Bioindicators & Biomonitors

Principles, Concepts and Applications

Trace Metals and other Contaminants in the Environment 6

Series Editor: Jerome O. Nriagu
Department of Environmental and Industrial Health
School of Public Health
University of Michigan
Ann Arbor, Michigan 48109-2029
USA

Other volumes in this series:

Volume 1: Heavy Metals in the Environment, edited by J.-P. Vernet
Volume 2: Impact of Heavy Metals on the Environment, edited by J.-P. Vernet
Volume 3: Photocatalytic Purification and Treatment of Water and Air,
edited by D.F. Ollis and H. Al-Ekabi
Volume 4: Trace Elements – Their Distribution and Effects in the
Environment, edited by B. Markert and K. Friese
Volume 5: Metals, Metalloids and Radionuclides in the Baltic Sea Ecosystem,
P. Szefer

Cover illustration:

Multiple bioindicators can be used for obtaining integrated data on quality of
the environment (with special courtesy to S. Wuenschmann for preparing the
illustration. Further details please find in Wuenschmann et al., 2001: Possible
use of wild-living rats (Rattus norvegicus) as bioindicators for heavy metal
pollution, UWSF – Z Umwetchem Ökotox 13 (5) 259–265.

Bioindicators & Biomonitors

Principles, Concepts and Applications

Edited by

B.A. Markert
Internationales Hochschulinstitut Zittau
Zittau, Germany

A.M. Breure
Laboratory for Ecological Risk Assessment
National Institute for Public Health and the Environment,
Bilthoven, The Netherlands

and

H.G. Zechmeister
Institute of Ecology and Conservation Biology, University of Vienna,
Vienna, Austria

ELSEVIER

Amsterdam – Boston – Heidelberg – London – New York – Oxford
Paris – San Diego – San Francisco – Singapore – Sydney – Tokyo

ELSEVIER B.V.
Sara Burgerhartstraat 25
P.O.Box 211, 1000 AE
Amsterdam, The Netherlands

ELSEVIER Inc.
525 B Street, Suite 1900
San Diego, CA 92101-4495
USA

ELSEVIER Ltd
The Boulevard, Langford Lane
Kidlington, Oxford OX5 1GB
UK

ELSEVIER Ltd
84 Theobalds Road
London WC1X 8RR
UK

First edition 2003
Second impression 2004

Library of Congress Cataloging in Publication Data
A catalog record is available from the Library of Congress.

British Library Cataloguing in Publication Data
A catalogue record is available from the British Library.

ISBN: 0-08-044177-7

♾ The paper used in this publication meets the requirements of ANSI/NISO Z39.48-1992 (Permanence of Paper).
Printed in The Netherlands.

Contents

Contributors

Dr Rolf Altenburger UFZ-Umweltforschungszentrum Leipzig-Halle, Sektion Chemische Ökotoxikologie, Leipzig, Germany
E-mail: ra@uoe.ufz.de

Mark J. Bagley, Ph.D. US Environmental Protection Agency, Office of Research and Development, National Exposure Research Laboratory, Ecological Exposure Research Division, Molecular Ecology Research Branch, Cincinnati, Ohio, USA
E-mail: bagley.mark@epa.gov

Prof. Dr Peter H. Becker Institut für Vogelforschung, Vogelwarte Helgoland, Wilhelmshaven, Germany
E-mail: ifv@terramare.de

Dr Jaap Bloem Alterra, Wageningen, The Netherlands
E-mail: J.Bloem@alterra.wag-ur.nl

Dr Peter Bosch European Environment Agency, Copenhagen, Denmark
E-mail: peter.bosch@eea.eu.int

Dr Anton M. Breure Laboratory for Ecological Risk Assessment, National Institute for Public Health and the Environment (RIVM), Bilthoven, The Netherlands
E-mail: ton.breure@rivm.nl

Barbara S. Brown US Environmental Protection Agency, Office of Research and Development, National Health and Environmental Effects Research Laboratory, Ecological Response Branch, Atlantic Ecology Division, Narragansett, Rhode Island, USA
E-mail: brown.barbara@epa.gov

Dr Andreas Chovanec Umweltbundesamt Wien, Abt. Aquatische Ökologie, Vienna, Austria
E-mail: chovanec@ubavie.gv.at

Dr Michael Cleuvers Dept. of Biology II (Zoology), Aachen University of Technology, Aachen, Germany

Dr Wim A.M. Didden Landbouw Universiteit Wageningen, Vakgroep Milieuwetenschappen, Bodembiologie Groep, Wageningen, The Netherlands
E-mail: wim.didden@bb.benp.wau.nl

Univ.-Prof. Dr Martin Dokulil   Institute for Limnology, Austrian Academy of Science, Mondsee, Austria
E-mail: martin.dokulil@oeaw.ac.at

Virginia D. Engle   US Environmental Protection Agency, Office of Research and Development, National Health and Environmental Effects Research Laboratory, Gulf Ecology Division, Gulf Breeze, Florida, USA
E-mail: engle.virginia@epa.gov

Prof. Wilfried H.O. Ernst   Vrije Universiteit, Dept of Ecology and Ecotoxicology, Faculty of Biology, Amsterdam, The Netherlands
E-mail: wernst@bio.vu.nl

Prof. Dr Otto Fränzle   Universität Kiel, Geographisches Institut, Kiel, Germany
E-mail: fraenzle@geographie.uni-kiel.de

Dr Maria do Carmo Freitas   Instituto Tecnológico e Nuclear, Sacavém, Portugal
E-mail: cfreitas@itn.itn.pt

Prof. Dr Jacob Garty   Department of Plant Sciences, University Campus/Ramat Aviv, Tel Aviv, Israel
E-mail: garty@post.tau.ac.il

Dr Bettina Götz   Federal Environmental Agency, Vienna, Austria
E-mail: goetz@ubavie.gv.at

Michael B. Griffith, Ph.D.   US Environmental Protection Agency, Office of Research and Development, National Center for Environmental Assessment, Cincinnati, Ohio, USA
E-mail: griffith.michael@epa.gov

Prof. Dr Krystyna Grodzińska   Polish Academy of Sciences, W. Szafer Institute of Botany, Krakow, Poland
E-mail: grodzin@ib-pan.krakow.pl

Dr Josef Hackl   Federal Environmental Agency, Vienna, Austria
E-mail: hackl@ubavie.gv.at

Dr Monika Hammers-Wirtz   RWTH Aachen, Lehrstuhl für Biologie V, Aachen, Germany
E-mail: monika.hammers-wirtz@bio5.rwth-aachen.de

Univ.-Prof. Dr Peter-Diedrich Hansen   Institut für Ökologie, Fachgebiet Ökotoxikologie, Technische Universität Berlin, Berlin, Germany
E-mail: pd.hansen@tu-berlin.de/E-mail: hanse32@ibm.net

Brian H. Hill, Ph.D. US Environmental Protection Agency, Office of Research and Development, National Health and Environmental Effects Research Laboratory, Mid-Continent Ecology Division, Watershed Diagnositic Research Branch, Duluth, Minnesota, USA
E-mail: hill.brian@epa.gov

Dr Rudolf Hofer Institute of Zoology and Limnology, Department Ecophysiology, University of Innsbruck, Innsbruck, Austria
E-mail: rudolf.hofer@uibk.ac.at

Dr Sebastian Höss ECOSSA (Ecological Sediment & Soil Assessment), München, Germany
E-mail: hoess@ecossa.de

Prof. Dr Antonius Kettrup GSF-Forschungszentrum für Umwelt u. Gesundheit GmbH, Institut für Ökologische Chemie, Neuherberg, Germany
E-mail: kettrup@gsf.de

Dr Karl Kienzl Federal Environmental Agency, Vienna, Austria
E-mail: kienzl@ubavie.gv.at

Prof. Dr Horst Kierdorf Universität Hildesheim, Institut für Biologie und Chemie, Abt. für Biologie, Hildesheim, Germany
E-mail: Kierdorf@rz.uni.hildesheim.de

David J. Lattier, Ph.D. US Environmental Protection Agency, National Exposure Research Laboratory, Office of Research and Development, Ecological Exposure Research Division, Molecular Ecology Research Branch, Cincinnati, Ohio, USA
E-mail: lattier.dave@epa.gov

Dr James M. Lazorchak US Environmental Protection Agency, Office of Research and Development, National Exposure Research Laboratory, Ecological Exposure Research Laboratory, Molecular Ecology Research Division, Cincinnati, Ohio, USA
E-mail: lazorchak.jim@epa.gov

Dr Carolin M. Lorenz Witteveen en Bos, The Hague, The Netherlands
E-mail: c.lorenz@witbo.nl

Anthony Maciorowski, Ph.D. US Environmental Protection Agency, Office of Water, Washington, DC, USA
E-mail: maciorowski.anthony@epa.gov

Univ.-Prof. Dr Bernd Markert Internationales Hochschulinstitut Zittau, Zittau, Germany
E-mail: markert@ihi-zittau.de

Dr Peter Matthiessen   Centre for Environment, Fisheries and Aquaculture Science (CEFAS), Fisheries Laboratory, Burnham-on-Crouch, Essex, UK
E-mail: P.Matthiessen@cefas.co.uk

Frank H. McCormick, Ph.D.   US Environmental Protection Agency, Office of Research and Development, National Exposure Research Laboratory, Ecological Exposure Research Division, Ecosystems Research Branch, Cincinnati, Ohio, USA
E-mail: mccormick.frank@epa.gov

Dr Christian Mulder   Laboratory for Ecological Risk Assessment, National Institute for Public Health and the Environment (RIVM), Bilthoven, The Netherlands
E-mail: christian.mulder@rivm.nl

Dr Ivo Offenthaler   Federal Environmental Agency, Vienna, Austria
E-mail: offenthaler@ubavie.gv.at

Dr Richard Öhlinger   Federal Office of Agrobiology, Linz, Austria
E-mail: oehlinger@agrobio.bmlf.gv.at

Univ.-Prof. Dr Jörg Oehlmann   Johann Wolfgang Goethe University Frankfurt, Faculty of Biology and Informatics, Department of Ecology and Evolution – Ecotoxicology, Frankfurt/Main, Germany
E-mail: oehlmann@zoology.uni-frankfurt.de

Dr Kevin Parris   OECD, Policies and Environment Division, Directorate for Food, Agriculture and Fisheries, Paris Cedex 16, France
E-mail: kevin.parris@oecd.org

Dr Ulla Pinborg   European Environment Agency, Copenhagen, Denmark
E-mail: ulla.pinborg@eea.eu.int

Dr Hans Toni Ratte   RWTH Aachen, Lehrstuhl für Biologie V (Ecology, Ecotoxicology, Ecochemistry), Aachen, Germany
E-mail: ratte@rwth-aachen.de

Dr Miguel dos Reis   Instituto Tecnológico e Nuclear, Sacavém, Portugal
E-mail: mareis@itn1.itn.pt

Dr Alarich Riss   Federal Environmental Agency, Vienna, Austria
E-mail: riss@ubavie.gv.at

Prof. Dr Fritz Schiemer   University of Vienna, Institute of Ecology and Conservation Biology, Vienna, Austria
E-mail: friedrich.schiemer@univie.ac.at

Dr Mechthild Schmitt-Jansen   UFZ – Umweltforschungszentrum, Leipzig-Halle,
Sektion, Chemische Ökotoxicologies, Leipzig, Germany
E-mail: mechthild.schmitt@uoe.ufz.de

Dr Ulrike Schulte-Oehlmann   Johann Wolfgang Goethe University Frankfurt,
Faculty of Biology and Informatics, Department of Ecology and Evolution –
Ecotoxicology, Frankfurt/Main, Germany
E-mail: oehlmann@zoology.uni-frankfurt.de

Dr Borut Smodiš   Section for Nutritional and Health-Related Environmental
Studies, IAEA, Vienna, Austria
E-mail: b.smodis@iaea.org

Dr Grazyna Szarek-Łukaszewska   Polish Academy of Sciences, W. Szafer Institute
of Botany, Krakow, Poland
E-mail: szarek@ib-pan.krakow.pl

Univ.-Prof. Dr Frieda Tataruch   Forschungsinstitut für Wildtierkunde und
Ökologie, Veterinärmedizinische Universität, Vienna, Austria
E-mail: frieda.tataruch@vu-wien.ac.at

Greg Toth, Ph.D.   US Environmental Protection Agency, Office of Research and
Development, National Exposure Research Laboratory, Ecological Exposure
Research Division, Molecular Ecology Research Branch, Cincinnati, Ohio, USA
E-mail: toth.greg@epa.gov

Prof. Dr Walter Traunspurger   Fakultät für Biologie, Universität Bielefeld,
Bielefeld, Germany
E-mail: traunspurger@biologie.uni-bielefeld.de

Dr Willhelm Vogel   Federal Environmental Agency, Vienna, Austria
E-mail: vogel@ubavie.gv.at

Dr Peter Weiss   Federal Environmental Agency, Vienna, Austria
E-mail: weiss@ubavie.gv.at

Dr Johann Wimmer   Attorney for the Environment, Upper Austria, Linz, Austria
E-mail: johann.wimmer@ooe.gv.at

Dr Bert Th. Wolterbeek   Interfaculty Reactor Institute, Delft University of
Technology, Delft, The Netherlands
E-mail: wolterbeek@iri.tudelft.nl

Univ.-Prof. Dr Harald G. Zechmeister   University of Vienna, Institute of Ecology
& Conservation Biology, Dept of Conservation Biology, Vegetation- and Landscape
Ecology, Vienna, Austria
E-mail: harald.zechmeister@univie.ac.at

About the editors

Bernd A. Markert, PhD, is professor of Environmental High Technology and Director of the International Graduate School (IHI) in Zittau, Germany.

His professional interests include the biogeochemistry of trace substances in the water/soil/plant system, instrumental analysis of chemical elements, eco- and human-toxicological aspects of hazardous substances, pollution control by use of bioindicators and technologies for waste management, environmental restoration and remedial action on soils.

Dr Anton M. Breure is a senior scientific staff member of the National Institute of Public Health and the Environment (RIVM) in Bilthoven, The Netherlands, where he is head of the department of Ecosystem Quality of the Laboratory for Ecological Risk Assessment. He works on the development of multi-stress models and indicators to quantify effects of contamination and other human impacts on the composition and functioning of ecosystems and on the bioavailability and biodegradation of organic pollutants. He holds an MSc in Chemistry and a PhD in Microbial Ecology from the University of Amsterdam and worked as an investigator at the Department of Microbiology of this university before joining the Institute. He has (co-)authored 100 open literature publications, reports and book chapters.

Prof. Dr. Harald G. Zechmeister is a senior scientist, working as a freelance at the Institute of Ecology and Conservation Biology of the University of Vienna. He finished his Masters degree as a high-school biology teacher and his PhD in Botany. Beside research activities he gives lectures on bioindication/biomonitoring and plant ecology. He has been working in several fields of vegetation ecology and biomonitoring. His work has an emphasis on lower plants, especially bryophytes. In his research he mainly deals with the impact of agricultural land use on biodiversity and the biomonitoring of atmospheric pollutants. His practical work is accompanied by a wide range of publications.

Preface

Our world undergoes rapid changes and is faced with an increasing number of known and unknown pollutants which combine with climate change and losses of biodiversity to threaten almost all ecosystems. This complex system of interactions and interrelations requires intensified efforts to provide integrated information on the status and development of environmental quality. Bioindicators and biomonitors have proven to be excellent tools in many of these cases and could provide information which cannot be derived from technical measurements alone.

Bioindicators and biomonitors yield extensive information. Thus an increasing knowledge of ecology gave way to the insight that organisms, cells and subcellular compounds likewise can be used as indicators for ecosystem qualities and for assessment of the impact of environmental stress on the composition and functioning of ecosystems. Indicators can be used to assess (environmental) quality, but also to investigate trends, e.g. by monitoring systems with measurements repeated in time. There are many interactions among biotic and abiotic components of ecosystems and the presence and abundance of specific species in ecosystems. The latter in turn depend on abiotic conditions, their role in the ecosystem and the presence of other organisms. Accordingly, management of ecosystems can be evaluated by biomonitoring.

This book *Bioindicators and Biomonitors* aims at giving a state-of-the-art overview of the current knowledge of how and where organisms can be used to assess environmental and ecosystems qualities. The principles of deriving indicators from presence and performance of organisms in ecosystems, and definitions that are used are given. A very important question to be answered in the book deals with the kind of information that can be obtained from biological indicators, and the advantages and disadvantages of their applications. Biological indicators can be used when easy to measure and allow for distinguishing between different states of the ecosystem by their discriminative power. Furthermore there must be a relationship between the presence and performance of the (group of) organisms used in the indicator and the kind of information to be obtained about the ecosystem. They can be used for assessment of ecosystem and environmental quality and for evaluation of management measures and environmental stress. The use of bioindicators/biomonitors has been developed from linear assessment (e.g. concentration of an element within an organism to deposition of this element) to integrated thinking on ecosystem levels. Therefore new indicators are being developed.

There are several chapters on integrative approaches of bioindicators for policy and legislature, company management and ecosystem management. These are followed by chapters on the use of single species indicators for assessment and predictions of ecosystem quality. Special attention is paid to biomarkers, subcellular systems such as enzymes that can be used to detect effects of compounds on the performance of organisms. Further standard laboratory tests are described, using standardized laboratory organisms to determine effects of toxic compounds or other stressors. Physicochemical methods used for purposes of identification of the presence and abundance of environmental stressors such as toxic compounds, "anomal" pH, or an excess or shortage of nutrients (nitrogen, phosphorus, sulphate, oxygen, trace metals) are described, giving indications on possible effects.

However, only assessment of the presence, abundance and performance of organisms in the field gives insight in the joint effects of the combination of stresses present in the ecosystem. Therefore the uses of different types of organisms (micro-organisms, lower plants, higher plants, invertebrates, vertebrates) as indicators are discussed in a series of subsequent chapters.

Important concepts and future developments for applications of biological indicators and uses of biomonitoring in national and international (monitoring) programmes complete the book. Up to date there is no other comprehensive review of these programmes and obviously this is a major advantage of this volume. Authors from the scientific world, each of them an authority on his or her subject, as well as national and international policymakers were invited to give their points of view on the use of bioindication for their respective purposes.

The book is a comprehensive overview of the present knowledge and developments and fills a wide gap on the textbook level and is written for conservationists, ecologists, ecotoxicologists, farmers, physiologists, policy makers, stewards of nature reserves and students.

We tried to be both as comprehensive and as thorough as possible with regard to this overwhelming scientific field of bioindication/biomonitoring. We invited contributions by leading persons in bioindication/biomonitoring from around the world reflecting the broad scope of current thinking and research, making this book essential reading for informed professionals as well as students. We tried to provide a comprehensive single source coverage of the entire field of bioindication/biomonitoring, from the ecological basics to the effects of chemicals on the indicators and monitors and the latest test strategies.

Of course there still remain shortcomings within this volume, for which only we as the editors can be blamed. E.g. we could not achieve the goal to give all bigger groups of species and organisms the state-of-the-art-discussion they deserve to get for being important ecosystem members. This is partly due to the limited number of pages and our not being successful in obtaining high-level manuscripts (e.g. on the use of arthropods). Possibly we may fill these gaps in a forthcoming edition once this volume is sold out (which will hopefully happen very rapidly). More complicated to overcome is the still remaining use of different definitions in terms of bioindication and biomonitoring. We as editors tried to use the qualitative/quantitative approach for clear-cut statements generally. You will find it in our introductory chapter.

On the other hand it seems to be evident that the rapid and dynamic international increase of use of these biotechniques did not save time and space for a more harmonized strategic development of techniques already successfully running. Therefore you will realize that other definitions are still scientifically attractive as long as they provide realistic perspectives in strategic and scientific planning. Therefore all definitions which make scientific sense should stand alternatively side by side unless one definition has developed such an impressive logistic and impetus to replace others automatically. A higher degree of standardization which could be accepted also in court as a protocol suited for judicial decision-making seems a mere dream for the future for a lot of biomonitoring techniques right now (exception: biotests). Possibly the integrated biomonitoring proposed in this book by several authors will open new innovative ways for progress in this direction.

As all of you are probably aware, such an endeavour to write some "standard" book needs quite a number of positive influences without which it could not be developed. The most pronounced influence came from the authors themselves. We are extremely grateful to these colleagues for a lot of constructive criticisms and ideas which of course made the work not easier for us as editors, but ended up in a product which highlighted the current state of the art. This way of intensive discussion, done mainly by means of our electronic computer systems, gave us as editors a partially new way of thinking in this field over the last years. Therefore the authors have acted throughout the preparation of this volume as teachers. And this reflects exactly that what *Bioindicators and Biomonitors* is meant to be: a guide and assistant to the world of different bioindicators, strategies and concepts. The international business and the intercultural exchange all over the globe which took place during preparation of this volume did show us once more how much science can benefit from a multi-face way of thinking. For these wonderful experiences we would like to give you a 1000-fold thank you.

In the same way the authors supported us by their scientific inputs we were fully dependent on our staff members at IHI Zittau. It would have been outright impossible to handle 1500 MS pages without their day and night work done with patience and care. Especially Ms Simone Wuenschmann and Ms Angelica Pedina (both Ph.D. candidates of Bernd Markert) have given all their time, vigour and energy for doing this arduous task. We all should be aware that especially such young, enthusiastic people are essential for fruitful further development of intelligent bioindicator systems. We thank both of you for having been our "backbones" and motivators during those times while we as editors had lost control of the material which came in. Thanks also to Dr Stefan Fraenzle (IHI Zittau) for his work on the language of both this preface and translations of other parts of the volume.

Of course, the help of the excellent computer equipment and infrastructure of IHI Zittau gave us the possibility in working to an internationally superb standard. Very often our network administrators at IHI, Bac le Trung and Hartmut Paetzold, have done a great job in overcoming technical problems we encountered during e-mail communications, graphical line drawings or with any other kind of technical hardships. Thank you.

And we would thank all colleagues and persons around us for their tolerance with our trembling nerves, minds and bodies during "high noon" situations. Thank you.

Eventually we would like to thank the staff members of Elsevier, especially Doris Funke, Sarah Moore and Peter Henn who created an excellent atmosphere of co-operative hand-in-hand work guaranteeing optimum working conditions for gathering expertise. Of course, Jerome Nriagu is thanked for opening his trace metals book series to a broader scope of topics and for guiding us through the North American continent.

We hope that this volume will intensify the discussion in between developers and users of bioindication methods. We hope that this volume will be distributed as far as the authors came from, which is to say all over the world. Please be invited to criticize all what would help to fit this volume. We hope that bioindication will be one just one resource for clever and intelligent bio-techniques in the new century, bringing us not only new scientific and practical insights, but in the same way intercultural and international exchange in between our multifaced nations and regions for a peaceful world.

Bernd Markert, Anton Breure and Harald Zechmeister
Zittau, Bilthoven and Vienna, June 2002

I

General aspects and integrative approaches

Bioindicators and biomonitors
B.A. Markert, A.M. Breure, H.G. Zechmeister, editors

Chapter 1

Definitions, strategies and principles for bioindication/biomonitoring of the environment

Bernd A. Markert, Anton M. Breure and Harald G. Zechmeister

Abstract

In the context of environmental monitoring studies bioindicators reflect organisms (or parts of organisms or communities of organisms) that contain information on quality of the environment (or a part of the environment). Biomonitors, on the other hand, are organisms (or parts of organisms or communities of organisms) that contain informations on the quantitative aspects of quality of the environment. When data and information obtained by bioindication are moved up to the level of knowledge the subjectivity of interpretation increases with the complexity and dynamics of a system ("staircase of knowing").

In this article clearcut definitions are attempted for most terms used in environmental monitoring studies. From there a comparison of instrumental measurements with the use of bioindicators/biomonitors with respect to harmonisation and quality control will be drawn. Precision, accuracy, calibration and harmonisation in between national standards and international routines seem to be the leading goals in quality studies of international working groups dealing with biomonitoring throughout the world. Common strategies and concepts will fill the gap in between single source results and integrated approaches related either for human health aspects or environmental protection purposes, f.e. via biodiversity monitoring. Here we report on well established monitoring programmes like Environmental Specimen Banking (ESB) or newly developed strategies as the Multi-Markered Bioindication Concept (MMBC) with its functional and integrated windows of prophylactic health care.

Keywords: Bioindicator(s)/biomonitor(s), definitions, information, quality control, precision, accuracy, calibration, harmonisation, integrated approaches, biodiversity monitoring, Environmental Specimen Banking (ESB), Multi-Markered Bioindicator Concept (MMBC)

1. Introduction

Organisms, populations, biocoenoses and ultimately whole ecosystems are naturally influenced by numerous biotic and abiotic stress factors such as fluctuations in climate, varying radiation and food supply, predator-prey relationships, parasites, diseases, and competition within and between species. This stress situation is vital at every level of biological organisation. Consequently, the ability to react to stressors is an important characteristic of all living systems, and conversely no development of the species and the ecosystem as a whole is possible without such natural stressors (Schüürmann and Markert, 1998). Stress is the locomotive of evolution. But within evolutive epochs the range of variation of the stressors is generally fairly constant and allows the species to adjust to changing environmental conditions.

In recent centuries these changes have reached a new dimension in terms of both quality and quantity. Through human activity the environment has been confronted with totally new substances that did not previously exist (xenobiotics, many radionuclides) and potentially harmful substances released in quantities unthinkable in the past (heavy metals, natural radionuclides). What is more, these new stressors usually have a multiplying effect, i.e. they are added to the effects of natural systems, or they themselves act in combination, with the result that the "tolerance level" of the organisms' ability to cope or to adjust to them is exceeded (Oehlmann and Markert, 1999).

In the last 200 years the increase in the world's population and the resulting global rise in energy consumption have led to a dramatic change in the natural basis of our lives. According to estimates by Schneider (1992), around 2500 square miles of tropical rainforest are being lost at a rate of 1.5 acres per second. Over 70 square miles of arable land are being irreversibly transformed into desert through mismanagement, overgrazing and over-population. Ten to 100 species of plants and animals may disappear from the planet each day, and an even bigger problem is that we do not know whether the number is 10 or 100. Nor we do know whether the total number of species on earth is one million or 10 million (Schneider, 1992). This means we have a problem of information on numbers in general.

Moreover, the situation will become more dramatic in the future. Each day we will have 250,000 more people on earth than the day before, and more than 40% of them will live their lives in abject poverty. Each day we will add 2700 more tons of chlorofluorocarbons and 51 million tons of carbon dioxide to our atmosphere and the earth will be a little warmer and a little more acidic, and more ultraviolet radiation will pass through our atmosphere to the surface of the earth (Schneider, 1992). A large amount of (chlorinated) pesticides will be used in the developed countries, and an even greater but unknown amount of such compounds will be applied in the less developed countries of the world. Analysis of most of these compounds in many countries will be too sporadic to allow definite conclusions about the toxic effects on man and the environment.

Chemical substances constitute the greatest pollutant burden on natural ecosystems. Table 1 gives examples of pollutants and burdening substances in different sectors of the environment. The number of known substances has risen rapidly since the development of modern chemistry. Between 1970 and 1993, for example, the number of substances registered with the Chemical Abstract Service of the American Chemical Society increased sixfold, from 2 million to 12 million. Assuming a linear course, this corresponds to a synthesis of nearly 1200 new substances every day (Markert et al., 2000). But only a fraction of these compounds are produced and used in such quantities that measurable effects on the environment are to be expected. Of the 73,000 chemicals that were on the market in Germany in 1985, 312 were produced in volumes of over 10,000 t/a. 2200 compounds were produced in quantities of 10–100 t/a. About 90% of the total annual production of the chemical industry world-wide consists of about 3000 substances. A knowledge of a) the chemical diversity (quality of pollution) and b) the amount (quantity) of the stressor per unit of time is necessary for a study of the ecological and human impact of pollution.

An inventory of the presence of chemicals on the European market resulted in the list of "Existing Chemicals" (those substances which were deemed to be on the

Table 1. Pollutant and burdening substances in different sectors of the environment (adapted from Markert et al., 2000).

Environmental compartment	Burdening substances	Examples	Places of origin
Water	Degradable organic compounds	Faeces, tensides, solvents, pesticides, industrial process materials, fats, oils, soluble animal and vegetable residues, basic chemicals, intermediate and end products	Towns and villages, households, agriculture, textile industry, metalworking, paintshops, food industry, chemical industry, paper industry, landfills
	Persistent organic compounds	Tensides, solvents, pesticides, industrial process materials, basic chemicals, intermediate and end products	Agriculture, textile industry, metal working, paintshops, chemical industry, paper industry, landfills
	Inorganic compounds	Heavy metals, salts, cyanide, chromate, fertilizers	Metal working, mining, leather production, towns and villages, agriculture, landfills
Soil	Degradable organic compounds	Faeces, pesticides, animal and vegetable residues, basic chemicals, intermediate and end products, sewage sludge, compost	Agriculture, landfills for domestic waste and waste requiring special monitoring and disposal facilities
	Persistent organic compounds	Tensides, solvents, pesticides, industrial process materials, basic chemicals, intermediate and end products	Landfills for waste requiring monitoring and for industrial waste
	Inorganic compounds	Heavy-metal compounds, salts, ash, slag	Landfills, incinerators, metal production
Air	Organic gases	Solvents, hydrocarbons, volatile pesticides, volatile industrial chemicals	Paintshops, etc., refineries, tank farms, agriculture, industry
	Inorganic gases	Carbon monoxide, hydrochloric and sulphuric acid, nitrogen oxides (ozone), metal vapours, carbon dioxide, ammonia	Firing systems, incinerators, engines, industry
	Dust and smoke	Metal oxides, PAH, soot	Metal production, waste incineration, firing systems in general

European market before September 18, 1981) and are listed in the EINECS inventory (European Inventory of Existing Commercial Chemical Substances). EINECS contains 100,195 substances (ECB, 2001). In Europe 2604 different chemicals are produced or marketed in quantities of more than 1000 t/a (IUCLID, 2000).

The number of species world-wide is thought to be 13 million, although only about 1.6 million species have been identified (Heywood and Watson, 1995). Some of them will never be identified. Pollution, habitat fragmentation and loss, intensification of agriculture and population pressure are leading to dramatic changes in biodiversity (McNeelay et al., 1995). The alarming loss of biological diversity within the last decades represents a major challenge to the scientific community and demands the development of appropriate strategies for land management and proper tools for monitoring. Besides having ecological consequences, this loss of species diversity may also affect economic processes. The prognosis for climate change (Mitchell et al., 1990; Watson et al., 1996; ICC, 2001) will change the viability of populations, the number and distribution of species and the structure, composition and functioning of ecosystems (Grabherr et al., 1994; Arft et al., 1999; Kappelle et al., 1999).

An objective of prophylactic environmental protection must be to obtain and evaluate reliable information on the past, present and future situation of the environment. Besides the classic global observation systems such as satellites and instrumental measuring techniques like trace gas and on-line water monitoring, increasing use should be made of bioindicative systems that provide integrated information permitting prophylactic care of the environment and human health. In the last 20 years, bioindicators have shown themselves to be particularly interesting and intelligent measuring systems. As long ago as 1980, Müller considered the "bioindicative source of information" one of the pillars of modern environmental monitoring, since "bioindication is the breakdown of the information content of biosystems, making it possible to evaluate whole areas".

2. Information on the environment – "old" and "new" ecology

Environmental chemicals affect biological systems at different levels of organisation, from individual enzyme systems through cells, organs, single organisms and populations to entire ecosystems. As a rule, the latter do not just react to single substances or parameters; they show species-specific and situation-specific sensitivity to the whole constellation of factors and parameters acting on them at their location. Information on the sensitivity and specificity of such reactions provides a basis for planning the use and evaluating the results of effect-related biological measuring techniques (Wagner, 1992).

The acquisition of information and a knowledge of our environment or environmental conditions, and the natural and anthropogenic changes these are undergoing, can be divided historically into three stages of development (although these are arbitrary and do not claim to be precise). They are (a) descriptive, observational biology up to the middle of the last century (up to 1950); (b) development of the environmental sciences in the second half of the last century (1950–2000); (c) the present synthesis of "old" and "new" ecology which takes the principle of sustainability as its scientific objective and includes use of the latest information and communication techniques and biotechnology.

Besides names such as Aristotle, Darwin and Linné that are associated with great scientific discoveries, Haeckel defined the "balance" of nature. In Europe, especially, approaches to animal and plant ecology then evolved that supplied the necessary methods for the second and probably most decisive phase in the development of the environmental sciences (both pure and applied) up to the end of the 20th century. During this time, ecology was "spun off" from general biology as a scientific field in its own right; in the course of its development it made increasing use of sophisticated techniques taken from molecular biology and industrial and computer-assisted methods and models for tackling unsolved problems. The 1950s saw the definition of applied aspects of environmental protection as focal points of modern ecological research in addition to basic research. Through the integrated, reciprocal effect of the principle of sustainability (Bundesministerium für Umwelt, Naturschutz und Reaktorsicherheit, 1992) and the latest biological, computing and communication technologies the present development of the "New Ecology" has generated scientific working methods that will in future permit totally new systems and system descriptions. The creation of virtual systems constitutes an important enlargement of the range of available test methods, too, since these systems will in future be used as a substitute in some areas (e.g. for experiments on vertebrates). Here the focus is on interdisciplinary approaches involving the natural, economic and social sciences and directed towards integrating the prophylactic and sustainable health and life of man and the world in which he lives. Governments are showing a growing interest in instruments for predicting the future performance of ecosystems that are or may be influenced by human activities. A keyword in this respect is "sustainable development" as defined by Brundtland's World Commission on Environment and Development (WCED, 1987). This also holds for "sustainable use of biodiversity", as adopted by the United Nations Conference on Environment and Development in Rio de Janeiro in Agenda 21 (UNCED, 1992). Sustainability describes the worldwide goal of all future efforts towards development. It chiefly means lastingly preventing the over-exploitation of:

- *natural resources*, especially soil and water;
- genetic diversity and the functional stability of the climate;
- *the social basis for a livelihood* and chances of development, especially in the poor sections of the population,

while making the most efficient use of available economic resources and ensuring that existing economic systems continue to function (Federal Ministry of the Environment, Nature Conservation and Reactor Safety 1992). Any serious consideration of these topics makes constant monitoring of environmental parameters and their trends an absolute necessity.

2.1. Descriptive, observational biology and the effects of environmental pollutants (up to 1950)

The observational, descriptive biology of the past two millennia was largely made up of chance observations of changes in the phenomena of the world around us caused by human activity. In his "Historiae naturalis", Pliny the Elder (23–79 AD) describes damage to coniferous trees around places where iron sulphide was converted to iron oxide by roasting; from the iron oxide it was then easier to separate off the ferrous metal so essential for daily use and the requirements of war. The sulphur dioxide released in this process caused acute damage to the needles of the surrounding conifers and gave rise to the first – greatly simplified – description of a connection between atmospheric pollution and damage to trees 2000 years ago.

Observations of fish kills in the Rhine and Thames in the late 17th and early 18th centuries quickly drew the attention of the inhabitants of major conurbations to the connection between the pollution of rivers and detrimental effects on fish. But of course it was not possible to foresee the tremendous, sometimes dramatic significance this connection would have in the future.

In the late Middle Ages people became aware of hazards to biodiversity and permanent damage to the landscape, although their intentions were often different from those of today. Examples are the Forestry Regulations issued by King Ferdinand I in 1535 to protect the forests planted as a barrier against avalanches. Without these forests it would scarcely have been possible to live in the valleys below (cf. Grabherr, 1991). Other regulations were concerned with sustainable hunting and fishing and intended to prevent the extermination of certain species.

There are many more examples of "chance" observations of the problem of pollutant substances in ecosystems and non-sustainable use of the land. What they all have in common is their random nature; they are phenomena that were scarcely heeded by the people living and working in those times, and which were not considered particularly risky. The unspecific way of looking at them reflects the prevailing "careless" attitude to nature.

2.2. Development of the environmental sciences – general and applied ecology (the "old" ecology, 1950–2000)

In the mid 20th-century the whole realm of ecological research acquired a scientific and therefore systematic structure. The 1950s saw an increasing interest in environmental phenomena and the connections between them, and with it the development of scientific research in this field, especially by the Odum family in the USA and a large number of scientists on the European continent. The focus was on understanding ecosystems as a whole and their various components. Motivating forces stimulated by UNESCO's "Man and the Biosphere Programme" were fundamental questions of how ecosystems and their components work, issues concerning the use and conservation of resources and the production and distribution of energy, and also forward-looking approaches to the responsible treatment of nature and its diversity. From the early 1950s to the late 1960s, especially, ecological research was devoted chiefly to understanding the basic characteristics of ecosystems in the classic sense, since the question of energy production seemed to have been answered by the advent of nuclear power, and questions of the availability of materials were regarded as a matter of logistics and distribution rather than as a problem of resources. Only gradually did it become clear that destruction of the environment meant elimination of the basic conditions for human life, both simultaneously and in the long term. A keystone in this respect was

Silent Spring by Rachel Carson (1962), who realised that the survival of man is very much dependent on nature, and that however artificial his dwelling, he cannot with impunity allow the natural environment of living things to be destroyed. Technologies like nuclear power, television and the telephone seemed to have made nature and its interaction with human social systems "controllable", at least for the time being ("unrestricted belief in technology"). The emergence of the computer sciences, the publication of the first net primary productivity maps, even the globalisation of ecological research seemed to be useful "by-products" of technical development. The landing of Apollo 11 on the moon in 1969 was regarded as the ultimate proof that nature could be controlled by man and his technologies.

The dream was brought to a sudden end by the reactor accident in Chernobyl in 1986. The unimpeded trans-boundary spread of the radioactive fallout from Chernobyl in Central and Northern Europe, the helplessness of the political decision-makers and their patently inadequate technical means of implementing a solution, brought the apparently controllable processes of nature back into the centre of attention. In the years that followed they led scientists to seek a more and more integrated approach to the objectives of environmental policy, an approach that necessitates an interdisciplinary view from the angle of the natural, economic and social sciences. A characteristic feature is the idea of "sustainable development", the objectives of which are stated in the Rio Protocol of 1992. The goals of environmental policy it contains, which have been moving more and more in the direction of global, prophylactic care of health and the environment since the end of the last century, are only just beginning to be acknowledged. The risk of epidemics and microbiological hazards such as the scarcely controllable cross bordering spread of AIDS, BSE and similar "plagues" of the 21st century make us aware of the difficult tasks of integrated, prophylactic care of the environment that lie ahead of us.

2.3. Present and future developments ("new" ecology)

In addition to social and economic developments, the last few decades have seen the advent of more precise and extensive environmental measurement programmes on the national and international level. UNESCO's "Man and the Biosphere Programme" mentioned above was superseded by the "International Geosphere/Biosphere Programme" (IGBP) that currently aims to co-ordinate international ecological research and generate global questions in respect of the details. In the field of environmental monitoring it has become very plain, in recent decades, that more and more precise analytical methods permit increasingly sophisticated frames of reference. In the early 1960s, classic chemical methods were still in general use for measuring environmental pollution, but with the development of atomic absorption spectrometry (AAS) and other analytical techniques it became possible to penetrate into the ppm and ppb range of pollutant research, and a huge amount of data was generated. Parallel to the development of extremely sensitive trace analysis down to the nanogram range, ecotoxicology began to turn its attention more and more to the damaging effects of trace substances such as Cd or dioxins on ecosystems. The data pool thus created initially contained a large number of dubious results, since the material was not subjected to any real quality control. In the early 1980s this problem was adequately solved by the introduction of standard reference materials on the instrumental side of measuring techniques, specific digestion and ashing methods, and the formulation of sampling programmes (Markert, 1996).

Besides numerous other tasks it is the recording of damage to organisms and populations in the field and prospective risk analysis before new chemicals are allowed onto the market that constitute the main field of work of ecotoxicology. By definition, ecotoxicology is the study of the scientific principles and methods that make it possible to identify and evaluate interference caused by substances introduced through human activity. Its objective is to recognise such interference and prevent possible damage, or to make suggestions for remedial action (Oehlmann and Markert, 1999).

Within a short time the relatively young sciences of bioindication and biomonitoring (definitions are given in sections of this article) have brought about numerous developments in method that have made it possible to meet the public's demand for protection of the environment against an increasing number and volume of environmental stress factors. The following are just a few such studies as an introduction and as examples, especially of the international context and the possible plant matrices; they reflect only a fraction of what has been described in thousands of works (Bamford et al., 2001; Bargagli, 1995, 1998; Bargagli et al., 1999; Brooks, 1998; Carreras et al., 1998; Conte et al., 1998; Conzales and Pignata, 1997; Conzales et al., 1998; Cortes et al., 2001; Djingova and Kuleff, 2000; Fraenzle and Markert, 2002; Freitas et al., 1999; Frontasyeva and Steinnes, 1995; Fytianos et al., 1999; Garty, 1998; Harada and Hatanaka, 1998; Herpin et al., 2001; Klumpp et al., 2000; Knauer et al., 1998; Kostka-Rick et al., 2001; Loppi et al., 1996; Markert, 1993; Markert and Weckert, 1993; Martinez-Cortizas et al., 1999; Michelot et al., 1999; Normandin et al., 1999; Rodushkin et al., 1999; Saiki et al., 2001; Schubert, 1991; Smodiš, 2002 (this book); Vutchkov, 2001; Wagner, 1987; Wappelhorst et al., 2000a/b; Wappelhorst et al., 2002; Watmough et al., 1999; Winter et al., 2000; Wolterbeek et al., 1995).

A recognised range of bioassay methods is now available for the provisional evaluation of individual substances (German Chemicals Act). Moreover, there is a great deal more data available on pesticides and also on water-polluting substances. Much has been achieved in the field of pollution protection and the development and establishment of cadasters of ecological effects. But in spite of this success there are a number of points to criticise, which plainly indicate deficits in ecotoxicology and therefore in bioindication and biomonitoring (Oehlmann and Markert, 1999). For example, there is a general problem with the rating of effects on the level of the individual or even on the suborganismic level for higher stages of complexity (populations, biocoenoses, ecosystems). The obvious gap between objectives and achievements in ecotoxicology and therefore in bioindication has been called the "dilemma" of this discipline by various authors. For ecotoxicology and bioindication do not merely claim to represent toxicology for just one or another animal or plant species. Integrated bioindication and biomonitoring should in future follow a comprehensive approach; i.e. besides determining effects on the level of the individual or species they should enable conclusions on the ecosystems level. Furthermore, modern research into bioindication and biomonitoring should do more to ensure the comparability of effects determined in the laboratory and in the field. In all the cases investigated there are definite signs of different threshold concentrations, sensitivities and extents of reaction in the laboratory and in the field. In bioassays, especially, it is usual to use genetically homogeneous plant and animal organisms in experimental work, whereas in nature genetic diversity is a typical characteristic of undisturbed populations. The multiple and simultaneous effects of several stressors on one or several groups of organisms increase complexity virtually to infinity, for it does not seem practicable at present to restrict experimentation to a limited combination of very different parameters. However, there is a steady development in risk assessment methods based on statistical interpretation of data obtained with individual species and single toxicants (Altenburger and Schmitt-Jansen, 2002 (this book); Posthuma et al., 2001).

Bioindication is essentially a tool of traditional conservation biology too. Indicator taxa are used to elucidate the effects of environmental change such as habitat alteration, fragmentation and climate change (McGeoch, 1998) on a spatial and temporal scale. Indicator species may act as surrogates for other groups of organisms or for larger communities (Meffe and Carrol, 1994).

Bioindicators have been used as indicators of biodiversity. In this field of research the species richness of one group of taxa is taken as an estimate of the species richness of another taxon irrespectively of its level (e.g. species, genus, order, etc.) or functional type (e.g. Shugart, 1997; Woodward and Kelly, 1997). Moreover, by using a multi-species approach it is possible to predict the species-richness of a whole landscape (e.g. 'shopping basket approach', Hammond, 1994; Sauberer et al., 2002).

Biodiversity indicators sometimes correlate closely with other variables that are important in terms of conservation issues, like genetic variables (e.g. allelic diversity, inbreeding, heterocygosity) or factors on the population or landscape level (e.g. demography, population and metapopulation dynamics, patch size, fragmentation, connectivity of habitats).

Climate change research often focuses on certain target species which react sensitively in the observed parameters (e.g. rise in precipitation or temperature). Again it is possible to make predictions on the ecosystem level, such as the 'northward movement' of plant communities (e.g. Gignac et al., 1998).

Ultimately, "new ecology" is just as difficult to define as "old ecology". Because of the ever-increasing complexity and dynamics of the subject, something new is constantly developing and being compared to what went before. And the new knowledge generated very soon becomes obsolete and is again replaced by something new. . . .

Apart from psychological aspects the "new ecology" of the present may be characterised as follows:

We are faced with a superexponential increase resulting from knowledge and understanding of the complexity of individual systems (complexity criterion) in conjunction with fast development and interlinking made possible by the computer technologies (dynamics). We have to demand that greater attention be given to global issues such as protection of the climate, AIDS, BSE, etc. and to the rapid integration of measures to protect both health and the environment (overall, global approach). At the same time we must continue specific investigations into individual local and regional processes; in their nanotechnological experimental approach these have to meet the most sophisticated requirements in respect of molecular, genetic and information technology (specific, regional approach). The numerous detailed regional models resulting from this are the most important basis for decision-making on global ecological policy.

Biomonitoring/bioindication does not only focus on the concentration and effects of heavy metals and other contaminants in the environment and particularly in the organisms living in the environment. Other environmental stresses, such as desiccation, acidification or eutrophication, and management practices such as agriculture also have a strong influence on ecology. The effects of such stresses can also be indicated by ecological monitoring. A major problem here is the causality between the effect observed and the stresses present.

In Germany, Ellenberg et al. (1992) have developed a very elegant system of associating the effects of pH, nitrogen, water content, light and salt with the presence of specific plant species in central Europe. Modern statistical techniques have recently made it possible to indicate the relative contribution of one stress factor to the integral effect of a mixture of stresses.

According to Costanza (1992), ecosystem health is a bottom-line normative concept. It represents a desired endpoint of environmental management, but the concept has been difficult to use because of the complex, hierarchical nature of ecological and economic systems. When developing and using (new) indicators, the following points have to be taken into consideration:

1. The need for pluralism. Multiple views are necessary to form an adequate picture of complex systems, but the multiple views require integration.
2. The need for integration: across space and time, across disciplines and sub-disciplines and across interest groups, in order to arrive at measures improving overall system performance.
3. These integrated measures may be called system health. Health is difficult to measure, cannot be quantified precisely; but it is a necessary concept.

The most impressive work in dimensions of this kind seems to be that of Costanza et al. (1997) on "the value of the world's ecosystem services and natural capital", in which the current value of 17 ecosystem services for 16 biomes is estimated. For the entire biosphere the value is calculated to be in the range of 16–54 trillion (10^{12}) US dollars p.a., the average being 33 trillion US dollars p.a. (more than the gross national product of the whole world).

When the data and information obtained by bioindication are moved up to the level of knowledge the subjectivity of interpretation increases with the complexity and dynamics of a system. A good way to illustrate the increase in subjectivity accompanying an increase in knowledge is the "staircase of knowing" (Roots, 1996). On this staircase (Fig. 1), observations and measurements, when verified according to agreed standards, become data. Data, properly selected, tested and related to subject areas can become information; information, organised and interpreted or applied to areas of interest or concern, can become knowledge; knowledge, if assimilated and subjected to mental assessment and enrichment, so that it is comprehended and integrated into a base of facts and impressions already assimilated, leads to understanding. And understanding, put into perspective with judgement according to certain values, can become wisdom. In general, by moving up the staircase, the material and ideas become increasingly subjective, with increasing human value added (Roots, 1996).

With respect to the age of "information technologies", Lieth (1998) tries to find a strategy for making the "digitalised bit world" more efficient for ecosystem research.

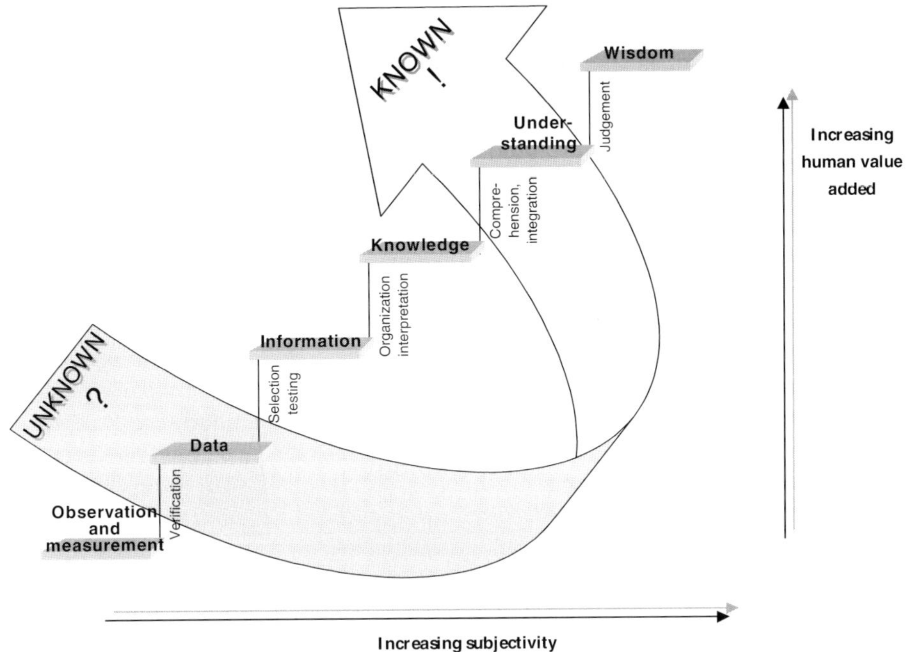

Figure 1. The staircase of "knowing", modified after Roots (1992). Explanations are given in the text.

According to Lieth we have to ask ourselves: what is the point of ecosystem research at all? What kind of information does an ecosystem offer? The information content of all the parts of the system elevates the ecosystem to the level of an intelligent system. Many toxicological implications involve the flow of information as the cause of significant changes in material fluxes and energy fluxes in the system. Plants may produce chemicals to protect themselves against animal grazing. Animals may produce toxic chemicals as weapons; humans may produce toxic chemicals to kill each other. Each process is controlled by "bits of information" which flow from one point in the ecosystem to another. A detailed description for further study of this straightforward concept is given in Lieth (1998).

2.4. Environmental medicine and ecological medicine

Finally we have to ask ourselves about the consequences of specific environmental burdens for man, i.e. we need to search for interactions between human beings and the environment in the pathogenic sense (Mersch-Sundermann, 1999). Recent medical history has seen the development of environmental medicine and ecological medicine. Environmental medicine and its methods (Section 4) tend to take the form of an individual approach (involving empirical research), whereas ecological medicine has more to do with basic research into causes together with the environmental sciences (Fig. 2).

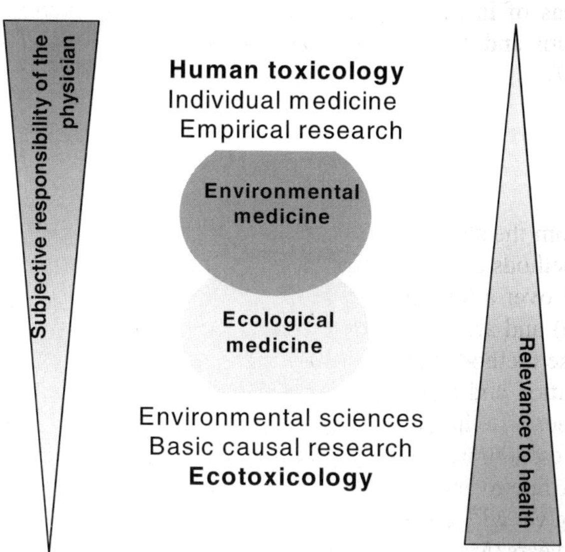

Figure 2. Differentiation between the terms "environmental medicine" and "ecological medicine" in respect of objectives and content. Definitions in the text (adapted from Mersch-Sundermann 1999).

According to Mersch-Sundermann (1999), environmental medicine is that branch of medicine that is concerned with identifying, investigating, diagnosing and preventing impairment of health and well-being and with identifying, investigating, assessing and minimising risks caused by definable spheres of interaction between man and the environment. The definable spheres of interaction between man and the environment are direct and indirect anthropogenic influences of a physical, chemical, biological, socio-psychological and perceptive nature. "Environment" is the totality of all processes and areas in which interaction between nature and civilisation takes place. In the context of this definition, environmental medicine deals with impairment of health and risks caused by definable anthropogenic influences on the environment. It therefore constitutes the link between health (as a state of equilibrium and adequate stability of essential measurements and values) and the environment (as processes and areas of interaction between civilisation and nature).

Ecological medicine is really an extension of environmental medicine, which centres on the patient, to supra-individual factors relating to health or superordinate risks resulting from interaction between man and the environment or between civilisation and nature (Mersch-Sundermann, 1999). There is no sharp dividing line between environmental medicine and ecological medicine, but environmental medicine takes a chiefly anthropocentric view of effects and risks from the environment, whereas ecological medicine analyses the characteristics of systems – i.e. biological, sociological and ecological factors – underlying these effects and risks. So whereas environmental medicine reflects the medical *effects* of interaction between man and the environment, ecological medicine is concerned with the *causes*. Following the ecosystem approach, ecological medicine constitutes a link between observations of the environment (as processes and areas of interaction between civilisation and nature) and health (as a state of equilibrium and stability of essential measurements and values) (Mersch-Sundermann, 1999).

3. Definitions

It seemed clear from the start that bioindication and biomonitoring are promising (and possibly cheap) methods of observing the impact of external factors on ecosystems and their development over a long period, or of differentiating between one location (e.g. an unpolluted site) and another (polluted site). The overwhelming enthusiasm shown in developing these methods has resulted in a problem that is still unsolved: the definitions of bioindication and biomonitoring respectively, and therefore the expectations associated with these methods, have never led to a common approach by the international scientific community, so that different definitions (and expectations!) now exist simultaneously. A fine overview of the various definitions is given by Wittig (1993).

In the following we will give some definitions that have been developed and used by us over the last 20 years (Markert et al., 1999), since we feel that they differentiate clearly between bioindication and biomonitoring using the qualitative/quantitative approach to chemical substances in the environment. This makes bioindicators directly comparable to instrumental measuring systems. From that angle it is possible to distinguish clearly between active and passive bioindication (biomonitoring). Especially where the bioindication of metals is concerned, the literature often makes a distinction between "accumulation indicators" and "effect indicators" in respect of the reaction of the indicator/ monitor to changes in environmental conditions. Here we should bear in mind that this differentiation does not imply a pair of opposites; it merely reflects two aspects of analysis. As the accumulation of a substance by an organism already constitutes a reaction to exposure to this substance which – at least in the case of high accumulation factors – is measurably reflected in at least one of the parameters used in defining the term "effect indicator/monitor" (e.g. morphological changes at the cellular level; formation of metal-containing intracellular granules in many invertebrates after metal accumulation), we should discuss whether it is worthwhile distinguishing between accumulation and effect indicators or whether both terms fall under the more general expression "reaction indicator". Often, too, it is not until a substance has been accumulated in organisms that intercellular or intracellular concentrations are attained that produce effects which are then analysed in the context of effect and impact monitoring (Fig 3).

From these preliminaries we come to the following definitions, given in Markert et al., 1997 and 1999:

A bioindicator is an organism (or part of an organism or a community of organisms) that contains information on the quality of the environment (or a part of the environment). A biomonitor, on the other hand, is an organism (or a part of an organism or a community of organisms) that contains information on the quantitative aspects of the quality of the environment. A biomonitor is always a bioindicator as well, but a bioindicator does not necessarily meet the requirements for a biomonitor.

We speak of active bioindication (biomonitoring) when bioindicators (biomonitors) bred in laboratories are exposed in a standardised form in the field for a defined period

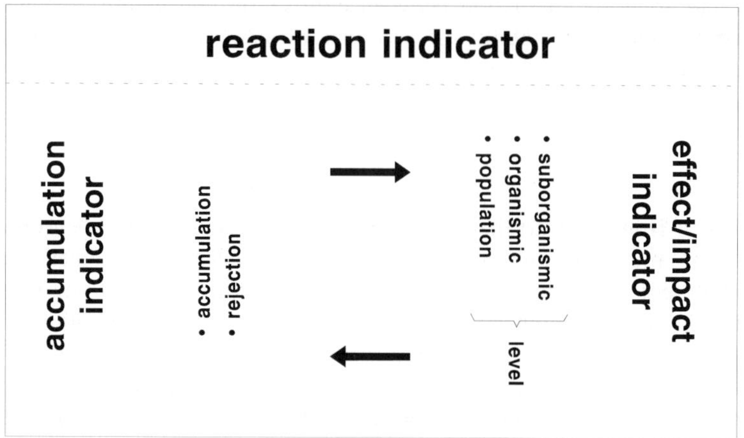

Figure 3. Illustration of the terms reaction, accumulation and effect/impact indicator (Markert et al., 1997). Explanations are given in the text.

of time. At the end of this exposure time the reactions provoked are recorded or the xenobiotics taken up by the organism are analysed. In the case of passive biomonitoring, organisms already occurring naturally in the ecosystem are examined for their reactions. This classification of organisms (or communities of these) is according to their "origin".

A classification of organisms (or communities of these) according to their "mode of action" (Fig. 3) is as follows: Accumulation indicators/monitors are organisms that accumulate one or more elements and/or compounds from their environment. Effect or impact indicators/monitors are organisms that demonstrate specific or unspecific effects in response to exposure to a certain element or compound or a number of substances. Such effects may include changes in their morphological, histological or cellular structure, their metabolic-biochemical processes, their behaviour or their population structure. In general the term "reaction indicator" also includes accumulation indicators/monitors and effect or impact indicators/monitors as described above.

When studying accumulation processes it would seem useful to distinguish between the paths by which organisms take up elements/compounds. Various mechanisms contribute to overall accumulation (bioaccumulation), depending on the species-related interactions between the indicators/monitors and their biotic and abiotic environment. Biomagnification is the term used for absorption of the substances from nutrients via the epithelia of the intestines. It is therefore limited to heterotrophic organisms and is the most significant contamination pathway for many land animals except in the case of metals that form highly volatile compounds (e.g. Hg, As) and are taken up through the respiratory organs, (e.g. trachea, lungs). Bioconcentration means the direct uptake of the substances concerned from the surrounding media, i.e. the physical environment, through tissues or organs (including the respiratory organs). Besides plants, that can only take up substances in this way (mainly through roots or leaves), bioconcentration plays a major role in aquatic animals. The same may also apply to soil invertebrates with a low degree of solarisation when they come into contact with the water in the soil.

Besides the classic floristic, faunal and biocoenotic investigations that primarily record rather unspecific reactions to pollutant exposure at higher organisational levels of the biological system, various newer methods have been introduced as instruments of bioindication. Most of these are biomarkers and biosensors.

Biomarkers are measurable biological parameters at the suborganismic (genetic, enzymatic, physiological, morphological) level in which structural or functional changes indicate environmental influences in general and the action of pollutants in particular in qualitative and sometimes also in quantitative terms. Examples: enzyme or substrate induction of cytochrome P-450 and other Phase I enzymes by various halogenated hydrocarbons; the incidence of forms of industrial melanism as markers for air pollution; tanning of the human skin caused by UV radiation; changes in the morphological, histological or ultra-structure of organisms or monitor organs (e.g. liver, thymus, testicles) following exposure to pollutants.

A biosensor is a measuring device that produces a signal in proportion to the concentration of a defined group of substances through a suitable combination of a selective biological system, e.g. enzyme, antibody, membrane, organelle, cell or tissue, and a physical transmission device (e.g. potentiometric or amperometric electrode, optical or optoelectronic receiver). Examples: toxiguard bacterial toximeter; EuCyano bacterial electrode. Biotest (bioassay): routine toxicological-pharmacological procedure for testing the effects of agents (environmental chemicals, pharmaceuticals) on organisms, usually in the laboratory but occasionally in the field, under standardised conditions (with respect to biotic or abiotic factors). In the broader sense this definition covers cell and tissue cultures when used for testing purposes, enzyme tests and tests using microorganisms, plants and animals in the form of single-species or multi-species procedures in model ecological systems (e.g. microcosms and mesocosms). In the narrower sense the term only covers single-species and model system tests, while the other procedures may be called suborganismic tests. Bioassays use certain biomarkers or – less often – specific biosensors and can be used in bioindication or biomonitoring.

In conservation biology several terms for bioindicators have been established but have been used in various ways. In our sense indicator taxa (syn. ecological indicators) are species which are known to be sensitive to processes or pollutants that lead to a change in biodiversity and are taken as surrogates for larger communities and act as a gauge for the condition of a particular habitat, community or ecosystem (in the meaning of Meffe and Carroll, 1994; McGeoch, 1998). In the field of biodiversity research these species or groups of species are often taken as a surrogate for the diversity of another group which might be more difficult to identify (Sauberer et al., 2002). Such species are also termed biodiversity indicators (for review see McGeoch, 1998).

In the field of vegetation ecology, especially, a broad discussion has evolved concerning the term 'monitoring'. Different words are used to define the various ways of observing vegetation changes (monitoring processes). Some terms are clearly defined, although different ideas are sometimes covered by the same terms (for review of the various terms see also Traxler and Zechmeister, 1997). The most important definitions are given here too:

Survey: Qualitative and quantitative observations made by standardised procedures without any regard to repetition.

Surveillance: An extended programme of surveys, undertaken in order to provide a
  time series, to ascertain the variability and/or range of states or values which might
  be encountered over time (but without preconceptions of what these might be;
  Hellawell, 1991).
Monitoring: Biological monitoring is the regular, systematic use of organisms to determine environmental quality (Cairns, 1979). This is an easily applicable definition
  which can be used in all fields of bioindication/biomontoring.

Spellerberg (1991) restricts his definition of monitoring to a specific problem designed
to provide information on the characteristics of the problem and changes in these over
the course of time. The most outstanding definition of monitoring is given by Hellawell
(1991). According to him, monitoring is an intermittent (regular or irregular) surveillance carried out in order to ascertain the extent of compliance with a predetermined
standard or the degree of deviation from an expected norm. This is in line with many
methods in ecotoxicology, especially the methods in which biosensors are used.

With regard to genetic and non-genetic adaptation of organisms and communities
to environmental stress we have to differentiate between the terms tolerance, resistance and sensitivity.

Tolerance (Oehlmann and Markert, 1997): desired resistance of an organism or
  community to unfavourable abiotic (climate, radiation, pollutants) or biotic factors
  (parasites, pathogens), where adaptive physiological changes (e.g. enzyme induction, immune response) can be observed.
Resistance, unlike tolerance, is a genetically derived ability to withstand stress
  (Oehlmann and Markert, 1997). This means that all tolerant organisms are resistant, but not all resistant organisms are tolerant. However, in ecotoxicology the
  dividing line between tolerance and resistance is not always so clear. For example,
  the phenomenon of PICT (pollution induced community tolerance) is described as
  the phenomenon of community shifts towards more tolerant communities when
  contaminants are present. It can occur as a result of genetic or physiological adaptation within species or populations, or through the replacement of sensitive
  organisms by more resistant organisms (Blanck et al., 1988; Rutgers et al., 1998).
Sensitivity of an organism or a community means its susceptibility to biotic or abiotic
  change. Sensitivity is low if the tolerance or resistance to an environmental stressor
  is high, and sensitivity is high if the tolerance or resistance is low.

4. From environmental monitoring to human health (environmental medicine)

Bioindication and biomonitoring must supply information on the degree of pollution
or degradation of ecosystems. Figure 4 explains the dilemma of integrated bioindication (high level of both complexity and dynamics). For integrative approaches
bioindication is not an "environmental monitoring machine" for a specific constellation of factors; ideally, it is an integrated consideration of various bioindicative test
systems which attempts, in conjunction with other environmental parameters, to

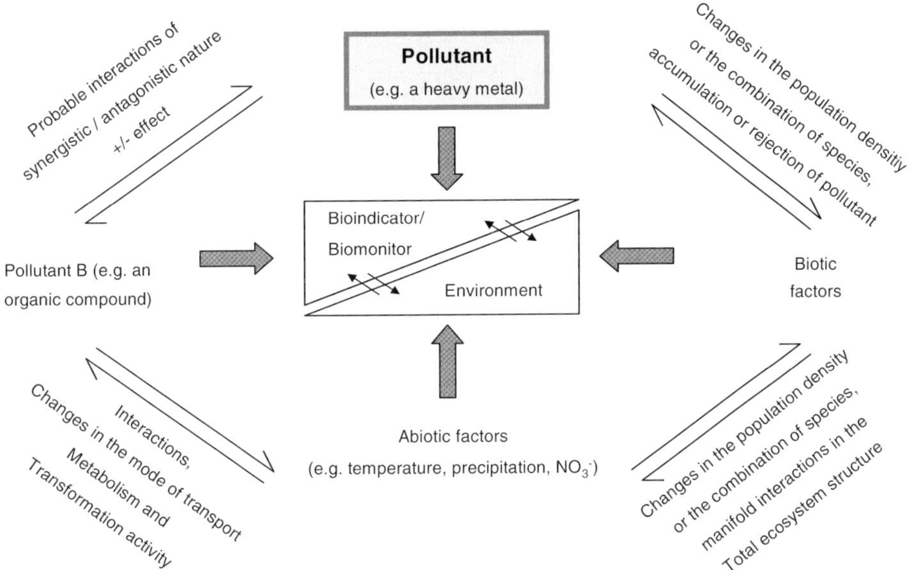

Figure 4. Simplified representation of complex (eco-)system interrelations with regard to a pollutant, and consequences for bioindication and biomonitoring (from Markert, 1996). As a rule, it is assumed that a pollutant affects an organism (bioindicator/biomonitor). Both the organism and the pollutant interact closely with other ecosystem compartments. The life activity of the organism is therefore influenced by a great number of abiotic and biotic factors and may often be subject to the action of several pollutants, especially under "natural" field conditions. With regard to the interpretation of the "information" given by the bioindicator/biomonitor, the problem often arises as to where the change observed or measured in the bioindicator/biomonitor really originates. Even a combined multifunctional and multi-structural view of the various ecosystem compartments has often left individual operative mechanisms unexplained . What makes matters even more difficult is that the pollutant to be monitored is in close competition with all other environmental compartments. So it is by no means certain, although rather probable, that pollutant A does not interact synergistically or antagonistically with pollutant B. Moreover, the absorption, location and metabolism of both have not yet been adequately described. However, pollutant A may also affect other biota, which may react even more sensitively to A than the bioindicator itself. If this sensitivity results in a change in the population density of a more sensitive organism, the occurrence of the bioindicator itself may also be affected, at least if the former is in direct or indirect competition with the latter. The question remains as to whether it is possible at all to make a statement about the current condition of the ecosystem as a whole by examining a single bioindicator.

produce a definite picture of a pollution situation and its development in the interests of prophylactic care of health and the environment.

Figure 5 is a diagram of a complete dynamic environmental monitoring system supported by bioindication. It can re-combine its measurement parameters according to the particular system to be monitored or the scientific frame of reference. The two main subjects of investigation – man and the environment – and the disciplines human toxicology and ecotoxicology derived from them are associated with various "toolboxes" and sets of tests ("tools", e.g. bioassays) for integrated environmental monitoring. The system shown in Figure 5 consists of six toolboxes. The first two are derived

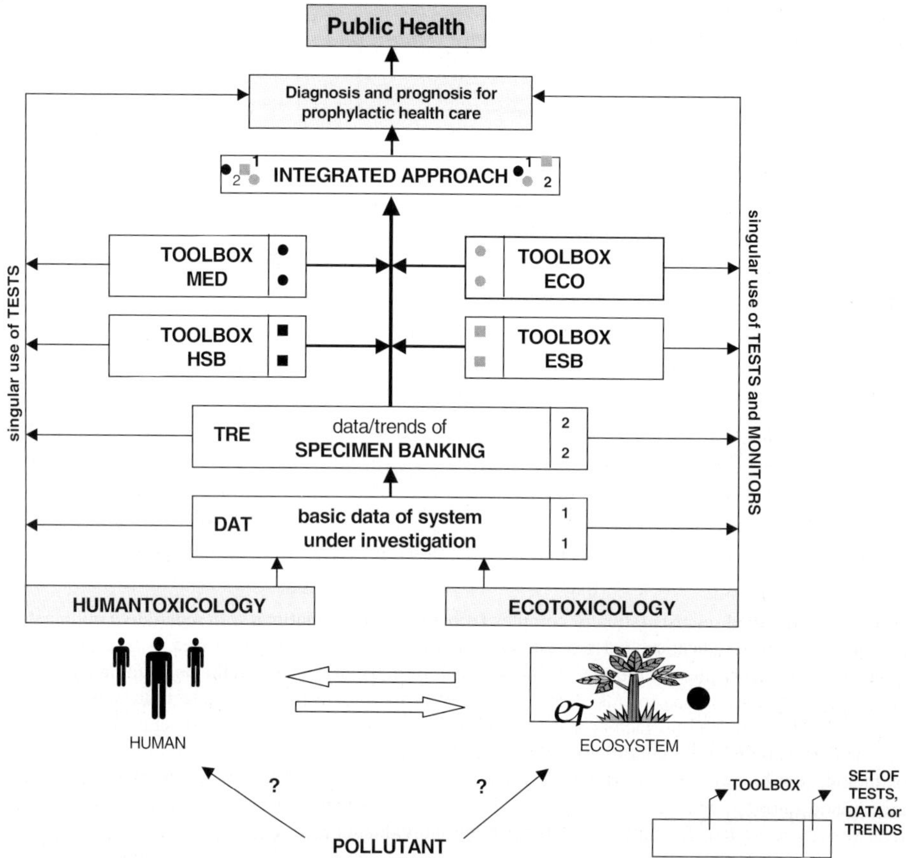

Figure 5. Possible hierarchical structure of a bioindicative toolbox model for integrative approaches in human- and ecotoxicology. The toolboxes MED and ECO contain single sets of tests that can be combined functionally to allow an integrated approach to the particular frame of reference or a specific scientific problem. The toolboxes HSB (human specimen banking) and ESB (environmental specimen banking) represent years of results from international environmental sample banks specialising in environmental and human toxicology; in addition to MED and ECO they provide important information on the ecotoxicological and human-toxicological behaviour of environmental chemicals. In the integrated approach, all the results obtained singly are substantiated by existing basic data available from (eco-)systems research, toxicology and environmental sample banks. The parameter constellations necessary for this are taken from the toolboxes TRE and DAT (from Markert et al., 2002, in preparation).

mainly from environmental research: DAT (for data) and TRE (for trend). DAT contains, as a set, all the data available from the (eco-)system under investigation, i.e. including data acquired by purely instrumental means, for example from the meteorological sphere. DAT also contains maximum permissible concentrations of substances in drinking water, food or air at the workplace and the data for the relevant ADI ("acceptable daily intake") and NO(A)EL ("no observed (adverse) effect level"). The toolbox TRE contains data on trends; these have been compiled mainly from years of investigations by national environmental sample banks, or information available from long-term national and international studies (e.g. Duvigneaud and Denayer-De Smet, 1973; Ellenberg et al., 1986; Likens et al., 1977). Specific conclusions and trend forecasts can then be prepared using the subsequent toolboxes HSB (human specimen banking) and ESB (environmental specimen banking) (see also Kettrup, 2002 in this book). The toolbox MED (medicine) contains all the usual methods employed in haematological and chemical clinical investigations of subchronic and chronic toxicity, whereas ECO is largely made up of all the bioindicative testing systems and monitors relevant to ecosystems which may be combined to suit the particular situation to be monitored.

The data from all the toolboxes must interact with each other in such a way that it is possible to assess the average health risk for specific groups of the population or determine a future upper limit of risk from pollutants by forming networks. This risk assessment ultimately makes use of all the toxicological limits that take the nature of the effect and dose-effect relationships into account according to the current status of scientific knowledge. Since toxicological experiments cannot be carried out on human beings, recourse has to be made to experience at the workplace and cases of poisoning in order to permit an evaluation and risk assessment. Besides examining reports on individual cases, greater efforts must be made to reveal the effects of substances as a cause of disease by means of epidemiological surveys with exposed groups as compared to a control group. The development and use of simulation models supported by information technology, taking all the data collected into account, will play an important role here, since a large number of parameters that do not interact directly have to be combined. They include various data from the field of epidemiology, from mutagenicity studies, toxicokinetics, metabolism research and structure-effect relationships.

Rapport (1992) suggests what he calls "ecomedical" indicators. For the ecosystem medicine approach, efforts should be made to apply systematic diagnostic protocols from human medicine to questions of ecosystem health. Beginning with the observation that medical practice has always relied on a suite of indicators for assessing human health, this broad approach is also required in screening ecosystems for possible pathologies. Further, in ecosystems, as in human health, no single indicator is likely to prove efficient as an early warning symptom, a diagnostic measure, and an integrated measure of the health of the entire system.

5. Comparison of instrumental measurement and the use of bioindicators/biomonitors with respect to harmonisation and quality control

The strong similarity in terms between instrumental chemical analysis (qualitative and quantitative measurements) and the field of bioindicators (as a qualitative approach to pollution control) and biomonitors (as a quantitative approach) makes it necessary to compare the two techniques.

5.1. Instruments and bioindicators

The more technical details of instrumental analysis are shown in Figure 6, which represent typical procedures for measuring chemical substances, enzyme activity or other ecosystem-relevant parameters by spectrometers or photometers. In a large number of

　　　　　　　　B.A. Markert, A.M. Breure, H.G. Zechmeister

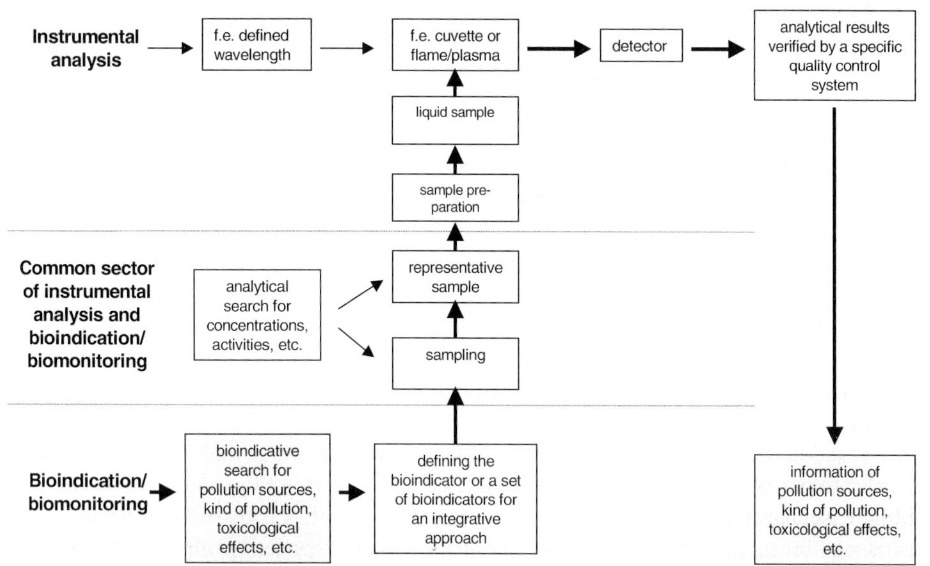

Figure 6. Comparison of measurements performed by spectrometers and bioindicators/biomonitors. In practice, instrumental measurements are often an integral part of bioindication (from Markert et al., 2002). A full instrumental flow chart for instrumental chemical analysis of environmental samples can be found in Markert (1996).

spectrometric methods a specific wavelength is used to obtain a signal by analysing a sample placed in a cuvette (photometer), flame (AAS), graphite furnace (AAS), plasma (ICP/MS or ICP/OES), supported by photomultipliers, amplifiers and other equipment and finally evaluated by detector systems. Quality control of the instrumental measurement is carried out with standard reference materials, for example. The main sources of error are the sampling procedure (up to 1000%) and sample preparation (up to 300%). A detailed discussion of typical errors in orders of magnitude is given by Markert (1996).

The direct comparison with a biological measuring device (bioindicator) in Figure 6 shows that the whole process of instrumental measurement is very often integrated into the procedure of bioindication, at least when samples have to be analysed for chemical compounds. This means that laboratory work on bioindicators depends heavily on instrumental measuring equipment to obtain additional information from the bioindicator. So when the question "bioindication or direct instrumental measurement?" is asked it seems that this relationship has not been fully understood. The practical laboratory problems encountered in biomonitoring are often the same as in chemical analysis. Take, for example, paradigm 1 of the sampling process: "The samples collected must be representative for the scientific question under review" (Markert, 1996). The representative collection of samples for monitoring or/and instrumental measurement has to done with the greatest care. This prerequisite is mentioned and explained in numerous excellent articles and textbooks and is not discussed here in detail (Keith, 1988; Klein and Paulus, 1995; Markert, 1994; Rasemann and Markert, 1998; Wagner, 1992, etc).

5.2. Precision and accuracy

In addition to the similar need for highest representative quality of the sample to be analysed or to be used as a bioindicator, most general rules and prerequisites of quality control in chemical analysis have to be taken into account in biomonitoring activities. In the last 20 years a strict differentiation between the terms "precision" (reproducibility) and "accuracy" (the "true" value) has been established in chemical analytical research (Fig. 7). The practical application of this differentiation makes it possible to determine the "true" or real content of a substance "X" in a sample "Y". The purpose of determining the precision of the data by repeatedly measuring the analytical signal is to track down and eliminate errors which might be generated, for example, by insufficient long-term stability of the measuring device (device-specific misadjustment). If the analytical procedures are not too complex, the precision will be 1 to 5%, and for most analytical problems this can be considered sufficiently exact. However, the mere fact that a signal is readily reproducible does not permit any statement about its accuracy. Even highly precise data can diverge greatly from the "true" (e.g. element) content of a sample. Correct analytical results can only be obtained if the entire analytical process is subjected to targeted quality control, where every result is checked for its precision and accuracy. Basically, two methods are now used to check the accuracy of analytical results: (a) use of standard reference materials (commercially available samples with a certified content of the compound to be measured and

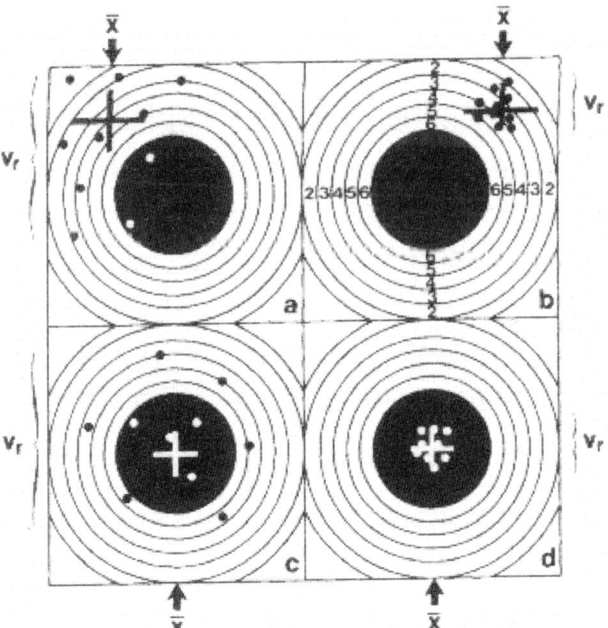

Figure 7. Illustration of the terms "precision" (reproducibility) and "accuracy" (the "true" value) in analytical chemistry (from Markert, 1996): a. Poor precision and poor accuracy, b. good precision and poor accuracy, c. poor precision and good accuracy, d. good precision and good accuracy, $\bar{x}$ = arithmetic mean, v_r = coefficient of variation.

a matrix similar to the original samples to be measured in the laboratory); and (b) use of independent analytical procedures.

With bioindicators we can, of course, carry out repeated sampling to get an idea of how "stable" the bioindicator under investigation is in respect of site and time variations. A more difficult problem is that of accuracy during the sampling procedure, for at present we have no "certified reference system" as a calibrator for accuracy in representative sampling. As a rule, "polluted" and "unpolluted" systems will be compared, but there is no way to be sure of working accurately. The only possible strategy here is that of "independent methods", when different research groups have the task of working in the same area with the same indicators, so that the data – obtained independently – can be compared. This is a very expensive method that can only be used in very special bioindication proposals where method development is of general concern, e.g. for EU or US directives.

5.3. Calibration

In general there is a considerable problem with bioindicators themselves, which does not usually arise with instrumental measurement techniques: the calibration of the biological system as such (Fig. 8). The limits within which organisms can indicate exposure become especially obvious in attempts to quantify environmental qualities, e.g. in biomonitoring in the stricter sense of the term (Markert et al., 1997). Although the number of potential bioindicators is virtually growing by the hour, it is difficult to find organisms (in nature) that meet the criteria of an active or passive biomonitor. For instance, the analysis of individual accumulation indicators for body burdens of certain substances does not necessarily permit conclusions about concentrations in the environment (Fig. 7). Many plants and animals display high accumulation factors for certain substances at low environmental concentrations, but the accumulation factors decrease sharply at higher environmental levels. The result is more or less a plateau curve for environmental concentration/body burdens (Fig. 8). On the other hand, many organisms succeed in keeping their uptake of toxic substances very low over a wide range of concentrations in the environment (Markert et al., 1997). Not until acutely toxic levels in the environment are exceeded do the regulatory mechanisms break down, resulting in a high degree of accumulation (Fig. 8). Exceptions are, of course, substances that are not taken up actively but enter the body by way of diffusion processes – doubtless rare in the case of inorganic metal compounds.

This often means that the relationship between the bioindicator/biomonitor and its environment in respect of the concentration of the compound to be accumulated is not linear but logarithmic. Even when linearity of the logarithmic function is achieved by mathematical conversion, the linear relationship between the two measurements is restricted to a small range. But organisms can only provide unequivocal information on their environment if a linear relationship exists which is comparable to the calibration line of measuring instruments.

Compared to spectrometric instrumental analytical methods, for example, where the linear calibration range normally covers several orders of magnitude, a linear range for bioindicators is more difficult to achieve since living organisms are constantly changing their "hardware" by biological, living processes. Standardisation of bio-

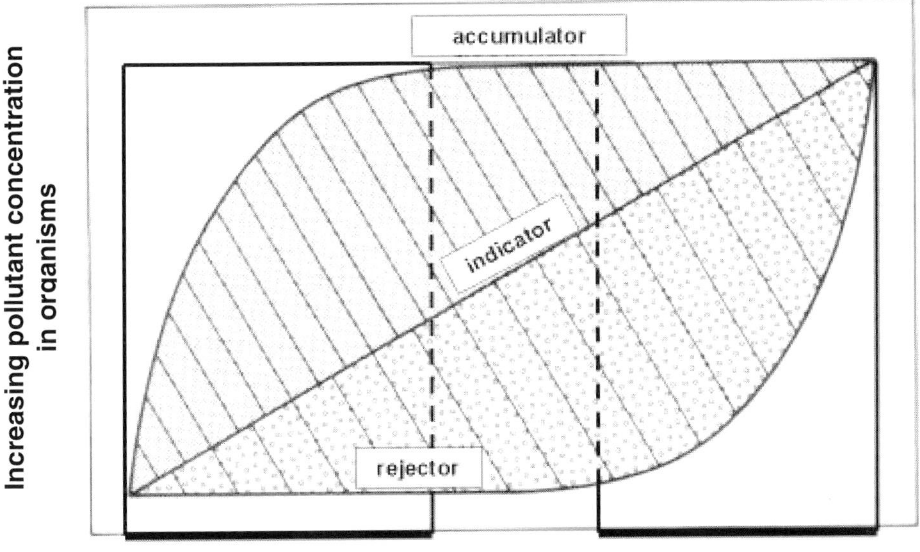

Figure 8.　Correlation between the environmental concentration of the pollutant to be monitored and the concentration in the organism. Linear ranges for calibration are very limited for accumulators and rejectors (from Markert et al., 1997).

indicators therefore seems unrealistic at the moment, which means that harmonisation between users of the same indicators is of specific and real concern for the future.

5.4. Harmonisation

Just as interlaboratory tests have for years enabled different laboratories to use real samples to optimise the quality of their own analyses in the field of analytical chemistry, greater attention must be given to harmonising the use of the same indicators in different places for the "calibration" of bioindicators. This is not so much true of work carried out in the laboratory, since bioassays as tests for chemicals, for example, are highly standardised and thus reproducible; it applies chiefly to all aspects of the use of bioindicators in the field. First of all, more cooperative planning in programme design seems absolutely necessary in order to compare results from individual working groups. On a regional and national level this is relatively easy to achieve, but on a global and intercontinental level the geographic distances between the research groups sometimes pose a problem. For example, the International Atomic Energy Agency (IAEA, 2001) tries to carry out biomonitoring of elements in different continents, and the high cost of personal meetings for an exchange of views has to be taken into account. Training and crash courses over a defined period of time (e.g. weeks) seem to be the first and best step towards harmonising scientific and (sometimes) cultural differences. And this should not be underestimated in a globalising world: bioindication in its different facets and on its different scientific levels can be performed by practically anybody, so that cross-border projects, especially, have a tremendous intercultural impact. We should beware of over-optimism, but "bioindication may be seen as a gateway to intercultural understanding and as a catalyst for peaceful international cooperation". Questions to be answered during this exchange of information might include how to relate observations of the same phenomena made by different techniques, such as remote sensing and on-site information (Smodiš, 2002 (this book); Roots, 1996). Scaling problems in space and in time are partly a matter of programme design. Programme design includes choice of measurements, sensors and recording methods and finally questions of information delivery and information technologies. Good examples of "questions in mind before starting the job" can be found in numerous national and international sampling campaigns for environmental observation and in literature dealing specifically with these harmonisation steps (e.g. Schroeder et al., 1996; Parris, 2002 (this book); Matthiessen, 2002 (this book); Bosch and Pinborg, 2002 (this book); Lazorchak et al., 2002 (this book)).

6. Strategies and concepts

The following reflects only a very small part of the overall existing and proposed strategies and concepts for bioindication. A great many more details on specific programmes are given, for example, by the Environmental Protection Agency (EPA, US), the OECD and the EEA. Further international and national organisations (the International Standards Organisation (ISO), CH), the European Union (EU, Belgium), especially in its section on "Measurement and Testing" (the former Bureau Community of Reference (BCR, Belgium)), Deutsches Institut für Normung (DIN, FRG) and others have elaborated various programmes for environmental control, observation and protection which are available on request via literature search or (more effective) via the internet.

The future development and coordination of bioindication methods should follow a two-levelled (A and B) parallel line:

– Level A optimises the development and harmonisation of existing and new indicators to make them suitable for practical use in risk management.
– Level B, already discussed in detail in Figure 4, represents a strongly integrated approach with environmental and health indicators to fill the gap between environmental biomonitoring and human health aspects.

A few concepts and examples of an integrated approach to bioindication based on forward-looking strategies are described below.

6.1. The Multi-Markered Bioindicator Concept (MMBC)

As we have already explained in detail in Section 4, the dilemma of bioindication lies in the fact that conclusions about the "overall condition" of an ecosystem have to be drawn from observations of a few representative indicator species. So because of the demands made on bioindication we have to ensure that the use of bioindicators is not carried *ad absurdum,* for its own sake, as a result of the extreme complexity of systems in conjunction with a high level of dynamic development. In future, simplifications – i.e. the reduction of a great diversity of species to a few (representative) bioindicators – should be carried out in a less isolated manner.

Besides increasing the specificity of bioindicators it is essential to place more emphasis on examining their functional interactions and interdependence, as we have already explained in Section 4 (Fig. 4).

Figure 9 gives a summary of the Multi-Markered Bioindication (MMB) Concept. The sole objective of this concept is to relate toxicological effects on a system to a potential hazard to human health. As described in detail in Section 4, the aim is to combine ecotoxicological data with data from environmental medicine by means of a toolbox model and the integrated use of various instrumental and bioindicative methods. As Figure 9 shows, possible mass balances for a particular pollutant are initially established by means of bioindicators and instrumental measurement methods; subsequently, they are traced back qualitatively and quantitatively to their probable sources, using the toolboxes ECO, ESB and TRE to facilitate the method (left side of Figure 9). At the same time, data on human toxicology from the fields of both environmental and ecological medicine are compiled with the aid of the toolboxes MED, HSB and TRE (right side of Figure 9). The MMB Concept is an attempt to combine data from human toxicology and ecotoxicology via "windows" in the context of an

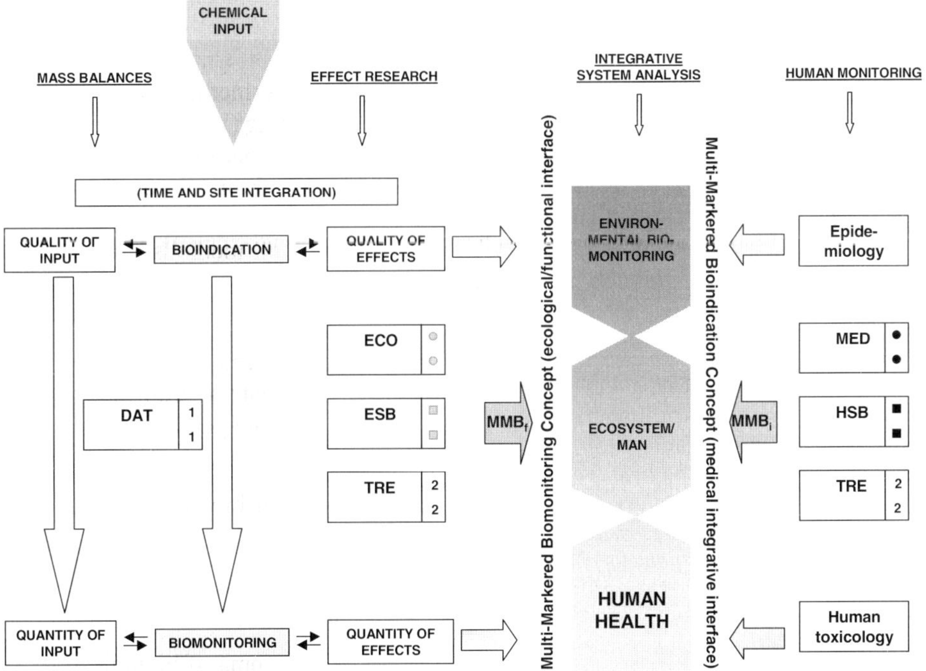

Figure 9. The Multi-Markered Bioindication Concept (MMBC) with its functional and integrated windows of prophylactic health care (from Markert et al., 2002). Explanations in the text. DAT, ECO, ESB, TRE, MED and HSB designate individual toolboxes and their test sets as shown in Fig. 5.

integrated system analysis in order to permit health care of a prophylactic and predictable nature. Intelligent calculation methods are required to take both functional (MMB_f) and integrated (MMB_i) aspects into account. Some of these methods have yet to be developed by basic research, since there is too little knowledge of certain functional and integrated connections.

6.2. Environmental sample banks

The purpose of environmental sample banks is to acquire samples capable of providing ecotoxicological information and to store them without change over long periods to permit retrospective analysis and evaluation of pollution of the environment with substances that could not be analysed, or did not seem relevant, at the time the samples were taken (Wagner, 1992). Individual aspects and background have been given in detail in Chapter 20 (Kettrup, 2002, this book) of this volume. The tasks and objectives of environmental sample banks may be outlined as follows (Klein, 1999):

- to determine the concentrations of substances that had not been identified as pollutants at the time the samples were stored, or which could not be analysed with sufficient accuracy (retrospective monitoring);
- to check the success or failure of current and future prohibitions and restrictions in the environmental sector;
- regular monitoring of the concentrations of pollutants already identified by systematic characterisation of the samples before archiving;
- prediction of trends in local, regional and global pollution;
- description of standardised sampling methods;
- documentation of the conditions under which the sample material is stored as a requirement for obtaining comparable results.

Table 2. Sample species from the German Federal Environmental Sample Bank (from Klein, 1999).

Sample species	Target compartment
Spruce (*Picea abies*)/pine (*Pinus sylvestris*)	Annual shoots
Red beech (*Fagus sylvatica*)/Lombardy Poplar (*Populus nigra "Italica"*)	Leaves
Domestic pigeon (*Columba livia f. domestica*)	Eggs
Roe deer (*Capreolus capreolus*)	Liver (kidneys)
Earthworm (*Lumbricus terrestris/Aporrectodea longa*)	Worm body without gut contents
Zebra mussel (*Dreissena polymorpha*)	Soft parts
Bream (*Abramis brama*)	Muscle tissue and liver
Brown algae (*Fucus vesiculosus*)	Thallus
Edible mussel (*Mytilus edulis*)	Soft parts
Blenny (*Zoarces viviparus*)	Muscle tissue and liver
Herring gull (*Latus argentatus*)	Eggs
Lugworm (*Arenicola marina*)	Worm body without gut contents

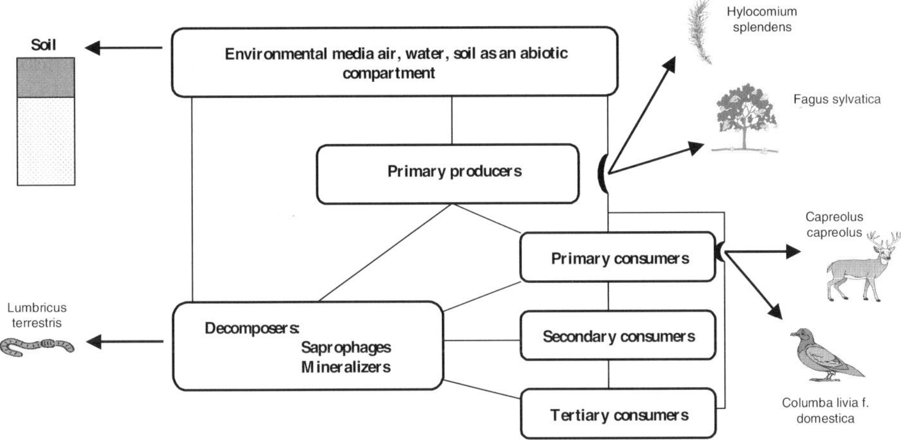

Figure 10. Selected sets of sample species (Hylocomium splendens is not included in any ESB-programme) at the ecosystem level for the German Federal Environmental Sample Bank (derived from Klein, 1999).

The German sample bank strategy also assumes that pollution at a particular location cannot be demonstrated by one bioindicator alone because of the different degree of exposure of the organisms in an ecosystem to pollutants and their different genetic predeterminants (Klein, 1999). Only a set of suitable bioindicators is capable of reflecting the pollutants present in the ecosystem.

Table 2 shows the bioindicators available at the German Federal Environmental Sample Bank. The criteria for choice of the sample species are discussed in detail in Klein and Paulus (1995). The expected functional connections between ecosystems are shown in Figure 10.

A problem posed by the environmental samples, which are carefully stored and refrigerated under liquid nitrogen, is the very high operating cost of the facility. There is also a certain lack of flexibility in taking in or handing out a bioindicator organism that has been analysed previously and over a period of years. The highly specific sampling guidelines often make it difficult to carry out comparisons with "normal" sampling protocols. These problems could be solved by integrating the results from the Environmental Sample Bank with other bioindication studies. In the MMBC this is shown by integrating the toolboxes ECO and MED with ESB and HSB in Figure 5.

6.3. *Example of integrated monitoring in the Euroregion Neisse (CZ, PL, D)*

By quantifying 12 chemical elements in the organ systems of rats (*Rattus norvegicus*) living wild in Zittau Zoo (Saxony) it was aimed to investigate the suitability of this species as a passive bioindicator (Wuenschmann et al., 2001, 2002). Besides determining "background concentrations" the emphasis was on sex and age specific accumulation of individual elements in the organ system of *Rattus norvegicus*. Individual elements were found to show an affinity for certain tissues and organs. In particular the

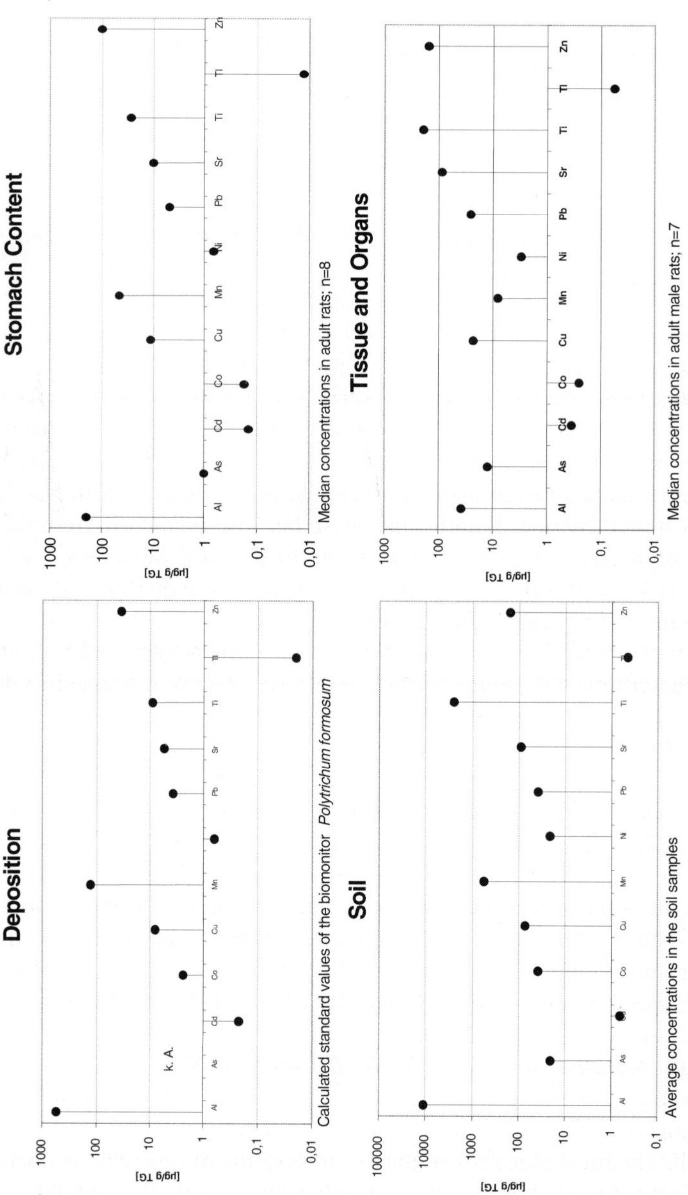

Figure 11. Integrated comparison of element distributions in the media air (deposition), soil, stomach content and tissue and organs in the study area. The highest median concentrations shown in the figure "Tissue and Organs" were measured in the following tissues and organs: Al, Ni, Pb, Sr, Ti and Zn in bones; Cd, Co and Te in the kidneys; Cu and As in the heart; Mn in the liver. All concentrations are stated in µg/g dry weight; k.A.: no information (from Wuenschmann et al., 2001, 2002).

sex and age specific characteristics found to exist for individual elements make it essential to prepare a detailed sampling strategy for later use of the rats as passive bioindicators.

Besides permitting an isolated view of individual elements in the animal's organ system, *Rattus norvegicus* is particularly suitable as an integrative bioindicator from the ecotoxicological point of view since it is affected indirectly by all the environmental media and directly via the food chain. But in order to ascertain such connections it is necessary to have study areas for which an adequate volume of additional ecotoxicological data with relevance to prophylactic health care has been acquired. In the Euroregion Neisse we are in the fortunate position of having data on both atmospheric deposition (from moss analyses) and soil data from years of research work.

Figure 11 is a comparison of the element concentrations from deposition, soil analyses and stomach content and the highest median concentrations revealed by tissue and organ analyses. The stomach content of the rats did not show unusually high levels of individual elements. This is surprising in that high arsenic concentrations were found in the environmental medium "soil", and an examination of the arsenic levels in the organ system of the rats revealed arsenic levels well above those of the stomach content. Using the calculation from the body-burden method it was possible to show that some tissues and organs have typical depot characteristics. In our investigations, for example, the elements Ni, Pb, Sr and Ti showed an increased affinity for bone tissue, whereas Cd and Tl tended to choose the kidneys as a depot organ. The tissue and organ concentrations shown here may therefore be regarded as possible initial background values for moderately polluted regions. The considerable natural fluctuations of individual elements according to organs, sex and age which are described in this study make it essential to devise a detailed sampling strategy if *Rattus norvegicus* is to be used successfully as a passive bioindicator (Wuenschmann et al., 2001, 2002).

6.4. Time- and site integration

The chief objective of biomonitoring is to permit statements about pollution and changes in biodiversity on various spatial and temporal scales. The site dependency of bioindicators/biomonitors is often affected by different biotopes which are characterised by different population structures and climatic, soil and food conditions. The latter can be delimited fairly easily by sampling the bioindicator from various locations at the same time. For this Wagner (1992) developed a system (Table 3) for fitting the sampling network to the quality of pollution control to be expected from the selected bioindicators (biomonitors) in use.

Compared with parameters resulting from the site, however, the behaviour of the bioindicator (biomonitor) along the time axis is much more difficult to determine. Especially in temperate climates, the great variation of seasonal effects causes variations of the pollutant concentration in one and the same bioindicator organism. For example, the seasonal fall in most of the heavy metal concentrations in spring (northern hemisphere) can be explained by the dilution effect of the first biomass of the year (Markert and Weckert, 1993). In particular a comparison of data obtained by different working groups using the same bioindicator has to be carefully checked with site-dependent and especially time-dependent parameters.

Table 3. Types of environmental monitoring networks used in ecological observation in Germany (from Wagner, 1992).

Types of monitoring network	Objectives	Characteristics of the network	Methods, examples
Permanent measuring stations/permanent observation sites, including ecosystem approaches	Reference and background data; time lines; integrated pollution and effect surveys; basis for comparison for environmental quality standards	Strictly according to regional statistics, avoiding local sources of interference; selected measuring points or sites to be observed	Widest possible range of methods as a reference basis, e.g. "Integrated Monitoring", DUFI. Baden Württemberg, also UBA monitoring network, ecosystem research + UPB, DWD
Monitoring networks for individual states	Overview of regional statistics; background data	Coordinate-based, wide-meshed networks (10-max. 50 km, avoiding local sources of interference)	Preferably passive biomonitoring, e.g. Bavarian moss and spruce monitoring network, Saarland poplar/spruce network
Regional monitoring networks	Screening (identification and delimitation of polluted areas or zones); integrated effects of complex or unknown types of pollution	Usually regular, relatively close-meshed measuring networks (approx. 1–10 km) limited in size (e.g. rural district, county, "polluted area")	Active and passive biomonitoring, effect cadaster in polluted and "clean air" regions, without reference to specific emitters
Emitter-related monitoring networks	To determine the extent of spread of pollution and the pollutant effects of an emitter	Usually close-meshed, often radial or linear networks or transects (<1–10 km between measuring points)	Primarily active or experimental methods geared to specific emitters or pollutants
Environmental impact analyses	To determine the degree of existing pollution and maximum tolerated burden before planned measures take effect (preservation of evidence)	As above	As above Possibly additional unspecific methods + UPB as preservation of ecotoxicological evidence

6.5. *Example of an integrated approach to bioindication of the biodiversity of a region and the influences acting on it*

A question much discussed internationally is that of the correlations in the biodiversity of different groups of organisms and those of the prime movers behind such connections. In a cultivated landscape, anthropogenic impacts naturally have to be taken into account in addition to natural parameters.

In a joint project carried out in the context of the extremely extensive study "Cultivated Landscape Research in Austria", over 30 research workers from eight institutions took 10 random samples from each of 41 square sampling sites with a side length of 600 m. The sites were chosen by means of a random number generator according to totally objective criteria. The exact documentation of positions naturally makes it possible to repeat the procedure at any time to permit monitoring. The manner of choosing sites, especially, has been unsatisfactory (i.e. subjective) in many previous bioindication studies. Greater attention should in future be given to this topic in general in the interests of proper statistical evaluation.

Data on the following organisms were collected at all 410 sampling points: ants; grasshoppers; ferns and spermatophytes; lichens; mosses and liverworts; ground beetles; mammals; snails and slugs; spiders and birds. These groups were chosen according to ecological/functional criteria. The objective was to determine the correlation between the various groups and the resulting indicative function of the individual groups of organisms in respect of others (e.g. biodiversity indicators; shopping basket approach; Sauberer et al., 2002).

A simultaneous analysis was made of the connections between the variability of physical, chemical and biochemical soil parameters and the abundance of species. Links were further established between various net primary production parameters (real NPP, potential NPP, NPP after deduction of the harvest, etc.) and biodiversity parameters.

In a subsequent step, links were determined between parameters of landscape morphology, location and biogeography (including the history of vegetation) and biodiversity. Information on the landscape was incorporated in a GIS (Geographical Information System) by means of digitised aerial photographs. This made it possible to relate geographic structural data to other parameters (e.g. biodiversity) via complex links (e.g. Fragstats; McGarigal and Marks, 1995; Moser et al., 2002). Finally, structures defined in terms of landscape ecology (Forman and Godron, 1986) are also a suitable means of determining basic properties of the areas under review in order to establish or test ecological theories. It is hoped that the establishment of links between basic patterns of landscape ecology (e.g. density, distribution of corridors) and biodiversity on the same sampled areas as were selected for the closer biodiversity analysis will make it possible to understand the effects of landscape on patterns of biodiversity and interpret them in the light of theoretical concepts.

Parameters relevant to cultivated landscapes have also been viewed in relation to biodiversity patterns, with the inclusion of socioeconomic data, and connections established that permit forecasts of how the biodiversity of the country will alter according to various scenarios of change in the cultivated landscape.

In this study the future development of various anthropogenic activities is forecast by individual disciplines in conjunction with specific frames of reference. When developing the scenarios it generally has to be taken into account that the individual anthropogenic impacts have different dynamics. For example, certain structural and functional impacts (e.g. drainage) develop over long periods of more than a century, whereas the release of pesticides and the effects of road traffic can be observed in periods of only decades (Braun, 1985; Nentwig, 1995). Further technological developments (such as genetic engineering) may have ecological effects that are not yet known.

It is very much in the interests of integrated monitoring to encourage an interdisciplinary staffing of research groups in future surveys too. This would permit rapid and flexible adjustment of the working groups to the particular frame of reference and enable a quick exchange of information between the individual disciplines.

References

Altenburger, R., Schmitt-Jansen, M., 2002 (2003). Predicting toxic effects of contaminants in ecosystems using single species investigations. In: Markert, B.A., Breure, A.M., Zechmeister, H.G. (Eds), Bioindicators and Biomonitors. Elsevier, Oxford, pp. 153–198.

Arft, A.M., Walker, M.D., Gurevitch, J., Alatalo, J.M., Bret-Harte, M.S., Dale, M., Diemer, M., Gugerli, F., Henry, G.H.R., Jones, M.,H., Hollister, R.D., Jónsdóttir, I. S., Laine, K., Lévesque, E., Marion, G.M., Molau, U., Mølgaard, P., Nordenhäll, U., Raszhivin, V., Robinson, C.H., 1999. Responses of tundra plants to experimental warming: meta analysis of the International Tundra Experiment. Ecological Monographs 69, 491–511.

Bamford, S., Osae E.K., Serfor-Armah, Y., Nyarko, B., Ofosu, F., Aboh, I.J., Odamtten, G.T., 2001. Biomonitoring of air pollution through trace element analysis. In: Co-ordinated Research Project on Validation and Application of Plants as Biomonitors of Trace Element Atmospheric Pollution, Analyzed by Nuclear and Related Techniques. IAEA, NAHRES-63, Vienna.

Bargagli, R., 1995. The element composition of vegetation and the possible incidence of soil contamination of samples. The Science of the Total Environment 176, 121–128.

Bargagli, R., (Ed.) 1998. Trace Elements in Terrestrial Plants – an Ecophysiological Approach to Biomonitoring and Biorecovery. Springer-Verlag, Berlin, Heidelberg.

Bargagli, R., Sanchez-Hernandez, J.C., Monaci, F., 1999. Baseline concentrations of elements in the Antarctic macrolichen. Chemosphere, 38 (3), 475–487.

Blanck H., Wängberg, S.A., Molander, S., 1988. Pollution-induced community tolerance. A new ecotoxicological tool. In: Cairns, J.J., Pratt, J.R. (Eds), Functional Testing of Aquatic Biota for Estimating Hazards of Chemicals. ASTM STP 988, American Society for Testing and Materials, Philadelphia, pp. 219–230.

Bosch, P., Pinborg, U., 2002 (2003). Bioindicators and the indicator approach of the EEA. In: Markert, B.A., Breure, A.M., Zechmeister, H.G. (Eds), Bioindicators and Biomonitors. Elsevier, Oxford, pp. 903–916.

Braun, A., 1985. Agrarökologie im Spannungsfeld des Umweltschutzes. Ulf Pedersen Braunschweig.

Brooks, R.R., 1998. Plants that Hyperaccumulate Heavy Metals. Their Role in Phytoremediation, Microbiology, Archaeology, Mineral Exploration and Phytomining. CAB International, Wallingford, New York.

Bundesministerium für Umwelt, Naturschutz und Reaktorsicherheit (BUNR), 1992. Umweltpolitik. Konferenz der Vereinten Nationen für Umwelt und Entwicklung im Juni 1992 in Rio de Janeiro – Dokumente – Agenda 21. Köllen Druck + Verlag GmbH, Bonn, Germany.

Cairns, J., 1979. Biological monitoring – concept and scope. In: Cairns, J., Patil, G.P., Waters, W.E. (Eds), Environmental Biomonitoring, Assessment, Prediction and Management. International Cooperative Publishing House, Maryland, pp. 3–20.

Carreras, H.A., Gudino, G.L., Pignata, M.L., 1998. Comparative biomonitoring of atmospheric quality in five zones of Cordoba city (Argentina) employing the transplanted lichen Usnea sp. Environmental Pollution 103, 317–325.

Carson, R., 1962. Silent Spring. Penguin Books, Harmondsworth.

Conte, C., Mutti, I., Puglisi, P., Ferrarini, A., Regina, G., Maestri, E., Marmiroli, N., 1998. DNA fingerprinting analysis by a PCR based method for monitoring the genotoxic effects of heavy metals pollution, Chemosphere 37 (14–15), 2739–2749.

Conzales, C.M., Pignata, M.L., 1997. Effect of pollutants emitted by different urban-industrial sources on the chemical response of the transplanted ramalina ecklonii (spreng.) mey. & flot. Toxicological and Environmental Chemistry 69, 61–73.

Conzales, C.M., Orellana, L.C., Casanovas, S.S., Pignata, M.L., 1998. Environmental conditions and chemical response of a transplanted lichen to an urban area. Journal of Environmental Management, 53, 73–81.

Cortes, E., Gras, N., Pereira, I., Andonie, O., Sepulveda, S., 2001. Study of air pollution in Chile using biomonitor. In: Co-ordinated Research Project on Validation and Application of Plants as Biomonitors of Trace Element Atmospheric Pollution, Analyzed by Nuclear and Related Techniques. IAEA, NAHRES-63, Vienna.

Costanza, R., 1992. Ecological economic issues and considerations in indicator development, selection, and use: towards an operational definition of system health. In: McKenzie, D., Hyatt, D., McDonald, V. (Eds), Ecological Indicators, Vol. 2. Elsevier Applied Science, London and New York, pp. 1491–1502.

Costanza, R., D'Arge, R., De Groot, R., Farber, S., Grasso, M., Hannon, B., Limburg, K., Naeem, S., O'Neill, R., Paruelo, J., Raskin, R., Sutton, P., Van Den Belt, M., 1997. The value of the world's ecosystem services and natural capital. Nature, 387, 253–260.

Djingova, R., Kuleff, I., 2000. Instrumental techniques for trace analysis. In: Markert, B., Friese, K. (Eds), Trace Elements, Their Distribution and Effects in the Environment. Elsevier, Amsterdam, pp. 137–185.

Duvigneaud, P., Denayer-De Smet, S., 1973. Biological cycling of minerals in temperate deciduous forests. Ecol. Stud. 1, 199–225.

ECB (European Chemicals Bureau), 2001. http://ecb.ei.jrc.it/

Ellenberg, H., Mayer, R., Schauermann, J., 1986. Ökosystemforschung, Ergebnisse des Solling Projektes. Ulmer, Stuttgart.

Ellenberg, H., Weber, H.E., Düll, R., Wirth, V., Werner, W., Paulißen, D., 1992. Indicator values of plants in central Europe. Scripta Geobotanica, Vol. 18, 2nd edn. Lehrstuhl für Geobotanik der Universität Göttingen.

Forman, R.T., Godron, M., 1986. Landscape Ecology. John Wiley, New York.

Fraenzle, S., Markert, B., 2002. The Biological System of the Elements (BSE) – a brief introduction to historical and applied aspects with special reference to "ecotoxicological identity cards" for different element species (f.e. As and Sn). Environmental Pollution, 120, 27–45.

Frontasyeva, M.V., Steinnes, E., 1995. Epithermal neutron activation analysis of mosses used to monitor heavy metal deposition around an iron smelter complex. The Analyst, 120, 1437–1440.

Freitas, M.C., Reis, M.A., Alves, L.C., Wolterbeek, H.T., 1999. Distribution in Portugal of some pollutants in the lichen Parmelia sulcata. Environmental Pollution, 106, 229–235.

Fytianos, K., Evgemodou, E., Zachariadis, G., 1999. Use of macroalgae as biological indicators of heavy metal pollution in Thermaikos Gulf, Greece. Bulletin of Environmental Contamination and Toxicology, 62, 630–637.

Garty, J., 1998. Airborne elements, cell membranes, and chlorophyll in transplanted lichens. Journal of Environmental Quality, 27, 973–979.

Gignac, L.D., Nicholson, B.J., Bayley, S.E., 1998. The utilization of bryophytes in bioclimatic modelling: predicted northward migration of peatlands in the Mackenzie River Basin, Canada, as a result of global warming. The Bryologist, 101, 572–587.

Grabherr G., 1991. Natur und Landschaft in Vorarlberg. In: Broggi, M., Grabherr, G., Alge, R. (Eds), Biotope in Vorarlberg. Natur und Landschaft in Vorarlberg 4. Vorarlberger Verlagsanstalt, Bregenz, pp. 36–76.

Grabherr, G., Gottfried, M., Pauli, H., 1994. Climate effects on mountain plants. Nature, 369, 448.

Hammond, P.M., 1994. Practical approaches to the estimation of the extent of biodiversity in speciose groups. Philosophical Transactions of the Royal Society London B 345, 119–136.

Harada, H., Hatanaka, T., 1998. Natural background levels of trace elements in wild plants. Soil Science Plant Nutrition, 44 (3), 443–452.

Hellawell, J. M., 1991. Development of a rationale for monitoring. In: Goldsmith, F.B. (Ed.), Monitoring for Conservation and Ecology, Chapman & Hall, London, pp. 1–14.

Herpin, U., Siewers, U., Kreimes, K., Markert, B., 2001. Biomonitoring – evaluation and assessment of heavy metal concentrations from two German moss surveys. In: Burga, C.A., Kratochwil, A. (Eds), General and Applied Aspects on Regional and Global Scales. Kluwer Academic, Dordrecht, Tasks for Vegetation Science, 35, pp. 73–95.

Heywood, V.H., Watson, R.T., 1995. Global biodiversity assessment. UNEP, Press Syndicate, Cambridge.

IAEA (International Atomic Energy Agency), 2001. Co-ordinated Research Project on Validation and Application of Plants as Biomonitors of Trace Element Atmospheric Pollution, Analyzed by Nuclear and Related Techniques. IAEA, NAHRES-63, Vienna.

IUCLID (International Uniform Chemical Information Database) CD-rom, 2000. Office for Official Publications of the European Communities, L-2985 Luxembourg.

ICC (International Climate Change), 2001. The Scientific Basis. Contribution of Working Group I to the Third Assessment Report of the Intergovernmental Panel on Climate Change (IPCC). Houghton, J.T., Ding, Y., Griggs, D.J., Noguer, M., van der Linden, P.J., Xiaosu, D. (Eds), Cambridge University Press, Cambridge.

Kappelle, M., VanVuuren, M.I., Baas, P., 1999. Effects of climate change on biodiversity: a review and identification of key research issues. Biodiversity and Conservation, 8, 1383–1397.

Keith, L.H. (Ed.), 1988. Principles of Environmental Sampling. ACS Professional Reference Book, American Chemical Society,Washington, DC.

Kettrup, A. 2002 (2003). Environmental Specimen Banking. In: Markert, B.A., Breure, A.M., Zechmeister, H.G. (Eds), Bioindicators and Biomonitors. Elsevier, Oxford, pp. 775–796.

Klein, R., Paulus, M. (Eds), 1995. Umweltproben für die Schadstoffanalytik im Biomonitoring. Gustav Fischer Verlag Jena, Stuttgart.

Klein, R., 1999. Retrospektive Wirkungsforschung mit lagerfähigen Umweltproben. In: Oehlmann, J., Markert, B. (Eds), Ökotoxikologie – Ökosystemare Ansätze und Methoden, Ecomed, Landsberg, pp. 285–293.

Klumpp, A., Domingos, M., Pignata, M.L., 2000. Air Pollution and vegetation damage in South America – state of knowledge and perspectives. In: Agrawal, S.B., Agrawal, M.A. (Eds), Environmental Pollution and Plant Responses. Lewis Publishers, Boca Raton.

Knauer, K., Ahner, B., Xue, H., Sigg, L., 1998. Metal and phytochelatin content in phytoplankton from freshwater lakes with different metal concentrations. Environmental Toxicology and Chemistry, 12, 2444–2452.

Kostka-Rick, R., Leffler, U.S., Markert, B., Herpin, U., Lusche, M., Lehrke, J., 2001. Biomonitoring zur wirkungsbezogenen Ermittlung der Schadstoffbelastung in terrestrischen Ökosystemen – Konzeption, Durchführung und Beurteilungsmaßstäbe im Rahmen von Genehmigungsverfahren, UWSF-Z. Umweltchemie und Ökotoxihologie, 12 (1), 5–12.

Lazorchak, J., Hill, B.H., Brown, B.S., McCormick, F.H., Engle, V., Lattier, M.J., Griffith, M.B., Maciorowski, A.F., Toth, G.P., 2002 (2003). U.S. EPA monitoring and bioindicator concepts needed to evaluate the biological integrity of aquatic systems. In: Markert, B.A., Breure, A.M., Zechmeister, H.G. (Eds), Bioindicators and Biomonitors. Elsevier, Oxford, pp. 837–873.

Lieth, H., 1998. Ecosystem principles for ecotoxicological analyses. In: Schüürmann, G., Markert, B. (Eds), Ecotoxicology – Ecological Fundamentals, Chemical Exposure and Biological Effects. John Wiley/Spectrum Akademischer Verlag, New York and Stuttgart, pp. 17–73.

Likens, G.E., Bormann, F.H., Pierce, R.S., Eaton, J.S., Johnson, N.M., 1977. Bio-geochemistry of a Forested Ecosystem. Springer, Berlin.

Loppi, S., Nelli, L., Aancora, S., Bargagli, R., 1996. Passive monitoring of trace elements by means of tree leaves, epiphytic lichens and bark substrate. Environmental Monitoring and Assessment 45, 81–88.

Markert, B. (Ed.), 1993. Plants as Biomonitors – Indicators for Heavy Metals in the Terrestrial Environment. VCH-Publisher, Weinheim.

Markert, B. (Ed.), 1994. Environmental Sampling for Trace Analysis. VCH-Publisher, Weinheim.

Markert, B., 1996. Instrumental Element and Multi-Element Analysis of Plant Samples. Wiley/VCH-Publisher, Chichester.

Markert, B., 2002. From biomonitoring to the Environmental Specimen Bank. In: Müller, P. (Ed.), Environmental Specimen Banking. In: Hutzinger, O. (Ed.), The Handbook of Environmental Chemistry. Springer-Verlag, Heidelberg, in preparation.

Markert, B., Weckert, V., 1993. Time-and-site integrated long-term biomonitoring of chemical elements by means of mosses. Toxicological Environmental Chemistry 40, 43–56.

Markert, B., Oehlmann, J., Roth, M., 1997. General aspects of heavy metal monitoring by plants and animals. In: Subramanian, G., Iyengar, V. (Eds), Environmental biomonitoring – exposure assessment and specimen banking. ACS Symp. Ser. 654, American Chemical Society, Washington, DC.

Markert, B., Wappelhorst, O., Weckert, V., Herpin, U., Siewers, U., Friese, K., Breulmann, G., 1999. The use of bioindicators for monitoring the heavy-metal status of the environment. Journal of Radioanalytical and Nuclear Chemistry, 240, 425–429.

Markert, B., Kayser, G., Korhammer, S., Oehlmann, J., 2000. Distribution and effects of trace substances in soils, plants and animals. In: Markert, B., Friese, K. (Eds), Trace Elements. Trace Metals in the Environment, Vol. 4. Elsevier, Amsterdam, pp. 3–31.

Markert, B., Fraenzle, S., Fomin, A., 2002. From the biological system of elements to biomonitoring. In: Merian, E., Anke, M., Ihnat, M., Stoeppler, M., Elements and Their Compounds in the Environment. Wiley/VCH, Chichester, in preparation.

Martinez-Cortizas, A., Pontevedra-Pombal, X., Garcia-Rodeja, E., Novoa-Munoz, J.C., Shotyk, W., 1999. Mercury in a Spanish peat bog: archive of climate change and atmospheric metal deposition. American Association for the Advancement of Science, 284, 939–942.

Matthiessen, P., 2002 (2003). Critical assessment of international marine monitoring programmes for biological effects of contaminants in the North-East Atlantic area. In: Markert, B.A., Breure, A.M., Zechmeister, H.G. (Eds), Bioindicators and Biomonitors. Elsevier, Oxford, pp. 917–939.

McGarigal, K., Marks, B.J., 1995. FRAGSTATS: spatial pattern analysis program for quantifying landscape structure. US Forest Service General Technical Report PNV 351.

McGeoch, M., 1998. The selection, testing and application of terrestrial insects as bioindicators. Biological Review 73, 181–201.

McNeely, J.A., Gadgil, M., Leveque, C., Padoch, C., Redford, K., 1995. Human influences on biodiversity. In: Heywood, V.H., Watson, R.T. (Eds), Global Biodiversity Assessment. UNEP, Press Syndicate, Cambridge, pp. 711–822.

Meffe, G.K., Carrol, C.R., 1994. Principles of Conservation Biology. Sinauer, Sunderland.

Mersch-Sundermann, V. (Ed.), 1999. Umweltmedizin, klinische Umweltmedizin, ökologische Medizin. Georg Thieme Verlag, Stuttgart.

Michelot, D., Poirer, F., Melendez-Howell, L.M., 1999. Metal content profiles in mushrooms collected in primary forests of Latin America. Archives of Environmental Contamination and Toxicology, 36, 256–263.

Mitchell, J.F.B., Manabe, S., Meleshko, I. V., Toroka, T., 1990. Equilibrium climate change and its implications for the future. In: Houghton, J.T., Jenkins, G.J., Ephraums, J.J. (Eds), Climate Change. The ICC Scientific Assessment, Cambridge University Press, Cambridge, pp. 134–172.

Moser, D., Zechmeister, H.G., Plutzar, C., Sauberer, N., Grabherr, G., 2002. Landscape shape complexity as an effective measure for plant species richness in rural landscapes. Landscape Ecology, in press.

Müller, P., 1980. Biogeographie. UTB, Ulmer-Verlag, Stuttgart.

Nentwig, W., 1995. Humanökologie. Springer, Berlin.

Normandin, L., Kennedy, G., Zayed, J., 1999. Potential of dandelion (Taraxacum officinale) as a bioindicator of manganese arising from the use of methylcyclopentadienyl manganese tricarbonyl in unleaded gasoline. The Science of the Total Environment, 239, 165–171.

Oehlmann, J., Markert, B., 1997. Humantoxikologie. Eine Einführung für Apotheker, Ärzte, Natur- und Ingenieurwissenschaftler. Wissenschaftliche Verlagsgesellschaft mbH Stuttgart.

Oehlmann, J., Markert, B. (Eds), 1999. Ökotoxikologie – Ökosystemare Ansätze und Methoden. Ecomed, Landsberg.

Parris, K., 2002 (2003). Assessing the environmental performance of agriculture: recent progress and future developments for OECD countries. In: Markert, B.A., Breure, A.M., Zechmeister, H.G. (Eds), Bioindicators and Biomonitors. Elsevier, Oxford, pp. 797–829.

Posthuma, L., Suter II, G.W., Traas, T.P., 2001. Species sensitivity distributions in ecotoxicology. CRC/Lewis Publishers, Boca Raton.

Rapport, D., 1992. Evolution of indicators of ecosystem health. In: McKenzie, D., Hyatt, D., McDonald, V. (Eds), Ecological Indicators, Vol. 1. Elsevier Applied Science, London.

Rasemann, W., Markert, B., 1998. Industrial waste dumps – sampling and analysis. In: Meyers, R.A. (Ed.), Encyclopedia of Environmental Analysis and Remediation, Vol. 4, John Wiley, 2356–2373.

Rodushkin, I., Ödman, F., Holmström, H., 1999. Multielement analysis of wild berries from northern Sweden by ICP techniques. The Science of the Total Environment, 231, 53–65.

Roots, E.F., 1992. Environmental information – a step to knowledge and understanding. Environmental Monitoring and Assessment, 50 (4), 87–94.

Roots, E.F., 1996. Environmental information – autobahn or maze? In: Schroeder, W., Fraenzle, O., Keune, H., Mandy, P. (Eds), Global Monitoring of Terrestrial Ecosystem. Ernst & Sohn Verlag für Architektur und technische Wissenschaften GmbH, Berlin, pp. 3–31.

Rutgers, M., Van 't Verlaat, I., Wind, B., Posthuma, L., Breure, A.M., 1998. Rapid method for assessing pollution-induced community tolerance in contaminated soil. Environmental Toxicology Chemistry 17, 2210–2213.

Saiki, M., Horimoto, L., Vasconcellos, M., Marcelli, M., Sumita, N., Saldiva, P., 2001. Determination of trace elements in lichen samples by instrumental neutron activation analysis. In: Co-ordinated Research Project on Validation and Application of Plants as Biomonitors of Trace Element Atmospheric Pollution, Analyzed by Nuclear and Related Techniques. IAEA, NAHRES-63, Vienna.

Sauberer N., Zulka, K.P., Abensperg-Traun, M., Berg, H.M., Bieringer, G., Milasowszky, N., Moser, D., Plutzar, C., Storch, C., Tröstl, R., Zechmeister, H.G., Grabherr, G., 2002. Biodiversity indicators in agricultural landscapes. Conservation Biology, in press.

Schneider, E.D., 1992. Global monitoring scales. In: McKenzie, D.H., Hyatt, D.E., McDonald, V.J (Eds), Ecological Indicators. Elsevier Applied Science, London, pp. 1009–1011.

Schroeder, W., Fraenzle, O., Keune, H., Mandy, P. (Eds), 1996. Global Monitoring of Terrestrial Ecosystems. Ernst & Sohn Verlag, Berlin.

Schubert, R., 1991. Possibilities and limitations in bioindication on landscape monitoring scales. In: McKenzie, D.H., Hyatt, D.E., McDonald, V.J. (Eds), Ecological Indicators. Elsevier Applied Science, London, pp. 1009–1011.

Schüürmann, G., Markert, B. (Eds), 1998. Ecotoxicology – Ecological Fundamentals, Chemical Exposure and Biological Effects. John Wiley, New York, and Spectrum Akademischer Verlag, Stuttgart.

Shugart, H.H., 1997. Plant and ecosystem functional types. In: Smith, T.M., Shugart, H.H., Woodward, F.I. (Eds), Plant Functional Types. Cambridge University Press, Cambridge, pp. 20–43.

Smodiš, B., 2002 (2003). IAEA approaches to assessment of chemical elements in atmosphere. In: Markert, B.A., Breure, A.M., Zechmeister, H.G. (Eds), Bioindicators and Biomonitors. Elsevier, Oxford, pp. 875–902.

Spellerberg, I.F., 1991. Monitoring Ecological Change. Cambridge University Press, New York.

Traxler A., Zechmeister, H.G., 1997. Definitionen und Begriffsabklärung des Monitorings. In: Traxler A. (Ed.), Handbuch des Vegetationsökologischen Monitorings. Monographien des Umweltbundesamtes 89a, pp. 4–12.

United Nations Conference on Environment and Development (UNCED), 1992. Agenda 21. Rio de Janeiro, June.

Vutchkov, M., 2001. Biomonitoring of air pollution in Jamaica through trace-element analysis of epiphytic plants using nuclear and related analytical techniques. In: Co-ordinated Research Project on Validation and Application of Plants as Biomonitors of Trace Element Atmospheric Pollution, Analyzed by Nuclear and Related Techniques. IAEA, NAHRES-63, Vienna.

Wagner, G, 1987. Entwicklung einer Methode zur großräumigen Überwachung mittels standardisierter Pappelblattproben von Pyramidenpappeln (Populus nigra Italica) am Beispiel von Blei, Cadmium und Zink. In: Stoeppler, M., Dürbeck, H.W. (Eds), Beiträge zur Umweltprobenbank, No. 5, Jülich Spezial, 412.

Wagner, G., 1992. Einsatzstrategien und Meßnetze für die Bioindikation im Umweltmonitoring. Ecoinforma, 1–8.

Wappelhorst, O., Korhammer, S., Leffler, U.S., Markert, B., 2000a. Ein Moosmonitoring zur Ermittlung atmosphärischer Elementeinträge in die Euroregion Neiße (D, PL, CZ), UWSF-Z. Umweltchemie und Ökotoxihologie 12 (4), 191–200.

Wappelhorst, O., Kuehn, I., Oehlmann, J., Markert, B., 2000b. Deposition and disease: a moss monitoring project as an approach to ascertaining potential connections. The Science of the Total Environment, 249, 243–256.

Wappelhorst, O., Kuehn, I., Heidenreich, H., Markert, B., 2002. Transfer of Selected Elements from Food into Human Milk (Ag, Ce, Co, Cr, Ga, La, Mo, Nb, Ru, Sb, Th, Ti and U). Nutrition, 18, 316–322.

Watmough, S., Hughes, R., Hutchinson, T., 1999. $^{206}Pb/^{207}Pb$ ratios in tree rings as monitors of environmental change. Environmental Science & Technology 33, 670–673.

Watson, R.T., Zinyowera, M.C., Moss, R.H., Dokken, D.J. (Eds), 1996. Climate Change 1995. Impacts, Adaptations and Mitigation of Climate Change: Scientific-technical Analysis. IPCC, Cambridge Univ. Press, Cambridge.

Winter, S., Wappelhorst, O., Markert, B., 2000. Löwenzahn Taraxacum officinale Web. als (städtischer) Bioindikator, UWSF-Z. Umweltchemie und Ökotoxihologie 12 (6), 311–321.

Wittig, R., 1993. General aspects of biomonitoring heavy metals by plants. In: Markert, B. (Ed.), Plants as Biomonitors – Indicators for Heavy Metals in the Terrestrial Environment. VCH-Publisher, Weinheim, pp. 3–27.

Wolterbeek, H.T., Kuik, P., Verburg, T.G., Herpin, U., Markert, B., Thöni, L., 1995. Moss interspecies comparisons in trace element concentrations. Environmental Monitoring Assessment 35, 263–286.

Woodward, F.I., Kelly, C.K., 1997. Plant functional types: towards a definition by environmental constraints. In: Smith, T.M., Shugart, H.H., Woodward, F.I. (Eds), Plant Functional Types. Cambridge University Press, Cambridge, pp. 47–65.

World Commission on Environment and Development (WCED), 1987. Our Common Future. Oxford University Press, New York.

Wuenschmann, S., Oehlmann, J., Delakowitz, B., Markert, B., 2001. Untersuchungen zur Eignung wildlebender Wanderratten (Rattus norvegicus) als Indikatoren der Schwermetallbelastung, Teil1: Alters- und geschlechtsspezifische Quantifizierung der Verteilung von Al, As, Cd, Co, Cu, Mn, Ni, Pb, Sr, Ti, Tl und Zn in den Organen Herz, Leber, Lunge, Niere, Muskulatur, Gehirn und Knochen. UWSF-Z. Umweltchemie und Ökotoxihologie 13 (5), 259–265.

Wuenschmann, S., Oehlmann, J., Delakowitz, B., Markert, B., 2002. Untersuchungen zur Eignung wildlebender Wanderratten (Rattus norvegicus) als Indikatoren der Schwermetallbelastung, Teil 2: Die Anwendung des Körperlast-Verfahrens von Depotkompartimenten. UWSF-Z. Umweltchemie und Ökotoxihologie 14(2), 96–103.

41

Chapter 2

Bioindicators and environmental stress assessment

O. Fränzle

Abstract

Bioindication is the analysis of the informational structure of living systems, ranging from single organisms to complex ecosystems, in order to define environmental quality or assess environmental hazards and risks. The indicative qualities of biotic systems are determined by inherent ecophysiological properties, population dynamics, and stress reactions with regard to physical and chemical changes in site conditions as described in the first section of the present contribution. Bioindication involves active and passive approaches which span the dimensional scale from single-species bioassays over micro and mesocosms to biocoenoses and ecosystems. In view of the mostly unspecific stress reactions of organisms the primary task of bioindicators is the general determination of physiological effects rather than the direct measurement of environmental concentrations of stressors. Thus, in early recognition perspective the lack of specificity has the advantage of a broad-based caveat, inducive of subsequent systematic search for quantitative causal interrelationships.

The second section provides a systematic review of both the typology and rational selection of bioindicators at the species, population, biocenotic and ecosystem levels. It is to show that, in addition to the above aspect, a further advantage of biomonitoring results from their low costs in comparison to those of instrumental measurements, even in the case of active monitoring networks. Another important point is the integrative recording character of both effect and accumulation bioindicators which always reflects the total exposure time, while an instrument can only produce a set of singular data within the framework of a given temporal resolution. In contrast to these positive aspects of bioindicator use, however, an essential deficiency results from the highly variable susceptibility of the different species exposed to stressors, which leads to difficulties in comparing specific effect data. But in these cases fuzzy logic approaches provide highly commendable operations for processing such data in order to more precisely define their physiological or toxicological information.

Active and passive biomonitoring techniques based on specimen exposure and observation yield spatially valid data only on condition the underlying measuring or sampling networks are implemented in compliance with basic geostatistical requirements or the corresponding test methodology such as variogram analysis and subsequent kriging procedures, respectively. By analogy, also the selection of complex bioindicators such as biocoenoses or ecosystems must be based on rigid criteria of spatial and temporal representativeness whose fulfillment should be corroborated by means of both traditional frequency statistics and complex geostatistical procedures as described in Section 2.3. The following section them is a critical comparative appraisal of the problems involved in biomonitoring, which leads to a set of suggestions for improving both the technical practicability and data quality of the various test procedures discussed.

Keywords: typology of bioindicators, stress reactions, representative measurements, biomonitoring networks, geostatistics

Hans Selye's discovery of the stress syndrome opened a new perspective in medicine and biology. Before Selye (1936), specificity was the fundamental question of pathology, and consequently in pathological diagnosis the so-called pathognomistic signs (Virchow, 1854) were considered the most important parameters. Selye was the first to emphasize the nonspecific, common symptoms of diseases which he summarized under the term "stress". Owing to its very general character the stress concept is used here to introduce a unified perspective into the discussion of bioindicators in order to more precisely define their scale-dependent realm of applicability and the reqirements for an adequate selection of such indicators and for the rational evaluation of biomonitoring data.

1. Stress and nonspecific resistance

Stress is the state of a biotic or abiotic system under the conditions of a "force"' applied, *strain* is the response to the stress, i.e. its expression before damage occurs, while *damage* is the result of too high a stress that can no longer be compensated for (cf. Csermely, 1998). Focusing on biological systems, it is indicated for reasons of terminological clarity not to apply the term "stress" to fast readjustments of metabolic fluxes, photosynthetic or transpiration rates induced by fluctuations in the photon flux density, slight changes in temperature, or rapid variations in air humidity. Plants are acclimated, i.e. usually respond flexibly to such steadily occurring normal changes of cell metabolism induced by variable environmental conditions. The same applies to the diurnal fluctuations in metabolic activities, growth patterns, and in cell division and differentiation processes. Besides such fast acclimations, plants can also respond to environmental changes by means of somewhat longer-term adaptations such as modifications of size and thickness of leaves, number and density of stomata, ultra-structure and function of the chloroplasts by raising the levels of photoprotecting enzymes and of stress metabolites.

When subjected to a stress, an organism is in a state of strain. As long as the strain is completely reversible, it is said to be elastic; beyond this point or threshold, the strain will be only partially reversible, and the irreversible part is called the permanent set or plastic strain. Unlike elastic strains, plastic strains are not constant for specific stresses, since they may eventually lead to disintegration of the system (organ, organism, population, biocenosis, ecosystem) affected. Since plastic strains may be dependent on the time exposed to the stress, the time factor must be measured whenever the resistance of biological systems to plastic strains is determined. Thus, elastic resistance is a measure of the system's ability to prevent reversible or elastic strains (physical or chemical changes) when exposed to a specific environmental stress, while plastic resistance is a measure of its ability to prevent irreversible or plastic strains and, therefore, injurious physical or chemical changes (Levitt, 1980).

Stress resistance has two main components, namely the innate internal properties of an organism which oppose the production of a strain by a specific stress on the one hand, and the repair system which reverses the strain on the other. Both the elastic and plastic resistances of an organism to a specific stress may be subject to changes depending on its adaptive potential. The adaptation may be either stable, having arisen over a large number of generations, or unstable, depending on the developmental stage of the organism and the environmental factors to which it has been exposed. This adaptation is important for both elastic and plastic strains. Plastic strains are by definition injurious. Therefore the adaptation leading to increased plastic resistance, i.e., resistance adaptation *sensu* Precht et al. (1955), will prevent (or at least reduce) injury by a stress which would injure (more intensely) the unadapted organism. Although elastic strains are reversible by removal of the stress and therefore *per definitionem* non-injurious, it must be realized that they may also lead to injury and even death if they are maintained for a long enough period. This may, for example, be due to the inability of the organism successfully to compete with others that undergo less elastic strain when subjected to the same stress (e.g., mesophiles versus psychrophiles at low temperatures). An elastic strain may also eventually injure the organism even in the absence of competition, for example, due to a disturbance of the metabolic balance. Thus, a low-temperature stress may simply decrease the rates of all metabolic processes reversibly, but to different degrees. As a consequence, if the stress is maintained for a long enough period, the resultant strain may conceivably lead to an accumulation of toxic substances or to a deficiency of essential intermediates. In either case, a sufficiently long exposure to the stress may injure or kill the organism (Levitt, 1980). An adapted organism, by way of contrast, may complete its life cycle, and regenerate in the presence of the stress. It displays capacity adaptation *sensu* Precht (1967), while resistance adaptation may not permit growth and may merely prevent the plastic strain and therefore the injury until the stress is removed or decreased to a level permitting growth and completion of the life cycle.

In the case of animals, elastic adaptation has been intensively studied in the past, while plastic adaptation has been largely ignored (Precht, 1967); for plants the reverse holds. Considering instantaneous response, there is an increase in elastic (reversible) strain with increasing stress up to the yield point; i.e., strain is proportional to stress. Beyond the yield point, however, a plastic (irreversible) strain occurs, and the strain increases more rapidly than the stress. In regard of the time scale involved, a further distinction is necessary. If a small stress with a corresponding elastic strain is maintained for some time, two kinds of adaptation may occur: (1) the strain may eventually decrease to a constant low value, leading to elastic ("capacity") adaptation, or (2) the resultant strain may remain constant. In this case, secondary changes induced in the organism may lead to a plastic ("resistance") adaptation.

Before stress exposure, the organism will be in a certain standard situation of physiology that is relatively optimum within the limits of the respective site or habitat factors, e.g., light, water, nutrient supply in the case of plants. Individual stressors or complex stress events will then lead to a series of strain reactions which can be subdivided into three phases. Considering plants by way of example, they respond at the beginning of a stress event (alarm phase) with a decline of one or several physiological functions, for example, the performance of photosynthesis, transport of metabolites, and uptake and translocation of ions. Thus, the plants deviate from their normal physiological standard, and as a consequence their vitality declines. Under these circumstances acute damage and senescence will occur rapidly in plants with low stress tolerance mechanisms or low resistance minimum, respectively. Normally, at the end of this phase plants begin to activate their stress-coping mechanisms such as acclimation of metabolic fluxes, activation of repair processes, and long-term metabolic and morphological adaptations. In the following restitution phase this leads to a hardening of the plants which attain their maximum resistance by establishing new physiological standards (Levitt, 1980; Lichtenthaler, 1984). Under conditions of long-term stress and stress intensities exceeding the plants' stress-coping mechanisms, however, the stage of exhaustion follows, when physiological activity and vitality are progressively reduced, which causes severe damage and finally death. However, when the stressors are removed in time, i.e., before senescence processes become dominant, the plants regenerate and develop new physiological standards (cf. Larcher, 1987; Lichtenthaler, 1998).

Usually, several stress factors act simultaneously, such as the combined heat, water and high-light stresses during high-pressure situations on plants in summer. In addition, the influence of primary stressors such as air pollution or drought which can reduce vitality considerably, may be followed by secondary stressors, e.g., bark beetles or particular fungi which further decrease vitality and finally lead to the death of the organism. In any case the stressors and stress constraints which have the character of external signals in terms of information theory, need to be registered by the organisms affected in order to respond by appropriate stress and strain reactions. There are multiple means and forms of such signal perception and transduction in an organism and its organs which will lead to direct metabolic responses on the one hand and to the activation of gene expression, enzyme formation, synthesis of stress proteins, stress metabolites, stress hormones, etc., on the other. The latter then further modify the metabolic responses under stress and control the stress resistance maximum and minimum of the organism. From a bioindicative point of view these reactions and the substances formed are of considerable importance, in particular in the framework of early recognition approaches for planning-related environmental assessment purposes.

2. Fields of bioindication

Knowledge of the existence of an environmental stress situation is the prerequisite for its solution or amelioration. In view of the different time scales of the resultant strains and their complex nature, early recognition of such situations is necessary before changes and damages become wide-spread and obvious. In this connection, habitual predictive assessment methods, such as pre-market testing of chemicals and environmental impact assessment of new technologies can never yield a sufficiently reliable forecast with regard to future exposure situations or marketing effects. This is due to the enormous number of existing chemicals, the complexity of related use patterns and environmental pathways, and the bewildering possibilities of antagonistic and synergistic effects (Schmidt-Bleek et al., 1987; Fränzle, 1993). Illustrative examples of the problems encountered are phenomena like forest dieback, atmospheric consequences from the use of fluoro-chloro-hydrocarbons, or problems arising from the application of thawing salt and the use of PCBs.

2.1. General typology of bioindicators

In a general ecological sense bioindicators are organisms or groups of organisms suited to determine qualitatively or quantitatively the state of the environment, in the narrower sense of the term the designation frequently refers to the organismic indication of anthropogenic environmental stressors. Bioindicators should respond to early stages in either exposure or effects conditions without disclosing cause-effect relationships. The determination of the latter is the necessary follow-up action after significant environmental changes have been detected so that preventive or corrective measures can be initiated.

From the pragmatic point of view three groups of bioindicators can be defined, although a clear-cut distinction is not always possible. Test organisms are used in toxicological test systems (bioassays) in order to quantitatively determine ecological effects of industrial chemicals. Individual species represent an intermediate level of biological organization between subcellular functions and community/ecosystem interactions. Many single species tests, comprising organisms of a very wide spectrum ranging from bacteria and plants over monocellular animals to metazoa, including mammals, are considered state-of-the-art and have correlated well with actual chemical impacts (Verschueren, 1983; Steinberg et al., 1995; Fränzle, 1999).

Indicator species are organisms that are used as qualitative indicators of specific natural or anthropogenic stress phenomena including in particular those caused by the release of industrial chemicals or their metabolites into the environment, the increase in radiation (Tietz and Weser, 2001) and sensible heat fluxes (cf. Hörmann, 1995) or structural interferences within the landscape. The presence or absence of indicator species, the habit and physiological constitution of the organisms are indicative of specific physical or chemical activities in, or aspects of the physical and chemical composition of, biotic systems (communities, ecosystems). It ensues from the introductory stress considerations that stenopotent (stenoecious) organisms or populations with their (partly very) limited adaptive capacity and correspondingly low stress-coping potential have principally a better indicator quality than eurypotent (euryoecious) ones which tolerate a relatively broad range of ecological conditions or amphioecious communities with their variable tolerance to habitat conditions as reflected in the presence of clines and subspecies.

Monitoring species are organisms which permit, on the basis of networks, to determine the impact of air and water pollutants on ecosystems; they are of particular importance in the framework of environmental protection measures. Criteria for assessing stress effects are: accumulation of xenobiotics at different organizational levels of the organism affected, disturbance of intracellular physiological and biochemical processes, submicroscopic and morphological modifications of organelles and organs, and changes in form and functioning of the whole organism. Stress response at the organismic level is also reflected in the structure of biocoenoses and even ecosystems (cf. Section 2.3).

Bioindicators are habitually subdivided into the groups of effect (or reaction) and accumulation indicators. The former respond (relatively) quickly and in an observable or measurable manner to physical or chemical stress, and it ensues from the above characterization of stress and strain resistance that effect indicators should be low-resistance systems with low adaptive potential, i.e. stenopotent organisms. Accumulation indicators, in contrast, must dispose of a fairly high amount of strain resistance, which enables them to incorporate for a considerable time, depending on the uptake-excretion ratio, potentially toxic substances without injury. Either type of indicators

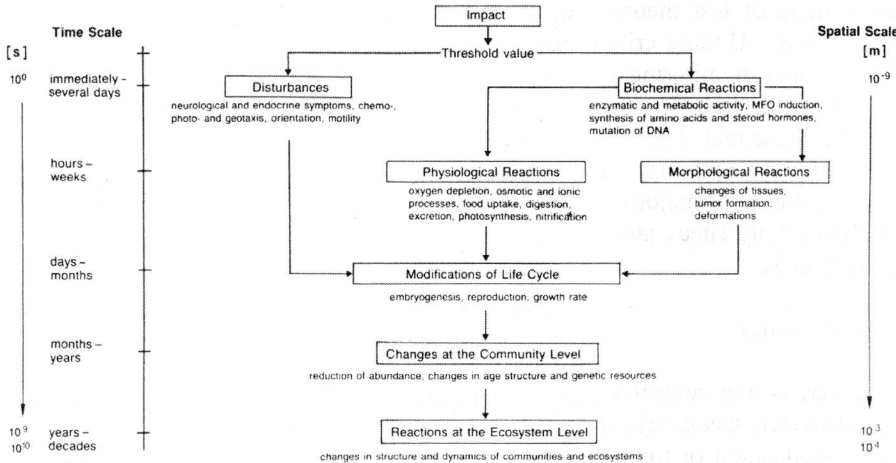

Figure 1. Average stress response times of biotic systems as related to size and complexity (after Korte, 1987, modified).

comprises a hierarchy of biotic systems ranging from the subcellular level to biocoenoses and ecosystems whose stress response time generally increases with the structural complexity and areal extent of the system.

From a more practical point of view the distinction between active and passive bioindication appears indicated. Active bioindication involves the intentional exposure of cultivated organisms, cells or organelles to segments of the environment for certain periods of time. Such active bioindicators should provide for wide taxonomic representation and include a range of biological processes, and ideally they can be combined with recording units in the form of *bioprobes* (Peichl et al., 1987; Schmidt-Bleek et al., 1987). Active biomonitoring has the commendable advantage of permitting rigid geostatistical controls when developing pertinent exposure networks (Fränzle, 1994; Fränzle et al., 1995; cf. Section 3). Passive bioindication is based on the comparative evaluation of stress reactions in selected components of the existing biocoenoses and ecosystems of a study area involving geostatistics and the application of geographical information systems. The methodology comprises a set of different approaches, ranging from the evaluation of more or less specific strain reactions of individual organisms (of one or several species) over the analysis of phyto-cenotic variations (composition, structure, biomass; release of phytoalexines, polyamines, ethene, stress proteins, etc.) to changes in structure and functioning of ecosystems.

2.2. Selection of bioindicators at the species and population levels

The quality of bioindicative data essentially depends (1) on the representativeness of the observed objects which constitute a random sample, (2) on the selection of object attributes relevant for hypotheses testing and (3) on the degree of compliance with the basic criteria of test theory, i.e., objectivity, reliability and validity (Schröder and Fränzle, 1996). If these criteria are not fulfilled at the lowest spatio-temporal level of environmental monitoring, this will affect any other level of data aggregation. Therefore efforts have to focus on these criteria, and quality assurance and quality control are essential. They comprise: (i) the selection of representative objects and the decision for the optimum assessment strategy, (ii) routine laboratory quality control (and, in the case of major monitoring networks, inter-laboratory quality control), and (iii) a plausibilty check and the evaluation of the spatio-temporal validity of the data (cf. Section 3).

2.2.1. Bioassays

A bioassay is any controlled, reproducible test to quantitatively determine the presence, character, specificity, or strength (potency or concentration) of a biological agent (e.g., a medication or toxicant) by measuring specified effects, i.e., stress responses (death, reproductive and/or behavioural dysfunction, and impairment of growth and development) upon a living organism, or on isolated tissue. In a wider sense the term also applies to any test of the effects of environmental variables on a biological system.

Such tests and test organisms should be selected on the basis of taxonomic, ecological, toxicological, and chemical exposure critera, among which sensitivity and type of stress reaction of the organisms are particularly important. Marine and freshwater invertebrates, molluscs, arthropods, and vertebrates are generally more susceptible to the influence of xenobiotics in their environment than the majority of terrestrial animals. This is due to various physiological characteristics: The epidermis is totally or partially (e.g. gills) hydrophilous; protective structures against dessication are largely lacking; cutaneous respiration plays a major role; parenteral uptake of organic and inorganic substance is common among invertebrates. In the light of these characteristics the following species proved suited for acute tests in freshwater systems and may be (potentially) useful as effect indicators under field conditions, too: *Entosiphon sulcatum, Chilomonas paramecium* (flagellates); *Colpoda maupasi, Paramecium caudatum, Uronema parduczi* (ciliates); *Daphnia magna, D. pulex, D. pulicaria* (daphnids); *Gammarus lacustris, G. fasciatus, G. pseudolimnaeus* (amphipods); *Cambarus spec., Oronectes spec., Procambarus spec.* (crayfish); *Baetis spec., Ephemerella spec., Hexagenia limbata* (mayflies); *Physa integra, P. heterostropha* (snails); *Dreissena polymorpha* (mussel); *Chironomus spec.* (midges). Marine and estuarine invertebrates used for acute laboratory (and predictive) tests are: *Acartia tonsa, A. clausi* (copepods); *Crangon crangon, C. septemspinosa, C. nigricauda, Mysidopsis bahia, Palaemonetes intermedius, P. pugio, Pandalus jordani, Penaeus duorarum, P. setiferus* (shrimps); *Callinectes sapidus, Carcinus maenas, Hemigrapsus spec., Pachygrapsus spec., Uca spec.* (crabs); *Crassostrea gigas, C. virginica* (oysters); *Capitella capitata* (polychaete) (Bick, 1972; Corliss, 1979; Nusch, 1982; Verschueren, 1983; Müller and Wagner, 1988).

Plants representing primary producers in aquatic test systems are the algae *Chlorella pyrenoidosa, C. vulgaris, Scenedesmus subspicatus, S. quadricauda, Selenastrum*

capricornutum, Skeletonema costatum (Grade et al., 1999) and among the higher plants
the duckweeds *Lemna minor, L. gibba* (Dolgerloh, 1999; Eberius and Vandenhirtz,
1999). Frequently used representatives of the decomposer subsystem in aquatic food
webs are the bacteria *Escherichia coli, Photobacterium phosphoreum, Pseudomonas
fluorescens,* and *P. putida.*

Favourite test organisms for the repesentation of terrestrial ecosystems are, for
instance, the edaphic algae *Chlamydomonas chlorococcoides, Klebsormidium dissec-
tum, Stichococcus bacillaris, Xanthonema montanum, X. tribonematoides* (Burhenne
et al., 1999) and the higher plants *Avena sativa, Brassica rapa,* and *Pisum sativum*
(Kalsch and Römpke, 1999). Among the animals earthworms and birds are of particular importance (cf. Section 2.2.4 for details).

Depending on the length of exposure three types of toxicological bioassays can be
distinguished. Acute tests, the simplest toxicity tests, determine whether a single exposure to a given chemical can produce a critical effect in a test organism. Depending
on the mode of exposure the following test approaches can be distinguished with
animals: acute oral, acute dermal, acute inhalation, primary skin irritation, and primary
eye irritation (cf. Ratte et al., 2002). The amount of test material required to kill 50%
of the animals with a 95% degree of confidence under a stated set of conditions is
called the LD_{50}, i.e. the lethal dose for 50% of the test population. It is normally
expressed in milligrams of chemical per kilogramme of body weight of animal. This
value does not permit, however, a direct weight-related extrapolation to other groups
of organisms, since it can (and normally does) vary with the kind of test animal and
even with the subspecies within a given species.

Most subchronic toxicity studies begin with information generated in the acute tests.
Fractional amounts of the acute LD_{50} (normally 20, 10, and 5%) are used in 14 to 90
days range-finding experiments to establish the long-term dosage level to be used in
chronic tests. Another purpose of subchronic testing is to identify target organs in
metazoa and early cumulative effects of toxic substances. Acute and subchronic testing
approaches differ in the fact that the latter always include a nontreated group of test
organisms, serving as a control and insuring that the observed stress reactions are treat-
ment-related.

The basic aim of chronic toxicity testing which normally involves comparatively
large populations of test organisms is to determine what happens after a lifetime of exposure to a chemical. Thus, tumorigenic or oncogenic studies are chronic by definition.
Reproductive tests include teratological experiments to provide information on embryonic effects, and three-generation protocols to determine long-term reproductive effects.

In current toxicological practice, mutagenicity tests are considered an essential part
of the test procedures carried out on a substance. The most widely used bacterial test
of this type is the one developed by Ames, using *Salmonella typhimurium* deprived of
the (natural) ability to synthesize the essential amino acid histidine. Mutagenic agents
mutate back these specially bred organisms to the natural state so that they no longer
require the addition of histidine to the growth medium for survival. This or basically
analogous tests (e.g., *Escherichia coli* test) are in general attributed a good predictability value, but they cannot be interpreted in isolation; false positive results have
been recorded with compounds known to be free of carcinogenic activity and false
negative results were obtained with some carcinogens (Verschueren, 1983).

Toxicological test systems on the basis of the above organisms are carried out on different levels of increasing structural complexity. The elementary forms comprise separate single-species tests and a combination of such tests with different species, which has the appreciable advantage of relative simplicity and a fairly high amount of reproducibility, but poses also manifold problems as to rationally and reliably extrapolating to the "real-world" situation. Therefore tests with different species tested simultaneously or compartment tests with spatially separated sections and different species appear more appropriate. In the normal case of simultaneous or sequential additive testing the organisms selected should be representatives of the main functional components of an ecosystem, i.e., primary producers, primary and secondary consumers, and decomposers. Such a combination could comprise *Scenedesmus subspicatus*, *Daphnia magna*, *Leuciscus idus*, and *Pseudomonas putida*, for example. Nevertheless the validity of the resultant data for assessing adverse chemical effects in ecosystems remains limited, because neither matrix effects which are essential for the bioavailability of substances are considered nor is the structural and functional complexity of natural systems sufficiently accounted for. Therefore, extrapolations from the bioassay to the ecosystem level involve safety factors, defined in dependence on the length of exposure of the test organisms and reconsidered in the light of comparative micro and mesocosm experiments (cf. ECETOC, 1997; Fränzle, 1999).

In compliance with national and international legal regulations the predicted no-effect concentration (PNEC) of an enviromental chemical, normally calculated from single-species acute or chronic laboratory tests, is one basis of ecotoxicological hazard assessments. The other is the predicted environmental concentration (PEC) which is derived from measurements or distribution models. Introducing the above safety factors, it is assumed that, where the PEC value exceeds the PNEC value, there could be a potential for adverse environmental effects; the corresponding risk is the system-specific probability of such a hazard (ECETOC 1993).

Hitherto more than 500 test species have been used in bioassays. A representative analysis of 137 BUA (GDCh-Advisory Committee on Existing Chemicals of Environmental Relevance) reports shows, however, that actually 25 predominate in testing (Figure 2).

The following exemplary matrix (Figure 3) summarizes the results of a systematic compilation of the base sets of test data which underlie the above BUA reports. It illustrates the considerable differences in toxicological information on the fate and behaviour of environmental chemicals, which poses serious problems in the framework of comparative evaluations. They can, at least to a considerable extent, be overcome by means of fuzzy logic approaches which are based on an extension of the classical meaning of the term "set" and formulate specific logical and arithmetical operations for processing imprecise or uncertain information (Salski et al., 1996; Friedrichs, 1999).

2.2.2. Autecological bioindicaton by plants

The major field of application for bioindicative plant species is air pollution monitoring. Although specific reactive patterns at the biochemical, physiological and morphological levels may be subject to systematic observation and measurement,

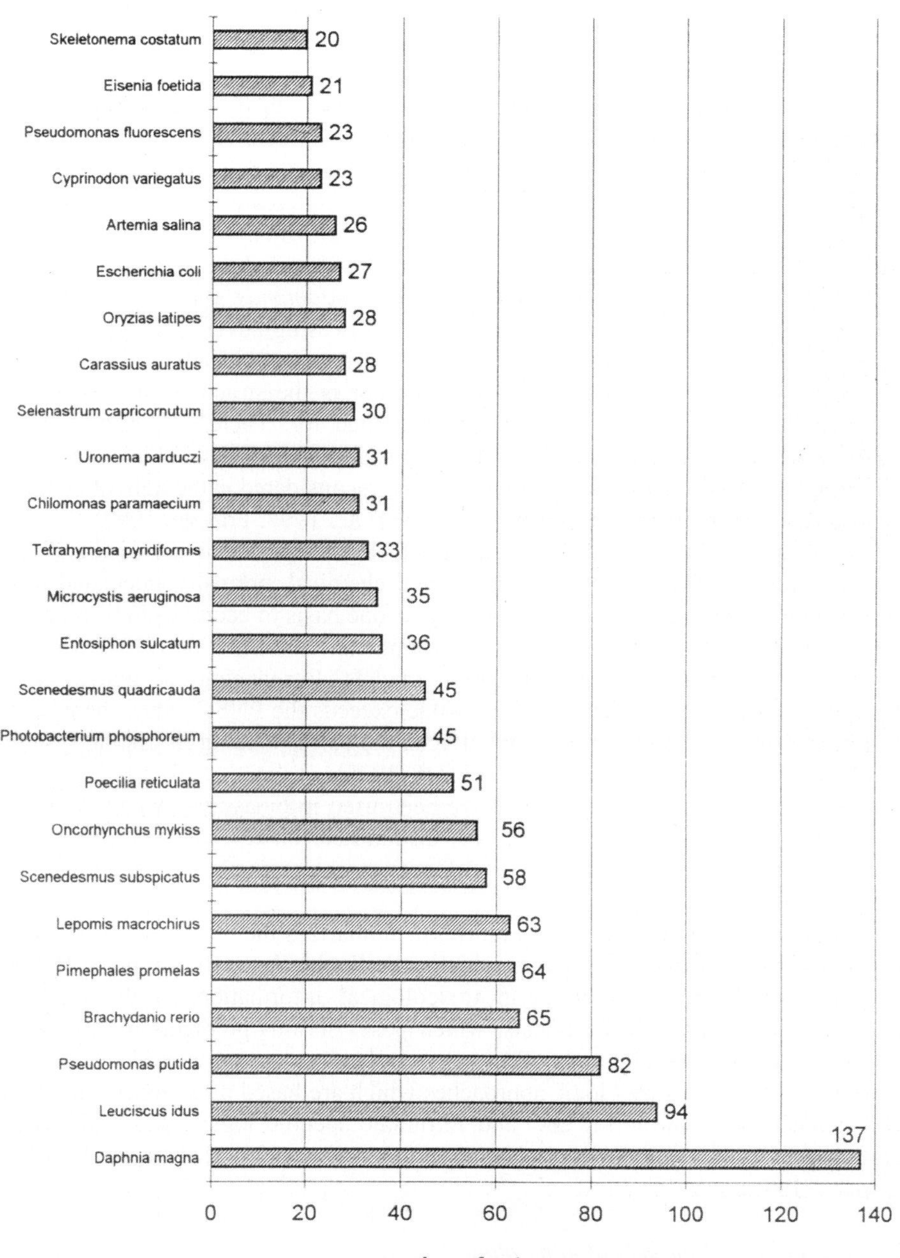

Figure 2.   Frequency distribution of favourite test organisms in toxicological bioassays, derived from the analysis of 137 selected BUA reports.

Species	1,1,1-Trichlorethan	1,1,2,2-Tetrachlorethan	1,1,2-Trichlorethan	1,1-Dichlorethen	1,2,3-Trichlorpropan	1,2,4,5-Tetrachlorbenzol	1,2,4-Trichlorbenzol	1,2-Dibromethan	1,2-Dichlor-3-nitrobenzol	1,2-Dichlor-4-nitrobenzol	1,2-Dichlorethan	1,2-Dichlorpropan	1,3,5-Trichlorbenzol	1,3,5-Trimethylbenzol	1,3-Dichlor-4-nitrobenzol	1,4-Dichlor-2-nitrobenzol	1,4-Dioxan	1,6-Hexandiol	1-Chlor-2,4-dinitrobenzol	1-Chlor-2-methyl-3-nitrobenzol	1-Methylnaphthalin	2,2'-Dithiobisbenzothiazol(MBTS)	2,3-Xylidin
Daphnia magna	4	3	26	3	5	4	29		9	4	30	9	1	4	9	4	6	2	8	9	2	1	3
Leuciscus idus	3						4			9	12		7			8	6		4	7			
Pseudomonas putida				1						1	2				2	1	1	2			1		
Brachydanio rerio	8			1			2		3	2							2						
Pimephales promelas	19	7	8	2	6		7				8	5	1				4				5		
Lepomis macrochirus	7	4	4	1	4	2	3	3			10	3					1					1	
Scenedesmus subspicatus	2			2							1							3				1	2
Oncorhynchus mykiss					8	2	4				7		3										
Poecilia reticulata	1	1	9			1	1	1			1		1		1	1			1	1	1		
Scenedesmus quadricauda											1			1									
Photobacterium phosphoreum	1				1	3			2		1		1		2	2				6			
Entosiphon sulcatum											1						1						
Microcystis aeruginosa											1						1				1		
Tetrahymena pyridiformis							1							1							1		1
Chilomonas paramaecium											1						1						
Uronema parduczi											1						1						
Selenastrum capricornutum	1	1				2	3				1										5	1	1
Carassius auratus													1									1	
Oryzias latipes							1																
Escherichia coli																							
Artemia salina			9								4			4	1						2		
Pseudomonas fluorescens																							
Cyprinodon variegatus	9	3		2		9	4				2												
Eisenia foetida	1		1	1			5				1	2										1	1
Skeletonema costatum	2	1				2	3	1			1	1											
Chlorella pyrenoidosa			1							1					1	1							
Chlorella vulgaris	1				1						1	1											
Crangon crangon			2								1	1											
Mysidopsis bahia	5	1				1	1				5	2											1
Number of species	14	8	8	8	7	9	14	3	5	5	21	8	7	3	5	7	10	3	3	5	7	6	4
Number of files	64	21	60	13	26	26	68	5	18	17	92	24	18	6	15	19	23	7	13	24	17	6	7

Species	2,4-Dichloranilin	2,4-Dichlorbenzol	2,4-Dinitrotoluol	Propylenglykol	Schwefelkohlenstoff	Styrol	TEL-CB	Terephthalsäure	Tetrachlorethen(PER)	Tetrachlormethan	Tetraethylblei	Tetramethylblei	TML-CB	Tributylamin	Tributylphosphat	Tributylzinnoxid	Trichlorethan	Triethanolamin	Triethylblei	Triethylentetramin	Triethylphosphat	Triisopropanolamin	Trimethylblei	Tris(2-chlorethyl)-phosphat	Vinylacetat	Vinylchlorid	Number of substances	Number of files
Daphnia magna	20	23	12	5	1	9			13	6				3	17	4		9		5	12	2			6		137	1046
Leuciscus idus		6	3		6	6			1	9				6	3		3	1		2	4	7			3	3	94	306
Pseudomonas putida		1		1		1			2	1					1		1	1			1	1			1	1	82	107
Brachydanio rerio	1	3	2		1			1	20	1					5			1							27		65	194
Pimephales promelas		14	7	4		12									4			1							3	1	64	361
Lepomis macrochirus		3	6			3			2									3							3	1	63	181
Scenedesmus subspicatus	4								1									6		3	2	3					58	209
Oncorhynchus mykiss		3	5	1		4		3	6	1					8	2									3		56	203
Poecilia reticulata	7	2		1	1	3			1																3		51	115
Scenedesmus quadricauda		1	1			1				1	1	1	1	1				1							1		45	49
Photobacterium phosphoreum	3	6		3		2			1					4		2	1								1		45	101
Entosiphon sulcatum		1	1			1				1	1	1						1							1		36	36
Microcystis aeruginosa	1	1	13			1				1	1	1						1							1		35	49
Tetrahymena pyridiformis	1									2					1			1						1	3		33	43
Chilomonas paramaecium		1	7	1						2								1							1		31	31
Uronema parduczi	1									1					1			1							3		31	31
Selenastrum capricornutum	1		2			2	1								1			1				1			1		30	98
Carassius auratus	1		1			2	1								1			1							1		28	49
Oryzias latipes		3				2									3												28	56
Escherichia coli										2								2									27	38
Artemia salina	1	1													1						1						26	55
Pseudomonas fluorescens				2														1	1	1					1		23	24
Cyprinodon variegatus			1			3												3									23	77
Eisenia foetida	4														2			2									21	40
Skeletonema costatum		1							2						1												20	44
Chlorella pyrenoidosa	1														1												17	20
Chlorella vulgaris			4						1	1	4														1		15	27
Crangon crangon										2																	15	31
Mysidopsis bahia								1	1	4					2	1		1				1				1	14	38
Number of species	11	15	11	10	5	15	2	2	13	10	2	3	2	4	16	7	8	14	3	5	7	4	3	3	14	1		
Number of files	44	67	58	20	13	50	5	4	56	25	3	3	5	15	53	15	11	29	3	12	22	13	3	5	53	1		

Figure 3. Toxicity data files of 137 selected BUA reports, illustrating differences in both kind and extent of test procedures.

nonspecific bioindicative approaches are generally more important. They are based on the hypothesis that environmental stresses in plants are geared to circadian rhythms in most of the biochemical and physiological processes, ranging from seconds to annual seasons. Based on data pertaining to salinity stress, activities of antioxidant substances, and enzymes (glutathione reductase, catalase, peroxidase, superoxide dismutase), and finding that environmental stress altered rhythmic responses to UV-B tolerance, K^+ uptake, stomatal movements, nutrition, and resistance to iron stress, it is assumed that duration, and more often amplitude, of rhythmic parameters are influenced by extreme environmental factors (Leshem and Kuiper, 1996). For instance, stomatal sensitivity to leaf-to-air vapour pressure deficit, i.e. the stress response reduction in stomatal aperture in dry air proved negatively correlated with the sum of photosynthetic photon flux density of the previous three weeks, but positively correlated with the mean vapour pressure deficit of the preceding month (Kutsch et al., 2001 a). The underlying mechanism is the activity of a light-stimulated plasma membrane proton pump, which enables the stomatal guard cells to accumulate such osmotic agents as potassium and chloride ions which increase turgor pressure, resulting in increased stomatal opening. Thus, chloroplasts exert, besides the traditional carbon fixation, a regulatory effect on the activity of the light-stimulated plasma membrane proton pump in such a way that at low atmospheric CO_2 pressure the proton pump is activated, while at high pressure CO_2 fixation is preferred. These findings and corresponding converging data may be interpreted in the sense of a general adaptation syndrome whereby different types of stress evoke identical coping mechanisms, which implies a co-stress response with one type of stress resistance imparting co-resistance to others. Common coping denominators may be physiological or morphological. The former include ion exchange, oxy-free radical scavenging, osmoregulation, abscicic acid, jasmonates, chaperones, heat shock proteins, and phytochelatins, and among these abscicic acid seems to be the panacea for plant stress, since it plays a key role in co-stress manifestations in the conversion of stressful environmental signals to gene expression (Chandler and Robertson, 1994). The latter include leaf pubescence, movements and stance, and rooting characteristics.

Lichens are sensitive, unspecific indicators of air pollution. SO_2 disturbs the stomatal regulation and causes a general stress situation which affects the whole metabolism of the plants. Consequently, unspecific alterations of enzyme and membrane activities occur which could also be induced by many other environmental stresses like water or temperature stress and pollutants like HCl, HF, NO_x, O_3, and PAN (peroxiacylnitrate). Lichens can therefore be used as pollution indicators for both active and passive biomonitoring purposes, which permits to a certain extent also the determination of pollution levels.

The standardized lichen exposure test with *Hypogymnia physodes* is described in detail in VDI Guideline 3799. It is based on 4 cm samples taken from the bark of oak trees, acclimated under low-stress conditions and exposed in autumn for one year. The end point of measurement is the visible damage (bleaching) of the thallus, determined by means of colour slides taken at the beginning and end of the exposure period (Larcher, 1994).

Lichens as perennial plants are also well suited for the bioindication of radionuclides. Lacking any excretion mechanisms, they may accumulate radioisotopes like

[137]Cs for comparatively long times, which permits valuable retrospective analyses. Thus, the radioactive outfall of the nuclear weapon tests of the 1960s can be traced quite reliably. With regard to the Chernobyl nuclear catastrophe in April 1986 the knowledge of the nuclide spectrum of the primary emission allows, on the basis of lichen samples, to estimate the regional deposition pattern due to concomitant fallout, rainout and washout processes. The 1992 decay rate of the radionuclides trapped (predominantly ^{137}Cs) is approximately 7% of the total primary deposition (Feige et al., 1990).

Mosses are very sensitive bioindicators of heavy metal contamination, but have also commendable advantages for organic xenobiotics assessment, since they can accumulate large amounts of these elements or compounds in their tissues. Their indicator quality is due to the following features: (1) Many species have a vast geographical distribution and they grow abundantly not only in various natural habitats but also in urban–industrial agglomerations. (2) The cell walls of mosses are easily penetrable for metal ions since they have neither epidermis nor cuticle. (3) Mosses have no organs for uptake of minerals from the substrate, but obtain them mainly from precipitation and dry deposition. (4) Some species of the genera *Hylocomium* and *Thuidium* have layer structure and annually produced organic matter forms distinct patterns. (5) Transport of minerals between the segments is very poor due to the lack of vascular tissues. (6) Mosses accumulate metals in a passive way acting as ion exchangers.

Owing to these qualities mosses are used to estimate the actual contamination of the environment at different spatial scales and to assess selected pollution levels of the historical past by means of specimens from herbaria or other sources. In the 1970s Rühling and Tyler (1973) measured the concentration of Cd, Co, Cu, Fe, Mn, Pb, and Zn in *Hylocomium splendens* samples collected at several dozen localities in Finland, Norway and Sweden. The concentration of lead and cadmium was ten times higher in the southern part of the survey area than in the northern. The corresponding increase for chromium and iron was three times as high in the south as in the north and for copper twice as high. Herrmann (1976) took samples of *Hypnum cupressiforme* ssp. *filiforme* from the regularly distributed nodes of two grids laid randomly over the Federal Republic of Germany. Maps of the regional variation of Be, Cd, Cu, Ni, Pb and Zn concentrations in the plants, as well as their interconnections disclosed by a principal component analysis, reveal high intercorrelations of Cd, Cu, Pb and Zn, whereas Ni and Be behave each in a different manner. A multidimensional discriminant analysis shows that the distribution pattern of the trace metals may be explained by the distance to the next nonferrous metal smelter and variables describing the economic structure. Thus, the highest deposits occur in the Rhenish industrial area and the northern Harz Mountains, while the lowest are recorded in Lower Bavaria and parts of northern Germany. A comprehensive sampling campaign for trace metals was conducted in 1991/92 in Germany, Poland, the Czech and Slovakian Republics by Markert et al. (1996). The moss samples taken at 831 sites comprised the species *Dicranium scoparium, Hylocomium splendens, Hypnum cupressiforme, Pleurozium schreberi, Polytrichum formosum* and *Sceropodium purum* which were analysed for Cd, Cr, Cu, Ni, Pb, and Zn. The results, presented in isopleth form, are indicative of a general W-E gradient of the heavy metal stress with the maximum strain located in Slovakia. With regard to lead, for example, the gradient is caused by the predominance of unleaded fuel in Germany. In smaller-scale inspection the nickel concentrations proved excessively increased in the neighbourhood of refineries and petrochemical factories.

Thomas and Herrmann (1980) also used *Hypnum cupressiforme* to determine the contamination with organic environmental chemicals along a larger-scale transect from Amsterdam to Munich and recorded the highest concentration of HCH in highly productive agricultural landscapes, the maximum PCB-60 content in the key areas of chemical industry, and Benzoperylene in the centres of heavy industry and in large urban agglomerations. In an analogous smaller-scale investigation Herrmann (1984) analysed the atmospheric transport and regional distribution of micropollutants in NE-Bavaria. Turbulent transport near the ground causes a regional distribution pattern characterized by a decrease of pollutant concentration in mosses within a short distance from the emission source. This pattern interferes with another one originating from long-distance transport. Therefore, trace metals and polycyclic aromatic hydrocarbons display comparable spatial patterns with high concentrations in densely populated areas and increasing concentrations along mountain ridges; singular emission sources, e.g. for lead, can be easily detected. In comparison to these groups of micropollutants, organo-chlorine pesticides have distinctly less delimitable distribution patterns, which is likely to be due to their more diffuse emission characteristics (Thomas, 1981).

Considering the results of xenobiotic monitoring by means of mosses shows that the concentration levels found depend on many factors. The most important among these are: specific strain reactions of the taxa exposed, organotropic accumulation mechanisms in, and age of, the specimens selected, type of sample (indigenous mosses, transplanted mosses, moss bags), type of sampling plots (i.e., open or forest area), date (season) of sampling, macro and microclimatic boundary conditions of deposition, and analytical procedures (cf. Grodzinska, 1982). In the light of these influential features a standardized methodology for moss monitoring purposes is indicated in order to reduce the extent of biassed estimates. In addition, geostatistical reqirements must be met and criteria of representativeness fulfilled (Section 3.2).

Like lichens and mosses also (higher) fungi have developed nonspecific stress-coping accumulation mechanisms with regard to heavy metals whose uptake from the soil is distinctly higher than from atmospheric deposition. This is due to the rapid growth of the fruiting bodies which are frequently characterized by a wide surface/volume ratio, and the extensive formation of mushroom hyphae in the topsoil. Their bioindicative value is still rather limited, however, for various reasons: (1) There are large inter- and intraspecific differences in uptake rates which frequently attain several orders of magnitude at one site and whose causes need to be explored in greater detail. (2) For several heavy metals there is no uptake from the soil matrix or solution; consequently a valid correlation between the concentration in soil and in the fruiting body cannot be established. (3) Knowledge of the volume of the hyphae system is frequently too limited to permit a reliable estimate of its role in heavy metal uptake. (4) Synergistic or antagonistic stress effects remain to be elucidated.

Higher plants are liable to xenobiotic impact by uptake from the soil or by atmospheric deposition (Fränzle et al., 1985). Considering the first pathway involves a sufficiently detailed account of a great number of chemicals interactions with soil as a three-phase system, among which dispersion, sorption and ion exchange, bioavailability and microbial biodegradation are the most important. Deposition rates of gases and particles are controlled by both transport to, and capture at, vegetation and soil surfaces and these sets of processes act, in terms of modelling, like resistances in series in an electric circuit. Some gases, such as HF and HNO_3, are so reactive that they penetrate most natural surfaces, including the waxy cuticle of leaves. Surface resistance is correspondingly small, and consequently deposition is largely determined by the resistance to transport through the air. The uptake of other gases, such as SO_2, O_3 or NO_2 with subsequent rapid interior absorption, is controlled by stomatal mechanisms which allow gas exchange through the cuticle (Lee et al., 1978; Ernst, 1993). The ratio of surface resistance for uptake of a gas to that for water vapour is then equal to the inverse ratio of the molecular diffusion coefficients in air for the two molecular species. Gases with a marked affinity to water, e.g., SO_2 and NH_3 in particular, are preferably deposited to water surfaces and to moist soils, while ozone is much more retained by dry soil. The uptake rate of gases which are consumed in metabolic or co-metabolic processes is on the one hand limited by the demand of the relevant physiological process, and physical parameters such as water solubility and diffusion coefficients on the other (Heagle et al., 1973; Fränzle, 1993).

In more or less explicit consideration of these problems a considerable number of higher plants has been recognized as bioindicators (Funke et al., 1993). A well known example of active biomonitoring is the application of standardized grass cultures with *Lolium multiflorum* ssp. *italicum* grown on unit pot soil with automatic water supply as developed by Scholl (1971). This plant has a comparatively high strain resistance and is consequently well suited as an accumulation indicator. On the basis of a differential analysis of the amounts of xenobiotics incorporated a hazard assessment of other plants exposed to the same chemical stressors is possible, provided the respective dose-effect relationships are known from extrapolatable laboratory experiments or field measurements. Such data can also, and with appropriate reservation, be interpreted with regard to ecological magnification which defines the increase in concentration of a substance in a food web when passing from a lower trophic level to a higher one (Korte, 1987; Fränzle, 1993). Comparable bioconcentration capacities are found with a considerable number of higher plants (cf. Weiss et al., 2002). Thus, for example, *Brassica oleracea cv. acephala* is frequently used for monitoring polycyclic aromatics (Nobel, 1987). A considerable accumulation potential is also found with numerous aquatic plant species, in particular with regard to heavy metal accumulation (Kohler 1982).

In contrast to accumulation indicators effect indicators are stenopotent plants with correspondingly low stress resistance and high reactivity. Thus, for ozone assessment purposes *Nicotiana tabacum* is used all over the world, in particular the most sensitive variety *Bel W3* which shows a quick ozone response in form of speckle necroses on the upper side of the leaves. Good PAN (peroxyacetylnitrate) indicators are *Urtica urens* and *Poa annua* (Posthumus, 1977). Chlorotic effects on leaves of the *Phaseolus vulgaris* cultivar "Pinto" are indicative of NO_2. Boron deposition on *Acer platanoides* leads to chlorotic and necrotic effects of the foliage. *Populus nigra, P. tremula* and *Fagus sylvatica* are sensitive to SO_2 exposure, while cloned poplar hybrids are sensitive to both SO_2 and O_3. Among the group of conifers *Abies alba, Pinus sylvestris* and *Picea excelsa* are SO_2 effect indicators, while *Pinus strobus* is O_3 sensitive. A comprehensive review of the bioindication of photooxidants is provided by Guderian et al. (1985).

It ensues from the foregoing that plants can also be used as indicators of specific site qualities which may take on the character of more or less pronounced stressors. Acidophytes or basiphytes are, in principle, indicative of acid or basic soils; but it must be taken into account that in reality this indication quite frequently relates to secondary soil parameters like texture or pedoclimate (Kreeb et al., 1990; Funke et al., 1993; Ellenberg, 1996). So quite a few of the central European basiphytes occur on acid and neutral soils in warmer climates, too; therefore the basic indicator quality seems to be rather related to the thermal capacity of the soil which co-variates for textural reasons with basicity under the conditions of a temperate macroclimate. Consequently some basiphytes (for example, species of the genera *Ophrys* and *Medicago*) are basically indicators of a warmer or drier soil climate. Metallophytes (chalcophytes) are characteristic of sites with heavy metal concentrations in soil above the normal background values of the micronutrients Fe, Mn, Zn, Cu, Co, Mo, Ni, V and the trace metals Cd, As, U, Pb, Tl, Cr, and Hg (cf. Lieth and Markert, 1990). A classic Zn indicator is *Viola calaminaria* which occurs in association with the heavy metal ecotypes of *Minuartia verna* and *Silene cucubalus*. The specific adaptation strategies developed by these plants are stress avoidance or stress tolerance (Levitt, 1980). Possibilities for trace metal tolerance are (i) binding to pectin residues and carboxyl groups in cell walls and (ii) complexing to organic acids in the cytosol followed by removal to the vacuole (De Knecht et al., 1994). From the viewpoint of a general adaptation syndrome detoxification of trace metals, metal-binding peptides, phytochelatins and membrane-located pumping mechanisms appear to be most important (Leshem et al., 1998). Contrary to these mechanisms, a recently discovered reaction which accounts for Al tolerance in transgenic tobacco and papaya appears to have a specific character. It is secretion of citric acid by roots which binds Al in the soil, thus preventing it from entering and damaging the plant (Fuente et al., 1997).

2.2.3. Bioindicator sets and phytocenoses

The uncertainty resulting from the above nonspecific stress reactions can be reduced by combining several indicator species of different sensitivity; thus, it is possible to more precisely determine individual stressors and the corresponding effective doses. In the Netherlands such sets of effect indicators are used for decades already in the framework of a national deposition assessment scheme (Posthumus, 1976). A further improvement of the indicator quality can be brought about by virtue of the comparative exposure of single or combined indicator plants to filtered and unfiltered ambient air. van Haut (1972) developed transportable test chambers which permit the selective determination of air pollutants and an estimate of the resultant hazard potentials. Further developments in this field are "open-top-chambers" (Heagle et al., 1973; Mandl et al., 1973) and the "zonal air pollution system" (Lee et al., 1978), the application of which is described by Kreeb (1990).

More frequent practice than active biomonitoring with selected lichen species is pollution mapping based on the distribution of lichen communities which has been carried out in many parts of the world during the past forty years. It has supplied important basic information for urban planning and the location of industrial plants (cf. Arndt et al., 1987; Zierdt, 1997). The essence of this approach is the comparative analysis of the abundance and luxuriance of diverse lichen species on selected trees which is interpreted in terms of the so-called "index of atmospheric purity" (IAP):

$$\text{IAP} = \sum_{i=1}^{n} (Q_i * f_i), \tag{1}$$

with Q = toxicological tolerance of species i (derived from the mean number of concomitant species on all sampling spots), f = frequency of species i (derived from dominance and abundance), and n = number of species. The higher the IAP value, the lower is the pollution stress of the lichen community. Provided, the geostatistical requirements described in Section 3.2 are met, isopleths of air quality can be deduced from the spatial IAP pattern.

Although the procedure is generally considered valuable in the estimation of air pollutant levels, several points merit attention: (1) The lichens substrate should occur consistently in the survey area, i.e., epiphytic lichens should be investigated on one (or a limited number of) tree species under consideration of micro-environmental conditions. (2) The trees selected for pollution mapping purposes should be free-standing, since in dense woodland a lower light intensity often results in a poorer development of lichens on tree trunks. Also trees in areas of intensive agriculture are problematic because of stress effects resulting from the application of organic and inorganic nutrients, herbicides and pesticides. The same applies to roadside trees which are usually subjected to the influence of exhaust gases and dust. (3) The buffer capacity of the tree bark exerts an important influence on the distribution of lichens, even in only slightly polluted areas. Trees with acidic bark are commonly unsuitable for mapping approaches, because their buffer capacity is (too) low. Thus, lichens disappear in Salzburg on *Picea abies* bark with the exception of *Scoliosporum chlorococcum*. In contrast, both *Aesculus hippocastanum* and *Fraxinus excelsior* support a rich lichen flora throughout the major part of the city, even under conditions of higher pollution stress, which makes these species equally unsuitable for comparative mapping purposes (Türk, 1982). (4) The different water capacities of tree barks and the nature of stemflow tracks must be appropriately analysed with regard to micro-environmental differentiations, and trees exhibiting irregularities due to these site factors should be excluded from surveys (Wirth and Türk, 1975). (5) Owing to the unspecific character of stress reactions the impoverishment of lichen communities may also be largely influenced by the different development of propagules and interspecific competition. (6) A change in the lichen flora as a result of increased pollutant emissions can be assessed after a comparatively short time; however, a reduction or cessation of detrimental emissions can be determined only after a certain time lag amounting to several years, depending on hysteresis effects in bark quality. For instance, Hafellner and Grill (1980) reported a reinvasion of lichens into the Leoben-Hinterberg area (Austria) some four to six years after a polluting factory had been shut down.

Already in the 1950s and 1960s extensive use was made of tree bark as a bioindicator of environmental acidity (cf. Barkman, 1958; Stäxang, 1969). In the following decade, for instance, Grodzinska (1978), Härtel and Grill (1972) and Zdanowska (1976) studied the level of tree bark acidity in industrial areas, Johnsen and Søchting (1973) and O'Hare (1974) in large urban agglomerations. It ensues from these and a number of comparable investigations that acidity is influenced by: tree species, age and health of trees, soil conditions of site, thickness of bark samples, length of storage prior to analysis, and analytical procedures applied. Generally the bark of deciduous tree species appears less acidified by nature, so it proved to be more sensitive an indicator than bark of conifers. With regard to analytical procedures it is worth mentioning that electrical conductivity of bark extracts proved to be a better indicative parameter for acidification measurements than pH, because it increases very markedly with an increase of the SO_2 concentration in the ambient air.

Up to the 1980s the relevant investigations were focused on individual plants or populations of mostly economically important plants; only thereafter a major interest was devoted to reactions of plant communities to air pollutants and other stressors (cf., e.g., Knabe, 1981). The observations resulted in a pragmatic distinction of high, medium and low impact levels. The effect of high pollutant concentrations is characterized by a more or less visible breakdown of the phytocenosis. In forests, for instance, first the tree layer of sensitive species is damaged by acute and chronical injury, then the less protected bushes, herbs and mosses; eventually barren ground may result. If the loads are not too high, the secondary succession as adaptive stress response leads to the formation of new stable structures of lesser complexity characterized by few species of high abundance. Also symbiotic interactions may be affected, e.g., the formation of nodules by rhizobia on leguminous plants and the development of mycorrhizae under the influence of photooxidants on some tree species (Letchworth and Blum, 1977). The effects of low xenobiotic loads on vegetation stands are within the marginal area between the level of normal (e.g. seasonal) fluctuations of unimpaired phytocenoses on the one hand and significant changes on the other. As described in Section 1, these effects range from growth stimulation on the one hand to marked reduction in vitality and reproductive potential on the other, depending on the relevant chemical stressor, its level of concentration, and the duration and temporal variability of the impact. If the stress persists for a long enough period, the resultant strain may conceivably lead to an accumulation of toxic substances or to a deficiency of essential intermediates, which may injure and eventually even kill a variable proportion of the community in dependence on the specific strain resistance of the individual plants and populations affected.

The essential principles of stress-related population dynamics can be described by the logistic (Verhulst-Pearl) equation

$$\frac{dN}{dt} = rN\,[(K - N)] \tag{2}$$

where r is the intrinsic rate of increase, N the number of individuals, and K the carrying capacity supporting the number of individuals in a given environment.

In the real world regulatory effects necessarily operate with a certain time delay, the characteristic magnitude of which may be denoted by T. Its incorporation into Equation (2) leads to a generalization of the logistic

$$\frac{dN}{dt} = rN\,[1 - N(t - T)/K\,] \tag{3}$$

If the time delay T is short in comparison to the characteristic return time T_R, disturbances will be damped monotonically back. As T approaches T_R, there is a tendency for the inherent regulatory (stress-coping) mechanisms to produce overshoot and overcompensation. Finally, as T becomes significantly larger than T_R, the pattern of overcompensation leads to self-sustaining stable cycles. The amplitude and period of the oscillations of population density $N(t)$ are determined uniquely by the parameters in Equation (3); as such, they may play a quantitative role in the bioindicative interpretation of population behaviour under stress.

Especially in this perspective it should be noted that difference equations admit of more realistic complications than the above differential equations. First, they permit to more appropriately describe the fact that for many plant and animal species generations overlap; second, they show that the regular pattern of stable cycles can give way to apparently chaotic fluctuations if the nonlinearities are sufficiently severe. For further variations on the theme of relative time scales the reader is referred to May (1981).

An extension of the preceding model helps to elucidate some basic features of two populations interacting with variable intensity as prey-predator, competitors, or mutualists. Modelling, for instance, the competition aspect in the behaviour of a two-population community leads to extensions of the above single-species logistic equation:

$$\frac{dN_1}{dt} = r_1 N_1 [1 - (N_1 + \alpha_{12} N_2)/K_1] \tag{4.1}$$

$$\frac{dN_2}{dt} = r_2 N_2 [1 - (N_2 + \alpha_{21} N_1)/K_2] \tag{4.2}$$

K_1 and K_2 here denote the carrying capacities of the environment, as perceived by the species 1 and 2, respectively; r_1 and r_2 are the corresponding intrinsic growth rates; α_{12} is a competition coefficient measuring the extent to which species 2 presses upon the resources exploited by species 1, α_{21} is the corresponding coefficient for the effect of species 1 on species 2. Characterizing the sensitivity of such a two-species community in terms of stability, the solution of Equations (4.1) and (4.2) indicates the possibilty of a stable equilibrium if intraspecific competition is stronger than interspecific competition; if the reverse holds, no stable coexistence is possible. The same applies in the case when the two species use the resources in an identical manner.

Multispecies generalizations of the above considerations lead to more realistic models of community behaviour and nondemographic measures of community sensitivity, which in turn permits to more precisely define the role of disturbances. They are, however, necessarily much more complicated if analogously formulated as systems of simultaneous differential equations, since already three-dimensional systems frequently display a rich dynamical complexity, which finds its appropriate reflection in strange attractors. Irrespective of the theoretical interest the approach has, its considerable formal complications often make a description of the stepwise transformation of a community under the impact of stressors with other techniques preferable.

In addition to the facilities of a description of stress-induced community modifications by means of geostatistic analyses of the spatial patterns of biocenotic injuries (cf. Section 3) or chronosequences of geographical data matrices also various methods of time series analysis are worthwhile. Principally their application involves the observation of a complete set of state variables, which may constitute a major drawback in empirical studies of communities or biocoenoses whose inherent dynamics is not known a priori. Under these circumstances, however, use can be made of the fact that in dynamical systems with a finite number of state variables the information on the momentary value of all state variables can be substituted by information on the recent history of a part of the variables. Thus, Grossmann et al. (1984) applied time series analysis in combination with geographic information systems (GIS) to 462 forest stands in Bavaria in order to develop a scenario method (POLLAPSE) which describes forest damage in sequential form by means of maps. These maps can be compared with the actual development so that deviations become readily discernible.

Analyses of the spatial correlation pattern of site qualities and damage levels by means of cross-tabulation techniques (Fränzle et al., 1985; Schröder et al., 1986) corroborate the importance of pedogenic nutrient supply and acidification processes for forest decline. The standardized damage level (i.e. proportion of injured/unaffected forested area) correlates best with pedogenic nutrient supply, frequency of fog situations and elevation a.s.l., while the correlation with other site factors is also significant on the 99.9% level although of little statistical relevance. Thus, stands on members of the ranker, cambisol and podzol groups with a pH < 4, marked nutrient deficiency and concomitant reduction of the buffering capacity (i.e. predominance of Al and Fe buffering systems) are most liable to dieback. Among the air-borne pollutants, SO_x species are particularly important. In combination with other pollutants their effect on both vegetation and soil is enhanced by fog. The inherent complexity of the dieback syndrome implies the recognition that critical levels for tolerable pollutant concentrations can be adequately defined only with regard to the whole set of concomitant stress factors affecting forests (cf. Altenbuger and Schmitt-Jansen, 2002; Ernst, 2002; Mulder and Breure, 2002).

2.2.4. Animals as bioindicators

In comparison to plants animals have generally developed a greater arsenal of stress-coping mechanisms; in addition, non-sessile animals can avoid a certain number of threatening environmental or anthropogenic stressors by virtue of their mobility or motility. With the exception of single-species laboratory systems, therefore the great potential for elastic and plastic adaptation makes stress analysis in general and the quantitative determination of individual stressors in particular a difficult task under field conditions (cf. Csermely, 1998; Iwama, 1998). Progress can only be achieved on the basis of very detailed inquiries into the nature and susceptibility of terrestrial and aquatic organisms of different taxonomic status and trophic position in food webs (see Chapters 11 and 12).

Owing to the generally higher sensitivity of aquatic animals to xenobiotics in comparison to that of terrestrial organisms they play a major role in acute, subchronic and chronic laboratory tests (cf. Section 2.2.1). Under field conditions inquiries into distribution patterns and the different propensities for organotropic accumulation of xenobiotics are the major fields of interest in various bioindicative approaches. In the latter respect a very general classification of higher taxa (i.e. phyla and classes) in terms of increasing accumulation potential for residues can be given: protozoans, worms, molluscs, annelids, insects, crustaceans, fishes, amphibians, reptiles, birds, mammals (Verschueren, 1983). Since the indicative qualities of the essential part of these animal groups are described in detail in Didden, 2002, Oehlmann and Schulte-Oehlmann, 2002, Chovanec et al, 2002, Becker, 2002 and Tataruch and Kierdorf, 2002, some explanatory remarks on the representative class of fishes may suffice.

Frequently used for test purposes are the following species: *Abramis brama* (bream), *Alburnus alburnus* (bleak), *Ameiurus melas, A. nebulosus* (black bullhead, American catfish), *Brachydanio rerio, Carassius auratus, C. carassius* (goldfish, crucian carp), *Clupea herengus* (herring), *Cyprinodon variegatus* (sheep head minnow), *Cyprinus carpio* (carp), *Esox lucius* (northern pike), *Gadus morrhua, G. pollachius* (cod, pollack), *Lepomis humilis, L. macrochirus* (common sunfish, bluegill sunfish), *Leuciscus leuciscus, L. idus* (dace), *Micropterus salmoides* (largemouth bass), *Oncorhynchus mykiss* (rainbow trout), *Orycias latipes, Perca fluviatilis* (perch), *Phoxinus phoxinus* (minnow), *Pimephales promelas* (fathead minnow), *Poecilia reticulata, Rutilus rutilus* (roach), *Salmo trutta* (brown trout), *Salvelinus salvelinus* (trout).

The traditional ichthyological characterization of water bodies is basically a biogeographical approach. It defines fish populations and communities in relation to their origins and the influence of past and present geographic and environmental factors that have shaped their distributions, habitats, and relationships at present or in historic times in an evolutionary framework (see, for example, Gulland, 1978; Johnson and Odada, 1996; Menting, 2001). Bioenergetic considerations open additional ways to more precisely determine habitat qualities, e.g. the trophic status of water bodies in terms of fish motility. In this connection individual-based modelling is particularly indicated to compare responses of different fish species to environmental change (cf. van Winkle et al., 1993). The comparative analysis of the *Rutilus* and *Abramis* populations of the eutrophic Lake Belau in Schleswig-Holstein (Germany), for instance, and the application of this methodology permitted on the one hand to simulate the behaviour of individual fishes and its variability in time and space, and these activity patterns displayed a close correlation with the results of respirometric measurements. On the other hand possible causes of phenotypic variability at the population level could be determined and the consequences of such variability for the development of populations assessed (Hölker, 2000).

For active biomonitoring purposes fish can be exposed in cages to monitor river waters or aqueous domestic and industrial wastes for harmful effects. In the framework of passive monitoring fish can be used as accumulation indicators, since the uptake of a compound from water into living tissue affects the movement, distribution and toxicity of chemicals in the environment. A substance that bioconcentrates may influence life far from the initial points of environmental release and may furthermore alter ecological processes at concentrations much lower than predicted from acute and subacute (subchronical) test results (cf. Section 2.2.1). Bioconcentration is the first step in the process of food chain biomagnification. Results of bioconcentration studies are useful in assessing environmental hazards especially if the substances of interest are highly lipid soluble (e.g., compounds with an octanol/water partition coefficient >1000), and do not undergo rapid chemical or biological transformation.

The functional importance of many invertebrates and vertebrates in terrestrial ecosystems makes them potentially suited for bioindication in general and toxicity testing in particular. Representatives of the above-mentioned phyla and classes are briefly described in their ecological roles, in order to illustrate their suitability as bioindicators or test organisms (cf. Section 2.2.1).

Earthworms, e.g. *Eisenia foetida, Lumbricus rubellus* and *L. terrestris* live in upper soil horizons and feed on decaying organic matter. They are ecologically important as soil mixers, aerators and drainers, and they serve as food for a great number of higher animals, e.g. robins, woodcock, mice, shrews. The diversity and wide distribution of worms make them equally desirable test species and accumulation indicators for monitoring purposes (cf. Friesel et al., 1984; Müller, 1984). Owing to their limited motility, they incorporate substances only from soil volumes in the order of magnitude of ten cubic metres, which permits a high-resolution soil monitoring with regard to many organic and inorganic chemicals (Didden, 2002). Terrestrial snails and slugs are primary consumers and feed on a varied diet of plant materials. Very widely distributed, they are a food source for larger insectivores. The accumulative potential of the bigger species, e.g. *Limax maximus, Arion rufus* or *Helix pomatia* and *H. aspersus* is considerable; owing to their considerable motility these gastropods may be integrative bioindicators for areas of several hectares (cf. Oehlmann and Schulte-Oehlmann, 2002).

The phylum Arthropoda comprises the classes Arachnida and Insecta. Ecologically important members of the Arachnida are spiders, mites and ticks, scorpions and harvestmen. Mites and ticks are parasitic on plants and animals, taking their diet directly from the fluids of their hosts. Spiders are carnivorous invertebrates whose food consists entirely of small animals, primarily insects, which makes them potentially good test and monitoring organisms. All arachnids are potential food sources for insectivores. Among the insect order Orthoptera, praying mantids (*Mantis* spec.) which are strictly carnivorous and rely heavily on insects for food, might accumulate certain chemicals or be markedly exposed to target animals. The order Hymenoptera contains ants, sawflies, ichneumons, chalcids, wasps, and bees. Many of these are important as pollinators and as parasites on other insects. They feed on pollen, plant juices, and many on other liquid foods, which permits to use some of them, in particular bees, as accumulation indicators, integrating information about acreages of hectares to square kilometres.

Generally speaking, primary consumers among the above groups are better suited as accumulation indicators than secondary consumers because they live on relatively low-energy diet. Consequently they have to take up considerable quantities of food, hereby incorporating also potentially toxic substances in substantial quantities. By way of contrast, secondary consumers normally live on high-energy food; thus, consuming only relatively moderate quantities, they can accumulate distinctly lesser amounts of toxic substances only. This applies in particular to short-lived species (Funke et al., 1993).

Birds are primary and secondary consumers, feeding on plants, invertebrates and vertebrates alike. They in turn are food for mammalian predators, a few amphibians and reptiles, and a few species of birds. Owing to their functional importance in ecosystems many avian species, for example pigeons (*Columba livia domestica, C. palumbus, C. oenas*) and goshawk (*Accipiter gentilis*) are good indicators of environmental quality. In consideration of their position as top predators the latter and other common Falconidae species are particularly interesting for biomagnification studies on organic chemicals and, because of marked organotropic accumulation effects in pinions, for heavy metal monitoring purposes. For example, the successive loss of the primary wing feathers of the female goshawk which are thrown in early summer during the normal annual moult is correlated with the reproductive (and hence feeding) behaviour, which in turn leads to a differential accumulation of lead and cadmium in the primaries (Müller, 1984; Becker, 2002). Also for toxicity testing birds proved useful (cf. Ratte et al., 2002). The objective of a quail dietary test, for instance, is to provide preliminary indication of potentially harmful effects of a chemical on terrestrial birds. To this end the bobwhite quail (*Colinus virginianus*) is particularly appropriate since it is easily and economically reared, widely available, and generally more sensitive to many hazardous substances than other common test species. In addition it is worth mentioning that in the course of the last forty years migrant birds have proved to be very sensitive indicators of environmental change, because they use to adapt their specific passage behaviour by virtue of genetically based extremely rapid selection processes (Berthold, 2000).

Among the mammals rabbits (*Oryctolagus cuniculus*), rats (*Rattus norvegicus*) and mice (*Mus musculus*) are readily available and easy to use in toxicity testing. However, because the albino strains of rats and mice have been selectively bred for laboratory purposes and long removed from the genetic influence of wild types, their relationship to the natural fauna is indefinable, and consequently the results of laboratory studies may not be applicable to wild populations. Also cats (*Felis domestica*) and dogs (*Canis familiaris)* are most suitable subjects for experimental work, e.g. for inhalation studies, but again it may be asked how suitable they are for toxicity testing, since their gene pools have been manipulated by man and, in some important ways, have been free for many years from natural selection pressures. This does not apply to roe deer (*Capreolus capreolus*) which is a favourite big game (e.g., >600 000 animals per year in Germany); consequently organs suited as accumulation indicators (e.g. livers) could be easily available for spatially valid monitoring purposes in sufficient quantities (cf. Tataruch and Kierdorf, 2002).

2.3. Bioindication on the basis of biocoenoses and ecosystems

The susceptibility of ecosystems to disturbances depends on the structure and size of the system and on the nature of the disturbances or stressors affecting it. Thus, the integrative stress reaction of a biocenosis or an ecosystem may be defined as a measurable alteration of the state of the community-forming organisms and their life-supporting substrates, which renders the individual, the population, or community more vulnerable to further injurious physical or chemical impacts.

2.3.1. Reactions of aquatic ecosystems to stress

Like terrestrial ecosystems, aquatic ecosystems under stress undergo changes in both structure and function. Changes in structure are manifested by modifications of the composition of the various biocoenoses and the related physical and chemical characteristics of the ambient water body; changes in function are reflected in differences in the organic matter production of the system and in the rates of utilization and release of different gases and nutrients.

According to Cairns and Niederlehner (1993) indicative stress reactions of aquatic ecosystems are the following:

1. Community respiration increases
2. Productivity/respiration ratio becomes unbalanced
3. Productivity/biomass ratio increases as energy is diverted from growth and reproduction into acclimation and compensation
4. Importance of auxiliary energy increases, i.e., import becomes necessary
5. Export of primary productivity increases
6. Nutrient turnover rates and losses increase
7. One-way transport increases, while internal cycling decreases
8. Lifespan decreases, turnover of organisms increases
9. Trophic dynamics shifts, food chains shorten, functional diversity declines
10. Efficiency of resources use decreases
11. Condition declines.

Many reports have shown a greater relative sensitivity for structural than functional variables. For example, Schindler (1987) found in his comparative studies of Canadian Shield lakes no significant changes in decomposition or nutrient cycling in acidified lakes, but the species composition of phytoplankton was among the earliest indicators of change. Crumby et al. (1990) studied the biological reaction of the Roaring River in Tennessee to stress caused by various constructions around the river and by inadequate agricultural practices in the watershed. Changes in species composition were reflected in a general decline in numbers of intolerant (stenoecious) species and a simultaneous increase of tolerant ones. Comparable results were obtained in two succesive analyses of the naturally brackish Jade Bay (Lower Saxony), where 40% of the species have changed within a 40 year span as a consequence of an increase in salinity and possibly of additional toxic effects due to xenobiotics, while the total number of species showed a slight increase (Michaelis, 1987; Zauke et al., 1987). Gnauck (1982) summarized structural and functional changes in aquatic ecosystems and gave examples of experiments equivalent to those of van Voris et al. (1980) which in turn correspond to findings of Uhlmann et al. (1978). These authors analysed the variability of species biomasses and some functional variables (expressed in terms of a relative index of instability) during a 50-day experiment with an artificial purification pond. The highest instability, equivalent to the highest sensitivity or bioindicative quality, results for the biomass of individual species (*Brachionus angularis* > Zooflagellates > *Liotonotus lamella* > *Chilodonella* spec. > *Ankistrodesmus falcatus*). In contrast, the functional variables such as oxygen concentration, primary production, turbidity, and the global biomass of functional groups of organisms (phytoplankton, zooplankton) or chlorophyll-a levels display a distinctly lesser sensitivity. The least sensitive appeared to be the organic carbon elimination capacity of the pond which characterizes, from the human point of view, its most important function.

However, there are also reports of a greater sensitivity of functional variables of aquatic ecosystems. Rodgers et al. (1980) found that process rate changes were more sensitive than biomass or chlorophyll concentration in detecting the effects of diverse chemicals on the periphyton in artificial streams. Crossey et al. (1988) stated in impaired rivers that production and respiration measures were more variable than the composition of macroinvertebrate communities. When dealing with enrichment, functional measures are also frequently a good warning indicator.

Once the stress has ceased, two possibilities exist: either reversible changes are induced or the changes prove irreversible (cf. Section 1). For chemical stress, the recovery of aquatic ecosystems depends mainly on the degree of accumulation of the respective chemical in the environment and the rate of flow, which can corroborate the findings in lakes with those in rivers. The recovery of eutrophic or highly polluted lakes after removal or reduction of the sources of organic pollution and phosphorus is very slow and takes up to 10–12 years (Straškraba 1995). The retarded reaction is due to enormous quantities of decomposable organic matter and phosphorus stored in the bottom mud. The oxygen at the bottom is consumed during the decomposition of organic matter, and in such conditions phosphorus is periodically (during the summer stagnation phase of holomictic-dimictic lakes) or continuously released. This phenomenon is called internal P load, which indicates that after termination of external loads internal release plays a major role in continuing eutrophication (Zeiler, 1996; Naujokat, 1997). In comparison to lakes, the recovery of rivers with high flushing may take place within about two years after a chemical stress event (Yount and Niemi, 1990).

2.3.2. Terrestrial ecosystems as integrative stress indicators

From a very general point of view stress reactions of a terrestrial ecosystem are analogous to those of an aquatic one. Considering various types of strain in the major compartments soil and biocenosis of such a system in greater detail, however, manifests far-reaching differences which are mainly related to functional aspects. Functions can be measured by estimating a "capacity" of the system or, particularly for carrier functions, its potential for use (Hanssen et al., 1991). For most of the functions a putative maximum value exists; depending on the stress-coping potential or resilience of the compartments affected, over-use or over-exploitation may lead to a decrease in capacity and ultimately to complete exhaustion or decline. With regard to the degree of naturalness of an ecosystem two types of functions can be specified. The first are so-called natural or, more precisely, ecological functions; and what is known as evaluation or valuation assessment of conservation values can be considered a measure of these ecological functions, in particular information and regulation functions. The second group of functions, particularly the carrier, storage and most of the production functions, may be called socio-economic functions. Obviously some functions are transitional in character, since they are bound to agricultural environments; consequently, and following van der Maarel (1978), they may be designated as rural functions comprising agricultural production, biotic production (e.g. forestry, production of industrial raw materials), rural carrying functions, and recreation.

Soil is one of the principal regulatory compartments of all terrestrial (and benthic) ecosystems. In a bioindicative context its susceptibility to disturbances should therefore be defined in terms of essential ecological functions such as regulation (comprising filtering, buffering and transformation processes), site (habitat), and productivity functions. Other functions which may attain importance in the framework of planning-related evaluations, e.g. subsoil as raw material for building purposes or as a substrate for waste deposits, soil as a geohistorical archive, are left out of account. Soil sensitivity to chemical impact is a highly variable property which can only be determined with a reasonable amount of practical accuracy when related to the ecological soil functions, past and current pedogenic processes, and agricultural or silvicultural use patterns. Chemical element speciation plays no less a role, since different species exhibit different mobilities in soils and sediments, have different plant availability, and different toxicity for organisms.

Thus, the use of ecological soil functions as indicator variables for chemical stress involves the regionalized determination of geogenic and anthropogenic background concentrations of potentially hazardous substances and the assessment of adverse chemical (and possibly concomitant physical) effects on soils and the related biota (Fränzle, 1998 a; Fränzle et al., 1993; Wiegmann, 1999). In terms of stress indication the definition of three threshold values of soil quality is commendable: (1) A stress-free soil is characterized by multifunctionality (the ecological functions are in the optimum range), an optimum conservation of species, site-specific quasi-natural climax communities, and absence of detrimental fluxes of chemicals. Under these circumstances any form of ecologically indicated land use is possible. (2) First low-level adverse effects on ecological soil functions are reflected in an initial loss of soil organisms, but there are not yet detrimental fluxes of chemicals into neighbouring ecosystems. (3) Long-lasting or permanent disturbances of essential ecological functions are coupled with a high-grade species deletion and injurious fluxes of hazardous chemicals into neighbouring ecosystems, which brings about a marked reduction of land use possibilities. Among the biological extrapolation methods for determining threshold values for these critical ranges of stress-induced disturbances of soil functions the van Straalen and Denneman (1989) approach appears appropriate. With regard to chemical stress it defines a hazard concentration (HC_p) which ensures the protection of a certain number of soil-living animals, e.g. 95% (i.e. $p = 5$) in the case of low-level effects and 50% ($p = 50$) in the case of permanent disturbances. The quality of this approach could be improved by defining p-values in terms of functionally relevant species.

An integrative approach to assessing ecosystem integrity or disturbances is related to the activity of soil microbial communities, which in turn is reflected in both the ATP content (when excluding the contribution of metazoic species) and the respiration rates of soil (Kokke and Winteringham, 1980). The measurement of soil-borne CO_2 as an essential component of the total carbon cycle of a system can be automated (infrared gas analyser), providing data which are also suitable for prospective interpretation in the framework of environmental change considerations (Kappen et al., 1992; Kappen, 1997). Comparative analyses carried out in Schleswig-Holstein on two fields on Cambic Arenosols and Podzoli-Cambic Arenosols, one cultivated as a maize monoculture with addition of cattle slurry and fertilizer, the other worked in a normal crop rotation system, permitted to define the essential factors of soil respiration at different time scales. Diurnal variations are highly positively correlated with soil temperature, while the annual fluctuations are due to the combined influence of temperature, soil moisture, rooting and the input of organic matter (crop remnants, manure). Thus, soil respiration constitutes a good indicator of ecosystem functioning and the influence of different agricultural and silvicultural mangement practices on systems evolution (Kutsch, 1996).

A more detailed insight into the transformation function of soil is provided by the study of the metabolic activity of its microbial communities. Using specific and adapted cultures of bacterial communities which play a key role in the carbon and nitrogen cycles of forest, grassland, and field ecosystems in Schleswig-Holstein, Bach (1996) found essential differences in the microbial transformation potentials due to the different intensity of agrarian and silvicultural land use on the one hand, and the hydro-geological and pedogenic site qualities on the other. On the *fields* under maize monoculture and crop rotation a marked reduction in both number and strain or species diversity is observed, coupled with a tendency to promoting the development of r-strategists which capitalize on the mineralisation of easily decomposable organic matter. The *grassland* communities are characterized by distinctly higher metabolic potentials than those of the fields, but have a similar prevalence of r-strategists. The nitrification potential appears impeded with a resultant tendency to the conservation of ammonium. *Wetlands* with a marked tendency to storage of organic matter have a high microbial transformation potential, coupled with a high biodiversity of the facul-tatively anaerobic microflora rich in K-strategists. Owing to acid soil reaction in the litter and Ah horizons and low N_{min} contents the *beech forest* soils have reduced trans-formation potentials with predominantly heterotrophic nitrification and reduced biodiversity. Irrespective of the high soil moisture, in the *alder carr* a bacterial microflora similar to that of the field and grassland sites has developed, specialized on the mineralisation of easily decomposable organic matter (Bach, 1996). Summing up, it may be said that bacterial populations are valuable integrative indicators of site qual-ities and the intensity of transformation processes in soil. In order to achieve a more comprehensive characterization of the regulation and habitat functions of soil the analysis of microbial populations could be methodologically combined with the above van Straalen and Denneman approach. Furthermore the inclusion of other components of the decomposer group appears feasible, but has to consider that the highest trophic level (predaceous Microarthopoda and Nematodes) seems to play a minor role in mineralisation than the lower levels (Schröter, 2001).

Vulnerability or susceptibility analysis of ecosystems looks at specific effects as caused by multiple, and possibly synergistic factors, while standard impact analysis addresses the multiple effects of a single causal factor; it seeks to identify a range of factors that may reduce response capacity and adaptation to stressors. Consequently the analysis involves inquiries into the scale-dependent sensitivity of the essential interrelated system components and their responses to both regular and singular stresses, or combination of stresses. Furthermore an inspection of critical response potentials in relevant exposure units due to intrinsic thresholds or positive feedbacks is necessary, coupled with an analysis of the adaptive capacity as a function of major entity characteristics like diversity and connectivity of system components (Clark et al., 2000; Moss et al., 2000).

Already in the 1970s van der Maarel (1978) developed tentative susceptibility estimates for Dutch ecosystems with five-point scales for stress situations due to eutrophication, desiccation (particularly lowering the groundwater table), and treading. Generally the susceptibility depends on the nutritional status of the systems (which makes oligotrophic variants most susceptible), the soil moisture conditions and the status of natural environmental dynamics. Here also the disposition may be important, i.e. how easily a potentially injurious impact can really reach a system. Furthermore the estimation of regeneration potentials or the resilience of the ecosystems (cf. Holling, 1976; Walker, 1995; Peterson et al., 1998) is of importance. Ellenberg (1972) suggested as a first rough approach a formula to estimate the "Belastbarkeit" (B) or load capacity which may be considered equivalent to vulnerability; it expresses to which extent a system (or its components) can be burdened before it definitely changes.

$$B = [(100 - D*L)R]*10^{-1} \tag{5}$$

where D = disposition, L = susceptibility, R = regeneration, all in 10-point estimation scales.

This and subsequent similar approaches are static and deterministic. Therefore, approaches are needed which are both dynamic (appropriately reflecting the time evolution of perturbation, sensitivity and adaptation) and statistical (defining probability distributions in order to calculate expected values). Among the existing or emerging formal (or formalizable) methodologies there are a number of suitable candidates for vulnerability analysis and assessment. Exemplary options comprise advanced versions of game and criticality theories, re-analysis of historical records, extreme-value statistics and non-linear dynamics, semi-quantitative typologies (e.g. degradation syndromes) and complex indicator approaches.

Among the latter the following merit particular attention (DFG 1983; Steinberg et al., 1995; Jørgensen and Müller, 2000):

- fluxes of energy through, and entropy production of, ecosystems;
- fluxes of selected macro and micronutrients such as K, Ca, Mg, P, S and Mn, Fe, Cu, Zn;
- duration of biogeochemical cycles;
- biomarkers, e.g. stress proteins, phytoalexines;
- changes in biodiversity of vegetation stands and faunal assemblages;
- dynamics of selected populations;
- changes in competitive behaviour of functionally important species;
- modifications of food web structure.

Some of the methodological problems involved in the determination of these and related integrative indicators are referred to in the following example of a novel multi-dimensional combination of integrative bioindicators, developed in the framework of the comprehensive ecosystem research programme in the Bornhöved Lake District (Fränzle, 1998b). Indicator construction is based on the following premise: The greater

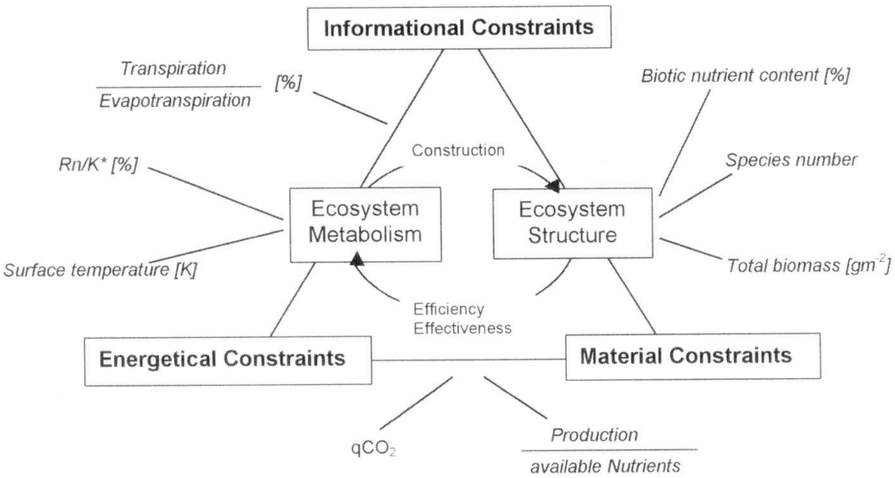

Figure 4. Indicators of ecological integrity derived from a concept of biological self-organization (after Kutsch et al., 2001b, modified).

an ecosystem's capacity for biological self-organization, the more likely is it for the system to be a reliable source of important ecosystem services on a long-term scale and in the face of unspecific ecological risks or stress situations, respectively. The essential interrelationships are summarized in the Figure 4.

The indicators were tested by comprehensive field data comprising the carbon, water, and energy budgets of adjacent crop field and beech forest systems which are edaphically and climatically similar but considerably different with regard to the intensity of human interference. In terms of the indicative parameters biomass storage, biologically bound nitrogen and phosphorus, species number, total ecosystem respiration per total biomass (qCO$_2$), total ecosystem assimilation per available nutrients, and transpiration per total evapotranspiration, there are clear differences between the systems. By way of contrast, ecosystem surface temperature and the R_n/K^* ratio (with R_n = net radiation and K^* = short wave radiation balance) were of limited usefulness for characterizing the two systems. For a detailed review of the metrological methodology and the results obtained the reader is referred to Kutsch et al. (2001b).

Thus, the degree of self-organization is in the first place indicated by structural patterns, such as biomass, intrabiotic nitrogen and phosphorus storage, and species number which all showed clear differences between the systems. Also *exergy storage*, i.e., the available work of the system (Jørgensen and Nielsen, 1998), is a suitable integrative indicator of ecosystem development since it expresses the distance from thermodynamic equilibrium and covers therefore both the size of the organized structure and its content of thermodynamic information. (Its determination, however, is faced with practical problems.) Species numbers, or *biodiversity* indicators derived therefrom (cf. Magurran, 1988), can only be a preliminary indicator of the capacity of ecological systems to self-organize. It is certain that biological information represents a constraint to self-organization processes, but there is no simple correlation between diversity and the actual organizational level attained. Genetic diversity, however, finds its predominant expression in species richness which provides ecological systems with the ability to adapt to changing environmental conditions. For the maintenance of processes under variable conditions, the most important effect of biological diversity is the provision of functionally redundant processors with varying ecological amplitudes (cf. Steedman and Haider, 1993; Walker, 1995; Peterson et al., 1998).

In the second place metabolic quotients reflecting functional or efficiency aspects are used to define system organization. *Transpiration* reflects the organizational capacity at the ecosystem level because it has a constructive function in addition to its dissipative character. The beech forest, where transpiration accounts for 63% of the total annual evapotranspiration, can be clearly distinguished from the maize field, with only 34%. In addition, the beech forest gained more biomass per available nutrients and proved able to maintain this biomass due to lower biomass-related respiration. These results confirm that the *transpiration/evapotranspiration ratio* (qCO_2), and the *production/available nutrient ratio* are valuable indicators of the degree of ecosystem self-organization.

More comprehensive in character, but necessarily more difficult to determine, are the entropy balance and ascendency of ecosystems. Ascendency (Ulanowicz, 1986, Ulanowicz and Norden, 1990) is assumed to be a measure of ecosystem growth and development, where growth is represented by an increase in the energy throughput of a system, while development is reflected in an increase in the information content of the flows. The underlying idea is that systems have a propensity to adapt to perturbations that possess regularities in time or space by adjusting their flow distributions so as to achieve higher values of ascendency. A peculiar formal property of the ascendency is that, even when it is calculated on a static network, clues to the dynamics behind the network pattern are built into the index of status. This contrasts remarkably with the situation in conventional dynamics, where information on static configurations tells nothing about the system's dynamical behaviour (Ulanowicz, 2001). It should be noted, however, that ascendency as presently calculated reflects much more growth (throughput) than the information content of flows as an indicator of development (cf. Christensen, 1994).

The entropy balance of ecosystems, i.e. the relationship between production and export of entropy, is another integrative parameter of ecosystem functioning, since all living systems and the entire ecosphere possess the essential thermodynamic characteristic of being able to create and maintain a high state of internal organization or a condition of low entropy, which is achieved by a continuous dissipation of energy of high utility (e.g. light or food) to energy of low utility (e.g. heat). For an ecosystem the entropy changes during a time interval can be decomposed into the entropy flux due to exchanges with the environment, and the entropy production due to irreversible processes inside the system such as diffusion, heat conduction and chemical reaction. An example of the application of the entropy balance concept to ecosystems of the Bornhöved Lake District (Schleswig-Holstein) is provided by Steinborn (2000); a general review of the concept is given by Svirezhev (1998).

The exergy concept which is related to the preceding one (but probably easier to conceive) was introduced by Evans (1966) and is based upon a classification of energy: energy which is useful and can do work, that is exergy, and energy which cannot do work, e.g. heat without a temperature gradient. By measuring the energy that can do work, exergy expresses energy with a built-in measure of quality, e.g. the chemical energy in biomass. Thus, exergy of a system can be defined as the amount of work (entropy-free energy) a system can perform, when it is brought into thermodynamic equilibrium with its environment. Loss of exergy and production of entropy are two different descriptions of the same thermodynamic reality, namely that all processes in open systems are irreversible.

The above indicators represent basic properties of the capacity of ecological systems for self-organization, which may be considered as the major protective strategy against uncertain long-term hazards to natural life-support systems. In order to depict the whole indicator set, a so-called amoeba diagram has the commendable advantage of illustrative clearness (Figure 5). The relative (or absolute) value of each single indicator is plotted on an axis of its own; the connection of these points then forms the "amoeba". In the present case the indicator values of a beech forest were used as reference values (benchmarks) for reasons of better comparability; it is not intended to convey the notion that the beech forest represents an optimum state of ecological integrity.

3. Geostatistic fundamentals of biomonitoring

Like many other spatially differentiated phenomena the components of vegetation and fauna of a study area exhibit such a variability that only a systematic statistical analysis prior to active monitoring activities or following the tentative steps of passive monitoring approaches can ensure the representativeness of data in general and the validity of areal extrapolation procedures based on primarily punctiform measurement data. In this context the term "representative" firstly means reproducing adequately the properties of sets of phenomena in terms of characteristic frequency distributions, and secondly it relates to specific spatial patterns. The latter aspect merits particular attention, when complex entities such as biocoenoses or ecosystems are considered which are not discrete independent and unambiguously identifiable objects, as ensues from their epistemological characterization; consequently the habitual statistical procedures must be supplemented by geostatistical analyses. The specific problems relating to areal data like mapping units on thematic maps, e.g. soil associations or ecosystem types, "concern (1) the arbitrariness involved in defining a [complex] geographical individual, (2) the effects of variation in size and shape of the individual areal units, (3) the nature and measurement of location" (Mather, 1972, p. 305).

3.1. Geostatistical measures of representativeness

Difficulties encountered in separating individual areal units from a continuum like soil or vegetation cover are most frequently, and at least partially, overcome by the selection of grid squares as the basic units, geographical characteristics being averaged out for each grid square. Since grid squares are all of the same shape and size their use

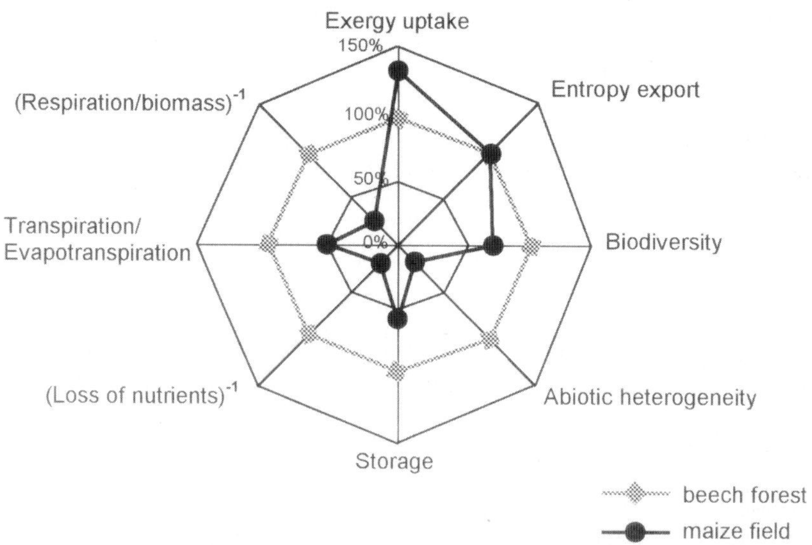

Figure 5. Integral graph ("amoeba" form) of an indicator set describing a beech and a crop field ecosystem of the Bornhöved Lake District (Schleswig-Holstein) (after Baumann, 2001).

eliminates variability in these properties and thus solves the second problem. The most commendable solution of the third problem is to make relative location as measured by spatial contiguity the dominant variable of analysis. It can be accomplished by means of geographical diversity analyses (Fränzle, 1978) or regionalization procedures based on comprehensive data matrices whose elements are derived from the digital evaluation of ecology or ecotope-related topographic, geological, hydrological, geobotanical, etc. base maps. The scale varies according to the size of the area to be investigated, i.e. normally between the 1:1 000 000 and 1:25 000 (or 5 000) scales (Fränzle, 1994).

Thus, for instance, the selection of representative core areas for comparative ecosystem research in the framework of the German surveillance concept (Ellenberg et al., 1978) was in the first step based on a specially developed algorithm (Fränzle et al., 1987). To this end the (old) Federal Republic of Germany was subdivided into a 12,706 mesh grid with an average grid square size of 21 km², each 10-dimensionally characterized by ecological indicator variables derived from the digitalization of relevant base maps. In the next evaluative step every square was compared with all others with regard to the variables, equality in a variable being labelled "1", and inequaliy "0". Averaging the number of comparisons (i.e., about $8.06*10^{10}$) the similarity of two grid squares then is characterized by a figure ranging from "0" (complete inequality) to "1" (equality in terms of indicator variables). Thereafter the vectorial distances of all of the squares were summarized in form of (virtual) histograms which define the representativeness of every square by means of the degree of right-skewed asymmetry. The transformation of these histograms into a 12,706-line matrix finally yields a gradation of (weighted) representativeness indices which, in turn, form the basis of a clustering procedure. It groups the matrix elements into clusters of decreasing representativeness.

In compliance with the second geostatistical requirement the localization of spatially representative grid squares out of the elements of these clusters was based on neighbourhood analysis. The methodology basically consists in determining the individual nearest-neighbourhood relationships of each grid square, i.e., their positive or negative autocorrelation which is a distance-weighted measure for each point in relation to its neighbours. The resultant data matrix permits to define average association frequencies as a basis for comparison of the individual autocorrelation status with the cluster averages. In terms of spatial structure it ensues that those grid squares or the 10-dimensionally defined ecotope complexes which they depict, are the most representative which differ least in their neighbourhood relationships from the average association pattern of the respective ecotope complex. The exact locations of study areas (typically comprising two or more of such complexes in the case of comparative ecosystem research) were eventually more precisely determined by applying the same geostatistical procedure to large-scale maps of these areas and their immediate surroundings, the results of which were finally corroborated by visual inspection in the field.

Two other methods (which require only IBM-compatible PCs) for reproducibly defining representative sampling or measuring locations are MUNAR (multidimensional neighbourhood analysis of representativeness) and CHAID (chi-square automatic interaction detection). MUNAR defines objects, for instance sites, not only multidimensionally by means of a set of characteristic qualities, but also in consideration of their neighbourhood relationships. Thus, it is a regionalized parameter for multistratigraphic binary data sets, e.g. k different thematic maps with m discrete features like soils or vegetation units (Vetter, 1989; Vetter and Maas, 1994). CHAID subdivides the statistical base set of indicative nominal, ordinal or metric data by means of likelihood ratio statistics for n two-way cross tables (Maas and Vetter, 1994; Schröder et al., 1992).

Mentioning different procedures to define representative objects for spatially valid biomonitoring purposes implies the understanding, ensuing also from elementary epistemological reasons, that representativeness is not an absolute but a relative term. The results of geostatistical analyses of categorized spatial data depend essentially on the procedure adopted, which requires a careful consideration in the light of the general research purposes envisaged.

3.2. Variogram analysis and kriging procedures

Values of a measured variable, for example a stress reaction of a bioindicator or a diagnostic soil property, are usually punctiform, i.e., the result of point measurements performed at selected locations. These data are then used to assess values at unobserved locations. Customary mathematical functions (e.g., linear or nonlinear interpolation approaches) are insufficient to give an adequate representation of such a regionalized variable because of its high degree of complexity and (frequently) small-scale variability or various correlations between neighbouring points. A useful statistical method would therefore have to inform in particular about the following problems: (1) Is it possible to decide upon the existence of a spatial distribution function on the basis of the available random samples? (2) To which spatial extent is a random sample representative within the limits of an imputed distribution function?

There are different geostatistical interpolation methods which can be used. Ordinary kriging is the best known among these; cokriging and external drift kriging are very efficient for the consideration of additional variables in the estimation (cf. Wackernagel, 1997). The problem with these techniques is, however, that they require numerical additional variables, and the relationships between the observed and the additional variables is supposed to be linear. If additional informations are available in categorical form only the Bayes-Markov kriging may be used (cf. Bárdossy et al., 1996).

The statistical hypothesis underlying ordinary kriging is the intrinsic assumption, i.e. it is supposed that the expected value of the variable Z is constant in the whole domain:

$$E[Z(x)] = m \tag{6}$$

and for all x

$$0.5 \, \text{Var} \, [Z(x + h) - Z(x)] = 0.5E\{[Z(x + h) - Z(x)]^2\} = \gamma(h) \tag{7}$$

where $\gamma(h)$, the semivariogram (or simply variogram), depends only on the vector h and not on the locations x and $x + h$ (Matheron, 1971). The linear estimator

$$Z^*(x) = \sum_{i=1}^{n} \lambda_i Z(x_i) \tag{8}$$

that minimizes the estimation variance can be found by solving the kriging system

$$\sum_{i=1}^{n} \lambda_i \gamma(x_j - x_i) + \mu = \gamma(x_j - x) \tag{9}$$

with $j = 1, \ldots, n$ and

$$\sum_{i=1}^{n} \lambda_i = 1 \tag{10}$$

Very low indicator values are difficult to interpolate. In this case it is commendable to estimate the exceedence probabilities of the classified data and apply the indicator kriging technique (Journel, 1983). Cokriging and external drift kriging can consider additional information, but only in a numerical form, assuming a linear relationship. An estimation method to cope also with non-linear relationships is the above Bayes-Markov kriging (Journel and Zhu, 1990) which, even in the simplified form of Bayes-Markov updating, proved superior to ordinary kriging (Lehmann, 1995).

3.3. Conclusions

An analysis of relevant literature in the light of the above geostatistical methodology shows that biomonitoring, geobotanical or geozoological studies sometimes neglect to indicate the limits of their statistical populations, and also fail to assess the degree of variability of the often undefined populations. It is clear that entirely spurious relationships can result between, for instance, stress-inducing chemical substances and strain reactions of individual organisms, populations or biocoenoses if they are derived from an insufficient amount of samples or measurements and where sample means obtained are not representative of population mean values.

Therefore, the selection of statistically relevant data sets for bioindicative purposes should be based on the following five-level approach:

- Approximate (preliminary) definition of representative areas for sampling or biomonitoring purposes on the basis of small-scale site-related maps (soil and vegetation maps, satellite images, etc.).
- Idem, on the basis of large-scale maps, stereocouples, etc., yielding a putative set of more precisely defined sites within the larger areas of the preceding step.
- Visual inspection and high-resolution mapping of sites, sampling, analyses of samples.
- Variogram analysis of data and, if necessary, renewed sampling on the basis of a denser reference grid.
- Kriging or related procedures in order to obtain valid spatial means of relevant parameters.

Both the importance and costs of many environmental assessment projects urgently recommend such a sequential approach in order to obtain spatially valid data.

4. Appraisal

Bioindication means unravelling the information structure of living systems, ranging from single organisms to complex ecosystems, in order to determine environmental quality or assess environmental hazards and risks. Biotic systems have an indicative quality on condition their ecophysiology and population dynamics are sufficiently well known and their stress reactions with regard to physical and chemical changes in site qualities can be measured against normal ecological boundary conditions and endogenous biorhythms.

- Environmental observation techniques which focus on individual and pre-selected physical or chemical stressors are not suited for the recognition of unexpected environmental changes in view of the enormous number of anthropogenic stressors, in particular chemicals, the complexity of environmental pathways and conversion products, and the multiple possibilities of synergistic and antagonistic effects. The demand for a timely observation of deleterious effects and altered environmental exposure situations before changes and damages become wide-spread and obvious can be satisfied by biomonitoring techniques. They involve active and passive approaches and span the dimensional scale from single-species bioassays over micro and mesocosms to biocoenoses and ecosystems. In view of the mostly unspecific stress reactions of organisms the primary task of bioindicators is the general determination of physiological effects rather than the direct measurement of environmental concentrations of stressors. In early recognition perspective the lack of specificity has the advantage of a broad-based *caveat*, inducive to subsequent systematic search for quantitative causal interrelationships.

  A further advantage is the comparatively low costs of biomonitoring approaches in comparison to those of instrumental measurements; even in the case of active monitoring networks the total costs would not amount to more than a few per cent of the latter. Another important point is the integrative recording character of both effect and accumulation bioindicators which always reflects the total exposure time, while an instrument can only produce a set of singular data within the framework of a given temporal resolution. The latter, however, is to a considerable extent responsible for the price of the instruments and the running costs involved.

In contrast to these positive aspects of bioindicator use, an essential disadvantage results from the highly variable susceptibility of the multitude of species exposed to stressors, which leads to difficulties in comparing specific effect data. Nevertheless, fuzzy logic approaches provide very useful operations for processing such data in order to more precisely define their physiological or toxicological information (cf. Section 2.2.1).

- In general the structurally simplest type of bioindicators are biomarkers and biotests (bioassays). The field of biomarkers has evolved rapidly in the last twenty years. Considering the substantially differing connotations of the term, in general a biomarker may be considered as a biological response to a chemical or chemicals that provides a measure of exposure and sometimes also of toxic effects. The development of new indices for defining exposure, effective dose, responses and susceptibility has been brought about by the application of new analytical techniques, mainly based on the tools of molecular biology, and the use of more detailed physiological and molecular-biological models. With regard to exposure monitoring, therefore, biomarkers have the advantage of reacting rapidly to exposure and (more or less) quantifying only the biologically available pollutants, and as effect monitors they are able to integrate the effects of multiple stressors (Peakall, 1994). Nevertheless, a better assessment of specific stress reactions can only be achieved by exposing sets of organisms of several susceptible taxa in the framework of active monitoring approaches. Theoretically feasible, but technically distinctly more difficult, would be the parallel exposure of one such set to "all" of the ambiental stressors and another (control set) to natural stresses such as drought, heat, etc. only. Thus, the step from measuring a selective response in a single (provably) representative organism (or a corresponding test population) and extrapolating from this selective biochemical change to estimate the ecological significance of a slight increase or decrease of a measured parameter is the major problem.

- Environmental rather than medical biomarkers may be used to investigate and more precisely define uncertainties that arise in preliminary hazard assessments, which involves establishing better links between exposure and biological effects. The fuzzy set theory offers commendable possibilities to grapple with these problems on the basis of existing data sets (Friedrichs, 1999; Salski et al., 1996). Models developed to this end must be refined in order to be better consistent with the spatial and temporal scales under consideration and should be based on a hierarchy of operationally defined verification, calibration and validation procedures. This implies that the realm of validity of structurally simpler models can only be reliably determined on the basis of, and in comparison with, more complex ones. For example, inquiries into the sensitivity of soil to environmental chemicals require the following hierarchy of test procedures: (1) badge experiments with different soil suspensions, (2) soil column experiments in a set of lysimeters, micro and mesocosms of systematically increased size, (3) controlled exposure of experimental plots of different biodiversity status to single and different combinations of chemicals (Fränzle, 1982, 1987).

- A distinctly greater obstacle to interpreting results of studies at higher levels of organization is the difficulty in determining cause/effect relationships between stress levels, e.g. ambient concentrations of environmental chemicals, and the degree of change in community and ecosystem level endpoints. In direct contrast to the field situation, in laboratory test sytems one or more (not necessarily field-relevant or functionally representative) species, with all individuals intentionally of the same age and health, are exposed to one or a few physical or chemical stressors. In ecosystems many species, each represented by several age classes with correspondingly great variations in physical condition due to a temporally and spatially varying mixture of "forces" (competition, predation, disturbances, etc.), will be exposed under fluctuating environmental conditions to chronic, possibly heterogeneously distributed anthropogenic stresses. In particular chemical stressors are numerous, and additive or synergistic effects are common; furthermore a high recalcitrance of chemicals may lead to a continuous increase in environmental concentration. Additionally, the bioavailability of potentially toxic substances may differ essentially from that in laboratory tests (cf., e.g., Spurgeon and Hopkin, 1995), and the susceptibility of organisms may be considerably increased owing to further stress factors, such as predation, competition or habitat stress (Bayne et al., 1985).

- The current state of research does not yet permit to decide whether the structure or the functioning of ecosystems is in general more sensitive to various stressors. One opinion has it that functional variables, especially those that are substrate-limited, will always be less sensitive than structural measures because there is functional redundancy in communities. So any loss of functional capacity by one organism will be immediately compensated by increased activity of another. Other researchers hold that functional capacity can be affected before compensatory mechanisms operate, especially when such mechanisms are themselves adversely affected by the stress or when they operate on a more lengthy time scale relative to the functional measure.

- An intersystemic comparison of biocenotic reactions must take into account the spatial and temporal variability of ecosystems, whose communities are organized by competition, predation and disturbances, with competition and predation being presumably less important in more disturbed environments. Consequently, there is no such thing as a single stability, susceptibility or vulnerability measure for a community or a whole ecosystem; these properties would rather be more appropriately expressed in multidimensional form such as "amoeba" diagrams of indicator variables or fuzzy clusters. The relative weight of the variables varies with the aspect of the system under study, which is basically an expression of the uniqueness and probabilistic character of ecosystems. Thus, simulations of systems behaviour under stress can only yield deterministic prognoses under precise, i.e. empirically defined boundary conditions (Breckling, 1990); predominantly, models are to provide formalized information to formulate hypotheses on causal chains, stress and strain mechanisms, or sensitive system properties. The epistemological relativity of the ecosystem concept implies that in dependence on the underlying model structure (i.e., elements and static and functional relationships defined) either possible damages may remain unidentified or unrealistic stress effects are postulated (Barnthouse, 1998).

- Active and passive biomonitoring techniques on the basis of single species and specimen exposure yield spatially valid data only if the underlying measuring or sampling networks are established in compliance with geostatistical requirements and corresponding test procedures such as variogram analysis and various kriging procedures. The selection of complex bioindicators such as biocoenoses or ecosystems must be based on rigid criteria of spatial and temporal representativeness whose fulfillment should be corroborated by means of traditional frequency statistics and, more specifically, by means of complex geostatistical procedures (cf. Kienzl et al., 2002).

References

Alterburger, R., Schmitt-Jansen, M., 2002 (2003). Predicting toxic effects of contaminants in ecosystems using single species investigations. In: Markert, B.A., Breure, A.M., Zechmeister, H.G. (Eds), Bioindicators and Biomonitors. Elsevier, Oxford, pp. 153–198.

Arndt, U., Nobel, W., Schweizer, B., 1987. Bioindikatoren – Möglichkeiten, Grenzen und neue Erkenntnisse. Ulmer, Stuttgart.

Bach, H.-J., 1996. Bakterielle Populationen und Stoffumsatzpotentiale in Acker-, Grünland- und Waldböden einer Jungmoränenlandschaft in Schleswig-Holstein. EcoSys – Beiträge zur Ökosystemforschung Suppl. Bd. 15. Verein zur Förderung der Ökosystemforschung zu Kiel, Kiel

Bárdossy, A., Haberlandt, U., Grimm-Strele, J., 1996. Regional scales of groundwater quality parameters and their dependence on geology and land use. In: Kobus, H. et al. (Eds), Groundwater and Subsurface Remediation. Springer, Berlin, pp. 195–203.

Barkman, J.J., 1958. Phytosociology and Ecology of Cryptogamic Epiphytes. Van Gorcum, Assen.

Barnthouse, L.W., 1998. Modeling ecological risks of pesticides: a review of available approaches. In: Schüürmann, G., Markert, B. (Eds), Ecotoxicology. Wiley, New York, and Spektrum, Heidelberg, pp. 769–798.

Baumann, R., 2001. Indikation der Selbstorganisationsfähigkeit terrestrischer Ökosysteme. Doctoral Thesis, University of Kiel.

Bayne, B.L., Brown, D.A., Burns, K., Dixon, D.R., Ivanovici, A., Livingstone, D.R., Lowe, D.M., Moore, M.N., Stebbing, A.R.D., Widdows, J., 1985. The Effects of Stress and Pollution on Marine Animals. Praeger, New York.

Becker, P.H., 2002 (2003). Biomonitoring with birds. In: Markert, B.A., Breure, A.M., Zechmeister, H.G. (Eds), Bioindicators and Biomonitors. Elsevier, Oxford, pp. 677–736.

Berthold, P., 2000. Vogelzug. Wissenschaftliche Buchgesellschaft, Darmstadt.

Bick, H., 1972. Ciliated Protozoa. An Illustrated Guide to the Species Used as Biological Indicators in Freshwater Biology. World Health Organization, Geneva.

Breckling, B., 1990. Singularität und Reproduzierbarkeit der Modellierung ökologischer Systeme. Doctoral Dissertation, University of Bremen.

Burhenne, M., Deml, G., Steinberg, C., 1999. Ein Biotestsystem mit verschiedenen Bodenalgen zur ökotoxikologischen Bewertung von Schwermetallen und Pflanzenschautzmitteln. In: Oehlmann, J., Markert, B. (Eds), Ökotoxikologie. Ecomed, Landsberg, pp. 88–99.

Cairns Jr., J., Niederlehner, B.R., 1993. Ecological function and resilience: neglected criteria for environmental impact assessment and ecological risk analysis. Environ. Professional 15, 116–124.

Chandler, P.M., Robertson, M., 1994. Gene expression regulated by abscicic acid and its relation to stress tolerance. Annual Rev. Plant Physiol. Plant Mol. Biol. 45, 113–141.

Christensen, V., 1994. On the behavior of some proposed goal functions for ecosystem development. Ecol. Modelling 75/76, 37–49.

Chovanec, A., Hofer, R., Schiemer, F., 2002 (2003). Fish as bioindicators. In: Markert, B.A., Breure, A.M., Zechmeister, H.G. (Eds), Bioindicators and Biomonitors. Elsevier, Oxford, pp. 639–676.

Clark, W.C., Jäger, J., Corell, R., 2000. Assessing vulnerability to global environmental risks. Report of the Workshop on Vulnerability to Global Environmental Change: Challenges for Research, Assessment and Decision Making. Warrenton, Virginia, 22–25 May 2000. Belfer Center for Science and International Affairs, John F. Kennedy School of Government, Harvard University, Cambridge.

Corliss, J.O., 1979. The Ciliated Protozoa. Characterization, Classification and Guide to the Literature. Pergamon Press, Oxford.

Crossey, M.J., LaPoint, T.W., 1988. A comparison of periphyton community structural and functional responses to heavy metals. Hydrobiologia 162, 109–121.

Crumby, W.D., Webb, M.A., Bulow, F.J., Cathey, H.J., 1990. Changes in biotic integrity of a river in North-Central Tennessee. Trans. Am. Fish. Soc. 119, 885–893.

Csermely, P. (Ed.) 1998. Stress of Life from Molecules to Man. Ann. New York Academy of Sciences, Vol. 851, New York.

De Knecht, J., van Dillen, M., Koevoets, P.L.M., Schat, H., Verkleij, J.A.C., Ernst, W.H.O., 1994. Phytochelatins in cadmium-sensitive and cadmium-tolerant Silene vulgaris. Plant Physiol. 104, 225–261.

DFG, Deutsche Forschungsgemeinschaft (Ed.), 1983. Ökosystemforschung als Beitrag zur Beurteilung der Umweltwirksamkeit von Chemikalien. Verlag Chemie, Weinheim.

Didden, W., 2002 (2003). Oligochaeta. In: Markert, B.A., Breure, A.M., Zechmeister, H.G. (Eds), Bioindicators and Biomonitors. Elsevier, Oxford, pp. 555–576.

Dolgerloh, M., 1999. Bildanalytische Auswertung des Wachstums von Lemna gibba G3 in Laborstudien. In: Oehlmann, J., Markert, B. (Eds), Ökotoxikologie. Ecomed, Landsberg, pp. 164–166.

Eberius, M., Vandenhirtz, D., 1999. Einsatz eines speziellen Bildanalysesystems zur ökotoxikologisch umfassenden und kosteneffizienten Auswertung des Wasserlinsentests. In: Oehlmann, J., Markert, B. (Eds), Ökotoxikologie. Ecomed, Landsberg, pp. 167–170.

ECETOC (European Chemical Industry Ecology and Toxicology Centre), 1993. Environmental Hazard Assessment of Substances. Technical Report 51, Brussels.

ECETOC, 1997. The Value of Aquatic Model Ecosystem Studies in Ecotoxicology. Technical Report 73, Brussels.

Ellenberg, H., 1972. Belastung und Belastbarkeit von Ökosystemen. Verh. Ges. Ökol. 1, 9–26.

Ellenberg, H., 1996. Vegetation Mitteleuropas mit den Alpen. Ulmer, Stuttgart.

Ellenberg, H., Fränzle, O., Müller, P., 1978. Ökosystemforschung im Hinblick auf Umweltpolitik und Umweltplanung, Umweltforschungsplan des Bundesministers des Innern. Ökologie Forschungsbericht 78 101 04/005, Bonn.

Ernst, W.H.O., 1993. Ecological aspects of sulfur in higher plants: the impact of SO_2 and the evolution of the biosynthesis of organic sulfur compounds on populations and ecosystems. In: De Kok, L.J., Stulen, I., Rennenberg, H., Brunold, C., Rauser, W.E. (Eds), Sulfur Nutrition and Assimilation in Higher Plants. SPB Academic, The Hague, pp. 295–313.

Ernst, W.H.O., 2002 (2003). The use of higher plants as bioindicators. In: Markert, B.A., Breure, A.M., Zechmeister, H.G. (Eds), Bioindicators and Biomonitors. Elsevier, Oxford, pp. 423–463.

Evans, R.B., 1966. A proof that exergy is the only consistent measure of potential work. Ph.D. Thesis, Dartmouth College, Hannover, NH.

Feige, G.B., Niemann, L., Jahnke, S., 1990. Lichens and mosses. Silent chronists of the Chernobyl accident. Bibl. Lichenol. 38, 63–77.

Fränzle, O., 1978. The structure of soil associations and cenozoic morphogeny of Southeast Africa. In: Nagl, H. (Ed.), Beiträge zur Quartar- und Landschaftsforschung. Hirt, Wien, pp. 159–176.

Fränzle, O., 1982. Modellversuche über die Passage von Umweltchemikalien durch die ungesättigte Zone natürlicher Bodenprofile sowie durch Bodenschlämme in Laborlysimetern und in Freiland. Umweltforschungsplan des Bundesministers des Innern. Forschungsbericht 106 02 005/02, Kiel.

Fränzle, O., 1987. Sensitivity of European soils related to pollutants. In: Barth, H., L'Hermite, P. (Eds), Scientific Basis for Soil Protection in the European Community. Elsevier, London, pp. 123–145.

Fränzle, O., 1993. Contaminants in Terrestrial Environments. Springer, Berlin.

Fränzle, O., 1994. Representative soil sampling. In: Markert, B. (Ed.), Environmental Sampling for Trace Analysis. VCH, Weinheim, pp. 305–320.

Fränzle, O., 1998a. Sensitivity of ecosystems and ecotones. In: Schüürmann, G., Markert, B. (Eds), Ecotoxicology. Wiley, New York, and Spektrum, Heidelberg, pp. 75–115.

Fränzle, O., 1998b. Ökosystemforschung im Bereich der Bornhöveder Seenkette. In: Fränzle, O., Müller, F., Schröder, W. (Eds), Handbuch der Umweltwissenschaften, Kap. V-4.3 (36 pp.). Ecomed, Landsberg.

Fränzle, O., 1999. Ökosystemare Toxikologie aus der Sicht des Ökologen. In: Oehlmann, J., Markert, B. (Eds), Ökotoxikologie. Ecomed, Landsberg, pp. 23–48.

Fränzle, O., Schröder, W., Vetter, L., 1985. Synoptische Darstellung möglicher Ursachen des Waldsterbens. Umweltforschungsplan des Bundesministers des Innern. Forschungsbericht 106 07 046/13, Kiel.

Fränzle, O., Kuhnt, D., Kuhnt, G., Zölitz, R., 1987. Auswahl der Hauptforschungsräume für das Ökosystemforschungsprogramm der Bundesrepublik Deutschland. Umweltforschungsplan des Bundesministers des Innern. Forschungsbericht 101 04 043/02, Kiel.

Fränzle, O., Bruhm, I., Grünberg, K.-U., Jensen-Huß, K., Kuhnt, D., Kuhnt, G., Mich, K., Müller, F., Reiche, E.-W., 1989. Darstellung der Vorhersagemöglichkeiten der Bodenbelastung durch Umweltchemikalien. Texte des Umweltbundesamtes 34/89, Umweltbundesamt, Berlin.

Fränzle, O., Jensen-Huss, K., Daschkeit, A., Hertling, T., Lüschow, R., Schröder, W., 1993. Grundlagen zur Bewertung der Belastung und Belastbarkeit von Böden als Teilen von Ökosystemen. Texte des Umweltbundesamtes 59/93, Umweltbundesamt, Berlin.

Fränzle, O., Krinitz, J., Schmotz, W., Delschen, T., Leisner-Saber, J., 1995. Harmonisierung der Untersuchungsverfahren und Bewertungsmaßstäbe für den Bodenschutz mit derRussischen Föderation. Texte des Umweltbundesamtes 60/95, Umweltbundesamt, Berlin.

Friedrichs, M., 1999. Entwicklung ökotoxikologischer Prüfverfahren mit Hilfe der Fuzzy-Clusteranalyse. Doctoral Dissertation, Kiel.

Friesel, P., Hansen, P.D., Kühn, R., Trénel, J., 1984. Überprüfung der Durchführbarkeit von Prüfungsvorschriften und der Aussagekraft der Stufe 1 und 2 des Chemikaliengesetzes – Teil VI. Umweltforschungsplan des Bundesministers des Innern. Forschungsbericht 106 04 011/08, Bonn.

Fuente, J.M. de la, Ramirez-Rodriguez, V., Cabrera-Ponce, J.L., Herrera-Estrella, L.H., 1997. Aluminum tolerance to alteration of citrate synthesis. Science 276, 1566–1568.

Funke, W., Feige, G. B., Jahnke, S., Reidl, K., 1993. Bioindikatoren. In: Kuttler, W. (Ed.), Handbuch zur Ökologie. Analytica, Berlin, pp. 60–68.

Gnauck, A., 1982. Strukturelle und funktionelle Änderungen in aquatischen Ökosystemen. Kongreß- und Tagungsberichte der Martin-Luther-Universität Halle-Wittenberg, Wittenberg, pp. 335–344.

Grade, R., Gonzalez-Valero, J., Höcht, P., Pfeifle, V.U., 1999. Ein "Higher Tier" Durchfluß-System im Fließgleichgewicht zur Bestimmmung der Toxizität von Pflanzenschutzmitteln gegenüber der Grünalge Selenastrum capricornutum. In: Oehlmann, J., Markert, B. (Eds), Ökotoxikologie. Ecomed, Landsberg, pp. 146–150.

Grodzinska, K., 1978. Acidity of tree bark as a bioindicator of forest pollution in southern Poland. Water, Air, Soil Poll. 7, 3–7.

Grodzinska, K., 1982. Monitoring of air pollutants by mosses and tree bark. In: Steubing, L., Jäger, H.-J. (Eds), Monitoring of Air Pollutants by Plants – Methods and Problems. Junk, The Hague, pp. 33–42.

Grossmann, W.-D., Schaller, J., Sittard, M., 1984. "Zeitkarten": eine neue Methodik zum Testen von Hypothesen und Gegenmaßnahmen bei Waldschäden. Allg. Forstz. 38, 837–843.

Guderian, R., Tingey, D.T., Rabe, R., 1985. Effects of photochemical oxidants on plants. In: Guderian, R. (Ed.), Air Pollution by Photochemical Oxidants. Formation, Transport, Control and Effects on Plants (Ecol. Studies 52). Springer, Berlin, pp. 127–346.

Gulland, J.A. (Ed.), 1978. Fish Population Dynamics. Wiley, Chichester.

Hafellner, J., Grill, D., 1980. Die Wiedereinwanderung von epiphytischen Flechten in den Raum Leoben-Hinterberg nach Stillegung des Hauptemittenten. Mitt. Forstl. Bundes-Versuchsanstalt Wien 131, 83–87.

Hanssen, U., Hingst, R., Irmler, U., Ritter, D., Schrautzer, J., 1991. Ökosystemforschung im Bereich der Bornhöveder Seenkette: Biozönotische Komplexe im Hauptforschungsraum des Bornhöveder Ökosystemforschungsprojektes. Verh. Ges. Ökol. 20, 127–136.

Härtel, O., Grill, D., 1972. Die Leitfähigkeit von Fichtenborken-Extrakten als empfindlicher Indikator für Luftverunreinigungen. Eur. J. Forest Path. 2, 205–213.

Heagle, A.S., Body, D.E., Heck, W.W., 1973. An open top chamber to assess the impact of air pollution on plants. J. Environ. Qual. 2, 365–368.

Herrmann, R., 1976. Modellvorstellungen zur räumlichen Verteilung von Spurenmetallverunreinigungen in der Bundesprepublik Deutschland, angezeigt durch den Metallgehalt in epiphytischen Moosen. Erdkunde 30, 241–253.

Herrmann, R., 1984. Atmosphärische Transporte und raumzeitliche Verteilung von Mikroschadstoffen (Spurenmetalle, Organochlorpestizide, polyzyklische aromatische Kohlenwasserstoffe) in Nordostbayern. Erdkunde 38, 55–63.

Hölker, F., 2000. Bioenergetik dominanter Fischarten (Abramis brama L. und Rutilus rutilus L.) in einem eutrophen See Schleswig-Holsteins – Ökophysiologie und Individuen-basierte Modellierung. EcoSys – Beiträge zur Ökosystemforschung Suppl. Bd. 32. Verein zur Förderung der Ökosystemforschung zu Kiel, Kiel.

Holling, C.S., 1976. Resilience and stability of ecosystems. In: Jantsch, E., Waddington, C.H. (Eds), Evolution and Consciousness. Addison-Wesley, Reading, pp. 73–92.

Hörmann, G. (Ed.), 1995. Auswirkungen einer Temperaturerhöhung auf die Ökosysteme der Bornhöveder Seenkette. EcoSys – Beiträge zur Ökosystemforschung Bd. 2. Verein zur Förderung der Ökosystemforschung zu Kiel, Kiel.

Iwama, G.K., 1998. Stress in fish. In: Csermely, P. (Ed.), Stress of Life from Molecules to Man. Ann. New York Academy of Sciences, Vol. 851, pp. 304–310.

Jørgensen, S.E., Nielsen, S.N., 1998. Thermodynamic orientors: exergy as a goal function in ecological modeling and as an ecological indicator for the description of ecosystem development. In: Müller, F., Leupelt, M. (Eds), Ecotargets, Goal Functions and Orientors. Springer, Berlin, pp. 63–86.

Johnsen, I., Søchting, U., 1973. Influence of air pollution on the epiphytic lichen vegetation and bark properties of deciduous trees in the Copenhagen area. Oikos 24, 344–351.

Johnson, T.C., Odada, E.O. (Eds), 1996. The Limnology, Climatology and Palaeoclimatology of the East African Lakes. Gordon and Breach, Amsterdam.

Journel, A.G., 1983. Non-parametric estimation of spatial distributions. Math. Geol. 15, 445–468.

Journel, A.G., Zhu, H., 1990. Integrating soft seismic data: Bayes-Markov updating, an alternative to cokriging and traditional regression. Report 3, Stanford Center for Reservoir Forecasting, Stanford.

Kalsch, W., Römbke, J., 1999. Zur chronischen Wirkung von TNT auf die Stoppelrübe Brassica rapa im Labortest. In: Oehlmann, J., Markert, B. (Eds), Ökotoxikologie. Ecomed, Landsberg, pp. 100–105.

Kappen, L. (Ed.), 1997. Übertragbarkeit ökophysiologischer Meßergebnisse von Pflanzenteilen auf Pflanzenbestände. EcoSys – Beiträge zur Ökosystemforschung Suppl. Bd. 20. Verein zur Förderung der Ökosystemforschung zu Kiel, Kiel.

Kappen, L., Gaedke, U., Geller, W., Kutsch, W., Tilzer, M.M., 1992. Strategien zur Untersuchung und Modellierung des Kohlenstoffhaushaltes in terrestrischen und aquatischen Ökosystemen. In: Erdmann, K.-H., Nauber, J. (Eds), Beiträge zur Ökosystemforschung und Umwelterziehung. MAB-Mitteilungen 36, 76–81.

Kienzl, K., Riss, A., Vogel, W., Hackl, J., Götz, B., 2002 (2003). Bioindicators and biomonitors for policy, legislation and administration. In: Markert, B.A., Breure, A.M., Zechmeister, H.G. (Eds), Bioindicators and Biomonitors. Elsevier, Oxford, pp. 85–122.

Knabe, W., 1981. Immissionsökologische Waldzustandserfassung in Nordrhein-Westfalen. Allge. Forstzeitsch. 36, 641–643.

Kohler, A., 1982. Wasserpflanzen als Bioindikatoren. Decheniana-Beihefte 26, 31–42.

Kokke, R., Winteringham, F.P.W., 1980. Labelled substrate techniques as indicators of agrochemical residue-biota interactivity in soil and aquatic ecosystems. IAEA Panel Proc. Ser. STI/PUB/548, Vienna, pp. 23–33.

Korte, F. (Ed.), 1987. Lehrbuch der Ökologischen Chemie. Thieme, Stuttgart.

Kreeb, K.H. (Ed.), 1990. Methoden zur Pflanzenökologie und Bioindikation. Fischer, Stuttgart.

Kreeb, K.H., Müller, J., Schneider, K., 1990. Kennzeichnung von Standortfaktoren durch Zeigerpflanzen. In: Kreeb, K.H. (Ed.), Methoden zur Pflanzenökologie und Bioindikation. Fischer, Stuttgart.

Kutsch, W.L., 1996. Untersuchungen zur Bodenatmung zweier Ackerstandorte im Bereich der Bornhöveder Seenkette. EcoSys – Beiträge zur Ökosystemforschung Suppl. Bd. 16. Verein zur Förderung der Ökosystemforschung zu Kiel, Kiel.

Kutsch, W.L., Herbst M., Vanselow, R., Hummelshøj, P., Jensen, N.O., Kappen, L., 2001a. Stomatal acclimation influences water and carbon fluxes of a beech canopy in northern Germany. Basic Appl. Ecol. 2, 265–281.

Kutsch, W.L., Steinborn, W., Herbst, M., Baumann, R., Barkmann, J., Kappen, L., 2001b. Environmental indication: a field test of an ecosystem approach to quantify biological self-organization. Ecosystems 4, 49–66.

Larcher, W., 1987. Streß bei Pflanzen. Naturwissenschaften 74, 158–167.

Larcher, W., 1994. Ökophysiologie der Pflanzen. Ulmer, Stuttgart.

Lee, J.J., Preston, E.M., Lewis, R.A., 1978. A system for the experimental evaluation of the ecological effects of sulfur dioxide. Proceedings 4th Joint Conference on Sensing of Environmental Pollutants. American Chem. Soc., pp. 49–53.

Lehmann, W., 1995. Anwendung geostatistischer Verfahren auf die Bodenfeuchte in ländlichen Einzugsgebieten. Mitt. Inst. für Hydrologie und Wasserwirtschaft der Univ. Karlsruhe, No. 52. Karlsruhe.

Leshem, Y.Y., Kuiper, P.J.C. 1996. Is there a GAS (general adaptation) response to various types of environmental stress? Biol. Plant. 38, 1–18.

Leshem, Y.Y., Kuiper, P.J.C., Erdei, L., Lurie, S., Perl-Treves, R., 1998. Do Selye's mammalian "GAS" concept and "co-stress" response exist in plants? In: Csermely, P. (Ed.), Stress of Life from Molecules to Man. Ann. New York Academy of Sciences, Vol. 851, New York; pp. 199–208.

Letchworth, M.B., Blum, U., 1977. Effects of acute ozone exposure on growth, nodulation and nitrogen content of Ladino clover. Environ. Pollut. 14, 303–311.

Levitt, J., 1980. Responses of Plants to Environmental Stresses. Academic Press, New York.

Lichtenthaler, H.K., 1984. Differences in morphology and chemical composition of leaves grown at different light intensities and qualities. In: Baker, N.R., Davies, W.J., Ong, K.C. (Eds), Control of Leaf Growth. Cambridge University Press, Cambridge, pp. 201–222.

Lichtenthaler, H.K., 1998. The stress concept in plants: an introduction. In: Csermely, P. (Ed.), Stress of Life from Molecules to Man. Ann. New York Academy of Sciences, Vol. 851, New York, pp. 187–198.

Lieth, H., Markert, B. (Eds), 1990. Element Concentration Cadasters in Ecosystems. VCH, Weinheim.

Maas, R., Vetter, L., 1994. CHAID – Chisquare automatic interaction detection. In: Schröder, W., Vetter, L., Fränzle, O. (Eds), Neuere statistische Verfahren und Modellbildung in der Geoökologie. Vieweg, Braunschweig, pp. 95–101.

Magurran, A.E., 1988. Ecological Diversity and its Measurement. Cambridge University Press, Cambridge.

Mandl, R.H., Weinstein, L.H., McCune, D.C., Keveny, M., 1973. A cylindrical open top chamber for the exposure of plants to air pollutants in the field. J. Environ. Qual. 2, 371–376.

Markert, B., Herpin, U., Berlekamp, J., Oehlmann, J., Grodzinska, K., Mankovska, B., Suchara, I., Sievers, U., Weckert, V., Lieth, H., 1996. A comparison of heavy metal deposition in selected eastern European countries using the moss monitoring method, with special emphasis on the "Black Triangle". Science of the Total Environment 193, 85–100.

Mather, P.M., 1972. Areal classification in geomorphology. In: Chorley, R.J. (Ed.), Spatial Analysis in Geomorphology. Methuen, London, pp. 305–322.

Matheron, G., 1971. The Theory of Regionalized Variables and its Application. Les Cahiers du Centre de Morphologie Mathématique, Fasc. 5. Fontainebleau.

May, R.M., 1981. Theoretical Ecology. Blackwell Scientific, Oxford.

Menting, G., 2001. Explosive Artbildung bei ostafrikanischen Buntbarschen. Naturw. Rundsch. 54, 401–410.

Michaelis, H., 1987. Strukturveränderungen der Wattenfauna am Beispiel des Jadebusens. In: Niedersächsisches Umweltministerium (Ed.), Umweltvorsorge Nordsee – Belastungen, Gütesituation und Maßnahmen, Niedersächs Umweltministerium, Hannover, pp. 151–157.

Moss, R., Brenkert, A., Malone, E.L., 2000. Measuring vulnerability: a trial indicator set. Pacific Northwest National Laboratory, Richland WA.

Mulder, Ch., Breure, A.M., 2002 (2003). Plant biodiversity and environmental stress. In: Markert, B.A., Breure, A.M., Zechmeister, H.G. (Eds), Bioindicators and Biomonitors. Elsevier, Oxford, pp. 501–525.

Müller, P., 1984. Experimental biomonitoring, food web monitoring and specimen banking. In: Lewis, R.A., Stein, N., Lewis, C.W. (Eds), Environmental Specimen Banking and Monitoring as Related to Banking. Martinus Nijhoff, Boston, pp. 180–199.

Müller, P., Wagner, G., 1988. Probenahme und Charakterisierung von repräsentativen Umweltproben. In: BMFT, Bundesministerium für Forschung u. Technologie (Ed.), Umweltprobenbank: Bericht und Bewertung der Pilotphase. Springer, Berlin, pp. 27–36.

Naujokat, D., 1997. Nährstoffbelastung und Eutrophierung stehender Gewässer – Möglichkeiten und Grenzen ökosystemarer Entlastungsstrategien am Beispiel der Bornhöveder Seenkette. Dissertations Druck Darmstadt.

Nobel, W., 1987. Einsatz von Bioindikatoren als Bestandteil einer kommunalen Luftreinhaltestrategie. In: Ökologische Probleme in Verdichtungsgebieten (= Hohenheimer Arbeiten). Ulmer, Stuttgart, pp. 203–213.

Nusch, E.A., 1982. Prüfung der biologischen Schadwirkungen von Wasserinhaltsstoffen mit Hilfe von Protozoentests. Decheniana-Beihefte 26, 87–98.

Oehlmannn, J., Schulte-Oehlmann, U. 2002 (2003). Molluscs as bioindicators. In: Markert, B.A., Breure, A.M., Zechmeister, H.G. (Eds), Bioindicators and Biomonitors. Elsevier, Oxford, pp. 577–635.

O'Hare, G.P., 1974. Lichens and bark acidification as indicators of air pollution in west central Scotland. J. Biogeogr. 1, 135–146.

Peakall, D.B., 1994. The role of biomarkers in environmental assessment. Ecotoxicology 3, 157–160.

Peichl, L., Reiml, D., Ritzl, I., Schmidt-Bleek, F., 1987. Übersicht biologischer Wirkungs-Testsysteme zur Beobachtung unerwarteter Umweltveränderungen – Biosonden. GSF-Bericht 28/87. Gesellschaft für Strahlen- und Umweltforschung München, Neuherberg.

Peterson, G., Allen, C.R., Holling, C.S., 1998. Ecological resilience, biodiversity and scale. Ecosystems 1, 6–18.

Posthumus, A.C., 1976. The use of higher plants as indicators for air pollutants in the Netherlands. In: Karenlampi, L. (Ed.), Proceedings Kuopio Meeting on Plant Damages by Air Pollutants, pp. 115–120.

Posthumus, A.C., 1977. Experimentelle Untersuchungen der Wirkung von Ozon und Peroxyacetylnitrat (PAN) auf Pflanzen. VDI-Berichte 270, 153–161.

Precht, H., Christophersen, J., Hensel, H., 1955. Temuratur und Leben. Springer, Heidelberg.

Precht, H., 1967. A survey of experiments on resistance adaptation. In: Troshin, A.S. (Ed.), The Cell and Environmental Temperature. Pergamon Press, Oxford, pp. 307–321.

Ratte, H.T., Hammers-Wirtz, M., Cleuvers, M., 2002 (2003). Ecotoxicity testing. In: Markert, B.A., Breure, A.M., Zechmeister, H.G. (Eds), Bioindicators and Biomonitors. Elsevier, Oxford, pp. 221–256.

Renkonen, O., 1938. Statistisch-ökologische Untersuchungen über die terrestrische Käferwelt der finnischen Bruchmoore. Ann. Zool. Soc. Zool.-Bot. Fenn. 6, 1–231.

Rodgers Jr., J.H., Dickson, K.L., Cairns, J., 1980. In: Wetzel, R.G. (Ed.), American Society for Testing and Materials. American Society for Testing and Materials, Philadelphia, pp. 142–167.

Rühling, A., Tyler, G., 1973. Heavy metal deposition in Scandinavia. Water, Air, Soil Pollut. 2, 445–455.

Salski, A., Fränzle, O., Kandzia, P. (Eds), 1996. Fuzzy logic in ecological modelling. Ecol. Model. 85 (1), Special Issue.

Schindler, D.W., 1987. Detecting ecosystem responses to anthropogenic stress. Com. J. Fish. Aquat. Sci. 44, suppl. 1, 6–25

Schmidt-Bleek, F., Peichl, L., Behling, G., Müller, K.W., Reiml, D., 1987. A Concept for Early Recognition and Assessment of Environmental Changes. GSF-Bericht 21/87. Gesellschaft für Strahlen- und Umweltforschung München, Neuherberg.

Scholl, G., 1971. Ein biologisches Verfahren zur Bestimmung der Herkunft und Verbreitung von Fluorverbindungen in der Luft. Landwirtsch. Forsch. 26, 1. Sonderheft, 29–35.

Schröder, W., Fränzle, O., Vetter, L., 1986. Ist eine synoptische Darstellung von standörtlichen Rahmenbedingungen der Waldschäden möglich? Allg. Forstz. 22, 543–544.

Schröder, W., Vetter, L., Fränzle, O., 1992. Einfluß statistischer Verfahren auf die Bestimmung repräsentativer Standorte für Umweltuntersuchungen. Petermanns Geogr. Mitt. 136, 309–318.

Schröder, W., Fränzle, O., 1996. Disparities in sampling, parameters and metadata: environmental monitoring and assesment as a unifying basis. In: Schröder, W., Fränzle, O., Keune, H., Mandy, P. (Eds), Global Monitoring of Terrestrial Ecosystems. Ernst and Sohn, Berlin, pp. 57–66

Schröter, D., 2001. Structure and function of decomposer food webs along a European north-south transect with special focus on testate amoebae. Doctoral Dissertation, University of Gießen.

Selye, H., 1936. A syndrome produced by various nocuous agents. Nature 138, 32 34.

Spurgeon, D.J., Hopkin, S.P., 1995. Extrapolation of the laboratory-based OECD earthworm toxicity test to metal-contaminated field sites. Ecotoxicology 4, 190–205.

Steedman, R., Haider, W., 1993. Applying notions of ecological integrity. In: Woodley, S., Kay, J., Francis, G. (Eds), Ecological Integrity and the Management of Ecosystems. St. Lucie Press, Ottawa, pp. 47–60.

Steinberg, C., Klein, J., Brüggemann, R., 1995. Ökotoxikologische Testverfahren. Ecomed, Landsberg.

Steinborn, W., 2000. Quantifizierung von Ökosystemeigenschaften als Grundlage für die Umweltbewertung. Doctoral Dissertation, Kiel.

Straškraba, M., 1995. Nutrient cycles of terrestrial and aquatic ecosystems. Ullmann's Encyclopedia of Industrial Chemistry, Vol. B7. VCH-Verlagsgesellschaft, Weinheim, pp. 40–54.

Stäxang, B., 1969. Acidification of bark of some deciduous trees. Oikos 20, 224–230.

Svirezhev, Y., 1998. Thermodynamic orientors: how to use thermodynamic concepts in ecology. In: Müller, F., Leupelt, M. (Eds), Ecotargets, Goal Functions and Orientors. Springer, Berlin, pp. 102–122.

Tataruch, F., Kierdorf, H., 2002 (2003). Mammals as bioindicators. In: Markert, B.A., Breure, A.M., Zechmeister, H.G. (Eds), Bioindicators and Biomonitors. Elsevier, Oxford, pp. 737–772.

Thomas, W., 1981. Entwicklung eines Immissionsmeßsystems für PCA, Chlorkohlenwasserstoffe und Spurenmetalle mittels epiphytischer Moose – angewandt auf den Raum Bayern. Bayreuther Geowiss. Arb. 3, Bayreuth.

Thomas, W., Herrmann, R., 1980. Nachweis von Chlorpestiziden, PCB, PCA und Schwermetallen mittels epiphytischer Moose als Biofilter entlang eines Profils durch Mitteleuropa. Staub – Reinhaltung der Luft 40, 440–444.

Tietz, A., Weser, L., 2001. Wirkung schwacher magnetischer Felder auf Pflanzen. Naturwiss. Rundsch. 54, 354–358.

Türk, R., 1982. Monitoring air pollutants by lichen mapping. In: Steubing, L., Jäger, H.-J. (Eds), Monitoring of Air Pollutants by Plants. Junk, The Hague, pp. 25–27.

Uhlmann, D., Mihan, H., Gnauck, A., 1978. Schwankungen des Sauerstoffhaushalts und der biologischen Struktur extrem nährstoffreicher Gewässer unter gleichbleibenden Umweltbedingungen (Modellversuche). Acta Hydrochim. Hydrobiol. 6, 421–444.

Ulanowicz, R.E., 1986. Growth and Development: Ecosystems Phenomenology. Springer, New York.

Ulanowicz, R.E., 2001. Information theory in ecology. Computers and Chemistry 25, 393–399.

Ulanowicz, R.E., Norden, J.S., 1990. Symmetrical overhead in flow and networks. Int. J. Systems Sci. 21, 429–437.

van Haut, H., 1972. Testkammerverfahren zum Nachweis phytotoxischer Immissionskomponenten. Environ. Pollut. 3, 123–132.

van der Maarel, E., 1978. Ecological principles for physical planning. In: Holdgate, M.W., Woodman, M.J. (Eds), The Breakdown and Restoration of Ecosystems. Plenum Press, New York, pp. 413–449.

van Straalen, N.M., Denneman, C.A.J., 1989. Ecotoxicological evaluation of soil quality criteria. Ecotox. Environ. Saf. 18, 241–251.

van Voris, P., O'Neill, R.V., Emanuel, R.W., Shugart, H.H., 1980. Functional complexity and ecosystem stability. Ecology 6, 1352–1360.

van Winkle, W., Rose, K.A., Winemiller, K.O., DeAngelis, D.L., Christensen, S.W., 1993. Linking life history theory, environmental setting, and individual-based modeling to compare responses of different fish species to environmental change. Trans. American Fisheries Soc. 122, 459–466.

Verschueren, K., 1983. Handbook of Environmental Data on Organic Chemicals. Van Nostrand Reinhold, New York.

Vetter, L., 1989. Evaluierung und Entwicklung statistischer Verfahren zur Auswahl von repräsentativen Untersuchungsobjekten für ökotxikologische Problemstellungen. Doctoral Dissertation, Kiel.

Vetter, L., Maas, R., 1994. Nachbarschaftsanalytische Verfahren. In: Schröder, W., Vetter, L., Fränzle, O. (Eds), Neuere statistische Verfahren und Modellbildung in der Geoökologie. Vieweg, Braunschweig, pp. 103–107.

Virchow, R., 1854. Handbuch der speziellen Pathologie und Therapie. Enke, Erlangen.

Wackernagel, H., 1997. Multivariate Geostatistics. Springer, Berlin.

Walker, B., 1995. Conserving biological diversity through ecosystem resilience. Conserv. Biol. 6, 18–23.

Wiegmann, S., 1999. Natürliche Schwermetallgehalte als planungs- und umweltrechtsrelevante Bewertungsgrundlage der Belastung norddeutscher Ackerböden. Cuvillier, Göttingen.

Weiss, P., Offenthaler, I., Öhlinger, R., Wimmer, J., 2002 (2003). Higher plants as accumulative bioindicators. In: Markert, B.A., Breure, A.M., Zechmeister, H.G. (Eds), Bioindicators and Biomonitors. Elsevier, Oxford, pp. 465–500.

Wirth, V., Türk, R., 1975. Über die SO_2-Resistenz und die mit ihr interferierenden Faktoren. In: Müller, P. (Ed.), Verh. Ges. f. Ökol., Erlangen 1974. Junk, The Hague, pp. 173–179.

Yount, J.D., Niemi, G.J., 1990. Recovery of lotic communities and ecosystems from disturbance: theory and applications. Environ. Managem. 14, 132–137.

Zauke, G.-P., Meurs, H.-G., Todeskino, D., Kunze, S., Bäumer, H.-P., Butte, W., 1987. Untersuchungen zur Verwendung von Bioindikatoren für die Umweltüberwachung im Ästuarbereich der Elbe, Weser und Ems. Teil 3: Zum Monitoring von Cadmium, Blei, Nickel, Kupfer und Zink in Balaniden (Cirripedia: Crustacea), Gammariden (Amphipoda: Crustacea) und Enteromorpha (Ulvales: Chlorophyta). Umweltforschungsplan des Bundesministers für Umwelt, Naturschutz und Reaktorsicherheit. Forschungsbericht 102 05 209. Oldenburg.

Zdanowska, D., 1976. Wplyw huty Warszawa na zakwaszenie srodowiska Puszczy Kampinoskiej. Frag. Faunistica 20, 353–367.

Zeiler, M., 1996. Nähr- und Spurenelementkreislauf in einem eutrophen Hartwassersee mit saisonal anoxischem Hypolimnion (Belauer See, Schleswig-Holstein). EcoSys – Beiträge zur Ökosystemforschung Suppl. Bd. 11. Verein zur Förderung der Ökosystemforschung zu Kiel, Kiel.

Zierdt, M., 1997. Umweltmonitoring mit natürlichen Indikatoren. Springer, Berlin.

Chapter 3

Bioindicators and biomonitors for policy, legislation and administration

K. Kienzl, A. Riss, W. Vogel, J. Hackl and B. Götz

Abstract

The political framework of bioindicators and biomonitors is shown, especially their recent use and potential to picture the environmental pillar of sustainable development.

Policy and legislature started to make use of bioindicators for monitoring chemical pollution and establishing chemical target values for certain media, especially air and water. Nowadays human biomonitoring as a part of environmental monitoring will get more important for political decision making processes. Especially for xenobiotics evaluation of human exposure for assessments of different sources and exposure paths will be necessary. The relevance of bioindicators in the field of target values and risk assessment is discussed, laying big emphasis on society and especially politicians, who should benchmark, how much risk they are willing to accept.

The advantage of using bioindicators in policy making is that they give insight into the joint effects of the combination of environmental stresses. The equipment is relatively cheap and mainly does not involve sophisticated high tech elements, which is a crucial aspect for administration. On the other hand, the investigations are often time-consuming and have a longer time horizon than an election period, which may be important for political decisions. Using bioindicators and biomonitors in political discussion should serve to bridge the gap in communication between science, policy makers and the public, which is highly correlated with social values. Bioindicators as living organism are sometimes closer to the emotional perception of people and raise more awareness than figures, e.g. of chemical analyses, on the other hand some people rely more on "hard figures". Therefore a combination of chemical and physical analyses and the reaction of a bioindicator is considered as the most powerful tool for interpretation and political reaction.

Some case studies for using bioindicators and biomonitors for policy and legislature on local, national and international levels are shown and differences of environmental indicators and bioindicators are emphasised.

Keywords: sustainable development, target value, risk assessment, political decisions, public, administration, environmental indicators.

1. The political framework of sustainable development

In 1992, the members of the United Nations and thus also the members of the European Union met at the World Conference on Environment and Development in Rio de Janeiro, Brazil. The Earth Summit was convened to address urgent problems of environmental protection and socio-economic development. The assembled leaders adopted Agenda 21, a 300-page plan for achieving sustainable development in the 21st century. Sustainable development implies a development that concerns ecological, economic as well as social aspects of all public domains and that has to be determined on an international and national as well as on regional and at local level. It may be defined as development which meets the needs of the present without compromising the ability of future generations to meet their own needs.

In the European Union a Commission proposal to the Gothenburg European Council in June 2001 about a European Union Strategy for Sustainable Development has been developed (COM (2001) 264 final, 2001). Two important principles are covered: the welfare of both present and future generations and the holistic assessment of the links and synergies between the economic, social and environmental dimensions of policies.

The quality of our environment is one pillar of sustainable development and an important factor of health and welfare of the citizens as well as a sign of culture and education and therefore also important for politicians and administration. Especially in the industrialised world huge sums of money are spent regularly for the protection of the environment to safeguard human health and to protect fauna, flora and landscape, e.g. for its recreational, commercial and intrinsic value. Environmental protection is a political factor and often addressed in election campaigns and it is a factor in the expenditures of the administration.

Politicians and administrations need information on the quality of the environment. And they need information on trends in order to use this knowledge either as an early warning system to detect damage at an early stage or to use it for monitoring the success of remediation measures. The information needed refers to chemical pollution as well as to structural changes of the environment. Chemical pollution can be quantified by chemical analysis of the different media. Other changes such as damage or loss of biotopes are more difficult to quantify. Organisms inhabit the environment and they reflect changes of the environmental conditions, either chemical or structural, at a highly integrated level. This reaction of the organisms, based on all aspects of interaction with their environment including chemical and structural changes, can be used to provide the crucial information which politicians and administrations need. Therefore bioindicators and biomonitors can be used to picture the environmental pillar of sustainable development.

2. Monitoring chemical pollution

2.1. A short history of politically relevant steps

When mankind started to use agricultural pesticides at a larger scale, these chemicals were considered to be safe for humans, wildlife and other non-target organisms. But in the 1950s and 1960s, when some agricultural pesticides were found to affect wildlife, widespread concern about possible ecological effects developed. It is probably Rachel Carson's *Silent Spring* (Carson, 1962) that describes this "period of alarm" in the most impressive way. Later, other groups of chemicals followed: Heavy metals and – after the Seveso accident in 1976 with the release of the accidentally produced 2,3,7,8 tetrachlorodibenzo-p-dioxine – dioxins and furans led to a new culmination of public awareness.

During recent decades the quantity and number of substances released into the environment have dramatically increased. Currently, about 5 million different chemical substances are known and some 80,000 are in use, with 500 to 1,000 new substances being added each year (Fent, 1998). About 600 pesticides are in use worldwide, with

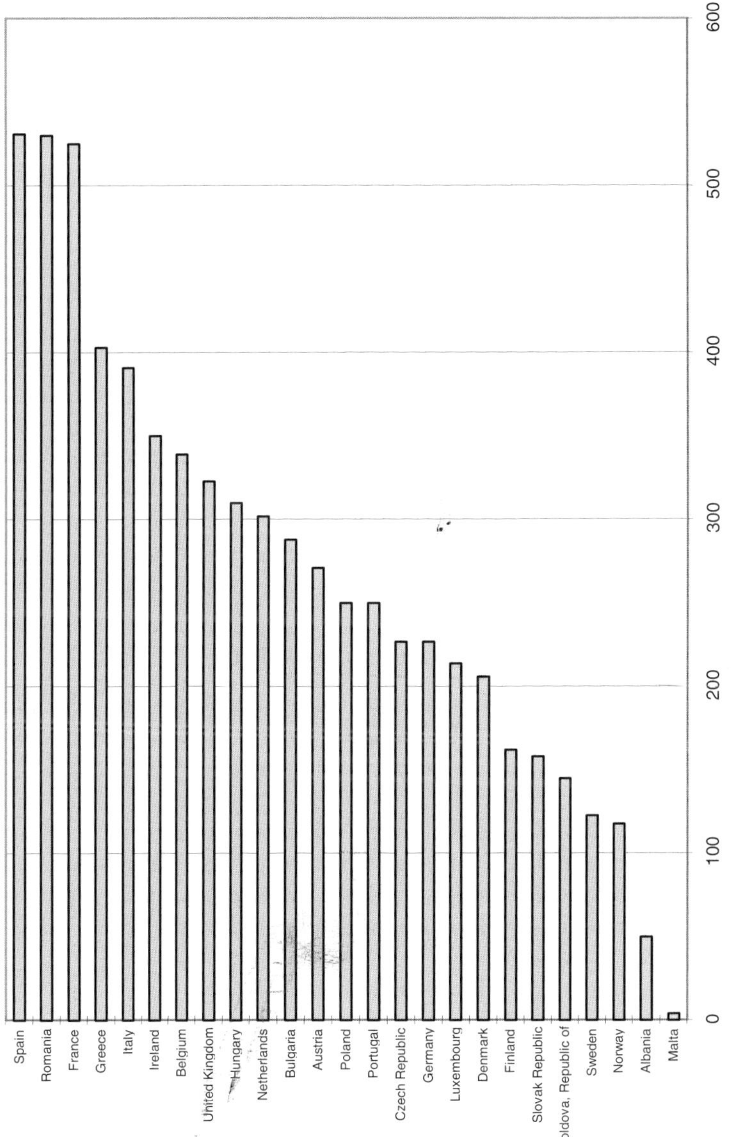

Figure 1. Number of pesticides (active ingredients) in European countries (Vogel and Grath, 1998).

a wide range of improved substances per country. About 500 different agents are approved in Spain, France and Romania, about 300 in the Netherlands, about 200 in Germany and about 100 in Sweden and Norway (see Fig. 1, Vogel and Grath, 1998). Those pesticides are intentionally released into the environment. Other air or water-borne pollutants enter the environment as an undesirable side effect of production and use, either during the normal working process or in case of accidents, like it is the case for PCBs. PCBs are substances produced intentionally for use in closed circuits. But nevertheless, PCBs have been found nearly everywhere in the environment since the 1950s. E.g. sediments of lake Ontario show increasing concentrations of PCBs beginning in 1950 and culminating about 1970 clearly reflecting the increasing sale of PCB in the US. Other substances are not produced intentionally, such as dioxins. Chlorodibenzo-p-dioxines and – furans emerge as unwanted (and often unknown) by-products of industrial chemical processes or in case of accidents.

Increasing public awareness has led to the need for reliable and comparable information on the state of the environment. In most European States specialised environmental agencies were established in the 1980s to collect this information. In the European Union the European Environment Agency provides this information on a European level. Within the EU countries free access to environmental information is guaranteed according to the information directive 90/313/EEC (see EEC, 1990).

The increasing scientific knowledge and public awareness led to stricter control of emissions into air and water as well as to the development of sometimes sophisticated programmes monitoring the environment. To determine chemicals in the environment chemical analysis of the environmental compartments as water, air and soil seems to be the first and most logical choice. But chemical analysis of those media has its limits and shortcomings.

2.2. *The basis for chemical target values: bioindication*

Chemical target values exist for environmental media such as air, soil and water as well as for food, especially in the field of drinking water. Even if the target values are of chemical nature, in most cases the basis of developing these values is bioindication.

For example, for drinking water target values for a large number of chemical substances have been stipulated, either as guidance values or as maximum admissible concentrations. Most of these values are based on recommendations provided by the World Health Organisation (WHO). The principle used by the WHO is the following (WHO, 1993):

For most kinds of toxicity, it is generally believed that there is a dose below which no adverse effects will occur. For chemicals causing toxic effects, a tolerable daily intake (TDI) can be derived as follows:

$$TDI = \frac{NOAEL\ or\ LOAEL}{UF}$$

where   NOAEL is the no-observed-adverse effect level,
         LOAEL is the lowest-observed-adverse effect level and
         UF is the uncertainty factor.

The TDI is an estimate of a substance in food or drinking water, expressed on a body weight basis (mg or μg per kg body weight) that can be ingested daily over a lifetime without appreciable health risk. The NOAEL or, if the NOAEL is not available, the LOAEL is based on bioindication using different species like guinea pigs, rats, mice, pigs, etc.

The uncertainty factor reflects our knowledge. A factor between 1 and 10 is applied for the following four sources of uncertainty: interspecies variation (animal to humans), intraspecies variations (individual variations), adequacy of studies or database and for nature and severity of effects. These factors give a maximum value (maximum uncertainty) of 10,000 in case 10 is applied for all four aspects.

Based on this TDI guideline values are calculated for drinking water using an average body weight (60 kg for adults, 10 kg for children, 5 kg for infants), an assumption of the fraction of the TDI allocated to drinking water (theoretically between 0 and 1) and the daily drinking water consumption (2 litres for adults, 1 litre for children, 0.75 litres for infants).

This principle of calculating ecotoxicological target values provides the basis for most national regulations (as well as for the EU regulations) by taking into account a variety of geographical, socio-economic, dietary and other conditions affecting potential exposure.

For ecotoxicological target values such as for ambient water quality and soil quality the same principle is applied but the organisms used for bioindication of adverse effects are different. For foodstuffs and drinking water organisms which from a physiological point of view are relatively similar to humans are chosen. In order to establish target values to protect ecosystems, the most sensitive parts of these ecosystems have to be selected. Daphnia are well known bioindicators as well as fishes or certain bacteria. To be on the safe side, tests have to be carried out with different types of organisms. Results are published in various journals and collected in series such as "the Rippen" (Rippen, 1988). But for the protection of the environment, however, there is no commonly accepted algorithm for the calculation of guidance values.

2.3. The nature of target values or "how safe is safe enough?"

Science can provide information on risk. Risk assessment is now a standard tool in the management of chemicals and therefore a lot of experience has been gained in recent decades. Nevertheless, "How safe is safe enough?" is a political question, which has to be answered by society itself. Science in general and risk assessment in particular can quantify the risk by providing information on the severity and likelihood of contamination and other effects, but it cannot answer the question of how much risk a society is willing to accept. Therefore, limiting values or other target values are the product of a, hopefully science based, political decision.

Risk assessment is based on existing information and prospective modelling. As described before uncertainties are dealt with by introducing factors. Although one of these uncertainty factors depends on the quality of information, there always remains a subjective dimension. A difference of a factor of 10 – an order of magnitude often reached in estimations by different experts – in many cases means a decision between "all clear" or need for action.

On the other hand the Precautionary Principle approach, meanwhile included in the constitutional Treaties of the European Union of Maastricht and Amsterdam and currently discussed for interpretation and application in different European institutions, concentrates on unknown but potential risks, uncertainties and not determined situations of fate and effects. Many examples in environmental history have shown that risk assessment based on actual knowledge strongly underestimates the environmental risks. This is why the public (the people and therefore also many politicians) gives preference to a Precautionary Principle over a science-based risk assessment concept.

According to Bro-Rasmussen (1999) "It is too late to be cautious when damage has occurred – But it is never too late to be cautious" it is evident that applying the Precautionary Principle is of special importance when irreversible processes are changing natural conditions. This concerns, for example, the release of xenobiotic substances to the environment, especially when they are persistent, have a potential to accumulate and/or are toxic to man or other species. In some cases bioindication can be one tool to recognise environmental effects early, sometimes even with the possibility to quantify health or environmental risks.

Figure 2 illustrates the different approaches of risk assessment and application of the Precautionary Principle for setting standards for food and drinking water. For environmental purposes additional reasons to prefer the Precautionary Principle approach are broader uncertainties and often irreversibility of processes both leading to an extended "grey-zone".

When we ask "How safe is safe enough?" we also have to ask "Safe enough for whom?" There are differences between the individual risk and the risk for society. For instance if we look at the low collective dose of radionuclides after the accident of the nuclear power plant in Chernobyl: Only 0.5% of the whole release of radionuclides had been deposited in Austria, although Austria was one of the most affected countries in Europe (UBA, 2001a). On the other hand if we make calculations on the basis of individual doses we have additional 1.5 cases of cancer per year which is not 'safe enough' from the individual point of view.

Another example are the emissions of gases which are responsible for climate change: We cannot identify a single case of individual harm but climate change may have a severe effect on society as a whole comprising various effects on individuals.

2.4. Human biomonitoring

Prediction of human exposure to xenobiotics on the basis of environmental data, allows the assessment of different scenarios, such as average or worst case exposure. As a part of environmental monitoring the way to evaluate human exposure is human biomonitoring. Results evaluate assessments of different sources of the pollutant and the role of different exposure paths.

Human biomonitoring measures the exposure of individuals or groups of persons to certain chemicals (monitoring of exposure) or to a distinct environmental situation (monitoring of effects). In the first case, concentration of chemicals is determined in easily available biological materials from humans (blood, urine, saliva, hair, deciduous teeth) while effect monitoring is based on functional parameters, behaviour or intellectual and physical capabilities.

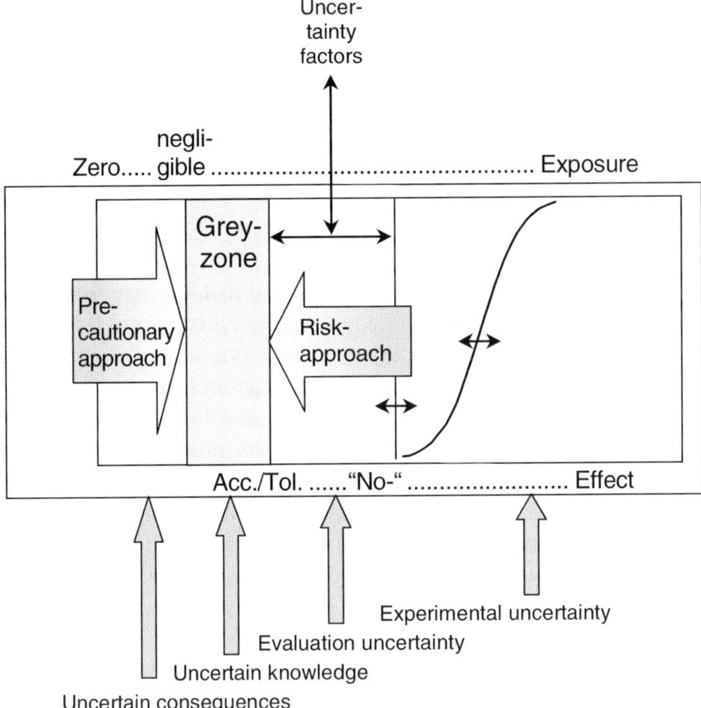

Figure 2. Comparison of individual limits for acceptable/tolerable concentrations of chemicals in food which are established via risk assessment and the common limit for negligible pesticide residues in drinking water which refers to the overriding Precautionary Principle. (Danish EPA, 1999, cit. from Bro-Rasmussen, 1999). For handling environmental risks the "Grey-zone" has to be extended because of more uncertainties and irreversibility of pollution.

Actual results underline human exposure to heavy metals and persistent organic pollutants, used as chemicals for different purposes such as technical use, ingredients in consumption goods, pesticides, pharmaceuticals or odorous substances (anthropogenic moshus compounds) or unintentionally produced as byproduct like polyaromatic hydrocarbons or dioxins.

3. Monitoring structural changes

It is not only chemical pollution that threatens our environment. With the reduction of emissions to air, water and soil the aspect of structural alteration becomes even more significant. For instance due to the establishment of biological waste water treatment plants and the application of strict emission limits for industrial plants chemical pollution of many rivers is no longer the major obstacle to the development of a biocoenosis relatively similar to the ones found in an anthropogenically undisturbed river of the same type. In the industrialised countries terrestrial ecosystems showing natural conditions are rare and restricted to remote areas. But the anthropogenic influence on the structure of the landscape is not only an unwanted byproduct like the chemical pollution of environmental media. Structural changes are undertaken deliberately. For example, rivers are modified in order to protect areas of human settlement, industrial agricultural areas or traffic routes from floods. Terrestrial ecosystems, such as agroecosystems and forests, are altered and formed in order to meet human needs e.g. food, energy supply, recreation and so on.

Most man-made structural changes in the landscape are carried out in order to meet the needs of the people. Nevertheless, the impact is not always limited to the necessary extent. It becomes more and more a common understanding that flood protection does not always imply the necessity of canalisation of a river and that, e.g. in forestry, the cultivation of only one or a few species of trees does not seem any more to be the only applicable way. Agricultural areas can be structured in a more or less environmentally sound way and even areas intensively used for human settlement show a certain potential to be the habitat for a high diversity of species.

Abundance of typical species and the establishment of a well-defined biocoenosis can characterise such areas. Here too, the method of bioindication provides a valuable tool to quantify the human impact and to give guidance on how to combine the goal of meeting human needs with the protection of species and ecosystems.

The structure of agricultural landscapes for instance consists of various environmental features (flora, fauna, habitats and ecosystems), land use patterns and distributions (e.g. crop types and systems of cultivation) and man-made objects (e.g. hedges, farm buildings). The structural landscape components provide the basis for the appearance of landscape and are closely connected with functions and values of landscape (OECD, 1997, 1999). Hedges and trees as well as grass strips, walls and fences are important field boundary features for biodiversity.

An example of bioindicators used to characterise the structure of agricultural landscape are bird species using farmland as habitat, such as they are used e.g. in the UK (UK Department of Environment, 1996). Use of pesticides and changing land use patterns in agriculture, especially the loss of extensive grazing land, causes unfavourable impacts on bird population (OECD, 1999). Within farmland habitats the decline in numbers of bird species was higher on cultivated arable land than on grazing land (MAFF, 2000). Structural changes of agricultural landscape covered pasture – a good source of invertebrate food – which has been lost from the arable areas, hedgerow removal and the loss of other uncropped habitats – all together led to reduction of nesting and feeding opportunities for some bird species. Other examples for key species for agricultural land used in surveys by the United Kingdom Mammal Society are voles (*Arvicola*), shrews (*Sorex*) and field mice (*Apodemus sylvaticus*). A decline in these species since the 1970s has been attributed to the loss of rough grazing land and small habitat features on farmland (such as ditches, hedges, etc.) as well as to the removal of field margins, by ploughing as close to field edges as possible, which are feeding areas for mammals and other wild (OECD, 1999).

Also ground beetles (*Carabidae*) and spiders show differences in distribution and species numbers because of indirect effects of fertilisers, different crop rotations and different numbers of prey available, which can be due to the landscape structure (Idinger et al., 1994).

To sum up certain species can be used as bioindicators for structural changes in agricultural landscape, but interpretation of these indicators is not straightforward and caution is required in relating reductions or increases of indicator species to agricultural practice. Other external factors, such as changes in the weather or populations of predators may have an important influence. It is always difficult to define a "natural" baseline for comparing recent numbers of species in agricultural ecosystems with future ones. In most cases the only practical baseline will be the first year of the monitoring programme. A significant constraint is that surveys of species populations can be very expensive and may require highly specialised skills. Methods for cost-effective and statistically reliable sampling have yet to be established for many groups.

4. Use of bioindicators for policy making and administration

4.1. *Bioindicators for bridging the gap between science and policy*

"Bridging the Gap" was the title of a conference in London in 1998 (Environment Agency for England and Wales, 1998). Politicians asked for indicators for the state of the environment and wanted headline indicators easily understandable for journalists and the public. The aim was to bridge the gap between science, journalists, politicians and the public, therefore indicators were claimed which can easily show complex inter-linkages.

In principle we have the same situation with bioindicators for policy making. We have to try to find bioindicators which can easily bridge the gap between necessary action in terms of sustainability for nature and man and maybe unpopular measures by politicians who want to be elected again.

Politicians can use bioindicators only if they can show and describe things very easily or even emotionally. Some examples are given in Section 5. Of course, basic research will always be necessary but scientists should also have a look on the practical use. It is clear that a single bioindicator cannot show every detail of complex situations but it should sharpen the eye for the necessity of political measures.

Therefore scientists should sometimes try to put themselves in the place of politicians who want to do the best for the environment and people, but only have a few years to show that they care more than other politicians. Within the context of bioindicators this could be very difficult because nature sometimes needs years or decades or even longer to show that something went wrong.

4.2. *Questions the administration needs to be answered*

What are the most important questions the administration of a country needs to be answered concerning environmental pollution? Most of them are like the following: Is the water safe to drink? Is the air safe to breathe? Is the soil safe to grow crops? Are the effluents or the air emissions of a industrial plant dangerous for the flora and fauna in its vicinity? Is fauna and flora in the environment unaffected by pollutants? These are the crucial questions. If the answer is yes, further investigations are of mainly academic interest. But if the answer is no, further action is needed.

If case of non-compliance with target values or if non-compliance seems to be likely (or possible) in the future more differentiated information is needed to see how the situation develops. If sanitation measures are applied, the effect of those measures has to be monitored.

4.3. Is bioindication the tool to provide the answer?

Bioindication can often answer the crucial questions about the risk without the need of elaborate chemical analysis. Fishes can be used as a "warning system" for drinking water. As long as a sensitive species living in the water which is to be tested shows no adverse effect we can consider the water to be more or less safe, i.e. free from acutely toxic substances in higher concentrations. As long as selected species of different plants show no adverse effect after being exposed to the air that is to be tested, we consider the air to be at least of a minimum quality. As long as selected sensitive plants (such as cress) can be grown in a certain soil we consider this soil as unpolluted, and as long as animals and plants typical for a certain biotope live in this biotope in expected diversity and abundance we consider this biotope as more or less unaffected.

In many cases it is not the reaction of an organism that is interpreted. Organisms often accumulate problematic substances making them available for chemical analysis or allowing comparable results. Spruce needles grown in the vicinity of an industrial plant provide an excellent matrix for analysing airborne pollutants leading to data integrating the status of air quality over a longer period of time. Mosses have been used as matrices for heavy metal analyses leading to air pollution maps of Europe showing hot spots and clean areas for the different elements.

It is not possible to analyse regarding all chemical substances. In most European countries about 400 different ingredients of agricultural pesticides are approved by the authorities. In practice, even chemical analysis covering all representatives of this relatively small group is hardly possible. Chemical analysis is extremely expensive if we do not know what to look for. Therefore, sometimes, a combination of chemical analysis and the reaction of a bioindicator is the most powerful tool: e.g. chemical analysis for substances suspected to be found and sensitive bioindicators indicating the absence (or presence) of toxic substances in very general terms. This approach is used in the licensing system for waste water emissions in many countries, e.g. in Austria using fishes, crustaceans or bacteria as indicator organisms.

4.4. Pros and cons of using bioindication for policy making

The advantage of using bioindicators, which give insight into the joint effects of the combination of stresses, is the direct approach: "what is to be protected is measured". No harmful substance is excluded and antagonistic and synergistic effects are measured as they occur. In general, the equipment for this type of bioindication is relatively cheap and does not involve sophisticated high tech elements. This is an important aspect for all countries and a crucial one for the administration in countries with very limited budgets, such as developing countries or states in transition. On the other hand, the investigations are often time-consuming and experienced staff is needed. Interpretation is sometimes difficult and, what causes even more problems, results are often difficult to communicate due to the fact that some people still consider results from chemical or physical analysis to be more precise and therefore more reliable.

5. Bioindicators – answers to political questions: some case studies

The following case studies shall give some impressions of how bioindication has been used for political decision making. They have been chosen because they were successful in one way or the other in supporting environmental politics or are actually used for this purpose and cover a broad spectrum of aspects from the local to the international level.

On the local level bioindication is used to give information on the nearby environment and on recent major environmental events, e.g. emissions of industrial plants or local structural changes. Using comparable methods bioindication can be applied on a national and international level as well, serving as a monitoring instrument for national and international environmental policy issues, enabling comparisons between countries and tracing of pollutants across borderlines.

5.1. Bioindication on a local and regional level

5.1.1. Case study: bioindication for identification of a local environmental problem, for acceptance of measures for solution and monitoring the success of environmental protecting measures

In this chapter we report about the identification of local effects of dioxin emissions from a copper reclamation plant on the environment and the local population in an inneralpine region, the establishment of environment and health protecting measures (especially to avoid further emissions) and the permanent monitoring of the success of the measures set by environmental politicians by carrying out control investigations especially by monitoring a whole food chain important for human nutrition.

Dioxins (polychlorinated dibenzo-dioxins and -furans, PCDD/F) are a group of persistent organic pollutants (POPs) which are heavily toxic for humans, accumulate in animal and human food chains, are spread ubiquitously because of long-range transport especially by air. Most of them reaching human food are originating from air deposition to plant surfaces. Therefore they got compounds of high public interest during the last two decades and were included to the list of POPs regulated in the POP-protocol of the UN-ECE Convention on Long-Range Transboundary Air Pollution (CLRTAP) of the UN-Economic Commission for Europe. In the meantime also a global regulation, the Stockholm Convention on Persistent Organic Pollutants prepared by UNEP was signed on 23rd May 2001.

The copper reclamation plant working for some decades and centuries before as a copperore smelter produced strong heavy metal pollution in its vicinity. It is situated in a village in an inneralpine valley in Tyrol/Austria with mainly grassland farming and milk production.

When in the late 1980 it became evident that metal reclamation plants could produce high amounts of PCDD/F, environmental investigations were carried out with bioindication methods as the most important element. After toxicological sensitivity analyses revealing the food chain air–fodder grass–cow's milk–humans to be the most important path for human body burden, the following steps became necessary:

5.1.1.1. Definition of the environmental problem, risk assessment and immediate steps

Chemical analyses on PCDD/F of soil (for long term deposition) and fodder grass (for actual deposition) were carried out to detect the spatial extent of the affected area. For toxicological risk assessment food produced from the affected region, especially cow's milk, and some samples of human blood and mother's milk from inhabitants were analysed. With regard to the few possibilities of chemical analysis with sufficiently low detection limits and their high costs in the late 1980s best efficiency of investigation design was necessary. Results of 20 soil and five fodder grass analyses determined the size of the affected area where measures for health protection had to be taken. Six analyses of cows' milk were necessary to cover all farms with elevated PCDD/F concentrations in milk fat to prevent further release of contaminated milk into the human food chain. Analyses of five samples of blood of exposed people (farmers and members of their families) showed in one case a more than ten-fold and in another case an about five-fold elevation of the PCDD/F load. Mother's milk (five samples) showed no deviations from background levels (Riss et al. 1990a,b; Riss 1993a). These results promoted analyses of PCDD/F in stack gas of the copper plant, which showed the release of high amounts of these contaminants to ambient air.

Figure 3 shows the PCDD/F load in different environmental compartments and in the food chain fodder grass – cow's milk in the year 1988, prior to the implementation of environmental protection measures.

The most important immediate measures were drastic emission control at the copper plant, feeding cows with fodder grass from other regions and withdrawing milk from the market for a period until tolerable concentrations in milk were reached.

5.1.1.2. Evaluation of the success of health and environmental protection measures

After reconstruction measures at the copper plant to reduce air emissions drastically the effect on the environment was monitored with simple bioindication methods. By periodically analysing some fodder grass and cows' milk samples the development of contamination could be monitored over successive seasons. The PCDD/F contamination of fodder grass and cows' milk did not decrease with the expected rate. The validity and comprehensibility of the results made further expensive emission-reducing measures at the plant possible, such as reduction of diffuse emissions from the smelter building, clean handling methods of materials at the plant area or general dust avoidance.

5.1.1.3. Bioindication for permanent monitoring of PCDD/F release

A bioindication system with high practical relevance and low costs for the permanent control of PCDD/F was developed.

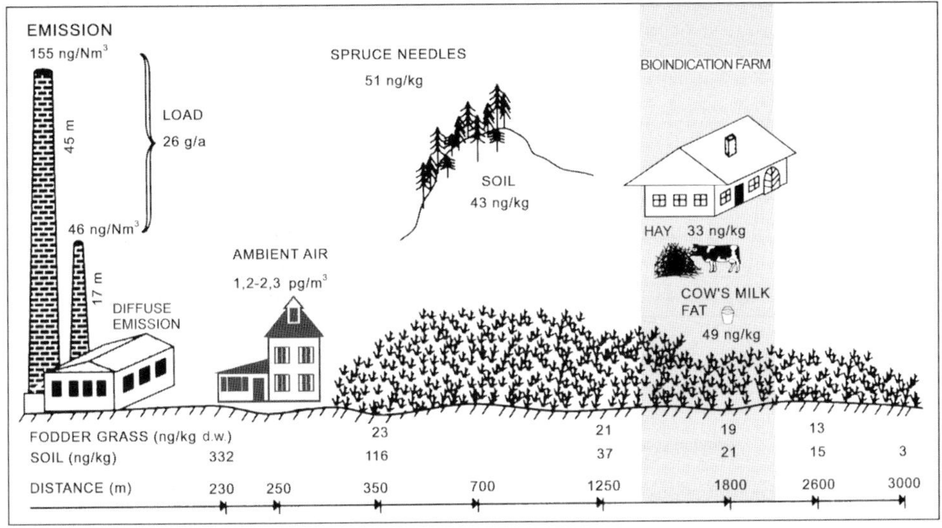

Figure 3.   Dioxin levels (in International Toxic Equivalents) in different environmental compartments and in a human food chain in the main wind direction prior to implementation of environmental protection measures. (Riss et al., 1990a; modified)

The first element is a farm producing cow's milk mainly from fodder grass. This "bioindication farm " was selected using the following criteria:

- The whole farm area should be situated in the potentially affected area; and
- nearly all part of the fodder for the milk cows should be produced at the farm area.

A representative integrated sampling strategy was developed, which included analyses of

- three grass weighted mixed (over the whole farm area) samples every season corresponding to usual grass cuttings in the region; and
- analysis of one cows' milk sample every year, representing the whole farm. The sampling time is spring, when all the hay from the last season has been fed during the winter and a steady state of PCDD/F contents between fodder, body burden and milk fat has been established. The contamination of the winter fodder can be calculated from the results obtained from the previous vegetation period.

The second element of the permanent bioindication system is a simple and cheap integrated monitoring of ambient air pollution with PCDD/F during one year. Plants, especially spruce needles, are widely used for bioindication of POPs in ambient air because of their ability to accumulate lipophilic compounds in their surface waxes.

For our bioindication purposes no naturally grown spruces located at the site with expected potentially highest ambient air pollution originating from the reclamation plant were available. Therefore it was decided to use young spruces cultivated in containers in a clean air area and to expose them every year at the required site

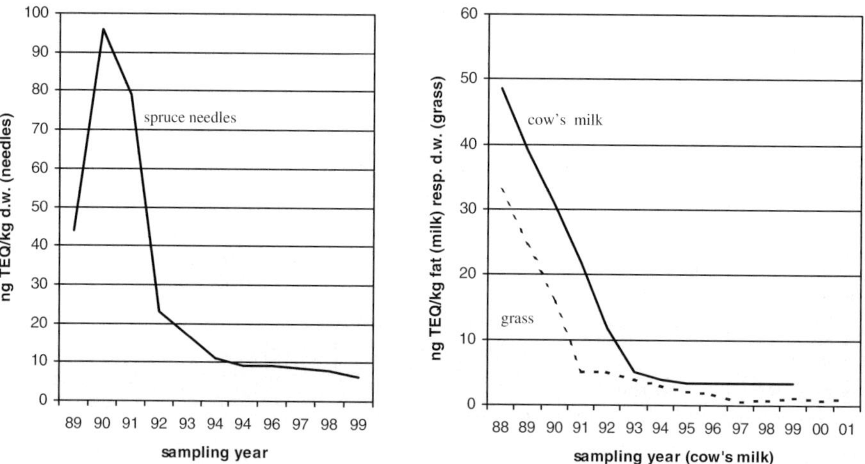

Figure 4. Dioxin concentrations (in International Toxic Equivalents) using spruce needles as bioindicators for ambient air pollution (left) and a "bioindication farm" for effects on an important human food chain. Sampling years of cow's milk, and corresponding fodder grass, sampled during the vegetation period in the previous year (left). Data: Amt der Tiroler Landesregierung, 1997–2000, Amt der Tiroler Landesregierung, 2000, Riss, 1993a, UBA, 1993–1995.

from May when new needles begin to develop for the following April. After one year of exposure one sample of these one year old needles is harvested for PCDD/F analysis. Some advantages of this method are in obtaining results representing ambient air contamination integrated over periods from a whole year, yearly uniform test trees without damage from sampling, and low costs.

5.1.1.4. Results

The results of bioindication of the PCDD/F load in the ambient air of the copper reclamation plant using spruce needles as a bioindicator for time-integrated reflection of ambient air pollution and the bioindication with special consideration of the local agricultural practice and food production over a period of more than a decade are shown in Figure 4. The development of the PCDD/F contamination of ambient air, fodder grass and cows' milk from the "bioindication farm" from the starting point of the investigations, after setting emission reduction measures and during the current environmental monitoring reflects the effectiveness of environmental protection measures.

All results show a strong decrease of contamination during the first years, and a stabilisation for many years on a level close to background levels, especially on the farm. The term "bioindication farm" is used to reflect the local state of the environment from a sensitive ("accumulation") and practical ("food chain of an important food") point of view and is more effect-related, while the "spruce needle method" is focused more on air emissions from the possible source. The combination of both methods allows an accurate description of a local state of the environment and, if necessary, the development and establishment of environmental and health protection measures.

5.1.1.5. Conclusions

A local environmental problem caused by a potentially strong emission source of substances harmful to health and environment had to be handled. At that time lack of experience of the fate of dioxins in the environment and to a great extent the lack of environmental and health reference data, uncertainties in risk assessment and poor chemical analytical capacities (with sufficient detection limits causing high costs) made it necessary to develop an investigation design aimed at getting a maximum of significant results using a minimum of samples. This was possible by combining analyses of the relevant environmental media and bioindication methods.

The spatial extension of the area affected for a long time could be delineated by analysing about 20 soil samples including reference sites.

The whole area affected regarding to risk assessment could be identified by analysis of eight fodder grass samples and 10 samples of cows' milk. Based on these results the fodder problem, the food chain enrichment and the relevance of food contamination could be assessed.

Analysis of 10 human samples (blood and mothers' milk) showed elevated concentrations of dioxins in two blood samples, which underlined the urgency to set environmental and health protection measures.

The first evaluation of the effectiveness of emission control measures at the polluter was made on the basis of results from analyses of four samples of cows' milk and 11 samples of fodder grass. The results showed reduced contamination in the food chain not at the expected rate. Therefore it was concluded that first measures limiting dioxin concentrations in stack gas were not sufficient enough and additional emission reduction measures had to be implemented.

In the described case the federal authority demands only one annual stack gas investigation for dioxins. Currently, the local state of the environment is monitored by an annual investigation of three fodder grass samples, one cows' milk sample and one sample of spruce needles.

For the implementation of environment protection measures it was necessary to establish an investigation design largely based on bioindication methods. General conclusions regarding the establishment of a successful investigation design could be as follows:

- The design has to be simple to understand and as far as possible based on daily experience. This makes it possible to convey investigation results and conclusions from it to all people involved, being inhabitants of the region, representatives of industry and their employees, representatives of agriculture, tourism and so on, local and federal authorities and administrations, politicians and, last but not least, journalists. It is evident that different social groups pursue different interests.

- To implement effective solutions for an environmental problem, co-operation of representatives of all involved interest groups is necessary but often difficult. In the reported case of environment contamination the provincial authorities dealing with environmental issues co-operated closely with a federal environmental institution not suspected to be influenced by regional interests. In particular the excellent and problem oriented support of the provincial farmer representation, at that time led by the present European commissioner for agriculture, allowed an effective sampling and carrying out the investigation programme.

- The lack of limit values for these "new" substances for food, agriculture and emissions made it difficult to set environmental and health protection measures.
- The validity and conclusiveness of the results of the investigations finally enabled the necessary steps to be taken by the permission authority for the industrial plant and the technical retrofitting to minimise further emissions from the enterprise.
- Current information from citizens, many of them economically dependent on the enterprise or affected as farmers, was an essential condition for finding acceptable solutions regarding the environmental impacts.
- By reducing the dioxin emissions from this industrial plant to nearly zero, total dioxin emissions estimated for Austria at that time could be reduced by a quarter (Riss and Aichinger, 1993).

5.1.2. Case study: human biomonitoring of lead exposure for monitoring effectiveness of environmental protection measures at an industrial site

In the south of Carinthia (Austria) lead and zinc ores and secondary materials were smelted for many decades. Metal processing, especially of lead, has a tradition of several centuries in this region. The industrial complex is situated close to small a little town.

Heavy metal pollution in the vicinity of the industrial site has been investigated since the late 1970s and during the 1980s (Halbwachs, 1982) and showed high lead concentrations especially in agricultural and forest ecosystems. Some soil samples showed lead contents up to some thousands mg/kg). In 1991 an investigation of soils from domestic gardens, playgrounds and sports grounds showed high pollution with lead and cadmium of the inhabited area especially up to a distance of about 1 km from the industrial complex (Kasperowski, 1993). Air quality and deposition studies underlined the long-term and actual exposure of the population living in this area to heavy metals, especially lead (Amt der Kärntner Landesregierung, 1999).

As a consequence local population was informed about health risks and their individual exposure due to their gardens, about how to behave to avoid further exposure. In addition general measures were developed in collaboration with the local authorities (Riss, 1993b). In the given environmental pollution situation, the necessity of carrying out human exposure studies was evident. Raised awareness of the people due to credible information and collaboration between authorities, environmental and medical experts and inhabitants allowed a human biomonitoring of more than a thousand people (investigation of lead in human blood and in a subsample, cadmium in human blood and lead in the deciduous teeth of children, Eisenmenger and Drasch, 1992).

Collaboration of persons and groups affected – especially inhabitants, local and federal authorities, politicians and representatives of industries (these groups not always pursuing the same interests) – was also necessary to develop and implement measures to minimise and avoid further risks to human health and to the environment.

The key measures to minimise human exposure were:

- Reorganisation of the activities at the industrial complex (closing the lead- and the zinc-smelter, restructuring measures at the industrial site, establishing low emission industries).
- Replacement of polluted soils at exposed sites in domestic gardens on a voluntary basis and at some playgrounds in the polluted area.
- General recommendations to the inhabitants concerning domestic vegetable cultivation.
- Recommendations for taking care with cleaning after soil or dust contact, especially for children.

One important consequence of the results obtained so far was to monitor the effects of the environment protection measures by air quality and deposition monitoring and by human biomonitoring.

In order to monitor the effectiveness of environmental and health measures to reduce the exposure of humans, the human biomonitoring investigations from 1992 were repeated in 1998 (Drasch et al., 2000).

A general decrease by more than 50% of lead concentrations in the blood of people through all age groups could be achieved as an encouraging result of successful environmental policy. Figure 5 shows one of the results, the intra-individual change of concentration of lead in blood of 40 children between 1992 and 1998. In all cases a decrease can be established, which is markedly larger in children with higher concentrations in 1992. In 1998 none of these cases the Human Biomonitoring Level I (HBW I) of 10 µg Pb/dl blood established in Kommission Human-Biomonitoring (1996) was exceeded.

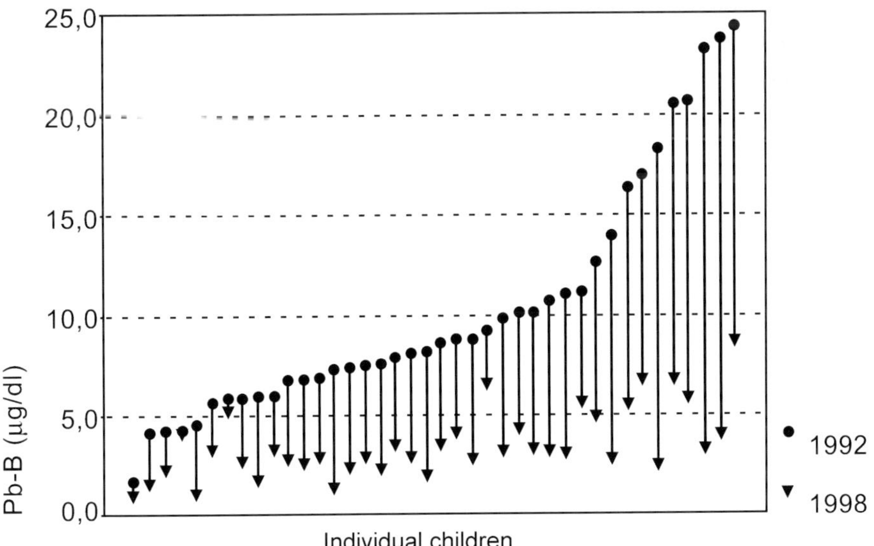

Figure 5.   Evaluation of environmental policy by using human biomonitoring: decrease of concentration of lead in blood of 40 children between 1992 and 1998 (Drasch et al., 2000).

5.1.3. Needle surface characteristics and element contents of Norway spruce needles – bioindicators for environmental control

The wax layer of spruce needles can be used for the assessment of effects of air pollution on forests. This method was developed at the Federal Environment Agency, Vienna. The wax layer is a protective barrier against any kind of environmental influence and excessive evaporative water loss. Especially in the epistomatal area these epicuticular waxes show very fine-reticulated and sensitive microstructures. The degradation of needle waxes is not pollutant-specific. However, investigations of epicuticular wax structures are suggested to be a sensitive and suitable bioindication tool for detecting incipient damage of trees from air pollution (Trimbacher and Weiss, 1999). Air pollutants and airborne particles may alter wax microstructure and lead to an increased ageing of the wax structures, causing premature senescence and shedding of the needles. Investigations are carried out by means of scanning electron microscopy and usually comprise wax quality, covering with dust and/or microorganisms. It is a suitable means for detecting incipient air pollution effects and has therefore been already used as a bioindication method for several years. The above mentioned micromorphological characteristics of Norway spruce needles are also called "needle surface characteristics".

The method is particularly suitable for complementing existing monitoring networks assessing air pollution and investigations of element contents of Norway spruce needles, representing essential data to characterise the nutritional status of trees. In some cases these data may help assess the ambient air pollution level for single elements and thus support the identification of polluting sources.

Although applicable for large scale assessments, the method is nowadays predominantly used at a local level to establish proof of air pollution in the vicinity of industrial plants and other local emission sources. The main reason for limited application at a large scale are the costs. However, at a local level the method worked in several cases like a joker. In other words it was the decisive factor and thus successfully supported implementation of measures for the reduction of polluting air emissions. It can be expected that this bioindicator may gain importance in the field of environmental control and for environmental impact statements and assessments.

The example shows that the political importance of a bioindicator is decisively influenced by its costs. New methods also are handicapped due to being unknown by political decision makers. It is the well established bioindicators (or other suitable methods) that are normally used for detection of impacts. In general these are the cheaper ones. The example also shows that there is little demand for specific bioindication that is less common, more sophisticated and more expensive, e.g. due to the needed employment of high technology. The use of such bioindication is (and will be) limited to solve specific problems. None the less it can contribute information of high political importance especially in delicate situations.

5.2. National level

5.2.1. Saprobiological index indicating river quality

Treatment of municipal and industrial waste water is expensive and the implementation of waste water reducing technologies (closed circuit technologies, etc.) is expensive too. Therefore, for politicians as well as for the administration it was (and still is) of crucial importance to be able to show the positive results of these investments in order to justify the money spent. In Austria, water quality maps have been used since 1962. On these maps which show the main rivers and streams of Austria, water quality is visualised with the colours blue, green, yellow and red indicating the range from high water quality (I, blue) to bad water quality (IV, red). With the steps in between (I-II, II-III, III-IV) this system has seven water quality classes.

The technical basis for these maps is bioindication. The fauna of e.g. a fast running, clean, and oxygen rich water body in the alpine region is totally different from the fauna of a slow running muddy river with a high organic load. The fauna of our first example might be dominated by larvae of ephemerides (may flies), plecopteres (stoneflies) and trichopteres (caddis flies), the fauna of the latter example might consist only of tubifex and chironimidae larvae (both are widely known as food for aquarium fish). Each species has the environment it prefers. Some with more tolerance to variation, some with less. This fact is used to judge the water by its inhabitants. Each organism has its indicator value indicating water quality from I to IV. The quality of the river is calculated from the type of species and their abundance, sometimes additional information is used, such as chemical parameters.

A water quality map clearly shows the water quality at a given time. Comparing water quality maps from different years gives a clear picture of the development over time. Comparing a water quality map from the late 80s with one from the early 90s reflects the rapid change of water quality especially in this period of time, triggered by changes in the pulp and paper industries (closed circuit technologies, biological waste water treatment) and an increasing part of municipal waste water being treated by biological waste water treatment. Red and yellow river parts have nearly vanished from all over Austria with the exception of the north eastern regions, where scattered settlements and intensive agriculture are combined with very low precipitation rates. These maps visualise the success of Austrian water management and show where problems still occur.

5.2.2. Use of bioindicators in the licensing of waste water emissions and as a basis for waste water taxes

According to the Austrian Water Act (Wasserrechtsgesetz 1959 (Federal Law Gazette No. 215/1959) according to the current version), about 60 ordinances were passed providing limit values (concentrations and in some cases loads) for waste water originating from different industrial branches as well as from communities. These values are based on the state of the art and give the framework for individual licences, which may be stricter than the ordinances e.g. in the case of a small and sensitive receiving water body. The limiting values vary from branch to branch and consist of two lists of physical and chemical parameters such as temperature, pH, BOD, COD, AOX, heavy metals, etc. One list is used for direct discharges, the other in case of discharge into a municipal sewerage system. In case of direct discharge the parameter list includes toxicity limits. In most cases limits are given for fish but in some cases additionally for daphnia and (luminescent) bacteria. The reason for this is that the composition of the waste water is sometimes very complex and usually not entirely known. To be on the safe side, in addition to the most important parameters of the branch, toxicity tests are stipulated in order to avoid the emission of toxic quantities of a substance, not listed in the chemical parameter list due to limited knowledge of the chemical processes involved. The figure given in the ordinance is the degree of dilution of waste water necessary to avoid toxic effects to the target organism. In the case of fish the stipulated limiting value (dilution factor) is generally <2 to <4.

In Germany, the emission of waste water is financially charged according to its chemical characteristics (Roth, 1997). For the most typical chemicals a price is listed in this ordinance (DM per kilogramme or tonne). In order not to exclude toxic chemicals from this system, a bioindication system is included. In addition to the chemical load, the cubic metre of waste water is charged according to its toxicity to fish. The more toxic the waste water, the higher the costs. This system is intended to act as an incentive to reduce the toxicity of waste water.

The costs for one emission unit range from 12 DM to more than 70 DM (since 1997). One unit is e.g. 50 kg COD, 3 kg phosphorous, 25 kg nitrogen, 2 kg AOX, 100 g cadmium and 20 g mercury. In case of the bioindicator "fish" (parameter: "fish toxicity"), one unit is 3000 cubic metres of waste water divided by the dilution factor, which indicates the dilution needed to avoid toxic effects.

This instrument acts as an incentive measure in order to motivate enterprises to apply advanced technics in order to minimize water pollution.

5.2.3. *Fishes protecting drinking water quality*

Drinking water from public waterworks is subject to frequent chemical control. If drinking water is abstracted from river water, changes in quality may occur very quickly due to accidents upstream of the abstraction point. In the case of an accident it cannot be foreseen which chemical will threaten the drinking water supply. To handle this risk monitoring systems have been established using bioindication as the basic concept. Fish (trout) were held in a tube with the abstracted water swimming against the current and indicating, as long as they swim, that the water is not acutely toxic. As soon as a fish dies the animal is transported with the water current touching a switch at the downstream end of the tube and setting an alarm. This system can be combined with an automatic sampling system (to allow chemical analyses in order to find the reason for the toxic effect) and with an automatic switching off of the abstraction. With a sophisticated computerised analysis of the movement of the fish even very minor changes of water quality can be detected triggering further analysis to safeguard drinking water quality.

Therefore, by using biomonitoring, the compliance of drinking water quality with the legal obligation of being safe for the consumers can be secured.

5.2.4. Bioindicators for biodiversity?

During recent decades nature protection was characterised by protecting single species, especially those which are attractive for people like the Giant Panda (*Alluropoda melanoleuca*). Within the European Union the first directive in the field of nature protection was the Bird Directive in 1978 (79/409/EEC) (EEC, 1979). This shows that politicians can protect nice looking birds more easily than reptiles or insects.

Red Lists of endangered species have been developed in many countries (Zulka et al., 2001) and have become popular over the years. More and more they have been used in Environmental Impact Assessments and when a new project is designed in an area which hosts endangered species.

For scientists it was clear that it is not only necessary to protect single species but also the habitats they live in. After the Earth Summit in Rio in 1992 (Convention on Biological Diversity) biodiversity was discussed a lot although it is used in very different senses. With the Council Directive on the conservation of natural habitats and of wild fauna and flora (EEC, 1992) the European Union got an basic instrument for protection of biodiversity. Some European countries are going to develop a list of endangered habitats (UBA, 2001b).

But it is still necessary to show people that man is only a part of nature. He also is the peak of the food pyramid and every impact of man on nature is an impact on the web he is part of.

5.3. International level

5.3.1. Crown condition survey and assessment of forest condition in Europe

During the 1980s the so called forest decline dominated the headlines of environmental news in Europe. It was assumed that the forests were suffering from air pollution to an extent that even larger forests could die as a consequence. In smaller regions this has been found to be true (e.g. in the northern part of former Czechoslovakia). Forests are of special interest for the public. It is not only their function to serve as a basis for the production of wood. Forests also have other values which is why people look after and protect them. The fact that forests have been affected by air pollution so drastically has shocked the public and led to a growing sensitivity of people to environmental matters. So politicians urgently needed information on the state of the forests to develop suitable measures for adequate reaction to this problem. Awareness was focused on visible symptoms on tree crowns and soon it was laid down that the parameter "crown condition" could provide enough information for describing the state of forests' health. The assessment of the "crown condition" is based on the degree of defoliation (and changes in ramification especially when using Colour Infrared Photography) and the discoloration of needles and leaves.

The knowledge of the principally negative impact of some airborne pollutants on trees and on the crown condition particularly led to relatively quick political reactions at national and international level. Some countries started national inventories on the state of the forests based on the parameter of the crown condition as early as at the beginning of the 1980s. As it was clear from the beginning that the problem was not limited to national borders, efforts were soon started to establish international and harmonised assessments, which could provide comparable data and an overall picture of the condition of forests in Europe. In the year 1985 the United Nations Economic Commission for Europe (UN-ECE) established the International Co-operative Programme on Assessment and Monitoring of Air Pollution Effects on Forests (ICP Forests) within the framework of the UN/ECE Convention on Long-Range Transboundary Air Pollution (CLRTAP). In Council Regulation (EEC) No. 3528/86, the European Union Scheme of the Protection of Forests against Atmospheric Pollution was laid down to contribute to the protection of forests. On the European scale, the spatial and temporal variation of forest condition is assessed by means of systematic large-scale monitoring, called "Level I" (UN/ECE and EC, 2000). It is based on the annual assessment of the crown condition in a transnational grid (16 $\times$ 16 km) and in national grids of individual scale.

Shortcomings of the bioindicator "crown condition" include the fact that crown condition is a non-specific indicator. The causes of different crown conditions can be manifold and comprise – apart from genetic variability and site conditions – a broad variety of biotic and abiotic stress factors. Whereas e.g. age and site conditions can be distinguished to a certain extent from other causes by assessing the crown conditions against reference trees, no differentiation is possible between changes in the crown condition due to air pollution (be it a consequence of soil acidification caused by the deposition of derivatives of sulphur dioxide and/or nitrogen compounds possibly followed by imbalances of nutrition, or the direct effects of ozone) (EC-UN/ECE, 1997; UN/ECE and EC, 2000; Augustin et al., 1997) and changes caused by other influences.

Being aware of these deficiencies, scientists have been calling for additional parameters to gather more information on the cause–effect relationships and to be able to better interpret forest condition. Some scientists even call for quitting the Level I crown condition survey in favour of comprehensive ecosystem-based analyses of the condition of forests (Deutscher Bundestag, 2000). Whether this will happen or not will be decided at a political level. Some of the ideas for ecosystem-oriented assessments have been implemented already. A reasonable number of ecological parameters is assessed at 864 selected plots in about 30 European countries by the Pan-European Programme for Intensive Monitoring of Forest Ecosystems, which was established in 1994 within the framework of the UN-ECE ICP Forest (UN/ECE and EC, 2000). This programme aims at gaining a better understanding of the effects of air pollution and other stress factors on forest ecosystems. In addition the implementation of the multi-disciplinary Integrated Monitoring Programme ICP IM (UN/ECE Convention on Long-range Transboundary Air Pollution, 2001), which is part of the effects-monitoring strategy under the CLRTAP can be seen as a valuable tool to complement the knowledge required.

Looking back it can be said that the bioindicator "crown condition" has played an important role in environmental policy. The fact that since the beginning of the public discussion forest decline has mainly been ascribed to air pollution has a decisive impact on the whole environmental discussion and for successful argumentation on the promotion of measures against air pollution.

The assessments of crown condition under the UN-ECE-ICP Forests have been carried out in some countries for about 15 years. Thirty-eight nations participate in this programme, which is part of several international initiatives dedicated to the protection of forests and the environment, respectively (e.g. the Ministerial Conference on the Protection of Forests in Europe). Quite a considerable number of scientific projects were initialised, which led to a considerable amount of knowledge on the effects of air pollution on forests. The shortcomings of the bioindicator "crown condition" associated with its scientific correctness and significance for the relationship between air pollution and forest condition may have been outweighed by the general profits, including the realisation that the consequences of human actions for ecosystems are much more complex than assumed. There is another aspect that may not be neglected: Lehmann (2001) states that forests are the most important metaphor for nature. This may hold true for a considerable part of the public, at least in Central Europe. Forest die-back is something that moves people emotionally. The interpretation of the condition of forests assessed by the (scientifically vague) crown condition has risen awareness for nature. Public awareness has remained at a high level, which has served environmental policy. This may lead to the conclusion that the usefulness of bioindication for (environmental) politicians is closely linked with the (emotional) perception of the bioindicator in the public.

To strengthen the significance of bioindicators which should describe the condition of forests (such as the crown condition) the combination of various investigation programmes is indispensable. These programmes should comprise analytical assessments of e.g. the deposition of pollutants as well as the assessment of integrative bioindicators. This should help optimise the application of precise but expensive analytical assessments on the one hand and of integrative and less precise but relatively cheap bioindication methods on the other (Mirtl, 2001).

5.3.2. Mosses as biomonitors for assessing heavy metal deposition in a European scale

Especially in the 1960s and 1970s the Nordic countries of Europe were affected by deposition of airborne acidifying substances joined with heavy metals transmitted primarily from Western and Central European countries. Sensitive ecosystems, especially non-buffered lakes, were strongly affected as fish showed unacceptable high heavy metal contents making them inedible and leading to the death of the whole ecosystem.

Therefore action was needed at international level to reduce emissions and to gather scientific information to quantify the spatial spreading and temporal development of heavy metal depositions (Nordic Council of Ministers, 1998).

The use of mosses as biomonitors for heavy metal deposition has some essential advantages. Compared to technical deposition sampling methods it is cheap and it can be standardised to be applicable over a wide geographic and climatic range. Therefore the establishment of a measuring grid with a high density of sampling sites is possible, allowing to draw pollution maps covering large areas.

When analysing moss shoots that are three years old, the results show the deposition of heavy metals integrated over this time period, a period useful to answer questions regarding general temporal trends of the input persistent pollutants into ecosystems.

The demands upon the sampling strategy are closely connected with the aims of the monitoring system. For monitoring of the long-range transboundary transport of airborne heavy metals the sampling strategy has to disregard effects of local emissions. Standards of demands upon sampling sites were developed: unaffected deposition (ideal: clearing in a forest), minimum distance to settlements and traffic routes, sampling sites in valleys two to four hundred metres above the valley floor). By considering these criteria impacts of settlements, traffic and precipitation elevated with altitude can be minimised.

During the years 1990 and 1995 the mapping of heavy metal deposition using mosses as biomonitors was carried out over wide parts of Europe and interpolated maps were established. For the sampling period 2000 the investigated area could again be extended. The aim of the Heavy Metal Protocol under the UN/ECE Convention on Long-Range Transboundary Air Pollution (CLRTAP) of the Economic Commission for Europe (UN-ECE) from 1998 is the reduction of the releases of lead, cadmium and mercury to the environment. For the other air borne pollutants under the convention (sulphur and nitrogen compounds) networks for monitoring the success of emission avoiding strategies are established.

So far no heavy metal deposition monitoring networks have been established giving comparable information on an international scale. It is evident that methods measuring the effectiveness of measures are needed.

Programmes to measure the ambient air concentrations of some heavy metals in the framework of the convention are to be launched at some sites. It is obvious that for a spatial monitoring of heavy metal deposition the moss method is appropriate fulfilling main demands for a monitoring network. The method is useful for the elements lead, cadmium and mercury dealt with in the protocol.

As a result the European-wide co-ordination and evaluation of the moss biomonitoring project was established at the International Cooperative Programme on Effects of Air Pollution on Natural Vegetation and Crops (ICP Vegetation) under the Working Group of Effects in the framework of the CLRTAP in 2000.

Results of the European-wide monitoring programme carried out in 1990 and 1995 showed a wide range of concentrations of heavy metals in mosses corresponding to emission centres, but also to regional deposition conditions depending on precipitation or barriers. Temporal developments between the years 1990 and 1995 show decreasing depositions of some elements such as lead and cadmium due to the introduction of unleaded petrol but also to restructuring and economic developments, especially in Central Europe.

A map of Europe shows a wide range of concentrations with regional hot spots of heavy metal deposition especially in the heavily industrialised regions in Eastern Europe. In countries with better environmental standards deposition of heavy metals is in general low. Nevertheless national maps of these countries with higher resolution show certain differences and regional characteristics. For example, the map of Austria reveals certain regional aspects (Zechmeister, 1994, 1997): In addition to an increase of some elements in the eastern region, mainly due to long-range transport, for special elements some industrial sites, as well as road traffic related depositions along international traffic routes through inneralpine valleys, the barrier effect of the Alps connected with elevated precipitation and geogenic characteristics for some elements are recognisable.

Biomonitoring of heavy metal deposition by means of the moss method provides a three dimensional picture of the state and the development of the environment: Grid density of sampling sites can be high because of low costs, making a regional and supraregional mapping with adequate precision possible. The third dimension is the monitoring of temporal developments.

The repeating of the investigation programme every five years is an optimum between environmental politician's information demand and methodological precision.

The need for a European-wide deposition measuring network monitoring the efficiency of emission reduction strategies of the Heavy Metal Protocol under the CLRTAP is evident. A cheap and efficient method is the biomonitoring of heavy metal deposition with mosses. Unfortunately, the method has not yet been officially implemented as an instrument of the convention.

Up to now decreasing depositions of some elements could be shown in some Western European countries. For some emission affected areas and especially for countries with comparably lower environmental standards, especially in the East and Central European countries, the necessity of emission reduction measures could be demonstrated by heavy metal deposition maps using the moss method. It can be an argument for a European wide environmental policy and environmental standards under the perspective of Enlargement of the European Union.

5.3.3. *European water framework directive*

In 2000 the European Union established with Directive 2000/60/EC a framework for community action in the field of water policy. The basic goal of this instrument is to reach at least a good status for surface water, which means a status achieved by a surface water body when both its ecological and its chemical status are at least good. The ecological status is an expression of the quality of structure and functioning of aquatic ecosystems associated with surface waters, which has to be classified according to Annex V of the Directive. According to this Annex the definition of a high, good and moderate ecological quality of a river, lake, transitional and coastal water is based on the composition and abundance of the basic biological quality elements such as phytoplankton, macrophytes, phytobenthos, benthic invertebrates and fishes.

The underlying concept of this classification is bioindication. Composition and abundance of the basic quality elements are to be compared with the ones expected and with those normally associated with the type of water under undisturbed conditions. If there is no or only very little evidence of distortion the water is to be classified as of high status. If "the values of the biological quality elements for the surface water body type show low levels of distortion resulting from human activity, but deviate only slightly from those normally associated with the surface water body type under undisturbed conditions" the water is to be classified as of good status. If those values deviate moderately the status is to be classified as moderate. If the biological communities deviate substantially from those normally associated with the surface water body type under undisturbed conditions the water is to be classified as of poor status, in the case of the absence of large portions of these communities as of bad status.

According to this directive "heavily modified water bodies " are bodies of surface water, which as a result of physical alterations by human activity are substantially changed in character, such as rivers used for the generation of electricity. In this case and in the case of artificial water bodies "good ecological status" is replaced by "good ecological potential". In this case the reference condition is the undisturbed condition associated with the surface water body type most similar to the artificial or heavily modified water body concerned. For instance, when a former running water is changed to a more or less lake-like water body by a dam in order to produce electricity, the reference conditions might be an undisturbed lake typical of this area.

If a surface water body is not of at least good status or of good ecological potential, action is needed with the aim of achieving good status or good ecological potential at the latest 15 years after the date of entry into force of the directive, which was in the year 2000.

5.3.4. *Changes in vegetation and global warming*

Recent public environmental discussion has been dominated by concerns about climate change. The Intergovernmental Panel on Climate Change points out that average temperatures could rise during the 21st century by about 1.4 to 5.9 degrees. Although this was stated by a team of experts well-renowned political efforts at the international level have not yet been successful in implementing strict measures to reduce greenhouse gas emissions. Uncertainties in the proposed scenarios due to limitations of calculation can be seen as one reason for the reluctance at the political level. Besides scenarios based on mathematical modelling the undoubtedly most important indicators are economic ones associated with the potential consequences of global warming. Regarding the potential ecological consequences bioindicators may play an important role in the upcoming political decision-making processes.

It may be assumed that for the public changes in vegetation are more impressive than a rise of the average temperature by 1.4 to 5.9 degrees within a century. This allows the conclusion that bioindicators could tip the balance in public opinion. One should not forget, however, that bioindicators can be very ambiguous. Grabherr et al. (1994) have shown the sensitivity of high alpine ecosystems and that even moderate warming induces migration processes, which are assumed to be under way already. They also point out that this might lead to disastrous plant extinction in these environments. Particularly the summit flora may be affected. On the other hand, some people may interpret the appearance of tree species above the former tree line, caused by higher temperatures in boreal zones, as a chance for better growth conditions for crops or increasing wood increment and for this reason a desirable goal. The political dimension of bioindicators in this field is evident.

Even now forest owners are made insecure by the scenarios related to forest development under the changes of temperatures expected. Whether they manage their forests by favouring or planting one or the other tree species can be decisive for the forests' stability in future. So it is of utmost importance for forest and environmental policy to gather information from nature itself as much as possible. To provide information that may give politicians a good means for decision making upon measures to be taken to reduce emissions of greenhouse gases science is challenged to develop bioindication methods that are able to give an impressive sign of potential negative consequences due to global warming. Bioindication for real changes in vegetation caused by global warming may add decisive value to statistical scenarios of consequences of climate change and therefore be of high political relevance.

5.3.5. *The role of bioindicators as a tool for proving sustainable forest management*

The disastrous speed at which tropical forests were destroyed during the last decades of the 20th century has led to a world-wide discussion on how to overcome this huge environmental problem. Together with the problems associated with forest dieback in Europe, this issue has highlighted a global forest problem. In many discussions the need for adequate strategies for the protection and sustainable use of forests world-wide has been stressed.

The World Summit in Rio de Janeiro in 1992 at the United Nations Conference on Environment and Development was one of several initiatives and processes encouraging countries to make commitments towards a more sustainable use of natural resources. These activities have also expressed concern related to the manifold causes threatening forests at a global scale. In the framework of several programmes established to cope with this issue, the creation of guidelines for sustainable forest management in tropical regions in 1993 by the International Tropical Timber Organisation (ITTO, 1998) was one of the most important measures. In addition, other regional programmes have been launched, such as the Ministerial Conference on the Protection of Forests in Europe (MCPFE) in 1990 (Ministerial Conference on the Protection of Forests in Europe, 2000). These high level political commitments comprise basic tools for the evaluation of the successful implementation of the proposed measures. Some of these tools are based on bioindicators. For example the MCPFE has established a set of six criteria for sustainable forest management. The evaluation of whether these criteria are met is facilitated by 27 quantitative indicators. Collecting data on these indicators shall make visible changes for each criterion over a period of time. One hundred and one descriptive indicators will help illustrate national political developments concerning the sustainable management of forests. Similar tools (sets of criteria and indicators) have been developed in other regions, such as the temperate and boreal zone outside of Europe (Montreal Process) and the already mentioned guidelines of the ITTO.

Apart from these governmental initiatives environmental NGOs have been playing a very active role in the context of proving sustainable use of forests. For example, the WWF promoted in the "Forests for Life" campaign the certification of sustainable forest management. In the meantime this issue has got a high political profile in environmental debates. Especially the certification schemes that are promoted by environmental NGOs often go beyond legal obligations as far as standards of sustainable forest management are concerned.

Again some bioindicators are part of the systems used for the overall assessment of the sustainable use of forests. There are two specific fields where bioindication is used, namely biological diversity and health and vitality. Some examples are given in Table 1.

It has to be mentioned that these bioindicators are not used exactly the same way. This is due to differing wordings and their adaptations to regional or local conditions.

Table 1. Examples of fields in which bioindicators are used for monitoring sustainable forest management by various governmental and non-governmental initiatives.

	Governmental initiatives	NGOs' initiatives
Extent of protected forest area	x	x
Forest area damaged by biotic or abiotic agents	x	x
Number of endangered species dependent on forests (condition)	x	x
Amount of standing dead wood		x
Percentage of forest stands with a natural mixture of tree species		x

> ***Excursus****: The elaboration of a set of criteria and indicators for SFM is difficult, especially when it comes to the definition of threshold values. These can only be developed for clearly defined areas and environmental conditions. Broad participation of the stakeholders often complicates the definition process but must be ensured. In some cases target values may only be established in an indirect way. For example, the regeneration of a forest stand depends on various variables. The browsing by ungulate game can regionally be the most dangerous factor and often causes severe and controversial discussions between stakeholders. To determine the influence of ungulate game many times fences are put up around defined areas in the forests. Thus a comparison of the composition and growth of plants in protected areas and in areas open for browsing is possible. The resulting ratio reveals the impact of game and the necessity of measures to regulate the game population.*

Rametsteiner (2000) found out that there are differing views on how certification may influence the ecological statuses of forests. However, especially the intensive debates on how to prove sustainable forest management have strengthened the role of indicators and involved bioindication in a new field of political action. Additionally, these discussions have illustrated the fact that bioindication is an important and necessary tool for a comprehensive monitoring of environmental changes caused by human intervention.

6. Using bioindicators as environmental indicators

Bioindicators are defined as organisms, cells and subcellular compounds that can be used to assess environmental and ecosystem quality, as well as the impact of environmental stress on the composition and functioning of ecosystems. Therefore the term bioindicator differs from the concept of environmental indicators. Bioindicators can be used as environmental indicators, but environmental indicators cover more than the organism aspect. The concept of environmental indicators builds upon a broader definition: they can be defined as parameters or values derived from parameters which point to/provide information about/describe the state of a phenomenon/environment/ area with a significance extending beyond that directly associated with a parameter value (OECD, 1993, 1994). They can be developed on the basis of

- data from environmental monitoring; and
- data from statistical systems.

Although the concept of environmental indicators is broad in terms of the possible raw data that can be used as environmental indicators, they are specific in another way: Environmental indicators must be related and give answer to a specific question, mainly driven by the political debate. It is always necessary to make a compromise between the issues at the centre of the political debate and the availability of data, because data availability is often one of the main constraints of indicator development.

Another important aspect is that environmental indicators shall be that easy to interpret that politicians can pick them up and communicate them to the public. Therefore it is necessary that indicators have a signal effect, raise attention and lead to a message which can be underlined by scientific results and assessments.

Indicators need – quite from the beginning of their development – underlying environmentally relevant questions and interpretation and assessment in the end. This is particularly important to avoid misinterpretation of shortened messages or illustrating pictures, such as the smiley faces used by the European Environment Agency (EEA) to give a concise assessment of indicators (see EEA; 2000a):

- {smile}positive trend of the indicator, moving towards target
- {face} some positive development, but either insufficient to reach target or mixed trends within the indicator;
- {sad} unfavourable trend.

It was the OECD which started developing environmental indicator sets and, as a first step, worked out criteria for indicator selection, which show the broader concept of environmental indicators compared to that of bioindicators (OECD, 1994, 1998). To be of political relevance and utility for users an environmental indicator should:

- provide a representative picture of environmental conditions, pressures on the environment or society's responses;
- be simple, easy to interpret and able to show trends over time;
- be responsive to changes in the environment and related human activities;
- provide a basis for international comparisons;
- be either national in scope or applicable to regional environmental issues of national significance;
- have a threshold or reference value against which to compare it so that users are able to assess the significance of the values associated with it.

Furthermore indicators should be analytically sound, which means that they should be theoretically well-founded both in technical and scientific terms, they should be based on international standards and international consensus about their validity and they should lend themselves to being linked to economic models, forecasting and information systems. Criteria for their measurability are that the underlying data are readily available or can be made available at a reasonable cost/benefit ratio. Furthermore, the data should be adequately documented and of known quality; they should be updated at regular intervals in accordance with reliable procedures (OECD, 1998).

Following these criteria bioindicators are in principle able to fulfil these claims and therefore can be used as environmental indicators. Beyond that, these criteria can also be met by statistical data, such as livestock numbers, data for gross energy consumption, amounts of waste or water use, when the figures are related to a certain reference unit and allow an assessment of the state of the environment, of pressures or political/social response to changes of the state of the environment. Furthermore, also other environmental information meets these criteria and is used for environmental indicator development, such as emission data, area of national parks or water quality data. Looking through current environmental indicator sets, today only few indicators are used as bioindicators. This can be due to many reasons, including

- expenditure and efforts of biomonitoring;
- too short time series of available biomonitoring information;
- too long time intervals between measurements and changes in bioindicator's behaviour, condition, growth, etc.;
- no or difficult comparability of biomonitoring data within a country and between different countries;
- difficulties in interpreting biomonitoring data, because changing biological reactions are often not linear to changing environmental pressures. In addition to that, changes in organisms, cells and subcellular compounds can have many other causes apart from additional environmental pressures from human beings, such as climate, substrate and habitat conditions and impacts of other organism (competitors, predators and prey relationships, etc.).

Environmental indicators have become more and more important in the last 10 years. They are mainly used to inform policy makers and the public about the state and development of environmental issues. On the international level many institutions work on the development and publication of environmental indicators, for instance:

- The OECD: On OECD-level, the demand for environmental indicators originated from an OECD Council meeting at ministerial level in 1989, which called for a more systematic and effective integration of environmental and economic decision-making. The OECD publishes regular statistical compendia, indicator reports and environmental performance reports based on the OECD core set of indicators. The OECD worked out the pressure–state–response framework to develop and organise environmental indicators: human activities exert pressures on the environment (described by *pressure indicators*) and change its quality and the quantity of natural resources (described by *state indicators*) (OECD, 1994). Society responds to these changes through environmental, general economic and sectoral policies (described by *response indicators*).
- The UN: On the basis of Agenda 21 the United Nations Commission on Sustainable Development (CSD) worked out a set of Indicators of Sustainable Development in 1996 and invited Member States for national testing of this set, which meant review and adaptation according to the national decision-making processes (UN, 1996).

The main goal of this indicator programme was to create a viable and flexible system for monitoring the progress of sustainable development strategies, policies and activities. The indicators of sustainable development cover social, economic, environmental and institutional aspects to arrive at a broader, more complete picture of societal development. Thus environmental indicators are one component of sustainable development indicators. The United Nations used the indicator framework of *"Driving Force – State – Response"* – indicators (DF-S-R), linking their concept with the work of the OECD, but introduced the term "Driving Force" instead of "Pressure". This was necessary because indicators of sustainable development also include – in addition to environmental pressures – economic and social factors, which exert positive or negative influence on sustainable development. Indicators for monitoring progress towards sustainable development are needed in order to assist decision-makers and policy-makers at all levels and to increase focus on sustainable development.

- The EEA: The European Environment Agency has developed the driving forces–pressure–state–impact–response–assessing and reporting framework (D-P-S-I-R) for environmental indicators and refers to this in its yearly indicator-based reports (EEA, 2000a), as well as in its co-operation reports with the Commission about sectoral integration indicators (for transport, energy, agriculture, e.g. EEA, 2000b). According to this system analysis view, social and economic developments (*driving forces*) exert *pressure* on the environment and, as a consequence, the *state* of the environment changes, such as the provision of adequate conditions for health, resources availability and biodiversity. Finally, this leads to *impacts* on human health, ecosystems and materials that may elicit a societal *response* that feeds back on the driving forces, or on the state or impacts directly, through adaptation or curative action (EEA, 1999).

- EUROSTAT: The statistical office of the European Communities has developed environmental *pressure* indicators (EUROSTAT, 1999a,b) on the basis of the results of two surveys of environmental experts carried out by EUROSTAT. Starting point for this initiative was a Communication from the Commission to the Council and the European Parliament on "Directions for the EU on Environmental Indicators and Green National Accounts" (COM(94)670). The work resulted in a publication of 60 environmental pressure indicators (EUROSTAT, 2000).

- The European Commission (DG Environment and EUROSTAT) together with Member States and the EEA: A joint report from the EU Member States, the European Commission (DG Environment and EUROSTAT) and the European Environment Agency "Towards a European Set of Environmental Headline Indicators" was published (EC and EEA, 2000). A set of a few (about ten) headline indicators is intended to fill the gap between one single index, such as the Gross Domestic Product (GDP) in economy, and a large set of environmental indicators addressing specific issues.

- European Council: The European Council at Cardiff in June 1998 in its conclusions requested the identification of indicators as a basis for monitoring progress with the environmental integration strategies to be developed for different sectors. The Transport, Energy and Agriculture Councils were invited to start this process. The Vienna European Council specifically called on the Commission to present a

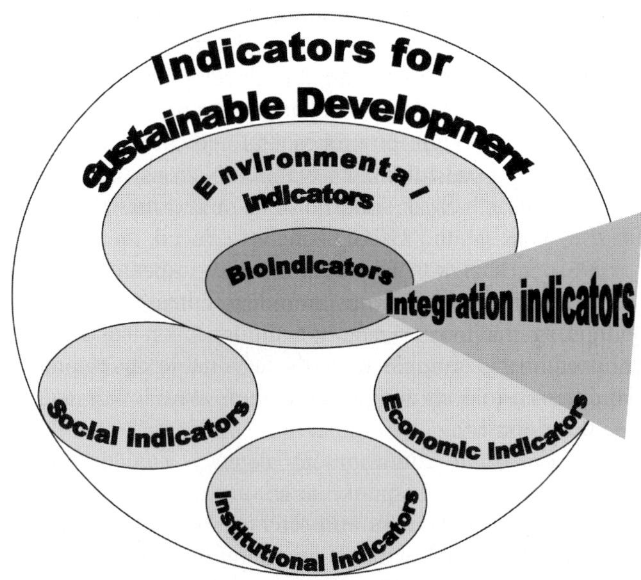

Figure 6.   Scheme for placing bioindicators within international indicator sets.

report on environmental and integration indicators, which was published 1999 (EC, 1999). These integration indicators should link environmental concerns with the (mainly economic) activities in the sectors.

An overview of these indicator levels is given in Figure 6. It shows that bioindicators can be used for environmental indicator sets, mainly to reflect on the state of the environment, but also as indicators for the environmental aspects of sustainable development or as integration indicators, when they are used to show interlinkages between economic and environmental issues.

Looking through the environmental indicator sets of international institutions only a few bioindicators are included at present. They are all used as environmental state indicators according to the D-P-S-I-R-framework or the DF-S-R-concept.

- *OECD Core set of environmental indicators*
  In 1994, the OECD for the first time published a core set of environmental indicators (OECD, 1994), which already included some bioindicators: In the chapter "Toxic contamination" the indicator describing the state of the environment was "Concentration of heavy metals and organic compounds in environmental media and living species", but this indicator was not elaborated further.

  In the chapter "Biodiversity/landscape" the "Threatened or extinct species as a share of total species known" are used as indicators. Data for this indicator cover mammals, birds, fish, reptiles, amphibians and vascular plants. Other groups such as invertebrates or fungi are not covered. The OECD pointed out that when interpreting this indicator it should be kept in mind that
  - the number of species known does not always accurately reflect the number of species in existence and that

Table 2. Bioindicators included in the set of UN-indicators of sustainable development.

Chapters of Agenda 21	Driving force indicators	State indicators	Response indicators
Chapter 18: Protection of the quality and supply of freshwater resources		Concentration of faecal coliform in freshwater bodies	
Chapter 17: Protection of the oceans, all kind of seas and coastal areas		Algae index	
Chapter 12: Managing fragile ecosystems: Combating desertification and drought		Satellite derived vegetation index value	
Chapter 15: Conservation of biological diversity		Threatened species as a percentage of total native species	
Chapter 19: Environmentally sound management of toxic chemicals		Chemically induced acute poisonings of inhabitants	

- the definitions are applied with varying degrees of rigour in Member countries, although international organisations such as the IUCN are promoting standard-isation.

- *UN set of indicators of sustainable development – environmental aspects*
  On a UN-level some bioindicators are used as environmental indicators describing the state of the environment according to the DF-S-R-concept, which is applied to the indicators of sustainable development. Bacteria, algae, vegetation cover, threatened species and man are used as bioindicators. The concrete indicators and related chapters of Agenda 21 are given in Table 2.

- *EEA indicator report: Environmental Signals 2000 (EEA, 2000a)*
  In the first EEA indicator report the sensitivity of certain organism is used in a way that exceedances of limit values and critical levels for human beings, certain crops and forests are shown as environmental indicators, for instance to describe air pollution. This is the case for tropospheric ozone concentration, but also exposure of human beings (exceedances of threshold values) to PM10 concentrations (fine particulate matter less than 10 mm in diameter) is used as an indicator for the state of the air in urban areas.
  In the chapter "wetlands" wintering waterbirds are used as environmental state indicators. Changes in the number of bird species and bird population are used as signals of general changes in the condition of and pressure on these aquatic ecosystems. Therefore data on 23 open-water species in 12 European countries (e.g. swans, ducks and the common coot) has been combined in an index. This index shows a slight increase in north-western Europe since 1980, but in some bird popu-

lations these increases are interpreted as being linked to the milder nature of the winters. This shows the difficulty of interpretation of bioindicators – the causing effects of changes in population dynamics are manifold and complex, so it is difficult to relate them to certain environmental pressures. An increasing number of wintering waterbirds might indicate that they are recovering due to favourable wintering conditions, and not because of less water pollution and less eutrophication or restoration of regulated rivers and habitat structures.

- *EUROSTAT: Pressure indicators*
  In 2000 EUROSTAT published a report which aims to give a comprehensive description of the most important human activities that have a negative impact on the environment (EUROSTAT, 2000). This excludes bioindicators by definition, environmental pressures not being shown by bioindicators, which reflect environmental pressures, but by data for the pressures themselves, like emission data, nutrient balances of soil, consumption of toxic chemicals, energy consumption and so on. For the issue "loss of biodiversity" indicators describing the loss of protected areas, wetland loss, fragmentation of forests and landscapes by roads, clearance of natural and semi-natural forested areas, etc. are applied.

- *European Commission and EEA: Environmental Headline Indicators (Working Document ENV/01/25)*
  The only way, in which bioindicators are used indirectly as environmental headline indicators, is using exceedances of critical loads for acidification and eutrophication as indicators for "air quality". These critical loads are derived from different degrees of sensitivity of organisms. But the main headline indicators for this issue are emission data.

  For the issue of biodiversity it is proposed for long term work that the ideal headline indicator would be a "bio-index" based on species, genes and habitats. In the medium term the current headline indicator for "Special Protection Areas" according to the Birds-directive should be accompanied by an indicator showing the development in absolute numbers of threatened species in EU 15 (number of Red List species).

7. Conclusions

The reason why bioindicators and biomonitors are able and suitable to provide crucial environmental information for politicians and administrations to reach decisions is that organisms inhabit the environment and reflect changes in the environment as a whole, both chemical and structural, at a highly integrated level.

One of the advantages of bioindication lies in its low costs, because the equipment does not comprise sophisticated high-tech elements. This is an important aspect for all countries and a crucial one for the administration in countries with very limited budgets such as developing countries or states in transition. Using e.g. mosses as biomonitors for heavy metal deposition is a cheap method compared to technical deposition sampling methods and can be standardised to be applicable over a wide geographic and climatic range. This makes the establishment of a measuring grid with a high density of sampling sites possible, allowing to create pollution maps for large areas. In addition to that bioindicators reflect rising awareness for environmental issues. Sometimes bioindicators are closer to the emotional perception of people than figures of chemical analyses (e.g. the disappearing of a lichen species is more tangible for many people than "parts per trillion" of a chemical substance). The interpretation of the condition of forests by means of the (scientifically vague) bioindicator crown condition has raised awareness for nature. Public awareness stabilised at a high level, which served environmental policy makers. This may lead to the conclusion that the usefulness of bioindication for (environmental) policy making is closely linked with the (emotional) perception of bioindicators in the public .

Bioindicators also have shortcomings. Investigations are often time-consuming and experienced staff is needed. Unlike precise analytical measurements interpretation of biomonitoring results is sometimes difficult. Moreover results are often difficult to communicate due to the fact that some people still consider results from chemical or physical analysis to be more precise and therefore more reliable. The aim is to bridge the gap in communication between science and policy makers, which can be reached by mutual dialogues and understanding.

Bioindication has to fulfil certain criteria to be successful in supporting environmental policy. Besides being scientifically well-founded, bioindication should be practicable, cost efficient and transparent (easily understandable) with a high degree of integration and positive perception of the people involved. The latter is highly correlated with social values. To implement effective solutions for an environmental problem, co-operation of representatives of all involved interest groups is necessary but often difficult. Based on the results of bioindication investigations necessary steps can be taken to obtain accepted solutions to reduce environmental impacts.

Bioindicators are used in many ways in political discussions, one important issue being risk assessment and the development of target values. To establish ecotoxicological target values in order to protect species and ecosystems, the most sensitive parts of these ecosystems have to be selected as bioindicators. Concerning risk assessment the environmentally sensitive public prefers the Precautionary Principle rather than a science-based risk concept, because the former takes into account broader uncertainties and the irreversibility of processes. The Precautionary Principle concentrates on unknown but potential risks, uncertainties and not determined situations of fate and effects. Bioindication can be a tool to identify environmental effects very early, sometimes even with the possibility of quantifying health or environmental risks. Quantification should differentiate between individual risk and the risk for society, which implies different levels of safety and scale.

For monitoring the structural changes of habitats bioindication provides a valuable tool to quantify human impact and to give guidance on how to combine the goal of meeting human needs with the protection of species and ecosystems. To get politically relevant messages in most cases the only practical baseline to show the development of the number of species will be the first year of a monitoring programme.

It can be concluded that a combination of chemical analyses and the reaction of a bioindicator is the most powerful tool for interpretation and political reaction. Chemical analysis for those substances suspected to be found together with sensitive bioindicators indicating the absence (or presence) of toxic substances as well as structural changes and changes in environmental condition in very general terms are the best basics to derive political measures. This approach allows policy makers and administrations to be on the safe side and it is already used e.g. in the licensing system for waste water emissions in Austria – using fish, crustaceans or bacteria as indicator organisms. It should help to optimise the application of precise but expensive analytical assessments on the one hand, and integrative and less precise but relatively cheap bioindication methods on the other.

For the future of bioindication and its use in policy making and administration, methods for cost-effective and statistically reliable sampling still have to be established for many groups of species. To strengthen the role of bioindication, continuous development and standardisation of techniques is necessary. This could help safeguard the advantage of a prosperous cost-benefit ratio which is an indispensable prerequisite for its application as a tool for environmental decision making.

References

Amt der Kärntner Landesregierung, 1999. Kärntner Umweltbericht 1999. Klagenfurt.

Amt der Tiroler Landesregierung, 1997–2000. Zustand der Tiroler Wälder. Innsbruck.

Amt der Tiroler Landesregierung, 2000. Bodennutzungs- und Bodenbelastungskataster Brixlegg. Berichte an den Arbeitskreis (unpublished).

Augustin, S., Degen, B., Lorenz, M., Schall, P., Schmieden, U., Schweizer, B., 1997. Auswertung der Waldschadensergebnisse (1982–1992) zur Aufklärung komplexer Ursache-Wirkungsbeziehungen mit Hilfe systemanalytischer Methoden. Umweltbundesamt. Berichte 6/97. Erich Schmidt Verlag, Berlin.

Bro-Rasmussen, F., 1999. Precautionary principle and/or risk assessment. A penitence in contemporary political culture. ESPR – Environ. Sci. & Pollut. Res. 6 (4), 188–192.

Carson, R., 1962. Silent Spring. Houghton Mifflin, Boston.

COM, 2001. A Sustainable Europe for a Better World: A European Union Strategy for Sustainable Development (Commission's proposal to the Gothenburg European Council).

Danish EPA, 1999. Government report on consequences of full or partial out-phasing of pesticide uses in Denmark (in Danish).

Deutscher Bundestag, 2000. Unterrichtung durch die Bundesregierung – Umweltgutachten 2000 des Rates von Sachverständigen für Umweltfragen – Schritte ins nächste Jahrtausend. Drucksache 14/3363.

Drasch, G., Roider, G., Bose-o'Reilly, S., Feenstra, O., Sampl, H., 2000. Rückgang der Bleibelastung der Bevölkerung im Umkreis einer Bleihütte in Arnoldstein/Kärnten durch erfolgreiche Sanierungsmaßnahmen. Umweltmedizin in Forschung und Praxis 5 (3), 233–237.

Eisenmenger, W., Drasch, G., 1992. Blutuntersuchungen in Arnoldstein. In: E. Kasperowski, 1993. Schwermetalle in Böden im Raum Arnoldstein. Umweltbundesamt, Monographien Band 33. Wien.

Environment Agency for England and Wales, 1998. Bridging the gap – proceedings of a conference held at Nelson's Dock in London. Bristol (Contact: alastair.ferguson@environment-agency.gov.uk).

EC (European Commission), 1999. Commission Working document SEC (1999) 1942. Report in Environmental and Integration Indicators to Helsinki Summit.

EC (European Commission) (DG Environment & EUROSTAT) & EEA (European Environment Agency), 2000. Towards Environmental Headline Indicators. Draft, 18 July 2000. EEA (European Environment Agency) (1999): Environmental indicators: typology and overview. Technical report No 25. Copenhagen.

EC-UN/ECE, 1997. Müller-Edzards, Ch., Erisman, J.W., de Vries, W., Dobbertin, M., Ghosh, S., 1997. Ten years of monitoring forest condition in Europe. Studies on temporal development, spatial distribution and impacts of natural and annthropogenic stress factors. Overview report. EC and UN/ECE, Brussels, Geneva.

EC-UN/ECE 2000. De Vries, W., Reinds, G.J., Kerkvoorde, M., Hendriks, C.M.A., Leeters, E.E.J.M., Gros, C.P. Voogd, J.C.H. Vel, E.M., 2000. Intensive monitoring of forest ecosystems in Europe. Technical report 2000. EC and UN/ECE, Brussels, Geneva.

EEA (European Environment Agency), 1999. Environmental indicators: typology and overview. Technical report. European Environment Agency, Copenhagen.

EEA, 2000a. Are we moving in the right direction? Indicators on transport and environment integration in the EU – TERM 2000. Copenhagen.

EEA, 2000b. Environmental signals 2000. European Environment Agency regular indicator report. Environmental assessment report No 6. Copenhagen.

EEC, 1979. Council Directive 79/409/EEC of 2 April 1979 on the conservation of wild birds. OJ L 103, 25.04.1979, p. 1.

EEC, 1990. Council Directive 90/313/EEC of 7 June 1990 on the freedom of access to information on the environment.

EEC, 1992. Council Directive 92/43/EEC of 21 May 1992 on the conservation of natural habitats and of wild fauna and flora. OJ L 206, 22.07.1992, p. 7.

EUROSTAT, 1999a. Toward environmental pressure indicators for the EU: indicator definition, 1999. Luxembourg

EUROSTAT, 1999b. Toward environmental pressure indicators for the EU: an examination of the sectors. Luxembourg.

EUROSTAT, 2000. Toward environmental pressure indicators for the EU. Luxembourg.

Fent, K., 1998. Ökotoxikologie. Georg Thieme Verlag, Stuttgart.

Grabherr, G., Gottfried, M., Pauli, H., 1994. Climate effects on mountain plants. Nature 369, 448.

Halbwachs, G. (Ed.), 1982. Das Immissionsökologische Projekt Arnoldstein. Carinthia II, 39. Sonderheft. Klagenfurt.

Idinger, J., Kromp, B., Steinberger, K.-H., 1994. Ground photoeclector evaluation of the numbers of carabid beetles and spiders found in and around grain fields treated with either inorganic or compost fertilisers. Arthropod natural enemies in arable land. Survival, reproduction and enhancement of beneficial predators and parasitoids in agroecosystems. Proceedings of the 2nd EU-workshop on Wageningen, 1–3 Dec. Acta Jutlandica.

ITTO, 1998. Annual Report 1998, Document GI-7/98 ITTO. Yokohama, Japan.

Kasperowski, E., 1993. Schwermetalle in Böden im Raum Arnoldstein. Umweltbundesamt, Monographien Band 33, Wien.

Kommission Human-Biomonitoring, 1996. Stoffmonographie Blei – Referenz- und Human-Biomonitoring-Werte (HBM). Bundesgesundheitsblatt 39 (6), 236–241.

Lehmann, A., 2001. Waldbewusstsein. Zur Analyse eines Kulturthemas in der Gegenwart. Forests and their perception by the general public. On the analysis of a present-day cultural subject. Forstwissenschaftliches Centralblatt. Jahrgang 120 (1), 37–49.

MAFF, 2000. Towards sustainable agriculture – a pilot set of indicators. Ministry of Agriculture, Fisheries and Food. London, United Kingdom. Available at the MAFF web-site at: http:/www.maff.gov.uk/farm/sustain.htm.

Ministerial Conference on the Protection of Forests in Europe, 2000. Ten Years of Commitment to European Forests – The Ministerial Conference on the Protection of Forests in Europe.

Mirtl, M., 2001. Integrated Monitoring – Langzeitmonitoring der Wirkung von Umweltstress auf Ökosysteme. In: UBA (Umweltbundesamt), 2001a, pp. 329–335.

Nordic Council of Ministers, 1998. Atmospheric heavy metal deposition in Europe 1995–1996. Nord (15). Copenhagen.

OECD (Organisation for Economic Co-operation and Development), 1993. OECD core set of indicators for environmental performance reviews. OECD Environment Monographs No. 83, Paris.

OECD (Organisation for Economic Co-operation and Development), 1994. Environmental indicators – OECD core set. Paris.

OECD (Organisation for Economic Co-operation and Development), 1997. Environmental indicators for agriculture. Paris.

OECD (Organisation for Economic Co-operation and Development), 1998. Environmental indicators – towards sustainable development. Paris.

OECD (Organisation for Economic Co-operation and Development), 1999. Environmental indicators for agriculture: methods and results – the stocktaking report – biodiversity. com/agr/ca/env/epoc(99)132/rev1, Paris. Concerning UK mammal population trends information can be found at: http://www.abdn.ac.uk/mammal/.

Rametsteiner, E., 2000. Sustainable forest management certification. frame conditions, system designs and impact assessment. Ministerial Conference on the Protection of Forests in Europe – Liaison Unit Vienna. Vienna.

Rippen, G., 1988 (ongoing supplementation) Handbuch Umweltchemikalien. Stoffdaten, Prüfverfahren, Vorschriften. Ecomed.

Riss, A., 1993a. Impact of PCDD/PCDF emissions of a copper reclamation plant: five years of experience with environmental monitoring. In: Fiedler, H., Frank, H., Hutzinger, O., Parzefall, W., Riss, A., Safe, S. (Eds), Dioxin '93. Organohalogen Compounds, Vol. 14. Vienna, pp. 23–26.

Riss, A., 1993b. Beurteilung der Ergebnisse. In: Kasperowski, E., Schwermetalle in Böden im Raum Arnoldstein. Umweltbundesamt, Monographien Band 33, Wien.

Riss, A., Aichinger, H., 1993. Reduction of dioxin emissions and regulatory measures in Austria. In: Fiedler, H., Frank, H., Hutzinger, O., Parzefall, W., Riss, A., Safe, S. (Eds), Dioxin '93. Organohalogen Compounds, Vol 14. Vienna, pp. 341–344.

Riss, A., Hagenmaier, H., Rotard, W., 1990a. Wirkungen von Dioxinimmissionen auf Boden, Grünlandaufwuchs und Kuhmilch – Fallstudie anhand einer Metallrückgewinnungsanlage in Österreich. In: Kommission Reinhaltung der Luft im VDI und DIN. Wirkungen von Luftverunreinigungen auf Böden, Einträge, Bewertung, Regelungen. VDI-Berichte 837. Düsseldorf.

Riss, A., Hagenmaier, H., Weberruß, U., Schlatter, C., Wacker, R., 1990b. Comparison of PCDD/PCDF levels in soil, grass, cow's milk, human blood and spruce needles in an area of PCDD/PCDF contamination through emissions from a metal reclamation plant. Chemosphere 21 (12), 1451–1456.

Roth, L., 1997. Gesetz über Ababe für das Einleiten von Abwasser in Gewässer (Abwasserabgabegesetz – AbwG) Wassergefährdende Stoffe 31. Erg. Lfg 10/97.

Trimbacher, C., Weiss, P., 1999. Needle surface characteristics and element contents of Norway spruce in relation to the distance of emission sources. Environmental Pollution 105, 111–119.

UBA (Umweltbundesamt), 1993–1995. Unpublished data from an environmental control programme. Wien.

UBA (Umweltbundesamt), 2001a. Umweltsituation in Österreich. 6. Umweltkontrollbericht des Bundesministers für Land- und Forstwirtschaft, Umwelt und Wasserwirtschaft. Wien.

UBA (Umweltbundesamt), 2001b: Rote Liste gefährdeter Biotoptypen Österreichs. Report in preparation, Wien.

UK Department of Environment, 1996. Indicators of sustainable development for the United Kingdom, London.

UN, 1996. Indicators of sustainable development: framework and methodologies. United Nations, New York.

UN/ECE and EC, 2000. Forest condition in Europe. Results of the 1999 crown condition survey. Technical Report, Geneva and Brussels.

UN/ECE Convention on Long-range Transboundary Air Pollution, 2001. International cooperative programme on integrated monitoring of air pollution effects on ecosystems. http://www.vyh.fi/eng/intcoop/projects/icp_im/im.htm.

Vogel, W.R., Grath, J., 1998. Groundwater in Europe – state and trends in quality and quantity. How to cope with degrading groundwater quality in Europe. International Workshop at Johannesberg, Sweden in October 1997. FRN-Report 98, 4, Sweden, pp. 24–37.

WHO, 1993. Guidelines for Drinking-Water Quality, Vol. 1, Recommendations. Geneva.

Zechmeister, H., 1994. Biomonitoring der Schwermetalldeposition mittels Moosen in Österreich. Umweltbundesamt. Monographien Band 42, Wien.

Zechmeister, H., 1997. Schwermetalldeposition in Österreich erfaßt durch Biomonitoring mit Moosen (Aufsammlung 1995). Umweltbundesamt. Monographien Band 94, Wien.

Zulka, K. P., Eder, E., Höttinger, H., Weigand, E., 2001. Grundlagen zur Fortschreibung der Roten Listen gefährdeter Tiere Österreichs. Umweltbundesamt. Monographien Band 135, Wien.

Chapter 4

Bioindicators for ecosystem management, with special reference to freshwater systems

C.M. Lorenz

Abstract

Bioindicators for ecosystem management provide information on the occurrence of ecological processes and structures. Furthermore, bioindicators provide information on the ecosystem condition by comparing the ecosystem with a reference level of good ecological functioning and on cause-effect relationships within an ecosystem. This chapter aims to develop bioindicators for two different aquatic ecosystems: A transboundary and impacted river, such as the Rhine and a eutrophicated shallow lake, such as the Loosdrechtse plassen in the Netherlands.

The bioindicators are developed in four steps: In the first step ecosystem functioning is described on the basis of ecological theories and concepts. For the river ecosystem the theoretical concepts reviewed are zonation, river continuum, stream hydraulics, resource spiralling, serial discontinuity, flood pulse, riverine productivity and catchment hierarchy. For the shallow lake theories on the eutrophication and rehabilitation of lakes are described. In the second step the dominant processes and structures of ecosystem functioning are defined. The river ecosystem is steered by abiotic processes, leading to longitudinal and lateral fluxes of matter, which in turn causes a spatial distribution of species. In a shallow lake the predominant effect of increased nutrient loads into lakes is eutrophication, which involves a cascade of direct and indirect effects. This cascade of effects can lead to one of the two equilibrium states of shallow lakes; a turbid, phytoplankton dominated lake in a meso- or eutrophic state. The other equilibrium state is a clear and macrophyte dominated lake, which is in a meso- or oligotrophic state. The balance between these two states depends on a bottom up force determined by producers and a top down force determined by consumers. In the third step bioindicators are selected, which provide information on ecosystem productivity and structure. The algal biomass is selected as indicator for river and lake productivity. For ecosystem structure, the indicators for river and lake differ. For the river ecosystem the bioindicators describe the diversity and occurrence of life cycles along the longitudinal and lateral dimension. For the lake ecosystem two bioindicators are selected that are characteristic for the switch between the two equilibrium states: (1) the area and biomass of macrophytes, indicating that turbidity has reduced to the extent that macrophytes can grow and (2) the ratio between preyfish and predatory fish, which has to be 1:1 to 2:1 to guarantee a long-term stable clear lake. In the fourth step abiotic steering factors and relating human pressures are defined to describe cause–effect relationships. For the river natural dynamics, habitat diversity, connectivity and the water quality are described. For the lake the preconditions for the two forces are selected as indicators: nutrient loading for the bottom-up force productivity and habitat area for predatory fish for the top-down force of consumption. Finally, reference levels for the bioindicator are discussed in order to assess ecosystem condition. A number of references for ecosystem assessment are presently used. The first type of references are based on an "undisturbed" river or lake having authentic hydrological, geomorphological and ecological characteristics comparable to the ecosystem, which is to be assessed. The second type of references relate to a historical analysis of river characteristics in a pre-"disturbance" phase (natural background water quality, species occurrence, hydro-geomorphological characteristics. The third type are effect reference levels based on the risk of ecological impact. The recognition of the irreversibility of human impact increasingly attracts attention to the return of ecosystem processes as the starting point for ecological assessment (such as flood frequency, sedimentation patterns, succession). Finally, references can be based on policy goals or a reference year for policy evaluation. Which of the abovementioned reference levels is selected for the indicators depends on the goal of the assessment (effectiveness of policy, assessment of ecosystem condition) and the availability of (data on) the reference levels.

Keywords: bioindicators, ecosystem management, river ecosystem, river concepts, shallow lake, eutrophication.

1. Introduction

Ecosystem management is management at the scale of a whole ecosystem. What the scale (or its boundary) of an ecosystem is, has been subject to discussion (see the thesis of F. Klijn, 1997 for more information). In this chapter an ecosystem is defined, as a system of structurally related abiotic and biotic components, which are also functionally related by physical, chemical and biological processes (after Tansley, 1935; Chorley and Kennedy, 1971; Van der Maarel and Dauvellier, 1978; Odum, 1983).

Ecosystems can be characterized as complex systems, because of:

- Their scale. Ecosystems, such as rivers, oceans, deserts, tropical forests are very large.
- The interaction of abiotic processes, biological processes and a large number of different species having their own habitat requirements, behaviour and population dynamics.
- The spatial and temporal dynamics of these interactions and processes.

Human impact has added an extra dimension to this complexity. Humans use the goods and services provided by ecosystems, such as the supply of space, fish, water and wood. The direct use of these goods and services (resource extraction) or the changes made to ecosystems in order to improve the supply (i.e. embankment, drainage) have impacted their functioning. An example of the complexity of processes and interactions can be seen in Figure 1, which gives an overview of the cause-effect relationships of human use of a river ecosystem.

The major aim of ecosystem management is to improve the functioning ecosystem and to reduce negative impacts on ecosystems. Therefore, ecosystem management needs information on (1) the ecosystem condition, (2) possible management actions to improve that functioning and (3) the effectiveness of applied measures.

Bioindicators can contribute to the information need of ecosystem management. An bioindicator is defined as an organism (or a part of an organism or a community of organisms) that contains information on the quality of the environment (or a part of the environment) (Markert et al., 1999). The organism has significance beyond what is actually measured; in addition to the information of its presence and abundance, it provides information on the occurrence of ecological processes. For example, the

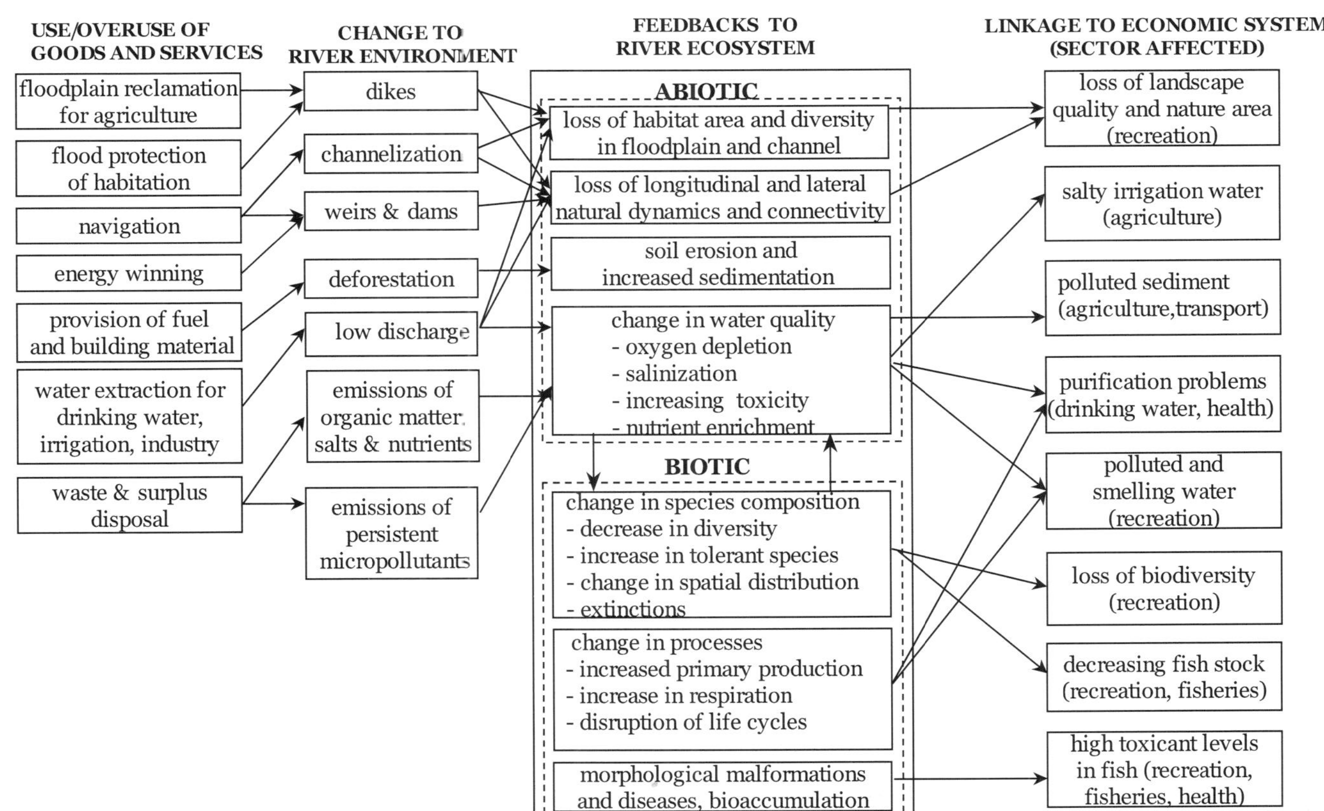

Figure 1. Overview of human use, changes made and effects on the river ecosystem and on socio-economic sectors.

presence of migratory fish implies the possibility of migration in a river. The occurrence and abundance of predators species indicate that the food web functions sufficiently to provide food for the predators.

Bioindicators can provide the following information for ecosystem management:

- A description of ecosystem processes and structures; both on general ecosystem processes, such as foodweb relations and biomass production and on more specific processes, such as fish migration in rivers or eutrophication processes in lakes.
- The ecosystem condition by comparing the ecosystem with a reference level of good ecological functioning.
- Cause–effect relationships within an ecosystem. Knowledge on these relationships is needed to define effective rehabilitation measures.

This chapter aims to develop bioindicators for two different aquatic ecosystems: A transboundary and impacted river, such as the Rhine and an eutrophicated shallow lake, such as the Loosdrechtse plassen in the Netherlands.

The river and lake are both aquatic ecosystems, but they differ in a number of ways. The most important difference is their scale: a transboundary river and its basin is most of the times much larger than a shallow lake. The Rhine basin has a surface of 185,000 km^2 and a length of 1320 km (Van Breukel, 1993; KHR, 1993), whereas the Loosdrechtse plassen have a surface of 981 ha, a depth of ±2 m and an average water residence time of 0.7 years (Janse et al., 1992). This difference in scale has implications for the complexity, knowledge and management of the systems (Lorenz, 1999).

In the first place, the large size of river basins and the numerous physical, chemical and biological processes, which are dynamic in time and space, make a specification of the cause(s) of any one effect very difficult. Cause–effect relationships have a spatial dimension, because of the unidirectional flow of water and matter from the catchment to the river mouth. A variety of upstream uses can culminate into downstream ecological effects.

Secondly, our knowledge on the functioning on transboundary river ecosystems is limited. Compared with other (semi-)aquatic systems, such as lakes, research on large rivers has been constrained by tradition, by methodological problems, and by the large geographic scale of these ecosystems. Moreover, natural, un-impacted large rivers no longer exist, and so there is no reference to assess the functioning of these river ecosystems. This lack of knowledge complicates not only assessment, but also restoration of rivers.

Thirdly, flow of water from upstream to downstream means that upstream users do not experience the negative effects of their actions and so are not directly motivated to adjust their practices. Moreover, river management depends on a large number of institutions with statutory authorities spread over different spatial scales (e.g. local, regional, national and international) and sectors (e.g. transport, agriculture, recreation, environment). Management of transboundary river basins under these conditions is complex and may easily become dominated by particular interests.

The shallow lakes in the Netherlands are smaller and have, more than rivers, the form of an uniform waterbody. The dominant impacts on shallow Dutch lakes are eutrophication due to nutrient emissions, destruction of natural banks and marshes and an unnatural waterlevel management (the waterlevel opposite from a natural water-level, namely high in summer and low in winter), disturbance by recreation and a direct impact on the foodweb by commercial fisheries.

The bioindicators for ecosystem management will be developed according to the following steps (Lorenz, 1999):

1. Description of ecosystem functioning on the basis of ecological theories and concepts.
2. Definition of dominant processes and structures of ecosystem.
3. Selection of bioindicators, providing information on the dominant ecological processes and structures.
4. Definition of the abiotic steering factors and relating human pressures to describe cause-effect relationships.
5. Selection of reference levels for the bioindicator (species diversity, species abundance) in order to assess its condition.

Ecosystem functioning will be described on the basis of ecological theories and concepts. For the river ecosystem a number of theoretical concepts will be used, which will be reviewed in the following section. In Section 3 the information from the river concepts will be integrated and the dominant processes and structures will be defined. For lakes theories on eutrophication processes and rehabilitation of lakes will be described in Section 5. On the basis of the reviews, indicators will be selected in Section 4 for the river ecosystem and Section 6 for the lake. Section 7 discusses possible reference levels for assessment. This chapter ends with some discussion points and conclusions.

2. Review of river concepts

In recent decades several river ecosystem concepts have been developed to describe the functioning and structure of natural, undisturbed rivers. These concepts identify dominant river ecosystem processes and structures and essential ecological character-istics (Lorenz et al., 1997). Dominant characteristics of the river concepts reviewed are summarised in Table 1. The table describes the type of river and dimension to which a concept applies. The essence of each concept is described by abiotic steering variables, functional and structural ecosystem characteristics.

In earlier times, ecological research on rivers focused on the description of biolog-ical communities in small streams (Cummins et al., 1995; Minshall, 1988). Research on large rivers was limited partly by tradition and partly by methodological problems, considering the large geographic scale of these ecosystems. The first step towards a development of more holistic concepts was the recognition that stream biota were influenced by the surrounding landscape (Ross, 1963; Hynes, 1975).

2.1. Zonation concept

The first attempt to describe the ecosystem of an entire river was the zonation concept. The zonation concept divides a river into zones characterised by fish communities

Table 1. Summary of river concept characteristics, the type of river and dimension that a concept applies to, and the abiotic steering variables determining the functional and structural river ecosystem characteristics according to the concept.

Concept	Type of river	Dimension	Abiotic steering variable	Functional ecosystem characteristics	Structural ecosystem characteristics
Zonation (Huet, 1954; Illies & Botosaneanu, 1963)	naturally undisturbed river	longitudinal	flow velocity temperature	adaptation of fishes and benthic fauna to temperature and flow velocity	zones of fish and benthic fauna
River Continuum Concept (Vannote et al., 1980)	naturally undisturbed river no floodplain river temperate climatic zone	longitudinal	stream size energy source: allochtonous OM or light	OM processing P/R ratio	shifts in functional feeding groups
Stream hydraulics (Statzner & Higler, 1986)	naturally undisturbed river	longitudinal	flow velocity water depth substrate roughness surface slope	adaptation of benthic fauna to hydraulic stress	zones of benthic fauna
Resource spiralling (Wallace et al., 1977; Newbold et al., 1981; Elwood et al., 1983)	naturally undisturbed river	longitudinal	flow velocity physical retention mechanisms nutrient limitation	recurrent nutrient and OM cycles along the length of the river	biological community (food web)

Table 1. (continued)

Concept	Type of river	Dimension	Abiotic steering variable	Functional ecosystem characteristics	Structural ecosystem characteristics
Serial discontinuity (Ward and Stanford, 1983a)	empounded river or floodplain river	longitudinal	position of dam	OM processing and P/R ratio shifts up- or downstream	functional feeding group ratio shifts up- or downstream
Flood pulse (Junk et al., 1989)	large flood plain river	lateral	flood pulse: duration frequency timing predictability amplitude water quality size and characteristics of floodplain	increased biological productivity and nutrient recycling in floodplain exchange of: nutrients and sediment OM biota	shifts in aquatic-terrestrial phases in floodplain high habitat and species diversity
Riverine productivity shifts in functional (Thorp and Delong, 1994)	large constricted river with well developed riparian zone	lateral	type and density of riparian zone retention structures flow velocity near riparian zone riparian zone		primary production feeding groups processing of OM from
Catchment concepts species distribution (Frissel, 1986; Gardiner, 1991; Naiman et al., 1992; Petts, 1994; Townsend, 1996)	whole catchment	longitudinal lateral vertical temporal	spatial and temporal dimensions and scales of abiotic variables	on a catchment scale	nutrient cycling on a catchment scale

(Huet, 1954) or macroinvertebrate communities (Illies & Botosaneanu, 1963). The zonation reflects differences in water temperature and flow velocity.

2.2. River Continuum Concept

The development of the River Continuum Concept (RCC) (Vannote et al., 1980) was an important step in river ecology, as it was the first attempt to describe both structural and functional characteristics of stream communities along the entire length of a river. The concept has been developed specifically in reference to natural river ecosystems in North America. The RCC argues that the biotic stream community adapts its structural and functional characteristics to the abiotic environment, which presents a continuous gradient from headwaters to river mouth. This is expressed by the distribution of organic matter and macroinvertebrate functional feeding groups.

In general, rivers can be divided in three parts based on stream size: headwaters (stream orders 1–3), medium-sized streams (orders 4–6) and large rivers (orders greater than 6). The headwaters of rivers are strongly influenced by riparian vegetation. Primary production is low because of shading and the vegetation contributes large amounts of allochtonous detritus. Thus, the ratio of gross primary productivity (P) to respiration (R) of the aquatic community is small (P/R < 1). The size of particulate organic matter in the water is rather large, consisting mainly of dead leaves and woody debris (Coarse Particulate Organic Matter (CPOM), >1 mm).

The influence of the riparian zone diminishes moving downstream; both the importance of terrestrial organic input and degree of shading decreases, whereas primary production and transport of organic matter from upstream increase. This is reflected by an increase in the P/R ratio (from P/R < 1 to P/R > 1). The size of suspended organic matter decreases to Fine Particulate Organic Matter (FPOM, 50 μm–1 mm) and Ultrafine Particulate Organic Matter (UPOM, 0.5–50 μm). Large rivers receive organic matter mainly from upstream, which has already been processed to a small size. Primary production is often limited by depth and turbidity. So, the P/R ratio decreases again (P/R < 1).

Changes in the size of organic matter along the length of the river are reflected in the distribution of functional feeding groups of invertebrates. In the headwaters shredders are co-dominant with collectors. Shredders process CPOM, such as leaf litter and the associated biomass. Collectors obtain their food by filtering them out of the water or gathering from the sediments FPOM and UPOM, which has been processed from CPOM by shredders. Collectors and grazers (or scrapers), which shear attached algae from surfaces, dominate the middle part of the river. In the lower reaches, the invertebrate assemblage consists mainly of collectors.

Since its development the applicability of the RCC has been tested on various river systems. Several field observations agreed with the RCC (Cushing et al., 1983; Minshall et al., 1983; Naiman, 1983; Conners and Naiman, 1984). In the cases where the concept did not comply with field observations, they could be explained by the dominant role of tributaries, climate, geology, local conditions, water quality or human disturbance (Winterbourne et al., 1981; Minshall et al., 1983, 1985; Bruns et al., 1984; Magdych, 1984; Conners and Naiman, 1984; Sedell and Frogatt, 1984; Cummins et al., 1984; Cummins et al., 1995). Minshall et al. (1985) stress the intention of the

RCC as a standard for natural, unperturbed lotic systems, in which general conditions and relationships can be identified and used to study and compare existing streams. It provides a framework for understanding the ecology of streams and rivers and is not intended as a description of biological components of all rivers individually.

2.3. Stream hydraulics concept

The RCC contrasts strongly to the zonation concept by emphasising gradients. An intermediate is the theory of stream hydraulics (Statzner and Higler, 1986). This theory distinguishes a zonation pattern of benthic fauna in which the distinct changes in species assemblages are linked to transitions in stream hydraulics. Stream hydraulics are determined by geomorphological and hydrological characteristics of the river and described by parameters, such as current velocity, depth, substrate roughness and surface slope. The pattern of velocity variation in space and time has a strong influence on biota, especially benthic invertebrates and fluvial algae (Statzner and Higler, 1986; Petts, 1994).

2.4. Resource Spiralling Concept

The Resource Spiralling Concept extends on the RCC by elaborating the processing of organic matter along the length of the river. The downstream flow of rivers adds a spatial dimension to resource cycles in stream ecosystems by downstream displacement of material. This results in partially open cycles or "spiralling" (Wallace et al., 1977; Newbold et al., 1981; Newbold et al., 1982a,b; Elwood et al., 1983; Newbold, 1992). Spiralling can be measured with the unit "spiralling length" (S), defined as the average distance along which the river flows during one cycle of a nutrient element, such as carbon. The shorter the spiralling length, the more efficiently the nutrient is utilised, i.e. the more times a nutrient is recycled within a given reach of stream (Newbold et al., 1981; Newbold et al., 1982a). Spiralling is a function of both downstream transport rate and retention processes (Minshall et al., 1983). A high transport rate, determined largely by water flow, will increase the spiralling length, whereas retention mechanisms, such as physical storage (e.g. wood debris, boulders, macrophyte beds, sedimentation) and biological uptake and storage will decrease the spiralling length. This biological retention is the recycling of nutrients in the ecosystem, which is largely controlled by environmental conditions, such as oxygen, temperature, nutrient availability and the structure and species composition of the food web (Minshall et al., 1983).

In general the spiralling length increases with stream size. Forested headwaters tend to conserve or store resources, because of their high biological activity and high retention of organic matter because of debris dams. In downstream parts of rivers, flow velocities increase and organic carbon is processed into successively smaller (and more transportable) particle sizes (Newbold et al., 1982a; Johnson et al., 1995). In side channels and floodplains the spiralling length may decrease, because of a high retention, both physically (e.g. sedimentation, woody debris, riparian vegetation) and biologically (e.g. high productivity) (Pinay et al., 1990).

2.5. Serial discontinuity concept

Another concept associated with the RCC is the serial discontinuity concept (Ward and Stanford, 1983a). This addresses the effects of dams on rivers. Dams disrupt the continuum and cause upstream-downstream shifts in abiotic and biotic parameters and processes. The effect is related to the position of the dam along the continuum. The serial discontinuum concept defines two parameters that may be used to evaluate the relative impact of a dam on riverine ecosystem structure and function (Ward and Stanford, 1983a). Firstly, the discontinuity distance, that is the distance over which the expected value of a physical or biological variable is shifted in downstream or upstream direction as a consequence of the discontinuity introduced by the dam. Secondly, the intensity, that is the absolute change in the variable as a consequence of regulation (i.e. the difference between the unregulated versus the regulated response at the same point along the longitudinal profile).

In general, dams increase the homogeneity of a variable between two discontinuities (Ward and Stanford, 1995). In the lentic water from a reservoir, temperature is more uniformly distributed than in flowing water. The transport of large organic matter particles, such as CPOM, will be blocked, whereas smaller particles (FPOM) can pass the dam more easily. Further, dams reduce the ecological connectivity between the mainstream and the riparian zone. In the first place, the blocking of CPOM originating from the riparian vegetation decouples the linkage between allochtonous inputs upstream and processing of organic matter downstream. In the second place, dam building is associated frequently with river regulation, which isolates river channels from their floodplain and riparian forest. Tests of the serial discontinuity concept in regulated rivers showed discontinuities in abiotic (temperature) (Stanford et al., 1988) and biotic variables (hydropsychid Trichoptera) (Hauer and Stanford, 1982; Stanford et al., 1988).

2.6. Flood Pulse Concept

The RCC predicts a diminishing influence of the riparian zone from headwaters to the downstream river. This, however, holds only for large rivers, which are confined to the river bed. Large floodplain rivers are significantly influenced by regular floods of the main stream into the bordering floodplains. The Flood Pulse Concept (FPC) (Junk et al. 1989) describes the effects of floods on both the river channel and its floodplain in an unmodified, large river-floodplain system.

Floodplains tend to establish their own nutrient cycles since organisms and environmental conditions differ considerably from the main channel. Nutrients originate mainly from river water. Release and storage of nutrients in the floodplain depend on the flood cycle, vegetation cover and, in temperate regions the growth cycle of the vegetation. During floods a layer of sediment, composed of nutrients and particulate organic and inorganic matter, is deposited on the floodplain. The quality of the sediment determines the fertility of the floodplain.

The carbon exchange between floodplain and main channel will depend on three factors; the presence of retention mechanisms keeping carbon in the floodplain and reducing leakage to the river channel (e.g. sedimentation, uptake by organisms, and retention by macrophytes and terrestrial vegetation); the duration and flushing rate of the flood; and the growth cycle of floodplain vegetation in temperate regions. Pinay et al. (1990) report that annual direct litter inputs to large rivers ranges between 10–40 g/m^2 of water surface.

With regard to biological productivity, a high P/R ratio is predicted for a large river-floodplain system, because of a high production in the floodplain and low import of organic matter from upstream. The flood pulse affects the primary production and respiration in the floodplain by determining occurrences, life cycles, and abundance of organisms. Furthermore, the change between terrestrial and aquatic phase accelerates the decomposition of organic material.

Life cycles of biota using floodplain habitats are related to the flood pulse in terms of its annual timing, duration, and rate of rise and fall. The floodplain is used for food supply, spawning and shelter. The main channel is used by fish as a migration route, for spawning, and as a refuge during for example droughts or for hibernation. Plant communities in the floodplain grow along a gradient of annual flooding, in which every plant has its optimum position.

River-floodplain systems show a high diversity of habitats. Sediments, deposited in the floodplain, form bars, levees, oxbows, backwaters and side channels. Differences in the duration of flooding, in soil structure and in vegetation result in many different small-scale habitats and physico-chemical conditions of oxygen, temperature, dissolved and suspended matter. These habitats can be unstable due to changing water levels, sediment deposition and erosion. Extensive studies on the geomorphological processes underlying the habitat diversity in floodplains have been carried out on the Rhône system by Amoros et al. (1987) and Bravard et al. (1986). As a consequence of the large variation in habitats, species diversity in river-floodplain systems is high. The flood pulse is expected to limit sedentary terrestrial and aquatic species, because of the physical stress of the flood, but will favour mobile organisms, such as fish and non-aquatic birds, which are able to avoid this stress.

Much research has been published on the importance and functions of floodplains in the river ecosystem (Pinay et al., 1990; Sparks et al., 1990; Bayley, 1991, 1995; Brunet et al., 1994; Sparks, 1995; Special issue of Reg. Rivers 11(1), 1995). In addition to the Flood Pulse Concept, functions of river-floodplain interaction, reported in the literature above, are:

- reduction of flood effects and the buffering of short-term water-level fluctuations, because flood peaks are absorbed in the floodplain (Bayley, 1991; Pinay et al. 1990);
- stimulation of the flood pulse advantage of fish yield, which is the amount by which fish yield per unit mean water area is increased by a natural predictable flood pulse compared with a system with a constant water level (Bayley, 1991);
- the natural filtering of nutrients and suspended matter by riparian forests against diffuse pollution from the floodplain (Pinay et al., 1990; Brunet et al., 1994).

Sedell et al. (1989) connected the RCC with the flood pulse concept by comparing a constrained river with a river-floodplain system. They concluded that the productivity of the floodplain vegetation greatly modified the longitudinal pattern of ecosystem processes predicted by the RCC.

Ward and Stanford (1995) extended their serial discontinuity concept for flood plain rivers. The increased interaction of the channel with the riparian zone in a floodplain river affects the river ecosystem in the lower reaches of the river considerably as described in the flood pulse concept. Consequently, damming and river regulation in a floodplain river has a marked effect on the channel stability, ecological connectivity between channel and riparian zone, thermal heterogeneity and species diversity.

2.7. Riverine productivity model

Thorp and Delong (1994) state that previous concepts rely too much on data from low order streams, floodplain rivers and collector-dominated river habitats; they have introduced an alternative hypothesis, the riverine productivity model (RPM). This concept states that carbon in constrained large rivers does not originate solely from downstream transport (as stressed by the RCC) but also from local autochthonous production and inputs from the riparian zone. The community composition and secondary production will differ among sites within a large river in response to both the physical characteristics of each habitat and the types of organic matter present. In general, high invertebrate densities are found in riparian zones due to their large habitat diversity and their role in retaining organic matter. The different sources of organic matter affect the composition of riverine food webs. Whereas the RCC predicts a dominance of collectors (filterers and gatherers) in the macroinvertebrate community which use FPOM transported from upstream sources, the RPM states that collectors are co-dominant with grazers (Thorp, 1992). Furthermore, phytoplankton productivity measurements in large rivers support the riverine productivity model that instream primary production is an important energy source in the downstream part of a the river (Reynolds, 1988, 1994; De Ruyter van Steveninck et al., 1990, 1992).

2.8. Catchment concepts

Finally, a number of authors have argued for a catchment-oriented approach. Frissell et al. (1986) proposed a hierarchical framework for stream habitat classification, which emphasises a stream's relationship to its watershed across a wide range of scales in space and time, from the entire channel network to pools, riffles, and microhabitats. Gardiner (1991) developed a manual for an holistic appraisal of river works on a catchment scale. Naiman et al. (1992) reviewed a number of specific classification systems for streams and showed consensus with regard to the ability to encompass broad spatial and temporal scales, to integrate functional and structural characteristics under various disturbance regimes and to convey information about the underlying mechanisms. Petts (1994) condensed the state-of-the-art research on the functioning of river systems (Calow and Petts, 1992) into five principles. Rivers are:

1. three dimensional systems;
2. driven by hydrology and fluvial geomorphology;
3. structured by food webs;
4. characterised by spiralling processes;
5. dependent upon change – changing flows, moving sediments and shifting channels.

Townsend (1996) argues in his catchment hierarchy approach for an integration of existing river concepts with the patch dynamics concept (White and Pickett, 1985; Pringle et al., 1988). An hierarchical framework of both river and patch dynamics concepts on the scale of the river catchment enables the prediction of spatial and temporal patterns of ecological variables in the river basin. For example, the dominant source of organic matter, such as transport from upstream, lateral input or instream production is predicted in different parts of the river basin. The temporal dimension is important in a dynamic environment, such as a river in which disturbance affects ecosystem structure and functioning (e.g. variable discharges, flood pulse).

3. From river concepts to indicators

The river concepts describe a natural undisturbed river system. Assessment of the impact of human activities on river ecosystems requires indicators relating cause to effect. Therefore, cause–effect chains have to be distinguished whereby human disturbance changes abiotic steering variables, which in turn affect the biotic structural and functional characteristics of the river ecosystem. An overview of human use and impacts on rivers is given in Figure 1. Below the dominant processes and structures will be defined by integrating the information from the review of river concepts. The effect of human changes to the river ecosystem will be taken into account.

A natural river ecosystem is steered by the abiotic environment. The boundary conditions are formed by the climate of the region (e.g. precipitation patterns, temperature ranges) and the geology of the basin (e.g. soil characteristics, fall between the up- and downstream part of the basin), which determine the hydrology (e.g. discharge, stream velocity), geomorphology (e.g. sedimentation and erosion, channel form and substrate) and water quality of the river (e.g. temperature, pH, oxygen, substance concentration). From up- to downstream an abiotic gradient occurs of increasing discharge, increasing channel size and decreasing substrate size. Temporal variation occurs due to daily and seasonal cycles and inter-annual variation of climate.

Most human impacts, except from fisheries or introduction of exotic species, go via the change of the abiotic environment (see also Figure 2.). Emissions have affected water quality (e.g. increase in temperature and substance concentrations, decrease of oxygen). The building of dams and river regulation affected predominantly the hydrological and geomorphological variation and dynamics and reduced the habitat area in channel and terrestrial surroundings.

The productivity in a natural river is determined by the fluxes of matter between the river and its terrestrial surroundings and the resulting resource cycling processes in the river. Nutrients and organic matter enter the river via direct input from the riparian zones and floodplains or via erosion and run-off from the basin. In the river, biomass can be produced via primary production of algae and macrophytes or degraded via respiration. Which biological process will dominate, primary production or respiration, depends on the nutrient and organic matter concentration and other abiotic steering parameters of these processes, such as temperature, turbidity, flow velocity. The resultant of these processes will be a certain algal concentration and water quality (e.g. oxygen content, turbidity, dissolved organic matter, mineral nutrients). Nutrient

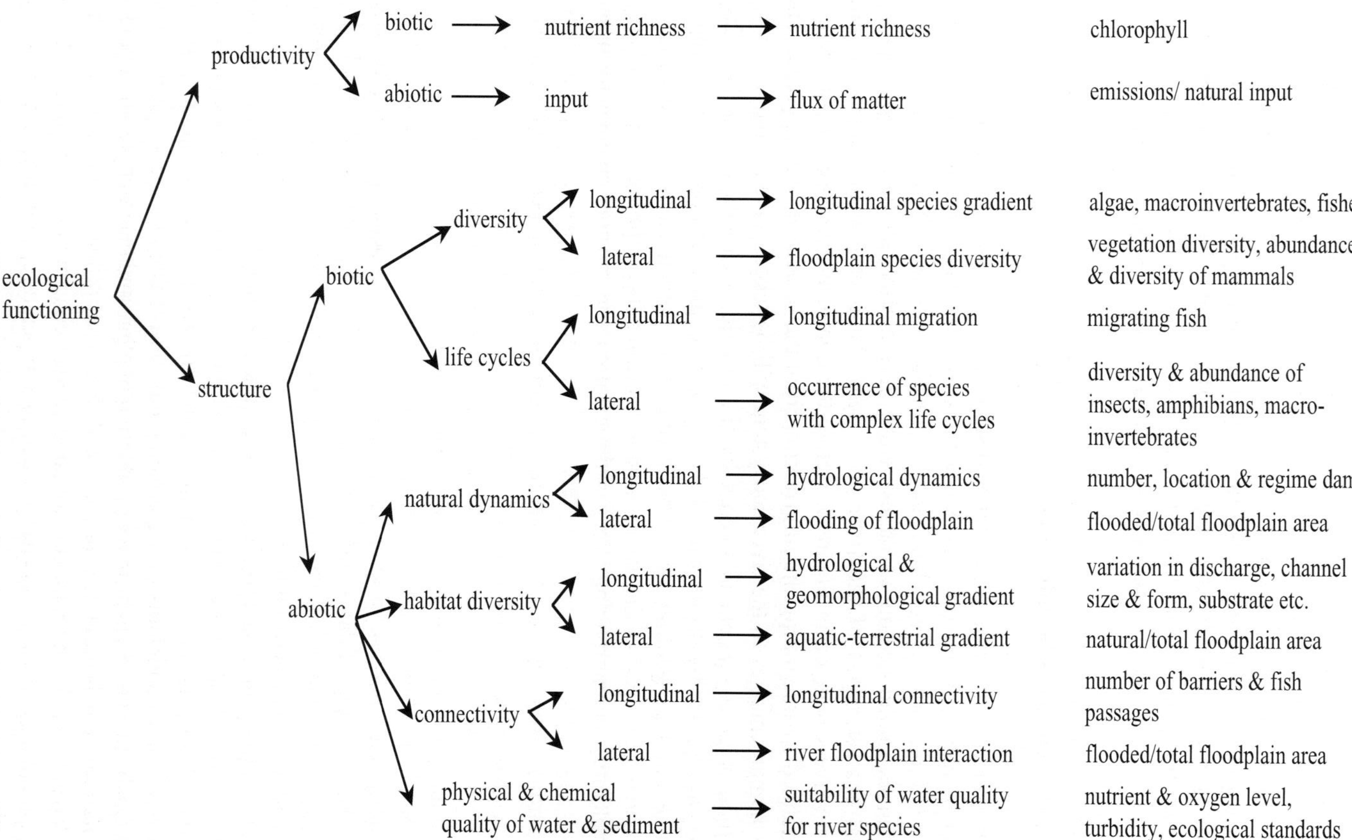

Figure 2. Overview of elements of environmental quality and the proposed indicators and variables for the river Rhine.

levels, turbidity and the retention time of the water are important determinants for the growth of primary producers, such as river plankton and macrophytes growing in the channel and on the floodplains. Oxygen and temperature are the crucial water quality parameter for the survival of river fauna, such as macroinvertebrates and fish. Nutrient emissions can lead to a higher primary production of algae. The embankment of floodplains and riparian zones has reduced the input of coarse particles of organic matter (e.g. CPOM). Dams will block the transport of large CPOM, whereas smaller particles (e.g. FPOM) can pass the dam more easily. This decouples the linkage between allochtonous inputs upstream and processing of organic matter downsteam (Ward and Stanford, 1983b, 1995). Due to dams, lentic parts occur in a normally lotic environment, leading to eutrophication, increased sedimentation and changes in water quality (e.g. turbidity, oxygen, nutrients, suspended matter).

The structural characteristics of the riverecosystem is described by the diversity, abundance and spatial distribution of species. The biological diversity in a natural river is high compared to other ecosystems. This is because of the high habitat diversity due to abiotic longitudinal, lateral and vertical gradients (Junk et al., 1989; Ward, 1989, 1998). Furthermore, the intermediate level of disturbance in rivers cause a maximum species richness (Ward and Stanford, 1983b). This is explained by the fact that disturbance creates niches for species that could not persist in the absence of disturbance due to competition. A too-high disturbance will however lead to the elimination of resident species and the dominance of colonising species. At the intermediate level the species richness is the highest, because (1) no competitive exclusion occurs, as the recurrence interval of disturbance events is shorter than the time necessary for competitive or predator-prey interaction leading to the elimination of species; (2) the river is populated by both resident species and colonizing species, exploiting the disturbed areas.

The gradients in the river are longitudinal from upstream headwaters to downstream large channels and lateral from the aquatic river channel to the terrestrial floodplain. The longitudinal gradient of stream velocity, discharge, turbidity, temperature and size of organic matter leads to shifts in spatial distribution of species, such as algae (e.g. diatoms, green and cyanobacteria) (Whitton, 1980), macroinvertebrate functional feeding groups (e.g. shredders, grazers, filter feeders) (Vannote et al., 1980) and fish (e.g. trout zone (*Salmo trutta*), grayling zone (*Thymallus thymallus*), barbel zone (*Barbus barbus*) and bream zone (*Abramis brama*) (Huet, 1954; Illies and Botosaneanu, 1963)).

In the lateral dimension the gradient from river to terrestrial surrounding (e.g. riparian zone, floodplain) is important because of its habitat diversity and area. Natural floodplains with a regular flood pulse have a high species diversity due to their diverse and highly dynamic habitat structure (Junk et al., 1989). The land-water gradient of moisture, oxygen content, groundwater quality, nutrient richness and dynamics leads to a gradient in vegetation communities with different species diversities and biomass production (Wassen, 1990).

Furthermore, the longitudinal and lateral dimensions in rivers contribute also to the completion of life-cycles of species. Floodplains are used for spawning, feeding and resting. Rivers have an ecological migration function within the river or between different rivers. Migration can be part of the life-cycle of a species or triggered by a

                                    C.M. Lorenz

changing environment. For migration the connectivity within the river channel (for spawning of migrating fish) is important, as well as connections between the channel and terrestrial surroundings (for spawning, feeding and resting) and connections between terrestrial surroundings (stepping stone mechanism for recolonization after a disturbance, or feeding and resting place during migration).

The building of dams has disrupted the longitudinal continuum and caused shifts in abiotic and biotic processes and characteristics (Ward and Stanford, 1983a) and has reduced the connectivity for species between up- and downstream. Migrating fish, such as salmon and trout, became extinct, as their life and reproduction cycle has been disturbed by barriers in the channel (Lelek, 1989). The installation of fish passages aims to increase the migration possibilities for anadromous fishes. However, the effectiveness of these facilities is still unclear and is presently under investigation (Mueller et al., 1994; IKSR, 1996). The embankment of floodplains has reduced the connectivity between channel and floodplains, reduced diversity of habitats and affected flooding, erosion and sedimentation processes. Channelization has reduced the diversity in channel sizes and forms, and affected erosion and sedimentation in the channel. A reduction of habitat diversity and dynamics (reducing the level of disturbance) will lead to lower species diversity. A disruption of life cycles will lead to the extinction of species depending on these life cycles.

The dominant processes and structures are the basis for potential indicators. In the following section a number of bioindicators and abiotic indicators are selected for river ecosystems.

4. Indicators for river ecosystems

In this section bioindicators for productivity and structure and abiotic indicators will be presented (see for an overview Figure 2). The indicators are derived from the summary of river functioning in Section 3.

4.1. Bioindicator for ecosystem productivity

Productivity is determined by the resource cycling processes, such as the input, processing and retention of organic matter. I propose the indicator algal biomass to describe productivity. The algal biomass is the result of the input and processing of nutrients and is indicated by the chlorophyll concentration. On the basis of the annual mean and maximum chlorophyll concentrations, nutrient richness is classified in trophic categories of ultra-oligo-, oligo-, meso-, eu-, and hypertrophy (OECD, 1982).

4.2. Bioindicator for ecosystem structure

Structure describes species diversity, abundance and spatial distribution of species in a river ecosystem. Since human influence affects the ecosystem primarily through changes of the abiotic environment, management actions have to focus on the rehabilitation of the abiotic environment as a prerequisite for better ecological functioning. Structure is described by biotic indicator species and abiotic environment indicators for the longitudinal or lateral dimension (except water and sediment quality).

The species diversity from up to downstream is expressed by the longitudinal species gradient, describing the changes in abundance and composition of algae, macroinvertebrate functional feeding groups and fish from up- to downstream. In the lateral direction the floodplain species diversity is based on: (1) the area and diversity of vegetation. A high vegetation diversity is considered a precondition for a high diversity of insects and amphibians in the floodplain; (2) mammals, as they need relatively large areas of natural floodplains for survival. Furthermore, they are predators and indicate that the food web functions sufficiently to provide food for the predators.

The connectivity within the river and the interaction between the channel and terrestrial surroundings is important for life cycles of typical river species. An indicator for the connectivity in the river channel is the longitudinal migration, measured by the abundance of migrating fish species, such as salmon and trout.

An indicator for river floodplain interaction are the occurrence of terrestrial species with aquatic life stages. They require different habitats during different stages of development, including a transition from wet to dry environment (such as insects and amphibians).

4.3. Abiotic indicators as precondition for bioindicators

The potential for productivity and species diversity, abundance and health is largely determined by the abiotic preconditions, namely hydrological and morphological dynamics, habitat diversity and connectivity, the nutrient input and the physical and chemical quality of water and sediment. Information on these abiotic preconditions is important for ecosystem management, as it provides starting points for the definition of rehabilitation measures for ecosystem improvement.

The flux of matter indicator describes the ratio between the human caused emissions of N and P and the natural input of N and P. The natural flux of nutrients comes from the erosion of rocks and soil and from vegetation (e.g. forests, wetlands) in the basin (Scholte Ubing, 1980; Meybeck and Helmer, 1989). The flux of matter is calculated both for nitrogen and phosphate.

$$flux\ of\ matter = \frac{human\ caused\ emissions}{natural\ input} \tag{1}$$

Emissions into the river and their biochemical processing will result in a certain physical and chemical quality of the water and sediment. The water quality can be assessed by using the models of the Habitat Evaluation Procedure (HEP), in which water quality requirements of river species are defined (Duel et al., 1994) or ecological standards based on the survival of 95% of the species of an ecosystem (Ministrie of VROM, 1994; Van Straalen and Denneman, 1989) or the difference with the natural background situation (CUWVO, 1988). Nutrient concentrations and turbidity will determine the growth of primary producers, such as macrophytes and algae. For higher trophic levels (macroinvertebrates, fish) oxygen concentrations and toxic compounds are crucial water quality parameters.

In the longitudinal dimension dams have changed hydrological dynamics. The longitudinal continuum in hydrology, geomorphology and water quality is disrupted, which lead to shifts in abiotic and biotic processes and structures. Ward and Stanford (1983) have defined two parameters to measure the effect of a dam; the discontinuity distance and the intensity (see serial discontinuity concept in Section 4.2). To determine these parameters data are necessary of abiotic and biotic variables before and after the building of the dam. These data are most of the times not available. Therefore, the characteristics of the dam, that affect the shift in continuum will be expressed in the indicator, namely the number and position of the dams and the discharge regime, determining the frequency and volume of water released by the dam.

$$longitudinal\ dynamics =$$

$$river\ reach\ between\ two\ dams\ *\ \frac{frequency\ of\ water\ release}{volume\ of\ water\ released} \tag{2}$$

In the lateral direction the dynamics are indicated by the flooded floodplain ratio, measured by the ratio between the regularly flooded and total floodplain area per river(reach).

$$flooded\ floodplain\ ratio = \frac{flooded\ floodplain\ area}{total\ floodplain\ area} \tag{3}$$

Habitat diversity in rivers is dependent on gradients. In the longitudinal direction, gradients in hydrology (e.g. stream order, discharge, stream velocity) and geomorphology (e.g. channel form, sediment, riparian bank type) determine the habitat diversity over the river length. The hydrological and geomorphological gradient can be described by the change in hydrological and geomorphological characteristics from up- to downstream.

$$longitudinal\ gradient = \frac{dy}{dx} \tag{4}$$

where

$x =$ river length
$y =$ hydrological or geomorphological characteristic:
- discharge
- stream velocity
- stream order
- channel size
- channel depth
- substrate particle size
- number of meanders.

In the lateral direction the gradient of land to water determines the diversity in moisture, nutrient and mineral richness, temperature, light conditions, sediment or soil surface. This gradient will be the most diverse for natural floodplains, as natural processes of flooding, sedimentation, erosion and succession can occur, leading to a mosaic of water, marshes, reed lands, floodplain forests. The lateral habitat diversity is indicated by the naturalness of floodplains, and measured by the ratio of the area of natural and semi-natural floodplains divided by the total area.

$$natural\ floodplain\ ratio = \frac{area\ of\ natural\ floodplain}{total\ floodplain\ area} \tag{5}$$

Longitudinal connectivity is measured by the ratio between the length of the river or river reach divided by the number of barriers and their effectiveness as fish passages. Longitudinal connectivity decreases with more dams and increases with a higher effectiveness of fish passages.

$$longitudinal\ connectivity =$$

$$\frac{length\ (river\ reach)}{(1+dams\ without\ fish\ pass.)+(1-effectivity\ fish\ pass.)*dams\ with\ fish\ pass.} \tag{6}$$

Lateral connectivity is indicated by the interaction of the river with the terrestrial surroundings. This depends on the area of flooded floodplains, therefore the flooded floodplain ratio (indicator for lateral dynamics) is also used as a proxy for the lateral connectivity.

5. Eutrophication in shallow lakes

This section summarises the ecological theory of eutrophication of shallow lakes. The predominant effect of increased nutrient loads into lakes is eutrophication, which involves a cascade of direct and indirect effects (Klinge et al., 1995; Hosper, 1997; Scheffer, 1998). This cascade of effects can lead to one of the two equilibrium states of shallow lakes: a turbid, phytoplankton dominated lake in a meso- or eutrophic state. The other equilibrium state is a clear and macrophyte dominated lake, which is in a meso- or oligotrophic state. The balance between these two states depends on two forces (Klinge et al., 1995):

- A bottom up force determined by producers, such as algae, zooplankton and plankti- and benthivorous fish. The potential biomass at different trophic levels, including the production of preyfish, is determined by the nutrient richness of the system.
- A top down force determined by consumers, such as predatory fish. The actual biomass and community structure are considered to depend mainly on consumer-control.

In shallow lakes (mean depth 1–4 m) high nutrient levels can lead to increased primary production of phytoplankton. The higher algal biomass increases the turbidity of the water and increases oxygen demand of the sediment due to decomposing algae. The turbidity can lead to a decline of submerged macrophytes. Macrophytes have clearing effects on the water, because they (1) provide refuge to grazing zooplankton; (2) protect the sediment from resuspension and (3) provide a habitat and hiding place for predatory fish, feeding on plankti- and benthivore fish. Turbid lakes devoid of vegetation have few predatory fish and have a high abundance of plankti- and benthivorous fish. Abundant planktivore fish control the zooplankton, resulting in low grazing of algae. Benthivore fish cause resuspension of sediment, increasing both turbidity and phosphate release from the sediment.

The biomass of predatory fish depends heavily on the morphometric conditions of the lake. These morphometric conditions determine the availability of suitable habitat for the predatory fish, such as submerged vegetation and emergent plants. The area of suitable habitat determines the maximum biomass of predatory fish and thereby the top down force of preyfish consumption. This area of suitable habitat has declined in the Netherlands due to a number of causes. Next to eutrophication, leading to turbidity and the disappearance of macrophytes, the application of artificial (high in summer and low in winter) or stable waterlevels, embankment of natural banks and floodplains have led to a reduction of submerged and emergent vegetation.

The transition between these states is characterised by a so-called hysteresis effect (Fig. 3). The hysteresis effect implies that there are different threshold nutrient loadings for transitions between the two equilibrium states, namely the critical nutrient loading for transition is lower, if the lake is in an eutrophic state than in a meso or oligotrophic state. This is illustrated by L1 and L2 in Figure 3. Hysteresis is caused by a number of physico-chemical (e.g. nutrient release from sediment, wind-induced resuspension of sediment in plant free lake) and ecological processes (e.g. increase of turbidity and nutrient release due to sediment perturbation by benthivore fish, dominance of non-consumable blue green algae at high turbidity) (Scheffer, 1998; Hosper, 1997). These processes cause a certain resistance to change with increasing and decreasing nutrient loading untill the loading is high or low enough to cause a switch to another equilibrium state (Hosper, 1997). The difference in threshold nutrient loading between oligotrophication and eutrophication depends on lake characteristics and the application of restoration measures. Lake characteristics are the depth of the lake, its retention time and sediment characteristics. Restoration measures can change the lake characteristics (e.g. flushing of lakes, lowering water level) or suppress the hysteresis effect (e.g. removing white fish and introducing predatory fish, dredging sediment).

6. Indicators for eutrophication in shallow lakes

Translating these eutrophication processes into the components of ecological functioning, productivity is described by the primary production of algae and the resulting nutrient richness of the lake. Ecosystem structure is described by changes within and between trophic levels of the food web, namely shifts between phytoplankton, macrophytes (both primary producers), zooplankton, zoobenthos (both primary consumers), planktivorous fish, benthivorous fish (secondary consumers) and predator or piscivorous fish (tertiairy consumers). Next to shifts among trophic groups, the total species diversity in turbid lakes decreases, as algae and a few fish species become dominating.

The effect of eutrophication on the environmental quality of a shallow Dutch lake will be indicated by the following indicators (see Fig. 4).

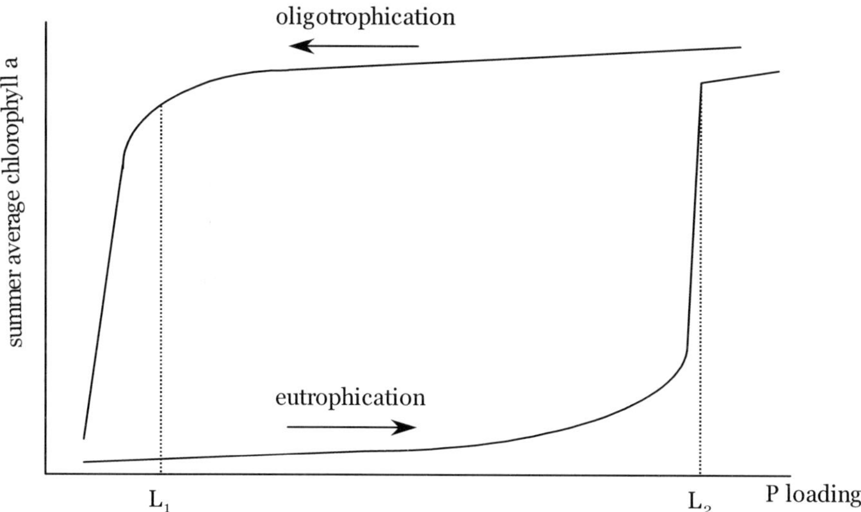

Figure 3. A graphic representation of the hysteresis phenomenon in a lake, in which the eutrophication process has a different relationship of chlorophyll versus phosphate loading compared to the oligotrophication process (Figure adapted from Hosper, 1997 and Scheffer, 1998).

6.1. Productivity

The productivity of a lake is described by the algal biomass. The nutrient richness is determined on the basis of annual mean and maximum chlorophyll concentrations. The algal biomass increases the turbidity of the lake, which next to chlorophyll indicates the productivity as well.

6.2. Structure

Eutrophication in a lake leads to food web changes. A clear lake has a higher species diversity than a turbid lake (Scheffer, 1998). When a clear lake turns into a turbid one, the invertebrates associated with the vegetation disappear and the birds and fishes that feed on the invertebrates or plants. Large zooplankton, using the vegetation as a refuge against predation, will strongly reduce. Furthermore, predatory fish will decline and preyfish will dominate. Indicators for structure are:

- The area and biomass of macrophytes, indicating that turbidity has reduced to the extent that macrophytes can grow. As stated above, macrophytes stabilize the clear equilibrium through a number of mechanisms.
- The ratio between preyfish and predatory fish, which has to be 1:1 to 2:1 to guarantee a long-term stable clear lake. The increased productivity leads to an increase of preyfish biomass and to a reduction of suitable predatory habitat, as turbidity increases, macrophytes disappear and the size of the preyfish is too large to eat for the predatory fish.

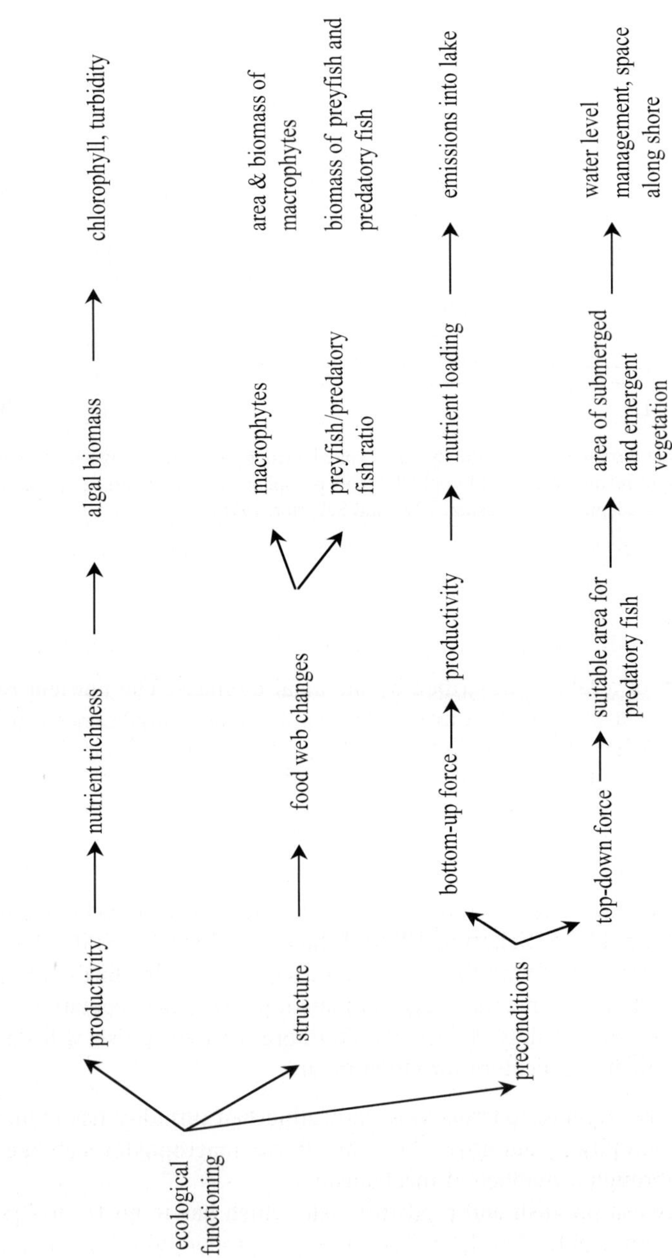

Figure 4. Overview of elements of environmental quality and the proposed indicators and underlying variables for the Dutch shallow lake

6.3. Indicators as precondition for the productivity and structure bioindicators

Indicators describing the abiotic preconditions for the process of eutrophication are the factors behind the balance between a clear and turbid lake:

- The nutrient loading determining the bottom-up force, namely the productivity of the lake.
- The area of suitable habitat of submerged and emergent vegetation for predatory fish, determining the top-down force. The precondition for emergent vegetation are natural water levels and space to grow along the shores. This space can be realized by creating extra shore length with sloping banks in small lakes or by creating (managed) marshes in larger lakes (Klinge et al., 1995)

7. Assessment and aggregation of indicator values

To come from bioindicator values to ecosystem condition the values of the different bioindicators have to be assessed and aggregated into one value. This section discusses possible reference levels for assessment and the aggregation procedure for ecological indicators.

7.1 Assessment

Value functions are used to assess indicator values. Value functions provide an explicit link between factual information (e.g. the indicator value) and human value judgement ("good" or "bad") in the form of a mathematical representation (Beinat, 1995). Reference conditions are useful to relate the indicator value to a value judgement. Reference conditions are the anchors for interpreting the meaning of a value score. For example, for an assessment of the water quality of a river the oxygen concentration as indicator for organic pollution can be valued one for the natural background value and zero for the worst case representing the situation of heaviest human impact measured (e.g. oxygen level in the 1970s) and 0.5 for the present situation. The form of the value function depends on the change in value judgement with the increasing (or decreasing) indicator value. Common forms are block functions, linear, sigmoid, convex or concave curves.

Value functions are often used in environmental management, without explicitly being called like that. In the Dutch AMOEBE indicator, the reference condition is the abundance of a set of indicator species around 1900. The relationship between the present and reference condition is considered linear (Ten Brink et al., 1991). Another example are the environmental standards in Dutch policy (Ministrie of VROM, 1994; Van Straalen and Denneman, 1989). They are based on the relationship between the concentration of a toxic compound and the No Observed Effect Concentration of a number of species of different trophic levels in the food web. The curve has a sigmoid form. The maximum acceptable concentration is set at the level of a protection of 95% of the species. This value function shows a mix of facts (e.g. NOEC of a compound for a species) and value judgements, namely the representation of the ecosystem by a number of species from different trophic levels (which is expert judgement) and the setting of the maximum acceptable concentration when 95% of the species are protected (which is a political decision).

Reference levels can be absolute or relative and lead to an absolute or relative judgement. For example, functional standards enable an absolute judgement on the suitability of a resource for a certain good or supply. No Observed Effect Concentrations are a threshold for absolute ecological effects to species in an ecosystem. Relative references enable only a relative valuation, relative in time, in space or relative to policy aims. The selection for an absolute or relative reference depends on the aim of the assessment and on the availability of data on reference levels. Often relative references are used, when no absolute references are available.

A number of references for ecosystem assessment are presently used. The first type of references are based on an 'undisturbed' river or lake having authentic hydrological, geomorphological and ecological characteristics comparable to the ecosystem, which is to be assessed. Finding a similar, but undisturbed large transboundary river as reference is difficult, as all large rivers in Europe and North America have been impacted. Only for smaller rivers can an undisturbed reference river be found (Boon, 1992; Wassen, 1990; Hooijer, 1996). With regard to lakes, small shallow undisturbed lakes can be found in Great Britain and the Scandinavian countries. However, these lakes have not the same history as Dutch lakes. Dutch lakes are man-made and are created due to peat extraction of marshes.

The second type of references relate to a historical analysis of river or lake characteristics in a pre-"disturbance" phase (natural background water quality, species occurrence, hydro-geomorphological characteristics) (Ten Brink, 1991). A historical reference has the disadvantage that it may turn out to be an unreachable goal, because many human effects are irreversible. The reference value in the AMOEBE pre-dates major changes and so can be equated to a relatively undisturbed state (Ten Brink, 1991; Ten Brink et al., 1991). However, the exotic species that have invaded the Rhine are not included in the AMOEBE of the Rhine, although they dominate the macroinvertebrate population (Van den Brink et al., 1991; Bij de Vaate, 1993; Bij de Vaate and Greijdanus-Klaas, 1995; Rajagopal et al., 1998). Therefore, the outcome of the AMOEBE can be questioned. Another example is the natural background water quality, showing the full extent of human impact. However, a certain level of human emissions will be inevitable and it might be more relevant to know the possible effects of increased concentrations.

The third type are effect reference levels based on the risk of ecological impact. Examples are ecotoxicological parameters (EC50, LC50, NOEC) as reference for ecological effects of toxic substances (Van Meent et al., 1990) or the threshold nutrient level in shallow lakes that mark the transition between clear and turbid states (Hosper, 1997; Scheffer, 1998, see Section 5). In the Netherlands, environmental standards for pollutants are based on a protection of 95% of the species in aquatic or terrestrial ecosystems (maximum acceptable risk level). This value is divided by an application factor to arrive at negligible risk level (Ministrie of VROM, 1994; Van Straalen and Denneman, 1989).

The recognition of the irreversibility of human impact increasingly attracts attention to the return of ecosystem processes as the starting point for ecological assessment, such as flood frequency, sedimentation patterns and succession (Nienhuis and Leuven,

1998; Pedroli et al., 1996; World Wildlife Fund, 1996) Therefore, theoretical knowledge on ecosystem functioning is needed, as described in Sections 2 and 5.

Finally, references can be based on policy goals or a reference year for policy evaluation. For example, the Rhine Action Plan has set the year 1985 as a reference year to assess the effectiveness of measures. Furthermore, the Rhine Action Plan had as policy goal the return of the salmon in the Rhine in the year 2000 (IRC, 1987).

Which of the abovementioned reference levels is selected for the indicators depends on the goal of the assessment (effctiveness of policy, assessment of ecosystem condition) and the availability of (data on) the reference levels.

7.2. Aggregation

Aggregation of a number of variables into one value for ecological functioning implies steps of selection, weighting (valuation), scaling (transforming indicators into dimensionless measures), and mathematical manipulation. If the aggregation method of weighted averages and weighted summation (see formula below) is used than no double counting and no interdependencies between the indicators may occur (Beinat, 1995). So, the final set of indicators has to be analysed on these two characteristics before aggregation. This is not an easy task, as a fundamental characteristic of ecosystems is that they are strongly interdependent. As there exist not one overall variable or indicator to describe and assess ecological functioning, different ecosystem elements (e.g. productivity, structure and resilience) have to be described. They are interdependent, as processes produce and affect species and species processes. If the indicators are interdependent, only the indicators that relate the most directly to the aim of the assessment should be aggregated. In our case the bioindicators relate more directly to ecosystem condition than the abiotic indicators.

8. Discussion

The chapter has developed bioindicators for a transboundary and impacted river, such as the Rhine and an eutrophicated shallow lake, such as the Loosdrechtse plassen in the Netherlands. The selection of bioindicators is based on organisms that provide information on the dominant processes and structures in the ecosystem, such as migrating fish on the connectivity of the river or algal biomass on the productivity. This chapter proposes organism groups, such migrating fish or predators as bioindicators instead of specific species. The application of organism groups as bioindicators is broader, as the occurrence of species can differ per river, whereas the dominant processes and the organism group indicating the process are the same for Western European rivers. On the basis of the proposed organism groups suitable indicator species can be selected for distinct rivers, such as the Rhine, Elbe or Meuse. There exist a number of advantages and disadvantages with regard to the use of bioindicators for ecosystem management. The advantages are:

- Bioindicators provide aggregated information on ecosystem functioning, as they provide information on structural (the occurrence and abundance of species) and functional aspects (species indicating ecological processes).

- Species are more appealing for policymakers and the public than the occurrence of ecological processes. For rehabilitation projects the political acceptance of investments with public means (e.g. building of fish traps) and measures applied to sectors (e.g. emission reduction by industry and agriculture) by the public is important. Examples of rehabilitation project in which a target species is chosen to increase the acceptance are numerous, for example the Salmon is symbol for a cleaner Rhine and the beaver for floodplain restoration projects in the Netherlands.

The disadvantages are:

- Measurement of the abundance of bioindicators can be difficult and time consuming than the monitoring of abiotic indicators. This disadvantage applies especially to mobile species as predatory fish, mammals, macroinvertebrates and insects
- Risk that management is focusing on the conservation of a small number of species instead of the functioning of the ecosystem. To avoid this problem the indicators presented in this chapter focus on the selection of organism groups (migrating fish, mammals, predatory fish) instead of distinct species. The diversity and abundance of these organism groups provides information on the processes in an ecosystem.

One of the major aims of bioindicators is to indicate the condition of the ecosystem. However, a quantitative assessment of ecosystem functioning on the basis of bioindicators can be complicated by the absence of clear references. For the shallow lake the objective to be achieved is clear; a clear lake with macrophytes and an equilibrium between predatory and preyfish. For the river the definition of a reference is more difficult due to numerous impacts, the complexity of processes, the disappearance of unimpacted reference rivers and a lack of knowledge. In Section 6 a number of reference levels are proposed. However, there is a need of quantitative information on reference levels, such as historical data, natural background values, data of a geographical reference ecosystem.

Next to the measurement of bioindicators, it is also important to monitor abiotic indicators. Times series on a combination of bioindicators and abiotic indicators will increase the knowledge on cause-effect relationships and on the effectiveness of rehabilitation measures.

The presented indicators for rivers and lakes are also relevant for the Europan Water Framework Directive (Commission of the European Communities, 2000). According to this directive the member states have to prepare river basin plans every six years including: (1) the results of the monitoring and ecological assessment of the water bodies in a river basin including rivers and lakes; (2) the measures to be taken to reach a "good ecological status" within 15 years. This means that river basin managers need information on the ecological functioning, the causes of ecological effects and the effectiveness of measures of the waterbodies in their catchment in order to comply with the directive. The monitoring of bioindicators in combination of the abiotic indicators can increase the knowledge on cause-effect relationships and on the effectiveness of rehabilitation measures.

References

Amoros, C., Roux, A.L., Reygrobellet, J.L., Bravard, J.P., Pautou, G., 1987. A method for applied ecological studies of fluvial hydrosystems. Reg. Rivers: Res. Mgmt. 1, 17–36.

Bayley, P.B., 1991. The flood-pulse advantage and the restoration of river-floodplain systems. Reg. Rivers: Res. Mgmt. 6, 75–86.

Bayley, P.B., 1995. Understanding large river-floodplain ecosystems. Bioscience 45, 153–158.

Beinat, E., 1995. Multiattribute value functions for environmental management. Ph.D. Thesis. Vrije Universiteit Amsterdam, December.

Bij de Vaate, A., 1993. Exotic aquatic macroinvertebrates in the Dutch part of the River Rhine: causes and effects. In: van Dijk G.M., Marteijn, E.C.L. (Eds), Ecological Rehabilitation of the River Rhine, The Netherlands Research Summary Report (1988–1992). Report of the project "Ecological Rehabilitation of the rivers Rhine and Meuse", Report no. 50, pp. 27–30.

Bij de Vaate, A., Greijdanus-Klaas,M., 1995. Macrofauna. In: Noordhuis, R. (Ed.), Biologische monitoring zoete rijkswateren: jaarrapportage 1993. RIZA nota nr. 95.002. Rijksinstituut voor Zoetwater en Afvalwaterzuivering, Lelystad, Nederland, pp. 24–27.

Boon, P.J., 1992. Essential elements in the case for river conservation. In: Boon, P.J. Calow, P., Petts, G.E. (Eds), River Conservation and Management. John Wiley, Chichester, pp. 11–34.

Bravard, J.P. Amoros, C., Pautou, G., 1986. Impact of civil engineering works on the successions of communities in a fluvial system. A methodological and predictive approach applied to the section of the Upper Rhône River, France. Oikos 47, 92–111.

Brunet, R.C., Pinay, G., Gazelle, F., Roques, L., 1994. Role of floodplain and riparian zone in suspended matter and nitrogen retention in the Adour River, south-west France. Reg. Rivers: Res. Mgmt. 9, 55–63.

Bruns, D.A., Minshall, G.W., Cushing, C.E., Cummins, K.W. Brock, J.T., Vannote, R.L., 1984. Tributaries as modifiers of the river continuum concept: analysis by polar ordination and regression models. Arch. Hydrobiol. 99, 208–220.

Calow, P., Petts, G.E. (Eds), 1992. The River Handbook. Hydrological and Ecological Principles, Vol. 1. Blackwell Scientific, Oxford, p. 526.

Chorley, R.J., Kennedy, B.A., 1971. Physical Geography. A Systems Approach, Prentice-Hall, London.

Commission of the European Communities, 2000. Council directive establishing a framework for community action in the field of water policy. European Commission publication of 20 July 2000.

Conners, M.E., Naiman, R.J. 1984. Particulate allochtonous inputs: relationships with stream size in an undisturbed watershed. Can. J. Fish. Aquat. Sci. 41, 1473–1484.

Cummins, K.W., Cushing, C.E., Minshall, G.W. 1995. Introduction: an overview of stream ecosystems. In: Cushing, C.E., Cummins, K.W., Minshall, G.W. (Eds), River and Stream Ecosystems. Ecosystems of the World 22. Elsevier Science, Amsterdam, pp. 1–8.

Cummins, K.W., Minshall, G.W., Sedell, J.R., Cushing, C.E., Petersen, R.C., 1984. Stream ecology theory. Verh. Int. Verein. Limnol. 22, 1818–1827.

Cushing, C.E., McIntire, C.D., Cummins, K.W., Minshall, G.W., Petersen, R.C., Sedell, J.R., Vannote, R.L., 1983. Relationships among chemical, physical and biological indices along river continua based on multivariate analyses. Arch. Hydrobiol. 98, 317–326.

CUWVO, 1988. Ecologische normdoelstellingen voor Nederlandse oppervlaktewateren. Coordinatiecommissie Uitvoering Wet Verontreiniging Oppervlaktewateren. The Hague, The Netherlands.

De Ruyter van Steveninck, E.D., Admiraal, W., Breebaart, L., Tubbing, G.M.J., van Zanten, B., 1992. Plankton in the River Rhine: structural and functional changes observed during downstream transport. J. Plankton Res. 14, 1351–1368.

De Ruyter van Steveninck, E.D., van Zanten, B., Admiraal, W., 1990. Phases in the development of riverine plankton: examples from the rivers Rhine and Meuse. Hydrobiol. Bull. 24, 47–55.

Duel, H., B.P.M. Specken, W.D. Denneman, Kwakernaak, C., 1994. The habitat evaluation procedure as a tool for ecological rehabilitation of wetlands in the Netherlands. Water Science Technol. 31, 387–391.

Elwood, J.W., Newbold, J.D. O'Neill, R.V., Van Winkle, W., 1983. Resource spiralling: an operational paradigm for analyzing lotic ecosystems. In: Fontaine, T.D., Bartell, S.M. (Eds), The Dynamics of Lotic Ecosystems. Ann Arbor Science, Ann Arbor, pp. 3–27.

Frissell, C.A., Liss, W.J., Warren, C.E., Hurley, M.D., 1986. A hierarchical framework for stream habitat classification: viewing streams in a watershed context. Environ. Manage. 10, 199–214.

Gardiner, J.L. (Ed.), 1991. River projects and conservation – a manual for holistic appraissal. Wiley, Chichester, p. 236.

Hauer, F.R., Stanford, J.A. 1982. Ecological responses of hydropsychid caddisflies to stream regulation. Can. J. Fish. Aquat. Sci. 39, 1235–1242.

Hooijer, A., 1996. Floodplain hydrology: an ecologically oriented study of the Shannon Callows, Ireland. Ph.D. Thesis, Vrije Universiteit, Amsterdam.

Hosper, H., 1997. Clearing lakes, an ecosystem approach to the restoration and management of shallow lakes in the Netherlands. Thesis, Agricultural University of Wageningen, Wageningen, The Netherlands, May 1997.

Huet, M., 1954. Biologie, profils en long et en travers des eaux courantes. Bulletin Français de Pisciculture 175, 41–53.

Hynes, H.B.N., 1975. The stream and its valley. Verh. Int. Verein. Limnol. 19, 1–15.

IKSR, 1996. Taetigkeitsbericht 1996. Internationale Kommission zum Schutze des Rheins.

Illies, J., Botosaneanu, L., 1963. Problèmes et méthodes de la classification et de la zonation écologiques des eaux courantes, considerées sutout du point de vue faunistique. Mitt. Internat. Verein. Limnol. 12, 1–57.

IRC, 1987. Rhine Action Plan. International Rhine Committee. Koblenz, Germany.

Johnson, B.L, Richardson, W.B., Naimo, T.J., 1995. Past, present and future concepts in large river ecology. BioScience 45, 134–141.

Janse, J., Aldenberg, T., Kramer, P.R.G., 1992. A mathematical model of the phosphorus cycle in Lake Loosdrecht and simulation of additional measures. Hydrobiologia 233, 119–136.

Junk, J.W., Bayley, P.B., Sparks, R.E., 1989. The flood pulse concept in river-floodplain systems. In: Dodge, D.P. (Ed.), Proceedings of the International Large River Symposium. Can. Spec. Publ. Fish. & Aqu. Sci. 106, 110–127.

KHR, 1993. Der Rhein unter der Auswirkung des Menschen- Ausbau, Schiffahrt, Wasserwirtschaft. International Kommission für die Hydrologie des Rheingebietes. Bericht nr I-11. Lelystad, The Netherlands.

Klijn, F., 1997. A hierarchical approach to ecosystems and its implications for ecological land classification. Thesis, University of Leiden.

Klinge, M., Grimm, M.P., Hosper, S.H., 1995. Eutrophication and ecological rehabilitation of Dutch lakes: presentation of a new conceptual framework. Water Science Technol. 31 (8), 207–218.

Lelek, A., 1989. The Rhine river and some of its tributaries under human impact in the last two centuries. In: Dodge. D.P. (Ed.), Proceedings of the International Large River Symposium. Can. Spec. Publ. Fish. & Aqu. Sci. 106, 469–487

Lorenz, C.M., 1999. Indicators for sustainable management of rivers. Thesis, Free University of Amsterdam.

Lorenz, C.M., G.M. van Dijk, A.G.M. van Hattum, W.P. Cofino, 1997. Concepts in river ecology: implications for indicator development. Reg. Rivers: Res. & Mgmt. 13, 501–516.

Magdych, W.P., 1984. Salinity stress along a complex river continuum: effects on mayfly (ephemeroptera) distributions. Ecology 65, 1662–1672.

Markert, B., Wappelhorst, O., Weckert, V., Herpin, U., Siewers, U., Friese, K., Breulmann, G., 1999. The use of bioindicators for monitoring the heavy-metal status of the environment. J. Rad. Nuc. Chem. 240 (2), 425–429.

Meybeck, M., Helmer, R., 1989. The quality of rivers: from pristine stage to global pollution. Paleogeogr. Paleoclimatol. Paleoecol. (global and planetary change section) 75, 283–309.

Minshall, G.W., 1988. Stream ecosystem theory: a global perspective. J. North Am. Benthol. Soc. 7, 263–288.

Minshall, G.W., Cummins, K.W., Petersen, R.C., Cushing, C.E., Bruns, D.A., Sedell, J.R., Vannote, R.L., 1985. Developments in stream ecosystem theory. Can. J. Fish. Aquat. Sci. 42, 1045–1055.

Minshall, G.W., Petersen, R.C., Cummins, K.W., Bott, T.L., Sedell, J.R., Cushing, C.E., Vannote, R.L., 1983. Interbiome comparison of stream ecosystem dynamics. Ecol. Monogr. 53, 1–25.

Ministrie of VROM, 1994. Environmental quality objectives in the Netherlands. Ministry of Housing, Spatial Planning and the Environment, The Hague, The Netherlands.

Mueller, R., Doenni, W., Zeh, M., 1994. Untersuchungen zur Verteilung und Wanderung der Fische in einer Staustufe des Hochrheins. In: Weidmann H., Meder, H. (Eds), Sandoz Rheinfonds. Sandoz International AG, Basel, pp. 138–146

Naiman, R.J., 1983. The annual pattern and spatial distribution of aquatic oxygen metabolism in boreal forest watersheds. Ecol. Monogr. 53, 73–94.

Naiman, R.J., Lonzarich, D.G., Beechie, T.J., Ralph, S.C., 1992. General principles of classification and the assessment of conservation potential in rivers. In: Boon, P.J., Calow, P., Petts, G.E. (Eds), River Conservation and Management. John Wiley, Chichester, pp. 93–123.

Newbold, J.D., 1992. Cycles and spirals of nutrients. In: Calow, P., Petts, G.E. (Eds), The River Handbook. Hydrological and Ecological Principles, Vol. 1. Blackwell Scientific, Oxford. pp. 379–408.

Newbold, J.D., Elwood, J.W., O'Neill, R.V., Van Winkle, W., 1981. Nutrient spiralling in streams: the concept and its field measurement. Can. J. Fish. Aquat. Sci. 38, 860–863.

Newbold, J.D., Mulholland, P.J., Elwood, J.W., O'Neill, R.V., 1982a. Organic spiralling in stream ecosystems. Oikos 38, 266–272.

Newbold, J.D., O'Neill, R.V., Elwood, J.W., Van Winkle, W., 1982b. Nutrient spiralling in streams: implications for nutrient limitation and invertebrate activity. Am. Naturalist 120, 628–652.

Nienhuis, P.H., Leuven, R.S.E.W., 1998. Ecological concepts for the sustainable management of lowland river basins: a review. In: Nienhuis, P.H., Leuven, R.S.E.W., Ragas, A.M.J. (Eds), New Concepts for Sustainable Management of River Basins. Backhuys, Leiden, pp. 7–34.

Odum, H.T., 1983. Systems Ecology, An Introduction. John Wiley, New York.

OECD, 1982. Eutrophication of Waters, Monitoring, Assessment and Control. OECD, Paris.

Pedroli, B., Postma, R., Rademakers, J., Kerkhofs, S., 1996. Welke natuur hoort bij de rivier. Landschap 13 (2), 97–113.

Petts, G.E., 1994. Rivers: dynamic components of catchment ecosystems. In Calow, P., Petts, G.E. (Eds), The River Handbook. Hydrological and Ecological Principles, Vol. 2. Blackwell Scientific, Oxford, pp. 3–22.

Pinay, G., Decamps, H., Chauvet, E., Fustec, E., 1990. Functions of ecotones in fluvial systems. In: Naiman, R.J., Decamps, H. (Eds), The Ecology and Management of Aquatic-Terrestrial Ecotones. Unesco, Paris, pp. 141–169.

Pringle, C.M, Naiman, R.J., Bretschko, G., Karr, J.R., Oswood, M.W., Webster, J.R., Welcomme, R.L., Winterbourne, M.J., 1988. Patch dynamics in lotic systems: the stream as a mosaic. J. North Am. Benthol. Soc. 7, 503–524.

Rajagopal, S., Van der Velde, G., Paffen B.G.P., Bij de Vaate A., 1998. Ecology and impact of the exotic amphipod Corophium curvispinum Sars, 1895 (Crustacea: Amphipoda), in the River Rhine and Meuse. Report no. 75-1998 of the project "Ecological rehabilitation of the Rivers Rhine and Meuse". RIZA, RIVM, IBN-DLO, RIVO-DLO, SCO-DLO, The Netherlands.

Regulated Rivers: Research and Management, 1995. International Conference: Sustaining the ecological integrity of large floodplain rivers. Reg. Rivers: Res. Mgmt. 11, 1–136.

Reynolds, C.S., 1988. Potamoplankton: Paradigms, Paradoxes and Prognoses. In: Round, F.E. (Ed.), Algae and the Aquatic Environment. Biopress, Bristol, pp. 285–311.

Reynolds, C.S., 1994. The long, the short and the stalled: on the attributes of phytoplankton selected by physical mixing in lakes and rivers. Hydrobiologia 289, 9–21.

Ross, H.H., 1963. Stream communities and terrestrial biomes. Arch. Hydrobiol. 59, 235–242.

Scheffer, M., 1998. The Ecology of Shallow Lakes. Chapman & Hall, London.

Scholte Ubing, D.W., 1980. Nutrienten in de Rijn. H$_2$O 5, 97–104.

Sedell, J.R., Frogatt, J.L., 1984. Importance of streamside forests to large rivers: the isolation of the Williamette River, Oregon, USA from its floodplain by snagging and streamside forest removal. Verh. Int. Verein. Limnol. 22, 1828–1834.

Sedell, J.R., Richey, J.E., Swanson, F.J., 1989. The river continuum concept: a basis for the expected ecosystem behaviour of large rivers. In: Dodge, D.P. (Ed.), Proceedings of the International Large River Symposium. Can. Spec. Publ. Fish. & Aqu. Sci. 106, 49–55.

Sparks, R.E, Bayley, P.B., Kohler, S.L., Osborne, L.L., 1990. Disturbance and recovery of large floodplain rivers. Environ. Manag. 14, 699–709.

Sparks, R.E., 1995. Need for ecosystem management of large rivers and their floodplains. BioScience 45, 168–182.

Stanford, J.V., Hauer, F.R., Ward, J.V., 1988. Serial discontinuity in a large river system. Verh. Int. Verein. Limnol. 23, 1114–1118.

Statzner, B., Higler, B. 1986. Stream hydraulics as a major determinant of benthic invertebrate zonation patterns. Freshwater Biol. 16, 127–139.

Tansley, 1935. The use and abuse of vegetational concepts and terms. Ecology 16, 284–307.

Ten Brink, B.J.E., 1991. The AMOEBE approach as a useful tool for establishing sustainable development. In: Kuik, O., Verbruggen, H. (Eds), In Search of Indicators for Sustainable Development. Kluwer Academic, Dordrecht, pp. 71–88.

Ten Brink, B.J.E., Hosper, S.H., Colijn, F., 1991. A quantitative method for description and assessment of ecosystems: the AMOEBE approach. Mar. Pollut. Bull. 23, 265–270.

Thorp, J.H., 1992. Linkage between islands and benthos in the Ohio River, with implications for riverine management. Can. J. Fish. Aquat. Sci. 49, 1873–1882.

Thorp, J.H., Delong, M.D., 1994. The riverine productivity model: an heuristic view of carbon sources and organic processing in large river ecosystems. Oikos 70, 305–308.

Townsend, C.R., 1996. Concepts in river ecology: pattern and process in the catchment hierarchy. Algol. Stud. 113, 3–24.

Van Breukel, R.M.A., 1993. De Rijn en Rijntakken, verleden, heden en toekomst. RIZA nota 93.004, Rijksinstituut voor Zoetwater en Afvalwaterzuivering. Lelystad, February 1995.

Van den Brink, F.W.B., van der Velde, G., Bij de Vaate, A., 1991. Amphipod invasion on the Rhine. Nature 352, 576.

Van der Maarel, E., Dauvellier, P.L., 1978. Naar een globaal ecologisch model voor de ruimtelijke ontwikkeling van Nederland. Staatsuitgeverij, Den Haag.

Van Meent, D., Aldenberg, T., Canton J.H., Van Gestel, C.A.M., Slooff, W., 1990. Streven naar normen. Rijkinstituut voor Volksgezondheid en Milieuhygiene. Bilthoven, Nederland.

Van Straalen, N.M., Denneman C.A.J., 1989. Ecotoxicological evaluation of soil quality criteria. Ecotoxicol. Environ. Saf. 18, 241–251.

Vannote, R.L., Minshall, G.W., Cummins, K.W., Sedell, J.R., Cushing, C.E., 1980. The river continuum concept. Can. J. Fish. Aquat. Sci. 37, 130–137.

Wallace, J.B., Webster, J.R., Woodall, W.R., 1977. The role of filter-feeders in flowing waters. Arch. Hydrobiol. 79, 506–532.

Ward, J.V., 1989. The four-dimensional nature of lotic ecosystems. J. N. Am. Benthol. Soc. 8 (1), 2–8.

Ward, J.V., 1998. Riverine landscapes: biodiversity patterns, disturbance regimes and aquatic conservation. Biol. Conserv. 83 (3), 269–278.

Ward, J.V., Stanford, J.A., 1983a. The serial discontinuity concept of lotic ecosystems. In: Fontaine, T.D., Bartell, S.M. (Eds), Dynamics of Lotic Ecosystems. Ann Arbor Science, Ann Arbor, MI.

Ward, J.V., Stanford, J.A., 1983b. The intermediate disturbance hypothesis: an explanation for biotic diversity patterns in lotic ecosystems. In: Fontaine, T.D. and Bartell, S.M. (Eds), Dynamics of Lotic Ecosystems. Ann Arbor Science, Michigan, pp. 347–356.

Ward, J.V., Stanford, J.A., 1995. The serial discontinuity concept: extending the model to floodplain rivers. Reg. Rivers: Res. Mgmt. 10, 159–168.

Wassen, M., 1990. Hydro-ecology of the river Biebrza. Ph.D. Thesis, Utrecht University, Utrecht.

White, P.S., Pickett, S.T.A., 1985. Natural disturbance and patch dynamics: an introduction. In: Pickett, S.T.A., White, P.S. (Eds), The Ecology of Natural Disturbance and Patch Dynamics. Academic Press, Orlando, Florida, pp. 3–16

Whitton, B.A., 1980. Algae. In: B.A. Whitton (Ed.), River Ecology. Blackwell Scientific, Oxford, pp. 81–105

Winterbourn, M.J., Rounick, J.S., Cowie, B., 1981. Are New Zealand stream ecosystems really different? N. Z. J. of Mar. and Freshwat. Res. 15, 321–328.

World Wildlife Fund, 1993. Levende Rivieren. World Wildlife Fund, Zeist, The Netherlands.

Chapter 5

Predicting toxic effects of contaminants in ecosystems using single species investigations

Rolf Altenburger and Mechthild Schmitt-Jansen

Abstract

The usefulness of information gained from investigations of single species for predicting adverse effects of chemical contamination on aquatic ecosystems is discussed in this contribution. The frame for prediction efforts is thereby provided by the type and number of chemicals considered, the time and spatial scale of pollution and the criteria for an assessment. Further, any attempt to predict ecotoxicological effects on the basis of single species observation requires instrumentalisation of current understanding of ecosystems and biological action of compounds for a specific assessment task.

Various methods and tools that link chemical and biological types of information specifically to establish a functional relationship between exposure concentration and effect, model biological responses of long-term exposure, identify components of ecotoxicological concern in complex contaminated samples, calculate expected combined effects for mixtures of pollutants, understand modes of ecotoxic action, and predict biological activities from compounds structural properties are reviewed. The methodologies introduced all have their specific drawbacks concerning the scope to assess and predict ecosystem responses. Various approaches addressing extrapolation problems like in situ-toxicity testing, species sensitivity distributions, comparative studies using laboratory assays and micro- and mesocosm studies are additionally considered.

It is concluded from the review of current literature that all evidence so far shows, that single species data on the toxicity of pollutants can be used to predict the potential of adverse effects in ecosystems. There is no evidence that complex model ecosystems are systematically more or less sensitive to toxicants than single species tests. Principal limitations for extrapolation emerge when longer time scales are of concern or when structures or processes above the level of populations are affected. Additionally, ecological issues considered from a recovery perspective like recolonisation or functional replacement of species might modify assessment views.

Keywords: mode of action, QSAR, combined effect, mixture toxicity, time-response relationship, concentration-response relationship, single species sensitivity distribution, bioassay-directed fractionation, biotest, in situ testing

1. Introducing the context

The protection of the environment has become an ultimate political goal and social value in industrialised countries in the early 1970s as documented with the United Nations Conference on the human environment held in Stockholm, Sweden in 1972 (Halpern, 1993). This process developed in scope from a mere conservationist point of view to a sustainability-oriented approach with the Rio Declaration in 1992 (Anon, 1992). Ever since the beginning raise in popular awareness that human activities and in particular the management of material flows may impair the human environment, sciences have been asked to assess and predict the consequences of releasing all sorts of chemicals into the environment. This perception can be traced e.g. when studying legal requirements regarding the protection of the environment that have considerably increased throughout the last three decades. They almost uniformly at one stage require scientific or expert judgement of anticipated deleterious effects. Various lines of discussion in policy defined legislative activities may be distinguished. For a historical perspective on this the reader may refer to e.g. Milles (1989, 1991). From a decision process perspective one may label the different purposes as hazard identification, hazard assessment, risk characterisation, risk assessment and risk management (OECD, 1995). The reader has to be aware, however, that there is no uniform understanding of any of the terms employed. Key wording regarding the scope of intended environmental protection as defined in specific laws are e.g. "no harmful effect on human or animal health, directly or indirectly (e.g. through drinking water, food or feed) or on groundwater; [. . .] no unacceptable influence on the environment" (EEC, 1991).

Transforming political and juridical concepts into regulatory and administrative work is a challenge in itself. Various scopes like predictive assessments required for industrial chemicals or active ingredients of drugs or pesticides are to be separated from retrospective judgements on the effects of effluent discharges into rivers or emissions to the air. Site- and time-specific evaluations like evaluating run-off from waste disposal sites may be separated from utilisation specific approaches. As examples for the latter the formulation of water quality objectives for the protection of fishing stocks or the preservation of aquatic biocoenosis may be named. To support the setting of procedures and standards that are likely to hold even in cases of legal controversy and challenge, whole groups of experts at various national and international fora (like DIN, AFNOR, SIS, BSO, ASTM, OECD, CEN, ISO, SETAC) discuss the definition of protocols for any assessment to be made.

When sciences were being ascribed to deliver rational approaches to assess and predict adverse effects of chemicals on the environment, a whole new branch called ecotoxicology emerged from the collaboration of several biological and chemical subdisciplines. Like in the medical sciences when considering the developments in pharmacology and toxicology, several lines of reasoning developed in ecotoxicology. They range from questions of identifying mechanisms of action, to understanding of translation of effects from molecular interactions to responses in the structure or function of an ecosystem. Also, directions of research vary greatly covering more academic questions of generic principles of interactions or rather applied aspects like the identification of remediation priorities or the management of a waste dump site.

This brief outline of the various activities of different stakeholders in the field hopefully enlightened the complex setting of expectations when writing about assessment and prediction for ecosystems. In no way a pure academic thinking on understanding ecosystems structure and function may prevail nor is a mere orientation on pragmatic issues like defining rules for discharge fees appropriate. The assessment of chemical effects in the environment is a demand from the general public linked with frameworks and values concerning what to consider and what to protect derived from various political and economic backgrounds. However, the development of rational methodology and procedures for performance of those societal choices is the task of scientists. Not surprisingly these goals are often too ambitious to be met by simple and universal solutions. Instead pragmatic tools, rational choices and refined methodologies are being developed in response to specific tasks. In order to make good use of available knowledge and instruments, it is therefore vital to understand the scope and limitations of our poor trials.

Assessment and prediction of effects of contamination on ecosystems commonly relies on consideration of chemical exposure i.e. identifying targets at risk, determining an exposure concentration with respect to the bioavailability of pollutants in a specific environmental milieu and assessing biological responses. The scope of this article within the context of this book is focused to highlight and reflect the differences in scientific approaches using single species as qualitative and quantitative indicators for predicting adverse effects of chemical contamination on aquatic ecosystems.

2. Frame for prediction activities

Apart from the chemical and biological issues of how to perform an assessment and predict effects of pollutants for ecosystems which will be considered in the subsequent parts of this contribution, there are factors that define a frame for any kind of prediction exercises, which have to be reflected. In particular, we have to deal with (i) a multitude of chemicals being present in the environment, (ii) time and scale of a pollution situation, (iii) definition of criteria for an assessment.

2.1. Which chemicals to address?

The chemical abstract service (CAS) in 2001 counted over 18 million organic and inorganic substances, of which more than 2 million were commercially available chemicals (http://info.cas.org/cgi-bin/regreport.pl). There is no way to handle these compounds on a one by one basis in any kind of hazard assessment. In view of this, many national and international authorities have begun to build different types of chemical inventories (e.g. European Inventory of Existing Commercial Chemical Substances – EINECS, or the Toxic Release Inventory – TRI) to select compounds for prior assessment. One approach in several OECD countries was to identify and list so-called high production volume chemicals (HPVCs), namely chemicals that are produced or traded in amounts exceeding 1000 t/a. The EEC regulation (EEC, 1990) counted about 2000 of such substances regarded as priority compounds for environmental risk assessment.

The US-Environmental Protection Agency (EPA) published a survey (Betts, 1998) on the availability of basic toxicological data for such long existing and traded chemicals of high production volumes (Fig. 1). It shows that for the majority even of priority chemicals there is a great lack of basic information typically needed for any normative hazard assessment.

Focusing on HPVCs means however, that whole chemical groups of high public concern like dioxins or pesticides are not covered. Moreover, the underlying problem is that for any given amount of resources for hazard and risk assessment activities, decisions have to be taken on the allocation of efforts which often leads to the question, do we want to focus on many compounds with little depth of knowledge or on just a few chemicals with great depth of understanding?

2.2. What are adequate time and spatial scales?

Occurrence of contaminants in the environment may vary greatly with respect to range, sequence and duration. Also in terms of management options exposure may be considered as accidental or unavoidable, as point source-related or diffuse. Similarly, type, extension in space and development of biosystems at risk show high variability. The scope for predictions is thus very much dependent on fixing borderlines in time and space. These in turn very often derive from specific demands for an assessment. A prospective evaluation of the potential of a new pesticide for e.g., investigated for admission, to affect invertebrate life by spraydrift to riverine systems, is to be distinguished from a retrospective assessment such as contaminants of a sediment investigated for suitability to be used as land fill material.

In assessing chemicals in the environment, discrete events (short-term) have to be distinguished form long-term exposure. Exposure over longer periods may cause a change of the abiotic conditions and may lead to adaptation processes in the ecosystem. Both exposure scenarios (long-term; short-term) interfere at different levels of an ecosystem and must be evaluated on different scales of time and space.

Because of the natural heterogeneity of an ecosystem, small scale effects are more difficult to detect than large scale events. On the other hand, short and "small" events are probably more frequent than disasters with large scale effects. E.g. the Sandoz catastrophe in the river Rhine provoked a total break down of the ecosystem (fish kill) but small but frequent effluents of a toxicant into a stream may remain unidentified because of the small scale of effects, e.g. changes in the microbenthic community. These small scale effects need more precise investigation methods to detect changes in the ecosystem.

Further the scale of abiotic ecosystem features must be taken into consideration, when assessing effects of a contamination in this ecosystem. E.g., aquatic ecosystems are characterised by a great variation of residence time of water, influencing persistence, degradation or accumulation of chemicals in the ecosystem. In running waters, toxic effluents are transported downstream, so a spatial discrepancy of the point of release and observed effects may be found. In standing waters, the toxicants may rest in the water body for a long time, translocated and bound to the sediments. These may be periodically resuspended e.g. by annual circulations of a mictic lake, so effects may be detected for a longer time period.

The mobility or space of colonisation of a population is further of importance, when assessing the scale of a contamination. For example in streams and rivers, macro-zoobenthic organisms, exposed to a short pulse of a toxicant can escape into the hyporheic interstitial, an ecotone, which exists in the loose sediments between the stream bed and the aquifer. From this refugium, they can recolonise the stream quite quickly. Smaller and immobile organisms, e.g. the meiobenthos, organisms of the biofilm, or macrophytes are not able to avoid the exposure. Disturbances may be

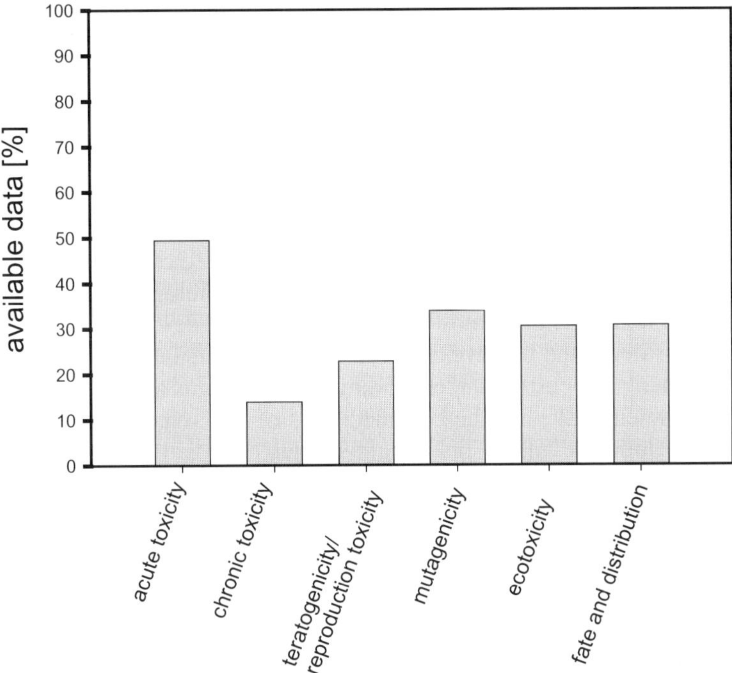

Figure 1. Availability of data for environmental risk assessment of priority chemicals (modified after Betts, 1998).

detectable much better within these groups of organisms than by investigating species with the potential of migration and recolonisation.

1.3. What are appropriate criteria for assessment?

Imagine a regionally contaminated aquifer is to be remediated. Ecotoxicologists are requested to provide tools that allow assessments of the treated groundwater and comparisons of different technological options (e.g. the SAFIRA project, http://safira. pro.ufz.de/) with respect to the protection of ecosystems. Commonly, faced with this question one would call for terms of reference by asking back: What type of bio-systems are to be protected?; What is to be regarded as a significant adverse effect?; And what in turn seems to be an acceptable hazard? An elaboration of these issues may be found in Forbes and Forbes (1994) and essentially they are a reflection that criteria for ecotoxicological assessment are *eo ipso* not scientific. Once the terms of reference have been defined scientific rationales may proceed. In the above example we may wish to conserve the fish population in the nearby freshwater for recreational purposes or we want to protect the microbial functions of the linked waste water treatment plant as a biological service. Another goal could be to avoid adverse effects on the interstitial faunistic community with stygophile and stygobite life forms in the adjacent groundwater as a biodiversity protection measure. Given the necessary resources are available, each of the protection goals can be addressed using specific models, biotests and expertise. However, none of the specific prediction instruments developed would allow extrapolative use for any of the others. Furthermore, additional requirements may have to be met in an assessment exercise, like that the assessment has to be made very fast to allow adequate intervention, or it may be asked for data that hold in a legal challenge.

3. Methodological considerations

Once the context of an ecotoxicological assessment or prediction exercise is defined, biological thinking prevails. Developing or utilising biological tools for specific purposes should regard what is known with respect to the description of structure and function of ecosystems and how we describe modes of toxic interference. Both aspects strive to instrumentalise current understanding of ecosystems and biological action for a specific assessment task. For the considerations of this overview, the methodological questions behind the aim to link single species observation to ecosystem effects basically are: What can be put under observation? What can be modelled in most simplistic ways?

3.1. Ecosystem description

The major challenge in ecotoxicology in the context considered here is to link measured endpoints of a single species under conditions as strictly defined as possible to ecological assessment endpoints, including communities and ecosystem structure and function.

There are various parameters that may be put under observation to characterise survival, growth, behaviour or development of single species or population dynamics (growth rate, death rate, density, etc.). Effects of toxicants can be quantified in the laboratory for these parameters, and used to assess direct effects of toxicants on the organism. Parameters of the population dynamics of a species may even be inserted in models, which try to simulate (predict) effects on a multispecies or ecosystem level.

On the other hand, all parameters, belonging specifically to a community level (species composition, species distribution, intra- and interspecific interactions) cannot be derived from a single species level. Effects of toxicants on these endpoints, which are essential in regulating community answers to toxicants cannot be observed in such test approaches.

Further, ecosystems are strongly influenced by abiotic factors (temperature, pH, matrix effects, etc.). These parameters also influence speciation, bioavailability and so the effects of a substance on a community in an exposed ecosystem. These interactions between pollutants, environmental milieu factors and biosystems may be investigated in a laboratory single species test in a restricted way. In a simple reaction chain, represented by one or a very few abiotic factors, substances and species, some causal connections can be derived and eventually modelled. The whole complexity of an ecosystem, however, can hardly be described.

Another aspect is the composition and diversity of species to be found in a community and its relation to chemical effects. There is no hierarchy in sensitivity of species, which could be easily generalised. However, it is not possible to test all species of an

Table 1. Biological levels of organisation and toxicological observation of specific interferences.

Level	Test system	Endpoints	Information
Organism	Single species test	Survival, growth, behaviour, physiological parameters, scope for growth	Direct impact on the organism, species sensitivity, mode of action
Population	Single species test	Growth/death rate, density, distribution	Parameters of population dynamics, intraspecific interaction
2–4 species	Multispecies test	Scope for growth, grazing rates (loss rates), competition	Interspecific interaction
Community	Community test	Species composition/ distribution, diversity, succession	Interspecific interaction, tolerance, adaptation, invasion, exclusion of species
Ecosystem	Microcosm, mesocosm, enclosures, field studies	Energy flow, food web, distribution of the toxicant	Impact of abiotic ecosystem parameters ecosystem structure, function bioavailability

(exposed) ecosystem for each chemical. Established single species test systems are mainly composed of ubiquists, which have turned out to be suitable for a good (reproducible) test situation and are easy to cultivate. In the ecosystem, specialists (stenoic species) are adapted to their environment, which characterise the ecosystem. These species are often rare (lists of endangered species) and sometimes characterised by complicated life cycles, tightly connected to the ecosystem characteristics. These species could be more sensitive towards chemical exposure and in many cases their elimination will be more difficult to overcome than for robust species. The niche of stenoic species will subsequently be occupied by an ubiquist resulting in a loss of diversity. These mechanisms might be more drastic in a sensitive ecosystem (e.g. bogs, springs) with many stenoeic species than in anthropogenic landscapes (e.g. agricultural landscape), which are disturbed by other factors already (structural changes etc.) and characterised by a small variety of ubiquists. Table 1 lists characteristics unique to different levels of biological complexity and parameters that may be accessible for toxicological consideration.

In summary, the above arguments show that in order to make useful contributions to an ecological assessment and prediction of chemical effects on the basis of single species testing, information regarding alterations with respect to behaviour, individual growth, mortality and reproductive success is needed using species with a well-understood ecology.

3.2. Effect and mode of action analysis

The second methodological consideration concerns the description of biological effects and the understanding of modes of toxic action. Knowledge of the mode of action of a chemical means to understand the interrelations of all observable effects provoked by the defined amount of a chemical in a biosystem. This of course is not much easier than describing an ecosystem in the first place. The most crucial point for our purpose therefore is the reflection of the effect to be put under observation. It might be helpful to distinguish between three aspects: firstly, the effect parameters as the biological structure or function which gains the focus like reproduction; secondly, the observation technique which is employed, like the photometric measurement of a suspension's light scattering, under a defined protocol, and thirdly, the derived effect descriptors like an ECx which is the estimated concentration (or dilution respectively) of a chemical that is predicted to elicit a certain response.

Using single organisms instead of whole ecosystems as instruments for assessing the status of the environment or pollution effects has a tradition in itself. The first indicator systems were established to assess the nutrient status of organically polluted water bodies. In order to obtain an indication of the nutrient status of a site of interest, the observation of the occurrence and abundance of species at sampling sites were linked to knowledge on the ecophysiological characteristics of indicator species. From there bioindication of pollutant effects that altered the occurrence or abundance of indicator species could be derived when relating site-specific observations to 'unpolluted' reference sites. Classical work has been performed by Kolkwitz and Marsson (1902) and Kolkwitz (1950) who invented the system of an index of "Saprobie" in running waters. Assessment of air pollution effects in industrial landscapes using lichens (Kreeb, 1990) and of readily decomposable organic water pollution using macroinvertebrates in streams (Diamond and Daley, 2000) are well known current examples of these approaches. When the observation of single species was extended to regard the performance of individuals, biomonitoring and biotesting of adverse effects of chemicals became established. Both fields rapidly developed various techniques and applications as it was possible to perform most work in laboratories and thus applying methods that have been developed in physiological or biochemical research. An overview of the various biomonitoring strategies is provided by de Zwart (1995), who categorises different fields such as toxicity monitoring of effluents, ambient toxicity monitoring, continuous biological monitoring, and ecosystem biomonitoring. In biotesting, environmental pollution is reduced to an environmental sample to be tested. Nusch (1992) gives an early account of the various demands that can be raised and specifically addressed in biotesting. After two decades of bioindicator, biotest and biomonitor development and use, a few rationales can be distinguished that may claim a consented status regarding the principles of effect assessment based on experimental biological data:

- *pars pro toto*-principle, i.e. test protocols are used employing definite species which than act as representative of whole taxa or trophic levels;
- use of biotest batteries instead of a single test organism realising that there is no such thing as a most sensitive species;
- bioassays with optimised signal to noise ratio should be used, thus allowing only the chemical to impose a constraint on the effect parameter under observation while providing optimum for all other factors.

When prediction is the goal in effect assessment, e.g. as it is the case in chemical hazard assessment, categorisation and modelling efforts become important in addition to the above described effect description tools. The methodologies developed for these purposes be it, e.g. quantitative structure-activity relationships or physiologically based pharmacokinetic modelling, do need some principal understanding regarding the interaction of pollutants with biosystems (Escher and Hermens, submitted). Principal understanding of toxic effects may derive from identification of primary molecular targets, biochemical studies of primary actions, or physiological and histological description of the following alterations. While the primary interaction may be referred to as mechanisms of action, toxic action is a process requiring the translation of functional or structural effects to response levels relevant for organismic performance, which is often referred to as mode of action.

In essence, when trying to predict ecosystem effects on basis of single species information the challenge of biodiversity translates into the effort to sufficiently represent different effect qualities that might be evoked from contamination of ecosystems with chemicals at high enough sensitivity. Finally, technical issues such as how to generate most precise and accurate information by regarding at the various sources of errors are discussed in the literature.

4. Methods and tools

Whenever the adverse effects of chemical contamination of ecosystems is to be predicted or to be assessed the challenge is to combine chemical and biological information, i.e. to link analytical data on occurrence, identity and quantity of xenobiotics to information on adverse effects on biota. Very often these two types of information are produced in separated monitoring and surveillance efforts and are then handled as completely independent type of information. This leaves the prediction or assessment job in trouble of either speculating about the hazard potential of a compound that has been analysed on a particular date, for whatever reason or with little clue about the causes of an observed impairment of a biological function. In the following we will therefore place special emphasis on methods and tools that link chemical and biological type of information to specifically:

- establish a functional relationship between exposure concentration and effect;
- model biological responses of long-term exposure;
- identify components of ecotoxicological concern in a complex contaminated sample;
- calculate expected combined effects for mixtures of pollutants;
- understand modes of ecotoxic action; and
- predict biological activities from compounds structural properties.

4.1. Concentration-response relationships

For prediction purposes, exposure concentrations of considered pollutants or dicharges that are to be regarded as non-toxic for ecosystems are required in order to provide references for management activities. Risk management procedures for chemicals commonly rely on effect assessments based on single substance evaluations and on fixing of threshold values i.e. no effect concentrations (NECs) as a borderline between an acceptable and an unacceptable risk. When asking scientific communities, this perception will immediately be translated into no observed effect concentrations (NOECs) or no observed adverse effect levels (NOAELs). To derive such values, statistical tests are used to compare the variances of a control and a treated situation and identify the highest test concentration that proves to be of no statistical difference from the control. For the last forty years this concept has been a basis for regulating various chemicals, and it is still enshrined in various guidelines, standards and norms. However, as it is difficult for the experimentator to observe "no effects" and with acknowledging furthermore that there are severe drawbacks from a statistical point of view (details of the discussion e.g. in Laskowski, 1995, Chapman et al., 1996, Moore and Caux, 1997) there seems now consensus reached to move away from this predictive approach (for review see OECD, 1998).

Instead an approach is favoured that focuses on a standing paradigm in toxicological research, namely that contaminant exposure and biological responses are functionally related. The objectives of determining such concentration-response relationships using quantitative models are to allow

- reproducible derivation of characteristic values used in chemical risk assessment procedures like an EC50 (effect concentration at which 50% of a specified effect is estimated to be evoked) or a LID (Lowest inhibitory dilution which produces a specified effect regarded as significant given a fixed dilution series of an environmental sample);
- comparison of compound properties in terms of intrinsic activity and effectiveness i.e. position and slope of a concentration-response curve;
- statistically valid predictions of low effect concentrations which are typical for many environmental contamination patterns.

Establishing a functional relationship between exposure concentrations of pollutants and biological effects requires experimentation using varying dilution often as geometric series in an appropriate range to observe varying responses of the effect put under observation. Figure 2 provides an example for the effect of various concentrations of the polyaromatic hydrocarbon naphthalene and its inhibitory effect on the reproduction of unicellular algae growing as a synchronous culture and being exposed for one generation cycle of 24 hrs (Walter at al., 2002).

The experimentally determined data are then fed into appropriate biometrical models that by iterative procedures do calculate estimates for the model parameters. Plotting the estimated function against observed data (Fig. 2) or performing residue analysis allows to assess the fit of the chosen function to the observed data. In this case using a probit model of the form:

$$\text{Effect} = \frac{1}{2\pi} \int_{-\infty}^{k(\text{Conc.})} \exp(-u^2/2)\, du,\ k(\text{Conc.}) = \theta_1 + \theta_2 \frac{\text{Conc.}^{\theta_3} - 1}{\theta_3} \tag{1}$$

$$= \text{Probit}\left(\theta_1 + \theta_2 \frac{\text{Conc.}^{\theta_3} - 1}{\theta_3} \right)$$

provides a good fit of the experimental data.

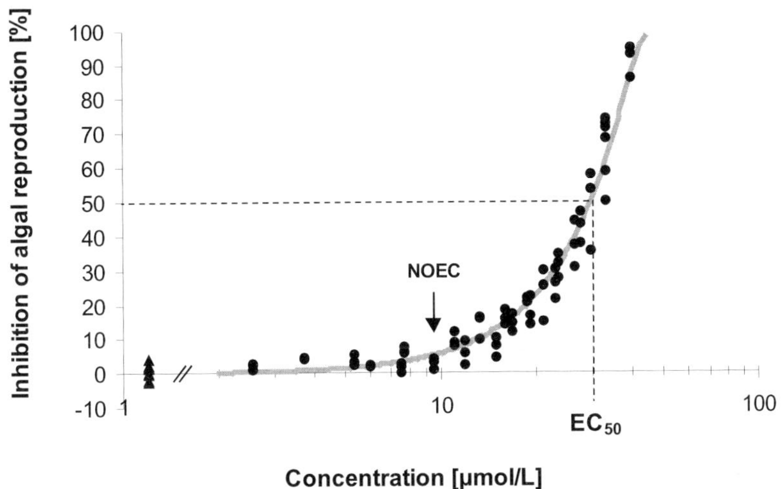

Figure 2.   Concentration-response model and data for the naphthalene induced inhibition of reproduction of the green alga *Scenedesmus vacuolatus* (from Walter et al., 2002).

The parameter estimates for this example were $\theta_1 = -85.44$, $\theta_2 = 52.67$, and $\theta_3 = 0.087$. There are various models to choose from, most of which deliver differences only when regarding effect estimations for high or low effects (Christensen 1984, Moore and Caux, 1997, Shukla et al., 2000, Scholze et al., 2001).

Once a concentration–response relationship has been established it may be used to derive parameters like ECx values for various purposes. Active ingredients, purposefully released to the environment may be assessed comparatively concerning their unwanted effects. Lützhoft et al. (1999) provide an example, comparing the phytotoxicity of seven antibacterial drugs applied in Danish fish farming using a cyanobacterium and two eucaryotic algal species as test organisms. Using estimated effect concentrations they were able to rank the different compounds according to their relative phytotoxicities. Further, they showed that the cyanobacterium *Microcystis aeruginosa* responded several orders of magnitude more sensitive compared with the eucaryotic plant species. This is easily understood considering the mode of action of the concerned compounds, which tend to be specific for interaction with procaryotic growth and reproduction processes.

An established concentration-response relationship may also be utilised to assess whether chemically detected amounts of a given pollutant sufficiently explain observable effects on organisms. Figure 3 provides an example for contaminated groundwater from the Bitterfeld area in Germany. Mass balances based on GC/MS-Screening showed that monochlorobenzene is the dominant contaminant for most ground water probes analysed in this particular area at the quarternary water table. Based on experimentally determined concentration-response functions for monochlorobenzene Figure 3 depicts the expected effects in *Vibrio fischeri* for chemically detected concentrations of chlorobenzene in different wells of the contaminated area. This expected effect is then compared with the observed toxicity of groundwater probes for the same

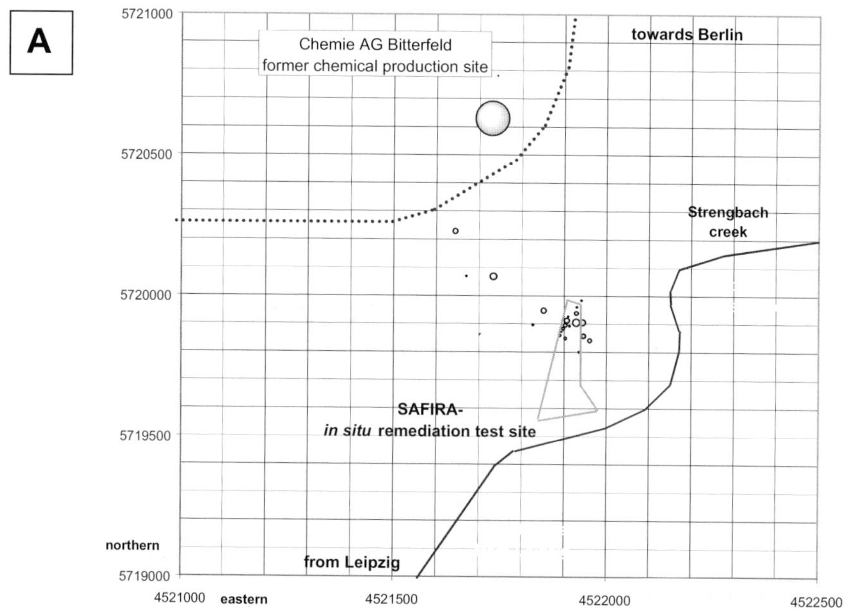

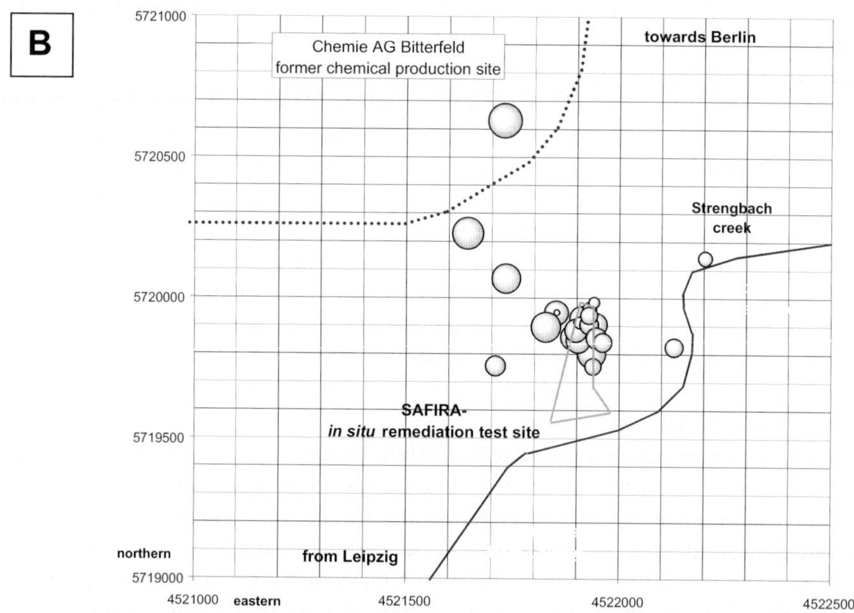

Figure 3. Map of expected (A) short-term bacterial toxicity in groundwater probes of a contaminated area, using analytically determined amounts and concentration-response relationships for chlorobenzene. (B) depicts the same area with observed responses of groundwater probes. The bubble sizes on the maps indicate the degree of response.

organism. Obviously, chlorobenzene though present in high amounts, is not responsible for the observable toxicity. This can also be shown for the alga *Scenedesmus vacuolatus* and the crustacean *Daphnia magna*. The modelling of the toxicity of different probes for the contaminated area, however, provides a consistent description. Thus other yet unidentified components yet unidentified might add in mixture to the observable toxic effects.

Concerning the predictive scope of a concentration–response analysis there are several issues to be taken in mind. The duration of exposure as well as the time for effect propagation do have to be regarded with respect to the chosen effect parameter. A chronic effect e.g. has to be seen in relation to the life span of a particular organism. Most unfortunate is the discussion of sensitive endpoints that can very often be found in bioassay discussions. Bearing the objective of most investigations in mind, that there has to be some assessment or prediction for an ecosystem the requirements should be pretty evident. If there is no biological argument for a definite time of observation, like the completion of a generation cycle for instance, than there has to be a consideration of the time-response relationship (see Section 4.2). Furthermore, effects to be observed in experiments that are meant to relate to ecosystem assessments should strive to link observations to life table parameters or growth to allow ecological considerations of effects at the population level. Finally, an environmental concentration of a pollutant may not easily translate into an effective dose for an organism. The understanding of the often dynamic relationships between contaminant exposure and bioavailable concentrations are again research topics in their own right.

There are various techniques available to cope with dose estimations for unknown or fluctuating exposure situations which utilise cumulative responses, exposure history data, flow through exposure systems, or bioaccumulation biomonitors. The concepts of lethal body burdens (LBBs) and physiologically based pharmokinetic modelling (PBPK) (e.g. Yang et al. 2000) provide scope for refined dose estimations in organism-based hazard assessment. The functional description of concentration-response relationships though often employed offers many untapped potentials for the prediction of effects beyond cut-off values. This includes contributions to the identification of modes of action (e.g. Altenburger et al., 1995) or the relevance of environmental milieu factors like pH for observable effects (Fahl et al., 1995).

4.2. Time-response relationships

The prediction of pollutant effects as a goal evidently is a time-related enterprise. Time in biological systems is an important variable e.g. regarding endogenous rhythms or different developmental stages of an individual; the age composition of a population or the succession state of an ecosystem to name a few. All these biological events in time have been shown to influence responses to chemical stress. Investigations addressing this most trivial fact explicitly, however, are not mainstream and a so-called endpoint discussion prevails instead. Even parameters intended to include time aspects like growth rates are often one point estimates. So whenever biological responses are not an end in themselves like a generation cycle, there is a need to identify the relevance of a measured number within the time scale, especially when regarding this measurements in the context of an ecological time frame.

Given the fact that the wealth of experimental ecotoxicity information exists for short-term standardised bioassays, and that for most chemicals this toxicity information is the only available, there is a long-lasting debate on the possibility to extrapolate chronic toxicity values from short-term toxicity data. Under the keyword acute to chronic ratio (ACR) several authors have generated experimental information for specific compounds using short-term and long-term exposure designs in distinct species (e.g. Morton et al., 1997). Others undertook to derive extrapolation factors from review of literature data for specific chemicals (e.g. Ford, 2001) or groups of chemicals (Länge et al., 1998, Roex et al., 2000) combining evidence from independent studies. Länge and co-workers (1998) e.g. compared ratios of EC50 values from acute studies to NOEC values from chronic toxicity studies for 71 substances. For that purpose, they drew their data for calculation from the factual database of the European Centre for Ecotoxicology and Toxicology of Chemicals (ECETOC) which puts special emphasis on the quality of the reviewed data with respect to the verification of toxicant concentrations in the studies to be included. Table 2 shows a descriptive statistics as an overview of their findings. Using the 90th percentile as a descriptor of the distribution of ACR values for different groups of chemicals for which the analysis could be performed the ratio between reported acute and chronic toxicity values varied from a factor of 16 to almost 200. The median value of the acute to chronic ratio for all chemical is 8.6 which is in good agreement with an analysis based on species sensitivity distributions for 89 pairs of acute and chronic toxicity descriptions for information from 3 to 262 species (de Zwart, 2002). Looking again at Table 2, it seems striking that metals and specifically acting pesticides are the chemical classes with the higher ratios, though the wide distribution of data as seen in the minimum and maximum values can be taken as a clear warning sign against applying generalisations to individual cases. The notion that the mode of action rather than the structure of a particular chemical plays an important role in explaining different ACR values has also been brought forward by Roex et al. (2000), who also show that the smallest variation in ACRs can be seen for nonpolar chemicals with an anticipated narcotic type of action.

Table 2. Acute EC50 to chronic NOEC ratios (ACRs) for fish and daphnid toxicity data for groups of substances (taken from Länge et al., 1998).

Substance group	No. of substances	Acute EC50:chronic EC50			90th percentile
		Min	50th percentile	Max	
All chemicals	71	0.13	8.63	1290	72.9
Pesticide a.i.	26	1.33	12.2	371	83.7
Other organics	26	0.13	3.91	27.5	15.9
Other organics, but at defined periods of exposure	19	1.25	3.60	28.3	24.5
Metals and organo-metals	14	0.30	28.0	1290	192
Other inorganics	7	2.92	8.39	69.3	20.1

Building acute to chronic toxicity ratios is, however, nothing more than trying to find rationales for extrapolation factors and thus dealing with the misery of regulatory biotesting using standardised protocols only, with arbitrarily chosen time periods for endpoint estimates. The alternative is available in form of functional descriptions of concentration-time relationships. For many years of pharmacological and toxicological research this is an issue of thought (Rand et al., 1995).

A canonic approach is to start with visualising concentration-time response surfaces. An illustrative example for comparing the metal toxicity against *Daphnia magna* clones has been provided by Barata et al. (1999). Figure 4 is taken from their work and illustrates that by simple transformation of the response scale according to a normal distribution, i.e. calculating probits a linear plane describes the experimental data in the three dimensional space already pretty well. Thus a multiple linear regression model of the form

$$E = a + b \ln (\text{Conc}) + c \left[1/\ln (T) \right] \tag{2}$$

fits the data (with E, effect in probits; Conc, concentration of the toxicant; T, time; and a, b, and c, linear regression parameters). More sophisticated functional descriptions of concentration-response surfaces may be found in the pharmacological literature (e.g. Levasseur et al., 1998).

Simplifications of such an approach, undertake to reduce information to a two-dimensional plot in order to derive characteristic values such as median lethal/effective time or incipient lethal/effective concentration (Rand et al., 1995). As in concentration-response analysis scale transformations are performed to allow simple regression techniques to estimate parameters of interest.

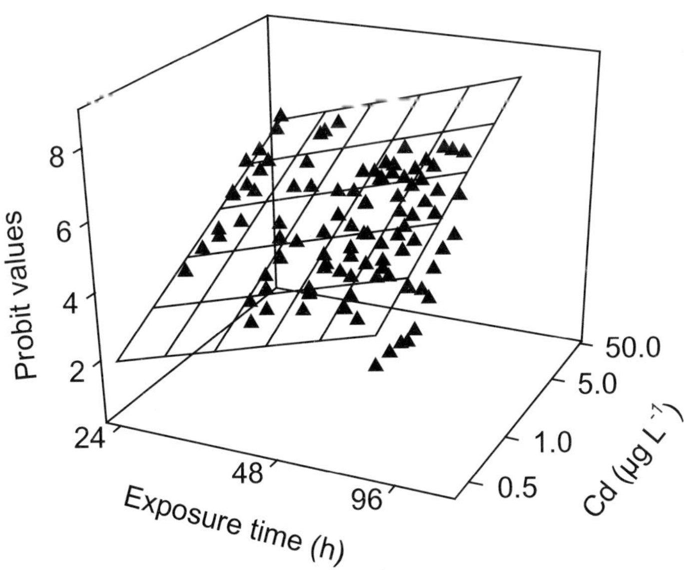

Figure 4. Concentration-time-response surface for the effect of Cd on *Daphnia magna* (immobilisation) for 96 h checked in 12 h intervals (from Barata et al., 1999).

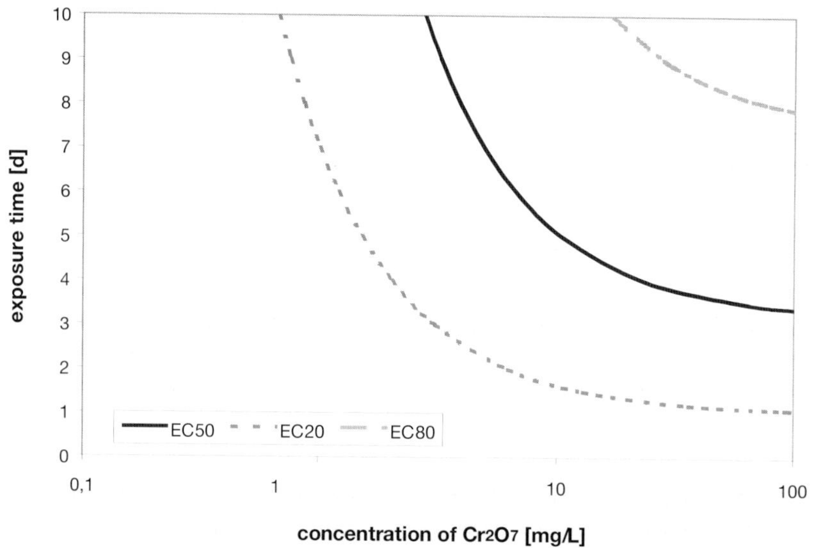

Figure 5. Time-response relationship for the effect of potassium dichromate on the frond growth of *Lemna minor* as calculated using the DEBtox-model of Kooijman and Bedaux (1996). Depicted are lines of equi-effects on a concentration-time scale.

An ecologically oriented theory that strives to describe toxic effects as process per-tubations is the dynamic energy budget theory, which is formalised as to the so-called DEBtox-model (Kooijman and Bedaux, 1996). On the basis of time series toxicity data for standard bioassays the DEBtox-model derives estimates for median effects like common concentration-response models. But in addition it generates no effect con-centrations and time dependent toxicities. The modelling works modular, assuming dif-ferent kinetics and effect propagation concerning costs for growth, maintenance and reproduction for the different test organisms used in standard biotest protocols like fish or daphnids. The calculus of the DEBtox-software package relies on solving sets of differential equations for the kinetics and dynamics of the compounds.

Implicit to most time-response modelling efforts is the assumption that effect prop-agation following an exposure to a toxicant is a steady process. This is probably a reasonable assumption for many unspecifically acting compounds like solvents or other industrial chemicals. It has also been shown to hold for a group of organophosphorus pesticides, which irreversibly bind specifically with their metabolised oxon analogues to acetylcholinesterase (Legierse et al., 1999). For such cases a so-called critical target occupation model has been proposed that describes the concentration in an organism at the time of death as a product of the area under the time-target tissue concentration and a constant, which can be derived from bioconcentration models and standard toxi-city estimates (Legierse et al., 1999). However, with specifically acting compounds one has to consider the possibility that primary interactions occur with processes at certain stages in the development of individuals and that such mechanisms of action will show a sensitive window in a life cycle. Examples for this are the sexual devel-opment of fish and its vulnerability to endocrine disruption (Segner et al., 2001) or the

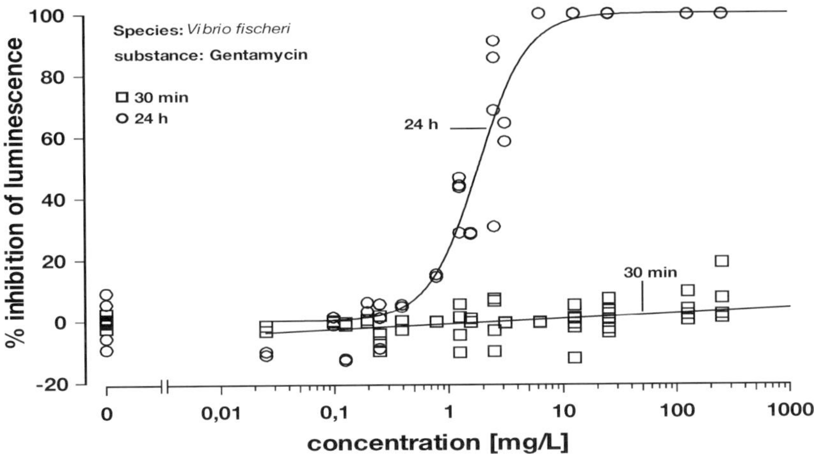

Figure 6. Effect of the antibiotic gentamycin on *Vibrio fischeri* after 30 min and 24 h of exposure (modified after Altenburger and Backhaus, 2000).

action of antibiotics on microorganisms (Backhaus et al., 1997). An example of the latter which can be interpreted in terms of mode and mechanism of action is provided in Figure 6. Gentamycin, an antibiotic known to specifically bind to bacterial 30S-RNA and thus interrupting protein biosynthesis, does not show any effect in water soluble concentrations in standard luminescence assays of 5–30 min exposure. However, if the test protocol is modified to allow a full cell cycle to take place the antibiotic potency of the compounds is easily demonstrated.

4.3. *Bioassay-directed fractionation and identification of toxicants*

The function of this approach for an assessment of ecosystems is to identify compounds of toxicological potency in complex contaminated environmental samples and to establish a causal link between occurrence of contaminants and possible adverse effects on biota. Very often in assessment of site-specific contaminations this knowledge cannot be adequately derived from existing emission information. In principle, one starts with defining the relevant toxic effect in the original sample. The original sample is than fractionated according to physico-chemical properties e.g. volatility, lipophilicity, or molecular size. The clue is to perform separation or clean-up in a way that allows subsequent testing of the remaining toxicity in the samples. The principle is illustrated in Figure 7.

For those fractions, that recover most of the toxicity of the original sample, this procedure may be continued in an iterative process, employing different separation principles. This process is continued until a chemical analysis concerning the elucidation of structures and their amounts due to sufficient clean-up can be performed. Chemical analysis is thus restricted to those fractions that retain toxicity. Also, it is no longer guesswork as to which identification technique might be adequate, due to the information gained from the fractionation techniques employed.

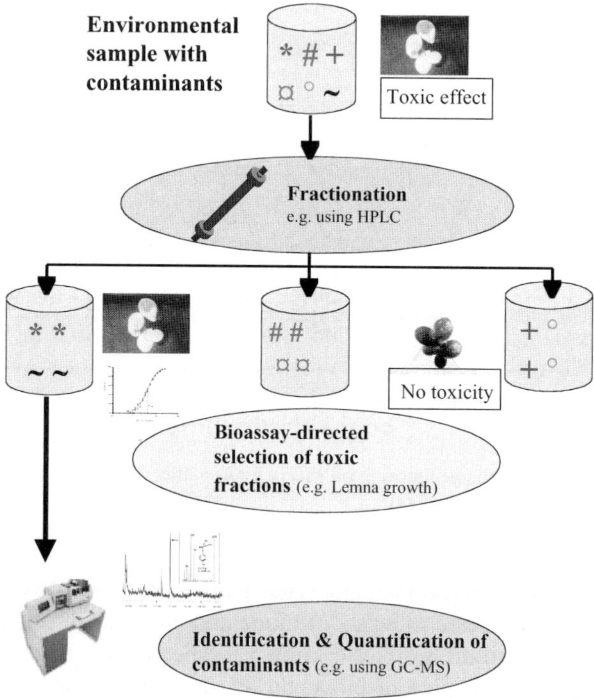

Figure 7.   Illustration of the principle of bioassay oriented fractionation and identification of toxicants in complex contaminated environmental samples.

As one example of this approach we consider a study by Brack and co-workers (Brack et al., 1999) who investigated a highly contaminated sediment in the riverine Spittelwasser, Germany, which flows into the river Mulde, a tributary to the Elbe stream. The Spittelwasser drains the areas of Bitterfeld and Wolfen, two of the major chemical production sites in Europe for over a hundred years. After political change and close-down of most production facilities, this riverine is to be remediated now. Though information on various contaminants exist, knowledge of the priority toxicants, however, is lacking. As production ceased the water body is no longer considered a priority problem, in contrast to the still heavily contaminated sediment.

To elucidate the composition of contaminants with potential ecotoxic effects, sediment samples from the creek were taken and Soxhlet-extracted with acetone. Such an extraction procedure allows to recover organic compounds of medium polarity to high lipophilicity. Metals and/or highly polar organic contaminants, however, will be lost. These extracts were then fractionated and biotested in a stepwise procedure till components could be identified using gas chromatography with mass selective detection (GC/MSD). Three biotests representing different life forms and types of response were used as effect detectors, namely: cellular reproduction of the unicellular green algae *Scenedesmus vacuolatus*, mobility of the water flea *Daphnia magna* and bioluminescence of the bacterium *Vibrio fischeri*. The so-called confirmation step comprises testing pure compounds that have been analytically identified in the effective subfractions and comparing effect concentrations determined for the pure compounds with the effect dilution for the fractions for which compound quantities can be estimated. Besides well-known and expectable toxicants, like organotin compounds and several polyaromatic hydrocarbons (PAHs) this study revealed effective concentrations of active ingredients of pesticides like prometryn (used in herbicides) and methyl parathion (used in insecticides) as well as a completely unexpected toxicant, namely N-phenyl-β-naphthalene amine. While a compound like parathion would not have been expected due to its rapid degradability in this system, the high phytotoxicity of N-phenyl-β-naphthalene amine was first identified in this study. Obviously, all bioassays employed detected different toxicants, thus proving, that the use of biotest batteries is necessary whenever there is no pre-defined focus for a specific toxic effect. Thus any remedial action considered, could now use criteria to assess biological efficiency of remediation activities in addition to purely chemically defined ones.

A second example where a defined effect quality was considered is provided by Purdom et al. (1994) and Desbrow et al. (1998), who undertook an elegant work to identify the causes of the previously reported estrogenic potency of effluents from sewage-treatment plants in British rivers. Sewage treatment plants, coping with industrial and domestic waste release highly complex effluents. Particularly, the non-ionic surfactant group of alkylphenols from household detergents are suspected to be responsible for the estrogenic potency of effluents, due to *in vitro* evidence. Fractionation of crude effluents of several sewage treatment plants into sub-samples containing volatiles, particulates and dissolved compounds in a first separation step using an in vitro yeast-based screen for oestrogenic activity rendered the dissolved phase as the only fraction containing any bioactivity. Three further fractionation steps, separating compounds according to lipophilicity of components on C18-solid phase extraction cartridges and subsequently on C18-HPLC columns left but a few active fractions. GC-MS analysis of these purified fractions identified estrone, 17β-estradiol and 17(-ethynylestradiol as the principal components. While the former two are supposedly of natural, human origin the latter compound is the main estrogenic component of the combined oral contraceptive pill. In a supplementary paper, the allocated effect quality of estrogenic responses was further validated for 17β-estradiol, estrone and an octylphenol (Routledge et al., 1998). In *in vivo* tank trial experiments, adult male rainbow trout (*Oncorhynchus mykiss)* and adult raoch (*Rutilus rutilus*) were exposed for 3 weeks to environmentally relevant concentrations of these compounds and the vitellogenin (VTG) content of blood samples was determined. The stimulation of the production of the female egg yolk protein VTG was used as a biomarker of response indicating oestrogenic contamination. All compounds investigated elucidated similar responses, however the natural steroidal oestrogens were three orders of magnitudes more potent as compared to octylphenol and furthermore they showed potencies at environmentally relevant concentrations.

Major limitations for using results from bioassay-directed fractionation to predict ecosystem effects result from the high demand on testing capacity and the lack of bioavailability information. The first point is a technical aspect, namely the limited use of laborious techniques to determine e.g. long-term, chronic effects which in turn preselect effect parameters and considered targets at risk ready for observation. In this respect, intelligent experimentation techniques such as molecular biomarkers or high-troughput devices are needed. The second issue, namely the assessment of bioavailability of compounds in the ecosystem context can only be addressed when complementing studies of site-specific toxicant identification, with investigations to determine the bioavailability *in situ*. An interesting approach not only to determine the bioavailability of organic pollutants in aquatic systems but additionally verify their toxic potentials has been provided by a combination of using semipermeable membrane devices as bio-mimetic passive sampler and employing bioassays on extracts derived from them (Sabliunas, 1999).

4.4. Effect analysis

The purposes of studies that focus on effect analysis in order to gain understanding of the mode of action of substances are in this context:

- to understand the relation between a substance and biological responses in order to identify sensitive taxa and processes within ecosystems;
- to clarify the scope for inference of effect assessments between different species;
- to assess the relevance of specific effects for ecosystem well-being and predict whether found environmental concentrations of contaminants may be linked to adverse effects.

Much of the variation seen in different organisms in response to toxicants has been attributed to the mode of action of chemicals (Vaal et al., 1997a). Moreover, while non-reactive organic chemicals, that act via unspecific so-called narcotic action reveal relatively small sensitivity distributions of acute toxicity data and commonly are straightforward to model regarding their acute to chronic toxicity relationships, the contrary seems true for reactive or specifically acting compounds (Figure 8) (Vaal et al., 1997b).

For specifically acting compounds the interference with taxa-specific processes or targets is of course the background for utilisation of compounds as active ingredients of drugs like antibiotics, herbicides, etc. A current review of the understanding of unspecific toxicity can be found in Caisukant et al. (1999), reviews of specifically acting compounds may be found in text books on phytopharmacology and human and veterinary drugs.

For most chemicals in the environment, however, the mode or even mechanism of action is unknown. When we aim to protect ecosystems, we have to acknowledge their property to consist of assemblies of various life forms and strategies. Commonly, the way to handle the resulting information gap on the toxicity potential for all organisms present in a specific ecosystem, biotest batteries consisting of selected species are constructed. These can be built on rationales like representing (i) different trophic levels of a food chain, (ii) various levels of biological complexity, (ii) different life strategies, or (iv) endangered species. Reviews on the ecotoxicity potential of specific compounds like chlorate (Van Wijk and Hutchinson, 1995) or 2,3,7,8-tetrachlorodi-benzo-*p*-dioxin (Boening, 1998) illustrate this thinking to identify most sensitive taxa. The challenge is to obtain information for specifically sensitive organisms, as in the example of chlorate the toxicity to aquatic organisms and ecosystems is relatively low for most species with values in the higher mg/L-range but very high for several species

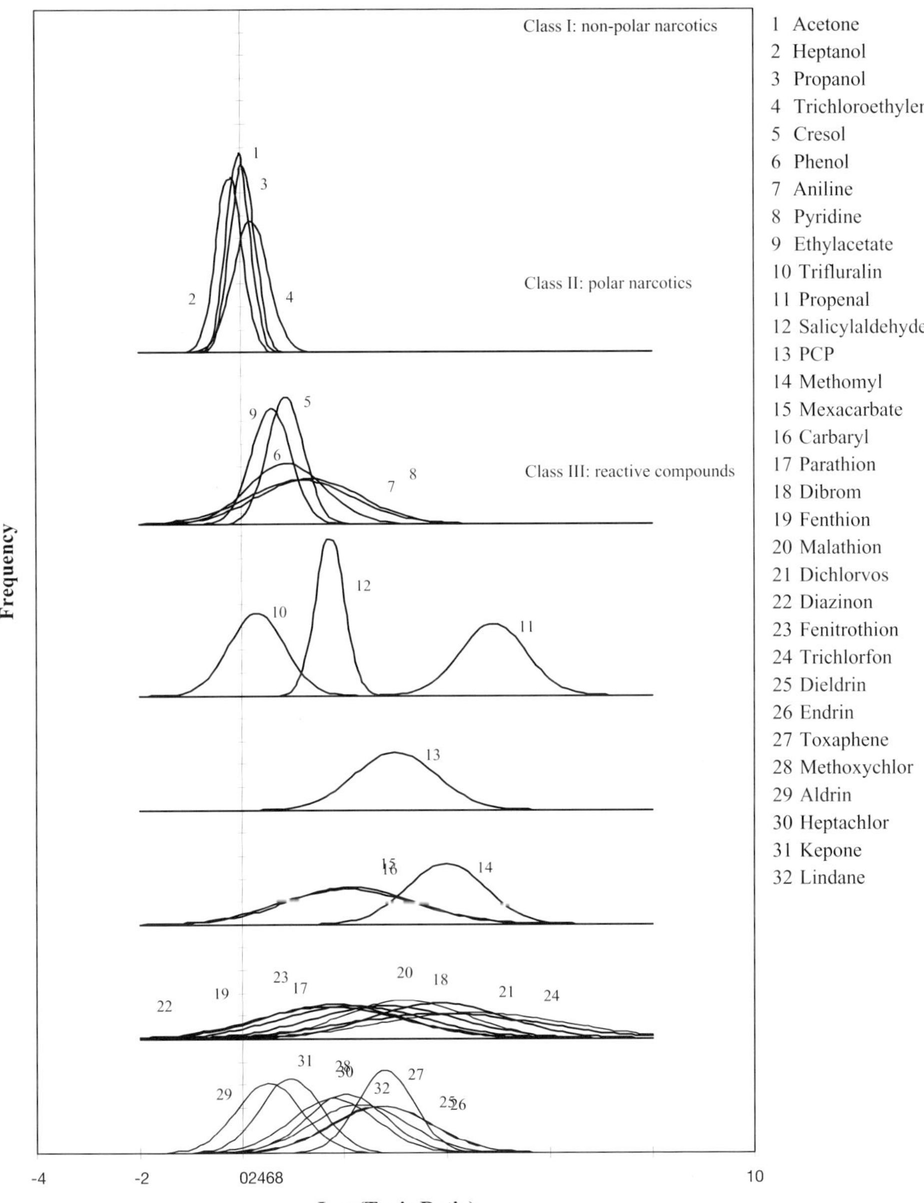

Figure 8. Species sensitivity distribution of acute toxicity data (Vaal et al., 1997b).

of marine macroalgae of the genus *Fucus* and for micro-phytoplankton communities with values in the μg/L-range (Van Wijk and Hutchinson, 1995). More often, the available information on the toxicity of chemicals towards different species is strongly biased towards species of particular economic impact like fish or bees and assessment efforts are thus hampered (Boening, 1998).

A different perspective comes from approaches assembling bioassays that represent different physiological competence and thus interaction potentials or that utilise batteries of biochemical and subcellular assays which may go as molecular as specific receptor-binding assays. The former approach is typically realised by constructing multi-parameter test-systems for single species like the observation of fish acute toxicity syndromes (McKim et al., 1987), the differentiation of lethal effects on egg and adult stages and non-lethal effects on food acquisition and production rates over time in *Daphnia* females (Barata and Baird, 2000), or the observation of different structural and functional parameters in the synchronised life cycle of unicellular green algae (Grimme et al., 1993). The latter strategy may combine *in vitro* assays that reflect different modes of action like the inhibition of acetylcholinesterase through organo-phosphorous compounds or the decoupling of oxidative phosphorylation through specific phenolic compounds (Wenzel et al., 1997). These approaches may correctly assess or predict the quality of the toxic potential of specific chemicals, however care must be taken regarding the quantification of the concentration–response relationships as *in vitro* tests are often less sensitive compared to organismic responses (Wenzel et al., 1997). A second problem in the prediction of the toxicity potential for ecosystems from a mode of action approach is the discrepancy of the recognition of a limited number of 7–10 modes of actions as discussed in the ecotoxicological literature (Nendza and Müller, 2000; Schüürmann, 1998, Wenzel at al., 1997), and the knowledge and utilisation of many more specific targets in drug application. Thus, Faust et al. (2000) could extract 40 different specific mechanisms of actions for herbicides and Backhaus et al. (2000) 32 mechanisms of action for antibiotics using textbook knowledge only. Also, the young history of ecotoxicology is a history of surprises regarding the discovery of new effect qualities like the current debate on endocrine disruption shows (Matthiessen, 2000).

A serious problem regarding the predictive scope for mode of action information apart from the concentration argument derives form the necessity to understand effect propagation from a molecular level of biosystem-pollutant interaction to interference with population and community relevant parameters. The question when we do understand the mode of action of a specific chemical is, how does it feed through the different levels of biological organisation towards a response that is to be seen on the population level. The linkage of parameters relevant at the population level like behaviour, growth, reproduction and mortality to physiological observations is often tried using energy budget considerations as observable in short-term experimentation (Kooijman and Bedaux, 1996, Barata and Baird 2000, Knops et al., 2001). An example for this is provided in Figure 9, where the results of an exposure of *Daphnia magna* females for eight days against a cationic surfactant (CTAB) and the metals copper and cadmium on scope for growth, dry weight increase and egg production are displayed (Knops et al., 2001). While scope for growth calculations based on oxygen consumption and food (algae) intake measurements showed a good correlation to achieved dry

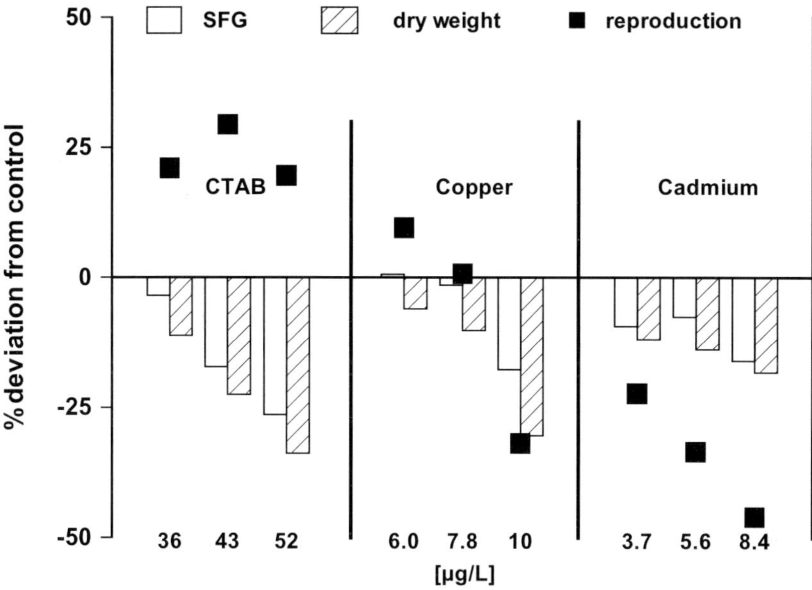

Figure 9. Effect propagation from physiological levels to population relevant responses (adapted from Knops et al., 2001). Scope for growth (SFG) was modelled on the basis of oxygen consumption and food intake measures.

weight increases, the parameter most relevant for population performance namely egg production shows a compound specific response. This demonstrates that there are no simple deterministic links between the parameters but that resource allocation is flexible and effect propagation a process in itself.

Typically, the effect assessment of compounds or contaminated environmental samples is based on single species studies that are performed under physiologically optimised conditions. This is done for the good reason, that a maximised signal to noise ratio is appropriate in effect quantification as discussed in various monographs on biotest development (Wells et al., 1998, Steinhäuser and Hansen, 1992). Regarding the prediction of effects based on these type of data, one has to consider that organisms outdoor have to cope with various environmental constraints, that may affect the sensitivity against exposure with pollutants. Examples have been provided demonstrating that the interaction due to density effects in a Daphnia population may alter responses to chemical stress (Goser, 1997). It has also been shown that interaction of environmental factors and chemical stress might affect responses of organisms. Thus, UV exposure and food ration increased the sensitivity of the amphipod *Paramoera walkeri* against copper exposure (Liess et al., 2001).

4.5. Combined effect analysis

Contamination of the environment is rarely a matter of single chemicals but rather of mixtures of components. This raises the question of the occurrence and relevance of combined effects for individual species as well as for communities or ecosystems. A differentiated methodology based on two pharmacologically founded concepts has been developed that comprise not only the possibility of developing rational and experimentally accessible approaches to this field but that also offers the opportunity to overcome the terminological confusion close to anarchy which still prevails in the field (Altenburger et al., 1993; Greco et al., 1995; Kortenkamp and Altenburger, 1998).

There is a vast body of literature studying how binary or multiple mixtures affect various biological responses usually observed in individuals of single species (Altenburger et al., 1993; Greco et al., 1995; Kortenkamp and Altenburger, 1999). The discussion employs terms like synergism or antagonism to qualify the observed effects. Synergism or antagonism are commonly taken to mean that the observed effect of a mixture was more or less than what had been expected. Thus the central question for any assessment of the effects provoked by mixtures is: What is a reasonable expectation for combination effects (Berenbaum, 1981, 1985, 1989)? Very early in the 20th century two different concepts that can be based on pharmacodynamic assumptions namely Concentration Addition and Independent Action were formulated that allow the calculation of expectable combined effects on the basis of information on the efficacy of the single components (Berenbaum, 1981; Greco et al., 1995). There are main differences between these concepts: Concentration Addition is based on the idea that one substance may act as an equitoxic dilution of another or in pharmacodynamic thinking calculates combined effects for substances that have a similar mechanism of action. Independent Action in contrast, regards effects of components as statistically independent and is thus thought to be valid for situations were the mixture components show dissimilar mechanisms of action (Grimme et al., 1996).

When moving from a pharmacodynamic level of molecular interaction to the assessment of mixture toxicities, a first question is whether simple ideas about combined effects at the level of molecular receptors translate into meaningful expectations at the level of intact organisms. Using two photosynthesis inhibitors with a known identical molecular binding site, Altenburger et al. (1990) studied mixture responses at different levels of plant responses employing Hill reaction measurements to quantify interaction with photosynthetic electron transport in isolated chloroplasts, photosynthetic oxygen production of algae after 15 minutes of exposure, cell volume growth performance after one growth phase and finally reproductive success after one generation. They were able to show that indeed concentration addition is a suitable concept for assessing combined effects on different levels of biological responses for this case of compound mixture with an identical mode of action. In subsequent studies the same group demonstrated for over a hundred different binary mixtures of pesticides and surfactants using an algae reproduction assay that indeed both concepts Concentration Addition and Independent Action provided quantitatively reasonable reference values for combined effect assessments (Faust et al., 1994; Altenburger et al., 1996). The case of multiple mixtures of compounds with unspecific modes of action like industrial chemicals such as solvents has been addressed in a series of studies by Könemann (1980, 1981), and by Hermens and coworkers (1982, 1984a,b, 1985a,b,c) using fish, daphnia and bacterial toxicity parameters. Even for cases where the concentrations of the individual chemicals were as low as 0.02 of their individual EC50 values significant combined effects were observable and close to what would be expected from concentration

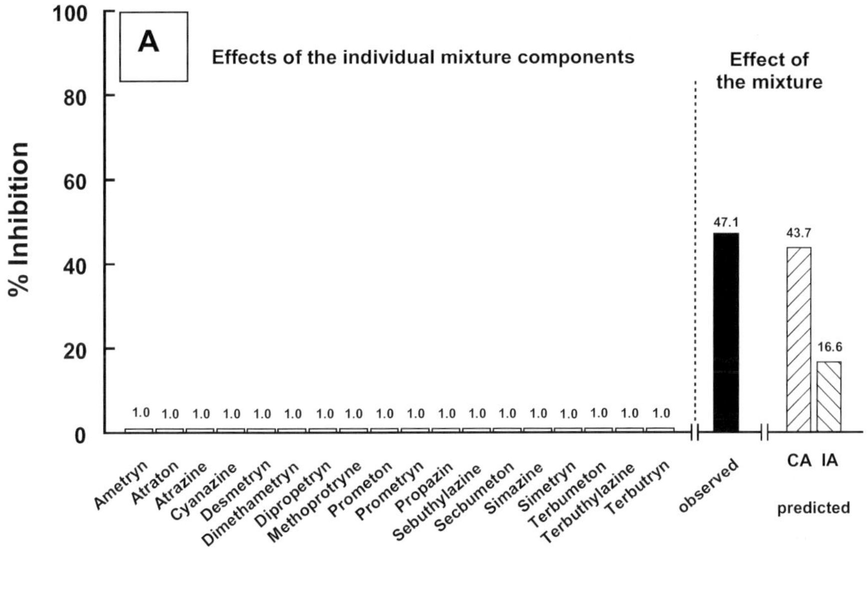

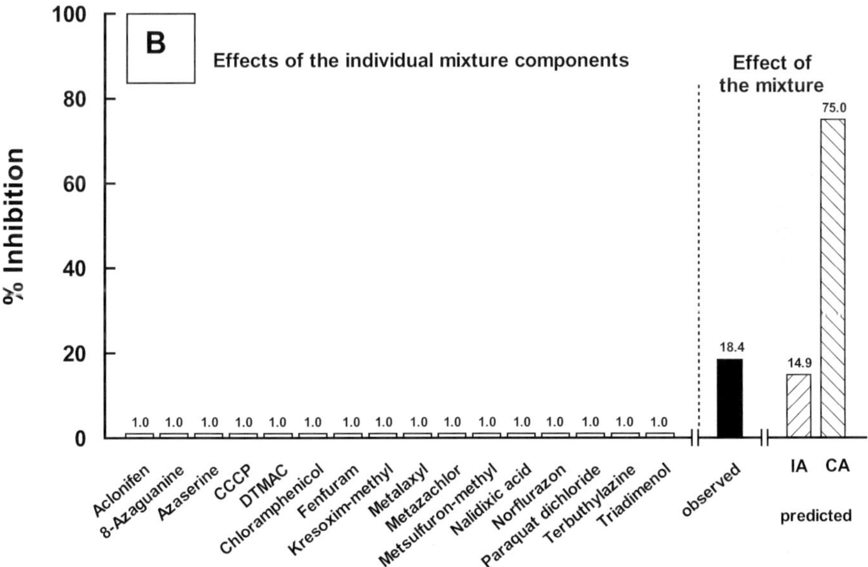

Figure 10. The mechanism of action provides a means to accurately calculate mixture toxicities for similarly (A) or dissimilarly (B) acting compounds on algal reproduction (adapted from Faust et al., 2001, 2002).

additive behaviour. The thus anticipated dispute as to whether concentration addition might be an universal model for higher levels of biological responses (Berenbaum, 1985; Pöch, 1993) became experimentally addressed in a series of papers by Grimme and colleagues (1996). They deliberately designed multiple target specific mixtures of drugs with well known molecular modes and mechanisms of action and investigated them in a bacterial and an algae toxicity assay (Altenburger et al., 2000; Backhaus et al., 2000; Faust et al., 2000). Figure 10 shows one of their striking findings, namely that indeed the knowledge about the mechanism of action provides a means to accurately calculate mixture toxicities for similarly or dissimilarly acting compounds (Faust et al., 2001, 2002).

The question whether the tools and understandings developed and validated in single species investigations are transferable to higher biological hierarchies has been addressed by Blanck and coworkers. In a first study an investigation on the combined effect of tri-n-butyl-tin and diuron on marine periphyton communities detected as pollution-induced community tolerance (PICT) was made (Molander et al., 1992). The authors were indeed able to detect not only single compound activities but also combination effects. From response surface modelling they concluded, that for the investigated mixture, the observed PICT-responses could be interpreted as occurrence of co-tolerance. Subsequent work performed with marine periphyton and epipsamon communities (communities, grown on sand) showed that while responses show higher variability as compared to lab-based single species algal toxicity testing, the type of mixture toxicities to be predicted and observed might be very similar to what has been described above (Vighi et al., in press).

The relevance of these recent advances in the understanding of mixture toxicity with regard to assessment schemes based on PNEC values for single compounds is currently discussed (Faust et al., 2001, 2002; Walter et al., 2002). Of course not all mixtures of contaminants to be found in the environment just behave as expected on the basis of a simplistic pharmacological reasoning. There are well documented cases of unexpected high combined effects on population and organismic levels of responses (Johnston et al., 1994, Babu et al., 2001). And there are of course investigations as to the mechanistic understanding of such interactions, which either focus on processes related to energy transducing membranes (Escher et al., 2001, Schweigert et al., 2000, 2001) or on interactions via altered internal contaminant concentration due to interference with biotransfomation enzymes (Johnston et al., 1994).

With regard to predictions of mixture toxicities to ecosystems apart from problems of correlated responses like co-tolerance occurrence or interference from indirect effects when considering mixtures with dissimilarly acting components, the definition of the type and ratio of mixtures to be assessed in terms of what is the actual exposure situation pose major challenges for the future.

4.6. Quantitative structure-activity relationships (QSAR)

As illustrated above there is the tremendous lack of even the most basic experimental data in ecotoxicology for most chemicals in everyday use. Considering a number like the about 100,000 chemicals which are listed in the European Inventory of Existing Commercial Chemical Substances (EINECS), it is apparent that neither resources nor the will may be allocated to change this picture in due course. The study of quantitative structure-activity relationships offers an approach to tackle the problem of lacking experimental data on biological effects. The scope for investigations of structure-activity relationships is to provide estimates of compound intrinsic biological activity properties in a systematic manner, i.e. to provide a generic toxicity profile. The basic principle lies in the comparison of several compounds of a similar structure (congenerity criteria) with respect to a defined biological activity at a fixed effect level. Various structural parameters may than be used to describe and analyse the observed effectiveness of the components of training and validation sets. Employing statistical methods e.g. regression techniques, predictive tools applicable for untested compounds can be derived. Good accounts of the principles of this methodology as utilised in ecotoxicology may be found in Nendza and Hermens (1995) and Schüürmann (1998). This approach has played a major role in reviving the concept of narcotic action of substances, QSARs derived from the correlating toxicities of non-polar, non-reactive organic compounds with compounds lipophilicity i.e. $logK_{ow}$ allow to estimate a baseline toxicity for almost any given organic chemical (Lipnick, 1989). Such values for an expectable minimum toxicity can be used immediately as a prediction in effect assessment, but may also serve as a reference to judge the existence of more specific interactions with biological systems.

Unlike its application in pharmacological studies where the preselection of an effect of interest is often highly appropriate, QSAR approaches to be used in ecotoxicology, have to cope with various possibilities of interactions of chemicals with biological systems. Thus the reflection of the compound selection is a crucial issue. Pioneering work of Verhaar et al. (1992) proposed a scheme based on earlier work to classify various organic chemicals into one of the following four classes: inert chemicals, less inert chemicals, reactive chemicals and specifically acting chemicals. Applying this scheme to 2000 chemicals labelled by the OECD as so-called high production volume chemicals (HPVCs) allows consideration of already 44% of these chemicals (Bol et al., 1993). For the compounds classified as inerts, QSAR equations to predict the short-term median aquatic toxicity values EC50 for fish, daphnia and algae based on a narcotic mode of action were calculated using the compounds octanol/water partioning coefficient $logK_{OW}$ as sole structure parameter (Verhaar et al., 1992). For the other groups that are expected to show somewhat higher toxicities due to interactions other than mere unspecific membrane disturbance, group specific empirical factors multiplied with the baseline toxicity value were proposed and used (Bol et al., 1993, Verhaar et al., 1992). This work became extended using various QSAR estimates for the toxicity for other organisms and endpoints like NOECs and deriving quality criteria for aquatic ecosystems based on species sensitivity distribution functions that were generated from the 19 estimated toxicity values (van Leeuwen et al., 1992). For risk assessment and management purposes this type of work allows coping with many chemicals that have yet not been assessed by regulators (Verhaar et al., 1992, Bol et al., 1993). However, it requires that hazard predictions based on suspicion rather than on numbers derived from base set routine test data become acceptable to risk regulators.

Inert chemicals evoking baseline toxicity, however, are to be among the least toxic substances, while higher toxicities of up to several orders of magnitude due to reactive or specifically acting compounds may be more crucial for ecosystems. Thus the challenge to discriminate further modes of action rather than just narcosis and allocate adequate QSARs has been taken up by various groups (Escher et al., 1999; Kapur et al., 2000; Marchini et al., 1999; Niculescu et al., 2000; Parkerton and Konkel, 2000; Vaal et al., 2000) of which two will be highlighted for illustration. A recent study by Basak and co-workers (Basak et al., 1998) employed molecular similarity, neural networks and discriminant analysis to assign the mode of action out of seven different types using acute fish toxicity data. For a set of 283 chemicals for which information as to the mode of action was available a correct assignment of 65 to 95% of these chemicals was possible. Similar outcomes were achieved when allocating 115 test chemicals to nine modes of action using quantum chemical descriptors and principal component analysis (Nendza and Müller, 2000).

Other tasks of QSAR studies regarding the predictive scope of this approach for ecosystems are oriented towards understanding the structural determinants of compounds to elicit effects of ecotoxic relevance. The recently heavily debated potency of various structurally unrelated pollutants to infere with the endocrine systems of heterotrophic organisms (Matthiesen, 2000; Jobling et al., 1995; Tyler et al., 1998) has launched activities to understand the structural determinants of estrogen receptor binding (e.g. Xing et al., 1999; Tong et al., 1998). For a group of nitroaromatic compounds that QSAR based effect assessment as proposed by the US-EPA would consider as merely narcotically acting and thus predicting a baseline toxicity, Schmitt et al. (2000) showed for algal reproduction toxicity data that these compounds are in general more toxic than nonpolar narcotics. Moreover, additional inclusion of quantum chemical electronic parameters like the energy of the lowest unoccupied orbital (E_{LUMO}) gained a consistent quantitative structure-activity relationship for all nitroaromatic compounds. In turn, employment of these structural parameters allowed to suggest additional modes of action in the organisms such as oxidative stress evoked from redox cycling of some of the compounds and toxicity from metabolites due to biotransformation (Fig. 11).

Finally, QSAR studies have occasionally treated the problem of chemical mixtures. The study of the mixtures of contaminants has been addressed using QSAR techniques to predict altered compound properties like modified solubilities and combined effects. Particularly, QSAR approaches have been used in the analysis of joint toxicity of chemicals to provide evidence for similar mode of action and thus concentration-additive mixture toxicity (see Section 4.5); to predict effect concentrations of untested components; to describe specific mixture effects deviating from expected responses; to discriminate between congeneric structures of dissimilar reactivity; to model exposure concentrations; and to derive mixture properties for prediction of joint toxicity (Altenburger et al., submitted).

Whatever the successes may be in filling data gaps by using quantitative structure-activities relationships, the major drawback regarding inference from this approach to ecosystems is the necessary preselection of effect parameters and targets at risk. This limitation has at least two aspects: firstly, the understanding of toxic action of pollutants in biosystems as a process involving pharmacokinetics and -dynamics as well as effect translation from mechanisms of interaction to modes of toxicity is ignored. Secondly, as QSAR studies perform and require comparisons of many compounds,

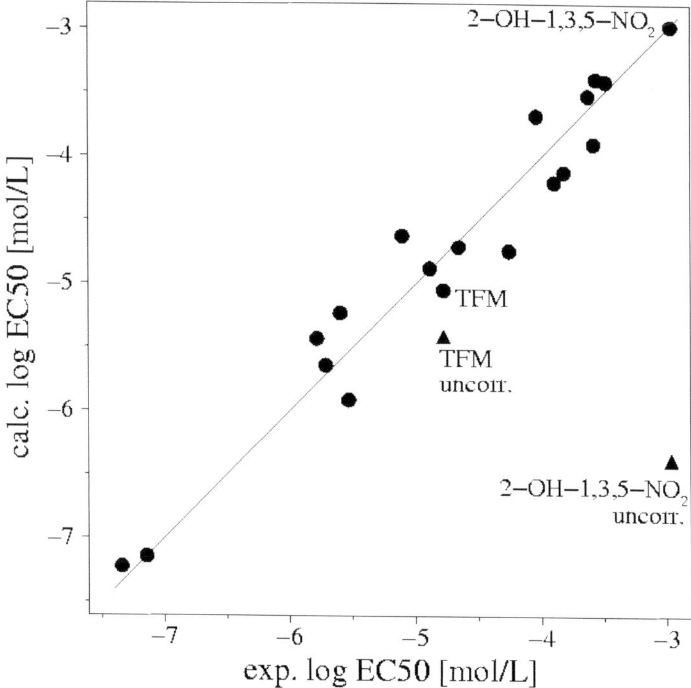

Figure 11. Calculated versus experimental log EC50 for inhibition of alga reproduction by nitrobenzenes using a three-variable regression model of the form log EC50 = -0.55 log D_{OW} + 1.69 E_{LUMO} − 34.3 $q_{nitro-N}$ + 18.4 (from Schmitt et al., 2000). For TFM and picric acid the triangles indicate their predicted log EC50 when using log K_{OW} instead of log D_{OW}. Abbr. D_{OW}, partition coefficient between octanol and water for the un-ionised species; E_{LUMO}, energy of the lowest unoccupied orbital; $q_{nitro-N}$, net atomic charge at the nitro nitrogen.

though in principle they may consider any type of effect in practice, this places high demand on testing capacity and limits use of laborius techniques. One therefore, will hardly find data allowing a QSAR analysis say for chronic toxicities not to speak of population or community level effects.

5. Scope for predictions

The methodologies introduced in the above sections all have their specific drawbacks concerning the scope to assess and predict ecosystem responses. The validity of quantitative hazard estimations of compounds or effluents based on laboratory investigations using single species is easily challenged on reasons of site-specific bioavailability of pollutants, intra- and interspecific variability, altered responses of organisms in the community context, relevance of environmental factors like feeding status influencing population sensitivities, and undetected indirect effects. Various approaches address these problems in extrapolating from isolated single species tests to higher levels of complexity. In the following we will briefly consider:

- *in situ*-toxicity testing;
- species sensitivity distributions;
- comparative studies using laboratory assays and micro- and mesocosm studies; and
- ecological considerations.

5.1. In situ-toxicity testing

A straightforward methodology to examine the influence of site-specific environmental milieu factors on estimations of effect concentrations is to perform toxicity testing *in situ*. Representing site-specific milieu factors in toxicity testing may mean including specific sediments or water bodies in static systems like tanks, ponds or ditches, flow-through systems like artificial stream experimentation (Debus et al., 1996; Drent and Kersting, 1993; Girling et al., 2000; Rand and Clark, 2000a,b), using bypass channels (Liess and Schulz, 1999) or caging of fish or macroinvertebrates directly on site (De la Torre et al., 2000; Ireland et al., 1996; Pereira et al., 1999; Pyle et al., 2001). A review on various existing designs for these kind of outdoor studies can be found in Caquet et al., (2000). Numerous parameters can thus be monitored and a major challenge becomes the task to reduce data to meaningful information (Girling et al., 2000; van Wijngaarden et al., 1996).

The environmental impacts of acid mine drainage (AMD) was investigated in a case study at the Puckett's Creek watershed in Virginia, USA using benthic macroinvertebrate sampling, *in situ*-toxicity testing with Asian clams (*Corbicula fluminea*), water column toxicity testing with the cladoceran *Ceriodaphnia dubia* and sediment toxicity testing with the cladoceran *Daphnia magna* and the midge larvae of *Chironomus tentans* (Soucek et al., 2000). Comparison of the different biological parameters investigated for 21 different sampling sites categorised for different AMD impacts revealed a fairly consistent pattern of biological responses for the sites exposed to acidic and neutral mining drainage. The water column testing for short-term survival of *Ceriodaphnia dubia* not only proved to be very sensitive in terms of distinction between different sites but correlated significantly with different indices describing the sampled benthic microinvertebrate community (r – values ranging from 0.49 to 0.81) (Soucek et al., 2000). The testing with clams (*Corbicula fluminea*) showed an almost identical response pattern regarding survival after 31 days of *in situ*-exposure compared to *Ceriodaphnia dubia*. A similar study performed on a long-abandoned mining site located in south-eastern Portugal compared laboratory test results for water column and solid phase samples with caged *in situ*-testing using the cladocerans *Ceriodaphnia dubia* and *Daphnia magna* (Pereira et al., 1999). Apart from a general good agreement in the observable short-term toxicities for 8 different sites at four sampling periods covering all seasons of the year, the mortality tended to be slightly higher for the bioassays, performed *in situ* and more similar to the solid phase tests (Pereira et al., 1999).

In a study with pyridaben, an active ingredient of a pesticide used as insecticide and acaricide, Rand and Clark (2000a,b) compared short-term toxicity findings for bluegill sunfish (*Lepomis macrochirus*) and mysids (*Mysidopsis bahia*) from laboratory studies using standard protocols with outdoor tank studies and employing natural photoperiod, single-pulse exposure and tanks filled with a defined sediment and specified water. They found that estimated LC50 values after 96 h exposure increased from laboratory conditions to outdoor studies for both organisms by about 1.5 orders of magnitude (Rand and Clark, 2000b). This significant decrease in toxicity is not too surprising considering that the actual concentration of pyridaben was halved in the tank studies about 8 h after application and regarding the low water solubility of pyridaben of about 12 µg/L as well as the high lipophilicity as characterised by an octanol water partition coefficient log K_{OW} of about 6.4. The environmental behaviour of this compound would thus be expected to strongly favour partioning from the water column and sorption onto organic particles and sediment (Rand and Clark 2000a). The degree to which these factors alter observable toxic effects and possibly compound assessment, however, will always be site-specific.

Liess and Schulz (1999) tried to link rainfall-induced surface runoff from arable land contaminated with several insecticides and subsequent exposure of the macroinvertebrate community in adjacent streams with the abundance of several macroinvertebrate species. They employed a runoff-triggered sampler to follow insecticide contamination after rainfall-induced runoff and were able to quantify parathion and fenvalerate exposure via the water column and suspended particles after several such events. In order to distinguish between stress factors accompanying runoff events in the stream like increase in current velocity and insecticide exposure, the authors used parallel bypass microcosms to isolate effects of contamination on survival and emergence of trichoptera larvae of *Limnephilus lunatus* and on survival of the amphipod *Gammerus pulex*. For a rainfall event where 6 µg/L of parathion could be detected in the swelling stream water for about one hour, significant decreases of the abundance of the populations of both organisms could be detected. These reactions are well in accordance with effect concentration data from several laboratory assays describing the short-term toxicity of parathion in *Gammerus* spp. (EPA-databank ECOTOX, http://www.epa.gov/ecotox/). Surprisingly, a short exposure period of only one hour suffices to reproduce these effects so exactly, that one might interpret the pharmacology of parathion on the basis of these findings as being very fast in uptake and provocation of mortality.

Maltby et al. (2000) investigated the biological impact of a point source discharge downstream from a bleaching work. Whole effluent testing in the laboratory predicted an acute toxicity for neonates of *Daphnia magna* that varied slightly in time regarding the dilution that proved to immobilise most neonates after 48 h of exposure. *In situ* toxicity tests with the same species and test regime confirmed this picture: while caged daphnids employed upstream of the discharge had little failure regarding survival, downstream of the effluent discharge all animals died within the 48 h of exposure. In a separate fractionation step (see above) the authors were able to attribute most of the observed toxicity to chlorine as the principal toxicant in the effluent.

Diamond and Daley (2000) were able to differentiate the picture on the predictive capability of whole effluent testing (WET) by reviewing data available for acute and chronic fish and daphnia whole effluent test data for the USA and relating these to assessments based on benthic macroinvertebrate inventories in various streams. The capacity of lab WET testing for predicting macroinvertebrate assemblages increased with frequency of WET tests being performed, with contribution of the discharger to the receiving water in terms of volume ratios, and when several types of tests were included in the assessments.

5.2. *Single species sensitivity distributions (SSDs)*

Instead of focusing on selected single numbers of single species to derive a prediction or assessment on the hazard imposed by a contamination, approaches have been developed that use more of the available information in probabilistic ways, i.e. basically regarding exposure and effect information as probability distributions. The biological reasoning for this derives from the view that the biological components of ecosystems might be regarded as assemblages of different species. Furthermore, it is widely assumed that there is no single most sensitive species regarding responsiveness to toxicants. Instead regarding responses of different species to toxicant exposure by modelling distribution of species sensitivities to a given chemical has been brought forward by Kooijman (1987) and van Straalen and Denneman (1989). Subsequently, several teams considered specific aspects of the occurrence of biological variances in response to toxicant exposure (Behra et al., 1999, Boutin and Rogers, 2000, Okkerman et al., 1991, McDaniel and Snell, 1999, Wagner and Lokke, 1991).

Two examples of the methodology are shown for the toxicity of the metal Cd using NOEC data for various soil organisms and for the toxicity of the insecticide lindane using NOEC values for aquatic organisms (Fig. 12, Traas et al., 2002). It can be seen that the unspecific toxicity of the metal results in a continuous distribution along the log concentration scale, which may be easily modelled by e.g. a logistic distribution function. Lindane on the other hand, as a specifically acting insecticide, produces jumps in the distribution, as non-target organisms will show clearly higher effect concentrations compared to arthropods. Depending on the goal one may even model this situation by employing different distribution functions for different taxa. An even clearer picture may be derived for herbicides such as atrazine (de Zwart, 2002). Two characteristic values suitable for risk assessment or prediction are also shown in Figure 12. When there is agreement on an intended level of protection in this example 95% of the species are to be protected (HC5 – hazardous concentration for 5% of the species in panel A) it is straightforward to estimate the corresponding concentration from the functional description. Vice versa, if an environmental concentration of a contaminant in the environment is known, a potentially affected fraction of species (PAF in panel B) may be derived. The various uses of the species sensitivity distributions in ecotoxicological risk assessment have just been compiled in a monograph by Posthuma et al. (2002).

While there is widespread acceptance of the achievements of this type of probabilistic approach to chemical hazard assessment, one has to be aware of different exposure profiles on a landscape scale due to different feeding strategies, discussed for example for bumblebees as compared to the standard test organism honeybee (Thompson and Hunt, 1999). This aspect is not to be modelled by species sensitivity distributions but matters e.g. in pesticide non-target assessments. A major technical drawback of the SSD concept is the requirement for sufficient sets of available experimental data for different species for a given toxin. Also, there are issues raised

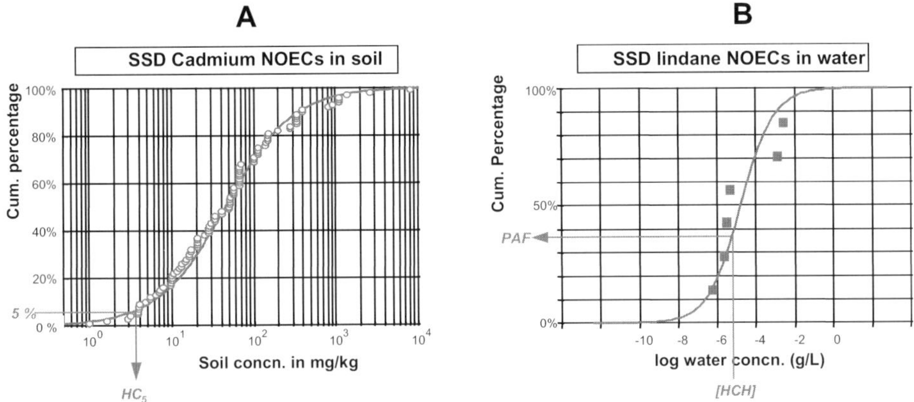

Figure 12. The use of species sensitivity distributions (SSDs) for deriving risk limits such as a certain hazardous concentration (HC$_5$ in panel A) and for risk assessment using potentially affected fractions of species (PAF in Panel B) (from Traas et al., 2002).

reflecting the fact that species sensitivity distributions are modelled on the basis of individual level effect parameters which only for populations with growth rates of about one will more or less mirror effects on populations (Forbes et al., 2001). Thus, including information on population dynamics could be an issue of further development in order to avoid undue over- or underestimations of risks.

5.3. Comparison of responses from single and multispecies testing

Several studies were performed to compare the effects of toxicants in single species tests and micro- or mesocosms directly, of which a few will be cited here in their main conclusions.

Traunspurger et al. (1996) evaluated the effects of the herbicide isoproturon in single species tests and mesocosms. They noticed a higher sensitivity of the laboratory tests and concluded, that theses tests are sensitive instruments for screening possible effects, though no effect concentrations in the mesocosms could be detected in this study. Jak et al. (1996) added a strain of *Daphnia magna* from the laboratory to a lake mesocosm study, to evaluate the effects of metals. This strain showed the same EC$_{50}$ in the mesocosms as evaluated in a single species test. On the other hand they noticed that species from the ecosystem were more sensitive and concluded that accurate safety factors must be considered to avoid the replacement of sensitive species and shifts in ecosystem function and structure. Rand and Clark (2000a,b) used three approaches for a risk assessment of the compound pyridaben. They estimated the environmental concentration (EEC) of the chemical and performed acute and chronic single species tests. By combining this data in a species sensitivity distribution, they evaluated, that there was a high risk for the most sensitive species from laboratory tests (based on EC5 level). On the other hand their outdoor studies showed a weaker sensitivity than the laboratory studies, resulting in a water-effect ratio of 18–24. They concluded that abiotic factors (photodegradation) reduced the bioavailability in the mesocosm study. Lampert et al. (1991) tested the effects of atrazine in systems of different complexity. Next to single species tests, artificial food chains and enclosure experiments were established. They showed, that the natural communities were the most sensitive and concluded, that the sensitivity of the systems increased with increasing complexity and that non-target-organisms (here daphnids) could be even more affected than the target organisms of the toxicant.

One important factor, changing the sensitivity of laboratory species tests and field populations is the development of tolerance. Ivorra et al. (2002) demonstrated, that a strain of the benthic diatom *Gomphonema parvulum* isolated from a stream, chronically subjected to high Zn (and Cd) contamination was more tolerant to Zn in the laboratory than the strain from an unpolluted stream. This tolerance was persistent for 2 years, suggesting a genetic based difference in tolerance. They concluded, that next to genetic adaptation tolerance might be related to different uptake rates or different intracellular pools of phytochelatins or glutathione. Barata and Baird (2000) compared the life history responses of field and laboratory populations of *Daphnia magna*, exposed to Cd and ethyl parathion. The results showed that the field population have a similar or greater tolerance to cadmium and ethyl parathion than the laboratory populations but the breadth of the tolerance distribution was higher. The authors concluded, that tolerance is strongly influenced by genetic factors; the use of genetically homogeneous laboratory populations has limited relevance in predicting long-term responses of field populations to toxic chemicals, however, short-term responses seem better predictable.

A comprehensive review of critically evaluated literature on model ecosystems studies for assessing substances deleterious effects on biocoenosis and a comparison with data from a single species database (Länge et al., 1998) has been provided by ECETOC (1997). Data for 34 chemicals were evaluated using marine, static freshwater and flowing freshwater ecosystem models, with a bias towards the latter. These data

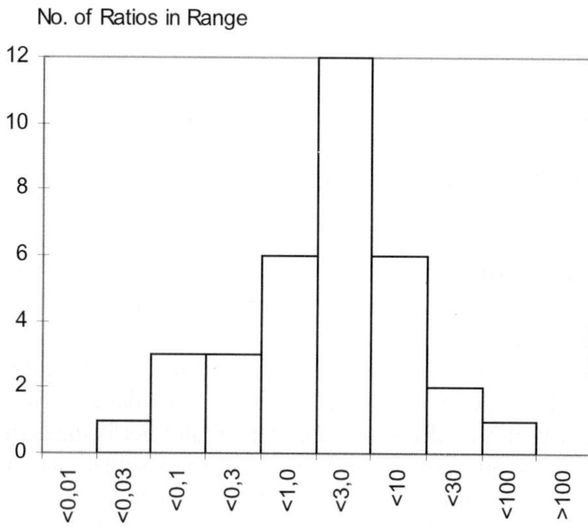

Figure 13.   Ratios of single species NOECs to model ecosystem NOECs for 34 chemicals of different chemical classes (from ECETOC, 1997).

were extracted out of 1108 original literature references selected for provision of NOEC and LOEC data. The main findings are aggregated in Figure 13. The median ratio for NOEC values derived from single species studies to model ecosystem investigations was 1.5 and 8.1 for the 90th percentile for all 34 chemicals thereby always using the most sensitive species or endpoints respectively. This clearly demonstrates that regulatory procedures using data for most sensitive species plus an additional assessment factor can be regarded as rational with respect to the available evidence.

5.4. Ecological considerations

Using effect data generated for specific chemicals employing single species for assessing or predicting effects for ecosystems commonly relies on comparison of predicted exposure concentrations for a specific media and lowest estimated no effect concentrations for a specified set of biotests. Many regulatory schemes use this procedure of a hazard quotient approach directly in form of so-called PEC/PNEC (predicted environmental concentration/predicted no effect concentration) or TER (toxicity/exposure) ratios to use derived values in decision rules of tier one risk assessment. Ecological thinking regarding this assessment strategy now reflects the plausibility of the assumed exposure and effect concentrations taken as basis for assessment. Mainly dissipation processes in reducing exposure, ecotoxic relevance of specific effects, and the importance of recolonisation of habitats are discussed as issue modifying assessments based on single species considerations in higher tier risk assessment schemes (Campbell et al., 1999; Heger et al., 2001).

Huber provides an early example reviewing and assessing the ecotoxicological relevance of atrazine in aquatic systems (Huber, 1993). Concerning the exposure estimations for atrazine he puts emphasis on the possibility of organisms to metabolise atrazine along different pathways, and highlights the biased sampling strategies underlying occurrence reports of atrazine in aquatic media. On the effect assessment side he raises the issue of ecosystem recovery, which he judges to occur very fast after exposure, so that his assessment of a relevant damage to aquatic ecosystems ends about one order of magnitude higher than provided on the basis of single species results (Huber, 1993). In view of its many years of application as a herbicide the author comes to conclude that for atrazine "the residual risk appear to be relatively low and easy to predict" (Huber, 1993).

Suter et al. (1999) undertook a site specific ecological assessment concerning the risk for a given fish population in the Clinch River, exposed to a variety of metals and PCB contaminations mainly due to activities of the US Department of Energy on its Oak Ridge Reservation in Tennessee. Biological survey data for different sites revealed in comparison to an ecological similar and relatively uncontaminated creek that all sites showed lower fish species richness and abundance. Multiple lines of evidence were then followed to investigate whether these community impairments could be detected and assessed unambiguously by other methods. Histopathological and reproductive bioindicators, ambient water toxicity estimated in development tests with fish eggs of medaka and redbreast channel catfish as well as short-term toxicity tests with *Ceriodaphnia* and fathead minnow, lethal body burdens measurements, and analysis of selected metals in the aquatic medium were used for that purpose. None of these lines used separately, provided unambiguous explanation for a significant risk, though all were indicative for at least periodic events of toxic contamination at one particular site, the Poplar creek embayment. Weighing the evidence subsequently in what could be called an eco-epidemiological approach, made plausible that the community survey results were consistent with a significant toxic effect at the Poplar creek embayment site, though habitat influence could not be completely discarded. Similar conclusions about the necessity to connect independent lines of evidence, i.e. combine chemical and toxicological information to identify sites with ecotoxic levels of contamination seen at population level, were drawn from the investigations of the benthic invertebrates in the same area (Jones et al., 1999).

A different line of ecological reasoning looks at indirect effects of contaminants on communities in ecosystems. If such effects occur in the sense that the effect is truly dependent on the interaction between species e.g. in a food web, there is no point in attempting to predict this by single species considerations. Other effects are, however related to differential sensitivities against toxicants in communities. These in turn may affect community composition by selecting for more tolerant species. Altered composition structures and functions may be detected by suitable methodologies such as for example the pollution induced community tolerance (PICT) (Blanck and Dahl, 1996, Blanck et al., 1988, Rutgers and Breure, 1999). In an early example Blanck and Dahl were able to demonstrate that shifts in marine periphyton community tolerance against exposure to TBT from ship antifouling paints reflected altered community compositions. Moreover, these effects were detectable at concentrations below those which could be predicted by surrogate species testing (Blanck and Dahl, 1996). The notion that changes in community structure are to be assessed as deleterious has been challenged on the basis that functional replacement of one species by another in a community might be regarded as ecologically acceptable (Heger et al., 2001). This perception of functional redundancy as a recovery potential is however disputed (Rutgers and Breure, 1999, Blanck et al. 1988).

The principal limitations of any ecologically oriented assessment are manifold: Ecosystems are usually unique, i.e. assessments made for one contaminant in a specific system cannot easily be inferred to other systems. Ecological considerations like functional replacement or recolonisation potencies are not consented criteria in chemical risk assessment unlike lethality or reproductive disturbances for individuals or populations. To develop these will be a long process as can be learned from the debate on the relevance of adverse effects of relevance in human toxicology. Finally, any statements on acceptability or negligibility of residual risks are not scientifically based. It needs simple thought to see that the residual risk is what we do not consider and therefore we cannot make statements as to its quantity or predictability. Of course, one may speculate or provide political judgements, but this should be clearly stated.

Besides widespread empirical evidence, there are theoretical considerations on the possibility of effect predictions for ecosystems. One important application of single species tests, used to assess and predict effects of chemicals on ecosystems are their implementation into models of the population dynamics in ecosystems. Model parameters are e.g. the growth rate, birth rate, respiration rate or death rate of a species, exposed to a chemical or the grazing pressure of a predator. These parameters, derived from laboratory tests for several chemicals could be inserted in a model, simulating species interaction or simple ecosystem features, e.g. seasonal temperature effects. Such models could then calculate direct and synergistic effects in the ecosystem, derived from species interaction of different trophic levels.

When looking for effects at larger scales in time or space, experimental approaches are often not practicable. Especially in complex ecosystems or when diffuse contamination at low doses but longer time periods have to be considered, disturbances often could first be detected after long time periods. Scales of investigation should then be considered in the context of generation times of the species under observation or of vegetation periods or abiotic cycles of the ecosystem. These time scales, often comprising several years, cannot be recorded with a mesocosm approach. Corresponding field studies are even more time consuming and often allow only a retrospection. Modelling effects of chemicals on ecosystems offer one practicable and often the only opportunity, to make prognoses on longer time scales.

In ecosystems, exposed to greater seasonal oscillations in contamination, the timing of a contamination could be of importance, when predicting xenobiotic effects. Such time courses could be e.g. an annual temperature effect or migration events of a population. In simulation approaches, the sensitive periods of the ecosystem could be detected, improving the efficiency of monitoring programmes.

On the other hand, the complexity of most ecosystems requires reductionistic approaches, not considering all parameters of relevance in one simulation. So results of simulations must be taken with caution for deterministic predictions. They can be used as a prognostic tool that reveals possible reactions of the ecosystem and offer the opportunity, to expose effects, which have not been considered, yet. It is their advantage to work out, which abiotic parameters or which level of organisms will be important for predictions. They could give hints, about what should be investigated and in which scales of time and space observations should be made, so being a useful tool, when planning new experimental designs. Further models, based on one set of data, cannot implicate all possible accidents. Computer based multiple simulations or the use of stochastic models are necessary to calculate the probability of effects, that could be expected. One possible tool, to consider the uncertainty of the model parameters, are Monte-Carlo-simulations. They do not base on fixed model parameters but on the probability of these values.

As an example Seitz and Ratte (1991) and Seitz and Poethke, (1995) developed a simulation model for pelagic systems of a deep dimictic lake. The aim of the study was, to derive a prognostic model, which reveals the potential reactions of this ecosystem to xenobiotics on the level of organisms. They reduced their system to two groups of algae and zooplankton and one fish group. As abiotic parameters, the seasonal variations of temperature, light and the nutrients phosphate and nitrate were implicated. As input data to the model, stress reactions of the organisms, caused by toxicants (e.g. increased respiration rates of cladoracea; decreased photosynthetic rates of algae) were considered. Their investigations showed several generalisable results, in which case models of population dynamics in ecosystems could help to understand effects of xenobiotics. The expectations of Seitz and Poethke (1995) were confirmed in that herbicides reduced the primary production of the phytoplankton and insecticides increased the respiration rates of zooplankton and, as an indirect effect the biomass of algae. These results were clear, when the simulation period was one year (one vegetation period). However, simulating over 10 years revealed contradicting results, where the population dynamics of the observed species developed against this hypothesis, and approached a chaotic system. This demonstrated, that conventional test systems e.g. micro- or mesocosms are limited in their scope to detect long-term effects.

5.5. What is the question?

Summarising, what has been laid out above, one may conclude from the review of current literature that all evidence so far shows, that single species data on the toxicity of pollutants can be used to predict the potential of adverse effects in ecosystems. There is no evidence that complex model ecosystems are systematically more or less sensitive to toxicants than single species tests. Higher variation in ecosystem as opposed to single species studies and the question of appropriate observation parameters for comparing different systems may cause technical difficulties in determining low effect concentrations. Additional problems arise in determining the correct concentration scale, as the environmental milieu effective in ecosystems may greatly alter the bioavailability of toxicants and may therefore show apparent lower toxicity. Also, as single species investigation commonly employ physiologically optimised conditions, effects of environmental factors on the sensitivity of species responses are easily overlooked. Principal limitations for extrapolation emerge when longer time scales are of concern or when structures or processes above the level of populations are affected. Additionally, ecological issues considered from a recovery perspective like recolonisation or functional replacement of species might modify assessment views.

The major challenge for an appropriate use of the many techniques available to gain single species information and the derivation of consistent assessments for ecosystems is to develop the right question for a predictive effort. This concern is illustrated in Table 3 which tries to distinguish commonly found goals for prediction in studies on contaminants effect and allocates priorities to various criteria regarding the suitability of a given biomonitor for an anticipated purpose.

Three thoughts for future perspectives to improve and refine the use of single species investigations for the assessment and prediction of adverse effects of chemicals in ecosystems shall conclude this chapter. (i) Laboratory studies using single species need to be more precise in what they want to predict in terms of ecosystem structure and function which hopefully leads to more focused instrumentation and approaches. (ii) Prediction and assessment of the effects of contaminants on ecosystems should always be regarded as a process. There is no end in itself, thus new evidence or theoretical considerations should be incorporated for adjustment or improvement. Practically speaking, when there is a requirement to assess a specific chemical at a given time, say in a pesticide admission process, we should not trust our prediction efforts in risk assessment more than a weather forecast, but instead acknowledge our ignorance accordingly in risk management and install appropriate monitoring tools to correct for false positive evidences coming up. (iii) Intelligent experimentation should replace too much standardised protocol testing.

Table 3. Matrix of application goals and suitability criteria for the utilisation of single species biotest systems.

Goals for prediction or assessment	Suitability criteria for selection of a bioindicator, biomonitor, or biotest							
	Fast and easy	High reproduci-bility	Accuracy	Precision	High detection specificity	High detection sensitivity	Scope for inference	Validation of effect quality
Remediation need	0	−	+	−	+	0	−	0
Remediation success	−	+	+	0	0	+	+	−
Identification of causes of ecotoxicity	0	0	+	−	+	−	0	+
Comparison of chemicals	+	+	0	+	−	−	0	−
Mode of toxic action	−	−	+	−	+	0	+	+
Identification of vulnerable ecosystem structure or function	−	−	+	−	0	0	+	+
Effects of low contaminant concentration/chronic impact assessment	0	0	+	0	−	+	0	0
Complex sample contamination/combined effects	0	0	0	+	0	0	0	+
Routine surveillance	+	0	−	+	0	+	0	−

+ important, 0 valuable, − less important

Acknowledgements

We acknowledge the constructive criticism of the manuscript provided by Matthias Liess, as well as by the anonymous reviewers.

This contribution draws from experience of several projects of which explicitly the following are gratefully acknowledged: Ökotoxikologische Testbatterie (DBU, 40–111422), PREDICT (*ENV4–CT96–0319;* http://www-user.uni-bremen.de/~predict/), BEAM (*EVK1–1999–00055;* http://www.aquatox.uni-bremen.de/beam/), and SAFIRA (BMBF, O2WT9949/3, http://safira.pro.ufz.de/).

References

Altenburger, R., Backhaus, T., Boedeker, W., Faust, M., Scholze, M., Grimme, L.H., 2000. Predictability of the toxicity of multiple chemical mixtures to *Vibrio fischeri*: mixtures composed of similarly acting chemicals. Environ. Toxicol. Chem. 19, 2341–2347.

Altenburger, R., Backhaus, T., 2000. Der Faktor Zeit bei der Beurteilung von biologischen Wirkungen. In: Mücke W., Link, W. (Eds), Biotests in der Praxis. Institut für Toxikologie und Umwelthygiene, Technische Universität München, pp. 61–74.

Altenburger, R., Bödeker, W., Faust, M., Grimme, L.H., 1990. Evaluation of the isobologram method for the assessment of mixtures of chemicals. Combination effect studies with pesticides in algal biotests. Ecotoxicol. Environ. Saf. 20, 98–114.

Altenburger, R., Bödeker, W., Faust, M., Grimme, L.H., 1993. Aquatic toxicology, analysis of combination effects. In: Corn, M. (Ed.), Handbook of Hazardous Materials. Academic Press, San Diego, pp. 15–27.

Altenburger, R., Bödeker, W., Faust, M., Grimme, L.H., 1996. Regulations for combined effects of pollutants: consequences from risk assessment in aquatic toxicology. Food and Chemical Toxicology 34, 1155–1157.

Altenburger, R., Callies, R., Grimme, L.H., Mayer A., Leibfritz, D., 1995. The mode of action of glufosinate in algae: the role of uptake and ammonia assimilation pathways. Pesticide Science 45, 305–310.

Altenburger, R., Nendza, M., Schüürmann, G. submitted. Mixture toxicity and its modeling by quantitative structure-activity relationships. Environ. Toxicol. Chem.

Anon., 1992. Report of the United Nations Conference on the Human Environment, Stockholm, 5–16 June 1972 (United Nations publication, Sales No. E.73.II.A.14 and corrigendum).

Babu, T.S., Marder, J.B., Tripuranthakam, S., Dixon, D.G., Greenberg, B.M., 2001. Synergistic effects of a photooxidized polycyclic aromatic hydrocarbon and copper on photosynthesis and plant growth: evidence that in vivo formation of reactive oxygen species is a mechanism of copper toxicity. Environ. Toxicol. Chem. 20, 1351–1358.

Backhaus, T., Altenburger, R., Bödeker, W., Faust, M., Scholze, M., Grimme, L.H., 2000. Predictability of the toxicity of a multiple mixture of dissimilarily acting chemicals to *Vibrio fischeri*. Environ. Toxicol. Chem. 19, 2348–2356.

Backhaus, T., Froehner, K., Altenburger, R., Grimme, L.H., 1997. Toxicity testing with *Vibrio fischeri*: a comparison between the long term (24 h) and short term (30 min) bioassay. Chemosphere 35, 2925–2938.

Barata, C., Baird, D.J., 2000. Determining the ecotoxicological mode of action from measurements made on individuals: results from instar-based tests with *Daphnia magna* Straus. Aquatic Toxicol. 48, 195–209.

Barata, C., Baird, D.J., Markich, S.J., 1999. Comparing metal toxicity among *Daphnia magna* clones: an approach using concentration-time-response surfaces. Arch. Environ. Contam. Toxicol. 37, 326–331.

Basak, S.C., Grunwald, G.D., Host, G.E., Niemi, G.J., Bradbury, S.P., 1998. A comparative study of molecular similarity, statistical, and neural methods for predicting toxic modes of action. Environ. Toxicol. Chem. 17, 1056–64.

Behra, R., Genomi, G.P., Joseph, A.L., 1999. Effect of atrazine on growth, photosynthesis, and between-strain variability in *Scenedesmus subspicatus* (Chlorophyceae). Arch. Environ. Contam. Toxicol., 37, 36–41.

Berenbaum, M.C., 1981. Criteria for analysing interactions between biologically active agents. Adv. Cancer Res. 35, 269–335.

Berenbaum, M.C., 1985. The expected effect of a combination of agents: the general solution. J. Theor. Biol. 114, 413–431.

Berenbaum, M.C., 1989. What is synergy? Pharmacol. Rev. 41, 93–141.

Betts, K.S., 1998. Chemical industry pressured to test high-production volume chemicals. Environ. Sci. Technol. 32, 251A.

Blanck, H., Dahl, B., 1996. Pollution-induced community tolerance (PICT) in marine periphyton in a gradient of tri-n-butyltin (TBT) contamination. Aquatic Toxicol. 35, 59–77.

Blanck, H., Wängberg, S.A., Molander, S., 1988. Pollution-induced community tolerance – a new eco-toxicological tool. In: Cairns, J.J., Pratt, J.R. (Eds), Functional Testing of Aquatic Biota for Estimating Hazards of Chemicals. ASTM STP 988, American Society for Testing Materials, Philadelphia, pp. 219–230.

Boening, D.W., 1998. Toxicity of 2,3,7,8-tetrachlorodibenzo-p-dioxin to several ecological receptor groups: a short review. Ecotoxicol. Environ. Safety 39, 155–163.

Bol, J., Verhaar, H.J.M., van Leeuwen, C.J., Hermens, J.L.M., 1993. Predictions of the aquatic toxicity of high-production-volume-chemicals. Part A. Introduction and methodology. The Hague, Ministry of Housing, Physical Planing and Environment, Report 1993/9A.

Boutin, C., Rogers, C.A., 2000. Pattern of sensitivity of plant species to various herbicides – an analysis with two databases. Ecotoxicology 9, 255–271.

Brack, W., Altenburger, R., Ensenbach, U., Möder, M., Segner, H., Schüürmann, G., 1999. Bioassay-directed identification of organic toxicants in river sediment in the industrial region of Bitterfeld (Germany) – a contribution to hazard assessment. Arch. Environ. Contam. Toxicol. 37, 164–174.

Caisukant, Y., Yu, Q., Connell, D.W., 1999. The internal critical level concept of nonspecific toxicity. Res. Environ. Contam. Toxicol. 162, 1–41.

Campbell, P.J., Arnold, D.J.S., Brock, T.C.M., Grandy, N.J., Heger, W., Heimbach, F., Maund, S.J., Streloke, M., (Eds), 1999. Guidance document on higher-tier aquatic risk assessment for pesticides (HARAP). SETAC, Brussels.

Caquet, T., Lagadic, L., Sheffield, S.R., 2000. Mesocosms in ecotoxicology (1): outdoor aquatic systems. Rev. Environ. Contam. Toxicol. 165, 1–38.

Chapman, P.M., Cadwell, R.S., Chapman, P.F., 1996. A warning: NOECs are inappropriate for regulatory use. Environ. Toxicol. Chem. 15, 77–79.

Christensen, E.R., 1984. Dose-response functions in aquatic toxicity testing and the Weibull model. Water Res. 18, 213–221.

De la Torre, F., Ferrari, L., Salibán, A., 2000. Long-term in situ toxicity bioassays of the Reconquista river (Argentina) water with *Cyprinus carpio* as sentinel organism. Water, Air and Soil Pollut. 121, 205–215.

Debus, R., Fliedner, A., Schäfers, C., 1996. An artificial mesocosm to simulate fate and effect of chemicals: technical data and initial experience with the biocenosis. Chemosphere 32, 1813–1822.

Desbrow, C., Routledge, E.J., Brighty, G.C., Sumpter, J.P, Waldock, M., 1998. Identification of estrogenic chemicals in STW effluents. 1. Chemical fractionation and in vitro biological screening. Environ. Sci. Technol. 32, 1549–1558.

de Zwart, D., 1995. Monitoring water quality in the future, Vol. 3, Biomonitoring. The Hague, VROM, 83.

de Zwart, D., 2002. Observed regularities in species sensitivity distributions for aquatic species. In: Posthuma, L., Suter, G.W., Traas, T.P. (Eds), The Use of Species Sensitivity Distributions (SSD) in Ecotoxicology. CRC Press, Boca Raton, pp. 133–154 (2002).

Diamond, J., Daley, C., 2000. What is the relationship between effluent toxicity and instream biological condition? Environ. Toxicol. Chem. 19, 158–168.

Drent, J., Kersting, K.,1993. Experimental ditches for research under natural conditions. Water Res. 27, 1497–1500.

ECETOC (European Centre for Ecotoxicology and Toxicology of Chemicals), 1997. The value of aquatic model ecosystem studies in ecotoxicology. Technical report no. 73. ECETOC, Brussels.

EEC (European Economic Community), 1990. Council Regulation on the evaluation and the control of the environmental risks of existing substances. COM (90) 227-final-syn 276.

EEC (European Economic Community), 1991. Council Directive 91/414/EEC of 15 July 1991 concerning the placing of plant protection products on the market. Official Journal L 230, 19 August 1991 pp. 1–32.

Escher, B., Hunziker, R.W., Schwarzenbach, R., 2001. Interaction of phenolic uncouplers in binary mixtures: concentration additive and synergistic effects. Environ. Sci. Technol., in press.

Escher, B., Hunziker, R., Schwarzenbach, R., 1999. Kinetic model to describe the intrinsic uncoupling activity of substituted phenols in energy transducing membranes. Environ. Sci. Technol. 33, 560–570.

Escher, B., Hermens, J., 2002. Modes of action in their role in ecotoxicology: body burdens, species selectivity, QSARs, and mixture effects. Environ. Sci. Technol. 36, 4201–4217.

Fahl, G.M., Kreft, L., Altenburger, R., Faust, M., Bödeker, W., Grimme, L.H., 1995. pH-dependent sorption, bioaccumulation and algal toxicity of sulfonylurea herbicides. Aquatic Toxicology 31, 175–187.

Faust, M., Altenburger, R., Backhaus, T., Blanck, H., Bödeker, W., Grammatica, P., Hamer, V., Scholze, M., Vighi, M., Grimme, L.H., 2001. Predicting the joint algal toxicity of multi-component s-triazine mixtures at low-effect concentrations of individual toxicants. Aquatic Toxicol. 56, 13–32.

Faust, M., Altenburger, R., Backhaus, T., Blanck, H., Bödeker, W., Grammatica, P., Hamer, V., Scholze, M., Vighi, M., Grimme, L.H., 2002. Joint algal toxicity of 16 dissimilarly acting chemicals is predictable by the concept of independent action. Aquat. Toxicol. 1–21.

Faust, M., Altenburger, R., Backhaus, T., Bödeker, W., Scholze, M., Grimme, L.H., 2000. Predictive assessment of the aquatic toxicity of multiple chemical mixtures. J. Eur. Qual. 29, 1063–1068.

Faust, M., Altenburger, R., Bödeker, W., Grimme, L. H., 1994. Algal toxicity of binary combinations of pesticides. Bull. Environ. Contam. Toxicol. 53, 134–141.

Forbes, V.E., Calow, P., Sibly, R.M., 2001. Are current species extrapolation models a good basis for ecological risk assessment? Environ. Toxicol. Chem. 20, 442–447.

Forbes, V.E., Forbes, T.L., 1994. Ecotoxicology in Theory and Practice. Chapman & Hall, London.

Ford, L., 2001. Development of chronic aquatic water quality criteria and standards for silver. Water Environ. Res. 73, 248–253.

Girling, A.E., Tattersfield, L.J., Mitchell, G.C., Pearson, N., Woodbridge, A.P., Bennett, D., 2000. Development of methods to assess the effects of xenobiotics in outdoor artificial streams. Ecotoxicol. Environ. Safety 45, 1–26.

Goser, B., 1997. Dichteabhängige Änderung der Entwicklung und reproduktion bei Cladoceren. Ursachen und ökologische Bedeutung. Shaker Verlag, Aachen. p. 210.

Greco, W., Bravo, G., Parsons, J.C., 1995. The search for synergy: a critical review from a response surface perspective. Pharmacol. Rev. 47, 331–385.

Grimme, L.H., Faust, M., Bödeker, W., Altenburger, R., 1996. Aquatic toxicity of chemical substances in combination: still a matter of controversy. Human Ecol. Risk Assess. 2, 426–433.

Grimme, L.H., Rieß, M.H., Manthey, M., Faust, M., Altenburger, R., 1993. Cell physiological parameters to detect ecotoxicological risks. Sci. Total Environ. Suppl. 741–748.

Halpern, S., 1993. United Nations Conference on Environment and Development: Process and Documentation. Academic Council for the United Nations System (ACUNS) Reports and Papers, no. 2. Providence, RI, Academic Council for the United Nations System (ACUNS).

Heger, W., Brock, T.C.M., Giddings, J., Heimbach, F., Maund, S.J., Norman, S., Ratte, H.-T., Schäfers, C., Streloke, M., 2001. Proceedings of the CLASSIC Workshop (Community Level Aquatic System Studies Interpretation Criteria). SETAC US Publication, in press.

Hermens, J., Canton, H., Janssen, P., de Jong, R., 1984a. Quantitative structure–activity relationships and toxicity studies of mixtures of chemicals with anaesthetic potency: acute lethal and sublethal toxicity to *Daphnia magna*. Aquat. Toxicol. 5, 143–154.

Hermens, J., Canton, H., Steyger, N., Wegman, R., 1984b. Joint effects of a mixture of 14 chemicals on mortality and inhibition of reproduction of *Daphnia magna*. Aquat. Toxicol. 5, 315–322.

Hermens, J., Broekhuyzen, E., Canton, H., Wegman, R., 1985a. Quantitative structure–activity relationships and mixture toxicity studies of alcohols and chlorohydrocarbons: effects on growth of *Daphnia magna*. Aquat. Toxicol. 6, 209–217.

Hermens, J., Busser, F., Leeuwangh, P., Musch, A., 1985b. Quantitative structure–activity relationships and mixture toxicity studies of organic chemicals in Photobacterium *phosphoreum*: the Microtox test. Ecotoxicol. Environ. Saf. 9, 17–25.

Hermens, J., Leeuwangh, P., Musch, A., 1985c. Joint toxicity of mixtures of groups of organic aquatic pollutants to the guppy (Poecilia reticulata). Ecotoxicol. Environ. Saf. 9, 321–326.

Hermens, J., Leeuwangh, P., 1982. Joint toxicity of mixtures of 8 and 24 chemicals to the guppy (Poecilia reticulata). Ecotoxicol. Environ. Saf. 6, 302–310.

Huber, W., 1993. Ecotoxicological relevance of atrazine in aquatic systems. Environ. Toxicol. Chem. 12, 1865–1881.

Ireland, D.S., Burton, G.A., Heiss, G.G., 1996. In situ toxicity evaluations of turbidity and photoinduction of polycyclic aromatic hydrocarbons. Environ. Toxicol. Chem. 15, 574–581.

Ivorra, N., Barranguet, C., Jonker, M., Kraak, M.H.S., Admiraal, W., 2002. Differences in Zn tolerance in strains of the freshwater microbenthic diatom *Gomphonema pavulum* (Bacillariophycede). Envir. Poll. 116 (1), 147–157.

Jak, R.G., Maas, J.L., Scholten, M.C.Th., 1996. Evaluation of laboratory derived toxic effect concentrations of a mixture of metals by testing fresh water plantkon communities in enclosures. Water Res. 30, 1215–1227.

Jobling, S., Reynolds, T., White, R., Parker, M.G., Sumpter, J.P., 1995. A variety of environmental persistent chemicals, including some phthalate plasticizers, are weakly estrogenic. Environ. Health Perspec. 103, 582–587.

Johnston, G., Walker, C.H., Dawson, A., 1994. Interactive effects of prochloraz and malathion in pigeon, starling and hybrid red-legged partridge. Environ. Toxicol. Chem. 13, 115–120.

Jones, D.S., Barnthouse, L.W., Suter G.W., Efrroymson, R.A., Field, J.M., Beauchamp, J.J., 1999. Ecological assessment in a large river-reservoir: 3. benthic invertebrates. Environ. Toxicol. Chem. 18, 599–609.

Kapur, S., Shusterman, A., Verma, R.P., Hansch, C., Selassie, C.D., 2000. Toxicology of benzyl alcohols: a QSAR analysis. Chemosphere 41, 1643–1649.

Knops, M., Altenburger, R., Segner, H., 2001. Alterations of physiological energetics, growth and reproduction of *Daphnia magna* under toxicant stress. Aquatic Toxicol. 53, 79–90.

Kolkwitz, R., 1950. Ökologie der Saprobien. Über die Beziehungen der Wasserorganismen zur Umwelt. Schriftenreihe des Verbandes für Wasser-, Boden- und Lufthygiene. 4, 1–64.

Kolkwitz, R., Marsson, M., 1902. Grundsätze für die biologische Beurteilung des Wassers nach seiner Flora und Fauna. Mitteilungen der königlichen Prüfanstalt Wasserversorgung Abwasserbeseitigung Berlin-Dahlem 1, 33–72.

Könemann, H., 1980. Structure–activity relationships and additivity in fish toxicities of environmental pollutants. Ecotoxicol. Environ. Saf. 4, 415–421.

Könemann, H., 1981. Fish toxicity tests with mixtures of more than two chemicals: a proposal for a quantitative approach and experimental results. Toxicology 19, 229–238.

Kooijman, S.A.L.M., 1987. A safety factor for LC50 values allowing for differences among species. Water Res. 21, 269–276.

Kooijman, S.A.L.M., Bedaux, J.J.M., 1996. The analysis of aquatic toxicity data. Amsterdam, VU University Press, p. 149.

Kortenkamp, A., Altenburger, R., 1998. Synergisms with mixtures of xenoestrogens – a reevaluation using the method of isoboles. Sci. Total Environ. 221, 59–73.

Kortenkamp, A., Altenburger, R., 1999. Approaches to assessing combination effects of oestrogenic pollutants. Sci. Total Environ. 233, 131–140.

Kreeb, K.H. (Ed.), 1990. Methoden der Pflanzenökologie und Bioindikation. G. Fischer Verlag, Jena.

Länge, R., Hutchinson, T.H., Scholz, N., Solbé, J., 1998. Analysis of the ECOTOC aquatic toxicity (EAT) database II – Comparison of acute to chronic ratios for various aquatic organisms and chemical substances. Chemosphere 36, 115–127.

Lampert, W., Fleckner, W., Pott, E., Schober, U., Störkel, K.U., 1991. Herbicide effects on planktonic systems of different complexity. Hydrobiologia 188/189, 415–424.

Laskowski, R., 1995. Some good reasons to ban the use of NOEC, LOEC and related concepts in ecotoxicology. Oikos, 73, 140–143.

Legierse, K.C.H.M., Verhaar, H.J.M., Vaes, W.H.J., de Bruin, J.H.M., Hermens, J.L.M., 1999. Analysis of time-dependent acute toxicity of organophosphorus pesticides: The critical target occupation model. Environ. Sci. Technol. 33, 917–925.

Levasseur, L.M., Slocum, H.K., Rustum, Y.M., Greco, W.R., 1998. Modelling of time-dependency of in vitro drug cytotoxicity and resistence. Cancer Res. 58, 5749–5761.

Liess, M., Champeau, O., Riddle, M., Schulz, R., Duquesne, S., 2001. Combined effects of ultraviolett-B radiation and food shortage on the sensitivity of the Antarctic amphipod Paramoera walkeri to copper. Environ. Toxicol. Chem. 20, 2088–2092.

Liess, M., Schulz, R., 1999. Linking insecticide contaminationn and population response in an agricultural stream. Environ. Toxicol. Chem. 18, 1948–1955.

Lipnick, R.L., 1989. Narcosis, electrophile and proelectrophile toxicity mechanisms: application of SAR and QSAR. Environ. Toxicol. Chem. 8, 1–12.

Lützhoft, H.C.H., Halling-Sorensen, B., Jorgensen, S.E., 1999. Algal toxicity of antibacterial agents applied in Danish fish farming. Arch. Environ. Contam. Toxicol. 36, 1–6.

Maltby, L., Clayton S.A., Yu, H., McLoughlin, N., Wood, R.M., Ying, D., 2000. Using single-species toxicity tests, community-level responses, and toxicity identification evaluations to investigate effluents impacts. Environ. Toxicol. Chem. 19, 151–157.

Marchini, S., Passerini, L., Hoglund, M.D., Pino, A., Nendza, M., 1999. Toxicity of aryl- and benzhalides to *Daphnia magna* and classification of their mode of action based on quantitative structure–activity relationship. Environ. Toxicol. Chem. 18, 2759–2766.

Matthiessen, P., 2000. Is endocrine disruption a significant ecological issue? Ecotoxicology 9, 21–24.

McDaniel, M., Snell, T.W., 1999. Probability distributions of toxicant sensitivity for freshwater rotifer species. Environ. Toxicol. 14, 361–366.

McKim, J.M., Bradbury, S.P., Niemi, G.J., 1987. Fish acute toxicity syndromes and their use in the QSAR approach to hazard assessment. Environ. Health Perspec. 71, 171–186.

Milles, D., 1989. Grenzen natürlicher Selbstreinigung – Zur Geschichte medizinischer Grenzwertkonzepte. In: Kortenkamo, K., Grahl, B., Grimme, L.H., (Eds), Die Grenzenlosigkeit der Grenzwerte. C.F. Müller, Karlsruhe, pp. 197–219.

Milles, D., 1991. Von Schädlingen und Schädigungen – Zur Geschichte der Pestizidzulassung. In: Rehbinder, E. (Ed.), Bremer Kolloquium über Pflanzenschutz (Tagungsband). Werner, Düsseldorf, pp. 17–43.

Molander, S., Dahl, B., Blanck, H., Jonsson, J., Sjöström, M., 1992. Combined effects of tri-n-butyl tin (TBT) and diuron on marine periphyton communities detected as Pollution induced community tolerance. Arch. Environ. Contam. Toxicol. 22, 419–427.

Moore, D.R.J., Caux, P-Y., 1997. Estimating low toxic effects. Environ. Toxicol. Chem. 16, 794–801.

Morton, M.G., Mayer F.L., Dickson, K.L., Waller, W.T., Moore, J.C., 1997. Acute and chronic toxicity of azinphos-methyl to two estuarine species, Mysidopsis bahia and Cyprinodon variegatus. Arch. Environ. Contam. Toxicol. 32, 436–441.

Nendza, M., Hermens, J., 1995, Properties of chemicals and estimation methodologies. In: van Leeuwen C.J., Hermens, J.L.M. (Eds), Risk Assessment of Chemicals: An Introduction. Kluwer, Dordrecht, pp. 239–292.

Nendza, M., Müller, M., 2000. Discriminating toxicant classes by mode of action: 2. Physico-chemical descriptors. Quant. Struct.-Act. Relat. 19, 581–598.

Niculescu, S.P., Kaiser, K.L.E., Schultz, T.W., 2000. Modeling the toxicity of chemicals to *Tetrahymena pyriformis* using molecular fragment descriptors and probabilistic neural networks. Arch. Environ. Contam. Toxicol. 39, 289–298.

Nusch, E.A., 1992. Grundsätzliche Vorbemerkungen zur Planung, Durchführung und Auswertung biologischer und ökotoxikologischer Testverfahren. In: Steinhäuser, K.G., Hansen, P.-D. (Eds), Biologische Testverfahren. Schr. Reihe Verein WaBoLu 89. Gustav Fischer Verlag, Stuttgart, pp. 35–48.

OECD (Organisation for Economic Co-operation and Development), 1995. Report of the OECD Workshop on Environmental Hazard/Risk Assessment. OECD Environment Monographs No. 105, Paris.

OECD (Organisation for Economic Co-operation and Development), 1998. Report of the OECD Workshop on Statistical Analysis of Aquatic Toxicity Data. OECD Series on Testing and Assessment, No. 10, Paris.

Okkerman; P.C., van der Plassche, E.J., Sloof, W., van Leeuwen, C.J., Canton, J.H., 1991. Ecotoxicological effects assessment: a comparison of several extrapolation procedures. Ecotoxicol. Environ. Safety 21, 182–193.

Parkerton, T.F., Konkel, W.J., 2000. Application of quantitative structure–activity relationships for assessing the aquatic toxicity of phthalate esters. Environ. Ecotoxicol. Safety 45, 61–78.

Pereira, A.M.M., da Maia Soares, A.M.V., Goncalves, F., Ribeiro, R., 1999. Test chambers and test procedures for in situ toxicity testing with zooplankton. Environ. Toxicol. Chem. 18, 1956–1964.

Pöch, G., 1993. Combined Effects of Drugs and Toxic Agents. Modern Evaluation in Theory and Practice. Springer Verlag, Wien.

Posthuma, L., Suter, G.W., Traas, T.P. (Eds), 2002. The Use of Species Sensitivity Distributions (SSD) in Ecotoxicology. CRC Press, Boca Raton.

Purdom, C.E., Hardiman, P.A., Bye, V.J., Eno, E.N., Tyler, C.R., Sumpter, J.P., 1994. Estrogenic effects of effluents from sewage treatment works. Chem. Ecol. 8, 275–285.

Pyle, G.G., Swanson, S.M., Lehmkuhl, D.M., 2001. Toxicity of uranium mine-receiving waters to caged fathead minnows, *Pimephales promelas*. Ecotoxicol. Environ. Safety 48, 202–214.

Rand, G., Clark, J.R., 2000a. Hazard/risk assessment of pyridaben: I. Aquatic toxicity and environmental chemistry. Ecotoxicology 9, 157–168.

Rand, G., Clark, J.R., 2000b. Hazard/risk assessment of pyridaben: II. Outdoor aquatic toxicity studies and the water-effect ratio. Ecotoxicology 9, 169–177.

Rand, G.M., Wells P.G., Mc Carty L.S., 1995. Introduction to aquatic toxicology. In: Rand G.M. (Ed.), Fundamentals of Aquatic Toxicology, 2nd edn. Taylor & Francis, New York, pp. 3–67.

Roex, E.W.M., van Gestel, C.A.M., van Wezel, A.P, van Straalen, N.M., 2000. Ratios between acute aquatic toxicity and effects on population growth rates in relation to toxicant mode of action. Environ. Toxicol. Chem. 19, 685–693.

Routledge, E.J., Shean, D., Desbrow, C., Brighty, G.C., Waldock, M., Sumpter, J.P., 1998. Identification of estrogenic chemicals in STW effluents. 2. In vivo responses in trout and roach. Environ. Sci. Technol. 32, 1559–1565.

Rutgers, M., Breure, A.M., 1999. Risk assessment, microbial communities, and pollution-induced community tolerance. Human Ecol. Risk Assess. 5, 661–670.

Sabliunas, D., 1999. Semipermeable membrane devices in monitoring of organic pollutants in the aquatic environment. Ph.D. Thesis, Lund University.

Schmitt, H., Altenburger, R., Jastorff, B., Schüürmann, G., 2000. Quantitative structure – activity analysis of the algae toxicity of nitroaromatic compounds. Chem. Res. Toxicol. 13, 441–450.

Scholze, M., Boedeker, W., Faust, M., Backhaus, T., Altenburger, R., Grimme, L.H., 2001. A general best-fit method for concentration-response curves and the estimation of low-effect concentrations. Environ. Toxicol. Chem. 20, 448–457.

Schüürmann, G., 1998. Ecotoxic modes of action of chemical substances. In: Schüürmann, G., Markert, B. (Eds), Ecotoxicology. Wiley, New York, and Spektrum, Heidelberg, pp. 665–749.

Schweigert, N., Acero, J.L., von Gunten, U., Canonica, S., Zehnder, A.J.B., Eggen, R.I.L., 2000. DNA degradation by the mixture of copper and catechol is caused by DNA-copper-hydroperoxo complexes, probably DNA-Cu(I)OOH. Environ. Mol. Mutagenesis 36, 5–12

Schweigert, N., Hunziker, R.W., Escher, B.I., Eggen, R.I.L., 2001. Acute toxicity of (chloro-)catechols and (chloro-)catechol-copper combinations in Escherichia coli corresponds to their membrane toxicity in vitro. Environ. Toxicol. Chem. 20, 239–247.

Segner, H., Wenzel, A., Janssen, C.R., Pascoe, D., 2001. Identification of endocrine disrupting effects in aquatic organisms (IDEA). Final report to the EU, project ENV4-CT97-0509.

Seitz, A., Poethke, H.J., 1995. Strukturanalyse und Modellierung von Zoopolankton-Fisch-Freilandsystemen zur Bewertung von Fremdstoffwirkungen in aquatischen Ökosystemen. In: Kirchner, M., Bauer, H. (Eds), Proceedings des Statusseminars zum Förderschwerpunkt Ökotoxikologie des BMBF.

Seitz, A., Ratte, H.T., 1991. Aquatic ecotoxicology: on the problems of extrapolation from laboratory experiments with individuals and populations to community effects on the field. Comp. Biochem. Physiol. 100, 301–304.

Shukla, R., Wang Q., Fulk, F., Deng, C. Denton, D., 2000. Bioequivalence approach for whole effluent toxicity testing. Environ. Toxicol. Chem. 19, 169–174.

Soucek, D.J., Cherry, D.S., Currie, R.J., Latimer, H.A., Trent, G.C., 2000. Laboratory to field validation in an integrated assessment of an acid mine drainage-impacted watershed. Environ. Toxicol. Chem. 19, 1036–1043.

Steinhäuser, K.G., Hansen, P.-D. (Eds), 1992. Biologische Testverfahren. Schr. Reihe Verein WaBoLu 89. Gustav Fischer Verlag, Stuttgart, p. 879.

Suter, G.W., Barnthouse, L.W., Efroymson, R.A., Jager, H., 1999. Ecological assessment in a large river-reservoir: 2. Fish community. Environ. Toxicol. Chem. 18, 589–598.

Thomson, H.M., Hunt, L.V., 1999. Extrapolating from honeybees to bumblebees in pesticide risk assessment. Ecotoxicology 8, 147–166.

Tong, W., Lowis, D.R., Perkins, R., Chen, Y., Welsh, W.J., Godette, D.W., Heritage, T.W., Sheehan, D.M., 1998. Evaluation of quantitative structure–activity relationship methods for large-scale prediction of chemicals binding to the estrogen receptor. J. Chem. Information Comp. Sci. 38, 669–677.

Traunspurger, W., Schäfer, H., Remde, A., 1996. Comparative investigation on the effect of a herbicide on aquatic organisms in single species tests and aquatic microcosms. Chemosphere 33, 1129–1141.

Traas, T.P., van de Meent, D., Posthuma, L., Hamers, T.H.M., Kater, B.J., de Zwart, D., Aldenberg, T., 2002. The potentially affected fraction as a measure of ecological risk. In: Posthuma, L., Suter, G.W., Traas, T.P. (Eds), The Use of Species Sensitivity Distributions (SSD) in Ecotoxicology. CRC Press, Boca Raton, pp. 315–344.

Tyler, C.R., Jobling, S., Sumpter, S.P., 1998. Endocrine disruption in wildlife: a critical review of the evidence. Critical Rev. Toxicol. 28, 319–361.

Vaal, M., van der Wal, J.T., Hermens, J., Hoekstra, J., 1997a. Pattern analysis of the variation in the sensitivity of aquatic species to toxicants. Chemosphere 35, 1291–1309.

Vaal, M., van der Wal, J.T., Hoekstra, J., Hermens, J., 1997b. Variation in the sensitivity of aquatic species in relation to the classification of environmental pollutants. Chemosphere 35, 1311–1327.

Vaal, M.A., van Leeuwen, C.J., Hoekstra, J.A., Hermens, J.L.M., 2000. Variation in sensitivity of aquatic species to toxicants: practical consequences for effect assessment of chemical substances. Environ. Management 25, 415–423.

van Leeuwen, C.J., van der Zandt, P.T.J., Aldenberg, T., Verhaar, H.J.M., Hermens, J.L.M., 1992. Application of QSARs, extrapolation and equilibrium partitioning in aquatic effect assessment. I. Narcotic industrial pollutants. Environ. Toxicol. Chem. 11, 267–282.

van Straalen, N.M., Denneman, C.A.J., 1989. Ecotoxicological evaluation of soil quality criteria. Ecotoxicol. Environ. Safety 18, 241–251.

Van Wijk, D.J., Hutchinson, T.H., 1995. The ecotoxicity of chlorate to aquatic organisms: a critical review. Ecotoxicol. Environ. Safety 32, 244–253.

van Wijngaarden, R.P.A., Van der Brink, P.J., Crum, S.J.H., Voshaar, J.H.O., Brock, T.C.M., Leeuwangh, P., 1996. Effect of the insecticide Dursban® 4E (active ingredient chlopyrifos) in outdoor experimental ditches: I. Comparison of short-term toxicity between the laboratory and the field. Environ. Toxicol. Chem. 15, 1133–1142.

Verhaar, H.J.M., van Leeuwen, C.J., Hermens, J., 1992. Classifying environmental pollutants. 1: Structure–activity relationships for prediction of aquatic toxicity. Chemosphere 25, 471–491.

Vighi, M., Altenburger, R., Arrhenius, Å., Backhaus, T., Bödeker, W., Blanck, H., Consolaro, F., Faust, M., Finizio, A., Froehner, K., Gramatica, P., Grimme, L.H., Grönvall, F., Hamer, V., Scholze, M., Walter, H., in press. Water quality objectives for mixtures of toxic chemicals: problems and perspectives. Ecotoxicol. Environ. Safety.

Wagner, C., Lokke, H., 1991. Estimation of ecotoxicological protection levels from NOEC toxicity data. Water Res. 25, 1237–1242.

Walter, H., Consolaro, F., Gramatica, P., Scholze, M., Altenburger, R., 2002. Mixture toxicity of priority pollutants at no observed effect concentrations (NOECs). Ecotoxicology 11, 299–310.

Wells, P.G., Lee, K., Blaise, C. (Eds), 1998. Microscale testing in aquatic toxicology. CRC Press, Boca Raton, p. 679.

Wenzel, A., Nendza, M., Hartmann, P., Kanne, R., 1997. Test battery for the assessment of aquatic toxicity. Chemosphere 35, 307–322.

Xing, L., Welsh, W.J., Tong, W., Perkins, R., Sheehan, D.M., 1999. Comparison of estrogen receptor alpha and beta suntypes based on comparative molecular field analysis (CoMFA). SAR QSAR Environ. Res. 10, 215–230.

Yang, R., Thurston, V., Neuman, J., 2000. A physiological model to predict xenobiotic concentration in fish. Aquatic Toxicology 48, 109–117.

II

**Bioindicators
in use**

IIa
Standard tests

Chapter 6

Biomarkers

P.-D. Hansen

Abstract

To understand the complexity of the structure of populations and processes behind the health of individual organisms living in a specific environment, populations and ecosystem health, we have to direct our efforts to promote rapid and cost-effective biochemical response parameters for early recognition of contaminants. One way of achieving that goal is to use biomarkers as a recognition system for the induced variation in cellular or biochemical components or processes, structures, or functions. The biomarkers achieve high sensitivities in a minimum of measuring time. New emerging effect related parameters are the genotoxic, immunotoxic and the endocrine potential of surface waters, effluents, groundwater and coastal waters. The biomarkers will become relevant for legislative framework but they have to meet the standardization procedures under e.g. ISO (ISO = International Standards Organisation), otherwise they will not be accepted by the regulatory authorities and industry. There are already biomarker-standards in the regulations for "genotoxicity" by ISO 13829; ISO/CD 16240; ISO/WD 21427 and in the process to become a ISO Standard for biotransformation and detoxification the ISO/NWI: EROD (EROD = Ethoxyresorufin-O-Deethylase) enzyme activity. For transitional and marine waters the biomarkers are already considered in connection with the implementation of the EU Water Frame Work.

Keywords: biomarker, ecosystem health, early recognition, DNA alterations, cholinesterase inhibition, vitellogenin, phagocytosis, genotoxicity, neurotoxicity, immunotoxicity, endocrine effects

Biochemical responses or biomarkers in organisms provide us with signals of potential damage in ecosystems due to environmental stress. These responses, if perceived early enough may prevent eventual damage in ecosystems. On the other hand once ecosystem damage has occurred, remedial action for their recovery could be expensive and pose certain logistical problems. Prevention of ecosystem deterioration is always better than curing a damaged one. Ideally, "early warning signals" in ecosystems using biomarkers would not only tell us the initial levels of damage, but could provide answers to develop control strategies and precautionary measures.

1. Introduction

To understand the complexity of the structure and processes behind the health of populations, communities and ecosystems, we have to direct our efforts to promote newly emerging, rapid and cost-effective parameters of ecological health. Newly emerging parameters are biochemical (biomarker) related parameters in the field of immuno-toxicity and endocrine effects. Environmental effects such as genotoxicity and clastogenicity have been detected in organisms from such "hot spots". Vital fluorescence tests are one way to allow us to unmask genetic alterations in field-collected animals or in situ-exposed organisms by a caging technique. Newly emerging ecosystem health parameters are linked closely to the biomarkers of organisms exposed to the monitored areas. One problem is to find the relevant interpretation and risk assessment tools to the environment, especially in the context of the new EU Water Directive and river basin management strategies. In some case studies biomarkers were very helpful in promoting an environmentally sensitive and sustainable use of ecosystems e.g. coastal zones (Bresler et al., 1999). A promising tool is the scale classification based on biomarkers in mussels. This scale classification, called Biomarker-Index, is applied for pollution monitoring along the European coastline (Narbonne et al., 1999). Currently available biomarkers or biochemical responses are used as biomonitoring tools to assess information on early responses of living organisms to environmental stressors, and to deliver signals on ecosystem damage and pathology due to both man-made and natural pollutants. Emphasis is made in the use of recently developed approaches for assessment and prediction of environmental pathology and alterations of ecosystem structure.

Traditional ecotoxicological monitoring includes a set of various field tests and laboratory bioassays which allow the detection of biological effects of pollutants and to assess the expression of these effects; e.g. level of pollution. The expression is calculated as LC50 (LC = lethal concentration) or NOEC (NOEC = no observed effect concentration). However, LC50 values depend on numerous environmental factors such as temperature, pH-values, ionic strength, DOC and changes in the toxicological profile due to the presence of humic acids. The NOEC-data are in some ways a biological paradox, because living organisms respond to any environmental alteration. The detection of such responses is a methodological problem. Using adequate methodologies, like the examination of inherent blue fluorescence of intracellular NADH, the responses can be detected easily (Bresler et al., 1999; Bresler and Yanko, 1995). Thus we must pass from ecotoxicological monitoring to diagnostics of environmental health, i.e. examination of selected main parameters which characterized the state of ecosystem health (see Table 1). One way of achieving that goal is to use relevant biomarkers in the context of environmental health.

2. Definitions

There are many definitions of biomarkers e.g.: "A biomarker is a xenobiotically-induced variation in cellular or biochemical components or processes, structures, or functions that is measurable in a biological system or sample" (National Research Council, 1987). The IPSC (International Programme on Chemical Safety of the WHO) has three classes of biomarkers identified: – biomarker of exposure – biomarker of effect and biomarker of susceptibility (International Programme on Chemical Safety, 1993). How can this very complex definition and biomarker classes be handled for understanding and documentation of ecosystem health and by governmental regulations? How can this definition be useful in an operational sense for governmental decision making in collaboration with industry and governmental authorities? But even if there are limitations there is already an initiative by ISO (Internationational Standards Organisation) to standardise biomarkers for governmental regulations. Environmental management and regulative requirements need a number of tools for risk assessment and risk minimisation and this can best be achieved by a standardised, highly sensitive, reaction-specific and widely applicable suite of biomarkers. In the area of precautionary measures for the protection of aquatic life, biomarkers still provide a most important aid in the fulfilling of actions after the precautionary principle.

3. Sampling and biomarkers

Organisms like mussels have to be collected by grab and scuba divers (in water depths of up to 40 m) and fish by netting or professional fishing on pollution gradients of river inputs in estuaries, hot spots or emission inputs, etc. Another very common way to evaluate polluted sites by biomarkers is to expose, by caging, fish or mussels (Gagné et al., 2001). The caging of organisms has the advantage of the so-called in-situ bioassay. But to have an "early warning" or alarm it is very it is difficult on the one side to handle these alarms and on the other side (see Table 2) there will be always a delay due to incubation time and valid response in the context of the used biomarkers. After sampling of the organisms many tissues require snap freezing in liquid nitrogen (e.g. liver tissue for EROD), other tissues require freezing at $-20°C$ (e.g. liver or brain for cholinesterase activity) while for other biomarkers such as MXRT (multi-drug resistance transport, Bresler et al., 1999) measurements must be carried out with freshly sampled tissues. Ideally, "early warning signals" (Fig. 1) in ecosystems using biomarkers and biosensors (Hansen, 1992; Hansen, 1993) would not only tell us the initial levels of damage, but these signals could also provide answers to develop control strategies and precautionary measures. Information on these initial signals can be obtained by processes such as detoxification, genotoxicity, immuno-suppression and the ability of the organism to reproduce. Presently, knowledge necessary to generate on-line signals is not available. Therefore biosensor detection approaches would be useful for generating information on early warning signals (initial onset) of environmental deterioration processes. Biosensors are defined here as selective biological systems (enzymes, antibodies, organelles, cells) combined with a transducer (thermistor, potentiometric and amperiometric electrode, piezoelectric and optical receivers) which generate on-line information from the investigated environment. The directions for effective biosensors at this time are the immunoassays. Progress has been already made in using enzymes and a bacterial bioluminescence detection system. Once the information from using biosensor approaches on geno-toxic potentials and immunotoxic potentials in ecosystems is available, effective control measures can be easily applied due to the response signals for coastal zone management.

4. Potential and limitations of biomarkers

One important tool for the acceptance of biomarkers in science, technology and governmental legislature are so-called inter-laboratory comparison studies. Laboratory studies have established a strong causal link between exposure of fish to PAHs and co-planar PCBs and the expression of cytochrome P4501A1 and its associated ethoxyre-sorufin-O-deethylase (EROD) activity. The induction of EROD activity in fish liver has been used extensively as a biomarker for the effects of these organic contaminants in several intercalibration exercises (see Stagg and Addison, 1995). The EROD induction is a classical biomarker and well established for MFOs (mixed function oxygenases) and biotransformation in ecotoxicology. But so far only phase I of the MFOs has been investigated and not the phase II reactions with conjugation which is the real detoxification process. Beside MFOs the other commonly used biomarker is the cholinesterase inhibition assay. The basic concept is that organophosphorus and carbamate pesticides inhibit the cholinesterase at different levels (Hansen, 1996; Baumard et al., 1999). For the quantification of neurotoxicity there are two well known cholinesterases, acetyl-cholinesterase and butrylcholinesterase, and the methology in principle is standardised by DIN (German Institute for Norming: DIN 38415-T1). The biomarkers already applied routinely to the water and organic matrices includes genotoxicity and the newly emerging biomarkers such as endocrine effects and immunotoxicity. But there are even more principles associated with the different scales of the biochemical processes (biomarker) relating to Ecosystem Health, these very complex parameters are shown in Table 1 after Bresler and Yanko (1995) and Bresler et al. (1999).

The effect related parameters or biochemical responses are very complex however they give a clear picture of the Health Status of a system under investigation. "Ecosystem Health" is defined as being synonymous with "environmental integrity", from which it follows that the scope of Ecosystem Health (EH) research encompasses all the tools and approaches which are efficacious in increasing the cognitive, curative, and preventive knowledge which has as its goal the preservation of environmental integrity. Ecosystem Health research thus directs its attention to the prediction of reversible and irreversible insults which human or other activities could potentially inflict on the environment. For the assessment of ecosystem health very promising biomarker approaches are centred on quantifying biochemical effects in organisms and populations.

In Table 2 the timescale responses by biomarkers in biological system are demonstrated. It is shown that there is a delay from the molecular level to the ecosystem. The principles associated with the different scales of the biochemical responses depends on the kinetics of the biological systems. In order to understand the complex ecosystem interaction, it is necessary to break down the functional and structural components to their respective parts. For the assessment at the organism level, in the context with the reproduction, incubation periods of 20 to 120 days would be required. During this period there could be the beginning of an adaptation process in the organism. This would reflect changes finally in the ecosystem. The adaptation processes can be monitored and quantified by the biomarker test batteries as well. This assessment would also include processes such as reorganization, redevelopment of the structure of the system. In Figure 1 the induction, inhibition and adaptation of enzymes

Table 1. Selected parameters for the assessment of environmental health of organisms and the biomarker methods used for their examination (Bresler and Yanko, 1995; Bresler et al., 2001).

Method	Characteristic of health
1. Measurement of blue and green fluorescence of NADH and FAD in living tissues	1. Metabolic state of mitochondria, cells or tissues respiration and glycolisis
2. Quantitative fluorescent cytochemistry	2. DNA, RNA, proteins and lipids content
3. Using permeable fluorogenic substrates of enzymes, specific inhibitors, and kinetic analysis	3. Enzyme activity in living cells in situ: a. Non-specific esterases b. Detoxifying enzymes c. Marker enzymes
4. Using special fluorescent anionic markers	4. Alterations of permeability of plasma membranes, epithelial layers and histohematic barriers
5. Using specific fluorescent transport substrates, inhibitors and kinetic analysis	5. State of carrier-mediated transport system for xenobiotics elimination
6. Using of fluorescent xenobiotics or fluorescent analog of the xenobiotics	6. Xenobiotics distribution, extra- and intracellular accumulation and storage
7. Using of special fluorescent xenobiotics or fluorescent analogs of xenobiotics	7. State and function of xenobiotic-binding proteins
8. Vital tests with Acridine Orange or Neutral Red	8. State of lysosomes and cell viability
9. Metachromatic fluorescence of intercalated or bound Acridine Orange, 590/530 nm Microfluorometr	9. Functional rate of nuclear chromatin, DNA denaturation
10. Complete cyto- and histopathological examination	10. Early pathological alterations and signs of environmental pathology
11. Electron microscopy	11. Cell structures and organoids
12. Cytogenetic examinations	12. Detection of environmental genotoxicity and clastogenicity

Table 2.　Timescale responses in the biological systems and ecotoxicological biomarker related endpoints ("signals").

Biomarker responses and reactions in biological systems

[> 2 years]　　　*Ecosystem Level*
- Alteration in ecosystem structure – redevelopment of the system's elements and structure. Pathology and ecosystem health decrease.

[0.5–1 year]　　　*Population Level:*
- *Alterations in population dynamics and structure*
  – self organization – reorganization

[1–2 month]
- Change in growth and adaptation. Pathology and population health decrease

[20–120 days]　　　*Organismic Level (exposure)*
- *Change in growth, reproduction, ant-xenobiotic defence and immunological defence. Environmental pathology*

[1–3 days]　　　*Cellular Level*
- Intercellular anti-xenobiotic defence like MXRtr (MXRtr = Multi Xenobiotic Resistancetransporter) and SATOA (SATOA = System of Active Transport of Organic Anions)
- Intercellular immunological defence
  genetic damage – repair of genetic damage
  reaction with macromolecules – cell pathology

　　　Molecular Level
　　　DNA adducts formation, Mutagenicity, DNA strand breaks
　　　Micronucleus formation, Metaphase chromosomal abnormality

[10 min.]
- "Biosensors", on-line – Early Warning Systems
  Biosensors available for the endpoints ("signals"):
  toxicity – genotoxic potential – immunosuppression

↑

Input of pollutants
(physical and chemical level)

---

or DNA-damage in a functional system is listed in terms of early recognition (alarm systems), adaptation and exhaustion.

The listed biochemical responses (biomarker) and selected parameters for the assessment of environmental health of organisms in Table 1 were difficult to form into operational effect related parameters for environmental standards (e.g. water quality criteria). In Figure 1 it is demonstrated how to organise the biochemical responses in an operational way to result in an environmental endpoint. At the same time there is

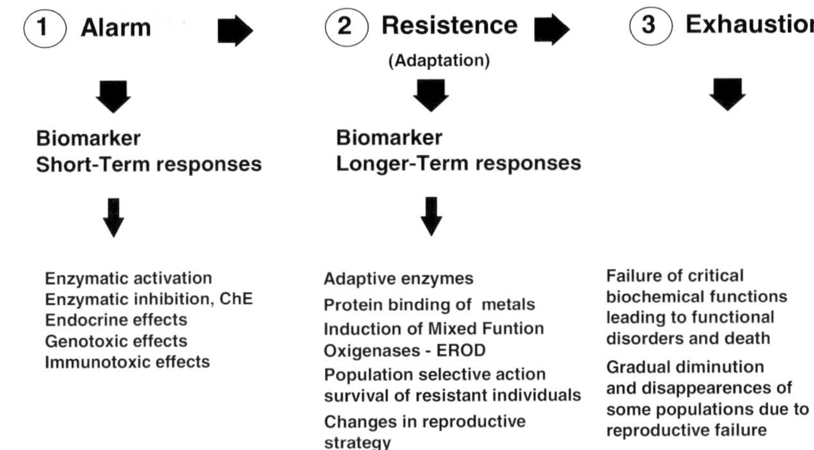

Figure 1. Environmental diagnosis by biomarker: operational effect related response.

need for a strong scientific background. Beside the new emerging parameters and biomarkers in the field of immunotoxicity and endocrine effects there are biomarkers already standardised by ISO in the field of genotoxicity and very soon biomarkers in the classical field of biotransformation-biomarkers (e.g. phase 1 = EROD [7-ethoxy-resorufin-O-deethylase]) will be standardised by ISO.

Acute toxicity results in organism selection, genotoxicity results in mutagenicity and physiological impairment (genetic disease syndromes), induction of MFO (biotransformation and detoxification) tells us that fish are induced with elevated detoxification levels and provides us with information on the effects of specific chemical species (warning signals). However, they do not have high ecological relevance like information from immunosuppression (phagocytosis). Genotoxic damage end points have high ecological significance as they relate to the ability of the organisms to reproduce. Stress responses at population levels have direct ecological implications even though they exhibit low specificity. Therefore, there should be a holistic ecosystem approach where the overall information (high specificity to low specificity) should be considered for proper "ecosystem health" management. In order to obtain information on the overall ecosystem assessment at least a two year cycle of vegetation is required, where we obtain initial signals on these events. These signals could be critical and information on such signals in relation to time and space (site specific) would determine the ecosystem pulses. Such information would be helpful in the proper understanding, management and restoration of ecosystem health. Genotoxicity is a good example for investigating an operational application of a relevant biomarker. Extended exposure of organisms to environmental genotoxins would result in several physiological disorders such as reproductive impairment and other related abnormalities. The response measurements to genotoxicity in the context of reproductive toxicity is essential for assessing the effects of anthropogenic stressors. The consequences of DNA damage (genotoxic potential) are demonstrated from the molecular to the ecosystem level of the biological system in Table 3.

The consequences of DNA damage of different organisational levels shown in Table 3 are very promising concerning the relevant endpoints. As long as the interpretation of the genotoxic effects is discussed at the molecular, tissue or organ level the results are quite clear but the more the discussion is at the organism, population and ecosystem level the more difficult it is to find relevant genotoxic effects in the environment resulting in reduction of population size, extinction or reduction of species diversity. There are catalogues of reference substances and their genotoxic, mutagenic and cancerogenic effects in well established assays (umu-assay, AMES-assay, COMET-assay, DNA-Unwinding, etc.) but finally the question is how these substances act in the real environment in the context of biomarkers.

The effects of genotoxic substances are not adequately considered in aquatic ecological hazard assessment, and comparatively little has been done to develop methods for predicting these effects. Several research programmes clearly show results and effects with model- or reference-substances, but in the field the genotoxic responses are for some assays less and for others more sensitive. This is especially true for biomarkers where the bioavailability and the matrix of exposure plays a key role. In general, the effects of genotoxic substances on reproductive processes in aquatic organisms have been ignored. In Figure 2 the genotoxic effects of effluents are shown in exposed trout to a relevant dilution row comparable to the water recipients. In the figures 2–5 biomarker effects (genotoxicity, neurotoxicity and endocrine effects [Vitellogenin]) will be demonstrated on the same matrix: effluents.

In Figure 2 there are clear genotoxic responses measured by the DNA-unwinding assay (Dizer et al., 2002). The arrow shows the threshold of the genotoxic "NOEC-response" and the lack of a clear relationship between effluent concentration and DNA-damage shows that repairing of the DNA-damage is already underway. This is a general problem with biomarkers in field studies. Repair and adaptation of the system to changing environmental conditions makes the interpretation of results difficult.

Table 3. Consequences of DNA damage of different organisational levels in biological systems.

Level of biological organisation	Effects
DNA	Mutations
Cell	Cell death
	Disordered proliferation and differentiation
	Neoplastic transformation
Tissue/Organ	Functional defects
	Malformations
	Tumours
Organism	Reduced viability
	Reduced fertility
Population	Reduction of population size
	Extinction
Ecosystem	Reduction of species diversity

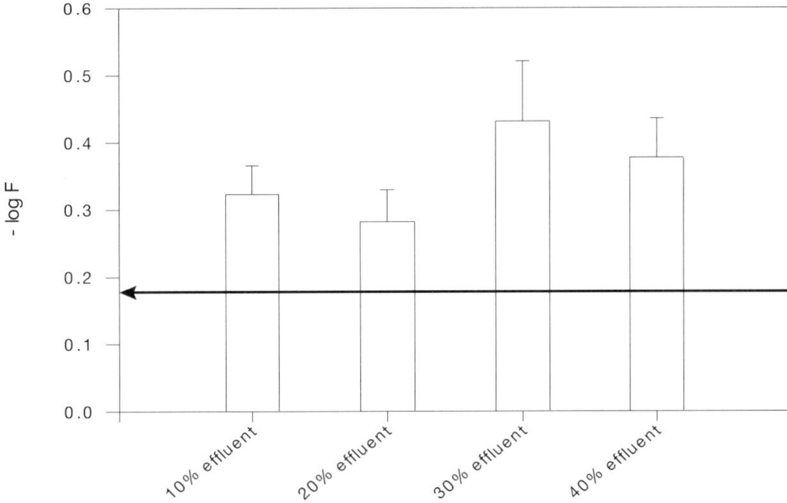

Figure 2. Genotoxic effects: DNA-damage in trout (liver) after seven days' exposure – 10%, 20%, 30% and 40% treated secondary effluent in the receiving water way (Havel River), $N=50$.

Other biomarkers, like cholinesterase (ChE) activity, in the context of neurotoxicity are more convincing and clear (see Fig. 3).

In Figure 3 the neurotoxic biomarker, cholinesterase activity, gives a clear response to the effluent exposure. These are interesting results, but again the question is, is there a relevance concerning the ecosystem? It is obvious that there are at least three potential genotoxicity related effects of exposure to genotoxic substances. First, reduced fertility (see Table 3) may occur if genetic damage induces cell death in dividing gametes. Second, reproductive success may be impaired if dominant – and recessive – lethal mutations are induced, causing embryo mortality or abnormality. Third, exposure to genotoxic contaminants may cause cancer. It is difficult to prove these genotoxic effects in the polluted environment. The influence of uv-B radiation on pelagic fish embryos and DNA-damage (Dethlefsen et al., 2001) is one of the relevant genotoxic effects in the field. The new emerging parameters for immunotoxicity and endocrine effects are also a field for newly developed biomarkers. For immunotoxicity or immunosuppression, phagocytosis is a relevant biomarker and standardised protocols are already available (Dizer et al., 2001).

For endocrine effects several receptor systems (ELRA, YES, etc.) are under discussion. One common biomarker concerning endocrine effects is the vitellogenin york protein. In Figure 4 with the same experimental set up the biomarker vitellogenin is demonstrated in its environmental response to endocrine disrupting compounds (EDCs).

In Figure 4 it is demonstrated that male fish exposed to effluents can, thus, be employed to monitor endocrine disruptions through multiple measurements of vitellogenin production, easily detected in their blood serum. The determination of vitellogenin is accomplished by means of a non-competitive enzymatic immunoassay (EIA) using monoclonal antibodies (Hansen et al., 1998). For the exposure experiments of fish

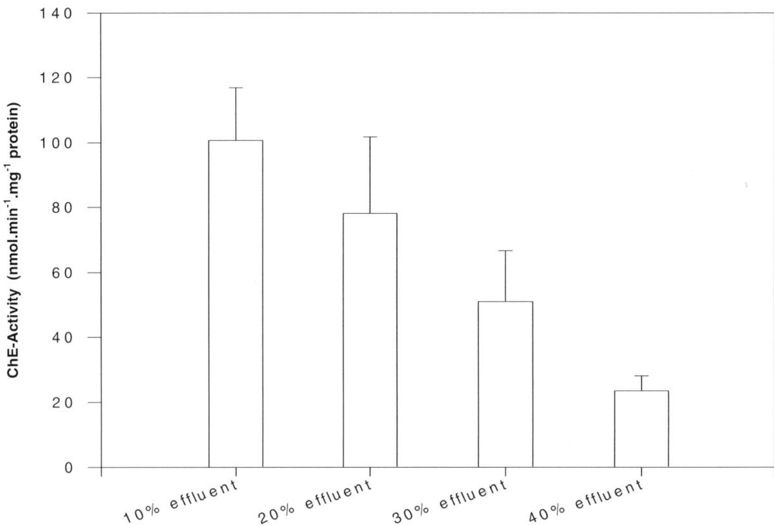

Figure 3.   Neurotoxic effects: ChE-Activity in trout muscle tissues after seven days' exposure – 10%, 20%, 30% and 40% treated secondary effluent in the receiving water way (Havel River), N=50.

in effluents the fish are exposed in the exposure tanks of a so called "on site WaBoLu-Aquatox Monitoring system". The fish are exposed to a mixture of definite amounts of effluent and dilution water in the on-line system. The dilution steps (10%, 20% 30% and 40% effluent) are relevant in relation to the effluent loading of the Berlin waterways during the seasons of the year. The effluent loading of 10%, 20%, 30% and 40% represents the water situation in the Berlin waterways at winter time (high water flow = app. 100 m^3/s) with 10% effluent, at spring and autumn (medium water flow = app. 30 m^3/s) with 20% respectively 30% effluent and at summer time (low water flow = app. 5 m^3/s) with 40 % effluent. Similar caging experiments with mussels shows comparable effects (Gagné et al., 2001; Blaise et al., 1999). All these investigations with the biomarker vitellogenin or vitellogenin like proteins (mussels) show a clear correlation to the exposure against positive (female) and negative (male) references. There was a clear response in vitellogenin production in the effluent exposed male fish (see Fig 4) fish according to the effluent concentrations. There is already a remarkable increase of the vitellogenin in the serum of the fish exposed to >20% effluent. In parallel the vitellogenin data of the non effluent exposed female fish (positive controls) and male fish (negative controls) were measured as references. But to validate this data field exposure has to be checked by laboratory experiments to show the kinetics of the production of vitellogenin. It could be demonstrated by induction experiments with 17β-Estradiol. The vitellogenin was induced by injection of 17β Estradiol into the peritoneum of the male fish. The results confirmed that during an incubation time of 5 days there was already a remarkable increase of the vitellogenin in the serum of the male fish and confirmed the relevance of this biomarker. The on-site exposure experiments with fish in the effluents of the Berlin-Ruhleben sewage plant and cause-effect studies with selected contaminants of the

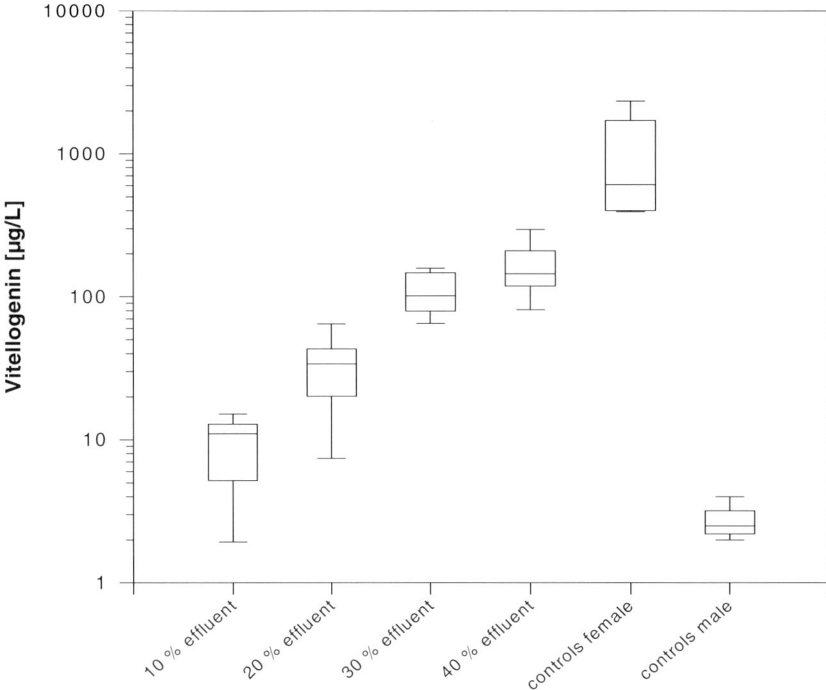

Figure 4.   Vitellogenin synthesis in the plasma of trout exposed to effluents for six months (N = 30).

effluents give rise to the further question of whether the increased vitellogenin synthesis in male fish (Figs 4 and 5) is responsible for the sex ratio problem in the water ways. To answer this question it is not possible just to discuss other additional biomarkers like steroids and receptor assays. Reproduction experiments with hatching eggs from parental fish exposed to effluents and the determination of the sex ratio in the F1 and F2 Generation will solve this problem. In Figure 5 the results in the F2 Generations are demonstrated. Beside the female and male fish there are also intersex oriented fish. This is very often observed in laboratory studies or effluent exposure but not in field surveys.

Only very few intersex fish were caught in the rivers. But now a days there are in some rivers in England already increasing numbers of intersex fish.

It can be shown in Figure 5 that in the effluent exposed F2-Generation a sex ratio with an increasing number of female fish will be developed. The exposure of the eggs to the effluents started with the pigmented eye stage of the fish embryos. After the "point of no return" (external food uptake) the F2- fish larvae were again exposed in the effluent concentrations of the F1-generation (10%; 20%; 30%; 40% effluent). After a growth and exposure for 6 and 12 months the juvenile fish (15–20 cm in length) were sexed again by epifocal fluorescence microscopy. The vitellogenin as biomarker and the interpretation of the consequences to the ecosystem is a good example for a screening tool though one should conduct histopathological studies or hatching experiments

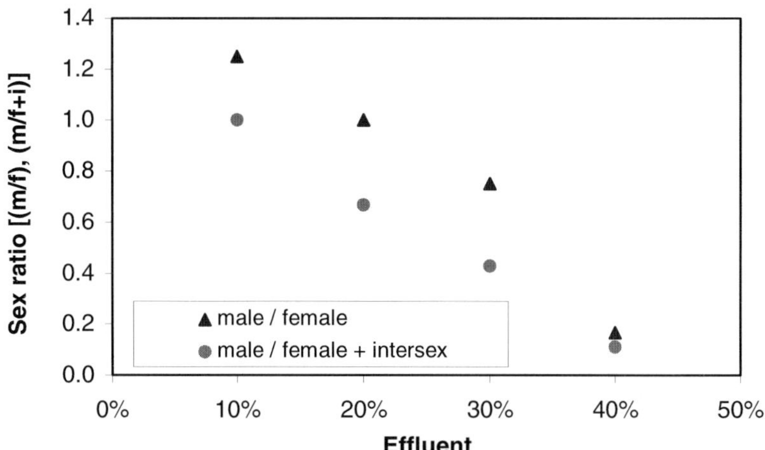

Figure 5.   Sex ratio of juvenile fish (N = 80) after hatching and growth in effluents for 12 months.

in context with reproduction to confirm the biomarker results. In Figure 6 the well standardised and already applied biomarkers are demonstrated already in use for the marine conventions (OSPARCOM, HELCOM and ICES) to protect the Sea and under discussion and progress in the freshwater area (ISO = International Standards Organisation).

The biomarker EROD (MFO- Phase 1–7-ethoxi-O-deethylase induction) is already srtandardised and intercalibrated (Stagg and Addision, 1995). Many historical data are available but still the interpretation of this biomarker is not that easy. The advantage of the extended EROD data base is the comparability of the data and the long-term monitoring of hot spots. The related biomarkers P450 and P450 1A1 are, from the point of scientific background, of interest but there is nearly no chance that these parameters, will become relevant for governmental regulations or long-term monitoring. For single new chemicals and drugs the P450 measurements are widely in use for biotransformation studies in the context of drugs and human health. The biomarker Cholinesterase (ChE) activity (inhibition) is a very well established biomarker for quantification of organophosphorus acid esters and carbamates. Numerous cholin esterases exist widely distributed in animals, partly specific to butyl or to acetyl groups as substrates (Obst et al., 1998; Sturm and Hansen, 1999; Sturm et al., 1999a; Sturm et al., 1999b; Dizer et al., 2002). Especially pesticides like organophosphorus acid esters and carbamates can inhibit the different species of cholinesterase. There are tables for the inhibition constants of acetylcholinesterase which can be used for the quantification of phosphoric acid esters by German Standard Methods (DIN 38415 T1, 1995). There are many other enzymes which can be used in the context of biochemical responses like Urease Inhibition (Wittekindt et al. 1996). Again the question is how specific and relevant are these biomarker responses for environmental effects monitoring to be accepted by the governmental authorities and industry? In this context biomarkers for genotoxicity are very successful. For example, in this field there are already standards for the umu assay (Hansen et al., 1998) by the German Institute for Norming (umu-

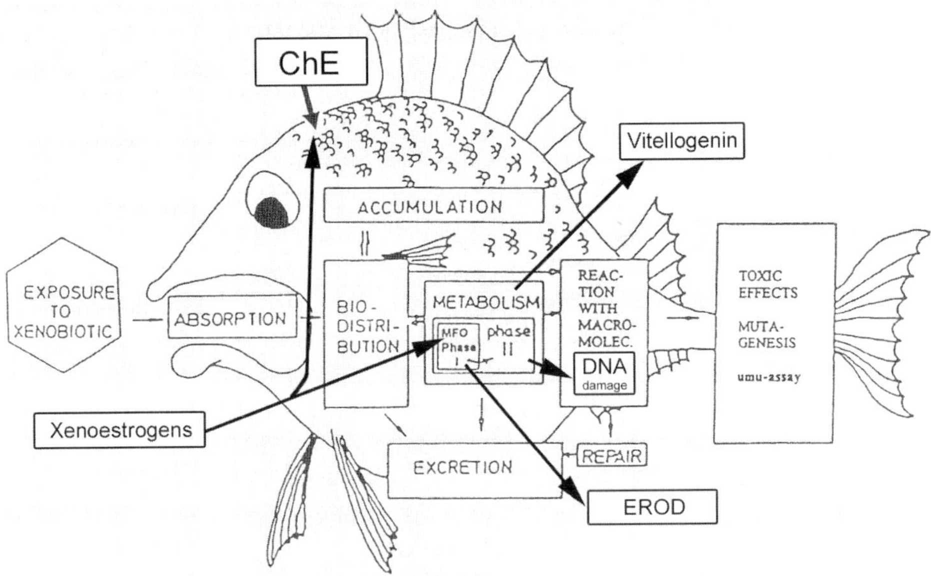

Figure 6. Biomarker responses of the biological system: neurotoxicity, Biotransformation (MFO-EROD), genotoxicity (DNA-damage), mutagenesis (DNA-damage and repair).

assay: DIN 38415-T3, AMES-assay: DIN 38415-T4) and by ISO (umu-assay: ISO 13829). There is, especially for the regulatory sector, a need for genotoxicity tests with eukaryotic organisms. The DNA-Unwinding assay (Herbert and Hansen, 1998; Rao et al., 1996a; Dizer et al., 2002) is a widely used and well standardised assay. As an alternative to the DNA-Unwinding assay as an assay with eukaryotic organisms, only the micronucleus assay with cellines, mussels and fish is of great relevance (Rao et al., 1996b). The genotoxicity as a sensitive parameter for ecosystem health and public health is also often used for the classification of surface water quality. An extended study in the context of the classification of the river Elbe upstream between the City of Hamburg and Schmilka by the DNA-Unwinding assay was done by Wittekindt et al. (2000). The genotoxic effects were monitored with exposed mussels (*Dreissena polymorpha*) at international monitoring stations (IKSE = International Commission for the River Elbe) along the River Elbe. The direction of the future is to classify rivers or coastal areas by biomarkers. An example for coastal areas is given by Narbonne et al. (1999) in the context of the "Biomarker Index" and by Baumard et al. (1999) concerning the pollution gradients of harbours along the Baltic Sea Coast. From the new emerging biomarkers the immunotoxicity and the endocrine effects plays a key role in effects monitoring. For the immunotoxicity the applied biomarker is the phagocytosis (Hansen, 1993; Hansen et al., 1991; Dizer et al., 2001). The phagocytosis of foreign particles by hemocytes of mussels is well standardised and is applicable in fresh and marine waters. For the quantification of biomarkers in the field of endocrine effects there are a lot of receptor assays (e.g. YES assay, YAS assay, ELRA assay, Seifert et al., 1999) which have been standardised. A number of comparative research studies

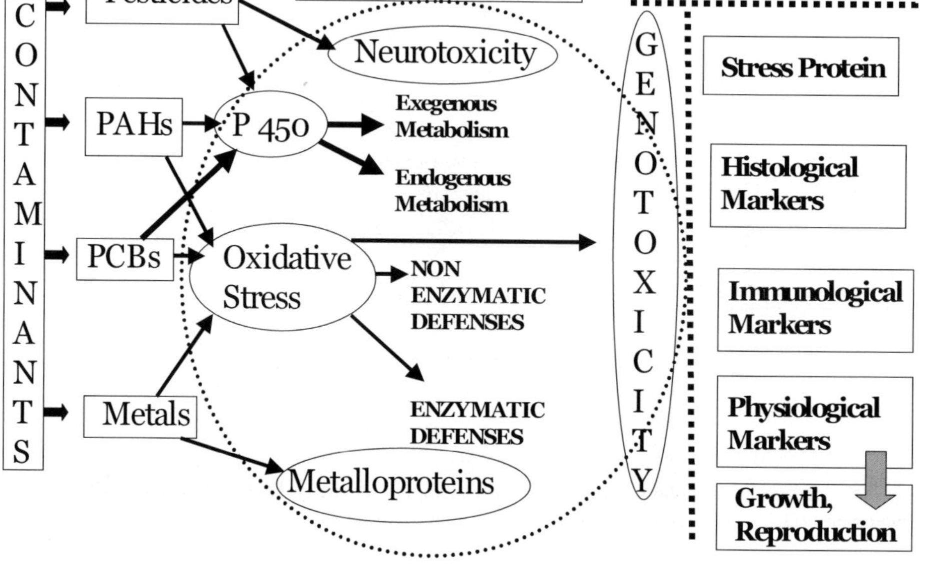

Figure 7.   Relationship between main contaminants and biomarkers: modified after Michel (1993).

were done with indicator and enzyme linked receptor assays for oestrogens and xeno-oestrogens under the EU-Projects COMREHEND (www.ife.ac.uk/comprehend/) and SANDRINE (www.sandrine-wwc.de). Beside the receptor assays and the proliferation assay with the MCF7 cell line, biomarkers like vitellogenin (see Fig. 5) were also investigated for endocrine effects in fish (Hansen et al., 1998) and mussels (Blaise et al., 1999). The vitellogenin assay has many advantages as a biomarker and the detection system by immunoassays and monoclonal antibodies (Marx et al., 2001). For the vitellogenin detection in fish blood recently a biosensor called "Vitello" was developed to enhance the measurements in the field. Figure 7 is an overview of the targets and the biomarkers available.

Beside these promising and well standardised biomarkers there are many other biomarkers of importance like the cytochrome P450 monooxygenase system or cytochrome P450 CDNAS (CYP1A1 and CYP4T2) as biomarker of organic pollution, metallothionein as a biomarker for heavy metals, DNA adduct and oxyradicals (ROD) detection (Fig. 7). But from the standpoint of practical approaches in environmental monitoring there are limitations concerning the fact that the biotransformation and MFO assays in use still only consider the Phase 1 reactions and not the phase 2 conjugation reactions (Sturm et al., 1999c). The biomarkers demonstrated in Figure 7 are reported by Garrigues et al. (2001) and will be checked and intercalibrated again in marine ecosystem under the EU-Projects BEQUALM and BEEP (www.lptc.u-bordeaux.fr/beep). The

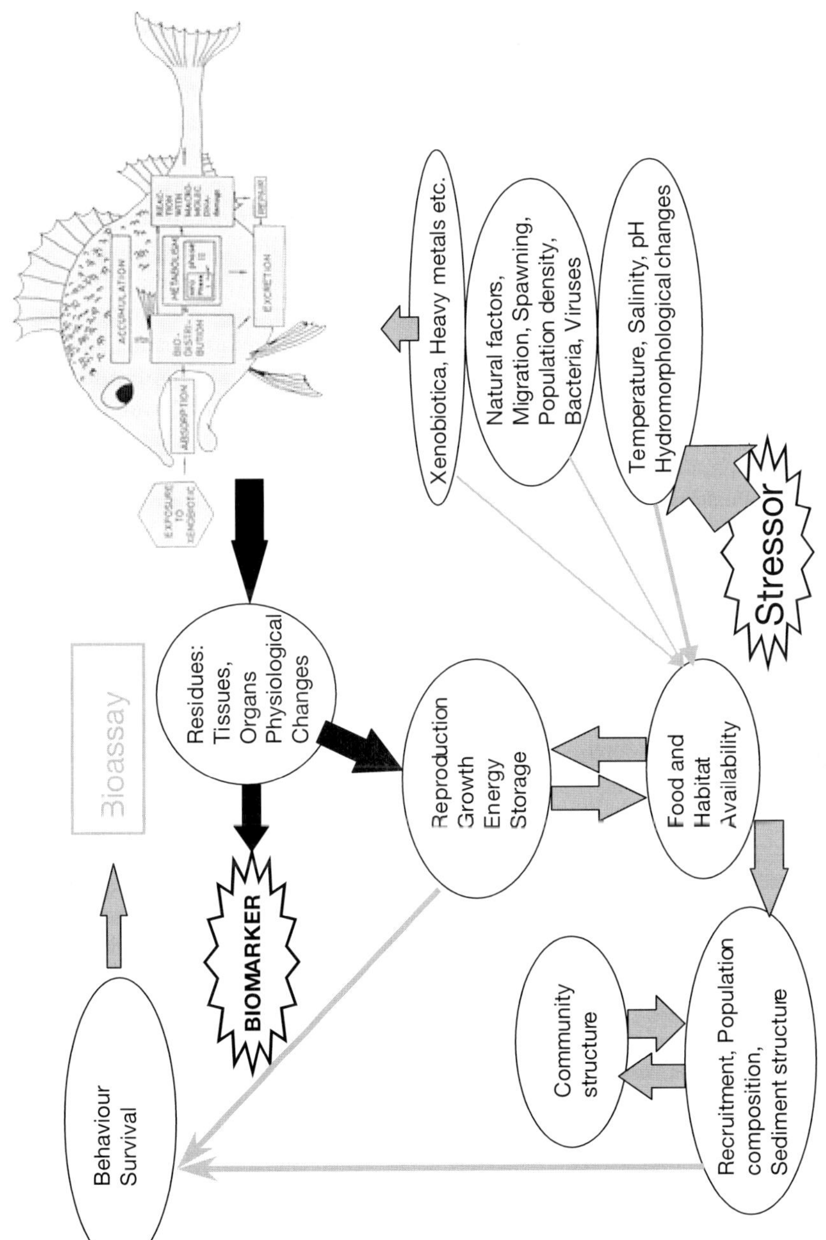

Figure 8. Relationship between ecological, bioassays and biomarker approaches.

positioning and relationship of the biomarkers and the interactions in the ecosystem is shown in Figure 8.

In Figure 8 it is demonstrated that considering the impact of either natural stress or man made stress always it has to be countered against detoxification, disease defence, regulation and adaptation processes. This situation makes the assessment approach by biomarkers rather complicated. On the other hand symptoms analysis including functional (behaviour, activity and metabolism) and structural changes in organism (cellular, tissue and organs), mean that biomarkers do have significant ecological assessment potential. In summary the biomarkers are in a sense multi-arrays, complementary to chemical methods (see Fig. 7) since they can alert as to the presence of toxic compounds that require further instrumental analysis or "bioresponse-linked instrumental analysis". Good examples of "bioresponse-linked instrumental analysis" are the biomarker cholinesterase inhibition (ChE), DNA binding of contaminants, the endocrine effects in the context of proteomics. For regulatory significance biomarkers have to meet the requirements by ISO or other international standardisation agreements. Biomarkers must be validated before they are applied as promising tools in the risk assessment process.

References

Baumard, P., Budzinski, H., Garrigues, P., Dizer, H., Hansen, P.D., 1999. Polycyclic aromatic hydrocarbons in recent sediments and mussels (*Mytilus edulis*) from the Western Baltic Sea: occurrence bioavailability and seasonal variation. Marine Environmental Research 47, 17–47.

Blaise, C., Gagné, F., Pellerin, J., Hansen, P.D., 1999. Measurement of vitellogenin-like protein in the hemolymph of Mya arenaria (Sagenay Fjord, Canada): a potential biomarker for endocrine disruption. Environmental Toxicology and Water Quality 14, 455–465.

Bresler, V., Yanko, V., 1995. Acute toxicity of heavy metals for benthic epiphytic foraminifera *Pararotalia spinigera* and influence of seaweed-derived DOC. Environmental Toxicology and Chemistry 14/10, 1687.

Bresler, V., Bissinger, V., Abelson, A., Dizer, H., Sturm, A., Krätke, R., Fishelson, L., Hansen, P.D., 2001. Marine molluscs and fish as biomarkers of pollution stress in littoral regions of the Red Sea, Mediterranean Sea and North Sea. Helgoland Marine Research 53, 3–4.

Dizer, H., Fischer, B., Harabawy, A.S.A., Hennion, M.C., Hansen, P.D., 2001. Toxicity of domoic acid in the marine mussel *Mytilus edulis*. Aquatic Toxicology 55,149–156.

Dizer, H., Wittekindt, E., Fischer, B., Hansen, P.D., 2002. The cytotoxic and genotoxic potential of surface water and wastewater effluents as determined by bioluminescence, *umu*-assays and selected biomarkers. Chemosphere 46, 225–233.

Dethlefsen, V., Westernhagen, H., Tüg, H., Hansen, P.D., 2001. Influence of solar ultraviolet-B on pelagic fish embryos: osmolality, mortality and viable hatch. Helgoland Marine Research 55, 45–55.

Gagné, F.C., Blaise, C., Salazar, M., Salazar, S., Hansen, P.D., 2001. Evaluation of estrogenic effects of municipal effluents to the freshwater mussel *Elliptio complanata*. Comparative Biochemistry and Physiology, Part C, 128, 213–223.

Garrigues, P., Barth, H., Walker, H., Narbonne, J.F., 2001. Biomarkers in Marine Organisms: A Practical Approach. Elsevier, Amsterdam, p. 550.

German standard methods for the examination of water, waste and sludge, DIN UA 7 subanimal testing DIN 38415 T1, 1995. Determination of cholinesterase inhibiting organophosphorus and carbamate pesticides (cholinesterase inhibition test). VCH Verlagsgesellschaft, Weinheim

Hansen, P.D., 1992. On-line Monitoring mit Biosensoren am Gewässer zur ereignisgesteuerten Probenahme. Acta hydrochimica hydrobiologica 20 (2), 92–95.

Hansen, P.D., 1993. Regulatory significance of toxicological monitoring by summarizing effect parameters. In: Richardson, M. (Ed.), Ecotoxicology Monitoring. VCH, New York, p. 384.

Hansen, P.D., 1996. Bioassays on sediment toxicity. In: Calmano, W. and Förstner, U. (Eds), Sediments and Toxic Substances. Springer, Heidelberg, p. 335.

Hansen, P.D., 1998. Small-scale in vitro genotoxicity tests for bacteria and invertebrates. In: Wells, P.G., Lee, K., Blaise, C. (Eds), Microscale Testing in Aquatic Toxicology. CRC Press, Boca Raton, p. 679.

Hansen, P.D., Bock, R., Brauer, F., 1991. Investigations of phagocytosis concerning the immunological defence mechanism of *Mytilus edulis* using a sublethal luminescent bacterial assay (Photobacterium phosphoreum). Comparative Biochemistry and Physiology 100C (1–2), 129–132.

Hansen, P.D., Dizer, H., Hock, B., Marx, A., Sherry, J., McMaster, M., Blaise, C., 1998. Vitellogenina – biomarker for endcrine disruptors. Trends in Analytical Chemistry 17, 448–451.

Hansen, P.D., Herbert, A. 1998. Small-scale in vitro genotoxicity tests for bacteria and invertebrates. In: Wells, P.G., Lee, K., Blaise, C. (Eds), Microsale Testing in Aquatic Toxicology. CRC Press, Boca Raton, pp. 237–252.

Herbert, A., Hansen, P.D., 1998. Genotoxicity in fish embryos. In: Wells, P.G., Lee, K., Blaise, C. (Eds), Microscale Testing in Aquatic Toxicology. CRC Press, Boca Raton, pp. 491–505.

International Programme on Chemical Safety – IPCS, 1993. Environmental Health Criteria 155, Biomarker and Risk Assessment: Concepts and Principles. World Health Organization, Geneva, p. 82.

ISO 13829 – ISO standard methods for the examination of water and waste water sludge, 1999. Water quality – Determination of the genotoxicity of water and waste water using the *umu*-test. ISO Secretariat TC 147/SC5/WG9. – DIN 39415-T3, VCH Verlagsgesellschaft Weinheim.

ISO/CD 16240 – ISO standard methods for the examination of water and waste water, 2002. Water quality – Determination of the genotoxicity of water and waste water using the Salmonella/microsome test (AMES-Test). ISO Secretariat TC 147/SC5/WG9. – DIN 38415-T4, VCH Verlagsgesellschaft Weinheim.

ISO/WD 21427 – ISO standard methods for the examination of water and waste water, 2002. Water quality – Evaluation of genotoxicity using Amphibia larvae (Xenopus laevis, Pleurodeles waltl). ISO Secretariat TC 147/SC5/WG9.

ISO/NWI – EROD – ISO standard methods for the examination of water and waste water, 2002. Water quality – Method for measuring the EROD enzyme activity in freshwater and saltwater fish. ISO Secretariat TC 147/SC5/WG3.

Marx, A., Sherry, J., Hansen, P.D., Hock, B., 2001. A new monoclonal antibody against vitellogenin from rainbow trout (*Oncorhynchus mykiss*). Chemosphere 44, 393–399.

Michel, X.R., 1993. Contributation a lètude des interactions entre les contaminants chimiques organiques et les organisms marins: bases moleculaires et applications a la biosurveilance de l'ènvironnement cotiers. These presentée a L'Université de Bordeaux, p. 234.

Narbonne, J.F., Daubèze, M., Clérandeau C., Garrigues, P., 1999. Scale of classification based on biochecmical markers in mussels: application to pollution monitoring in European coasts. Biomarkers, 4 (6), 415–424.

National Research Council – NRC, 1987. Committee on biological markers. Environmental Health Perspective 74, 3.

Obst, U., Wessler, A., Wiegand-Rosinus, M., 1998. Enzyme inhibition for examination of toxic effects in aquatic systems. In: Wells, P.G., Lee, K., Blaise, C. (Eds), Microscale Testing in Aquatic Toxicology. CRC Press, Boca Raton, p. 679.

Rao, S.S., Neheli, T.A., Carey, J.H., Herbert, H., Hansen, P.D., 1996a. DNA alkaline unwinding assay for monitoring the impact of environmental genotoxins. Environmental Toxicology and Water Quality: An International Journal 11, 351–354.

Rao, S.S., Neheli, T., Metcalfe, C.D., 1996b. Hepatic micronucleus assay for the assesment of genotoxic responses in fish. Environmental Toxicology and Water Quality: An International Journal 11, 167–170.

Seifert, M., Haindl, S., Hock, B., 1999. Development of an enzyme linked receptor assay (ELRA) for estrogens and xenoestrogens. Analytica Chimica Acta 386, 191–199.

Stagg, R.M., Addision, R.F., 1995. An inter-laboratory comparison of measurements of ethoxyresorufin O-de-ethylase activity in Danb (*Limanda limanda*) liver. Marine Environmental Research 40 (1), 93.

Sturm, A., Hansen, P.D., 1999. Altered cholinesterases and monooxygenase levels in daphnia magna and chironimus riparius exposed to environmental pollutants. Ecotoxicology and Environmental Safety 42, 9–15.

Sturm, A., da Silva de Assis, H.C., Hansen, P.D., 1999a. Cholinesterases of marine teleost fish: enzymological characterisation and potential use in the monitoring of neurotoxic contamination. Marine Environmental Research 47, 389–398.

Sturm, A., Hansen, P.D., Wolfram, J., Liess, M., 1999b. Potential use of cholinesterase in monitoring low levels of organophosphates in small streams: natural variability in three-spined stickleback (*Gasteroteus aculeatus*) and relation to pollution. Environmental Toxicology and Chemistry 18 (2), 194–200.

Sturm, A., Hodson, P.V., Carey, J.H., Hansen, P.D., 1999c. Hepatic UDP-glucuronosyltransferase in rainbow trout (*Oncorhynchus mykis*) and preliminary assesment of response to pulp mill cooking liquor. Bulletin Environmental Contamination and Toxicology 62, 608–615.

Wittekindt, E., Werner, M., Reinicke, A., Herbert, A., Hansen, P.D., 1996. A microtiter-plate urease inhibition assay – sensitive, rapid and cost-effective screening for mercury and other heavy metals in water. Environmental Technology 17, 597–603.

Wittekindt, E., Mathess, C., Gaumert, T., Hansen, P.D., 2000. Die gentoxische Gewässergüte-Klassifizierung der Elbe – entwickelt mit Hilfe des DNA-Aufwindungstests mit der Dreikantmuschel. Hydrobiologie und Wasserbewirtschaftung 44 (3), 131–144.

Chapter 7

Ecotoxicity testing

Hans Toni Ratte, Monika Hammers-Wirtz
and Michael Cleuvers

Abstract

Ecotoxicity tests are required in the context of national legislation and regulation of pesticides, other chemicals and environmental quality. Among the OECD (Organisation for Economic Cooperation and Development) countries the requirements differ only marginally as do also the current guidelines for the conduct of ecotoxicity tests. Ecotoxicity testing can be divided in lower-tier and higher-tier testing. In lower-tier tests single cohorts of a test species from different trophic levels and environmental compartments are examined in the laboratory under standardised conditions. International standards and guidelines have been developed by CEN (Comité Européen de Normalisation), ISO (International Organization for Standardization) and OECD. Depending on the selected test, the effect of the test material on one to several endpoints is investigated, mainly variables of survival, reproduction, growth and metabolism. The special characteristics, endpoints considered and problems of current lower-tier tests are described and discussed. Mainly two statistical designs are performed: the point-estimation approach to compute the EC_x (effective concentration for an effect of x%) as toxicity parameter and the hypothesis-testing approach to determine the NOEC (no-observed effect concentration). Sometimes a so-called limit test is conducted, in which the effect of only one high concentration is investigated and statistically tested. Hypothesis testing is problematic since there is often the danger of a false-negative result, i.e. statistical test reveals "no effect" but there is one. Therefore, the hypothesis-testing approach has been increasingly criticised during the past decade, so that OECD and ISO decided on a phasing-out of the NOEC as toxicity parameter, which at this point appears to be not followed unequivocally

In higher-tier testing more flexible and complex ecotoxicity tests are performed, a standardisation of which is impossible. They are required if a chemical substance fails to meet the trigger values in lower-tier risk assessment. These tests range from experiments with additional species, populations, small communities in the laboratory to larger-scaled outdoor mesocosms. The general aim is to research fate and effects of a test substance under more realistic environmental conditions, i.e. fate processes can take place and the tested populations of the test community undergo biotic and abiotic interactions.

Keywords: Ecotoxicity tests, lower-tier, higher-tier, authorisation of chemicals, guidelines, statistical design

1. Introduction

Ecotoxicity tests are biological experiments with various test species under presence of chemical substances or environmental samples. They are required for risk assessment of new and existing chemicals as well as for monitoring the environmental quality (e.g. of effluents, sediment and soil samples). In each test the response of one or more characteristic endpoints are measured. The term "endpoint" (also called "response variable") refers to the biological parameter measured/observed, e.g. survival, number of eggs, size or weight, enzyme level. The terms "effect" and "response" describe the change in the endpoint considered. For quantal endpoints (e.g., mortality; see "Statistical Treatment" section) an effect is defined in terms of a change in the number of animals affected, for continuous endpoints (e.g., growth rate) it is defined in terms of a percentage change in the average level of the endpoint, for both quantal and continuous endpoints in comparison to the controls.

An important requirement for ecotoxicity tests is their reproducibility and repeatability, because the results have to be justiciable. Therefore, the tests are run under the principles of Good Laboratory Practice (GLP; OECD 1998b) and using internationally accepted guidelines and standards, such as developed within the OECD (Organisation of Economic Cooperation and Development), CEN (Comité Européen de Normalisation) and ISO (International Organization for Standardization).

To bring new chemical substances onto the market, in Europe ecotoxicity testing is prescribed by national and European directives (e.g., 91/414/EEC, 1991). According to the EEC directive, a tiered approach for identifying potential risks for aquatic species is performed. Effects on biota are being characterised by acute and chronic laboratory studies (so-called Lower-tier Tests) using various aquatic organisms, e.g. fish, invertebrates, algae and macrophytes. As effect parameters effect concentrations or threshold concentrations (e.g., EC50s and NOECs, respectively) from these studies are compared with predicted environmental concentrations (PECs) to obtain toxicity exposure ratios (TERs). In Annex II (91/414/EEC, 1991), trigger values for these TERs of 100 and 10 are applied to acute and chronic studies, respectively. If the ratio for an active ingredient exceeds these trigger values, no further biotesting is required. If the ratio is below these trigger values, so-called Higher-tier studies are required for a refined risk assessment. If a chemical substance fails to meet the trigger values, a range of higher-tier studies have been recommended by the HARAP workshop (Campbell et al. 1999). The results from these studies can be used to demonstrate acceptability of the chemical substance. In addition, Annex II of Directive 91/414 EEC (1991) suggests aquatic microcosm or mesocosm studies as higher-tier tests.

With respect to monitoring of environmental samples only lower-tier tests are used to determine a dilution which is regarded as tolerable. The toxicity parameters determined are either a low EC_x (e.g., EC20, EC10) or the Lowest Ineffective Dilution (LID). In some countries these sort of tests forms the base for the legal control of effluents from e.g. industrial waste water treatment plants (e.g., in Germany according to the "Wasserhaushaltsgesetz (WHG)" and "Abwasserabgabengesetz (AbwAG)"). Depending on the results companies can be sentenced by court to pay for compensation measures.

Compared with the more than five million species on earth only very few species are used as test organisms in ecological risk assessment to assess effects on the ecosystem. Important criteria for the selection of ecotoxicological test species are sensitivity, representation of the ecosystem that may receive the impact, abundance and availability of the species, the ecological importance, as well as practical aspects like easy handling and culturing of species in the laboratory.

Furthermore background information on a species (i.e., its physiology, genetics, and behaviour) is helpful for interpreting the test results. Because of the variation in sensitivity between different species, the test species used has a great impact on the assessment of a chemical. The generally most sensitive test species does not exist because the sensitivity of a species depends on the chemical exposed to and the test parameter recorded. Therefore it appears to be important to investigate several species exhibiting different sensitivities.

The use of test species belonging to different trophic levels has been established in risk assessment. Representative species traditionally used for risk assessment of chemicals are various freshwater species representing algae, invertebrates and fishes. Some green algae like *Scenedesmus subspicatus*, *Chlorella vulgaris* or *Pseudokirchneriella subcapitata* are in use as standard test organisms representing primary producers, while *Daphnia magna* or other daphnids are traditionally used as representatives for the primary consumers or arthropods in general. Furthermore different fish species represent the trophic level of the secondary consumers or top-predators.

The oldest and widely used test species are freshwater species, while species representing soil organisms are currently becoming more important in ecotoxicology. Up until now, among the terrestrial species mainly used in risk assessment of chemicals are some higher plants and the earthworm *Eisenia fetida*.

2. Lower-tier testing

2.1. Ecotoxicological parameters and investigated level of organisation

We distinguish between two types of ecotoxicological biotests: the acute tests and the prolonged tests. In acute tests normally quantal responses like mortality or immobilisation are observed. The prolonged tests focus on the effects on reproduction, growth or other physiological processes. Almost in all of the acute tests organisms are investigated, while in the prolonged studies different levels of organisation (organisms, populations) are considered. In the algal growth inhibition test the population growth is determined by measuring the density of algal cells; here several generations of algae are involved during the 72 h test duration. In contrast in the Daphnia reproduction test the effect on reproduction is measured on organism level rather than on population level, since the offspring number is only measured in one generation.

The reasons for choosing the organism level in ecotoxicological tests are mostly the duration of the test and the handling of those tests. There are first steps in developing tests on sub-organism level, which are helpful tools in screening and monitoring the environmental quality. However, their ecological relevance is more problematic than with tests on organism or population level.

Even if the tests are conducted on organism level, we have to accept that not all population-relevant aspects are considered. In the Daphnia reproduction test for example the number of offspring is the parameter used for risk assessment. In this test however the quality of this offspring is not considered although it can also be affected by the test material and is of great importance for daphnids population growth (Hammers-Wirtz and Ratte, 2000). The parameters investigated in standardised laboratory tests are always a compromise between practicability and ecological relevance.

But we have to be aware which level of organisation is investigated and which conclusions can be drawn from this level for the population in the field.

2.2. Standardisation

Ecotoxicological tests are used in the legal control of chemicals and the quality assessment of effluents. Therefore an ecotoxicological test has to give approximately the same result in different laboratories and in the same laboratory at different times, i.e. the results of such biotests have to be reproducible and repeatable. To reach a high repeatability and reproducibility the test conditions need to be strictly regulated. Thus since the 1980s various organisations (e.g. OECD, CEN, ISO, ASTM) are developing guidelines for the relevant ecotoxicity tests in order to standardise the test and evaluation methods.

Of course, when a test will be repeated several times or will be conducted at one time in different places, the results won't be exactly the same, due to biological variability of the test organisms involved. The major potential sources of variation in ecotoxicological tests are genetic and environmental ones. In the tests with the water-flea *Daphnia magna* the genetic variation can be eliminated because the daphnids normally produce genetically identical neonates due to parthenogenetic reproduction. Hence these test organisms provide an opportunity to distinguish between genetic variability and environmental variability. Different studies with several clones of *Daphnia magna* showed that some genotypes were more tolerant to a toxicant than others, but the results depended on the toxicant applied (Baird and Barata 1999, Baird et al. 1991, Baird et al. 1990, Soares and Calow 1993). There exist no generally sensitive or tolerant genotypes. Thus, when we use a clone in ecotoxicological studies we can reduce the genetic variability, but we have to be cognisant that this clone is possibly not as sensitive as another genotype of this species in the field.

The environmental variability can be reduced by strictly regulating the ambient conditions. Some environmental factors can easily be controlled such as temperature, and lighting regimes. But there are many factors, first of all the supply of live food of approximately the same quality, which is difficult to control and has great impact on the development of the test organisms and therefore possibly on the toxicant's effect (Soares and Calow 1993). In most of the test systems the variability measured is a combination of both genetic and environmental variation. To determine the level of variation in reproducibility and repeatability so-called "ring tests" were performed where different laboratories investigate one or several toxic compounds several times. The more complex the test system is and the less the number of test organisms involved, the higher is usually the variability of the test system. We need a high reproducibility and a high repeatability in toxicological tests, but on the other hand in view of the practitioner and those paying for them, a test should be practicable, preferably little labour-intensive and cheap (Bödeker et al. 1992). Furthermore, the test organisms have to be easy to rear and nevertheless highly sensitive and representative for the biotic community because only few species are tested as representatives for the biocenosis.

From the above considerations it becomes clear that any ecotoxicity test is a compromise between practicability and cost efficiency on the one hand and the representativeness and sensitivity of the species investigated on the other.

2.3. Test species

A multitude of different species is used worldwide in scientific studies evaluating the risk of chemicals or effluents, but only a few of this species have been established as standard test organisms of worldwide use. This chapter will mainly focus on standard test organisms in ecotoxicology more than on species tested for special aspects. In the following standard test species and the principle of the tests with these species will be presented. Furthermore shortcomings of these standard tests will be discussed. Other test species of less importance are only shortly mentioned. An overview of the standardised test guidelines for the different trophic levels is given in Table 1.

2.3.1. Bacteria

Bacteria play a decisive role in the ecosystem as decomposer of organic material. Furthermore they are of great importance for the biodegradation of organic compounds in wastewater treatment plants, sewage sludge and soils. Bacteria are commonly used test organisms in the Respiration inhibition test (OECD, 1984f) with activated sludge used to assess effects of chemicals or effluents on the aerobic bicoenosis of microorganisms in sewage treatment plants.

Another bacterium used as test species is the marine luminescent bacterium *Vibrio fischeri*, formerly known as *Photobacterium phosphoreum*. Luminescent bacteria transmit a bioluminescence as metabolism product. The enzyme luciferase catalyses the oxidation of reduced riboflavin phosphate which is accompanied by emission of light (Kaiser, 1998). This process is linked with the microbial metabolism, and therefore a reduction in the natural bioluminescence of *Vibrio fischeri* is an indication of a toxic effect. In the bioassay with *Vibrio fischeri* the inhibitory effect of chemicals or effluents on the light emission is determined. The bacteria are incubated with the toxicant or effluent for 30 minutes. Thus the duration of this bioassay is very short in contrast to other acute bioassays with a duration of 24 to 96 h. The bacteria used for the test can be taken from freshly prepared, liquid-dried or freeze-dried stock cultures which have to be prepared differently. For this reason the ISO standard 11348 consists of three parts. Due to the possibility that the bacteria stock culture can be kept frozen and easily prepared this bioassay is less labour-intensive than other biotests where a stock culture has permanently to be maintained. The field of main application of this test is the testing of wastewater.

2.3.2. Algae and plants

2.3.2.1. Alga growth inhibition test

The ecological importance of aquatic plants can hardly be overestimated. For example, approximately 70% of the world's atmospheric oxygen is generated by phytoplankton (Reynolds, 1984). Furthermore, phytoplankton as well as benthic and epiphytic attached microalgae are the primary energy source for a lot of aquatic ecosystems. Another crucial function, which is currently discussed regarding the greenhouse effect and the "global warming" phenomenon, is the ability of algae to bind carbon dioxide and thus reducing the amount of atmospheric green house gases.

Table 1. Existing test guidelines (OECD) and standards (ISO) for the different trophic levels; please find the references of actual versions under http://www.oecd.org and http://iso.org, respectively.

Trophic level	Species	Test guidelines
Decomposer	Marine luminescent bacterium *Vibrio fischeri* (formerly known as Photobacterium phosphoreum)	ISO 11348–1/-2 /-3 (1998)
	Activated sludge microorganisms	ISO 15522 (1999)
Primary producer	Green algae *Scenedesmus subspicatus* *Chlorella vulgaris* *Pseudokirchneriella subcapitata* (= Selenastrum capricornutum)	OECD 201: Algal Growth Inhibition Test (Updated Guideline, adopted 1984) ISO 8692: Algal Growth Inhibition Test (1989)
	Marine algae *Skeletonema costatum* *Phaeodactylum tricornutum*	ISO 10253: Algal Growth Inhibition Test (1995)
	Duckweeds *Lemna minor* *Lemna gibba*	OECD 221 (New Guideline, 2000): *Lemna* growth inhibition test
	Terrestrial plants	OECD 208: Growth test (Original Guideline, adopted 1984) OECD 208 A: Seedling Emergence and Seedling and Seedling Growth (Draft Updated Guideline, 2000) OECD 208 B: Vegetative Vigour Test (Draft Updated Guideline, 2000)
Primary consumer	Freshwater crustacean *Daphnia magna*	OECD 202: Acute Immobilisation Test (Updated Guideline, adopted 1984) ISO 6341: Inhibition of the mobility (1996) OECD 211: Reproduction Test (Original Guideline, adopted 1998) ISO 10706: Long term toxicity (2000)

Table 1. (continued)

Trophic level	Species	Test guidelines
	Marine crustaceans *Acartia tonsa* *Tisbe battagliai* *Nitocra spinipes*	ISO 14669: Acute lethal Toxicity (1999)
	Chironomids *Chironomus tentans* *Chironomus riparius*	OECD 218: Sediment-Water Chironomid Toxicity Test (using spiked sediment)/OECD 219 (using spiked water) (Draft New Guideline, 2000)
	Earthworm *Eisenia fetida*	OECD 207: Acute Toxicity Test (Original Guideline, adopted 1984) ISO 11268–1: Acute Toxicity (1993)
	Eisenia fetida/andrei *Eisenia fetida*	OECD: Earthworm Reproduction Test (Draft New Guideline, 2000) ISO 11268–2: Reproduction Test (1998)
	Enchytraeid worm *Enchytraeus sp.*	OECD 220: Reproduction Test (Draft New Guideline, 2000)
Secondary consumer	Fishes e.g. *Danio rerio*	OECD 203: Acute Toxicity Test (Updated Guideline, 1992) ISO 7346–1/-3: Acute Toxicity (1996)
	e.g. *Danio rerio,* *Oncorhynchus mykiss,* *Pimephales promelas*	OECD No. 212 (Short-term Toxicity Test on Embryo and Sac-Fry Stages) (Original Guideline, 1998)
	e.g. *Danio rerio,* *Oncorhynchus mykiss,* *Pimephales promelas*	OECD No. 210 (Early-Life Stage Toxicity Test) (Original Guideline, 1992)
	e.g. *Danio rerio* *Oncorhynchus mykiss*	OECD 204: Prolonged Toxicity Test (Original Guideline, 1984) ISO 10229: Prolonged Toxicity Test (1994)

Various parameters of natural algal communities have been monitored to estimate the quality of their habitat. Shubert (1984) has presented a summary about the use of algae as ecological indicators in environmental studies. Due to their significance it was a logical step to use algae routinely in toxicity tests designed to provide information on the environmental safety of chemicals and effluents. The first standard methods, the marine and freshwater bottle tests, which were developed in the early 1970s (US EPA 1971, 1974), have been followed by more additional methods published by standard writing organisations and various regulatory agencies (Payne and Hall 1979, FDA 1982, Horning and Weber 1985, APHA 1985). The most important international guidelines currently in use are the OECD guideline 201 (OECD 1984), the ISO standard 8692 (ISO 1989) and the guideline published by the European Commission (EEC 1993), which differ in form and content only in minor aspects.

Test species: Regarding the algal test species the long lasting expert-knowledge of Scandinavian (e.g. Skulberg 1967), American (e.g. Miller et al. 1976) and German (e.g. Bringmann and Kühn 1956) scientists was useful during the development of test protocols. Unicellular protococcale green algae like *Chlorella, Scenedesmus* and *Selenastrum* proved to be particularly suitable. *Scenedesmus subspicatus* (new name: *Desmodesmus subspicatus*) and *Selenastrum capricornutum* (= *Pseudokirchneriella subcapitata*) are the most common test species for algal growth inhibition tests (ISO 8692).

Standard test design: The base of each evaluation of algal tests is the growth curve of the population in a defined test duration (normally 72 hours). Algae are grown in a test medium with a defined nutrient content and different concentrations of the test substance and in the control samples (test medium without test substance). The initial cell density should be 10,000 cells/ml. The algae came from an exponentially growing pre-culture, which has been set up 72 hours before the start of the test under the same conditions as in the test. For each tested concentration, three replicates and six replicates for the control are incubated. Under the prescribed light and temperature conditions the algae population grows exponentially during the test (at least in the control).

Mean cell densities are measured after 24, 48 and 72 hours. Plotting the mean cell density versus the test duration leads to the growth curves of the population in the control as well as in the various test material concentrations.

The inhibition of the algal growth caused by the toxicity of a test substance can be derived from different variables. There is an ongoing debate on the whether the cell number, biomass integral or growth rate should be used in risk assessment. Some prefer the cell number (or the derived variable "biomass integral", which is the area under the growth curve (Nusch, 1982, 1983)), while the other favour the average growth rate (Nyholm, 1985, 1990, 1994; Dorgerloh, 1997). The pros and cons of the various response variables are discussed thoroughly in Ratte et al. (1998).

Surrogate toxicity data-sensitivity: In the past it was proposed to extrapolate phytotoxicity from data derived with fishes or daphnia (Stephan et al., 1985), because algae were considered less sensitive against chemicals than animal species, as some studies indicated (Kenaga and Moolenar, 1979; Kenaga, 1982). But many publications came to a different result (Patrick et al., 1968, Shehata and Nawar, 1979; Bringmann and Kühn, 1980; Walsh et al., 1980, 1982; Giddings et al., 1983; Sloff et al., 1983; Adema et al., 1983; Walsh and Merrill, 1984; Karpinska-Smulikovska, 1984; Miller et al., 1985), with algae being more sensitive than fish or daphnids in many cases. Generally, there is no universally sensitive test species or group of species, since the sensitive was found to be chemical- and species-specific. Two studies which have used different data-bases for notifications of substances came to the congruent result that algae were more sensitive than animal species in 50–60% and less sensitive in about 30% (Benenati, 1990, Weyers et al., 2000). Thus, it can be seen as confirmed that algal toxicity is not predictable from animal test species (Lewis, 1990, 1995) and that phytotoxicity data are essential for a reliable risk assessment. But in any case it would be reasonable to use more than the two standard test species *Scenedesmus* and *Selenastrum* because the sensitivity to chemicals and effluents was found to be strongly species specific (Bringmann and Kühn, 1978; Sloff et al., 1983; Stratton, 1987; Swanson et al., 1991; Lewis, 1995) and can differ clearly even between different test strains or geographical races of the same species (Venkatamaran and Rajyalakshmi, 1972; Millie and Hersh, 1987; Riedel, 1989). Consequently, it was demanded on various occasions to use more algal species and also cyanophytes in a test battery (Blanck et al., 1984; Wängberg and Blanck, 1988; Swanson et al., 1991; Cleuvers, 2001).

Difficult substances: Difficult substances (ISO, 1997, OECD, 1999) are e.g. poorly soluble materials, volatile compounds, adsorbing or complexing substances, substances that degrade in the test system and coloured substances like dyes. While the former have an effect on generally all test species, in the latter case primarily algae are affected. Coloured substances are a sizeable part of substances applied for notification of new substances in the European Union. The difficulty for interpretation results from the fact, that dyes are able to inhibit the growth of algae in two basically different ways. First, coloured substances absorb light with the result that the light intensity util-isable for algae is diminished and additionally also the spectral quality of light is changed. If light saturation is not achieved, the growth of algae will be inhibited due to a physical property of the tested substance. Second, the substance can act toxically in a narrower sense, due to its chemical effect on algae and the determination of this toxic action is the important point, while the European Commission stipulates that inhi-bition due to the shading effect of the coloured substance shall not be mentioned. Because both effect occur concomitantly, is it a priori not possible to make a state-ment about the extent of the single effects. Some attempts were made to clear this point (Memmert et al., 1994; Comber et al., 1995) and recently this topic was thor-oughly discussed by Cleuvers and Ratte (2002), who showed that by increasing the light intensity to a level above saturation and reducing the culture volume of algae to shorten the light path the shading effect of different dyes even at very high concen-trations could be completely eliminated.

2.3.2.2. Macrophytes – Lemna growth inhibition test
Macrophytes are used less frequently than algae in toxicity tests (Lewis, 1995). In one literature survey, only 7% of 528 reported phytotoxicity tests used macrophytic species

(Blanck et al., 1984). When macrophytes have been used, the duckweeds (*Lemna* spp.) have been the species of choice and they are often used as a representative species for other vascular plants. Lemnaceans are monocotyledon, angiosperm aquatic plants which belong within the subclass Arecidae to the Arales. It is a fast growing higher plant which is widespread worldwide in lenthic waters from the tropics to the temperate and arctic zone. The propagation takes place mainly vegetatively by the creation of daughter fronds (fronds = structure analogous to leaves). Like algae and cyanophytes as primary producers lemnaceans belong to the base of the food web.

From the family of lemnaceans are mainly *Lemna*-species like *L. minor* (Wallbridge, 1979; Wang, 1986; Taraldsen and Norberg-King, 1990; Jenner and Janssen-Mommen, 1993; Lomagin and Ulyanova, 1993; Weltje et al., 1997) and *L. gibba* (Holst and Ellwanger, 1982; Hughes et al., 1988; Cowgill and Milazzo, 1989; Wang, 1990; Cowgill et al., 1991; Lakatos et al., 1993; Day and Hodge, 1996), and less often also *L. perpusilla* and *Spirodela polyrhizza* (APHA et al., 1989) in use for toxicity testing. *L. triscula* was recommended too because of its easy culture and rapid growth (Huebert et al., 1990).

Various national guidelines are currently in use (AFNOR, 1990; ASTM, 1991; APHA et al., 1995; SIS, 1995; US EPA, 1996; EC, 1998; DIN, 2001), which differ regarding the preferred test species, the used culture medium, the test duration and, most important, the observed endpoint. Thus, the comparability of results obtained with different test protocols is restricted. In 2000, an international guideline and standard was drafted by OECD and ISO, respectively. In both drafts the test duration is set to seven days. All tests start with a definite number of fronds in each vessel, normally ten or twelve, whereas only young plants with two or three fronds are used. The plants were inspected in a defined time scheme, at day 0, 3, 5 and 7. Beneath the number of fronds either the total frond area, or, at the end of the test, the dry weight or the fresh weight must be measured. Furthermore, changes in the development of the plants ought to be noted, like the occurrence of chlorosis, necrosis or changes in the length of the rhizoid. A useful tool in this context is the digital image analysis, which facilitates the evaluation of test results clearly and enables the investigator to record further parameters, e.g. the distribution pattern of fronds in different size classes. The choice of the endpoint has a strong influence on the toxicity data; Cleuvers (2001), e.g. found differences of more than factor 300 between EC_{10}'s calculated from inhibition of the average growth rate on the one hand and either total frond area or dry weight on the other hand.

Regarding the sensitivity one could guess, that *Lemna* is less sensitive than algae, because it is often used to remove toxicants and nutrients from sewage waters and it shows indeed a considerable tolerance against metals (Tripathi and Chandra, 1991). But a general trend does not exist (Lewis 1995). In a study with 16 herbicides Fairchild et al. (1997) found *Lemna* in eight cases to be more sensitive and in eight cases to be less sensitive than the green algae *Selenastrum capricornutum*. Peterson et al. (1997) confirmed the findings, that green algae are not generally more or less sensitive than *Lemna*. In a study with eight pesticides, Lemna was either as sensitive as or more sensitive than the tested green algae (Grossman et al., 1992).

In any case, a *Lemna* growth inhibition test is very useful as an additional source of information about phytotoxicity, whereas generally a extension of test species, e.g.

by using other macrophytes, was frequently recommended (Swanson et al., 1991; Lewis, 1995; Fairchild et al., 1997).

2.3.3. Aquatic invertebrates

2.3.3.1. Daphnia

The freshwater cladoceran *Daphnia magna* is one of the oldest and widely used test organisms in aquatic toxicology (Baudo, 1987; Enserink, 1995). The genus *Daphnia* is an important link in freshwater trophic chains as dominant consumer of primary producers and as food for both invertebrate and vertebrate predators. The species of the genus *Daphnia* differ in their world distribution. *Daphnia magna* is a pond species with limited geographical range (Eurasia), whereas *Daphnia pulex* is a more widespread species (North and South America, Greenland and Europe) (Baudo, 1987; Flößner, 2000). Therefore in the USA the species *D. pulex* and another cladoceran *Ceriodaphnia dubia* are also used as test organisms representing the filter-feeding zooplankton (Mark and Solbe, 1998).

The choice of *Daphnia magna* as standard test species was influenced by several advantageous characteristics. It is of small size (compared to fish, molluscs or macrocrustaceans) and easy to culture in the laboratory. Its parthenogenetic reproduction under non-stressed conditions allows the testing of clones, which enhances the reproducibility and repeatability of the test results. Furthermore *Daphnia magna* is relatively sensitive to chemicals compared with other freshwater invertebrates (Mark and Solbe, 1998; Baird et al., 1989; Radix et al., 1999; Versteeg et al., 1997) and its relatively short life-span and reproductive cycle are favourable for the chronic testing. The ecology of *Daphnia* has long been studied so that we have a lot of background information on its biology and ecology.

Thus *Daphnia magna* is the most commonly tested freshwater species in acute as well as in chronic tests (Baird et al., 1989; Mark and Solbe, 1998; Baird and Barata, 1999).

Acute test: In acute tests with *Daphnia magna* young daphnids, aged less than 24 hours, are exposed to different concentrations of a test substance in a static system for a period of 48 hours. After 24 and 48 hours the immobilisation of the test animals is recorded. Those animals are considered as immobilised that are not able to swim or to move the appendages or the postabdomen after gentle agitation of the liquid. During the acute test the daphnids were not fed. With the percentage of immobilised daphnids an EC50 at 24 h and 48 h should be calculated. The test procedure is highly standardised. There exist different test methods that differ mainly in age of test organisms and test medium used. A comparison of the commonly international guidelines for testing the acute toxicity of Daphnia is given in Versteeg et al. (1997).

In acute tests with *Daphnia magna* usually clones are tested in order to reduce the variability due to genetic differences. An important source of variation for both acute and chronic tests with *Daphnia magna* are the food supply and the culture conditions of the daphnids in the stock culture (e.g. Baird and Barata, 1999). The feeding of the mothers and their density in the stock culture have a great impact on the size, weight and fat content of their neonates (Cowgill et al., 1985; Enserink et al., 1990; Goser,

1997), and the "quality" of the neonates introduced in the test influences the sensitivity of these test organisms (Enserink et al., 1990; Goser, 1997). The lower the food supply of the mothers, the larger and fatter is the offspring (Boersma, 1997a, 1997b; Enserink et al., 1995; Cowgill et al., 1985; Guisande and Gliwicz, 1992; Glazier, 1998; Sokull-Klüttgen, 1998) and higher densities in the culture lead to larger and fatter neonates (Cleuvers et al., 1997; Goser, 1997). In acute tests larger neonates have been shown to be less sensitive than smaller ones exposed to cadmium (Enserink, et al., 1990), bromide (Naylor et al., 1992), or dichloraniline (Baird et al., 1989; Naylor et al., 1992; Goser, 1997). Although the culture conditions influence the test results, the culture conditions particularly the feeding and the density of the culture are up to now not regulated in the test guidelines. Here it is only stated that the culture conditions shall be similar to those in the test and that the daphnids shall be from a healthy stock showing no signs of stress such as high mortality, presence of males or ephippia, discoloured animals or delay in first reproduction. To get neonates of similar quality the culture conditions have to be specified in detail.

Another factor affecting the results of acute tests is the age of the test animals at the start of the test. The test animals inserted shall be younger than 24 h at the beginning of the test. Klein (2000) showed recently that also in this range of 0 to 24 h age-dependent differences in sensitivity can occur. Animals with an age of 20 to 24 h were up to 4-fold more sensitive to potassium dichromate than younger animals. But for other chemicals this trend was not confirmed (Bögi, 1998). Furthermore the age-dependent sensitivity was different at exposure times of 24 and 48 h. Thus, on the basis of these facts it seems not to be useful to change the age of the test animals.

Daphnia reproduction test: The chronic test with *Daphnia magna* is conducted to assess the effect of a test material on the reproductive output of *Daphnia magna*. Young daphnids less than 24 h old are exposed for a period of 21 days to a test substance, industrial or sewage effluents. The test can be performed in a static or a flow-through system. The flow-through design has to be used if the test substance is volatile or not stable over a period of three days, i.e. the concentration of the test substance falls below 80% of the initial measured concentration within three days. The effect of a test material on the reproduction of *Daphnia magna* can be influenced by various biotic and abiotic factors. The abiotic conditions during the test are highly standardised, while the food quantity and quality fed to the test organisms is not exactly regulated. Different green algae *(Chlorella ssp., Pseudokirchneriella subcapitata, Scenedesmus subspicatus)* can be used as food for the daphnids and the quantity of food shall be between 0.1 and 0.2 mg C per animal and day. The food quantity and quality are known to have a great impact on the reproduction of *Daphnia magna* (Green, 1954; Cowgill et al., 1985; Boersma, 1997a, 1997b; Kilham et al., 1997). Therefore a more exact regulation of feeding would reduce that variation in reproducibility and repeatability.

The parameters recorded in the chronic test are the survival of the females together with the number of living offspring per live female at the end of the test. The mean number of live offspring per female in each exposure concentration is compared to the control mean by statistical tests in order to determine the LOEC and NOEC. Additionally the EC50 can be calculated.

In the Daphnia reproduction test the number of offspring in the treatments is not always reduced but can also be enhanced compared to control. Such increasing effects normally induced by low concentrations of some chemicals, so called hormetic effects, occur in various species and different test endpoints (Calabrese and Baldwin, 1998). In the Daphnia reproduction test increased offspring number compared to control have been observed several times induced by different chemicals (Biesinger and Christensen, 1972; Francis et al., 1986; Van Leeuwen et al., 1987; Van der Hoeven, 1990; Baldwin et al., 1995; Klüttgen et al., 1996; Brown et al., 1998; Hammers-Wirtz and Ratte, 2000; Wollenberger et al., 2000). Those increases in offspring number also occurred in solvent controls induced by organic solvent added to the test medium (Shurin and Dodson, 1997; Brown et al., 1998; Sanchez et al., 2000). Up to now those "positive" effects on reproduction are not considered in risk assessment. But one experiment showed definitely that the increase in offspring number was coupled with a reduction in offspring quality (Hammers-Wirtz and Ratte, 2000). A reduction of offspring quality has a great impact on the beginning and extent of reproduction, thus on the fitness of the population, which is the one that should be protected. Therefore it must be questioned if increases in offspring number can further be ignored in risk assessment, because the increases in offspring numbers can be coupled with a decrease in neonate fitness.

Beyond it several chronic studies with *Daphnia magna* showed that the quality of neonates was affected by chemicals even if the offspring number was reduced. Although the offspring size is not routinely recorded in chronic tests with daphnids several authors observed a reduction in neonate size induced by several chemicals like copper (Flickinger et al., 1982), cadmium (Bodar et al., 1990), lead (Enserink et al., 1995), the pesticide carbaryl (Hanazato and Dodson, 1995), and a dispersant (Hammers-Wirtz and Ratte, 2000). These facts clearly indicate that the measurement of offspring number alone is not sufficient to detect ecological relevant effects on population level, the fitness of the neonates also has to be considered. Measuring the body size of the neonates can indicate their fitness, but body size is not always correlated to the fitness. To determine the neonates fitness either an additional test with these neonates is necessary or instead of the chronic test an population test has to be carried out where the chemical effect on subsequent generations is recorded. If only the offspring number is considered in risk assessment the effect on population level can be underestimated (Hammers-Wirtz and Ratte, 2000).

2.3.3.2. Further aquatic invertebrates
Another test species used as representative for the zooplankton is the cladoceran *Ceriodaphnia dubia*. This test species is commonly used in the USA, but of minor interest in Europe. The important advantage of *Ceriodaphnia* compared to *Daphnia* is the short generation time submitting the reduction of the chronic test duration to seven days in contrast to 21 days in the Daphnia reproduction test. A comparative study of toxicity tests with *Daphnia magna* and Ceriodaphnia dubia demonstrate the *C. dubia* are on average more sensitive than *Daphnia magna* in acute toxicity tests and of similar sensitivity in chronic toxicity tests (Versteeg et al., 1997).

The rotifer *Brachionus calyciflorus* is a further invertebrate of interest as ecotoxicological test species because of its short generation time. *Brachionus calyciflorus* reproduces by parthenogenesis and is able to hatch several times during 48 h under favourable conditions. These animals are primary consumers and serve as prey for several invertebrates and fishes. In the test with *Brachionus calyciflorus* the population growth over a period of 48 h is determined. This test is not yet standardized but there exist a Working Draft of ISO. An advantage of this test is its short duration of only two days especially for testing of instable effluents. But first investigations with Brachionus showed that this test is on average three times less sensitive than the Daphnia reproduction test (Radix et al., 1999).

2.3.4. Terrestrial invertebrates

2.3.4.1. Earthworm

The earthworm *Eisenia fetida* is one of the mostly used test organism for the terrestrial field. This terrestrial species representing the soil fauna is used in acute tests as well as in reproduction tests. The acute toxicity of a chemical on the earthworm can be determined in two different ways. Either the mortality is determined after the earthworms were exposed for hours to an filter spiked with the test substance or the mortality is recorded after 7 or 14 days rearing in an artificial soil merged with the chemical. In the reproduction test adult worms were exposed to a chemical and lethal and sublethal effects were recorded over a period of 8 weeks. After four weeks growth and mortality effects were recorded and then adults were removed from the soil. The effects on reproduction were assessed after further four weeks by counting the number of offspring present in the soil. For the reproductive output an NOEC (no observed effect concentration) and, if possible, an EC50 is calculated.

2.3.4.2. Enchytraeids

Although the earthworm *Eisenia fetida /E. andrei* is recommended in various guidelines as standard test organism for the terrestrial environment (Römbke and Moser 1998), the use of enchytraeids as standard test organisms is dicussed in the recent years. Soil-inhabiting enchytraeids are more ecologically relevant than *E. fetida/ E. andrei* occurring almost in compost heaps. Furthermore enchytraeids are important members of the soil biocoenosis in many different habitats, especially where earthworms are rare (Didden 1993). The test handling with enchytraeids is more practicable than with earthworms because the enchytraeus species are easy to handle and breed, their generation time is significantly shorter than for earthworms (test duration only 4 to 6 weeks) and the volume of the test vessels is significantly smaller than for the earthworm test (amount of soil: 20 g vs. 500 g).

In the enchytraeid reproduction test 10 adult worms of the test species *Enchytraeus albidus* or *enchytraeus sp.* are exposed to chemical spiked artificial soil (artificial soil according to OECD (1984)). The tests are conducted at 20°C and the worms fed weekly with rolled oats. After 3 weeks the adult worms are removed and after 6 weeks the juveniles hatched counted. The variables recorded in this test are mortality and behaviour of the adult worms (after 3 weeks) and the number of juveniles hatched (after 6 weeks).

Table 2. Listing of the current test guidelines with fishes; please find the references of actual versions under http://www.oecd.org and http://iso.org, respectively.

Guideline	Title
OECD 203	Fish, Acute Toxicity Test (Updated Guideline, adopted 1992)
OECD 204	Fish, Prolonged Toxicity Test: 14-Day Study (Original Guideline 1984)
OECD 210	Fish, Early-Life Stage Toxicity Test (Original Guideline, adopted 1992)
OECD 212	Fish, Short-term Toxicity Test on Embryo and Sac-Fry Stages (Original Guideline, adopted 1998)
ISO 7346 (1996)	Water quality – Determination of the acute lethal toxicity of substances to a freshwater fish (*Brachydanio rerio* Hamilton-Buchanan (Teleostei, Cyprinidae)) Part 1: Static method Part 2: Semi-static method Part 3: Flow-through method
ISO 10229 (1994)	Water quality – Determination of the prolonged toxicity of substances to freshwater fish – Method for evaluating the effects of substances on the growth rate of rainbow trout (*Oncorhynchus mykiss* Walbaum (Teleostei, Salmonidae))
ISO 12890 (1999)	Water quality – Determination of toxicity to embryos and larvae of fresh water fish – Semi-static method.

2.3.4.3. Collembola

Another invertebrate used as test organism for the terrestrial environment is the collembola *Folsomia candida*. Collembola are representing the detrivorous consumers. *Folsomia candida* shows different characteristics advantageous for test organisms: it is easy to breed, has a short generation time and a high reproduction rate and reproduces pathenogenetically. This collembola species has been used for a long time in risk assessment of pesticides in acute tests as well as in reproduction tests.

2.3.5 Fish

Fish are used as test organisms in aquatic toxicology because of their top-position in the trophic chain and their role as food for humans. The acute and chronic fish tests are used to assess effects of chemicals or effluents on the survival or sub-lethal parameters. An overview about the guidelines with fish as test organism is shown in Table 2.

Different test species are used as test organisms representing the aquatic vertebrates. Commonly used test species are the rainbow trout *Oncorrhynchus mykiss*, the Zebrafish *Danio rerio*, the Common carp *Cyprinus carpio*, the fathead minnow *Pimephales promelas*, and the guppy *Poecilia reticulata*. In principle, also salt water fish species can be used in these tests assumed the species can be held under satisfactory conditions in the laboratory. Up to now the main database of fish toxicity data exist for freshwater species. Comparative studies of toxicity data of freshwater fish and marine species show that marine species were more sensitive than freshwater species for the majority of substances tested, but for 91% of all substances the sensitivity ratios between freshwater and saltwater fish were within a factor of 10 (Hutchinson et al., 1998). Thus an extrapolation from freshwater fish to marine fish seems to be possible, the more so as the sensitivity between different freshwater species may vary by orders of magnitude (Nagel and Isberner, 1998).

2.3.5.1. Acute tests

In acute fish tests usually juvenile or adult fish are exposed to chemicals or effluents preferably for a period of 96 hours (e.g. OECD guideline 203). After 24 h, 48 h or 96 h the mortality in the treatments is recorded in order to calculate the LC50. The test species used for the test can be chosen by the testing laboratory, but the choice may be focus on practical criteria (availability throughout the year, ease of mainte-nance, convenience for testing) as well as relevant economic, biological an ecological factors. Several species are recommended as test species in the OECD guideline. The temperatures used in the tests and further test conditions depend on the test species chosen. All fish must be obtained and held in the laboratory for at least 12 days before they are used for testing. During this time the fishes were fed at least three times a week, but during the test period the fish were not fed. At least seven fish must be used at each test concentration and in the controls. The validity criteria for the test are less than 10% mortality in the controls and a dissolved oxygen concentration higher than 60 per cent of the air saturation value throughout the test.

Since acute fish toxicity tests are in conflict with current Animal Rights Welfare legislation possible alternative methods on sub-organism level were developed. Possible alternatives of acute fish test might be acute toxicity tests with embryo of zebrafish *Danio rerio* and cytotoxicity tests with fish cells. In the acute embryo test of zebrafish fertilized eggs were exposed to a test substance for 48 h. After 48 h of static exposure different lethal and sublethal parameters were recorded (Schulte and Nagel 1994). Different comparative studies demonstrate that the toxicity data derived from the embryo test are in good accordance with data from corresponding acute toxi-city tests with juvenile or adult fish (Nagel and Isberner 1998). Tests investigating the effect of several industrial effluents showed that the embryo toxicity test was as sensi-tive or even more sensitive than the conventional acute fish test (Friccius et al. 1995). Thus the embryo test is a promising alternative to conventional acute fish toxicity tests in routine waste water control (Nagel and Isberner, 1998). For the testing of effluents this test recently has been standardized as German DIN-Norm DIN 38415–6.

In cytotoxicity tests with fish cells the permanent cell line RTG-2 derived from rainbow trout gonads was commonly used. Comparative studies with different chem-ical compounds showed that in most cases the zebrafish embryo test was more sensitive than both the acute toxicity test and the RTG-2 cell test (Nagel and Isberner, 1998).

2.3.5.2. Short-term toxicity test on embryo and sac-fry stages

This test is a short-term test in which the life stages from the newly fertilized egg to the end of the sac-fry stage are exposed to a chemical. This embryo and sac-fry test can be performed as semi-static or as flow-through test. Several fish species can be used for this tests, but most experience has been with the freshwater fish *Danio rerio*. Test conditions and test duration will depend on test species used. In this test lethal, and to a limited extent, sub-lethal effects are investigated to determine the lowest observed effect concentration and the no observed effect concentration, or to estimate LC/EC_x values. The parameters observed in this type of test are survival, hatching, abnormalities in appearance and behaviour, as well as length and weight at the end of the test.

This embryo and sac-fry test is a link between lethal and sub-lethal fish tests. It can be used as screening test for the full Early Life Stage test or for chronic toxicity tests.

2.3.5.3. Chronic fish tests

In chronic fish test lethal and sub-lethal effects are investigated for a period of 14 to 28 days. The parameters recorded are changes in morphology (in colour), food uptake and swimming behaviour.

2.3.5.4. Prolonged toxicity tests

In the prolonged toxicity tests fish are exposed to several concentrations of a chemical for at least 14 days, but can be extended by one or two weeks. This test can be performed under semi-static conditions or in flow-through systems. Several test species are recommended for this test as in the acute fish test. During this test fish were fed at least daily. Parameters observed in this test beside mortality are any other than lethal effects like effects on behaviour (swimming behaviour, food intake), changes in appearance of fish or changes in length or body weight. The results are used to determine the lowest observed effect concentration (LOEC) and the no observed effect concentration (NOEC). Unfortunately, in most prolonged toxicity tests mortality was the crucial test endpoint for the determination of LOEC and NOEC (Nagel and Isberner, 1998).

2.3.5.5. Early life stage toxicity test

In the early life stage toxicity test different life stages of fishes like fertilized eggs, embryo stages, and sac fry stages are tested, because these stages arc highly sensitive to chemicals and other test material. Fertilised fish eggs are exposed to a range of test concentrations preferably under flow-through conditions at least until all the control fish are free-feeding. There are different species recommended for this test. The test conditions and test duration as well as feeding and handling of the different life stages are depending on the test species chosen for this test.

Several parameters are recorded during this test: observations on hatching and survival, abnormalities of body form in larvae or fish, abnormalities in behaviour (e.g. hyperventilation, uncoordinated swimming, atypical quiescence, atypical feeding behaviour). Furthermore the weight and length of all surviving fish are recorded at the end of the test. Lethal and sub-lethal effects are assessed in order to determine the lowest observed effect concentration and the no observed effect concentration.

In comparative studies the early life stage test proved to be more sensitive than the prolonged fish test (28 days) for the majority of the chemicals tested (Nagel and Isberner, 1998). Due to its sensitivity and its higher ecological relevance the early life stage test represents a promising alternative to the conventional prolonged fish test (Nagel and Isberner, 1998).

2.3.5.6. Life cycle tests

In life cycle tests sub-lethal effects on fish were recorded over the whole life cycle in order to assess effects on reproduction and offspring quality. For complete life cycle tests only small, rapidly growing warm water fish such as zebrafish may be used to complete this test within a reasonable period of time. Life cycle tests with zebrafish usually start with fertilised eggs until the larvae of the second generation have been developed. This type of test requires a lot of time (25 weeks) and money. As alternative to complete life cycle tests partial life cycle tests, early life stage tests and extrapolation from acute toxicity data to chronic data have been discussed (Nagel and Isberner, 1998). Partial life cycle tests as well as early life stage tests are no adequate alternatives for the full life cycle test since in these test effects on the second generation which are often the most sensitive endpoint can not be considered. The method of acute-to-chronic ratio (ACR) is no suitable alternative because this ratio shows a high variability for different chemical compounds and might only carefully be extrapolated to other species (Nagel and Isberner, 1998). Furthermore in chronic studies the effects on populations and not on individuals should be detected. Thus if we want to study effects on population level complete life cycle tests have to be carried out.

2.4. Statistical analysis

The type of statistical methods applied in ecotoxicity tests is of prime importance for the interpretation of results and conclusions from these tests, and consequently for the associated policy decisions. The statistical treatment of biotest data aims to generate parameters that adequately describe the toxicity of a test material. During the last decade an intense discussion arised on the use of appropriate statistical methods (Pack, 1993; SETAC, 1995; Hoekstra and Van Ewijk, 1993; Kooijman and Bedaux, 1996; Laskowkj, 1995; Chapman et al., 1996; OECD, 1998c; ASTM, 2000; Newman, 1994; Sparks, 2000). Many of the authors claimed the lack of statistical guidance given in guidelines and standards. As a consequence, a combined ISO (ISO TC147/SC5/WG 10) and OECD expert group has just started to work on an agreed general guidance document on the statistical treatment of ecotoxicity test results. The current section does not intend to anticipate the results and conclusions of this guidance document, but some general aspects of the statistical evaluation will be dealt with below.

A first important point is that the selection of an appropriate statistical method is determined by the type of response variable measured or observed in an ecotoxicity test. Sokal and Rohlf (1981) divide variables as follows:

- Measurement variables (quantitative variables)
  - Continuous variables (metric variables, unbounded; e.g. weight, length, experimental time).
  - Discontinuous variables (discrete variables, unbounded; e.g. count data such as number of offspring, number of eggs).

- Ranked variables (discrete variables, unbounded)
  These data are ordered or ranked by their magnitude; differences in ranks cannot be interpreted as metric distances as it can be done with measurement variables. These types of data are rarely found in ecotoxicity tests; however, measurement

variables are sometimes transformed into ranked variables, in case rank-based statistical tests are to be applied.

- Attributes (nominal variables, discrete variables, bounded)
  Among the relevant variables for ecotoxicity tests are mortality (two states: dead or alive), fertility (two states: fertile or non-fertile), emergence (two states: emerged or not emerged). These variables are mostly termed "quantal" or "incidence data", i.e. "a number out of another number". So far as we can see, variables with more than two states are not used in laboratory ecotoxicity testing.

- Derived variables
  These are mainly either ratios or rates and can be based on a measurement variable or an attribute (percentages, growth rate, metabolic rate, etc.).

A second important aspect is that generally two types of statistical methods are existing to determine toxic effect concentrations: hypothesis testing and point estimation.

2.4.1. Point estimation

A point estimate of a toxicity parameter is a concentration value obtained from interpolation or extrapolation of a concentration/response relationship, such as an EC_x, LC_x, ET_x. EC_x is the concentration of the test material in water (e.g. in mg/l) or soil or sediment (e.g. in mg/kg) that is estimated to cause some defined toxic effect to $x\%$ of the test organisms. The duration of exposure must be specified, as EC_x is a function of time. ET_x is the time at which an effect of $x\%$ is expected when the test organisms are exposed to a given concentration of test material (in water or sediment or soil). ET_x is a function of the concentration of exposure which therefore has to be specified. EC_x (ET_x) is termed LC_x (LT_x) in case the observed variable is survival.

To obtain a sound point estimate, a data set showing a clear concentration-effect relation, a concentration-response model (with ET_x a time-response model) and an appropriate fitting method is needed. Currently, the most used dose/response functions are the normal sigmoidal (probit), logistic (logit) and Weibull together with a weighted-maximum-likelihood regression approach as described in Finney (1978). This approach compensates for the heterogeneity of variances in quantal variables. An advantage of this method appears to be the determination of only two parameters (slope, intercept) and easy computing of the confidence interval around the fitted function (and the EC_x).

The aforementioned functions are applied for both the quantitative and qualitative variables, being transformed to ratios relative to the control and relative frequencies, respectively. In case of qualitative variables (e.g., mortality) at least three treatments should result in partial kills. Problems arise with less than three partial kills. In these cases, the effect concentrations are often determined by means of the Moving Averages (Thompson, 1947) or Trimmed Spearman-Kärber method (Hamilton et al., 1977).

With quantitative variables it is often disregarded that with these functions a modified weighting in regression is required because of a differing variance structure of quantitative variables (for details e.g. Christensen, 1984 and Christensen and Nyholm, 1984) and in many cases fitting of these functions is problematic, as hormesis effects (e.g. promotion of growth at lower concentrations of the test material) are sometimes occurring and the test material undergoes some alteration during the prolonged test period. Hence, deviations from the normal sigmoidal (probit), logistic (logit) and Weibull are often observed. In these cases the application of more flexible concentration/response functions with more parameters are recommended.

Among the most recommended functions is the four-parameter logistic, which forms also the base for modified equations to account for hormetic effects (Brain and Cousens, 1989, Van Ewijk and Hoekstra, 1993). In a recent paper Scholze et al. (2001) describe and validate a general best-fit method for the estimation of effect concentrations and recommend the use of a pool of 10 different sigmoidal regression functions for continuous toxicity data. Due to heterogeneous variabilities in replicated data (i.e., heteroscedasticity), the concept of generalized least squares is used for the estimation of the model parameters in combination with smoothed variance spline function describing the heteroscedasticity. The best-fit model is chosen individually for each set of data. A bootstrap method is applied for constructing confidence intervals for the estimated effect concentrations. At present, this approach appears to be the method of choice for concentration/response modelling with ecotoxicity tests.

2.4.2. Hypothesis testing

Statistical testing is mainly used to determine the LOEC/NOEC out of a series of test concentrations. In case of a so-called limit test, only one high concentration is compared with a control by a pairwise-test, whereas the LOEC is determined using a multiple statistical test, i.e. a test which keeps the experiment-wise error at the chosen significance level (e.g., $\alpha = 0.05$). With continuous measurement variables the powerful ANOVA procedures are applied, if the data distribution does not substantially deviates from a normal distribution and variances are homogeneous. In these cases the multiple t-tests of Dunnett (Dunnett, 1964; Dunnett and Tamhane, 1991; Tamhane et al., 1996) or Williams (Williams, 1971, 1972) belong to the most powerful ones. If the aforementioned requirements are not fulfilled and cannot be established by an appropriate data transformation (e.g., log, square root, reciprocal), a rank-based test replaces the multiple t-tests, such as the Bonferroni-U-test after Holm (Holm, 1979). With rank-based tests, measurement data are ordered by increasing size and the test uses their ranks rather than their metric values. Hence, there is some loss of information and consequently of statistical power. If a limit-test was conducted, the Student t-test or the Mann-Whitney U-test is applied. So far as we can see, NOECs are not determined for quantal responses.

The NOEC concept has increasingly been criticised in recent years (e.g. Moore and Caux, 1997; Bruce and Versteeg, 1992; Chapman et al., 1996; Hoekstra and Van Ewijk, 1993; Laskowskj, 1995). The NOEC is seen as problematic as it is determined from hypothesis testing, which often cannot be performed with sufficient statistical power. It has to be admitted that the current design of the majority of ecotoxicity tests is determined by optimising cost-effectiveness rather than statistical power. The NOEC depends on the choice of a significance level, the statistical test selected, and the experimental design. Poor experimental design, such as small sample size, improper concentration spacing, and large experimental variability, is rewarded, since it tends to increase the NOEC. In addition, information about the concentration-effect relationship is disregarded and confidence intervals cannot be calculated. The NOEC suggests that there is no or only a minimum effect. However, in practice this does not hold. Even in the alga growth inhibition test, where the variability is relatively low, it was shown that on average (38 tests) the NOEC was in the range of the EC20; in single case it exceeded the EC50 (Ratte et al., 1998). In other words, it is to be expected that false-negative results ("no-effect" is assumed but there is one) are frequent.

As a consequence, there has been an ISO resolution (ISO TC147/SC5/WG10 Antalya 3) as well as an OECD recommendation (OECD, 1998c) that the NOEC should be phased out from international standard. However, up until now NOECs are used by regulatory authorities and probably will be used also in future, since it appears that these decisions will not be followed strictly.

If the NOEC is intended to be used further, measures have to be taken to lessen the aforementioned problems. Among these are: setting of a reasonable effect threshold that should be detected (e.g. 20% reduction relative to control), reducing variability, controlling the statistical power (e.g. $\beta \leq 0.2$). The inherent biological variability of test cohorts can hardly influenced by the experimenter, but in some cases might be reduced to some extent by selecting more homogeneous stems of test animals. Also the careful control of the environmental factors during the experiment and selection of more homogeneous test substrate can reduce variability. The most powerful measure, however, appears to be to increase the replication to the necessary level (for formulas see Horn and Vollandt 2001).

3. Higher-tier tests

If the preliminary assessments from lower-tier ecotoxicity tests give rise to concerns, further evaluation of the potential risks is required to determine impacts under more environmentally realistic conditions. Directive 91/414/EEC (1991) states that if TERs are below the trigger values "no authorisation shall be granted, unless it can be shown that under field conditions there is no unacceptable impact".

In order to address concerns arising from preliminary assessments and to demonstrate that the impact under field conditions may be acceptable, in the past field or field-like studies have been performed such as farm pond and monitoring studies, large-scale mesocosm studies and, more recently, outdoor microcosm studies. Over the past few years, however, other approaches have been discussed such as additional, expert-designed laboratory tests, which are seen as useful alternatives, or at least intermediates between standard toxicity tests and field experiments.

An international workshop on "Higher-tier Aquatic Risk Assessment for Pesticides" (HARAP, Campbell et al., 1999) was organised under the auspices of the EC, SETAC-Europe and the OECD, on which pesticide regulators and registrants evaluated various approaches and developed guidance on what aspects each approach could address, which methods are suitable to determine potential risks, and how the data generated by these alternative approaches can be used in a risk assessment.

Based on the current EU Authorisations Directive for Plant Protection Products (91/414/EEC, 1991), ways were discussed to refine estimates of effect concentrations, which are broadly applicable to the aquatic risk assessment of pesticides in general. Depending on the properties of the compound in question and its use pattern, the following approaches/techniques were recommended:

- Interrogation use of core (tier 1) data;
- further single-species studies;
- laboratory multi-species tests;
- field studies.

The results from such higher-tier studies are seen to improve the risk characterisation and to reduce the uncertainty associated with preliminary risk characterisation (e.g., by reducing uncertainty relating to species sensitivity).

3.1. Interrogation of core data

Core pesticide data packages, which often contain more useful information than the relatively limited number of end-points typically used in a preliminary risk assessment, can be used in an initial refinement of a preliminary assessment and may provide information useful for defining the scope of any further risk assessment that may be required (for details see Campbell et al., 1999).

3.2. Further single-species studies

The HARAP recommendation includes three possible approaches to further single-species studies:

- tests with additional species;
- modified exposure studies;
- population-level studies.

Data from these additional single-species toxicity tests can be used to reduce uncertainty in an effects assessment by providing information on the distribution of species sensitivity. Results from such tests may justify a reduction in the regulatory triggers by up to an order of magnitude, depending on the number and appropriateness of the additional species tested. Additional species data may also be used to develop probabilistic effects distributions, as being used in probabilistic approaches (e.g., SETAC, 1994; Solomon et al., 1996).

Modified exposure studies are performed, in case evidence can be obtained that the environmental fate of a pesticide can be an important factor in risk mitigation. One approach is to alter the test system to allow fate processes to take place, e.g., by the addition of sediment (e.g., Hamer et al., 1992). Another approach is for example to simulate the dissipation of the chemical in a toxicity test by using a variable dosing system during a flow through test, which allows to produce any desired duration and shape of exposure (Hosmer et al., 1998). The toxicity endpoints generated by such studies can then be used to re-evaluate the conclusions of a preliminary risk characterisation by modifying the worst-case exposure assumptions associated with the core data package.

Population-level studies can be used to identify an ecologically-acceptable concentration (EAC). The EAC is the concentration at which the ecological function and community structure is not adversely affected (temporary effects on certain taxa are considered to be acceptable). Population-level studies are conducted mainly with plants and invertebrate species and allow conclusions to be made about potential impacts at the population level and subsequent recovery, rather than on individual life-stages. They may be simple or complex, multilife-stage studies which attempt to simulate "natural" population dynamics and evaluate population growth rates over one or several life cycles (e.g., Taylor et al., 1992; Maund et al., 1992; Hammers-Wirtz and Ratte, 2000). The assessment of effects on population growth is described in, e.g., Van Straalen and Kammenga (1998).

Also models of the effects of pesticides on populations including those of *Daphnia* and fish have been developed (e.g., Kooijman and Metz, 1984; Barnthouse et al., 1987; Gurney et al., 1990; De Angelis et al., 1991; Hommen et al., 1993; Ratte et al., 1994; Hallam and Lassiter, 1994; Acevedo et al., 1995; Sibly, 1996; Maund et al., 1997; Calow et al., 1997). The advantage of models is seen in the wide range of scenarios which can be evaluated.

3.3. Indoor multi-species tests

HARAP divided the indoor multi-species tests as follows:

1. *Simple indoor multi-species* tests to study specific interactions/processes (e.g., bioavailability, population responses).
2. *Indoor defined microcosm tests* comprising well-described assemblages of organisms at different trophic levels to assess critical ecological threshold levels.
3. *Indoor semi-realistic microcosms* comprising complex natural assemblages which can be used to define directly an EAC. The uncertainty factor to be applied to such an EAC needs to be assessed on a case-by-case basis, based on the uncertainty and acceptability of the test.

3.3.1. Simple indoor multi-species tests

A few selected species are studied with regard to their interactions that are suspected to be of importance under realistic field conditions, in order to illustrate the influence of particular biotic factors on pesticide effects (e.g., the impact of the presence of macrophyte or algal biomass on the bioavailability and toxicity of an insecticide to fish or *Daphnia;* transfer of the pesticide via the food). Some examples of simple indoor multi-species tests are described in Day and Kaushik (1987), Gomez et al. (1997), Hamers and Krogh (1997) and Klüttgen et al. (1996).

3.3.2. Indoor defined microcosm tests

Relatively small test systems are inoculated with a well-described assemblage of small organisms but characterised by several trophic levels (e.g., primary producers, consumers, decomposers). These microcosms may be seeded with organisms from stock cultures or field or even standard test species (e.g., Taub, 1969, 1974; Leffler, 1981; Kersting 1991).

For a number of pesticides, both defined microcosms tests and outdoor microcosm/mesocosm tests have been performed (for references see Campbell et al., 1999) showing that the ecological threshold levels (i.e., the concentration above which effects on population densities and functional endpoints become apparent) of pesticides observed in indoor defined microcosms may be comparable to those measured in complex outdoor experimental ecosystems.

3.3.3. Indoor semi-realistic microcosms tests

This type of system intends to represent natural assemblages of organisms characterised by several trophic levels. They are constructed directly with samples of natural ecosystems. Species covering a wide range of sensitivities and biological diversity can be included. In general, indoor semi-realistic microcosms can include populations of microorganisms, planktonic, periphytic and benthic algae, zooplankton, meiofauna, macroinvertebrates and, when large enough, also macrophytes (for examples see Breneman and Pontasch, 1994; Fliedner et al., 1997; Van den Brink et al., 1997).

Among the advantages of indoor semi-realistic laboratory microcosm tests over outdoor field tests are experimenting throughout the year, easier control over the experimental conditions, and less set-up costs. In contrast, disadvantages are seen in unrealistic population densities of large organisms (e.g., fish, newts, frogs and nymphs of larger insects), disturbance by larger organisms, lack of natural fluctuations, difficulty to investigate long-term effects and recovery of species with complex life cycles, limited number of microhabitats, and possibility of disturbing certain populations by sampling (e.g., macroinvertebrates and macrophytes).

3.4. Field studies

If laboratory studies (lower- and higher-tier) indicate potential risks, field studies (micro-/mesocosm) are required, in order to test specific hypotheses about ecological effects, i.e. population-level and community-level effects. The aim is to derive an EAC. This means that for certain taxa or end-points, effects observed in a field study may be considered acceptable, if with appropriate expert ecological judgement, it is considered that they would not pose significant ecological risks to natural aquatic ecosystems. However, if a keystone, indicator or protected species is substantially affected, this is considered as not acceptable. HARAP recommends that the results of field studies should be accompanied by clear explanations as to why a given observed effect should be considered ecologically significant or acceptable when they are presented to regulatory authorities. Furthermore, wherever possible, such studies should be reviewed by an expert panel to provide the least-biased advice.

Every field study is designed with a specific purpose and is unique in at least some respects. Therefore, it has to be designed accordingly. Guidance for conducting field studies was developed during several workshops held at Monks Wood (SETAC, 1991), Wintergreen (SETAC/RESOLVE, 1991), and Potsdam (SETAC, 1992), resulting in an OECD (1998a) draft guidance document for field studies.

The test system is usually a "naturally" developed aquatic community maintained under outdoor conditions which usually contains naturalised sediment and appropriate organisms such as zoo- and phytoplankton, pelagic and benthic macroinvertebrates, and macrophytes (inclusion of fish is not generally recommended). It may not be necessary to include macrophytes if the study objectives only concern phytoplankton and zooplankton. Organisms are identified to "the lowest practical taxon" (arthropods: usually species or genus; algae: sometimes only to class). For a powerful statistical analysis of community structure, species-level identification is preferable (e.g., use of Principal Response Curves and Similarity Analysis; see e.g., Van den Brink and Ter Braak, 1999).

It was already recognised at the HARAP workshop, that aquatic microcosms and mesocosms are highly complex test systems, for which the ecological interpretation and use in risk assessment needed further discussion and guidance. Therefore, the CLASSIC Workshop (Community Level Aquatic System Studies – Interpretation Criteria; Giddings et al., 2002) was organised to provide such a forum. Among the most important recommendations of CLASSIC were: an exposure-response experimental design as the treatment regime (including the maximum PEC if possible; based on the expected effects); preference for the "toxicological approach" (chemical is uniformly dosed into the water) over the "simulation approach" (dosing simulates field exposure such as due to run-off or spray drift; application of the test substance in spring (generally higher species richness and abundance; potentially sufficient time to study recovery); taxonomic resolution as high as scientifically justified or practically feasible (special efforts for those groups that are identified in laboratory studies as potentially the most sensitive); higher aquatic plants (macrophytes) and fish in general not the principal endpoints; univariate statistical methods in analysis of population-level effects, multivariate methods for community-level effects; EAC as toxicity parameter (highest concentration with no *ecologically significant* effects, some *statistically significant* effects at the EAC possible if considered as ecologically insignificant); structural and functional endpoints of the same importance (species structure the principal protection aim, functional endpoints alone not considered as appropriate); with full population recovery an initial effect is not regarded as ecologically significant; with incomplete recovery requirement of additional tools (e.g. further laboratory studies) to address the remaining uncertainty; development of ecological extrapolation models for extrapolation recommended; EACs from a well-designed and appropriately performed microcosm or mesocosm study representative for common field situations in isolated, static water bodies (databases on the abiotic and biotic conditions of surface water needed to aid interpretation and extrapolation between different waters and regions); landscape ecology important for evaluating the uncertainty of mesocosm results (water bodies in agricultural landscapes often not isolated and/or completely exposed); application of an uncertainty factor to the EAC only in exceptional cases; long-term persistence of populations of non-target organisms derivable from mesocosm data in regulatory risk assessment; additional guidance, training and tools needed by those conducting and evaluating microcosm or mesocosm studies.

Examples of micro/mesocosm studies can be found in Hill et al. (1994), Jak et al. (1994), Graney et al. (1994), Leeuwangh et al. (1994), and Van den Brink and Ter Braak (1999).

4. Concluding remarks

In recent years, the design of higher-tier toxicity tests, the statistical analysis and the interpretation of the results was the subject of a series of workshops which carefully produced guidance for the experimenter and the risk assessor as well. At this point, practical experience is developed, which can be used in future to refine the recommendations and guidance on the conduct and interpretation of higher-tier toxicity tests. Because the measurement endpoints in these tests (population and community data) are much closer to the assessment endpoint (persistence of the community structure and function) than in lower-tier test, uncertainty about the real effects in the field can be substantially reduced.

In contrast, in view of the authors the concept behind lower-tier testing should be carefully reconsidered. A large number of tests have been developed in the past, for which guidelines and standards have been established. However, it appears that one of the current "philosophies" in lower-tier testing, that representative species of the trophic levels are subjected to toxicity testing and the results can be extrapolated to other members of the same trophic level (OECD, 1993; Fent, 1998; Shaw and Chatwick, 1999), can be dangerous and not protective to the community, if the current risk assessment practices are applied. Ratte and Hammers-Wirtz (2001) give examples which point to severe shortcomings in the current approach: In outdoor mesocosm experiments, by which two fungicides were studied, *Daphnia magna*, the "backbone" of testing trophic level 2 in aquatic systems, was less and similarly affected by the fungicides. However, tremendous sensitivity differences between *Daphnia magna* and some rotifers were observed. With Fungicide 1, the EC_{50} of the most sensitive rotifer (*Brachionus* spec.) was nearly two orders of magnitude lower than in *Daphnia magna*, whereas with Fungicide 2 the difference between the highly sensitive *Keratella quadrata* and *Daphnia magna* was even about three orders of magnitude. According to OECD (1992) a safety factor of 1000 is used, if data from one or two acute tests are available, it is 100, in case the acute LC_{50} or EC_{50} are available for the base set (alga, daphnid, fish) and 10, if the chronic NOEC is available for the complete base set. For the above example, the magnitude of sensitivity difference cannot be balanced even by the highest safety factor.

As a consequence, risk assessment cannot be based on *Daphnia magna* as the lone representative of the herbivorous trophic compartment in stagnant freshwater systems. The example supports a change from the trophic approach (i.e. choosing the test organisms according to the trophic level) to the *taxonomic* approach (i.e. inclusion of enough representatives from the dominant species of a community). The rationale behind is that the sensitivity of a species depends on its physiological properties, being more related to the taxonomic rather than the trophic position of a species. Therefore, we recommend to include more (not all) species from different taxonomic groups, which play a major role in the considered community, into toxicity testing, such as representatives from rotifers, molluscs, oligochaetes.

Also the design of lower-tier toxicity tests and the selection of appropriate endpoint needs to be reconsidered. Besides the fact that the statistical design of many tests is probably poor, the suite of endpoints considerably differs among the various test and there is a lack of endpoints for the population level. The population in the field rather than the individual organism is the real assessment endpoint, hence more population endpoints should be measured in a test. Even in the *Daphnia* reproduction test, which is commonly seen as a real life-cycle test, important effects on the next generation are disregarded, since the quality of offspring produced by the treated mothers is generally not assessed. It could be shown that this can lead to a strong underestimation of the effects on the population (Hammers-Wirtz and Ratte, 2000; Ratte and Hammers-Wirtz, 2001).

Some of these shortcomings probably could have been avoided, if there would have been a similar supervising by expert workshops as has been done in higher-tier testing. Together with the development of an improved theoretical concept behind lower-tier testing we recommend also the use of powerful simulation tools, by which the optimal endpoints and statistical design can be found and the applied statistical methods can be verified (for examples see Ratte, 1996; Ratte et al., 1998; Scholze et al., 2001). Such type of computer simulations should be applied *before* ring-tests are conducted and a guideline is developed.

References

91/414/EWG, 1991. Council Directive 91/414/EEC of 15 July 1991 concerning the placing of plant protection products on the market. Commission of the European Communities Directorate General, Brussels.

Acevedo, M.F., Waller, W.T., Smith, D.P., Poage, D.W., McIntyre, P.B., 1995. Modelling cladoceran population to stress with particular reference to sexual reproduction. Nonlinear World 2, 97–129.

Adema, D.M.M., Kuiper, J., Hanstveit, A.O., Canton, H.H., 1983. Consecutive tests for assessment of the effects of chemical agents in the aquatic environment. In: Miyamoto, J. (Ed.), IUPAC Pesticide Chemistry: Human Welfare and the Environment. Pergamon Press, Elmsford, NY, pp. 537–544.

AFNOR (Association Francaise de Normalisation), 1990. Determination of the inhibitory effect on the growth of *Lemna minor* XP T 90–337, 10 pp.

APHA, AWWA, WPCF (American Public Health Association, American Water Works Association and Water Pollution Control Federation), 1989. Standard methods for the examination of water and wastewater, 18th edn, Section 8220. Washington, DC.

APHA, AWWA, WEF (American Public Health Association, American Water Works Association and Water Environment Federation), 1995. Toxicity, Part 8000, p. 8 40 8 42. In: Standard Methods for the Examination of Water and Wastewater, 19th edn. American Public Health Association, Washington, DC.

APHA, AWWA, WPCF (American Public Health Association, American Water Works Association and Water Pollution Control Federation), 1985. Toxicity testing with phytoplankton (tentative). In: Standard Methods for the Examination of Water and Wastewater, 16th edn. American Public Health Association, Washington, DC, pp. 735–739.

ASTM (American Society for Testing and Materials), 1991. Standard guide for conducting static toxicity tests with *Lemna gibba* G3, E 1415–91. In: Book of ASTM Standards, Philadelphia, pp. 1–10.

ASTM, 2000. E1847–96. Standard Practice for Statistical Analysis of Toxicity Tests Conducted under ASTM Guidelines. In: ASTM Annual Book of Standards, Vol. 11.05. ASTM, West Conshohocken, PA.

Baird, D.J., Barata, C., 1999. Genetic variation in the response of *Daphnia* to toxic substances: implications for risk assessment. In: Forbes, V.E. (Ed.), Genetics and Ecotoxicology. Taylor and Francis, Washington, DC, pp. 207–221.

Baird, D.J., Barber, I., Calow, P., 1990. Clonal variation in general responses of *Daphnia magna* Straus to toxic stress. I. Chronic life-history effects. Functional Ecology 4, 399–407.

Baird, D.J., Barber, I., Bradley, M., Calow, P., Soares, A.M.V.M., 1989. The *Daphnia* bioassay: a critique. Hydrobiologia 188/189, 403–406.

Baird, D.J., Barber, I., Bradley, M., Soares, A.M.V.M., Calow, P., 1991. A comparative study of genotype sensitivity to acute toxic stress using clones of *Daphnia magna* Straus. Ecotoxicology and Environmental Safety 21, 1–9.

Baldwin, W.S., Milam, D.L., LeBlanc, G.A., 1995. Physiological and biochemical pertubations in *Daphnia magna* following exposure to the model environmental estrogen diethylstilbestrol. Environmental Toxicology and Chemistry 14, 945–952.

Barnthouse, L.W., G.W. Suter II, A.E. Rosen, J.J. Beauchamp. 1987. Estimating responses of fish populations to toxic contaminants. Environ. Toxicol. Chem. 6, 811–824.

Baudo, R., 1987. Ecotoxicological testing with *Daphnia*. In: Peters, R.H., de Bernardi, R. (Eds), Daphnia. Mem. Ist. Ital. Idrobiol. 45, 461–482.

Benenati, F., 1990. Keynote address: plants – keystone to risk assessment. In: Wang, W., Gorsuch, J.W., Lower W.R. (Eds), Plants for toxicity assessment. ASTM STP 1091, American Society for Testing and Materials, Philadelphia, pp. 5–13.

Biesinger, K.E., Christensen, G.M., 1972. Effects of various metals on survival, growth, reproduction, and metabolism of *Daphnia magna*. Journal Fisheries Research Board of Canada 29, 1691–1700.

Blanck, H., Wallin, G., Wängberg, S., 1984. Species dependent variation in algal sensitivity to chemical compounds. Ecotoxicology and Environmental Safety 8, 339–351.

Bodar, C.W.M., Voogt, P.A., Zandee, D.I., 1990. Ecdysteroids in *Daphnia magna*: their role in moulting and reproduction and their levels upon exposure to cadmium. Aquatic Toxicology 17, 339–350.

Bödeker, W., Altenburger, R., Faust, M., Grimme, L.H., 1992. Biometrische Verfahren zur Auswertung von Biotests. Schr.-Reihe Verein WaBoLu 89. Gustav-Fischer Verlag, Stuttgart.

Boersma, M., 1997a. Offspring size and parental fitness in *Daphnia magna*. Evolutionary Ecology 11, 439–450.

Boersma, M., 1997b. Offspring size in *Daphnia*: does it pay to be overweight? Hydrobiologia 360, 79–88.

Bögi, C., 1998. Einfluß des Alters von *Daphnia magna* Straus (Cladocera, Crustacea) auf die Empfindlichkeit im akuten Toxizitätstest. Diplomarbeit, Ruprecht-Karls-Universität, Heidelberg.

Brain, P., Cousens, R., 1989. An equation to describe dose responses where there is stimulation of growth at low doses. Water Research 29, 93–96

Breneman, D.H., Pontasch, K.W., 1994. Stream microcosm toxicity tests: predicting the effects of fenvalerate on riffle insect communities. Environmental Toxicology and Chemistry 13, 381–387.

Bringmann, G., Kühn, R., 1956. Der Algentiter als Maßstab der Eutrophierung von Wasser und Schlamm. Gesundheitsingenieur 77, 374–381.

Bringmann, G., Kühn, R., 1978. Testing of substances for their toxicity threshold: model organisms Microcystis aeruginosa and Scenedesmus quadricauda. Mitteilungen der Internationalen Vereinigung für Limnologie 21, 275–284.

Bringmann, G., Kühn, R., 1980. Comparison of toxicity threshold of water pollutants to bacteria, algae, and protozoa in the cell multiplication test. Water Research 14, 231–241.

Brown, D., Croudace, C.P., Williams, N.J., Shearing, J.M., Johnson, P.A., 1998. The effect of phthalate ester plasticisers tested as surfactant stabilised dispersions on the reproduction of *Daphnia magna*. Chemosphere 36, 1367–1379.

Bruce, R.D., Versteeg, D.J., 1992. A statistical procedure for modelling continuous toxicity data. Environmental Toxicology and Chemistry 11, 1485–1494.

Calabrese, E.J., Baldwin, L.A., 1998. Hormesis as a biological hypothesis. Environmental Health Perspectives 106, 357–362.

Calow, P., Sibly, R.M., Forbes, V.E., 1997. Risk assessment on the basis of simplified life-history scenarios. Environmental Toxicology and Chemistry 16, 1983–1989.

Campbell, P.J., Arnold, D.J.S., Brock, T.C.M., Grandy, N.J., Heger, W., Heimbach, F., Maund, S.J., Streloke, M., 1999. Guidance document on higher-tier aquatic risk assessment for pesticides (HARAP). SETAC-Europe, Brussels.

Chapman, P.M., Caldwell, R.S., Chapman, P.F., 1996. A warning: NOECs are inappropriate for regulatory use. Environmental Toxicology and Chemistry 15 (2), 77–79.

Christensen, E.R., 1984. Dose-response functions in aquatic toxicity testing and the Weibull model. Water Research 18, 213–221.

Christensen, E.R., Nyholm, N., 1984. Ecotoxicological assays with algae: Weibull dose-response curves. Environmental Science and Technology 18, 713–718.

Cleuvers, M., 2001. Analyse und Bewertung des phytotoxischen Potentials farbiger Xenobiotika – Modifikationen des Algenhemmtestes mit *Scenedesmus subspicatus* und Erfahrungen mit alternativen Testspecies. Dissertation RWTH Aachen, Shaker Verlag, Aachen.

Cleuvers, M., Goser, B., Ratte, H.T., 1997. Life-strategy shift by intraspecific interaction in *Daphnia magna*: change in reproduction from quantity to quality. Oecologia 110, 337–345.

Cleuvers, M., Ratte, H.T., 2002. The importance of light intensity in algal tests with coloured substances. Water Research, in press.

Comber, M.H.I., Smyth, D.V., Thompson, R.S., 1995. Assessment of the toxicity to algae of coloured substances. Bulletin of Environmental Contamination and Toxicology 55, 922–928.

Cowgill, U.M., Hopkius, D.L., Applegath, S.L., Takashi, I.T., Brooks, S.D., Milazzo, D.P., 1985. Brood size and neonate weight of *Daphnia magna* produced by nine diets. In: Bahner, R.C., Hansen, D.J. (Eds), Aquatic Toxicology and Hazard Assessment 8th Symposium, ASTM STP 891, American Society for Testing and Materials, Philadelphia, pp. 233–244.

Cowgill, U.M., Milazzo, D.P., 1989. The culturing and testing of two species of duckweed. In: Cowgill, U.M., Williams, L.R. (Eds), Aquatic Toxicology and Hazard Assessment. Vol. 12, ASTM STP 1027, American Society for Testing and Materials, Philadelphia, pp. 379–391.

Cowgill, U.M., Milazzo, D.P., Landenberger, B.D., 1991. The sensitivity of *Lemna gibba* G-3 and four clones of *Lemna minor* to eight common chemicals using a 7-day test. Research Journal WPCF 63.

Day, K., Kaushik, N.K., 1987. The adsorption of fenvalerate to laboratory glassware and the alga *Chlamydomonas reinhardii*, and its effects on uptake of the pesticide by *Daphnia galeata* mendotae. Aquatic Toxicology 10, 131–142.

Day, K.E., Hodge, V., 1996. The toxicity of the herbicide metolachlor, some transformation products and a commercial safener to an alga (*Selenastrum capricornutum*), a cyanophyte (*Anabaena cylindrica*) and a macrophyte (*Lemna gibba*). Water Quality Research Journal of Canada 31, 197–214.

De Angelis, D., Godbout, L.L., Shuter, B.J., 1991. An individual-based approach to predicting density-dependent compensation in smallmouth bass populations. Ecological Modelling 57, 91–115.

Didden, W.A.M., 1993. Ecology of terrestrial enchytraeidae. Pedobiologia 37, 2–29.

DIN, 2001. Deutsches Institut für Normung: DIN-Arbeitskreis "Bioteste". Vorlage für einen ISO/CEN Entwurf zum Lemnatest (ISO/WD 20079). Water quality – duckweed growth inhibition; determination of the toxic effect of water constituents and waste water to duckweed (*Lemna minor*).

Dorgerloh, M., 1997. Labor-Algentest: Bedeutung der toxikologischen Endpunkte. UWSF – Zeitschrift für Umweltchemie und Ökotoxikologie 9, 222–224.

Dunnett, C.W., 1964. New tables for multiple comparisons with a control. Biometrics 20, 482–491.

Dunnett, C.W., Tamhane, A.C., 1991. Step-down multiple tests for comparing treatments with a control in unbalanced one-way layouts. Statistics in Medicine 20, 939–947.

EC (Environment Canada), 1998. Biological test method: a test measuring the inhibition of growth using the freshwater macrophyte *Lemna minor*. Report EPS 1/RM/37.

EEC, 1993. Commission of the European Communities: Methods for determination of ecotoxicity; Annex V, C3, Algal inhibition test. L383A, 179 186. EEC Directive 92/69/EEC.

Enserink, E.L., 1995. Food mediated life history strategies in *Daphnia magna*: their relevance to ecotoxicological evaluations. Thesis, Landbouw Universiteit, Wageningen.

Enserink, E.L., Kerkhofs, M.J.J., Baltus, C.A.M., Koeman, J.H., 1995. Influence of food quantity and lead exposure on maturation in *Daphnia magna*; evidence for a trade-off mechanism. Functional Ecology 9, 175–185.

Enserink, E.L., Luttmer, W., Maas-Diepeveen, H., 1990. Reproductive strategy of *Daphnia magna* affects the sensitivity of its progeny in acute toxicity tests. Aquatic Toxicology 17, 15–25.

Fairchild, J.F., Ruessler, D.S., Haverland, P.S., Carlson, A.R., 1997. Comparative sensitivity of *Selenastrum capricornutum* and *Lemna minor* to sixteen herbicides. Archives of Environmental Contamination and Toxicology 32, 353–357.

FDA (Food and Drug Administration), 1982. Algal assay test. In: Environmental Assessment Technical Guide. Bureau of Veterinary Medicine and Bureau of Foods. Washington, DC.

Fent, K., 1998 Ökotoxikologie. Thieme, Stuttgart, pp. 288.

Finney, D.J., 1978. Statistical Method in Biological Assay. 3rd edn, Cambridge University Press, London.

Flickinger, A.L., Bruins, R.F.J., Winner, R.W., Skillings, J.H., 1982. Filtration and phototactic behavior as indices of chronic copper stress in *Daphnia magna* Straus. Archives Environmental Contamination and Toxicology 11, 457–463.

Fliedner, A., Remde, A., Niemann, R., Schäfers, C., 1997. Effects of the organotin pesticide azocyclotin in aquatic microcosms. Chemosphere 35, 209–222

Flößner, D., 2000. Die Halopoda und Cladocera (ohne Bosminidae) Mitteleuropas. Backhuys, Leiden.

Francis, P.C., Grothe, D.W., Scheuring, J.C., 1986. Chronic toxicity of 4-Nitrophenol to *Daphnia magna* Straus under static-renewal and flow-through conditions. Bulletin of Environmental Contamination and Toxicology 36, 730–737.

Friccius, T., Schulte, C., Ensenbach, U., Seel, P., Nagel, R., 1995. Der Embryotest mit dem Zebrabärbling – eine neue Möglichkeit zur Prüfung und Bewertung der Toxizität von Abwasserproben. Vom Wasser 84, 407–418.

Giddings, G., O'Neill, A., Gardner, R., 1983. An efficient algal bioassay based on short-term photosynthetic response. In: Bishop, W.E., Caldwell, R.D., Heidolph, B. (Eds), Aquatic Toxicology and Hazard Assessment. STP 802. American Society for Testing and Materials, Philadelphia, pp. 445–459.

Giddings, J.M., Brock, T.C.M., Heger, W., Heimbach F., Maund, S.J., Norman, S.M., Ratte H.T., Schäfers, C., Streloke, M., 2002. Community level aquatic system studies – interpretation criteria. Society of Environmental Toxicology and Chemistry (SETAC), p. 43.

Glazier, D.S., 1998. Does body storage act as food-availability cue for adaptive adjustment of egg size and number in *Daphnia magna*? Freshwater Biology 40, 87–92.

Gomez, A., Ceccine, G., Snell, T.W., 1997. Effects of pentachlorophenol on predator-prey interactions of two rotifers. Aquatic Toxicology 37, 271–282.

Goser, B., 1997. Dichteabhängige Änderungen der Entwicklung und Reproduktion bei Cladoceren – Ursachen und ökologische Bedeutung. Dissertation RWTH Aachen, Shaker Verlag, Aachen.

Graney, R.L., Kennedy, J.H., Rodgers Jr., J.H., 1994. Aquatic Mesocosm Studies in Ecological Risk Assessment. Lewis, London.

Green, J., 1954. Size and reproduction in *Daphnia magna* (Crustacea: Cladocera). Proceedings of the Zoological Society of London 126, 535–545.

Grossman, K., Berghaus, R., Retzlaff, G., 1992. Heterotrophic plant cell suspension cultures for monitoring biological activity in agrochemicals research. Comparison with screens using algae, germination seeds and whole plants. Pesticide Science 35, 283–289.

Guisande, C., Gliwicz, Z.M., 1992. Egg size and clutch size in two *Daphnia* species grown at different food levels. Journal of Plankton Research 14, 997–1006.

Gurney, W.S.C., McCauley, E., Nisbet, R.M., Murdoch, W.W., 1990. The physiological ecology of *Daphnia*: a dynamic model of growth and reproduction. Ecology 71, 716–732.

Hallam, T.G., Lassiter, R.R., 1994. Individual-based mathematical modelling approaches in ecotoxicology: a promising direction for aquatic population and community ecological risk assessment. In: Kendall, R.J., Lacher, T.E. (Eds), Wildlife Toxicology and Population Modelling. Lewis, Boca Raton.

Hamer, M.J., Maund, S.J., Hill, I.R., 1992. Laboratory methods for evaluating the impact of pesticides on water/sediment organisms. Proceedings, British Crop Protection Council Conference (Pests and diseases), Brighton, United Kingdom, 6A-4, pp. 487–496.

Hamers, T., Krogh, P.H., 1997. Predator-prey relationships in a two-species toxicity test system. Ecotoxicology and Environmental Safety 37, 202–212.

Hamilton, M.A., Russo, R.C., Thurston, R.V., 1977. Trimmed Spearman-Karber method for estimating median lethal concentrations in toxicity bioassays. Environmental Science and Technology 11, 714–719

Hammers-Wirtz, M., Ratte, H.T., 2000. Offspring fitness in *Daphnia*: is the daphnia reproduction test appropriate for extrapolating effects on the population level? Environmental Toxicology and Chemistry 19, 1856–1866.

Hanazato, T., Dodson, S.I., 1995. Synergistic effects of low oxygen concentration, predator kairomone, and a pesticide on the cladoceran *Daphnia pulex*. Limnology Oceanography 40 (4), 700–709.

Heger, W., Brock, T.C.M., Giddings, J.M., Heimbach, F., Maund, S.J., Norman, S., Schäfers, C., Streloke, M., 2000. Proceedings of the CLASSIC Workshop (Community Level Aquatic System Studies – Interpretation Criteria). Fraunhofer Institute, Schmallenberg, Germany, 30 May – 2 June 1999; SETAC Final Draft, August 2000.

Hill, I.R., Heimbach, F., Leeuwangh, P., Matthiessen, P., 1994. Freshwater Field Tests for Hazard Assessment of Chemicals. Lewis, London.

Hoekstra, J.A., Van Ewijk, P.H., 1993. Alternatives for the no-observed-effect level. Environmental Toxicology and Chemistry 12 (2), 187–194.

Holm, S., 1979. A simple sequentially rejective multiple test procedure. Scandinavian Journal of Statistics 6, 65–70.

Holst, R.W., Ellwanger, T.C., 1982. Pesticide Assessment Guidelines. Subdivision J. Hazard Evaluation: Non-target plants. EPA-54019–82–020US EPA, Washington, DC.

Hommen, U., Poethke, H.J., Dülmer,U., Ratte, H.T., 1993. Simulation models to predict ecological risk of toxins in freshwater systems. ICES Journal of Marine Sciences 50, 337–347.

Horn, M., Vollandt, R., 2001. A manual for the determination of sample sizes for multiple comparisons – formulas and tables. Informatik, Biometrie und Epidemiologie in Medizin und Biologie 32, 1–28.

Horning, W., Weber, C., 1985. Short-term methods for estimating the chronic toxicity of effluents and receiving waters to freshwater organisms. EPA 600/4–85/014. US Environmental Protection Agency, Environmental Monitoring and Support Laboratory, Cincinatti.

Hosmer, A.J., Warren, L.W., Ward, T.J., 1998. Chronic toxicity of pulsed-dosed fenoxycarb to *Daphnia magna* exposed to environmentally realistic concentrations. Environmental Toxicology and Chemistry 17, 1860–1866.

Huebert, D.B., McIlraith, A.L., Shay, J.M., Robinson, G.G.C., 1990. Short communication: axenic cultures of *Lemna triscula*. Aquatic Botany 38, 295–301.

Hughes, J.S., Alexander, M.M., Balu, K., 1988. An evaluation of appropriate expressions of toxicity in aquatic plant bioassays as demonstrated by the effects of atrazine of algae and duckweed. In: Adams, W.J., Chapman, A., Landis, W.G. (Eds), Aquatic Toxicology and Hazard Assessment, Vol. 10, ASTM STP 971, American Society of Testing and Materials, Philadelphia, pp. 531–547.

Hutchinson, T.H., Scholz, N., Guhl W., 1998. Analysis of the ECETOC aquatic toxicity (EAT) database. IV. Comparative toxicity of chemical substances to freshwater versus saltwater organisms. Chemosphere 36, 143–154.

ISO, 1989. ISO 8692. Water quality: Fresh water algal growth inhibition test with *Scenedesmus subspicatus* and *Selenastrum capricornutum*. International Organization for Standardization.

ISO, 1993. ISO 11268–1. Soil quality: effects of pollutants on earthworms (Eisenia fetida). Part 1: Determination of acute toxicity using artificial soil substrate. ISO.

ISO, 1994. ISO 10229. Water quality: determination of the prolonged toxicity of substances to freshwater fish – Method for evaluating the effects of substances on the growth rate of rainbow trout (Oncorhynchus mykiss Walbaum (Teleostei, Salmonidae)). ISO.

ISO, 1995. ISO 10253. Water quality: marine algal growth inhibition test with Skeltonema costatum and Phaeodactylum tricornutum. ISO.

ISO, 1996. ISO 6341. Water quality: determination of the inhibition of the mobility of Daphnia magna Straus (Cladocera, Crustacea). Acute toxicity test. ISO.

ISO, 1996. ISO 7346–1/–2/–3. Water quality: determination of the acute lethal toxicity of substances to a freshwater fish [Brachydanio rerio Hamilton-Buchanan (Teleostei, Cyprinidae)]. Part 1: static method; Part 2. semi-static method; Part 3: flowthrough method. ISO.

ISO, 1997. Revised Committee draft of ISO/CD 14442 Guidance for algal growth inhibition tests with poorly soluble materials, volatile compounds, metals and waste water. ISO/TC 147/SC 5/WG 5. Toxicity to algae. ISO.

ISO, 1998. ISO 11268–2. Soil quality: effects of pollutants on earthworms (Eisenia fetida) – Part 2: determination of effects on reproduction. ISO.

ISO, 1998. ISO 11348–1/–2/–3. Water quality: determination of the inhibitory effect of water samples on the light emission of Vibrio fischeri (Luminescent bacteria test). Part 1: method using freshly prepared bacteria; Part 2: method using liquid-dried bacteria; Part 3: method using freeze-dried bacteria. ISO.

ISO, 1999. ISO 12890. Water quality: determination of toxicity to embryos and larvae of freshwater fish. Semi-static method. ISO.

ISO, 1999. ISO 14669. Water quality: determination of acute lethal toxicity to marine copepods (Copepoda, Crustacea). ISO.

ISO, 1999. ISO 15522. Water quality: determination of the inhibitory effect of water constituents on the growth of activated sludge microorganisms. ISO.

ISO, 2000. ISO 10706: Water quality: determination of long term toxicity of substances to Daphnia magna Straus (Cladocera, Crustacea). ISO.

Jak, R.G., Schobben, H.P.M., Scholten, M.C.Th., Karman, C.C., 1994. A comparison of the ecotoxicological effects measured in mesocosms and laboratory single-species tests (in Dutch). TNO-94/139. Report. Netherlands Organisation for Applied Scientific Reasearch (TNO), Delft, The Netherlands.

Jenner, H.A., Janssen-Mommen, J.P.M., 1993. Duckweed *Lemna minor* as a tool for testing toxicity of coal residues and polluted sediments. Environmental Contamination and Toxicology 25, 3–11.

Kaiser, K.L.E., 1998. Correlations of *Vibrio fischeri* bacteria test data with bioassay data for other organisms. Environmental Health Perspectives 106(2), 583–591.

Karpinska-Smulikovska, J., 1984. Studies on the relationship between composition and molecular mass of non-ionic surfactants of the pluronic type and their biotoxic activity. Tenside Determination 21, 243–246.

Kenaga, E., 1982. Hazard Assessment. Review: the use of environmental toxicology and chemistry data in hazard assessment: progress, needs, challenges. Environmental Toxicology and Chemistry 1, 69–79.

Kenaga, E., Moolenar, R., 1979. Fish and *Daphnia* toxicity as surrogates for aquatic and vascular plants and algae. Environmental Science and Technology 13, 1479–1480.

Kersting, K., 1991. Microecosystem state and its response to the introduction of a pesticide. Internationale Vereinigung für theoretische und angewandte Limnologie – Verhandlungen 23, 1641–1646.

Kilham, S.S., Kreeger, D.A., Goulden, C.E., Lynn, S.G., 1997. Effects of algal food quality on fecundity and population growth rates of *Daphnia*. Freshwater Biology 38, 639–647.

Klein, B., 2000. Age as a factor influencing results in the acute daphnid test with *Daphnia magna* Straus. Water Research 34, 1419–1424.

Kooijman S.A.L.M., Bedaux, J.J.M., 1996. The analysis of aquatic ecotoxicity data. Vrije Universiteit University Press, Amsterdam.

Kooijman, S.A.L.M., Metz, J.A.J., 1984. On the dynamics of chemically additional stressed populations: the deduction of population consequences from the effects on individuals. Ecotoxicology and Environmental Safety 8, 254–274.

Lakatos, G., Meszaros, I., Bohatka, S., Szabo, S., Makadi, M., Csatlos, M., Langer, G., 1993. Application of Lemna species in ecotoxicological studies of heavy metals and organic biocides, The Science of the Total Environment, supplement 1993, Elsevier Science, Amsterdam.

Laskowskj, R., 1995. Some good reasons to ban the use of NOEC, LOEC and related concepts in ecotoxicology. Oikos 73 (1), 140–144.

Leeuwangh, P., Brock, T.C.M., Kersting, K., 1994. An evaluation of four types of freshwater model ecosystem for assessing the hazard of pesticides. Human Experimental Toxicology 13, 888–899.

Leffler, J.W., 1981. Aquatic microcosms and stress criteria for assessing environmental impact of organic chemicals. Subcontract No T64 11(7197) 025, Report, United States Environmental Protection Agency, Office of Pesticides and Toxic Substances, Washington DC.

Lewis, M.A., 1990. Are laboratory-derived toxicity data for freshwater algae worth the effort? Environmental Toxicology and Chemistry 9, 1279–1284.

Lewis, M.A., 1995. Use of freshwater plants for phytotoxicity testing – a review. Environmental Pollution 87, 319–336.

Lomagin, A.G., Ulyanova, L.V., 1993. A new test for water pollution using duckweed *Lemna minor*. Russian Plant Physiology 40, 302–303.

Mark, U., Solbe, J., 1998. Analysis of the ECETOC aquatic toxicity (EAT) database V – the relevance of *Daphnia magna* as a representative test species. Chemosphere 36, 155–166.

Maund, S.J., Taylor, E.J., Pascoe, D., 1992. Population responses of the freshwater amphipod crustacean *Gammarus pulex* (L.) to copper. Freshwater Biology 28, 29–36.

Maund, S.J., Sherratt, T.N., Stickland, T., Biggs, J., Williams, P., Shillabeer, N., Jepson, P., 1997. Ecological considerations in risk assessment for pesticides in aquatic ecosystems. Pesticide Science 49, 185–190.

Memmert, U., Motschi, H., Inauen, J., Wüthrich, W., 1994. Inhibition of algal growth caused by coloured test substances, ETAD Project E 3023. Ecological and Toxicological Association of Dyes and Organic Pigments Manufacturers, Basel.

Miller, W.E., Greene, J.C., Shiroyama, T., 1976. Application of algal assays to define the effects of wastewater effluents upon algal growth in multiple use river systems. In: Middlebrooks et al. (Eds), Biostimulation and nutrient assessment. Ann Arbor Science, Ann Arbor, pp. 77–91.

Miller, W.E., Peterson, S.A., Greene, J.C., Callahan, C.A., 1985. Comparative toxicology of laboratory organisms for assessing hazardous waste sites. Journal of Environmental Quality 14, 569–574.

Millie, D.F., Hersh, C.M., 1987. Statistical characterization of the atrazine induced photosynthetic inhibition of *Cyclotella menghiniana* (Bacillariophyta). Aquatic Toxicology 10, 239–249.

Moore, D.R.J., Caux, P.-Y., 1997. Estimating low toxic effects. Environmental Toxicology and Chemistry 16, 794–801.

Nagel, R., Isberner, K., 1998. Testing of chemicals with fish – a critical evaluation of tests with special regard to zebrafish. In: Braunbeck, T., Hinton, D.E., Streit, B. (Eds), Fish ecotoxicology. Birkhäuser Verlag, Basel, pp. 337–352.

Naylor, C., Bradley, M.C., Calow, P., 1992. Effect of algal ration – quality and method of quantification – on growth and reproduction of *Daphnia magna*. Archiv für Hydrobiologie 125, 311–321.

Newman, M.C., 1994. Quantitative Methods in Aquatic Ecotoxicology. Lewis, London.

Nusch, E.A., 1982. Evaluation of growth curves in bioassays. ISO/TC 147/SC 5/WG5 N62. Nederlands Normalisatie-Instituut, Delft.

Nusch, E.A., 1983. ISO document ISO/TC 147/SC 5/WG5 N76. Nederlands Normalisatie-Instituut, Delft.

Nyholm, N., 1975. Kinetic studies of phosphate limited algal growth, Ph.D. Thesis (in Danish), Technical University of Denmark, Lyngby.

Nyholm, N., 1985. Response variable in algal growth inhibition test – biomass or growth rate. Water Research 19, 273–279.

Nyholm, N., 1990. Expression of results from growth inhibition toxicity tests with algae. Archives of Environmental Contamination and Toxicology 19, 518–522.

Nyholm, N., 1994. Comments on test duration and selection of response variable in algal growth inhibition test. ISO TC 147/SC 5/WG N 158, "Algae". International Standards Organization, Delft.

OECD (Organisation for Economic Cooperation and Development and Growth), 1984a. Alga, Growth Inhibition Test, No. 201. OECD Guidelines for Testing of Chemicals, Paris.

OECD, 1984b. Guideline for the Testing of Chemicals: Daphnia sp. Acute Immobilisation Test and Reproduction Test. OECD Guideline 202 (Updated Guideline), Paris, France.

OECD, 1984c. Guideline for the Testing of Chemicals: Fish, Prolonged Toxicity Test: 14-Day Study. OECD Guideline 204 (Original Guideline), Paris, France.

OECD, 1984d. Guideline for the Testing of Chemicals: Earthworm, Acute Toxicity Tests. OECD Guideline 207 (Original Guideline), Paris, France.

OECD, 1984e. Guideline for the Testing of Chemicals: Terrestrial Plants, Growth Test. OECD Guideline 208 (Original Guideline), Paris, France.

OECD, 1984f. Guideline for the Testing of Chemicals: Activated Sludge, Respiration Inhibition Test. OECD Guideline 209 (Original Guideline), Paris, France.

OECD, 1992. Guidline for the Testing of Chemicals: Fish, Acute Toxicity Test. OECD Guidline 203 (Updated Guidline), Paris, France.

OECD, 1992a. Guideline for the Testing of Chemicals: Fish, Early-Life Stage Toxicity Test. OECD Guideline 210 (Original Guideline), Paris, France.

OECD, 1992b. Report of the OECD workshop on the extrapolation of laboratory toxicity data to the real environment. OECD Environment Monographs No. 59, Paris, France.

OECD, 1993. Guidelines for Testing of Chemicals. OECD, Paris, France.

OECD, 1994. Data Requirements for Pesticide Registration in OECD Member Countries: Survey Results. Report. Organisation for Economic Co-operation and Development, Series on Pesticides No. 1, Paris.

OECD, 1995. Environment Monographs No. 105. Report. Organisation for Economic Co-operation and Development, Environment Directorate Workshop on Environmental Hazard/Risk Assessment, Paris.

OECD, 1997. Guideline for Testing Chemicals: *Daphnia magna* Reproduction Test. OECD Guideline 211 (revised draft), Paris.

OECD, 1998a. Guidance Document for Freshwater Lentic Field Tests. Draft report. Organisation for Economic Co-operation and Development, Environment Directorate Test Guidelines Programme, Paris.

OECD, 1998b. Series on Principles of Good Laboratory Practice and Compliance Monitoring, No. 1. ENV/MC/CHEM (98) 17, Paris.

OECD, 1998c. Report on the OECD Workshop on Statistical Analysis of Aquatic Toxicity Data. Series on Testing and Assessment, No 10. Environmental Health and Safety Publications. Series on Testing and Assesment. ENV/MC/CHEM(98)18, Paris, France.

OECD, 1998d. Guideline for the Testing of Chemicals: Daphnia magna Reproduction Test. OECD Guideline 211 (Original Guideline), Paris, France.

OECD, 1998e. Guideline for the Testing of Chemicals: Fish, Short-term Toxicity Test on Embryo and Sac-Fry Stages. OECD Guideline 212 (Original Guideline), Paris, France.

OECD, 1999. Draft Guidance Document on Aquatic Toxicity Testing of Difficult Substances, Paris, France.

OECD, 2000. *Lemna* Growth Inhibition Test, Draft Guideline. OECD Guidelines for the Testing of Chemicals, Paris, France.

OECD, 2000a. Guideline for the Testing of Chemicals: Seedling Emergence and Seedling and Seedling Growth. OECD Guideline 208A (Draft Updated Guideline), Paris, France.

OECD, 2000b. Guideline for the Testing of Chemicals: Vegetative Vigour Test. OECD Guideline 208B (Draft Updated Guideline), Paris, France.

OECD, 2000c. Guideline for the Testing of Chemicals: Earthworm Reproduction Test (Eisenia fetida/andrei). OECD Guideline 222 (Draft New Guideline), Paris, France.

OECD, 2000d. Guideline for the Testing of Chemicals: Enchytraeidae Reproduction Test. OECD Guideline 220 (Draft New Guideline), Paris, France.

OECD, 2001a. Guideline for the Testing of Chemicals: Sediment-Water Chironomid Toxicity Test Using Spiked Sediment. OECD Guideline 218 (Draft New Guideline), Paris, France.

OECD, 2001b. Guideline for the Testing of Chemicals: Sediment-Water Chironomid Toxicity Test Using Spiked Water. OECD Guideline 219 (Draft New Guideline), Paris, France.

OECD, 2002. Guideline for the Testing of Chemicals: Lemna sp., Growth Inhibition Test. OECD Guideline 221 (Draft Revised Guideline), Paris, France.

Pack, S., 1993. A Review of Statistical Data Analysis and Experimental Design in OECD Aquatic Toxicology Test Guidelines. Shell International Research, Paris.

Patrick, R., Cairns, J., Scheir, A., 1968. The relative sensitivity of diatoms, snails and fish to twenty common constituents of industrial wastes. Progressive Fish-Culturist 30, 137–140.

Payne, A.G., Hall, R.H., 1979. A method for measuring algal toxicity and its application to the safety assessment of new chemicals. In: Marking, L.L., Kimerle, R.A. (Eds), Aquatic Toxicology. STP 667, American Society for Testing and Materials, Philadelphia.

Peterson, H.G., Boutin, C., Freemark, K.E., Martin, P.A., 1997. Toxicity of hexazinone and diquat to green algae, diatoms, Cyanobacteria and duckweed. Aquatic Toxicology 39, 111–134.

Radix, P., Leonard, M., Papantoniou, C., Roman, G., Saouter, E., Gallotti-Schmitt, S., Thiebaud, H., Vasseur, P., 1999. Comparison of Brachionus calyciflorus 2-d and Microtox chronic 22-h tests with Daphnia magna 21-d test for the chronic toxicity assessment of chemicals. Environmental Toxicology and Chemistry 18 (10), 2178–2185.

Ratte, H.T., 1996. Statistical implications of end-point selection and inspection interval in the Daphnia reproduction test – a simulation study. Environmental Toxicology and Chemistry 15, 1831–1843.

Ratte, H.T., Hammers-Wirtz, M., 2001. Does risk mitigation needs also modified approaches in toxicity testing? Mitteilungen der Biologische Bundesanst alt für Land- und Forstwirtschaft 383, 21–24.

Ratte, H.T., Hammers-Wirtz, M., Cleuvers, M., 1998. Influence of the growth pattern on the EC50 of cell number, biomass integral and growth rate in the algae growth inhibition test. Umweltbundesamt Project Report No. 360030 10, Berlin, Germany.

Ratte, H.T., Poethke, H.J., Dülmer, U., Hommen, U., 1994. Modelling aquatic field tests for hazard assessment. In: Hill, I.R., Heimbach, F., Leeuwangh, P., Matthiessen, P. (Eds), Freshwater Field Tests for Hazard Assessment of Chemicals. Lewis, Michigan, pp. 399–423.

Reynolds, C.S., 1984. The Ecology of Freshwater Phytoplankton. Cambridge University Press, Massachusetts.

Riedel, G.F., 1989. Interspecific and geographical variation of the chromium sensitivity of algae. In: Suetr, G.W., Lewis, M.A. (Eds), Aquatic Toxicology and Environmental Fate, Vol. II, ASTM STP 1007. American Society of Testing and Materials, Philadelphia.

Römbke, J., Moser, Th., 1998. Organisation and performance of an international ringtest for the validation of the enchytraeid reproduction test. Research and development project of the German Federal Environmental Agency, draft report.

Sanchez, M., Ferrando, M.D., Sancho, E., Andreu, E., 2000. Physiological pertubations in several generations of Daphnia magna Straus exposed to diazinon. Ecotoxicology and Environmental Safety 46, 87–94.

Scholze, M., Boedeker, W., Faust, M., Backhaus, T., Altenburger, R., Grimme, L.H., 2001. A general best-fit method for concentration-response curves and the estimation of low-effect concentration. Environmental Toxicology and Chemistry 20, 448–457.

Schulte, C., Nagel, R., 1994. Testing acute toxicity in the embryo of zebrafish, Brachydanio rerio, as an alternative to the acute fish test: preliminary results. ATLA 22, 12–19.

SETAC, 1991 (Society of Environmental Toxicology and Chemistry – Europe). Guidance document on testing procedures for pesticides in freshwater static mesocosms. Report. Society of Environmental Toxicology and Chemistry – Europe, Workshop at Monks Wood Experimental Station, Huntingdon, United Kingdom.

SETAC, 1992 (Society of Environmental Toxicology and Chemistry – Europe). Guidance document on testing procedures for pesticides in freshwater static ecosystems. Report. Society of Environmental Toxicology and Chemistry – Europe, Brussels.

SETAC, 1994 (Society of Environmental Toxicology and Chemistry). Pesticide risk and mitigation: final report of the aquatic risk assessment and mitigation dialog group. Report. Society of Environmental Toxicology and Chemistry, Foundation for Environmental Education, Pensacola.

SETAC, 1995 (Society of Environmental Toxicology and Chemistry – Europe). Asking the right questions: ecotoxicology and statistics. In: Report of a Workshop Held at Royal Holloway College, University of London.

SETAC/RESOLVE, 1991 (Society of Environmental Toxicology and Chemistry/RESOLVE). Workshop on aquatic microcosms for ecological assessment of pesticides. Report from a meeting held in Wintergreen, Virginia, 11 October 1991.

Shaw, I.C., Chatwick, J., 1999. Principles of Environmental Toxicology. Taylor & Francis, London, pp. 216.

Shehata, S.A., Nawar, S.S., 1979. Toxicity effect of anti-germ 50 to algae and fish. Zeitschrift für Wasser-Abwasser-Forschung 12, 226–229.

Shubert, L.E., 1984. Algae as Ecological Indicators. Academic Press, London.

Shurin, J.B., Dodson, S.I., 1997. Sublethal toxic effects of cyanobacteria and nonoylphenol on environmental sex determination and development in *Daphnia*. Environmental Toxicology and Chemistry 16, 1269–1276.

Sibly, R. M., 1996. Effects of pollutants on individual life histories and population growth rates. In: Newman, M.C., Jagoe, C.H. (Eds), Ecotoxicology: a hierarchical approach. Lewis, Boca Raton.

SIS (Swedish Standards Institute), 1995. Water quality determination of growth inhibition (7-d) *Lemna minor*, duckweed, Svensk Standars SS 02 82 13.

Skulberg, D.M., 1967. Algal cultures as means to assess the fertilizing influence of pollution. Advances in Water Pollution Research 1, 113–138.

Sloff, W., Canton, J.H., Hermens, J.L.M., 1983. Comparison of the susceptibility of 22 freshwater species to 15 chemical compounds: I. (Sub)acute toxicity tests. Aquatic Toxicology 4, 113–128.

Soares, M.V.M., Calow, P., 1993. Progress in Standardization of Aquatic Toxicity Tests. Lewis, Boca Raton.

Sokal R.R., Rohlf, F.J., 1981. Biometry, 2nd edn, Freeman, San Francisco, p. 859.

Sokull-Klüttgen, B., 1998. Die kombinierte Wirkung von Nahrungsangebot und 3,4-Dichloranilin auf die Lebensdaten von zwei nahverwandten Cladocerenarten, *Daphnia magna* und *Ceriodaphnia quadrangula*. Dissertation RWTH Aachen, Shaker Verlag, Aachen.

Solomon, K.R., Baker, D.B., Richards, R.P., Dixon, D.R., Klaine, S.J., LaPoint, T.W., Kendall, R.J., Weiskopf, C.P., Giddings, J.M., Giesy, P., Hall, L.W., Williams, W.M., 1996. Ecological risk assessment of atrazine in North American surface waters. Environmental Toxicology and Chemistry 15, 31–74.

Sparks, T., 2000. Statistics in Ecotoxicology. John Wiley, Chichester.

Stephan, C., Mount, D., Hansen, D., Gentile, G., Brungs, W., 1985. Guidelines for derivation national water quality criteria for the protection of aquatic organisms and their uses. US Environmental Protection Agency, Office of Water Regulation and Standards, PB85–227049. National Technical Information Service, Springfield, Virginia.

Stratton, G.W., 1987. The effects of pesticides and heavy metals towards phototrophic microorganisms. In: Hodgson, E. (Ed.), Review in Environmental Toxicology, Vol. 3. Elsevier, New York, pp. 71–147.

Swanson, S.M., Rickard, C.P., Freemark, K.E., MacQuarrie, P., 1991. Testing for pesticide toxicity to aquatic plants: recommendations for test species. ASTM Special Technical Publication 1115: Plants for Toxicity Assessment, pp. 77–98.

Tamhane, A.C., Hochberg, Y., Dunnett, C.W., 1996. Multiple test procedures for dose finding. Biometrics 52, 21–37.

Taraldsen, J.E., Norberg-King, T.J., 1990. New method for determining effluent toxicity using duckweed (*Lemna minor*). Environmental Toxicology and Chemistry 9, 761–767.

Taub, F.B., 1969. A biological model of a freshwater community: a gnotobiotic ecosystem. Limnology and Oceanography 14, 136–142.

Taub, F.E., 1974. Closed ecological systems. Ann. Rev. Ecology and Systematics. 5, 139–160.

Taylor, E.J., S.J. Blockwell, S.J. Maund, Pascoe, D., 1992. Effects of lindane on the life cycle of a freshwater invertebrate *Chironomus riparius* Meigen (Insecta: Diptera). Archives of Environmental Contamination Toxicology 24, 145–150.

Thompson, W.R., 1947. Use of moving averages and interpolation to estimate median-effective dose. I. Fundamental formulas, estimation of error, and relation to other methods. Bacteria Reviews 11, 115–145.

Tripathi, B.D., Chandra, P., 1991. Chromium uptake by *Spirodela polyrhiza* (L.) in relation to metal chelators and pH. Bulletin of Environmental Contamination and Toxicology 47, 764–769.

US EPA (United States Environmental Protection Agency), 1971. Algal assay procedure: Bottle test. United States Environmental Protection Agency, Corvallis, Oregon, USA.

US EPA (United States Environmental Protection Agency), 1974. Marine algal assay procedure: bottle test. Eutrophication and Lake Restoration Branch, National Environmental Research Center, Corvallis, Oregon.

US EPA (United States Environmental Protection Agency), 1996. Aquatic plant toxicity test using Lemna ssp., Tiers I and II. Public draft. Ecological effects test guidelines OPPTS 850, 4400, United States Environmental Protection Agency, Pesticides and Toxic Substances (7101) EPA 712-C-96-156.

Van den Brink, P.J., Ter Braak, C.J.F., 1999. Principal response curves: analysis of time-dependent multivariate responses of biological community to stress. Environmental Toxicology and Chemistry 18, 138–148.

Van den Brink, P.J, Hartgers, E.M., Fettweis, U., Crum, S.J.H., Van Donk, E.,Brock, T.C.M., 1997. Sensitivity of macrophyte-dominated freshwater microcosms to chronic levels of the herbicide linuron. I. Primary producers. Ecotoxicology and Environmental Safety 38, 13–24.

Van der Hoeven, N., 1990. Effect of 3,4-Dichloranilin and Metavanadate on Daphnia populations. Ecotoxicology and Environmental Safety 20, 53–70.

Van Ewijk, P.H., Hoekstra, J.A., 1993. Calculation of the EC50 and its confidence interval when subtoxic stimulus is present. Ecotoxicology and Environmental Safety 25, 25–32.

Van Leeuwen, C.J., Niebeek, G., Rijkeboer, M., 1987. Effects of chemical stress on the population dynamics of *Daphnia magna*: a comparison of two test procedures. Ecotoxicology and Environmental Safety 14, 1–11.

Van Straalen, N.M., Kammenga, J.E., 1998. Assessment of ecotoxicity at the population level using demographic parameters. In: Schuurmann, G., Markert, B. (Eds), Ecotoxicology. John Wiley, New York.

Venkataraman, G.S., Rajyalakshmi, B., 1972. Relative tolerance of nitrogen-fixing blue-green alga to pesticides. Indian Journal of Science 42, 119–121.

Versteeg, D.J., Stalmans, M., Dyer, S.D., Janssen, C., 1997. *Ceriodaphnia* and *Daphnia*: a comparison of their sensitivity to xenobiotics and utility as a test species. Chemosphere 34 (4), 869–892.

Wallbridge, C.T., 1979. A flow through testing procedure with duckweed (*Lemna minor*). US Environmental Protection Agency, Duluth, Minnesota. EPA-600/3-77-108, Washington, DC.

Walsh, G.E., Bahner, L., Horning, W., 1980. Toxicity of textile mill effluents to freshwater and estuarine algae, crustaceans and fishes. Environmental Pollution Series A21, 169–179.

Walsh, G.E., Duke, K.M., Foster, R.B., 1982. Algae and crustaceans as indicators of bioactivity of industrial wastes. Water Research 16, 879–883.

Walsh, G.E., Merrill, R.G., 1984. Algal bioassays of industrial and energy process effluents. In: Schubert, L.E. (Ed.), Algae as Ecological Indicators. Academic Press, London, pp. 329–360.

Wang, W., 1986. Toxicity tests of aquatic pollutants by using common duckweed. Environmental Pollution 11, 1–14.

Wängberg, S.A., Blanck, H., 1988. Multivariate patterns of algal sensitivity to chemicals in relation to phylogeny. Ecotoxicology and Environmental Safety 16, 72–82.

Weltje, L., Hattink, J., Harms, A.V., 1997. Toxicity, uptake and transformation of Technetium (^{99}TC) in duckweed (*Lemna minor* L.). TU Delft, Delft.

Weyers, A., Sokull-Klüttgen, B., Baraibar-Fentanes, J., Vollmer, G., 2000. Acute toxicity data: a comprehensive comparison of results of fish, *Daphnia* and algae tests with new substances notified in the EU. Environmental Toxicology and Chemistry 19, 1931–1933.

Williams, D.A., 1971. A test for differences between treatment means when several dose levels are compared with a zero dose control. Biometrics 27, 103–117.

Williams, D.A., 1972. The comparison of several dose levels with a zero dose control. Biometrics 28, 519–531.

Wollenberger, L., Halling-Soerensen, B., Kusk, K.O., 2000. Acute and chronic toxicity of veterinary antibiotics to *Daphnia magna*. Chemosphere 40, 723–730.

IIb

Microbial indicators

Bioindicators and biomonitors
B.A. Markert, A.M. Breure, H.G. Zechmeister, editors

Chapter 8

Microbial indicators

Jaap Bloem and Anton M. Breure

Abstract

Micro-organisms are present in high amounts in all kinds of environments and play key roles in important ecosystem functions such as decomposition and nutrient cycling. Their small size and high surface to volume ratio causes a high affinity for very low concentrations of substances. Because of the intimate contact and interaction with the environment, microbes are very sensitive and respond quickly to contamination and other types of environmental stress. Microbiological indicators can therefore serve as early warnings. Techniques to determine microbial biomass, activity and diversity are summarised. Examples of results from contaminated sites and from a soil quality monitoring network are given. In copper-contaminated soils bacterial growth rate and diversity were significantly reduced or changed already at field concentrations below the current EU limit. At higher metal concentrations also biomass, respiration and mineralisation were strongly reduced. Remediation resulted in recovery to normal levels. Significant differences between soil types and land-use forms in the soil quality monitoring network demonstrate reproducibility and discriminative power of the microbial indicators. Biological and extensive farms showed higher C and N mineralisation rates than intensively managed farms. It is important to use a set of various indicators. Some may be more sensitive to pollution, others to agricultural management and soil fertility.

Keywords: microbial biomass, microbial activity, microbial diversity, microbial indicator, soil quality

1. Introduction

1.1. Role of micro-organisms in food webs and nutrient cycling

Micro-organisms are unicellular or multicellular organisms, which are only visible through a microscope. Their sizes range from about 0.2 to 200 μm (<0.2 mm). They include fungi, bacteria, protozoa, and algae. Micro-organisms are present in many different places on earth: in soil and water, high in the air and more than 1000 metres below the surface, in high temperature vents and in arctic ice, in aerobic and in anaerobic environments. Micro-organisms have adapted to life in almost all types of environment. They may obtain their energy from light or from the oxidation of chemical compounds. The electron donors may be organic or inorganic compounds and the carbon source may be CO_2 or complex organic compounds. Table 1 gives a general overview of the different types of metabolism of bacteria, the most abundant and diverse group of micro-organisms.

Microscopic algae and cyanobacteria (phytoplankton) are the major primary producers of biomass in aquatic ecosystems. In lakes and oceans, mainly bacteria

Table 1.   Different types of metabolism in bacteria.

Electron donor		Carbon source	
Inorganic	*Organic*	*CO_2*	*Organic*
Energy source: Light			
Photolithotrophy	Photo-organotrophy	Photolithoautotrophs	Photo-organoheterotrophs
H_2O, H_2S, S, H_2	Succinate, acetate	Cyanobacteria, green and purple sulphur bacteria	Rhodospirillaceae
Energy source: Chemical compounds			
Chemolithotrophy	Chemo-organotrophy	Chemolithoautotrophs	Chemo-organo heterotrophs
H_2, H_2S, NH_3, Fe^{2+}, NO_2	Many organic compounds	Hydrogen bacteria, colourless sulphur bacteria, nitrifying bacteria, iron bacteria, methanogens, methanotrophs	Most bacteria

decompose phytoplankton exudates and cell residues. The bacterial secondary production is on average about 20% of the phytoplankton primary production (Cole et al., 1988; Schwaerter et al., 1988). Although heterotrophic bacteria have been found to sustain high growth rates, the variation in bacterial numbers is relatively small. Azam et al. (1983) recognised that actively growing marine bacteria (size 0.2–2 µm) are kept below a density of about a million bacteria ml^{-1} by protozoa. In water, the main bacterivores (bacteria eating organisms) are heterotrophic flagellates (size 2–20 µm), which reach abundances of about thousand cells ml^{-1}. Flagellates are in turn consumed by bigger protozoa, such as ciliates, of the same size range as the phytoplankton (20–200 µm). Thus energy released as organic matter by phytoplankton is returned to the main food chain through a microbial loop of bacteria and protozoa.

In terrestrial ecosystems higher plants are the major primary producers of biomass. Carbon and energy are released into the soil by root exudates and plant residues. In soil, bacteria and fungi are the primary decomposers of dead organic matter such as plant residues, root exudates, decaying micro-organisms and animal manure. Microbes are the food source of microbivores, such as protozoa and nematodes, and play a key role in food webs and nutrient cycling (Bouwman et al., 1994; Bloem et al., 1997). A food web can be defined as a network of consumer-resource interactions among different functional groups of organisms. The breakdown of complex biopolymers into CO_2, H_2O, mineral nitrogen (N), phosphorus (P) and other mineral elements is called mineralisation. Mineralisation is performed not only by microbes but also by microbivores (grazers) and predators, which decompose microbes and other organisms. Mineral nutrients released by decomposers are available for uptake by plants and microbes for production of new biomass. Thus, nutrients are cycled through ecosystems.

An agricultural soil (0–25 cm depth) contains about 3000 kg (fresh weight) biomass per hectare. This is equivalent to about four cows or 60 sheep. The biomass and activity in soil is strongly dominated by micro-organisms, which often contribute more than 80% to the total (De Ruiter et al., 1993; Bloem et al., 1994; Velvis, 1997). In an arable soil the microbial food web decomposes about 5000 kg carbon (C) ha^{-1} year^{-1}, and releases about 100 kg nitrogen (N) ha^{-1} year^{-1} in mineral form. This is about 50% of the N-requirement for growth of an arable crop in high production agriculture in western Europe. Besides in decomposition and nutrient cycling, soil micro-organisms and microbivores are also involved in other important ecosystem functions like the formation and preservation of soil structure (Paul and Clark, 1989; Guggenberger et al., 1999).

1.2. Why microbial indicators?

Micro-organisms are useful indicators for environmental monitoring and ecological risk assessment because they are present in high amounts in all kinds of environments and play key roles in food webs and element cycles, i.e. of nitrogen, carbon, sulphur, and phosphorus (Domsch, 1977; Bloem et al., 1997). Thus, micro-organisms are indispensable for the life of higher organisms. The small size and high surface to volume ratio of microbes cause a high affinity for very low concentrations of substances. Because of the intimate contact and interaction with the environment, microbes are very sensitive and respond quickly to contamination and other types of environmental stress (Brookes, 1995; Giller et al., 1998). The microbial activity reflects the sum of all physical, chemical and biological factors regulating the decomposition and transformation of nutrients (Elliot, 1997; Stenberg, 1999). Microbiological indicators can therefore serve as early warnings in monitoring programmes (Jordan et al., 1995). A considerable amount of the total biomass and an enormous amount of biodiversity is present in microbes (Torsvik et al., 1990; Bloem et al., 1994). Genetic differences between plant and animal species are often much smaller than differences within microbial communities (Karp et al., 1998). Because micro-organisms are the most abundant and smallest biological entities in the ecosystem, the problem of geometric scale does not exist. The sampling strategy directly determines the scale for the interpretation of the results.

Environmental risk analysis in cases of water, sediment and soil pollution is generally based on chemical analysis of a selection of xenobiotic compounds. Measured total concentrations are used to predict the degree of pollution and the potential risks. However, at a given total concentration the bioavailability of contaminants varies strongly with the properties of the environment. E.g. in soils and sediments the availability and toxicity of heavy metals is inversely related to pH, organic matter content and clay content (Peijnenburg et al., 1997; Wuertz and Mergeay, 1997). Despite a partial understanding of the main factors, which affect contaminant availability, it is still hard to assess their effects accurately. Many extrapolation methods to estimate effects of the presence of contaminants are in use. However, ecotoxicological laboratory testing generally focuses on toxic effects of pure chemicals on single species. When results of such studies are extrapolated to the field situation in order to predict ecological effects, interactions between populations and communities are not taken into account and food chain effects are neglected (Posthuma et al., 2001; Admiraal et al., 2000). In addition, toxicity tests generally have (sub)lethal toxicity endpoints and are based on relatively short-term test periods. Their results are therefore not necessarily ecologically relevant. In laboratory bioassays acute toxicity (disturbance) is determined, whereas in monitoring of field sites responses to long-term chronic toxicity (stress) is measured (Giller et al., 1998). Much less effects may be expected in field situations where pollution has been present for decades already. There are two reasons for this phenomenon: most contaminants may have been adsorbed to soil organic matter and mineral soil particles, and soil organisms may have been adapted to the heavy metals (e.g. Alexander, 1995; Boivin et al., 2002).

Furthermore, often a complex mixture of contaminants is present in the field, making estimation of effects even more difficult due to lack of adequate models for mixture toxicity (Van der Geest et al., 2000). The last complicating factor is that pollution is often not the only environmental stress. Other important stresses on ecosystems may be eutrophication, desiccation, acidification or human management practices. One pragmatic way to overcome that problem in field situations is the application of direct toxicity assessment in incubated field samples using bioassays with test organisms in the lab (e.g. Van Gestel et al., 2001). That, however, is not the subject of this paper. Here we focus on measurement of field communities.

Ecological effects of environmental stress also include loss of biodiversity and impairment of life support functions such as decomposition and nutrient cycling. It is difficult to measure decomposition and mineralisation processes directly in situ. Often long incubations are needed and the methods available are subject to artefacts. Moreover it may take years before effects of contamination and disturbance on these processes become apparent, e.g. by accumulation of plant residues and organic matter due to a reduced decomposition (Giller et al., 1998). Therefore, sensitive biological indicators are needed to detect changes in ecosystems.

Of course there are problems. The species concept has developed towards maturity in plant and animal ecology, but is not easily applicable in microbial ecology. The definition of a species is not fully clear and the question of redundancy is not solved (Naeem, 1998). There is an incredible amount of microbial species (genotypes) in the environment. For soil, estimates range from 10^4–10^5 different species per gram (Torsvik et al., 1990; Dykhuizen, 1998). Thus, it is impossible to include all microbial diversity in any assessment. Instead, current techniques of community analysis include the dominant microbial populations. In general, the complexity of this kind of analysis is corresponding to 10 to 100 different species, which is comparable to ecological field studies on animals and plants. Using specific primers for multiplication of DNA by the polymerase chain reaction (PCR), it is possible to zoom in at specific populations with very low abundances. Yet ecologically relevant parts of the microbial community may be missed.

2. Microbiological methods applied in monitoring

Different types of techniques can be applied to determine microbiological parameters. Three types of information may be obtained: the amount of biomass, the activity, and the diversity of the microbial community. For all three types of information different techniques are used. New methods for the analysis of microbial communities have become available due to developments including analytical chemistry, molecular biology, computerised equipment, image analysis, and improved data handling.

In principle determination of microbiological parameters can be performed with bacteria from all kinds of habitats: water, soil, and sediment. Many of the methods mentioned further on, e.g. to determine microbial biomass (microscopy and image analysis), growth rate (^{3}H-thymidine incorporation) and diversity (DGGE and Biolog) originate from aquatic ecology (Sieracki et al., 1985; Moriarty, 1986; Garland and Mills, 1991; Muyzer et al., 1993) and have been applied in soil ecology since about 1990 (Bååth and Johansson, 1990; Zak et al., 1994; Bloem et al., 1995a; Heuer and Smalla, 1997; Engelen et al., 1998; Paul et al., 1999).

Micro-organisms have been used already for a long time as indicators of water quality. Total and faecal coliform bacteria are commonly used to assess potential contamination of drinking and swimming water with pathogenic bacteria of intestinal origin. The degree of eutrophication and ecological water quality have been assessed for decades by counting numbers and relative abundances of different species of algae and cyanobacteria and measurement of their activities in photosynthesis (oxygen production) and respiration (biological oxygen demand). The characteristics of these phytoplankton organisms are treated in Chapter 9 by Dokulil (this volume).

The use of microbial indicators in soil is less common and started more recently. Since about 1980 there has been increasing concern about the effects of contamination and intensive farming practices on soil quality (Domsch et al., 1983; Bååth, 1989; Doran and Parkin, 1994). Routine monitoring of ecological soil quality has been initiated in several countries since about 1990 (Stenberg, 1999). In most countries one or more microbiological indicators have been included in such monitoring schemes. Alef and Nannipieri (1995) have given details of many methods in applied soil microbiology. Here we focus on the main methods which are already in use for monitoring programmes.

Bacteria are more often used as indicators than fungi and protozoa. Although the latter two functional groups are certainly important for the functioning of ecosystems, they are more difficult to measure. Molecular techniques to analyse fungal DNA in environmental samples are under development, but are not yet being used in monitoring (Smit et al., 1999). Also palynological techniques are under development at the moment to be used as a way to quantify fungal diversity, but these also are not yet ready for large scale routine monitoring (Mulder, 2001; Mulder et al., 2002).

2.1. Determination of the amount of microbial biomass

The classical way to determine the amount of living micro-organisms is by plating them on a solidified culture medium in a glass dish as described first by Petri (1887) and modified by many others. The actively growing bacteria form colonies on the plates. The number of colony forming units (CFU) is dependent on the type of medium used in the plates. About 1% of the total number of bacteria present in an environmental sample can be determined in this way.

Using direct microscopy and automatic image analysis, it is possible to obtain information about the total number, biomass and morphological characteristics of the microbial community (Bloem et al., 1995a,b; Bloem, 1995; Paul et al., 1999). Total number and cell volumes of fluorescently stained bacteria and fungi are used to calculate the biomass. Fungal biomass is estimated from the biovolume using a specific carbon content of 1.3×10^{-13} g C μm^{-3}. Because smaller cells tend to have a higher density, for bacteria a higher specific carbon content of 3.1×10^{-13} g C μm^{-3} is used (Fry, 1990). Besides bacteria and fungi, protozoa and nematodes may contribute significantly to the total microbial biomass (Bloem et al., 1994). Their specific carbon content is similar to that of the fungi. Protozoa (flagellates, amoebae, ciliates) are usually counted by the most probable number method using dilution series in a growth medium with a food bacterium (Darbyshire et al., 1974). Numbers are estimated from the dilution where no protozoa are found anymore. Nematodes are elutriated from the soil and also counted under the microscope (Boon et al., 1998).

Another commonly used measure of total biomass in soil is the increase in extractable organic carbon (and nitrogen) after fumigation of the soil with chloroform. Chloroform dissolves cell membranes and thus releases cell constituents (Brookes et al., 1985; Vance et al., 1987). The amount of organic C (and N) in an unfumigated sample (control) is subtracted from the amount in the fumigated sample. Because not all biomass can be extracted from soil (roughly 50%) the measured biomass is multiplied by a correction factor.

2.2. Determination of microbial activity

Microbial activity can be determined relatively easy by measuring soil respiration under standardised conditions in the laboratory. This requires relatively long incubations (2–6 weeks) in most soils where the availability of easily decomposable substrates is limited.

Another technique is substrate-induced respiration (SIR) (Anderson and Domsch 1978), which is used in Germany for monitoring purposes (Höper, 1999). In this method increased respiration (CO_2-evolution) in the first hours after addition of an easily degradable substrate (glucose) to the soil, is used as a measure of the (responsive) microbial biomass.

Bacterial growth rate may be determined as the incorporation rate of ^{3}H-thymidine and ^{14}C-leucine into bacterial DNA and proteins during a short incubation of 1 h (Michel and Bloem, 1993). Using this dual label approach both parameters are measured in a single assay. Because the bacterial DNA content is more constant than the protein content, thymidine incorporation is more proportional to growth rate than leucine incorporation. Since cells synthesise more proteins than DNA, leucine incorporation is an order of magnitude higher than thymidine incorporation (e.g. Fig. 1). Therefore measurements of leucine incorporation are more accurate, especially at low growth rates. Only bacteria incorporate thymidine, but not all bacteria can incorporate ^{3}H-thymidine. All bacteria incorporate leucine, but leucine can be incorporated

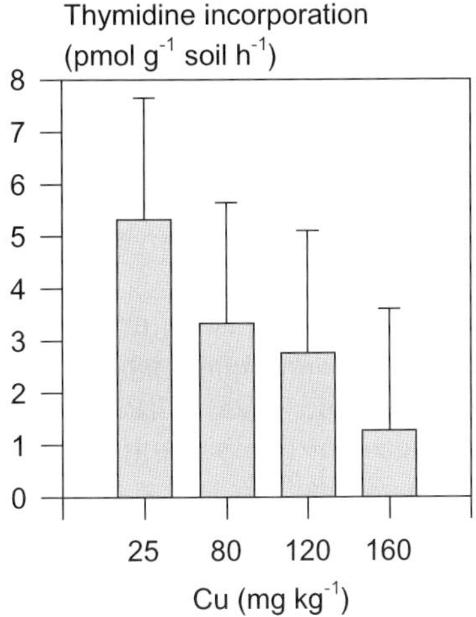

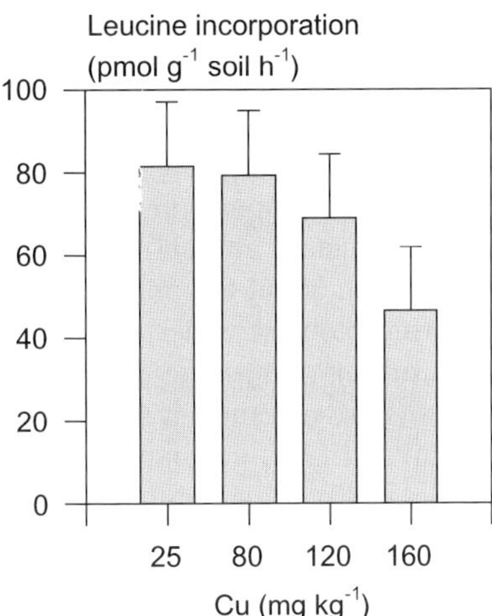

Figure 1.   Reduced bacterial growth rate at increasing levels of copper contamination in soil. Growth rate was measured as incorporation of thymidine and leucine into DNA and proteins. Error bars indicate the LSD (least significant difference at $P < 0.05$).

also by other organisms. As both methods have limitations they are used simultaneously.

The potential nitrogen mineralisation rate and the potential carbon mineralisation rate in soil can be determined simultaneously by incubation for six weeks at 20°C and 50% water holding capacity (WHC) (Bloem et al., 1994). N mineralisation rate is calculated from the increase in mineral N concentration between week six and week one. Results of the first week are not used to avoid disturbance effects of sample handling. The potential carbon mineralisation rate in soil is calculated from the CO_2 evolution (respiration) between week six and week one. In some soils of marine origin, which contain high amounts of $CaCO_3$, CO_2 evolution is not reliable. In such soils O_2 consumption is used assuming that 1 Mol of O_2 consumed corresponds to 1 Mol of CO_2 evolved. Any O_2 used for nitrification is subtracted (Bouwman et al., 1994).

2.3. Chemical techniques for determination of microbial diversity

The signature lipid biomarker (SLB) method is a molecular approach that has become widely used to study microbial communities. Lipids in microorganisms are found primarily in the cell wall but also as storage materials. It has been argued that SLBs in their most sophisticated and extensive analysis can provide both taxonomic and physiological information about microbial communities (White and Ringelberg, 1998). Lipids from microorganisms can be extracted, fractionated into various classes and analysed by gas chromatography. It is rare, however, to find studies in which all classes of lipids have been quantified, as this is time consuming. One of these classes, the phospholipids, are essential membrane components of microorganisms but are not found in storage materials or in dead cells. Most studies have used a simplified phospholipid fatty acid (PLFA) analysis to investigate microbial community structure

(Zelles, 1999). Although individual PLFAs are often not specific to an organism, or even groups of organisms, they can be used as biochemical markers because they predominate in certain taxonomic groups and are relatively conservative in their concentrations within them (White, 1993). Measuring the concentrations of different PLFAs extracted from soils can, therefore, provide a biochemical fingerprint of the soil microbial community (Tunlid and White, 1991). The PLFA profiles reflect the community structure and show which groups are dominant. PLFA give no quantitative information about the number of species.

Measuring biomass either by the total quantity of PLFAs present, or as the quantity of a subset of specific PLFAs, has been shown to correlate well with other techniques (Bardgett and McAlister, 1999; Frostegard et al., 1993). PLFA profiles are affected by soil type, vegetation, climate and management (Bossio and Scow, 1998; Yao et al., 2000). In general, PLFA analysis compares well in terms of sensitivity, often detecting change due to pollutant stress when other conventional measures do not (Bååth et al., 1998b; Pennanen, 2001; Bundy et al., 2001).

PLFA analysis is potentially a good method for environmental monitoring because of its relative simplicity, speed and its potential for semi-automation and standardisation. The overriding needs however, are to establish the interpretation of results within an ecotoxicological framework with respect to normal spatial and temporal variation and background data, to correlate observed changes with potential harm to the environment and to further compare the method with traditional monitoring methods.

2.4. Genetical techniques for determination of microbial diversity

To determine the genetic diversity different molecular techniques are available. DNA is extracted from environmental samples (soil, water, or sediment) (Van Elsas and Smalla, 1995) and multiplied by the polymerase chain reaction using a general probe for bacterial 16S-ribosomal DNA. When bacterial DNA is analysed using denaturing gradient gel electrophoresis (DGGE) (Muyzer et al., 1993; Karp et al., 1998; Griffiths et al., 2000), DNA fragments of equal length are loaded on a gel which contains a gradient of a denaturing agent. During electrophoresis the DNA fragments are running through an increasing concentration of denaturant. Depending on the strength (composition) of the DNA, the fragments start to melt and form a band at a specific denaturant concentration. This technique yields a banding pattern where the number of DNA bands reflects the number of "species" (genotypes) of abundant bacteria, and the band-intensity reflects the relative abundance of the species. The banding patterns are analysed and quantified by image analysis. Also other, related techniques are in use to determine genetic diversity of microbial biomass, such as temperature gradient gel electrophoresis (TGGE), where there is a temperature gradient in the polyacrylamide gel to attain denaturation of the DNA molecules. A slightly different technique is called amplified ribosomal DNA restriction analysis (ARDRA), a DNA fingerprinting technique based on PCR amplification of 16S ribosomal DNA using primers for conserved regions, followed by restriction enzyme digestion and agarose gel electrophoresis (Karp et al., 1998; Smit et al., 1997). ARDRA fingerprints show qualitative differences between communities, but do not reflect the number of species.

2.5. *Physiological techniques for determination of microbial diversity*

Physiological approaches have been developed based on the respiration of specific carbon compounds (Degens and Harris, 1997) or the ability of the community to metabolise specific carbon substrates in Biolog plates (Winding, 1994; Garland and Mills, 1991; Garland, 1997). Testing the ability of microbial communities to utilise a range of (31 or 95) sole-carbon-source substrates in Biolog™ multiwell plates characterises functional diversity. Garland and Mills (1991) proposed the application of Biolog plates for characterisation of microbial communities. Because of the inoculum density dependent response, Garland (1997) proposed standardisation of the inoculum, i.e. the number of colony forming units added, to permit comparison of results obtained with communities derived from different spots. We developed a procedure that indicates the distribution of activities in a microbial community independent of inoculum density (Breure and Rutgers, 2000; Rutgers and Breure, 1999). The aim of our study was to develop an identification method for microbial communities to be applied in monitoring schemes. Utilisation of a specific carbon source is indicated by colour development of a redox indicator dye, which indicates respiration in a specific well. Changes in the overall patterns of carbon utilisation (community level physiological profiles, CLPP) are assessed by multivariate statistics.

2.6 *Reference values*

Microbial characteristics are effected not only by anthropogenic stress but also by soil characteristics such as pH, clay, organic matter content, and the quality of available organic substrates. Therefore, to establish effects of stress an uncontaminated control with the same soil characteristics is required. In cases where no valid control can be found, it may be helpful to use quotients like the specific respiration (= metabolic quotient = amount of CO_2 evolved per unit of biomass) (Anderson and Domsch, 1986; Dahlin et al., 1997), or the biomass C to organic C ratio. Such ratios constitute a kind of internal standard, but do not solve all reference problems (Wardle and Ghani, 1995). Therefore, reference values have to be deduced from many observations with sufficient replicates per soil type (see Discussion).

2.7. *Determination of causal relationships between pollution and field observations*

The techniques mentioned above give the opportunity to measure a quality aspect of the ecosystem. A problem still is to determine the cause of the quality. There is not necessarily a causal relationship between the measured values of the microbial indicators and the environmental stress. Such a relationship can be determined only by use of statistical techniques, given that sufficient data are available. This is a main problem in the extrapolation of laboratory toxicity data to field effects.

One of the possible effects of pollution is that community characteristics in an ecosystem shift towards increased tolerance as a result of exposure to contaminants. Communities may develop increased overall tolerance by various mechanisms:

physiological or genetic adaptation of the local populations, loss of the most sensitive part of populations, and recolonisation of the site by tolerant species.

Blanck et al. (1988) recognised that this phenomenon (PICT, pollution induced community tolerance) could be used to determine field effects of pollutants. The following issues are recognised:

First, regarding the question on the role of suspect compounds in causing ecological effects, the PICT concept covers the issue of causality better than classical ecological community response parameters like species densities or species diversity indices, as the suspect compound (or compounds) causing the observed effect can be deduced with relatively clear inference from artificial exposure experiments.

Second, PICT directly addresses a level of biological integration (the community), the level of concern for many ecological risk assessment methods. Other methods for risk assessment, like toxicity testing or bioassays, focus on individual or population-level effects, and extrapolate these data to the higher level of ecological integration. Such an extrapolation step may pose problems regarding validity.

Using the PICT approach a causal relationship between the presence of a pollutant and an ecological effect may be determined. This can be done in laboratory dose response measurements with the organisms sampled in the field and the suspect contaminant. For this approach different types of activity measurements may be performed such as acetate mineralisation (Van Beelen et al., 2001), application of Biolog plates (Rutgers et al., 1998a,b) or ^{3}H-thymidine incorporation (Bååth et al., 1998a). A decreased sensitivity of a microbial community derived from a polluted site indicates adaptation to the pollutant and therewith, that the pollutant is present in a bioavailable form and exerts an effect on the microbial community in the field.

3. Results

Using the techniques mentioned above effects of environmental stress are investigated. The reasons for their application are mainly to answer the following questions:

(a)   What is the influence of contaminants?
(b)   What is the integral effect of pollution, management practices and other environmental stress on the microbial community and ecosystem processes?

3.1. *Effects of heavy metals on microbial communities*

Using different techniques it can be shown that heavy metals affect microbial communities. Respiration appears to be sensitive to metal contamination in forest soils (Bååth, 1989), but for agricultural soils the results are conflicting (Giller et al., 1998). Respiration by itself is not a sensitive indicator of contamination because it appears to be unaffected at heavy metal concentrations at around current EU mandatory limits (Brookes, 1995). However, the specific respiration or metabolic quotient has been shown to be a more sensitive indicator of stress (Anderson and Domsch, 1986; Dahlin et al., 1997). In general ^{3}H-thymidine incorporation has been found to be more sensitive to contamination than biomass and respiration rate both in water and soil (Jones et al., 1984; Bååth, 1992).

The results with respect to the sensitivity of fungi for heavy metals are conflicting. Nordgren et al. (1986) and Bååth (1992) showed a lower sensitivity and Pennanen et al. (1996) found the fungal part of the microbial biomass to be more sensitive to heavy metals.

We studied the effects of heavy-metal contamination on the microbial community in several field and laboratory studies. To determine the effects of copper increasing amounts of copper sulphate were added to clean sandy soil (pH 5.0) in microcosms. Two days after amendment of copper the bacterial growth rate (^{3}H-thymidine incorporation) was significantly reduced already at a very low copper concentration (10 mg Cu kg^{-1}). The bacterial biomass and the respiration rate were reduced at much higher copper concentrations of 100–1000 mg Cu kg^{-1}. Thus, microbial growth rate appeared to be a more sensitive indicator of metal-stress than biomass or respiration.

Further we determined ecological parameters at three different field sites with sandy soils. One was contaminated with copper (maximum 160 mg Cu kg^{-1}), one with nickel (Ni) and chromium (Cr) (2800 and 430 mg kg^{-1}, respectively), and the third was polluted with Zn (10,000 mg kg^{-1}). The copper contaminated site consisted of arable field plots which have been exposed to different levels of copper (70–160 mg kg^{-1}) and pH (-KCl) (4–6), in a factorial design, for more than 10 years (Korthals et al., 1996). The soil contained 4% organic matter and 3% clay. The nickel and chromium contaminated site is a grassland near a galvanising company. Samples were taken at the most polluted spot around a former basin, and at distances of 10 and 50 m (unpolluted control). Soil characteristics were pH 6.0, 4% organic matter and 5% clay. The Zn-contaminated soil was obtained from a nature area around a former zinc-smelter, which has caused an extremely high contamination with in particular Zn, but also with Cd, Cu, and Pb. This has resulted in a complete disappearance of the natural vegetation in an area of 135 hectares. In an experiment established in 1990, three hectares of heavily contaminated bare soil were rehabilitated using metal immobilising cyclonic ashes, also called beringite (Vangronsveld, 1995). This was combined with addition of municipal waste and sowing of zinc tolerant grass. Consequently, the pH of the pore water increased from 5.5 to 7.5. This resulted in a drastic reduction of the solubility of all metals. Furthermore, the municipal waste provided the poor soil with nutrients for plant growth and increased the water holding capacity of the soil.

In the slightly to moderately Cu contaminated soil, both the rates of ^{3}H-thymidine incorporation (-76%) and ^{14}C-leucine incorporation (-43%) indicated a significantly reduced bacterial growth rate, in comparison with the unpolluted control where the Cu concentration was at a normal background level of 25 mg Cu kg^{-1} soil (Fig. 1). The incorporation of ^{3}H-thymidine was reduced already at field concentrations below the current EU limit of 140 mg Cu kg^{-1} soil. Most other parameters, i.e. the bacterial and fungal biomass, protozoa, and nematodes, the soil respiration rate and N-mineralisation rate also tended to be lower at higher copper concentrations but these differences were not statistically significant (Table 2). Most of the parameters that were not affected by copper, were significantly affected by pH. Thus, like in the microcosm experiment, a reduced bacterial growth rate appeared to be the most sensitive indicator of copper stress.

In the soil more severely contaminated with Ni and Cr, not only the bacterial growth rate but also the bacterial biomass decreased with increasing Cr and Ni concentration

(Fig. 2, Table 2). Like in the copper contaminated soil, ^{3}H-thymidine incorporation (-82%) was reduced more than ^{14}C-leucine incorporation (-66%). In our experience, heavy metals inhibit DNA synthesis (^{3}H-thymidine incorporation) more than protein synthesis (^{14}C-leucine incorporation). The decrease in bacterial growth rate and biomass was reflected in decreased densities of nematodes (-90%). In the extremely contaminated Zn desert all parameters, the micro-organisms, the microbivores and the C-mineralisation rate were extremely low, at least 85% lower than in the plots, which had been remediated with the metal immobilising beringite and metal tolerant grass (Table 2). In the remediated plots the microbes and microbivores had recovered to levels similar to those in the unpolluted soils. The soil microbial parameters were measured six years after the remediation.

[3]H-Thymidine incorporation appears to be more sensitive to contamination than biomass and respiration rates. A plausible explanation for a reduced growth rate in contaminated environments is that micro-organisms under stress divert energy from growth to cell maintenance functions (Killham, 1985; Giller et al., 1998). Physiological processes required for detoxification carry an additional energy burden to micro-organisms. Thus, less energy is available for synthesis of new biomass (growth), and a higher proportion of the available substrate is respired and converted to CO_2. Finally this will result in reduced biomass and increased specific respiration.

At all three sites differences in DNA profiles indicated differences in microbial community structure between the metal contaminated soil and the uncontaminated control (Table 2, Fig. 3). In the Cu contaminated experimental plots the largest differences in the DNA banding patterns were found between the uncontaminated soil at neutral pH (6.1), and the most contaminated soil at the lowest pH (4.0), which

Table 2. Ecological parameters in metal contaminated soils. Reduction or change is expressed as percentage of the value in the uncontaminated control.

Parameter	Change in metal contaminated soil (% of control)		
	Cu 160 mg/kg	Ni/Cr 2800/430 mg/kg	Zn 10,000 mg/kg
Bacteria			
Growth rate	-76*	-82*	-89*
Biomass	-4	-66*	-95*
No of DNA bands	-16	-2	-38*
DNA profile	23*	18 *	33*
Fungal hyphae	-34	ND	-81*
Protozoa	-9	ND	-98*
Nematodes	-19	-92*	-99*
Respiration	-33	ND	-99*
N-mineralisation	-10	ND	ND

* Significant difference ($P < 0.05$)

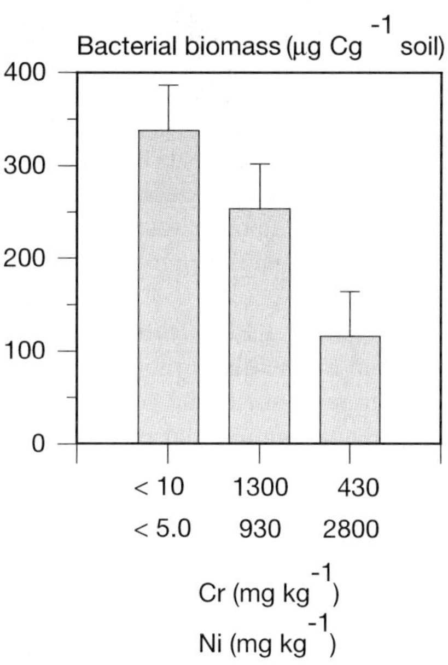

Figure 2. Reduced bacterial biomass in soil contaminated with nickel and chromium. Bacterial biomass was measured automatically by confocal laser-scanning microscopy and image analysis. Error bars indicate the LSD.

contained 125 mg Cu kg^{-1}. At the lowest pH, Cu mobility and toxicity were highest. Here, the number of DNA bands was 42 ± 2 (mean $\pm$ SE, $n = 2$) versus 50 ± 0 in the unpolluted soil. The DNA profiles were normalised so that equal amounts of DNA (integrated grey level) are compared. Differences between DNA profiles were quantified by integrating differences in absolute terms, irrespective of increase or reduction in grey level (Engelen et al., 1998). Thus, theoretically a maximum difference of 100% would be obtained if there would be no overlap between two profiles. The DNA profiles are highly reproducible. The difference between replicates and the mean of each treatment (pH+Cu) was on average 7% (SE = 0.5%, $n = 12$). There was a much larger and statistically significant difference of 23% between the Cu polluted and the unpolluted soil (P = 0.03, LSD = 11.6%) (Table 2). This difference can be due to both differences in Cu content (125 versus 25 mg kg^{-1}) and pH (4.0 versus 6.1). Therefore we also analysed DNA banding patterns of different combinations of Cu and pH. In plots with equal pH (4.7) but different Cu content (25 versus 152 mg kg^{-1}) the difference in DNA profile was 18% (P = 0.02, LSD = 6.5%). In unpolluted soil with the same Cu content (25 mg kg^{-1}) but a different pH of 6.1 versus 4.7, the difference in DNA profile was 17% (P = 0.02, LSD = 6.8%). Thus, both Cu and pH caused differences in DNA profiles and tended to decrease the number of DNA bands. Using another DNA-technique (ARDRA), also Smit et al. (1997) found shifts in microbial community structure in the same copper contaminated soil. In the Ni and Cr polluted soil the number of DNA bands (48.5 ± 0.5) was not significantly reduced compared to the unpolluted soil (47.5 ± 0.5), but the DNA profiles were significantly different. The difference was 18% (P = 0.034, LSD = 8.5%). The largest difference of 33%

272   *J. Bloem, A.M. Breure*

(P < 0.001, LSD = 3.1%) was found between the DNA profiles of the heavily Zn polluted soil and the remediated control where the pollution had been immobilised (Fig. 3). In the Zn polluted soil the number of DNA bands was 31 ± 1 versus 50 ± 1 in the remediated soil (P = 0.006, LSD = 6.1). Thus, in all three contaminated soils microbial DNA profiles (community structure) were different, but the diversity (number of DNA bands) was not necessarily reduced. Changed microbial community structure in metal contaminated soils has been found with various DNA techniques and with phospholipid fatty acid analysis (Frostegård et al., 1996; Griffiths et al., 1997; Bååth et al., 1998b; Sandaa et al., 1999).

In Figure 4 the results are given of a PICT measurement on sandy soil samples from a gradient of Zn pollution in the neighbourhood of a zinc smelter using Biolog plates (Rutgers et al., 1998a). The presence of the contaminant in the soil has resulted in a Zn-tolerance of the microbial community, that is proportional with the Zn-concentrations in the field. The shape of the Gauss curves in Fig 4B indicates that the functional diversity of the population at the most polluted side of the gradient is smaller than on the other side of the gradient. These are strong indications, that Zn may cause a decrease in biodiversity of soil organisms.

3.2. *Routine monitoring of soil quality*

Routine monitoring of soil quality has been initiated in several countries since about 1992. In Germany basal respiration and substrate-induced respiration are used (Höper, 1999). In Switzerland the chloroform fumigation extraction method, basal respiration

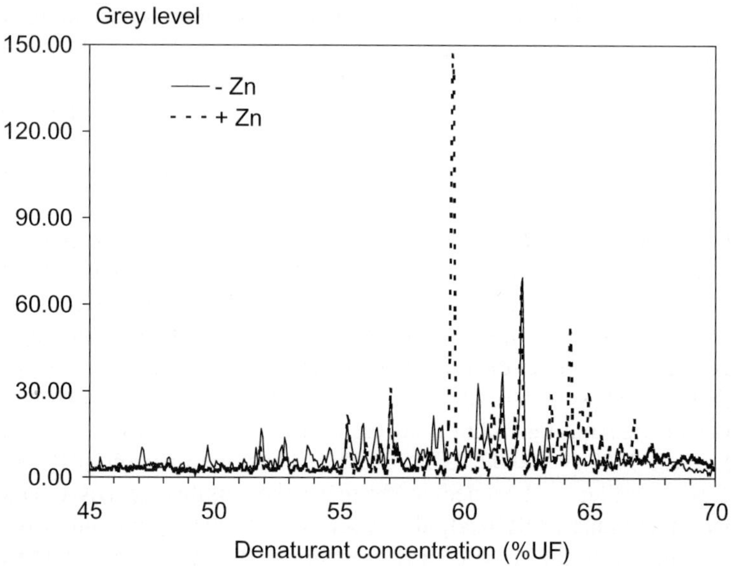

Figure 3.   Reduced genetic diversity in zinc contaminated soil (10,000 mg Zn kg⁻¹). The number of DNA bands (peaks), obtained by DGGE, is 31 in the contaminated soil (+ Zn) and 50 in the remediated soil (− Zn).

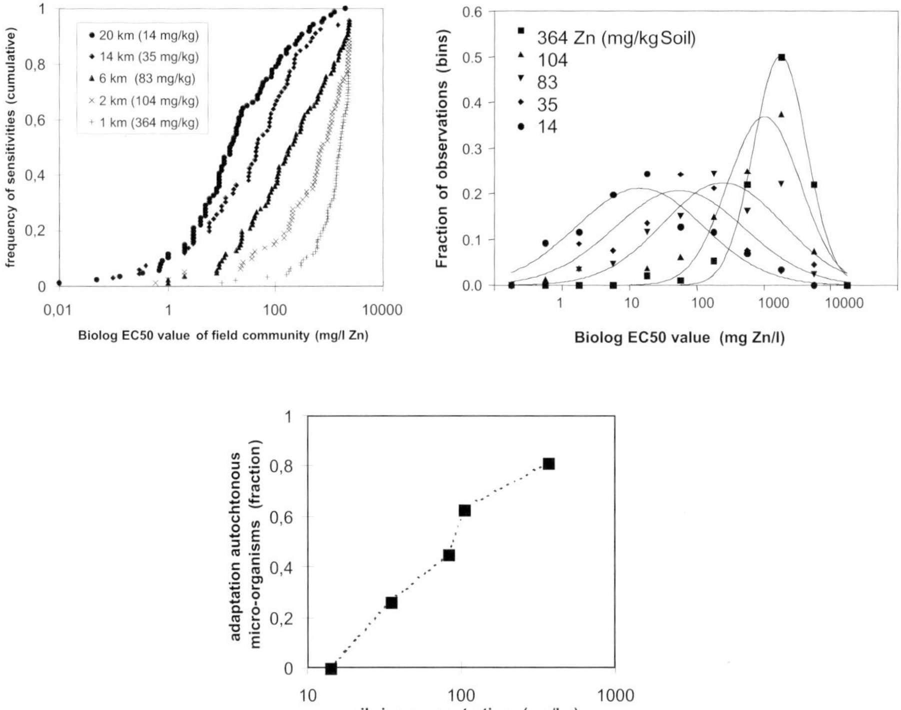

Figure 4. Results from a PICT experiment using samples from a zinc contaminated area near a zinc smelter (Budel, The Netherlands):

A. Frequency distribution of all calculated Biolog EC50 values of five communities around the zinc smelter. All markers in one curve indicate different Biolog substrate conversions. The zinc total concentrations in the samples and the distance to the smelter are indicated in the legend. B. Fitted community densities of the microbial communities (Gauss curves; non-linear regression; fixed area) based on the frequency distribution of Biolog EC50 values. C. Summarised results from PICT measurement of case 1 showing the shift of community in the density of sensitivity values relative to the reference site (calculated from B). Data taken from Rutgers et al., 1998a.

and potential N-mineralisation are used for monitoring since 1999 (Maurer-Troxler, 1999). For a Swedish monitoring programme the following microbiological indicators have been proposed: basal respiration, microbial biomass by substrate induced respiration, potential N-mineralisation, potential ammonium oxidation, and potential denitrification (Stenberg, 1999).

In the Netherlands an integrated biological indicator for soil quality (BISQ) is being developed within the already existing infrastructure of the Dutch Soil Quality Network (Schouten et al., 2000a,b). This network started in 1993, initially to obtain policy information on abiotic soil status. The integrated biological indicator consists of several indicatory variables, which have been selected on the basis of their role in life support functions and underlying ecological processes. Microbial indicators currently used are bacterial biomass, growth rate and diversity. These indicators showed clear effects in polluted soils. In addition potential nitrogen and carbon mineralisation (respiration) are determined.

The network covers the main types of soil and land-use in the Netherlands. It consists of 10 combinations of soil type and land-use with 20 replicates per type, making a total of 200 sites. The replicates are mainly conventional farms. In addition 50 to 100 sites from outside this network are sampled, for instance biological farms or polluted areas which are supposed to be "good" and "bad" references, respectively. Each year two types of soil and land-use are sampled (40 sites plus reference sites). Thus it takes five years to complete one round of monitoring the whole network plus references. The Dutch network started in 1993 to obtain policy information on abiotic soil status. The aim was to measure changes over time and finally to evaluate the actual soil quality. A set of biological indicators has been included since 1997.

In 1997, as a pilot study, dairy farms with grassland on clay and horticultural farms (vegetables and flower bulbs) were sampled. Both categories are intensively managed. The programme was continued in 1999 when grassland farms on sand were sampled with a range of management intensities. Large differences in bacterial biomass and growth rate were found between the different categories (Fig. 5). Both bacterial growth rate and biomass were high in grassland on clay. The bacterial growth rate was also high at the horticultural farms, but here the biomass was low. Bacterial biomass was fairly high but bacterial growth rate was low in grassland on sand. Thus the specific growth rate was much lower in grassland on sand than in grassland on clay. The Community Level Physiological Profiles obtained by the use of Biolog plates as described by Breure and Rutgers (2000) were clearly different between the land use types investigated. Use of ordination techniques (principal component analysis, PCA) showed very clear differences between the soil microbial communities in the two types of farms (Fig. 6) (Schouten et al., 2000a).

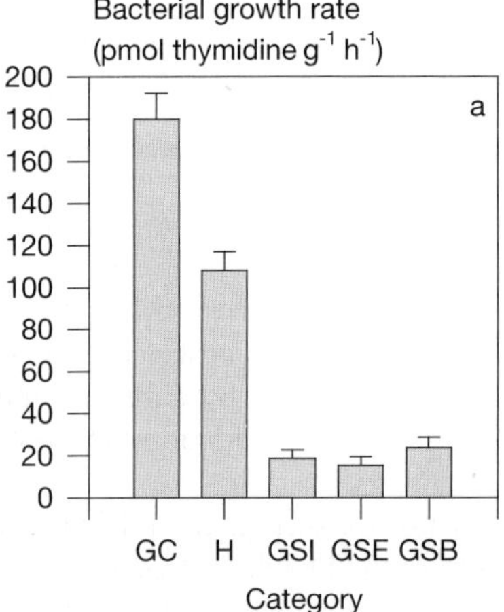

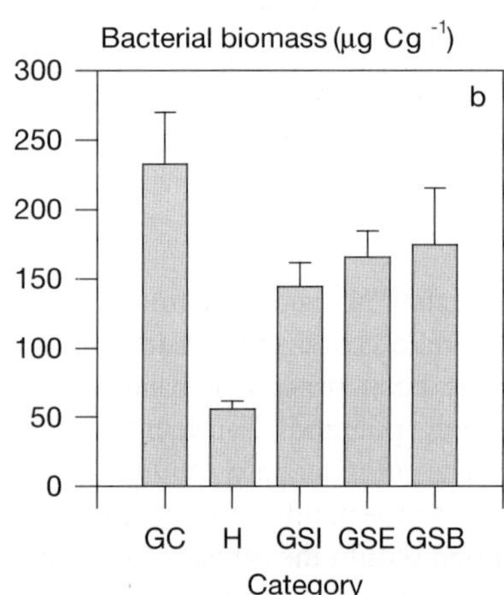

Figure 5. Bacterial growth rate (a) and biomass (b) in different categories of soil and land-use of the Dutch Soil Quality Network: grassland on clay (GC, $n = 20$), horticulture (H, $n = 17$), intensive grassland on sand (GSI, $n = 20$), extensive grassland on sand (GSE, $n = 19$) and biological grassland on sand (GSB, $n = 10$). Error bars indicate SE.

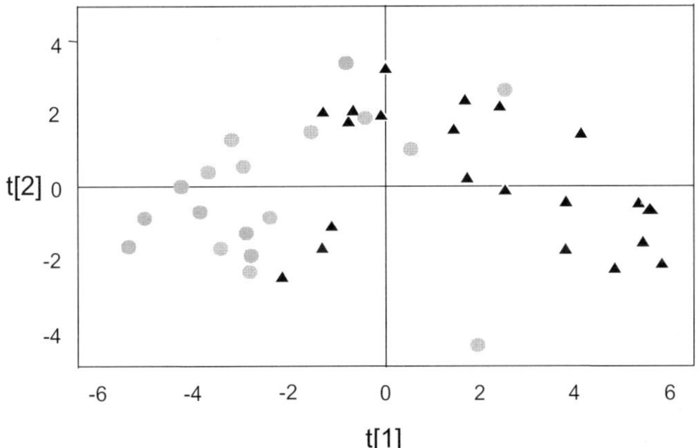

Figure 6. Principal component analysis of community level physiological profiles derived in the soil monitoring network (from Schouten et al., 2000a).

The significant differences between the investigated soil types and land-use forms demonstrate reproducibility and discriminative power of the microbial indicators. This is not trivial because the scale of sampling (whole farms spread over the country, long sample handling time) is much larger than usual in experiments or at polluted sites. Variance is caused not only by the method of analysis but also by differences within and between farms of the same category. The coefficient of variation was 30% for bacterial growth rate and 60% for bacterial biomass measurements. Using 20 replicates (farms) the discriminative power was sufficient to establish statistically significant differences between categories.

In the first year genetic diversity was measured only at one farm of both categories. Analyses were performed in June, September, November and March to investigate seasonal variability. The number of genotypes was rather constant during the year. The mean number of DNA bands was 50.13 $\pm$ 2.34 ($\pm$ SD, $n = 8$). There was no significant difference between the dairy farm on clay and the horticultural farm.

Soil type (clay or sand) strongly influences the microbial biomass and the microbial activity. In 1999 at a single soil type (sand) three different land-use intensities were sampled. The differences in bacterial biomass and growth rate between intensively, extensively and biologically managed grassland on sand were smaller than the differences between grassland on clay and horticultural farms observed in 1997. The average bacterial growth rate (thymidine incorporation) was highest at the biological farms (23.6 pmol g^{-1} h^{-1}), 27% higher than at the intensively managed farms (18.6 pmol g^{-1} h^{-1}) (Fig. 5). The bacterial biomass was highest at the biological farms (174 μg C g^{-1} soil) and lowest at the intensively managed farms (144 μg C g^{-1} soil). The difference was 21%. Also genetic diversity was highest at the biological farms

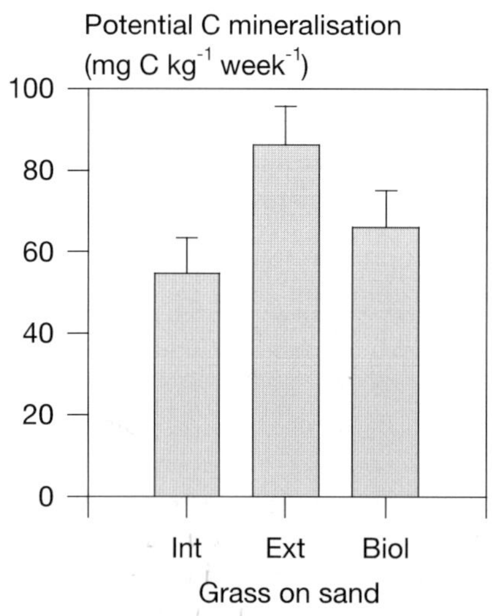

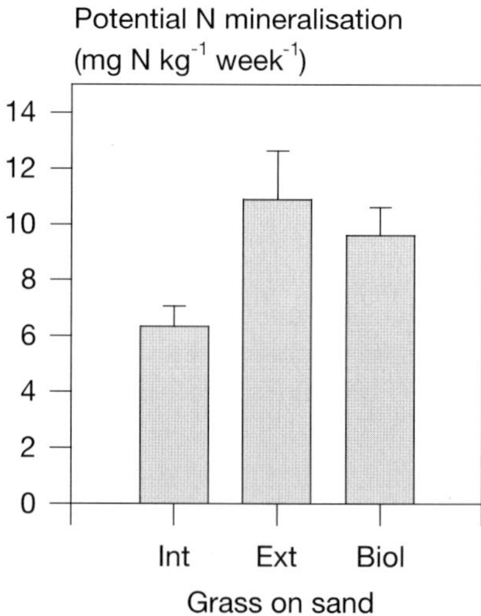

Figure 7. Potential carbon mineralisation (a) and nitrogen mineralisation (b) in intensive ($n = 20$), extensive ($n = 19$) and biological ($n = 10$) grassland on sand. Error bars indicate SE.

(49.4 ± 2.85 genotypes or DNA bands, mean $\pm$ SE) and lowest at the intensive farms (47.8 ± 1.45 DNA bands) although the differences were very small (3%). Thus microbiological indicators tended to be higher at biological farms and lower at intensive farms. This could be expected if extensive and biological farming would promote microbial life in soil. However, the differences between the three categories of dairy farms on sandy soil were not statistically significant as for bacterial biomass, growth rate and diversity. Larger and significant differences were found for potential C-mineralisation and N-mineralisation (Fig. 7). C-mineralisation (soil respiration) was 58 and 21% higher at the extensively managed and biological farms, respectively, than at the intensively managed farms. The difference between extensively managed and intensively managed farms was significant ($P = 0.013$). N-mineralisation was 71 and 51% higher at the extensive and biological farms, respectively, than at the intensive farms ($P = 0.004$) (Fig. 7). The higher mineralisation rates indicate higher soil fertility at extensively managed and biological farms. The higher mineralisation rates corresponded with higher soil organic matter contents, which were 69% and 53% higher at the extensive and biological farms, respectively, than at the intensive farms (7.8 and 8.6 versus 5.1% organic matter).

4. Discussion

Micro-organisms play key roles in ecosystems and are useful indicators for environmental monitoring. Microbial biomass, activity and diversity are significantly reduced or changed by contaminants such as heavy metals. The results illustrate that it is important to use a set of various indicators, and not a few a priori selected indicators which are supposed to be the most sensitive. It is possible that some indicators, such as bacterial growth rate and diversity, are more sensitive to pollution while others, such as N mineralisation, are more sensitive to differences in soil fertility and agricultural management. Ecosystems are complex and their functioning can not be monitored meaningfully with a few simple tools or criteria. Many different aspects need to be measured (Lancaster, 2000). Soil quality has been defined as "the capacity of a soil to function within ecosystem boundaries to sustain biological productivity, maintain environmental quality, and promote plant and animal health" (Stenberg, 1999; Doran and Parkin, 1994). "Within ecosystem boundaries" implies that each soil is different. There are no absolute quality estimates and each soil must be evaluated in relation to natural differences such as soil type, land-use and climate.

A major problem remains to decide what is good and what is bad soil quality (Lancaster, 2000). In pollution-gradients it is possible to use a local unpolluted control. However, in many cases such a reference is not available. Generally the value of an indicator is affected not only by stress factors, but also by soil type, land-use and vegetation. Therefore, reference values for specific soil types have to be deduced from many observations, e.g. 20 replicates per type. The choice of a desired reference is a political issue rather than a scientific issue. For a specific soil and land-use type the reference could be the current average of 20 conventional farms, or the average of 20 biological farms. Soils showing very low or very high indicator values may be suspect and need further examination. Monitoring changes of indicators over time can reduce the importance of (subjective) reference values. Such changes may be easier to interpret than momentary values. Spatially extensive and long-term monitoring may not be ideal, but it is probably the most realistic approach to obtain objective information on differences between, temporal changes within, and human impact on ecosystems.

An additional problem is the causal relation between environmental stress and observed effects in the field. Until now only statistical techniques are capable to couple effects and stress. Only the PICT approach bears somewhat more causality than the other approaches. The use of this approach indicates, that microbial communities adapt to pollutants and they shift towards tolerance. However, such shifts are not only positive as there are strong indications that such shifts may lead to loss of diversity (Rutgers et al., 1998a) or even lead to function loss of the community (Boivin et al., 2002).

Acknowledgements

We thank P.R. Bolhuis, L.A. Bouwman, M. Rutgers, A.J. Schouten, A. Vos and M.R. Veninga for their contribution to the results. We thank C.D. Campbell (Macaulay Land Use Research Institute, Aberdeen) for comments on the manuscript and especially for supplying the section on the signature lipid biomarker (SLB) method.

References

Admiraal, W., Barranguet, C., Van Beusekom, S.A.M., Bleeker, E.A.J., Van den Ende, F.P., Van der Geest, H.G., Groenendijk, D., Ivorra, N., Kraak, M.H.S., Stuijfzand, S.C., 2000. Linking ecological and ecotoxicological techniques to support river rehabilitation. Chemosphere 41, 289–295.

Alef, K., Nannipieri, P. (Eds), 1995. Methods in Applied Soil Microbiology and Biochemistry. Academic Press, London.

Alexander, M., 1995. How toxic are toxic chemicals in soil. Environ. Sci. Technol. 29, 2713–2717.

Anderson, J.P.E., Domsch, K.H., 1978. A physiological method for the quantitative measurement of microbial biomass in soil. Soil Boil. Biochem. 10, 215–221.

Anderson, T.-H., Domsch, K.H., 1986. Carbon link between microbial biomass and soil organic matter. In: Megusar, F., Ganter, M. (Eds), Proceedings of the Fourth International Symposium on Microbial Ecology. Slovene Society for Microbiology, Ljubljana, Jugoslavia, pp. 467–471.

Azam, F., Fenchel, T., Field, J. G., Gray, J. S., Meyer-Reil, L.-A., Thingstad, F., 1983. The ecological role of water column microbes in the sea. Mar. Ecol. Prog. Ser. 10, 257–263.

Bååth, E., 1989. Effects of heavy metals in soil on microbial processes and populations (a review). Water Air Soil Poll. 47, 335–379.

Bååth, E., 1992. Measurement of heavy metal tolerance of soil bacteria using thymidine incorporation into bacteria after homogenization-centrifugation. Soil Biol. Biochem. 24, 1167–1172.

Bååth, E., Díaz-Ravina, M., Frostegård, Å, Campbell, C. 1998a. Effect of metal-rich sludge amendments on the soil microbial community. Appl. Environ. Microbiol. 64, 238–245.

Bååth, E., Frostegård, Å., Díaz-Ravina, M., Tunlid, A., 1998b. Microbial community-based measurements to estimate heavy-metal effects in soil: the use of phospholipid fatty acid patterns and bacterial community tolerance. AMBIO 27, 58–61.

Bååth, E., Johansson, T., 1990. Measurement of bacterial growth rates on the rhizoplane using ^{3}H-thymidine incorporation into DNA. Plant Soil 126, 133–139.

Bardgett, R.D., McAlister, E., 1999. The measurement of soil fungal: bacterial biomass ratios as an indicator of ecosystem self-regulation in temperate meadow grasslands. Biol. Fertil. Soils 29, 282–290.

Blanck, H., Wängberg, S.A., Molander, S., 1988. Pollution-induced community tolerance. A new eco-toxicological tool. In: Cairns, J.J., Pratt, J.R. (Eds), Functional Testing of Aquatic Biota for Estimating Hazards of Chemicals, ASTM STP 988, 1986, American Society for Testing and Materials, Philadelphia, pp. 219–230.

Bloem, J., 1995. Fluorescent staining of microbes for total direct counts. In: Akkermans, A.D.L., Van Elsas, J.D., De Bruijn F.J. (Eds), Molecular Microbial Ecology Manual. Kluwer Academic, Dordrecht, pp. 4.1.8: 1–12.

Bloem, J., Bolhuis, P.R., Veninga, M.R., Wieringa, J., 1995a. Microscopic methods for counting bacteria and fungi in soil. In: Alef, K., Nannipieri, P. (Eds), Methods in Applied Soil Microbiology and Biochemistry. Academic Press, London, pp. 162–173.

Bloem, J., De Ruiter, P.C., Bouwman, L.A., 1997. Food webs and nutrient cycling in agro-ecosystems. In: Van Elsas, J.D., Trevors, J.T., Wellington, E. (Eds), Modern Soil Microbiology. Marcel Dekker, New York, pp. 245–278.

Bloem, J., Lebbink, G.M., Zwart, K.B., Bouwman, L.A., Burgers, S.L.G.E., De Vos, J.A., De Ruiter, P.C., 1994. Dynamics of microorganisms, microbivores and nitrogen mineralisation in winter wheat fields under conventional and integrated management. Agric. Ecosys. Environ. 51, 129–143.

Bloem, J., Veninga, M., Shepherd, J., 1995b. Fully automatic determination of soil bacterium numbers, cell volumes and frequencies of dividing cells by confocal laser scanning microscopy and image analysis. Appl. Environ. Microbiol. 61, 926–936.

Boivin, M.-E.Y., Breure, A.M., Posthuma, L., Rutgers, M., 2002. Determination of field effects of contaminants. The importance of pollution-induced community tolerance. Human Ecol. Risk Assess. 8, 1035–1055.

Boon, G.T., Bouwman, L.A., Bloem, J., Römkens, P.F.A.M., 1998. Effects of a copper tolerant grass (*Agrostis capillaris*) on the ecosystem of a copper-contaminated arable soil. Environ. Toxicol. Chem. 17, 1964–1971.

Bossio, D.A., Scow, K.M., 1998. Impacts of carbon and flooding on soil microbial communities: phospholipid fatty acid profiles and substrate utilization patterns. Microb. Ecol. 35, 265–278.

Bouwman, L.A., Bloem, J., Van den Boogert, P.H.J.F., Bremer, F., Hoenderboom, G.H.J., De Ruiter, P.C., 1994. Short-term and long-term effects of bacterivorous nematodes and nematophagous fungi on carbon and nitrogen mineralization in microcosms. Biol. Fertil. Soils 17, 249–256.

Breure, A.M., Rutgers, M., 2000. The application of Biolog plates to characterise microbial communities. In: Benedetti, A., Tittarelli, F., De Bertoldi, S., Pinzari, F. (Eds), Biotechnology of Soil: Monitoring, Conservation and Bioremediation, Proceedings of the COST Action 831 Joint Working Group Meeting, 10–11 December 1998, Rome (EUR 19548), pp. 179–185.

Brookes, P.C., 1995. The use of microbial parameters in monitoring soil pollution by heavy metals. Biol. Fertil. Soils 19, 269–279.

Brookes, P.C., Landman, A., Pruden, G., Jenkinson, D.S., 1985. Chloroform fumigation and the release of soil nitrogen: a rapid direct extraction method to measure microbial biomass nitrogen in soil. Soil Biol. Biochem. 17, 837–842.

Bundy, J.G., Paton, G.I., Campbell, C.D., 2001. Microbial communities in different soil types do not converge after diesel contamination. J. Appl. Microbiol. 91, 1–13.

Cole, J.J., Findlay, S., Pace, M.L., 1988. Bacterial production in fresh and saltwater ecosystems: a cross-system overview. Mar. Ecol. Prog. Ser. 43, 1–10.

Dahlin, S., Witter, E., Mårtensson, A., Turner, A., Bååth, E., 1997. Where's the limit? Changes in the microbiological properties of agricultural soils at low levels of metal contamination. Soil Biol. Biochem. 29, 1405–1415.

Darbyshire, J.F., Wheatley, R.E., Greaves, M.P., Inkson, R.H.E., 1974. A rapid micromethod for estimating bacterial and protozoan populations in soil. Rev. Écol. Biol. Soil. 11, 465–475.

De Ruiter, P.C., Van Veen, J.A., Moore, J.C., Brussaard, L., Hunt, H.W., 1993. Simulation of nitrogen mineralization in soil food webs. Plant Soil 157, 263–273.

Degens, B.P., Harris, J., 1997. Development of a physiological approach to measuring the catabolic diversity of soil microbial communities. Soil Biol. Biochem. 29, 1309–1320.

Dokulil, M.T., 2002 (2003). Algae as ecological bioindictators. In: Markert, B.A., Breure, A.M., Zechmeister, H.G. (Eds), Bioindicators and Biomonitors. Elsevier, Oxford, pp. 285–327.

Domsch, K.H., 1977. Biological aspects of soil fertility. In: Proceedings of the International Seminar on Soil Environment and Fertility Management in Intensive Agriculture, Tokyo, pp. 737–743.

Domsch, K.H., Jagnow, G., Anderson, T.-H., 1983. An ecological concept for the assessment of side-effects of agrochemicals on soil organisms. Residu Reviews 86, 65–105.

Doran, J.W., Parkin, T.B., 1994. Defining and assessing soil quality. In: Doran, J.W., Coleman, D.C., Bezdicek, D.F., Stewart, B.A. (Eds), Defining Soil Quality for a Sustainable Environment, 35. American Society of Agronomy Special Publication, Madison, pp. 3–21.

Dykhuizen, D.E., 1998. Santa Rosalia revisited: why are there so many species of bacteria. Antonie van Leeuwenhoek 73, 25–33.

Elliot, E.T., 1997. Rationale for developing bioindicators of soil health. In: Pankhurst, C.E., Doube, B.M., Gupta, V.V.S.R. (Eds), Biological Indicators of Soil Health. CAB International, Wallingford, pp. 49–78.

Engelen, B., Meinken, K., Von Wintzingerode, F., Heuer, H., Malkomes, H.-P., Backhaus, H., 1998. Monitoring impact of a pesticide treatment on bacterial soil communities by metabolic and genetic fingerprinting in addition to conventional testing procedures. Appl. Environ. Microbiol. 64, 2814–2821.

Frostegård, Å., Tunlid, A., Bååth, E., 1993. Phospholipid fatty acid composition, biomass and activity of microbial communities from two soil types experimentally exposed to different heavy metals. Appl. Environ. Microbiol. 59, 3605–3617.

Frostegård, Å., Tunlid, A., Bååth, E., 1996. Changes in microbial community structure during long-term incubation in two soils experimentally contaminated with metals. Soil Biol. Biochem. 28, 55–63.

Fry, J.C., 1990. Direct methods and biomass estimation. In: Grigorova, R., Norris, J.R. (Eds), Methods in Microbiology, Vol 22, Techniques in Microbial Ecology. Academic Press, London, pp. 41–85.

Garland, J.L., 1997. Analysis and interpretation of community-level physiological profiles in microbial ecology. FEMS Microbiol. Ecol. 24, 289–300.

Garland, J.L., Mills, A.L., 1991. Classification and characterization of heterotrophic microbial communities on the basis of patterns of community level sole-carbon-source utilization. Appl. Environ. Microbiol. 57, 2351–2359.

Giller, K.E., Witter, E., McGrath, S.P., 1998. Toxicity of heavy metals to microorganisms and microbial processes in agricultural soils: a review. Soil Biol. Biochem. 20, 1389–1414.

Griffiths, B.S., Díaz-Ravina, M., Ritz, K., Mcnicol, J.W., Ebblewhite, N., Bååth, E., 1997. Community DNA hybridisation and %G+C profiles of microbial communities from heavy metal polluted soils. FEMS Microbiol. Ecol. 24, 103–112.

Griffiths, B.S., Ritz, K., Bardgett, R.D., Cook, R., Christensen, S., Ekelund, R., Sörensen, S.J., Bååth, E., Bloem, J., De Ruiter, P.C., Dolfing, J, Nicolardot, B., 2000. Ecosystem response of pasture soil communities to fumigation-induced microbial diversity reductions: an examination of the biodiversity-ecosystem function relationship. Oikos 90, 279–294.

Guggenberger, G., Elliott, E.T., Frey, S.D., Six, J., Paustian, K., 1999. Microbial contributions to the aggregation of cultivated grassland soil amended with starch. Soil Biol. Biochem. 31, 407–419.

Heuer, H., Smalla, K., 1997. Application of denaturing gradient gel electrophoresis and temperature gradient gel electrophoresis for studying soil microbial communities. In: Van Elsas, J.D., Trevors, J.T., Wellington, E. (Eds), Modern Soil Microbiology. Marcel Dekker, New York, pp. 353–373.

Höper, H., 1999. Bodenmikrobiologische Untersuchungen in der Bodendauerbeobachtung in Deutschland. VBB-Bulletin 3, 13–14. Arbeitsgruppe Vollzug Bodenbiologie. FiBL, CH-5070 Frick, Switzerland (in German).

Jones, R.B., Gilmour, C.C., Stoner, D.L., Weir, M.M., Tuttle, J.H., 1984. Comparison of methods to measure acute metal and organometal toxicity to natural aquatic microbial communities. Appl. Environ. Microbiol. 47, 1005–1011.

Jordan, D., Kremer, R.J., Bergfield, W.A., Kim, K.Y., Cacnio, V.N., 1995. Evaluation of microbial methods as potential indicators of soil quality in historical agricultural fields. Biol. Fertil. Soils 19, 297–302.

Karp, A., Isaac, P.G. Ingram, D.S., 1998. Molecular Tools for Screening Biodiversity – Plants and Animals, 1st edn. Chapman & Hall, London.

Killham, K., 1985. A physiological determination of the impact of environmental stress on the activity of microbial biomass. Environ. Poll. 38, 283–294.

Korthals, G.W., Alexiev, A.D., Lexmond, T.M., Kammenga, J.E., Bongers, T., 1996. Long-term effects of copper and pH on the nematode community in an agroecosystem. Environ. Toxicol. Chem. 15, 979–985.

Lancaster, J., 2000. The ridiculous notion of assessing ecological health and identifying the useful concepts underneath. Human Ecol. Risk Assess. 6, 213–222.

Maurer-Troxler, C., 1999. Einsatz bodenbiologischer Parameter in der langfristigen Bodenbeobachtung des Kantons Bern. VBB-Bulletin 3, 11–13. Arbeitsgruppe Vollzug Bodenbiologie. FiBL, CH-5070 Frick, Switzerland (in German).

Michel, P.H., Bloem, J., 1993. Conversion factors for estimation of cell production rates of soil bacteria from thymidine and leucine incorporation. Soil Biol. Biochem. 25, 943–950.

Moriarty, D.J.W., 1986. Measurement of bacterial growth rates in aquatic systems from rates of nucleic acid synthesis. Adv. Microbial Ecol. 9, 245–292.

Mulder, Ch., 2001. Quantitative correlations between mycoflora and landscape ecological parameters from a contemporary wet heathland in The Netherlands. In: Goodman, D.K., Clarke, R.T. (Eds), Proceedings of the IX International Palynological Congress, Houston, Texas (1996); American Association of Stratigraphic Palynologists Foundation, pp. 549–555.

Mulder, Ch., Breure, A.M., Joosten, J.H.J, 2002. Fungal functional diversity inferred along Ellenberg's abiotic gradients: palynological evidence from different soil biota. Grana 41, in press.

Mulder, Ch., Janssen, C.R., 1999. Occurrence of pollen and spores in relation to present-day vegetation in a Dutch heathland area. J. Veg. Sci. 10, 87–100.

Muyzer, G., De Waal, E.C., Uitterlinden, A.G., 1993. Profiling of complex microbial populations by denaturing gradient gel electrophoresis analysis of polymerase chain reaction-amplified genes coding for 16S rRNA. Appl. Environ. Microbiol. 59, 695–700.

Naeem, S., 1998. Species redundancy and ecosystem reliability. Conserv. Biol. 12, 39–45.

Nordgren, A., Kauri, T., Bååth, E., Söderström, B., 1986. Soil microbial activity, mycelial lengths and physiological groups of bacteria in a heavy metal polluted area. Environ. Poll. 41, 89–100.

Paul, E.A., Clark, F.E., 1989. Soil Microbiology and Biochemistry. Academic Press, San Diego.

Paul, E.A., Harris, D., Klug, M., Ruess, R., 1999. The determination of microbial biomass. In: Robertson, G.P., Coleman, D.C., Bledsoe, C.S., Sollins, P. (Eds), Standard Soil Methods for Long-Term Ecological Research. Oxford University Press, New York, pp. 291–317.

Peijnenburg, W.G.J.M., Posthuma, L., Eijsackers, H.J.P., Allen, H.E., 1997. A conceptual framework for implementation of bioavailability of metals for environmental management purposes. Ecotox. Environ. Saf. 37, 163–172.

Pennanen, T., 2001. Microbial communities in boreal coniferous forest humus exposed to heavy metals and changes in soil pH – a summary of the use of phospholipid fatty acids, Biolog (R) and H-3-thymidine incorporation methods in field studies. Geoderma 100, 91–126.

Pennanen, T., Frostegård, Å., Fritze, H., Bååth, E., 1996. Phospholipid fatty acid composition and heavy metal tolerance of soil microbial communities along two heavy metal-polluted gradients in coniferous forests. Appl. Environ. Microbiol. 62, 420–428.

Petri, R.J., 1887. Eine kleine Modifikation des Koch'schen Plattenverfahrens. Centralblatt für Bakteriologie und Parasitenkunde 1, 279–280.

Posthuma, L., Schouten, T., Van Beelen, P., Rutgers, M., 2001. Forecasting the effects of toxicants at the community level: four case studies comparing observed community effects of zinc with forecasts from a generic ecotoxicological risk assessment method. In: Rainbow, P.S., Hopkin, S.P., Crane, M. (Eds), Forecasting the Environmental Fate and Effects of Chemicals. John Wiley, Chichester, pp. 151–176.

Rutgers, M., Breure, A.M., 1999. Risk assessment, microbial communities, and pollution induced community tolerance. Human Ecol. Risk Assess. 5, 661–670.

Rutgers, M., Sweegers, B.M.C., Wind, B., Veen, R.P.M. van, Folkerts, A.J., Posthuma, L., Breure, A.M. 1998a. Pollution induced community tolerance in soil microbial populations. In: Contaminated Soil '98, Vol. I. Thomas Telford, London, pp. 337–343.

Rutgers, M., Van 't Verlaat, I., Wind, B., Posthuma, L., Breure, A.M., 1998b. Rapid method for assessing pollution-induced community tolerance in contaminated soil. Environ. Toxicol. Chem. 17, 2210–2213.

Sandaa, R.-A., Torsvik, V., Enger, Ø., Daae, F.L., Castberg, T., Hahn, D., 1999. Analysis of bacterial communities in heavy metal-contaminated soils at different levels of resolution. FEMS Microbiol. Ecol. 30, 237–251.

Schouten, A.J., Bloem, J., Breure, A.M., Didden, W.A.M., Esbroek, M. van, Ruiter, P.C. de, Rutgers, M., Siepel, H., Velvis, H., 2000b. Pilotproject Bodembiologische Indicator voor Life Support Functies van de bodem. RIVM report 607604001 (in Dutch with English summary).

Schouten, A.J., Bloem, J., Didden, W.A.M., Rutgers, M., Siepel H., Posthuma, L., Breure A.M., 2000a. Development of a biological indicator for soil quality. SETAC Globe 1 (4), 30–32.

Schwaerter, S., Søndergaard, M., Riemann, B., Jensen, L.M., 1988. Respiration in eutrophic lakes: the contribution of bacterioplankton and bacterial growth yield. J. Plankt. Res. 10, 515–531.

Sieracki, M.E., Johnson, P.W., Sieburth, J.M., 1985. Detection, enumeration, and sizing of planktonic bacteria by image-analyzed epifluorescence microscopy. Appl. Environ. Microbiol. 49, 799–810.

Smit, E., Leeflang, P., Glandorf, B., Van Elsas, J.D., Wernars, K., 1999. Analysis of fungal diversity in the wheat rhizosphere by sequencing of cloned PCR-amplified genes encoding 18S rRNA and temperature gradient gel electrophoresis. Appl. Environ. Microbiol. 65, 2614–2621.

Smit, E., Leeflang, P., Wernars, K., 1997. Detection of shifts in microbial community structure and diversity in soil caused by copper contamination using amplified ribosomal DNA restriction analysis. FEMS Microbiol. Ecol. 23, 249–261.

Stenberg, B., 1999. Monitoring soil quality of arable land: microbiological indicators. Acta Agric. Cand., Sect B, Soil and Plant Sci. 49, 1–24

Torsvik, V., Goksøyr, J., Daae, F.L., 1990. High diversity of DNA of soil bacteria. Appl. Environ. Microbiol. 56, 782–787.

Tunlid, A., White, D.C., 1991. Biochemical analysis of biomass, community structure, nutritional status and metabolic activity of microbial communities in soil. In: Bollag, J.-M., Stotzky, G. (Eds), Soil Biochemistry. Marcel Dekker, New York, pp. 229–262.

Van Beelen, P., Fleuren-Kemilä, A.K., Aldenberg, T., 2001. The relation between extrapolated risk, expressed as potentially affected fraction, and community effects, expressed as pollution-induced community tolerance. Environ. Toxicol. Chem. 20, 1133–1140.

Van der Geest, H.G., Greve, G.D., Boivin, M.E., Kraak, M.H.S., Van Gestel, C.A.M., 2000. Mixture toxicity of copper and diazinon to larvae of the mayfly (*Ephoron virgo*) judging additivity at different effect levels. Environ. Toxicol. Chem. 19, 2900–2905.

Van Elsas, J.D., Smalla, K., 1995. Extraction of microbial community DNA from soils. In: Akkermans, A.D.L., Van Elsas, J.D., De Bruijn, F.J. (Eds), Molecular Microbial Ecology Manual. Kluwer Academic, Dordrecht, pp. 1.3.3.1–1.3.3.11.

Van Gestel, C.A.M., Van der Waarde, J.J., Derksen, J.G.M., Van der Hoek, E.E., Veul, M.F.X.W., Bouwens, S., Rusch, B., Kronenburg, R., Stokman, G.N.M., 2001. The use of acute and chronic bioassays to determine the ecological risk and bioremediation efficiency of oil-polluted soils. Environ. Toxicol. Chem. 20, 1438–1449.

Vance, E.D., Brookes, P.C., Jenkinson, D.S., 1987. An extraction method for measuring soil microbial biomass C. Soil Biol. Biochem. 19, 703–707.

Vangronsveld, J., Van Assche, F., Clijsters, H., 1995. Reclamation of a bare industrial area contaminated by non-ferrous metals: in situ metal immobilization and revegetation. Environ. Pollut. 87, 51–59.

Velvis, H., 1997. Evaluation of the selective respiratory inhibition method for measuring the ratio of fungal:bacterial activity in acid agricultural soils. Biol. Fertil. Soils 25, 354–360.

Wardle, D.A., Ghani, A., 1995. A critique of the microbial metabolic quotient (qCO_2) as a bioindicator of disturbance and ecosystem development. Soil Biol. Biochem. 27, 1601–1610.White, D.C., 1993. In-situ measurement of microbial biomass, community structure and nutritional status. Philosophical Transactions of the Royal Society of London Series A – Mathematical Physical and Engineering Sciences 344, 59–67.

White, D.C., Ringelberg, D.B., 1998. Signature lipid biomarker analysis. In: Burlage, R.S., Atlas, R.M., Stahl, D.A., Geesey, G., Sayler, G.S. (Eds), Techniques in Microbial Ecology. Oxford University Press, New York, pp. 255–272.

Winding, A., 1994. Finger printing bacterial soil communities using Biolog micro titer plates. In: Ritz, K., Dighton, J., Giller, K.E. (Eds), Beyond the Biomass. Wiley-Sayce, London, pp. 85–94.

Wuertz, S., Mergeay, M., 1997. Impact of heavy metals on soil microbial communities and their activities. In: Van Elsas, J.D., Trevors, J.T., Wellington, E. (Eds), Modern Soil Microbiology. Marcel Dekker, New York, pp. 607–642.

Yao, H., He, Z., Wilson, M.J., Campbell, C.D., 2000. Microbial biomass and community structure in a sequence of soils with increasing fertility and changing land use. Microbial Ecol. 40, 223–237.

Zak, J.C., Willig, M.R., Moorhead, D.L., Wildman, H.G., 1994. Functional diversity of microbial communities: a quantitative approach. Soil Biol. Biochem. 26, 1101–1108.

Zelles, L., 1999. Fatty acid patterns of phospholipids and lipopolysaccharides in the characterisation of microbial communities in soil: a review. Biol. Fertil. Soils 29, 111–129.

IIc

Lower plants

Chapter 9

Algae as ecological bio-indicators

M.T. Dokulil

Abstract

The value of algae as bio-monitors and bio-indicators has already been recognised in the mid 19th century: The first concept which has been developed was the system of saprobity. It was mainly designed for organic pollution of streams and rivers. This system was altered, modified and expanded over the years by several authors. Because saprobity is defined as the intensity of heterotrophic activity, all photoautotrophic species were finally excluded from the saprobic system when problems of inorganic nutrient load to rivers became increasingly important. Parallel to the progress in running waters, the trophic classification system has been developed for lakes which is based on inorganic nutrients and their loading from the catchments. The basis of both this concepts and other systems is the belief that the presence, absence or abundance of species or species assemblage readily reflects the character of the habitat within which they are found. Those species are usually identified as bio-indicators. This concept of indicators can be extended beyond presence/absence by relating abundance, biomass or growth of algal species to environmental impacts in general or specific stress symptoms in particular. The indicator species then becomes a 'bio-sensor' for the bioassay of environmental contamination. Another concept associates indicator species with organisms accumulating substances from the surrounding environment so as to reflect natural levels and exposure to these substances Such species are 'bio-accumulators' which are especially useful when concentrating very low levels of a substance.

In this chapter, principles of algal bio-indication and bio-monitoring in the environment is outlined for streams and rivers, lakes and reservoirs, as well as for marine ecosystems. Both pelagic and benthic algal groups and species are considered. Field and laboratory bioassay procedures and techniques are described and discussed for both natural assemblages and laboratory cultures. Aspects of sediment testing are included. Since environmental contamination and pollution has severely expanded in the recent past, ecotoxicological methods became increasingly important. More integrative new approaches such as 'ecosystem health' and 'environmental integrity' are briefly discussed.

Keywords: algal bio-indicators, rivers, lakes, ecotoxicology, bioassay, saprobity, trophic-system

1. Introduction

The value of algae as bio-monitors for fresh waters has already been recognised in the mid 19th century (Cohn, 1853). The first attempt to classify aquatic organisms as indicators of water quality was made by Cohn (1870), later modified by Mez (1898). The relation of organisms to the quality of water was more clearly defined by Kolkwitz and Marsson (1902, 1908, 1909) who also created the name '*saprobic organisms*'.

The system of saprobity, a term to describe the biotope introduced by Šrámek-Hušek (1956), was further developed and revised by Kolkwitz (1950) and Liebmann (1962). Several simple and more elaborated definitions of saprobity by various authors are listed by Sládeček on page 27 in his comprehensive overview of 1973. Because saprobity is defined as the intensity of heterotrophic activity, all photoautotrophic species were excluded from the saprobic system in a revision (Friedrich, 1990) to avoid over-lapping with trophic indication In short, the presence, absence or abundance of species or species assemblage readily reflects the character of the habitat within which they are found. Those species are usually identified as bio-indicators.

Since then, various elaborated systems deducing water quality from observations of indicator organisms have been developed, evolved, and diversified both in the field and in the laboratory as bioassay. The terms "indicator", "bio-indicator" or "indicator species" may be used and understood, however in several different ways. As a pre-requisite for an alga to become an indicator we need to know the requirements of that species with regard to one or more environmental variables. The presence of such a species in a given habitat will then indicate that one or more parameters are within the tolerance limits of that species. This concept of indicators can be extended beyond presence/absence by relating abundance, biomass or growth of algal species to environmental impacts in general or specific stress symptoms in particular. The indicator species then becomes a "bio-sensor" for the bioassay of an environmental contamination. Another concept associates indicator species with organisms accumulating substances from the surrounding environment so as to reflect natural levels and exposure to these substances Such species are "bio-accumulators" which are especially useful when concentrating very low levels of a substance.

Algae are most useful as indicators in the context of eutrophication but have been employed as well to detect organic pollution because of their well documented tolerance (e.g. Palmer, 1969). Their value as bio-accumulators of e.g. pesticides or heavy metals is limited. Some species, such as *Selenastrum capricornutum*, are used as bio-sensors in laboratory bioassays while natural phytoplankton assemblages are often used for *in situ* bioassays (Schelske, 1984). Ecotoxicology is another field in which algae have been applied. Some common phytoplankton bioassay techniques mentioned later in the text are summarised in Figure 1.

In a wider concept, organisms are seen as fundamental sensors that respond to any stress affecting the system in which they live (Loeb and Spacie, 1994). Any stress, physical, chemical or biological, imposed on an aquatic system manifests its impact on the organisms living within that ecosystem through their health. The health of an aquatic ecosystem is affected when its capacity to absorb stress is exceeded. The concept proposes that the environmental health of aquatic ecosystems can be assessed by biological monitoring using organisms as diagnostic tools.

2. Bio-indication and bio-monitoring in the environment

Field assessment of environmental quality usually uses algae which are either plank-tonic or attached to surfaces. True planktonic forms are confined to lakes and large,

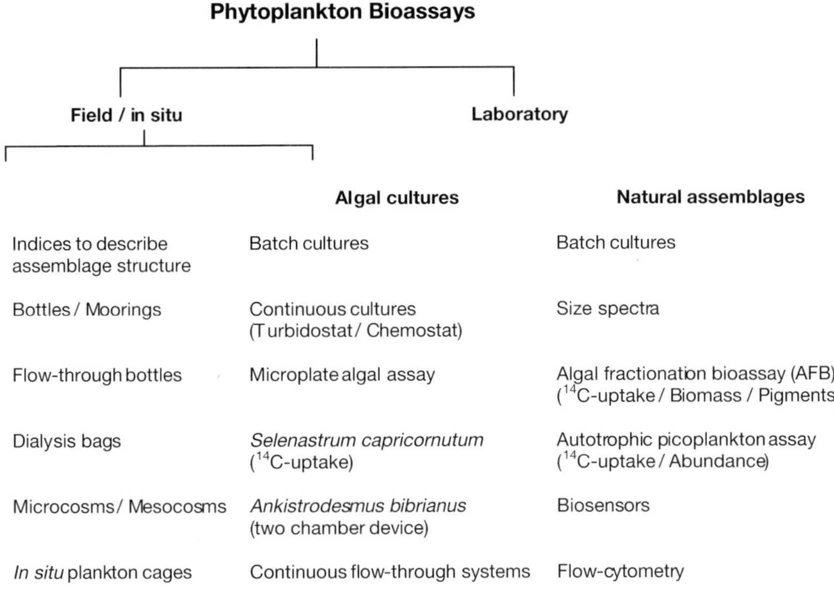

Figure 1.   Phytoplankton bioassays commonly employed in laboratory and field/in situ situations (modified from Munawar et al., 1989.

slow flowing rivers. They are simple and inexpensive to collect by conventional limnological water-samplers. Attached algae may be removed in the field by scraping or brushing of definite areas when quantification is attempted. In order to overcome several deficiencies of detachment techniques, artificial substrates of various kinds have been tried with some success (Hellawell, 1986). Detailed description of methodology and statistical evaluation of benthic algae used as bio-monitors is included in Lowe & Pan (1996).

2.1. Rivers

Historically the concept of saprobity included autotrophic and heterotrophic organisms (Sládeček 1977). Therefore several indices have been developed and used for both. As organic pollution in rivers decreased due to restoration measures, trophic problems became more pronounced resulting in the development of separate trophic bio-indices (Table 3). Ultimately all autotrophic organisms were treated separately from the saprobic system (Friedrich, 1990). General indices of water pollution can be found in Abel (1989).

Trophic classification of rivers from phytoplankton and periphyton structure and abundance is now included in the EC-Water Framework Directive (2000) permitting biology based quality estimation. While the usage of benthic algae for the classification of running waters has quite a long tradition, phytoplankton has widely been neglected which is quite the opposite situation compared to lakes.

Table 1. Criteria for trophic classification of plankton-dominated rivers (modified from Schmitt, 1998).

Trophic classification		Primary productivity	Chlorophyll-a 90-percentiles [μg l^{-1}]	Chlorophyll-a Average [μg l^{-1}]
I	Oligotrophic	Very low	3–8	<1–4
I–II	Mesotrophic	Low to moderate	8–30	3–8
II	Eutrophic	Moderate	20–100	7–30
II–III	Eu- to polytrophic	Moderate to high	70–150	25–50
III	Polytrophic	High	120–250	50–100
III–IV	Poly- to saprotrophic	Very high	200–400	>100
IV	Saprotrophic	Extremely high		>400

2.1.1. Phytoplankton

Long-term changes in rivers using plankton biocoenoses are relatively easy to detect because methods are similar to lakes but standard protocols for surveys are jet to be developed. River plankton assemblages are most often dominated by diatoms. Green algae and Cryptophyceae appear in summer (Dokulil, 1991, 1996). At reduced flow rates Cyanobacteria can grow and sometimes produce short-lived blooms because of improved light conditions and less turbulence (Steinberg and Hartmann, 1988). In general, the plankton flora of rivers is far less diverse than those of lakes and is often dominated by centric diatoms (Rojo et al., 1994; Reynolds and Descy, 1996). Species which might be used as indicators are *Aulacoseira granulata, Actinocyclus normanii, Stephanodiscus neoastrea, Cyclotella meneghiniana* among many others. The usefulness of river plankton for bio-indication is, however hampered because of the wide ecological tolerance of most species (Lange-Bertalot, 1978, 1979).

A general trophic classification of plankton-dominated rivers (Schmitt, 1998) uses the 90th percentiles of the chlorophyll-a concentrations from the growing season, March–October (Table 1). Peak and average values may be used in addition. Ranges indicate the changing chlorophyll content of algal biomass with varying algal composition which increases according to Behrendt and Opitz (1996) from Bacillariophyceae (diatoms) to Cyanobacteria (Cyanoprokaryota) to Chlorophyta (green-algae).

A similar trophic system, including algal abundances and primary production, has been published by Felföldy (1987) for Hungarian rivers and lakes (Table 2). This system unifies trophic categorization for all types of surface waters as discussed by Hamm (1996).

2.1.2. Phytobenthos

In contrast to plankton species, the algal periphyton in running waters include many of the requirements for an excellent monitoring system because they occur ubiquitously from clean springs to highly polluted river sections (Patrick, 1994). Macro-

Table 2. Parameters for the trophic characterization of rivers and lakes (modified from Felföldy, 1987).

Trophic level	Algal abundance [10^6cells l^{-1}]	Chlorophyll-a [μg l^{-1}]	Primary production [mg C m^{-2} d^{-1}]	Primary production [g C m^{-2} a^{-1}]
1 Ultra-oligotrophic	<0.01	<1	<5β	<10
2 Oligotrophic	0.01–0.05	1–3	50–125	10–25
3 Oligo-mesotrophic	0.05–0.1	3–10	125–250	25–50
4 Mesotrophic	0.1–0.5	10–20	250–500	50–100
5 Meso-eutrophic	0.5–1.0	20–50	500–900	100–175
6 Eutrophic	1–10	50–100	900–1,500	175–300
7 Eu-polytrophic	10–100	100–200	1,500–2,500	300–500
8 Polytrophic	100–500	200–800	2,500–4,000	500–800
9 Hypertrophic	>500	>800	> 4,000	>800

scopic conglomerations of algae are sampled and evaluated from transects or squares. Evaluation of microscopic periphyton is done after scraping off from the natural habitat or uses artificial substrata such as glass slides, styrofoam, plexiglass or tiles.

In principle, the presence of any species whose environmental limits are clearly understood could be used as an indicator. In practice, its ecological range is often to broad or too little is known to be of any use. Healthy growth of a species often is a much better indicator than just its presence. *Cladophora glomerata*, for example appears in almost all streams but large growth is only found when nutrient levels are high (Whitton, 1979). Several lists of individual species do exist classifying species according to their reaction to one or the other type of pollution or contaminant (Mauch, 1976; Rott et al., 1997, 1999).

The most objective accounts of the tolerance of individual species have been made for diatoms. Continuous long-time monitoring of rivers using diatometers (Patrick and Hohn, 1956) clearly show that algal assemblages on glass-slides reflect well perturbations such as increase in pollution, building of dams or small amounts of toxic pollution (Patrick, 1976). Winter and Duthie (2000) evaluate for instance in-stream nutrient concentration from patterns of epilithic diatom distribution. From a combination of chlorophyll-a measurements and an analysis of benthic algal assemblages Biggs (2000) constructed a nomograph relating nutrient concentrations and days of accrual to trophic conditions (Fig. 2). These and many other observations and elaborations have led to the formulation of many different indices for algal communities, especially diatoms by a large variety of authors, summarised in Table 3. Besides saprobic and trophic indication, diatoms are used for many other indications such as salinity, acidity, pH-value, Al-concentration, dissolved organic carbon (DOC) and humic substances (Schönfelder, 2000). In palaeolimnology, water-temperatures, pH-values or phosphorus concentrations from the past are reconstructed from the analysis of diatom frustuls found in the sediments.

Some of the more recently developed diatom indices are often based on experimental investigations such as those by Reimann and Hamm (1996), who analysed

Table 3.   Indices for the assessment of running waters based on algal biocoenoses of the natural environment (updated from Ghetti & Ravera, 1994 and DePauw et al., 1992).

Indices	Communities	References
Saprobic indices		
Biol. Effect of Org. Load (BEOL)	PA	Knöpp 1954
Relative Purity	PA	Knöpp 1954
Saprobic Index (S)	PA	Pantle and Buck 1955, DIN 38-410
Saprobic Index (S)	PA	Zelinka and Marvan 1991
Saprobic Index (S)	D	Sladecek 1986
Saprobic Index (SI)	D	Kobayasi and Mayama 1989
Saprobic Index (SI_{MI})	AD	Rott et al. 1997
Saprobic quotient (SQ)	P	Dresscher and Van der Mark 1976
Biotic indices		
Cemagref diatom Index (IDC)	PAD	Cemagref 1984
Diatom Index (IDD)	AD	Descy 1979
Diatom Index (IILB)	AD	Lange-Bertalot 1979
Diatom Index (IPS)	AD	Cemagref 1982
Diatom Index (ILM)	AD	Leclerq and Maquet 1987
Diatom Index (CEC)	AD	Descy and Coste 1991
Diatom assembl. Index (DAIpo)	D	Watanabe et al. 1986
Generic diatom Index (GDI)	AD	Rumeaux and Coste 1988
Median diatomic Index (MI)	AD	Bazerque et al. 1989
Trophic diatom Index (TDI)	D	Schiefele and Kohmann 1993,
Trophic diatom Index (TDI)	D	Kelly and Whitton 1995; Kelly 1996
Eutrophic Pollution Index (E/P-I)	D	Dell 'Uomo 1996
Trophic Index (BRB)	D	Schönfelder 1997
Trophic Diatom Index (TDI)	D	Coring et al. 1999
Trophic Index (TI and TI_{DIA})	AD	Rott et al. 1999
Specific diversity indices		
Equitability	D	Lloyd and Ghelardi 1964
Log-normal distribution	D	Preston 1948
Number of individuals per taxon	PA	Helawell 1986, Plafkin et al. 1989
Sequential Comparative Index (SCI)	A	Cairns et al. 1968
Taxa richness (S)	PA	Helawell 1986, Plafkin et al. 1989
Total number of individuals (N)	PA	Helawell 1986, Plafkin et al. 1989
Comparative indices		
Fluctuation Index (D)	D	Dubois 1973

P = Plankton, A = Periphyton (Aufwuchs), D = Diatoms.

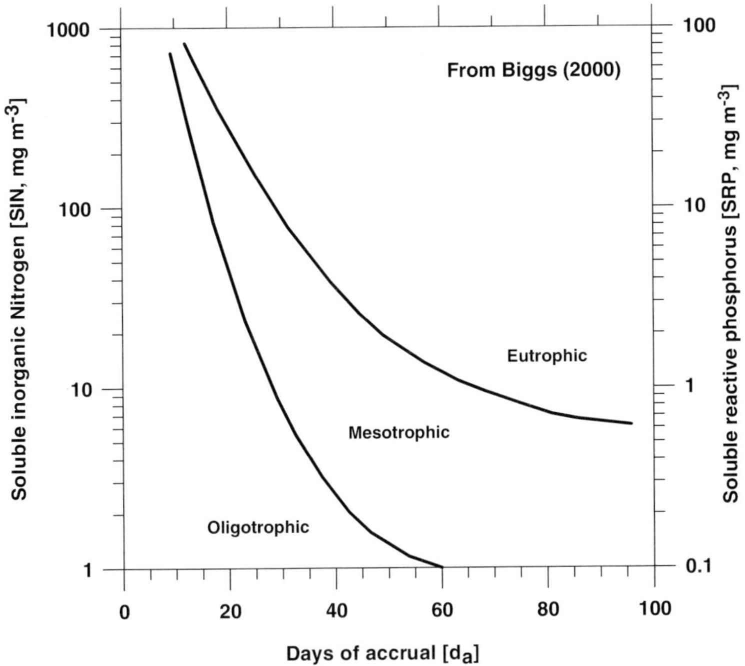

Figure 2. Nomograph predicting maximum benthic algal biomass as chlorophyll-a indicative of oligo-trophic, mesotrophic, and eutrophic conditions from mean monthly soluble inorganic nitrogen (SIN, left *y*-axis), soluble reactive phosphorus SRP, right *y*-axis) and days of accrual (d$_a$, *x*-axis). Boundaries were set to 60 mg m^{-2} chlorophyll-a to separate oligotrophic from mesotrophic, and to 200 mg m^{-2} for mesotrophic to eutrophic (modified from Biggs, 2000).

periphytic diatoms in artificial field and laboratory mesocosms. Based on the concept of differentiating species (see below) and intensive field investigations throughout Germany Schiefele and Kohmann (1993) developed a weighted trophic diatom index (TDI):

$$TDI_{SS} = \frac{\sum_{i=1}^{n} Y_i\, TDI_i W}{\sum_{i=1}^{n} Y_i W_i}$$

with TDI_{SS} = Trophic diatom index for sampling site (*SS*)
  Y_i = relative abundance of species $i = 1$ to n at the sampling site
  TDI_i = Index based on either phosphorus or phosphorus and nitrogen for species 1 to n
  W_i = Weight $i = 1$ to n for species 1 to n

This formula is mathematically equivalent to the saprobic index by Pantle and Buck (1955) or Zelinka and Marvan (1961).

                                                          M.T. Dokulil

Table 4. Relation of trophic levels to the Trophic Diatom Index (TDI), trophic condition and nutrient load (modified from Schiefele and Kohmann, 1993).

Trophic level	TDI	Trophic condition	Nutrient load
1	1.0–1.4	oligotrophic (o)	natural
1.5	1.5–1.8	oligo-mesotrophic (om)	low
2	1.9–2.2	mesotrophic (m)	moderate
2.5	2.3–2.7	meso-eutrophic (me)	critical
3	2.8–3.1	eutrophic (e)	significant
3.5	3.2–3.5	eu-hypertrophic (eh)	high
4	3.6–4.0	hypertrophic (h)	very high

The Trophic Diatom Index (TDI) characterises the trophic level of streams and rivers using seven levels similar to the saprobic system (Table 4). Both classifications are independent and their levels unequal. The value of the TDI lies mainly in its ability better to classify nutrient loads than some of the commonly used saprobic indices It is best applied to neutral or slightly alkaline, meso- to hypertrophic waters The weighted trophic diatom index of Schiefele and Kohmann (1993) for running waters is similar and comparable in methodology to the index developed for lakes by Hofmann (1993, see below Section 2.2).

Besides indices, several authors have attempted to describe community structure with elaborated differential concepts (Lange-Bertalot, 1978, 1979; Schiefele, 1987; Steinberg and Schiefele, 1988). These authors finally defined five groups of different tolerance levels against pollution and two groups describing nutrient conditions:

- most tolerant species (mt); reproduce even in polysaprobic areas
- highly tolerant species (ht) which occur up to the α-meso-polysaprobic level
- tolerant species (t) tolerating α-meso-saprobic conditions
- sensitive species (s) which are sensible against pollution but tolerating β- to α-meso-saprobic situations
- highly sensitive species (hs) which avoid saprobities greater than β-meso-saprobic
- oligotraphentic species (ol) indicating low nutrient concentrations
- eutraphentic species (eu) preferring high nutrient levels.

Using this system, running waters can be classified into three levels of pollution and four classes of trophy from relative species abundances according to the scheme given in Table 5. Examples of applications to running waters of various types include e.g. Dokulil et al. (1997) and Pipp and Rott (1994).

Other authors have used ordination techniques and weighted-averaging regression-calibration models for inferring stream water conditions from diatom patterns versus nutrients (e.g. Schönfelder, 2000; Winter and Duthie, 2000).

In some cases, phytosociological techniques were applied to bioindication of water quality using planktonic and benthic algal species (Möller and Pankow, 1981; Täuscher, 1999).

Table 5. Bioindication of trophy and pollution according to the system of differential diatom species (from Steinberg and Schiefele, 1988). ol = oligotraphentic species, eu = eutraphentic species, hs = highly sensitive species, s = sensitive species, t = tolerant species, ht = highly tolerant species, mt = most tolerant species. For more details refer to the text.

Relative abundances of the differentiating algal groups

Trophic level

I	ol $\geqslant$ 50%	hs $\geqslant$ 10%	eu < 10%	t + ht + mt + s < 10%
II	ol $\geqslant$ 10%	hs $\geqslant$ 10%	eu < 50%	t + ht + mt + s < 10%
III	ol < 10%	hs $\geqslant$ 10%	eu $\geqslant$ 50%	t + ht + mt + s < 10%
IV	ol < 10%	hs < 10%	eu $\geqslant$ 50%	

Pollution class

1	ol + hs <10%	eu < 50%	t + ht + mt + s $\geqslant$ 10%
2	ol + hs <10%	eu < 50%	t + ht + mt + s $\geqslant$ 50%
3	ol + hs <10%	eu < 10%	t + ht + mt + s $\geqslant$ 50%

Attached algae other than diatoms are additional valuable indicators of conditions in flowing waters (e.g. Backhaus, 1973). Macroscopic and microscopic sessile green algae, although difficult to identify, are often the most common species present in river beds during summer. John and Johnson (1991) developed a field and laboratory protocol to enable the use of these species for detection of response to heavy metals, nutrient enrichment and other types of pollution. In conjunction with the EC Directive on the ecological quality of waters, many countries develop protocols, standards and lists of indicator species for the assessment of river water quality (e.g. Jarlman et al., 1996; Rott et al., 1997, 1999).

The indices developed by Rott el al. (1997, 1999) include saprobic as well as trophic indices. The latter is either based on all algal classes or solely on diatoms. Mathematically it is similar to the trophic index of Schiefele and Kohmann (1993):

$$TI = \frac{\sum_{i=1}^{n} TW_i G_i H_i}{\sum_{i=1}^{n} G_i H_i}$$

with TI = trophic index of all algal groups or diatoms only (TI_{DIA})
$\quad TW_i$ = trophic value of species i (tabulated in Rott et al., 1999)
$\quad G_i$ = Weight given to species i (tabulated in Rott et al., 1999)
$\quad H_i$ = relative abundance of species i in %

Biotic integrity of rivers is estimated with a new index of biological integrity (PIBI) developed from periphyton assemblages using a wide variety of estimators such as algal genera richness, relative abundance of diatoms, various types of acidophilic, eutraphentic or motile dominant diatom genera, cyanobacteria, chlorophyll and ash-free dry biomass (Hill et al., 2000).

Biomass per unit area or ratios of different components have sometimes been used as indicators of water quality. For instance, Weber and McFarland (1969), quoted from Whitton (1979), proposed an index of ash-free dry weight of periphyton to their respective chlorophyll-a content, both in g m^{-2}. This index should be higher in polluted areas that contain a larger proportion of heterotrophic organisms.

The response of photosynthesis and respiration to factors such as nutrient enrichment or a pollutant can be used to evaluate water quality. Of especial importance is the P/R-ratio which is: <1 in septic zones, increases rapidly and reaches values >1 in the recovery zone. Further downstream the P/R ratio approaches one. River primary production is often estimated from continuous upstream-downstream recordings of oxygen and other parameters. Water quality is deduced from these measurements (Kelly et al., 1976).

2.2. Lakes and reservoirs

Several techniques, indices and indicator species have been proposed by a variety of authors for the trophic classification of lakes and reservoirs with natural phytoplankton assemblages. The phytobenthos (periphyton, Aufwuchs) in lakes has attracted much less attention, especially when compared to river benthos. In some cases differences at higher taxonomic levels (algal groups) were used to characterise trophic levels of lakes.

2.2.1. Phytoplankton

2.2.1.1. Indices using algal groups

- Chlorococcal – Desmid Quotient (Thunmark, 1945)

Trophic levels are characterised by the relationship of the number of species found in a sample according to

$$Q = \frac{\text{Chlorococcal species number}}{\text{Desmid species number}}$$

Oligotrophic lakes have values <1, usually between 0.2 and 0.7; eutrophic waters are characterised by Q $\geqslant$ 1 (1–3); hypertrophic lakes may reach values as high as 14.

Other authors could not validate this quotient and reported high variability.

- Algal quotients according to Nygaard (1949)

In addition to Thunmark's index, Nygaard developed further indices based on various algal groups:

$$\text{Myxophyceae Quotient} = \frac{\text{Myxophyceae}}{\text{Desmids}}$$

$$\text{Diatom Quotient} = \frac{\text{Centrales}}{\text{Pennales}}$$

$$\text{Euglenophyceae Quotient} = \frac{\text{Euglenophyceae}}{\text{Myxophyceae} - \text{Chlorococcal greens}}$$

Compound Quotient =

$$\frac{\text{Myxophyceae} - \text{Chlorococcales} + \text{Centrales} + \text{Euglenophyceae}}{\text{Desmids}}$$

Characteristic values of trophic levels are:

Dystrophic	0–0.3
Oligotrophic	<1
Mesotrophic	1–2.5
Eutrophic	3–5
Hypertrophic	5–20
Polytrophic	10–43

Again, the compound index, as all others proved to be of rather limited value.

- E:O und EV:EO ratios according to Järnefelt et al. (1963 cit. acc. to Heinonen, 1980)

The ratio of eutraphentic to oligotraphentic species (E:O) and the quotient of the total biomass of eutraphentic species to the biomass of oligotraphentic species EV:OV) is defined on species level:

$$\frac{E}{O} = \frac{\text{Number of eutraphentic species}}{\text{Number of oligotraphentic species}}$$

$$\frac{EV}{OV} = \frac{\text{Total biomass of eutraphentic species}}{\text{Total biomass of oligotraphentic species}}$$

According to Heinonen (1980) the E/O index fits better at higher trophic levels while the biomass based quotient is very variable. Moreover, application is restricted due to the limited number of oligotraphent indicator species.

- Algal quotient according to Stockner (1971)

The index is based on the ratio of the two diatom groups Araphidineae/Centrales. Originally, it was developed for diatom frustules in recent sediments. The author proposed the following classification:

A/C ratio	Lake type
0.0–1.0	Oligotrophic
1.0–2.0	Mesotrophic
> 2.0	Eutrophic

2.2.1.2. Classification based on indicator species

Some of the above mentioned authors as well as several others have tried to classify lakes using the indicator species concept. A pre-requisite for defining indicators is a good knowledge of the algal species specific taxonomy and their related environmental requirements (e.g. Teubner, 1995).

- Indicative species according to Thunmark (1945), Nygaard (1949), Järnefelt (1952) or Teiling (1955)

All these authors proposed various lists of algal species which are either indicative for specific trophic situations, are indifferent or have no indicator value for lakes. For more details one must consult the original reference because the listings are voluminous. These approaches are of limited regional importance because most information originated from Scandinavian countries.

- Dominant limnetic algae according to Rawson (1956)

The author proposed a list for Western Canada in which the dominant algal species are placed in approximate sequence from oligotrophic to eutrophic occurrence (Table 6). Dominance is defined as a high percentage of the species in phytoplankton counts over much of the summer season. It should be made clear that this list shall only be used in Canada. Lakes in different regions of the world may need different species list (see further down).

- Qualitative characterisation according to Heinonen (1980)

Classification was based on qualitative phytoplankton analyses and on a differentiation of lakes based on their total plankton biomass. Lakes with a biomass ranging from

Table 6. Approximate trophic distribution of dominant algae in lakes of Western Canada (from Rawson, 1956).

Oligotrophic	*Asterionella formosa*
	Aulacoseira islandica
	Tabellaria flocculosa var. fenestrata
	Tabellaria flocculosa
	Dinobryon divergens
	Fragilaria capucina
	Stephanodiscus niagarae
	Staurastrum spp.
	Aulacoseira granulata
Mesotrophic	*Fragilaria crotonensis*
	Ceratium hirundinella
	Pediastrum boryanum
	Pediastrum duplex
	Coelospherium naegelianum
	Anabaena spp.
	Aphanizomenon flos-aquae
	Microcystis aeruginosa
Eutrophic	*Microcystis flos-aquae*

0.01–0.50 mg l^{-1} are considered oligotrophic those with B >2.5 mg l^{-1} are called eutrophic. Indicator species dominant in one or the other lake type are listed. Comparison to the species listed by Järnefelt et al. (1963, cited in Heinonen 1980) substantiated the indicative value of species such as *Arthrodesmus incus, Dinobryon cylindricum* and *Mallomonas borgei*. Many of the so called oligotraphentic species, however were often found in eutrophic lakes.

- Trophic Lake Index (Hörnström, 1981)

Hörnström postulates that the composition of the phytoplankton reacts more slowly to changing trophic conditions (>1 year) while total biomass readily reflects the nutrient situation. Based on these assumptions, he devleoped a Trophic Lake Index (I_L) which is calculated from

$$I_L = \frac{\sum (f * I_S)}{\sum f}$$

with I_S = Trophic Index of the species (range 0–100)
       f = frequency as log Biovolume in μm^3 ml^{-1} (modification by Tremel, 1996).

The indicator valence of an algal species, ranging from 0 to 100 with increasing trophic state, is estimated from calculating median biomass for all the lakes in which a species occurs relative to the highest median observed (Fig. 3). This index is interesting for classification because it is based on relative frequencies which should remain more stable than absolute occurrence in case of zooplankton grazing.

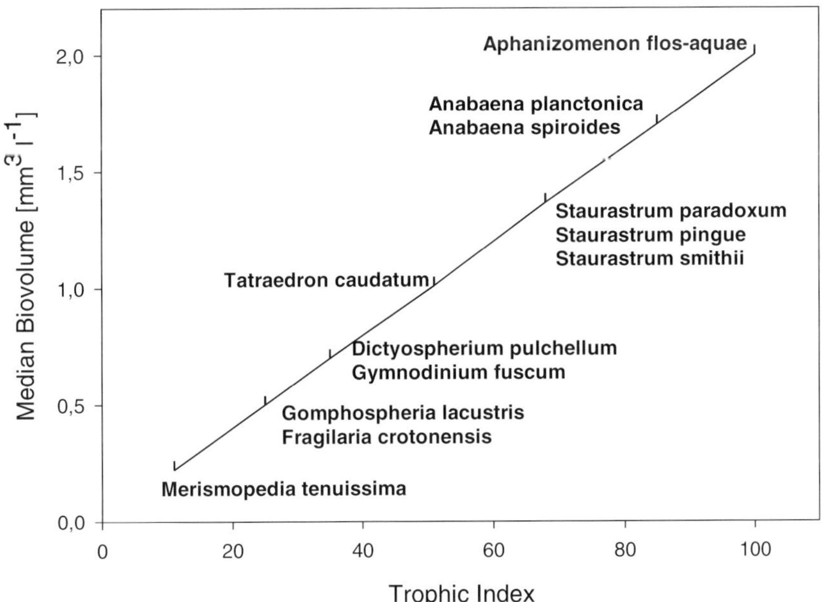

Figure 3. Relation between median volume and trophic index of phytoplankton species (modified from Hörnström, 1981).

Table 7. Algal bioindicators for trophic levels (from Kümmerlin, 1990). Species which are common but have no indicative value are listed under 'eutraphent'.

Trophic level	Algal group	Algal species
Oligotrophic	Bacillariophyceae	*Cyclotella bodanica*
	Chrysophyceae	*Chromulina erkensis*
		Chromulina rosanoffii
	Xanthophyceae	*Istmochloron trispinatum*
	Cryptophyceae	*Cryptomonas obovata*
Oligo-mesotrophic	Cyanophyceae	*Microcystis wesenbergii*
	Cryptophyceae	*Cryptaulax vulgaris*
Mesotrophic	Bacillariophyceae	*Tabellaria fenestrata*
Eutrophic	Cyanophyceae	*Microcystis aeruginosa*
		Aphanizomenon flos-aqae
		Anabaena planctonica
	Bacillariophyceae	*Stephanodiscus hantzschii*
		St. astrea
		St. binderanus
	Conjugatophyceae	*Mougeotia thylespora*
Eutraphent (euryök)	Bacillariophyceae	*Asterionella formosa*
		Cyclotella radiosa
	Dinophyceae	*Ceratium hirundinella*
	Cryptophyceae	*Rhodomonas minuta*
		Cryptomonas ovata

- Indicator species and indicator group study by Rosén (1981)

From a large data-set of medium and small sized Swedish lakes, algal species with clear environmental characteristics were defined from distribution functions. Results indicate that blue-green and green planktonic algae, besides well defined eutrophic species, comprise types indicative of clear lakes or low or high humic content. Chrysophyceans often dominate in nutrient poor waters. Diatoms are absent from ultra-oligotrophic lakes. Dinoflagellates and Cryptophyceans are confined to certain lake types. Within the Chloromonadophyceans, *Gonyostomum semen* is an excellent indicator for humic lakes. The study contains detailed lists of the various species and their indictor value with respect to several limnological important variables.

- Algal Bioindicators according to Kümmerlin (1990)

Indicator species are deduced from long-term observations on Lake Constance, Germany (Table 7).

- Algal Bioindicators and Trophic Index by Brettum (1989)

The system used by Brettum (1989) is an extension and elaboration of the method earlier developed by Hörnström (1981, see above). More than 120 species are assigned to seven trophic categories (ultra-oligotrophic to hyper-eutrophic) according to the probability of their highest appearance calculate from

$$p = \frac{n_i}{N_i} \times V_i$$

with N_i = total number of algal species within a trophic class
n_i = number of a specific species (i) per group
V_i = percentage contribution of species i to total biovolume

These values are normalised to the interval at which the species contributes most (=100) which results in a numeric distribution of all species in the seven trophic classes which are summarised in Brettum's study. A compound index is finally calculated from the individual species indices:

$$I_T = \frac{\sum (v_i * i_{iT})}{\sum v_i}$$

with I_T = index for the trophic level T
v_i = total biovolume of species i
i_{iT} = index value of species i for the trophic category T

This index has the advantage that it uses volumes rather than relative abundances. Similar to the study by Rosén (1981), distribution of species is related to several environmental variables

- BRB-Index (Schönfelder, 1997, 2000)

Bioindication with the BRB-index was developed for *b*icarbonate-*r*ich waters in *B*randenburg, Germany and is therefore restricted to this and similar types of waters but can be used for plankton and benthic diatoms both in streams and lakes. The concept of Schönfelder (1997) is based, similar to many other approaches, on the optimum and the tolerance range of diatom species to total phosphorus concentrations which are calculated from:

$$\ln TP - \text{Optimum}_k = \frac{\sum_{i=1}^{S} \ln TP_i * d_{k,i}}{\sum_{i=1}^{S} d_{k,i}}$$

with $d_{k,i} = n_{k,i}/n_i$

where TP = total phosphorus
k = taxon for which TP is estimated
i = sample number
s = number of samples
$d_{k,i}$ = dominance of taxon k in sample i
$n_{k,i}$ = abundance of taxon k in sample i
n_i = abundance of all species in sample i

The tolerance range of the individual species is estimated from the standard deviation of ln *TP*:

$$t_{\ln\,TP,k} = \left(\frac{\sum_{i=1}^{S} d_{k,i} * (\ln\,TP - \text{Optimum} - \ln\,TP_i)^2}{\sum_{i=1}^{S} d_{k,i}} \right)^{1/2}$$

These tolerance values are then converted to integer TP-factors from:

$$SF_{TP,k} = 3.4999 - 3.333\,t_{TP,k}$$

Mathematically negative results are considered to be zero. These factors are inversely proportional to the indication of TP by the species ($SF_{TP,k} = 0$ equals wide ecological tolerance, $SF_{TP,k} = 3$ little tolerance).

The trophic index is finally calculated from the dominance, the TP-factors and TP-optima of all the species (*m*) from:

$$\text{BRB} - \text{Index} = \frac{\sum_{k=1}^{m_i} d_{k,i} * SF_k * \ln\,TP - \text{Optimum}_k}{\sum_{k=i}^{m} d_{k,i} * SF_k}$$

The calculated index is calibrated against the natural-logarithms of the measured TP-concentrations which are related to 11 trophic conditions (Table 8). The TP factors for a large number of benthic and planktonic species can be found in Schönfelder (1997).

● Phytoplankton Indicators (Lepistö and Rosenström, 1998; Lepistö, 1999)

Most recent collection of extensive lists of indicator species for various types of trophic conditions. Indication based on an evaluation of references and own observations.

2.2.1.3. Classification from biomass or biovolume

Phytoplankton biovolume or biomass has been used by several authors for the trophic classification of lakes. The systems of Rosén (1981) and Rott (1984) are identical. The Norwegian (Brettum, 1989) and the Swedish classifications (Willén, 2000) are both based on either mean or maximum values. Four systems are compared in Table 9 from which it becomes evident that greatest discrepancies among delineation by authors are between Heinonen (1980) and all others. The main differences lie in the number of trophic categories considered. Brettum's classification is the most differentiated one within this comparison while those of Rosén (1981) and Rott (1984) have overlapping values but consider only three trophic levels. Categorisation and delineation using algal biomass by different authors is graphically summarised in Figure 4.

2.2.1.4. Classification based on seasonal phytoplankton associations

Detailed analyses of phytoplankton succession and seasonal development culminated in the description of 26 provisional associations by Reynolds (1997) which in his view

Table 8. Concentrations of total phosphorus (TP) for the trophic categories as defined by Schönfelder (1997).

Trophic status	Range of TP [μg l^{-1}]
Ultraoligotrophic	<4.3
Ultra- to oligotrophic	4.3–7.0
Oligotrophic	7.0–11.6
Oligo- to mesotrophic	11.6–19.1
Mesotrophic	19.1–31.5
Meso- to eutrophic	31.5–51.9
Eutrophic	51.9–85.6
Eu- to polytrophic	85.6–141.2
Polytrophic	141.2–232.8
Poly- to hypertrophic	232.8–383.8
Hypertrophic	>383.8

Table 9. Comparison of trophic delineation from phytoplankton fresh-weight biomass according to various authors.

Trophy	Average fresh weight biomass [mg l^{-1}]			
	Heinonen (1980)	Rosén (1981), Rott (1984)	Brettum (1989)	Willén (2000)
Ultra-oligotrophic	<0.2		<0.12	<0.1
Oligotrophic	0.21–0.5	0.1–1.0	0.12–0.40	0.1–0.5
Oligo-mesotrophic	0.5 –1.0		0.4 –0.6	
Mesotrophic	1.0 –2.5	0.5–5.0	0.6 –1.5	0.5–1.5
Meso-eutrophic			1.5 –2.0	1.5–2.5
Eutrophic	2.5 –10	>2	2.0 –5.0	2.5–5.0
Polytrophic			2.0 –5.0	
Hypertrophic	>10		>5	>5

are different vegetation types recognisable within freshwater phytoplankton (Table 10). The number of entries and the species associated with are seen as an open, changeable system by Reynolds. Associations are not defined via species but through functional algal groups. Adaptations to limiting factors can result in the preference of certain morphotypes, such as colonial or filamentous forms when grazing pressure increases, independent from their taxonomical position. Although Reynolds (1997) never uses the term 'indicator' about 60% of the species and genera mentioned in the broad description of his associations can be assigned to different trophic categories.

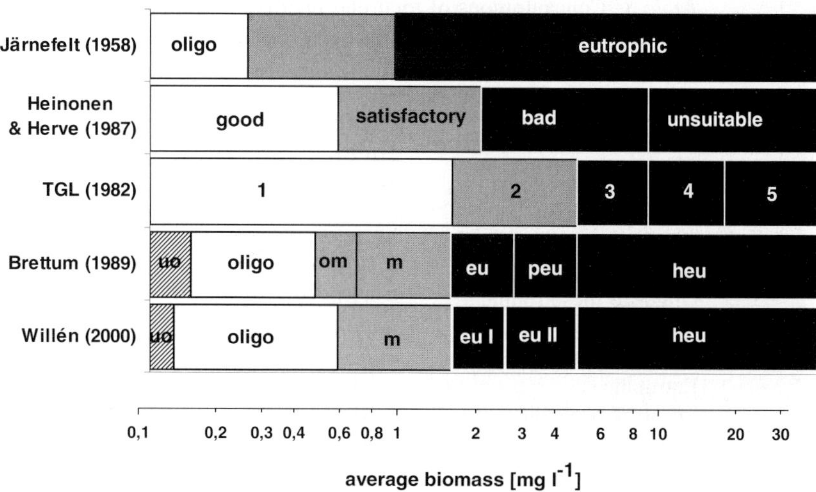

Figure 4. Comparison of five classification systems for lakes using average phytoplankton biomass.

2.2.1.5. Other Approaches

Palaeolimnological investigations can provide background information of the recent past of lake ecosystems to be monitored (Simola et al., 1996). This approach is discussed in more detail in e.g. Charles et al. (1994) which came to the conclusion that palaeolimnological investigations can significantly enhance the usefulness and applicability of monitoring data by extending the temporal record of ecosystem conditions for a considerable time into the past, and providing a context for evaluating more recent measurements.

Environmental changes over longer time periods can be monitored using algal microfossils preserved in freshwater sediments (Dixit et al., 1992). Among many other potential indicators, the morphological remains of diatom frustules, chrysophyte scales and cysts are usually abundant in lake sediments, and they often form essential parts of palaeolomnological studies. Correlation of diatom and chrysophyte changes with specific lake water variables allow acidity, trophic and salinity reconstruction. The most effective way of studying algal populations with respect to lake water quality is to analyse the microfossils present in surface sediments from a set of lakes with known water chemistry. These diatom and chrysophyte training or calibration sets, integration over time and space, contain enough autecological and synecological information to enable deduction of environmental conditions from species composition of samples (Wunsam and Schmidt, 1995; Kamenik and Schmidt, 2001). Recent developments in ordination analysis has greatly improved our understanding of the relations between species distribution and environmental variables. Moreover, as a result of recent refinements in methodology of sediment coring and sectioning procedures detection of lake water quality changes of the last 5 to 10 years is now possible. The rapidly increasing data sets suggest that a number of environmental variables such as Secchi-depth, conductivity, several chemical elements etc. can be monitored using sedimentary algal remains.

Table 10. Associations of freshwater phytoplankton from Reynolds (1997). Trophic status and description of assemblages added from the text there-in.

	Trophic status	Typical taxonomic units	Description of plankton assemblage
A	Oligotrophic	*Urosolenia spp., Cyclotella comensis*	Oligotraphentic diatom dominated
B	Oigotrophic	*Asterionella spp., Aulacoseira italica*	Oligotraphentic spring-diatom s
C	Etrophic	*Asterionella, Stephanodiscus rotula, Aulacoseira ambigua*	Oligotraphentic spring-diatom
D	Hertrophic	*Stephanodiscus hantzschii*	Diatoms in hypertrophic shallow lakes
E	Mesotrophic	*Dinobryon, Chrysosphaerella*	Mesotraphentic Chrysophycean dominated
F	Oligotrophic	*Sphaerocystis, Botryococcus*	Oligotraphentic green algae
G	Eutrophic	*Eudorina, Pandorina*	Eutraphentic green algae
H	Eutrophic	*Anabaena*	N-fixing blue-green aggregates
J	Eutrophic	*Pediastrum, Scenedesmus, Oocystis borgei*	Eutraphentic green algae
K		*Aphanocapsa, Aphanothece*	Small sized blue-green aggregates
L_0	Oligo-mesotrophic	*Ceratium, Peridinium inconspicuum, Gomphosphaeria*	Oligotraphentic Dinoflagellates
L_M	Meso-eutrophic	*Ceratium, Microcystis*	Eutraphentic Dinoflagellates
M		*Microcystis*	*Microcystis*-dominated
N	Oligo-mesotrophic	*Cosmarium, Tabellaria*	Oligotraphentic desmid-diatom plankton
P	Eutrophic	*Staurastrum, Fragilaria.*	Eutraphentic plankton
R		*Planktothrix rubescens/mougeotii*	Deep-living blue-green algae
S	Eutrophic	*Planktothrix agardhii / Limnothrix redekei*	Eutraphentic, filamentous blue-green algae at low transparency
S_N	Eutrophic	*Cylindrospermopsis*	N-fixing, filamentous blue-green algae
T		*Geminella, Binuclearia, Tribonema*	Filamentous algae at high mixing
U		*Uroglena*	Early summer plankton at very low phosphorus concentrations
V		Phototrophic bacteria	Phototrophic bacterio-plankton
W		*Euglena, Synura, Gonium*	Plankton in ponds at high organic load
Xl	Eutrophic	*Chlorella, Ankyra, Monoraphidium*	Eutraphentic nanoplankton
X2	Eutrophic	*Rhodomonas, Chrysochromulina*	Eutraphentic mobile nanoplankton
X3	Oligotrophic	*Koliella, Chrysococcus*	Oligotraphentes nanoplankton
V	Meso-eutrophic	*Cryptomonas spp.*	Eutraphentic
Z	Oligotrophic	*Synechococcus, Chlorella minutissima*	Oligotraphentic picoplankton

In some cases algal-based models can help to predict trophic level changes (e.g. Dokulil and Frisk, 1993; Jørgensen, 1992.).

2.2.2. Phytobenthos

Periphyton is an important component of the litoral zone of lakes and reservoirs. Among the many algal groups which have been tried as bio-indicators for the lake litoral by several authors (e.g. Kann, 1978, 1986), the diatoms attracted particular attention because of their widespread distribution, high sensibility, good preservation and well developed indication techniques for both saprobity and trophy.

Most of the many investigations that deal with litoral diatoms are either systematically orientated or are interested in correlations to variables other than those responsible for eutrophication. As a consequence, information on species distribution and their environmental requirements within the trophic spectrum is limited (Lowe, 1974; Whitmore, 1989).

The trophic diatom index by Hofmann (1993, 1999) is one of the few examples of bio-indication using lake litoral diatoms. About 400 algal species from the epilithon, epiphyton and from artificial substrates were analysed for their requirements with special emphasis on total phosphorus because of its relevance for the trophic state. The organisms were assigned into five categories: a group containing all the ubiquitous, tolerant species found at all trophic levels and four indicative classes (oligotraphentic, oligo-mesotraphentic, meso-eutraphentic, and eutraphentic taxa). These four levels were combined with three weight-factors. The index is then calculated from the formula of Zelinka and Marvan (1961) with an equation similar to one of those already shown above. Classification is bases on five trophic classes from oligo- to hyper-trophic.

2.3. Marine ecosystems

Most what has been outlined above for freshwater indicators and bioassays, equally applies to bio-indicators in the marine environment. Approaches and protocols for marine phytoplankton may be found in Maestrini et al. (1984). The most common approach for assessing the relation between phytoplankton and the nutritional environment is the classical descriptive one of drawing information from (i) concentrations of nutrients, (ii) phytoplankton biomass or (iii) phytoplankton biochemical activity. The complex nature of natural assemblages, however does not allow unequivocal answers. Nitrogen additions to coastal waters often result in increased phytoplankton productivity and perhaps biomass (Costa et al., 1992). Red and brown tides or the appearance of toxic algae is often linked to water quality (Watanabe 1983) but may also be considered as early indication of climatic change (Hinckley and Tierney, 1992).

Advantage is gained therefore from experimental approaches which use the response of organisms in bioassays. By definition, bioassays include a multitude of methods and techniques, such as in situ versus in vitro bioassays using either natural assemblages or unialgal cultures (for details refer to Maestrini et al., 1984, and Appendix in Maestrini et al., 1984).

Macroalgae or seaweeds in coastal marine waters are far more important as bio-monitors than macrophytic algae in fresh-waters. Seaweeds have several intrinsic advantages as organisms for monitoring environmental impacts. Because of their sessile nature they are easily collected in abundance and can be used to characterise locations over time. Accumulation of compounds from the surrounding water make them idle as bioaccumulators. Sometimes the extent of benthic macroalgal distribution can be quantified by aerial mapping (Costa et al., 1992).

Investigations at the community level are time consuming and require expertise. Interpretation of results is difficult and detection of impacts often requires long-term studies. Alternatively, the responce of individual species to environmental conditions, measured as growth or productivity, can be used for biodetection under laboratory bioassay or field deployment conditions (Levine, 1984).

The diverse life history types among the algae offer a wide variety of approaches. Annual species reflect conditions over a well-defined period while perenniating algae integrate the milieu of several years past. Within-species assays are possible with the various stages in algal life cycles which can be expected to have different susceptibilities. The reproductive cells are of particular interest because they are usually most vulnerable.

Brown algae (Phaeophyceae), often dominating seaweed communities in the littoral and sub-littoral, are frequently employed for coastal monitoring. Members of the Fucales (*Fucus, Ascophyllum*) and Laminariales (*Laminaria, Macrocystis*) received most attention. Pollution assessment studies have also used red algae (Rodophyceae) which often make up the major biomass in subtidal communities. The life cycle of these algae involves three plant types: gametophytes, carposporophytes and tetrasporophytes. Any or all of these stages can be used for monitoring pollution effects.

Among the green seaweeds (Chlorophyceae), the genera *Ulva* and *Enteromorpha* have attracted considerable attention as biomonitors. The ability of gametes to develop parthenogenetically in *Ulva* is of particular interest, offering genotypically identical plants minimizing effects of genetic differences between experimental organisms.

2.3.1. Size measurements

All these plants grow by cell division and thallus elongation. The magnitude and rate of these processes can be measured simplest by the increase in dimensions. Alternative approaches include the determination of biomass and the estimation of rates of primary production.

In seaweeds, size is the most frequently used measurement for impact assessments. Some seaweeds have localized meristematic regions which can easily be used for growth determination. In the kelps, for instance, growth is primarily restricted to the base of the blades. Growth measurements are taken by punching a hole into the blade at a predetermined distance from the meristematic zone. After a certain period of growth, another hole is punched at the same location. The distance between holes is an estimation of growth. This approach has also been used in continuous flow culture systems of *Laminaria saccharina*. In addition, early developmental stages have been used by several authors to test effects of toxicants on scawccds (for references ref. to Levine, 1984). Similar or more complicated techniques were developed for growth studies using a large number of red algal species, fucoid algae and *Ulva*.

Because sewage pollution in coastal waters is correlated with abundant *Ulva* growth several in situ and laboratory bioassays have been developed using discs cut from the thalli of *Ulva* or deploying genotypically identical *Ulva* plantlets (Levine, 1984).

2.3.2. Biomass

Another way of estimating growth and productivity in seaweed is through the determination of biomass. A complete over-view on sampling and quantitative procedures is given by Gonor and Kamp (1978). Bellamy et al. (1973) developed an approach to determine productivity from biomass estimations.

2.3.3. Photosynthesis and respiration rates

Seaweed productivity can be estimated from photosynthetic and respiration rates. Techniques commonly employed include CO_2-detection by infrared absorption, the oxygen evolution/consumption method and the carbon-14 technique. Rate measurements in situ or under laboratory conditions were used to detect effects of oil coating, sewage effluents, iron-ore dust, etc. on various types of marine algae used as bio-indicators.

2.3.4. Reproduction

The reproductive processes of seaweeds offers yet another way to investigate effects of pollutants or toxicants. Meiosis is a particularly sensitive phase in the life cycle of most organisms. Tropic responses, motility of reproductive cells and sexual processes offer a wide variety of possible test alternatives for pollutants such as oil, petroleum, iron-ore dust, detergents and other toxicants.

2.3.5. Bioaccumulation

Accumulation of polluting substances by marine macro-algae have received much attention because these attached plants reflect environmental conditions over prolonged time periods. Moreover, their sessile nature enables relatively easy collection. Seaweeds accumulate heavy metals, hydrocarbons, pesticides, PCBs, radionuclides and numerous other compounds from the water. The accumulation and release of these different compounds largely depends on their chemical properties and concentration in the environment but is strongly modified by several circumstances. The position of the plant on the shore or in the water column affects the degree of contamination since some contaminants are restricted to the surface layers while others sink relative rapidly to deeper layers or become associated with the sediment.

The principal mode for accumulation of substances by algae appears to be the process of adsorption. Frequently, the uptake involves two stages: (1) an initial passive accumulation by adsorption to the exterior surface, followed by (2) a slower uptake mediated by metabolic processes which in turn depend on external physical variables. The environmental regime at different localities, regions or seasons can therefore significantly influence the interpretation of bioaccumulation data.

In addition, structural differentiation must be considered. Many marine macroalgae accumulate substance with virtually every cell while in some of the more differentiated seaweeds uptake and accumulation varies in different portions of the plant. Generally the older plant parts have accumulated higher concentrations of pollutants. Many complications are involved, however, in the interpretation of such data (Levine, 1984).

2.3.6. Mutagenicity assays

Many man-made chemicals introduced into the environment are suspect to induce cancer. These compounds can be tested for cancerogenesis by their ability to induce mutations in microbial DNA, since carcinogenic and mutagenic chemicals are highly correlated. Application to water samples, however, is restricted owing to dilution and the large volumes needed therefore. This procedural bottleneck can be circumvented by using extracts derived from bioaccumulator species. Mutagens were detected within tissue extracts derived from *Porphyra umbilicalis, Fucus vesiculosus,* and *Enteromorpha* spp. by Barnes (1980). Due to its parenchymatous nature, lowest mutagenic activity was found in Fucus.

Mutagenic substances are either endogenously produced by seaweeds or accumulated from the external environment. For coastal monitoring, the character of mutagenic agents is of prime importance. Many compounds are present in all marine waters as a result of both natural production an anthropogenic inputs. Tissues of bioaccumulator species integrating environmental regimes in conjunction with mutagenicity assays provide reliable screening procedures for hazardous chemicals in marine ecosystems.

3. Bioassays

Theoretical principles, selection of organisms and their pre-cultivation in algal bioassays are extensively discussed in Marvan et al. (1979). Biotests with algae are carried out either in the natural environment or, under more 'standardised' conditions, in the laboratory with single cultivated algal strains, mixtures of them or natural assemblages. In general algal bioassays can be an important tool for the assessment of present or potential deterioration of water quality (Bellinger, 1979). A disadvantage of many algal bioassays, however, is that they do not consider nutrient recycling in the water-body. Results obtained therefor often refer to maximum attainable algal biomass rather than to total algal growth. Since such algal-bioassays only show what happens in the short term, longer-term bioassays become increasingly important.

3.1. Field approaches and in situ techniques

Field or *in situ* approaches do not use "standard conditions" of growth for incubations, but rely on available nutrient supply, temperature and light conditions prevailing at the time. The disadvantage is the infinite range of variables in nature. Field incubations of algae, usually phytoplankton or cultivated algal species used in the bioassay, are carried out using one of a variety of chambers or enclosures. There is an almost endless variety of systems either closed or open, a multitude of experimental designs, species used and measurement parameters. The individual investigator must decide and choose which technique is the best for the problem to be solved (Trainor, 1984).

3.1.1. Types of enclosures

In situ enclosures may be divided into five types: bottles, bags, tubes, curtains, and specialised types of flow-through studies.

Approaches using *bottles* are probably the oldest. As early as 1891, Regnard used bottles to expose cress and radish seedlings to the underwater light gradient in an attempt to explore the growth and chlorophyll regulation (cit. acc. to Talling, 1984). Capacities range from 125 ml bottles to 'carboys' of 20 litres or more. Strickland and Terhune (1961) were the among the first to use *bags* or *spheres* to study the dynamics of phytoplankton. The size of these bags can range from about 120 m^3 to as small as 1 or 2 litres. *Tubes* have two important differences to spheres; they have a more or less fixed diameter and they can be open at the top or on both ends. When open, the tubes do not isolate the enclosed water mass from the atmosphere or the bottom sediment. A large range in size among different types of tubes has been employed for ecological studies. The first small *in situ* enclosures were probably the 'plankton-test-lots' used by Thomas (1964) for nutrient enrichment experiments. Other enclosures range from diameters of about 1 m through medium-sized 'limnocorrals' to the 45 m diameter 'Lund tube' (Lund, 1972). The largest enclosures are generally those used in marine studies (Schelske, 1984). Water masses within an aquatic system may be separated from one another by *curtains* extending from the surface to the bottom sediments effectively restricting the exchange of water between the two parts (Schindler, 1974). This type has since been used by many investigators. Specialised types of field enclosures include the *in situ* chemostat (deNoyelles and O'Brien, 1974), sacs constructed from dialysis membranes and devices using either membrane or similar filters or dialysis membranes to separate cultures or natural assemblages from the surrounding water.

There is no obvious advantage of one single type of enclosure. All types have shortcomings. Application to a particular experiment will depend on several factors such as the number of independent variables

3.1.2. Algal fluorescence

Similar to standardised laboratory bioassays (see below), the toxicity of chemicals, mixtures of chemicals or polluted waters is measured as their inhibitory effect on the photosynthesis of natural algal assemblages. Measurements continuously monitor chlorophyll-a fluorescence signals. Additional parameters include cell numbers and turbidity (Sayk and Schmidt, 1983; Noack, 1987). Other 'on line' systems use algal cultures (*Scenedesmus subspicatus, Chlamydomonas reinhardtii* or *Microcystis sp.*) which are added to the water as test organisms. The delayed fluorescence signal from darkened cells (DF-algal test) is measured every 30 minutes (Gerhardt and Putzger, 1992). Both techniques are cost intensive on installation, require regular servicing but offer quasi-permanent control abilities. Correct interpretation of results is sometimes difficult. In combination with other (semi)-automatic tests these techniques can be used for alert systems. Further on line systems under development are summarised in Gunkel (1994, p. 356ff).

3.1.3. Oxygen-production and respiration rates in situ

Standard test: DIN 38412, L13, 1983a
Organisms: Natural phytoplankton
Oxygen concentrations are measured chemically or amperometrically in light and dark bottles at the beginning and at the end of an incubation period of up to 24 hours. Because of the rapid light attenuation in natural waters this test is usually performed at several different depths.

O_2 in light – O_2 in dark = Gross oxygen-production
Initial O_2 – O_2 in dark$_{24h}$ = Respiration

Sensitive, cost effective test with high ecological relevance because of multi-species approach. Not used in ecotoxicology. Methodological comparison to other techniques see e.g. Sakamoto et al. (1984).

3.1.4. Bioassays using picoplankton

A number of studies (Munawar et al., 1987; Munawar and Weisse, 1989; Weisse, 1991) indicate that photoautotrophic picoplankton are useful indicators of contaminant stress in marine and freshwater ecosystems. They are often a dominant component of oligotrophic, pelagic ecosystems (Weisse, 1993; Stockner et al., 2000). From recent *in situ* nutrient enrichment bioassays with picoplankton Schallenberg and Burns (2001) concluded that picocynaobacteria in oligotrophic lakes are sensitive to extremely small changes in nutrient availability and therefore highly useful as early indicators of nutrient enrichment.

3.1.5. Assessment of micro-'Aufwuchs' biocoenoses

Not standardised
Organisms: Aufwuchs-biocoenoses in running waters
The adaptation of the Aufwuchs organisms to ecological conditions are assessed through microscopical analysis, dry weight estimation, chlorophyll analysis or measurement of the oxygen production potential (see below) after an incubation period of 3–4 weeks.

Sensitive, ecologically relevant and integrative assessment at moderate costs. Large time lags and interpretation problems when specific parameters must be monitored.

Techniques for *in situ* algal assays with periphyton using bottles or artificial substrates are discussed by Sládečková (1979).

3.2. Laboratory tests and bioassays

In several countries tests have been developed to determine the algal growth potential, the AGP, of a water. Basically the experiment determines the maximum concentration

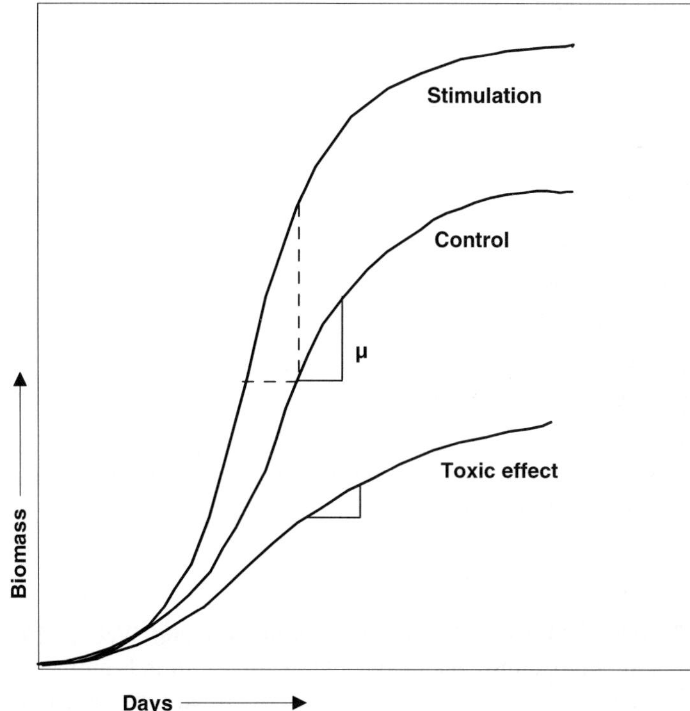

Figure 5. Schematic growth curve of algae in batch culture compared to stimulation and toxic effects of a test substance. μ = specific growth rate coefficient.

of algae that can grow in a water sample under standardized conditions. These tests are often used to judge:

- the degree of eutrophication of surface water (Thomas, 1953; Skulberg, 1964)
- the eutrophication potential of the effluent of sewage treatment plants (Forsberg, 1972)
- the possible effects of environmental measures on the degree of eutrophication of water systems (van der Does and Klapwijk, 1987).

Growth curves obtained by such tests or bioassays schematically are displayed in Figure 5 together with schematic effect-curves of a stimulating and a toxic substance. In some cases, these growth curves show a second exponential phase after a retardation phase. depending on the culture medium used (Bolier and Donze, 1989).

Growth tests are performed in batch cultures under defined nutrient, CO_2, pH and light conditions. Biomass development over time is estimated from microscopic or electronic cell counts, chlorophyll-a concentration, ATP-, DNA-content or similar parameters. Turbidity measurements may overestimate biomass when bacterial contamination is high. Test results are analysed from algal growth curves against an untreated control (Gunkel, 1994). Laboratory tests used in eco-toxicology are summarised in Steinberg et al. (1995).

3.2.1. Algal growth inhibition test

Standard test: OECD Guideline for testing of chemicals 201, 7.1984
Organisms: unicellular green algae (*Ankistrodesmus bibraianum, Scenedesmus subspicatus* or *Chlorella vulgaris*)
Dilution series between NOEC and concentration >LC 50. Growth is followed by cell counting for 72 hours.
End point: EC 50 and NOEC

3.2.2. Inhibition of green-alga by water contaminants (Scenedesmus growth-inhibition test)

Standard test: DIN 38 412, L9, 1989 (ISO/DIS 6862:06.87).
Organism: *Scenedesmus subspicatus* CHODAT, a unicellular green alga
Dilution series of the substance or water to be tested run in 100 ml Erlenmayr-flasks with culture media at 23°C, 8 000 Lux continuous light for 3 days.
Biomass must at least be estimated after 24, 48 and 72 hours.
End point: EC 10 and EC 50 after 72 hours

3.2.2. Measurement of non-toxic effects of water contaminants on green algae (Scenedesmus-chlorophyll-fluorescence test) in dilutions

Standard test: DIN 38 412, L33, 1991
Organism: *Scenedesmus subspicatus* CHODAT, a unicellular green alga
Dilution series are incubated as above. Fluorescence is measured at the end at 685 nm from all dilutions relative to the untreated control. Toxic effects are present if fluorescence is inhibited by 20%.

All the above mentioned tests can largely be automated.

Typical results of algal growth potential (AGP) tests are shown in Figure 6 from an intensive study in the Netherlands (Klapwijk et al., 1989) and from a deep alpine lake (Dokulil, unpublished). Both observations indicate that phosphorus was the substance most likely limiting growth. Seasonal varying nutrient limitation was found in English lakes using laboratory bioassays with *Asterionella formosa* and *Rhodomonas lacustris* as test organisms (Barbosa, 1989). Phosphate was the major element limiting both species throughout the year, except during spring diatom development when dissolved silica became limiting. Chelated iron increased growth, particularly in combination with phosphate. Comparison of AGP-bioassay and phosphate uptake kinetics with natural phytoplankton, however gave somewhat inconclusive results as reported by Van Donk et al. (1989).

3.2.3. Bioassays with macroalgae

Macroalgae may also be used as test-organisms. Inhibition of trichom movement in the blue-green *Phormidium autumnale* (cyanobacteria) is used to screen for toxic substances (Noll and Bauer, 1973; Breitig and Tümpling, 1982). Håglund et al. (1990) use the red-alga *Gracilaria tenuistipitata* (Rodophyta) to test marine and brackish waters for toxic pollutants. Effects of tributyltin (TBT) on community metabolism

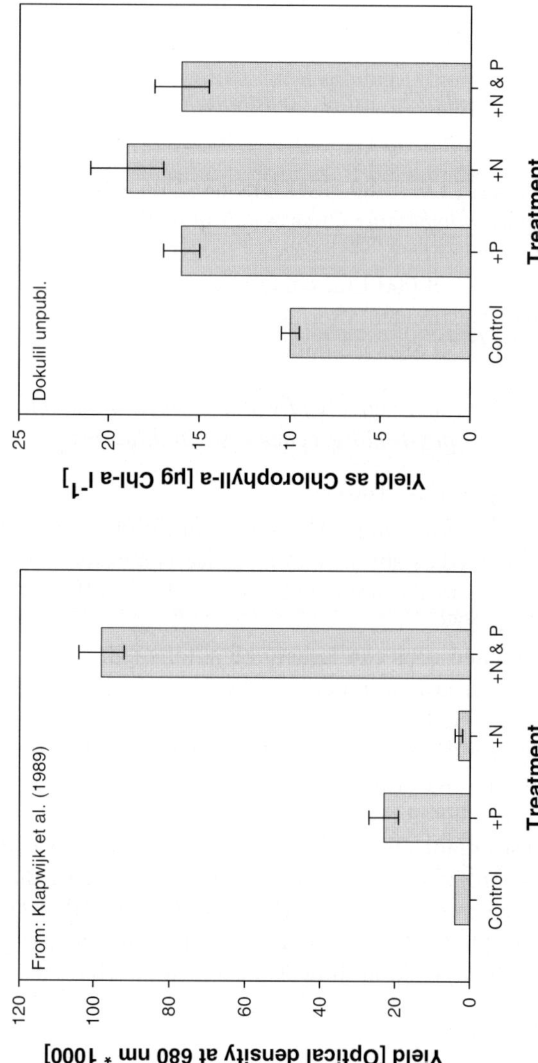

Figure 6. Two typical examples of algal growth potential (AGP) bioassay experiments with nitrogen and phosphorus enrichment. The left hand panel, modified from Klapwijk et al. (1989), shows yields (± st. dev.) as optical density for laboratory batch experiments with water from the Reeuwijk lakes in The Netherlands using the alga *Scenedesmus quadricauda* as test organism. In the right hand panel. yields are expressed as chlorophyll-a for an in situ AGP bioassay using natural phytoplankton from Mondsee, Austria (Dokulil, 1989). Both experiments were presumably limited by phosphorus.

dominated by *Fucus vesiculosus* were measured with a portable continuous flow-through system in a study by Lindblad et al. (1989).

3.2.4. Bioprobes

Bio-electrodes are currently developed which use the electron flow produced by photosynthesis or respiration of the cyanobacterium *Synechococcus*, embedded in alginate and fixed to the tip of the electrode, as a measurement signal (Steinberg et al., 1992).

3.2.5. Ataxonomic bioindication

Because size distribution of pelagic organisms is continuous in undisturbed systems (Gaedke, 1992), biomass spectra may be used to evaluate environmental disturbances. Optimal size spectra can most easily by obtained by modified flow-cytometry (Steinberg et al., 1999).

3.2.6. Oxygen-production potential – light-dark bottle laboratory test (DIN 38 412, L14, 1983b)

This test estimates the production potential as well as the respiration of freshwaters. Both rates are essential for the overall oxygen budget.

Six oxygen bottles are filled with the test-water. Two bottles are used to estimate the initial oxygen concentration, two are incubated under constant light in the laboratory at 20°C for 24 hours, and the remaining two are completely darkened. At the end of the incubation period the oxygen levels in the bottles are measured. Results are expressed as:

$$O_2 \text{ in light}_{24h} - O_2 \text{ in dark}_{24h} = \text{Gross-potential production}$$
$$\text{Initial } O_2 - O_2 \text{ in dark}_{24h} = \text{Respiration-potential}$$

Cheap, sensitive and easy assessment of potential multi-species production and destruction rates. Rarely used for effect monitoring of contaminants.

3.3. Sediments

Many pollutants are associated with sediments in aquatic systems. Hazard assessment to establish sediment quality criteria require rapid, inexpensive screening test. Direct bioassay with algae have proven to be very sensitive indicators of contaminant stress.

Numerous methods are available for the assessment of environmental impacts of sediment-associated contaminants Ahlf and Munawar (1988). Effects of sediment elutriates on algae are measured as the amount of inhibition in photosynthetic ^{14}C assimilation of e.g. *Selenastrum capricornutum* under laboratory conditions (Ross et al., 1988). Approaches using natural phytoplankton were developed by Munawar and Munawar (1987). Effects of increasing concentrations of sediment elutriates on photosynthetic rates of natural phytoplankton under controlled laboratory conditions are tested in the short-term "algal fractionation bioassay" (AFB). Carbon-uptake of size fractions (>20 μm – netplankton ; 5–20 μm – nanoplankton ; 1–5 μm – ultraplankton ; < 1 μm – picoplankton) is estimated against an untreated control. Similarly, long-term effects of sediment elutriates can be evaluated from 4 day bioassays in 5-litre bottles using natural phytoplankton or a mixed culture of *Ankistrodesmus braunii* (Naeg.) Brunnthaler and *Chlorella vulgaris* Beyerinck. This test may be expanded to include the solid phase of the sediment allowing differential bioassay of the effects of both solid- and liquid-phase of sediment contaminants. In this case, the sediment compartment may be separated from the water/organism part by a membrane allowing exchange of substances but prevent mixing (Ahlf, 1985). The use of sediments directly in bioassays with algae is recommended over elutriates because a large number of toxic chemicals can not be extracted with water (Ongley et al., 1988).

All these bioassay techniques integrate the response of test organisms to contaminants and nutrients. They often give best results when combined with other assessment methods (Ahlf et al., 1989; Gregor and Munawar, 1989).

4. Ecotoxicology

In ecotoxicology biomonitoring is the accumulation of contaminants in cells or tissues of organisms without severe damage or even death. The contaminant and its quantity can only be evaluated after chemical analysis (*exposure-monitoring*). Effect-monitoring estimates the quality and quantity of a contaminant through analysis of the population structure (bioindication) because populations or assemblages change characteristically when impacted by polluting substances.

A contaminating substance (xenobiotica) must be biologically available to be of environmental relevance and hence be taken up by organisms in one way or the other. The ability of many plants and animals to accumulate exotic substances makes them idle for biomonitoring.

Criteria for effective biomonitors for organic contaminants include the following:

1. The organisms must accumulate the xenobiotic substances without being affected by environmental relevant concentrations
2. The organisms should preferably be sessile to be representative for the investigated area.
3. The organisms should either live everywhere in the area investigated or be tolerant to exposure in chambers, cages, etc.
4. The organisms shall be long lived to act as integrators of contaminations.
5. The organisms shall be of such a size that enough tissue for chemical analysis is available.
6. Collection and handling of the organisms should be easy.
7. A simple correlation should exist between the mean concentration of the contaminant in the environment and the content in the organism.
8. All individuals of a species used in biomonitoring must, under all circumstances, have the same relation to the concentration of the contaminant.

4.1. Measurement techniques

Acute toxicity is usually estimated from 72 hour growth tests using the green-algae *Scenedesmus subspicatus*. The no observed effect concentration (NOEC) is defined as the concentration at which less than 20% of the organisms ($<EC_{20}$) are affected (Steinberg et al., 1999). A similar growth test is used to estimate the quality of sediments.

Alternatively, acute aquatic toxicity can be assessed by a practical and cost-efficient micro-bioassay using microplates with *Selenastrum capricornutum* as test organism following the standardized protocol developed by Blaise (1986) and Blaise et al. (1986).

Long-term sublethal toxicity is much more difficult to assess. Rhee (1989) used a two-stage continuous algal culture bioassay to investigate steady-state responses of a diatom (*Fragilaria crotonensis*), a green-alga (*Ankistrodesmus falcatus*) and a cyano-bacterium (*Microcystis sp.*) to organic pollutants (PCBs). Results showed a variety of sublethal effects such as enhancement of growth, photosynthesis and P-uptake as well as their inhibition, growth rebound and development of resistance.

4.2. Effects of inorganic nitrogen substances

A summary of acute and chronicle toxicity effects of ammonium and nitrite on algae, invertebrates and fish is provided by Schwoerbel et al. (1991). Phytoplankton species such as *Chlorella vulgaris* show acute toxic effects at concentrations (LC-50/5d) of 8.55 mg l^{-1} NH_3. Chronicle effects on CO_2-uptake of five species of green and blue-green algae were only observed at ammonia concentrations much higher than those commonly observed in running waters.

4.3. Uptake of organic contaminants by algae

Uptake and accumulation of a contaminant by algae follow a saturation curve where saturation must be seen as an equilibrium between adsorption and de-sorption. Mathematically these behaviours can be described by Langmuir's isotherms, the Michaelis-Menten equation or as equilibrium distribution according to Nernst.

Uptake of contaminants in algae is mediated by a variety of cellular processes which results in non-constant bio-concentration factors (BCF's). Quantitative correlation of accumulation in algae with storage products other than lipids or oils deviate from animals in the slope of the regression equation.

An example of the variability of the BCF's is shown in Table 11 for the bio-accumulation of atrazin by the green coccoid algae *Scenedesmus acutus*.

The sorption capacity increases with increasing concentrations in the surrounding medium according to

$$BCF_F = Sk/c_w \tag{1}$$

where BCF_F is the bio-concentration factor related to fresh-weight, Sk is the sorption capacity and c_w is the concentration of the xenobiotic substance in the medium.

The uptake of herbicides by algae has two steps: a *protein-specific binding* which follows a saturation function and a *unspecific binding* where distribution-equilibria with the lipid phase are important.

Steinberg et al. (1992) summarise BCFs by the green algae *Chlorella fusca* for more than 100 selected organic chemicals. These BCFs span several orders of magnitude

Table 11. Dependence of the bio-concentration factors (BCF) of *Scenedesmus acutus* from external concentrations of Atrazin (after Steinberg et al., 1992).

Atrazin concentrations [mg l^{-1}]	BCF-value [volume related]	Sorption coefficient (Sk) [mg kg^{-1} dry weight]
0.0012	51	0.36
0.012	27	1.97
0.100	10	6.80
1.100	6	44.20

from 10 for 2,6-dichlorbenzamid to 28,000 for methanol which is incorporated and metabolised like acetate or urea. Condensed aromatic substances have intermediate BCFs.

Eco-toxicological effects may be estimated from physico-chemical attributes of the (untested) substance by the Quantitative Structure Activity Relationship (QSAR). If the concentrating phase in the organism are lipids then the BCFs should have a simple relationship to the octanol/water partition coefficient K_{OW}. For green algae, such as *Chlorella fusca* or *Ankistrodesmus bibraianum,* these correlations deviate significantly from those observed for various animals (cit acc. to Steinberg et al., 1992). Uptake by algae, picoplankton and bacteria is a two-step process involving lipid independent adsorption to surfaces and later incorporation into the lipid phase (see Falkner and Simonis, 1982 and Steinberg et al., 1992 for further discussion).

Bioaccumulation and -magnification within the food chain depends on the partition coefficient K_{OW} of the contaminant. At values of log K_{OW} less than 5 accumulation is not important. Substances with values between 5 and 7 are strongly magnified within the food chain. At log K_{OW} > 7, effects will largely depend on assimilation and accumulation by phytoplankton (Thomann, 1989).

4.4. Heavy metals

Several algal species accumulate considerable amounts of metals and can thus be used as monitors for elements such as cadmium, copper or lead (Hellawell, 1986; Whitton, 1984). Both field and laboratory populations have been used with success. A detailed description of toxicity effects of various metals can be found in Moore and Ramamoorthy (1984). Most metals are slightly to highly toxic to algae, arsenic, copper, mercury and zinc having the greatest effect. Impacts on algae in natural waters is highly variable. Cyanobacterial strains reacted more sensitive to heavy metals in a comparative laboratory growth inhibition test than green algae (Kusel-Fetzmann et al., 1989). Because of its widespread occurrence, the filamentous green alga *Cladophora* has been assessed more than any other species except perhaps for *Chlorella* in the laboratory. Both species concentrate various metals proportionally to ambient concentrations. An example of pH-dependent Zn uptake by *Cladophora glomerata* (L.) Kütz.

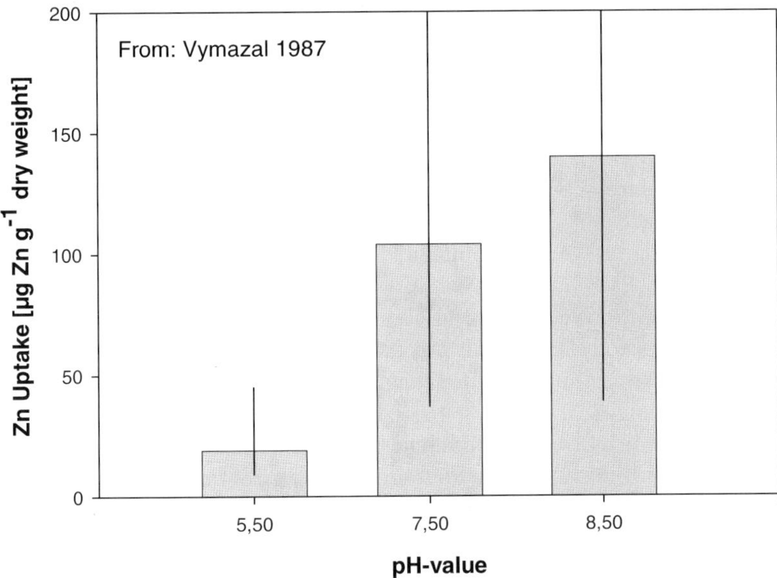

Figure 7. Relationship between Zn uptake in *Cladophora glomerata* and pH. Bars indicate maximum and minimum values (modified from Vymazal, 1987).

is given in Figure 7 (Vymazal, 1987). In contrast, enrichment ratios of the red-alga *Lemanea fluviatilis* decreased with increasing aqueous concentration (Fig. 8, Whitton, 1984).

4.5. Polychlorinated biphenyls (PCBs)

Polychlorinated biphenyls (PCBs) are usually mixtures of isomeres marketed under a variety of names. They are non-ionic, non-flammable compounds with extremely low water solubility but are highly lipophilic, and hence of significance to biota.

Algae as indicators of PCB-pollution are advantageous because they represent organism at the basis of the food chain. Marine phytoplankton for instance has an enormous capacity for accumulating organohalogen compounds such as polychlorinated biphenyls (Ramade, 1987). Concentrations of PCB in the range of 11 to 111 µg l⁻¹ were reported to inhibit growth and photosynthesis in green algal species (see Steinberg et al., 1992 or Hellawell, 1986).

4.6. Pesticides

Several of the many different groups of pesticides can not be biomonitored with algae mainly because of their low bioaccumulation (e.g. urea-based pesticides, comp. Steinberg et al., 1992, p. 178 ff). Although nitrogen-based herbicides such as e.g. atrazin are strongly accumulated by phytoplankton and the coccoid green-algal species *Scenedesmus acutus* and *Chlorella fusca,* biomonitoring is not possible because of the

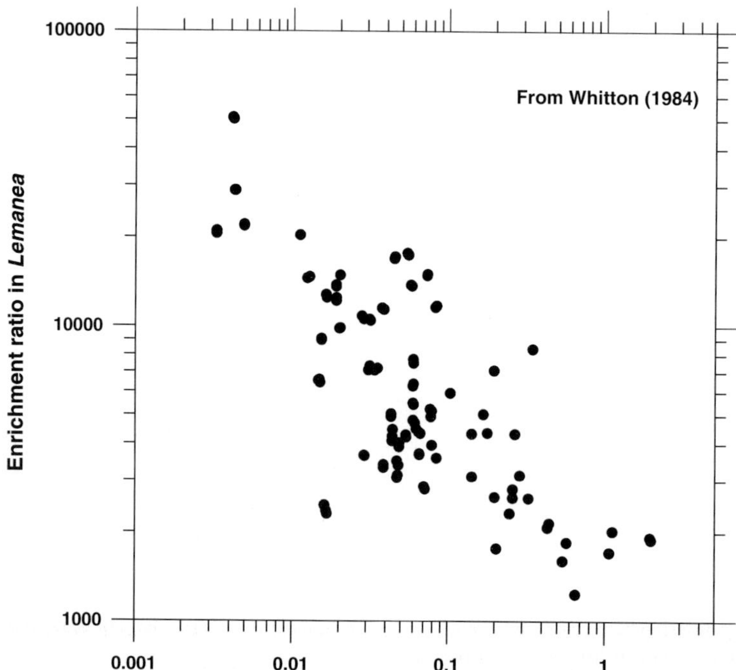

Figure 8. Relationship between enrichment ratio for zinc in *Lemanea fluviatilis* and total zinc concentration in water of stream and rivers ($r = -0.84$), modified from Whitton (1984)

high variability (Kusel-Fetzmann et al., 1989) and discrepancy of measured and calculated BCF-values. Side-effects of atrazin on aquatic ecosystems are reported, however for single species, communities and food chains (Lampert et al., 1989). In running water experiments, the composition and quantity of periphytic algae, especially *Rhopalodia, Phormidium* and *Cladophora* are affected at atrazin-levels of 1 mg l^{-1}. Adverse effects on diatoms become visible already at concentrations of 0.01 mg l^{-1}. Pre-incubation with the herbizide did not result in adaptation (Kosinski, 1984; Kosinski and Merkle, 1984). Tabulated data on accumulation and toxicity of selected pesticides by algal species are summarized in Steinberg et al. (1992).

The specific diversity of phytoplankton and biomass estimations via chlorophyll-a in ponds were used by Goacolou and Echaubard (1987) to evaluate in situ pesticide contamination. The biocoenotic structure, species richness and chlorophyll levels were significantly altered in ponds affected by pesticides.

4.7. Tensides

Toxicity from various tenside classes vary by four orders of magnitude within a single algal species. In general, however, kationic tensides are far more effective to algae than anionic or non-ionic tensides. The sensitivity of different algae to a single tenside varies by three orders of magnitude, depending on the species used, their physiology

Table 12. Mean concentration factors for various radioisotopes in marine algae (from Ramade, 1987).

Radionuclides	Conc. factor	Radionuclides	Conc. factor
^{3}H	0.9	^{89,90}Sr	50
^{7}Be	250	^{90,91}Y	500
^{14}C	4000	^{95}Zr, ^{95}Nb	1500
^{24}Na	1	^{103,105}Ru, ^{106}Rh	400
^{32}P	10^4	^{131}I	5000
^{45}Ca	2	^{133}Xe	1
^{45}Sc	1200	^{137}Cs	15
^{51}Cr	2000	^{140}Ba, ^{140}La	25
^{54,56}Mn	3000	^{141,144}Ce	700
^{55,59}Fe	2×10^4	^{183,187}W	5
^{57,58,60}Co	500	^{203,210}Pb	700
^{65}Zn	10^3	^{210}Po	1000
^{85}Kr	1	^{226}Ra	1000
		^{239}Pu	1300

and the overall test conditions. The cyanobacterium *Microcystis aeruginosa* for instance, is 10-times more sensible than the green *Ankistrodesmus bibraianum*. Toxicity values for tensids are tabulated by Lewis (1990).

4.8. Radioisotopes

According to the summary by Ramade (1987) the average concentration factor of radioisotopes by algae varies considerably depending on the isotope (Table 12).

References

Abel, P.D., 1989. Water Pollution Biology. John Wiley, New York.

Ahlf, W., 1985. Behaviour of sediment-bound heavy metals in a bioassay with algae: bioaccumulation and toxicity. Vom Wasser 65, 183–188.

Ahlf, W., Calmano, W., Erhard, J., Förstner, U., 1989. Comparison of five bioassay techniques for assessing sediment-bound contaminants. Hydrobiologia 188/189, 285–289.

Ahlf, W., Munawar, M., 1988. Biological assessment of environmental impact of dredged material. In: Salomons, W., Förstner, U. (Eds), Chemistry and Biology of Solid Waste. Springer Verlag, Heidelberg, pp. 127–142.

Backhaus, D., 1973. Fliesswasseralgen und ihre Verwendbarkeit als Bioindikatoren. Verhandlungen der Gesellschaft für Ökologie, Saarbrücken 1973, 149–168.

Barbosa, F.A.R., 1989. Evidence from algal bioassays of seasonal nutrient limitations in two English lakes. Hydrobiologia 188/189, 211–228.

Barnes, W.S., 1980. Assays for dispersed mutagens in marine environments using extracts of bioconcentrators. Considerations, problems, and applications. Ph.D. Dissertation, Botany Department, University of Massachusetts, Amherst.

Bazerque, M.F., Laville, H., Brouquet, Y., 1989. Biological quality assessment in two rivers of the northern plain of France (Picardie) with special reference to chironomid and diatom indices. Acta Biologicae Debrecen, Oecologia Hungarica 3, 29–39.

Behrendt, H., Opitz, D., 1996. Ableitung einer Klassifizierung für die Gewässergüte von plankton-dominierten Fließgewässern und Flussseen im Berliner Raum. Berichte des Instituts für Gewässer-ökologie und Binnenfischerei (IGB) 1, 1–26.

Bellamy, D.J., Witteck, A., John, D.M., Jones, D.J., 1973. A method for the determination of seaweed production. In: A Guide to the Measurement of Marine Primary Productivity under Some Special Conditions. UNESCO, Paris, pp. 27–33.

Bellinger, E.G., 1979. The response of algal populations to changes in lake water quality. In: James, A. and Evison, L. (Eds), Biological Indicators of Water Quality. John Wiley, Chichester, pp. 9-1–9-25.

Biggs, B.J.F., 2000. Eutrophication of streams and rivers: dissolved nutrient-chlorophyll relationships for benthic algae. Journal of the North-American Benthological Society 19, 17–31.

Blaise, C., 1986. Micromethod for acutee aquatic toxicity assessment using the green alga *Selenastrum capricornutum*. Toxicity Assessment 1, 377–385.

Blaise, C., Legault, R., Bermingham, N., van Coillie, R., Vasseur, P., 1986. A simple micoplate algal assay technique for aquatic toxicity assessment. Toxicity Assessment 1, 261–281.

Bolier, G., Donze, M., 1989. On the accuracy and interpretation of growth curves of planktonic algae. Hydrobiologia 188/189, 175–179.

Breitig, G., Tümpling, W. v., 1982. Ausgewählte Methoden der Wasseruntersuchung. Bd. II. Biologische, mikrobiologische und toxikologische Methoden. 2. Aufl., G. Fischer Verlag, Jena.

Brettum, P., 1989. Alger som indikatorer på Vannkvalitet i norske innsjøer Planteplankton. NIVA, Blindern, Oslo.

Cairns, J., Albaugh, D.W., Busey, F., Chaney, M.D., 1968. The sequential comparison index. A simplified method to estimate relative differences in biological diversity in stream pollution studies. Journal of the Water Pollution Control Federation 40, 1607–1613.

Cemagref, 1982. Etude des méthodes biologiques d'appreciation quantitative de la qualité des eaux. Rapport Q.E.Lyon-A.F- Bassin Rhône-Méditerraneé-Corse.

Charles, D.F., Smol, J.P., Engstrom, D.R., 1994. Paleolimnological approaches to biological monitoring. In: Loeb, S.L., Spacie, A. (Eds), Biological Monitoring of Aquatic Systems. Lewis Publishers, Boca Raton, pp. 233–293.

Cohn, F., 1853. Über lebende Organismen im Trinkwasser. Z. klein. Medizin 4, 229–237.

Cohn, F., 1870. Über den Brunnenfaden (*Crenothrix polyspora*) mit Bemerkungen über die mikroskopische Analyse des Brunnenwassers. Cohn's Beiträge zur Biologie der Pflanzen 3, 1–108.

Coring, E., Schneider, S., Hamm, A., Hofmann, G., 1999. Durchgehendes Trophiesystem auf der Grundlage der Trophieindikation mit Kieselalgen. DVWK Materialien 6, 1–219.

Costa, J.E., Howes, B.L., Giblin, A.E., Valiela, I., 1992. Monitoring nitrogen and indicators of nitrogen loading to support management action in Buzzards bay. In: McKenzie, D.H., Hyatt, D.E., McDonald, V.J. (Eds), Ecological Indicators. Elsevier Applied Science, London, pp. 499–531.

Dell' Uomo, A., 1996. Assessment of water quality of an apennine river as a pilot study for diatom-based monitoring of Italian watercourses. In: Whitton, B.A., Rott, E., (Eds), Use of Algae for Monitoring Rivers II. Proceedings of an International Symposium, University of Innsbruck, Austria, pp. 65–71.

DeNoyelles, F.Jr., O'Brien, W.J., 1974. The in situ chemostat – a self contained continuous culturing and water sampling system. Limnology Oceanography 19, 326–331.

DePauw, N., Ghetti, P.F., Manzini, D.P., Spaggiari, D.P., 1992. Biological assessment methods for running water. In: Newman, P.J., Piavaux, M.A., Sweeting, R.A. (Eds), River Water Quality. Ecological Assessment and Control. European Communities Commission, Luxembourg, pp. 217–248.

Descy, J.-P., 1979. A new approach to water quality estimation using diatoms. Nova Hedwigia, Beiheft 64, 305–323.

Descy, J.-P., Coste, M., 1991. A test of methods for assessing water quality based on diatoms. Verhandlungen der Internationalen Vereinigung für Limnologie 24, 2112–2116.

DIN (Deutsche Einheitsverfahren zur Wasser-, Abwasser- und Schlammuntersuchung), L13, 1983a. Bestimmung von Sauerstoffproduktion und Sauerstoffverbrauch im Gewässer mit der Hell-Dunkel-flaschen-Methode SPG und SPV (Biogene Belüftungsrate). VCH, Weinheim.

DIN (Deutsche Einheitsverfahren zur Wasser-, Abwasser- und Schlammuntersuchung), L14, 1983b. Bestimmung der Sauerstoffproduktion mit der Hell-Dunkelflaschen-Methode unter Laborbedingungen SPL (Sauerstoffproduktionspotential). VCH, Weinheim.

DIN (Deutsche Einheitsverfahren zur Wasser-, Abwasser- und Schlammuntersuchung), L9, 1989. Bestimmung der Hemmwirkung von Wasserinhaltsstoffen auf Grünalgen (*Scenedesmus*-Zellvermehrungs-Hemmtest). VCH, Weinheim.

DIN (Deutsche Einheitsverfahren zur Wasser-, Abwasser- und Schlammuntersuchung), L33, 1991. Bestimmung der nicht-giftigen Wirkung von Abwasser gegenüber Grünalgen (*Scenedesmus*-Chlorophyll-Fluoreszenztest) über Verdünnungsstufen. VCH, Weinheim.

Dixit, S.S., Cumming, B.F., Smol, J.P., Kingston, J.C., 1992. Monitoring environmental changes in lakes using algal microfossils. In: McKenzie, D.H., Hyatt, D.E., McDonald, V.J. (Eds), Ecological Indicators. Elsevier Applied Science, London, pp. 1135–1155.

Dokulil, M.T., 1989. Bioassay experiments with natural phytoplankton. Unpublished draft manuscript.

Dokulil, M.T., 1991. Review on recent activities, measurements and techniques concerning phytoplankton algae of large rivers in Austria. In: Whitton, B.A., Rott, E., Friedrich, G. (Eds), Use of Algae for Monitoring Rivers. Institut für Botanik, University of Innsbruck, pp. 53–57.

Dokulil, M.T., 1996. Evaluation of eutrophication potential in rivers: the Danube example, a review. In: Whitton, B.A., Rott, E. (Eds), Use of Algae for Monitoring Rivers II. Institut für Botanik, University of Innsbruck, pp. 173–178.

Dokulil, M.T., Frisk, T., 1993. Long-term trophic dynamics and predictive model of a flood-water control system of the river Danube in Vienna, Austria. International Symposium Transboundary River Basin Management and Sustainable Development, RBA Centre, Delft University of Technology, 18–22 May 1992. Proceedings, Vol. II, UNESCO, Paris, pp. 191–195.

Dokulil, M.T., Schmidt, R., Kofler, S., 1997. Benthid diatom assemblages as indicators of water quality in an urban flood-water impoundment, Neue Donau, Vienna, Austria. Nova Hedwigia 65, 273–283.

Dresscher, T.G.N., Van der Mark, H., 1976. A simplified method for the biological assessment of the quality of fresh and slightly brackish water. Hydrobiologia 48, 199–201.

EC Water Framework Directive, 2000. *Official Journal* (*OJL* 327), 22 December.

Falkner, R., Simonis, W., 1982. Polychlorierte Biphenyle (PCB) im Lebensraum Wasser (Aufnahme und Anreicherung durch Organismen – Probleme der Weitergabe in der Nahrungspyramide). Archiv für Hydrobiologie, Ergebnisse der Limnologie, Beiheft 17, 74S.

Felföldy, L., 1987. A biológiai vízminosítés (Biological water quality evaluation). Vízügyi Hidrobiológia 16. VGI, Budapest.

Forsberg, C., 1972. Algal assay procedure. Journal Water Pollution Control Federation 44, 1623–1628.

Friedrich, G., 1990. Eine Revision des Saprobiensystems. Zeitschrift für Wasser- und Abwasser-Forschung 23, 141–152.

Gaedke, U., 1992. The size distribution of plankton biomass in a large lake and its seasonal variability. Limnology Oceanography 37, 1202–1220.

Gerhardt, V., Putzger, J., 1992. Ein Biotest zur Gewässerüberwachung auf der Grundlage der verzögerten Fluoreszenz von Algen. Schriftenreihe des Vereins Wasser-Boden-Luft 89, 277–292.

Ghetti, P.F., Ravera, O., 1994. European perspective on biological monitoring. In: Loeb, S.L., Spacie, A., (Eds), Biological Monitoring of Aquatic Systems. Lewis Publishers, Boca Raton, pp. 31–46.

Goacolou, J., Echaubard, M., 1987. Influence des traitements phytosanitaires sur les biocoenoses du phytoplancon limnique. Hydrobiologia 148, 269–280.

Gonor, J.J., Kemp, P.F., 1978. Procedures for quantitative ecological assessment in intertidal environments. EPA-600/3-78-087.

Gregor, D.J., Munawar, M., 1989. Assessing toxicity of Lake Diefenbaker (Saskatchewan, Canada) sediments using algal and nematode bioassays. Hydrobiologia 188/189, 291–300.

Gunkel, G., 1994. Bioindikation in aquatischen Ökosystemen. G. Fischer Verlag, Jena, Stuttgart.

Håglund, K., Törnquist, L., Pedersén, M., 1990. Toxicity tests of marine and brackish pollutants by using the macroalga *Gracilaria tenuistipitata* (Rhodophyta). Proceedings of the International Symposium on Ecotoxicology, GSF-Bericht 1/92, 349–355.

Hamm, A., 1996. Möglichkeiten und Probleme einer durchgehenden Trophiebewertung. Deutsche Gesellschaft für Limnologie (DGL) – Tagungsberichte 1995 (Berlin), 11–15.

Heinonen, P., 1980. Quantity and composition of phytoplankton in Finish inland waters. Publications of the Water Research Institute 37, Vesihallitus-National Board of Waters, Finland.

Heinonen, P., Herve, S., 1987. Water quality clasification of inland waters in Finland. Aqua Fennica 17, 147–156.

Hellawell, J.M., 1986. Biological indicators of freshwater pollution and environmental management. Elsevier Applied Science, London.

Hill, B.H., Herlihy, A.T., Kaufmann, P.R., Stevenson, R.J., McCormick, F.H., Burch Johnson C., 2000. Use of periphyton assemblage data as an index of biotic integrity. Journal of the North-American Benthological Society 19, 50–67.

Hinckley, D., Tierney, G., 1992. Early ecological indicators of climate change. In: McKenzie, D.H., Hyatt, D.E., McDonald, V.J. (Eds), Ecological Indicators. Elsevier Applied Science, London, pp. 1157–1164.

Hofmann, G., 1993. Aufwuchs – Diatomeen in Seen und ihre Eignung als Indikatoren der Trophie. Dissertation. J.W. Goethe-Universität, Frankfurt am Main.

Hofmann, G., 1999. Trophiebwertung von Seen anhand von Aufwuchsdiatomeen. In: Von Tümpling, W., Friedrich, G. (Eds), Biologische Gewässeruntersuchung, Methoden der biologischen Wasseruntersuchung. Gustav Fischer Verlag, Stuttgart, 2, 319–333.

Hörnström, E., 1981. Trophic characterisation of lakes by means of qualitative phytoplankton analysis. Limnologica 13, 249–361.

Jarlman, A., Lindstrøm, E.-A., Eloranta, P., Bengtsson, R., 1996. Nordic standard for assessment of environmental quality in running water. In: Whitton, B.A., Rott, E., Friedrich, G. (Eds), Use of Algae for Monitoring Rivers. Institut für Botanik, University of Innsbruck, pp. 17–28.

Järnefelt, H., 1952. Plankton als Indiktor der Trophiegruppen der Seen. Annales Academia Scientiarum Fennica A IV, Biol. 18, 1–29.

John, D.M., Johnson, L.R., 1991. Green microphytic algae as river water quality monitors, In: Whitton, B.A., Rott, E., Friedrich, G. (Eds), Use of Algae for Monitoring Rivers. Institut für Botanik, University of Innsbruck, pp. 41–47.

Jørgensen, S.E., 1992. Ecological indicators and ecological modelling. In: McKenzie, D.H., Hyatt, D.E., McDonald, V.J. (Eds), Ecological Indicators. Elsevier Applied Science, London, pp. 201–209.

Kamenik, C., Schmidt, R., 2001. The impact of temperature on recent and fossil chrysophyte stomatocyst composition in an high alpine lake (Gossenköllesee, Tyrol). Nova Hedwigia, Beihefte, 122, 1–22.

Kann, E., 1978. Typification of Austrian streams concerning algae. Verhandlungen der internationalen Vereinigung für Limnologie 20, 1523–1526.

Kann, E., 1986. Können benthische Algen zur Wassergütebestimmung herangezogen werden. Archiv für Hydrobiologie 73, 405–423.

Kelly, M.G., 1996. The trophic diatom index. Bowburn Consultancy, R&D Technical Report E2, 1–148.

Kelly, M.G., Hornberger, G.M., Cosby, B.J., 1976. Automated measurement of river productivity for eutrophication prediction. In: Cairns, J., Dickson, K.L. Westlake, G.F. (Eds), Biological Monitoring of Water and Effluent Quality. American Society for Testing and Materials, pp. 133–146.

Kelly, M.G., Whitton, B.A., 1995. The trophic diatom index: a new index for monitoring eutrophication in rivers. Journal of Applied Phycology 7, 433–444.

Klapwijk, S.P., Bolier, G., Van der Does, J., 1989. The application of algal growth potential tests (AGP) to the canals and lakes of western Netherlands. Hydrobiologia 188/189, 189–199.

Knöpp, H., 1954. Ein neuer Weg zur Darstellung biologischer Vorfluteruntersuchungen, erläutert an einem Gütelängsschnitt des Mains. Wasserwirtschaft, 45, 9–15.

Kobayasi, H., Mayama, S., 1989. Evaluation of river water quality by diatoms. Korean Journal of Phycology 4 (2), 121–133.

Kolkwitz, R., 1950. Ökologie der Saprobien. Schriftenreihe des Vereins für Wasser-, Boden- und Lufthygiene 4, 1–64.

Kolkwitz, R., Marsson, M., 1902. Grundsätze für die biologische Beurteilung des Wassers nach seiner Flora und Fauna, Mitteilungen der Prüfungsanstalt für Wasserversorgung und Abwasserreinigung 1, 33–72.

Kolkwitz, R., Marsson, M., 1908. Ökologie der pflanzlichen Saprobien. Berichte der Deutschen Botanischen Gesellschaft 26A, 505–519.

Kolkwitz, R., Marsson, M., 1909. Ökologie der tierischen Saprobien. Internationale Revue der gesamten Hydrobiologie 2, 126–152.

Kosinski, R.J., 1984. The effect of terrestrial herbicides on the community structure of stream periphyton. Environmental Pollution A36, 165–189.

Kosinski, R.J., Merkle, M.G., 1984. The effect of four terrestrial herbicides on the productivity of artificial stream algal communities. Journal of Environmental Quality 13, 75–82.

Kümmerlin, R., 1980. Plnakton-Gemeinschaften als bioindikatoren für Stehgewässer. Ökologie & Naturschutz 3, 227–241.

Kusel-Fetzmann, E., Latif, M., Zach, B., 1989. Vergleichende Toxizitätsbestimmungen ausgewählter Schadstoffe mittels Algen als Indikatororganismen. Wasserwirtschaft-Wasservorsorge. Forschungsarbeiten. Bundesministerium Land- und Forstwirtschaft, Wien.

Lampert, W., Fleckner, W, Pott, E., Schober, U., Störkel, K.-U., 1989. Herbicide effects on planktonic systems of different complexity. Hydrobiologia 188/189, 415–424.

Lange-Bertalot, H., 1978. Diatomeendifferentialarten anstelle von Leitformen: ein geeigneteres Kriterium der Gewässerbelastung. Archiv für Hydrobiologie, Supplement 51, 393–427.

Lange-Bertalot, H., 1979. Pollution tolerance of diatoms as a criterion for water quality estimation. Nova Hedwigia, Beiheft 64, 285–304.

Leclercq, L., Maquet, B., 1987. Deux nouveaux indices chimique et diatomique de qualité d'eau courante. Application au Samson et à ses affluents (bassin de la Meuse belge). Comparaison avec d'autres indices chimiques, biocénotiques et diatomiqués. Institut Royal des Sciences Naturelles de Belgique, document de travail 28, 1–113.

Lepistö, L., 1999. Phytoplankton assemblages reflecting the ecological status of lakes in Finland. Monographs of the Boreal Environment Research 16, Tammer Paino Oy, Tampere.

Lepistö, L., Rosenström, U., 1998. The most typical phytoplankton taxa in four types of boreal lakes. Hydrobiologia 369/370, 89–97.

Levine, H.G., 1984. The use of seaweeds for monitoring coastal waters.. In: Shubert, L.E. (Ed.), Algae as Ecological Indicators. Academic Press, London, pp. 189–210.

Lewis, M.A., 1990. Chronic toxicities of surfactants and detergents builders to algae: a review and risk assessment. Ecotoxicology and Environmental Safety 20, 123–140.

Liebmann, H., 1962. Handbuch der Frischwasser- und Abwasserbiologie. Bd. I, 2. Aufl., G. Fischer Verlag, Jena.

Lindblad, C., Kautsky, U., Andrè, C., Kautsky, N., Tendengren, M., 1989. Functional response of *Fucus vesiculosus* communities to tributyltin measured in an *in situ* flow-through system. Hydrobiologia 188/189, 277–283.

Loeb, S.L., Spacie, A. (Eds), 1994. Biological Monitoring of Aquatic Systems. Lewis Publishers, Boca Raton.

Lowe, R.L., 1974. Environmental requirements and pollution tolerance of freshwater diatoms. US Environmental Monitoring Series. EPA 670/4–74–005, 1–334.

Lowe, R.L., Pan, Y., 1996. Benthic algal communities as biological monitors. In: Stevenson, R J, Rothwell, M.L., Lowe, R L., 1996. Algal Ecology. Freshwater Benthic Ecosystems. Academic Press, London, pp. 705–739.

Lund, J.W.G., 1972. Preliminary observations on the use of large experimental tubes in lakes. Verhandlungen der Internationalen Vereinigung für Limnologie 18, 71–77.

Maestrini, S.Y., Bonin, D.J., Droop, M.R., 1984a. Phytoplankton as indicators of sea water quality: bioassay approaches and protocols. In: Shubert, L.E. (Ed.), Algae as Ecological Indicators. Academic Press, London, pp. 71–132.

Maestrini, S.Y., Droop, M.R., Bonin, D.J., 1984b. Test algae as indicators of sea water quality: prospects. In: Shubert, L.E. (Ed.), Algae as Ecological Indicators. Academic Press, London, pp. 133–188.

Marvan, P., Přibil, S., Lhotský, O. (Eds), 1979. Algal Assays and Monitoring Eutrophication. E. Schweizerbart'sche Verlagsbuchhandlung, Stuttgart.

Mauch, E., 1976. Leitformen der Saprobität für die biologische Gewässeranalyse. Courier Forschungs Institut Senckenberg 21, Vol. 1–5.

Mez, C., 1898. Mikroskopische Wasseranalyse. Springer Verlag, Berlin.

Möller, B., Pankow, H., 1981. Algensoziologische und saprobiologische Untersuchungen an Vorflutern der Elbe. Limnologica 13, 291–350.

Moore, J.W., Ramamoorthy, S., 1984. Heavy Metals in Natural Waters. Applied Monitoring and Impact Assessment. Springer Verlag, New York.

Munawar, M., Munawar, I.F., 1987. Phytoplankton bioassays for evaluating toxicity of in situ sediment contaminants. Hydrobiologia 149, 87–105.

Munawar, M., Munawar, I.F., Leppard, G.G., 1989. Early warning assays: an overview of toxicity testing with phytoplankton in the North American Great Lakes. Hydrobiologia 188/189, 237–246.

Munawar, M., Munawar, I.F., Norwood, W.P., Mayfield, C.J., 1987. Significance of autotrophic picoplankton in the Great Lakes and their use as early indicators of contaminant stress. Archiv für Hydrobiologie, Ergebnisse der Limnologie 25, 141–155.

Munawar, M., Weisse, T., 1989. Is the 'microbial loop' an early warning indicator of anthropogenic stress? Hydrobiologia 188/189, 163–174.

Noack, U., 1987. Quasi-kontinuierliche Chlorophyllfluoreszenz-Messung in der Gewässerüberwachung. Archiv für Hydrobiologie, Ergebnisse der Limnologie, Beiheft 29, 99–105.

Noll, M., Bauer, U., 1973. *Phormidium autumnale* als Indikatororganismus für algizide Substanzen im Wasser. Verhandlungen der Gesellschaft für Ökologie, Saarbrücken 1973, 169–173.

Nygaard, G., 1949. Hydrobiological studies on some Danish ponds and lakes. Part II: The quotient hypothesis and some little known plankton organisms. Vidensk Danske Selsk. Biologica Skripta 7, 1–293.

Ongley, E.D., Birkholz, D.A., Carey, J.H., Samoiloff, M.R., 1988. Is water a relevant sampling medium for toxic chemicals? An alternative environmental sensing strategy. Journal of Environmental Quality 17, 391–401.

Palmer, C.M., 1969. A composite rating of algae tolerating organic pollution. Journal of Phycology 5, 78–82.

Pantle, R., Buck, H., 1955. Die biologische Überwachung der Gewässer und die Darstellung der Ergebnisse. Gas und Wasserfach 96, 604.

Patrick, R., 1976. The importance of monitoring change, In: Cairns, J., Dickson, K.L. Westlake, G.F. (Eds), Biological Monitoring of Water and Effluent Quality. American Society for Testing and Materials, pp. 157–190.

Patrick, R., 1994. What are the requirements for an effective biomonitor? In: Loeb, S.L., Spacie, A. (Eds), Biological Monitoring of Aquatic Systems. Lewis Publishers, Boca Raton, pp. 23–29.

Patrick, R., Hohn, M.H., 1956. The diatometer – a method for determining the pattern of the diatom flora. Proceedings of the American Petroleum Institute 36, 332–339.

Pipp, E., Rott, E., 1994 Classification of running water sites in Austria based on benthic algal community structure. Verhabdlungen der Interantionalen Vereinigung für Limnologie 25, 1610–1613.

Plafkin, J.L., Bardour, M.T., Porter, K.D., Gross, S.K., Hughes, R.M., 1989. Rapid bioassessment protocols for use in streams and rivers. Benthic macoinvertrebrates and fish. Report No. EPA/444/4–89–001. US EPA Office of Water (WH-553), Washington DC 20460.

Preston, F.W., 1948. The commonness and rarity of species. Ecology 29, 254–283.

Ramade, F., 1987. Ecotoxicology. John Wiley, Chichester.

Rawson, D.S., 1956. Algal indicators of trophic lake types. Limnology & Oceanography 1, 18–25.

Reimann, I., Hamm, A., 1996. Experimentelle Untersuchungen zur Trophieindikation anhand von Aufwuchsdiatomeen. Informationsberichte des Bayerischen Landesamtes für Wasserwirtschaft 3/96, 1–108.

Reynolds, C.S., 1997. Vegetation processes in the pelagic. A model for ecosystem theory. In: Kinne, O. (Ed.), Excellence in Ecology 9. Ecology Institute, Oldenburg, pp. 1–371.

Reynolds, C.S., Descy, J.-P., 1996. Production, biomass and structure of phytoplankton in large rivers. Archiv für Hydrobiologie Supplemente (Large Rivers) 113, 161–187.

Rhee, G.-Y., 1989. Continuous culture algal bioassays for organic pollutants in aquatic ecosystems Hydrobiologia 188/189: 247–257.

Rojo, C., Cobelas, M.A., Arauzo, M., 1994. An elementary, structural analysis of river phytoplankton. Hydrobiologia 289, 43–55.

Rosén, G., 1981. Phytoplankton indicators and their relations to certain chemical and physical factors. Limnologica (Berlin) 13, 263–290.

Ross, P., Jarry, V., Sloterdijk, H., 1988. A rapid bioassay using the green algal Selenastrum capricornutum to screen for toxicity in St. Lawrence River sediment elutriates. In: Cairns, J. Jr., Pratt, J.R. (Eds), Functional Testing of Aquatic Chemicals. ASTM STP 1988, American Society for Testing and Materials, Philadelphia.

Rott, E., 1984. Phytoplankton as biological parameters for the trophic characterization of lakes. Verhandlungen der Internationalen Vereinigung für Limnologie 22, 1078–1085.

Rott, E., Hofmann, G., Pall, K., Pfister, P., Pipp, E., 1997. Indikatorlisten für Aufwuchsalgen. Teil 1: Saprobielle Indikation. Wasserwirtsc-haftskataster BMLF, Wien.

Rott, E., Pfister, P., Van Dam, H., Pall, K., Binder, N., Pipp, E., Ortler, K., 1999. Indikatorlisten für Aufwuchsalgen. Teil 2: Trophieindikation und autökologische Anmerkungen. Wasserwirtschaftskataster. BMLF, Wien.

Rumeau, A., Coste, M., 1988. Initiation à la systématique des diatomées d'eau douce pour l'utilisation pratique d'un indice diatomique générique. Bulletin de French Pêche Pisciculture 309, 1–69.

Sakamoto, M., Tilzer, M.M., Gächter, R., Rai, H., Collos, Y., Tschumi, P., Berner, P., Zbaren, D., Zbaren, J., Dokulil, M., Bossard, P., Uehlinger, U., Nusch, E.A., 1984. Joint field experiments for comparison of measuring methods of photosynthetic production. Journal of Plankton Research 6, 365–383.

Sayk, D., Schmidt, D., 1986. Algen-Floureszenztest, ein vollautomatischer Biotest. Zeitschrift Wasser und Abwasser-Forschung 19, 182–184.

Schallenberg, M., Burns, C.W., 2001. Tests of autotropjh picoplankton as early indicators of nutrient enrichment in an ultra-oliogorophic lake. Freshwater Biology 46, 27–38.

Schelske, C.L., 1984. *In situ* and natural phytoplankton assemblage bioassays. In: Shubert, L.E. (Ed.), Algae as Ecological Indicators. Academic Press, London, pp. 15–47.

Schiefele, S., 1987. Indikationswert benthischer Diatomeen in der Isar zwischen Mittenwald und Landshut. Diplomarbeit Ludwig-Maximilians-Universität München.

Schiefele, S., Kohmann, E., 1993. Bioindikation der Trophie in Fließgewässern. Bayerisches Landesamt für Wasserwirtschaft. Forschungsbericht Nr. 10201504.

Schindler, D.W., 1974. Eutrophication and recovery in experimental lakes: implications for lake management. Science 184, 897–899.

Schmitt, A., 1998. Trophiebewertung planktondominierter Fließgewässer – Konzept und erste Erfahrungen. Münchener Beiträge zur Abwasser-, Fischerei- und Flussbiologie 51, 394–411.

Schönfelder, I., 1997. Eine Phosphor-Diatomeen-Relation für alkalische Seen und Flüsse Brandenburgs und ihre Anwendung für die paläolimnologische Analyse von Auensedimenten der unteren Havel. Dissertationes Botanicae 283, J. Cramer, Berlin.

Schönfelder, I., 2000. Indikation der Gewässerbeschaffenheit durch Diatomeen. In: Steinberg, C., Clamano, W., Klapper, H., Wilken, R.-D., (Eds), Handbuch Angewandte Limnologie. 2, VIII-7.2, 9. Erganzungs Lieferung 4/00, Ecomed, Landsberg.

Schwoerbel, J., Gaumert, D., Hamm, A., Hansen, P.D., Nusch, E.A., Schilling, N., Schindele, X., 1991. Akute und chronische Toxizität von anorganischen Stickstoffverbindungen unter besonderer Berücksichtigung des Ökosystemschutzes im aquatischen Bereich. In: Hamm, A. (Ed.), Studie über Wirkungen und Qualitätsziele von Nährstoffen in Fließgewässern, Academia Verlag, Sankt Augustin, pp. 112–205.

Simola, H., Meriläinen, J.J., Sandman, O., Martilla, V., Karjalainen, H., Kukkonen, M., Julkunen-Tiitto, R., Hakulinen, J., 1996. Palaeolimnological analyses as information source for large lake biomonitoring. Hydrobiologia 322, 283–292.

Skulberg, O.M., 1964. Algal problems related to the eutrophication of European water supplies, and a bioassay method to assess fertilizing influences of pollution on inland waters. In: Jackson, D.F. (Ed.), Algae and Man. Plenum Press, New York, pp. 262–299.

Sládeček, V., 1973. System of water quality from the biological point of view. Archiv für Hydrobiologie, Beihefte Ergebnisse der Limnologie 7, 1–218.

Sládeček, V. (Ed.), 1977. Symposium on Saprobiology. Archiv für Hydrobiologie, Beihefte Ergebnisse der Limnologie 9. 1–245.

Sládeček, V., 1986. Diatoms as indicators of organic pollution. Acta Hydrochimica et Hydrobiologica 14, 555–566.

Sládečková, A., 1979. Periphyton assays in situ. In: Marvan, P., Přibil, S., Lhotský, O. (Eds), Algal Assays and Monitoring Eutrophication. E. Schweizerbart'sche Verlagsbuchhandlung, Stuttgart, pp. 205–209.

Šrámek-Hušek, R.,1956. Zur biologischen Charakterisierung der höheren Saprobitätsstufen. Archiv für Hydrobiologie 51, 376–390.

Steinberg, C., Brüggemann, R., Traunspurger, W. 1999. Öko(toxi)kologische Beurteilung von Gewässerverunreinigungen. In: Frimmel, F.H. (Ed.), Wasser und Gewässer. Spektrum Akad. Verlag, Berlin, pp. 447–505.

Steinberg, C., Hartmann, H., 1988. Planktische blütenbildende Cyanobakterien (Blaualgen) und die Eutrophierung von Seen und Flüssen. Vom Wasser 70, 1–10.

Steinberg, Ch., Kern, J., Pitzen, G., Traunspurger, W., Geyer, H., 1992. Biomonitoring in Binnengewässern. Grundlagen der biologischen Überwachung organischer Schadstoffe für die Praxis des Gewässerschutzes. Ecomed, Landsberg.

Steinberg, C., Klein, J., Brüggemann, R., 1995. Ökotoxikologische Testverfahren. Übersicht über bestehende Testverfahren. Modellierung in der Ökotoxikologie. Empfehlungen für Normung und Forschung. Ecomed, Landsberg.

Steinberg, C., Schiefele, S., 1988. Biological indication of trophy and pollution of running waters. Zeitschrift für Wasser- und Abwasser-Forschung 21, 227–234.

Stockner, J.G., 1971. Preliminary characterization of lakes in the Experimental Lakes Areas, northwestern Ontario using diatiom occurrence in sediments. Journal of the Fisheries Research Board Canada 28, 265–275.

Stockner, J.G., Callieri, C., Cronberg, G., 2000. Picoplankton and other non-bloom forming cyanobacteria in lakes. In: Whitton, B.H., Potts, M. (Eds), Ecology of Cyanobacteria. Their Diversity in Time and Space. Kluwer Academic, Norwell, pp. 195–231.

Strickland, J.D.H., Terhune, L.D.B., 1961. The study of in-situ marine photosynthesis using a large plastic bag. Limnology Oceanography 6, 93–96.

Talling, J.F., 1984. Past and contemporary trends and attitudes in work on primary productivity. Journal of Plankton Research 6, 203–217.

Täuscher, L., 1999. Planktic and benthic diatom assemblages as indicators of water quality in the floodplains of the middle area of the River Elbe and lower area of the River Havel (Brandenburg, Sachsen-Anhalt, Germany). Berichte de Instituts für Gewässerökologie und Binnenfischerei (IGB) 7 (Sonderheft II), 82–85.

Teiling, E., 1955. Some mesotrophic phytoplankton indicators. Verhandlungen der Internationalen Vereinigung für Limnologie 12, 212–215.

Teubner, K., 1995. A light microscopical investigation and multivariate statistical analysis of heterovalvar cells of *Cyclotella*-species (Bacillariophyceae) from lakes of the Berlin-Brandenburg region. Diatom Research 10, 191–205.

TGL 27885/01, 1982. Fachbereichstandard der DDR. Nutzung und Schutz der Gewässer. Stehende Binnengewässer. Berlin, DDR, PP. 1–6.

Thomann, R.V., 1989. Bioaccumulation model of organic chemical distribution in aquatic food chains. Environmental Science and Technology 23, 699–707.

Thomas, E.A., 1953. Zur Bekämpfung des See-Eutrophierung: Empirische und experimentelle Untersuchungen zur Kenntnis der Minimunistoffen in 46 Seen der Schweiz und angrenzender Gebiete. Monatsbull. Schweiz. Ver. Gas-Wasserfachm. 33, 25–32, 71–79.

Thomas, E.A., 1964. Nährstoffexperimente in Plankton-Test-Loten. Verhandlungen der Internationalen Vereinigung für Limnologie 15, 342–351.

Thunmark, S., 1945. Zur Soziologie des Süßwasserplanktons. Eine methodisch-ökologische Studie. Folia Limnologica Scandinavia 3, 1–66.

Trainor, F.R., 1984. Indicator algal assays: laboratory and field approaches. In: Shubert, L.E. (Ed.), Algae as Ecological Indicators. Academic Press, London, pp. 3–14.

Tremel, B., 1996. Determination of the trophic state by qualitative and quantitative phytoplankton analysis in two gravel pit lakes. Hydrobiologia 323, 97–105.

Van der Does, J., Klapwijk, S.P., 1987. Effects of phosphorus removal on the maximal algal growth in bioassay experiments with water from four Dutch lakes. Internationale Revue der gesamten Hydrobiologie 72, 27–39.

van der Does, T. Klapwijk, I.S.P., 1987. Effects of phosphorus removal on the maximal algal growth in bioassay experiments with water from four Dutch lakes. Int. Revue ges. Hydrobiol. 72, 27–39.

Van Donk, E., Mur, L.R., Ringelberg, J., 1989. A study of phosphate limitation in Lake Maarsseveen: phosphate uptake kinetics versus bioassays. Hydrobiologia 188/189, 201–209.

Vymazal, J., 1987. Zn uptake by *Cladophora glomerata*. Hydrobiologia 148, 97–101.

Watanabe, M., 1983. The modelling of red tide blooms. In: Jørgensen, S.E. (Ed.), Application of Ecological Modeling in Environmental Management. Part A. Elsevier, Amsterdam.

Watanabe, T., Asai, K., Houki, A., 1986. Numerical estimation to organic pollution of flowing water by using the epilithic diatom assemblage index (DAIpo). Science of the Total Environment 55, 209–218.

Weisse, T., 1991. The microbial food web and its sensitivity to eutrophication and contaminant enrichment: a cross-system overview. Internationale Revue der gesamten Hydrobiologie 76, 327–337.

Weisse, T., 1993. Dynamics of autotrophic picoplankton in marine and freshwater ecosystems. Advances in Microbial Ecology 13, 327–370.

Whitmore, T.J., 1989. Florida diatom assemblages as indicators of trophic stage and pH. Limnology Oceanography 34, 882–895.

Whitton, B.A., 1979. Plants as indicators of river water quality. In: James, A. and Evison, L. (Eds), Biological Indicators of Water Quality. John Wiley, Chichester, pp. 5-1–5-34.

Whitton, B.A., 1984. Algae as monitors of heavy metals in freshwaters. In: In: Shubert, L.E. (Ed.), Algae as Ecological Indicators. Academic Press, London, pp. 257–280.

Willén, E., 2000. Phytoplankton in water quality assessment – an indicator concept. In: Heinonen, P., Ziglio, G., Van der Beken, A. (Eds), Hydrological and Limnological Aspects of Lake Monitoring. John Wiley, Chichester, pp. 58–80.

Winter, J.G., Duthie, H.C., 2000. Epilithic diatoms as indicators of stream total N and total P concentration. Journal of the North-American Benthological Society 19, 32–49.

Wunsam, S., Schmidt, R., 1995. A diatom-phosphorus transfer function for Alpine and pre-alpine lakes. Memorie dell' Instituto Italiano di Idrobiologia 53, 85–99.

Zelinka, M., Marvan, P., 1961. Zur Präzisierung der biologischen Klassifikation der Reinheit fliessender Gewässer. Archiv für Hydrobiologie 57, 389–407.

329

Chapter 10

Bryophytes

Harald G. Zechmeister, Krystyna Grodzińska and
Grazyna Szarek-Łukaszewska

Abstract

*The use of bryophytes as bioindicators and biomonitors in terrestrial and aquatic habitats is
reviewed in this article.*

*Bryophytes are excellent indicators for a wide range of contaminants. This is in conse-
quence of a series of morphological and physiological properties like the lack of a cuticle or
the existence of large cationic exchange properties within the cell wall. Mosses have mainly
been used as accumulation indicators especially for heavy metals, radionucleides and for toxic
organic compounds. Reviewing a wide range of investigations on this topic, advantages and
further needs for research are discussed. Sulphurous and nitrogen depositions can hardly be
analysed by methods in the field of accumulation monitoring but by investigating the frequency,
distribution, fertility and vitality of bryophyte species and populations. Similar methods are
targeted by global change research, especially for the analysis of climate warming and the
influence of land-use intensity on biodiversity.*

Keywords: Bryophytes, bioindicators, biomonitors, heavy metals, sulphur, nitrogen, toxic
organic compounds, radionuclides, global change, terrestrial and aquatic habitats.

1. Introduction

Bryophytes are autotrophic cryptogames comprising approximately 25,000 species.
Taxonomically, they are divided into four classes, the hornworts (Anthocerotopsida),
two classes of the liverworts (Marchantiopsida, Jungermanniopsida) and the mosses
(Bryopsida).

The life history of bryophytes involves an alternation between sporophytic and
gametophytic generations that differ in form and function. The actual plant is represented by the gametophytic generation, which is the most evolved haploid generation
in the whole plant kingdom.

The spores germinate to form a branched or thallose protonema which resembles
green algae. The germination as well as its growth is very sensitive to all kinds
of natural and human influences which exceeds by far the sensitivity of the green
gametophore. Therefore, in many cases the resistance of the protonema against eco-
toxicologically relevant substances is the main limiting factor for the distribution of a
species (e.g. Gilbert, 1968). The green gametophore produces the sex organs. After
successful pollination, a sporophytic generation evolves which remains attached to
the green plant and is nourished by the gametophore. Depending on size, most of the

spores released by the sporangium are dispersed by wind, by rising up to several thousand metres (Longton, 1997) they can be transported over large distances. Many species also produce vegetative reproductive units which enables them to a less energy demanding propagation or the survival of unfavourable conditions.

Bryophytes are generally small (less than 5 cm) but some can grow up to a length of 70 cm (e.g. *Polytrichum, Dawsonia*). In contrast to vascular plants, they rarely grow as single stems but in groups, which form turfs, cushions, wefts or other growth forms (e.g. Mägdefrau, 1982). Bryophytes show a limited range of anatomical or morphological features but a wide range of physiological and dispersal adaptations to stress caused by natural or anthropogenic disturbance (e.g. Smith, 1982; Bates and Farmer, 1992). A great variety of life traits can be found mainly in short lived habitats (e.g. During, 1979; Proctor, 1990). In Grime's (1979) triangular scheme they are mainly ruderals and/or stress tolerants (Grime et al., 1990). The various modes of reproduction play an important role in the life cycles of mosses especially in stands with a high disturbance (e.g. During, 1997; Longton, 1997; Zechmeister and Moser, 2001).

Bryophytes thrive in humid climates, but can be found all over the world, even in arid regions. As a consequence of slow evolution (e.g. Szweykowski, 1984), many dominant species can be found all over the world, or show at least a circumpolar distribution. Their biomass production is important in subarctic ecosystems and mountainous tropical rain forests only (Longton, 1988; Pòcs, 1980), but they are a significant ecological factor in a variety of habitats (e.g. bogs, water springs, alpine grasslands). On the other hand, a wide range of species can grow in areas unable to be colonised by any other plant which is significant for many aspects in bioindication also. Bryophytes colonise nearly every kind of terrestrial substrate (e.g. bare stones, bark, skeletons, etc.) and grow in freshwater but are absent from saline waterbodies (salt lakes, oceans).

2. The physiological basis for the use of bryophytes as indicators

The use of bryophytes in an increasing number of monitoring programmes is based on a wide range of remarkable anatomical and physiological properties, which are briefly reviewed. Further information on this topic are given in extensive reviews by Bates (1992), Brown (1984), Brown and Bates (1990), Proctor (1982, 1990), Sveinbjörnson and Oechel (1992), Tyler (1990) or Onianwa (2001).

2.1. Water relations

Bryophytes are poikilohydric species, but among them there is a diversity of means of water and mineral uptake. As most of the bryophytes are small and the leaves of many mosses as well as those of folious liverworts are built up by only one cell layer, the surface/volume ratio is high. According to their small size, the micro-environment in some climates is often much more important than the macro-environmental conditions.

Most of the bryophytes are ectohydric species, which means that most of the species receive water as well as mineral nutrients predominantly by atmospheric depositions.

They are well adapted to this strategy since they have no or only very small vacuoles, and beside some surface wax structures (e.g. papillae) there is no continuous water-repellent cuticle. Some species obtain additional water and soluble nutrients from the substrate. In a few cases (e.g. Polytrichales) bryophytes have water conducting tissues (endohydric species). Additionally, the physical and chemical properties of the substrate is essential for the establishment and survival of the green plants. Beside species adapted to wet habitats, which hold most water within large cells, external capillary conducting systems retain the major part of the water content. They are diverse and can be found on surfaces of leaves, between various parts of the gametophore (e.g. rhizoid wefts, auricles between stem and leaves) or between single stems. The latter leads to a wide range of growth forms in dependency of habitat conditions. Desiccation tolerance is mainly based on physiological adaptations and enables some species to grow in dry and hot environments.

2.2. Mineral requirements

Mineral requirements are similar to those in vascular plants. Mineral uptake by the cell is controlled by a semipermeable membrane. The protonema and the early gametophyte is attached to the substrate and significant stocks of nutrients may be accumulated from the surface at this stage. Later on, many pleurocarpous species leave the close contact to the substrate and it is generally assumed that the main source of minerals for these species are atmospheric sources (e.g. Tamm, 1953; Bates and Bakken, 1998), though some elements (e.g. Ca, K, P) seem to be derived further via the substrate (e.g. Bates, 1992; Wells and Boddy, 1995; Brown and Brūmelis, 1996; Brūmelis and Brown, 1997).

Elements associated with well developed gametophores can be attributed to four possible locations (Brown and Bates, 1990; Bates, 1992): trapped particulate matter, intercellular soluble, extracellular, bound to cell wall on charged exchange sites, or intracellular. Particulate matter and intercellular elements are unbound ions in the water free space and can easily be removed by washing or mechanic treatment. Exchangeable cations are bound to positively charged exchange properties of the cell wall and are fixed by a process mainly depending on physico-chemical processes and are not physiologically active, whereas intracellular elements fulfil a physiological function.

2.2.1. Extracellular uptake

The very high cation exchange capacity is related mainly to unesterified polyuronic acid molecules (Clymo, 1963). In *Sphagnum* galacturonic acid and in some liverworts mannuronic acid is reported to be present in the cell walls too (Brown, 1984). Binding conforms to strict physico-chemicals rules. External uptake is rapid and occurs within the first few minutes during rainfall (Gjengedahl and Steinnes, 1990). The uptake depends on the nature of the elements only, irrespective of the physiological condition of the plant. Monovalent cations (e.g. K, Na) show less affinity for anionic sites and divalent cations with Class B characteristics (e.g. Pb, Cu) show greater affinity than Class A metals (eg. Cd, Mg, Zn; Brown and Brown, 1990). The uptake of heavy metals is slightly influenced by pH of the precipitation and air temperature. Laboratory studies have shown that the presence of Na^+, Mg^{2+} and Cl^- in coastal precipitation reduces the uptake of Zn and Cu (e.g. Berg et al. 1995).

Several investigations have shown that uptake efficiencies seem to follow the order Pb >> Co = Cr > Cu = Cd = Mo = Ni = V = Sb > Zn > As (e.g. Steinnes, 1985; Berg et al., 1995; Thöni et al., 1996; Berg and Steinnes, 1997). Slightly different results are given by Rühling and Tyler (1970) and Ross (1990) who showed a markedly higher retention capacity for Cu. Čeburnis et al. (1999) found accumulations as follows: Ni < V < Cr < Zn. Čeburnis and Valiulis (1999) who compared concentrations in bulk depositions with throughfall under moss obtained an absolute uptake efficiency of approximately 60% but did not find differences for the various metals.

The total metal binding is determined by the number of available exchange sites and morphological structures of the bryophytes, which differs from species to species. Therefore, most species have different uptake capacities. There are no significant differences in the accumulation of a wide range of trace metals (e.g. Pb, Cd, Cu, V) between the mosses *Pleurozium schreberi* (Brid.) Mitt. and *Hylocomium splendens (Hedw.)* B.S.G., (Herpin et al., 1994; Zechmeister, 1994; Berg and Steinnes, 1997; Halleraker et al., 1998). These mosses have been used in a range of investigations (see Section 4.2). Differences occur mainly in metals, which are either not well retained (like As) or have high background levels in the moss (Zn; Rühling and Steinnes, 1998). Interspecific calibration is advised if data are compared between different species (e.g. Zechmeister, 1998).

Younger parts of the plants show higher amounts of monovalent cations and nutrient anions than older parts. Divalent cations, especially heavy metals, show the reverse distribution. Dead tissues retain polyvalent cations more effectively still (e.g. Rühling and Tyler, 1970; Pakarinen and Rinne, 1979).

2.2.2. Intracellular uptake

In contrast to the extracellular uptake mechanism, intracellular uptake is influenced by various aspects of plant metabolism. Entry to the cell plasma is determined e.g. by the affinity for an appropriate carrier, competitive elements, gradients in element concentration or the energy status. Elements located within the cell influence cell metabolism. Uptake rates are in general much lower than at the extracellular sites.

As shown by various authors (e.g. Pickering and Puia, 1969; Burton and Peterson, 1979; Wells and Brown, 1987), non-physiological elements like heavy metals also pass the limiting plasma membrane of the cell and affect cell metabolism. In consequence heavy metals induce the production of thiol-containing peptides such glutathiones which therefore can be used as biomarker for heavy-metal pollution (e.g. Bruns et al., 1999, 2001). Nevertheless, the cell wall is an efficient barrier against the penetration of heavy metals into the protoplasma of the bryophyte cell (Shimwell and Laurie, 1972; Skaar et al., 1973). Young shoots tend to have a more effective barrier than older ones (Lüttge and Bauer, 1968).

Some bryophyte species (e.g. copper-mosses) also tolerate elevated levels of toxic elements on a physiological level (e.g. Url, 1959; Shaw, 1987).

2.3. Sources of elements

Mineral cations as well as anionic nutrients derive mainly from atmospheric depositions (e.g. Rieley et al., 1979; Okland et al., 1999; Reimann et al., 2001). Positive correlations exist between the quantity of rainfall and nutrient concentrations in bryophytes, which can vary with plant growth (Brown and Bates, 1990). Nevertheless, some elements (e.g. Ca, Mg, K) depend also on the uptake from the substrate, especially in endohydric species as well as in mosses growing in the form of turfs, cushions or covers (e.g. Büscher and Koedam, 1979; Brown and Bates, 1990; Bates, 1992; Okland et al., 1999) although there are contrasting results (Brūmelis et al., 2000).

The amount of occult depositions (fog, mist, clouds) seems to be high in some areas. There are few investigations on the percentage of dry in relation to total depositions (e.g. Fowler et al., 1993). The estimation is difficult as there are also dry depositions during rainfall. The results show a wide range for the percentage of the dry depositions : 20 % (Ross, 1990; Ruijgrook et al., 1993), 33% (Svensson and Lidèn, 1965) and 50% (Galloway et al., 1982). There are fairly large differences for the various metals and there is a strong correlation with annual precipitation and the distance to emission sources (Brown and (Brūmelis, 1996; Zechmeister, 1997).

Internal translocation of elements within the bryophytes seems to play an important role in their nutrient cycles, although most bryophytes lack recognisable conducting tissues. Acropetal movement of photoassimilates and essential elements has been reported by several authors (e.g. Brown and Bates, 1990; Wells and Boddy, 1995; Brown and Brūmelis, 1996; Bates and Bakken, 1998) though further investigations are needed.

3. General reactions on pollutants and areas of applications

Under the influence of human induced changes of the environment, bryophytes respond sensitively and this can effectively be used for monitoring purposes.

Mostly they show a decline in vitality, which, for example, can be detected by changes in colour following damages in the chloroplast structure (e.g. Martĩnez-Abaigar and Núñez-Olivera, 1998), or less vigorous growth by individuals or populations (e.g. Bengtson et al., 1982; Callaghan et al., 1997). Sometimes there is also a shift in the reproduction mode, favouring asexual reproduction under stress (Rao, 1982; Otnyukova, 1995; Zechmeister and Moser, 2001). The ultimate response is population loss and finally extinction.

There is a wide range from toxitolerant species to extremely sensitive ones. Tolerances vary not only from species to species but also with the type of pollutants. Additionally, it must be considered that climatic conditions are much more influential for the survival of poikilohydric organisms than e.g for flowering plants.

On the other hand, bryophytes are very resistant against a series of substances which are highly toxic for other plants (e.g. heavy metals, radionucleides, various toxic organic compounds). As a consequence of their nutrient cycling and uptake mechanisms (see above) they even tend to accumulate these pollutants.

Regarding both their accumulation capacity and their sensibility to various toxic compounds, bryophytes can be used either as accumulation indicators or as sensitive indicators for a series of human influences, mainly polluting substances.

4. Bryophytes as indicators and monitors in terrestrial habitats

4.1. Heavy metals

Heavy metals originate from both natural and anthropogenic sources in the environment. In the atmosphere, natural sources of these elements are volcanic eruptions, cosmic and terrestrial dusts, vegetation fires, and salt spray from the oceans. Anthropogenic sources include emissions from different industrial plants (steel and non-ferrous metallurgy, smelters, alloying plants, petrochemical industry, fertiliser plants, coal power plants, industrial and home furnaces) and motor traffic. The amount of heavy metals originating from natural sources in the atmosphere is small as compared with the anthropogenic flux of these elements.

Airborne heavy metals enter the ecosystems where they circulate and, depending on their concentration and toxicity, pose a greater or smaller threat to the components of these ecosystems. The accumulation of heavy metals in the soil and living organisms may have a damaging effect on the environment (Lieth and Markert, 1990; Markert, 1993; Herpin et al., 1996).

In the 1950s and 1960s, the quick development of industry and motor transport caused a dramatic increase in dust emissions containing heavy metals. It is only natural that ecologists focused their attention on threats posed by heavy metals to the biotic and abiotic environment. They began to look for sensitive and, cheap biological methods for assessing the environmental level of heavy metals, above all the most toxic ones (Cd, Pb, Hg).

In the late 1960s, two Swedish ecologists, Rühling and Tyler (1968; 1969), first used mosses as indicators of heavy metals pollution. They recognised these plants to have many features of good bioindicators.

The suitability of bryophytes for the indication of heavy metal depositions is based on their accumulation which is a result of a series of morphological and physiological properties which have already been given in Section 1 (e.g. cationic exchange properties).

Additionally in certain species of feather mosses (e.g., *Hylocomium splendens* (Hedw.) B.S.G.) and in *Sphagnum* spp. it is possible to recognise and separate annual growth increments, facilitating determination of the age and exposure time of the material to be used in monitoring. This is one of the most important advantages of using mosses for the estimation of heavy metal depositions.

Several species are very abundant and widespread (cosmopolitan or circumpolar) in defined habitats.

Owing to the longevity of most bryophytes they may be used to integrate the deposition over a considerable time, usually 2–5 years, depending on the species and sampling methods.

Moss methods in heavy metals monitoring work are inexpensive and simple.

Since the time of Rühling and Tyler's pioneer research (1968; 1969; 1970; 1973), mosses have become commonly used indicators of heavy metals contamination of the environment (e.g., Groet, 1976; Grodzińska, 1978; Rinne and Barclay-Estrup, 1980; Herpin et al., 1994; Liiv et al., 1994; Zechmeister, 1994; Kuik and Wolterbeek, 1995; Herpin et al., 1996; Markert et al., 1996; Äyräs et al., 1997; Berg and Steinnes, 1997; Čeburnis et al., 1997; 1999; Mankovska, 1997; Alliksaar and Punning, 1998; Pott and Turpin, 1998; Sucharová and Suchara, 1998; Galsomies et al., 1999; Grodzińska et al., 1999; Davis et al., 2001; Figueira et al., 2002). Onianwa (2001) gives an extensive overview on this topic.

Mosses have been used for evaluating (1) the present state of environmental contamination by heavy metals in both extensive and small areas and (2) the pollution level in the past. Some examples of these studies are presented below.

4.1.1. Assessment of the current heavy metals levels of the environment

4.1.1.1. Indigenous species

Moss technique was first used for large – scale monitoring purposes in the Scandinavian countries: Sweden, Norway, Finland and Denmark (Gydesen et al., 1983; Rühling et al., 1987). National monitoring programmes were co-ordinated by the Nordic Group of Heavy Metal Deposition. In the late 1980s, European countries were invited to the European Heavy Metal Deposition Programme by the Environmental Monitoring and Data Group in the Nordic countries. The 1990 survey was extended to 21 European countries (Rühling, 1994), and the 1995 survey to 28 European countries (Rühling et al., 1996; Rühling and Steinnes, 1998). According to the European programme guidelines, moss monitoring should be carried out at five-year intervals. Thus the next sampling of mosses fell in the year 2000.

Some examples of monitoring studies carried out on the European continent using mosses as bioindicators are given below.

The concentrations of heavy metals (As, Cd, Cr, Cu, Fe, Pb, Ni, V, Zn) in moss samples (*Hylocomium splendens* (Hedw.) B.S.G., *Pleurozium schreberi* (Brid.) Mitt.) collected in north European countries (Norway, Sweden, Finland, Denmark, Greenland, Iceland) during the 1970s and 1980s showed considerable spatial differentiation (Rühling and Tyler, 1973; Steinnes, 1977; Gydesen et al., 1983). The highest levels of heavy metals were found in mosses from the southern part of the Scandinavian Peninsula (Fig. 1), which was under the influence of heavy metals emissions from Western European countries. The metal content of mosses decreased markedly towards the north, reaching the lowest values in Iceland and Spitsbergen (Gydesen et al., 1983; Rühling et al., 1987). This decreasing gradient of heavy metals levels in mosses from south to north was in conformity with the spatial distribution of metal concentrations in the atmosphere.

The results of studies conducted within the framework of European joint moss research in 1990–1995 (Rühling, 1994; Rühling and Steinnes, 1998) using three moss species (*Pleurozium schreberi* (Brid.) Mitt., *Hylocomium splendens* (Hedw.) B.S.G. and *Hypnum cupressiforme* agg.) confirmed regional variations in the deposition of heavy metals in Europe, as noticed in earlier decades. They showed a sharply decreasing gradient from Central Europe to northern Scandinavia for Cd, Pb and V

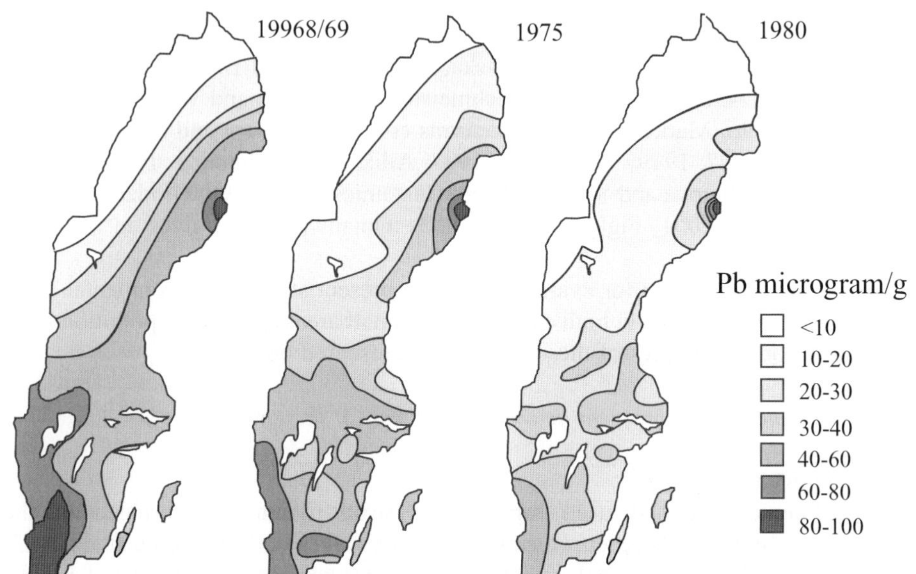

Figure 1. Concentrations of Pb in forest moss in Sweden (Rühling and Tyler, 1984).

(Fig. 2). A weaker gradient was obtained for Cr, Cu, Fe, Ni and Zn, and particularly for Ni and Cu in the far north, in the western part of the Kola Peninsula where there are big smelting plants.

Continental and national studies on heavy metals using the moss technique indicate a general decrease in the concentrations of most metals during the last 15 years (Gydesen et al., 1983; Rühling et al., 1987; Rühling, 1992; Rühling, 1994; Rühling and Steinnes, 1998, Kunert et al., 1999). This decrease, which has taken place over large areas, is due principally to better emissions control legislation, better filter technique and the closure of old polluting industrial plants. For lead the decrease is due to reduction in the use of leaded petrol.

Assessment of heavy metals contamination using mosses has also been carried out on a regional scale (e.g. Šoltès, 1992; Kauneliene, 1995; Markert et al., 1996; Äyräs et al., 1997; Halleraker et al., 1998; Gerdol et al., 2000; Grodzińska and Szarek-Łukaszewska, 2001). Because for these studies a denser network of moss sampling sites was established than for the large-scale monitoring studies, spatial variations in deposition could be examined more thoroughly.

The heavy metals concentrations in *Pleurozium schreberi* (Brid.) Mitt. collected in two industrial regions in southern Poland (Kraków-Silesia region and Legnica-Głogów Copper Basin) showed evident spatial diversification. The highest concentrations of the elements occurred in places with a high number of plants and smelters, hence the high correspondence between the volume of production and industrial emissions and the levels of contaminants in mosses (Fig. 3).

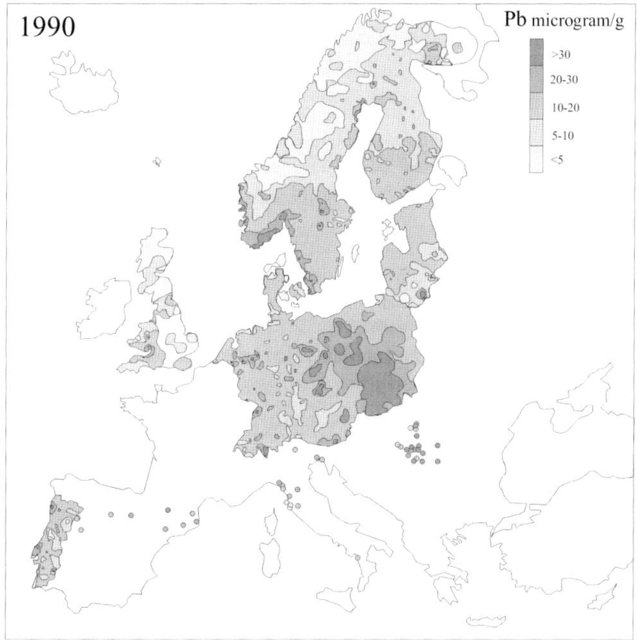

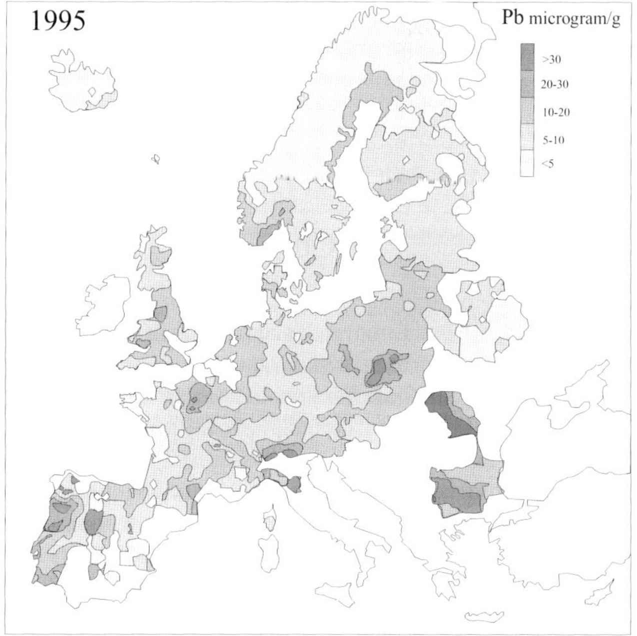

Figure 2. (a) Concentration of Pb in forest moss in Europe in 1990 (Rühling, 1994). (b) Concentration of Pb in forest moss in Europe in 1995 (Rühling and Steinnes, 1998).

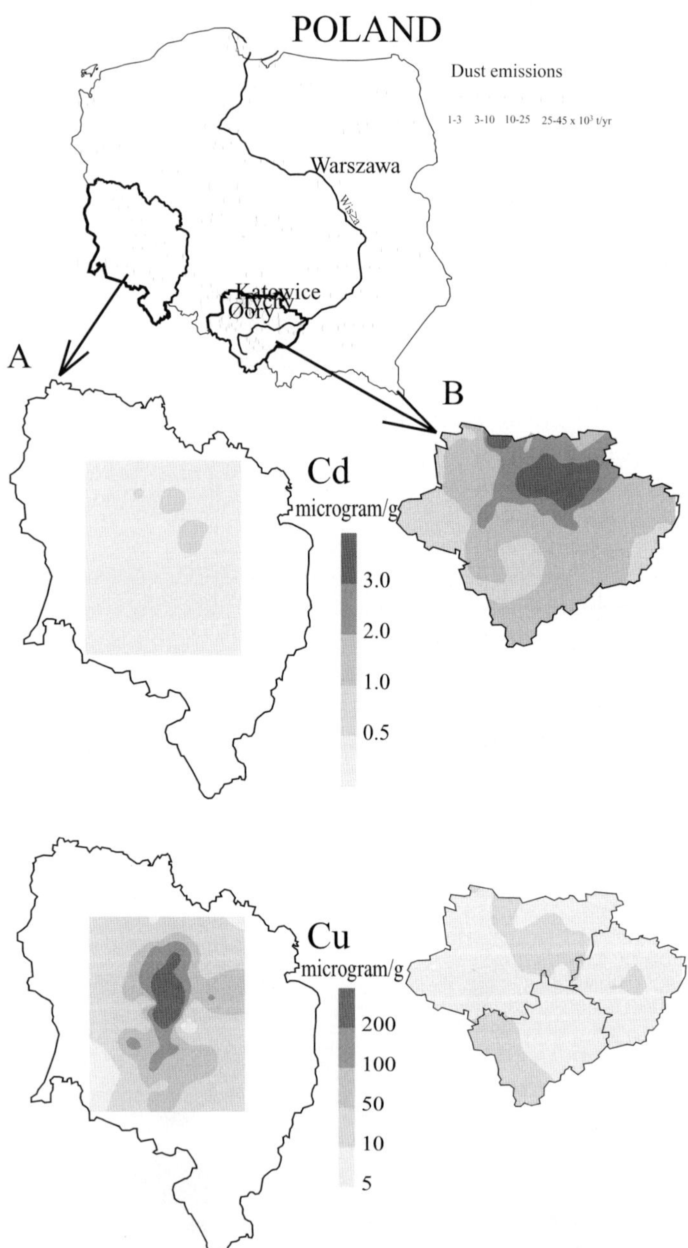

Figure 3. Concentrations of Cd and Cu in *Pleurozium schreberi* in two industrial regions in southern Poland. A – Legnica – Głogów Copper Basin; B – Kraków – Silesia Region (Szarek, Klich, Grodzińska unpublished data).

The extremely high concentrations of copper and nickel in moss (*Hylocomium splendens* (Hedw.) B.S.G.) samples from the western part of the Kola Peninsula correlated significantly with the distribution of emission sources (nickel mining and smelting) in this Arctic area (Äyräs et al., 1997).

The moss technique is also very useful for assessing environmental contamination on a local scale, in the neighbourhood of industrial plants (Makomaska, 1978; Pilegaard, 1979; Barclay-Estrup and Rinne, 1979; Folkeson, 1981; Mäkinen, 1983, 1994a), in towns (Lötschert et al., 1975; Grodzińska and Kaźmierczakowa, 1977; Hertz et al., 1984; Mäkinen, 1987), and in the vicinity of highways (Rühling and Tyler, 1968; Markert and Weckert, 1994). It was found that the concentrations of heavy metals decreased markedly with increasing distance from an industrial plant, a town centre (Fig. 4) or highway (Fig. 5).

Folkeson (1981) assessed the concentrations of 10 heavy metals in *Pleurozium schreberi* (Brid.) Mitt. along a 0.06–10 km transect from a peat-fired power plant. A considerable decrease in the levels of heavy metals in mosses with increasing distance from the power plant was observed.

Grodzińska and Kaźmierczakowa (1977) determined the accumulation of Cd, Pb and Fe in moss samples collected from urban parks in the urban-industrial agglomeration of Kraków, located some 5–17 km from a steelworks. In moss samples from the park farthest from the works the concentration of Fe was 2.5 times lower than in mosses from the park closest to the emissions source. The differences in Cd and Pb levels were small. This is evidence that in the investigated urban agglomeration the

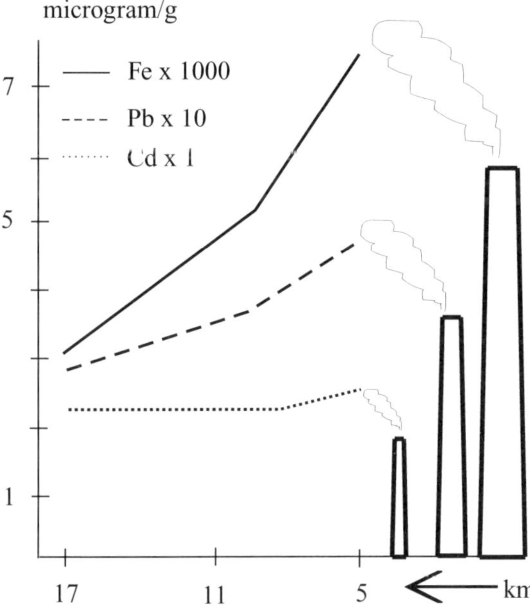

Figure 4.   Concentrations of Fe, Pb and Cd in moss in parks of Kraków city (southern Poland) (Grodzińska and Kaźmierczakowa, 1977).

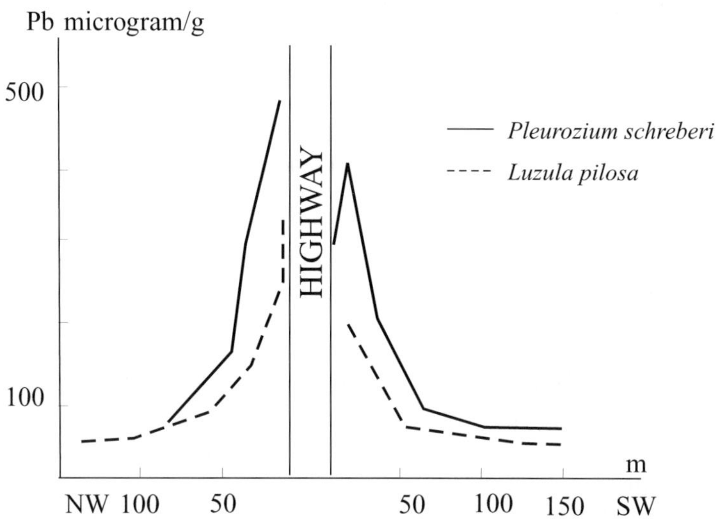

Figure 5.   Lead gradients in the road transect at Stavsjo (S. Sweden) (Rühling and Tyler, 1968).

most important source of Fe was the steelworks, while Pb and Cd are also emitted by other sources in the town itself (Fig. 4).

Rühling and Tyler (1968) showed that the lead concentrations in mosses growing near highways with heavy motor traffic in southern Sweden decreased markedly (nine-fold) over a distance of a mere 150 m from the highways (Fig. 5).

A first attempt to compare heavy metal concentrations in mosses with the incidence of various types of human diseases was peformed by Wappelhorst et al. (2000). They found a connection between respiratory diseases and levels of the elements of Ce, Fe, Ga and Ge. A correlation also existed between thallium concentrations and heart disease. Further investigations have been advised.

Estimation of deposition rates: In most studies of heavy metal pollution using the "moss-method", *concentrations* in mosses are compared with each other. This is satis-factory enough if mosses of the same species are compared or if intercalibration data are available for the species in use (See Section 4.2).

But for environmental protection measures that require political decisions, *deposi-tion* data (g/m^2, kg/ha etc.) are needed.

Sample strategies that are used in most of the national and international programmes (e.g. Rühling et al., 1987, Rühling and Steinnes, 1998) involve moss collection at sampling points without any interference of canopy and throughfall. Therefore, bulk depositions which facilitate the calculation of area-related depositions (e.g. Bergerhoff), have been compared with concentrations in mosses. The "efficiency factor" gives an estimation of the percentage of ions bound by the moss in relation to the absolute deposition. Following a series of studies (e.g. Steinnes, 1985; Ross, 1990; Gjengedahl and Steinnes, 1990; Berg et al., 1995; Thöni et al., 1996; Berg and Steinnes, 1997; Zechmeister, 1997) efficiency factors can be estimated as: As (25%),

Cd (60%) Co (60%), Cr (65%), Cu (35%), Fe (70%), Mo (60%), Ni (50%), Pb (100%), V (60%) and Zn (60%).

Whereas there are a series of parameters influencing uptake capacities (see also Section 2.2.1), the factors thus obtained seem good enough for overall estimations. Nevertheless, further studies on efficiency-factors are urgently needed.

Two methods have been used for the evaluation of deposition data from concentrations in mosses, both of which can be seen as additional advantages of the moss method.

Method A: Calculations following the formula (Rühling et al., 1987):

$$D = C * G / E$$

D: Deposition
C: Concentration analysed in mosses
G: Annual increase of biomass by area for the species in use
E: Efficiency factor

For this calculation, data on area-related bryophyte growth are needed. These data have been obtained by Tamm (1953), Rühling et al. (1987) and Zechmeister (1994, 1995a, 1998) for most of the species in use (e.g. *Hylocomium splendens* (Hedw.) B.S.G., *Pleurozium schreberi* (Brid.) Mitt., *Hypnum cupressiforme* Hedw. s. str. and *Abietinella abietina*(Hedw.) Fleisch.).

Method B: Calculation by regression equation
An alternative method is to calculate moss data versus atmospheric deposition values from analysis of precipitation samples by using linear regression. The regression equation can be used to transform moss concentration data directly to absolute deposition rates (Ross, 1990; Berg et al., 1995; Berg and Steinnes, 1997).

4.1.1.2. Live moss transplanting and moss bag technique
Two other assessment techniques are used, namely the transplantation method and the moss bag method. The former consists in transplanting living mosses together with their substratum from clean areas to areas under the influence of emissions. Mosses are exposed in polluted areas over a period of some weeks or months, and next examined for heavy metals concentrations (Goodman and Roberts, 1971; Pilegaard, 1979; Johnsen et al., 1983; Markert, 1993). For transplantation three moss species have been used: *Dicranoweissia cirrata* (Hedw.) Lindb., *Hypnum cupressiforme* agg. and *Tortula ruralis* agg. This method is not recommended as, in addition to atmospheric pollution, the transplanted plants experience stress induced by other habitat factors (e.g., changes in light, humidity) (Tyler, 1990; Čeburnis and Valiulis, 1999).

In the moss bag method, samples of dried or fresh mosses collected from clean areas are placed in nylon nets and exposed in a polluted area over different periods of time (two weeks, one month). Afterwards the concentration of heavy metals is measured in the samples. The most often used moss species are *Sphagnum* spp. and more rarely *Hypnum cupressiforme* agg, *Hylocomium splendens* (Hedw.) B.S.G. or *Pleurozium schreberi* (Brid.) Mitt. The moss bag technique has been used most often in urban areas and industrial agglomerations where indigenous mosses are absent (Goodman and Roberts, 1971; Little and Martin, 1974; Hynninen, 1986; Mäkinen, 1987; Yurukova and Ganeva, 1997; Čeburnis and Valiulis, 1999). The concentration of heavy metals in moss bags correlates with atmospheric heavy metals content. Tyler (1990) and Čeburnis and Valiulis (1999) recommend to use the moss bag technique in areas without indigenous moss flora. However, they also point to factors limiting the use of this method (type of bag, time of exposure).

4.1.1.3 Factors modifying the level of heavy metals in mosses

Many years of surveys carried out by researchers in different countries in and beyond Europe have shown that the level of heavy metals in mosses can be influenced by factors other than air pollution (climatic and edaphic factors, moss species, etc.)

In the following some factors most often mentioned by researchers are listed with references to publications where the reader can find detailed data. Researchers' opinions on the importance of particular factors vary and often are controversial. Some questions have already been discussed in Section 2. Most of the controversial results have to be a matter of research in the future to improve the method.

Beside local or global emissions factors modifying the heavy metals content of mosses are as follows:

1. Precipitation: intensity, frequency and duration can lead to an increase in concentrations as well as to a rinsing of already deposited metals (Rühling and Tyler, 1969; 1971; Ross, 1990; Tyler, 1990; Schmid-Grob et al., 1992; Rühling, 1992; 1994; Steinnes et al., 1992; Thöni et al., 1996).
2. Altitude: usually there is an increase in deposition with rising altitude (Groet, 1976; Zechmeister, 1994; 1995b; Šoltès, 1998), although there are controversial results (Schmid-Grob et al., 1992; Gerdol et al., 2002).
3. Mineral particles: mainly windblown dust from local soils, which make it difficult to distinguish between recent atmospheric depositions and waste from the past; several metals (e.g. Al, Ti) are used as indicators for the identification of the amount of windblown particles (Brown & Brown, 1990; Steinnes et al., 1992; Mäkinen, 1994b; Rühling, 1994; Steinnes, 1995; Čeburnis et al., 1999; Fernández and Carballeira, 2002; Figuera et al., 2002).
4. Natural cycling processes, in particular atmospheric transport from marine environments; this leads to a shift in element concentrations and changes in uptake efficiences (see also Section 2.2.1) (Rühling, 1992; Steinnes et al., 1992; Kuik and Wolterbeek, 1995; Čeburnis et al., 1999).
5. Root uptake in vascular plants from soil and subsequent transfer to mosses by leaching from living or dead plant tissues (Brown and Bates, 1990; Rühling, 1992, 1994; Steinnes et al., 1992; Kuik and Wolterbeek, 1995; see also Sections 2.2 and 2.3).
6. Moss species: *Pleurozium schreberi* (Brid.) Mitt. and *Hylocomium splendens* (Hedw.) B.S.G. are the most often used moss species (Rühling and Tyler, 1969; Grodzińska, 1978; Gydesen et al., 1983; Rühling, 1992; 1994; Herpin et al., 1994; Wolterbeek et al., 1995; Rühling and Steinnes, 1998, Galsomies et al., 1999; Sucharová and Suchara, 1998; Zechmeister, 1998; Grodzińska et al., 1999) but *Hypnum cupressiforme* agg. (Herpin et al., 1994; Zechmeister, 1994; Wolterbeek

et al., 1995) and *Scleropodium purum* (Hedw.) Limpr. (Markert et al., 1994; Sucharova and Suchara, 1998) somewhat rarely. Other moss species utilised as indicators of heavy metal pollution are *Dicranum scoparium* Hedw. (Markert et al., 1996), *Abietinella abietina* (Hedw.) Fleisch., *Rhytidium rugosum* (Hedw.) Kindb., *Ctenidium molluscum* (Hedw.) Mitt. (Zechmeister, 1994; 1998*), Polytrichum formosum* Hedw. (Markert et al., 1996), *Bryum argenteum* Hedw. (Thöni et al., 1993), *Thuidium tamariscinum* (Hedw.) B.S.G. (Galsomies et al., 1999), *Pohlia nutans* (Hedw.) Lindb. (Folkeson, 1979), *Leucobryum glaucum* Brid. (Groet, 1976), various species of *Sphagnum* and *Polytrichum* (Šoltès, 1992; 1998). The level of accumulation of heavy metals depends on the species. According to Folkeson (1979) the use of different monitor species in a heavy metal deposition survey require interspecies calibration. Thöni et al. (1996) suggest not to convert the data of one species to the other, even when there seem to be tendencies for different concentrations. A conversion may result in a bigger mistake. Wolterbeek et al. (1995) and Galsomies et al. (1999) suggest that the use of inter-species calibration from different countries may lead to unreliable results when concentration ranges are different.

7. Age of mosses – different exposure time (Pakarinen, 1977; Grodzińska, 1978; Makomaska, 1978; Brown and Brūmelis, 1996).

8. Part of mosses: some authors report about different accumulation rates in stems or leaves (Lötschert et al., 1975; Siebert et al., 1996).

9. Growth rate of species: variations in growth according to climatic conditions leads to different concentrations at comparable deposition rates (Zechmeister, 1995a; 1998).

10. Date of sampling (seasonal variation): there are results which indicate a variation of concentration according to sampling period as a result of moss-growth (Markert and Weckert, 1989; Zechmeister, 1994; Zechmeister et al., 2002a), others did not find any variations (e.g. Fernandez and Carballeira, 2002).

11. Type of moss samples (indigenous, transplanted, bag).

12. Sample preparation procedure and analytical methods (Steinnes et al., 1993; Markert et al., 1994; Wolterbeek and Bode, 1995).

Data evaluation and presentation has to be taken carefully and is critically analysed by several authors (e.g. Kostka-Rick et al., 2001; Herpin et al., 2001).

Despite these variations, when properly applied, moss methods are useful and often more useful than other biological methods in assessing atmospheric heavy metal pollution.

4.1.2. Heavy metal contamination of the environment in the past

Using mosses from old herbarium collections, one may assess the level of heavy metals contamination in the past (Rühling and Tyler, 1968, 1969, Herpin et al., 1997). According to Rühling and Tyler the concentrations of Pb, Ni and Cr in *Hypnum cupressiforme* agg. collected at the same localities in southern Sweden more than doubled over a period of 150 years. For Cu and Zn the estimated increase was much smaller. Although heavy metal concentrations were low in herbarial material of the years between 1845 and 1901, Herpin et al. (1997) found that the concentrations of As, Cd, Cr, Cu, Pb and Zn in mosses collected between 1845 and 1974 were generally higher than in mosses collected in 1991. They conclude that this was in consequence to improved methods of preventing air pollution in recent days than in earlier periods.

Peatbogs also well document the history of heavy metals contamination. According to Lee and Tallis (1973), a bog situated in the industrial part of England 'registered' a low level of Pb by the early 18th century, and next its gradual increase until the present time. Surveying ombrotrophic bogs in southern England, Jones and Hao (1993) demonstrated an increase in the Cd, Cu, Pb and Zn concentrations with the age of the bogs, and no increase in the level of Fe; they suggested that the history of heavy metals pollution cannot be reconstructed from profiles of bogs under hydrological influences. McKanzie et al. (1998) concluded that ombrotrophic, unsaturated peat bogs where heavy metals are immobilized may reflect the historical level of these elements, while minerotrophic and saturated bogs in which heavy metals are mobile elements cannot be treated as indicators. According to Tyler (1972), the patterns of heavy metals accumulation in ombrotrophic bogs in Sweden do not allow these bogs to be considered sensitive indicators of past changes in the concentrations of heavy metals.

Further information on historical heavy metal pollution records in peat bogs can be found in several papers (e.g. Martinez-Cortizas et al., 1997; Kuester and Rehfuess, 1997; West et al., 1997) published in the Proceedings of the workshop on "peat bog archives of atmospheric metal deposition".

4.2. Nitrogen compounds

The main natural sources of nitrogen emissions to the atmosphere are biomass burning, lightning, microbial soil processes, stratospheric input, and marine phytolotic and biological processes (Pitcairn and Fowler, 1995). The main sources of anthropogenic emissions are combustion of coal, particularly in power plants producing electricity, heating of buildings, industry and transport (diesel or petrol-driven vehicles), and production and spreading of animal manure (Europe's Environment, 1998). In the early 1980s the emissions of nitrogen compounds were approx. 35 million tonnes annually in Europe. In the mid-1990s they decreased to approx. 27 million tonnes (Europe's Environment, 1998). However, the rapid development of the transport sector can lead to the resurgence of nitrogen compound emissions in a short time.

Atmospheric deposition of fixed nitrogen as nitrate and ammonium in rain and by dry deposition of nitrogen dioxide, nitric acid and ammonia has increased throughout Europe during the last two decades of the 20th century, from 2–6 kg N/ha/year to 15–60 (80) kg N/ha/year (Sutton et al., 1993; Pitcairn and Fowler, 1995).

Nitrogen concentration in mosses can be a good indicator of the content of this element in the atmosphere, as exemplified by Baddley et al. (1994), Pitcairn and Fowler (1995) or Woolgrove and Woodin (1996). According to Baddley et al., nitrogen concentrations in *Racomitrium lanuginosum* (Hedw.) Brid. were clearly correlated with nitrogen deposition. In moss samples collected at the end of the 1980s the nitrogen concentration was 50% higher than in the 1950s and 2–3 times higher than in the 19th century. The other authors found that in 41 years (1950–1990) the nitrogen concentrations increased from 38 to 63% in some several moss species (among others *Rhytidiadelphus triquetrus* (Hedw.) Warnst., *Hypnum cupressiforme* agg, *Sphagnum* spp.) in regions of Great Britain where deposition was high (15–30 kg N/ha/yr); no such increase was noted in mosses growing in areas with low deposition (6 kg N/ha/yr). Similar effects were found by Woolgrove and Woodin (1996) in Scottish snowbed communities. They conclude that nitrogen depositions exceed by far what is recognised currently as critical loads.

Another appropriate biochemical marker for enhanced atmospheric nitrogen supply is the nitrate reductase activity. A series of studies which are reviewed by Lee et al. (1998) show that the activity of this enzyme reduces or get lost in polluted environments.

Press et al. (1986), Pitcairn et al. (1991), and Dirkse and Martakis (1992) found marked reduction in the growth of *Sphagnum cuspidatum* Hoffm em. Warnst., *Hylocomium splendens* (Hedw.) B.S.G. and *Pleurozium schreberi* (Brid.) Mitt. with an increase in nitrogen deposition (from 30 to 60 kg N/ha/yr). High atmospheric nitrogen input is partly responsible for the loss of certain bryophyte species and a decrease in bryophyte cover over the past 30–40 years (Pitcairn et al., 1991, Baddley et al., 1994), and declining health of *Sphagnum* communities (Press et al., 1986; Lee and Baxter, 1990; Woodin and Farmer, 1993; Gunnarsson 2000). Rao (1982) found losses of bryophyte species on Dutch chalk grasslands between 1953 and 1983, and the decline of *Rhytidiadelphus squarrosus* (Hedw.) Warnst. and *Scleropodium purum* (Hedw.) Limpr. in the Netherlands.

4.3. Sulphur compounds

Sulphur dioxide originates from both natural and anthropogenic sources in the environment. Natural sources of the sulphur in the atmosphere are volcanic activity, fires and bacterial activity, the sea and ocean. The main sources of anthropogenic emissions of sulphur are the energy sector and various factories where fossil fuels are used, especially coal and oil. For the northern hemisphere the natural sulphur input was 3.3 million t S/yr in the 1990s, while the anthropogenic input was 29 million t S/yr (Whelpdale et al., 1997); thus anthropogenic emissions constituted approx. 90% of the total sulphur input to the atmosphere.

Global sulphur emissions have been increasing steadily since 1850. At that time they were estimated at 1.2 million tonnes annually, while in the 1990s they reached 71.5 million tonnes (Elvingson, 2000). This steady growth of global sulphur input has been caused mainly by developing industries in Asian countries. In Europe, however, sulphur input has decreased markedly since the 1980s. In Europe at the end of the 20th century SO_2 emissions were by 53% lower than in the 1980s (Europe's Environment, 1998). The geographic distribution of SO_2 emissions is uneven in Europe; they are highest in Central and Eastern Europe (Europe's Environment, 1998).

Bryophytes are used for assessing contamination of the atmosphere with sulphur compounds by a wide range of methods, though quantitative estimations are much more rarely than for example assessing heavy metal pollution. They mainly respond by changes in their distribution and frequency of occurrence, and with changes in biomass, health, and structure of communities.

Mosses have been used as accumulators (Świeboda and Kalemba, 1987; Niskavaara and Äyräs, 1991; Grodzińska and Godzik, 1991; Szarek and Chrzanowska, 1991; Makinen, 1994b). Using a moss bag method, Świeboda and Kalemba (1987) exposed *Sphagnum recurvum* Pal. Beauv. for 6 weeks in the vicinity of the aluminium works and the power plant in Skawina (southern Poland) and then determined their sulphur content. Szarek and Chzanowaska (1991) determined the concentration of sulphur in *Pleurozium schreberi* (Brid.) Mitt. and *Hylocomium splendens* (Hedw.) B.S.G. collected in 14 national parks in Poland in 1975 and 1986. Niskavaara and Äyräs (1991) measured it in *Hylocomium splendens* (Hedw.) B.S.G. growing in forests surrounding the town of Rovaniemi in northern Finland, and Äryräs et al. (1997) in *Pleurozium schreberi* (Brid.) Mitt. and *Hylocomium splendens* (Hedw.) B.S.G. in northern Finland, Norway and Russia (Kola penninsula). Data on sulphur concentrations in some moss species, among others *Dicranum groenlandicum* Brid., *Hylocomium splendens* (Hedw.) B.S.G. and *Sanionia uncinata* (Hedw.) Loeske, growing in Spitsbergen (Horsund region) were given by Grodzińska and Godzik (1991). All these authors agree that accumulation of sulphur by mosses is not a good indicator of atmospheric SO_2 pollution. According to Äyräs et al. (1997) "there may be several reasons for this behaviour of the mosses: (1) being a nutrient sulphur has so high natural levels in mosses that no major additional uptake is possible, or (2) SO_2 as a mainly gaseous emission is transported high into the atmosphere and spread over large areas limiting the amount of sulphur offered for uptake in the immediate surroundings of the industrial plants or (3) sulphur is mostly deposited as sulphuric acid (acid rain) to which the uptake mechanisms of the moss do not respond favourably."

Pakarinen (1981) determined the concentration of sulphur in *Sphagnum fuscum* (Schimp.) Klinggr. and *Sph. balticum* (Russ.) C. Jens growing on bogs in southern Finland and found a clear correlation between the amount of sulphur in mosses and the level of atmospheric SO_4 deposition.

A large number of studies deal with the impact of sulphur depositions (wet and dry) on vitality and distribution of bryophyte species and populations. Bryophytes seem to be fairly sensitive to SO_2, even more than lichens (Türk and Wirth, 1975).

A series of investigations deal with the effect of SO_2 fumigation on bryophytes. Results clearly depend on (a) the concentrations, (b) time of exposure (c) air humidity (sensitivity increases with moisture) and (d) transformation of SO_2 into other substances, for example H_2SO_3 or H_2SO_4 (e.g. Gilbert, 1970 a,b; Frahm, 1977; Greven, 1992a,b; see also reviews in Winkler, 1977; Frahm, 1998).

Similar effects could be found by the results of experiments which transplanted bryophytes from unpolluted areas to regions with enhanced SO_2 concentrations like towns or the vicinities of point sources (Gilbert, 1968; Taoda, 1973; Greven, 1992a).

With increasing SO_2 concentrations bryophytes show a transformation of chlorophyll *a* into phaeophytin *a,* followed by a reduction or loss of chlorophyll content which finally could lead to extinction. Severe air pollution reduces sexual reproduction (e.g. Rao, 1982; Greven, 1992b; Otnyukova, 1995). Nash and Nash (1974) showed that bryophyte protonema is by a factor 10 less resistant than adult plants.

Several standardised methods have been involved to investigate the distribution of bryophytes within distinct areas or along a gradient of pollutants. In most of these cases the results reflect the overall air pollution of an area including several pollutants. Nevertheless sulphur and nitrogen pollutants are the most influential ones.

LeBlanc and De Sloover (1970) involved a method which estimated the 'Index of Atmospheric Purity' (IAP). It is based on the quantitative and qualitative distribution of epiphytic bryophytes (and lichens) in the investigated area.

$$\text{IAP} = \sum_{i=1}^{n} (Q_i \cdot f_i)$$

n = number of species at each sampling plot

i = index of each species

Q = the ecological index of each recorded species, calculated by a defined method and representing the overall resistance or sensibility against pollutants.

f = the coverage value at each sampling plot given in a defined scale.

The IAP index is calculated for each investigated sampling plot. In many cases contour maps with isolines of comparable I.A.P index which stands for similar pollution effects were drawn (e.g. LeBlanc et al., 1974; Zechmeister et al., 2002b, Fig. 6).

The method was used in a large number of investigation all over the world (e.g. Sergio, 1987; Inui and Yamaguchi, 1996; Palmieri et al., 1997).

Another method which originally is defined for lichens within the Association of Engineers standards list (VDI 1995, Richtlinie 3799) is in progress to be adapted for bryophytes too.

Derived from all these investigations many authors (e.g. Gilbert, 1970; Greven, 1992a,b; Frahm, 1998; Sauer, 2000; Zechmeister et al., 2002b) provided lists which showed the various tolerances of epiphytic bryophytes in regard to environmental pollution. The tolerances of species within these lists vary with respect to geographical areas. Habitat conditions (climate, bark structure etc.) strongly influence the distribution of bryophyte species and sometimes overlap pollution impacts.

A few examples for tolerances of epiphytic bryophytes in Central Europe are given:

- Resistant: *Bryum argenteum* Hedw., *Ceratodon purpureus* (Hedw.) Brid., *Dicranoweissia cirrata* (Hedw.) Lindb.
- Insensitive: *Amblystegium serpens* (Hedw.) B.S.G., *Brachythecium rutabulum* (Hedw.) B.S.G., *Hypnum cupressiforme* Hedw., *Orthotrichum diaphanum* Brid., *Plagiothecium laetum* B.S.G.
- Nearly insensitive: *Bryum capillare* agg., *Plagiothecium nemorale* (Mitt.) Jaeg., *Platygyrium repens* (Brid.) B.S.G.
- Moderate sensitive: *Metzgeria furcata* (L.) Dum., *Orthotrichum pumilum* Sw., *Orthotrichum obtusifolium* Brid., *Pylaisia polyantha* B.S.G., *Ulota crispa* (Hedw.) Brid.
- Sensitive: *Frullania dilatata* (L.) Dum., *Leucodon sciuroides* (Hedw.) Schwaegr., *Orthotrichum lyellii* Hook & Tayl., *Pterigynandrum filifome* Hedw.,
- Very sensitive: *Antitrichia curtipendula* (Hedw.) Brid., *Lejeunea cavifolia* (Ehrh.) Lindb., *Neckera pennata* Hedw., *Orthotrichum rogeri* Brid.

As a consequence to all these investigations we do not recommend the use of mosses as accumulator type – indicators of the level SO_2 in the air, but we believe

Figure 6. Map with isolines of comparable pollution derived from the mapping of epiphytic bryophytes in the heavy industrialised city of Linz (Austria; Zechmeister, unpublished data).

changes in the frequency and abundance of moss species, and changes in their health and community structure can be that good indicators of ambient SO_2 especially for monitoring at the ecosystem level.

4.4. Toxic organic compounds

The geochemistry of organic compounds is strongly connected with carbon circulation. On the one hand, they enter into the composition of the structures forming living organisms and they co-ordinate processes taking place in them. On the other hand, many organic compounds can be a potential source of environmental contamination. They often have toxic effects (carcinogenic, mutagenic, teratogenic) on living organisms. Organic compounds do not degrade quickly, so they persist in the environment for many years. The most frequent organic compounds occurring in the environment are aliphatic hydrocarbons (AH), mono- and polycyclic aromatic hydrocarbons (MAH, PAH), polychlorinated biphenyls (PCB) and chloro-organic pesticides. Many AH, MAH and PAH compounds are of natural origin, geological or biological, and these as a rule do not pose threats to the environment. More dangerous are compounds originating from anthropogenic sources connected with burning and processing caustobiolites and wastes. In the course of these processes toxic dioxins and furans are produced. Other organic compounds produced by man include solvents (e.g. chloroform), wood preservatives (e.g. pentachlorophenol), transformer (PCB), and pesticides.

The nature of these organic compounds is a major problem for their indication by bryophytes. Some of these substances are unstable and/or cannot be trapped by traditional methods, whereas some are beyond any detection limits (e.g. Umlauf et al., 1994).

Only a limited number of substances (mainly PAHs, PCB) have been indicated by mosses. A series of unanswered questions remains. Because of the lack of a cuticle, hydrophobic organic compounds do not show a special affinity to bryophyte surfaces. The accumulation of toxic organic compounds is not only dependent on atmospheric pollution levels but also on enrichment parameters, which describe physiological parameters as well as pollutant characteristics (Thomas, 1984; Strachan and Glooschenko, 1988). Drying of samples prior to storage might lead to secondary contamination by several organic pollutants. Lead et al. (1996) highlighted this fact for PCBs (especially tri- and tetrachlorinated groups) and concluded, therefore, that low-level samples should be analysed wet whenever possible.

Intensive investigations using mosses as indicators were performed by Thomas and Herrmann (1980) and Thomas (e.g. 1984; 1986), who studied a series of organic pollutants [α-HCH, γ-HCH (Lindan), DDT, PCB 60, Fluoranthen, Benzo(a)pyren] at 37 sites along a geographic profile through Central Europe. Epiphytic *Hypnum cupressiforme* L. ssp. *filiforme* was used as biomonitor, the amount of analysed moss was five gram dry weight per sample. A close correlation of pollutant concentration in moss and emissions mainly by agricultural sources was shown.

The enrichment factor of PAH's in mosses compared to concentrations measured in bulk precipitation was high (e.g. 500 for PCB, 2570 for 3.4 benzopyrene), but much lower if concentrations in mosses were compared to dry depositions (Thomas, 1984). It can be concluded that accumulation of PAH's in mosses depends also on the hydration of the moss.

The fluxes from the atmosphere to the ground surface of Benzo(a)pyren (as an example for PAH) were calculated by Milukaitė (1999). Three hundred samples of indigenous *Pleurozium schreberi* (Brid.) Mitt. and *Hylocomium splendens* (Hedw.) B.S.G. from Lithuania were analyzed. The concentration in mosses (average 54.7 μg kg^{-1}) were higher than in leaves of flowering plants, but lower than in needles or roots. Partitioning of BaP among the various parts of the moss, of different age did showed no marked differences. This was interpreted as a result of BaP stability in moss.

Sixteen polycyclic aromatic hydrocarbons were detected along a roadside by Viskari et al. (1997). They used moss bags filled with 20 gram of cleaned 2 cm tips of *Pleurozium schreberi* (Brid.) Mitt. and exposed the moss for seven weeks during summer. They emphasised that moss bags are an efficient collector of airborne PAHs.

Moss bags containing *Sphagnum* peat were used by Strachan and Glooschenko (1988) for the detection of PCB's and a series of organochlorines (á-HHC, Lindan, DDT, DDE, Dieldrin etc.). They highlight the various modes of distribution of these pollutants and that results therefore depend on exposure modes.

A series of PCB congers were analysed in northern and southern Norway. A general decline of PCBs in the investigated areas was provided with a less strong tendency for hexa- or heptachlorinated PCB congeners in the North, which might be evidence for the global fractionation hypothesis (Lead et al., 1996).

4.5. Radionuclides

Radioactive substances are often accumulated in bryophytes in great quantities. Svenson and Linden (1965) found that *Pleurozium schreberi* (Brid.) Mitt. absorbed Zr, B, Ba, La from fallouts after nuclear tests, Clymo (1978) and Oldfield et al. (1979) determined the concentrations of radiocaesium derived from the nuclear weapon tests of the 1960s in the top segment of peat mosses, Kwapuliński and Sarosiek (1988) estimated the ^{226}Ra/^{228}Ra ratio in dustfall, air and *Hypnum* species nearby and around a power station in Poland. Concentration of radioactive caesium (^{137}Cs) was determined in mosses (*Pleurozium schreberi* (Brid.) Mitt.) collected in Polish national parks in 1986 (Chernobyl accident) (Grodzińska, et al. 1993). The most contaminated with ^{137}Cs were parks localised in northeastern and southern Poland. It was correlated with the wind direction and precipitation in first days after the Chernobyl accident, and also with dust level in atmosphere in that time. Gerdol et al. (1994) determined the vertical distribution of ^{137}Cs in the uppermost layer of *Sphagnum* collected in Alps in 1988. They found that Chernobyl–radiocaesium peak was well distinguishable three years after the Chernobyl accident and disappeared three more years later. Cherchintsev et al. (2000) estimated several radionuclides incl. ^{124}Sb, ^{134}Cs, ^{131}Ba, ^{86}Rb in *Hylocomium splendens* (Hedw.) B.S.G. and *Pleurozium schreberi* (Brid.) Mitt. collected in Chelyabinsk region (Ural Mts.). They found extremely high concentration of Sb (2.3 ppm av. value, 29 ppm max.) in moss samples from a "steel town" (Magnitogarsk). These examples show that mosses are very valuable accumulator of radionuclides.

5. Bryophytes as indicators in aquatic habitats

5.1. General aspects

Bryophytes are often conspicuous elements of the macrophyte vegetation of aquatic fresh water habitats. In some of these habitats they are even more abundant than higher plants, for example in water spring vegetation (e.g. Zechmeister and Mucina, 1994), acidic lakes (e.g. Grahn, 1986) or fast running mountain streams (e.g. Geissler, 1976). They can not be found in marine ecosystems, though some terrestrial species on rocky coasts tolerate salt spray or can be found in brackish water (e.g. *Fontinalis dalecarlica* Br. Eur).

A wide range of threats affect aquatic habitats and some of them strongly reduce bryophyte populations. The increase of nutrients for example elevates algal growth and diminishes light for mosses and liverworts. Changes in the structure of aquatic ecosystems (e.g. bank reinforcements, river control) alter flow regimes as substrates and mostly lead to dramatic changes in the aquatic vegetation (e.g. Glime, 1992).

On the other hand, the decrease of pH caused by human activities like the deposition of airborne particles containing sulphur or the pollution by organic pollutants sometimes even benefit bryophytes. Pollutants suppress non-bryophyte aquatic macrophytes which are strong competitors for light, space and nutrients in unpolluted habitats. Vascular plants are less resistant than aquatic bryophytes against a series of substances and have less effective protective mechanisms against pollutant-induced changes of the environment (e.g. Grahn, 1977; Ek et al., 1995).

Whereas a wide range of species live close to permanent water reservoirs, only bryophytes submerged throughout the year can be used for biomonitoring (e.g. Tremp, 1992, 1999). The main advantages of using bryophytes as indictors in aquatic habitats are:

- There is a constant uptake of pollutants from water over the entire surface.
- Most aquatic bryophytes are fairly tolerant against a wide range of pollutants like heavy metals, which they tend to accumulate.
- Bryophytes react quickly to changes in water quality according to increases or decreases in nutrients or toxic substances.
- They form stable and homogeneous populations and they show green leaves and active metabolism throughout the year, which favours them over higher plants which lie dormant during the winter season, or algae which often show restricted life spans.
- There is only a limited number of submerged species in the northern hemisphere, which is in contrast to sometimes enormous biomass easily to identify in most of the cases.

Based on their ability either to accumulate pollutants or respond sensitively to changes in water quality, bryophytes are used either as accumulation indicators or the bryophyte species assemblages are investigated for indication of water quality (including the nutrient status) or changes in the pH.

5.2. Accumulation indicators

5.2.1. Heavy metals

The ability of bryophytes to accumulate heavy metals has already been described and there is little difference between aquatic and terrestrial mosses and liverworts regarding uptake mechanisms.

Of greater importance than in terrestrial habitats is the pH of the surrounding water. A low pH influences the uptake efficiencies of some metals negatively (Whitton et al., 1982; Say and Whitton, 1983; Claveri et al., 1995; Yoshimura et al., 1998; Carballeira et al., 2001). Gagnon et al. (1999) emphasise the importance of water hardness. Additionally, temperature clearly determines the uptake at least for copper (Glime, 1992).

Aquatic bryophytes generally have a high tolerance against a range of heavy metals like Pb, Zn or Fe (e.g. Glime and Keen, 1984). But with the exception of the "copper mosses", there is generally a fairly low resistance to copper (e.g. Glime and Keen, 1984; Tyler, 1990) which causes disturbance to photosynthesis (Sommer and Winkler, 1982). *Rhynchostegium riparoides* (Hedw.) Card. and *Scapania undulata* (L.) Dum. seem to be the most resistant species according to water pollution by heavy metals (e.g. Empain, 1976; Empain et al., 1980).

Like in terrestrial mosses the ability to accumulate heavy metals differs for each species and for each heavy metal. In general *Rhynchostegium riparoides* (Hedw.) Card. has the highest uptake rates compared to *Fontinalis antipyretica* Hedw., *Amblystegium riparium* (Hedw.) Lindb., *Cinclidotus danubicus* Schiffn. and Baumg., *Cinclidotus fontinaloides* (Hedw.) P. Beauv. or *Fissidens crassipes* Wils. (Wehr and Whitton, 1983a,b; Glime 1992; Mersch and Reichard, 1998; Gagnon et al., 1999).

There is a significant correlation between concentrations in water and moss tissues for some heavy metals, especially for Cd, Cu, Pb and Zn. (e.g. Whitton et al., 1982; Kelly and Whitton, 1989; Goncalves et al., 1994; Carter and Porter, 1997 [Cu, Zn]; Bruns et al., 1997), but there also exist contrasting reports for Cd and Pb (Say and Whitton, 1983; Carter and Porter, 1997).

As in terrestrial species the tips of the shoots show significantly lower concentrations than the whole plant (e.g. Whitton et al., 1982; Wehr et al., 1983), although there are differences between metals. For Cd, the differences in accumulation between the older and younger parts are in many cases greater than for the elements Cu, Pb and Zn in the same plant (Siebert et al., 1996).

The whole plant reflects the metals of the environment over a much longer period, whereas the tips should be used for monitoring short term changes, repeated surveys or the comparison of different river systems (Say and Whitton, 1983; Wehr and Whitton, 1983a).

Because of their high accumulation capacity, aquatic bryophytes have received increasing attention as bioindicators within the last three decades, especially in highly polluted water. Nevertheless, they can also be used in areas where the metal concentration in water-samples is beyond the detection limits (e.g. Jones, 1985). For Pb, an enrichment ratio of 3.5×10^5–1.2×10^6 in the liverwort *Scapania undulata* (L.) Dum. has been reported by Satake et al. (1989). Enrichment factors in *Rhynchostegium riparioides* (Hedw.) Card. as presented by Wehr and Whitton (1983b) are 9.4×10^5 for lead, and 2.3×10^6 for cadmium.

5.2.1.1. Methods

The considerable literature on aquatic bryophytes used as heavy metal accumulators is based on a variety of different practical methods. Benson-Evans and Williams (1976) give an overview on earlier studies. Wehr et al. (1983) tested a series of methods and gained sometimes markedly differing results. They emphasise the care which should be taken when comparing the metal compositions of aquatic plants from studies reported in the literature.

In many studies **indigenous plant material** was used to detect changes in pollution released by point sources or to evaluate the overall pollution of rivers in industrialised areas all over the world (e.g. Empain, 1976; Burton and Peterson, 1979, Nimis et al., 2002).

Transplants of mosses are used more often in aquatic than in terrestrial habitats (e.g. Mouvet, 1984; Carter and Porter, 1997; Mersch and Reichard, 1998, Sergio et al., 2000).

Moss bags are used as transplanting devices. Most of them are nylon bags with a size of 20 × 20 cm or 20 × 30 cm and with different mesh sizes (1 mesh cm^{-1}, 0.9 mesh cm^{-1}, 0.7 cm^{-1}, 0.07 cm^{-1}; e.g. Kelly et al., 1987). Additionally, Gimeno and Puche (1999) used garden mesh cylinders (1 mesh cm^{-1}) with a length of 25 cm length and 10 cm diameter. Anchored plastic tubes are used by Mouvet (1984). Mesh size appeared to have little effect upon metal accumulation by moss inside the bag, however, it is recommended that larger mesh size than 0.07 mesh cm^{-1} are used (Kelly et al., 1987). Various systems have been used for a stable anchoring of the bags (stones, steel stakes), which all must consider that transplants should be found even in cloudy waters and should be fixed according to the current velocity of the rivers (Benson-Evans and Williams, 1976).

Aquatic bryophytes are transferred mainly from unpolluted to polluted sites. A comparable water pH of the control and treatment study site is advised, as this influences heavy metal concentrations. A steep decline in concentrations of Fe and Al in mosses transplanted from acidic to neutral streams has been observed (e.g. Engleman and McDiffet, 1996).

The metal content of the autochthonous bryophytes is sometimes different from those of the transplanted mosses which implies different conclusions about the contamination level of the water. Adaptation to different pH conditions leading to physiological and structural specificity's could explain the different abilities of autochthonous and transplanted populations to accumulate metals in acidic surroundings (Claveri et al., 1995). Johansson (1995), who investigated the pollution gradient along the stream Smedbyan, north-east of Stockholm, found that the impact on the water quality as a consequence of increased urbanisation was more pronounced in the indigenous mosses than in the transplanted bryophytes.

As a **physiological parameter** regarding heavy metal contamination, the induction of thiol-containing peptides such glutathiones were investigated (e.g. Bruns et al., 1997; Doering et al., 1999; Bruns et al., 2001). A positive correlation was found between glutathione levels and Cd levels in the moss samples. These authors discuss the suitability of this biochemical response to stress as a biomarker for heavy-metal pollution at field locations.

Another method involves the identification of sites with different pollutant levels on the basis of a physiological stress criterion (the D665/D665a pigment index; Carballeira and Lopez, 1997). They also used this method to estimate background levels for each metal in each of the five investigated bryophytes (*Fontinalis antipyretica* Hedw., *Fissidens polyphyllus* Wils., *Brachythecium rivulare* Schimp., *Rhynchostegium riparioides* (Hedw.) Card, and *Scapania undulata* (L.) Dum.).

Experimenal attempts to use bryophytes for **biotechnical purification** of water have been performed by Samecka and Kempers (1996) using *Scapania undulata* (L.) Dum, and Ho et al. (1996) using *Sphagnum* ssp.

Regarding the total number of aquatic species investigations, the monitoring of heavy metal pollution has been restricted to only a very few species:

- *Amblystegium riparium* (Hedw.) Lindb. (e.g. Say et al., 1981; Wehr and Whitton, 1983b)
- *Brachythecium rivulare* Schimp. (e.g. McLean and Jones, 1975; Carballeira and Lopez, 1997)
- *Cinclidotus danubicus* Schiffn. and Baumg. (e.g. Empain, 1976; Mersch and Reichard, 1998)
- *Cinclidotus nigricans* (Brid.) Wijk and Marg. (e.g. Empain, 1976)
- *Fontinalis dalecarlica* Br. Eur. (e.g. Glime and Keen, 1984; Gagnon et al. 1999)
- *Fontinalis duriaei* Schimp. (e.g. Glime and Keen, 1984)
- *Fontinalis hypnoides* Hartm. (e.g. Gimeno and Puche, 1999)
- *Fontinalis squamosa* Hedw. (e.g. McLean and Jones, 1975; Say et al., 1981)
- *Fontinalis antipyretica* Hedw. (e.g. Pickering and Puia, 1969; Dietz, 1972; Kirchmann and Lambinon, 1973; Empain, 1976; Say et al., 1981; Say and Whitton, 1983; Wehr and Whitton, 1983a,b; Wehr et al., 1983; Mouvet, 1984; Glime and Keen, 1984; Kelly et al., 1987; Johansson, 1995; Siebert et al., 1996; Bruns et al., 1997; Carballeira and Lopez, 1997; Mersch and Reichard, 1998)
- *Fissidens polyphyllus* Wils. (e.g. Carballeira and Lopez, 1997)
- *Hygrohypnum ochraceum* (Wils.) Loeske (Claveri et al., 1995; Carter and Porter, 1997)
- *Pellia epiphylla* (L.) Corda (Claveri et al., 1995)
- *Rhynchostegium riparioides* (Hedw.) Card (e.g. Empain, 1976; Say et al., 1981; Say and Whitton, 1983; Wehr and Whitton, 1983a,b; Wehr et al., 1983; Glime and Keen, 1984; Kelly et al., 1987; Carballeira and Lopez, 1997; Mersch and Reichard, 1998; Gimeno and Puche, 1999; Gagnon et al., 1999) and
- *Scapania undulata* (L.) Dum.) (e.g. McLean and Jones, 1975; Burton and Peterson, 1979; Whitton et al., 1982; Wehr and Whitton, 1983a; Jones, 1985; Satake et al., 1989; Samecka and Kempers, 1996; Carballeira and Lopez, 1997; Yoshimura et al., 1999).

5.2.2. Radionuclides

A series of publications deals with the accumulation of radioactive elements derived from nature (e.g. Justyn and Stanek, 1974; Shacklette and Erdmann, 1982).

Recently, radionuclides derived from anthropogenic sources has been emphasised. Aquatic mosses were found to be useful monitoring organisms to detect effluents containing radionuclides derived from a power station in the river Meuse in France (Kirchmann and Lambinon, 1973; Lambinon et al., 1976). Compared to algae as well as higher plants, *Cinclidotus danubicus* Schiffn. and Baumg. had a much higher accumulation capacity for a range of elements (e.g. ^{60}Co, ^{134}Cs, ^{137}Cs) in that investigation.

Mersch and Kass (1994) detected the γ-radiation activity of artificial radionuclides (^{58}Co, ^{60}Co, ^{110}Ag, ^{124}Sb) by *Fontinalis antipyretica* Hedw. Their investigation recommends a monitoring system of the river Moselle downstream of a nuclear power plant near Cattenom.

5.2.3. Toxic organic compounds

Some of the organic pollutants (e.g. phenols) are quickly incorporated in tissues of aquatic bryophytes (Morrison and Wells, 1981). Depending on the concentrations, bryophytes either decompose and/or accumulate these substances, which lead to changes in the chlorophyll and phaeopigment concentrations (e.g. chl a, chl b, OD665/OD665a, Peñuelas, 1984; Martĭnez-Abaigar and Núñez-Olivera, 1998), plasmolysis (Auerbach et al., 1973) and to inhibition in growth (Glime, 1992). Tolerance against phenol or benzopyrenes varies within different species, and is high for *Rhynchostegium riparioides* (Hedw.) Card, *Amblystegium riparium* (Hedw.) Lindb. or *A. tenax* (Hedw.) Jens. and less for *Fontinalis antipyretica* Hedw.; liverworts seem to be the most sensitive group (Frahm, 1975; Peñuelas, 1984; Glime, 1992; Kosiba and Sarosiek, 1995; Vanderpoorten, 1999a).

5.2.3.1. Chlorinated organic compounds:

The spatial distribution of effluents from an insecticide producing factory in France containing hexachlorocyclohexanes (HCHs) and polychlorinated biphenyls – PCBs have been investigated by Mouvet et al. (1985). The moss *Cinclidotus danubicus* Schiffn. and Baumg. was exposed in moss bags. Chlorinated hydrocarbons as DDT or α-HCH accumulations reflected their concentration in the water. Hexachlorocyclohexan ($\acute{\alpha}$-, β-, γ- δ-, ϵ-), chlorophenols, hexachlorobenzene and DDTs were detected by Chovanec et al. (1994) in the rivers Danube and Traun along industrial areas in Austria. They found higher concentrations of these substances than in the underlying sediments. *Fontinalis antipyretica* Hedw. was used as indicator species in this investigation.

PCBs are accumulated in higher amounts than chlorinated hydrocarbons (Mouvet et al., 1985; Chovanec et al., 1994). From these investigations it can be concluded that *Cinclidotus danubicus* Schiffn. and Baumg. as well as *Fontinalis antipyretica* Hedw. can be used as bioindicator if the concentration of PCBs or HCHs are beyond detection limit in water-samples or sediments.

5.2.3.2. Polycyclic aromatic hydrocarbons

The concentrations of PAHs and sediments were compared by Chovanec et al. (1994). Whereas the concentrations in sediments were higher than in *Fontinalis antipyretica* Hedw., the moss proved to be a suitable indicator for a wide series PAHs (e.g. phenantrene, coronene).

The accumulation of PAHs and responses of antioxidant enzymes in *Fontinalis antipyretica* Hedw. transplanted around a Finnish city harbour were also studied by Roy et al. (1996). Glass fibre bags containing this species ("moss bags") were exposed for 35 days. This study introduced a new approach to investigate the cause-effect relationship between bioaccumulation of aquatic pollutants and the biochemical responses in organisms following exposure to such pollutants in a field setting.

5.2.3.3. Others

The distribution of tebufenozide (RH-5992, MIMIC) (N′-t-butyl-N′-(3,5-dimethylbenzoyl)-N-(4-ethyl-benzoyl) hydrazine) after spraying on forests in order to fight the spruce budworm (*Choristoneura fumiferana* Clem.) in Canadian forestry has been investigate using several indicators including the moss *Drepanocladus* sp. (Sundaram et al., 1996).

5.3. Indicators of water quality and acidification

Beside alterations in flow regime and substrates (Suren, 1996), changes in bryophyte species assemblages are mainly a result either of toxic substances (e.g. Cu, PCB's), shifting in the trophic status or the pH of the water. Depending on environmental changes, bryophytes can be favoured or disadvantaged against higher plants. Nutrient enrichment, shade effects, mechanic destruction, changes in water-pH cause different reactions by the various groups of organism. These factors correlate with each other and can often not be separated. Despite that, emphasis has been on two major topics involving bryophyte assemblages: water quality and acidification.

5.3.1. Water quality

Water quality mainly is a result of the trophic level and the quantity of toxic substances within the water as well as the natural ability of an aquatic habitat to regenerate from these influences. Trophic parameters mainly influence oxygen content, which is detected by a standard method for the evaluation of water quality, the 'saprobial-system' (e.g. Liebmann, 1962). Whereas several authors refer to this system, only a few bryophyte species have been considered as indicators within this system (e.g. Frahm, 1974; Papp and Rajczy, 1995). Oostendorp and Schmidt (1977) analysed the influence of effluents from a brewery on bryophyte biomass and calculated indices for the degree of saprobity for several species. Table 1 provides a list of species ranked within the various classes of the saprobial system, following the classifications of Frahm (1974, 1998), Papp and Rajczy (1995), Oostendorp and Schmidt (1977), Tremp (1999) and Zechmeister (unpublished data).

The system could be significantly improved by taking into account the results of a wide range of monitoring studies (e.g. Vanderporten, 1999a,b).

Bryophytes will be part of the evaluation system of the Water Framework Directive 2000 / 60 / EC of the European Union. They will play an important role mainly in alpine areas where vascular macrophytes are a less important part of river vegetation. The classification system has to be provided for each country.

A distinct change of bryophyte assemblages within the last years following changes in water quality has been reported for many rivers and lakes (e.g. Empain, 1973). After a period of a marked decline in species richness as a result of water pollution, an improvement in water quality during the last twenty years lead to an increase in species richness (e.g. in the Lower Rhine, Frahm and Abts, 1993). Furthermore, the raising of water temperatures within the last century seems to support the invasion of species (e.g. *Cinclidotus danubicus* Schiffn. and Baumg.) originating from milder climatic regions (Frahm, 1997). An increase in species richness as a result of a decrease in trophic levels and an increase in heavy metals is also reported by Vanderpoorten (1999b) for the rivers Meuse and Sambre.

Table 1. List of species ranked within the various classes of the saprobial system, following the classifications of Frahm (1974, 1998), Papp and Rajczy (1995), Oostendorp and Schmidt (1977), Tremp (2000), Zechmeister (unpublished). 0: xenosaprobic, 1: oligosaprobic, 2: α-mesosaprobic, 3 β-mesosaprobic, 4 eu-/poly-saprobic

Species	Sapobric class
Brachythecium rivulare (Hedwig) B.S.G.	1, 2, 3
Chiloscyphus polyanthus (L.) Corda	1, 2
Cinclidotus aquaticus (Hedw.) B.S.G.	1, 2, 3
Cinclidotus danubicus Schiffn. and Baumg.:	2, 3
Cinclidotus nigricans (Brid.) Wijk and Marg.:	2, 3
Cinclidotus fontinaloides (Hedw.) P. Beauv.	2, 3
Fissidens crassipes Wils.	2, 3
Fissidens arnoldii Ruthe	2
Fissidens fontanus (Pyl.) Steud.	2
Fontinalis squamosa Hedw.:	1
Fontinalis antipyretica Hedw.:	1, 2, 3, 4
Hygroamblystegium fluviatile (Hedw.) Loeske	2, 3
Hygroamblystegium tenax (Hedw.) Jenn.	0, 1, 2
Hygrohypnum ochraceum (Wils.) Loeske	0, 1, 2
Hygrohypnum luridum (Hedw.) Jenn.	2
Leptodictium riparium (Hedw.) Warnst.	2, 3, 4
Leskea polycarpa Ehrh.	2, 3
Rhynchostegium riparioides (Hedw.) Card:	0, 1, 2, 3, 4
Scapania undulata (L.) Dum.	0, 1
Schistidium alpicola (Hedw.) Limpr.	2

Frahm (1975) experimentally evaluated threshold values for several pollutants (e.g. NH_4^+, SO_4^{2-}, Cl^-), which are relevant for some aquatic bryophytes taken from the River Rhine.

Empain (1978) provided an early approach to quantifying the relationship between physico-chemical properties of water and the quantitative distribution of aquatic bryophytes as a result of species resistance to pollution in Belgium and northern France. Data synthesis lead to a water quality index. Vanderpoorten and Palm (1998) developed a method for predicting water quality by linear regression using aquatic bryophyte canonical variables as predictors and highlighted the importance of NH_4^+, NO_3^-, PO_4^{3-} and water temperature.

The photosynthetic pigment composition of aquatic bryophytes has been used as an indicator for water pollution in a series of investigations. The reduction of chlorophyll concentrations as a response to pollution has been reported (e.g. Glime, 1992). Different pigments exhibit different responses to pollution, sometimes associated with a decrease of the chlorophyll a/b ratio, although it has also been reported that this index might not be a proper tool for the indication of water pollution (Martínez-Abaigar and

Núñez-Olivera, 1998). Gimeno and Puche (1999) examined chlorophyll content and morphological changes in cellular structure induced by transplanting two species of aquatic bryophytes [*Rhynchostegium riparioides* (Hedw.) Card. and *Fontinalis hypnoides* Hartm.] in different sites of a river in Spain. Two transplanting devices were tested, a nylon mesh bag and a plastic net cylinder. They found that the changes in cellular structure mainly affect chloroplasts and follows a sequence of alterations which ends with cellular death.

Wehr and Whitton (1986) correlated a series of ecologiocal factors and pollutants with morphological variations in *Rhynchostegium riparioides* (Hedw.) C.Jens.

5.3.2. Acidification

Acidification is mainly a result of atmospheric depositions of acid effluents derived from industrial sources. Sources for acidification are SO_2, NO_x, and NH_3 emissions. NO_x orginate mainly from road traffic and NH_3 from cattle farming.

Several investigations underline the importance of water-pH as a decisive factor influencing species richness and causing changes in the floristic compositions of aquatic macrophytes (e.g. Roelofs, 1983; Arts, 1990; Frahm, 1992; 1998; Tremp and Kohler, 1995; Karttunen and Toivonen, 1995). Water-pH influences the availability and uptake of pollutants like heavy metals (5.2.1) or nutrients. Schuurkes et al. (1986) showed that acid tolerant species have an ammonium-dominated nitrogen utilization. Obviously bryophyte growth is strongly determined by these variables (e.g. Tremp, 1992).

Streams or lakes with a moderately lowered pH are rich in bryophyte species (Frahm, 1999; Satake et al., 1995), although a further decrease leads to a rapid loss in species (e.g. Yan et al. 1985; Ormerod and Wade, 1990).

Consequences from human induced acidification are more pronounced for bryophytes in oligotrophic lakes or streams with naturally reduced pH which developed in areas with old siliceous bedrock. The acidification of Scandinavian or Canadian lakes or streams and shifts in species assemblages has thus been reported by several authors (e.g. Grahn 1977).

During the last years, developments show reverse tendencies and indicate the reduction of environmental pollutants. Ek et al. (1995) ascribe the recent recovery of bryophytes or fishes in Swedish lakes to reduced deposition of sulphate. Brouwer et al. (1997) conclude from their experimental long term studies that the recovery of the water chemistry and vegetation of soft-water ecosystems is much slower after ammonium sulphate addition, compared with that of the recovery following sulphuric acid addition. Regarding the current nitrogen deposition levels in Europe, it is unlikely that recovery will occur within the next ten years.

Beside the mapping of species distributions measures of chlorophyll fluorescence was used as a measure for the tolerance of lowered water-pH (Tremp, 1993).

Reviews by Burton (1990), Glime (1992), Frahm (1998) or Tremp (2000) provide further information on this topic.

6. Indicators for global change

Based on their physiological and morphological features, bryophytes seem to be some of the plants most sensitive to changes in environmental conditions.

Bryophytes are appropriate indicators either in ecosystems which they dominate in terms of biomass or species richness (e.g. sub-Arctic), or in landscapes which are poor in species richness as a result of intensive human influence (e.g. arable fields).

6.1. Climate change

It must be expected that the increased release of greenhouse gases as CO_2, CH_4 or N_2O will lead to global warming. The predicted changes within the next fifty years give an increase between 1.5 to 6 °C (Mitchell et al., 1990; IPCC 2000). Warming will be pronounced in the Arctic, Antarctic and Alpine regions (Jacka and Budd, 1991; Maxwell, 1992; Watson et al., 1996). Additionally precipitation is also expected to increase in these areas. Both lead to a series of changes such as increased soil microbial activity, stimulated nutrient mineralization and enhanced emission of greenhouse gases (e.g. NH_4^+) by natural processes (e.g. Press et al., 1998). A shift in species composition of plants must be expected (e.g. Wookey et al., 1993; Grabherr et al. 1994; Kappelle et al., 1999; Arft et al., 1999).

In all Arctic and Alpine regions bryophytes are important in terms of biomass and ecosystem diversity (e.g. Longton, 1988; Russel, 1990; Gignac et al., 1998). They also favour growth of vascular plants in preventing erosion, intercepting pollutants or by nitrogen fixation (Press et al., 1998). Studies have been performed to analyse either growth responses or shifts in species diversity of bryophytes as a result of climate warming.

In a series of investigations the biomass production of populations and/or increments of single shoots of bryophytes have been investigated by environmental manipulation studies. By using Open Top Chambers (OTC) the influence of increased temperatures, precipitation and nitrogen depositions has been investigated in Arctic Europe and Canada. Additionally, studies have been performed which compare analogous variables at various latitudes or altitudes, including the Alps and Antarctic regions.

Whereas increased precipitation leads to greater growth rates in nearly all of the experimental studies, results for temperature and fertiliser application are controversial.

As reported by Callaghan et al. (1997), growth parameters were strongly correlated with early summer temperatures and the length of the growing seasons for *Hylocomium splendens* (Hedw.) B.S.G. (Hedw.) B.S.G. and *Polytrichum commune* Hedw. at seven circumpolar located arctic/subarctic sites. A positive correlation between temperature and bryophyte growth rates has also been reported by Zechmeister (1995a; 1998) for several species in Alpine environments. Potter et al. (1995) and Molau and Alatalo (1998) reported a negative or nil response of bryophyte biomass production on enhanced temperatures. This stands not only in contrast to the results given above but also to laboratory results obtained by several authors (e.g. Longton and Green, 1979; Furness and Grime, 1982).

Bryophyte species richness increased in poor heath communities at higher temperatures, but not in rich meadows (Molau and Alatalo, 1998).

Fertiliser application increased the general bryophyte cover in five-years experiments (Robinson et al., 1998), and similar results have been obtained for the cover of *Polytrichum commune* Brid. in long-term studies of Potter et al. (1995). Nevertheless, there were also negative responses to nitrogen application in some other studies (e.g. Jónsdóttir et al., 1995; Molau and Alatalo, 1998; see also Section 4.2).

Contrary to the parameters mentioned above there will be no response of bryophytes to enhanced CO_2, as plants already experience concentrations of atmospheric CO_2 predicted to occur over the next 50 years (Sonesson et al., 1992). This is due to their small size and therefore their growth just above soil surfaces (Sveinbjörnson and Oechel, 1992).

Because of large interannual climatic variations, changes in vegetation cover are difficult to estimate in the Antarctic. Changes seem to be more dramatic in the sub-Antarctic than in the continental Antarctic flora (Melick and Seppelt, 1997).

Modelling of changes of vegetation dominated by bryophytes which are based on climatic and geographic data have been performed by Gignac et al. (1998) for peatlands in the Mackenzie River Basin. They predict a movement of the southern boundary of peatland ecosystems 780 km north of today, but no changes in the current peatland species diversity.

Statistical modelling of bryophyte-environment relationships for several taxa of *Andraea* and *Racomitrium* are presented by Birks et al. (1998). They suggest models to predict the future geographical distribution of single species based on present-day climatic demands of the species. Similar models predicting the impact of climate change on species distributions have been presented for vascular plants in the Alps (Gottfried et al., 1998).

In regions with a temperate climate, changes are expected not to be remarkable. Nevertheless, it has been reported that 27 bryophyte species with a distribution mainly in the Mediterranean or Atlantic climate spread over central Europe within the last 12 years. This invasion is interpreted as a consequence of the increase of winter temperatures by 1.5 °C (Frahm and Klaus, 1997).

The results of experimental studies implicate a variation in bryophyte response to changing environmental conditions. As there is a lack of long-term data, further experimental studies will be needed if confident predictions can be made. The attempt to predict future changes by geographical modelling should be enforced. These models should also include data obtained by studies on the autecology of species as well as metapopulation data.

6.2. Land-use intensity

The alarming loss of biological diversity within the last decades, which is often caused by increasing land-use intensity represents a major challenge to the scientific community and demands the development of appropriate strategies of land management. Intensification of agriculture and forestry including fertilisation, irrigation and the use of pesticides are currently recognised as one of the major threats to biodiversity (European Commision, 1998; Hallingbäck, 1998; Matson et al., 1997).

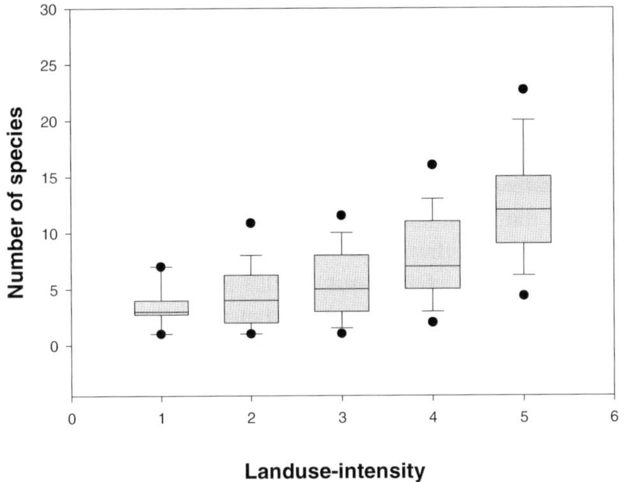

Figure 7. Box and whisker plots for the number of species within each state of land-use intensity at the habitat-scale. The boxes represent the median and 75th percentiles, the bars 90th percentiles; dots show the 95th percentiles for each group; classes of land-use from 1 (high intensity) to 5 (low intensity) (redrawn from Zechmeister and Moser, 2001).

Many investigations show a strong negative correlation between bryophyte diversity and land-use intensity both in landscapes used mainly by agriculture or forestry (e.g. Rydgren et al., 1998; Mensing et al., 1998; Jonsson and Jonsell, 1999; Zechmeister and Moser, 2001).

The most important variables explaining bryophyte diversity in boreal forests are the amount of dead wood (Ohlson et al., 1997), which correlates significantly with forest management intensity (Grabherr et al., 1998) and the age of a forest (Crites and Dale, 1998). In tropic forests as in boreal forests the total number of species on trees in old-growth coniferous forests is nearly twice that of species on trees in secondary-growth stands (e.g. Cooper-Ellis, 1998).

In agricultural landscapes correlations between species richness and land-use intensity are dependent upon disturbance intensity as well as geographic scale. Zechmeister and Moser (2001) showed that species richness on bare soils is at the highest at an intermediate disturbance regime, whereas at other substrates or geographic scales (e.g. habitat or landscape scale) species richness is negatively correlated with land-use intensity is (see also Fig. 7). There is also a correlation between land-use intensity and the number of endangered species (Zechmeister et al., 2002c). This is mainly a consequence of habitat destruction as many bryophytes are tolerant against a series of herbizides (Newsmaster et al., 1999).

Therefore, for further monitoring of changes in human land-use by means of bryophytes, either keystone indicators (e.g. hornworts; Bisang, 1995) or diversity indices including species richness can be used.

7. Conclusions and further prospects

Bryophytes have been an essential group in the field of bioindication for at least four decades. Most of the relevant research is reported in the above sections.

In the field of the estimation of atmospheric heavy metal pollution there is hardly any other group of organisms which give that suitable results, although there is still a need for research to improve the method (see Section 1.1.3).

As a consequence of the perfect results gained by this method there is an enormous demand for the integration of this method into the legislative procedure. Bryophytes should be acknowledged by law as indicator species for the setting and control of deposition limits for heavy metal imissions. A step in this direction was made by the integration of the European heavy metal programme (e.g. Rühling, 1998) into the UN/ECE Convention on Long-range Transboundary Air Pollution (ICPs Working Group on Effects) in the year 2000, although further steps are still missing.

The same goes for aquatic habitats, although in aquatic habitats a series of well investigated groups of organism (e.g. fish, see Dokulil, 2002; Chovanec et al., 2002; Oehlmann and Schulte-Oehlmann, 2002; Lorenz, 2002) might fit better for some questions, especially regarding pollutants in the food chain.

Bryophytes proved well in many fields of environmental control. Additionally, bryophytes seem to become an important group of species especially in the field of climate change research in the future (see also Section 6), which will be a major task in the next years.

Finally we like to say that bryophytes which are mostly small in size, are essentially for the integral understanding and control of the present state and future development of our environment.

References

Alliksaar, T., Punning, J.-M., 1998. The spatial distribution of characterized fly-ash particles and trace metals in lake sediments and catchment mosses, Estonia. Water, Air and Soil Pollution 106, 219–239.

Arft, A.M., Walker, M.D., Gurevitch, J., Alatalo, J.M., Bret-Harte, M.S., Dale, M., Diemer, M., Gugerli, F., Henry, G.H.R., Jones, M.H., Hollister, R.D., Jónsdóttir, I.S., Laine, K., Lévesque, E., Marion, G.M., Molau, U., Mølgaard, P., Nordenhäll, U., Raszhivin, V., Robinson, C.H., 1999. Responses of tundra plants to experimental warming: meta analysis of the International Tundra Experiment. Ecological Monographs 69, 491–511.

Arts, G.H.P., 1990. Aquatic bryophytes as indicators of water quality in shallow pools and lakes in the Netherlands. Annales Botanici Fennici 27, 19–32.

Auerbach, S., Prüfer, P., Weise, G., 1973. Gasstoffwechselphysiologische Schädigungskriterien bei submersen Makrophyten vom Typ *Fontinalis antipyretica* L. unter Einwirkung von Schwermetallen oder Phenol. Internationale Revue der Gesamten Hydrobiologie 58, 19–32.

Äyräs, M., Niskavaara, H., Bogatyrev, J., Chekushin, V., Pavlov, V., de Caritat, P., Halleraker, Jo.H.M, Finne, T.E., Kashulina, G., Reimann, C., 1997. Regional patterns of heavy metals (Co, Cr, Cu, Fe, Ni, Pb, V and Zn) and sulphur in terrestrial moss samples as indication of airborne pollution in a 188,000 km^2 area in northern Finland, Norway and Russia. Journal of Geochemical Exploration 53, 269–281.

Baddley, J.A., Thompson, D.B.A., Lee, J.A., 1994. Regional variation in the nitrogen content of *Racomitrium lanuginosum* in relation to atmospheric nitrogen deposition. Environmental Pollution 84, 189–196.

Barclay-Estrup, P., Rinne, R.J.K., 1979. Trace element accumulation in a feather moss and in soil near a kraft paper mill in Ontario. Bryologist 82, 599–602.

Bates, J.W., 1992. Mineral nutrition acquisition and retention by bryophytes. Journal of Bryology 17, 223–240.

Bates, J.W., Bakken, S., 1998. Nutrient retention, desiccation and redistribution in mosses. In: Bates, J.W., Ashton, N.W., Duckett, J.G. (Eds), Bryology in the Twenty-first Century. Maney Publishers and BBS, Leeds, pp. 293–304.

Bates J.W., Farmer, A., 1992. Bryophytes and lichens in a changing environment. Clarendon Press, Oxford.

Bengtson, C., Folkeson, L., Göransson, A., 1982. Growth reduction and branching frequency in *Hylocomium splendens* near a foundry emitting copper and zinc. Lindbergia 8, 129–138.

Benson-Evans, K., Williams, P.F., 1976. Transplanting aquatic bryophytes to assess river pollution. Journal of Bryology 9, 81–91.

Berg, T., Røset, O., Steinnes, E., 1995. Moss (*Hylocomium splendens*) used as biomonitor of atmospheric trace element deposition: estimation of uptake efficiencies. Atmospheric Environment 29, 353–360.

Berg, T., Steinnes, E., 1997. Use of mosses (*Hylocomium splendens* and *Pleurozium schreberi*) as biomonitors of heavy metal deposition: from relative to absolute deposition values. Environmental Pollution 98, 61–71.

Birks, H.J.B., Heegard, E., Birks, H.H., Jonsgard, B., 1998. Quantifying bryophyte-environment relationships. In: Bates, J.W., Ashton, N.W., Duckett, J.G. (Eds), Bryology in the Twenty-first Century. Maney Publishers and BBS, Leeds, pp. 305–319.

Bisang, I., 1995. The diaspore bank of hornworts (Anthocerotae, Bryophyta) and its role in the maintenance of populations in cultivated fields. Cryptogamica Helvetica 18, 107–116.

Brouwer, E., Bobbink, R., Meeuwesen, F., Roelofs, J.G.M., 1997. Recovery from acidification in aquatic mesocosms after reducing ammonium and sulphate deposition. Aquatic-Botany 56 (2), 119–130.

Brown, D.H., 1984. Uptake of mineral elements and their use in pollution monitoring. In: Dyer, A.F., Ducket, J.G. (Eds), The Experimental Biology of Bryophytes. Academic Press, London, pp. 55–62.

Brown, D.H., Bates J.W., 1990. Bryophytes and nutrient cycling. Botanical Journal of the Linnean Society 104, 129–147.

Brown, D.H., Brown, R.M., 1990. Reproducibility of sampling for element analysis using bryophytes. In: Lieth, H., Markert, B. (Eds), Element Concentration Cadasters in Ecosystems. VCH, Weinheim, pp. 55–62.

Brown, D.H., Brümelis, G., 1996. A biomonitoring method using the cellular distribution of metals in moss. The Science of the Total Environment 187, 153–161.

Brümelis, G., Brown, D.H., 1997. Movements of metals to new growing tissues on the moss Hylocomium splendens (Hedw.) BSG. Annales Botanici 79, 679–686.

Brümelis, G., Lapina, L., Tabors, G., 2000. Uptake of Ca, Mg and K during growth of annual segments of the moss *Hylocomium splendens* in the field. Journal of Bryology 22, 163–174.

Bruns, I., Friese, K., Markert, B., Krauss, G.J., 1997. The use of *Fontinalis antipyretica* L. ex Hedw. as a bioindicator for heavy metals. 2. Heavy metal accumulation and physiological reaction of *Fontinalis antipyretica* L. ex Hedw. in active biomonitoring in the River Elbe. Science of the Total Environment 204, 161–176.

Bruns, I., Friese, K., Markert, B., Krauss, G.J., 1999. Heavy metal inducible compounds from Fontinalis antipyretica reacting with Ellman's reagent are not phytochelatines. Science of the Total Environment 241, 215–216.

Bruns, I., Sutter, K., Menge, S., Neumann, D., Krauss, G.J., 2001. Cadmium lets increase the glutathione pool in bryophytes. Journal of Plant Physiology, 158, 79–89.

Burton, M.A.S., 1990. Terrestrial and aquatic bryophytes as monitors of environmental contaminants in urban and industrial areas. Botanical Journal of the Linnean Society 104, 267–280.

Burton, M.A.S., Peterson P.J., 1979. Studies on zinc localization in aquatic bryophytes. The Bryologist 82, 594–598.

Büscher, P., Koedam, N., 1979. Ecological aspects of cation absorption in several mosses and ferns. Bulletin Societé Royale de Botanique de Belgique. 112, 216–229.

Callaghan, T.V., Carlson, B.Å., Sonesson, M., Temesváry, A., 1997. Between-year variation in climate-related growth of circumarctic populations of the moss *Hylocomium splendens*. Functional Ecology 11, 157–165.

Carballeira, A., Lopez, J., 1997. Physiological and statistical methods to identify background levels of metals in aquatic bryophytes: dependence on lithology. Journal-of-Environmental-Quality 26, 980–988.

Carballeira, A., Vazquez, M.D., Lopez, J., 2001. Biomonitoring of sporadic acidification of rivers on the basis of release of preloaded cadmium from the aquatic bryophyte *Fontinalis antipyretica* Hedw. Environmental Pollution 111, 95–106.

Carter, L.F., Porter, S.D., 1997. Trace-element accumulation by *Hygrohypnum ochraceum* in the upper Rio Grande Basin, Colorado and New Mexico, USA. Environmental Toxicology and Chemistry 16, 2521–2528.

Čeburnis, D., Rühling, A., Kvietkus, K., 1997. Extended study of atmospheric heavy metal deposition in Lithuania based on moss analysis. Environmental Monitoring and Assessment 47, 135–152.

Čeburnis, D., Steinnes, E., Kvietkus, K., 1999. Estimation of metal uptake efficiencies from precipitation in mosses in Lithuania. Chemosphere 38, 445–455.

Čeburnis, D., Valiulis, D., 1999. Investigation of absolute metal uptake efficiency from precipitation in moss. The Science of the Total Environment 226, 247–253.

Cherchintsev, V.D., Frontasyeva, M.V., Lyapunov, S.M., Smirnov, L.J., 2000. Biomonitoring air pollution in Chelyabinsk region (Ural Mountains, Russia) through trace elements and radionuclides: temporal and spatial trends. Research Contract No. 9939/R2/Regular Budget Fund.

Chovanec, A., Vogel, W.R., Lorbeer, G., Hanus-Illnar, A., Seif, P., 1994. Chlorinated organic compounds, PAHs, and heavy metals in sediments and aquatic mosses of two Upper Austrian rivers. Chemosphere 29, 2117–2133.

Chovanec, A., Hofer, R., Schiemer, F., 2002 (2003). Fish as bioindicators. In: Markert, B.A., Breure, A.M., Zechmeister, H.G. (Eds), Bioindicators and Biomonitors. Elsevier, Oxford, pp. 639–676.

Claveri, B., Guerold, F., Pihan, J.C., 1995. Use of transplanted mosses and autochthonous liverworts to monitor trace metals in acidic and non-acidic headwater streams (Vosges mountains, France). Science of the Total Environment 175, 234–244.

Clymo, R.S., 1963. Ion exchange in Sphagnum and its relation to bog ecology. Annales Botanici 27, 309–324

Clymo, R.S., 1978. A model of peat bog growth. In: Heal, O.W., Perkins, D.F. (Eds), Production Ecology of British Moors and Montane Grasslands. Springer Verlag, Berlin, pp. 187–223.

Cooper-Ellis S., 1998. Bryophytes in old-growth forests of western Massachusetts. Journal of the Torrey Botanical Society 125, 117–132.

Crites, S., Dale, M.R.T., 1998. Diversity and abundance of bryophytes, lichens, and fungi in relation to wood substrate and successional stage in aspen mixedwood boreal forests. Canadian Journal of Botany 76, 641–651.

Davis, D.D., McClenahen, J.R., Hutnik, R.J., 2001. Use of an epiphytic moss to biomonitor pollutant levels in southwestern Pennsylvania. Northeastern Naturalist 8, 379–392.

Dietz, F., 1973. The enrichment of heavy metals in submerged plants. Advances in Water Pollution Research. Pergamon Press, Oxford, pp. 53–62.

Dirkse, G.M., Martakis, G.F.P., 1992. Effects of fertiliser on bryophytes in Swedish experiments on forest fertilisation. Biological Conservation 59, 155–161.

Doering, S., Oetken, M., Korhammer, S., Markert, B., 1999. Investigation of induction of phytochelatines in Fontinalis antipyretica and its use as a biomarker. Proceedings of the 9th Annual Meeting of SETAC-Europe, 25–29 May, Leipzig.

Dokulil, M.T., 2002 (2003). Algae as ecological bioindicators. In: Markert, B.A., Breure, A.M., Zechmeister, H.G. (Eds), Bioindicators and Biomonitors. Elsevier, Oxford, pp. 285–327.

During, H., 1997. Bryophyte diaspore banks. Advances in Bryology 6, 103–134.

During, H., 1979. Life strategies of bryophytes: a preliminary review. Lindbergia 5, 2–17.

Ek, A., Grahn, O., Hultberg, H., Renberg I., 1995. Recovery from acidification in Lake Orvattnet, Sweden. Water, Air and Soil Pollution 85, 1795–1800.

Elvingson, P., 2000. Global SO_2 emissions steady increase. Acid News 3, 26.

Empain, A., 1973. The aquatic and sub aquatic bryophytic vegetation of the Belgian part of the river Sambre related to ecological conditions and water pollution. Lejeunea 69, 1–58.

Empain, A., 1976. Les bryophytes aquatiques utilises comme traceurs de la contamination en métaux lourdes des eaux douces. Mémoires Société Royale Botanique Belgique 7, 141–156.

Empain, A., 1978. Quantitative relationships between populations of aquatic bryophytes and pollution of streams definition of an index of water quality. Hydrobiologia 60, 49–74.

Empain, A., Lambinon, J., Mouvet, C,, Kirchmann, R., 1980. Utilisation des bryophytes aquatiques et sub-aquatiques comme indicateurs biologiques de a la qualité des eaux courantes. In: Pesson, P. (Ed.), La pollution des eaux continentales. Gauthier-Villars, Paris, pp. 195–223.

Engleman, C.J., McDiffett, W.F., 1996. Accumulation of aluminum and iron by bryophytes in streams affected by acid-mine drainage. Environmental-Pollution 94, 67–74.

Europe's Environment, The Second Assessment, 1998. European Environment Agency, Elsevier Science, Oxford.

European Commission, 1998. First report on the implementation of the convention on biological diversity by the European Commission. DG XII, European Community, Luxembourg.

Fernandez, J.A., Carballeira, A., 2002. Biomonitoring metal deposition in Galicia (NW Spain) with mosses: factors affecting bioconcentration. Chemosphere 46, 535–542.

Figueira, R., Sergio, C., Sousa, A.J., 2002. Distribution of trace metals in moss biomonitors and assessment of contamination sources in Portugal. Environmental Pollution 118, 153–163.

Folkeson, L., 1979. Interspecies calibration of heavy metal concentrations in nine mosses and lichens – applicability to deposition measurements. Water, Air and Soil Pollution 11, 253–260.

Folkeson, L., 1981. Heavy metal accumulation in the moss *Pleurozium schreberi* in the surroundings of two peat-fired power plants in Finland. Annales Botanici Fennici 18, 245–253.

Fowler, D., Gallagher, M.W., Lovett G.M., 1993. Fog, cloud water and fog deposition. In: Lövblad, G., Erisman, J.W., Fowler D. (Eds), Models and Methods for the Quantification of Atmospheric Input to Ecosystems. Nordiske Seminar- og Arbejdsrapporter 1993, 573, Nordisk Council of Ministers, Copenhagen, pp. 51–73.

Frahm, J.P., 1974. Wassermoose als Indikatoren für die Gewässerverschmutzung am Beispiel des Niederrheins. Gewässer und Abwässer 53/54, 91–106.

Frahm, J.P., 1975. Toxitoleranzversuche an Wassermoosen. Gewässer und Abwässer 57/58, 59–66.

Frahm, J.P., 1977. Experimentelle Untersuchungen über Moose als Indikatoren für die Luftverschmutzung. Staub- Reinhaltung der Luft 37, 55–58.

Frahm, J.P., 1992. Ein Beitrag zur Wassermoosvegetation der Vogesen. Herzogia 9, 141–148.

Frahm, J.P., 1997. Zur Ausbreitung von Wassermoosen am Rhein (Deutschland) und seinen Nebenflüssen seit dem letzten Jahrhundert. Limnologica 27, 251–261.

Frahm, J.P., 1998. Moose als Bioindikatoren. Biologische Arbeitsbücher 57. Quelle, Meyer, Wiesbaden.

Frahm, J.P., Abts, U., 1993. Veränderungen in der Wassermoosflora des Niederheins 1972–1992. Limnologica 23, 123–130.

Frahm, J.P., Klaus, D., 1997. Moose als Indikatoren von Klimafluktuationen in Mitteleuropa. Erdkunde 51, 181–190.

Furness, S.B., Grime, J.P., 1982. Growth rate and temperature responses in bryophytes. Journal of Ecology 70, 525–536.

Gagnon, C., Vaillancourt, G., Pazdernik, L., 1999. Cadmium accumulation and elimination by two aquatic mosses, Fontinalis dalecarlica and Platyhypnidium riparioides: effect of Cd concentration, exposure time, water hardness and moss species. Revue des Sciences de l'Eau 12, 219–237.

Galloway, J.N., Thornton, J.D., Norton, S.A., Volchok, H.L., McLean, R.A.N., 1982. Trace metals in atmospheric deposition: a review and assessment. Atmospheric Environment 16, 1677–1700.

Galsomies, L., Letrouit, M.A., Deschamps, C., Savanne, D., Avnaim, M., 1999. Atmospheric metal deposition in France, initial results on moss calibration from the 1996 biomonitoring. The Science of the Total Environment 232, 39–47.

Geissler, P., 1976. Zur Vegetation alpiner Fliessgewässer. Beiträge zur Kryptogamenflora Schweiz, Teufen, 14, 1–52 + Table.

Gerdol, R., Bragazza, L., Marchesini, R., Aber, R., Bonetti, L., Lorenzoni, G., Achilli, M., Buffoni, A., De Marco, N., Franchi, M., Pison, S., Giaquinta, S., Palmieri, F., Spezzano, P., 2000. Monitoring of heavy metal deposition in Northern Italy by moss analysis. Environmental Pollution 108, 201–208.

Gerdol, R., Bragazza, L., Marchesini, R., 2002. Element concentrations in the forest moss Hylocomium splendens: variation associated with altitude, net primary production and soil chemistry. Environmental Pollution 116, 129–135.

Gerdol, R., Degetto, S., Mazzotta, D., Vecchiati, G., 1994. The vertical distribution of the Cs137 derived from Chernobyl fallout in the uppermost Sphagnum layer of two peatlands in the southern Alps (Italy). Water, Air and Soil Pollution 75, 93–106.

Gignac, L.D., Nicholson, B.J., Bayley, S.E., 1998. The utilization of bryophytes in bioclimatic modelling: predicted northward migration of peatlands in the Mackenzie River Basin, Canada, as a result of global warming. The Bryologist 101, 572–587.

Gilbert, O.L., 1968. Bryophytes as indicators of air pollution in the Tyne valley. New Phytologist 67, 15–30.

Gilbert, O.L., 1970a. A biological scale for the estimation of sulphur dioxide pollution. New Phytologist 69, 629–634.

Gilbert, O.L., 1970b. Further studies on the effect of sulphur dioxide on lichens and bryophytes. New Phytologist 69, 605–627.

Gimeno, C., Puche, F., 1999. Chlorophyll content and morphological changes in cellular structure of *Rhynchostegium riparioides* (Hedw.) Card. (Brachytheciaceae, Musci) and *Fontinalis hypnoides* Hartm. (Fontinalaceae, Musci) in response to water pollution and transplant containers on Palancia river (East, Spain). Nova-Hedwigia 68, 197–216.

Gjengedahl, E., Steinnes, E., 1990. Uptake of metal ions in moss from artificial precipitation. Environmental Monitoring and Assessment 14, 77–87.

Glime, J.M., 1992. Effects of pollutants on aquatic species. In: Bates, J., Farmer, A. (Eds), Bryophytes and Lichens in a Changing Environment. Clarendon Press, Oxford, pp. 333–361.

Glime, J.M., Keen, R.E., 1984. The importance of bryophytes in a man-centered world. Journal of the Hattori Botanical Laboratory 55, 133–146.

Goncalves, E.P.R., Soares, H.M.V.M., Boaventura, R.A.R., Machado, A.S.C., Esteves da Silva, J.C.G., 1994. Seasonal variations of heavy metals in sediment and aquatic mosses from the Cavado river basin (Portugal). Science of the Total Environment 142, 143–156.

Goodman, G.T., Roberts, T.M., 1971. Plants and soils as indicators of metals in the air. Nature 231, 287–292.

Gottfried, M., Pauli, H., Grabherr, G., 1998. Prediction of vegetation patterns at the limits of plant life: a new view on the alpine-nival ecotone. Arctic and Alpine Research 30, 207–221.

Grabherr, G., Koch, G., Kirchmeir, H., Reiter, K., 1998. Hemerobie österreichischer Waldsysteme. Veröff. des österr. MaB-Programms. Bd.17, Universitätsverlag Wagner, Innsbruck.

Grabherr, G., Gottfried, M., Pauli, H., 1994. Climate effects on mountain plants. Nature 369, 448.

Grahn, O., 1977. Macrophytes succession in Swedish lakes caused by deposition of airborne acid substances. Water, Air and Soil Pollution 7, 295–305.

Grahn, O., 1986. Vegetation structure and primary production in acidified lakes in southwestern Sweden. Experientia 42, 465–470.

Greven, H.C., 1992a. Changes in the Dutch bryophyte flora and air pollution. Dissertationes Botanici 194, 1–237.

Greven, H.C., 1992b. Changes in the moss flora of the Netherlands. Biological Conservation 59, 133–137.

Grime, J.P., 1979. Plant strategies and vegetation process. Wiley, Chichester.

Grime, J.P., Rincon, E.R., Wickerson, B.E., 1990. Bryophytes and plant strategy theory. Botanical Journal of the Linnean Society 104, 175–186

Grodzińska, K., 1978. Mosses as bioindicators of heavy metal pollution in Polish National Parks. Water, Air and Soil Pollution 9, 83–97.

Grodzińska, K., Godzik, B., 1991. Heavy metals and sulphur in mosses from southern Spitsbergen. Polar Research 9, 133–140.

Grodzińska, K., Godzik, B., Szarek, G., 1993. Heavy metals, sulphur and radioactive Cs137 pollution of Polish national parks. Prądnik, Prace Muzeum Szafera 7–8, 153–158 (in Polish).

Grodzińska, K., Kaźmierczakowa, R., 1977. Heavy metal content in the plants of Cracow parks. Bulletin de l'Acadèmie Polonaise des Sciences Cl. V, 25, 227–234.

Grodzińska, K., Szarek-Łukaszewska, G., 2001. Response of mosses to heavy metal deposition in Poland – an overview. Environmental Pollution 114, 443–451.

Grodzińska, K., Szarek-Łukaszewska, G., Godzik, B., 1999. Survey of heavy metal deposition in Poland using mosses as indicators. The Science of the Total Environment 229, 41–51.

Groet, S.S., 1976. Regional and local variations in heavy metal concentrations of bryophytes in the northeastern United States. Oikos 27, 445–456.

Gunnarsson, U., 2000. Vegetation changes on Swedish mires. Effects of raised temperature and increased nitrogen and sulphur influx. Acta Universitatis Upsaliensis. Comprehensive summaries of Uppsala dissertations from the faculty of science and technology, 561. Uppsala, 25 pp.

Gydesen, H., Pilegaard, K., Rasmmussen, L., Rühling, Å., 1983. Moss analyses used as a means of surveying the atmospheric heavy metal deposition in Sweden, Denmark and Greenland in 1980. Bulletin SNV PM 1670, 1–44.

Halleraker, J.H., Reimann, C., de Caritat, P., Finne, T.E., Kashulina, G., Niskaavaara, H., Bogatyrev, I., 1998. Reliability of moss (*Hylocomium splendens* and *Pleurozium schreberi)* as a bioindicator of atmospheric chemistry in the Barents region, interspecies and field duplicate variability. The Science of the Total Environment 218, 123–139.

Hallingbäck, T., 1998. Threats and protection of bryophytes in Sweden. Journal of the Hattori Botanical Laboratory 84, 175–185

Herpin, U., Berlekamp, J., Markert, B., Wolterbeek, B., Grodzińska, K., Siewers, U., Lieth, H., Weckert, V., 1996. The distribution of heavy metals in a transect of the three states the Netherlands, Germany and Poland, determined with the aid of moss monitoring. The Science of the Total Environment 187, 185–198.

Herpin, U., Markert, B., Siewers, U., Lieth, H., 1994. Monitoring der Schwernetallbelastung in der Bundesrepublik Deutschland mit Hilfe von Moosanalysen. Bundesministerium f. Umwelt, Naturschutz und Reaktorsicherheit, 1–161.

Herpin, U., Markert, B., Weckert, V., Berlekamp, J., Friese, K., Siewers, U., Lieth, H., 1997. Retrospective analysis of heavy metal concentration at selected locations in the Federal Republic of Germany using moss material from a herbarium. The Science of the Total Environment 205, 1–12.

Herpin, U., Siewers, U., Kreimes, K., Markert, U., 2001. Biomonitoring – evaluation and assessment of heavy metal concentrations from two German moss monitoring surveys. In: Burga, C.A., Kratochwil A. (Eds), Biomonitoring: general and applied aspects on regional and global scales. Kluwer, Dordrecht, pp. 73–95.

Hertz, J., Schmid, I., Thöni, L., 1984. Monitoring of heavy metals in airborne particles by using mosses collected from the city of Zurich. International Journal of Environmental Analytical Chemistry 17, 1–12.

Ho, Y.S., Wase, D.A.J., Forster, C.F., 1996. Removal of lead ions from aqueous solution using sphagnum moss peat as adsorbent. Water South Africa Pretoria 22, 219–224.

Hynninen, V., 1986. Monitoring of airborne metal pollution with moss bags near an industrial source at Haryavalta, southwest Finland. Annales Botanici Fennici 23, 83–90.

Inui, T., Yamaguchi, T., 1996. Epiphytic bryophytes in Naha City (subtropical urban area in Okinawa Island, southern Japan), with special reference to air pollution. Hikobia 12, 161–168.

IPCC, 2000. Summary for policymakers. A report of Working Group I of the intergovernmental panel on climate change. www.ipcc.ch/pub/tpbiodiv.pdf.

Jacka, T.H., Budd, W.F., 1991. Detection of temperature and sea-ice extension changes in the Antarctic and Southern Ocean. In: Weller, G., Wilson, C, Severin, B. (Eds), Proceedings of the International Conference on the Role of Polar Regions in Global Change. University of Alaska, pp. 63–70.

Johansson, L., 1995. Detection of metal contamination along a small river through transplantation of the aquatic moss *Fontinalis antipyretica*. Aqua-Fennica 25, 49–55.

Johnsen, I., Pilegaard, K., Nymand, E., 1983. Heavy metal uptake in transplanted and in situ yarrow (*Achillea millefolium)* and epiphytic cryptogams at rural, urban and industrial localities in Denmark. Environmental Monitoring and Assessment 3, 13–22.

Jones, J.M., Hao, J., 1993. Ombrotrophic peat as a medium for historical monitoring of heavy metal pollution. Environmental Geochemistry and Health 15, 67–74.

Jones, K.C., 1985. Gold, silver and other elements in aquatic bryophytes from a mineralised area of North Wales, UK. Journal of Geochemical Exploration 24, 237–246.

Jónsdóttir, I.S., Callaghan, T.V., Lee, J.A., 1995. Fate of added nitrogen in a moss-sedge Arctic community and effects of increased nitrogen deposition. The Science of the Total Environment 160/161, 677–685.

Jonsson, B.G., Jonsell, M., 1999. Exploring potential biodiversity indicators in boreal forests. Biodiversity and Conservation 8, 1417–1433.

Justyn, J., Stanek, Z., 1974. Accumulation of natural radio nuclides in the bottom sediments and by aquatic organisms of streams. Int. Rev. Gesamten Hydrobiologie 59, 593–609.

Kappelle, M., Van Vuuren, M.M.J., Baas, P., 1999. Effects of climate change on biodiversity: a review and identification of key research issues. Biodiversity and Conservation 8, 1383–1397.

Karttunen, K., Toivonen, H., 1995. Ecology of aquatic bryophyte assemblages in 54 small Finnish lakes, and their changes in 30 years. Annales-Botanici-Fennici 32 (2), 75–90.

Kauneliene, W., 1995. Atmospheric heavy metal deposition in Kaunas region monitored by moss analyses. Atmospheric Physics 17, 53–59.

Kelly, M.G., Girton, C., Whitton, B.A., 1987. Use of moss bags for monitoring heavy metals in rivers. Water Research 21, 1429–1435.

Kelly, M.G., Whitton, B.A., 1989. Interspecific differences in zinc, cadmium and lead accumulation by freshwater algae and bryophytes. Hydrobiologia 175, 1–12.

Kirchmann, R., Lambinon, J., 1973. Bioindicateurs végétaux de la contamination d'un cours d'eau par des effluents d'une centrale nucléaire a eau pressurisée. Bulletin Société Royale de Botanique de Belgique 106, 187–201.

Kosiba, P., Sarosiek, J., 1995. Disappearance of aquatic bryophytes resulting from water pollution by textile industry. Cryptogamica Helvetica 18, 85–93.

Kostka-Rick, R., Leffler, U.S., Markert, B., Herpin, U., Lusche, M., Lehrke, J., 2001. Biomonitoreing zur wirkungsbezogenen Ermittlung der Schadstoffbelastung in terrestrischen Ökosystemen. Zeitschrift Umweltchemie und Ökotoxikologie 12, 5–12.

Kuester, H., Rehfuess, K.E., 1997. Pb and Cd concentrations in a southern Bavarian bog profile and the history of vegetation as recorded by pollen analysis. Water, Air and Soil Pollution 100, 379–386.

Kuik, P., Wolterbeek, H.Th., 1995. Factor analysis of atmospheric trace-element deposition data in the Netherlands obtained by moss monitoring. Water, Air and Soil Pollution 84, 323–346.

Kunert, M., Friese, K., Weckert, V., Markert, B., 1999. Lead isotope systematics in Polytrichum formosum: an example from a biomonitoring field study with mosses. Environment Sci. Technology 33, 3502–3505.

Kwapuliński, J., Sarosiek, J., 1988. 226Ra/228Ra quotient in some species of mosses as a new method of estimation of the influence of a power station. In: Glime, J.M. (Ed.), Methods in Bryology. Workshop, Mainz, pp. 245–247. Hattori Botanical Laboratory, Nichiunan, Japan.

Lambinon, J., Kirchmann, R., Colard, J., 1976. Evolution récente de la contamination radioactive de écosystèmes aquatique et ripicole de la Meuse par les effluents de la Centrale nucléaire de la SENA (Chooz, Ardennes Françaises). Mémoires Société Royale de Botanique de Belgique 7, 157–175.

Lead, W.A., Steinnes, E., Jones, K.C., 1996. Atmospheric deposition of PCBs to moss (*Hylocomium splendens*) in Norway between 1977 and 1990. Environmental Science, Technology 30, 524–530.

LeBlanc, F., De Sloover, J., 1970. Relation between industrialization and the distribution and growth of epiphytic lichens and mosses in Montreal. Canadian Journal of Botany 48, 1485–1496.

LeBlanc, F., Robitaille, G., Rao, D., 1974. Biological responses to lichens and bryophytes to environmental pollution in the Murdochville Copper Mine area. Journal of the Hattori Botanical Lab. 38, 205–433.

Lee, J.A., Baxter, R., 1990. Responses of Sphagnum species to atmospheric nitrogen and sulphur deposition. Botanical Journal of the Linnean Society 104, 255–265.

Lee, J.A., Caporn, S.J.M., Carroll, J., Foot, J.P., Johnson, D., Potter, L., Taylor, A.F.S., 1998. Effects of ozone and atmospheric nitrogen deposition on bryophytes. In: Bates, J.W., Ashton, N.W., Duckett, J.G. (Eds), Bryology in the Twenty-first Century. Maney Publishers and BBS, Leeds, pp. 331–341.

Lee, J.A., Tallis, J.H., 1973. Regional and historical aspects of lead pollution in Britain. Nature 245, 216–218.

Liebmann, H., 1962. Handbuch der Frischwasser- und Abwasserbiologie. Band 1. Gustav Fischer Verlag, Jena.

Lieth, H., Markert, B. (Eds), 1990. Element concentration cadaster in ecosystems, methods of assessment and evaluation. VCH Verlagsgesellschaft mbH, Weinheim.

Liiv, S., Sander, E., Eensaar, A., 1994. Territorial distribution of heavy metals content in Estonian mosses. Methodological investigation. Journal of Ecological Chemistry 3, 101–110.

Lippo, H., Poikolainen, J., Kubin, E., 1995 The use of moss, lichen and pine bark in the nationwide monitoring of atmospheric heavy metal deposition in Finland. Water, Air and Soil Pollution 85, 2241–2246.

Little, P., Martin, M.H., 1974. A survey of zinc, lead and cadmium in soil and natural vegetation around a smelting complex. Environmental Pollution 3, 241–254.

Longton, R.E., 1988. The Biology of Polar Bryophytes and Lichens. Cambridge University Press, Cambridge.

Longton, R.E., 1997. Reproductive biology and life-history strategies. Advances in Bryology 6, 65–101.

Lorenz, C.M., 2002 (2003). Bioindicators in ecosystem management. In: Markert, B.A., Breure, A.M., Zechmeister, H.G. (Eds), Bioindicators and Biomonitors. Elsevier, Oxford, pp. 123–152.

Lötschert, W., Wandtner, R., Hiller, H., 1975. Schwermetallanreicherung bei Bodenmossen in Immisionsgebieten. Berichte der Deutschen Botanischen Gesellschaff 88, 419–431.

Lüttge, U., Bauer, K., 1968. Die Kinetik der Ionenaufnahme durch junge und alte Sprosse von *Mnium cuspidatum*. Planta 78, 310–320.

Mägdefrau, K., 1982. Life forms of bryophytes. In Smith, A.J.E. (Ed.), Bryophyte Ecology. Chapman & Hall, London, pp. 45–58.

Mäkinen, A., 1983. Use of *Hylocomium splendens* for regional an local heavy metal monitoring around a coal-fired power plant in Southern Finland. Symposia Biologica Hungarica 35, 777–794.

Mäkinen, A., 1987. *Sphagnum* moss bags in air pollution monitoring in the city of Helsinki. Symposia Biologica Hungarica 35, 755–775.

Mäkinen, A., 1994a. Heavy metal and arsenic concentrations of a woodland moss, *Hylocomiun splendens* (Hedw.) Br. Et Sch., growing around a coal-fired power plant in coastal southern Finland. Projekt KHM (Kol. Hälsa. Miljo). Teknisk Rapport 85, 1–85.

Mäkinen, A., 1994b. Biomonitoring of atmospheric deposition in the Kola Peninsula (Russia) and Finnish Lapland, based on the chemical analysis of mosses. Ministry of the Environment, Rapport 4, 1–83.

Makomaska, M., 1978. Heavy metal contamination of pinewoods in the Niepołomice Forest (Southern Poland). Bulletun de l'Acadèmie Polonaise des Sciences Cl. II, 26, 679–685.

Mankovska, B., 1997. Deposition of heavy metals in Slovakia. Assessment on the basis of moss and humus analyses. Ecology (Bratislava) 16, 433–442.

Markert, B. (Ed.), 1993. Plants as biomonitors – indicators for heavy metals in terrestrial environment. VCH Verlagsgesellschaft mbH, Weinheim.

Markert, B., Herpin, U., Berlekamp, J.,Oehlmann, J., Grodzińska, K., Mankovska, B., Suchara, I., Siewers, U., Weckert, V., Lieth, H., 1996. A comparison of heavy metal deposition in selected Eastern European countries using the moss monitoring method with a special emphasis on the "Black Triangle". The Science of the Total Environment 193, 85–100.

Markert, B., Reus, U., Herpin, U., 1994. The application of TXRF in instrumental multielement analyses of plants, demonstrated with species of moss. The Science of the Total Environment 152, 213–220.

Markert, B., Weckert, V., 1989. Fluctuations of element concentrations during the growing season of *Polytrichum formosum* (Hedw.). Water, Air and Soil Pollution 43, 177–189.

Markert, B., Weckert, V., 1994. Higher lead concentrations in the environment of former West Germany after the fall of the "Berlin Wall". The Science of the Total Environment 158, 93–96.

Martĩnez-Abaigar, J., Núñez-Olivera, E., 1998. Ecophysiology of photosynthetic pigments in aquatic bryophytes. In: Bates, J.W., Ashton, N.W., Duckett, J.G. (Eds), Bryology in the Twenty-first Century. Maney Publishers and BBS, Leeds, pp. 277–292.

Martinez-Cortizas, A., Pontevedra-Pomba, X., Nuova-Numoz, J.C., Garcia-Rodeja, E., 1997. Four thousand years of atmospheric Pb, Cd and Zn deposition recorded by the ombrotrophic peat bog of Penido Vello (northwestern Spain). Water, Air and Soil Pollution 100, 387–403.

Matson, P.A., Parton, W.J., Power, A.G., Swift, M.J., 1997. Agricultural intensification and ecosystem properties. Science 277, 504–509.

Maxwell, B., 1992. Arctic climate: potential for change under global warming. In: Chapin, F.S., Jefferies, R.L., Reynolds, J.F., Shaver, G.R., Svoboda, J. (Eds), Arctic Ecosystems in a Changing Climate. Academic Press, San Diego, pp. 11–34.

McKanzie, A.B., Logan, E.M., Cook, G.T., Pulford, I.D., 1998. Distributions, inventories and isotopic composition of lead in ^{210}Pb-dated peat cores from contrasting biogeochemical environments, implications for lead mobility. The Science of the Total Environment 223, 25–35.

McLean, R.O., Jones, A.K., 1975. Studies of tolerance to heavy metals in the flora of the river Ystwyth and Clarach, Wales. Freshwater Biology 5, 431–444.

Melick, D.R., Seppelt, R.D., 1997. Vegetation patterns in relation to climatic and endogenous changes in Wilkes Land, continental Antarctica. Journal of Ecology 85, 43–56.

Mensing, D.M., Galatowitsch, S.M., Tester, J.R., 1998. Anthropogenic effects on the biodiversity of riparian wetlands of a northern temperate landscape. Journal of Environmental Management 53, 349–377.

Mersch, J., Kass, M., 1994. La mousse aquatique *Fontinalis antipyretica* comme traceur de la contamination radioactive de la Moselle en aval de la centrale nuclèaire de Cattenom. Bull. Soc. Nat. Luxemb. 95, 109–117.

Mersch, J., Reichard, M., 1998. In situ investigation of trace metal availability in industrial effluents using transplanted aquatic mosses. Archives of Environmental Contamination and Toxicology 34, 336–342.

Milukaitė, A., 1999. Flux of benzo(a)pyrene to the ground surface and its distribution in the ecosystem. Water, Air, and Soil Pollution 105, 471–480.

Mitchell, J.F.B., Manabe, S., Meleshko, V., Toroka, T., 1990. Equilibrium climate change and its implications for the future. In Houghton, J.T., Jenkins, G.J., Ephraums, J.J. (Eds), Climate Change. The ICC Scientific Assessment, Cambridge University Press, Cambridge, pp. 134–172.

Molau, U., Alatalo, J.M., 1998. Responses of subarctic-alpine plant communities to simulated environmental change: bodiversity of bryophytes, lichens, and vascular plants. Ambio 27, 322–329.

Morrison, B.R.S., Wells, D.E., 1981. The fate of fenitrothion in a stream environment and its effect on the fauna, following aerial spraying of a Scottish forest. Science of the Total Environment 19, 233–239.

Mouvet, C., 1984. Accumulation of chromium and copper by the aquatic moss *Fontinalis antipyretica* L. ex Hedw. transplanted in a metal-contaminated river. Environmental Technological Letters 5, 541–548.

Mouvet, C., Galoux, M., Bernes, A., 1985. Monitoring of polychlorinated biphenyls (PCBs) and hexachlorocyclohexanes (HCH) in freshwater using the aquatic moss *Cinclidotus danubicus*. Science of the Total Environment 44, 253–267.

Nash, T.H., Nash, E.H., 1974. Sensitivity of mosses to sulphur dioxide. Oecologia 17, 257–263.

Newsmaster, S.G., Vitt, D.H, Bell, F.W., 1999. The effects of trichlopy and glyphosate on common bryophytes and lichens in Nortwestern Ontario. Canadian Journal of Forest Research 29, 1101–1111.

Nimis, P.L., Fumagalli, F., Bizzotto, A., Codogno, M., Skert, N., 2002. Bryophytes as indicators of trace metals pollution in the River Brenta (NE Italy). The Science of the Total Environment 286, 233–242.

Niskavaara, H., Äyräs, M., 1991. Sulphur and heavy metals in feather moss in Rovaniemi urban area. In: Pulkkinen, E. (Ed.), Geological Survey of Finland, Special Paper 9, pp. 213–222.

Oehlmann, J., Schulte-Oehlmann, U., 2002 (2003). Molluscs as bioindicators. In: Markert, B.A., Breure, A.M., Zechmeister, H.G. (Eds), Bioindicators and Biomonitors. Elsevier, Oxford, pp. 577–635.

Ohlson, M., Söderstrom, L., Hornberg, G., Zackrisson, O., Hermansson, J., 1997. Habitat qualities versus long-term continuity as determinants of biodiversity in boreal old-growth swamp forests. Biological Conservation 81, 221–231.

Okland, T., Okland, R., Steinnes, E., 1999. Element concentrations in the boreal forest moss *Hylocomium splendens*: variation related to gradients in vegetation and local environmental factors. Plant and Soil 209, 71–83.

Oldfield, F., Appleby, P.G., Cambray, R.S., Eakins, J.D., Barber, K.E., Battarbee, R.W., Person, G.R., Williams, J.M., 1979. 210Pb, 137Cs and 239Pu profiles in ombrotrophic peat. Oikos 33, 40–45.

Onianwa, P.C., 2001. Monitoring atmospheric metal pollution: a review of the use of mosses as indicators. Environmental Monitoring and Assessment 71, 13–50.

Oostendorp, W., Schmidt, E., 1977. Untersuchungen zur Biomasseverteilung submerser Bryophyten in der Selbstreinigungsstrecke eines Brauereivorfluters (Mettma, Hochschwarzwald). Gewässer, Abwässer 62/63, 85–96.

Ormerod, S.J., Wade, K.R., 1990. The role of acidity in the ecology of Welsh lakes and streams. In: Edwards, R.W., Gee, A.S., Stoner, J.H. (Eds), Acid Waters in Wales. Kluwer, Dordrecht, pp. 93–119.

Otnyukova, T., 1995. Sporophyte abnormalities as a cause for decline and disappearance of mosses in polluted areas. Cryptogamica Helvetica 18, 67–75.

Pakarinen, P., 1977. Element contents of *Sphagna*, variation and its sources. Bryophytorum Bibliotheca 13, 751–762.

Pakarinen, P., 1981. Regional variation of sulphur concentrations in *Sphagnum* mosses and *Cladonia* lichens in Finnish bogs. Annales Botanici Fennici 18, 275–279.

Pakarinen, P., Rinne, R.J.K., 1979. Growth rates and heavy metal concentrations of five moss species in paludified spruce forests. Lindbergia 5, 77–83.

Palmieri, F., Neri, R., Benco, C., Serracca, L., 1997. Lichens and moss as bioindicators and bioaccumulators in air pollution monitoring. Journal of Environmental Pathology, Toxicology and Oncology 16, 175–190.

Papp, B., Rajczy, M., 1995. Changes in the bryophyte vegetation and habitat conditions along a section of the river Danube in Hungary. Cryptogamica Helvetica 18, 95–105.

Peñuelas, J., 1984. Pigments of aquatic mosses of the river Muga, NE Spain, and their response to water pollution. Lindbergia 10, 127–132.

Pickering, D.C., Puia, I.L., 1969. Mechanism for the uptake of zinc by *Fontinalis antipyretica*. Physiologia Plantarum 22, 653–661.

Pilegaard, K., 1979. Heavy metals in bulk precipitation and transplanted *Hypogymnia physodes* and *Dicranoweisia cirrata* in the vicinity of a Danish steelworks. Water, Air and Soil Pollution 11, 79–91.

Pitcairn, C.E.R., Fowler, D., 1995. Deposition of fixed atmosheric nitrogen and foliar nitrogen content of bryophytes and *Calluna vulgaris* (L.) Hull. Environmental Pollution 88, 193–205.

Pitcairn, C.E.R., Fowler, D., Grace, J., 1991. Changes in species composition of semi-natural vegetation associated with the increase in atmospheric inputs of nitrogen. Report to Nature Conservancy Council. Institute of Terrestrial Ecology, Edinburgh.

Pòcs, T., 1982. Tropical forest bryophytes. In Smith, A.J.E. (Ed.), Bryophyte Ecology. Chapman & Hall, London, pp. 59–104.

Pott, U., Turpin, D.H., 1998. Assessment of atmospheric heavy metals by moss monitoring with Isothecium stoloniferum Brid in the Fraser Valley, BC, Canada. Water, Air and Soil Pollution 101, 25–44.

Potter, J.A., Press, M.C., Callaghan, T.V., Lee, J.A., 1995. Growth responses of *Polytrichum commune* and *Hylocomium splendens* to simulated environmental change in the sub-arctic. New Phytologist 131, 533–541.

Press, M.C., Callaghan, T.V., Lee, J.A., 1998. How will European Arctic ecosystems respond to projected global environmental change? Ambio 27, 306–311.

Press, M.C., Woodin, S.J., Lee, J.A., 1986. The potential importance of an increased atmospheric nitrogen supply to the growth of ombrotrophic *Sphagnum* species. New Phytologist 103, 45–55.

Proctor, M.C.F., 1982. Physiological ecology: water relations, light and temperature responses, carbon balance. In: Smith, A.J.E. (Ed.), Bryophyte Ecology. Chapman & Hall, London, pp. 333–381.

Proctor, M.C.F., 1990. The physiological basis of bryophyte production. Botanical Journal of the Linnean Society 104, 61–77.

Rao, D.N., 1982. Responses of bryophytes to air pollution. In. Smith, A.J.E. (Ed.), Bryophyte Ecology. Chapman & Hall, London, pp. 445–471.

Reimann, C., Niskavaara, H., Kashulina, G., Filzmoser, P., Boyd, R., Volden, T., Tomilina, O., Bogatyrev, I., 2001. Critical remarks on the use of terrestrial moss (*Hylocomium splendens* and *Pleurozium schreberi*) for monitoring of airborne pollution. Environmental Pollution 113, 41–57.

Rieley, J.O., Richards, P.W., Bebbington, A.D.L., 1979. The ecological role of bryophytes in a North Wales woodland. Journal of Ecology 67, 497–527.

Rinne, R.J.K., Barclay-Estrup, P., 1980. Heavy metals in a feather moss: *Pleurozium schreberi*, and in soils in NW Ontario, Canada. Oikos 34, 59–67.

Robinson, C.H., Wookey, P.A., Lee, J.A., Callaghan, T.V., Press, M.C., 1998. Plant community responses to simulated environmental change at a high arctic polar semi-desert. Ecology 79, 856–866.

Roelofs, J.G.M., 1983. Impact of acidification and eutrophication on macrophyte communities in soft waters in The Netherlands. I. Field studies. Aquatic Botany 17, 139–155.

Ross, H.B., 1990. On the use of mosses (*Hylocomium splendens* and *Pleurozium schreberi*) for estimating atmospheric trace metal deposition. Water, Air, Soil Pollution 50, 63–76.

Roy, S., Sen, C.K., Hanninen, O. 1996. Monitoring of polycyclic aromatic hydrocarbons using "moss bags": Bioaccumulation and responses of antioxidant enzymes in *Fontinalis antipyretica* Hedw. Chemosphere 32, 2305–2315.

Rühling Å. (Ed.), 1992. Atmospheric heavy metal deposition in northern Europe 1990. Nord 1992, 12. Nordic Council of Ministry, Copenhagen.

Rühling Å. (Ed.), 1994. Atmospheric heavy metal deposition in Europe – estimations based on moss analysis. Nord 1994, 9. Nordic Council of Ministry, Copenhagen.

Rühling, Å., Rasmmusen, L., Pilegaard, K., Mäkinen, A., Steinnes, E., 1987. Survey of atmospheric heavy metal deposition in the Nordic countries in 1985. Nord 1987, 21, Nordic Council of Ministry, Copenhagen.

Rühling Å., Steinnes, E. (Eds), 1998. Atmospheric heavy metal deposition in Europe 1995–1996. Nord 1998, 15. Nordic Council of Ministry, Copenhagen.

Rühling, Å., Steinnes, E., Berg, T., 1996. Atmospheric heavy metal deposition in northern Europe 1995. Nord 1996. Nordic Council of Ministry, Copenhagen.

Rühling, Å., Tyler G., 1973. Heavy metal deposition in Scandinavia. Water, Air and Soil Pollution 2, 445–455.

Rühling, Å., Tyler, G., 1969. Ecology of heavy metals – a regional and historical studies. Botaniska Notiser 22, 248–259.

Rühling, Å., Tyler, G., 1970. Sorption and retention of heavy metals in the woodland moss *Hylocomium splendens* (Hedw.) Br. et Sch. Oikos 21, 92–97.

Rühling, Å., Tyler, G., 1971. Regional differences in the deposition of heavy metals over Scandinavia. The Journal of Applied Ecology 8, 497–507.

Rühling, Å., Tyler, G., 1984. Recent changes in the deposition heavy metals in northern Europe. Water, Air and Soil Pollution 22, 173–180.

Rühling, Å., Tyler, G., 1968. An ecological approach to the lead problem. Botaniska Notiser 121, 321–342.

Ruijgrook, W., Nicholson, K.W., Davidson C.I., 1993. Dry deposition of particles. In: Lövblad, G., Erisman, J.W., Fowler, D. (Eds), Models and Methods for the Quantification of Atmospheric Input to Ecosystems. Nordiske Seminar- og Arbejdsrapporter 1993, 573. Nordisk Council of Ministers, Copenhagen, pp. 147–162.

Russel S., 1990. Bryophyte production and decomposition in tundra ecosystems. Botanical Journal of the Linnean Society 104, 3–22.

Rydgren K., Hestmark G., Økland R.H., 1998. Revegetation following experimental disturbance in a boreal old-growth Picea abies forest. Journal of Vegetation Science 9, 763–776.

Samecka, C., Kempers, A.J., 1996. Bioaccumulation of heavy metals by aquatic macrophytes around Wroclaw, Poland. Ecotoxicology and Environmental Safety 35, 242–247.

Satake, K. Oyagi, A., Iwao, Y., 1995. Natural acidification of lakes and rivers in Japan: the ecosystem of Lake Usoriko (pH 3.4–3.8). Water, Air and Soil Pollution 85, 511–516.

Satake, K., Takamatsu, T. Soma, M. Shibata, K. Nishikawa, M, Say, P.J., Whitton, B.A., 1989. Lead accumulation and location in the shoots of the liverwort *Scapania undulata* (L.) Dum. in stream water at Greenside Mine, England. Aquatic Botany 33.

Sauer, M., 2000. Moose als Bioindikatoren. In: Nebel, M., Philippi G. (Eds), Die Moose Baden-Würtembergs. Band 1., Eugen Ulmer, Stuttgart.

Say, P.J., Harding, J.P.C., Whitton, B.A., 1981. Aquatic mosses as monitors of heavy metal contamination in the River Etherow, Great Britain. Environmental Pollution (Ser. B.) 2, 295–307.

Say, P.J., Whitton, B.A., 1983. Accumulation of heavy metals by aquatic mosses: 1. *Fontinalis antipyretica* Hedw. Hydrobiologia 100, 245–260.

Schmid-Grob, I., Thőni, L., Hertz, J., 1992. Applicability of the moss *Hypnum cupressiforme* Hedw. S. 1. for biomonitoring of heavy metals, 2. A comparison with the moss *Hylocomium splendens* (Hedw.) Schimp. and the freights in bulk precipitation in Switzerland. Metal Compounds in the Environment and Life 4, 153–161.

Schuurkes, J.A.A.R., Kok, C.J., Den Hartog, C., 1986. Ammonium and nitrate uptake by aquatic plants from poorly buffered and acidified waters. Aquatic Botany 24, 131–146.

Sergio, C., 1987. Epiphytic bryophytes and air quality in the Tejo estuary. Symposia Biologica Hungarica 35, 795–814.

Sergio, C., Figueira, R., Crespo, A.M.V., 2000. Observations of heavy metal accumulation in the cell walls of Fontinalis antipyretica, in a Portuguese stream affected by mine effluent. Journal of Bryology 22, 251–255.

Shacklette, H.T., Erdmann, J.A., 1982. Uranium in spring water and bryophytes at Basin Creek in central Idaho. Journal of Geochemical Exploration 17, 221–236.

Shaw, J., 1987. Evolution of heavy metal tolerance in bryophytes. II: An ecological and experimental investigation of the "copper moss", *Scopelophila cataractae* (Pottiaceae). American Journal of Botany 74, 813–821.

Shimwell, D.W., Laurie, A.E., 1972. Lead and zinc contamination of vegetation in the Southern Pennines. Environmental Pollution 3, 291–301.

Siebert, A., Bruns, I., Kraus, G.-J., Miersch, J., Markert, B., 1996. The use of the aquatic moss *Fontinalis antipyretica* L. ex Hedw. as a bioindicator for heavy metals. The Science of the Total Environment 177, 137–144.

Skaar, H., Ophus, E., Gullvag, G.M., 1973. Lead accumulation within nuclei of moss leaf cells. Nature 241, 215–216.

Smith, A.J.E., 1982. Bryophyte ecology. Chapman and Hall, London.

Šoltés, R., 1992. Heavy metal concentrations in the mosses of the Tatra Mountains (Czecho-Slovakia): multivariate analysis. Oecologia Montana 1, 31–36.

Šoltés, R., 1998. Correlation between altitude and heavy metal deposition in the Tatra Mountains (Slovakia). Biologia 53, 85–90.

Sommer, C., Winkler, S., 1982. The effect of heavy metals on the rates of photosynthesis and respiration of *Fontinalis antipyretica*. Arch. Hydrobiolog. 93, 503–524.

Sonesson, M., Gehrke, C., Tjus, M., 1992. CO_2 environment, microclimate and photosynthetic characteristics of the moss *Hylocomium splendens* in a subarctic environment. Oecologia 92, 23–29.

Steinnes, E., 1977. Atmospheric deposition of trace elements in Norway studied by means of moss analysis. Kjeller Report 154.

Steinnes, E., 1985. Use of mosses in heavy metal deposition studies. EMEP/CCC Report 3/85, 161–170.

Steinnes, E., 1995. A critical evaluation of the use of naturally growing moss to monitor the deposition of atmospheric metals. Science of the Total Environment 160/161, 243–249.

Steinnes, E., Hanssen, J.E., Rambaek, J.P., Vogt, N.B., 1993. Atmospheric deposition of trace elements in Norway, temporal and spatial trends studied by moss analysis. Water, Air and Soil Pollution 69, 121–138.

Steinnes, E., Rambaek, J.P., Hanssen, J.E., 1992. Large scale multi-elements survey of atmospheric deposition using naturally growing moss as biomonitor. Chemosphere 25, 735–752.

Strachan, W.M.J., Glooschenko, W.A., 1988. Moss bags as monitors of organic contamination in the atmosphere. Bulletin of Environmental Conntamination Toxicology 40, 447–450.

Sucharová, J., Suchara, I., 1998. Atmospheric deposition of chosen elements in the Czech Republic determined in the framework of the International Bryomonitoring Programme 1995. The Science of the Total Environment 223, 37–52.

Sundaram, K.M.S., Nott, R., Curry, J., 1996. Deposition, persistence and fate of tebufenozide (RH-5992) in some terrestrial and aquatic components of a boreal forest environment after aerial application of MIMIC. Journal of Environmental Science and Health Part B Pesticides Food Contaminants and Agricultural Wastes 31, 699–750.

Suren, A., 1996. Bryophyte distribution patterns in relation to macro-, meso-, and micro-scale variables in South Island, New Zealand streams. New Zealand Journal of Marine and Freshwater Research 30, 501 523.

Sutton, M.A., Fowler, D. Moncrieff, J.B., 1993. Exchange of atmospheric ammonia with vegetation surfes, unfertilised vegetation. Quarterly Journal of the Royal Meteorological Society 119, 1023–1045.

Sveinbjörnson, B., Oechel, W.C., 1992. Controls on growth and productivity of bryophytes: environmental limitations under current and anticipated conditions, In: Bates, J.W., Farmer, A. (Eds), Bryophytes and Lichens in a Changing Environment. Clarendon Press, Oxford, pp. 77–102.

Svensson, G.K., Linden, K., 1965. The quantitative accumulation of 95Zr + 95Nb and 140Ba + 140La in carpets of forest moss. A field study. Health Physiology 11, 1033–1042.

Świeboda, A., Kalemba, A., 1987. The use of *Spahgnum recurvum* Pal. Beauv. As biological tests for determination of the level of pollution with fluorine compounds and sulphur dioxide in the environment. Acta Societatis Botanicorum Poloniae 50, 541–551.

Szarek, G., Chrzanowaska, E., 1991. The total sulphur content in the mosses of Polish National Parks – changes within the last 10 years. Acta Societatis Botanicorum Poloniae 60, 101–110.

Szweykowski, J., 1984. What do we know about the evolutionary process in bryophytes? Journal of the Hattori Botanical Lab. 55, 209–218.

Tamm, C.O., 1953. Growth, yield and nutrition in carpets of a forest moss (*Hylocomium splendens*). Meddelanden Fran Statens Skogsforskningsinstitut 43, 1–140.

Taoda, H., 1973. Effect of air pollution on bryophytes. I. SO2 tolerance of bryophytes. Hikobia 6, 238–250.

Thöni, L., Schnyder, N., Hertz, J., 1993. Schätzung der Schwermetalldeposition mit Hilfe des Mooses Bryum argenteum als Biomonitor. Staub – Reinhaltung der Luft 53, 319–325.

Thöni, L., Schnyder, N., Krieg, F., 1996. Comparison of metal concentrations in three species of mosses and metal freights in bulk precipitations. Fresenius Journal of Analytical Chemistry 354, 703–708.

Thomas, W., 1984. Statistical models for the accumulation of PAH, chlorinated hydrocarbons and trace metals in epiphytic *Hypnum cupressiforme*. Water, Air, and Soil Pollution 22, 351–371.

Thomas, W., 1986. Representativity of mosses as biomonitor organisms for the accumulation of environmental chemicals in plants and soils. Ecotoxicology and Environmental Safety 11, 339–346.

Thomas, W., Herrmann, R., 1980. Nachweis von Chlorpestiziden, PCB, PCA und Schwermetallen mittels epiphytischer Mosse als Biofilter entlang eines Profils durch Mitteleuropa. Staub-Reinhaltung der Luft 40, 440–444.

Thomas, W., Schunke, E., 1985. Polyaromatic hydrocarbons, chlorinated hydrocarbons and trace metals in moss samples from Iceland. Lindbergia 10, 27–32

Tremp, H., 1992. Einsatz submerser Bryophyten als Bioindikatoren in versauerten Fliessgewässern des Schwarzwaldes. In: Kohler, A., Arndt, U. (Eds), Bioindikatoren für Umweltbelastung. 24. Hohenheimer Umwelttagung. Verlag Josef Margraf, Weikersheim, pp. 143–158.

Tremp, H., 1993. Ein einfacher Nachweis der unterschiedlichen Säuretoleranz bei Wassermoosen – Visuelle und fluoreszenzoptische Bonitur. Ber. Inst. Landschafts-Planzenökologie. Univ. Hohenheim, Heft 2, 281–286.

Tremp, H., 1999. Submerged bryophytes in running waters, ecological characteristics and their use in biomonitoring. Environmental Science Forum 96, 233–242.

Tremp, H., Kohler, A., 1995. The usefulness of macrophyte monitoring systems, exemplified on eutrophication and acidification of running waters. Acta Botanica Gallica 142, 541–550.

Türk, R., Wirth, V., 1975. Über die SO_2-Empfindlichkeit einiger Moose. Bryologist 78, 187–193.

Tyler, G., 1972. Heavy metals pollute nature, may reduce productivity. Ambio 1, 57–59.

Tyler, G., 1990. Bryophytes and heavy metals, a literature review. Botanical Journal of the Linnean Society 104, 231–253.

Umlauf, G., Hauk, H., Reissinger, M., Hutzinger, O., 1994. Untersuchungen zur atmosphärischen Deposition lipophiler organischer Verbindungen auf Pflanzen am Beispiel *Picea abies*. In: Alef, K., Fiedler, H., Hutzinger, O. (Eds), Umweltmonitoring und Bioindikation. ECO-Informa-'94, Umweltbundesamt. Wien, pp. 129–145.

Url, W., 1959. Über Schwermetall-, zumal Kupferresistenz einiger Moose. Protoplasma 46, 768–793.

Vanderpoorten, A., 1999a. Correlative and experimental investigations on the segregation of aquatic bryophytes as a function of water chemistry in the Walloon hydrographic network. Lejeunia 159, 1–17.

Vanderpoorten, A., 1999b. Aquatic bryophytes for a spatio-temporal monitoring of the water pollution of the rivers Meuse and Sambre (Belgium). Environmental Pollution 104, 401–410.

Vanderpoorten, A., Palm, R., 1998. Canonical variables of aquatic bryophyte combinations for predicting water trophic level. Hydrobiologia 386, 85–93.

VDI, 1995. Richtlinie 3799. Blatt 1. Messen von Immissionswirkungen. Ermittlung und Beurteilung phytotoxischer Wirkungen von Immissionen mit Flechten; Flechtenkartierung zur Ermittlung des Luftgütewertes (LGW). VDI Handbuch Reinhaltung der Luft, Band 1, Düsseldorf.

Viskari, E.-L., Rekilä, R., Roy, S., Lehto, O., Ruuskanen, J., Kärenlampi, L., 1997. Airborne pollutants along a roadside: assessment using snow analyses and moss bags. Environmental Pollution 97, 153–160.

Wappelhorst, O., Kühn, I., Oehlmann, J., Markert, B., 2000. Deposition and disease: a moss monitoring project as an approach to ascertaining potential connections. Science of the Total Environment 249, 243–256.

Watson, R.T., Zinyowera, M.C., Moss, R.H., Dokken, D.J. (Eds), 1996. Climate change 1995. Impacts, adaptations and mitigation of climate change: scientific-technical analysis. IPCC, Cambridge University Press, Cambridge.

Wehr, J.D., Empain, A., Mouvet, C., Say, P.J., Whitton, B.A., 1983. Methods for processing aquatic mosses used as monitors of heavy metals. Water Research 17, 985–982.

Wehr, J.D., Whitton, B.A., 1983a. Accumulation of heavy metals by aquatic mosses. 2. *Ryhnchostegium riparoides*. Hydrobiologia 100, 261–284.

Wehr, J.D., Whitton, B.A., 1983b. Accumulation of heavy metals by aquatic mosses. 3. Seasonal changes. Hydrobiologia 100, 285–291.

Wehr, J.D., Whitton, B.A., 1986. Ecological factors relating to morphological variation in the aquatic moss *Rhynchostegium riparioides* (Hedw.) C. Jens. Journal of Bryology 14, 269–280.

Wells, J.M., Boddy, L., 1995. Phosphorus translocation by saprophytic basidiomycete mycelial cord systems on the floor of a mixed deciduous woodland. Mycological Research 99, 977–980.

Wells, J.M., Brown, D.H., 1987. Factors affecting the kinetics of intra- and extracellular Cadmium uptake by the moss *Rhytidiadelphus squarrosus*. New Phytologist 105, 123–137.

West, S., Charman, D.J., Grattan, J.P., Cherburkin, A.K., 1997. Heavy metals in Holocene peats from south west England: detecting mining impacts and atmospheric pollution. Water, Air and Soil Pollution 100, 343–353.

Whelpdale, D.M., Summers, P.W., Sanhueza, E., 1997. A global overview of atmospheric acid deposition fluxes. Environmental Monitoring and Assessment 48, 217–247.

White, N.E., Brooks, R.R., 1969. Aquatic bryophytes as indicators of uranium mineralization. The Bryologist 72, 501–507.

Whitton, B.A., Say, P.J., Jupp, B.P., 1982. Accumulation of zinc, cadmium, and lead by the aquatic liverwort *Scapania*. Environmental Pollution 3, 299–316.

Winkler, S., 1977. Flechten und Moose als Bioindikatoren. In, Frey, W., Hurka, H, Oberwinkler, F. (Eds), Beiträge zur Biologie der Niederen Pflanzen. G. Fischer, Stuttgart. pp. 155–176.

Winner, W.E., Bewley, J.D., 1978a. Contrasts between bryophyte and vascular plant synecological responses in an SO$_2$-stressed white spruce association in Central Alberta. Oecologia (Berl.) 33, 311–325.

Winner, W.E., Bewley, J.D., 1978b. Terrestrial mosses as bioidicators of SO$_2$ pollution stress. Oecologia (Berl.) 35, 212–230.

Wolterbeek, H.Th., Bode, P., 1995. Strategies in sampling and sample handling in the context of large-scale plant biomonitoring surveys of trace element air pollution. The Science of the Total Environment 176, 33–43.

Wolterbeek, H.Th., Kuik, P., Verburg, T.G, Herpin, U., Markert, B., Thőni, L., 1995. Moss interspecies comparisons in trace element concentrations. Environmental Monitoring and Assessment 35, 263–286.

Woodin, S.J., Farmer, A.M., 1993. Impacts of sulphur and nitrogen deposition on sites and species of nature conservation importance in Great Britain. Biological Conservation 63, 23–30.

Wookey, P.A., Parsons, A.N., Welker, J.M., Potter, J.A., Callaghan, T.V., Lee, J.A., Press, M.C., 1993. Comparative responses of phylogeny and reproductive development to simulated environmental change in sub-arctic and high arctic plants. Oikos 67, 490–502.

Woolgrove C.E., Woodin, S.J., 1996. Current and historical relationships between the tissue nitrogen content of a snowbed bryophyte and nitrogenous air pollution. Environmental Pollution 91, 283–288.

Yan, N.D., Miller, G.E., Wile, I., Hitchin, G.G., 1985. Richness of aquatic macrophyte floras of soft water lakes of differing pH and trace metal content in Ontario, Canada. Aquatic Botany 23, 27–40.

Yoshimura, E., Kitai, K., Nishizawa, N., Satake, K., Mori, S., Yamazaki, S., 1998. Accumulation of metals and cellular distribution of aluminium in the liverwort *Scapania undulata* in acidic and neutral streams in Japan. Journal of Environmental Science and Health (Part A) 33, 671–680.

Yurukova, L., Ganeva, A., 1997. Active biomonitoring of atmospheric element deposition with Sphagnum species around a copper smelter in Bulgaria. Journal of Applied Botany 71, 14–20.

Zechmeister, H.G., 1994. Biomonitoring der Schwermetalldepositionen mittels Moosen in Österreich. Monographien des Umweltbundesamtes Wien 42, 1–168

Zechmeister, H.G., 1995a. Growth rates of five pleurocarpous moss species under various climatic conditions. Journal of Bryology 18/3, 455–468.

Zechmeister, H.G., 1995b. Correlation between altitude and heavy metal deposition in the Alps. Environmental Pollution 89, 73–80.

Zechmeister, H.G., 1997. Schwermetalldepositionen in Österreich. Aufsammlung 1995. Monographien des Umweltbundesamtes Wien 94, 1–145.

Zechmeister, H.G., 1998. Annual growth of four pleurocarpous moss species and their applicability for biomonitoring heavy metals. Environmental Monitoring and Assessment 52, 441–451.

Zechmeister, H.G., Moser, D., 2001. The influence of agricultural land-use intensity on bryophyte species richness. Biodiversity and Conservation 10, 1609–1625.

Zechmeister H.G., Mucina L., 1994. High ranked syntaxa of the European water spring vegetation (Montio-Cardaminetea). Journal of Vegetation Science 5, 385–402.

Zechmeister, H.G., Hohenwallner, D., Riss, A., Hanus-Illnar, A. 2002a. Variations in heavy metal concentrations in the moss species *Abietinella abietina* (Hedw.) Fleisch. according to sampling time, within site variability and increase in biomass. Science of the Total Environment, in press.

Zechmeister, H.G., Tribsch, A., Hohenwallner, D., 2002b. Mooskartierung Linz mit bioindikatorischem Schwerpunkt. Naturkundliches Jahrbuch der Stadt Linz, in press.

Zechmeister H.G., Tribsch, A., Moser, D., Wrbka, T., 2002c. Distribution of endangered bryophytes in Austrian cultural landscapes. Biological Conservation 103, 173–182.

Bioindicators and biomonitors
B.A. Markert, A.M. Breure, H.G. Zechmeister, editors
© 2003 Elsevier Science Ltd. All rights reserved.

Chapter 11

Biomonitors in use: lichens and metal air pollution

H.T. Wolterbeek, J. Garty, M.A. Reis and M.C. Freitas

Abstract

Lichens are reviewed for their potential to reflect metal air pollution. They are discussed both as being symbiotic plants and as biomonitor organisms. The attention is focused on their biological performance, the dynamics of metal uptake and release, on particle interception and on (physiological) processes underlying metal accumulation within the thallus. Laboratory-linked detailed physiological assessments of lichen behaviour are recognized as are the more pragmatic approaches often adopted in field work. Lichen morphology and physiology are regarded as of similar relevance in overall metal accumulation. A simplified approach is discussed in which the lichen is seen as essentially "homogeneous", in the sense that no symbiont-dependent differentiations are regarded, but where the lichen's morphological characteristics are seen as of importance in initial interception of airborne particles. Modelled results from this approach relative to field survey assessments indicate the potential of this view to a lichen.

Discussed literature data on lichen physiology indicate that, in surveys, apart from lichen metal determinations, lichen behaviour may be critically viewed if not "normalized" by simultaneous analysis of selected lichen physiological parameters: apart from the assessment of the lichen's surface-to-mass ratio, a number of possible analyses are presented, among which the determination of leaching through damaged cell membranes, stress-ethylene, the rate of photosynthesis, the potential quantum yield of photosystem II, the lichen's spectral reflectance, or the lichen's chlorophyll content.

Keywords: Lichen, Metal Air Pollution, Morphology, Physiology, Quantification, Biomonitor, Modelling

1. Introduction

1.1. Metal air pollution

Longest standing attention for wide-spread air pollution stems from our concern over its ecological effects. The environmental impact of atmospheric deposition has been studied for more than a century: probably the first effect that was described on a scientific basis was the decline of epiphytic lichens in areas with high levels of atmospheric pollution. Ever since Nylander's (1866) classical report on the epiphytic lichens of Paris and its surroundings, extensive studies have been performed in many areas (Barkman, 1958; De Wit, 1976; Hawksworth, 1971). Although these research efforts have led to a greatly improved scientific understanding of the abiotic effects of atmospheric deposition, especially in the fields of atmospheric chemistry (Asman, 1987), soil chemistry (Mulder, 1988) and water chemistry (Van Dam et al., 1990), many of the biotic effects are still poorly understood, particularly in the terrestrial environment. Although many changes in vegetation are now generally attributed to atmospheric deposition, dose-effect relationships are usually poorly known (Heij et al., 1991). General information on monitoring, behaviour and impact of terrestrial trace element pollution can be found in various reviews (Adriano, 1986; Bowen, 1979; Farago, 1994; Lepp, 1981a,b; Markert, 1993; Martin and Coughtrey, 1982).

In addition to the ongoing concern for ecosystem performance as such, attention has been and becomes increasingly more directly focused on human health. This may be ascribed to the generally recognized impact of ecosystem performance on human well-being; furthermore, health-care has also been progressively developing towards approaches which include our nutrition and our social and environmental surroundings (Vandenbroucke and Hofman, 1993). As a result, throughout the world, epidemiological studies were set up on air pollution and mortality rates and respiratory health effects, initially mostly on air particulates, ozone, acid rain, NO_x and sulphur oxides (Castillejos et al., 2000; Dockery et al., 1993; Lebowitz, 1996; Spix et al., 1993), but today the attention includes contaminants such as heavy metals, polycyclic aromatic compounds and halogenated organics, which all differ widely with respect to their environmental and health impact properties (Badman and Jaffe, 1996; Carter et al., 1997; Heinrich et al., 1999; Laden et al., 1999; Roemer et al., 2000; Wappelhorst et al., 2000; Wiederkehr, 1991).

1.2. Quantitative assessment

The necessary *quantitative* information on metal air pollutants is generally obtained by modelling of the dispersion (which is a source-oriented approach, making use of a priori known information on emission sources), or by field measurements of the immission (which is a receptor-oriented approach). Immission measurements should be regarded as indispensable: they should be used to validate dispersion models, and the data obtained may indicate the presence of sources which are not known or registrated (Wolterbeek and Freitas, 1999). Immission measurements require long-term sampling at large numbers of sampling sites. Such measurements using technical equipment (e.g. air samplers) have been few, mainly due to the high costs, the risk of vandalism, and the lack of sufficiently sensitive and inexpensive techniques which permit the simultaneous measurement of many air contaminants (Puckett, 1988). It is here that biomonitoring comes in.

1.3. Biomonitoring

Biomonitoring, in a general sense, may be defined as the use of bio-organisms to obtain *quantitative* information on certain characteristics of the biosphere (see Garty, 1993; Markert et al., 1999; 2000; Puckett, 1988; or Wittig, 1993) for clear overviews of what is meant by the terms monitors, indicators, or collectors). The relevant information in biomonitoring may be deduced from changes in the behaviour of the organism (Herzig et al., 1990) (impact: occurrence of species, ecological performance, morphology), but most metal air pollution studies focus on the assessment of the concentrations of the relevant substances in the monitor tissues. With proper selection of organisms, the general advantage of the biomonitoring approach is related primarily to the permanent and common occurrence of the organism in the field, even in remote areas, the ease of sampling, and the absence of any necessary expensive technical treatment.

1.4. Lichens and their use in air pollution studies of SO_2 and metals

Apart from mosses, lichens may be considered as the most commonly applied biomonitor organisms in studies of metal air pollution. This is largely based on their lack of any roots comparable to higher plants. Although effects from substrata may occur (De Bruin and Hackenitz, 1986; Goyal and Seaward, 1981; Prussia and Killingbeck, 1991; Sloof and Wolterbeek, 1993), this absence of roots makes that lichens are generally thought to obtain their mineral supplies from aerial sources (Sloof and Wolterbeek, 1993). Further arguments supporting the use of lichens comprise their metal accumulative and time-integrative behaviour (Wolterbeek and Bode, 1995; Wolterbeek et al., 1996).

The use of lichens as biomonitors of metal air pollution has evolved from the long-standing obervations on the relationships between air pollution and lichen species occurrence. Ever since the first obervations by Nylander (1866) on the epiphytic lichens of Paris and its surroundings in relation with atmospheric pollution, many studies have been performed in many areas (Barkman, 1958; De Wit, 1976; Hawksworth, 1971), with special reference to atmospheric SO_2 and acidic precipitation. The decline of epiphytes has been ascribed to SO_2 (Hawksworth, 1971): the hypothesized working mechanisms are either a direct toxic effect of gaseous SO_2 (Barkman, 1958) or bark acidification caused by SO_2 deposition (Gilbert, 1970; Johnsen and Søchting, 1973; Van Dobben, 1993). Relationships between species richness and atmospheric SO_2 are shown as rather complex: decreases in atmospheric SO_2 were reported as most strongly reflected by recurrences of especially nitrophytic lichen species, whereas acidophytes such as *Lecanora conizaeoides*, *Hypogymnia physodes*, or *Pseudevernia furfuracea* showed decreases in mean abundance with decreasing atmospheric SO_2 (Van Dobben, 1993).

The effects of SO_2 on lichen physiology makes that the lichen's response to metal supply should be suspected to be affected by SO_2. Therefore, quantitative lichen-based metal air pollution studies (see Section 2.5) should include comparative assessment of lichen vitality (see Section 4.4.2; see also Section 5).

1.5. The quantification problem

The present chapter focuses on the *quantitative* assessment of metal air pollution. It implicitly addresses larger-scaled surveys, which makes that biomonitoring should be seen as a serious alternative and supplementing method for technical air sampling. The issue which is most specifically addressed in the present paper is the translation of the biomonitor data into quantitative information on metal air pollution. This makes that much of what will be reviewed has a more general relevance than being of importance for lichens only. However, the points raised will be seen most specifically in the context of the lichen's potential to meet requirements associated to its use as a biomonitor. Thus, the centre of attention, and most of the reviewed literature will be focused on the lichen: the lichen, which is generally suggested as "reflecting" its ambient atmospheric metal availability, while that very same metal availability may have its impact on lichen physiology. In other words, the lichen should be reviewed both for its biological performance and for its metal accumulation characteristics. Such a review can never give a complete representation of all existing information: when searching the keyword *"lichen"*, the Institute for Scientific Information (ISI)'s combined SCI, SSCI and HCI databases yielded some 7500 entrées over the last 12 years. Instead, the present paper gives a number of illustrative references for all points and concepts raised and introduced.

2. The lichen as a plant

2.1. General

Lichens are symbiotic organisms composed of a fungal partner (the mycobiont) and a green or blue-green algal partner (the photobiont) (Garty, 1993; Richardson, 1999) (Figs 1 and 2), and are important constituents of the vegetation of many ecosystems of the world. Lichens include about 13,500 species or more (Hawksworth, 1988). Details on lichen physiology, its bionts, and the initiation and nature of the symbiosis have been published throughout literature (Ahmadjian, 1973; Egan, 1982; Honegger, 1991; Jahns, 1993; Richardson, 1999; Smith, 1963). Symbiotic interactions are quite extensive and involve nitrogen metabolism, synthesis of secondary metabolites and the transfer of carbohydrates. Regarding lichen reproduction, many mycologists assume that once a lichen fungal ascospore has contacted a suitable photobiont, or once a soredium, isidium or lichen fragment has landed on an appropriate surface, a lichen thallus will develop into a mature lichen (Phillips and Stumpel-Rienks, 1980), which may survive for tens, if not hundreds or even thousands of years (Bowen, 1979; Richardson, 1999). It should be noted here that thallus fragments are reported to provide short-distance dispersion only (Heinken, 1999). For germination, the fungal spores of most lichens do not need the photobiont: they are reported to grow and encircle any spherical structures of suitable size, including glass beads or rods (Ahmadjian, 1973; 1990). In liquid cultures, however, the isolated fungus is reported to behave partly like a yeast, producing large amounts of single cells, whereas the isolated algae show thickened cell walls (Ahmadjian and Hale, 1973).

The fungal component may comprise some 75% of the total lichen mass (Goyal and Seaward, 1981; Richardson, 1999; Tyler, 1988), but hardly anything is known about the precise participations of the symbiotic algae and fungi in metal accumulation (Wastlhuber and Loos, 1996). Laboratory culturing of lichens and the synthesis of the symbiotic organism from its isolated bionts have been attempted for the last 50 years, but only certain species could be grown successfully (Ahmadjian, 1990, 1993; Jahns, 1993). In laboratory culture experiments with isolated lichen fungi and algae, component-specific uptake and accumulation behaviour was observed (Demon et al., 1988; 1989; Goldsmith et al., 1997), but the reported marked changes in physiology

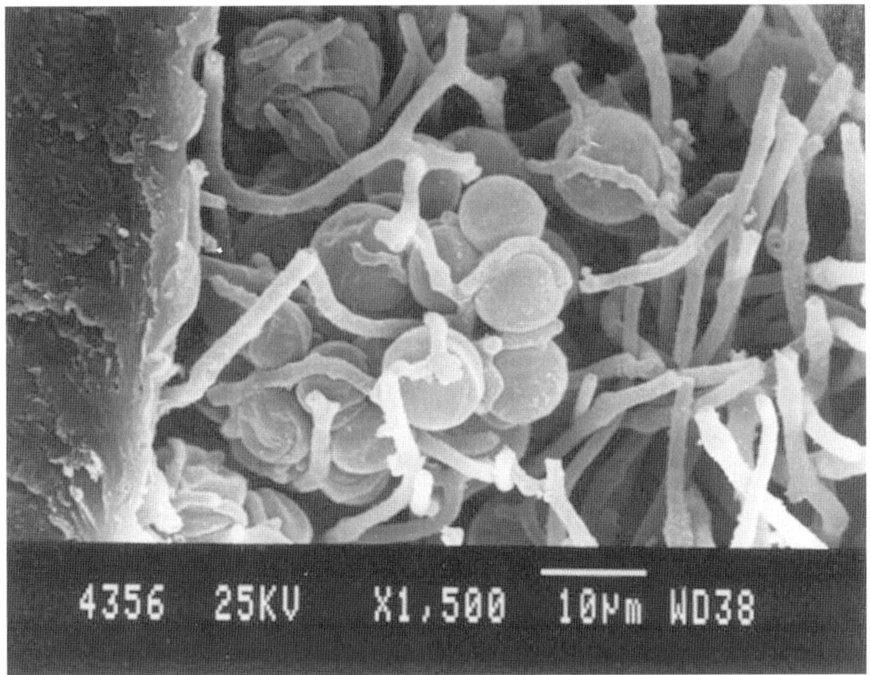

Figure 1. Cross-section of thallus of *Ramalina lacera* from HaZorea, an unpolluted site in Israel. Algal cells (*Trebouxia*) are seen among the fungal hyphae. Scale bar = 10 μm (Garty, Kunin, Delarea, unpublished). Reproduced by permission of Elsevier Science Ltd.

and morphology of lichen bionts, due to separation (Ahmadjian and Hale, 1973; Waslhuber and Loos, 1996), make it hard to draw any firm conclusions for the in-situ situations (Hickman, 1965; Romano, 1966; Turner, 1971). Of interest here is also the report by Wells and Brown (1987), which indicates that differences between two field populations of the moss Rhytidiadelphus squarrosus in Cd uptake characteristics, cation contents, gas exchange and dry weights per unit length were eliminated by laboratory culture.

2.2. Biological performance

Lichens may be regarded and analysed in terms of their morphology, histology, ecology and physiology, in short- or long-term periods of time (Ahmadjian and Hale, 1973). Measurements of growth rate, productivity, reproductive capacity, deformity, discoloration, chlorophyll content, membrane integrity, respiratory activity, ionic content, geographical occurrence, substrate-related distributional limitations, or water relations are all examples of the assessment of the lichen's biological performance (Barkman, 1958; De Wit, 1976; Garty et al., 1998; González and Pignata, 1999; Goyal and Seaward, 1982; Henderson-Sellers and Seaward, 1979; Seaward, 1974, 1976). However, as pointed out by Seaward (1976), in terms of biological scaling, the interpretation of "symptom mapping" is often rather difficult. The increase in size and

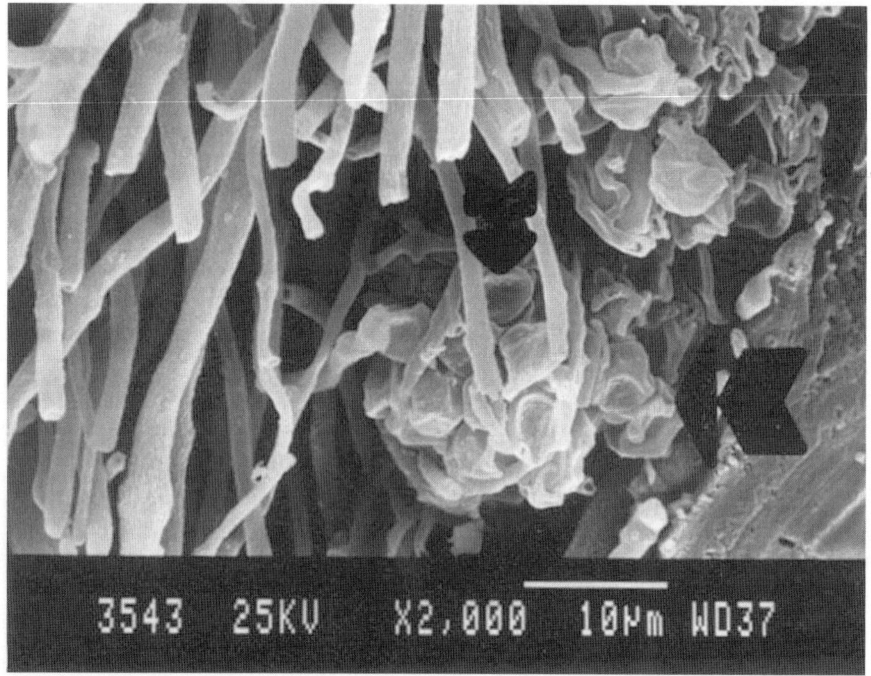

Figure 2. Cross-section of thallus of *Ramalina lacera* from HaZorea, collected in November 1994, transplanted to the vicinity of a chemical plant in Ashdod, SW Israel, and retrieved in August 1995. Note the shrunk *Trebouxia* cells. Scale bar = 10 μm (Garty, Kunin, Delarea, unpublished).

number of apothecia may reflect an increase in reproductive capacity but may also merely show a malformation of the central parts of the thallus. Discoloration may be associated to general senescence, but may also reflect injuries from insects, diseases or chemical sprays. As a last example, changes in the distribution of species may reflect changes in air pollution (Barkman, 1958; De Wit, 1976; Henderson-Sellers and Seaward, 1979), but may also be associated to other environmental parameters (Henderson-Sellers and Seaward, 1979), such as changes in humidity or changes in the availability of preferent substrates (De Wit, 1976; Manning and Feder, 1980). The strong-points of the approach, however, can be illustrated by a North-Italian study, showing the lichen biodiversity as strongly correlated with both anthropogenous pollutants such as SO_2, NO_3^-, dust and SO_4^{2-}, and the incidence of lung cancer mortality (Cislaghi and Nimis, 1997).

2.3. Metal accumulation

2.3.1. General

Plants are often classified as accumulators, indicators, or excluders (Farago, 1994; Farago and Mehra, 1992; Markert et al., 1999). Following Baker (1981) and Farago (1994a), accumulators tend to concentrate elements to levels higher than the ambient concentrations, indicators show concentrations similar to the ambient ones, and excluders maintain relatively low concentrations, up to a certain ambient level, after which the plant's exclusion mechanism breaks down and unrestricted uptake and toxicity sets in. It should be noted here that the term *indicator* is somewhat misleading: in the context of metal accumulation, Farago and Mehra (1992) suggested to better use the term *concentration indicator*. Relative to ambient atmospheric metal concentrations, lichens may be invariably classified as accumulators for the majority of metals considered.

Elements are often classified as being macronutrients or micronutrients, and as either essential or non-essential for the plant, whereas their concentrations are generally indicated as either deficient, sufficient or toxic. Plant nutrients yet identified and best known as essential are C, H, O, N, P, K, S, Ca, Mg (as macronutrients) and B, Cl, Co, Cu, Fe, Mn, Mo, Ni, Si, Na and Zn (as micronutrients) (Davies, 1994). Further general and more detailed data on metal essentiality, occurrence in soils and plants, toxicity and uptake forms are given by Bowen (1979) and Markert (1996). Due to the plant's metabolically controlled maintenance of required levels of essential elements, lichens may show rather high base-line concentrations for especially the essential elements under low atmospheric availability conditions. It should be realized here, however, that this control involves both growth and accumulation (Hale, 1974; Quispel and Stegwee, 1984).

2.3.2. Uptake and release

Lichens may accumulate metals from airborne particles or from dissolved and suspended material. The mechanisms of element uptake are not very well known, nor are the precise participations of the symbiotic algae and fungi. Trapping of particles, ion-exchange, electrolyte sorption, and metabolically controlled absorption are all likely to be involved in element accumulation, as indicated by uptake studies with intact lichens (Puckett et al., 1973; Richardson et al., 1984). Metal uptake by lichens has been extensively studied in the laboratory, mostly emphasizing the rapid, exchangeable process of metal binding to cell walls (Brown, 1976; Goyal and Seaward, 1982; Nieboer and Richardson, 1981; Nieboer et al., 1978). However, studies by Demon et al. (1988; 1989) with isolated bionts, and of Brown and co-workers (Beckett and Brown, 1984; Brown and Beckett, 1984; Brown and Buck, 1979) with intact lichens all suggested the presence of a metabolism-related, temperature-dependent component of the accumulation, numerically depending on both element and lichen biont. It should also be realized that, although uptake efficiencies may be strictly ordered for a number of metals (Rühling and Tyler, 1970), competition effects may significantly influence uptake and release. High sea salt input in moss showed effects even on the retention of strongly absorbed metals such as Pb and Cu (Berg and Steinnes, 1997; Berg et al., 1995; Gjengedal and Steinnes, 1990), in lichens both the extra- and intracellular uptake of Cd was shown to be reduced in the presence of a range of other metals (Beckett and Brown, 1984). In intact lichens, uptake is reported as pH dependent (Puckett and Burton, 1981; Puckett et al., 1973; Richardson et al., 1984); Demon et al. (1988; 1989) showed that metal uptake into algae was generally faster at pH 7, fungal uptake was higher at pH 5. Effects from temperature appear to depend on metal species: Beckett and Brown (1984) reported temperature effects of especially the intracellular uptake of Cd in *Peltigera*, Demon et al. (1988, 1989) suggested the absence of effects for Cu and La for both fungal and algal cells in short-term 4 h experiments, and reported temperature dependencies for uptake of As, W, Zn and Cd in the algal components.

As said earlier, here it should be noted that short term experiments on plant cellular metal uptake mostly highlight the rapid process of cell wall sorption (Sentenac and Grignon, 1981; Wolterbeek et al., 1995): insight in the subsequent slower but prolonged uptake asks for longer-term assessment (Sutcliffe, 1962). This makes that short-term comparisons are sometimes misleading: the *capacity* of extracellular sorption cannot be compared directly with the *rate* of intracellular uptake (Beckett and Brown, 1984; Puckett et al., 1973; Wolterbeek et al., 1995). In this respect, the process of total accumulation comprises a dynamic combination of cell wall sorption and uptake: the cell wall potential disturbs the ion distribution in its immediate vicinity, so that the local concentrations and *fluxes* through membranes are a complicated combination of the characteristics of both cell wall and membrane (Dalton, 1984; Helfferich, 1962; Thibaud et al., 1984; Wolterbeek, 1987).

It should be noted here that in accumulation kinetics, influx and efflux are of equal importance (Shipley and Clark, 1972; Wolterbeek et al., 1995): metal levels in lichens should therefore be regarded as the invariable resultance of both uptake and release. Losses of ammonia from lichens were greatest in living material, inferring the initial involvement of intercellular uptake (Miller and Brown, 1999). Also the Chernobyl accident has provided much information on both accumulation and release (Devell et al., 1986; De Vries and Van der Kooij, 1986; Martin and Coughtrey, 1982; Papastefanou et al., 1988, 1989; Raes et al., 1990; Sloof, 1993): fallout was measured in air particulate matter, deposition and biomonitors. Retention efficiencies and release rates could be calculated from the time course of radioactivity levels in biomonitors after the short-term atmospheric influx (Ellis and Smith, 1987; Sloof and Wolterbeek, 1992; Smith and Ellis, 1990). In addition, Déruelle (1984) showed that lead accumulated by lichens after transplantation to sites which were contaminated with automobile exhaust was lost within months after they were returned to their original uncontaminated sites. Furthermore, in the three years following the cessation of U bearing emissions from a mine near Elliot Lake, Ontario, Canada, nearby growing *Cladonia* species were reported as exhibiting substantially decreased uranium levels (Trembley et al., 1997). Apart from direct releases, metals may also be lost due to processes of biotransformation (e.g. redox reactions, methylation). The volatilization of As, Se, Hg and other metals is well documented (Bargagli et al., 1987; Challenger, 1945; Summers, 1978). Demon et al. (1988) suggested the decreases in algal/fungal As after initial uptake as probably due to the release of methylated As-compounds.

2.3.3. Particle interception

The interception by vegetation of atmospheric depositing particulates has been studied for decades: general principles and data up to the 1980s have been reviewed by Martin and Coughtrey (1982), a number of more recent reviews on dry deposition and particulate interactions with natural surfaces were published by Baumbauch (1996), Gallagher et al. (1997), and Davidson and Wu (1989). The most important processes involved in transfer of particulates are those of Brownian motion, sedimentation, impaction and interception (Martin and Coughtrey, 1982), their relative importances depend on a variety of factors such as windspeed, particle size, temperature, humidity and surface roughness (Chamberlain, 1970; Hicks et al., 1991). Relevant plant-related parameters are surface roughness, "stickyness" and surface-orientation (horizontal vs vertical). Here, the work of Schuepp (1984) should be mentioned, which suggested that porous lichen surfaces accumulate more sub-micron-sized aerosols by convective deposition than do flat surfaces.

To the best of our knowledge there are hardly any studies that relate the retention of particulates on plant surfaces to *wet* deposition fluxes from the atmosphere. Field studies are complicated since rainfall can be expected to remove at least part of the previously (dry) deposited particulates from surfaces. To further complicate the point, the efficiency of rainfall scavenging in atmospheric *deposition* is reported to depend on particle diameter, and is suggested as important for particles >10 μm (Menzel, 1967), but as very small for particles < 4 μm (Chamberlain, 1970; Wedding et al., 1975). Moreover, *washout from surfaces* is reported to increase with rainfall rate (Robig et al., 1978; Slinn, 1978). Thus, the fraction that is retained is a proportion of the intercepted fraction, and is a function of the time period involved between deposition and measurement. In terms of metal retention, it is also related to the chemistry

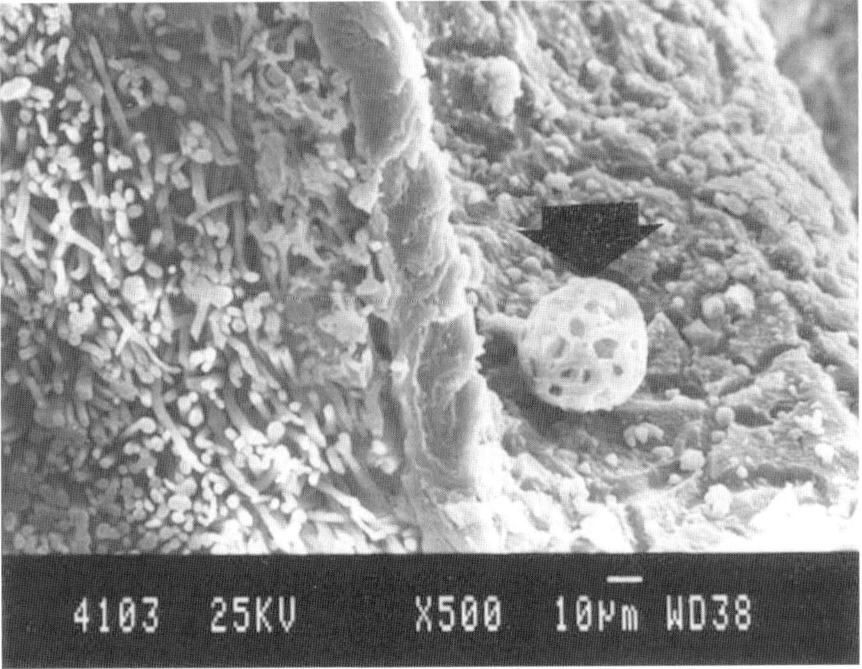

*Figure 3.*An airborne particle (arrow) derived from heavy-fuel combustion, detected on the thallus surface of *in situ Ramalina lacera* growing on *Faidherbia (Acacia) albida* trees in a nature reserve, north-east of an industrial town (Ashdod, SW Israel). The cross-section shows mostly fungal hyphae (left part of the figure). The thallial surfaces are shown in the right part of the figure. Scale bar = 10 μm (Garty, Kunin, Delarea, unpublished).

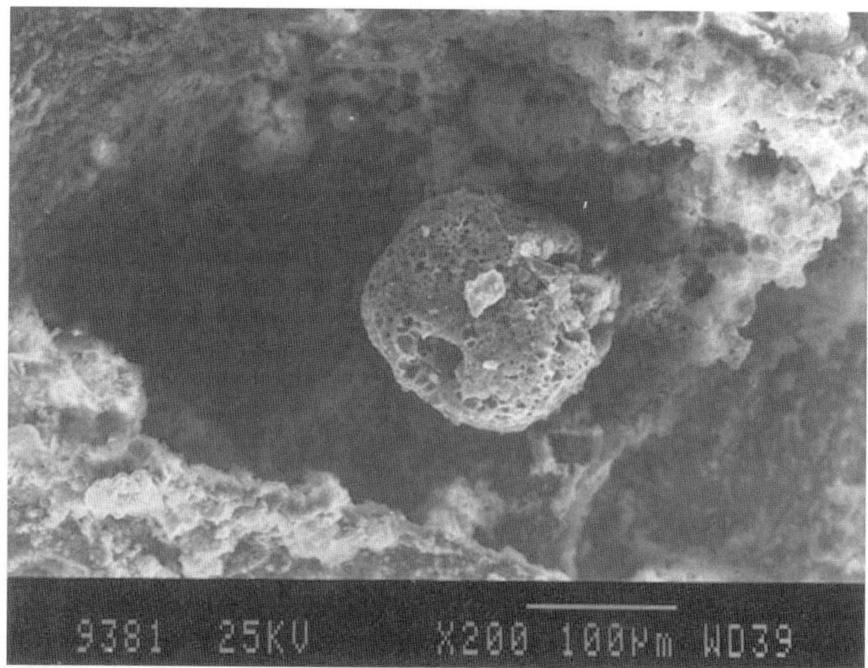

Figure 4. A very large airborne particle derived from heavy-fuel combustion, detected on the thallus surface of *in situ Ramalina lacera* growing on *Faidherbia (Acacia) albida* trees in a nature reserve near the town of Ashdod, SW Israel. Scale bar = 100 μm. (Garty, Kunin, Delarea, unpublished).

of the particulate concerned and the relative solubility and absorption of its chemical components into the plant surface. All this makes that longer-term assessment of deposition invariably encompasses both deposition and retention. The overall half-life for retention of deposited particulates is reported as about 14 days, but it is also reported as varying between 1 to 100 days (Garten, 1978; Martin and Coughtrey, 1982). *Resuspension* from soil to plant of previously deposited materials is commonly accounted for by estimation of the degree of soil contamination, *e.g.* by the use of the Fe:Ti ratios in soil and plant (Martin and Coughtrey, 1982; Nieboer et al., 1978), or by the application of factor-analytical data-processing (Wolterbeek et al., 1996).

For lichens, electron microprobe studies have demonstrated the presence of heavy metals in particles trapped within the thallus (Garty, 1993; Garty et al., 1979; Garty, 2000a) (Figs 3 and 4). Other particles (called pruina) observed in the lichen thallus are reported to contain only calcium oxalates (Wadsten and Moberg, 1985): the reasoning here is that the lichen probably needs to dispose of excess calcium (Wadsten and Moberg, 1985), or that the particles possibly represent material from the immediate substratum or local dust (Brown and Beckett, 1984). However, here the Purvis et al. (1987) data should also be mentioned: results indicated the occurrence of copper-oxalates in medulla tissue of *Lecidea lactea*. In addition, in more recent reports, Pb and Zn are reported to accumulate both through complexation to carboxylic groups of fungal cell walls and through enhanced synthesis of oxalates; toxic elements may precipitate as insoluble salts by the latter process (Sarret et al., 1998). Here, prevailing acidic conditions may (re)mobilize metals such as Pb and may eventualy result in Pb accumulation in especially fungal tissues (Purvis et al., 2000). In the above calcium reasoning, the formation of oxalate hydrates may also be seen as an expression of lichen physiology: the hydrates are suggested to function in the lichen water balance (Wadsten and Moberg, 1985). In a more general metal context, it should also be noted that intracellular synthesis and deposits of organic acids such as malates or citrates have been frequently indicated in general plant metal tolerance mechanisms (Thurman, 1981; Baker, 1987).

3. The lichen as a biomonitor

3.1. General

Considering lichens as biomonitors means that the plant-characteristics of lichens should be seen relative to the atmospheric metal availability. The inferred accumulation of the metals implies that both morphology and physiology of the lichen should be taken into account (Tretiach and Brown, 1995; Brown, 1991). Morphology and physiology are closely related (Nash and Gries, 1995; Sancho and Kappen, 1989): in a lichen study along a latitudinal gradient from the temperate to the Arctic climatic zone, Schipperges et al. (1995) reported simultaneous variations in CO_2 exchange, water relations and morphology, the lichen populations thereby representing different ecotypes rather than exhibiting plastic acclimation ability. Further illustrations of this point are a study on north- and south-facing rock-growing *Ramalina* populations (Pintado et al., 1997), where differences in both chlorophyll content, water retention capacity, and thalli thickness were observed, and a study by Máguas et al. (1997), who observed differences in both morphology and physiology between central and marginal thallus zones in large-thallied foliose lichens.

The impact of metal accumulation on lichen physiology has been reviewed by Puckett and Burton (1981) and, more recently, by Garty (2000, 2000a). It should be noted here that responses to heavy metal deposition may differ between the algal and fungal lichen components (Tarhanen, 1998). Lichen morphology in the context of metal accumulation has been discussed by Goyal and Seaward (1982) and Brown (1991): thallial size and thickness, rhizinal density and the relative thicknesses of algal and fungal tissue layers were all considered as affected by and/or determining metal accumulation.

The indicated effects should probably be generalized: the non-metal components of air pollution (e.g. ozone, acidity) are also reported to result in changes in lichen physiology, ultrastructure and morphology (Piervittori et al., 1997; Tarhanen et al., 1997); furthermore, lichen functioning (e.g. growth) will be ruled by general environmental factors such as (micro)climate or season (Jenkins, 1987; Richardson et al., 1982; Werner, 1990).

Largely following the considerations above, the following paragraphs give some thoughts on the lichen as a biomonitor, introduced by a generalized concept of the lichen functioning, and discussed by reviewing the existing information on the relationships between metal accumulation on the one hand and aspects of both lichen morphology and physiology on the other.

3.2. A view of a lichen

3.2.1. The surface-layer model: general considerations

Apart from the numerous lichen metal accumulation studies, which have focused on the role of various lichen components, the internal localization, chemical form and translocation of the metal Demon et al., 1988; 1989; Falnoga et al., 2000; Garty, 2000; Garty et al., 1979; Goyal and Seaward, 1981, 1982; Purvis et al., 1987; Sarret et al., 1998; Wadsten and Moberg, 1985), many other studies have implicitly modelled the lichen as an essentially homogeneous organism towards metal uptake and release (Ellis and Smith 1988; Papastefanou et al., 1988; 1989; Puckett et al., 1973; Raes et al. 1990; Ramelow et al., 1996; Schwartzman et al., 1991; Sloof, 1995; Sloof and Wolterbeek, 1992; Smith and Ellis, 1990; Tyler, 1988). Generally speaking, these two approaches may be regarded as representing laboratory findings and field-related results respectively. In a recent attempt to bridge these points of view, Reis et al. (1999) developed a physico-mathematical view to a lichen, which combined an interface and lichen interior, thereby modelling both interception and accumulation. In this view, the lichen is seen as essentially homogeneous, but its morphology is recognized as relevant in interception processes. Furthermore, the performance of the lichen interior may be regarded as an expression of the lichen's physiological status quo, comprising both algal and fungal contributions.

Figure 5 shows a simplified representation of this view of a lichen, giving an outsided interface (the lichen's surface S), enveloping the lichen's interior tissues I. Regarding all N notations as denoting the amounts of metal, and regarding the lichen's ambient environment A, the model implicates that the available amount of atmospheric metal N_A is deposited at a rate $D_S.N_A$, the rate constant D_S thereby expressing the first order characteristics of the deposition process. In analogy, the release from the surface is governed by $R_S.N_S$, the metal influx into the lichen interior by $I_I.N_S$, and its efflux by $E_I.N_I$, with rate constants R_S, I_I and E_I. Here, the rate constants D_S and R_S should be interpreted as representing surface/morphology characteristics, whereas the rate constants I_I and E_I should be seen as more intimately related to the lichen's physiology. The full mathematical evaluation of this view is beyond the scope of the present paper, but can be read in Reis et al. (1999).

Essentially, the lichen's performance is regarded as a dynamic one: both uptake and release operate simultaneously and continuously, their rates governed by the overall lichen's performance, the latter both in a physiological and a morphological sense. Figure 6 gives a basic illustration of the implications of the model, in terms of the lichen response to variations in the ambient metal availability. The left part of the graph shows the lichen in equilibrium with its ambient metal availability: the reflectance is given by the ratio L_1/A_1. A block-change in availability (A_2) results in a progressive change in lichen metal status, at an overall rate governed by all rate constants, until a new equilibrium is reached at lichen level L_2. A second block-change makes that the ambient metal availability returns to a level A_3: the lichen's response

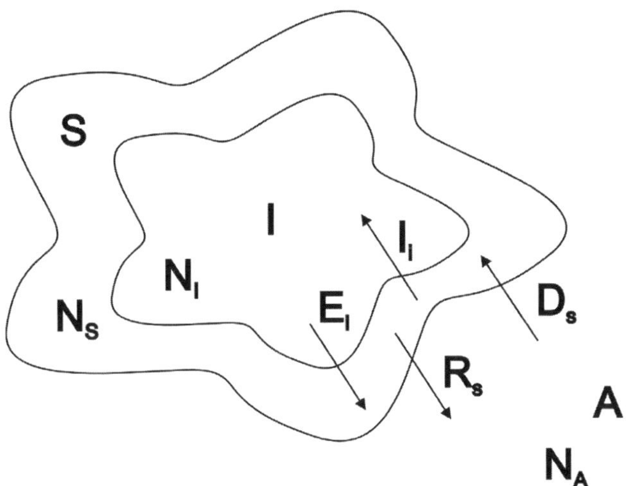

Figure 5. A view to a lichen. Simplified representation of a lichen. The Lichen consists of a Surface-layer (S) and an Interior (I), and is placed in its ambient atmospheric environment (A). The metal occurrence in A, S and I is represented by N_A, N_S and N_I respectively. Deposition (D) and release (R) onto and from the surface are governed by the rate constants D_S and R_S respectively, with associated element fluxes $D_S.N_A$ and $R_S.N_S$. In analogy, element influx and efflux are governed by rate constants I_I and E_I respectively, the associated fluxes should be written as $I_I.N_S$ and $E_I.N_I$. The respective rate constants will be influenced by lichen morphology, ambient conditions (rainfall, particle sizes, wind, etc.) and metal chemical forms (D_S, R_S), and by lichen physiological parameters (carrier affinity/specificity, membrane permeability, internal chemical and spatial compartimentalization, etc.) (I_I, E_I). The ambient (atmospheric) metal availability (N_A) may be expressed by (a combination of) total, wet or dry deposition, classed following particle sizes, aerosols, and assessed by air and deposition sampling. Although the lichen representation simplifies the contributions of the algal and fungal components, the view recognizes the relevance of both morphology and physiology in processes of metal accumulation and release.

implies a progressive change in metal status towards status L_3. The graph shows that the lichen needs time to "follow" changes in availability: the rates of the processes of uptake and release make that the lichen spends time in "forgetting" the old situation. Reis et al. (1999) have called this time the lichen's *remembrance* time, and have suggested mathematical approaches to a "memory loss" function. The model suggests that any change in rate constants D_S, R_S, I_I or E_I (Fig. 1) may lead to changes in both lichen equilibrium levels and remembrance times.

In an attempt to account for pronounced effects from rainfall (Chamberlain, 1970, 1970a; Garten, 1978; Hicks et al., 1991; Menzel, 1967; Robig et al., 1978; Slinn, 1978; Wedding et al., 1975), Reis (2001) expressed a phenomenological rainfall function, which has an initial positive effect on lichen's metal status, due to rainfall scavenging (Chamberlain, 1970; Menzel, 1967) at small rainfall rates, but which has progressively negative effects on the lichen's metal status at higher rates of rainfall, the latter due to increasing washout (Crittenden, 1983, 1985; Robig et al., 1978; Slinn, 1978). To illustrate the overall approach, Figure 7 shows the time course of Na in *Parmelia sulcata* transplants, both measured and model-fitted relative to the Na total deposition, over a one-year exposure period at Portugese field-stations (Reis, 2001).

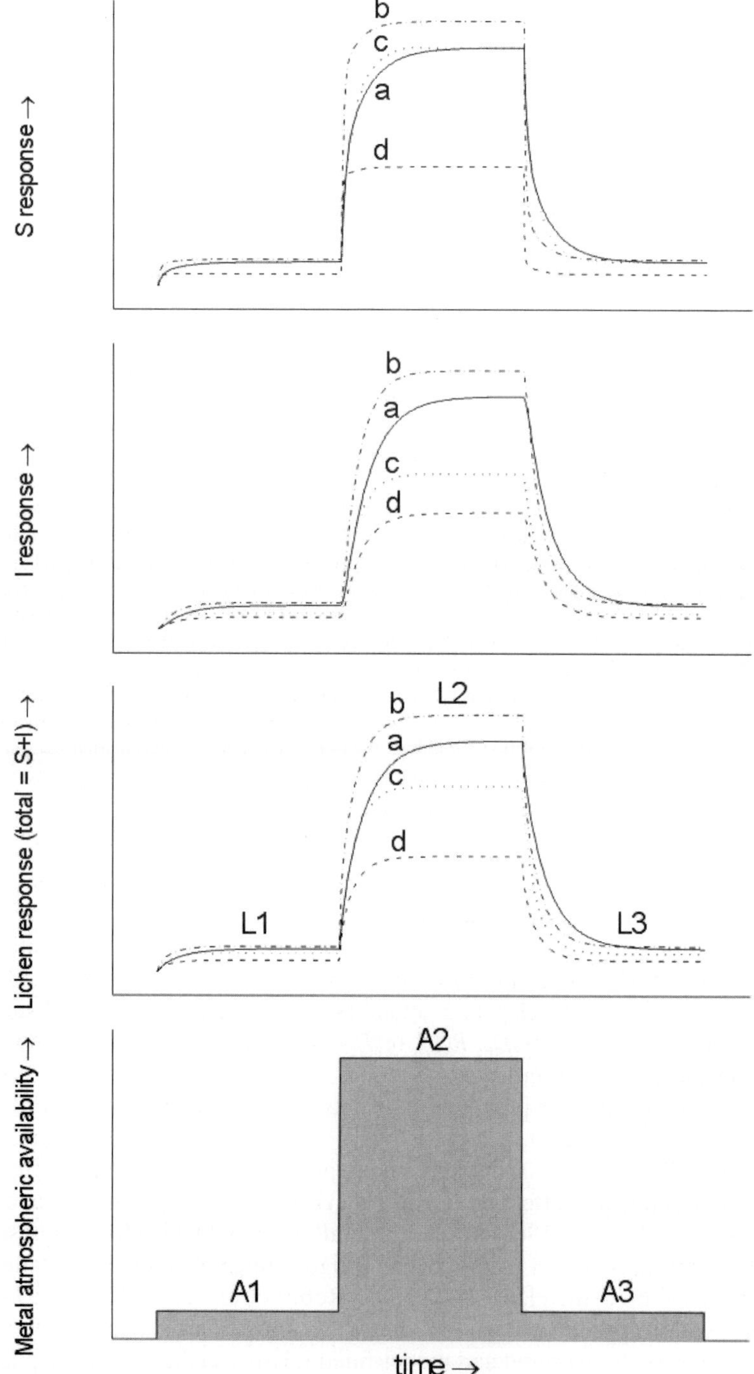

S response →
b
c
a
d
I response →
b
a
c
d
Lichen response (total = S+I) →
b
L2
a
c
d
L1
L3
Metal atmospheric availability →
A2
A1
A3
time →

3.2.2. *The surface-layer model: some implications*

The approach set out in Figure 5 may serve to illustrate that both lichen physiology and morphology operate simultaneously in metal accumulation. Figure 6 also shows that there may be clear-cut relationships between lichen metal levels L and metal availabilities A, *provided* that variable metal availabilities do not lead to variable rate constants D_S, R_S, I_I and E_I. As said above, any significant change in rate constants affects both lichen equilibrium levels and remembrance times (Fig. 6): this means that changes in physiology and morphology should be regarded as of similar relevance.

In reality, equilibria may never be reached: time/season/wind-directional variances in ambient metal availability (Ernst, 1990; Hickman, 1965; Markert and Weckert, 1989) may make that the lichen is subject to continuous changes in metal content, at rates which depend on velocity-parameters of both ambient metal-specific availability and metal-specific lichen properties. In turn, this makes that the availability *period* which is reflected by a metal-accumulating and releasing lichen should be regarded as essentially *metal-specific*. The presently followed general uptake-release concept, which is principally applicable for all living systems, implies that the metal a-specific approaches followed in the moss-deposition relationships (Berg and Steinnes, 1997; Berg et al., 1995; Ross, 1990; Rühling et al., 1987) may need a more critical metal-specific re-examination: the latter approach presumes an infinite remembrance coupled to a metal specific efficiency in moss uptake/retention processes.

In principle, the rate constants may be regarded to reflect lichen properties, but an exception should be made for D_S: this last rate constant relates availability to the lichen, and the question may be raised as to which ambient availability parameter(s) should be applied. Ambient availability may be expressed in terms of wet, dry or total deposition (see Fig. 7), but may also be related to atmospheric concentrations, to metal chemical forms or to metal occurrences in certain particle size classes (Milford and Davidson, 1985). Further studies should increase our insight in their relative

---

Figure 6. Simulation of the lichen response to changes in element availability. The bottom picture shows the ambient element availability, which block-changes in time from level A1 to A2 to A3. The lichen pictures (top three pictures) represent the lichen behaviour following the model shown in Figure 1. The solid lines in the top three pictures (a-lines) represent the lichen's response (total, surface, interior) based on a specific set of numerical values for the rate constants D_S, R_S, I_I and E_I (Figure 1). The a-lines indicate the initial equilibrium L1/A1, they show the time needed for the lichen to reach a new equilibrium L2/A2 in an accumulation period, and they indicate the time needed for the lichen to release elements towards a third equilibrium L3/A3. The accumulation and release periods towards new equilibria may be called the *lichen remembrance time*, which is element-specific, which may depend on ambient and lichen morphological/physiological conditions, and which gives an expression of the length of the foregoing metal availability period reflected by the lichen's elemental content.

The b-line indicates the lichen response when D_S and R_S are increased by 30% and 200% respectively, under constant I_I and E_I regime: the results indicate higher equilibrium values in the lichen, and a shorter remembrance time.

The c-lines indicate a situation where I_I and E_I are increased by a factor 2 and 3 respectively, under constant D_S and R_S regime. The results indicate hardly any changes in surface elements (top picture); but the total element content under equilibrium changes largely due to the decrease in accumulation.

The d-lines show the situation where only R_S is increased by a factor 2. The results show decreased equilibrium ratios, but also indicate, as did the b-lines, that high release rates lead to relatively short remembrance times.

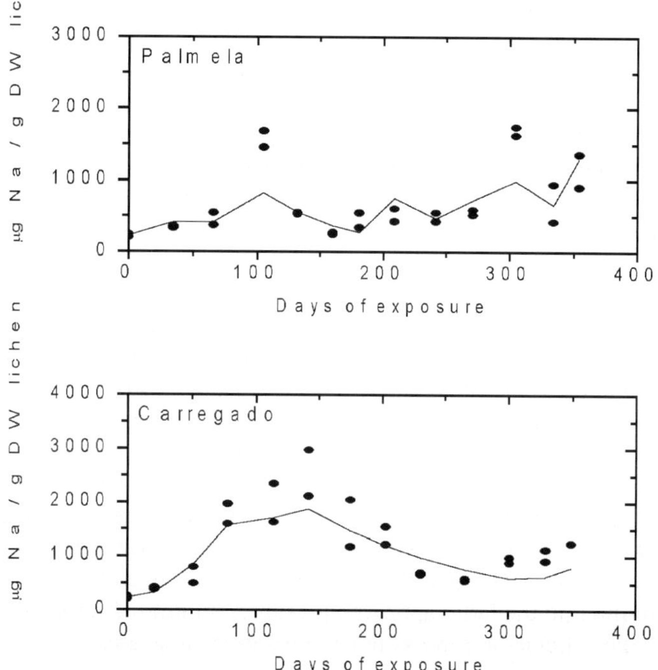

Figure 7. Response modelling of *Parmelia sulcata* lichen transplants. Modelling of the sodium (Na) concentration course in transplanted *Parmelia sulcata* lichens. Full details on the exposure experiments can be found in Reis (2001) and Reis et al. (1999). Illustrative results for the two Portugese experimental stations at sites called Palmela and Carregado are given. The elemental content in the lichens were modelled following the ideas put forward in Figure 5, thereby considering the effects from rainfall, and were seen relative to the locally assessed atmospheric total deposition ATD data. The black dots in the graphs denote the Na concentrations determined in the transplants by INAA analytical techniques (Reis, 2001; Reis et al., 1999). Experimental set-up, sampling and analysis were all in duplo: the individual observations show the variability of the lichen responses. The drawn lines are fitted from the ATD data, the lichen Na concentrations and the presumed behaviourial response equations (Figure 5). The eventual results show the potential of the approach, which is specifically suggested by the way the model reflects the fluctuations in the lichen Na content, which are sometimes fast and rapid if seen relative to the total length of exposure. The variability in the lichen Na content, expressed in both rapid decreases and increases strongly underlines the dynamics of the lichen behaviour, which comprises continuous and simultanuous elemental uptake and release (Figure 5).

importances, and to what extent metals may be grouped into general terms of different availability parameters.

A further general comment which may be made is related to the handling of the sampled lichen itself: Figure 6 shows a theoretical lichen response to ambient metal availability, expressed as the metal content of the *total* lichen, that is, the content of both interior and surface, and of the calculated individual surface- and interior-related elemental contents. Figure 7 gives model predictions, based on assumed removal of surface elements by 30-second rinsing in sample handling before analyses (Reis et al., 1999): this means that the model accounted for the total lichen as operating in accumulation, but that only the lichen interior (N_I) was considered as represented by the elemental data. This made that further mathematical formulations in data interpretation were adapted to the presumed analytical circumstances. The above illustrates that sample handling may have its consequences for the interpretation of the data; the generality of this discussion on sample cleaning is given below.

In literature, a generally posed question concerns the cleaning of plant material before chemical analyses. Ernst (1978), Wyttenbach et al. (1985), Porter (1986), and Markert (1993) all used and discussed various cleaning and washing procedures for various plant materials; all procedures were essentially meant to remove surface contamination. Djingova and Kuleff (2000) suggested a standardized short wash with tap/bidistilled water for the removal of mechanically surface-deposited dust and/or aerosols, if possible on fresh plants or practically in the field. Washing soon after collection is also recommended for both lichens and mosses by Buck and Brown (1977; 1979), to avoid ion leakage during washing induced by previous desiccation. For lichens used in survey contexts, no washing to very short laboratory washing procedures (5–30 sec) are in use or recommended (Boonpragob and Nash, 1990; IAEA-NAHRES, 1999; Reis et al., 1996; Wolterbeek et al., 1996). Boonpragob and Nash (1990), in field studies with lichens, estimated that their 30-second rinses removed short-term surface held dry deposition only. This last approach is in line with the practical assumptions adopted in experimental procedures followed by Reis et al. (1999), and is also argumented on basis of the prolonged rinsing and washing procedures reported as necessary by Brown and co-workers (Beckett and Brown, 1984; Brown and Beckett, 1985; Wells and Brown, 1987) to remove and discriminate between extracellular and intracellular lichen elements. It should be noted here, however, that metal-rich particles may reach the thallus interior (Brown, 1991; Galun et al., 1984; Garty et al., 1979; Olmez et al., 1985), thereby making releases by washing difficult.

The above discussion clearly emphasizes both the importance of dedicated, reliable and reproducible sample handling for adequate interpretation of data obtained, and the need for more study into what is meant by atmospheric elemental availability in the context of lichen metal accumulation.

3.3. Metal accumulation and morphology

In literature, general anatomical and morphological features are largely discussed in the context of the lichen's natural phenotypic variability on the one hand, and its adaptive potential under extreme environmental conditions on the other: Seaward (1976) reviewed *Lecanora muralis* as a highly polymorphic species in natural habitats, and as showing a remarkable consistency in morphological, histological and ecological performance when growing in an urban environment.

A large number of reports discuss morphology in relation with e.g. temperature (Ott and Sancho, 1993; Schipperges et al., 1995), light (Tretiach and Brown, 1995), gas exchange (Máguas et al., 1997; Tretiach and Brown, 1995; Valladares et al., 1993) or water use efficiency (Huiskes et al., 1997; Souza-Egipsy et al., 2000; Valladares, 1994; Valladares et al., 1993). A comparison between epiphytic and epilithic populations of *Parmelia pastillifera* showed the epilitic samples as having reduced photosynthetic rates, a lower chlorophyll content, an increased thickness of the upper cortex, and much more developed rhizinae: this combination of morphological and physiological features were all apparently meant to buffer the higher levels of irradiance and to increase the water storage capacity (Tretiach and Brown, 1995). In line, south-facing populations of rock-growing *Ramalina capitata* showed thicker thalli, mostly due to increased medulla thickness, and a higher water-retention capacity (Pintado et al., 1997). The medullary structure is indicated by Valladares (1994) as probably determining thallus porosity and water storage capacity. Similar reasoning is reported for *Xanthoria* species, where abundant extracellular water was observed between medullary hyphae (Souza-Egipsy et al., 2000).

Lichen morphology, and morphological performance, including growth (Barkman, 1958; De Wit, 1976; Seaward, 1976), have also been associated to atmospheric SO_2 and the incidence of acidic precipitation. It should be noted here, however, that lichen growth may virtually cease during the winter season: there may be a considerable difference in general lichen growth rates between summer and winter in temperate latitudes (Seaward, 1976). Apart from measurable growth effects, sulphur, but also fluoride and ozone have all been reported to cause ultrastructural changes (Palomäki et al., 1992; Piervittori et al., 1997; Tarhanen, 1998; Terhanen et al., 1997), the effects being specific for both lichen species and component. A more explicit illustration of morphological changes due to acidity can be found from the results by Piervittori et al. (1997): exposure of thalli of *Pseudevernia furfuracea* to simulated acid rain caused a progressive reduction of the surface amorphous layer, the latter being essentially formed by surface thickened and conglutinated hyphal walls, which, bonded together, form a more or less smooth outer surface (Anglesea et al., 1982). Variations in thalli thicknesses (Seaward, 1976) may have consequences for the expression of elemental data: Sloof and Wolterbeek (1992), considering three classes of fruiting body density, discriminated between (contour) surface area and lichen mass, and observed an almost constant ^{137}Cs activity per unit surface area for every density class. These results suggest that the regular expression of elemental data on a lichen mass basis may be subject to variations unless the surface to mass ratio is taken into account. Here it is also interesting to note that for *Cladonia cristatella*, fruiting of the fungus was observed as correlated to both acidity and drying (Ahmadjian, 1973).

Irrespective of its seasonality, growth should be regarded as an important lichen parameter. First, growth "dilutes" internal elemental concentrations by increase in mass (Markert and Weckert, 1989), and, second, growth dictates that the lichen consists of a continuum of old to new plant mass. The higher metal concentrations found in older lichen thalli parts (Brown, 1991) may be due to the longevity of the parts themselves (but note Brown (1991) and Garty (1993), who both argue that equilibration to new metal levels may be rapid, and see also Figure 6), or to the possibility that older parts have higher cation exchange capacities, the latter presumably due to the progressive process of cell senescence, in turn associated to the exposure of additional exchange sites (Brown, 1991). This last assumption is in line with results on ^{210}Pb and Pb uptake in *Flavoparmelia baltimorensis*: Schwartzman and co-workers (Schwartzman et al., 1991) observed consistently decreasing partition coefficients ($[Pb]_{water}/[Pb]_{lichen}$) with thallus age, and concluded that older thalli should carry a greater concentration of ion-exchange sites.

In a series of laboratory experiments on lichen metal accumulation, Goyal and

Seaward (Goyal and Seaward, 1981, 1982, 1982a) suggested the rhizinae as absorptive, accumulative and regulative in function. They also found differences in metal uptake capacity between the upper and the lower thallial surface. At high metal supply, the upper surface showed reduced capacity, whereas that of the lower one was enhanced: the results were discussed as indicating a greater accumulation and translocation capacity of the mycobiont relative to that of the phycobiont (Goyal and Seaward, 1982a). Furthermore, rhizinae were found to accumulate the highest metal concentrations under high metal supply conditions (Goyal and Seaward, 1981). For *Peltigera* species in metal polluted environments, Goyal and Seaward (1982) observed reduced thallial size and rhizinal length, a more dense rhizinal growth (related to a more profusely branched venation), and an increased thallial thickness, as well as the relative thicknesses of the algal and fungal tissue layers. These morphological modifications may be regarded as expressing the lichen's basis of metal tolerance and detoxification. The smaller thallial size would provide a reduced interception and absorption area, and an increase in the rhizinal density and thickness of the medulla would protect the phycobiont from direct exposure. Here, the absorption capability of the rhizinae may provide a natural method of detoxification (Goyal and Seaward, 1982a).

The overall results strongly indicate that the lichen morphological form may account for variances in elemental composition. Goyal and Seaward (1981, 1982, 1982a) suggested that the observed differences in tissue dimensions of *Peltigera* may have been induced by exposure to heavy metals, but, in the field, some of the reported changes may also reflect responses to other stresses, such as water (Huiskes et al., 1997; Pintado et al., 1997; Snelgar and Green, 1981; Souza-Egipsy et al., 2000; Valladares, 1994; Valladares et al., 1993) or acidity (Anglesa et al., 1982; Piervittori et al., 1997). Since the assessed lichen's elemental content is commonly expressed on a per unit of mass basis, in comparative surveys, as a initial and practical approach, it may be advisable to also assess the lichen's surface to mass ratio. This makes that for both assessments similar thallial parts should be used, the latter meant to minimize difficulties associated to the possible differences between contour- and actual lichen surfaces (Sloof and Wolterbeek, 1992).

As different lichen species in the same location contain varying amounts of metal, it is obvious that the amount of metal contained by a lichen is species-dependent. According to Upreti and Pandey (1994) the lichen *Umbilicaria decussata* collected in Antarctica contained more Pb and Fe than *U. aprina* in the same site. This difference was attributed to the thin thallus of *U. decussata* which makes for a high ratio of surface area to dry weight as compared with *U. aprina*.

Marked differences characterize the metal content of lichens belonging to different genera. The morphological determinant which has the greatest relevance is made evident by the greater capability of foliose lichens to accumulate metal-containing, airborne particles over and above fruticose lichens. The foliose lichens *Flavoparmelia baltimorensis* and *Xanthoparmelia conspersa* contained greater amounts of metal than the fruticose lichen *Cladonia subtenuis* (Lawrey and Hale, 1981), the foliose lichen *Hypogymnia enteromorpha* contained greater amounts of metal than different species of *Usnea* (Gough et al., 1988) and the foliose lichens *Parmelia subrudecta* and *P. sulcata* contained greater amounts of metal than the fruticose lichen *Anaptychia ciliaris* (Glenn et al., 1995). The foliose lichen *Hypogymnia physodes* was found to contain more Fe, Zn, Ca, Mg, Mn, Cu and Ni than the fruticose lichen *Usnea hirta*, both exposed in Oulu, N. Finland (Garty et al., 1996). Other elements, i.e. K, Na, Pb and Cd did not differentiate significantly between the two lichens. In contrast, Puckett and Finegan (1980) suggested that finely divided thalli of lichens such as *Usnea* or *Alectoria* have a greater affinity to particulate matter than *Umbilicaria* which has an undivided thallus. A rough, pitted, sticky or tomentose surface as compared with a smooth surface, assists the retention of particles (Boileau et al., 1982). The capability to retain water for longer periods (Larson, 1979) is probably an additional determinant of the efficiency of particulate entrapment (Garty et al., 1996). A gelatinous surface further enhances the capability to entrap particulate matter (Garty et al., 1996; Puckett and Finegan, 1980).

Certain studies suggested that foliose lichens depend to a great extent on atmospheric dryfall whereas fruticose lichens depend more on wetfall. According to Bosserman and Hagner (1981), increased amounts of K, Fe, B, Zn, Al, Si and P contained in the foliose lichen *Parmelia* sp., were related to dryfall, whereas higher amounts of Ca, Mg, Na and Sr in the fruticose lichen *Usnea* sp. were attributed to the dependence of this lichen on atmospheric wetfall.

Apart from their relevance to the uptake, retention and accumulation of airborne metal-containing particles, morphological features determine the replicability of results. A poor replicability of metal content in a fruticose lichen relative to the replicability of a foliose lichen was explained by the great variability of surface extent (Gailey and Lloyd, 1986) Foliose lichens in contrast have an equal sized surface, thus promoting an equal accumulation of metals.

A comparison of metal content, of different species of foliose lichens, e.g. *Xanthoparmelia conspersa* and *Peltigera canina* and a crustose lichen, *Lecanora subfusca*, followed the sequence: *P. conspersa* > *P. canina* > *L. subfusca* (Bartók, 1988).

3.4. *Metal accumulation and physiology*

3.4.1. General

Air pollution has been and still is widely associated with an abundance of lichens (Barkman, 1958; De Wit, 1976; Nylander, 1886). In a large body of literature reports, lichen abundance and species richness is associated to acidic precipitation, both in responses to increasing pollution levels and to improving atmospheric conditions (Cislaghi and Nimis, 1997; Henderson-Sellers and Seaward, 1979; Leblanc et al., 1974; Nash, 1988; Oksanen et al., 1990; Seaward, 1980). In two recent studies in The Netherlands, SO_4, NH_4, NO_3, 22 elements and acidity of bark substrates and associated lichen abundances were determined (Van Dobben et al., 2001; Wolterbeek et al., 1996): the results indicated that sulphate, ammonia and nitrate did not significantly affect bark metal retention, that acidity influenced bark retention of only Ca and Hg (Wolterbeek et al., 1996), and that lichen biodiversity was affected almost exclusively by atmospheric SO_2 and NO_2 (Van Dobben et al., 2001). In this respect, the reported synergistic inhibitory effects on lichen pigment status of combined SO_2 and NO_3 are worth mentioning (Balaguer and Manrique, 1991). These results indicate that, in the assessment of comparative lichen metal accumulation, insight in the lichen physiological responses to ambient environmental conditions is of paramount importance: these include those to substrates, season, climate, temperature, acidity, ozone or altitude (Jenkins, 1987; MacFarlane and Kershaw, 1980; Máguas et al., 1997; Piervittori et al., 1997; Pintado et al., 1997; Richardson et al., 1982; Schipperges et al., 1995; Terhanen et al., 1997; Werner, 1990; Zambrano et al., 1999). In addition, however, the impact of increased metal supply should be critically assessed.

The physiology of metal toxicity begins with an increased metal supply, and proceeds to the failure of an "essential to life" plant process (Baker, 1987; Puckett and Brown, 1981; Thurman, 1981). Metal accumulation may be regarded as the first stage in the physiology of toxicity. The metals are accumulated by several mechanisms: particulate entrapment, ion exchange, electrolyte sorption and processes mediated by metabolic energy. The accumulation of particulates, if the particles are not readily solubilized, may be of little metabolic significance and the particulates would be much less toxic when compared to the effects induced by an equivalent amount of metal in solution. In a number of studies ion exchange models are applied to account for lichen metal accumulation (Puckett et al., 1973; Ramelow et al., 1996; Schwartzman et al., 1991): underlying reasoning was the rapid passive process in soluble cation uptake, involving the reversible binding to negatively charged anion sites in the cell wall. Early hypotheses implied that the cell walls would play an important role in metal tolerance by preventing uptake into the cell interior (Lepp, 1981a; Puckett and Burton, 1981; Thurman, 1981), later studies, however, indicated that the cell wall Donnan potential largely results in changed local ion concentrations and associated changed ion fluxes across plasma membranes (Chang, 1981; Dalton, 1984; Helfferich, 1962; Sentenac and Grignon, 1981; Thibaud et al., 1984; Wolterbeek, 1987). In addition, Brown and Beckett (1984), experimenting with *Peltigera* from background and polluted sites, found that the population from the contaminated site displayed a much reduced rate of intracelular Cd uptake, without any significant difference between the two populations in extracellular Cd binding with respect to both incubation time or Cd concentration. The intracellular uptake across the cellular membranes may be by simple diffusion, by carrier-assisted facilitated diffusion or by active transport (Chang, 1981).

The latter transport involves the movement of substances across the membrane against a concentration gradient, it requires energy, and a carrier molecule for complexing the ion. The carrier molecule is such, that a cellular energy supply (*e.g.* ATP) is directly coupled to permit the active transport. Carrier-mediated transport generally displays Michaelis-Menten kinetics (Chang, 1981), and is referred to in a number of studies on cations by Brown and co-workers (Beckett and Brown, 1984; Brown and Beckett, 1984; 1985) and is also indicated for anions by Richardson et al. (1984). It is here that uptake is most directly coupled to cell physiology.

The overall process of metal accumulation, comprising particulate entrapment and both extracellular and intracellular uptake, may turn out to affect several processes at the cellular physiological and biochemical level, without any direct visual symptoms, such as transpiration, respiration or photosynthesis, but, at increased metal levels, may also result in visible symptoms such as stunted growth, decoloration or chlorosis (Van Grunsveld and Clijsters, 1994). Whatever the ultimate sites of action, the metals have to cross the plasma membrane before reaching the intracellular compartments. Therefore, this barrier (including the cell wall, and including its associated transport systems) can be considered as the first target for metal action. Only after passing the membrane, metals can interact with further cellular components and processes (Van Gronsveld and Clijsters (1994). In a review on metal phytotoxicity, Van Gronsveld and Clijsters (1994) recognized (in)direct effects of metals on the permeability of the plasma membrane, inhibition of enzymes, increases in enzyme capacities, interactions with nucleic acids, and a number of defence mechanisms, the later including anti-oxidative intermediates, phytochelatins, peroxidases etc. Most important, however, are the *diagnostic critera* for the assessment of phytotoxicity. As said earlier above, it should be noted here that in the context of the use of the lichen as a biomonitor, phyto-toxicity should be diagnosed essentially irrespective of its cause. Any cause which leads to changed physiology and morphology should be suspected to alter the characteristics of metal accumulation: this includes metal accumulation itself.

3.4.2. Assessment of lichen vitality by physiological parameters

Different experimental procedures were developed to determine the impact of metals on lichen vitality. Certain procedures relate to the entire thallus. Others refer to the photobiont part of the thallus. The exclusive response of the mycobiont was less documented. The greater part of the analytical procedures applied to assess lichen vitality are destructive as are many of the analytical methods applied to determine metal content. A limited number of procedures enables a non-destructive investigation of the alteration of the physiological status.

3.4.2.1. Response of the thallus as a whole

Leakage through damaged cell membranes: Metals may alter the plasma membrane permeability, leading to leakage of ions like potassium and other solutes (Van Grunsveld and Clijsters, 1994). Damage may be brought about by both the oxidation and crosslinking of membrane protein sulphydryls and by the induction of lipid peroxydation (De Vos, 1991): in experiments with the green alga *Chlorella* the extent of the damage could correlated to the strength of the metal-sulphydryl bond (Van Grunsveld and Clijsters, 1994). Notwithstanding experimental difficulties in data interpretation (Brown and Beckett, 1984), K leakage is used as a measure of membrane damage. Garty and co-workers (Garty et al., 1993, 1997, 1998) assessed both the specific leakage of K and the general electrolyte leakage (measured by the electric conductivity of lichen washing solutions) as an indication of membrane integrity. The results in transplants were attributed to the synergistic effects of both SO_2 and ambient element abundances (Garty et al., 1998).

Leachates of lichen thalli may contain large amounts of K^+ following exposure to metals. The exposure of lichen thalli to increasing concentrations induced a gradual loss of K^+ whereas other metals such as Ni, Co, Cd and Pb initiated an abrupt efflux at a specific concentration (Boonpragob and Nash, 1990). In polluted areas leachates contained, in addition to K^+, elements such as Ca, Na, Fe, Si, Cd and Ba, in higher concentrations than in unpolluted sites (Boonpragob and Nash, 1990).

The decrease of the total K content of the thallus is a well-known indicator of K leakage due to exposure to chemical pollutants. This decrease was demonstrated under field conditions (Garty et al., 1985, 1997, 1997a, 1998, 1998a, 2000; Tarhanen et al., 1999), or upon treatment with metals such as Ni (Hyvärinen et al., 2000; Tarhanen et al., 1999) and Cu (Tarhanen et al., 1999). In many cases, the decrease of K in the thallus was accompanied by an increase of electric conductivity of the leachate (Garty et al., 1997, 1997a, 1998, 1998a, 2000; Hyvärinen et al., 2000; Rope and Pearson, 1990; Silberstein et al., 1996; Tarhanen et al., 1999). Branquinho and co-workers provided essential knowledge of the uptake of heavy metals and their effect on cell membrane integrity in lichens (Branquinho et al., 1997). The lichens *Usnea* spp. and *Ramalina fastigiata* were analysed to establish the efficiency of Na_2EDTA as a chelating agent for extracellular Cu supplied to lichens under laboratory conditions, without causing cellular membrane injury. A sequential elution procedure was used to determine the cellular location of Cu in the lichens. The patterns of extracellular uptake versus time or concentration were anticipated by conventional kinetic studies of other organisms and heavy metals. Copper supplied in the laboratory replaced the naturally acquired extracellular Mg and Ca and induced alterations in the passage of K across the cell membrane. An additional study (Branquinho et al., 1997a) confirmed that Pb is capable of entering cells of different photosynthetic symbionts and various forms of thallus growth: *Lobaria pulmonaria, Ramalina farinacea, Parmelia caperata* and *Peltigera canina*. This study did not show, however, modifications of lichen membrane permeability, such as indicated by the loss of intracellular K and Mg. On the other hand this study confirmed that Pb displaced extracellular exchangeable Ca and Mg. The authors concluded that the deficiency of Mg and Ca, if crucial to physiological processes, may become apparent upon a long-term Pb contamination. The degree of electric conductivity and quantity of K^+ in leachates of the epiphytic lichens *Hypogymnia physodes* and *Bryoria fuscescens* were determined for samples collected in northern Finland and the Kola Peninsula by Tarhanen et al. (1996) who found that K^+ leakage was the most sensitive indicator of membrane disturbance. The damage was attributed to pollution derived from the industrial activity in the area which produces Cu, Ni and SO_2, among other contaminants. The exposure of *Bryoria fuscescens* to wet deposited metals and to simulated acid rain under controlled conditions caused an increase of the electric conductivity of the leachate, an enhanced concentration of K^+ in leachates and a decrease of total K content of the thallus (Tarhanen et al., 1999). A treatment with the metals Ni and Cu caused a reduction of the content of ergosterol in the lichen, indicating an inhibition of sterol biosynthesis that may lead to a disturbance of membrane functioning and an efflux of K^+. Ergosterol is a major sterol of the fungal plasma (Ekblad et al., 1998). It occurs also in green algae, including the unicellular green alga *Trebouxia*, a common photobiont in lichens, isolated from the lichen *Cladonia impexa* (Goodwin, 1974). It is possible that ergosterol is present in equal concentrations in the plasma membranes of both photobiont and mycobiont (Sundberg et al., 1999). Although many studies indicated a decrease of K content of lichens exposed to heavy metal pollution, low K levels should not be expected in all cases. Certain studies indicated that K-containing dust may reach the lichen thallus in the vicinity of fertilizer producers (Holopainen, 1983; Kauppi, 1976, 1984; Palomäki et al., 1992; Tynirinen et al., 1992), fertilizer terminals (Garty et al., 1998, 1998a) and apatite strip mines which emit dust containing elevated levels of K (Palomäki et al., 1992; Tynirinen et al., 1992). In these cases, a certain share of K lost through damaged cell membranes, may be replaced by airborne K deriving from the above mentioned anthropogenic sources. Another source of K which may compensate for K losses in thalli exposed to metal polution is the natural deposition of aerosols especially in arid and semi-arid zones. The chemical composition of settling particles from 23 dust storms in Israel has been determined by Ganor and co-workers (Ganor et al. (1991) over a 20-year period. The authors detected 8600 $\pm$ 3580 $\mu g\ g^{-1}$ K in samples of these settling aerosols.

Stress-ethylene and metal pollution: Lichens belonging to different systematic groups were proved to produce ethylene (Epstein et al., 1986; Lurie and Garty, 1991; Ott, 1993; Ott and Schieleit, 1994; Ott et al., 2000; Ott and Zwoch, 1992; Schieleit and Ott, 1996; 1997). The first indication of the impact of air pollution on the ethylene-producing system was provided by Epstein et al. (1986) who transplanted *Ramalina lacera* originating in a relative clean air site, to a busy road intersection in Israel for a period of one year. This site was proved to be contaminated by Pb (Garty and Fuchs, 1982; Garty et al., 1985; 1985a), Zn (Garty et al., 1985; Garty and Fuchs, 1985a) and Cu (Garty et al., 1985, 1985a). The production of stress-ethylene was measured in thalli of both sites and found to be relatively high in the polluted site. A contaminated industrial region had a similar impact on lichen transplants (Garty et al., 1993a). Relatively high levels of ethylene produced by transplants of *R. lacera* in the vicinity of industrial sites in Israel coincided with a high content of sulphate-S, Pb, Cu, Ni, V and Cr (Garty et al., 1997) and of Cu, Zn and Ba (Garty et al., 2000). Similarly, the enlarged amounts of ethylene produced by transplants of *Hypogymnia physodes* in an urban site with slow traffic and bad ventilation in Oulu, N. Finland, coincided with relative high levels of Fe, Mg, Zn, Pb and Cu (Garty et al., 1997). In comparison, transplants of *Usnea hirta* displayed under the same conditions an enlarged level of ethylene coinciding with high levels of Ca, Mg, Fe, Zn, Mn, Pb, Ni and Cu (Garty et al., 1997). The macroelements K, Ca, Na, Mg and Fe detected in transplants of *H. physodes* and *U. hirta* derive mainly from geochemical sources and are important constituents of dust. Other elements such as Pb, Zn and Cu are linked for the greater part to anthropogenic activity, including car traffic.

Controlled experiments exposing thalli of *R. lacera* to solutions containing metals indicated an enlarged level of stress-ethylene: thalli exposed to 20 mM of either $PbCl_2$, $ZnCl_2$, $MnCl_2$, $CuCl_2$ or $FeCl_2$ produced respectively 3.12, 2.64, 4.12, 7.09 and 17.86 nl $g^{-1}\ h^{-1}$ ethylene (Lurie and Garty, 1991). The application of Fe^{2+} or Fe^{3+} combined with Cl^- at pH 2.0, stimulated the release of ethylene over and above the combination of Fe^{3+} and NO_3^- or Fe^{2+} and SO_4^{2-}. The results point to a synergistic effect of Fe^{3+}, Fe^{2+} and Cl^- ions (Garty et al., 1995). An additional lichen species, *Cladonia stellaris*, produced large amounts of ethylene upon exposure to Fe salts at pH 3.5 and lesser amounts upon exposure to Cu and Zn salts (Kauppi et al., 1998). Combined treatments with H_2SO_4 or H_2SO_4 + HNO_3 followed by either $FeCl_2$ or $FeSO_4$ yielded higher concentrations of ethylene than the same treatment in a reversed order.

The production of enhanced amounts of ethylene by lichens exposed to metal-containing solutions is comparable with stress factors detected in other micro-organisms.

Maillard and co-workers (Maillard et al., 1993) found that the production of ethylene by the unicellular green alga *Haematococcus pluvialis* was stimulated by Co^{2+}, Mn^{2+} and Ag^{2+}. In fungi, such as *Mucor hiemalis*, the ethylene production was stimulated by metal solutions, especially by Fe^{2+} or Fe^{3+}, whereas Cu^{2+} and Mn^{2+} inhibited it (Lynch, 1974). Ferrous ions (Fe^{2+}) stimulated the rate of ethylene formation noticeably in the fungus *Fusarium oxysporum* f.sp. *tulipae* whereas Fe^{3+}, Cu^{2+} or Zn^{2+} had little or no effect (Hottiger and Boller, 1991).

3.4.2.2. Photobiont injuries

The rate of photosynthesis and the impact of metal uptake: Controlled experiments, designed to investigate the impact of metals, detected significant changes in the rate of photosynthesis in lichens. The degree of toxicity of different metals was related to the degree of injury endured by the photosynthetic system of *Umbilicaria muhlenbergii*. The relative toxicity was represented by the sequence Ag, Hg > Co >Cu, Cd > Pb, Ni for short-term exposures and Ag > Cu >> Pb, Co > Ni for extended exposures (Boonpragob and Nash, 1990). Additional studies referred to the limited impact of Ni on photosynthesis expressed as ^{14}C fixation (Nieboer et al., 1976). A decrease of 20% of ^{14}C fixation in *U. muhlenbergii* was caused by Pb-containing solutions whereas Cu-containing solutions produced a ca. 50% decrease (Richardson et al., 1979). An incubation of *Hypogymnia physodes* in $Pb(NO_3)_2$ for 18 h led to a decrease of 45% of photosynthesis whereas a mixed solution contining $Pb(NO_3)_2$ and NaCl caused a decrease of 70% (Punz, 1979).

Relative low concentrations of Zn and Cd (200 μM) were needed to depress photosynthesis in *Cladonia uncialis* and *Lasallia papulosa* (Nash III, 1975). Thalli of *C. uncialis* displayed a greater sensitivity to Zn and Cd than thalli of *L. papulosa* whereas a simultaneous application of Zn and Cd depressed the net photosynthesis to a lesser extent in *L. papulosa* than a separate application of these metals.

Zinc, Cd and Cu were found to inhibit photosynthesis in lichens containing cyanobacterial photobionts at lower concentrations than the concentrations needed to decrease photosynthesis in lichens containing green algae as photobionts (Brown and Beckett, 1985a). The authors found the reduction in gross photosynethesis to be 5% to 40% of controls in ten species of cyanobacteria-containing lichens of the genera *Collema*, *Lichina*, *Lobaria*, *Nephroma*, *Peltigera* and *Sticta*. The distinction made between lichens with cyanobacterial photobionts and lichens with chlorophycean photobionts was not, however, related to a differential content of Zn, Mg, Ca or K in the thallus, nor to the quantity of Zn taken up into intracellular sites. Another cyanobacterial lichen, *Peltigera membranacea* was collected by Beckett and Brown (1983) in heavy-metal contaminated areas to study the impact of heavy metal uptake and accumulation on photosynthesis apart from its tolerance of certain metals. A pretreatment with 12 μM $ZnSO_4$ followed by a H_2O treatment led to a percentage of 96% of photosynthesis in comparison with control material. A pretreatment with 100 μM $CdSO_4$ yielded a percentage of 43% whereas a pretreatment with 12 μM $ZnSO_4$ followed by a treatment with $CdSO_4$ yielded a percentage of 71%. These results suggested that a pretreatment with Zn increases the resistance to Cd in *P. membranacea*.

A comparison of the ^{14}C fixation and the Cd content of the chlorophycean lichen *Cladonia rangiformis* and the cyanobacterial lichen *Peltigera horizontalis*, was made by Brown and Beckett (1983). The cyanobacterial lichen was found to be more sensitive to Cd than the chlorophycean lichen.

The toxic effect of uranium uptake in lichens was studied by measurements of the photosynthesis of *Cladonia rangiferina* exposed to cationic, neutral and anionic forms of uranyl ions. The anionic complex of oxalate ($UO_2\ L_2^{2-}$) was more toxic than the uncomplexed cations UO_2^{2+} whereas no detrimental effect could be assigned to the neutral phtalate complex (UO_2L) (Boileau et al., 1985).

Certain lichen species were studied to assess the impact of field conditions on photosynthesis. Samples of *Evernia prunastri, Hypogymnia physodes* and *Flavoparmelia caperata* collected at a distance of 15 m off a motorway exhibited a considerable decrease of net photosynthesis: 74% in *H. physodes*, 45.5% in *F. caperata* and 42.7% in *E. prunastri*, relative to controls sampled at a distance of 600 m (Déruelle and Petit, 1983). The impact of car-exhaust pollutants on photosynthesis in comparison with artificially-supplied Pb was studied by Lemaistre (1985). In naturally polluted samples of *Flavoparmelia caperata* net photosynthesis increased relative to controls. In artificially-polluted samples net photosynthesis decreased. For *Cladonia portentosa* both natural and artificial pollution enhanced the net photosynthesis, although the increase was much smaller in the presence of artificial pollutants. These data accord to a certain extent with data referring to gas-exchange in the presence of a high Pb content reported by James (1973), Punz (1979), and Déruelle and Petit (1983).

A recent study designed to identify the relationship of physiological parameters of the photosynthetic system with the elemental content of the lichen *Ramalina lacera*, was performed by Garty and co-workers (Garty et al., 2001). Thalli of *R. lacera* were collected in an unpolluted site and transplanted to a national park and an industrial region in Israel for eight months. The net rate of CO_2 fixation per gram dry weight correlated inversely with the Ba, Cu, and Ni content. The rate of net CO_2 fixation calculated per mg chlorophyll of the photobiont correlated inversely with the V and Zn content.

Metals and the potential quantum yield of photosystem II (PSII): One of the non-destructive methods to assess changes associated with photosystem II (PSII) in lichens is the measurement of modulated chlorophyll *a* fluorescence. The potential quantum yield of PSII is measured by a pulse amplitude modulated (PAM) fluorometer. Many studies express the results as an F_v/F_m ratio, which represents the potential (or optimal) quantum yield of electron transfer through PSII. The F_v/F_m ratio is calculated for dark-adapted lichens by maximal fluorescence (F_m) minus minimal fluorescence (F_o), divided by F_m: ($F_mP - F_o)/F_m = F_v/F_m \cdot F_v/F_m$ ratios are frequently used to demonstrate the impact of stress on the photosynthetic apparatus (Schreiber and Bilger, 1993).

Analyses of lichens exposed to chemicals under controlled conditions were carried out by Branquinho and co-workers (Branquinho et al., 1997, 1997a) who investigated the impact of heavy metals on chlorophyll fluorescence. The authors detected a decrease following an uptake of Cu (Branquinho et al., 1997). *Usnea* spp. were found to be more sensitive to Cu than *Ramalina fastigiata*. The absence of certain lichens in an area adjacent to a copper mine, was ascribed to its metal emissions (Branquinho et al., 1999). A total inhibition of PSII photochemical reactions occurred under controlled conditions when intracellular Cu concentrations exceeded a threshold of approximately 2.0 μmol g^{-1}. The presence of the metal thus explained both the physiological change and the survival or absence of *R. fastigiata* in the surroundings of the copper mine. Comparable findings obtained by Odasz-Albrigtsen and co-workers (Odasz-Albrigtsen et al., 2000) referred to a decreased photosynthetic efficiency of lichens exposed to multiple airborne pollutants along the Russian–Norwegian border. Lichen species like *Hypogymnia physodes, Cladonia* spp. and *Parmelia olivacea* from the South-Varanger region at a distance of 7 to 45 km from metal smelters, located in Nikel and Zapolyarnij, Russia, demonstrated an inverse correlation of F_v/F_m values and Ni and Cu content.

A decrease of the ratio F_v/F_m expressing the impairment of PSII photochemical reactions, may be obtained by an exposure of lichens to Pb, particularly in cyanobiont lichens (Branquinho et al., 1997a). A decrease of the F_v/F_m ratio coincided with a relatively high content of Cu, Zn, Ba, S, Ni and V in thalli of the epiphytic lichen *Ramalina lacera* transplanted in the vicinity of oil refineries, a power station (Garty et al., 2000) and electrochemical industries in Israel, where thalli were enriched by Hg (Garty et al., 2001).

Metal content and the spectral reflectance response of the thallus: An additional non-destructive method to assess change in the vitality of the photobiont in lichens is based on the application of spectroscopic measurements of the reflectance response of the thallus. In the last three decades, remote sensing satellites were employed for investigations of the status and dynamics of vegetation. Studies of the spectral reflectance response of vegetation led to the development of vegetation indices (VIs) (Bannari et al., 1995; Gamon and Qiu, 1999). Different VIs were based on combinations of two or more spectral bands assuming that a multi-band analysis would enlarge the amount and quality of information. The multi-temporal nature of satellite imagery facilitates the investigation of the vegetative component, based on its typical reflectance, which is apparent for the greater part in the red (R, 600–700 nm) and near infrared (NIR, 700–1100 nm) bands of the electromagnetic spectrum (Tucker, 1979; Sellers, 1985). Vegetation indices were confirmed to correlate with other parameters of vegetation such as green biomass (Tucker, 1979), chlorophyll concentration (Buschman and Nagel, 1993; Peñuelas and Fillela, 1998), leaf area index (Asrar et al., 1984), foliar loss and damage (Vogelmann, 1990), photosynthetic activity (Sellers, 1985) and carbon fluxes (Tucker et al., 1986).

Ecological studies applied for the greater part the Normalized Difference Vegetation Index (NDVI) formulated as: NDVI = $(\rho NIR - \rho R) / (\rho NIR + \rho R)$ (Rouse et al., 1982). This index as well as its less popular modifications, is based on the difference between the maximum absorption of radiation in the red due to chlorophyll pigments and the maximum reflection of radiation in the NIR due to leaf cellular structure, apart from the fact that soil spectra, lacking these mechanisms, do not show a dramatic spectral difference.

As a non-destructive method, NDVI demonstrates the possibility to detect pollutant-induced stress in lichens referring to heavy metals and other substances. This parameter enables the detection of early signs of stress before changes in other physiological parameters become measurable.

The lichen *Ramalina lacera* was used to estimate the degree of stress induced by exposure to pollutants, as expressed by changes in the spectral reflectance of the thallus (Garty et al., 1997, 1997a, 1997b, 1997c, 2000, 2001, 2001a). The alteration of NDVI was compared with the amount of mineral elements in thalli of *R. lacera* collected in an unpolluted site and transplanted to industrial regions in Israel for different periods. The polluted air induced significant shifts in the spectral response of the lichen. Pearson correlation coefficients of NDVI and elemental content indicated an inverse correlation of NDVI and total S, sulphate-S, Ni, V, Pb, Mn, Al, Cr, Fe, Ba, Cu and Ti, and a direct correlation of NDVI, K and P. Not many studies applied the parameter of spectral response under controlled conditions, with the exception of Cox et al. (1991) who found that lichens exposed to Cu concentrations of > 20 μg g^{-1} displayed a significant shift of 2–3% of the spectral response. In a series of controlled experiments thalli of *Ramalina lacera* from an unpolluted site in Israel were treated either with double distilled water or with simulated acid rain (H_2SO_4:HNO_3, 2:1, pH 3.01±0.07) for a period of five-months (Garty, 2000a). Analyses of the NDVI ratios obtained in this experiment simulated the degradation of chlorophyll upon exposure to acid rain. The exposure of thalli of *R. lacera* to simulated acid rain caused also a significant decrease in the amount of Ca, K, Mg, Na, P, Mn, Sr and Ba, in comparison with the content of transplants left in the control site. On the other hand, the amounts of both total S and sulfate-S increased as expected in thalli treated with solutions of H_2SO_4:HNO_3.

Metals and the chlorophyll content and integrity of the photobiont: The bleaching of lichen thalli as a result of chlorophyll degradation is an obvious sign of damage to lichens affected by airborne pollutants. Lichens exposed to metal-containing solutions underwent a significant change of the chlorophyll absorption spectrum: according to Puckett (1988) the alterations were dependent on the specific ion and the duration of exposure. Marked deviations from the range of control spectra were produced in *Umbillicaria muhlenbergii* treated with solutions of Cu, Hg or Ag ions for 15 h. The application of Cd, Co, Ni and Pb, on the other hand, did not affect the chlorophyll in this lichen. Combined treatments of *Ramalina lacera* with each of the metal ions Fe^{2+}, Fe^{3+}, Mn^{2+}, Cu^{2+}, Pb^{2+} and Zn^{2+} and the ions SO_4^{2-}, NO_3^- or Cl^- under acidic conditions, were assumed to intensify the detrimental effect of heavily polluted sites. Treatments with Cu, Zn, Mn, Pb, Fe(II) and Fe(III) salts under acidic conditions were particularly effective (Garty et al., 1992). SO_4^{2-} combined with K^+ proved to be rather harmful to the photobiont chlorophyll, over and above Cl^- anions. A $CuCl_2$-treatment led to a severe chlorophyll degradation. The chlorophyll degradation in the above-mentioned study was measured by the Ronen and Galun method (Ronen and Galun, 1984). The extraction of chlorophyll is performed by dimethyl sulphoxide (DMSO). The ratio of optical density at wavelengths 435 and 415 nm was confirmed as a reliable parameter for an estimation of chlorophyll degradation. Its reliability derives from its high correlation with the acid factor used for the measurement of the ratio chlorophyll *a*/phaeophytin *a*.

Certain metals were linked directly with the disintegration of chlorophyll and/or with a decrease of chlorophyll synthesis. Evidence relating to the interference of Cu with the biosynthesis of chlorophyll in addition to its part in the process of lipid peroxidation in photosynthetic membranes, was provided by Chettri and co-workers (Chettri et al., 1998). Different lichens displayed a disparate response to Cu-treatments. A Cu content of up to 1600 μg g^{-1} dry weight had no effect on the total chlorophyll content of *Cladonia rangiformis* whereas a Cu content exceeding 175 μg g^{-1} led to a decrease of the total chlorophyll content of *Cladonia convoluta*.

Under field conditions the degradation of chlorophyll of lichens exposed to chemical pollution corresponds in many cases with enlarged amounts of metals and other elements contained in the thallus. The chlorophyll integrity of *Ramalina lacera* correlated negatively with Br, Pb, Fe and Ti content after one year of exposure in different biomonitoring sites, including sites polluted by vehicular and industrial activity (Garty et al., 1985). Another study detected negative correlations for chlorophyll integrity and Zn, Pb, and Cu for transplants of *Ramalina lacera* after one year of exposure (Garty et al., 1988). An exposure of 10 months in other study areas in Israel yielded negative correlations for chlorophyll integrity and Al, Fe, Ni, sulphate-S, Ti and V (Garty et al., 1997a), B, Cr, Fe, Mn, Ni and Pb (Garty et al., 1998a). An additional study held for a duration of eight months in a different region indicated negative correlations for chlorophyll integrity and Ba, Cu and Zn (Garty et al., 2000). A duration of 11 months in the same region yielded negative correlations for chlorophyll integrity and Ba, Cu, Ni, V and SO_4^{2-} (Garty et al., 2001a).

Enlarged amounts of Mg, Cr, Fe and Cd in transplants of *Xanthoria parietina* (Bartók et al., 1992) and of Cd in samples of *Anaptychia ciliaris, Lobaria pulmonaria* and *Ramalina farinacea* (Riga-Karandinos and Karandinos, 1998) coincided with symptoms of chlorophyll degradation. Negative correlations were obtained for chlorophyll integrity and S and Al content of samples of *Punctelia subrudecta* transplanted in Cordoba, Argentine (Gonzáles and Pignata, 1994).

A substantial decline of chlorophyll and net photosynthesis was detected in *Ramalina menziesii* in a polluted site in summer in Los Angeles by Boonpragob and Nash (Boonpragob and Nash, 1991). The percentage of phaeophytization was found to increase in proportion. Leachable Mg, Ca, P, Na and K were found to increase in this lichen in summer (Boonpragob and Nash, 1990). The authors assumed that at least a certain part of the elevated leachable Mg in *R. menziesii* derived from the degradation of chlorophyll algal cells of this lichen in air-contaminated sites. Comparable results were obtained for thalli of *Ramalina lacera* collected in an unpolluted site in Israel and transplanted for a period of 10 months in the vicinity of oil refineries and a steel smelter in the Haifa Bay that exhibited a definite degradation of chlorophyll (Garty et al., 1993). The seasonal variation of Mg percentages detected by scanning electron microscopy (SEM) and energy dispersive X-ray analysis (EDX) on/in the cortical cells in the lichen was compared with changes in the chlorophyll integrity of the *R. lacera* transplants. The amounts of Mg on/in the thallial cortex increased gradually in samples from the Haifa Bay whereas low Mg values were obtained for control thalli. The values obtained for Mg seem to represent a significant leakage from algal cells, indicated by a detectable accumulation on the lichen surface. As the control thalli left in the unpolluted site exhibited a different pattern of Mg accumulation under similar climatic conditions, it was assumed that its accumulation on/in the cortical cells is associated with a leakage of ions from internal parts of the thallus, followed by a deposition on surface cells.

4. The lichen in use: problems and practical perspectives

In considering lichens as biomonitors for metal air pollution, the present paper under-lines that the lichen should be recognized as the plant it is, and that clear views should be expressed as to what is meant by metal air pollution.

Presently, the lichen's general dynamics in elemental uptake and release were put central in the biomonitoring approaches. In principle, elemental uptake and release should be considered for both photobiont and mycobiont parts of the symbiotic lichen organism (Figs 1,2), but the simplified approach expressed by Figure 5 regards the lichen as a thoroughly mixed system of both components. This means that although any component-specific behavioural change may show up in the approach, lichen morphological heterogeneities which affect total behaviour, which are relevant in sampling, or which are of importance in the handling of the lichen samples, may remain obscured in the Figure 5 view of the lichen. Further study is necessary to clear these points of interest.

The approach recognizes the relevances of both morphology and physiology, thereby more or less bridging detailed physiological laboratory work and the prag-matic approaches which characterize most field work. Considering both physiology and morphology of the lichen also implicitly means that both physico-chemical/ mechanical and biochemical/physiological processes are regarded as of relevance in the lichen's behaviour and responses. Any consequences for field approaches, sampling strategies or sample handling procedures should be thoroughly discussed. Set in the context of overall lichen's elemental accumulation efficiency and retention, the above suggests that the comparability of lichen (sub)parts should be studied in terms of particle entrapment and uptake and release processes, the latter also because of the heterogeneities in internal distribution of myco- and photobionts.

It should be noted that lichen physiology and morphology are of paramount impor-tance in metal accumulation processes, but that a large variety of factors influence these very lichen's characteristics: apart from the influences by metals themselves, morphology and physiology may be affected, tuned and directed by light, temperature, (sub)climatic conditions, substrates, humidity and other environmental factors. This means that especially for lichen surveys on metal air pollution which are larger-scaled in a geographical sense, lichen materials should not only be analysed for their metal content, but also for a number of selected morphological/physiological characteristics, the latter exclusively meant to indicate the basic comparability of the lichen's responses. The present paper addressed the surface-to-mass ratio, rates of K-leaching, spectral reflectance, chlorophyll content, stress ethylene production, the rate of the photosynthesis processes, the potential quantum yield of photosystem II as possible points of attention. Further discussion should clarify these or any other approaches in terms of larger-scaled usability, sensitivity and practical interpretation in the context of accepting, rejecting or normalizing of the lichen sample's metal contents.

The present paper also addressed the metal atmospheric abundance. In many studies, variances in the lichen's metal content are interpreted as indicating geograph-ical or time differences in metal availability. The point of interest here is that in many cases we don't know what is meant by metal availability: is this to indicate total depo-sition, any fraction or particle size class of the total deposition, or is this also including metals in aerosols, in air filters, total or in any particle size class? Is this to indicate total metal abundance or is this mostly related to specific highly available metal chemical forms? Although the lichen may principally accumulate metals through the combined action of all processes and from a combination of all metal available fractions, one may speculate as to whether metals may be generally classified as belonging to groups which mostly associate to certain processes or sources: here the reasoning is that some metals may be accumulated mostly by morphology (surface) related processes, or that specific metals may be relevant for the lichen largely in small diameter particle size classes. This line of reasoning makes that further work in defining and analytical fractionation of metal atmospheric abundance should be seen as relevant as the study of the basic lichen responses themselves.

A point of interest which combines metal abundance and the lichen response is the lichen dynamics (Figs 5, 6, 7): any period of exposure should be seen relative to the lichen response dynamics, which means that metal abundances may not be averaged over periods of exposure without thoughts on the lichen's remembrance times: short-term high metal level abundances in periods which are low-level on average (or the other way around) may have impacts on lichen metal contents, the latter largely depending on the metal specifics of uptake and release.

Apart from the difficulties and associated challenges associated to the use of lichens as biomonitors of atmospheric metal pollution, the data illustrated in Figure 7 also indicate that progress is made: lichen responses may be modelled to the extent that observed short term fluctuations in metal abundance and lichen metal contents are reflected in the theoretical expectations. Furthermore, the lichen remembrance time may be calculated, which indicates the lengths of metal exposure which are reflected by metal content of the lichen. Here should be noted that this remembrance length combines characteristics of both lichen and ambience: it may be both metal- and conditions-specific, and it may be affected by chemical forms, competition effects, etc.

Further study should be focused on inversed methodologies: air metal abundances should not only be reflected by lichen metal contents, but the lichen metal contents should be used in inversed calculations of air metal abundance characteristics, thereby accounting for lichen dynamics.

References

Ahmadjian, V., 1973. Resynthesis of lichens. In: Hale, M.E., Ahmadjian, V. (Eds), The Lichens. Academic Press, New York, pp. 565–579.

Ahmadjian, V., 1980. Separation and artificial synthesis of lichens. In: Cook, C.B., Pappas, P.W., Rudolph, E.D. (Eds), Cellular Interactions in Symbiosis and Parasitism. Ohio State University Press, Columbus, pp. 3–30.

Ahmadjian, V., 1990. What have synthetic lichens told us about real lichens. Bibliothecha Lichenology 38, 3–12.

Ahmadjian, V., 1993. The lichen photobiont – what can it tell us about lichen systematics? Bryologist 96, 310–313.

Ahmadjian, V., Hale, M.E. (Eds), 1973. The Lichens. Academic Press, New York.

Adriano, D.C., 1986. Trace Elements in The Terrestrial Environment. Springer Verlag, New York.

Anglesea, D., Veltkamp, C., Greenhalgh, G.N., 1982. The upper cortex of *Parmelia saxatilis* and other lichen thalli. Lichenologist 14, 29–38.

Asman, W.A.H., 1987. Atmospheric behaviour of ammonia and ammonium. Ph.D. Thesis, University of Utrecht.

Asrar, G., Fuchs, M., Kanemasu, E.T., Hatfield, J.L., 1984. Estimating absorbed photosynthetic radiation and leaf area index from spectral reflectance in wheat. Agronomical Journal 76, 300–306.

Badman, D.G., Jaffe, E.R., 1996. Blood and air pollution: state of the knowledge and research needs. Otolaryngology – Head and Neck Surgery 114, 205–208.

Baker, A.J.M., 1981. Accumulators and excluders-strategies in the response of plants to heavy metals. Journal of Plant Nutrition 3, 643–654.

Baker, A.J.M., 1987. Metal tolerance. New Phytologist 106, 93–111.

Balaguer, L., Manrique, E., 1991. Interaction between sulfur dioxide and nitrate in some lichens. Environmental and Experimental Botany 31, 223–227.

Bannari, A., Morin, D., Bonn, F., 1995. A review of vegetation indices. Remote Sensing Reviews 13, 95–120.

Bargagli, R., Iosco, F.P., Barghigiani, C., 1987. Assessment of mercury dispersal in an abandoned mining area by soil and lichen analysis. Water, Air and Soil Pollution 36, 219–225.

Barkman, J.J., 1958. Phytosociology and Ecology of Cryptogamic Epiphytes. Van Gorcum, Assen.

Bartók, K. 1988. Heavy metal distribution in several lichen species in a polluted area. Rev. Roum. Biol. – Biologie Végétale 33, 127–134.

Bartók, K., Nicoará, A., Victor, B., Tibor, O., 1992. Biological responses in the lichen *Xanthoria parietina* transplanted in biomonitoring stations. Rev. Roum. Biol. – Biologie Végétale 37, 135–142.

Baumbauch, G., 1996. Air quality control. In: Foerstner, U., Rulkens, W.H. (Eds), Environmental Engineering Series. Springer-Verlag, Berlin.

Beckett, R.P., Brown, D.H., 1983. Natural and experimentally-induced zinc and copper resistance in the lichen genus *Peltigera*. Annals of Botany 52, 43–50.

Beckett, R.P., Brown, D.H., 1984. The control of cadmium uptake in the lichen genus *Peltigera*., Journal of Experimental Botany 35, 1071–1082.

Berg, T., Royset, O., Steinnes, E., 1995. Moss (*Hylocomium splendens*) used as biomonitor of atmospheric trace element deposition: estimation of uptake efficiencies. Atmospheric Environment 29, 353–360.

Berg, T., Steinnes, E., 1997. Use of mosses (*Hylocomium splendens* and *Pleurozium schreberi*) as biomonitors of heavy metal deposition: from relative to absolute values. Environmental Pollution 98, 61–71.

Boileau, L.J.R., Beckett, P.J., Lavoie, P., Richardson, D.H.S., 1982. Lichens and mosses as monitors of industrial activity associated with uranium mining in northern Ontario, Canada – Part 1: Field procedures, chemical analysis and inter-species comparisons. Environmental Pollution (Ser. B) 4, 69–84.

Boileau, L.J.R., Nieboer, E., Richardson, D.H.S., 1985. Uranium accumulation in the lichen *Cladonia rangiferina*. Part II. Toxic effects of cationic, neutral, and anionic forms of the uranyl ion. Canadian Journal of Botany 63, 390–397.

Boonpragob, K., Nash III, T.H., 1990. Seasonal variation of elemental status in the lichen *Ramalina menziesii* Tayl. from two sites in southern California: evidence for dry deposition accumulation. Environmental and Experimental Botany 30, 415–428.

Boonpragob, K., Nash III, T.H., 1991. Physiological responses of the lichen *Ramalina menziesii* Tayl. to the Los Angeles urban environment. Environmental and Experimental Botany 31, 229–238.

Bosserman, R.W., Hagner, J.E., 1981. Elemental composition of epiphytic lichens from Okefenokee Swamp. Bryologist 84, 48–58.

Bowen, H.J.M., 1979. Environmental Chemistry of the Elements. Academic Press, London.

Branquinho, C., Brown, D.H., Máguas, C., Catarino, F., 1997. Lead (Pb) uptake and its effects on membrane integrity and chlorophyll fluorescence in different lichen species. Environmental and Experimental Botany 37, 95–105.

Branquinho, C., Brown, D.H., Catarino, F., 1997a. The cellular location of Cu in lichens and its effects on membrane integrity and chlorophyll fluorescence. Environmental and Experimental Botany 38, 165–179.

Branquinho, C., Catarino, F., Brown, D.H., Perera, M.J., Soares, A., 1999. Improving the use of lichens as biomonitors of atmospheric metal pollution. The Science of the Total Environment 232, 67–77.

Brown, D.H., 1976. Mineral uptake by lichens. In: Brown, D.H., Hawskworth, D.L., Bailey, R.H. (Eds), Systematics Association Special Volume, No. 8, Lichenology: Progress and Problems. Academic Press, London, pp. 255–261.

Brown, D.H., 1991. Lichen mineral studies – currently clarified or confused. Symbiosis 11, 207–223.

Brown, D.H., Beckett, R.P., 1983. Differential sensitivity of lichens to heavy metals. Annals of Botany 52, 51–57.

Brown, D.H., Beckett, R.P., 1984. Uptake and effect of cations on lichen metabolism., Lichenologist 16, 173–188.

Brown, D.H., Beckett, R.P., 1985. Intracellular and extracellular uptake of cadmium by the moss *Rhytiadiadelphus squarrosus*. Annals of Botany 55, 179–188.

Brown, D.H., Beckett, R.P., 1985a. Minerals and lichens: acquisition, localisation and effect. In: Vicente, C., Brown, D.H., Legaz, M.E. (Eds), Surface Physiology of Lichens. Editorial de la Universidad Complutense de Madrid, Madrid, pp. 127–149.

Brown, D.H., Buck, G.W., 1979. Desiccation effects and cation distribution in bryophytes. New Phytologist 82, 115–125.

Buck, G.W., Brown, D.H., 1977. Cation analysis of bryophytes: the significance of water content and ion location. Bryophyta Bibliotheca 13, 735–750.

Buck, G.W., Brown, D.H., 1979. The effect of desiccation on cation location in lichens. Annals of Botany 44, 265–277.

Buschmann, C., Nagel, E., 1993. *In vivo* spectroscopy and internal optics of leaves as basis for remote sensing of vegetation. International Journal of Remote Sensing 14, 711–722.

Carter, J.D., Ghio, A.J., Devlin, R.B., 1997. Cytokine production by human airway epithelial cells after exposure to an air pollution particle is metal-dependent., Toxicology and Applied Pharmacology 146, 180–188.

Castillejos, M., Borja-Aburto, V.H., Dockery, D.W., Gold, D.R., Loomis, D., 2000. Airborne coarse particles and mortality. Inhalation Toxicology 12, 61–71.

Challenger, F., 1945. Biological methylation. Chemical Reviews 36, 315–361.

Chamberlain, A.C., 1970. Deposition and uptake by cattle of airborne particles. Nature 225, 99–100.

Chamberlain, A.C., 1970a. Interception and retention of radioactive aerosols by vegetation. Atmospheric Environment 4, 57–58.

Chang, R., 1981. Physical Chemistry with Applications to Biological Systems. Macmillan, New York.

Chettri, M.K., Cook, C.M., Vardaka, E., Sawidis, T., Lanaras, T., 1998. The effect of Cu, Zn and Pb on the chlorophyll content of the lichens *Cladonia convoluta* and *Cladonia rangiformis*. Environmental and Experimental Botany 39, 1–10.

Cislaghi, C., Nimis, P.L., 1997. Lichens, air pollution and lung cancer. Nature 387, 463–464.

Cox, J.D., Beckett, P.J., Courtin, G.M., 1991. Factors affecting spectral responses from lichens. In: The Eighth Thematic Conference on Geologic Remote Sensing, Vol. 2. Proceedings. Denver, Colorado, 30 April – 2 May 1991. Environmental Research Institute of Michigan, Ann Arbor, pp. 1125–1137.

Crittenden, P.D., 1983. The role of lichens in the nitrogen economy of subarctic woodlands: nitrogen loss from the nitrogen-fixing lichen *Stereocaulon paschale* during rainfall. In: Lee, J.A., McNeill, S., Rorison, I.H. (Eds), Nitrogen as an Ecological Factor. Blackwell, Oxford, pp. 43–68.

Crittenden, P.D., 1985. Nitrogen relations of mat-forming lichens. In: Boddy, L., Marchant, R., Read, D.J. (Eds), Nitrogen, phosphorus and sulphur utilization by fungi. Cambridge University Press, Cambridge, pp. 243–268.

Dalton, F.N., 1984. Dual pattern of potassium transport in plant cells: a physical artifact of a single uptake mechanism. Journal of Experimental Botany 35, 1723–1732.

Davidson, C.I., Wu, Y.L., 1989. Dry deposition of trace elements. In: Pacyna, J.M., Ottar, B. (Eds), Control and Fate of Atmospheric Trace Metals. NATO ASI Series, Kluwer Academic, Dordrecht, pp. 147–202.

Davies, B.E., 1994. Soil chemistry and bioavailability with special reference to trace elements. In: Farago, M.E. (Ed.), Plants and the Chemical Elements. Biochemistry, Uptake, Tolerance and Toxicity. VCH Verlagsgesellschaft, Weinheim, pp. 1–30.

De Bruin, M., Hackenitz, E., 1986. Trace element concentrations in epiphytic lichens and bark substrate. Environmental Pollution 11, 153–160.

De Vos, R., 1991. Copper-induced oxydative stress and free radical damage in roots of copper tolerant and sensitive *Silene cucubalus*. Ph.D. Thesis, Vrije University, Amsterdam.

De Vries, W., Van Der Kooij, A., 1986. Radioactivity measurements arising from Chernobyl. IRI-report 190–86–01. Interfaculty Reactor Institute, TU Delft, Delft.

De Wit, A., 1976. Epiphytic lichens and air pollution in The Netherlands. Bibliotheca Lichenologica 5. Cramer, Vaduz.

Demon, A., De Bruin, M., Wolterbeek, H.Th., 1988. The influence of pH on trace element uptake by an alga (*Scenedesmus pannonicus* subsp. Berlin) and fungus (*Aureobasidium Pullulans*). Environmental Monitoring and Assessment 10, 165–173.

Demon, A., De Bruin, M., Wolterbeek, H.Th., 1989. The influence of pre-treatment, temperature and calcium ions on trace element uptake by an alga (*Scenedesmus pannonicus*, subsp. Berlin) and fungus (*Aureobasidium pullulans*). Environmental Monitoring and Assessment 13, 21–33.

Déruelle, S., 1984. L'utilisation des lichens pour la detection et l'estimation de la pollution par le plomb. Bulletin Ecology 15, 1–6.

Déruelle, S., Petit, P.J.X., 1983. Preliminary studies on the net photosynthesis and respiration responses of some lichens to automobile pollution. Cryptogamie, Bryologie. Lichénologie 4, 269–278.

Devell, L., Tovedal, H., Bertgstrom, U., Appelgren, A., Chyssler, J., Andersson, L., 1986. Initial observations of fall-out from the reactor accident at Chernobyl. Nature 321, 192–193.

Djingova, R., Kuleff, I., 2000. Instrumental techniques for trace analysis. In: Markert, B., Friese, K. (Eds), Trace Elements. Their Distribution and Effects in the Environment. Elsevier Science, Oxford, pp. 137–185.

Dockery, D.W., Pope, C.A., Xu, X.P., Spengler, J.D., Ware, J.H., Fay, M.E., Ferris, B.G., Speizer, F.E., 1993. An association between air-pollution and mortality in 6 United States cities. New England Journal of Medicine 329, 1753–1759.

Egan, R.S., 1982. Recent literature on lichens 113. Bryologist 85, 342–348.

Ekblad, A., Wallander, H., Näsholm, T., 1998. Chitin and ergosterol combined to measure total and living biomass in ectomycorrhizas. New Phytologist 138, 143–149.

Ellis, K., Smith, J.N., 1987. Dynamic model for radionuclide uptake in lichen. Journal of Environmental Radioactivity 5, 185–208.

Epstein, E., Sagee, O., Cohen, J.D., Garty, J., 1986. Endogenous auxin and ethylene in the lichen *Ramalina duriaei*. Plant Physiology 82, 1122–1125.

Ernst, W.H.O., 1978. Ökologische Risiken, verursacht durch emittierte Feinstäube aus Verbrennungs-motoren. Folgerungen für die analytische Chemie. Chemische Rundschau 31, 337–370.

Ernst, W.H.O., 1990. Element allocation and (re)translocation in plants and its impact on representative sampling. In: Lieth, H., Markert, B. (Eds), Element Concentration Cadasters in Ecosystems. Methods of Assessment and Evaluation. VCH Verlagsgesellschaft, Weinheim, pp. 17–40.

Falnoga, I., Tusek-Znidaric, M., Jeran, Z., Jacimovic, R., Pacicevic, J., Duric, D., Scancar, J., 2000. Water soluble metalloproteins in *Hypogymnia physodes*. In: Abstract Book Second International Workshop on Biomonitoring of Atmospheric Pollution (BioMAP), P-15, Praia da Vitoria, Azores Islands, 28 Aug. – 3 Sept.

Farago, M. (Ed.), 1994. Plants and The Chemical Elements. VCH Verlagsgesellschaft mbH, Weinheim.

Farago, M.E., 1994a. Plants as indicators of mineralization and pollution. In: Farago, M.E. (Ed.), Plants and the Chemical Elements. Biochemistry, Uptake, Tolerance and Toxicity. VCH Verlagsgesellschaft, Weinheim, pp. 221–240.

Farago, M.E., Mehra, A., 1992. Uptake of elements by the copper tolerant plant *Armeria Maritima* In: Merian, E., Haerdi, W. (Eds), Metal Compounds in Environment and Life, 4: Interrelation between Chemistry and Biology. Science and Technology Letters, Northwood, pp. 163–169.

Gailey, F.A.Y., Lloyd, O.Ll., 1986. Methodological investigations into low technology monitoring of atmospheric metal pollution: Part 3 – the degree of replicability of metal concentrations. Environmental Pollution Series B12, 85–109.

Gallagher, M., Fontan, J., Wyers, P., Ruijgrok, W., Duyzer, J., Hummelshoj, P., Pilegaard, K., Fowler, D., 1997. Atmospheric particles and their interactions with natural surfaces. In: Slanina, S. (Ed.), Biosphere-atmosphere Exchange of Pollutants in the Troposphere, Vol. 4. Springer Verlag, Berlin.

Galun, M., Garty, J., Ronen, R., 1984. Lichens as bioindicators of air pollution. Webbia 38, 371–383.

Gamon, J.A., Qiu, H.-L., 1999. Ecological applications of remote sensing at multiple scales. In: Pugnaire, F.I., Valladares, F. (Eds), Handbook of Functional Plant Ecology. Marcel Dekker, New York, pp. 805–846.

Ganor, E., Foner, H.A., Brenner, S., Neeman, E., Lavi, N., 1991. The chemical composition of aerosols settling in Israel following dust storms. Atmospheric Environment 25, 2665–2670.

Garten, C.T., 1978. A review of the parameter values used to assess the transport of plutonium, uranium and thorium in terrestrial food chains. Environmental Research 17, 437–452.

Garty, J., 1993. Lichens as biomonitors for heavy metal pollution. In: B. Markert (Ed.), Plants as Biomonitors. Indicators for Heavy Metals in the Terrestrial Environment. VCH, Weinheim, pp. 193–264.

Garty, J., 2000. Environment and elemental content of lichens. In: Markert, B., Friese, K. (Eds), Trace Elements. Their Distribution and Effects in the Environment. Elsevier Science, Amsterdam, pp. 245–276.

Garty, J., 2000a. Trace metals, other chemical elements and lichen physiology: research in the nineties. In: Markert, B., Friese, K. (Eds), Trace Elements. Their Distribution and Effects in the Environment, Vol. 4, Trace Metals in the Environment Series. Elsevier Science, Oxford, pp. 277–322.

Garty, J., Cohen, Y., Kloog, N., 1998a. Airborne elements, cell membranes and chlorophyll in transplanted lichens. Journal of Environmental Quality 27, 973–979.

Garty, J., Cohen, Y., Kloog, N., Karnieli, A., 1997. Effects of air pollution on cell membrane integrity, spectral reflectance and metal and sulfur concentrations in lichens. Environmental Toxicology and Chemistry 16, 1396–1402.

Garty, J., Fuchs, C., 1982. Heavy metals in the lichen *Ramalina duriaei* transplanted in biomonitoring stations. Water, Air and Soil Pollution 17, 175–183.

Garty, J., Galun, M., Kessel, M., 1979. Localization of heavy metals and other elements accumulated in the lichen thallus. New Phytologist 82, 159–168.

Garty, J., Karary, Y., Harel, J., 1992. Effect of low pH, heavy metals and anions on chlorophyll degradation in the lichen *Ramalina duriaei* (De Not.) Bagl. Environmental and Experimental Botany 32, 229–241.

Garty, J., Karary, Y., Harel, J., 1993. The impact of air pollution on the integrity of cell membranes and chlorophyll in the lichen *Ramalina duriaei* (De Not.) Bagl. Transplanted to industrial sites in Israel. Archives of Environmental Contamination and Toxicology 24, 455–460.

Garty, J., Karary, Y., Harel, J., Lurie, S., 1993a. Temporal and spatial fluctuations of ethylene production and concentrations of sulfur, sodium, chlorine and iron on/in the thallus cortex in the lichen *Ramalina duriaei* (De Not.) Bagl. Environmental and Experimental Botany 33, 553–563.

Garty, J., Karary, Y., Harel, J., Lurie, S., 1995. The impact of heavy metals on lichens. In: Wilken, R.-D., Förstner, U., Knöchel, J. (Eds), Heavy Metals in the Environment, Vol. 1. Proceedings of the International Conference, Hamburg, CEP Consultants, Edinburgh, pp. 152–155.

Garty, J., Kardish, N., Hagemeyer, J., Ronen, R., 1988. Correlations between the concentration of adenosine triphosphate, chlorophyll degradation and the amounts of airborne heavy metals and sulphur in a transplanted lichen. Archives of Environmental Contamination and Toxicology 17, 601–611.

Garty, J., Karnieli, A., Wolfson, R., Kunin, P., Garty-Spitz, R., 1997a. Spectral reflectance and integrity of cell membranes and chlorophyll relative to the concentration of airborne mineral elements in a lichen. Physiologia Plantarum 101, 257–264.

Garty, J., Kauppi, M., Kauppi, A., 1996. Accumulation of airborne elements from vehicles in transplanted lichens in urban sites. Journal of Environmental Quality 25, 265–272.

Garty, J., Kauppi, M., Kauppi, A., 1997. The influence of air pollution on the concentration of airborne elements and on the production of stress-ethylene in the lichen *Usnea hirta* (L.) Weber em. Mot. transplanted in urban sites in Oulu, N. Finland. Archives of Environmental Contamination and Toxicology 32, 285–290.

Garty, J., Kauppi, M., Kauppi, A., 1997. The production of stress ethylene relative to concentration of heavy metals and other elements in the lichen *Hypogymnia physodes*. Environmental Toxicology and Chemistry 16, 2404–2408.

Garty, J., Kloog, N., Cohen, Y., 1998. Integrity of lichen cell membranes in relation to concentration of airborne elements. Archives of Environmental Contamination and Toxicology 34, 136–144.

Garty, J., Kloog, N., Wolfson, R., Cohen, Y., Karnieli, A., Avni, A., 1997b. The influence of air pollution on the concentration of mineral elements, on the spectral reflectance response and on the production of stress-ethylene in the lichen *Ramalina duriaei*. New Phytologist 137, 587–587.

Garty, J., Kloog, N., Cohen, Y., Wolfson, R., Karnieli, A., 1997c. The effect of air pollution on the integrity of chlorophyll, spectral reflectance response, and on concentrations of nickel, vanadium, and sulfur in the lichen *Ramalina duriaei* (De Not.) Bagl. Environmental Research 74, 174–187.

Garty, J., Ronen, R., Galun, M., 1985. Correlation between chlorophyll degradation and the amount of some elements in the lichen *Ramalina duriaei* (De Not.) Jatta. Environmental and Experimental Botany 25, 67–74.

Garty, J., Tamir, O., Hassid, I. Eshel, A., Cohen, Y., Karnieli, A., Orlovsky, L., 2001. Photosynthesis, chlorophyll integrity and spectral reflectance in lichens exposed to air pollution. Journal of Environmental Quality 30, 884–893.

Garty, J., Weissman, L., Cohen, Y., Karnieli, A., Orlovsky, L., 2001. Transplanted lichens in and around the Mount Carmel National Park and the Haifa Bay industrial region in Israel: physiological and chemical responses. Environmental Research 85A, 159–176.

Garty, J., Weissman, L., Tamir, O., Beer, S., Cohen, Y., Karnieli, A., Orlovsky, L., 2000a. Comparison of five physiological parameters to assess the vitality of the lichen *Ramalina lacera* exposed to air pollution. Physiologia Plantarum 109, 410–418.

Garty, J., Ziv, O., Eshel, A., 1985a. The effect of coating polymers on accumulation of airborne heavy metals by lichens. Environmental Pollution (Series A) 38, 213–220.

Gilbert, O.L., 1970. Further studies on the effect of sulphur dioxide on lichens and bryophytes. New Phytologist 69, 605–627.

Gjengedal, E., Steinnes, E., 1990. Uptake of metal ions in moss from artificial precipitation., Environmental Monitoring and Assessment 14, 77–87.

Glenn, M.G., Gomez-Bolea, A., Lobello, R., 1995. Metal content and community structure of cryptogam bioindicators in relation to vehicular traffic in Montseny Biosphere Reserve (Catalonia, Spain). Lichenologist 27, 291–304.

Goldsmith, S.J., Thomas, M.A., Gries, C., 1997. A new technique for photobiont culturing and manipulation. Lichenologist 29, 559–569.

González, C.M., Pignata, M.L., 1994. The influence of air pollution on soluble proteins, chlorophyll degradation, MDA, sulphur and heavy metals in a transplanted lichen. Chemical Ecology 9, 105–113.

González, C.M., Pignata, M.L., 1999. Effect of pollutants emitted by different urban-industrial sources on the chemical response of the transplanted *Ramalina ecklonii* (Spreng.) Mey. & Flot. Toxicological and Environmental Chemistry 69, 61–73.

Goodwin, T.W., 1974. Sterols. In: Stewart, W.D.P. (Ed.), Algal Physiology and Biochemistry. Blackwell Scientific, Oxford, pp. 266–280.

Gough, L.P., Jackson, L.L., Sacklin, J.A., 1988. Determining baseline element composition of lichens. II. *Hypogymnia enteromorpha* and *Usnea* spp. at Redwood National Park, California. Water, Air and Soil Pollution 38, 169–180.

Goyal, P., Seaward, M.R.D., 1981. Metal uptake in terricolous lichens. I. Metal localization within the thallus. New Phytologist 89, 631–645.

Goyal, R., Seaward, M.R.D., 1982. Metal uptake in terricolous lichens. II. Effects on the morphology of *Peltigera canina* and *Peltigera rufescens*. New Phytologist 90, 73–84.

Goyal, R., Seaward, M.R.D., 1982a. Metal uptake in terricolous lichens. III. Translocation in the thallus *of Peltigera canina*. New Phytologist 90, 85–98.

Hale, M.E., 1974. The Biology of Lichens. Edward Arnold, London.

Hawksworth, D.L., 1971. Lichens as a litmus for air pollution: a historical review. International Journal of Environmental Studies 1, 281–296.

Hawksworth, D.L., 1988. The Fungal Partner. In: Galun, M. (Ed.), Handbook of Lichenology, Vol. 1. CRC Press, Boca Raton, pp. 35–38.

Heij, G.J., De Vries, W., Posthumus, A.C., Mohren, G.M.J., 1991. Effects of air pollution and acid deposition on forests and forest soils. In: Heij, G.J., Schneider, T. (Eds), Acidification Research in The Netherlands, Final Report of the Dutch Priority Programme on Acidification. Elsevier, Amsterdam, pp. 97–137.

Heinken, T., 1999. Dispersal patterns of terricolous lichens by thallus fragments. Lichenologist 31, 603–612.

Heinrich, J., Wijst, M., Jacob, B., Hoelscher, B., Wichmann, H.E., 1999. Environmental epidemiological investigations in the area of Bitterfeld, Hettstadt and a control area 1992–1999. Allergologie 22, 3–13.

Helfferich, F., 1962. Ion Exchange. McGraw-Hill, New York.

Henderson-Sellers, A., Seaward, M.R.D., 1979. Ring lichen reinvasion of ameliorating environments. Environmental Pollution 19, 207–213.

Herzig, R., Urech, M., Liebendörfer, L., Ammann, K., Guecheva, M., Landolt, W., 1990. Lichens as biological indicators of air pollution in Switzerland: passive biomonitoring as a part of an integrated measuring system for monitoring air pollution. In: Lieth, H., Markert, B. (Eds), Element Concentration Cadasters in Ecosystems. CH Verlagsgesellschaft, Weinheim, pp. 317–332.

Hickman, C.J., 1965. Fungal structure and organisation. In: Ainsworth, G.C., Sussman, A.A. (Eds), The Fungi: An Advanced Treatise, Vol. I, The Fungal Cell. Academic Press, New York, pp. 21–48.

Hicks, B.B., Hosker, R.P., Meyers, T.P., Womack, J.D., 1991. Dry deposition inferential measurement techniques. I. Design and tests of a prototype meteorological and chemical system for determining dry deposition. Atmospheric Environment 25A, 2345–2359.

Holopainen, T., 1983. Ultrastructural changes in epiphytic lichens, *Bryoria capillaris* and *Hypogymnia physodes*, growing near a fertilizer plant and a pulp mill in central Finland. Annales Botanici Fennici 20, 169–185.

Holopainen, T., 1984. Cellular injuries in epiphytic lichens transplanted to air polluted areas. Nordic Journal of Botany 4, 393–408.

Honegger, R., 1991. Functional aspects of the lichen symbiosis. Annual Review of Plant Physiology and Plant Molecular Biology 42, 553–578.

Hottiger, T., Boller, T., 1991. Ethylene biosynthesis in *Fusarium oxysporum* f. *sp. tulipae* proceeds from glutamatate/2-oxoglutarate and requires oxygen and ferrous ions *in vivo*. Archives of Microbiology 157, 18–22.

Huiskes, A.H.L., Gremmen, N.J.M., Francke, J.W., 1997. Morphological effects on the water balance of Antarctic foloise and fruticose lichens. Antarctic Science 9, 36–42.

Hyvärinen, M., Roitto, M., Ohtonen, R., Markkola, A., 2000. Impact of wet deposited nickel on the cation content of a mat-forming lichen *Cladina stellaris*. Environmental and Experimental Botany 43, 211–218.

IAEA-NAHRES-43 Report. 1999. Co-ordinated research project on validation and application of plants as biomonitors of trace element atmospheric pollution, analysed by nuclear and related techniques. Report on the First RCM Meeting, Vienna, Austria.

Jahns, H.M., 1993. Culture experiments with lichens. Plant Systematics and Evolution 187, 145–174.

James, P.W., 1973. The effect of air pollutants other than hydrogen fluoride and sulphur dioxide on lichens. In: Ferry, B.W., Baddeley, M.S., Hawksworth, D.L. (Eds), Air Pollution and Lichens. Athlone Press, London, pp. 143–175.

Jenkins, J.A., 1987. Trace elements in saxicolous lichens. In: Coughtrey, P.J., Martin, M.H., Unsworth, M.H. (Eds), Pollutant Transport and its Fate in Ecosystems. British Ecological Society Special Publication No. 6, Blackwell Scientific, Oxford, pp. 249–266.

Johnsen, I., Søchting, U., 1973. Influence of air pollution on the epiphytic vegetation and bark properties of deciduous trees in the Copenhagen area. Oikos 24, 344–351.

Kauppi, M., 1976. Fruticose lichen transplant technique for air pollution experiments. Flora 165, 407–414.

Kauppi, M., Kauppi, A., Garty, J., 1998. Ethylene produced by the lichen *Cladina stellaris* exposed to sulphur and heavy-metal-containing solutions under acidic conditions. New Phytologist 139, 537–547.

Laden, F., Neas, L.M., Dockery, D.W., Schwartz, J., 1999. The association of elemental characteristics of fine particles with mortality in six cities. Epidemiology 10, 163.

Larson, D.W., 1979. Lichen water relations under drying conditions. New Phytologist 82, 713–731.

Lawrey, J.D., Hale Jr., M.E., 1981. Retrospective study of lichen lead accumulation in the northeastern United States. Bryologist 84, 449–456.

Leblanc, F., Robitaille, G., Rao, D.N., 1974. Biological response of lichens and bryophytes to environmental pollution in the Murdochville copper mine area, Quebec, Journal of the Hattori Botanical Laboratory 38, 405–433.

Lebowitz, M.D., 1996. Epidemiological studies of the respiratory effects of air pollution. European Respiratory Journal 9, 1029–1054.

Lemaistre, V., 1985. Influence of automobile exhaust and lead on the oxygen exchange of two lichens measured by a new oxygen electrode method. In: Brown, D.H. (Ed.), Lichen Physiology and Cell Biology. Plenum Press, New York, pp. 173–183.

Lepp, N.W. (Ed.), 1981a. Effect of Heavy Metal Pollution on Plants. Vol. 1, Effects of Trace Metals on Plant Function, Applied Science, London.

Lepp, N.W. (Ed.), 1981b. Effect of Heavy Metal Pollution on Plants, Vol. 2, Metals in the Environment. Applied Science, London.

Lurie, S., Garty, J., 1991. Ethylene production by the lichen *Ramalina duriaei*. Annals of Botany 68, 317–319.

Lynch, J.M., 1974. Mode of ethylene formation by *Mucor hiemalis*. Journal of General Microbiology 83, 407–411.

Macfarlane, J.D., Kershaw, K.A., 1980. Physiological-environmental interactions in lichens. IX. Thermal stress and lichen ecology. New Phytologist 84, 669–685.

Máguas, C., Valladares, F., Brugnoli, E., 1997. Effects of thallus size on morphology and physiology of foliose lichens: new findings with a new approach. Symbiosis 23, 149–164.

Maillard, P., Thepenier, C., Gudin, C., 1993. Determination of an ethylene biosynthesis pathway in the unicellular green alga, *Haematococcus pluvialis*. Relationship between growth and ethylene production. Journal of Applied Phycology 53, 93–98.

Manning, W.J., Feder, W.A., 1980. Biomonitoring air pollutants with plants. Applied Science, London.

Markert, B. (Ed.), 1993. Plants as Biomonitors. Indicators for Heavy Metals in the Terrestrial Environment. VCH Verlagsgesellschaft, Weinheim.

Markert, B., 1993. Instrumental analysis of plants. In: Markert, B. (Ed.), Plants as Biomonitors. Indicators for Heavy Metals in the Terrestrial Environment. VCH Verlagsgesellschaft, Weinheim, pp. 65–103.

Markert, B., 1996. Instrumental Element and Multi-element Analysis of Plant Samples. John Wiley, Chichester.

Markert, B., Oehlmann, J., Roth, M., 2000. Biomonitoring of heavy metals: definitions, possibilities and limitations. In: Proceedings of the International Workshop on Biomonitoring of Atmospheric Pollution (with emphasis on trace elements), BioMAP, Lisbon, 21–24 September 1997, IAEA TECDOC 1152, IAEA, Vienna, Austria.

Markert, B., Wappelhorst, O., Weckert, V., Herpin, U., Siewers, U., Friese, K., Breulmann, G., 1999. The use of bioindicators for monitoring the heavy-metal status of the environment. Journal of Radio-analytical and Nuclear Chemistry 240, 425–429.

Markert, B., Weckert, V., 1989. Fluctuations of element concentrations during the growing season of *Polytrichum formosum* (Hedw.). Water, Air and Soil Pollution 43, 177–189.

Martin, M.H., Coughtrey, P.J., 1982. Biological Monitoring of Heavy Metal Pollution. Land and Air. Applied Science, London.

Menzel, R.G., 1967. Airborne radionuclides and plants. In: Brady, N.C. (Ed.), Agriculture and the Quality of Our Environment. American Association Advances in Science Publication 85, pp. 57–75.

Milford, J.B., Davidson, C.I., 1985. The sizes of particulate trace elements in the atmosphere: a review. Journal of Air Pollution Control Association 35, 1249–1260.

Miller, J.E., Brown, D.H., 1999. Studies of ammonia uptake and loss by lichens. Lichenologist 31, 85–93.

Mulder, J., 1988. Impact of acid atmospheric deposition on soils: field monitoring and aluminium chemistry. Ph.D. Thesis, University of Wageningen.

Nash III, T.H., 1975. Influence of the effluents from a zinc factory on lichens. Ecol. Monogr. 45, 183–198.

Nash III, T.H., 1988. Correlating fumigation studies with field effects. Lichens, bryophytes and air quality. Bibl. Lichenol. 30, 201–216.

Nash, T.H., Gries, C., 1995. The response of lichens to atmospheric deposition with an emphasis on the Arctic. The Science of the Total Environment 161, 737–747.

Nieboer, E., Puckett, K.J., Grace, B., 1976. The uptake of nickel by *Umbilicaria muhlenbergii*: a physico-chemical process. Canadian Journal of Botany 54, 724–733.

Nieboer, E., Richardson, D.H.S., 1981. Lichens as monitors of atmospheric deposition. In: Eisenreich, S.J. (Ed.), Atmospheric Pollutants in Natural Waters. Ann Arbor, Michigan, pp. 339–388.

Nieboer, E., Richardson, D.H.S., Tomassini, F.D., 1978. Mineral uptake and release by lichens: an overview. Bryologist 81, 226–246.

Nylander, W., 1866. Les lichens du Jardin du Luxembourg, Bulletin de la Societé Botanique de France 13, 364–372.

Odasz-Albrigtsen, A.M., Tommervick, H., Murphy, P., 2000. Decreased photosynthetic efficiency in plant species exposed to multiple airborne pollutants along the Russian–Norwegian border. Canadian Journal of Botany 78, 1021–1023.

Oksanen, J., Tynnyrinen, S., Karenlampi, L., 1990. Testing for increased abundance of epiphytic lichens on a local pollution gradient. Annales Botanici Fennici 27, 301–307.

Olmez, I., Cetin-Gulovali, M., Gordon, G.E., 1985. Trace element concentrations in lichens near a coal-fired power plant. Atmospheric Environment 19, 1663–1669.

Ott, S., 1993. The influence of light on the ethylene production by lichens. In: Feige, G.B., Lumbsh, H.T. (Eds), Phytochemistry and Chemotaxonomy of Lichenized Ascomycetes – A Festschrift in Honour of Siegfried Huneck. Bibliothecha Lichenologica 53, Cramer, Berlin and Stuttgart, pp. 185–190.

Ott, S., Krieg, T., Spanier, U., Schieleit, P., 2000. Phytohormones in lichens with emphasis on ethylene biosynthesis and functional aspect on lichen symbiosis. Phyton 40, 83–94.

Ott, S., Sancho, L.G., 1993. Morphology and anatomy of *Caloplaca coralligera* (*Teloschistaceae*) as adaption to extreme environmental conditions in the maritime antarctic. Plant Systematics and Evolution 185, 123–132.

Ott, S., Schieleit, P., 1994. Influence of exogenous factors on the ethylene production by lichens. I. Influence of water content and water status conditions on ethylene production. Symbiosis 16, 187–201.

Ott, S., Zwoch, I., 1992. Ethylene production by lichens. Lichenologist 24, 73–80.

Palomäki, V., Tynnyrinen, S., Holopainen, T., 1992. Lichen transplantation in monitoring fluoride and sulfur deposition in the surroundings of a fertilizer plant and a strip mine at Siilinjarvi. Annales Botanici Fennici 29, 25–34.

Papastefanou, C., Manolopoulou, M., Charalambous, S., 1988. Radiation measurements and radioecological aspects of fall-out from the Chernobyl reactor accident. Journal of Environmental Radioactivity 7, 49–64.

Papastefanou, C., Manolopoulou, M., Sawidis, T., 1989. Lichens and mosses: biological monitors of radioactive fallout from the Chernobyl reactor accident. Journal of Environmental Radioactivity 9, 49–64.

Peñuelas, J., Fillela, I., 1998. Visible and near infrared reflectance techniques for diagnosing plant physiological status. Trends in Plant Science 3, 151–158.

Phillips, R., Stumpel-Rienks, S.E., 1980. Grassen, Varens en Korstmossen. Spectrum Natuurgids, Spectrum, Utrecht.

Piervittori, R., Usai, L., Alessio, F., Maffei, M., 1997. The effect of simulated acid rain on surface morphology and n-alkane composition of *Pseudevernia furfuracea*. Lichenologist 29, 191–198.

Pintado, A., Valladares, F., Sancho, L.G., 1997. Exploring phenotypic plasticity in the lichen *Ramalina capitata*: morphology, water relations and chlorophyll content in north- and south-facing populations. Annals of Botany 80, 345–353.

Porter, J.R., 1986. Evaluation of washing procedures for pollution analysis of *Ailanthus altissima* leaves. Environmental Pollution Series B 12, 195–202.

Prussia, C.M., Killingbeck, K.T., 1991. Concentrations of ten elements in two common foliose lichens: leachability, seasonality, and the influence of the rock and tree bark substrates. The Bryologist 94, 135–142.

Puckett, K.J. 1988. Bryophytes and lichens as monitors of metal deposition. In: Nash III, T.H., With, V. (Eds), Lichens, Bryophytes and Air Quality, Bibliotheca Lichenologica 30, Cramer, Berlin, pp. 231–267.

Puckett, K.J., Burton, M.A.S., 1981. The effect of trace elements on lower plants. In: Lepp, N.W. (Ed.), Effect of Heavy Metal Pollution on Plants, Vol. 2, Metals in the Environment, Applied Science, London, pp. 213–238.

Puckett, K.J., Finegan, E.J., 1980. An analysis of the element content of lichens from the Northwest Territories, Canada. Canadian Journal of Botany 58, 2073–2089.

Puckett, K.J., Nieboer, E., Gorzynski, M.J., Richardson, D.H.S., 1973. The uptake of metal ions by lichens: a modified ion-exchange process. New Phytologist 72, 329–342.

Punz, W., 1979. The effect of single and combined pollutants on lichen water content. Biologia Plantarum 21, 472–474.

Purvis, O.W., Elix, J.A., Broomhead, J.A., Jones, G.C., 1987. The occurrence of copper-norstictic acid in lichens from cupriferous substrata. Lichenologist 19, 193–203.

Purvis, O.W., Williamson, B.J., Bartok, K., Zoltani, N., 2000. Bioaccumulation of lead by the lichen *Acaspora smagardula* from smelter emissions. New Phytologist 147, 591–599.

Quispel, A., Stegwee, D. (Eds), 1984. Plantenfysiologie. 2nd edn. Bohn, Scheltema & Holkema, Utrecht.

Raes, F., Graziani, G., Stanners, D., Girardi, F., 1990. Radioactivity measurements in air over Europe after the Chernobyl accident. Atmospheric Environment 24A, 909–916.

Ramelow, U.S., Guidry, C.N., Fisk, S.D., 1996. A kinetic study of metal ion binding by biomass immobilized in polymers. Journal of Hazardous Materials 46, 37–55.

Reis, M.A., 2001. Biomonitoring and assessment of atmospheric trace elements in Portugal. Methods, response modeling and nuclear analytical techniques. Thesis, University of Delft, The Netherlands.

Reis, M.A., Alves, L.C., Freitas, M.C., Van Os, B., Wolterbeek, H.Th., 1999. Lichens (*Parmelia sulcata*) time response model to environmental elemental availability. The Science of the Total Environment 232, 105–115.

Reis, M.A., Alves, L.C., Freitas, M.C., Wolterbeek, H.Th., Verburg, T., Gouveia, M.A., 1996. Main atmospheric heavy metal sources in Portugal by biomonitor analysis. Nuclear Instruments and Methods in Physical Research B109/110, 493–497.

Richardson D.H.S., Nieboer, E., Lavoi, P., Padovan, D., 1979. The role of metal-ion binding in modifying the toxic effects of sulphur dioxide on the lichen *Umbilicaria muhlenbergii*. II. ^{14}C-fixation studies. New Phytologist 82, 633–643.

Richardson, D.H.S., 1995. Metal uptake in lichens. Symbiosis 18, 119–127.

Richardson, D.H.S., 1999. War in the world of lichens: parasitism and symbiosis as exemplified by lichens and lichenicolous fungi. Mycological Research 103, 641–650.

Richardson, D.H.S., Nieboer, E., Lavoie, P., Padovan, P., 1984. Anion accumulation by lichens. I. The characteristics and kinetics of arsenate uptake by *Umbilicaria muhlenbergii*. New Phytologist 96, 71–82.

Richardson, D.H.S., Nieboer, E., Van Dobben, H.F., 1982. Lichens, bioindicators of air quality. Natuur en Techniek 11, 870–889.

Riga-Karandinos, A.N., Karandinos, M.G., 1998. Assessment of air pollution from a lignite power plant in the plain of Megalopolis (Greece) using as biomonitors three species of lichens: impacts of some biochemical parameters of lichens. The Science of the Total Environment 215, 167–183.

Robig, G., Porstendorfer, J., Ahmed, A., 1978. Experimental investigations of the collection efficiency of spheres for submicron particles with regard to washout in the atmosphere. In: Benarie, M.M. (Ed.), Atmospheric Pollution. Proceedings 13th International Colloquium, Paris. Studies in Environmental Science, Vol. 1, Elsevier Science, Amsterdam, pp. 273–278.

Roemer, W., Hoek, G., Brunekreef, B., Clench-Aas, J., Forsberg, B., Pekkanen, J., Schultz, A., 2000. PM10 elemental composition and acute respiratory health effects in European children (PEACE project). European Respiratory Journal 15, 553–559.

Romano, A.H., 1966. Dimorphism. In: Ainsworth, G.C., Sussman, A.A. (Eds), The Fungi: An Advanced Treatise, Vol. II, The Fungal Organism. Academic Press, New York, pp. 181–209.

Ronen, R., Galun, M., 1984. Pigment extraction from lichens with dimethyl sulfoxide (DMSO) and estimation of chlorophyll degradation. Environmental and Experimental Botany 24, 239–245.

Rope, S.K., Pearson, L.C., 1990. Lichens as air pollution biomonitors in a semiarid environment in Idaho. Bryologist 93, 50–61.

Ross, H.B., 1990. On the use of mosses (*Hylocomium splendens* and *Pleurozium schreberi*) for estimating atmospheric trace metal deposition. Water, Air and Soil Pollution 50, 63–76.

Rouse, J.W., Haas, R.H., Schell, J.A. Deering, D.W., Harlan, J.C., 1982. Monitoring the Vernal Advancements and Retrogradation (Greenwave Effect) of Nature Vegetation. NASA/GSFC Final Report. NASA, Greenbelt.

Rühling, Å., Rasmussen, L., Pilegaard, K., Mäkinen, A., Steinnes, E., 1987. Survey of atmospheric heavy metal deposition in the Nordic countries in 1985 – monitored by moss analysis. NORD 1987, 21.

Rühling, Å., Tyler, G., 1970. Sorption and retention of heavy metals in the woodland moss *Hylocomium splendens* (Hedw.). Oikos 21, 92–97.

Sancho, L.G., Kappen, L., 1989. Photosynthesis and water relations and the role of anatomy in *Umbilicariaceae* (lichens) from central Spain. Oecologia 81, 473–480.

Sarret, G., Manceau, A., Cuny, D., Van Haluwyn, C., Deruelle, S., Hazemann, J.L., Soldo, Y., Eybert-Berard, L., Menthonnex, J.J., 1998. Mechanisms of lichen resistance to metallic pollution. Environmental Science and Technology 32, 3325–3330.

Schieleit, P., Ott, S., 1996. Ethylene production and 1-aminocyclopropane-1-1 carboxylic acid content of lichen bionts. Symbiosis 21, 223–231.

Schieleit, P., Ott, S., 1997. Ethylene production in lichens with respect to possible bacterial contamination. Lichenologist 29, 492–495.

Schipperges, B., Kappen, L., Sonesson, M., 1995. Intraspecific variations of morphology and physiology of temperate to arctic populations of *Cetraria nivalis*. Lichenologist 27, 517–529.

Schreiber, U., Bilger, W., 1993. Progress in chlorophyll fluorescence research: major developments during the past years in retrospect. In: Dietmar Behnke, H., Lüttge, U., Essen, K., Kadereit, W., Runge, M. (Eds), Progress in Botany. Springer Verlag, Berlin, pp. 151–173.

Schuepp, P.H., 1984. Observations on the use of analytical and numerical models for the description of transfer to porous surface vegetation such as lichen. Boundary-Layer Meteorology 29, 59–73.

Schwartzman, D.W., Stieff, L., Kasim, M., Kombe, E., Aung, S., Atekwana, E., Johnson, J., Schwartzman, K., 1991. An ion-exchange model of lead-210 and lead uptake in a foliose lichen; application to quantitative monitoring of airborne lead fallout. The Science of the Total Environment 100, 319–336.

Seaward, M.R.D., 1974. Some observations on heavy metal toxicity and tolerance in lichens. Lichenologist 6, 58–164.

Seaward, M.R.D., 1976. Performance of *Lecanora muralis* in an urban environment. In: Brown, D.H., Hawksworth, D.L., Bailey, R.H. (Eds), Systematics Association, Special Volume 8, Lichenology: Progress and Problems. Academic Press, London, pp. 323–357.

Seaward, M.R.D., 1980. The use and abuse of heavy metal bioassays of lichens for environmental monitoring. In: Spaleny, J. (Ed.), Proceedings 3rd International Conference Bioindicators Deteriorisationis Regionis, September 1977, Liblice, Czechoslakia, Academia, Praha, pp. 375–384.

Sellers, P.J., 1985. Canopy reflectance, photosynthesis and transpiration. International Journal of Remote Sensing 6, 1335–1372.

Sentenac, H., Grignon, C., 1981. A model for predicting ionic equilibrium concentrations in cell walls. Plant Physiology 68, 415–419.

Shipley, R.A., Clark, R.E., 1972. Tracer methods for in vivo kinetics. Theory and Applications. Academic Press, New York.

Silberstein, L., Siegel, B.Z., Siegel, S.M., Mukhtar, A., Galun, M., 1996. Comparative studies on *Xanthoria parietina*, a pollution-resistant lichen, and *Ramalina duriaei*, a sensitive species. I. Effect of air pollution on physiological processes. Lichenologist 28, 355–365.

Slinn, W.G.N., 1978. Parameterizations for resuspension and for wet and dry deposition of particles and gases for use in radiation dose calculations. Nuclear Safety 19, 205–219.

Sloof, J.E., 1993. Environmental Lichenology: Biomonitoring Trace Element Air Pollution. Thesis, Delft University of Technology, Delft, The Netherlands.

Sloof, J.E., 1995. Lichens as quantitative biomonitors for atmospheric trace-element deposition, using transplants. Atmospheric Environment 29, 11–20.

Sloof, J.E., Wolterbeek, H.Th., 1992. Lichens as biomonitors for radiocaesium following the Chernobyl accident. Journal of Environmental Radioactivity 16, 229–242.

Sloof, J.E., Wolterbeek, H.Th., 1993. Substrate influence on epiphytic lichens. Environmental Monitoring and Assessment, 25, 225–234.

Smith, D.C., 1963. Experimental studies of lichen physiology. Symbiotic Associations: Symposia of The Society for General Microbiology 13, 31–50.

Smith, J.N., Ellis, K., 1990. Time dependent transport of Chernobyl radioactivity between atmospheric and lichen phases in Eastern Canada. Journal of Environmental Radioactivity 11, 151–168.

Snelgar, W.P., Green, T.G.A., 1981. Ecologically-linked variation in morphology, acetylene reduction and water relations in *Pseudophellaria dissimilis*. New Phytologist 87, 403–411.

Souza-Egipsy, V., Valladares, F., Ascaso, C., 2000. Water distribution in foliose lichen species: interactions between method of hydration, lichen substances and thallus anatomy. Annals of Botany 86, 595–601.

Spix, C., Heinrich, J., Dockery, D.W., Schwartz, J., Volksch, G., Schwinkowski, K., Collen, C., Wichmann, H.E., 1993. Air pollution and daily mortality in Erfurt, East Germany, 1980–1989. Environmental Health Perspectives 101, 518–526.

Summers, A.O., 1978. Microbial transformations of metals. Ann. Rev. Microbiol. 32, 637–672.

Sundberg, B., Ekblad, A. Näsholm, T. Palmqvist, K., 1999. Lichen respiration in relation to active time, temperature, nitrogen and ergosterol concentrations. Functional Ecology 13, 119–125.

Sutcliffe, J.F., 1962. Mineral Salt Absorption in Plants. Pergamon Press, Oxford.

Tarhanen, S., 1998. Ultrastructural responses of the lichen *Bryoria fuscescens* to simulated acid rain and heavy metal deposition. Annals of Botany 82, 735–746.

Tarhanen, S., Holopainen, T., Poikolainen, J., Oksanen, J., 1996. Effect of industrial emission on membrane permeability of epiphytic lichens in northern Finland and the Kola peninsula industrial area. Water, Air and Soil Pollution 88, 189–201.

Tarhanen, S., Holopainen, T., Oksanen, J., 1997. Ultrastructural changes and electrolyte leakage from ozone fumigated epiphytic lichens. Annals of Botany 80, 611–621.

Tarhanen, S., Metsärine, S., Holopainen, T., Oksanen, J., 1999. Membrane permeability response of lichen *Bryoria fuscescens* to wet deposited heavy metals and acid rain. Environmental Pollution 104, 121–129.

Thibaud, J.B., Romieu, C., Gibrat, R., Grouzis, J.P., Grignon, C., 1984. Local ionic environment of plant membranes: effects on membrane functions. Zeitschrift fuer Pflanzenphysiologie 114, 207–213.

Thurman, D.A., 1981. Mechanism of metal tolerance in higher plants. In: Lepp, N.W. (Ed.), Effect of Heavy Metal Pollution on Plants, Vol. 2, Metals in the Environment. Applied Science, London, pp. 239–250.

Trembley, M.L., Fahselt, D., Madzia, S., 1997. Localization of uranium in *Cladonia rangiferina* and *Cladonia mitis* and removal by aqueous washing. Bryologist 100, 368–376.

Tretiach, M., Brown, D.H., 1995. Morphological and physiological differences between epilithic and epiphytic populations of the lichen *Parmelia-pastillifera*. Annals of Botany 75, 627–632.

Tucker, C.J., 1979. Red and photographic linear combinations for monitoring vegetation. Remote Sensing of the Environment 8, 127–150.

Tucker, C.J., Fung, I.Y., Keeling, C.D., Gammon, R.H., 1986. Relationship between atmospheric CO_2 variations and a satellite-derived vegetation index. Nature 319, 195–199.

Turner, W.B., 1971. Fungal Metabolites. Academic Press, London.

Tyler, G., 1988. Uptake, retention and toxicity of heavy metals in lichens. Water, Air and Soil Pollution 47, 321–333.

Tynnyrinen, S., Palomäki, V., Holopainen, T., Karenlampi, L., 1992. Comparison of several bioindicator methods in monitoring the effects on forest of a fertilizer plant and a strip mine. Annales Botanici Fennici 29, 11–24.

Upreti, D.K., Pandey, V., 1994. Heavy metals of Antarctic lichens. Feddes Repertorium 105, 197–199.

Valladares, F., 1994. Form-functional trends in Spanish *Umbilicariaceae* with special reference to water relations. Cryptogamie Bryologie Lichenologie 15, 117–127.

Valladares, F., Wierzchos, J., Ascaso, C., 1993. Porosimetric study of the lichen family *Umbilicariaceae* – anatomical interpretation and implications for water storage capacity of the thallus. American Journal of Botany 80, 263–272.

Van Dam, H., Houweling, H., Van Dobben, H.F., 1990. Water and ion mass budgets of seepage pools. Conference Abstracts International Conference on Acidic Deposition: Its Nature and Impacts, Glasgow, 16–21 September 1990, Royal Society of Edinburgh.

Van Dobben, H.F., 11993. Vegetation as a monitor for deposition of notrogen and acidity. PhD. Thesis, University of Utrecht, Utrecht, The Netherlands.

Van Dobben, H.F., Wolterbeek, H.Th., Wamelink, G.W.W., Ter Braak, C.J.F., 2001. Relationships between epiphytic lichens, trace elements and gaseous atmospheric pollutants. Environmental Pollution 112, 163–169.

Van Gronsveld, J., Clijsters, H., 1994. Toxic effects of metals. In: Farago, M.E. (Ed.), Plants and the Chemical Elements. VCH Verlagsgesellschaft, Weinheim, pp. 149–177.

Vandenbroucke, J.P., Hofman, A., 1993. Grondslagen der Epidemiologie. Wetenschappelijke Uitgeverij Bunge, Utrecht.

Vogelmann. J.E., 1990. Comparison between two vegetation indices for measuring different types of forest damage in the north-eastern United States. International Journal of Remote Sensing 11, 2281–2297.

Wadsten, T., Moberg, R., 1985. Calcium oxalate hydrates on the surface of lichens. Lichenologist 17, 239–245.

Wappelhorst, O., Kühn, I., Oehlmann, J., Markert, B., 2000. Deposition and Disease – a moss monitoring project as an approach to ascertaining potential connections. The Science of the Total Environment 249, 243–256.

Wastlhuber, R., Loos, E., 1996. Differences between cultured and freshly isolated cyanobiont from *Peltigera* – is there symbiosis-specific regulation of a glucose carrier? Lichenologist 28, 67–78.

Wedding, J.B., Carlson, R.W., Stukel, J.J., Bazzaz, F.A., 1975. Aerosol deposition on plant leaves. Environmental Science and Technolology 9, 151–153.

Wells, J.M., Brown, D.H., 1987. Factors affecting the kinetics of intra- and extracellular cadmium uptake by the moss *Rhytidiadelphus squarrosus*. New Phytologist 105, 123–137.

Werner, A., 1990. Lichen growth rates for the northwest coast of Spitsbergen, Svalbard. Arctic and Alpine Research 22, 129–140.

Wiederkehr, P., 1991. Control of hazardous air pollutants in OECD countries. In: Proceedings of the International Conference on Managing Hazardous Air Pollutants: State of the Art, Washington DC, November.

Wittig, R., 1993. General aspects of biomonitoring heavy metals by plants. In: Markert, B. (Ed.), Plants as Biomonitors. VCH Verlagsgesellschaft, Weinheim, pp. 3–27.

Wolterbeek, H.Th., 1987. Relationships between adsorption, chemical state and fluxes of cadmium applied as $Cd(NO_3)_2$ in isolated xylem cell walls of tomato. Journal of Experimental Botany 38, 419–432.

Wolterbeek, H.Th., Bode, P., 1995. Strategies in sampling and sample handling in the context of large-scale plant biomonitoring surveys of trace element air pollution. The Science of the Total Environment 176, 33–43.

Wolterbeek, H.Th., Bode, P., Verburg, T.G., 1996. Assessing the quality of biomonitoring via signal-to-noise ratio analysis. The Science of the Total Environment 180, 107–116.

Wolterbeek, H.Th., Freitas, M.C., 1999. Preface. The Science of the Total Environment 232, 1–2.

Wolterbeek, H.Th., Kuik, P., Verburg, T.G., Wamelink, G.W.W., Van Dobben, H., 1996. Relations between sulphate, ammonia, nitrate, acidity and trace element concentrations in tree bark in The Netherlands. Environmental Monitoring and Asessment 40, 185–201.

Wolterbeek, H.Th., Viragh, A., Sloof, J.E., Bolier, G., Van Der Veer, B., De Kok, J., 1995. On the uptake and release of zinc (^{65}Zn) in the growing alga *Selenastrum capricornutum* Printz. Environmental Pollution 88, 85–90.

Wyttenbach, A., Bajo, S., Tobler, L., Keller, T., 1985. Major and trace element concentrations in needles of *Picea abies*: levels, distribution functions, correlations and environmental influences. Plant and Soil 85, 313–325.

Zambrano, A., Nash, T.H., Gries, C., 1999. Physiological effects of the Mexico city atmosphere on lichen transplants on oaks. Journal of Environmental Quality 28, 1548–1555.

IId

Higher plants

Chapter 12

The use of higher plants as bioindicators

W.H.O. Ernst

Abstract

Various aspects of the (potential) use of higher plants as bioindicators are reviewed. Dependent on the scope of the enviromental analysis, many options for indicative as well as for causal identification of environmental disturbance and changes are possible at the level of individuals and populations of a species, at the level of plant communities and ecosystems. Ellenberg's indicator values give information on the qualitative relationship between the occurrence of a plants and its natural environment by highlighting several components of that environment, the indicator values for salinity and heavy metals are modified and improved. Quantitative analysis of such relations demands experimental testing. Bioindication has to consider the ecological differentiation and the selection of resistant ecotypes as a longterm natural process.

Anthropogenic changes of environmental conditions can roughly be indicated by the disappearance of species (Red lists) without identifying the reason for this process. Evolution of resistant ecotypes is a good strategy of a plant population with a high genetic potential to survive in changing environments, but it camouflages the loss of non-resistant populations.

Test plants can be used for localizing emission sources and effects of emissions or for aimed release of compounds either at acute or chronic exposure. Most tests, however, are developed for testing the acute toxicity of a compound, mostly without a well-defined selection and description of the cultivar, ecotype or variety of the plant species under investigation and without a proper conditioning of the accompanying environmental abiotic and biotic conditions, i.e. air humidity, nutrition, quantity and quality of radiation, temperature, and association with symbiotic organims. Endpoints of these tests range from germination up to a full lifecycle analysis in the laboratory or/and in the field partly accompanied with registration of visible symptoms.

Phytometer, i.e. planting model plants in existing vegetation with or without isolation, can help to identify environmental processes within one or a few growing seasons.

The pros and cons of a lot of approaches are highlighted.

Keywords: air quality, carbon dioxide, chronic exposure, ecotype, Ellenberg's indicator value, hazard assessment, heavy metals, red list, salinity, sulphur dioxide

1. Introduction

As long ago as Roman times it was known that certain plant species are good environmental indicators such as willows for a good drinking water quality and some other plant species for highly mineralized soils. Later, from the Middle Ages onwards up to modern times, this knowledge of bioindicators was especially applied to ore prospection (Ernst, 1993a). Von Linstow (1929) first summarized the relationship between plant occurrence and soils in his famous book on soil-indicating plants. Nearly 20 years later, Ellenberg (1950) tried to qualify the interaction between higher plants and their *agricultural* environment in Central Europe by classifying the observed responses of ruderal species to several environmental factors, the so-called 'indicator-values'. Predominance of one factor may suppress the reaction pattern to other factors. In an environment with moderate levels of many factors plant species will respond in a fine-tuned manner to the combination of factors.

Plant species or ecotypes of one species with comparable ecological demands aggregate to plant communities or vegetation types. The evaluation of changes in species combination may overcome the restrictions of single indicator species and thus enhance the quality of bioindication. At least in Europe the knowledge of plant communities is very comprehensive to enable such an indication (Grabherr and Mucina, 1993; Pott, 1996 for Germany; Schaminée et al., 1995–2000 for the Netherlandsp; Rodwell, 1998–2000 for UK). Bioindication of the change of environmental conditions can be enlarged to plant-animal and plant-microbial relationships because a change in the physiology of a host plant may change the number of herbivores and the effect of the micro-organisms, i.e. ecosystem evaluation.

Human activities can modify environmental conditions either to the benefit of a few species and ecosystems by improving the growth conditions for these species (eutrophication) or to a disadvantage by deterioration of the natural conditions for these species (diminishing water availability or increasing soil acidification), or by the intended release of pesticides or the unintended release of toxicants with effects at the level of acute and/or chronic exposure. Changes of the frequency of certain plant species in a region have been formalized in "Red lists" by classifying the degree of the change in a numerical system ranging from extinct (Class 0) to potentially endangered (class 4) and not endangered (in principal class 5) (Blab et al., 1984; Van der Meijden, 1996).

An evaluation of the causal relationship between environmental change and plant response demands observation of visible changes ranging from field observations of growth performance of individuals (Bergmann, 1983) to remote sensing of changes in ecosystems by satellites (Hoeks, 1972; Gossmann, 1989) or/and measurement of ecophysiological reaction patterns by analysing specific plant compounds (Ernst and Peterson, 1994).

As soon as members of a plant population can adapt to changing environmental conditions it will finally result in the selection of resistant ecotypes which will hamper the expression of visible effects. Changes of the genetic composition of plant species under adverse environmental conditions were first demonstrated by Prat (1934) for the impact of copper contamination on a population of *Melandrium rubrum*. Nearly 30 years later, Gregory and Bradshaw (1965) developed a technique to identify metal resistance of grasses by comparing the root growth of plants from contaminated and non-contaminated regions. Other experimental tests have later been developed to identify the population genetic changes in plant species exposed to air pollutants (Bell and Mudd, 1976) and herbicides (Radosevich and Appleby, 1973).

The exposure of standardized plants to an environment with potentially negative effects is a well-established procedure for biomonitoring. It was initiated already in 1911 by Sorauer to test the adverse effects of industrial emissions on plants. In the

1970s Schönbeck et al. (1970) and Scholl (1970) improved and standardized the methods of "active" biomonitoring. Another approach was followed by Heggestad and Menser (1962) selecting those cultivars or populations of plant species which were highly sensitive and thus very responsive to a specific environmental component. Further modifications of these principles were elaborated for testing the toxicity of substances to be released into the environment. Laboratory tests with very different endpoints of evalutation were developed or formalized by various environmental agencies such as ASTM, OECD, USFAD, USEPA and others.

A new approach in ecology is the use of phytometers, i.e. planting test plants from the same ecological background in an environmental gradient and analysing growth and concentration of mineral nutrients during one or a few growing seasons.

In this contribution I will highlight some aspects of the use of of higher plants as bioindicators and hope to stimulate the functional appraoch of the use of bioindicators in the future.

2. Ellenberg indicator values and bioindication

2.1. The scale of indicator values for different ecological parameters and conditions

The occurrence of a plant species in one or more communities and ecosystems is governed by the ability to germinate and finalize the life cycle by seeds under the environmental conditions exposed to. Ellenberg (1950) qualified, not quantified the relationship between *estimated* environmental components and the occurrence of higher plants in Central Europe by classifying the observed responses of higher plants in grasslands and arable fields as so-called indicator-values ("Zeigerwerte"). It was originally a scale of five. Later Ellenberg (1974) sophisticated the scale from originally 5 to 9 and an "x" for indifferent behaviour (thus effectively 10 reaction degrees) for responses light, temperature, soil acidity, mineral nitrogen and continentality, and with 12 plus one indifferent value for soil humidity (thus 13 reaction degrees). In the last edition Ellenberg et al. (1991) updated and corrected some classifications and extended the information for resistence of plants to salinity and heavy metals. Ellenberg et al. (1991) enlarged the indicator values for salinity from 6 in the original scale of Scherfose (1990) to a scale of 10, based on chloride concentration in the soil solution. In the case of heavy metals Ellenberg et al. (1991) introduced a scale of three, species with avoid metal-enriched environments, those with a moderate metal resistance, those with a moderate one (b) and those with a high metal resistance (B). Ellenberg's indicator value of a species is said to be based on an *expert judgement* and should give a first information of the position of a plant in an environmental gradient with regard to the above mentioned abiotic conditions. A calibration of Ellenberg indicator values by Ertsen et al. (1998) has shown that the data of soil moisture, acidity and chloride concentration in groundwater have a reasonable correlation with Ellenberg's indicator values. A comparison of the relation between soil pH and distribution of some plant species in Danish grassland with the Ellenberg's acidity

Table 1. Percentage distribution of some meadow species in pH soil classes (Olsen, 1923) and the soil acidity indicator value (R) given by Ellenberg et al. (1991) to these species.

Plant species	pH range									R-value
	3.5 to 3.9	4.0 to 4.4	4.5 to 4.9	5.0 to 5.4	5.5 to 5.9	6.0 to 6.4	6.5 to 6.9	7.0 to 7.4	7.5 to 7.9	
Avenella flexuosa	54	31	15	.	.	.	.	.	.	2
Calluna vulgaris	31	23	31	15	.	.	.	.	.	1
Molinia caerulea	25	25	25	11	.	4	4	6	.	x
Hieracium pilosella	.	.	20	30	30	30	.	.	.	x
Anthoxanthum odoratum	7	7	15	17	26	17	7	2	2	5
Deschampsia caespitose	.	.	.	3	30	27	12	15	12	x
Tussilago farfara	.	.	.	.	.	.	11	44	33	8
Scirpus sylvaticus	.	.	.	.	.	.	63	12	25	4

(R)-values should show one problem of indicator values (Table 1). The response of a plant to soil acidity is more than the reaction to a concentration of H^+-ions because pH has a strong impact on the mobility of many elements, often in different directions. The result of this comparison is as follows: the R-value of two species does not fit with ecological reality. The grass *Anthoxanthum odoratum* has not only in Denmark, but also in the Netherlands and Germany a wide ecological range, therefore Ellenberg's $R5$ value does not reflect the ecology of this grass and should be changed to "x". The acidity values ($R4$) for *Scirpus sylvaticus* ($R4$) is also a mismatch, at least in Denmark and has also to be changed to "x".

The weak correlation of Ellenberg's N-values with soil chemical variable as evaluated by Ertsen et al. (1998) is not surprising. A plant has a lot of options to adapt its physiological processes to the various combinations of bioavailable concentration of the 15 essential mineral nutrients, a few other beneficial and some tens of non-essential chemical elements with a strong impact on plant performance.

Another aspect of indicator values is the differential potential of species with narrow and broad ecological response. Plant species with a narrow ecological niche ("stenoecious") can relatively easily be associated with a certain value of an environmental factor. Plant species with a broad ecological spectrum for one or more of the above mentioned factors ("euryoecious" species) are indicated by Ellenberg et al. (1991) often by an "x", which means indifferent. The more ecological knowledge will be built up for so-called "stenoecious" species, the more a change from a number to the indifferent value "x" will occur, thus decreasing indicator quality of the species, but a better identification and characterization of ecoytpes will strongly improve the indicative potential. As shown below, a new solution have to be found be giving indicator values to well-defined ecotypes of a species differntiated for one or more environmental conditions.

2.2. Regional differences in indicator values and impact of ecological differentiation within a species

For a first look of consistency I will compare Ellenberg's indicator values for some plant species occurring in plant communities on calcareous and acidic sands in the Netherlands and in Germany (Table 1). It has to result in a high consistency. A moderate variation in the response between species is acceptable because each species has to find its niche in a vegetation, but inconsistency, i.e. more than three Ellenberg units, may indicate either regional adaptation, ecotypic differentiation or a wrong evaluation.

For the Phleo-Tortuletum on alkaline sand and the Spergulo-Corynephoretum on acidic sand it seems to be obvious that the light requirement of the plant species is high and that there is a good adaptation to a low amounts of major nutrients.

Values for soil moisture have to consider the precipitation pattern and the life cycle of plants which is important for annual plants. They are often differentiated in summer and/or winter annual populations, e.g. *Senecio sylvaticus* (Ernst, 1987), which demands a further differentiation of the indicator values. During late winter time and early spring plants have the highest growth performance and the soil moisture is high, whereas low soil moisture during summer is of very minor ecological relevance (Ernst, 1983; Rozijn et al., 1990). Thus the evaluation has *not* only to consider the water availability during the summer, but during winter and spring. Therefore the indicator value for soil moisture of many winter annuals such as *Phleum arenarium* (Ernst, 1981), *Erophila verna* and *Vicia lathyroides* (Rozijn et al., 1990) has to be upgraded from low values to moderate values (Table 2). In the present comparison a good consistency among indicator values of species of one plant community is also absent for the estimated response of species to soil acidity, the most important ecological difference between both plant communities. In the Netherlands *Carex arenaria* is not restricted to acidic soils, but occurs on both calcareous and acidic sands, thus its ranking with regard to soil acidity demands an "x". In contrast, *Vicia lathyroides* is not associated with acidic, but with calcareous soils, thus it demands an "R7". Also *Festuca ovina tenuifolia* has a great ecological range with many local ecotypes, thus it demands also an "x". In contrast, *Spergula morisonii* is not indifferent to soil acidity as indicated by Ellenberg et al. (1991), but strongly associated with acidic sands, thus an indicator value of *R*2.

Indicator values for nutrients, especially nitrogen and phosphorus availability, the "nitrogen-value" (Ellenberg et al., 1991) deliver even more problems. Both nutrients have a high turnover in the soil (Blair, 1988). Over a period of 15 years fertilization of Scots pine forest stands with N, P, K and treatment with lime and dulphuric acid there were treatment effects on the undergrowth of the pine forest, but Van Dobben et al. (1999) have to admit that Ellenberg's indicator values were not as sensitive as a redundancy analysis. In addition many species with a symbiosis with arbuscular-mycorrhizal fungi receive most of phosphorus and other nutrients and water from the fungi (Allen, 1991). Plant species can be so dependent on the mycorrhizal partner that the AM fungus determines their presence in a plant community (Ozinga et al., 1997; van der Heijden et al.,1998).

Ecophysiologists have introduced another parameter for evaluating the possibilities of plant species to handle the nutrient availability of a soil, i.e. nutrient use efficiency

Table 2. Ellenberg's indicator values of some plant species belonging to the Phleo-Tortuletum ruraliformis on calcareous dune sand and to the Spergulo-Corynephoretum on acidic dune sand.

Plant species	Environmental component			
	Radiation	Moisture	Soil acidity	Nutrients
Phleo-Tortuletum				
Carex arenaria	7	3	2*→x	2
Cerastium semidecandrum	8	3	6	x
Erodium glutinosum	8	4	7	2
Erophila verna	8	3*→5	x	2
Phleum arenarium	9	3*→5	7	3
Saxifraga tridactylites	8	2*→5	7	1
Vicia lathyroides	8	2*→5	3*→7	2
Viola curtisii	8	3	6	3
Spergulo-Corynephoretum				
Carex arenaria	7	3	2*→x	2
Corynephorus canescens	8	2	3	2
Festuca ovina tenuifolia	7	4	3*→x	2
Rumex acetosella	9	3	2	1
Spergula morisonii	9	3	x*→2	2

Notes
1–9: 1 = lowest, 9 = highest, x = indifferent)
* indicates inconsistency between the ecological behaviour in the Netherlands and in Germany, as explained in the text; → indicates the new indicator value.

(Aerts and Chapin III 2000). If the nutritional status of the plant is considered, then many plant species of so-called nutrient-poor sites have high nitrogen and phosphorus concentrations, as the above mentioned plant species of the Phleo-Tortuletum (Ernst, 1983), thus indicating a high nutrient use efficiency.

2.3. Ecotypic differentation and Ellenberg's indicator values

Plant species with a broad ecological amplitude (euryoecious species) are evaluated as indifferent by an "x". This evaluation ignores one of the most important ecological processes, i.e. differentiation of a species into highly specialized local and regional ecotypes. These ecotypes are often very specific indicators of a specific dominant (i.e. selective) environmental factor, but – unfortunately – most of these ecotypes cannot be distinguished by their morphology in nature, but only by their ecophysiological response after experimental testing. Ecotypes of euryoecious species have evolved for many abiotic conditions and biotic factors. *Solidago virgaurea* has populations highly adapted to sun or to shade (Björkman, 1968); therefore it has to be evaluated with L5 for the shade ecotype (Ellenberg et al., 1991), but with L9 for the sun ecotype, thus for the species as a whole a light value "x". Differences in drought resistance are known for ecotypes of *Dactylis glomerata* (Valero and Olivieri, 1985), demanding a change of the moisture value of F5 to "x".

The evolution of ecotypes for soil chemical conditions has received much attention. Plant species have evolved ecotypes with a specific adaptation to a low, moderate or high supply of major nutrients and minor nutrients in their habitat. Populations of *Senecio vulgaris* growing in dune sand poor in nitrogen and phosphorus have to be evaluated with $N = 2$–3, and not with $N = 8$ (Ellenberg et al. 1991) which is only correct for populations growing in fertilized agricultural and horticultural soils. As a consequence the N-value has to be given as indifferent, thus "x" instead of "8". *Silene nutans* has evolved populations with different responses to calcareous and siliceous soils (De Bilde and Lefèbvre 1990) so that Ellenberg's soil acidity indication has not only to be given as $R7$ for the calcareous, but also as $R3$ for the siliceous population, thus finally as "x" for the species.

It is already a progress that some of these ecological differentiations are added as footnotes to some euryoecious species, concerning responses to salinity and heavy metal (see below). But very good observation and experimental testing will detect more of such a differentiation of euryoecious species and will strongly improve bioindication.

2.3.1. Salinity indicator values and the ecological reality

Many plant species have evolved different degrees of resistance to a surplus of sodium chloride and sodium sulphate. Ellenberg et al. (1991) are only considering soil chloride as differentiating condition for species with populations near and far away from the coast (Table 3). Walter (1960) has already demonstrated the importance of the cation (Na, K, Ca) which determines the survival of the plant and distinguished between sulphate and chloride halophytes. The ecophysiological relevance of the cation is confirmed by unravelling the various cation transporters different for K, Na, and Ca (Fox and Guerinot, 1998).

Iversen (1936) had already established three halophyte classes for Denmark. Walter (1960) has extended the evaluation for plant species growing in brackish and saline sodium soils into five classes. Euryoecious plant species have differentiated in salt-sensitive and salt resistant ecotypes, partly recognized by Ellenberg et al. (1991). The difference in the degree of salinity responses between populations can be as small as three classes as in *Phragmites australis* or increase up to seven classes as in *Atriplex hastata*. However, not all salinity values fit well with field and experimental data. First of all, the salt sensitivity (class S0) has no physiological background. In contrast, all higher plants demand chloride for the water-splitting system of photosystem II (cf. Marschner, 1995). Therefore an indicator value of S0 is physiologically and ecologically impossible, and all plant species in this class has to be upgraded to class S1. In general, salinity classification has to be revised for many species. The salt sensitivity (S0) given to *Juncus alpino-articulatus* (Ellenberg et al., 1991) is too low for the populations characteristic for brackish water; its growth is only affected if the soil solution is surpassing 0.21% Cl, thus deserving an improvement of the indicator value by two classes to S2. Also the inland form of *Agrostis stolonifera* is more salt resistant than S0 is suggesting. The salinity value for the coastal ecotype of *Juncus bufonius* has to be upgraded from S2 to S3.

Table 3.   Salinity (chloride) indicator values of populations from inland and coastal sites in relation to the percentage chloride concentration of soil water (Ellenberg et al., 1991) compared with the salt resistance from ecological field data and from growth in hydroponics at chloride contrations resulting in 10% (EC^{10}) growth reduction (Rozema, 1978). Proposed changes of the S-indicator values are given by an arrow ($\rightarrow$).

Plant species	Salt number	Chloride in soil water(%)	Chloride resistance (% Cl at EC$_{10}$)
Agrostis stolonifera	0→2	0.0	0.21
ssp. maritime	6	0.9–1.2	0.53
Festuca rubra ssp. rubra	0→1	0.0	0.07
ssp. arenaria	1→3	0.0–0.1	0.32
ssp. litoralis	7	1.2–1.6	0.60
Juncus alpino-articulatus	0→2	0.0	0.25
Juncus bufonius			
inland ecotype	0→1	0.0	0.04
costal ecotype	2→3	0.05–0.3	0.43
Juncus gerardii	7	1.2–1.6	0.21
Juncus maritimus	6→7	0.9–1.2	1.24

The indicator potential of plants for salt-affected soils is proven for roadsides receiving de-icing salts in winter. Two salt-resistant plant species, *Cochlearia danica* (S4) and *Puccinellia distans* subsp. *distans* (S7?), are strongly expanding along highways and roads in Central and Western Europe populations (Beyschlag et al., 1992), thus nicely indicating anthropogenic salinization.

2.3.2. Indicator values for heavy metals

Plants take up many heavy metals, not only those which are essential for a plant's metabolism such as cobalt, copper, iron, manganese, nickel and zinc, but also elements which are not essential for plant growth such as aluminium, cadmium, lead and thallium. Many plant species have evolved ecotypes highly adapted to a surplus of one heavy metal or a combination of several heavy metals (Ernst, 1974). This adaptation of a population to the biologically available metal concentration of the soil is encoded only by a few genes specific for each metal (Schat et al., 1996). This genetic design results in a high ecological specialization of each ecotype with a strong impact on its performance on soils with another combination and concentration of heavy metals (Fig. 1), as shown for the Zn-Cd-ecotype from the metal-enriched soils near Plombière (Ernst and Nelissen, 2000 a,b).

Ellenberg and co-workers (1991) have given two values for plant species which are pronounced (B) and moderately resistant (b) to heavy metals – the letter [B,b] is derived from the German term "Blei" = lead. In contrast to salinity metal-sensitive species have not received an indicator value. Obviously the authors have recognized that they had to differentiate between the differential demand and accumulation of all

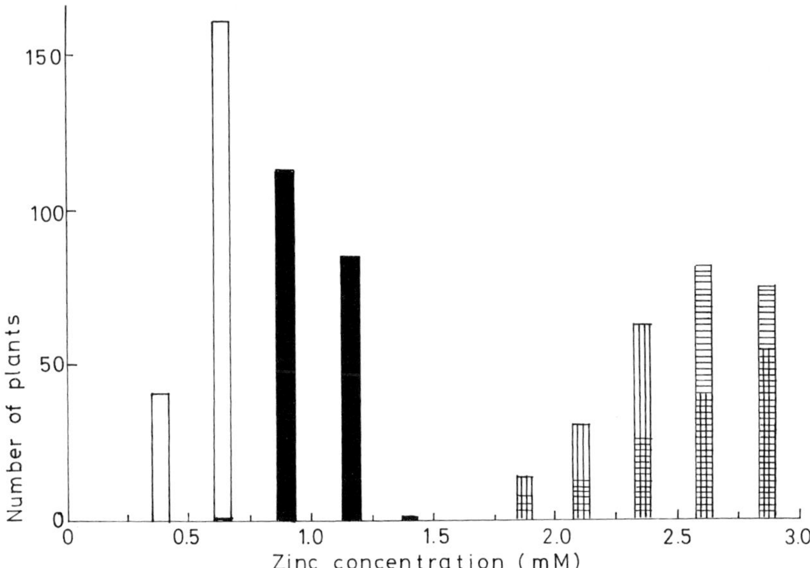

Figure 1. Absolute frequencies of individuals in ecological effect concentrations (EC_{100}-classes) of zinc for root growth, as established in a sequential test with 0.25 mM Zn concentration intervals. The ecotypes are derived from soils with a normal zinc supply at Amsterdam (open column) and Marsberg (black column) which is rich in copper, and from soils with high zinc concentrations at Blankenrode (vertical strips) and Harlingerode (horizontal strips). Overlap of both Zn-resistant populations appear as cross-hatched. Data from Schat et al. (1996).

higher plants for cobalt, copper, iron, manganese, nickel and zinc (see Table 6), and the different degrees of resistance to metabolically non-essential elements such as cadmium and lead. Indicator values for heavy metals were presented only for nine plant species with an "occurrence in habitats with high concentrations of lead, zinc and other heavy metals. Some other species evolve resistant races but occur predominantly on normal soils" (p. 70), but have not been considered in the evaluation.

What is the problem with the metal indicator values B and b. First of all, lead (B, b) is the least selective factor on heavy metal-enriched soils (Antosiewicz, 1995) due its low availability in contrast to cadmium, copper, manganese and zinc (cf. Ernst and Nelissen, 2000a). In Central Europe a surplus of copper and/or zinc is the most selective factor, thus an abbreviation of metal resistance with a letter "M" will be more appropriate. Secondly, Ellenberg et al. (1991) have insufficiently considered the publications on metal-resistance (especially zinc-) of plant species from heavy metal-enriched soils (Gries, 1966; Rüther, 1967; Ernst, 1974, 1976b). Ellenberg et al. (1991) underestimate the high gradation of heavy metal resistance in plants by stating (p. 19): "More precise gradation as well as differentiation of resistance against metal ions are presently not yet possible and will not be achieved for the practical application".

The concentration of the bioavailable metal(s) in the soil strongly governs the selection of metal resistance. Therefore the degree of metal resistance can vary up to a factor of thousand among ecotypes of plant species on metal-enriched soils (Table 4), independent if this resistance is tested with the method of comparative protoplasma-tology, with growth experiments in hydroponics or in metal-enriched soils. The indicator values for metal resistance should be classified in at least six classes. The highest resistance to zinc is present in ecotypes of *Arabidopsis halleri* (M5), *Armeria halleri* (M5), *Silene vulgaris* (M5) and *Thlaspi caerulescens* (M5). *S. vulgaris* has evolved ecotypes with one of the highest Zn- and/or Cu-resistance in angiosperms (Gries, 1966; Rüther, 1967; Ernst, 1974, 1976b; Schat et al., 1996; Ernst and Nelissen, 2000a,b) and consequently the underestimated metal indicator value ("b") has to be upgraded to "B". In the proposed new classification these ecotypes receive the highest value of M5. Less Zn- and Cu-resistant are ecotypes of the amycorrhizal *Minuartia verna* and the mycorrhizal grasses *Agrostis capillaris* and *Festuca ovina* (M4). On soils enriched with copper the fungicidal potential of copper hampers the development of arbuscular mycorrhizal fungi so that the grasses cannot rely on fungal metal-protection. The zinc violets are the least Zn-resistant species on heavy metal-enriched soils, as already shown by Rüther (1967). Therefore the indicator value has to be diminished from category "B" to "b", or to the new low class M1. The ability of zinc violets to grow on zinc-enriched soils is due to the high degree of mycorrhization with arbuscular mycorrhizal fungi which are diminishing the metal transfer from the soil to the zinc violet (Hildebrandt et al., 1999; Tonin et al., 2001). The Zn resistance of the zinc violets is as low as that of Zn-sensitive ecotypes of *Silene vulgaris*. The Zn-resis-tance of *Viola tricolor arvensis* is a factor of 5 lower than that of *V. tricolor guestphalica* (Kakes, 1981). The Zn-resistance of these ecotypes has to be classified in the same sensitivity class M1 as that of the model plant of molecular plant biolo-gists, *Arabidopsis thaliana* (Van der Zaal et al., 1999). Many grass species with a proven metal-resistance have to be given a M3 or even M4 value: *Agrostis canina, A. stolonifera, Anthoxanthum odoratum, Arrhenatherum elatius, Deschampsia caespi-tosa, Festuca rubra, Holcus lanatus* and *Molinia caerulea* (Ernst, 1974; Lefèbvre and Vernet, 1990). Also many herbs occurring on metal-enriched soils have at least moderate metal resistance to be evaluated at least with M3, for example *Biscutella laevigata, Campanula rotundifolia, Euphrasia stricta, Plantago lanceolata, Rumex acetosa, Silene dioica, S. nutans* and *Thymus serpyllum* (Prat, 1934; Rüther, 1967; Ernst, 1974, 1976b; Mathys, 1977; Godzik, 1991).

Also the frequency of metal resistant populations of euryoecious species is not correctly described by Ellenberg et al. (1991, p. 19): "Plants or ecotypes are *frequently* to be found on soils more or less enriched with *lead* and other heavy metals". With the exception of *Arabidopsis halleri, Armeria halleri* and the zinc violets, all other plant species are NOT frequently growing on soils enriched heavy metals and have a majority of ecotypes on normal soils (Table 4).

In conclusion, the evaluation of metal indicator values demands a new approach as explained above, and the Ellenberg values should be considered as not appropriate and deleted.

2.4. How to improve the reliability of Ellenberg's indicator values

After criticism by Walter and Breckle (1983), Ellenberg (in Ellenberg et al., 1991) emphasized that the "indicator-values" ("Zeigerwerte") should not be mistaken as

Table 4. Distribution frequency of species with metal-resistance in West-Germany (based on Haeupler and Schönfelder, 1988) and the metal resistance "B" (highly metal resistant) and "b" (moderately metal resistant) as judged by Ellenberg et al. (1991) compared with cell physiological Zn resistance (Gries, 1966; Rüther, 1967; Ernst, 1974) and root growth in metal solution (Ernst, 1974, 1976b; Mathys, 1977; Kakes, 1981; Schat et al., 1996; Ernst and Nelissen, 2000; Assuncao et al., 2001) resulting in the new proposed metal indicator classes M1 (low) to M5 (very high).

Plant species	Frequency	Ellenberg value	Metal resistance (mol Zn L^{-1}) of		New metal-value
			Cells[c]	Roots[r]	
Arabidopsis halleri heavy metal soil	51.9	B	0.0004–0.02[c]		→ M5
Arabidopsis thaliana				0.00003[r]	→ M1
Agrostis capillaris Zn-resistant	3.4	b		0.002[r]	→ M4
non-Zn-resistant				0.0002[r]	→ M3
Armeria halleri	100.0	B	0.0004–0.04[c]		→ M5
Armeria maritime coastal ecotype				0.00008[r]	→ M1
Festuca ovina Zn-resistant	3.1	b		0.008[r]	→ M4
non-Zn-resistant				0.00001[r]	→ M1
Minuartia verna	38.3	B	0.0004–	0.004[c]	→ M4
Silene vulgaris heavy metal soils	3.7	b			
Zn – Blankenrode			0.004–0.04[c]	0.002[r]	→ M5
Cu, Zn-Langelsheim				0.04[r]	→ M5
normal soils			0.00004[c]	0.0003[r]	→ M2
Thlaspi caerulescens heavy metal soils	1.7	B			
Zn- Aachen			0.04[c]		→ M5
Zn – Silberberg				>0.001[r]	→ M4
Zn – La Calamine				>0.001[r]	→ M4
normal soils				0.00005[r]	→ M1
Thymus serpylllum	0.8	–	0.0004[c]		→ M3
Viola calaminaria	100.0	B	0.00004[c]		→ M1
Viola tricolor guestphalica	100.0	B	0.00004[c]	0.00015[r]	→ M2
Viola tricolor arvensis				0.00003[r]	→ M1

measured values ("Messwerte"). But he stimulated this misinpretation of a quantitative value as a qualititative one by calculating means of plant communities, thus adding up the indicator value of all species of a specific plant community and dividing their sum by the number of species resulting in a community mean, disregarding the "x" value in the calculation. If the indicator values will have ecological relevance, it is urgent to quantify them by field measurement and/or experimental tests.

To increase the reliability of Ellenberg's indicator values the abiotic conditions of many sites of plants species have to be analysed and each of the environmental factors has to be correlated with the performance of that species, i.e. number of individuals, biomass, seed production and rejuvination rate. The reaction pattern of a plant population to *one* environmental component can be described by a response curve ranging from low productivity (or population size) via a maximum to a decline with increasing adverse conditions (see Fig. 3.1 in Ernst, 1996). In nature, the survival of an individual and the persistence of a population in a certain environment is based on the integrated reaction of a plant species to all environmental factors at that site, therefore the ecological optimum is often different from the physiological optimum (for a review: Ernst, 1978). The law of the relative habitat constance of a species (Walter, 1954) has to be the guiding principle in relating the occurrence of a plant species to its ecological reaction pattern. Therefore ranking the response of a plant population in the field demands a lot of ecological knowledge to identify the deviation of a plant from its optimum performance. Only then, Ellenberg's indicator values may be very helpful as bioindicators. Instead of oversophisticating a sound scale as that of salinity from original 6 (Scherfose, 1990) to 10 (Ellenberg et al., 1991), it would be more helpful to diminish the scale to the original range of 5, as was necessary for comparing six models of vegetation responses to the hydrology of the habitats (Olde Venterink and Wassen, 1997).

3. Are 'Red list' species reliable bioindicators?

'Red lists' of organisms are designed to give information about the 'welfare' of a species in its present environment in comparison to a reference point in the past. The categories range from "0" (extinct species in a region or country) up to "4" (potentially endangered species); not endangered species are not registered (Blab et al., 1984; Van der Meijden et al., 2000). The reasoning for a categorization of plant species in 'Red list' as endangered is mostly ignoring a species' history, especially the introduction of a species to a certain area by human activities in the past. Strictly spoken species have only to be categorized as endangered in a Red list if human activities are diminishing their *natural habitat* and thus reducing their natural population size, changing their ecological behaviour and genecological potentials. Although the genetic constitution of a population is very important for its vitality on the long term, it is only considered in 'Red lists' if genotypes can be distinguished by morphological traits (Van der Meijden et al., 2000).

A decrease in the range of suitable *natural* environments due to human activities are manifold: Examples of mechanical impacts are the reclamation of peat, destruction of rocks, disturbance of naturally metal-enriched environments due to mining, loss of habitats due to urbanization, flooding of valleys for the construction of dams, and lowering soil water level for improvement of the growth condition for agricultural crops or for for drinking water supply with desiccation of natural environments. Example for chemical changes of the environment are emission of compounds in toxic concentrations resulting in air, soil and water contamination and/or acidification and soil fertilization resulting in losses of the oligo- and mesotrophic conditions (eutrophication).

Species are in principle also not to be categorized as endangered if natural events in a geological time scale diminish their survival at a certain site or region. The disappearance of the tundra vegetation in Central Europe was caused by increasing temperature after the last glaciation and the replacement of subtropical woodlands by short grasslands or deserts in the Sahara was due to natural changes in precipitation amounts and patterns. Even climatic changes in shorter geologic periods can have strong effect, as the replacement of *Fagus grandiflora* by oak and pine in North America caused by rapid Little Ice Age cooling between 1400 and 1670 AD (Campbell and McAndrews, 1993). Freshly flushing beech leaves are prone to late frost in spring (Von Wuehlisch et al., 1995) which still limits the altitudinal and northern latitudinal expansion of *Fagus sylvatica* in Europe.

Most species in 'Red lists' and in the European Habitat Directive are selected due to changes of not only their *natural habitats*, but predominently on the criterion of anthropogenic impacts on their *man-made habitats*. The overall decrease of their occurrence in geographic grids is thus not differentiating between natural and man-made environments (RIVM, 1996). I have nothing against a conservation of those plant species which have invaded or strongly expanded by human activities, i.e. by historic events, but it has to be clearly defined as such. If the human activity does change, e.g. land-use, then of cause, plant species related to these activities will decrease or even disappear, thus reversing the historical expansion.

"Red lists" and the European Habitat Directive have a strong bias considering only the recent, one to two hundred-year-old history of a species. They overemphasize the species with narrow ecological niches and do not consider the highly specialized ecotypes of wide-spread species which may be lost by human impacts. The local or regional ecotypes of a wide-spread species are as important as indicators as those ("stenoecious") species which are restricted to one environmental complex. Ecologists and environmentalists are often not aware of the loss of important ecotypes of wide-spread plant species, thus neglecting evolutionary processes of tens to thousands of years. The identification of such highly specialized ecotypes of a wide-spread species demands experimental testing and can only partly be established in the field.

Insufficient credit is given to the regional history of the species, to its range in a geological time frame, to its phenology as seed or photosynthetically active plant.

3.1. Threatening due to reverse historic expansion

Some examples will be given to help to rationalize the discussion of "Red list" species which are not threatened *per se* if the prehistoric population size is considered. In principle, a protection of these species demands the reconstruction or maintenance of historic forms of human activities, such as the type of human settlement, agricultural activies and crop selection, and grazing activities of domesticated animals.

Surrounding human settlements with walls constructed from stones with mortar has let species expand from natural rocky areas into villages and cities. In addition to the destruction of these walls, the new construction material and methods such as blocks with armoured concrete has diminished the establishment and survival of many wall-related species. Strictly speaking species such as *Asplenium adiantum-nigrum, A. viride, Cystopteris fragilis* are not threatened, except that a conservation of their expansion in historic times is the scope of protection.

A similar argument holds for plant species ("weeds") accompanying agriculture. Introduction of agricultural crops was associated with the expansion and introduction of agricultural weeds as nearly all species belonging to the vegetation classes Chenopodietea and Secalietea. Weeds in an open crop have good access to sunlight (caused by sowing or planting in rows) and to nutrients. Intensifying the crop density and the crop height, the introduction of improved harvest and cleaning techniques, and the application of pesticides have strongly affected the crop-associated weeds. Therefore it is not surprising that fields on chalkstone has the highest numbers of endangered species in the Netherlands (Van der Meijden et al., 2000) and elsewhere if they cannot find ecological niches at the borderlines of fields or elsewhere. The strongest impact on the maintenance of a species will be on those weeds which are parasites of the agricultural crop, as the holoparasites *Cuscuta epilinum* (3 in D, 2 in NL) on flax (*Linum usitatissimum*) and *Orobanche ramosa* (3 in D; 0 in NL) on *Nicotiana tabacum* or *Cannabis sativa.*

Extension of domesticated animal herds has first damaged the understorey in and juvenation of forests. Later the etablishment of grasslands has destroyed the area of forests and opened the access to species formerly restricted to small open patches in the natural environments. A decrease of grazing activities will immediately initiate a succession of so-called half-natural grassland to woodlands and forests. Consequently populations of species with high light demands and less strong competition will decrease in numbers or will disappear, as indicated by the many "Red list" species of chalk grassland and heather.

3.2. Differences in expansion history

The ranking of a species in a Red list is not free of a national bias due to different histories of the same species in two neighbouring countries. A good example is the yellow-flowering zinc violet (*Viola calaminaria*), a character species of naturally metal-enriched soils in certain parts of Belgium and Germany (Ernst, 1974). In the Western part of Germany this species is correctly categorized as strongly endangered (category 2) because many sites of outcropping zinc ores are removed. In the Netherlands, however, the zinc violet together with *Thlaspi caerulescens*, another species of metal-enriched grassland in Western Europe, was introduced together with metal-contaminated sediments originating from Belgian mine sites. From the Middle Ages onwards these heavily contaminated sediments were deposited on the river banks of the Geul during a couple of centuries (Ernst, 1974; Leenaers, 1989) and let develop a heavy metal vegetation imported from Belgium. At the end of the 20th century, Belgian authorities have let remove much of the tailings and thus blocked the erosion of metal-contaminated soil into the Geul river. As a result, the river sediments were cleaned up so that the zinc contamination of the river banks of the Geul in the Netherlands was drastically diminished and consequently let decrease the population of the zinc violet. Environmentalists and Nature Conservation should be happy with such a decrease in soil contamination in this area and emphasize the disappearence of this plant species as a good bioindication of environmental improvement. By ignoring that reality, the zinc violet is categorized as highly threatended (category 1) on the Red List of the Netherlands!

3.3. Red lists and the seed bank

At the moment the status of a species in a Red list is nothing more than a very rough indication of a change of the occurrence of a species in a certain geographic region without indicating directly any environmental process. Categorizing a species in a Red list makes the assumption that a plant has to be present in a visible form above the soil level. Plant species can survive as seeds for decades and perhaps centuries in a soil as seed bank (Thompson et al., 1997). The seed is hampered to develop into a "visible" individual due to insufficient environmental conditions. The maintenance or break down of seed dormancy often demands a change of the red/far-red component on the soil surface by vegetation succession and mechanical soil disturbance. Such a species may be categorized as extinct or highly endangered although it sufficiently present in a dormant state. I will highlight this aspect for foxglove (*Digitalis purpurea*), a biennial herb characteristic for woodland and forest clearings which was ranked as highly endangered (value 2) in the Red list of the Netherlands (Van der Meijden, 1996). The seeds of foxglove germinate only if exposed to the full radiation spectrum of sunlight. After creation of a gap in a forest by the death of a huge tree or by wind, fire or clear cutting and simultaneous disturbance of the soil profile, the foxglove seeds germinate. But as soon as the vegetation of such a forest gap has grown to such a height that the leaves of taller plants modify the light quality by filtering especially the red component and increasing the far-red/red ratio, the seeds of the new generation fell dormant and remain vital at the site up to the next gap event (Van Baalen and Prins, 1983) which may occur once in hundred or more years as shown by pollen and macrofossil analysis (Pott, 1986). Recently, the recognition of this below-ground survival of foxglove seeds as essential parameter in the classification of this species has resulted in a removal from the recent Red list of the Netherlands, thus a jump from class 2 to class 5 (Van der Meijden et al., 2000).

The survival of a species in a seed bank is species-specific and has received growing scientific interest during the last decade with emphasis on species from pastures and natural grassland (Poschlod and Jackel, 1993; Eriksson and Eriksson, 1997; Bekker et al., 1998; Van der Valk et al., 1999; Bekker et al., 2001). The longevity of a seed in a seed bank can also conserve ecotypes which may be selected under other environmental conditions (Hurka and Neuffer, 1991). Seed banks are not only important for the enumeration of a species' status in Red lists, but have also consequences for the restoration of natural and semi-natural vegetation.

Relating the endangered category of a species to its occurrence in vegetation types as exemplified by Van der Meijden et al. (2000) for the Netherlands is a first step to put the classification in an environmental context.

4. Plant communities as bioindicators

The relation of plant communities with specific factors of their environment are well elaborated for several European countries and regions, e.g. Grabherr and Mucina (1993) for Austria, Oberdorfer (1977–1992) for Southern Germany, Pott (1996) for Germany, Schaminée et al.(1995–2000) for the Netherlands, Rodwell (1998–2000) for the United Kingdom. Changes in species composition of such plant communities may be indicators of changing environmental conditions, but it demands a long-standing data base. To elucidate the reason of the changing floristic composition, it is necessary not only to describe the species inventory (cf.for a review: Wittig, 1991), but to analyse also abiotic factors (Heil and Diemont, 1983; Aerts and Berendse, 1988; Roelofs, 1991; Van Wijnen and Bakker 1997). Many long-standing observations and measurements have sufficiently documented that changes in and of plant communities are very promising approaches in bioindication.

Changes in the floristic composition of communities may be caused by different processes:

1. Climatic changes on a geological time scale, i.e. long-term natural processes, as the change from a tundra vegetation to the broad-leaved forests in the present temperate zone of the Northern Hemisphere, as documented by pollen analysis and fossil remnants.
2. Medium-term processes resulting in succession with the predicting the development of plant communities, e.g. the vegetation development of the Wadden Islands based on a 40-year lasting data base (Westhoff and van Oosten, 1991).
3. Short-term man-made processes with a strong selective component may enhance the presence or frequency of certain plant communities, i.e. the disappearance of wet dune slacks due to decreasing the water level (Grootjans et al., 1998), the destruction of forests by acid precipitation (Smith, 1981), the eutrophication of oligotrophic terrestrial and aquatic plant communities by airborne nitrogen and by fertilization (Berendse and Aerts, 1984; Van Beckhoven, 1995; van den Berg et al., 1998) or the establishment of metal-resistant plant communities in the vicinity of smelters (Ernst, 1999b).
4. Often plant species in a plant community decline or disappear due to a change of land-use practices of half-natural ecosystems, such as grasslands and anthropogenic heathlands (Willems, 1983; Bakker, 1989; Stampfli and Zeiter, 1999). The species richness of these ecosystems is often maintained by mowing, grazing or even overgrazing so that strong competitors are hampered to increase.
5. Epidemics in animals can have a great impact on vegetation development. An anthrax outbreak in East Africa killed more than 90% of the impala population, and rinderpest some 20% of the buffalo population (Prins and Weyerhaeuser, 1987). These ruminant diseases in African savanna are responsible for the establishment of shrub- and tree-savannas with even-aged *Acacia* trees.

   Myxomatosis in European rabbits (*Oryctolagus cuniculus*) has resulted in a population crash by often more than 95% (Fenner and Myers, 1978). As a consequence shrubs and young trees are not destroyed in grassland ecosystems and the succession of grassland into shrubland is enhanced (Harper, 1969). Very recently the outbreak of foot and mouth disease and BSE has strongly reduced the cattle herds and hampered grazing due to sanity regulations, thus changing grassland use for one season with obviously no impact on plant populations except that the flowering and fruiting of some species such as *Cardamine pratensis* was very obvious, at least in the Netherlands. Not only animal diseases, but also plant diseases can have a strong impact on ecosystems. The breakdown of the *Ulmus* populations by the elm disease (*Ophiostoma ulmi*) around 3000 BC (Digerfeldt, 1997) has opened the forests in Europe and gave new options for other tree species, especially beech (*Fagus sylvatica*).

Chestnut blight caused by the parasitic fungus *Cryptonectria (Endothia) parasitica* has almost eliminated *Castanea dentata*, a dominant tree in large areas of North America, and favoured the growth of the co-occurring *Quercus prinus* (Ruffner and Abrams, 1998). The same parasite had less strong effects on *Castanea sativa* in Caucasus forests (Pridnya et al., 1996), perhaps due to long-term co-adaptation in its area of origin. The relatively recently detected involvement of mycoviruses in the effectiveness of such pathogens (Brasier, 1990) is a warning against oversimplification of data interpretation and opens new areas in bioindication.

6. Human industrial activities, resulting in acid precipitation had strong effects on biota, already recognized in 1850 near silver smelters in Saxony (Stöckhardt, 1850) and more thirty years later in the vicinity of Zn smelters in the Harz mountains (von Schroeder and Reuss, 1883). More than hundred years later, once more in a period of high SO_2 emission, governments started to act by either demanding the switch from coal to natural gas for electricity generation (1968 in the Netherlands) or the precipitating sulphur dioxide as calcium sulphate, but still nowadays many countries have still to act (ECE, 1987; RIVM, 1996) despite the impact on human health and the large-scale damage to forests and aquatic ecosystems (D'Itri, 1982).

5. Visible morphological and physiological changes as bioindicators of general metabolic disturbance

Visible injury of plants are the result of insufficient adaptation of a plant to changing environmental conditions, either by natural processes such as the eruption of volcanoes, salt spray or flooding, or by man-made processes, mostly caused by acute or chronic exposure to changes in concentration of natural environmental components or to the release of new chemical compounds.

Visible plant damage by changes of leaf colour (chlorosis, necrosis) and changes of the growth performance was first recognized by Stöckhardt (1850) in the Harz mountains. He related negative plant responses to high sulphur dioxide emissions.

Sorauer (1911) exposed sensitive plants of *Phaseolus vulgaris* in areas with adverse environmental conditions and evaluated later the damage at the macroscopic and microscopic level, followed by chemical analysis, thus a very integrated approach of bioindication. From the 1930s onwards visible symptoms of a disturbed plant metabolism are highly appreciated as diagnostic criteria for nutrient deficiencies in agricultural crops (Chapman, 1965; Bergmann, 1983). Timely bioindication and biomonitoring have used these symptoms in evaluation of forest health and of the disturbance of environmental processes.

5.1. Premature leaf losses

A severe impact on the metabolism of a plant will finally affect the longevity of plant or its organs. A quite common reaction pattern to hazardeous environmental conditions is the premature loss of plant leaves. Lack of needle classes in coniferous trees is a well-known example of the use of a plant organ as bioindicator for the evaluation of the health of coniferous forests (Langeweg, 1988). The factors causing leaf losses, however, may be quite different, ranging from exposure to air pollutants, e.g. SO_2 and cement-kiln dust (Lerman and Darley, 1975), and increasing soil acidity, to biota such as injury to ectomycorrhizal fungi, attact by parasitic fungi, e.g. *Chrysomyxa* (Crane and Hiratsuka, 2000) and *Lophodermium* species on coniferous trees (Muller and Hallaksela, 1998), and by caterpillars, e.g. *Thaumetopoea processiones* on oaks, the latter also with strong effects on human health (Hesler et al., 1999). Care, however, have to be taken that natural processes are not confounded with anthropogenic impacts. The low number of annual year classes on Scots pine in the Netherlands was first taken for the impact of acid rain, but part of the damage was caused by *Lophodermium pinastri* on accessions from Poland and planted in the 1930s in the country due to shortage of indigenous proveniances.

5.2. Malformation of plant organs

Plants which are insufficiently or not at all adapted to changing environmental conditions can express disturbance of the metabolism by changes in their morphology, often visible as dwarfed life forms (dwarfism, nanism) or gigantism of leaves. Growth performance of plants is very responsive to a lot of environmental factors so that the reason for anormalities has to be stated by (physico)chemical analysis, except if they are caused by animals, e.g. galls.

5.2.1. Malformation caused by radioactivity

Response to increased natural or industrial radioactivity (radiomorphosis, Savchenko, 1995) is indicated by anomalies of plant growth or by irregular formation of leaves and flowers. On soils naturally enriched by radionuclides (carnorite and monazite) in Brazil, India, Poland, Russia and the United States of America (Penna-Franka et al., 1965; Nair, 1961; Sarosiek and Leonowicz-Babiakowa, 1970; Kovalskii et al., 1967; Osborn, 1961). Dwarfism of shoots and misformation of flowers are frequent responses to enhanced doses of radionuclides. After the Chernobyl disaster the emitted radionuclides have caused much morphological deviation in leaves of oak and pine (Savchenko, 1995).

5.2.2. Malformation by a surplus of heavy metals.

On soils over lead-enriched outcrops *Papaver macrostomum* shows misformation of the petals (Maljuga et al., 1959) whereas dwarfism of plants and needles is frequently occurring in coniferous trees at exposure to increased levels of heavy metals, such as Zn (Ernst, 1985) and Ni (Kozlov and Niemela, 1999). Deficiency of zinc is also visible in dicots by stunted growth of the shoot, by shortening the internodes ('rosetting') and by a strong decrease in leaf size ('mottle leaves') (for a review: Baumeister and Ernst, 1978). Asymmetry of Scots pine needles is another indicator of the impact of metal contamination (Kozlov and Niemela, 1999).

5.2.3. Malformation by phosphorus deficiency.

Phosphorus deficiency may be expressed by a reduction of leaf surface areas in crop plants (Fredeen et al., 1989) and in wild plants. The needle length of Scots pine growing in the border line of fens is often half of that of healthy ones due to phosphate fixation by iron. In contrast a high degree of mycorrhizal colonization stimulates needle length (Timonen et al., 1997).

All malformations are good indicators of inadequate environmental conditions, but they demand a (bio)chemical analysis to identify the kind and the quantity of the stressor.

5.3. Modification of tree-ring width and wood biomass

Radial growth responses of trees have received a lot of attention in reconstructing palaeoclimatology (Fritts, 1976). It can also be used in the evaluation of the impact of other environmental factors. It is long known that biological processes can diminish the annual increment of ring width. An internal change of the allocation of carbon hydrates from wood to fruit production in high fecundity (mast) years of beech and oak strongly affected ring width (Rohmeder, 1967). A same effect can be caused by a mass development of defoliating insects diminishing the photosynthetic leaf areas (Varley and Gradwell, 1962).

Abiotic factors can have a negative or a positive impact on wood diameter growth. A growing season with low precipitation diminishes the annual increment (Fritts, 1976), but high precipitation stimulates it (Kozlowski, 1971). Similar contrasting responses can be evoked by high and low temperatures. The impact of these natural sources of variation can be accentuated by anthropogenic factors: Diminished ring width results from long-term exposure to a surplus of sulphur dioxide (Lux, 1965; Pollanschütz, 1971; Grill et al., 1979) and heavy metals (Carlson and Bazzaz, 1977). Recently, the increase of ring width in *Pinus cembra* is related to enhanced atmospheric CO_2-concentration (Nicolussi et al., 1995).

In the vicinity of a Cu and Ni smelter in Finland there was a sharp gradient in the annual increment of stem volume of Scots pine as a result of a combined effect of SO_2, Cu, Ni and Zn emission with only 7.7% at 0.5 km, 45.3% at 4 km compared to 100% at 5 km distance. The impact of these emissions on the height growth of the trees was less, with 39% and 10% reduction at 0.5 and 4 km distance from the emission source, respectively (Mälkönen et al., 1999).

Due to the many plant-internal and external factors which modify the physiological processes involved in radial growth of trees, however, it will be difficult to demonstrate a causality originating from one single environmental factor.

Table 5.   Chlorosis and necrosis in leaves of higher plants caused by deficiency or toxicity of chemical elements.

Leaf age	Type of disturbance	Prevailing symptom	Element
			Deficiency
Mature	Chlorosis	Uniform	N, S
		Interveinal or blotche	Mg, Mn
	Necrosis	Tip and marginal scorch	K
		Interveinal	Mg, Mn
Young	Chlorosis	Uniform	Fe, S
		Interveinal or blotched	Zn, Mn
	Necrosis		B,Ca, Cu
			Toxicity
Mature	Chlorosis	Uniform	Cd, Zn
	Necrosis	Marginal scorch	Cd,NaCl, Zn
Young	Chlorosis	Spots	Mn, B
		Uniform	Zn

Modified after Marschner (1995).

5.4. Chlorosis and necrosis

Disturbance of the chlorophyll synthesis often results in pale green to yellow leaves (lack of chlorophyll synthesis, chlorosis) or brown leaves (breakdown of chloroplast pigments, necrosis). Long-known is the lime chlorosis of plants suffering from a low iron supply on calcareous soils, not only in crops (Bergmann, 1983; Chen and Hadar, 1991), but also in wild plants (Hutchinson, 1968). A deficiency of other nutritional elements, e.g. boron, magnesium, nitrogen, sulphur and zinc causes also chlorosis (Marschner, 1995), often specifically expressed in leaves (Table 5). If the amount of heavy metals taken up by (even highly adapted) plants is surpassing the physiological regulation, chlorosis will also appear (Ernst, 1999a). Recently chlorosis in *Betula platyphylla* var. *japonica* is proposed as a bioindicator of soil acidification in Japan (Kitao et al., 2001).

Changes in the concentration of components of the ambient air can also cause chlorosis, as shown for *Picea abies* and *Pinus sylvestris* after exposure to enhanced concentration of volatile hydrocarbons (Schröder, 1998) or ozone (Utriainen and Holopainen, 1998). Selected sensitive cultivars, ecotypes or varieties of plants may develop specific types of chlorosis (spickled, homogeneous, along leaf veins) and necrosis (leaf margin, leaf tips). Spickled necrosis was caused by ozone in the tobacco cultivar Bel-W3 (Heggestad and Menser, 1962); this cultivar was later used in the Netherlands to analyse ozone damage on a national scale (Floor and Posthumus, 1977). Other cultivars with specific sensitivity to one air pollutant are the *Gladiolus* cultivars

Sneeuwprinses and the *Tulipa* variety Blue Parrot for fluorides, and *Trifolium pratense* cultivars for SO_2 (Floor and Posthumus, 1977).

The application of herbicides is another cause of chlorosis. In the target plant, protoporphyrinogen oxidase, the final enzyme in the tetrapyrrole biosynthesis pathway before it branches to chlorophyll or haem, is blocked by "bleaching" herbicides, e.g. acifluorfen (Devine and Preston, 2000) and benzoylisoxazole (Pallett, 2000) so that insufficient chlorophyll is synthesized.

Deficiency and surplus of chemical elements, but also exposure to enhanced levels of HCl, HF, and SO_2 and salt spray, can affect the plants in such a manner that parts of the plants or/and some plant organs die-off resulting in necrosis (Table 5).

Due to the various environmental factors causing chlorosis and necrosis, a chemical analysis of the affected plants is necessary to identify the reason for chlorosis and necrosis to avoid misinterpretation. On coastal dunes in the vicinity of an industrial complex leaves of *Sambucus nigra* show many leaf necroses. Asking participants of excursions for the source for this injury, all were pointing to the industrial complex, but never to the North Sea because they did not realize that salt spray, i.e. nature, can also injure plants.

6. Bioindicators of chronic exposure to changes in environmental compounds and processes

Now the main question arises: Can higher plants indicate chronic changes of environmental compounds and processes and help to elucidate the reasons for the losses indicated by Red list qualifications. The forest dye-off in many industrial areas and by long-distance transport of air pollutants also in remote areas of the world are the result of long-term (chronic) exposure to pollutants (cf. Smith, 1981). Some of the components of chronic exposure can be identified by a thorough analysis of several of the affected plant species.

6.1. Changes in air quality and radiation intensity

During the past 50 years human acitivities have changed the radiation intensity, especially in the UV-B range (Rozema et al., 1997), the chemical composition of the atmosphere and the chemistry of soils and waters by aerial fall-out. As a consequence, the genetics and physiology of organisms and the composition of biological communities were modified by the exposure to these changes, ranging in geographic scale from local to worldwide exposure. The greatest change in atmospheric chemistry has taken place worldwide with regard to carbon dioxide. In Europe other pollutants were formerly more important, such as sulphur dioxide and locally hydrogen sulphide. More recently the air quality was modified by an increase in traffic-based emissions of nitrogen oxides and by cattle-based emissions of ammonia. In addition to these changes in air chemistry, the decrease of stratospheric ozone has enhanced the penetration of more UV-B radiation to the earth surface. Can these changes be indicated by higher plants?

*6.1.1. Stomata density is not a reliable indicator of a change of atmospheric
CO_2 levels?*

From the last glacial period onwards the atmospheric CO_2 concentration has nearly
doubled with an enhanced increase during the past decades. But how important is this
increase if it is compared to the evolutionary time of angiosperms and to the essentially shorter time of anthropogenic impacts? CO_2 concentrations have shown a high
cyclic variation over geological time periods so that plant species may have an evolutionary memory. In historic times high CO_2 exposure has certainly occurred in the
vicinity of human settlements already during the Middle Ages; due a clean air act in
London in the 13th century high industrial activity was forbidden during the assem-
blation of the lords.

Therefore it is conceivable that many plant species have had already a long-term
history of enhanced CO_2 exposure, prior to the industrial revolution.

Woodward (1987) has proposed that an increase in atmospheric CO_2 concentration
will diminish stomata density and improve water economy by comparing the change
in stomata density in leaves from herbarium specimens and with those from the actual
vegetation. Palaeobotanists have taken this information for granted and try to use stomatal density as a proxy for palaeo-CO_2 levels (Royer, 2001) without considering the
relevant physiological and ecological processes affecting this morphological parameter. However, leaf material from herbaria is strongly biased by unknown sampling
procedures (Nicolussi et al., 1995) and information on the environment of the specimen
sampled.

The best relationship should be expected in sites with a natural elevated CO_2
concentration as in the vicinity of CO_2 vents with CO_2 concentration up to 2200 ppmv.
However, all investigations have shown that stomatal density varied with species
and/or season (Jones et al., 1995; Tognetti et al., 2000) or remained indifferent
(Miglietta and Raschi, 1993). A similar species-specific response was found in an
altitudinal analysis of coniferous species in the Rocky Mountains (Hultine and
Marshall, 2000).

Biotic and abiotic parameters may modify stomatal density in different directions:
In *Fagus sylvatica* sun leaves have higher stomatal densities when compared to shade
leaves (Lockheart et al. 1998). A 2.5-fold interleaf variation of stomatal density by
was present in *Alnus glutinosa* (Poole et al., 2000). Increasing number of chromosome
sets (polyploidy) let decrease stomatal density in cultivars of sugar beet (Bogaert
and Lemeur, 1994); if this phenomenon will be confirmed than strongly hybridizing
species with different degrees of polyploidy such as *Betula pendula* and *B. pubescens*
will never deliver conclusive results. Population differences were described for *Bromus
erectus* populations on calcareous grassland (Lauber and Körner, 1997), drought-
tolerant populations of *Pinus ponderosa* which have a lower stomatal density than
drought sensitive ones (Cregg, 1994) in contrast to a drought-tolerant variety of
Sorghum (Tsay et al., 1994). *Raphanus raphanistrum* has a heritable variation for
stomatal density among populations, which let conclude Case et al. (1998) that selection due to elevated $[CO_2]$ is not likely to act on these traits; their argument is based
on no detectable effect on lifetime fecundity. Another aspect of CO_2 exposure has to
consider the natural gradient of $[CO_2]$ in the environment. Small plants make profit

from the release of CO_2 from soil respiration, on a mean up to 3000 ppm CO_2. Most experiments, however, did not consider the natural exposure of plant species to this essential higher $[CO_2]$ in nature (Ernst, 1993c, 1998). Therefore small plants will suffer CO_2 deficiency when they are grown in a greenhouse lacking such additional soil CO_2. Each fumigation up to the above-mentioned CO_2 levels and the positive responses is therefore nothing more than the restoration of the natural conditions for these forms. Tall plant species, especially trees, experience a developmental change in the exposure to $[CO_2]$ during their development from seedling to mature plant. During their often long-lasting seedling phase they are exposed to high $[CO_2]$ derived from the soil due to decomposition of organic matter and the activity of soil organisms. Therefore it is not surprising that stomata density did not change in tree saplings of *Alnus glutinosa* (Poole et al., 2000) and *Quercus myrtifolia* (Lodge et al., 2001). When plants have passed the seedling and sapling stage they can no longer profit from the elevated $[CO_2]$ and will highly appreciate any additional CO_2 supply. To cope with developmental changes in the CO_2 exposure demands high phenotypic flexibility.

A lot of abiotic conditions affect stomatal density such as drought, UV-B radiation, a surplus of heavy metals in non-metal resistant plants and potassium supply decrease stomatal density (Losch et al., 1992; Elias, 1995; Paakkonen et al., 1998; Baryla et al., 2001; Kostina et al., 2001).

An interaction between $[CO_2]$ and temperature was used to calibrate the geographical distribution of C_3 and C_4 grasses and proposed to use their frequency as an indicator for the impact of enhance CO_2. C_4 grasses should have a competitive advantage under low CO_2 concentration even at cooler temperature, thus they should be dominant during the last glacial maximum (Ehleringer et al., 1997; Collatz et al., 1998). However, the evaluated data base excluded many cool-temperate and all tundra areas of the world. In addition the authors have disregarded the above mentioned high CO_2 levels near the soil surface and the precipitation (pluvial periods) during that time in the tropics. Furthermore, the ecological reality falsifies their hypothesis: wild C_4 grasses such as *Spartina* species at the coast and *Eragrostis* species on the European continent are just expanding over the past century; the agricultural succes of *Zea mays*, a well-known C_4 crop, is still increasing in cool-temperate areas of the world.

Another approach, i.e. experiments with subambient $[CO_2]$ could also not deliver consistent results (Malone et al., 1993).

Therefore a lot of long-term experiments have to be performed to establish a strong relation of stomatal density to increased $[CO_2]$, starting from the seedling in areas with high and low soil respiration, following the stomata density in the sapling by exposure to ambient and enhanced $[CO_2]$ and during its further development to a mature plant, by considering also the differences between sun and shade leaves, insertion and age of leaves, the degree of polyploidy and many abiotic factors. As Morison (2001) concluded from a survey of European forest species exposed for long-term to enhanced $[CO_2]$ (Medlyn et al., 2001) the present approaches and especially the present models are insufficient to support the hypothesis of a change of stomata density as response to high CO_2 levels.

6.1.2. *The response of plant species to the sulphur status of the environment*

The negative effects of SO_2 emissions are the most investigated aerial pollutants from the 1850s onwards on coniferous trees (Stöckhardt, 1850). When the concentrations of air pollutants pass a critical level, most plant species suffer visible injury or disappear which is best documented for lichens. During the period of high sulphur emission (1950–1968; Stuyfzand, 1993) many epiphytic lichens disappeared in the industrial areas of Europe (cf. Wotterbeek et al., 2002) and after the reduction of the SO_2 emission in Western Europe they are now strongly recovering, thus being good indicators of certain components of air quality.

How have higher plants reacted to such an SO_2 exposure? Many plant species were highly injured. The bleeching bark of Scots pine, the loss of needles, the above mentioned chlorosis and necrosis of leaves of many plant species in the vicinity of SO_2 emission sources were good visible symptoms of the affected metabolism (Mudd, 1975). But there was also another reaction pattern. The increase of atmospheric SO_2 has consequently enhanced the deposition of SO_4 and increased the sulphur concentration of soils. Because sulphur belongs to the major plant nutrients, positive effects may be expected in higher plants. Analysis of the floristic database in the Netherlands and Germany (Ernst, 1993b) has shown that indeed many plant species belonging to the family Brassicaceae have extended their range on a national scale in both countries (Haeupler and Schönfelder, 1988; Van der Meijden et al., 1989). Ernst (1993b) speculated that the expansion of species belonging to this plant family may be related to the enhanced sulphur demand, partly due to the synthesis of sulphur-demanding glucosinolates. By testing this hypothesis, it was obvious that *Arabidopsis thaliana*, one of the strongly expanding cruciferous species, proved to be not highly sulfur-demanding, but resistent to high SO_2 concentration (Van der Kooij et al., 1997). Another expanding Brassicaceae in the Netherlands and Germany, *Brassica nigra*, showed also an increased sulfur resistance (Ernst, 2000).

After the strong decrease of sulphur emission in Western Europe, from 1968 onwards, the sulphur supply to soils has dropped down to levels of the 1930s (Somhorst and Stolk, 1996; Zhao et al., 1997). As a consequence, the incidence of sulphur deficiency has increased in agricultural crops such as rape seed (*Brassica napus*) in various Western European countries due to a lack of aerial fallout of sulphur compounds and due to insufficient application of S-fertilizer which can compensate for the loss by harvests (Fieldsend and Milford 1994). S-deficiency in the yellow-flowering *B. napus* can be easily detected by a diminished concentration of flavonoids resulting in pale-yellow petals (Schnug and Haneklaus, 1994), whereas no obviously faintly coloured flowers were observed in wild Brassicaceae, obviously due to the high sulphur conservation in natural ecosystems (Ernst, 2000). The experiment with *B. nigra* supports the hypothesis that the sulphur demand of wild Brassicaceae is still sufficiently satisfied by the present sulphur deposition so that change in flower colour is not a good indicator of sulphur deficiency in the wild plant species belonging to the Brassicaceae.

Selective forces have enabled a third reaction pattern of plants to chronic SO_2 esxposure. Species of higher plants which can survive at SO_2-polluted sites will integrate the exposure to air pollutants over long periods. During the period of high SO_2 emission in the 1960s, several plant species highly exposed to SO_2 evolved SO_2-resistant populations. Examples are populations of *Geranium carolinianum, Lepidium virginicum, Lolium perenne*, and *Silene vulgaris* (Bell and Mudd, 1976; Taylor and Murdy, 1978; Murdy 1979; Ernst et al., 1985; Bell et al., 1991 for a review). The inhomogeneity of exposure concentration and exposure time is obviously the reason that resistance to air pollutants has less frequently evolved than that to heavy metals (Ernst, 1998b).

To enable or improve plant growth in air-polluted areas, plants from naturally or anthropogenic polluted areas were collected and/or further improved by breeding (Mejnartowicz, 1984). SO_2 resistant populations of *Pinus pumila* from volcanoes in Japan have been transfered in the early 1970s to the SO_2-polluted areas at Sudbury (Canada) and performed well whereas indigenous conifer species died. Planting resistant cultivars, however, will hamper the indication of a polluted environment.

6.1.3. *Exposure to enhanced levels of ammonia and nitrogen dioxide*

Concentrations of NH_4 are increasing in agricultural areas. On a local and regional scale a lot of ammonia is emitted into the atmosphere by the cattle-rearing industry in Belgium, Denmark, the Netherlands and Northern Germany. Certain nitrophilous lichen species, e.g. *Xanthoria parietina*, have shown a positive response to enhanced NH_4 concentrations (Van Herk, 1999). Higher plants do not only respond to the nitrogen status of the soil (Marschner, 1995), but also to ammonia-containing air. The latter can have serious impacts on individual plants, resulting in injured trees of *Pinus sylvestris* (Van der Eerden, 1992), and on oligotrophic vegetation by stimulating the growth of nitrogen-responsive species (see Section 4). The populations of the annual herb *Ceratocapnos (Corydalis) claviculata* has strongly increased its occurrence and frequency in many forest types in the vicinity of cattle-rearing areas in France, Germany and the Netherlands. Nowadays it is a very common understorey species in broad-leaved and coniferous forests and a good bioindicator of areas with high ammonia deposition (Schmidt, 1999; Decocq, 2000; Lethmate and Wendeler, 2000).

Enhanced concentrations of NO_x are related to the exhaust of running vehicles. Nitrogen dioxide at high concentrations may injure plants and select NO_2 resistant populations, as shown for *Lolium perenne* growing in the vicinity of a nitrogen fertilizer factory (Taylor and Bell, 1988). In the same study, it was shown that another perennial grass, *Dactylis glomerata,* has not this evolutionary ability. Up to now, a good bioindicator for NO_2-contamination has not (yet) been found perhaps due to the rapid turnover in the soil to NO_3.

6.1.4. *Can increased UV-B levels be monitored by plant responses?*

During the last decades the stratospheric ozone layer is decreasing in thickness, due to increased emission of reactive anthropogenic organic compounds such as chlorinated fluorocarbons (CFC's) which break down ozone (Herman et al., 1996). Thus anthropogenic pollution elevates the natural UV-B dose to which plants are exposed. High UV-B fluxes can damage biomembranes, the photosynthetic apparatus, proteins, and DNA by formation of DNA dimers (Jansen et al., 1998). Due to the evolution of adaptation mechanisms to prevent UV-B damage by enhanced DNA repair and synthesis of UV-B scavenging secondary metabolites (Bornman et al., 1997), the responses to enhance UV-B doses vary from nearly neutral (Van de Staaij et al., 1995) to injury in UV-B sensitive cultivars of cultivated plants (Teramura et al., 1990, 1991; Ziska et al., 1993). At present it will be difficult to monitor effects of enhanced UV-B radiation with higher plants in nature.

6.1.5. Element concentration for bioindication of changes in air quality

Only a few chemical elements are directly taken up via the leaves i.e. C as CO, CO_2, CH_4, volatile organic carbons (VOCs) such as PCBs (Buckley 1982), F as HF (Dässler, 1976), N as NH_4, NO and NO_2 (Wellburn, 1990), O as O_2, O_3 and peroxyacetylnitrate (Nouchi et al., 1984), and S as SO_2, H_2S and COS (Taylor et al., 1983). Many other chemical elements can be adsorbed to the leaf surface and can be analysed in unwashed plant material. If there is a pollution gradient the element concentration in the analysed plant part can help to identify and localize the emission source. However, the co-occurring contamination of the soil by aerial fall-out, the uptake of the emitted and deposited element by roots and translocation to the shoot will then be a combined indication of air and soil contamination load at the site. The identified contamination gradient can vary from a few metres as in plants and soils under copper high tension lines (Kraal and Ernst, 1976) up to more than ten kilometres as caused by metal smelters in the Kola Peninsula (Barcan et al., 1998).

6.2. Changes in soil quality

Diminishing the water table (desiccation), fertilization (eutrophication) and the deposition of chemicals (acidification, contamination) will change the environmental quality of a soil and consequently change the chemical composition of plants, the species composition and diversity of the vegetation or the genotypes in a population.

6.2.1. Changes in chemical composition of plants

Whereas air pollutants will pass along the plant and one part of it will evoke direct or indirect responses (Keller and Schwager, 1971), another part of air pollutants will be deposited on the soil, thus increasing chronic exposure, and interact with the soil and its organisms and afterwards with the higher plants. The high persistence of many chemical elements in the soils can be analysed even if the original contamination source has ceased to cause new contamination. After more than 5000 years smelting sites of heavy metals in the Bronze Age can still be identified by enhanced metal concentrations in plants (Repp, 1963), often together with changes in vegetation composition (Ernst and Nelissen, 2000). Prior to identify a contamination, a careful chemical background analysis of the plant species under investigation is necessary due to the very species-specific uptake and accumulation in leaves and other plant parts (Table 6). Biological enrichment of an element by its natural accumulation may be mistaken for environmental contamination. One of the pitfalls is still the natural high concentration of manganese in *Fagus sylvatica* and the high zinc concentration in *Betula, Populus* and *Salix*- species (cf. Table 6; Denaeyer-De Smet, 1970; Baumeister and Ernst, 1978; Ernst, 1984).

Table 6.   Mean heavy metal concentration in leaves of trees and shrubs growing in a sand dune in North Holland Dune Reserve. Highest value of each metal in all leaves is italicised.

Plant species	Concentration (μmol g^{-1} dry mass)			
	Fe	Mn	Zn	Cu
Acer pseudoplatanus	9.10	1.66	0.62	0.10
Betula pendula	2.23	1.90	3.80	0.14
Crataegus monogyna	16.70	1.97	2.48	*0.17*
Euonymus europaeus	15.05	1.64	0.47	0.08
Ligustrum vulgare	3.65	0.96	1.04	0.11
Pinus nigra	*29.80*	*2.97*	0.46	0.05
Populus alba	13.70	1.95	*4.28*	0.17
Populus nigra	3.47	0.59	1.75	0.06
Quercus robur, mature	9.74	1.22	0.52	0.12
Quercus robur, flushing	4.21	0.31	0.45	0.13
Salix repens	10.40	0.78	2.18	0.09
Sambucus nigra	8.40	0.97	1.04	0.11

Sampling date: 8 July 1994.

In addition to species-specific element concentration chemical analysis as a tool in bioindication has to consider plant age, plant parts, and species characters (see Ernst, 1990) and ecotypic differentiation (see 6.2.1).

6.2.2. Selection of ecotypes resistant to heavy metals

Long-term exposure to soil contaminants has a strong impact on the genetic composition of plant populations with many evidences for heavy metal contamination. As mentioned above, wide-spread plant species are differentiated in local and regional ecotypes with mostly a multitude of genotypes in the population. If the environmental conditions are changing and the selection pressure is high, genotypes with often marginal presence in the population can get their chance. Then selection in favour of the resistant ecotype takes place within a couple of years or decades finally resulting in ecotypes with a high resistance to the orginally adverse soil factor (Fig. 1).

In the vicinity of metal-processing industries, established far away from mineralized soils, the emission and deposition of heavy metals can exceed such levels that most plant species of the former clean sites cannot survive. Among members of the local population of wild plants metal-resistant genotypes are selected, especially in grasses, as demonstrated for species of *Agrostis* (Bradshaw, 1976; Ernst, 1976; Dueck et al., 1984; De Koe and Jaques, 1993; Archambault and Winterhalder, 1995) and *Festuca* (Brown and Brinkmann, 1992; Harrington et al., 1995) and other grass species such as *Agropyron repens* (Brej, 1998), *Deschampsia caespitosa* (Frenckell-Insam and Hutchinson, 1993), and *Holcus lanatus* (McNair and Cumbes, 1987). It may be surprising that this evolution in the vicinity of emission sources can occur between 5 and 40 years. In contrast to plant species on ore outcrops with a natural surplus of one or more metals, the fertility of the soils in the industrial area was originally high. After contamination by industrial metal fall-out, the selection process is restricted to one or a few metals although the resistance to each metal demands the involvement of one or two genes (Schat, 1999). All the resistant ecotypes can indicate the long-term exposure to metal contaminated environments and can be used for bioindication after laboratory tests, mostly implemented by the rooting test (Bradshaw, 1976; Schat and Ten Bookum, 1992). The roots and shoots of individuals adapted to the contaminated soil will grow happily, non-adapted individuals will fail.

6.3. Selection of ecotypes resistance to pesticides

Frequent application of the same herbicide to agricultural crops for a longer period let select within a decade herbicide-resistant weeds so that the efficiency of the herbicide is lost. Although it was assumed for long time that this evolution will not take place, there is now sufficient evidence that it has occurred and is still underway. Herbicide-resistant populations evolved after long-term application of triazine herbicides in many agricultural weeds, such as *Amaranthus retroflexus, Capsella bursa-pastoris, Chenopodim album, Poa annua, Senecio vulgaris,* and *Stellaria media* (for a review: Warwick, 1991). Triazine-resistant ecotypes of *Senecio vulgaris* had lower carbon assimilation and quantum yields than susceptible ones, finally resulting in diminished growth and productivity (McCloskey and Holt, 1990). Due to the high variability of these parameters in wild plant populations, a test of herbicide resistant can only be done in the laboratory. Recently the frequent application of glyphosate has forced the selection of glyphosate-resistant weeds, independently in *Lolium rigidum* in Australia (Powles et al., 1998) and *Eleusine indica* in Malaysia (Lee and Ngim, 2000) despite the recent statement by Baylis (2000) that it will not occur. Due to the long-persistent seedbank of many weeds, up to hundred years, the local weed population will consist of a mixture of ecotypes as shown for *Capsella bursa-pastoris*, a common weed in agricultural fields (Hurka and Neuffer, 1991). As a consequence, analysing the response of agricultural weeds to a certain environmental contaminant, will be biased on the long-term by a mixture of genotypes in the seedbank: seeds which were already present in the seedbank prior to the application of the contaminant (sensitive ones) and those which have evolved resistance to environmental contaminant(s) (resistant ones).

7. Test plant procedure in the laboratory for hazard assessment

For establishing the toxicity of compounds prior to their release into the environment, several toxicity tests have been developed for Regulatory Documents and Standard Test Procedures. The general principle of all these tests was the cost efficiency often hampering or strongly diminishing the biological quality, thus the relevance of the test. It is still surprising that in all of the plant tests proposed by ASTM (1994), OECD (1984), USEPA (1991) and USFDA (1987) the standardization of the plant material is far below an acceptable scientific level; the prescription gives only the species (cf. Kapustka, 1997) without defining their provenience, population, ecotype or variety. To enhance the reliability of all these tests will demand a good definition and selection of the provenience, cultivar, variety or ecotype of the species under consideration. Even then, the heterogeneity in a population will cause some minor variation in the outcome of the tests. The endpoints of the test, however, will strongly depend on the substrate. Artifical mixtures as those in the OECD (1984) procedure are not soils, but badly defined components of clay and peat without describing the conditions for an equilibrium between the test substance and the mixture, the pH of the mixture and the water holding capacity; there is no standardization of the plant-specific growth conditions such as quality and quantity of radiation, air humidity and air temperature, the addition of specific arbuscular mycorrhizal fungi and in the case of legumes as test plants *Rhizobium* bacteria.

7.1. Germination as endpoint of the test

Due to economic arguments, the first life phase of a plant, i.e. germination or seedling emergence is taken as endpoint of the test, not realizing that this life phase is characterized by very population- and species-specific responses. Analysing germination, will demand a very good definition of the endpoint. It should be the break-through of the radicula through the testa which does mean that this observation cannot be made in a soil as substrate. Another problem of the first life phase as endpoint is the impermeability of the seed coat to many chemicals and the precipitation or complexation of chemicals in the cell walls. The quantification of the endpoint is the percentage germination after a very short time, often no longer than 5 days and a 50% decrease of seed germination is used for the EC_{50} determination (USFDA, 1987; USEPA, 1985). Many cultivars of agricultural crops have indeed a very rapid germination, but it will be necessary that the maximum percentage of germination in the control has to be in the vicinity of 95%. As soon as plant species from arable fields, grasslands and other ecosystems are involved, the experiment has to be extended often for more than one year due to the genetically and environmentally determined long-term dormancy (Baskin and Baskin, 1998) to achieve a meaningful germination percentage. Germination of many plant species is governed by the radiation quantity and quality, especially the ratio of red/far-red and temperature. Therefore the procedure for germination test should clearly define all these abiotic test conditions.

7.2. Root elongation as endpoint of the test

As soon as the radicle is in contact with the environment, nutrients and other chemical compounds can be taken up and affect the metabolism. It depends on the seed reserve, how long the seedling will rely on its own sources and thus avoid the uptake of external material. Root elongation is often one of the endpoints of the emergence tests because it is long known that cell division and root elongation are often hampered at exposure to increased concentration of a chemical. There are many procedures for the rooting test. The more recent improvement is developed by Schat and Ten Bookum (1992) extending the experiment as long as root elongation does stop (EC_{100}) by dipping the root into a solution with active charcoal prior to the transfer to a sequential increase of the concentration. One pitfall of the root elongation test is the composition of the nutrient solution which has to be chosen as close as possible to the natural nutrient supply. Growth of roots only in the solution of the chemical under consideration will be disturbed by deficiency of major and minor nutrients and thus imbalancing the integrity of the biomembranes. A second pitfall of the root elongation test is the root morphology of a species. A plant species without a main (tap) root, but with a rapid development of side-roots, stops soon elongation of the main root to enhance the development of side-roots for a good exploration of the soil environments.

With the exception of plant species belonging to the family of Brassicaceae, Caryophyllaceae, Chenopodiaceae and Cyperaceae, the roots of most plant species are shortly after emergence infested by mycorrhizal fungi. These fungi can modify the toxicity of a substance; thus testing plant exposure to environmental compounds without addition of the appropriate mycorrhizal fungus to the substrate will give a result which is not very relevant for the response in a real environment (Joner et al., 2000). Although many root elongation experiments are carried out with lettuce (without defining the cultivar), none of them reports the involvement of a mycorrhizal fungus, thus the third pitfall.

7.3. Early seedling growth

The early seedling growth assay relies upon the development of the seedling into the C-autotrophic stage (photosynthesis). In this case it is necessary that the seedling is growing in either a nutrient solution well designed for the specific cultivar or in a soil which chemistry and water supply is relevant for the cultivar and species under consideration. The artificial mixture proposed in the OECD procedure (OECD, 1984) is not well defined (which type of peat? which pH?) and environmentally not relevant. Seedling growth does not only rely on the root environment, but depends strongly on radiation, air humidity, and temperature. So-called "room temperature" is an ambiguous term ranging from 15°C in cool temperate laboratories up to 30°C in tropical laboratories. Many plant species demand a day/night cycle of radiation and temperature for good growth performance. All these conditions have to be defined for a meaningful test procedure, but are lacking in all procedures.

7.4. Life cycle bioassay

All plant species have to produce seeds to ensure the survival of the population and the species. Therefore life cycle bioassays are the most biologically relevant procedures in testing toxicity of environmental components because the plants passes all stages, from the germination via the vegetative phase to reproduction. As shown for a Cd-Zn-resistant ecotype Plombières of the non-mycorrhizal herb *Silene vulgaris* (Fig. 2) the response to metal-enriched soils with different combination of Cd, Cu and Zn varied strongly between the various development stages of the plants (Ernst and Nelissen, 2000). The life cycle of plants vary from a month up to some hundreds of years. For economic reasons, life cycle bioassays are often restricted to short living plant species or those which can reproduce already in the first year after germination. There may be two endpoints of these assays, the total biomass and the amount of seeds. Reproduction of many plant species demands the transfer of pollen either from neighbouring flowers or from flowers of other individuals due to stigma incompatibility.

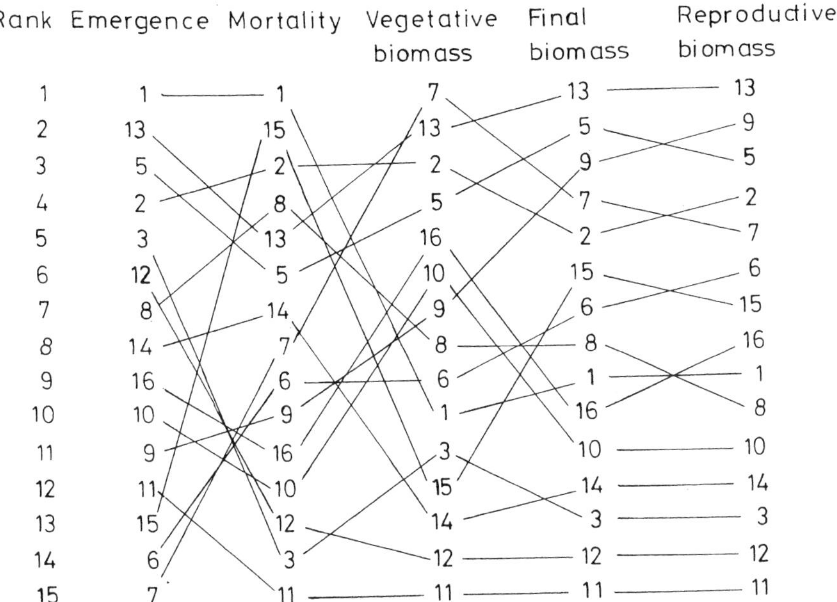

Figure 2. Ranking of the responses of the Zn-Cd resistant ecotype 'Plombières' of *Silene vulgaris* to the exposure on orogenic soils during various phases of the lifecycle. Rank 1 gives the highest mergence and the lowest mortality of the seedlings, and the highest biomass in all other phases of the lifecycle. The mean of three pots with the same soil is given per rank. A lot of changes of plant performance occur during one life cycle. On soil the moderately Zn and Cd-enriched soil 19 Langelsheim) emergences was the best and seedling mortality the lowest, but vegetative growth and the production of seeds was more hampered on this soil compared to soil 7 (Welfesholz, having only half of the Zn and one third of the Cd concentration of soil 1, but a three times higher Cu concentration), but finally the highest biomass production at the time of seed ripeness and seed harvest was realized on soil 13 (having half of the Cd and Zn concentration and the same Cu level as soil 7). The extremely Zn- and Cu-enriched soils of the Bronze Age smelting site near Langelsheim hampered the plants already early in their development and let not produce seeds. From Ernst and Nelissen (2000). Reproduced by permission of Elsevier Science Ltd.

Only self-compatible species such as the model plant of molecular botanists, *Arabidopsis thaliana*, are independent of pollen transfer by abiotic (wind, water) or biotic agents (insects, bats, birds). The selection of this model plant species has the advantage of a short lifecycle (Ratsch et al., 1986), but at the same time it is known that the various genotypes react quite differently to environmental factors, as demonstrated for the copper sensitivity of ten ecotypes (Murphy and Taiz, 1997) and the high resistance to sulphur dioxide (Van der Kooij et al., 1997) and hydrogen sulfide (Van der Kooij and De Kok, 1998). The latter adaptation has stimulated the expansion of this speicies during the period of high sulphur dioxide emission (Ernst, 1993b). Such a metabolic bias may diminish the reliability of this model plants for testing sulphur-containing contaminants.

In conclusion, plant species selected for life cycle bioassays has to consider the (eco)physiological background of the specific cultivar, ecotype or variety of that species and its association with mycorrhizal fungi and/or nitrogen-fixing bacteria and finally the quality of the produced seeds.

References

Aerts, R., Berendse, F., 1988. The effect of increased nutrient availability on vegetation dynamics in wet heathlands. Vegetatio 76, 63–69.

Aerts, R., Chapin III, F.S., 2000. The mineral nutrition of wild plants revisited: a re-evaluation of processes and patterns. Advances in Ecological Research 30, 2–67.

Allen, M.F., 1991. The Ecology of Mycorrhizae. Cambridge University Press, Cambridge.

Antosiewicz, D.M., 1995. The relationship between constitutional and inducible Pb-tolerance and tolerance to mineral deficits in *Biscutella laevigata* and *Silene inflata*. Environmental and Experimental Botany 35, 55–69.

Archambault, D.J., Winterhalder, K., 1995. Metal tolerance in *Agrostis scabra* from the Sudbury, Ontario, area. Canadian Journal of Botany 73, 766–775.

Assuncao, A.G.L., Da Costa Martins, P., De Folter, S., Vooijs, R., Schat, H., Aarts, M.G.M., 2001. Elevated expression of metal transporter genes in three accesssions of the metal hyperaccumulator *Thlaspi caerulescens*. Plant, Cell Environment 24, 217–226.

ASTM, 1994. Standard Practice for Conducting Early Seedling Growth Tests. IE 1598–94. American Society for Testing and Materials, Philadelphia.

Bakker, J.P., 1989. Nature Management by Grazing and Cutting. Kluwer, Dordrecht.

Barkan, V.Sh., Kovnatsky, E.F., Smetannikova, M.S., 1998. Absorption of heavy metals in wild berries and edible mushrooms in an area affected by smelter emission. Water, Air and Soil Pollution 103, 173–195.

Baryla, A., Carrier, P., Franck, F., Coulomb, C., Sahut, C., Havaux, M. 2001. Leaf chlorosis in oilseed rape plants (*Brassica napus*) grown on cadmium-polluted soil: causes and consequences for photosynthesis and growth. Planta 212, 696–709.

Baskin, C.C., Baskin, J.M., 1998. Seeds. Ecology, Biogeography, and Evolution of Dormancy and Germination. Academic Press, San Diego.

Baumeister, W., Ernst, W., 1978. Minerallstoffe und Pflanzenwachstum, 3rd edn. G. Fischer Verlag, Stuttgart.

Baylis, A.D., 2000. Why glyphosate is a global herbicide: strengths, weaknessess and prospects. Pest Management Science 56, 299–308.

Bekker, R., Ernst, W., De Vries, Y., 2001. Zaadvoorraad van duinvalleien. Bron of belemmering voor herstel? Landschap 18, 173–184.

Bekker, R.M., Bakker, J.P., Grandin, U., Kalamees, R., Milberg, P., Podschlod, P., Thompson, K., Willems, J.H., 1998. Seed size, shape and vvertical distribution in the soil: indicators for seed longvity. Functional Ecology 12, 834–842.

Bell, J.N.B., Ashmore, M.R., Wilson, G.B., 1991. Ecological genetics and chemical modifications of the atmosphere. In: Taylor, G.E., Pitelka, L.F., Clegg, M.T. (Eds), Ecological Genetics and Air Pollution. Springer Verlag, Berlin, pp. 33–59.

Bell, J.N.B., Mudd, C.H., 1976. Suphur dioxide resistance in plants: a case study of *Lolium perenne*. In: Mansfield, T.A. (Ed.), Effects of Air Pollutants on Plants. Cambridge University Press, Cambridge, pp. 87–103.

Berendse, F., Aerts, R., 1984. Competition between *Erica tetralix* L. and *Molinia caerulea* (L.) Moench as affected by the availability of nutrients. Oecologia Plantarum 5, 3–14.

Bergmann, W., 1983. Ernährungsstörungen bei Kulturpflanzen. Entstehung und Diagnose. G. Fischer Verlag, Stuttgart.

Beyschlag, W., Ryel, R.J., Ullmann, I., 1992. Experimental and modelling studies of competition for light in roadside grasses. Botanica Acta 105, 285–291.

Björkman, O., 1968. Further studies on differentiation of photosynthetic properties in sun and shade ecotypes of *Solidago virgaurea*. Physiologia Plantarum 21, 84–99.

Blab, J., Nowak, E., Trautmann, W., Sukopp, H., 1984. Rote Liste der gefährdeten Tiere und Pflanzen in der Bundesrepublik Deutschland. Kilda Verlag, Greven.

Blair, J.M., 1988. Nitrogen, sulfur and phosphorus dynamics in decomposing deciduous leaf litter in the southern Apalachians. Soil Biology and Chemistry 20, 693–701.

Bogaert, G., Lemeur, R., 1994. Ontogenetic effects on the stomatal apparatus of 4 Belgian sugar-beet (*Beta vulgaris* subsp. *vulgaris*) cultivars. Photosynthetica 30, 333–339.

Bornman, J.F., Reuber, S., Cen, Y.P., Weissenböck, G., 1997. Ultraviolet radiation as a stress factor and the role of protective pigments. In: Lumsden, P.L. (Ed.), Plants and UV-B: Responses to Environmental Change. Cambridge University Press, Cambridge, pp. 157–168.

Bradshaw, A.D., 1976. Pollution and evolution. In: Mansfield, T.A. (Ed.), Effects of Air Pollutants on Plants. Cambridge University Press, Cambridge, pp. 135–159.

Brasier, C.M., 1990. The unexpected element: mycovirus involvement in the outcome of two recent pandemics, Dutch elm disease and chestnut blight. In: Burdon, J.J., Leather, S.R. (Eds), Pests, Pathogen and Plant Communities. Blackwell Scientific, Oxford, pp. 289–307.

Brej, T., 1998. Heavy metal tolerance in *Agropyron repens* (L.) Beauv. populations from the Legnica copper smelter area, Lower Silesia. Acta Societas Botanicorum Poloniae 67, 325–333.

Brown, G., Brinkmann, K., 1992. Heavy metal tolerance in *Festuca ovina* L. from contaminated sites in the Eifel Mountains, Germany. Plant and Soil 143, 239–247.

Buckley, E.H., 1982. Accumulation of airborne polychorinated biphenyls in foliage. Science 216, 520–522.

Campbell, I.D., McAndrews, J.H. 1993. Forest disequilibrium caused by rapid Little Ice Age cooling. Nature 366, 336–338.

Carlson, R.W., Bazzaz, F.A., 1977. Growth reduction in American sycamore (*Platanus occidentalis* L.) caused by Pb-Cd interaction. Environmental Pollution 12, 243–253.

Case, A.L., Curtis, P.S., Snow, A.A., 1998. Heritable variation in stomatal responses to elevated CO_2 in wild radish, *Raphanus raphanistrum* (Brassiaceae). American Journal of Botany 85, 253–258.

Chapman, H.D. (Ed.), 1965. Diagnostic criteria for plants and soils. Riverside.

Chen, Y., Hadar, Y. (Eds), 1991. Iron Nutrition and Interactions in Plants. Kluwer Academic, Dordrecht.

Collatz, G.J., Berry, J.A., Clark, J.S., 1998. Effects of climate and atmospheric CO_2 partial pressure on the global distribution of C_4 grases: present, past, and future. Oecologia 114, 441–454.

Crane, P.E., Hiratsuka, Y., 2000. Evidence for environmental determination of uredinia and telia production in *Chrysomyxa pirolata* (inland spruce cone rust). Canadian Journal of Botany 78, 660–667.

Cregg, B.M., 1994. Carbon allocation, gas-exchange, and needle morphology of *Pinus ponderosa* genotypes known to differ in growth and survival under imposed drought. Tree Physiology 14, 883–898.

Dässler, H.G., 1976. Einfluss von Luftverunreinigungen auf die Vegetation. G. Fischer Verlag, Jena.

De Bilde, J., Lefèbvre, C., 1990. The strategy of *Silene nutans* on calcareous and siliceous soils. Acta Oecologia 11, 399–408.

Decocq, G., 2000. A case-report of forest phytogeography: the presence of *Ceratocapnos claviculata* (L.) Liden in northern France. Acta Botanica Gallica 147, 143–150.

De Koe, T., Jaques, N.M.M., 1993. Arsenate tolerance in *Agrostis castellana* and *Agrostis delicatula*. Plant and Soil 151, 185–191.

Denaeyer-De Smet, S., 1970. Consideration sur l'accumulation du zinc par les plantes poussant sur sols calaminaires. Bulletin de l'Institute royale de Science naturelle de Belgique 46 (11), 1–13.

Devine, M.D., Preston, C., 2000. The molecular basis of herbicide resistance. In: Cobb, A.H., Kirkwood, R.C. (Eds), Herbicides and Their Mechanisms of Action. Sheffield Academic Press, Sheffield, pp. 72–104.

Digerfeldt, G., 1997. Reconstruction of Holocene lake-level changes in Lake Kalvsjon, southern Sweden, with contribution to the local palaeohydrology at the Elm Decline. Vegetation History and Archaeobotany 6, 9–14.

D'Itri, F.M. (Ed.), 1982. Acid Precipitation. Effects on Ecological Systems. Ann Arbor Science, Ann Arbor.

Dueck, T.A., Ernst, W.H.O., Faber, J., Pasman, F., 1984. Heavy metal immission and genetic constitution of plant populations in the vicinity of two metal emission sources. Angewandte Botanik 58, 47–59.

ECE, 1987. National Strategies and Policies for Air Pollution Abatement. Convention on Long-Range Transboundery Air Pollution. United Nations, New York. ECE/EB.AIR/14.

Ehleringer, J.R., Cerling, T.E., Helliker, B.R., 1997. C_4 photosynthesis, atmospheric CO_2, and climate. Oecologia 112, 285–299.

Elias, P., 1995. Stomata density and size of apple-trees growing under irrigated and non-irigated conditions. Biologia 50, 115–118.

Ellenberg, H., 1950. Landwirtschaftliche Pflanzensoziologie. I. Unkrautgemeinschaften. Ulmer, Stuttgart.

Ellenberg, H., 1974. Zeigerwerte der Gefässpflanzen Mitteleuropas. Scripta Geobotanica 9, 1–97.

Ellenberg, H., Weber, H.E., Düll, R., Wirth, V., Werner, W., Paulissen, D., 1991. Zeigerwerte von Pflanzen in Mitteleuropa. Scripta Geobotanica 18, 1–148.

Eriksson, A., Eriksson, O, 1997. Seedling recruitment in semi-natural pastures: the effects of disturbance, seed size, phenology and seed bank. Nordic Journal of Botany 17, 469–482.

Ernst, W., 1974. Schwermetallvegetation der Erde. G. Fischer, Stuttgart.

Ernst, W., 1976a. Physiological and biochemical aspects of metal tolerance. In: Mansfield, T.A. (Ed.), Effects of Air Pollutants on Plants. Cambridge University Press, Cambridge, pp. 115–133.

Ernst, W., 1976b. Ökologische Grenze zwischen Violetum calaminariae und Gentiano-Koelerietum. Berichte der Deutschen Botanischen Gesellschaft 89, 381–390.

Ernst, W., 1978. Discrepancy between ecological and physiological optima in plant species. A re-interpreation. Oecologia Plantarum 13, 175–188.

Ernst, W.H.O., 1981. Ecological implication of fruit variability in *Phleum arenarium* L., an annual dune grass. Flora 171, 387–398.

Ernst, W.H.O., 1983. Element nutrition of two contrasted dune annuals. Journal of Ecology 71, 197–209.

Ernst, W.H.O., 1984. Indicatoren van een overmaaat aan zware metalen in terrestrische ecosystemen. In: Best, E.P.H., Haeck, J. (Eds), Ecologische Indicatoren voor de Kwaliteitsbeoordeling van Lucht, Water, Bodem en Ecosystemen. Pudoc, Wageningen, pp. 109–120.

Ernst, W.H.O., 1985. Schwermetallemisionen – ökophysiologische und populations-genetische Aspekte. Geobotanische Colloquium Düsseldorf 2, 43–57.

Ernst, W.H.O., 1987. Impact of the aphid *Aulacorthum solani* Kltb. on growth and reproduction of winter and summer annual life forms of *Senecio sylvaticus*. Acta Oecologica, Oecologia Generalis 8, 537–547.

Ernst, W.H.O., 1990. Element allocation and (re)translocation in plants and its impact on representative sampling. In: Lieth, H., Markert, B. (Eds), Element Concentration Cadasters in Ecosystems. VCH Verlagsgesellschaft, Weinheim, pp. 17–40.

Ernst, W.H.O., 1993a. Geobotanical and biogeochemical prospecting for heavy metal deposits in Europe and Africa. In: Markert, B. (Ed.), Plants as Biomarkers. Indicators for Heavy Metals in the Terrestrial Environment. VCH, Weinheim, pp. 107–126.

Ernst, W.H.O., 1993b. Ecological aspects of sulfur in higher plants: the impact of SO_2 and the evolution of the biosynthesis of organic sulfur compounds on populations and ecosystems. In: De Kok, L.J., Stulen, I., Rennenberg, H., Brunold, C., Rauser, W.E. (Eds), Sulfur Nutrition and Assimilation in Higher Plants: Regulatory, Agricultural and Environmental Aspects. SPB Academic Publishing, The Hague, pp. 295–313.

Ernst, W.H.O., 1993c. Population dynamics, evolution and environment: adaptation to environmental stress. In: Fowden, L, Mansfield, T., Stoddart, J. (Eds), Plant Adaptation to Environmental Stress. Chapman & Hall, London, pp. 19–44.

Ernst, W.H.O., 1996. Schwermetalle. In: Brunold, C., Rüegsegger, A., Brändle, R. (Eds), Stress bei Pflanzen. Haupt-Verlag, Bern, pp. 191–219.

Ernst, W.H.O., 1998a. Effects of heavy metals in plants at the cellular and organismic level. In: Schüürmann, G., Markert, B. (Eds), Ecotoxicology. Ecological Fundamentals, Chemical Exposure, and Biological Effects. John Wiley, Heidelberg, pp. 587–620.

Ernst, W.H.O., 1998b. Ecotypic variation and environmental adaptation to air pollution and global change. In: De Kok, L.J., Stulen, I. (Eds), Responses of Plant Metabolism to Air Pollution and Global Change. Backhuys, Leiden, pp. 217–232.

Ernst, W.H.O., 1999a. Biomarkers in plants. In: Peakall, D.B., Walker, C.H., Migula, P. (Eds), Biomarkers: A Pragmatic Basis for Remediation of Severe Pollution in Eastern Europe. Kluwer Academic, Dordrecht, pp. 135–151.

Ernst, W.H.O., 1999b. Evolution of plants on soils anthropogenically contaminated by heavy metals. In: Van Raamsdonk, L.W.D., Den Nijs, J.C.M. (Eds), Plant Evolution in Man-Made Habitats. Hugo de Vries Laboratory, University of Amsterdam, Amsterdam, pp. 13–27.

Ernst, W.H.O., 2000. Expansion of *Brassica nigra* populations is not due to sulfur demand, but sulfur resistance. Landbauforschung Völkenrode Sonderheft 218, 31–33.

Ernst, W.H.O., Nelissen, H.J.M., 2000. Life-cycle phases of a zinc- and cadmium-resistant ecotype of *Silene vulgaris* in risk assessment of polymetallic mine soils. Environmental Pollution 107, 329–338.

Ernst, W.H.O., Peterson, P.J., 1994. The role of biomarkers in environmental assessment (4). Terrestrial plants. Ecotoxicology 3, 180–192.

Ernst, W.H.O., Tonneijck, A.E.C., Pasman, F.J.M., 1985. Ecotypic response of *Silene cucubalus* to air pollutants (SO_2, O_3). Journal of Plant Physiology 118, 439–450.

Ertsen, A.C.D., Alkemade, J.R.M., Wassen, M.J., 1998. Calibrating Ellenberg indicator values for moisture, acidity, nutrient availability and salinity in the Netherlands. Plant Ecology 135, 113–124.

Fenner, F., Myers, K., 1978. Myxoma virus and myxomatosis in retrospect: the first quarter century of a new disease. In: Kurstak, E., Maramorosch, K. (Eds), Viruses and Environment. Academic Press, New York, pp. 539–570.

Fieldsend, J., Milford, G.F.J., 1994. Changes in glucosinolates during crop development in single- and double-low genotypes of winter oilseed rape (*Brassica napus*). I. Production and distribution in vegetative tissues and developing pods during development and potential role in the recycling of sulphur within the crop. Annals of Applied Biology 124, 531–542.

Floor, H., Posthumus, A.C., 1977. Biologische Erfassung von Ozon- und PAN-Immissionen in den Niederlanden 1973, 1974 und 1976. VDI-Berichte 270, 183–190.

Fox, T.C., Guerinot, M.L., 1998. Molecular biology of cation transport in plants. Annual Review of Plant Physiology and Plant Molecular Biology 49, 669–696.

Fredeen, A.L., Rao, I.M., Terry, N., 1989. Influence of phosphorus nutrition on growth and carbon partitioning in *Glycine max*. Plant Physiology 89, 225–230.

Frenckell-Insam, B.A.K., Hutchinson, T.C., 1993. Occurrence of heavy metal tolerance in *Deschampsia caespitosa* (L.) Beauv. from European and Canadian populations. New Phytologist 125, 555–564.

Fritts, H.C., 1976. Tree Rings and Climate. Academic Press, New York.

Godzik, B., 1991. Accumulation of heavy metals in *Biscutella laevigata* (Cruciferae) as a function of their concentration in substrate. Polish Botanical Studies 2, 241–246.

Gossmann, H., 1989. Satelliten-Fernerkundung. Geografische Rundschau 41, 674–680.

Grabherr, G., Mucina, L. (Eds), 1993. Die Pflanzengesellschaften österreichs. G. Fischer Verlag, Jena.

Gregory, R.P.G., Bradshaw, A.D., 1965. Heavy metal tolerance in populations of *Agrostis tenuis* Sibth. and other grasses. New Phytologist 64, 131–143.

Gries, B., 1966. Zellphysiologische Untersuchungen über die Zinkresistenz bei Galmeiökotypen und Normalformen von *Silene cucubalus* Wib. Flora B156, 271–290.

Grill, D., Liegl, E., Windisch, E., 1979. Holzanatomische Untersuchungen an abgasbelasteten Bäumen. Phytopathologische Zeitschrift 94, 335–342.

Grootjans, A.P., Ernst, W.H.O., Stuyfzand, P.J., 1998. European dune slacks: strong interactions of biology, pedogenesis and hydrology. Trends in Ecology and Evolution 13, 96–100.

Haeupler H., Schönfelder P., 1988. Atlas der Farn- und Blütenpflanzen der Bundesrepublik Deutschland. Ulmer Verlag, Stuttgart.

Harper, J.L., 1969. The role of predation in vegetational diversity. Brookhaven Symposia in Biology 22, 48–62.

Hartrington, C.F., Roberts, D.J., Nickless, G., 1995. The effect of cadmium, zinc, and copper on the growth, tolerance index, metal uptake, and production of malic acid in two strains of the grass *Festuca rubra*. Canadian Journal of Botany 74, 1742–1752.

Heggestad, H.E., Menser, H.A., 1962. Leaf spot-sensitive tobacco strain Bel W3, a biological indicator of the air pollutant ozone. Phytopathology 52, 735.

Heil, G.W., Diemont, W.H., 1983. Raised nutrient levels change heathland into grassland. Vegetatio 53, 113–120.

Herman, J.R., Bhartia, P.K., Ziemke, J., Ahmad, Z., Larko, D., 1996. UV-B increases (1979–1992) from decreases in total ozone. Geophysical Research Letters 23, 2117–2120.

Hesler, L.S., Logan, T.M., Benenson, M.W., Moser, C., 1999. Acute dermatitis from oak processionary caterpillars in a US military community in Germany. Military Medicine 164, 767–770.

Hildebrandt, U., Kaldorf, M., Bothe, H., 1999. The zinc violet and its colonization by arbuscular mycorrhizal fungi. Journal of Plant Physiology 154, 709–717.

Hoeks, J., 1972. Effect of leaking natural gas on soil and vegetation in urban areas. Agricultural Research Report 778. Pudoc, Wageningen.

Hultine, K.R., Marshall, J.D., 2000. Altitude trends in conifer leaf morphology and stable carbon isotope composition. Oecologia 123, 32–40.

Hurka, H., Neuffer, B., 1991. Colonizing success in plants: genetic variation and phenotypic plasticity in life history traits in *Capsella bursa-pastoris*. In: Esser, G., Overdieck, D. (Eds), Modern Ecology. Basic and Applied Aspects. Elsevier, Amsterdam, pp. 77–96.

Hutchinson, T.C., 1968. A physiological study of *Teucrium scorodonia* ecotypes which differ in their susceptibility to lime-induced chlorosis and iron-deficiency chlorosis. Plant and Soil 28, 111–136.

Iversen, J., 1936. Biologosche Pflanzentypen als Hilfsmittel in der Vegetationsforschung. Meddedelinger fra Skalling Laboratoriet, Copenhagen.

Jansen, M.A.K., Gaba, V., Greenberg, B.M., 1998. Higher plants and UV-B radiation: balancing damage, repair and acclimation. Trends in Plant Science 3, 131–135.

Joner, E.J., Briones, R., Leyval, C., 2000. Metal-binding capacity of arbuscular mycorrhizal mycelium. Plant and Soil 226, 227–234.

Jones, M.B., Brown, J.C., Raschi, A., Miglietta, F., 1995. The effect on *Arbutus unedo* L. of long-term exposure to elevated CO_2. Global Change Biology 1, 295–302.

Kakes, P., 1981. Genecological investigations on zinc plants. IV. Zinc tolerance of *Viola calaminaria* ssp. *westfalica* (Lej.) Ernst, *Viola arvensis* Murr. and their hybrids. Acta Oecologica, Oecologia Plantarum 2, 305–317.

Kapustka, L.A., 1997. Selection of phytotoxicity tests for use in ecological risk assessment. In: La Point, T.W. (Ed.), Environmental Toxicicology and Risk Assessment. CRC Press, Boca Raton, pp. 517–550.

Keller, Th., Schwager, H., 1971. Der Nachweis unsichtbarer ("physiologischer") Fluor-Immissions-schädigungen an Waldbäumen durch eine einfache kolorimetrische Bestimmung der Peroxidase-Aktivität. European Journal of Forest Pathology 1, 6–18.

Kitao, M., Lei, T.T., Nakamura, T., Koike, T., 2001. Manganese toxicity as indicated by visible foliar symptoms of Japanese white birch (*Betula platyphylla* var.*japonica*). Environmental Pollution 111, 89–94.

Kostina, E., Wulff, A., Julkunen-Tiitto, R., 2001. Growth, structure, stomatal responses and secondary metabolites of birch seedlings (*Betula pendula*) under elevated UV-B radiation in the field. Tree-Structure and Function 15, 483–491.

Kovalskii, V.V., Voronitskaja, J.E., Lekarev, V.S., 1967. Biogeochemical food chain of uranium in aquatic and terraneous organisms. Radio-Ecological Concentration Process. Proceedings of an International Conference Stockholm, Oxford.

Kozlov, M.V., Niemela, P, 1999. Difference in needle length – a new and ojective indicator of pollution impact on Scots pine (*Pinus sylvestris*). Water, Air and Soil Pollution 116, 365–370.

Kraal, H., Ernst, W., 1976. Influence of copper high tension lines on plants and soils. Environmental Pollution 11, 131–135.

Langeweg, F. (Ed.), 1988. Zorgen voor morgen. Nationale milieuverkenning 1985–2010. Tjeenk Willink, Alphen a.d. Rijn.

Lauber, W., Körner, C., 1997. In situ stomatal responses to long-term CO_2 enrichment in calcareous grass-land plants. Acta Oecologica 18, 221–229.

Lee, L.J., Ngim, J., 2000. A first report of glyphosate-resistant goosegrass (*Eleusine indica* (L.) Gaertn.) in Malaysia. Pest Management Science 56, 336–339.

Leenaers, H., 1989. The dispersal of metal mining wastes in the catchment of the river Geul (Belgium – The Netherlands). Netherlands Geographical Studies 102, 1–223.

Lefèbvre, C., Vernet, P., 1999. Microevolutioary processes on contaminated deposits. In: Shaw, A.J. (Ed.), Heavy Metal Tolerance in Plants: Evolutionary Aspects. CRC Press, Boca Raton, pp. 285–299.

Lerman, S.L., Darley, E.F., 1975. Particulates. In: Mudd, J.B., Kozlowski, T.T., 1975. Responses of Plants to Air Pollutants. Academic Press, New York, pp. 141–158.

Lethmate, J., Wendeler, M., 2000. Das chemische Klima des Riesenbecker Osning in den Messjahren 1988 and 1998. Osnabrücker Naturwissenschaftliche Mitteilungen 26, 121–133.

Lockheart, M.J., Poole, I., Van Bergen, P.F., Evershed, R.P., 1998. Leaf carbon isotope composition and stomatal characters: important considerations for palaeoclimate reconstruction. Organic Geochemistry 29l, 1003–1008.

Lodge, R.J., Dijkstra, P., Drake, B.G., Morison, J.I.L., 2001. Stomatal acclimation to increased CO_2 concentration in a Florida scrub oak species *Quercus myrtifolia* Willd. Plant, Cell and Environment 24, 77–88.

Losch, R., Jensen, C.R., Andersen, M.N., 1992. Diurnal courses and factorial dependencies of leaf conductance and transpiration of differently potassium fertilized and watered field-grown barley plants. Plant and Soil 140, 205–224.

Lux, H., 1965. Ergebnisse von Zuwachsuntersuchungen (Bohrspahnanalyse) im Rauchschadengebiet Dübener Heide. Archiv für Forstwesen 14, 433–442.

Mälkönen, E., Derome, J., Fritze H., Helmisaari, H.S., Kukkola, M., Kytö, M., Saarsalmi, A., Salemaa, M., 1999. Compensatory fertilization of Scots pine stands polluted by heavy metals. Nutrient Cycling in Agroecosystems 55, 239–268.

Maljuga, D.P., Malazeina, N.S., Makarova, A.I., 1959. Biogeochemical research in Kadzarane, Armenian SSR (in Russian). Geochimija 1959, 519–528.

Malone, S.R., Mayeux, H.S., Johnson, H.B., Polley, H.W., 1993. Stomatal density and aperture length in 4 plant species grown across a subambient CO_2 gradient. American Journal of Botany 80, 1413–1418.

Marschner, H., 1995. Mineral Nutrition of Higher Plants, 2nd edn. Academic Press, London.

Mathys, W., 1977. The role of malate, oxalate, and mustard oil glucosides in the evolution of zinc-resistance in herbage plants. Physiologia Plantarium 40, 130–136.

McCloskey, W.B., Holt, J.S., 1990. Triazine resistance in *Senecio vulgaris* parental and nearly isonuclear back crossed biotypes is correlated with reduced productivity. Plant Physiology 92, 954–962.

McNair, M.R., Cumbes, Q., 1987. Evidence that arsenic tolerance in *Holcus lanatus* L. is caused by an altered phosphate uptake system. New Phytologist 107, 387–394.

Medlyn, B.E., Barton, C.V.M., Broadmeadow, M.S.J., Ceulemans, R., De Angelis, P., Forstreuter, M., Freeman, M., Jackson, S.B., Kellomaki, S., Laitat, E., Rey, A., Robertnz, P., Sigurdsson, B.D., Strassemeyer, J., Wang, K., Curris, P.S., Jarvis, P.J., 2001. Stomatal conductance of European forest species after long-term exposure to elevated [CO_2]: a synthesis of experimental data. New Phytologist 149, 247–264.

Mejnartowicz, L., 1984. Enzymatic investigations on tolerance in forest trees. In: Koziol, M.J., Whatley, F.R. (Eds), Gaseous Air Pollutants and Plant Metabolism. Butterworth, London, pp. 381–398.

Miglietta, F., Raschi, A., 1993. Studying the effect of elevated CO_2 in the open in a naturally enriched environment in Central Italy. Vegetatio 104, 391–400.

Morison, J.I.L., 2001. Increasing atmospheric CO_2 and stomata. New Phytologist 149, 154–158.

Mudd, J.B., 1975. Sulfur dioxide. In: Mudd, J.B., Kozlowski, T.T. (Eds), Responses of Plants to Air Pollution. Academic Press, New York, pp. 9–22.

Muller, M.M., Hallaksela, A.M., 1998. Diversity of Norway spruce needle endophytes in various mixed and pure Norway spruce stands. Mycological Research 102, 1183–1189.

Murdy, W.H., 1979. Effect of SO_2 on sexual reproduction in *Lepidium virginicum* L. originating from regions with different SO_2 concentrations. Botanical Gazette 140, 299–303.

Murphy, A., Taiz, L., 1997. Correlation between potassium efflux and copper sensitivity in ten *Arabidopsis* ecotypes. New Phytologist 136, 211–222.

Nair, G.G., 1961. Floral study of the sterile *Crotalaria striata* L. observed in the radioactive monazite sand. Transaction of the Bose Research Institute 24, 67–72.

Nicolussi, K., Bortenschlager, S., Körner, C., 1995. Increase in tree-ring width in subalpine *Pinus cembra* from the central Alps that may be CO_2-related. Trees 9, 181–189.

Nouchi, I., Mayumi, H., Yamaoze, F., 1984. Foliar injury responses of petunia and kidney bean to simultaneous and alternate exposures to ozone and PAN. Atmospheric Environment 18, 453–460.

Oberdorfer, E., 1977–1992. Süddeutsche Pflanzengesellschaften, Teil I–IV, 2nd edn. G. Fischer Verlag, Stuttgart.

OECD, 1984. OECD Guideline for Testing Chemicals, 208. Terrestrial Plants, Growth Test. OECD, Paris.

Olde Venterink, H., Wassen, M.J., 1997. A comparison of six models predicting vegetation response to hydrological habitat change. Ecological Modelling 101, 347–361.

Olsen, C., 1923. Studies on the hydrogen ion concentration of the soil and its significance to the vegetation, especially to the natural distribution of plants. Compte rendu des travaux de Laboratoire de Carlsberg 15, 1–166.

Osborn, W.S., 1961. Variation in clones of *Penstemon* growing in natural areas of differing radioactivity. Science 134, 342–343.

Ozinga, W.A., Van Andel, J., McDonnell-Alexander, M.P., 1997. Nutritional soil heterogeneity and mycorrhiza as determinants of plant species diversity. Acta Botanica Neerlandica 46, 237–254.

Paakkonen, E., Vahala, J., Pohjolai, M., Holopainen, T., Karenlampi, L., 1998. Physiological, stomatal and ultrastructural ozone repsonses in birch (*Betula pendula* Roth.) are modified by water stress. Plant, Cell and Environment 21, 671–684.

Pallett, K.E., 2000. The mode of action of isoxaflutole: a case study of an emerging target site. In: Cobb, A.H., Kirkwood, R.C. (Eds), Herbicides and their Mechanisms of Action. Sheffield Academic Press, Sheffield, pp. 215–238.

Penna-Franca, E., Almeida, J.C., Beckker, J., Emmerich, M., Roser, F.X., Kegel, G., Hainsberger, L., Cullen, T.L., Petrouw, H., Drew, R.T., Eisenhut, M., 1965. Status of investigations in the Brasilian area of high natural radioactivity. Health Phsyics 11, 699–712.

Pollanschütz, J., 1971. Die ertragskundlichen Messmethoden zur Erkennung und Beurteilung von forstlichen Rauchschäden. Mitteilungen der forstlichen Bundesversuchsanstalt Wien, 93, 153–206.

Poole, I., Lawson, T., Weyers, J.D.B., Raven, J.A., 2000. Effect of elevated CO_2 on the stomatal distribution and leaf physiology of *Alnus glutinosa*. New Phytologist 145, 511–521.

Poschlod, P., Jackel, A.K., 1993. The dynamics of the generative diaspore bank of calcareous grassland plants. 1. Seasonal dynamics of diaspore rain and diaspore bank in 2 calcareous grassland sites in the Suebian-Alb. Flora 188, 49–71.

Pott, R., 1986. Der pollenanalytische Nachweis extensiver Waldbewirtschaftungen in den Haubergen des Siegerlandes. In: Behre, K.E. (Ed.), Anthropogenic Indicators in Pollendiagrams. Balkema, Rotterdam, pp. 125–134.

Pott, R., 1996. Die Pflanzengesellschaften Deutschlands, 2nd edn. E. Ulmer Verlag, Stuttgart.

Powles, S.B., Lorraine-Colwill, D.F., Dellow, J.J., Preston, C., 1998. Evolved resistance to glyphosate in rigid grass (*Lolium rigidum*) in Australia. Weed Science 46, 604–607.

Prat, S., 1934. Die Erblichkeit der Resistenz gegen Kupfer. Berichte der Deutschen Botanischen Gesellschaft 52, 65–67.

Pridnya, M.V., Cherpakov, V.V., Paillet, F.L., 1996. Ecology and pathology of European chestnut (*Castanea sativa*) in the deciduous forests of the Caucasus Mountains in southern Russia. Bulletin of the Torrey Botanical Club 123, 213–222.

Prins, H.H.T., Weyerhaeuser, F.J., 1987. Epidemics in populations of wild ruminants: anthrax and impala, rinderpest and buffalo in Lake Manyara National Park. Oikos 49, 28–38.

Radosevich, S.R., Appleby, A.P., 1973. Relative susceptibility of two common groundsel (*Senecio vulgaris* L.) biotypes to six s-triazines. Agronomy Journal 65, 553–555.

Ratsch, H.C., Johndro, D.J., McFarlane, J.C., 1986. Growth inhibition and morphological effects of several chemicals in *Arabidopsis thaliana* (L.) Heynh. Environmental Toxicology and Chemistry 5, 55–60.

Repp, G., 1963. Die Kupferresistanz des Protoplasmas höherer Pflanzen auf Kupfererzböden. Protoplasma 57, 643–659.

RIVM, 1996. Achtergronden bij Milieubalans 96. Rijksinstituut voor Volksgezondheid en Milieu. Tjeenk Willink, Alphen a.d. Rijn.

Rodwell, J.S. (Ed.), 1998–2000. British Plant Communities, Vols 1–5. Cambridge University Press, Cambridge.

Roelofs, J.G.M., 1983. Inlet of alkaline water into peaty lowlands: effects on water quality and *Stratiotes aloides* L. Aquatic Botany 39, 267–293.

Rohmeder, E., 1967. Beziehugen zwischen Frucht- bzw. Samenerzeugung und Holzerzeugung der Waldbäume. Allgemeine Forstzeitung 22, 33–39.

Royer, D.I., 2001. Stomatal density and stomatal indices as indicators of palaeoatmospheric CO_2 concentration. Review of Palaeobotany and Palynology 114, 1–28.

Rozema, J., Van der Staaij, J.W.M., Björn, L.O., Cadwell, M.M., 1997. UV-B as an environmental factor in plant life: stress and regulation. Trends in Ecology and Evolution 12, 22–28.

Rozijn, N.A.M.G., Ernst, W.H.O., Van Andel, J., Nelissen, H.J.M., 1990. Growth responses to different levels of soil fertility in four winter annual species during their life cycle. Flora 184, 302–312.

Rüther, F., 1967. Vergleichende physiologische Untersuchungen über die Resistenz von Schwermetallpflanzen. Protoplasma 64, 400–425.

Ruffner, C.M., Abrams, M.D., 1998. Relating land-use history and climate to the dendrology of a 326-year-old *Quercus prinus* talus slope forest. Canadian Journal of Forest Research 28, 347–358.

Sarosiek, J., Leonowics-Babiakowa, K., 1970. The effect of chronic gamma radiation upon *Smphytum officinale* L. under natural conditions. Ekologia Polska 18, 435–464.

Savchenko, V.K., 1995. The Ecology of the Chernobyl Catastrophe. UNESCO, Paris, and Parthenon, New York.

Schaminée, J.H.J., Stortelder, A.H.F., Westhoff, V., 1995–2000. De vegetatie van Nederland, Vols 1–5. Opulus, Uppsala.

Schat, H., 1999. Plant responses to inadequate and toxic micronutrient availability: general and nutrient-specific mechanisms. In: Gissel-Nielsen, G., Jensen, A. (Eds), Plant Nutrition – Molecular Biology and Genetics. Kluwer Academic, Dordrecht, pp. 311–326.

Schat, H., Ten Bookum, W.M., 1992. Genetic control of copper tolerance in *Silene vulgaris*. Heredity 68, 219–229.

Schat, H., Vooijs, R., Kuiper, E., 1996. Identical major gene loci for heavy metal tolerances that have independently evolved in different local populations and subspecies of *Silene vulgaris*. Evolution 50, 1888–1895.

Scherfose, V., 1990. Salz-zeigerwerte von gefässplanzen der salzmarschen, tideröhrichte und salzwassertümpel an der deutschen nord- und ostseeküste. Jahrbuch des Niedersächsischen Landesamtes für wasser und abfall, Forchungstelle Küste 39, 31–82.

Schmidt, W., 1999. Bioindiation und Monitoring von Pflanzengesellschaften. Konzepte, Ergebnisse, Anwendungen, dargestellt an Beispielen aus Wäldern. Berichte der Reinhold Tüxengesellschaft 11, 138–141.

Schnug, E., Haneklaus, S., 1994. Sulfur deficiency in *Brassica napus* – biochemistry, symptomatology, morphogenesis. Landbauforschung Völkenrode, FAL-Braunschweig, Sonderheft 144.

Schönbeck, H., Buck, M., van Haut, H., Scholl, G., 1970. Biologische Messverfahren für Luftverunreinigungen. VDI-Berichte 149, 225–236.

Scholl, G., 1970. Ein biologisches Verfahren zur Bestimmung der Herkunft und Verbreitung von Fluorverbindungen in der Luft. Landwirtschaftliche Forschung 26 (Sonderheft 1), 29–35.

Schröder, P., 1998. Halogenated air pollutants. In: De Kok, L.J., Stulen, I. (Eds), Responses of Plant Metabolism to Air Pollution and Global Change. Backhuys, Leiden, pp. 131–145.

Shaw, A.J. (Ed.), 1999. Heavy Metal Tolerance in Plants: Evolutionary Aspects. CRC Press, Boca Raton.

Smith, W.H., 1981. Air Pollution and Forests. Interactions between Air Contaminants and Forest Ecosystems. Springer Verlag, Berlin.

Somhorst, M.H.M., Stolk, A.P., 1996. Landelijk Meetnet Regenwatersamenstelling. Meetresultaten 1994. National Institute of Public Health and the Environment, Bilthoven, Report no. 723101027.

Sorauer, P., 1911. Die mikroskopische Analyse rauchgasgeschädigter Pflanzen. Abhandlungen Abgase und Rauchschäden 7, 1–58.

Stampfli, A., Zeiter, M., 1999. Plant species decline due to abandonment of meadows cannot easily be reversed by mowing. A case study from the southern Alps. Journal of Vegetation Science 10, 151–164.

Stöckhardt, A., 1850. Über die Einwirkung des Rauches von Silberhütten auf die benachbarte Vegetation. Polytechnisches Centralblatt 1850, 257.

Stuyfzand, P.J., 1993. Hydrochemistry and Hydrology of the Coastal Dune Area of the Western Netherlands. KIWA, Nieuwegein.

Taylor, G.E., Murdy, W.H., 1978. Population differentiation of an annual plant species, *Geranium carolinianum* L., in response to sulfur dioxide. Botanical Gazette 139, 362–368.

Taylor, G.E., McLaughlin, S.B., Shriner, D.S., Selvidge, W.J., 1983. The flux of sulfur containing gases to vegetation. Atmospheric Environment 17, 789–796.

Taylor, H.J., Bell, J.N.B., 1988. Studies on the tolerance of SO_2 of grass populations in polluted areas. V. Investigations into the development of tolerance to SO_2 and NO_2 in combination and NO_2 alone. New Phytologist 110, 327–338.

Teramura, A.H., Sullivan, J.H., Lydon, J., 1990. Effects of UV-B radiation on soybean yield and seed quality: a 6-year field study. Physiologia Plantarum 80, 5–11.

Teramura, A.H., Sullivan, J.H., Sztein, A.E., 1991. Changes in growth and photosynthetic capacity of rice with increased UV-B radiation. Physiologia Plantarum 83, 373–380.

Thompson, K., Bakker, J., Bekker, R., 1997. The Soil Seed Banks of North West Europe. Methodology, Density and Longevity. Cambridge University Press, Cambridge.

Timonen, S., Tammi, H., Sen, R., 1997. Outcome of interactions between genets of two *Suillus* spp. and different *Pinus sylvestris* genotype combinations: identity and distribution of ectomycorrhizas and effect on early seedling growth in N-limited nursery soil. New Phytologist 137, 691–702.

Ting, I.P., Dugger, W.M. Jr., 1971. Ozone resistance in tobacco plants: possible relationship to water balance. Atmospheric Environment 5, 147–150.

Tognetti, R., Minnocci, A., Penuelas, J., Raschi, A., Jones, M.B., 2000. Comparative field water relations of three Mediterranean shrub species co-occurring at a natural CO_2 vent. Journal of Experimental Botany 51, 1135–1146.

Tonin, C., Vandenkoornhuyse, P., Joner, E.J., Straczek, J., Leyval, C., 2001. Assessment of arbuscular mycorrhizal fungi diversity in the rhizosphere of *Viola calaminaria* and effect of these fungi on heavy metal uptake by clover. Mycorrhiza 10, 161–168.

Tsay, J.S., Wu, I.Y., Chen, C.Y., 1994. Responses of soybean, corn, and sorghum to water deficiency – effects of long term drought stress on abaxial epidermal cell and stomatal apparatus of soybean, corn and sorghum. Journal of the Agricultural Association of China 168, 49–62.

USEPA, 1991. Plant Tier Testing: A Workshop to Evaluate Nontarget Plant Testing in Subdivision J Pesticide Guidelines. EPA/600/9–91/041. US Environmental Protection Agency, Washington, DC.

USFDA, 1987. Seed germination and root elongation. In: Center for Food Safety and Applied Nutrition, Center for Veterinary Medicine (Eds), Environmental Assessment Technical Handbook 4.06. US Food and Drug Administration, Washington, DC.

Utriainen, J., Holopainen, T., 1998. Effects of elevated CO_2 and O_3 concentrations on dry matter partitioning, chlorophyll content and needle ultrastructure of Scots pine seedlings. In: De Kok, L.J., Stulen, I. (Eds), Responses of Plant Metabolism to Air Pollution and Global Change. Backhuys, Leiden, pp. 467–469.

Valero, M., Olivieri, I., 1985. Adjacent populations of cocksfoot (*Dactylis glomerata* L.): a detailed study of allozyme variation across contrasting habitats. In: Jacquard, H., Heim, G., Antonovics, J. (Eds), Genetic Differentiation and Dispersal in Plants. Springer Verlag, Berlin, pp. 339–354.

Van Baalen, J., Prins, E.G.M, 1983. Growth and reproduction of *Digitalis purpurea* in different stages of succession. Oecologia 58, 84–91.

Van Beckhoven, K., 1995. Rewetting of coastal dune slacks: effects on plant growth and soil processes. Doctorate Thesis, Vrije Universiteit, Amsterdam.

Van den Berg, M.S., Scheffer, M., Coops, H., Simons, J., 1998. The role of Characean algae in the management of eutrophic shallow lakes. Journal of Phycology 34, 150–157.

Van der Eerden, L.J., 1992. Fertilizing effects of atmospheric ammonia on semi-natural vegetations. Doctorate Thesis, Vrije Universiteit, Amsterdam.

Van der Heijden, M.G.A., Boller, T., Wiemken, A., Sanders, I.A., 1998. Different arbuscular mycorrhizal fungal species are potential determinants of plant community structure. Ecology 79, 2082–2091.

Van der Kooij, T.A.W., De Kok, L.J., Haneklaus S., Schnug E., 1997. Uptake and metabolism of sulphur dioxide by *Arabidopsis thaliana*. New Phytologist 135, 101–107.

Van der Kooij, T.A.W., De Kok, L.J., 1998. Kinetics of deposition of SO_2 and H_2S to shoots of *Arabidopsis thaliana* L. In: De Kok, L.J., Stulen, I. (Eds), Responses of Plant Metabolism to Air Pollutants and Global Change. Backhuys, Leiden, pp. 481–483.

Van der Meijden, R., 1996. Heukels' Flora van Nederland, 22nd edn. Wolters-Noordhoff, Groningen.

Van der Meijden, R., Odé, B., Groen, K.(C.)L.G., Witte, J.P.M., Bal, D., 2000. Bedreigde en kwetsbare vaatplanten in Nederland. Basisrapport met voorstel voor de Rode Lijst. Gorteria 26, 85–141.

Van der Meijden, R., Plate, C.L., Weeda, E.J., 1989. Atlas van de Nederlandse Flora. 3. Minder zeldzame en algemene soorten. Rijksherbarium/Hortus Botanicus, Leiden.

Van der Valk, A.G., Bremholm, T.L., Gordon, E., 1999. The restoration of sedge meadows: seed viability, seed germination requirements, and seedling growth of *Carex* species. Wetlands 19, 756–764.

Van der Zaal, B.J., Neuteboom, L.W., Pinas, J.E., Chardonnens, A.N., Schat, H., Verkleij, J.A.C., Hooykaas, P.J.J., 1999. Putative zinc transporter genes from animals can lead to enhanced zinc resistance and accumulation. Plant Physiology 119, 1047–1055.

Van de Staaij, J.W.M., Ernst, W.H.O., Hakvoort, J.W.J., Rozema, J., 1995. Ultraviolet-B (280–320 nm) absorbing pigments in the leaves of *Silene vulgaris*: the role in UV-B tolerance. Journal of Plant Physiology 147, 75–80.

Van Dobben, H.F., ter Braak, C.J.F., Dirkse, G.M., 1999. Undergrowth as a biomonitor for deposition of nitrogen and acidity in pine forest. Forest Ecology and Management 114, 83–95.

Van Herk, C.M., 1999. Mapping of ammonia pollution with epiphytic lichens in the Netherlands. Lichenologist 31, 9–20.

Van Wijnen, H.J., Bakker, J.P., 1997. Nitrogen accumulation and plant species replacement in three salt marsh systems in the Wadden Sea. Journal of Coastal Conservation 3, 19–26.

Varley, G.C., Gradwell, G.R., 1962. The effect of partial defoliation by caterpillars on the timber production of oak trees in England. Proceedings of the 11th International Congress of Entomology, Wien, Vol. 2, pp. 211–214.

Von Linstow, O., 1929. Bodenanzeigende Pflanzen. Abhandlungen der Preussisch-Geologischen Landesanstalt NF 114, 1–247.

Von Schroeder, J. Reuss, C., 1883. Die Beschädigung der Vegetation durch Rauch und die Oberharzer Hüttenrauchschäden. Verlag Paul Parey, Berlin.

Von Wuehlisch, G., Krusche, D., Muhs, H.J., 1995. Variation in temperature sum requirement for flushing of beech provenances. Silvae Genetica 44, 343–346.

Walter, H., 1954. Einführung in de Phytologie. III. Grundlagen der Pflanzenverbreitung. II. Teil Arealkunde. Ulmer Verlag, Stuttgart.

Walter, H., 1960. Einführung in die Phytologie. III. Grundlagen der Pflanzenverbreitung. I. Teil Standortslehre, 2nd edn. Ulmer Verlag, Stuttgart.

Walter, H., Breckle, S.W., 1983. Ökologie der Erde. I. Ökologische Grundlagen in globaler Sicht. G. Fischer Verlag, Stuttgart.

Warwick, S.I., 1991. Herbicide resistance in weedy plants: physiology and population biology. Annual Review of Ecology and Systematics 22, 95–114.

Wellburn, A.R., 1990. Why are atmospheric oxides of nitrogen usually phytotoxic and not alternative fertilizers. New Phytologist 15, 395–429.

Westhoff, V., Van Oosten, M.F., 1991. De plantengroei van de Waddeneilanden. Stichting Uitgeverij Koninklijke Nederlandse Natuurhistorische Vereniging, Utrecht.

Willems, J.H., 1983. Species composition and above ground phytomass in chalk grassland: the role of seed rain and soil seed bank. Vegetatio 52, 171–180.

Wittig, R., 1991. Veränderungen im Artenspektrum von Waldgesellschaften als Indikatoren erhöhter Säure- und Stickstoffeinträge. VDI Berichte 901, 407–418.

Wolterbeck, H.T., Garty, J., Reis, M.A., Freitas, M.C., 2002 (2003). Biomonitors in use: lichens and metal air pollution. In: Markert, B.A., Breure, A.M., Zechmeister, H.G. (Eds), Bioindicators and Biomonitors. Elsevier, Oxford, pp. 377–419.

Woodward, F.I., 1987. Stomatal numbers are sensitive to increases in CO_2 from pre-industrial levels. Nature 327, 617–618.

Zhao, F.J., Withers, P.J.A., Evans, E.J., Monaghan, J., Salmon, S.E., Shewry, P.R., McGrath, S.P., 1997. Sulphur nutrition: an important factor for the quality of wheat and rapeseed. Soil Science and Plant Nutrition 43, 1137–1142.

Ziska, L.H., Teramura, A.H., Sullivan, J.H., McCoy, A., 1993. Influence of ultraviolet-B (UV-B) radiation on photosynthetic and growth characteristics in field-grown cassava (*Manihot esculentum crantz*). Plant Cell and Environment 16, 73–79.

Bioindicators and biomonitors
B.A. Markert, A.M. Breure, H.G. Zechmeister, editors

Chapter 13

Higher plants as accumulative bioindicators

Peter Weiss, Ivo Offenthaler, Richard Öhlinger and
Johann Wimmer

Abstract

This chapter gives an introduction to accumulative biomonitoring with higher plants. The basic difference between active and passive biomonitoring is explained, including information about those species routinely used for each approach. The differences and possible advantages over more tech-orientated screening methods are discussed (in awareness of a recurring confusion between the terms "technic" and "scientific"). The reader is confronted with a variety of biotic and abiotic factors that influence accumulation processes, including soil conditions, deposition patterns, plant morphology and many more. A number of important methodical considerations is presented together with some caveats and common pitfalls. The necessity of adequate and comprehensive documentation is discussed as well the particular demands on statistical analysis arising from small sample sizes and ostensible "outlier" values. A summarizing lookup table offers quick orientation among different screening designs. Throughout the chapter, the practical relevance of accumulative biomonitoring is illustrated by numerous examples of field studies from different countries and impact situations.

Keywords: accumulative bioindication, higher plants, approaches, methods, screening design, inorganic, organic pollutants

1. Introduction

Higher plants and plant communities play a fundamental role for nutrition and life on earth. As non-mobile organisms they are always exposed to the environmental conditions, for instance to air pollutants, at their sites of growth. The properties of the aerial plant parts (e.g. surface roughness, passive pollutant diffusion through stomata, uptake and accumulation by the cuticle) and the nature of the pollutants are responsible for accumulation of several harmful compounds in the plants. Plants have high leaf area indices; i.e. the surface area of the plant is much higher than the area on ground covered by the aerial plant parts (Kimmins, 1987). As a consequence, the aerial plant parts effectively filter out air pollutants.

The investigation of higher plants as accumulative indicators for air pollution has in the meanwhile a long tradition. Already in the 19th century and at the beginning of the 20th century plant organs were chemically analysed to detect the impact of emitters (Portele, 1891, cit. in Stefan and Fürst, 1998; Swain and Harkins, 1908, cit. in Martin and Coughtrey, 1982). Particularly in the last decades of the 20th century a rapid increase in bioindication studies for pollutant loads in higher plants can be observed (e.g. Arndt et al., 1987, 1996; Markert, 1993a; Martin and Coughtrey, 1982; Mulgrew and Williams, 2000; Steubing and Jäger, 1982; VDI 1987, 1992). The reasons behind were the increasing emissions rates and their impact on the biosphere and ecosystems. As a consequence, there has been a great need for cost-saving monitoring methods as well as tools that allow a direct assessment of pollutant contamination in plants and of the pollutant exposure of their subsequent consumers, particularly man. Only bioindication techniques are able to fulfil both of these requirements. Meanwhile, several methods have been developed, tested, used and further refined, partly up to a high degree of standardisation (e.g. VDI 3792–1, VDI 3792–2, VDI 3792–3, VDI 3792–5, VDI 3957–3, VDI 3957–5). The instructions for sampling of tree leaves within international monitoring programmes under the UN-ECE "Convention on Long-range Transboundary Air Pollution" represent examples for internationally standardised sampling techniques (UN-ECE,1998; Stefan et al., 2000). In some countries the use of bioindication techniques was even object of legal regulations, for instance in the German federal regulation for the protection against harmful environmental impacts by air pollution (BIMSCHG, 1990) or in the Austrian second regulation against air pollution affecting forests (BGBL, 199/1984). The latter regulation also includes limit values for the sulphur, fluorine and chlorine contents of tree leaves. The new regulations for environmental impact assessments of major industrial or traffic projects, which were recently laid down in many countries, will very likely bring a new impetus for bioindication methods (Wimmer, 1998; Zimmermann et al., 1998).

The present chapter deals mainly with the use of higher plants as accumulative bioindicators for air pollution impact. In addition, the nutrient and pollutant uptake of plants via roots, its interference with the bio-accumulation signals from the uptake via atmosphere and the related use of plants for the biomonitoring of soil contamination will be discussed briefly. The small segment of environmental monitoring which is discussed in this chapter should not lead to a preference of too narrow approaches. Even the inclusion of few further parameters in biomonitoring programmes can broaden significantly the gain of knowledge and allow more comprehensive evaluations of the findings (e.g. Weiss, 2000a). Recently, comprehensive monitoring programmes were started (e.g. EC-UN/ECE, 1995; UN-ECE 2001), because narrow approaches left too much questions open with respect to the impact of environmental pollution.

2. Factors influencing pollutant concentrations in higher plants

2.1. Type of deposition

Airborne pollutants can reach plant surfaces through wet and occult deposition, dry gaseous and dry particulate matter deposition. The contribution of these different deposition types to the total pollutant flux to the plant surface is influenced (e.g. Bidleman, 1988; Kalina et al., 1998; Kömp and McLachlan, 1997a; McLachlan et al., 1995; McLachlan, 1999; Miller et al., 1993; Riederer, 1992):

- by the emission type;
- by the physical-chemical properties of the pollutants (e.g. their atmospheric behaviour and their gas phase and particle phase partitioning);
- by the atmospheric, climatic and site conditions (e.g. amount of particulate matter in the atmosphere, temperature, precipitation, altitude);
- and by the different susceptibility of plants and plant parts to pollutant deposition.

Innes (1995) showed significant correlations between needle nitrogen and sulphur contents of various coniferous trees and the atmospheric deposition of these compounds. From his results he assumed that direct air pollution had a greater effect on the needle contents than indirect pollution (wet deposition). Experiments by Umlauf et al. (1994) and Welsch-Pausch et al. (1995) demonstrated the dominant role of gaseous deposition for the concentrations of several semivolatile organic compounds in spruce needles and rye grass, as compared to wet deposition and deposition of particles. Relatively involatile organic compounds (e.g. the higher chlorinated dioxins and furans, Horstmann and McLachlan, 1998) and heavy metals (Mayer, 1981; except mercury, for which further research in this field is needed, Lin and Pehkonen, 1999) are mainly deposited as particulate matter through dry or wet deposition.

2.2. The uptake mechanisms

Air pollutants can be taken up via stomata, cuticle or indirectly by uptake via roots after deposition of the air pollutants to the soil. For a detailed description of these different uptake mechanisms the reader is kindly referred to general overviews (e.g. Heath, 1980; Marschner, 1995; Schönherr and Riederer, 1989). In the present chapter only a coarse overview on this issue is given, which should demonstrate the implications of these different uptake mechanisms for the monitoring of air pollution impact with higher plants.

Gaseous air pollutants like SO_2, NO_2, HF, HCl diffuse via stomata (which are active "valves" for the gas exchange through the cuticle) into the plant interior. The permeation of all gaseous compounds from the atmosphere to the needle interior is to a high degree dependent on the stomatal and mesophyllic resistances (Heath, 1980). The opening of the stomata and, hence, the stomatal diffusion of gaseous pollutants is affected by various factors, for instance light, air humidity, water supply, wind velocity, internal CO_2 concentration and the effect of pollutants on the stomatal movement (Ziegler, 1984). Various experiments gave evidence that the uptake of SO_2 and NO_2 via stomata is related to the atmospheric concentrations (e.g. Guderian, 1970a; Kaiser et al., 1993; Schätzle et al., 1990; Thoene et al., 1991).

Beside the uptake via stomata, gaseous pollutants may also permeate through the cuticle. In terrestrial higher plants, the cuticle covers all aboveground parts unless it is replaced with more robust protective layers during secondary growth. The cuticle is formed by a lipophilic membrane, which is composed of the biopolymer cutin and waxlike lipids embedded within the cutin matrix and/or covering the outer surface of the cuticle. In principle, the cuticle represents a transport-limiting barrier, which protects from excessive water loss and leaching of solutes and severely restricts the uptake of solutes. The uptake of substances across the cuticle is influenced by the presence of soluble cuticular lipids, the polarity of the molecules and the concentration gradient (Riederer, 1989). The fatty ingredients of the plant cuticles serve as a reservoir and are assumed to be also a source for the sustained contamination of the interior plant tissue with lipophilic organic compounds. The accumulation of lipophilic compounds in isolated cuticles has been extensively investigated (Schönherr and Riederer, 1989). Several studies under controlled and natural conditions identified a clear accumulation of semivolatile organic compounds (SOCs) in the cuticular lipids (e.g. Hauk et al., 1994; Reischl et al., 1987; Strachan et al., 1994). These studies suggest that the uptake via the cuticle is of significant relevance for the SOC concentrations in aerial plant parts. This passive uptake mechanism is one of the reasons why higher plants are well suited to detect the atmospheric impact with SOCs. Nevertheless, further studies are still required for SOCs to clarify the relative importance of the different uptake mechanisms into aerial plant parts. Also inorganic gaseous pollutants that are deposited on dry cuticles or into the water films on cuticles can enter the plant interior via the cuticle (Lendzian, 1987). However, it is assumed that for these pollutants the resistance of the cuticular pathway is very high and stomatal uptake far more important (Rennenberg and Gessler, 1999; Riederer, 1989; Wellburn, 1988).

Particle bound air pollutants like heavy metals and the less volatile organic compounds are deposited to the plant surfaces and may be adsorbed there, embedded or even taken up in the plant cuticles. Several studies gave evidence that the surface wax layer of leaves and needles works as an accumulator of particle bound air pollutants (compilation in Djingova and Kuleff, 1994). Through analytical scanning electron microscopy, X-ray microanalysis and mineralogical–geochemical phase analysis such adhering particles, their elemental composition and structural characterisation (incl. indications for their origin) can be determined and semi-quantitatively assessed (Mankovska, 1992; Neinave et al., 2000; Trimbacher and Weiss, 1999; Weiss and Trimbacher, 1998).

Removing the cuticle with organic solvents significantly alters the leaf concentration of various elements. This observation clearly demonstrates the effective accumulation of particle bound elements on plant surfaces and supports the principal suitability of higher plants to monitor the corresponding atmospheric load. Particularly for the following elements a significant contribution of the plant cuticle layer to the overall needle or leaf concentration was shown: Al, As, Br, Co, Cr, Cs, Cu, Fe, Hg, La, Mo, Na, Ni, Pb, Sc, Sb, Sn, Th, V (Bäumler et al., 1995; Keller et al., 1986; Krivan and Schaldach, 1986; Krivan et al., 1987; Lick and Dorfer, 1998; Mößnang, 1990; Steubing,, 1987; Wagner 1987; Wyttenbach et al., 1985, 1989). Water rinsing treatments or even heavy rainfall could not effectively remove such surface bound elements (Djingova and Kuleff 1994; Krivan and Schaldach, 1986; Krivan and Schäfer, 1989). These results give evidence that certain plant species exhibit long-term surface accumulation of selected airborne elements.

On the contrary, cuticular deposits usually do not contribute significantly to the overall needle or leaf concentration of Mg, K, Ca, Ba, Cd, Mn, Tl, Zn (see above cited references). These general statements, however, may be not valid for all species (see compilation and discussion in Djingova and Kuleff, 1994) and for very specific environmental conditions. Even for Mg and Ca specific emission situations exist (e.g., close to quarries, cement kilns and magnesite plants) in which surface deposits of these

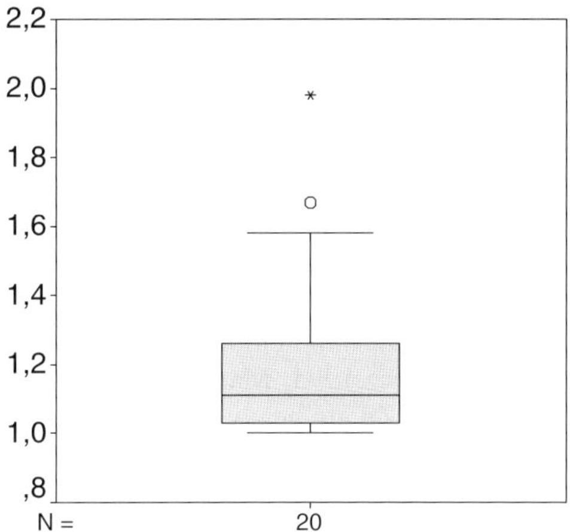

Cd [conc. unwashed/conc. washed]

Figure 1. Cd concentration quotient of unwashed Norway spruce needles and Norway spruce needles which were washed with an organic solvent to remove the cuticle. Samples origin from sites close to steel works (Trimbacher and Weiss, 2002, in prep.).

elements can be clearly detected and contribute to the overall needle concentrations (Trimbacher and Weiss, 1997, 1999, 2000). In a similar way, it has been shown that at some sites close to an emitter a significant contribution of Cd in the cuticle layer to the overall Cd needle concentration can be detected, although the results for most investigated sites are in line with the literature findings of an insignificant contribution of surface deposits to the overall Cd needle concentration (see outliers and wider range of the upper quartile in Figure 1, Trimbacher and Weiss, 2002, in preparation). These specific observations may demonstrate that an appropriate and more detailed approach of passive bioindication with higher plants (investigation of plants taken from the site of growth) allows to identify atmospheric impacts, even for elements usually not well suited for this method.

For semivolatile organic compounds the contribution of particle deposition to the pollutant concentration in aerial plant parts was tested by a few experiments. Even for relatively involatile organic compounds which are present in two phases in the atmosphere (gaseous and particle bound) evidence was given that cuticular uptake from the gas phase seems to be the prevailing source for the concentrations in aerial plant parts (Umlauf et al., 1994; Welsch-Pausch et al., 1995). Only for the mainly particle bound compounds with log octanol/air partition coefficients above 11, like the hepta- and octachlorinated dioxins and furans and the polycyclic aromatic compounds with more than five rings, particle deposition seems to be relevant for the concentrations in aerial plant parts (Böhme et al., 1999; Welsch-Pausch et al., 1995).

Wet deposited particle bound or dissolved compounds may also be retained and incorporated by aerial plant parts. However, for various pollutants direct uptake from wet deposition is assumed to be quantitatively less relevant than other deposition types and uptake mechanisms (see above).

These different uptake mechanisms via atmosphere and the related accumulation of air pollutants within aerial plant parts may be masked by the pollutant uptake via roots and shoots and the translocation to the investigated plant parts. Nutrient elements and some more mobile heavy metals are taken up to a significant amount from the soil and translocated to the aerial plant parts. In the case of nitrogen, the high natural contents in plants and the dynamic behaviour of the element within individual and ecosystem is an important reason for the scarce accumulative bioindication techniques to detect its atmospheric input (Hicks et al., 2000; Mohr, 1999). Sulphur, on the other hand, is a nutrient element for which – despite the significant uptake via soil – an analysis of conifer needles is often done with the aim of monitoring the atmospheric sulphur impact (e.g. Dmuchowski and Bytnerowicz,1995; Mankovska, 1997; Rudolph, 1987; Stefan and Fürst, 1998). Cd, Mn, Tl and Zn are heavy metals with a considerable mobility in soil and large soil/plant transfer coefficients. These properties complicate the use of any higher plants to detect the atmospheric input of these elements. The opposite, a comparatively very low uptake via roots holds true for the elements Co, Cr, F, Hg and Pb (Kloke et al., 1984; Sauerbeck, 1986; Scheffer and Schachtschabel, 1989). For these elements a low "noise-level" in the concentrations, which cannot be related to the uptake from the atmosphere, but to soil uptake, can be expected (see also above).

The physical–chemical properties of organic pollutants play an important role whether a root uptake and translocation of such compounds within the plant is likely or not. Matucha et al. (2001) and Sutinen et al. (1995) detected a significant root uptake and translocation to the upper plant parts of the phytotoxic trichloroacetic acid. An uptake by the roots, a translocation to the upper plant parts and a significant contribution to the concentrations in aerial plant parts is likely for non-ionised organic chemicals with octanol/water partition coefficients (log K_{OW}) between 0 and 4. For the more lipophilic organic pollutants, a significant contribution of the root uptake to the concentrations in aerial plant parts can be widely excluded (Briggs et al., 1982; Tiefenbach et al., 1983). The absence of edaphic influences on the concentrations of aerial plant parts is one of several reasons why higher plants are very well suited for the biomonitoring of lipophilic semivolatile organic compounds.

2.3. Accumulation, losses, degradation and metabolism of compounds

An accumulation of the pollutants in the plant is a requirement for their detection by chemical–analytical methods. Therefore reactive or rapidly metabolised compounds are not suitable for accumulative biomonitoring. Among them are pollutants of major environmental concern, like ozone, for which only effect related bioindication techniques exist (overview in Mulgrew and Williams, 2000). For a broad spectrum of air pollutants an accumulation in plant parts has been detected. Particularly significant accumulation rates can be assumed for

- numerous heavy metals (Deu and Kreeb, 1993; Kovacs et al., 1982; Wagner, 1990);
- other elements like S, F, Cl (Guderian, 1970a, 1970b; Kaiser et al., 1993; Kovacs et al., 1982; Kronberger et al., 1978; Schätzle et al., 1990; Vike and Habjorg, 1995);
- lipophilic semivolatile organic compounds (Bacci and Gaggi, 1987; Bacci et al., 1990; Böhme et al., 1999; Reischl et al., 1989; Strachan et al., 1994; Umlauf et al., 1994; Welsch-Pausch et al., 1995);

- and also some other halogenated and nitrated organic compounds, like trichloro-acetic acid and nitrophenols (Frank, 1991; Hinkel et al., 1989; Plümacher et al., 1993; Weiss, 2000b).

Besides this principal accumulation of certain compounds, the seasonal changes in element content (see detailed discussion in Djingova and Kuleff, 1994) illustrate the need for a certain and narrow sampling period when carrying out a biomonitoring study with higher plants (see Section 2.4). Sometimes in literature, an increase in the concentrations in plant over time is readily assumed to be an indication of atmospheric pollution. Such an approach neglects the fact that even in widely unpolluted environments an increase in the contents of certain elements (e.g. lead) over time can be detected (Ahrens, 1964; Bäumler et al., 1995; Deu and Kreeb, 1993; Ernst, 1990; Guha and Mitchell, 1966; Kovacs et al., 1982, see also Fig. 2). Therefore, the mere increase of elements with needle age does not allow any conclusion about atmospheric pollution.

For semivolatile organic compounds (SOCs) the accumulation rate is related to the physical–chemical properties of the compound, to the properties of the plant or plant community and to atmospheric characteristics like the pollutant concentration, temperature etc. (Böhme et al., 1999; Kömp and McLachlan, 1997a; Umlauf et al., 1995; Welsch-Pausch et al., 1995). For the less volatile SOCs, the cuticular storage capacities and the volume-specific surface area of the leaves/needles seem to influence the accumulation rate in a way that is specific for the particular plant (Böhme et al., 1999). The physical–chemical properties of the chemicals are relevant for the prevailing deposition mode and for the time needed to reach the temperature-dependent equilibrium partitioning between plant- and atmospheric concentration. For SOCs, a few months up to years (for the higher boiling compounds) have been reported to reach equilibrium partitioning (McLachlan et al., 1995; Paterson et al., 1991; Riederer, 1990; Umlauf et al., 1994). From the biomonitoring's perspective these long times are advantage and disadvantage in one. The advantage is that even a short lasting pulse of atmospheric SOC pollution during the exposure period of the plant is captured and

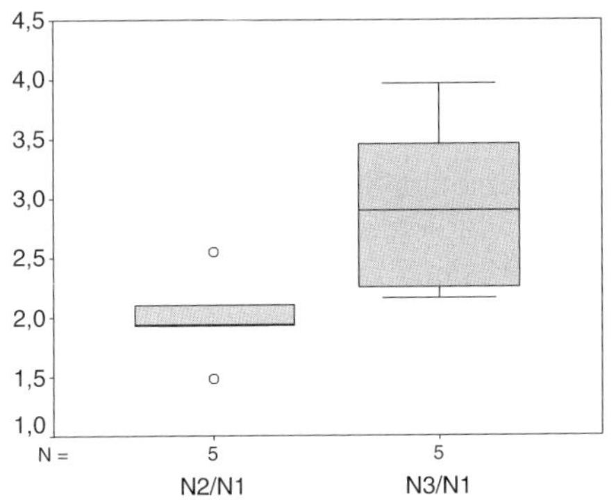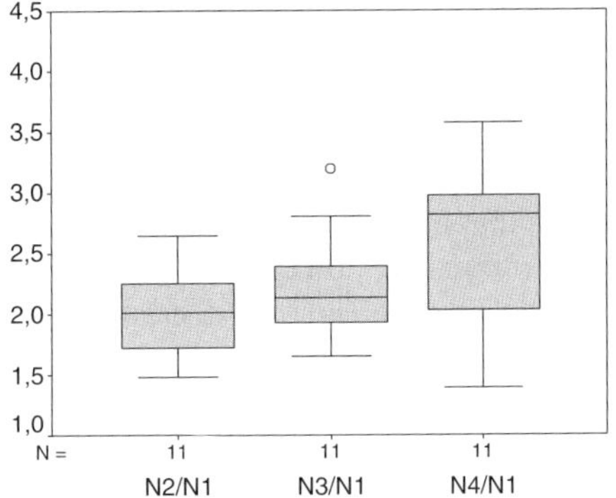

Figure 2. Lead concentration quotients for Norway spruce needles of different needle age (N1: ½-year old, N2: 1½-year old, N3: 2½-year old, N4: 3½-year old). Left: remote sites (calculated from Weiss, 1998), right: urban-industrial sites (calculated from Wimmer et al., 2002, in prep.)

reflected by the concentration in the plant part. This "memory" effect, however, represents also a disadvantage because the concentration in plant may be mainly caused by this short pulse of high atmospheric load, which is not typical for the average load at the site during the exposure time. However, this "SOC integration" by plants over time allows to draw more adequate conclusions about the impact of the occurring pollution on the plant or on its consumer than results from air quality measurements. Even if air SOC measurements are conducted long and frequently enough to trace such a single pollution peak (irrespective of the considerable financial and methodical efforts), the consequences for the surrounding vegetation or the herbivores would remain unclear.

For volatile organic compounds the equilibrium distribution between the plant and the surrounding atmosphere is reached within a few hours (Figge, 1990; Frank and Frank, 1986; Riederer, 1990). The accumulation of such compounds is followed by a quick process of clearance in clean air (re-volatilisation of the compound). Nevertheless, fumigation experiments with perchloroethylene (PER) by Figge (1990) showed that after 28 days of desorption 3.4% of the initial PER concentration was still in the needles as non-desorbable portion. The absolute amount of this portion was correlated with the atmospheric PER concentration and the length of the fumigation period. Since equilibrium partitioning for volatile organic compounds is reached fairly quickly, no differences between the concentrations in different needle age classes would be expected in the field. Under such conditions the plant-internal concentration would primarily change with temperature (Kömp and McLachlan, 1997a). This is confirmed by results of Plümacher et al. (1993) who took monthly needle samples over two years and found seasonal variations with higher concentrations of volatile halogenated organic compounds in the cold season. On the other hand, Weiss et al. (1998b) detected that among needles sampled on the same date, the concentrations of volatile halogenated organic compounds were significantly higher in the older needle age classes. Differences between needle age classes, regarding uptake, accumulation, revolatilisation and degradation, as well as some accumulation over time (as measured by Figge, 1990) were assumed to be responsible for this result. These somehow controversial findings suggest further investigations to clarify the underlying mechanisms. Anyhow, the quickly reachable equilibrium partitioning of volatile organic compounds underline that concentrations of these compounds in plants are mainly indicative for the short term atmospheric load with these compounds during the time of sampling. The temperature during the time of sampling plays a further crucial role in this context.

Apart from the different accumulation behaviour of pollutants and plants, other factors that may influence plant concentrations and will be discussed in the following, include: translocation to other plant parts, leaching, erosion of the cuticles and adhering particles, metabolism and degradation within the plants.

The redistribution and re-translocation of several nutrient elements within plant parts is a well known process to transfer these elements to plant parts of demand or to avoid excess nutrient losses with the litterfall (Fiedler et al., 1973; Marschner, 1995). A steady accumulation in the leaves has been shown for Ca, B, Fe, Sr, Ba, Si and almost all non-essential metals, while for N, P, K, Mg, Zn, Cu and Na a redistribution within the plant takes place. But even for some pollutants a re-translocation within the plant has been reported. Kronberger et al. (1978) and Kühnert et al. (1996) gave evidence for redistribution processes of F between needle age classes of Norway spruce.

A further influence on the concentration may be the leaching of elements out of the leaf. Particularly under conditions of acid rain a significant leaching of various elements has been shown. Experiments by Kreutzer and Bittersohl (1986) and Seufert (1990) showed a higher leaching of the nutrient elements Ca, Mg, K, Mn and Zn for Norway spruce trees, which were sprinkled with acid rain. Most of the studies by Asche (1985), Godt et al. (1985), Mayer (1981) and Zöttl (1985) showed a significantly higher total precipitation deposition of the elements Cd, Mn, Zn underneath the forest crown canopy compared to unstocked areas, whereas Fe, Co, Cr, Cu, Ni, Pb were in most cases not or only slightly elevated. Total ecosystem balances suggest a significant interception of the latter elements by the tree crowns (mainly dry deposition) and a quantitatively small removal of these intercepted elements by rainfall (Mayer, 1981). The higher deposition of Cd under the canopy layer is remarkable, if connected with the general findings in literature and own results that removal of the cuticles does not significantly change the Cd concentration of needles (see Section 2.2). Although there are no relations between these two types of investigation, the common observation of the findings of both suggests that leaching of Cd is the more important source for higher Cd deposition under canopies than wash-off or erosion of surface deposits. With respect to leaching and the significant uptake from the soil (see Section 2.2), the accumulation behaviour of Cd, Mn and Zn is unfavourable from the perspective of accumulative passive bioindication, and these elements' rapid ecosystem internal cycle adds further complication. While this does not principally exclude a detection of the atmospheric impact of Cd, Mn and Zn on plant parts, a sound survey will involve additional approaches (see Sections 2.2 and 2.4).

The lipophilic nature of semivolatile organic compounds makes a significant wash-off of these compounds less likely than their removal by cuticular erosion. Deposition measurements in forests and an adjacent clearing showed a significant higher SOC deposition under the crown cover. It was assumed that this higher deposition is partly related to the erosion of cuticle waxes (Horstmann et al., 1996, 1997; Horstmann and McLachlan, 1998). More specific investigation would be needed to elucidate this phenomenon.

Metabolisation and degradation may also contribute to the losses of chemicals in plant parts. For many inorganic air pollutants this causes no major biomonitoring problem as long as certain elements of the pollutants accumulate to an extent that can be identified as a signal of air pollution. For organic compounds, however, the situation is different: The concentrations are usually several orders of magnitude lower and, hence, the amount of an element released by the breakdown of its "carrier" pollutant would be negligibly small compared to the plant's natural content of the element. Plant metabolism of organic chemicals, particularly pesticides, has been described in several overviews (Coupland, 1991; Lamoureux and Rusness, 1989; Lamoureux et al., 1991; Sandermann, 1994; Sandermann et al., 1997; Shimabukuro and Walsh, 1978; Schröder et al., 1998). Particularly, many recently used agrochemicals are known to be quickly metabolised. Any biomonitoring of such compounds needs to target the original compound and its metabolites. This remains, however, a very difficult if not impossible task, the success of which depends on the kinetics of the compound's metabolism, the knowledge of its metabolites and the availability of methods to detect them.

2.4. Further abiotic and biotic influences on the pollutant concentrations in higher plants

In addition to the previously listed possible influences, several other biotic and abiotic factors may affect the pollutant concentrations in higher plants. These factors equally deserve attention when planning or carrying out a biomonitoring study.

2.4.1. Indicator-related accumulation behaviour

2.4.1.1. Interspecific variability
In Section 2.3 the uptake mechanisms of pollutants were discussed. These active and passive uptake mechanisms vary quantitatively from species to species. As a consequence, species usually differ in the concentrations of inorganic and organic pollutants (e.g. Böhme et al., 1999; Buckley, 1982; Kömp and McLachlan, 1997b; Kovacs et al., 1990, 1993; Thomas et al., 1985; Wagner, 1990). Therefore, a comparison of the pollutant loads of sites should be based on the results of one species. It is, however, not always possible or desired to confine the investigation to a single species. The heterogeneous ecological conditions of a region may not allow one to monitor the whole area with one ubiquitous species. In this case, the number of samples of the origin study design must be increased by the number of parallel samplings of both species per site which are needed for concentration calibrations or, alternatively, active biomonitoring methods may be used. A further approach could be to normalise the detected concentrations for each species to arrive at "normal" or "background" concentration ranges typical for the indicator species. These "unpolluted" concentration ranges could be taken from investigations of unpolluted sites or calculated according to Erhardt et al. (1996) who developed an iterative statistical approach to identify this range within a bi- (or poly-)modal distribution of biomonitoring data. If a sufficient number of well distributed sites for each species within a region is given, this "normal range" for each species can serve as a basis for classification and characterisation of the pollutant impact situation, even if data from two species have to be merged to combine different areas in one spatial map (see an example in Figure 4, Section 3).

2.4.1.2. Intraspecific variability
In addition to interspecific variation, the element concentrations differ between individuals of one species. Apart from external biotic and abiotic influences, which will be discussed later, genetic differences between the plants and their impact on the uptake of compounds, storage properties and the biochemical differences may be a reason for such intra-specific variation (Ernst, 1990; Heinze and Fiedler, 1992; Markert,1993b). Paulus et al. (1995) detected significant correlations between the concentrations of inorganic and organic compounds and the biometric differences of Norway spruce needles from one sampling site. If a significant influence of individual constitution is expected, it is necessary to increase the number of sampled plants per site, to select a less variable species or to use genetically identical material (e.g. clones of Norway spruce) together with active biomonitoring. The inclusion of selected biometric parameters allows testing whether these parameters have a significant influence on the detected concentration differences between sites (Weiss, 1998). Evidence

for an influence of a different plant age of perennials on the elemental concentration differences in plant parts between individuals is scarce (Heinze and Fiedler, 1992). It may be a wise and precautionary practice to avoid large variation in the age of the investigated perennials.

2.4.1.3. Variability within the individual

In a next step, the presence of concentration differences between plant parts of an individual must be taken into account for a proper monitoring design. Examples on such differences are given, among others, in Djingova and Kuleff (1994), Ernst (1990), Heinze and Fiedler (1992). They underline the need to define the sampled plant parts.

A further reason for sampling standardisation is given by the possible concentration variations due to different heights and exposures of the harvested plant parts (e.g. upper vs. lower or wind shaded vs. wind exposed part of the tree; Guha and Mitchell, 1966; Heinze and Fiedler, 1992; Knabe, 1982; Krivan and Schaldach, 1986; Markert, 1993b; Steubing, 1982; Strachan et al., 1994; Wagner, 1990).

The age of a plant part, its developmental stage and, as a consequence, the exposure period may also have a marked influence on the concentrations in plants. The concentrations of several nutrient elements show seasonal variations, as do many pollutants, among which a number of inorganic and semi-volatile organic compounds tend to increase during the growing season, at least in leaves of trees (Ahrens, 1964, Deu and Kreeb, 1993, Djingova and Kuleff, 1994, Ernst, 1990, Guha and Mitchell, 1966; Heinze and Fiedler, 1992; Knabe, 1982; Markert, 1993b; Martin and Coughtrey, 1982; Nakajima et al., 1995; Plümacher et al., 1993; Umlauf et al., 1994; Wagner, 1990). Furthermore, significant concentration differences between the age classes of conifer needles have been detected for several pollutants (Bäumler et al., 1995; Gaggi and Bacci, 1985; Raisch, 1983; Strachan et al., 1994; Wagner, 1990; Weiss, 1998, 2000b, see also Figure 2 in Section 2.3). Markert (1993b) showed that the seasonal changes in the contents of various elements can even exceed the variations from the analytical methods and the site-by-site differences. As a consequence of these findings the sampling period

Table 1. Significant increase of the concentrations (i, in % of the initial concentrations per day) of some elements in Norway spruce needles from an urban-industrial environment during a sampling period of seven weeks.

Element		i (% d^{-1})	s.e.	n
Cd	N2	0.987	0.317	11
Hg	N1	0.902	0.198	11
Hg	N2	1.069	0.234	11
Pb	N1	1.657	0.332	10
V	N1	1.325	0.457	10
V	N2	0.643	0.229	10

The daily accumulation rate was linearly interpolated from two sampling dates (11, 12 September; repeated sampling: 29, 31 October); N1: ½-year old needles, N2: 1½-year old needles; s.e.: standard error of i (Wimmer et al., 2002, in prep.).

should be kept as short as possible when carrying out a study to detect the site specific impact of air pollution on plants and the age classes of conifer needles must be separated. For some elements even a sampling duration of a few weeks may have a significant influence on the concentrations (Table 1). If a long sampling period cannot be avoided, a subsequent small-scale repetition allows the identification and removal of the influences from the long sampling period with appropriate statistical methods.

2.4.2. External biotic influences

Besides internal, many external biotic and abiotic factors have been identified to influence pollutant or element concentrations in higher plants. The interactions between pest infestations and the nutrient contents of plants are thoroughly described (Bergmann, 1993; Marschner, 1995). Such effects have been also detected for air pollutants. Wagner (1990) reports a significantly enhanced fixation of lead containing particles at surface of leaves sticky with aphid secretions. Observations with respect to pest infestations should be part of the sampling protocol. Differences in the mycorrhizal partners may also vary plant concentrations among individuals of one species (Ernst, 1985, 1990).

2.4.3. External abiotic influences

2.4.3.1. Meteorological conditions

The meteorological conditions during exposure time, before and during sampling can modify element and pollutant concentrations. Based on a dense grid of sites and long-time observations, Stefan and Gabler (1998) detected variations of the annual contents of several nutrients in Norway spruce needles connected with meteorological differences between the years. Wagner (1990) found a pronounced effect of a prolonged rainy period on the Zn and Cd contents of poplar leaves. On the other hand, Krivan and Schaldach (1986) and Krivan and Schäfer (1989) detected no clear influence of artificial water rinsing of needles or heavy rainfall on the concentrations of some heavy metals in Norway spruce needles. For the more volatile organic compounds temperature has a marked influence on the concentrations in plants, but for organic compounds with log octanol/air partition coefficients higher than eight (corresponding to compounds like the polychlorinated dioxins and furans, the higher chlorinated biphenyls and the heavier polycyclic aromatic hydrocarbons) such a temperature dependence of the plant/air partitioning is considered to be very unlikely (Kömp and McLachlan, 1997a). Nevertheless, an observation of the meteorological conditions during the time of exposure, the time immediately before and during the sampling campaign should be part of any biomonitoring project.

2.4.3.2. Soil conditions

Particularly for higher plants the soil conditions may have a marked influence on the element concentrations. This is especially the case for all elements which cycle rapidly in the ecosystems, like several nutrient elements (incl. also the heavy metals Zn, Mn) but also certain non-essential elements such as Cd, whose mobility and uptake via the soil varies considerably with the soil conditions. The main determining soil properties

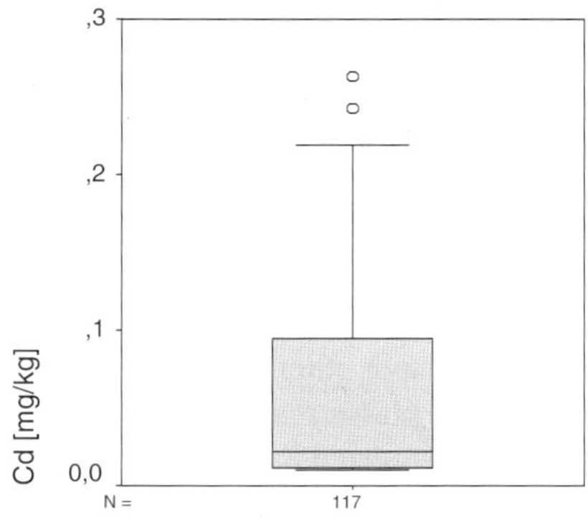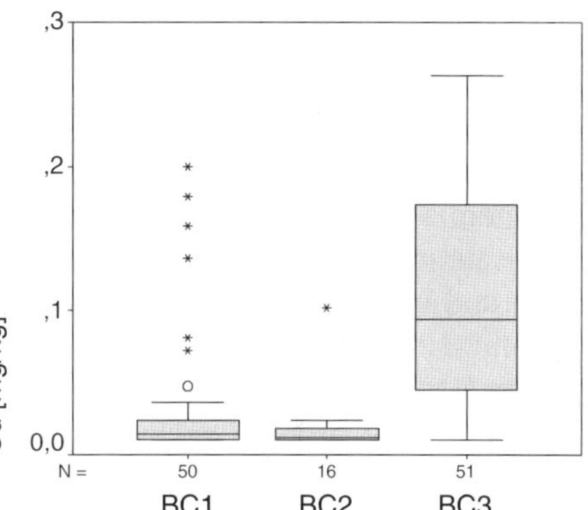

Figure 3. Cadmium concentrations in ½-year old Norway spruce needles from sites of an urban industrial area. Left: All sites pooled, Right: Sites grouped by soil buffer capacity (BC1: high buffer capacity, BC3: low buffer capacity; Wimmer et al., 2002, in prep.)

for the elemental concentrations in plants may be summarised as follows: total element concentration, concentration of the mobile and exchangeable fraction of the element, pH-value, organic matter content, cation exchange capacity, soil texture, soil water content and the presence or absence of competing ions (Bergmann, 1993; Marschner, 1995; Martin and Coughtrey, 1982).

In Figure 3 an illustrative example on the masking effect of the soil properties on the Cd concentrations in Norway spruce needles is given. The boxplot for all sites (left side of Figure 3) shows a non-normal distribution with a small share of sites with clearly higher needle Cd contents. Grouping the sites by soil buffer capacity, which is an indicative parameter for the mobility of Cd in the soils, yields a very different result. The group of sites with a low buffer capacity in soils ("BC3", higher Cd mobility) shows an approximately normal distribution and significant higher Cd concentrations than the groups "BC1" and "BC2" with a higher buffer capacity (right side of Figure 3). It is obvious that the significant higher Cd concentrations of BC3 are very likely to a large extent soil burden. An interpretation approach could incorporate this hypothesis to treat those sites with higher concentrations with regard to the generally increased Cd-levels in BC3. Remarkable are the outliers in BC1 and BC2 (see circles and asterisks in right side of Figure 3). Please note that they are completely obscured in the left side and ungrouped part of Figure 3. The Cd concentrations of these outlier sites in BC 1 and BC 2 seem to be related to other influences than to the soil conditions, probably to atmospheric input. Aiming at the biomonitoring of atmospheric impact, these outlier sites should gather particular attention during interpretation and further analysis. If a biomonitoring study should also identify atmospheric loads of such highly soil dependent elements like Cd, and the soil conditions of the investigation area vary considerably, it is necessary to include a soil survey (or results from such a survey), to use active methods of bioindication (see Section 4) or to use mosses as biomonitors, which are virtually independent from the local soil conditions (see Zechmeister et al., 2002). An approach as given in Figure 1 above (the comparison of washed and unwashed needles) allows to discriminate between atmospheric and uptake from the soil for such pollutants.

Most of this chapter focuses on the use of higher plants as indicators for atmospheric pollutant impact. The present section on the soil's influence gives a good opportunity to address the use of higher plants as monitors for soil contamination or for biogeochemical ore prospecting, which has a very long tradition (Cannon, 1960, several authors in Markert, 1993a; Martin and Coughtrey, 1982; Robinson and Edgington, 1945). Once more it should be emphasised that a premium advantage of bioindication in case *one* (bioindication of soil contamination) is the more effect related and integrative evidence provided with respect to the possible impact on the plant and herbivores. A less contaminated soil may still have other unfavourable soil properties, which in turn lead to higher concentrations in plants than a more contaminated soil. Case *two* (biogeochemical ore prospecting) represents a more practicable and cheaper approach than geochemical techniques, and may be even more successful in certain cases (e.g. deep rooting plants can reveal buried mineralised areas, which are not accessible by surface sampling of soil). It has to be admitted though, that the evidence gathered during an isolated use of one of these two bioindication techniques is limited without additional geochemical and soil analyses, respectively. One of the biggest problems both applications have in common, while being rather different by aim, approach and technique, is the significant dependency of the plant concentrations from a multitude of environmental parameters (see above). Linear relationships between the total soil content and the concentration in plants are more of an exception than a rule. Once again, the results in Figure 3 may serve as an illustrative example: "BC3" with worse conditions for Cd accumulation in the soil has higher Cd concentrations in the needles. The significant discrepancies between the element contents of plant parts with and without cuticles given above show that for the biomonitoring of soil contamination with aerial plant parts a removal of the cuticles may be needed. However, in contaminated environments, where the emitter has already been closed, the opposite approach may be more promising. In such a situation the analysis of unwashed aerial plant material may allow to detect the recent impact by the deposition of wind-eroded contaminated soil particles (Trimbacher and Weiss, 2000).

2.4.3.3. Location of the plant

The location of the plant within the plant community, its social status (predominant to suppressed), the structure and texture of the plant community but also the surroundings of solitary plants may influence the pollutant concentrations of the investigated plant parts. Godt et al. (1985) investigated the heavy metal deposition onto an unstocked area, within a forest and at the forest's edge. A clearly higher deposition of some elements (Fe, Cd, Pb, Zn) was detected at the edge. The social status within a plant community has an influence on the radiation, temperature, wind and deposition imposed on the plant parts as well as on their biometric, biochemical and morphological properties (Hutchinson and Hicks, 1985; Kimmins, 1987; Monteith 1976). As discussed above, all of these factors may have an influence on the concentrations in plants. With respect to the use of trees as accumulative biomonitors, the choice of dominant trees is recommended in monitoring programmes (Knabe, 1982; Stefan et al., 2000; VDI 3792–5). The structure and texture of the plant community around the investigated plants, the surroundings of solitary plants or exposed plant pots of active approaches influence the radiation, the temperature, the exposure to wind and deposition of the investigated plant. For instance, Steubing (1982, 1987) detected significant differences between the heavy metal contents of plants located in the wind shadow of hedges and exposed to the wind. It is therefore important to consider these factors when carrying-out a monitoring study and standardise them as far as possible.

2.4.3.4. Atmospheric pollution and pollution patterns

Last, but most relevant for the objectives of the present book: The degree, duration and variation of atmospheric pollution (short-term vs. long-term atmospheric input, chronic low dose vs. single high dose input) have in many cases a significant effect on the pollutant concentration in plants. These effects are used in accumulative bioindication techniques with higher plants (see Section 4). As already mentioned, the objective of the use of higher plants as accumulative bioindicators for air pollution is not to detect the instantaneous atmospheric pollutant concentration, but to measure the atmospheric impact on plants (i.e. the present concentration in the plant under the given environmental conditions including the amount, duration and variation of the atmospheric pollutant concentration). The various influences on the plant pollutant concentration as discussed above are responsible for the present lack of general valid relationships, which would allow to directly calculate the air concentration from a plant's concentration. Nevertheless, several field studies showed excellent correlations between the concentrations in higher plants and the atmospheric pollution with inorganic and organic compounds detected by air measurements (Guderian, 1970a, 1970c; Heidt et al., 1987; Horntvedt, 1995; Ikeda et al., 2001a, 2001b; Innes, 1995; Kosta-Rick, 1992; Manninen and Huttunen, 1995; Morosini et al., 1993; Nobel et al., 1992).

3. Methodical considerations and limitations

In the previous sections a multitude of possible influences on the pollutant concentrations in higher plants were discussed together with ways to circumvent these influences with respect to biomonitoring of air pollution. The abundance of information may hopefully not cause a frustrated switch back to methods of atmospheric measurement or a rejection of accumulative bioindication techniques with higher plants:

1. In many cases, the concentrations in plants allow a better risk assessment of the existing environmental pollution, particularly for those compounds, which are phytotoxic and/or mainly affect animals and humans via the consumption of plants.
2. For various harmful pollutants chemical or physical methods of atmospheric measurement either do not exist or are overly complicated, resource demanding, expensive or even less sensitive than the bioindication techniques.
3. Section 4 lists a variety of very successful approaches to quantitatively detect the air pollutant impact on higher plants.

The listing of numerous influences on pollutant concentrations in the previous section aims at demonstrating that careful planning and conduction of such a study can avoid artefacts and lead to results which allow a sound interpretation. While this section gives only a brief overview on the most relevant issues, in-depth advice in this field can be found in several comprehensive articles, books and guidelines (Djingova and Kuleff, 1994; Ernst, 1990, 1994; Markert, 1993b; Stefan et al., 2000; Steubing and Jäger,1982; Umweltbundesamt,1996; UN-ECE 1998, VDI 3792–1, VDI 3792–2, VDI 3792–5, VDI 3793–3, VDI 3957–3, VDI 3957–5; Wagner, 1995, 1997; Wagner and Klein, 1995; Zimmermann et al., 2000).

3.1. Selection of the appropriate bioindicator and bioindication technique

Like in every study, the specific problem, the questions to be answered and the objectives are the points to start from. Based on these, the appropriate bioindicator organism and bioindication techniques are selected. Specifying a generally valid approach for bioindication according to the demands of accumulative biomonitoring, the selected higher plant and method should: 1. be sensitive to the pollutants to be measured, 2. give accurate and reproducible signals, 3. be pollutant specific, 4. allow to identify the spatial and temporal resolution of the pollutant impact, 5. provide specific, representative and important results for the region and for the objectives of the study and 6. provide representative and transferable results with respect to other organisms (Halbwachs and Arndt,1992; Zimmermann et al., 2000). It becomes evident that these requirements rely on evidence from previous studies and experiments, advice from literature and/or guidelines as a basis for the selection procedure of a bioindicator and a bioindication method. As far as possible, already existing standardised and well developed methods should be used in biomonitoring studies. If no well established, validated and/or standardised method is available, preliminary experiments are needed to develop the method before carrying out a comprehensive monitoring programme. Otherwise, smaller scale studies, also including other monitoring tools (e.g. physical, chemical or better established methods of bioindication) or different monitoring tools should be chosen. All aspects which were discussed in Section 2 should undergo careful consideration during the planning stage and should be defined as far as possible, respectively. Even the choice of a commonly used or standardised method may be inapt for a specific situation (e.g. the plant of an active biomonitoring technique may not grow under the given climatic conditions of the study area). Therefore, a wise planning process would collect and evaluate all relevant information concerning the study area (e.g. the site conditions like the soil properties and meteorological parameters, relevant emitters in the region) as well as the compounds to be investigated and the methods and bioindicators chosen. The example in Figure 3 (Section 2.4) underlines the usefulness of such an approach in study areas with heterogeneous influences on the pollutant concentrations to be detected. If such information on important influencing factors is missing, preliminary investigations or the inclusion of an investigation on these factors in the biomonitoring programme may be required.

3.2. Selection and number of sites

One crucial question is the selection and number of sites within a biomonitoring programme. Again, a general valid advice cannot be given. Few sites with different distances to roads were needed to give evidence for the heavy metal impact of motor vehicle traffic on higher plants (e.g. Keller and Preis, 1967; Wäber et al., 1996). It seems that for the detection of the impact of particle bound air pollutants with short

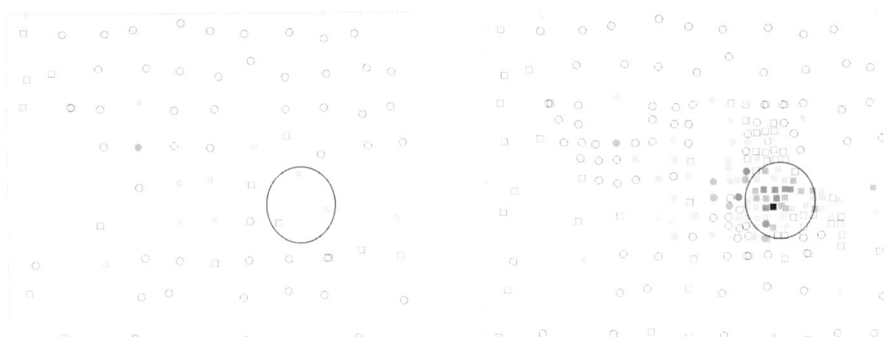

Figure 4.   Iron in ½-year old Norway spruce needles (circles) and poplar leaves (squares) of an urban-industrial environment (Wimmer et al., 2002, in prep.). Left: regional distribution of the Fe-concentrations as observed with a 2 × 2 km grid ($n = 91$). Right: enhanced spatial resolution due to an additional condensed grid (0.5 × 0.5 km, $n = 221$). The ellipse represents the area of steel works.

Concentration classes: Class 0, colourless: "urban background" or "normal" concentration range calculated individually for both tree species according to Erhardt et al. (1996) with a modified classification scheme for elevated concentrations; Spruce needles (condensed grid): 45.0–151.0 mg Fe/kg d.w. (Mean: 98.7 mg/kg d.w.), Poplar leaves (condensed grid): 75.0–259.2 mg Fe/kg d.w. (Mean: 163.2 mg/kg d.w.). Class 1, light grey: > class 0 and up to two times higher than the maximum of class 0. Class 2, dark grey: > class 1 and up to four times higher than the maximum of class 0. Class 3, black: > class 2 and up to eight times higher than the maximum of class 0.

transmission distances an approach with a few sites at different distances from the line source will be sufficient. This is only valid, however, if significant contributions of other emitters or other major disturbing influences to the pollutant concentrations can be excluded. For the biomonitoring of an isolated point source, Zimmermann et al. (2000) recommend a minimum of seven sites (in addition to the possibly required reference measurements), distributed according to a defined scheme which is based on the anticipated pollutant dispersion or the distribution of the wind directions. For the detection of a long-range pollutant impact at sites remote from sources valuable outcomes can be gained from a few carefully chosen sites only, if appropriate selection criteria are strictly obeyed (Weiss et al., 2000). Likewise, a small number of study sites are sufficient to explore time trends in the pollution data (Herman, 1998). These examples show that the use of reasonable selection criteria may help to reduce the number of sites without jeopardising the significance of the study.

A lot of sample points may be needed if information on the pollutant dispersion around a source is not available, if the emitter is located between other potential sources or if the environmental conditions in the study area are heterogeneous. Regular grids of sufficient density may be required to get representative, unbiased results for large study areas and large-scale monitoring programmes. In areas of particular interest (e.g. close to emitters) an additional condensed grid may help to study the local impact of or the small-scale variations around such sources (e.g. Stefan and Fürst, 1998). Figure 4 represents an example for this approach in a heterogeneous urban-industrial environment. The results in Figure 4 also point out the advantages of a generous sampling design, although – at first glance – they may exceed the actual objectives of the study: both, the origin and the intensification grid were sampled during one campaign. In a first step, the samples of the original coarse grid were analysed (left side of Figure 4). The second step, an additional analysis of the plant samples from the intensification sites allowed to resolve the Fe impact of the steel works (right side of Figure 4). Without that generous sampling design in the first place, a further study and undoubtedly more resources would have been needed for the same evidence. The planning of a biomonitoring study with higher plants should be guided by the rationale that the measurements cannot be repeated under the same conditions. This is another out of several reasons (see the implementation steps below) for an inclusion of reference plant material (e.g. from sites with a generally low pollution load, or at least without significant pollution by the investigated compounds).

3.3. Duration of exposure/sampling date

The decision about the time of exposure (active biomonitoring) or sampling (passive biomonitoring) of the plants is a further important point. With the exception of evergreen plants, like most conifers, these periods are restricted to the growing season. If the winter impact (e.g. by domestic heating systems, Köhler et al., 1995) has to be investigated, only evergreen plants can be employed. For passive biomonitoring methods the progressive change of the pollutant concentrations with the development of the investigated plant parts is one more point of consideration (see Section 2). The final conclusion on the sampling or exposure time is based on the objectives of the study and is decisive for the selection of the plant, plant part and method.

3.4. Guidelines

Before commencing the project, the method, from sampling design to the chemical analysis, should be clarified and laid down in unambiguous and reproducible guidelines, e.g. in form of standard operating procedures or a case-specific study plan (OECD, 1997). This is a prerequisite for sound, reproducible and comparable results, particularly in large-scale and long-term biomonitoring programmes involving a large number of personnel staff or institutes (e.g. UN-ECE, 1998; Stefan et al., 2000). Test phases and round robins for the inter- and intra-calibration and harmonisation of sampling, sample treatment and analysis are a further requirement, even more for large monitoring programmes, to avoid "the mere detection of the different bias of each individual team within a study". One of the most important reasons for "data cemeteries" (i.e. irreproducible data which cannot be used anymore) is the missing of any sound documentation concerning the applied methods, implementation of the study, and relevant "meta-information" (i.e. "obvious" personal expertise and know-how).

3.5. Adequate resources

Specifically all steps of a biomonitoring study which cannot be repeated (i.e. all steps from sampling to sample treatment before chemical analysis), should be allotted adequate labour-, material- and time resources (e.g. good qualification of the staff, sufficient time to avoid sloppy work, state-of-the-art material to avoid artefacts).

3.6. Sampling

Important sampling and processing criteria include the selection of the plant material (individual, plant part, location, sample volume), measures to avoid interfering contamination and pollutant losses during sampling, storage and sample treatment and others (further details are given in the method focused literature cited at the beginning of Section 3). The sampling, storage and sample treatment of plants which are analysed for organic compounds is more demanding than that required for the analysis of inorganic compounds, because volatilisation, degradation, further accumulation or contamination happens much easier (VDI 3957–3; Weiss et al., 1998b; Zimmermann, 1995). Prepared sampling forms featuring indexed fields for each collectable parameter (e.g. possible influencing factors), should be used. It may be the chronic underestimation of plant sampling and sample treatment as "unscientific" matters that these implementation steps are prone to introduce larger errors in biomonitoring studies (Markert, 1992).

3.7. Sample treatment

One frequent question is whether the plant sample should be cleaned and washed before chemical analysis. Moreover, the individual cleaning techniques (e.g. brushing, stripping techniques, water rinsing, washing with organic solvents) lead to incompatible results. Numerous papers compiled and discussed different cleaning procedures and their advantages and disadvantages with respect to the objectives of biomonitoring studies (e.g. Djingova and Kuleff, 1994; Krivan and Schaldach, 1986; Markert,1993b; Zimmermann et al., 2000). Seeking the advice in the literature with respect to this question will retrieve somehow controversial information. The proposals in Table 2 may help to make a decision whether and how the samples should be washed. As can be seen, it may become necessary to carry out all three proposed treatments if several of the objectives are valid for the study. There is no general answer to the question which washing treatment will serve the investigator's purposes best. Rather, the nature of the pollutant together with existing methodical experience will decide, if a clearer evidence with respect to the objectives can be given by the analysis of washed or untreated samples, or if it is even required to compare unwashed and cleaned material – like demonstrated in Figure 1 (Section 2.2). The issue is further complicated by the fact that the outcomes of an identical washing procedure vary between species (see overview in Djingova and Kuleff, 1994).

3.8. Chemical analysis

It is beyond the scope of the present chapter to discuss the various suitable techniques to analyse plant parts chemically. Two general considerations, however, may be given: 1. The methods of chemical analysis evolve and change very quickly. 2. The attempt to avoid variations between different methods, laboratories and personnel (see e.g. Stefan et al., 1997) should be one of the guiding principles in the planning process of a biomonitoring study. To minimise adverse consequences of 1 and 2 and ensure the long-term usefulness of a biomonitoring study, one should provide an adequate stock of reserve and reference samples, moreover round robins and a detailed documentation of the procedures, as is obligatory for accredited laboratories.

Table 2. Decision guidance for a selection of washing procedures in an accumulative biomonitoring study with higher plants (combination of these methods may be necessary).

Objective(s) of the study	No washing	Washing with water	Washing with organic solvents (i.e. removal of the cuticle)
Detection of atmospheric pollutant input	X[1]		
Evidence for an atmospheric impact of pollutants (if the approach above is not sufficient)	X[3,4]		X[3,4]
Detection of the impact through the soil			X[2,3]
Distinction between air borne and soil burden pollutant accumulation	X[2,3,4]		X[2,3,4]
Detection of the contents in interior plant parts (e.g. for physiological or nutritional considerations)			X
Impact for human plant consumers		X[5]	
Impact for animal plant consumers	X		
Studies on the pollutant transfer through the cuticle (e.g. for xenobiotics without soil uptake or stomatal uptake)	X[3,4]		X[3,4]
Compensating effects of intensive rain events during sample collection		X	

[1] Only valid for strongly differing air pollution impact situations and for passive methods, which are sensitive for the investigated compound, or for standardised active biomonitoring methods.

[2] Only valid for compounds which are also taken up from the soil.

[3] Only valid for air pollutants which are mainly taken up by and accumulated in the cuticle and not valid for pollutants which primarily enter through the stomata or rapidly permeate the cuticle.

[4] These objectives require two chemical analyses per sample: separate analysis of the plant part before and after washing.

[5] Simulation of the plant preparation in kitchen.

3.9. Documentation and statistical analysis

Nevertheless, neither the best preparation nor the most diligent execution of a monitoring study can exclude artefacts or the interference of unexpected influences at single sample points. Therefore it is good to have tools to identify and explain such conspicuous observations (e.g. quality control procedures and a thorough documentation of all relevant implementation steps, site conditions and other potentially important circumstances during the sampling). An illustrative example of how a sampling protocol helped to identify the reason for an apparently unexplainable concentration peak is given in Figure 5.

Even such untypical results applying to only a single sampling point must be duly evaluated unless the value has already been pinpointed to artefacts incurred

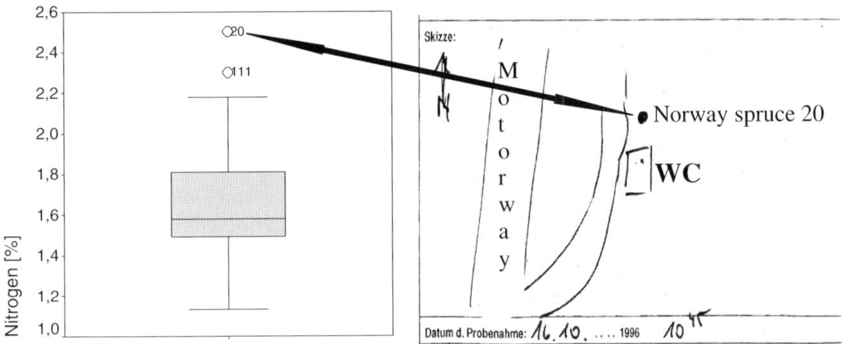

Figure 5. Explanation of the extremely high nitrogen value observed for one of 117 Norway spruce needle samples (½-year old) from an industrial urban area (Wimmer et al., 2002, in prep.). The telltale sketch-map of the site (right, translated), drawn as part of the sampling protocol, marks the position of the sample tree – and of a roadside public lavatory.

during sampling, sample preparation or analytical procedures. One of the advantages of biomonitoring is the direct detection of those concentrations occurring in the plant: for the plant as well as its consumer the pollutant content of the individual counts rather than the statistical classification of the respective site (which might swiftly be disregarded as a statistical outlier). Besides, closer investigations of such single outliers often revealed surprising or new information about environmental pollution or allowed improvements of monitoring techniques. With this example, the authors would like to stress the need for a careful analysis, interpretation and evaluation of the results, including also the statistical outliers which are often dismissed for their mere categorisation as statistical outliers (or even deleted uncritically). No need to mention that analysis and interpretation of the results require the appropriate statistical tools, and a foray through the relevant literature. Particularly bioindication studies, which still seem to have an ambivalent reputation in some circles, will benefit from this approach through an increased acceptance by the public and the decision-making authorities.

4. Examples for methods and studies

4.1. General remarks

A complete compilation of methods would be beyond the scope of most bioindication monographs. Therefore, the present section can only offer examples of frequently used methods with respect to the various existing questions. More detailed and comprehensive compilations on methods and plant species are given for instance in Arndt et al. (1987), Markert (1993a), Martin and Coughtrey (1982), Mulgrew and Williams (2000).

In biomonitoring, two techniques are distinguished: active and passive methods. Passive methods use biological material taken from the site of growth, while active methods use defined plant material that is grown and exposed in a defined way. Table 3 gives a comparison of these two methods which may help in decision finding, but should

Table 3.   A comparative outline of active and passive methods for accumulative bioindication with higher plants (X indicates the principally better suited approach for the criterion).

	Active biomonitoring	Passive biomonitoring
Standardisation (e.g. exposure time, soil conditions, plant material, location of the investigated plant)	X	
Comparability and reproducibility of results	X	
Pollutant to be monitored requires control of endogenous and exogenous influences	X	
Monitoring of elements with rapid biological turnover or of pollutants which are very sensitive to site conditions	X	
Heterogeneous investigation area (e.g. with respect to soil conditions, plant distribution)	X	
Highly artificial environments are part of the study area (e.g. industrial and urban areas without suitable plants as passive indicators)	X	
Independence from seasonal plant development required (e.g. monitoring of the seasonal change in pollution)	X	
Risk assessment and supply of ecotoxicological evidence		X
Integration of effects under natural conditions		X
Representative results and meaningful transfer of results on other organisms (e.g. plant consumers)		X
Retrospective analysis (e.g. the impact of an accident)		X
Limited infrastructural, labour or time resources		X
Costs		X
Large-scale monitoring networks		X

not be mistaken as a conclusive evaluation of both methods. The appropriate choice depends very much on the specific framework of the study.

A random evaluation of the relevant literature suggests that passive methods play still a quantitatively more important role than active ones. With the introduction of standardised active methods in recent years, however, the use of active methods has been steadily increasing. An important share of passive monitoring studies used trees, particularly conifers, while rye grass (*Lolium multiflorum* Lam.) is probably the most frequently used species in active biomonitoring. Besides trees, a great number of other higher plant species can be found in passive monitoring studies. The literature on active methods is restricted to very few species, mainly rye grass, kale (*Brassica oleracea*

L. *var. acephala*) and Norway spruce (*Picea abies* [L.] Karst.). In most studies, assimilating plant parts were monitored. To assess the impact on humans, however, fruits and other edible plant parts are investigated. Forage plants (e.g. grasses, corn) are used to examine the impact on livestock and subsequent consumers.

4.2. Passive biomonitoring

The European "ICP Forest" monitoring programme under the UN-ECE "Convention on Long-range Transboundary Air Pollution" and under the EU is a high ranking example for a large scale monitoring programme which includes also the investigation of tree leaves for their elemental contents. Nutrient elements in the leaves of spruce, pine, beech and oak species are analysed to control the nutritional status of the trees, but some noxious elements are also included in the observations (Stefan et al., 1997; Rademacher, 2001). The investigation methods are laid down in a detailed manual (Stefan et al., 2000). The "ICP Forest" programme integrates to some extent the results of national foliar forest condition surveys. Starting with the emerging debate on "novel forest decline" several large-scale surveys of accumulative biomonitoring with trees were initiated. They partly own a long tradition and specific designs according to the individual conditions, objectives and requirements of each country or province. Frequently, S, F, Cl and also some heavy metals in tree leaves are measured in these surveys (e.g. Gulder and Kölbel, 1993; Heidingsfeld, 1992; Knabe, 1983; Landolt et al., 1989; Stefan and Fürst, 1998).

Different from such large-scale monitoring programmes, intense forest ecosystem studies and other local forest monitoring programmes frequently include the measurement of pollutant accumulation in trees (but also in other plants like mosses and herbs). The studies of Mayer (1981) and Raisch (1983) may serve as examples. The "ICP Integrated Monitoring Programme" under the UN-ECE "Convention on Long-range Transboundary Air Pollution" represents a co-operative programme between long-term ecosystem monitoring sites in Europe (UN-ECE 2001). Local long-term monitoring studies are often well suited to detect trends of the air pollutant impact on higher plants. For instance, Grünhage and Jäger (1988) and Herman (1998) suggested gradual decrease of lead concentrations in tree leaves during the last decades, referring to long-term observations at local study sites. In addition, the results of Herman (1998) were used to develop a new scheme for the classification of the lead contents in Norway spruce needles with respect to the pollution impact.

Tree leaves are also a common indicator of the environmental impact of point and line sources, or urban areas. For this purpose, not only forest trees but also solitaire trees are investigated, sometimes utilising less conventional tree species. For instance, Wagner (1987) developed a standardised method for Lombardy poplar (*Populus nigra* "Italica") which has successfully been used in several other monitoring programmes and regions (Djingova et al., 1995, 1996, 1999; Wagner, 1993). Deu and Kreeb (1993) and Öhlinger (2000) tested the suitability of several fruit tree leaves as bioindicators for heavy metals. Based on comprehensive data sets Öhlinger (2000) derived reference values for the leaves of five fruit trees. Like in the above mentioned forest monitoring programmes, the frequently investigated tree species in environments close to emitters or in conurbations in Europe are:

- *Picea abies* Karst.: e.g. Lick and Dorfer, 1998; Keller and Preis, 1967; Keller et al., 1986; Mankowska, 1984, 1998; Mankovska and Steinnes, 1995; Mukherjee and Nuorteva, 1994; Tichy, 1996; Trimbacher and Weiss, 1997, 1999, 2000; Vogel and Riss, 1992; Wagner and Müller, 1979; Weiss and Trimbacher, 1998, (see compilation in Weiss, 1999),
- *Pinus spp.:* e.g. Dmuchowski and Bytnerowicz, 1995; Kratz, 1996; Mukherjee and Nuorteva, 1994; Rautio et al., 1998; Vike and Habjorg, 1995; Weißflog et al., 1994;
- *Fagus sylvatica* L. and *Quercus spp.:* e.g. Alfani et al., 2000; Keller et al., 1994; Mankovska, 1984, 1998; Mankovska and Steinnes, 1995; Monaci et al., 2000.

Most of the cited studies deal mainly with inorganic substances. However, the accumulative behaviour of needles from spruce and pine is equally useful to detect the atmospheric impact of semivolatile and some other organic compounds at:

- remote or rural sites (Calamari et al., 1994; Eriksson et al., 1989; Fiedler et al., 1995; Frank, 1991; Frank et al., 1992, 1994; Gaggi et al., 1985; Holoubek et al., 2000; Höpker, 1992; Köhler et al., 1995; Notarianni et al., 1998; Plümacher and Schröder, 1994; Reischl et al., 1990a, 1990b; Sinkkonen et al., 1995; Thompson and Treble, 1995; Tremolada et al., 1996; Weiss et al., 1998a, 1998b, 2000b; Weißflog et al., 1999a; Wenzel et al., 1997)
- and at sites close to various emission sources (Hinkel et al., 1989; Holoubek et al., 2000; Höpker, 1992; Ikeda et al., 2001b, 2001c; Juuti et al., 1995; Kolic et al., 1995; Kratz, 1996; Notarianni et al., 1998; Plümacher et al., 1993; Reischl et al., 1990b; Riss et al., 1990; Sinkkonen et al., 1996; Thompson and Treble, 1995; Umlauf et al., 1990; Weiss et al., 2001; Weißflog et al., 1999b; Wenzel et al., 1997).

Unfortunately, these various studies on organic pollutants show large methodical differences so that a sound comparison of data is seldom possible. Standardisation and harmonisation in this field is much required. The guidelines for other biomonitoring programmes which use trees (Stefan et al., 2000; VDI 3792–5) would be a starting point, however, these methodical instructions have to be adapted according to the specific sampling, storage and sample treatment requirements for organic compounds (VDI 3957–3; Weiss et al., 1998b; Zimmermann, 1995). In addition, the amount of plant material needed for the analysis of organic compounds is considerable higher.

Apart from the leaves, bark (overview in Walkenhorst et al., 1993), tree rings (overview in Hagemeyer, 1993) and pollen were used as bioindication tools for inorganic pollutants. The use of tree rings for trace metal pollution, however, was several times critically reassessed (Hagemeyer, 1993; Nabais et al., 1999).

As mentioned in Section 4.1, several other higher plants were used in response to the specific objectives and ecological conditions in various study areas. Suggestions for tropical regions are given in Arndt et al. (1987) and Franz-Gerstein (1996). In Northern America, also leaves of *Acer spp.* have frequently been analysed (Smith, 1972, 1973).

A widely used passive accumulator for inorganic and semivolatile organic compounds in agriculture is pasture (Fomin et al., 1996; Vogel and Riss, 1992). However, possible contamination by soil/plant transfer, soil particles or volatilisation of soil bound compounds has to be considered during the interpretation of such data.

Therefore, an analysis of grass samples should always include an investigation of the soil. In connection with animal bioindicators like cow's milk, the pollutant concentration in grass is used to assess the transfer risks and possible impacts along the food chain (e.g. Glawischnig and Halbwachs, 1987; Krause et al., 1993; Riss et al., 1990). National or regional limit values for harmful compounds in grass (or fodder) often allow a direct legal intervention in case of contamination. Öhlinger (2000, 2002) analysed various heavy metals and other inorganic pollutants in many grass samples of study areas with different ecological conditions, thus providing natural reference concentrations for these elements.

Fruits and vegetables are often included in biomonitoring programmes around sources to assess the alimentary load imposed on humans (Kasperowski,1993; Nobel et al., 1992; Vogel et al., 1991; Wickström et al., 1986; Woidich et al., 1981). Similar to fodder, limit values often facilitate the evaluation of pollutant concentration in agricultural products. Section 2.4 already dealt with the detection of soil contamination and the identification of mineralised soil layers in connection with ore prospecting, both of which are special fields of bioindication with higher plants. An example for the former purpose is the analysis of agricultural plants to control the soil/plant transfer of noxious compounds following the agricultural application of organic recycling material like sewage sludge (e.g. Kampe, 1987; Witte, 1989). Overviews of methods to use higher plants in ore prospecting are given in Markert (1993a) and Martin and Coughtrey (1982).

4.3. Studies which use active methods

In many biomonitoring programmes ryegrass cultures (*Lolium multiflorum* Lam.) are exposed to detect the atmospheric impact with inorganic and organic compounds. The plants are grown six to seven weeks in greenhouses and then exposed for 28 days at defined height (1.5 m above ground). Detailed descriptions on the methods and practical hints can be found in Erhardt et al. (1994), Kosta-Rick and Arndt (1987), Scholl (1987), VDI 3792–1, VDI 3792–2, VDI 3792–3, VDI 3957–2. Several inorganic pollutants like sulphuric or fluorine compounds and heavy metals are routinely measured with this method. Ryegrass cultures are successfully used close to line- and single point sources, but also in regional monitoring networks with heterogeneous environmental conditions (Andre, 1992; Bockholt, 1987; Fomin et al., 1996; Frauendorfer, 1987, 1992, 1996; Kosta-Rick, 1992; Nobel and Michenfelder, 1987; Nobel and Maier-Reiter, 1996; Öhlinger and Döberl, 1992; Öhlinger 2000, 2002; Peichl et al., 1994; Peichl, 2001; Scholl, 1992; Steubing, 1987; Wäber et al., 1996). The feasibility of this method for semivolatile organic pollutants has been demonstrated by several biomonitoring studies (Höpker, 1992; Nobel et al., 1992, 1993; Nobel and Maier-Reiter, 1996; Öhlinger, 2000; Peichl, 2001). The pollutant concentrations are frequently referenced against official limit values for green fodder.

In recent years the spectrum of indicator plants for active biomonitoring has been extended to kale (*Brassica oleracea acephala*) and Norway spruce (*Picea abies*). The corresponding methods are described in detail in Zimmermann and Baumann (1994) and VDI 3957–5 for spruce and in Radermacher and Rudolph (1994) and VDI 3957–3 for kale. Both species have a distinct waxy surface layer on the leaves/needles, which qualifies them for the monitoring of semivolatile organic compounds. In some studies, however, inorganic compounds were also analysed (Andre, 1992; Feist et al., 1995; Höpker,1992; Nobel et al., 1992, 1993; Nobel and Maier-Reiter, 1996; Zimmermann, 1990).

References

Alfani, A., Baldantoni, D., Maisto, G., Bartoli, G., Virzo De Santo, A., 2000. Temporal and spatial variation in C, N, S and trace element contents in the leaves of *Quercus ilex* within the urban area of Naples. Environ. Poll. 109, 119–129.

Ahrens, E., 1964. Untersuchungen über den Gehalt von Blättern und Nadeln verschiedener Baumarten an Kupfer, Zink, Bor, Molybdän und Mangan. Allg. Forst- u. Jagdzeitung 135, 8–16.

Andre, W., 1992. Biologische Meßverfahren mit Kulturpflanzen zur Bewertung von regionalen und anlagenbezogenen Immissionswirkungen – Beispiele angewandter Bioindikation. In: VDI (Ed.), Bioindikation – ein wirksames Instrument der Umweltkontrolle, Vol. 1. VDI-Verlag, Düsseldorf, VDI-Bericht 901, pp. 495–511.

Arndt, U., Fomin, A., Lorenz, S. (Eds), 1996. Bioindikation – Neue Entwicklungen, Nomenklatur, Synökologische Aspekte. Heimbach, Ostfildern.

Arndt, U., Nobel, W., Schweizer, B., 1987. Bioindikatoren – Möglichkeiten, Grenzen und neue Erkenntnisse. Ulmer, Stuttgart.

Asche, N., 1985. Komponenten des Schwermetallhaushaltes von zwei Waldökosystemen. VDI-Verlag, Düsseldorf, VDI-Bericht 560, pp. 357–386.

Bacci, E., Gaggi, C., 1987. Chlorinated hydrocarbon vapours and plant foliage: kinetics and application. Chemosphere 16, 2515–2522.

Bacci, E., Cerejeira, M.J., Gaggi, C., Chemello, G., Calamari, D., Vighi, M., 1990. Bioconcentration of organic chemical vapours in plant leaves: the Azalea model. Chemosphere 21, 525–535.

Bäumler, R., Goerttler, T., Zech, W., 1995. Nährelement- und Schwermetallgehalte in den Nadeln von Fichten und Tannen eines Bergmischwaldes auf Flysch (Tegernseer Alpen). Forstw. Cbl. 114, 30–39.

Bergmann, W., 1993. Ernährungsstörungen bei Kulturpflanzen, 3rd edn. Gustav Fischer Verlag, Jena.

BGBL, 1984. Verordnung des Bundesministers für Land- und Forstwirtschaft vom 24. April 1984 über forstschädliche Luftverunreinigungen. – 199. Verordnung: Zweite Verordnung gegen forstschädliche Luftverunreinigungen. Austria.

Bidleman, T.F., 1988. Atmospheric processes: wet and dry deposition of organic compounds are controlled by their vapour-particle partitioning. Environ. Sci. Technol. 22, 361–367.

BIMSCHG, 1990. Gesetz zum Schutz vor schädlichen Umwelteinwirkungen durch Luftverunreinigungen, Geräusche, Erschütterungen und ähnliche Vorgänge. BGBL 1.

Bockholt, B., 1987: Anlagenbezogene Ermittlung der räumlich-zeitlichen Fluor-Immissionsbelastung durch Anwendung des Verfahrens der standardisierten Graskultur. VDI (Ed.), Bioindikation – wirkungsbezogene Erhebungsverfahren für den Immissionsschutz. VDI-Verlag, Düsseldorf, VDI-Bericht 609, pp. 317–336.

Böhme, F., Welsch-Pausch, K., McLachlan, M., 1999. Uptake of airborne semivolatile organic compounds in agricultural plants: field measurements of interspecies variability. Environ. Sci. Technol. 33, 1805–1813.

Briggs, G.G., Bromilow, R.H., Evans, A.A., 1982. Relationships between lipophilicity and root uptake and translocation of non-ionised chemicals by barley. Pestic. Sci. 13, 495–504.

Buckley, E.H., 1982. Accumulation of airborne polychlorinated biphenyls in foliage. Science 216, 520–522.

Calamari, D., Tremolada, P., Di Guardo, A., Vighi, M., 1994. Chlorinated hydrocarbons in pine needles in Europe: fingerprint for the past and recent use. Environ. Sci. Technol. 28, 429–334.

Cannon, H.L., 1960. Botanical prospecting for ore deposits. Science 132, 591–598.

Coupland, D., 1991. The role of compartimentation of herbicides and their metabolites in resistance mechanisms. In: Caseley, J.C., Cussans, G.W., Atkin, R.K. (Eds), Herbicide resistance in weeds and crops. Butterworth, Oxford, pp. 263–278.

Deu, M., Kreeb, K.H., 1993. Seasonal variations of foliar metal content. In: Markert, B. (Ed.), Plants as Biomonitors. VCH-Verlag, Weinheim, pp. 577–600.

Djingova, R., Kuleff, I., 1994. On the sampling of vascular plants for monitoring of heavy metal pollution. In: Markert, B. (Ed.), Environmental Sampling for Trace Analysis. VCH, Weinheim, pp. 395–414.

Djingova, R., Wagner, G., Peshev, D., 1995. Heavy metal distribution in Bulgaria using Populus nigra "Italica" as a biomonitor. Sci. Total Environ. 172, 151–158.

Djingova, R., Wagner, G., Kuleff, I., Peshev, D., 1996. Investigations on the time-dependent variations in metal concentrations in the leaves of Populus nigra "Italica". Sci. Total Environ. 184, 197–202.

Djingova, R., Wagner, G., Kuleff, I., 1999. Screening of heavy metal pollution in Bulgaria using *Populus nigra* "Italica". Sci. Total Environ. 234, 175–184.

Dmuchowski, W., Bytnerowicz, A., 1995. Monitoring environmental pollution in Poland by chemical analysis of Scots Pine (*Pinus sylvestris* L.) needles. Environ. Poll. 87, 87–104.

EC-UN/ECE, 1995. European programme for the intensive monitoring of forest ecosystems. General information on the permanent observation plots in Europe (Level II). International Co-operative Programme on Assessment and Monitoring of Air Pollution Effects on Forests (ICP Forest). EC, UN-ECE, Brussels, Geneva.

Erhardt, W., Fischer, I., Wildenmann, K., 1994. Bioindikationsmethoden – Aktive Verfahren: Standardisierte Graskultur. UWSF-Z. Umweltchem. Ökotox. 6, 219–222.

Erhardt, W., Höpker, K.A., Fischer, I., 1996. Verfahren zur Bewertung von immissionsbedingten Stoffanreicherungen in standardisierten Graskulturen. UWSF-Z. Umweltchem. Ökotox. 8, 237–240.

Eriksson, G., Jensen, S., Kylin, H., Strachan, W., 1989. The pine needle as a monitor for atmospheric pollution. Nature 341, 42–44.

Ernst, W.H.O., 1985. Schwermetallimmissionen – ökophysiologische und populationsgenetische Aspekte. Düsseldorfer Geobot. Kolloq. 2, 43–57.

Ernst, W.H.O., 1990. Element allocation and (re)translocation in plants and its impact on representative sampling. In: Lieth, H., Markert, B. (Eds), Element Concentration Cadasters in Ecosystems. VCH, Weinheim, pp. 17–40.

Ernst, W.H.O., 1994. Sampling of plants for environmental trace analysis in terrestrial, semiterrestrial and aquatic environments. In: Markert, B. (Ed.), Environmental Sampling for Trace Analysis. VCH, Weinheim, pp. 381–394.

Feist, B., Thuß, U., Popp, P., 1995. Contents of PCDD/PCDF in soil kale and deposition in an area in southern Saxony-Anhalt. Organohal. Compds. 24, 333–336.

Fiedler, H.J., Nebe, W., Hoffmann, F., 1973. Forstliche Pflanzenernährung und Düngung. Gustav Fischer Verlag, Stuttgart.

Fiedler, H., Lau, C., Cooper, K., Andersson, R., Kulp, S.E., Rappe, C., Howell, F., Bonner, M., 1995. PCDD/PCDF in soil and pine needle samples in a rural area in the United States of America. Organohal. Compds. 24, 285–292.

Figge, K., 1990. Luftgetragene, organische Stoffe in Blattorganen – Vorgang der ad-/absorptiven Anreicherung. UWSF-Z. Umweltchem. Ökotox. 2, 200–207.

Fomin, A., Ansel, W., Arndt, U., 1996. Bioindikatormeßnetz Niederlausitz – Akkumulation von Fluor in Standardisierter Graskultur und in natürlich aufgewachsenen Gräsern. In. Arndt, U., Fomin, A., Lorenz, S. (Eds), Bioindikation – Neue Entwicklungen, Nomenklatur, Synökologische Aspekte. Heimbach, Ostfildern, pp. 213–220.

Frank, H., 1991. Airborne chlorocarbons, photooxidants and forest decline. Ambio 20, 13–18.

Frank, H., Frank, W., 1986. Photochemical activation of chloroethenes leading to destruction of photosynthetic pigments. Experientia 42, 1267–1269.

Frank, H., Scholl, H., Renschen, D., Rether, B., Laouedj, A., Norokorpi, Y., 1994. Haloacetic acids, phytotoxic secondary air pollutants. ESPR – Environ. Sci. & Pollut. Res. 1, 4–14.

Frank, H., Scholl, H., Sutinen, S., Norokorpi, Y., 1992. Trichloroacetic acid in conifer needles in Finland. Ann. Bot. Fennici 29, 263–267.

Franz-Gerstein, C., 1996. Entwicklung und Erprobung von Bioindikatoren für die Tropen und Subtropen. In: Arndt, U., Fomin, A., Lorenz, S. (Eds), Bioindikation – Neue Entwicklungen, Nomenklatur, Synökologische Aspekte. Heimbach, Ostfildern, pp. 79–97.

Frauendorfer, B., 1987. Das Bioindikatorenmeßprogramm (BIMP) in Hamburg. In: VDI (Ed.), Bioindikation – wirkungsbezogene Erhebungsverfahren für den Immissionsschutz. VDI-Verlag, Düsseldorf, VDI-Bericht 609, pp. 271–286.

Frauendorfer, B., 1992. Bioindikation von Luftschadstoffen in Hamburg – Ergebnisse des Meßprogramms BIMP 1980–1990. VDI (Ed.), Bioindikation – ein wirksames Instrument der Umweltkontrolle, Vol. 2. VDI-Verlag, Düsseldorf, VDI-Bericht 901, pp. 765–778.

Frauendorfer, B. 1996. Bioindikation von Luftverunreinigungen mit Lolium multiflorum Lam. – Erfahrungen im Dauerkulturbetrieb. In: Arndt, U., Fomin, A., Lorenz, S. (Eds), Bioindikation – Neue Entwicklungen, Nomenklatur, Synökologische Aspekte. Heimbach, Ostfildern, pp. 239–244.

Gaggi, C., Bacci, E., 1985. Accumulation of chlorinated hydrocarbon vapours in pine needles. Chemosphere 14, 451–456.

Gaggi, C., Bacci, E., Calamari, D., Fanelli, R., 1985. Chlorinated hydrocarbons in plant foliage: an indication of the tropospheric contamination level. Chemosphere 14, 1673–1686.

Glawischnig, E., Halbwachs, G., 1987. Das Rind als Indikator für Fluorimmissionen. In: VDI (Ed.), Bioindikation – wirkungsbezogene Erhebungsverfahren für den Immissionsschutz. VDI-Verlag, Düsseldorf, VDI-Bericht 609, pp. 123–131.

Godt, J., Mayer, R., Georgi, B., 1985. Die Interceptionsdeposition als wichtiger Faktor der Schwermetallbelastung von Waldökosystemen. VDI-Verlag, Düsseldorf, VDI-Bericht 560, pp. 333–356.

Grünhage, L., Jäger, H.J., 1988. Entwicklung der Nährstoff- und Schwermetallgehalte in Fichtennadeln aus dem Rhein-Main-Gebiet. Angew. Botanik 62, 85–91.

Guderian, R., 1970a. Untersuchungen über quantitative Beziehungen zwischen dem Schwefelgehalt von Pflanzen und dem Schwefeldioxidgehalt der Luft – I. Teil. PflKrankh. 4/5, 200–220.

Guderian, R., 1970b. Untersuchungen über quantitative Beziehungen zwischen dem Schwefelgehalt von Pflanzen und dem Schwefeldioxidgehalt der Luft – II. Teil. PflKrankh. 6, 289–308.

Guderian, R., 1970c. Untersuchungen über quantitative Beziehungen zwischen dem Schwefelgehalt von Pflanzen und dem Schwefeldioxidgehalt der Luft – III. Teil. PflKrankh. 7, 387–399.

Guha, M.M., Mitchell, R.L., 1966. The trace and major element composition of the leaves of some deciduous trees. Plant and Soil 24, 90–112.

Gulder, H.J., Kölbel, M., 1993. Waldbodeninventur in Bayern. Forstliche Forschungsberichte München 132, Schriftenreihe der Forstwissenschaftlichen Fakultät der Universität München und der Bayerischen Forstlichen Versuchs- und Forschungsanstalt.

Hagemeyer, J., 1994. Monitoring trace metal pollution with tree rings: a critical reassessment. In: Markert, B. (Ed.), Plants as Biomonitors. VCH-Verlag, Weinheim, pp. 541–563.

Halbwachs, G., Arndt, U., 1992. Möglichkeiten und Grenzen der Bioindikation. In: VDI (Ed.), Bioindikation – ein wirksames Instrument der Umweltkontrolle, Vol. 1. VDI-Verlag, Düsseldorf, VDI-Bericht 901, pp. 7–15.

Hauk, H., Umlauf, G., McLachlan, M., 1994. Uptake of gaseous DDE in spruce needles. Environ. Sci. Technol. 28, 2372–2379.

Heath, R.L., 1980. Initial events in injury to plants by air pollutants. Ann. Rev. Plant. Physiol. 31, 395–431.

Heidingsfeld, N., 1991. Nähr- und Schadstoffgehalte in Fichten- und Kiefernnadeln als Bioindikator im Rahmen großräumiger Waldzustandserhebungen. In: VDI (Ed.), Bioindikation – ein wirksames Instrument der Umweltkontrolle, Vol. 1. VDI-Verlag, Düsseldorf, VDI-Bericht 901, pp. 235–258.

Heidt, V., Kehlberger, S., Schulz, A., 1987. Die Verwendung von Bioindikatoren neben anderen Meßverfahren zur Erfassung der raum-zeitlichen Belastungsstruktur durch Luftschadstoffe. In: VDI (Ed.), Bioindikation – wirkungsbezogene Erhebungsverfahren für den Immissionsschutz. VDI-Verlag, Düsseldorf, VDI-Bericht 609, pp. 249–269.

Heinze, M., Fiedler, H.J., 1992. Ernährung der Gehölze. In: Lyr, H., Fiedler, H.J., Tranquillini, W. (Eds), Physiologie und Ökologie der Gehölze. Gustav Fischer Verlag, Jena, pp. 43–116.

Herman, F., 1998. Investigation of the lead content of spruce needles in remote and rural areas over a thirty-year period. Environ. Sci. & Pollut. Res., Special Issue No. 1, 70–74.

Hicks, W.K., Leith, I.D., Woodin, S.J., Fowler, D., 2000. Can the foliar nitrogen concentration of upland vegetation be used for predicting atmospheric nitrogen deposition? Evidence from field surveys. Environ. Poll. 107, 367–376.

Hinkel, M., Reischl, A., Schramm, K.W., Trautner, F., Reissinger, M., Hutzinger, O., 1989. Concentration levels of nitrated phenols in conifer needles. Chemosphere 18, 2433–2439.

Holoubek, I., Korinek, P., Seda, Z., Schneiderova, E., Holoubkova, I., Pacl, A., Triska, J., Cudlin P., Caslavsky, J., 2000. The use of mosses and pine needles at local and regional scales. Environ. Poll. 109, 283–292.

Höpker, K.A., 1992. Bioindikation organischer Luftschadstoffe – Erste Erfahrungen im Ökologischen Wirkungskataster Baden-Württemberg. In: VDI (Ed.), Bioindikation – ein wirksames Instrument der Umweltkontrolle, Vol. 2. VDI-Verlag, Düsseldorf, VDI-Bericht 901, pp. 827–836.

Horntvedt, R., 1995. Fluoride uptake in conifers related to emissions from aluminium smelters in Norway. Sci. Total Environ. 163, 35–37.

Horstmann, M., McLachlan, M.S., 1996. Evidence of a novel mechanism of semivolatile organic compound deposition in coniferous forests. Environ. Sci. Technol. 30, 1794–1796.

Horstmann, M., Bopp, U., McLachlan, M., 1997. Comparison of the bulk deposition of PCDD/Fs in a spruce forest and an adjacent clearing. Chemosphere 34, 1245–1254.

Horstmann, M., McLachlan, M.S., 1998. Atmospheric deposition of semivolatile organic compounds to two forest canopies. Atmos. Environ. 32, 1799–1809.

Hutchinson, B.A., Hicks, B.B. (Eds), 1985. The forest atmosphere interaction. Reidel, Dordrecht.

Ikeda, K., Aoyama, T., Takatori, A., Miyata, H., Pond, P., 2001a. Correlation of dioxin analogues concentrations between ambient air and pine needle in Japan 1. Organohal. Compds. 51, 84–87.

Ikeda, K., Aoyama, T., Takatori, A., Miyata, H., Pond, P., Brzic, B., Serwotka, C., 2001b. Correlation of dioxin analogues concentrations between ambient air and pine needle in Japan. 2. Case study in greater Tokyo area. Organohal. Compds. 51, 88–91.

Ikeda, K., Aoyama, T., Takatori, A., Kusaba, H., Miyata, H., Pond, P., 2001c. Correlation of dioxin analogues concentrations between ambient air and pine needle in Japan. 3. Trend and its estimated source. Organohal. Compds. 51, 5–8.

Innes, J.L., 1995. Influence of air pollution on the foliar nutrition on conifers in Great Britain. Environ. Poll. 88, 183–192.

Juuti S., Norokorpi, Y., Ruuskanen, J., 1995. Trichloroacetic acid (TCA) in pine needles caused by atmospheric emissions of kraft pulp mills. Chemosphere 30, 439–448.

Kaiser, W., Dittrich, A., Heber, U., 1993. Sulfate concentrations in Norway spruce needles in relation to atmospheric SO_2: a comparison of trees from various forests in Germany with trees fumigated with SO_2 in growth chambers. Tree Physiol. 12, 1–13.

Kalina, M.F., Zambo, E., Puxbaum, H., 1998. Assessment of wet, dry and occult deposition of sulphur and nitrogen at an alpine site. Environ. Sci. & Pollut. Res., Special Issue No. 1, 53–58.

Kampe, W., 1987. Organische Stoffe in Böden und Pflanzen nach langjährigen intensiven Klärschlammanwendungen. Korrespondenz Abwasser 8/87, 820–827.

Kasperowski, E., Frank, E., 1989. Boden- und Vegetationsuntersuchungen im Bereich der Scheitelstrecke der Tauernautobahn. Federal Environment Agency, Wien, M-15.

Kasperowski, E., 1993. Schermetalle in Böden im Raum Arnoldstein. Federal Environment Agency, Wien, M-33.

Keller, T., Bajo, S., Wyttenbach, A., 1986. Gehalte an einigen Elementen in den Ablagerungen auf Fichtennadeln als Nachweis der Luftverschmutzung. Allg. Forst- u. J.-Ztg. 157, 69–78.

Keller, T., Matyssek, R., Günthardt-Goerg, M.S., 1994. Beech foliage as a bioindicator of pollution near a waste incinerator. Environ. Poll. 85, 185–189.

Keller, T., Preis, H., 1967. Der Bleigehalt von Fichtennadeln als Indikator einer verkehrsbedingten Luftverunreinigung. Schweiz. Z. Forstwesen 118, 143–162.

Kimmins, J.P., 1987. Forest Ecology. Macmillan, New York.

Kloke, A., Sauerbeck, D., Vetter, H., 1984. The contamination of plants and soils with heavy metals and the transport of metals in terrestrial food chains. In: Nriagu, J.O. (Ed.), Changing Metal Cycles and Human Health. Springer, Berlin, pp. 113–141.

Knabe, W., 1982. Monitoring of air pollutants by wild life plants and plant exposure: suitable bioindicators for different immissions types. In: Steubing, L., Jäger, H.J. (Eds), Monitoring of Air Pollutants by Plants – Methods and Problems. Dr W. Junk, The Hague, pp. 59–72.

Knabe, W., 1983. Immissionsökologische Waldzustandserfassung in Nordrhein-Westfalen (IWE 1979) – Fichten und Flechten als Zeiger der Waldgefährdung durch Luftverunreinigungen. Landwirtschaftsverlag, Münster, Wissenschaftliche Berichte und Diskussionsbeiträge 37, 138 pp.

Köhler, J., Peichl, L., Dumler-Gradl, R., Thoma, H., Vierle, O., 1995. Monitoring of PCDD/F-levels with bioindicator plants. Organohal. Compds, 24, 205–208.

Kolic, T.M., MacPherson, K.A., Reiner, E.J., McIlveen W.D., 1995. Norwegian spruce needles: a monitoring technique for PCDD/Fs during and after a tire fire. Organohal. Compds. 24, 195–199.

Kömp, P., McLachlan, M.S., 1997a. Influence of temperature on the plant/air partitioning of semivolatile organic compounds. Environ. Sci. Technol. 31, 886–890.

Kömp, P., McLachlan, M.S., 1997b. Interspecies variability of the plant/air partitioning of polychlorinated biphenyls. Environ. Sci. Technol. 31, 2944–2948.

Kosta-Rick, R., 1992. Messung der Fluorid-Immissionskonzentrationen, -Immissionsraten (SAM) sowie aktives Biomonitoring in der Umgebung eines lokalen Emittenten. In: VDI (Ed.), Bioindikation – ein wirksames Instrument der Umweltkontrolle, Vol. 2. VDI-Verlag, Düsseldorf, VDI-Bericht 901, pp. 801–811.

Kosta-Rick, R., Arndt, U., 1987. Methodische Untersuchungen zur Optimierung des Verfahrens der standardisierten Graskultur. VDI, (Ed.), 1987. Bioindikation – wirkungsbezogene Erhebungsverfahren für den Immissionsschutz. VDI-Verlag, Düsseldorf, VDI-Bericht 609, pp. 301–316.

Kovacs, M., Opauszky, I., Klincsek, P., Podani, J., 1982. The leaves of city trees as accumulation indicators. In: Steubing, L., Jäger, H.J. (Eds), Monitoring of Air Pollutants by Plants – Methods and Problems. Dr W. Junk, The Hague, pp. 149–153.

Kovacs, M., Turcsanyi, G., Nagy, L., Koltay, A., Kaszab, L., Szöke, P., 1990. Element concentration cadasters in a Quercetum petraeae-cerris forest. In: Lieth, H., Markert, B. (Eds), Element Concentration Cadasters in Ecosystems. VCH, Weinheim, pp. 255–264.

Kovacs, M., Turcsanyi, G., Penksza, K., Kaszab, L., Szöke, P., 1993. Heavy metal accumulation by ruderal and cultivated plants in a heavily polluted district of Budapest. In: Markert, B. (Ed.), Plants as Biomonitors. VCH-Verlag, Weinheim, pp. 495–505.

Kratz, W., 1996. Ökotoxikologische Bioindikation – Schwermetalle, PAK und PCB in Kiefernnadeln. UWSF-Z. Umweltchem. Ökotox. 8, 130–137.

Krause, G.H.M., Delschen, T., Fürst, P., Hein, D., 1993. PCDD/F in Böden, Vegetation und Kuhmilch. UWSF-Z. Umweltchem. Ökotox. 5, 194–203.

Kreutzer, K., Bittersohl, J., 1986. Stoffauswaschung aus Fichtenkronen (Picea abies [L.] Karst.) durch saure Beregnung. Forstw. Cbl. 105, 357–363.

Krivan, V., Schäfer, F., 1989. Surface deposits on spruce needles as a possible indicator for the degree of heavy metal pollution of the atmosphere. Fresenius Z Anal. Chem. 333, 726.

Krivan, V., Schaldach, G., 1986. Untersuchungen zur Probenahme und -vorbehandlung von Baumnadeln zur Elementanalyse. Fresenius Z Anal. Chem. 324, 158–167.

Krivan, V., Schaldach, G., Hausbeck, R., 1987. Interpretation of element analyses of spruce-needle tissue falsified by atmospheric surface deposition. Naturwissenschaften 74, 242–245.

Kronberger, W., Halbwachs, G., Richter, H., 1978. Fluortranslokation in *Picea abies* (L.) Karst. Angew. Botanik 52, 149–154.

Kühnert, M., Halbwachs, G., Kühnertova, M., 1996. Bioindikation auf einem Altlastenstandort. In: Arndt, U., Fomin, A., Lorenz, S. (Eds), Bioindikation – Neue Entwicklungen, Nomenklatur, Synökologische Aspekte. Heimbach, Ostfildern, pp. 277–280.

Lamoureux, G.L., Rusness, D.G., 1989. The role of glutathione and glutathione S-transferases in pesticide metabolism, selectivity and mode of action in plants and insects. In: Dolphin, D., Poulson, R., Avramovic, O. (Eds), Glutathione: Chemical Biochemical and Medical Aspects, Vol IIIB, Series: Enzyme and Cofactors. Wiley, New York, pp. 153–189.

Lamoureux, G.L., Shimabukuro, R.H., Frear, D.S., 1991. Glutathione and glucoside formation in herbicide selectivity. In: Caseley, J.C., Cussans, G.W. Atkin, R.K. (Eds), Herbicide Resistance in Weeds and Crops. Butterworth, Oxford, pp. 227–262.

Landolt, W., Guecheva, M., Bucher, J.B., 1989. The spatial distribution of different elements in and on the foliage of Norway spruce growing in Switzerland. Environ. Poll. 56, 155–167.

Lendzian, K.J., 1987. Aufnahme und zellphysiologische Wirkung von Luftschadstoffen. Naturwissenschaften 74, 282–288.

Lick, H., Dorfer, A., 1998. Schwermetallbelastung der Wälder in der Steiermark. In: Amt der Steiermärkischen Landesregierung – Fachabteilung für das Forstwesen (Ed.), Der Zustand des Steirischen Waldes 1996/97. Amt der Steiermärkischen Landesregierung – Fachabteilung für das Forstwesen, Graz, pp. 34–52.

Lin, C.-J., Pehkonen, S.O., 1999. The chemistry of atmospheric mercury: a review. Atmos. Environ. 33, 2067–2079.

Mankovska, B., 1984. The effects of atmospheric emissions from the Krompachy, Nizna Slana, Rudnany iron ore mines on forest vegetation and soils. Ekologia (CSSR) 3, 331–344.

Mankovska, B. 1992. Chemical composition of solid particles on vegetative surface in Slovak forests. Ekologia (CSFR) 11, 205–214.

Mankovska, B., 1997. Variations in sulphur and nitrogen foliar concentration of deciduous and coniferous vegetation in Slovakia. Water, Air Soil Poll. 96, 329–345.

Mankovska, B., 1998: The chemical composition of spruce and beech foliage as an environmental indicator in Slovakia. Chemosphere 36, 949–953.

Mankovska, B., Steinnes, E., 1995. Effects of pollutants from an aluminium reduction plant on forest ecosystems. Sci. Total. Environ. 163, 11–23.

Manninen, S., Huttunen, S., 1995. Scots pine needles as bioindicators of sulphur deposition. Can. J. For. Res. 25, 1559–1569.

Markert, B. 1992. Multi-element analysis in plant materials. In: Adriano, D.C. (Ed.), Biogeochemistry of Trace Metals. Lewis Publishers, Boca Raton, pp. 401–428.

Markert, B. (Ed.), 1993a. Plants as Biomonitors. VCH-Verlag, Weinheim.

Markert, B., 1993b. Instrumental analysis of plants. In: Markert, B. (Ed.), Plants as Biomonitors. VCH-Verlag, Weinheim, pp. 65–103.

Marschner, H., 1995. Mineral Nutrition of Higher Plants. 2nd edn. Academic Press, London.

Martin, M.H., Coughtrey, P.J., 1982. Biological Monitoring of Heavy Metal Pollution. Applied Science, London.

Matucha, M., Uhlirova, H., Bubner, M., 2001. Investigation of uptake, translocation and fate of trichloroacetic acid in Norway spruce (*Picea abies*/L./Karst.) using ^{14}C-labelling. Chemosphere 44, 217–222.

Mayer, R., 1981. Natürliche und anthropogene Komponenten des Schwermetallhaushalts von Waldökosystemen. Göttinger Bodenkundliche Berichte 70.

McLachlan, M.S., Welsch-Pausch, K., Tolls, J., 1995. Field validation of a model of the uptake of gaseous SOC in Lolium multiflorum (Rye grass). Environ. Sci. Technol. 29, 1998–2004.

McLachlan, M.S., 1999. Framework for the interpretation of measurements of SOCs in plants. Environ. Sci. Technol. 33, 1799–1804.

Miller, E.K., Friedland, A.J., Arons, E.A., Mohnen, V.A., Battles, J.J., Panek, J.A., Kadlecek, J., Johnson, A.H., 1993. Atmospheric deposition to forests along an elevational gradient at Whiteface Mountain, NY, USA. Atmos. Environ. 27A, 2121–2136.

Mohr, K., 1999. Passives Monitoring von Stickstoffeinträgen in Kiefernforsten mit dem Rotstengelmoos (Pleurozium schreberi (Brid.) Mitt.). UWSF-Z. Umweltchem. Ökotox. 11, 267–274.

Monaci, F., Moni, F., Lanciotti, E., Grechi, D., Bargagli, R., 2000. Biomonitoring of airborne metals in urban environments: new tracers of vehicle emission, in place of lead. Environ. Poll. 107, 321–327.

Monteith, J.L. (Ed.), 1976. Vegetation and the atmosphere, Vol. 2. Academic Press, New York.

Morosini, M., Schreitmüller, J., Reuter, U., Ballschmiter, K., 1993. Correlation between C-6/C-14 chlorinated hydrocarbons levels in the vegetation and in the boundary layer of the troposphere. Environ. Sci. Technol. 27, 1517–1523.

Mößnang, M., 1990. Element contents of spruce needles (*P. abies* [L.] Karst.) along an altitudinal gradient in the Bavarian Alps. In: Zöttl, H.W., Hüttl, R.F. (Eds), Management of Nutrition in Forests under Stress. Kluwer Academic, Dordrecht, pp. 107–112.

Mukherjee, A.B., Nuorteva, P., 1994. Toxic metals in forest biota around the steel works of Rautaruukki Oy, Raahe, Finland. Sci. Total Environ. 151, 191–204.

Mulgrew, A., Williams, P., 2000. Biomonitoring of air quality using plants. WHO Collaborating Centre for Air Quality Management and Air Pollution Control at the Federal Environmental Agency Germany, Berlin, Report 10.

Nabais, C., Freitas, H., Hagemeyer, J., 1999. Dendroanalysis: a tool for biomonitoring environmental pollution? Sci. Total Environ. 232, 33–37.

Nakajima, D., Yoshida, Y., Suzuki, J., Suzuki, S., 1995. Seasonal changes in the concentration of polycyclic aromatic hydrocarbons in Azalea leaves and relationships to atmospheric concentration. Chemosphere 30, 409–418.

Neinave, H., Pirkl, H., Trimbacher, C., 2000. Herkunft und Charakteristik von Stäuben. Federal Environment Agency, Wien, BE-171.

Nobel, W., Maier-Reiter, W., 1996. 12 Jahre Bioindikation im Raum Esslingen/Altbach (1983 bis 1994). In: Arndt, U., Fomin, A., Lorenz, S. (Eds), Bioindikation – Neue Entwicklungen, Nomenklatur, Synökologische Aspekte. Heimbach, Ostfildern, pp. 31–50.

Nobel, W., Maier-Reiter, W., Finkbeiner, M., Frank, W., Sommer, B., Kosta-Rick., R., 1993. Levels of polychlorinated dioxins and furans in ambient air, plants and soil as influenced by emission sources and differences in land-use. Organohal. Compds. 12, 171–174.

Nobel, W., Maier-Reiter, W., Sommer, B., Finkbeiner, M., 1992. Biomonitoring organischer Luftschadstoffe, insbesondere Dioxine/Furane. In: VDI (Ed.), Bioindikation – ein wirksames Instrument der Umweltkontrolle, Vol. 2. VDI-Verlag, Düsseldorf, VDI-Bericht 901, pp. 813–826.

Nobel, W., Michenfelder, K., 1987. Routinemäßiger Einsatz von pflanzlichen Bioindikatoren im Rahmen immissionsschutzrechtlicher Genehmigungsverfahren. VDI (Ed.), Bioindikation – wirkungsbezogene Erhebungsverfahren für den Immissionsschutz. VDI-Verlag, Düsseldorf, VDI-Bericht 609, pp. 367–394.

Notarianni, V., Calliera, M., Tremolada, P., Finizio, A., Vighi, M., 1998. PCB distribution in soil and vegetation from different areas in northern Italy. Chemosphere 37, 2839–2845.

OECD, 1997. Revised OECD principles of good laboratory practice (GLP). Organisation for Economic Co-operation and Development, Paris.

Öhlinger, R., 2000. Biomonitoring von Luftschadstoffen und deren Bewertung aus landwirtschaftlicher Sicht. Veröff. Bundesamt für Agrarbiologie Linz/Donau 22, 115–136.

Öhlinger, R., 2002. Aktives und passives Biomonitoring – Richtwerte. www.agrobio.bmlf.gv.at.

Öhlinger, R., Döberl, H., 1992. Immissionskontrollen an standardisierten Pflanzen in Oberösterreich. In: VDI (Ed.), Bioindikation – ein wirksames Instrument der Umweltkontrolle, Vol. 1. VDI-Verlag, Düsseldorf, VDI-Bericht 901, pp. 513–529.

Paterson, S., Mackay, D., Bacci, E., Calamari, D., 1991. Correlation of the equilibrium and the kinetics of leaf-air exchange of hydrophobic organic chemicals. Environ. Sci. Technol. 25, 866.

Paulus, M., Klein, R., Zimmer, M., Jacob, J., Rossbach, M., 1995. Biomonitoring und Umweltprobenbank IV – Die Rolle der biometrischen Probencharakterisierung in der Umweltanalytik am Beispiel der Fichte (*Picea abies*). UWSF-Z. Umweltchem. Ökotox. 7, 236–244.

Peichl, L., 2001. Umweltindikatoren für Immissionswirkungen – Berechnung von Indices. UWSF-Z Umweltchem Ökotox 13, 130–138.

Peichl, L., Wäber, M., Reifenhäuser, W., 1994. Schwermetallmonitoring mit der Standardisierten Graskultur im Untersuchungsgebiet München – Kfz-Verkehr als Antimonquelle? UWSF-Z. Umweltchem. Ökotox. 6, 63–69.

Plümacher, J., Renner, I., Schröder, P., 1993. Volatile chlorinated hydrocarbons and trichloracetic acid in conifer needles. In: Schröder, P., Frank, H., Rether, B. (Eds), Volatile Organic Pollutants: Levels, Fate and Ecotoxicological Impacts. Wiss.-Verlag Maraun, Frankfurt, IFU Schriftenreihe 23/93, pp. 37–51.

Plümacher, J., Schröder, P., 1994. Accumulation and fate of C1/C2-chlorocarbons and trichloroacetic acid in spruce needles from an Austrian mountain site. Chemosphere 29, 2467–2476.

Portele, K., 1891. Über die Beschädigung von Fichtenwaldbeständen durch schwefelige Säure. Öst. Landwirt. Cbl. 1, 27–38.

Rademacher, P., 2001. Atmospheric heavy metals and forest ecosystems. UN/ECE, Geneva.

Radermacher, L., Rudolph, H., 1994. Bioindikationsmethoden – Aktive Verfahren. Grünkohl als Bioindikator. UWSF-Z. Umweltchem. Ökotox. 6, 384–386.

Raisch, W., 1983. Bioelementverteilungen in Fichtenökosystemen der Bärhalde. Freiburger Bodenkundliche Abhandlungen 11.

Rautio, P., Huttunen, S., Lamppu, J., 1998. Effects of sulphur and heavy metal deposition on foliar chemistry of Scots pine in Finnish Lapland and on the Kola peninsula. Chemosphere 36, 979–984.

Rennenberg, H., Gessler, A., 1999. Consequences of N deposition to forest ecosystems – Recent results and future research needs. Water, Air Soil Poll. 116, 47–64.

Reischl, A., Reissinger, M., Hutzinger, O., 1987. Occurrence and distribution of atmospheric organic micropollutants in conifer needles. Chemosphere 16, 2647–2652.

Reischl, A., Reissinger, M., Thoma, M., Hutzinger, O., 1989. Uptake and accumulation of PCDD/F in terrestrial plants: basic considerations. Chemosphere 19, 467–474.

Reischl, A., Reissinger, M., Thoma, H., Mücke, W., Hutzinger, O., 1990b. Biomonitoring of PCDD/F in Bavaria (Germany). Organohal. Compds. 4, 229–232.

Reischl, A., Zech, W., Reissinger, M., Lenoir, D., Schramm, K.W., Hutzinger, O., 1990a. Distribution of chlorinated aromatics in leaves, needles and two soils from the Fichtelgebirge (NE-Bavaria), FRG. Organohal. Compds. 4, 223–228.

Riederer, M., 1989. The cuticles of conifers. In: Schulze, E.-D., Lange, O.L., Oren, R. (Eds), Forest decline and air pollution. Springer, Berlin, pp. 157–192.

Riederer, M., 1990. Estimating partitioning and transport of organic chemicals in the foliage/atmosphere system: discussion of a fugacity based model. Environ. Sci. Technol. 24, 829–837.

Riederer, M., 1992. Uptake of organic chemicals in conifer needles. Surface adsorption and permeability of cuticles. Environ. Sci. Technol. 26, 153–159.

Riss, A., Hagenmaier, H., Weberruß, U., Schlatter, C., Wacker, R., 1990. Comparison of PCDD/PCDF levels in soil, grass, cow's milk, human blood and spruce needles in an area of PCDD/PCDF contamination through emissions from a metal reclamation plant. Chemosphere 21/12, 1451–1456.

Robinson, W.O., Edgington, G., 1945. Minor elements in plants and some accumulator plants. Soil Science 60, 15–28.

Rudolph, E., 1987. Passives und aktives Monitoring mit Fichten – Schwefelgehalte und deren zeitliche Entwicklung im Zeitraum von 1978–1984, dargestellt am Beispiel des passiven Monitoring. In: VDI (Ed.), Bioindikation – wirkungsbezogene Erhebungsverfahren für den Immissionsschutz. VDI-Verlag, Düsseldorf, VDI-Bericht 609, pp. 337–350.

Sandermann, H., 1994. Higher plant metabolism of xenobiotics: the 'green liver' concept. Pharmacogenetics 4, 225–241.

Sandermann, H., Haas, M., Messner, B., Pflugmacher, S., Schröder, P., Wetzel, A., 1997. The role of glucosyl and malonyl conjugation in herbicide selectivity. In: Hatzios, K.K. (Ed.), Regulation of Enzymatic Systems Detoxifying Xenobiotics in Plants. Kluwer Academic, Dordrecht, pp. 211–231.

Sauerbeck, D., 1986. Schadstoffeinträge in den Boden durch Industrie, Besiedelung, Verkehr und Landbewirtschaftung (anorg. Stoffe). In: VDLUFA (Ed.), Kongreßband 1985. VDLUFA-Verlag, Darmstadt, pp. 59–72.

Schätzle, H., Seufert, G., Bender, J., Groß, G., Arndt, U., Jäger, H.J., 1990. Mineral content in the soil and tree foliage. Environ. Poll. 68, 253–273.

Scheffer, F., Schachtschabel, P., 1989. Lehrbuch der Bodenkunde, 12th edn. Enke, Stuttgart.

Schönherr, J., Riederer, M., 1989. Foliar penetration and accumulation of organic chemicals in plant cuticles. Rev. Environ. Contam. Toxicol. 108, 1–70.

Scholl, G., 1987. Grundlagen des Verfahrens der standardisierten Graskultur. In: VDI (Ed.), Bioindikation – wirkungsbezogene Erhebungsverfahren für den Immissionsschutz. VDI-Verlag, Düsseldorf, VDI-Bericht 609, pp. 287–300.

Scholl, G., 1992. Untersuchungen über die Beteiligung verschiedener Emissionsquellen an überhöhten Bleigehalten im Weideaufwuchs. VDI (Ed.), Bioindikation – ein wirksames Instrument der Umweltkontrolle, Vol. 1. VDI-Verlag, Düsseldorf, VDI-Bericht 901, pp. 531–538.

Schröder, P., Messner, B., Weiss, P., 1998. Detoxification by conjugation and metabolism of xenobiotics in conifers. In: Weiss, P., Schröder, P., Rether, B., Keith, G., Collins, C., Bach, Th. (Eds), Organic Xenobiotics and Plants: Impact, Metabolism and Toxicology. Proceedings of the 4th IMTOX-Workshop. Federal Environment Agency, Wien, CP-24, pp. 139–153.

Seufert, G., 1990. Ions in percolating waters through a model ecosystem. Environ. Poll. 68, 231–252.

Shimabukuro, R.H., Walsh, W.C., 1978. Xenobiotic metabolism in plants: in vitro tissue, organ, and isolated cell techniques. In: Paulson, G.D., Frear, S.D., Marks, E.P. (Eds), Xenobiotic Metabolism: in vitro Methods. ACS Symposia, Series 97, pp. 3–34.

Sinkkonen, S., Raitio, H., Paasivirta, J., Rantio, T., Lahtiperä M., Mäkelä, R., 1995. Concentrations of persistent organochlorine compounds in spruce needles from western Finland. Chemosphere 30, 1415–1422.

Sinkkonen, S., Welling, L., Vattulainen, A., Lahti, L., Lahtiperä, M., Paasivirta, J., 1996. Short chain aliphatic halocarbons and polychlorinated biphenyls in pine needles: effects of metal scrap plant emissions. Chemosphere 32, 1971–1982.

Smith, W.H., 1972. Lead and mercury burden of urban woody plants. Science 176, 1237–1239.

Smith, W.H., 1973. Metal contamination of urban woody plants. Environ. Sci. Technol. 7, 631–636.

Stefan, K., Fürst, A., Hacker, R., Bartels, U., 1997. Forest foliar condition in Europe – results of large-scale foliar chemistry surveys (survey 1995 and data from previous years). Forest Foliar Co-ordinating Centre, EC, Brussels, UN/ECE, Geneva, FBVA, Vienna.

Stefan, K., Fürst, A., 1998. Indication of S and N inputs by means of needle analyses based on the Austrian Bio-Indicator Grid. ESPR – Environ. Sci. & Pollut. Res., Special Issue 1, 63–69.

Stefan, K., Gabler, K., 1998. Connections between climatic conditions and the nutritional status of spruce needles determined from the Austrian bio-indicator grid. ESPR – Environ. Sci. & Pollut. Res., Special Issue No. 1, 59–62.

Stefan, K., Raitio, H., Bartels, U, Fürst, A., 2000. Manual on methods and criteria for harmonized sampling, assessment, monitoring and analysis of the effects of air pollution on forests. Part IV: Sampling and analysis of needles and leaves. UN-ECE, ICP Forests, Geneva.

Steubing, L., 1982. Problems of bioindication and the necessity of standardisation. In: Steubing, L., Jäger, H.J. (Eds), Monitoring of Air Pollutants by Plants – Methods and Problems. Dr W. Junk, The Hague, pp. 19–24.

Steubing, L., 1987. Bioindikation von Schwermetallen in verschiedenen Ökosystemen. In: VDI (Ed.), Bioindikation – wirkungsbezogene Erhebungsverfahren für den Immissionsschutz. VDI-Verlag, Düsseldorf, VDI-Bericht 609, pp. 351–366.

Steubing, L., Jäger, H.J. (Eds), 1982. Monitoring of Air Pollutants by Plants – Methods and Problems. Dr W. Junk, The Hague.

Strachan, W.M.J., Eriksson, G., Kylin, H., Jensen, S., 1994. Organochlorine compounds in pine needles: methods and trends. Environ. Toxicol. Chem. 13, 443–451.

Sutinen, S., Juuti, S., Koivisto, S., Turunen, M., Ruuskanen, J., 1995. The uptake of and structural changes induced by trichloroacetic acid in the needles of Scots pine seedlings. J. Exp. Bot. 46, 1223–1231.

Swain, R.E., Harkins, W.D., 1908. Arsenic in vegetation exposed to smelter smoke. Am. Chem. Soc. J. 30, 915–928.

Thoene, B., Schröder, P., Papen, H., Egger, A., Rennenberg, H., 1991. Absorption of atmospheric NO_2 by spruce (*Picea abies* L. Karst.) trees. I. NO_2 influx and its correlation with nitrate reduction. New Phytol. 117, 575–585.

Thomas, W., Simon, H., Rühling, A., 1985. Classification of plant species by their organic (PAH; PCB; BHC) and inorganic (heavy metals) trace pollutant concentrations. Sci. Total Environ. 46, 83–94.

Thompson, T.S., Treble, R.G., 1995. Use of pine needles as an indicator of atmospheric contamination by pentachlorphenol. Chemosphere 31, 4387–4392.

Tichy, J., 1996. Impact of atmospheric deposition on the status of planted Norway spruce stands: a comparative study between sites in southern Sweden and the northeastern Czech Republic. Environ. Poll. 93, 302–312.

Tiefenbach, K., Tuschl, P., Woidich, H. 1983. Studium der Aufnahme von polycyclischen aromatischen Kohlenwasserstoffen durch höhere Pflanzen mit Hilfe chromatographischer Methoden und Isotopentechnik. Die Bodenkultur 34, 147–160.

Tremolada, P., Burnett, V., Calamari, D., Jones, K.C., 1996. A study of the spatial distribution of PCBs in the UK atmosphere using pine needles. Chemosphere 32, 2189–2203.

Trimbacher, C., Weiss, P., 1997. Wachsqualität, Nähr- und Schadstoffkonzentrationen von Fichtennadeln an belasteten und unbelasteten Standorten in Österreich – Untersuchungsergebnisse 1995. Federal Environment Agency, Wien, M-90.

Trimbacher, C., Weiss, P., 1999. Needle surface characteristics and element contents of Norway spruce in relation to the distance of emission sources. Environ. Poll. 105, 111–119.

Trimbacher, C., Weiss, P., 2000. Nadeloberflächenparameter und Elementgehalte von Fichtennadeln ausgewählter Industriestandorte – Gesamtergebnisse 1997. Federal Environment Agency, Wien, BE-174.

Trimbacher, C., Weiss, P., 2002 (in preparation). Federal Environment Agency, Wien, Report in preparation.

Umlauf, G., Reischl, A., Reissinger, M., Richartz, H., Hutzinger, O., Weissflog, L., Wenzel, K.D., Martinez D., 1990. Atmospärische Belastung in Nordbayern und im Ballungsraum Halle – Leipzig. UWSF-Z. Umweltchem. Ökotox. 2, 193–194.

Umlauf, G., Hauk, H., Reissinger, M., 1994. Deposition of semivolatile organic compounds to spruce needles. II. Experimental evaluation of the relative importance of different pathways. ESPR – Environ. Sci. & Pollut. Res. 1, 209–222.

Umweltbundesamt (Ed.), 1996. Umweltprobenbank des Bundes – Verfahrensrichtlinien für Probenahme, Transport, Lagerung und chemische Charakterisierung von Umwelt und Human-Organproben. Umweltbundesamt, E. Schmidt, Berlin.

UN-ECE, 1998. Manual for integrated monitoring. Finnish Environment Institute, Helsinki.

UN-ECE, 2001. International Cooperative Programme on Integrated Monitoring of Air Pollution Effects on Ecosystems. http://www.vyh.fi/eng/intcoop/projects/icp_im/im.htm.

VDI (Ed.), 1987. Bioindikation – wirkungsbezogene Erhebungsverfahren für den Immissionsschutz. VDI-Verlag, Düsseldorf, VDI-Bericht 609.

VDI (Ed.), 1992. Bioindikation – ein wirksames Instrument der Umweltkontrolle, Vols 1, 2. VDI-Verlag, Düsseldorf, VDI-Bericht 901.

VDI 3792–1, 1978. Messen der Wirkdosis – Verfahren der standardisierten Graskultur. Beuth, Berlin.

VDI 3792–2, 1982. Messen der Immissions-Wirkdosis – Messen der Immissions-Wirkdosis von gas- und staubförmigem Fluorid in Pflanzen mit dem Verfahren der standardisierten Graskultur. Beuth, Berlin.

VDI 3792–3, 1991. Measurement of the response dose – measurement of the response dose of ambient lead in plants with standardized grass cultures. Beuth, Berlin.

VDI 3792–5, 1991. Response dose determination – standardization of sampling of leaves and needles from trees at their natural site. Beuth, Berlin.

VDI 3957–2, 2001 (draft). Biologische Messverfahren zur Ermittlung und Beurteilung der Wirkung von Luftverunreinigungen auf Pflanzen (Bioindikation) – Verfahren der standardisierten Graskultur. Beuth, Berlin.

VDI 3957–3, 2000. Biological measuring techniques for the determination and evaluation of effects of air pollutants on plants (bioindication) – standardized exposure of green cabbage. Beuth, Berlin.

VDI 3957–5, 2001. Biological measuring techniques for the determination and evaluation of effects of air pollutants on plants (bioindication) – standardised exposure of spruce. Beuth, Berlin.

Vike, E., Habjorg, A., 1995. Variation in fluoride content and leaf injury on plants associated with three aluminium smelters in Norway. Sci. Total Environ. 163, 25–34.

Vogel, W., Kienzl, K., Riss, A., 1991. Die Treibacher Chemischen Werke – Wirkungen auf die Umwelt. Federal Environment Agency, Wien, M-26.

Vogel, W., Riss, A., 1992. Grünlandaufwuchs und Fichtennadeln als Akkumulationsindikatoren. In: VDI (Ed.), Bioindikation – ein wirksames Instrument der Umweltkontrolle, Vol. 1. VDI-Verlag, Düsseldorf, VDI-Bericht 901, pp. 323–335.

Wäber, M., Laschka, D., Peichl, L., 1996. Biomonitoring verkehrsbedingter Platin-Immissionen – Verfahren der standardisierten Graskultur im Untersuchungsgebiet München. UWSF-Z. Umweltchem. Ökotox. 8, 3–7.

Wagner, G., 1987. Entwicklung einer Methode zur großräumigen Überwachung der Umweltkontamination mittels standardisierter Pappelblattproben von Pyramidenpappeln (Populus nigra "Italica") am Beispiel von Blei, Cadmium und Zink. In: Stoeppler, M., Dürbeck, H.W. (Eds), Beiträge zur Umweltprobenbank Nr. 5. Kernforschungsanlage Jülich, Jül-Spez-412.

Wagner, G., 1990. Variability of element concentrations in tree leaves depending on sampling parameters. In: Lieth, H., Markert, B. (Eds), Element Concentration Cadasters in Ecosystems. VCH, Weinheim, pp. 41–54.

Wagner, G., 1993. Large-scale screening of heavy metal burdens in higher plants. In: Markert, B. (Ed.), Plants as Biomonitors. VCH-Verlag, Weinheim, pp. 425–434.

Wagner, G., 1995. Basic approaches and methods for quality assurance and quality control in sample collection and storage for environmental monitoring. Sci. Total Environ. 176, 63–71.

Wagner, G., 1997. Biological samples. In: Stoeppler, M. (Ed.), Sampling/Sample Preparation. Springer Lab Manual, Springer, pp. 88–107.

Wagner, G., Klein, R., 1995. Sampling strategy in environmental monitoring. In: Quevauviller, P. (Ed.), Quality Assurance in Environmental Monitoring from Sampling to Laboratory. VCH, Weinheim, pp. 25–50.

Wagner, G., Müller, P., 1979. Fichten als "Bioindikatoren" für die Immissionsbelastung urbaner Ökosysteme unter besonderer Berücksichtigung von Schwermetallen. Verhandlungen der Gesellschaft für Ökologie (Münster 1978), Vol. VII, pp. 307–314.

Walkenhorst, A., Hagemeyer, J., Breckle, W., 1993. Passive monitoring of airborne pollutants, particularly trace metals with tree bark. In: Markert, B. (Ed.), Plants as Biomonitors. VCH-Verlag, Weinheim, pp. 523–540.

Weiss, P., 1998. Persistente organische Schadstoffe in Hintergrund-Waldgebieten Österreichs. Federal Environment Agency, Wien, M-97.

Weiss, P., 1999. Elementgehalte von Fichtennadeln, Pappel- und Ahornblättern – Eine tabellarische Zusammenstellung von Literaturdaten. Federal Environment Agency, Wien, BE-143.

Weiss, P., 2000a. Vegetation/soil distribution of semivolatile organic compounds in relation to their physico-chemical properties. Environ. Sci. Technol. 34, 1707–1714.

Weiss, P., 2000b. Nitrophenole, leichtflüchtige halogenierte Kohlenwasserstoffe und Trichloressigsäure in Fichtennadeln. In: Umweltbundesamt (Ed.), Pflanzentoxische organische Schadstoffe und enzymatische Reaktionen in Fichten emittentenferner Waldstandorte Österreichs. Federal Environment Agency, Wien, M-123, Part 1, pp. 1–69.

Weiss, P., Lorbeer, G., Scharf, S., 1998a. Persistent organic pollutants in remote Austrian forests – altitude-related results. ESPR – Environ. Sci. & Pollut. Res., Special Issue No. 1, 46–52.

Weiss, P., Lorbeer, G., Scharf, S., 2000. Regional aspects and statistical characterisation of the load with semivolatile organic compounds at remote Austrian forest sites. Chemosphere 40, 1159–1172.

Weiss, P., Lorbeer, G., Stephan, C., Svabenicky, F., 1998b. Short chain aliphatic halocarbons, trichloroacetic acid and nitrophenols in spruce needles of Austrian background forest sites. In: Weiss, P., Schröder, P., Rether, B., Keith, G., Collins, C., Bach, Th. (Eds), Organic Xenobiotics and Plants: Impact, Metabolism and Toxicology. Proceedings of the 4th IMTOX-Workshop. Federal Environment Agency, Wien, CP-24, pp. 49–64.

Weiss, P., May, R., Schröder, P., 2001. Nitrophenole, halogenierte Kohlenwasserstoffe und enzymatische Reaktionen in Fichtennadeln emittentennaher Standorte Österreichs. Federal Environment Agency, Wien, M-151.

Weiss, P., Trimbacher, C., 1998. Nadeloberflächenparameter und Elementgehalte von Fichtennadeln ausgewählter Industriestandorte – Gesamtergebnisse 1996. Federal Environment Agency, Wien, R-154.

Weißflog, L., Wienhold, K., Wenzel, K.D., Schüürmann, G., 1994. Ökologische Situation der Region Leipzig-Halle – I. Immissionsmuster luftgetragener Schwermetalle und Bioelemente. UWSF-Z. Umweltchem. Ökotox. 6, 75–80.

Weißflog, L., Manz, M., Popp, P., Elansky, N., Arabov, A., Putz, E., Schüürmann, G., 1999a. Airborne trichloroacetic acid and its deposition in the catchment area of the Caspian Sea. Environ. Poll. 104, 359–364.

Weißflog, L., Wenzel, K.D., Manz, M., Kleint, F., Schüürmann, G., 1999b. Economic upheaval in 1990–93 and the ecological situation in central Germany. Environ. Poll. 105, 341–347.

Wellburn, A.R., 1988. Air pollution and acid rain: the biological impact. Longman Scientific & Technical, Burnt Mill.

Welsch-Pausch, K., McLachlan, M.S., Umlauf, G., 1995. Determination of the principal pathways of polychlorinated dibenzo-p-dioxins to *Lolium multiflorum* (Welsh Ray Grass). Environ. Sci. Technol. 29, 1090–1098.

Wenzel, K.D., Weißflog, L., Paladini, E., Gantuz, M., Guerreiro, P., Puliafito, C., Schüürmann, G., 1997. Immission patterns of airborne pollutants in Argentina and Germany. II. Biomonitoring of organochlorine compounds and polycyclic aromatics. Chemosphere 34, 2505–2518.

Wickström, K., Pyysalo, H., Plaami-Heikkilä, Tuominen, J., 1986. Polycyclic aromatic compounds (PAC) in leaf lettuce. Z Lebensm. Unters. Forsch. 183, 182–185.

Wimmer, J., 1998. Die Umweltverträglichkeitserklärung. In: Bergthaler, J., Weber, K., Wimmer, J. (Eds), Die Umweltverträglichkeitsprüfung – Praxishandbuch für Juristen und Sachverständige. Manz, Wien, pp. 203–275.

Wimmer, J., Offenthaler, I., Weiss, P., 2002. Biomonitoring von Schwermetallen und Nährelementen im Raum Linz. In: Umweltbundesamt (Ed.), Ergebnisse des Biomonitoringprogrammes im Raum Linz. Federal Environment Agency, Wien, Monograph (in preparation).

Witte, H., Langenohl,T., Offenbächer, G., 1989. Untersuchungen zum Eintrag von organischen Schadstoffen in Boden und Pflanze durch die landwirtschaftliche Klärschlammverwertung. Umweltbundesamt, Berlin, Texte 26/89.

Woidich, H., Pfannhauser, W., Tiefenbacher, K., 1981. Zur Standortfrage der Schadstoffbelastung von Nahrungspflanzen (Gemüse, Obst, Getreide) in Österreich. In: Bundesministerium für Gesundheit und Umweltschutz (Ed.), Rückstandsuntersuchungen in Lebensmitteln. Bundesministerium für Gesundheit und Umweltschutz, Wien.

Wyttenbach, A., Bajo, S., Tobler, L., Keller, Th., 1985. Major and trace element concentrations in needles of Picea abies: levels, distribution functions, correlations and environmental influences. Plant and Soil 85, 313–325.

Wyttenbach, A., Tobler, L., Bajo, S., 1989. Nadelinhaltsstoffe und Ablagerungen auf Nadeloberflächen von Fichten (Picea abies Karst.). Forstw. Cbl. 108, 233–243.

Zechmeister, H.G., Grodińska, K., Szarek-Łukaszewska, G., 2002 (2003). Bryophytes. In: Markert, B.A., Breure, A.M., Zechmeister, H.G. (Eds), Bioindicators and Biomonitors. Elsevier, Oxford, pp. 329–375.

Ziegler, H., 1984. Weg der Schadstoffe in der Pflanze. In: Hock, B., Elstner, E.F. (Eds), Pflanzentoxikologie. B.I.-Wissenschaftsverlag, Mannheim, pp. 35–46.

Zimmermann, R.D., 1990. Erste Ergebnisse des Klon-Fichten-Meßnetzes Baden-Württemberg. AFZ 11, 281–284.

Zimmermann, R.D., Baumann, R., 1994. Bioindikationsmethoden – Aktive Verfahren: Das Klon-Fichtenverfahren. UWSF-Z. Umweltchem. Ökotox. 6, 111–115.

Zimmermann, R.D., 1995. Pflanzen als Akkumulationsindikatoren. UWSF-Z. Umweltchem. Ökotox. 7, 187–189.

Zimmermann, R.D., Debus, R., Franzaring, J., Höpker, K.A., Maier, W., Reiml, D., Finck, M., 1998. Empfehlungen zum Einsatz von Bioindikationsverfahren im Rahmen des Umweltverträglichkeitsprüfungsgesetzes (UVPG). Gefahrst. Reinh. L. 58, 479–486.

Zimmermann, R.D., Wagner, G., Finck, M., 2000. Guidelines for the use of biological monitors in air pollution control (plants). Part I. WHO Collaborating Centre for Air Quality Management and Air Pollution Control at the Federal Environmental Agency Germany, Berlin, Report 12.

Zöttl, H.W., 1985: Schwermetalle im Stoffumsatz von Waldökosystemen. Bielefelder Ökol. Beitr. 1, 31–49.

501

Chapter 14

Plant biodiversity and environmental stress

Ch. Mulder and A.M. Breure

Abstract

This chapter describes relationships between environmental stress and the reaction of plants thereupon. Plants might react to environmental stress on various levels: on the biochemical, cellular, or morphological scale, and at species or population level. The characteristics of organisms change with their size (allometric relationships), with their life history (and related allocation strategies) and with their functional evolution. We discuss in this chapter, what kind of information may be derived from the composition of the vegetation (species diversity and species abundance). We provide insights in the effects of environmental stress on biodiversity., e.g. shifts in communities, the loss of species and appearance of new species. These insight may contribute to the validation of environmental standards. Another aspect we have dealt with is the use of plants to track climate and land-use changes in history. The use of plants as a bioaccumulative indicator is not discussed.

When using a bioindicator, it is particularly important to state clearly which stress factor it is supposed to be sensitive for. It is shown in this chapter, that the sensitivity of plants for, e.g., heavy metals is dependent on the combination of morphological, anatomical and phenological characteristics of the plant. Evergreen plants are far more sensitive than early spring green plants, while shrubs are more sensitive than trees and obligate annuals (therophytes) are more sensitive than facultative annuals. Dependent on the character states, plants, or plant parts, may be, and are, used as climatic indicators, e.g., life forms and leaf anatomy in relation to temperature, rainfall and atmospheric CO_2, or as reliable ecotoxicological tools (rootlet growth in testing solutions; and as time proxies (e.g. tree rings).

The composition of plant communities can deliver a lot of information about environmental conditions. Further the implications of the so-called Ellenberg approach are analysed. Changes in vegetation over time indicate changes in climate patterns, as the main biotic response to climate change is migration, to track optimal conditions for growth. The type of photosynthetic pathways (C_3 versus C_4 photosynthesis) and leaf anatomy of plants as indicators for climatic changes is discussed and coupled with atmospheric tele-connections like El Niño and its antagonist La Niña.

A very interesting way of bioindication by plants is used by palynologists. Their use of plant remains as proxies for vegetation history and human influence provides information about the development of the present society, also from periods where there were no written sources. They show us what the world looked like in historic and pre-historic era. This latter work can be seen as assessments of early human impacts on the environment.

Keywords: autecology, biodiversity, character state, climate change, ecotoxicology, effects, Ellenberg, flora, heavy metal, IEPS, insecticide, PAF, palynology, photosynthesis, synecology, vascular plants, vegetation

1. Introduction

The earth's environment is defined by three major natural patterns: climate, vegetation and soil. Due to their high diversity and abundance in the world's ecosystems and their overall importance in human economy, vascular plants have been used as bioindicators and biomonitors for many centuries. The flora of Europe alone encompasses 12,500 higher plant species (ECNC, 2001). Whether people are looking for ores, nutrient rich grounds or polluted soils, there are always plant species that may indicate the hidden with more or less accuracy. Human activities cause different types of environmental stress on plants, not only on a global scale (climate change, desertification etc.), but also directly at local scales (see Markert et al., 2002, this volume). Beside a large number of toxic chemicals, which are released in enormous amounts every day, the various practices in land-use are causes for a severe stress to the vegetation. Terrestrial plants are important receptors of airborne xenobiotics (Franzaring and van der Eerden, 2000) and soil contaminants (Sanità di Toppi and Gabbrielli, 1999).

When using a bioindicator, it is particularly important to state clearly which stress factor it is supposed to be sensitive for. On the one hand plants may be used to indicate environmental concentrations of pollutants by using their bioaccumulative properties. That aspect is not being dealt with in this chapter, but in the previous one (Weiss et al., 2002, this volume). On the other hand, plants might react to environmental stress on various levels: on the biochemical, cellular, or morphological scale, and at species or population level. The characteristics of organisms change with their size (allometric relationships), with their life history (and related allocation strategies) and with their functional evolution. This scaling of biological processes strongly dominates the community ecology and the environment resilience. The ecological importance of scaling and ranking of biological processes will be discussed in the following sections. Stuessy (1990) distinguishes two primary operations of classification in plant sciences. (1) The grouping operation, which suits floristic and vegetation studies. Selected characters within all aspects of the investigated individual organisms are measured for a comparison between character states (statistic, cladistic, genetic, etc.). The final aim of the grouping operation is the description of (the occurrence of) plant taxa. (2) The ranking operation, which selects appropriate character states for all investigated taxa to allow for an evaluation in physiological categories (like trophic groups). This latter operation of classification can be a powerful tool in ecological risk assessments. A major drawback in the identification of vascular plant as proxies or bioindicators remains that their optimal habitat changes not only in time (as expected) but also in space. In fact, primary niches are not independent (autecological) variables. The survival, reproduction and finally the success of any organism have to deal with competition and facilitation opportunities (selection).

Taxonomy cannot be static, as classifications rely upon an increasing number of morphological, cytological, genetic and physiological informations. These crucial characters are often debated, although superb character lists have been defined (e.g., Watson and Dallwitz, 1994). But also at higher hierarchical levels the natural vegetation of distinct regions is characterised by certain plant forms, i.e. by a given physiognomy. The identification of the variable that provides reliable information on a complex situation as a whole is a main goal in ecotoxicology.

In this contribution we want to give answers to the questions: What can you recognize in the field as a result of environmental stress and management practices? When you go outside, the world is green, although environmentalists tell you that there is severe stress. What are you supposed to look for?

It is our aim to give insights into the effects of environmental stress on biodiversity as seen in the field, e.g. shifts in communities, the loss of species and appearance

Table 1. Morphological, anatomical and phenological character states of plants suitable for quantitative environmental risk assessment.

Plant architecture	*Photosynthethic organs*	*Water requirement*
Renewal buds:	Bundle sheath cells:	Hydrophytic
Annual life form	Kranz anatomy (C$_4$)	Helophytic
Therophytes (weeds, herbs)	Non-Kranz anatomy	Mesophytic
	(C$_3$ and CAM)	Xerophytic
Biennials or perennials	Leaf characters:	
Geophytes/Helophytes	Size of leaves	*Belowground organs*
Hemicryptophytes	Length of leaves	
	Width of leaves	Root morphology:
Perennial life forms	Consistency	Root depth
Chamaephytes (shrubs)	Tomentosity	Root spread
Nanophanerophytes (shrubs)		Horizontal roots
Megaphanerophytes (trees)	*Temporal cycle*	Tap root system
		Vertical-horizontal roots
Periodically shed organs:	Longevity:	Hemispheric roots
Whole plant	Life duration of plant	Netted roots
Shoot	Life duration of leaves	
Branch	Life duration of stems	Root modifications:
Leaf		Fleshy roots
	Seasonality of:	Woody roots
Stem characters:	assimilating organs	Tubers
Consistency	shoot growth	Sucker bearing
Thickness	phenology:	Contractile
Spinescence	Flowers	None
	(seasonal flowering)	
Plant characters:	(pyrogenetic flowering)	Rootlet modification:
Height	Seeds	Mycorrhizal roots
Canopy		Root nodules
Crown (density + diameter)	*Pollination syndrome*	Others
		None
	Wind pollination	
Trophic types	Water pollination	Underground stems:
	Insect pollinated	Short internodes
Autrophic	Melittophily	Elongated internodes
N-fixing synthesisers	Cantharophily	Stem tubers
Semi-parasites	Myophily	Bulbs
Saprophytes	Phalaenophily	Corms
Parasites	Psychophily	None
Carnivorous	Other (Aves, Chiroptera	
	etc.)	

of new species. These insights may contribute to the validation of environmental standards, often derived from laboratory and model studies.

Another aspect we have dealt with is the use of plants to track climate and land-use changes in history.

2. Character states of species within a flora

Functional adaptations to environmental stress (aridity, temperature, overgrazing) within a taxon have always been a major topic in geobotany during the past century. Looking for universal relationships, the autecology of vascular plants has in fact been unravelled (e.g., Raunkiaer, 1910; Braun-Blanquet, 1951; Walter, 1951). This process had great consequences for the hierarchical classification of both taxa and communities. Major strategies in plant adaptation to environmental conditions have been recognised by Du Rietz (1931) and Raunkiaer (1934). Since density estimates depend upon the scale of measurements, botanists started to focus on typical characters that could easily be recognised in the field. Within Europe, the attention to these functional groups of plant species rapidly increased. The location of the perennating buds in a plant, the leaf seasonality and the mean plant longevity (i.e., the specimen history) is crucial not only in competition and survival of natural environmental stress, but is also – and probably especially – determinant in a successful response to pollutants. Table 1 provides a survey of a combination of morphological, anatomical and phenological characters suitable for a quantitative environmental risk assessment. An example of the important role of character states within the habitat–response relation of a given taxon can be inferred from the calculated multi-substance Potentially Affected Fraction (msPAF *sensu* Posthuma et al., 2002) for Cd, Cu and Zn in the Netherlands.

Species Sensitivity Distributions (SSDs) are used to take differences in toxicant sensitivity among taxa into account, and these results are used both in the derivation of Environmental Quality Criteria (EQCs) as well as in (Probabilistic) Environmental Risk Assessment (ERA) to quantify toxic stress (expressed as Potentially Affected Fraction, PAF) at contaminated sites and water bodies. Input data for SSDs consists of single-species laboratory toxicity data, and together these data are analysed using statistical distribution theories. The SSD itself is the statistical description of empirically observed differences between species with regard to their sensitivity to toxicants. The distribution is characterised by the mean value of the log (NOEC), α, and the standard deviation σ of log transformed NOECs. The advantage of the logistic density function is that it allows for analytical evaluation of the cumulative distribution for all species, $f(x)$, by integration:

$$f(x) = 1/\{1 + \exp\,[(\alpha - x)/\beta]\})$$

Where β is proportional to the standard deviation of the distribution, and is equal to approximately 0.5σ ($\beta = \sigma\,[(\sqrt{3})/x)]$) and x is the logarithm of the *chemical*, or *bioavailable*, concentration in the environment. It takes the form of a cumulative distribution function or as a probability density function.

PAF can be derived from SSDs, given a known local environmental chemical concentration. Given a concentration of a toxicant in water, soil or sediment, the proportion of species for which the NOEC is exceeded (PAF) can be derived according to:

$$PAF = 1/\{1 + \exp\,[(\alpha - x)/\beta]\}$$

Toxic stress often results from the presence of mixtures of toxicants, and multi-substance PAF values are calculated to assess the potential ecological effects of exposure to these mixtures. For compounds with similar modes of action, the multi-substance PAF is calculated assuming that the concentration is additive (which means that relative toxicities of different compounds may be added). Groups of compounds with different toxic modes of action are assumed to cause effects additively. For heavy metals the multi substance PAF can be derived from:

$$PAF_{hm} = 1 - \prod_i (1 - PAF_i)$$

Within a data set of 95,529 field relevés (690 species), mostly in open landscapes and grasslands, Bakkenes et al. (2002) show that the toxic stress by heavy metals is not a distinguishing variable to explain the occurrence of most plant species. Still, the toxic stress plays an important role at least in 191 species. In fact, if we take into account the character states of Table 1, the sensitivity to pollutants of these vascular plants shows a much more physiological constraint. Although the PAF_{hm} is low, the sensitivity distributions of these plant species in Figures 1 (upper right corner) and 2 (on the left) point out a much lower amount of PAF-sensitive species recorded within

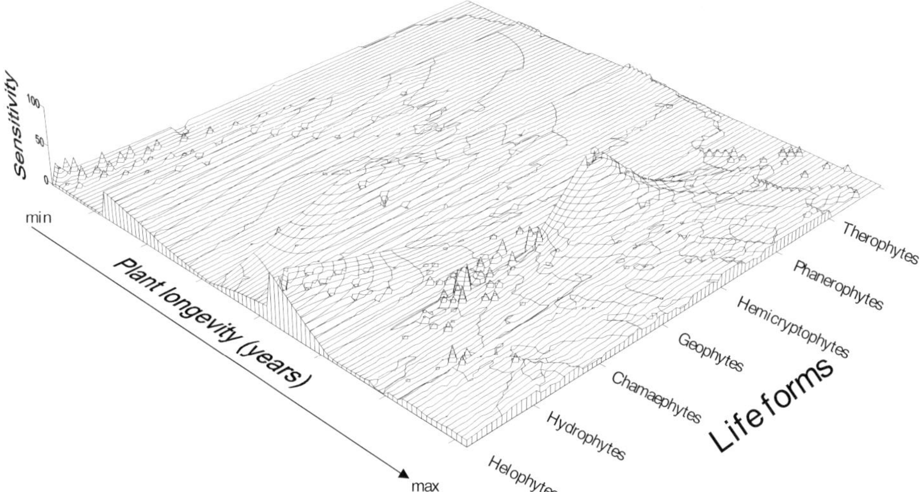

Figure 1. Density of PAF-sensitive vascular plants ranked according to their life form (*sensu* Raunkiaer, 1934) and longevity (mainly according to literature). Wireframe map modelling the radial basis function of the anisotropically re-scaled analysis of variance (ANOVA) of life form (*x*-axis), longevity (*y*-axis) and sensitivity (*z*-axis) of 191 vascular plants (raw data from Bakkenes et al., 2002). Multiquadratic function type, smoothing factor $R^2 = 0.0128$ (cf. Carlson and Foley, 1991).

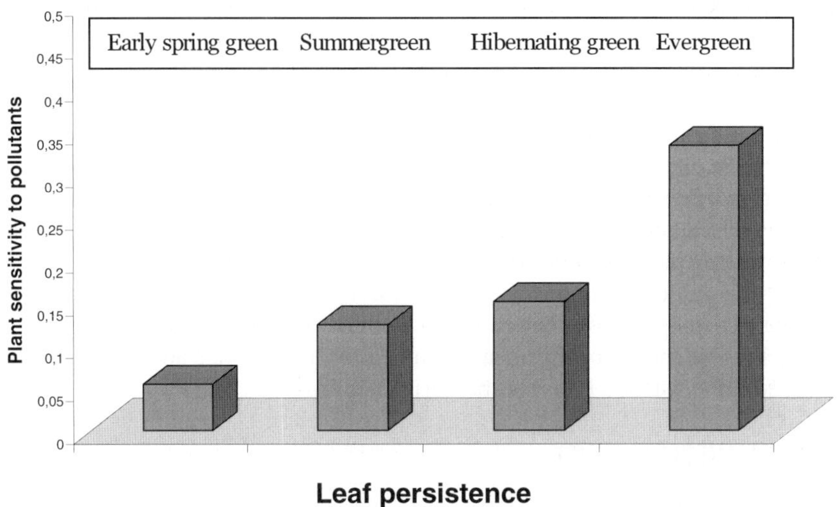

Figure 2. Mean sensitivity to heavy metals of Dutch grasslands according to the time of the year in which the leaves of the investigated vascular plants are green (categories as in Ellenberg et al., 1992).

wooden species whose leaves are green only from early spring until summer (early-spring green and summergreen).

Some conclusions can be drawn. (1) Although the average value of the sensitivity to pollutants of early-spring green species (mostly herbs, crops, weeds) is the lowest among the four categories of leaf persistence (*x*-axis of Figure 2), a widespread tolerance-sensitivity range to Cd, Cu and Zn is evident on the lower *y*-axis of Figure 1. (2) Figure 2 shows a maximal sensitivity of evergreen species for heavy metals, and although not many field records exist comparing the evergreen species with the early-spring green species, a significant correlation between plant longevity (and wooden life form) and [Cd, Cu and Zn]-msPAF exists. (3) A Gaussian-like sensitivity distribution of hibernating green species (with leaves remaining green during the winter) is suggested in Figure 1 by the kernel density of hemicryptophytes (hibernating buds near the soil surface). The target character states in a metal-polluted soil seem to be trees, wooden (and to a lesser extent herbaceous) shrubs and short-living species, passing the winter in the form of seeds (minimal plant longevity). These results confirm the recent study of Salemaa et al. (1999), where heavy metal induced shoot death in polluted environments, and rapid re-growth and plastic branching contribute to an increased resistance mechanism to heavy metals. As the mean age of the whole plant from phanerophytes through to therophytes forms a natural sequence from the K-selection (slow growth and smaller proportion of energy devoted to reproduction) to the r-selection (rapid growth and large proportion of energy devoted to reproduction, thus a high amount of relatively small seeds), these ecotoxicological results state that the energy devoted to reproduction is a key Life Support Function in heavy metal polluted soils. Although the decrease in the resilience to unfavourable conditions during the growing season from phanerophytes through to therophytes is less evident in temperate biomes, this generalised relationship can be invoked to explain shifts from obligate annuals to facultative annuals and from shrubs to trees and *vice versa*. Actually, the development of chamaephytes and phanerophytes can either proceed to further stages or remain restricted to earlier stages under certain environmental conditions (Schulze, 1982; Woodward, 1986). Eriksson (2000) has suggested only recently actual hypotheses on the hitherto neglected role of life-cycle characteristics in disturbance and increased ecosystem resilience. He stated that some character states, like creeping rhizomes, should enable the development of remnant populations. Since such plant populations are persistent, these species are especially suitable as bioindicators (or "ecosystem engineers" *sensu* Jones et al., 1994) because their activity would promote the availability of indispensable conditions for other organisms (Eriksson, 2000).

Many character states shown in Table 1 are currently used as climatic bioindicators e.g., life forms and leaf anatomy in relation to temperature, rainfall and atmospheric CO_2. Other applications are their use for ecotoxicological tests (e.g., rootlet growth of lettuce-sprouts in hydroponic solution), and as time proxies (e.g., tree rings in dendrochronological analyses; see later in this chapter). Not many studies considered so far the kind and the combination of character states which mark a given taxon not only as an evolutionary heritage of the investigated taxon, but also -or even especially- as changing fingerprints between different habitats. The co-existence of several character states within a plant community (increasing functional diversity) is supposed to reduce the impact of perturbations.

The survival and reproduction of a plant in a polluted environment remains probably one main assay. Aside from the seed germination, the growth period of a seedling until it develops in a flowering plant is crucial. Brej (1998) showed that in a polluted soil most seeds are not sensitive to heavy metals occurring in the study area (Cu, Zn, Pb, Cd and Ni). However, the subsequent development of seedlings was much less tolerant to metal pollution and the vegetative reproduction is critical even in unpolluted soils (slow development of rhizome buds). Also the root/shoot ratio is determinant in the plant tolerance to soil pollutants. Phytotoxicity assessments of the Ni and Pb content in (not standardised) soils show that the average NOEC of herbaceous plants (crops) is, depending of the lutum fraction of the soil, up to 20 times higher than the average NOEC of woody species, 58.8 vs. 3 mg/kg and 318.3 vs. 18.6 mg/kg, respectively (Lijzen et al., 2002). Surprisingly, Cd and Cu do not show such a dichotomy. Therefore, although most mechanistic whole-plant models focus on the key processes of nutrient fluxes and shoot-root allocation, the application of such models to ecotoxicological approaches remains disputed. Albeit the ecological significance of species-specific differences in C and N metabolism have been widely demonstrated and the effects of independent root and shoot competition are accepted (e.g., De Ruiter et al., 1993, Bijlma and Lambers, 2000), very little is known about the natural mechanisms involved in the uptake, retention and reallocation of heavy metals in the volume of soil influenced by the roots activity (rhizosphere). Pollutants are supposed to become stored or even immobilised in the mycelium, but only limited experimental data exist, and field empirical records are rare. Furthermore several factors contribute to make a comparison between different sites difficult, like surface water runoff and erosion which deeply affect the amount of mycorrhizal propagules in the soil, limiting the possibilities for fungal colonisation and plant regeneration. Yet, there is increasing evidence that within the rhizosphere the ecotoxicological tolerance of plants is improved by metal accumulation in the mycorrhizal fungi. These mycorrhizal symbionts would enable radionuclide accumulation and final removal by vascular plants (Entry et al., 1996; Delvaux et al., 2000; Steiner et al., 2002). The roots of most plant species are associated with fungal symbionts: external infection (ectomycorrhizas) and internal infection. The latter form of symbiosis is the most widespread mycorrhiza (vesicular-arbuscular or simply arbuscular mycorrhiza), although some other forms occur within the Ericales plant-hosts. The relevance of ectomycorrhizas for a possible limitation of heavy metal contamination in their plant hosts is unclear, since metal concentrations were usually not altered in infected trees although the same metals were found to accumulate in the extramatrical hyphae of macromycetes (Wilkins, 1991). Furthermore, the tolerance of ectomycorrhizal fungi to heavy metal varies (Kahn et al., 2000). On the contrary, internal infections of plant roots suggest a selective advantage for the infected plants. In particular, the importance of arbuscular mycorrhizas in plants growing on heavy metal contaminated sites is more evident, since mycorrhizas play a crucial role in protecting the plant roots (Galli et al., 1994). Ericoid mycorrhizal associations are widespread in acidic, nutrient-poor soils, where high levels of toxic organic acids produced by microbial conversion of fatty and phenolic residues are detoxified and assimilated by the fungus *Hymenoscyphus ericae* (Leake et al., 1989). Low pH and anaerobic conditions facilitate also the mobilisation of metals, but ericoid mycorrhizas seem to confer to their *Calluna* plant hosts adaptive resistance to AsO_4^{3-} and constitutive resistance to Cu^{2+} in a contaminated mine site (Sharples et al., 2001). *Calluna* heathlands in mycorrhizal symbiosis with *H. ericae* offers unique possibilities for environmental monitoring and site remediation. Not only the kind of fungal infection, but also the morphology of the root system (e.g., deep rooting plants like woody shrubs or trees, superficial roots in grass hummocks) will obviously affect the nutrients uptake. Also the acquisition and translocation of trace substances deserves more field studies (Marschner et al., 1996; Entry et al., 1999). Such a fundamental mechanism has to be taken into account for future dose assessment studies, since the final results would help to explain the impact of mycorrhizae on the root uptake and plant adsorption of trace substances.

Füll et al. (2000) claim the importance of a significant screening of test species (and consequently character states) to evaluate the effects of plant protection products on non-target plants. These authors suggest the standard inclusion of one leguminose species (as representative for N-fixing plants), one *Brassica* species (as representative for hemicryptophytes) and one *Avena* species (as representative for crops) in acute toxicity tests. For non-herbicide products six families (three monocots and three dicots) have to be used. However not only the character states of test plants have to be in line with the application method (drilling or spray), but also their synergy with insects. For instance, fumigation experiments allowed the quantification of a striking correlation between resistance of bean genotypes to aphids herbivory and NO_2 pollution (Masters and McNeill, 1996). Similar trends could be expected between pollinators and phenology. In fact reproducing plants depend on external vectors for the pollen dispersal and during this period they are extremely sensitive to disturbance. Alas, the correlation of land-use, pollination syndromes and biodiversity is much less debated. A powerful and urgent example is provided by the close interaction between pollinators and plant species (Fægri and van der Pijl, 1979). Everybody agrees that the loss of key non-target insect species may severely disrupt ecosystem functioning. According to the US Department of Agriculture, both wild and managed pollinators are disappearing at alarming rates (USDA-ARS, 1991). The impact of such decimation on the populations of bees and other insects is unknown, just as the impact of the reduced (or eliminated) population of pollinators on the occurrence and dispersal of wild and cultivated plant species. The amount of non-target insect pollinators necessary in a forest ecosystem is high, as we can see in Figure 3. The rate of insect pollinators in open landscapes is even higher (cf. Kastinger and Weber, 2001). Most plants are adapted to more than one pollinator, and their functional diversity is important for the resilience and the health of the aboveground community.

Still, the knowledge of the pollination syndrome of the occurring plant species may supply precious proxies for management of agroecosystems and conservation biology. In our forest example, a ratio between the vascular plant species (fern excluded) and their pollination syndrome can be proposed for a quantitative ecological risk assessment. Most insect-pollinated plants show two pollination syndromes. The values of Figure 3 present two thirds of the local flora with an Insecticide-Endangered Pollination Syndrome (403 plant taxa : 699 syndromes, with a ratio of 0.58). The IEPS-ratio can be used as indicator for a quantitative assessment of environmental risks of non-target effects of two modes of action, the cholinesterase inhibitors by organophosphorus pesticides and the modification in the functioning of sodium channels of nerve membranes by pyrethroids. Approaching the theoretical maximal value ($R = 1$), the ecosystem resilience to insecticide deposition from neighbouring cultivated areas is supposed to collapse with remarkable loss of plant diversity. In a patchy area, with

Insecticide-Endangered Pollination Syndromes

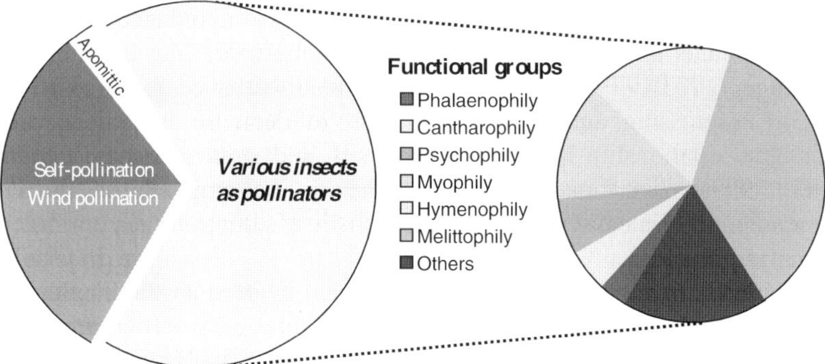

Figure 3. Spectrum of the pollination syndromes of a Bavarian managed forest. Floristic data from Mulder and Janssen (1998). The "risk" local flora (66% of the pollination within the study area involves various insects) can be used to quantify the ecosystem resilience to continuous use of pesticides in adjacent croplands.

several cultivated areas adjacent small, fragmented plant communities, the situation can become even worse. However, this ratio cannot be used directly in croplands (they have low biodiversity – most are monocultures – and low functional ecology – due to a high percentage of wind pollinated or even apomittic species). Even if the IEPS-ratio can approach the value one in such agroecosystems (monospecific croplands), the land management allows the maintenance of an ecosystem almost without resilience. Furthermore, a selective loss of higher plants implies severe consequences for the trophic chain, for instance due to disappearance of specialised butterflies (cf. Mathijssen-Spiekman and Wolters-Balk, 2001).

3. Character states of vegetation: species pool or species carousel?

Much attention has been paid to existing relationships between the occurrence of one plant species (or a certain life strategy, or a given vegetation physiognomy) and climatic conditions on micro- and mesoscales. Therewith the interplay between environmental changes in space and time of the local vegetation on both historical and geological scales was taken into account. Climate and edaphic properties determine the ability of a given taxon to become established or even competitive. Consequently, historically the land-use has been constrained by climate, technology, and economics. During several centuries, dramatic changes in land-use driven by human pressures have accelerated species extinctions, broken trophic links and exacerbated the fragmentation of (semi-)natural ecosystems. These have been occurring for at least several centuries. In addition to these world-wide problems, the superimposed effects of global change of the atmospheric composition show an unambiguous empirical evidence of a strict link between biodiversity, ecological structure and function and climate change.

The 1990s have seen a resurgence in botanical interest all over the world. This raising interest in plant-diversity issues started much before Rio de Janeiro, and culminated with the Report of the World Commission on Environment and Development (WCED, 1987) on the urgency of identifying bioindicators and instituting long-term monitoring networks to conserve biodiversity. Processes of avoidance, resistance and tolerance to pollutants are reflected in theoretical frameworks like the hump-backed model of Grime (1973,1979) and the intermediate disturbance hypothesis (Collins et al., 1995). Progressing activities finally lead to decrease in species richness. Nevertheless, the relationship between biodiversity and stress appears controversial (Waide et al., 1999), since most theoretical studies on the temporal variability of the stability or resilience of ecosystems rely on unrealistic assumptions (Cottingham et al., 2001).

Plant species sharing the same character states can be seen as functionally similar (cf. Grime, 1997; Aarssen, 2001). But to what extent does environmental information on a certain location rely upon the *occurrence* of a given species? Can the species-competition obscure the habitat-response relationship of a given species within a plant community? Actually, Humphry et al. (2001) show clearly that even the plant *abundance* has a strong effect on dose-response relationships like the ED_{50}. The field surveys (quantitative relevés according to empirical scales of plant cover and dominance) are

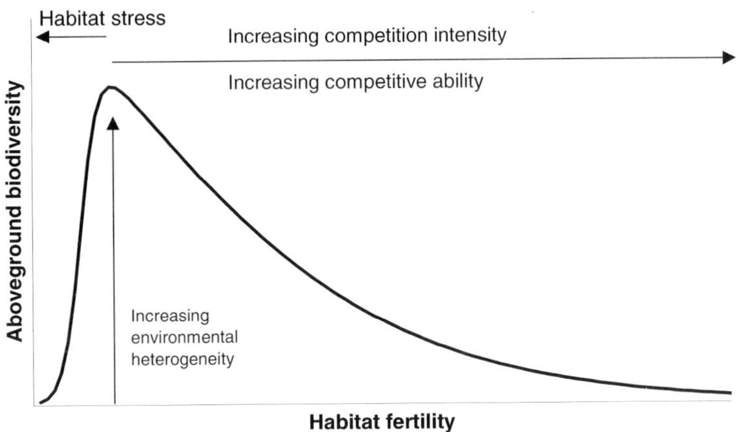

Figure 4. The hump-shaped relationship relies on empirical data observed across different habitats (Grime, 1973, 1979), whereas the recorded plant productivity is taken as proxy for habitat fertility.

supposed to provide a valuable tool to assess at the same time the occurrence and the dominance (plant density) of a given species within the local vegetation. The empirical scales necessary for this systematic method (the so-called Zürich-Montpellier phytosociological school) has been developed seventy five years ago for a quantitative study of the plants occurring in a mosaic of alpine vegetation units dominated by edaphic and microclimatic constraints (Braun-Blanquet and Jenny, 1926). Meanwhile it appears evident that although this approach is highly reliable in fragmented and man-disturbed landscapes, it fails in vast, continuous and homogeneous landscapes like most vegetation units outside Europe. In many countries in the world it remains extremely difficult to characterise a plant community by dominant, characteristic or simply "marker" species, forcing to deviate from the original method (e.g., Danin and Orshan, 1999).

Ernst (2002, this volume) discussed the use of the so-called Ellenberg indicator values to describe the environmental status with respect to eutrophication, moisture content of the soil, salinity, acidification, continentality, light status, and heavy metals (as defined in Ellenberg et al., 1992). From his discussion it is clear, that the use of this system needs good insights in ecological processes. Indicator values inferred from the occurrence of a single plant species within one region might not be applicable in other regions. Two of the most widespread species of the world, namely *Fagus sylvatica* (beech) and *Phragmites australis* (reed), provide fine examples of their migration flexibility and adaptation possibility. The present-day beech is one of the most common temperate trees in the Northern Hemisphere, but in the geological past this tree was a bioindicator for tropical, warm and wet forests (May, 1995). The still cosmopolitan, perennial reed is not only a bioindicator for wetlands (Gore, 1983, Walter, 1964), as *Phragmites* can easily grow with surprising speed in completely desiccated riverbeds of southern Africa (cf. Gibbs Russell et al., 1991). Yet, if in good hands, Ellenberg's approach is a very powerful tool to assess the environmental status, as will become clear in the following part.

According to Mueller-Dombois and Ellenberg (1974) and van der Maarel (1996), plant species would move freely around and across the community with different turn-around time according to the following factors: the present-day flora (species pool), the migration ability (vegetation history) and the ecological properties (like the kurtosis of their niche). Aside from species with wide ecological amplitudes like beech and reed, this "carousel model" implies that environmental assessments based on an as large as possible amount of species are reliable. Most indicator values increased the reliability of forecasting and ecological modelling, since their averages show significant correlations with field-measured data. Zobel (1997) described a ranking within "ecological groups" of species along an environmental gradient as a successful compromise between the field approach and the experimental approach. Although other semi-quantitative approaches became popular during the last decades for diagnostic and prognostic purposes (Landolt, 1977; Soó, 1980; Karrer and Kilian, 1990; Karrer, 1992; Borhidi, 1995; Horváth et al., 1997; Bassler et al., 2000), this methodology allowed for an independent environmental ranking of vegetation units in Central Europe. Ellenberg et al. (1992) stated that the ranking of the environmental factors in an artificial scale is mainly based upon species lists (an approach successfully tested even within regions outside Central Europe). This implies that the autecological assessment of the environmental sensitivity of a vascular plant (or the belowground tolerance of mycorrhizal fungi shown by Mulder et al., 2002) can not be misunderstood as synecological characterisation of any plant community, since the same indicator values are supposed to help us only in an ecological ranking of similar behaviours. Therefore very large metabases were assembled during the last decade.

Up to now, MOVE (MOdel for VEgetation) probably relies upon the largest amount of field-measured data for phytosociological purposes (Schaminée et al., 1995). 169,000 vegetation relevés enable a reliable evaluation of the empirical plant response to environmental factors in order to simulate and predict the change of occurrence of the local flora in the framework of acidification, eutrophication, and desiccation scenarios (e.g. Latour and Reiling, 1993; Alkemade et al., 1998; Wamelink et al., 1998). The obtained algorithms are extremely suitable for prognostic purposes. In the latest version (MOVE-3), 914 species occurring in the Netherlands have been taken into account and 14 independent dynamic models were running. Such a large amount of records demands care in statistical approaches. A too large number of environmental variables would support in fact any pattern and would give misleading evidence in multivariate analysis (McCune, 1997). To avoid this problem, a step-wise regression analysis and an independent test like the Hosmer-Lemeshow goodness-of-fit criterion ($\hat{C}$) are necessary. This criterion is a summary measure used when the number of predicted patterns is large relative to the sample size (Bio, 2000). The test has been performed as follows:

$$\hat{C} = \sum_{k=1}^{g} \frac{(o_k - n_k \bar{\pi}_k)^2}{n_k \bar{\pi}_k (1 - \bar{\pi}_k)}$$

Where n_k represents the number of records, o_k the number of observed presences (species occurrences), $\bar{\pi}_k$ the mean estimate of the likelihood in the k-group (average predicted probability, cf. Bio, 2000). After the calculation of the goodness-of-fit, 690 species showed to lie within the 95% confidence interval and have been selected to

Table 2. Environmental variables of MOVE-3.

Variable	Type	Range
Vegetation types	Discrete	5 categories
Physical/geomorphic region	Discrete	11 categories
Moisture	Continue	1–12
Open canopies	Continue	1–9
Mowing	Continue	1–9
Soil reaction	Continue	1–9
Nitrogen	Continue	1–9
Glycophyly (*)/Halophily	Continue	0*; 1–9
Combined potential affected fraction to soil heavy metals (Cd, Cu and Zn)	Continue	0.0–1.0

Redrawn from Bakkenes et al. (2002).

provide a model (Bakkenes et al., 2002). The independent environmental variables that have been taken into account are given in Table 2. For most plant species in the data set, the variables that have the fewest occurrences are the tolerances to salt and heavy metals, and the discrete categories that have been used to classify the Dutch geographic regions and vegetation types. This conclusion supports the importance of the character states of Table 1, since also variables with low occurrences become reliable in ecotoxicological studies as soon the dataset is weighted with the plant morphology and life form (Fig. 1).

4. Flora and vegetation in space and time

Vascular plants play a key role as monitor, motor and moderator of environmental changes. Different approaches aim for a quantitative assessment of the actual role of plants. Macroecology utilises inductive methods to study the emergent characteristics of large datasets of species distributed in space and time (Brown, 1995, 1999). Macroecology can therefore offer a reliable tool to outflank the difficulties of recognising the present-day ecological weight of species. The main biotic response to changing climatic patterns remains in fact migration of both plants and animals, allowing for individual species to track the best climatic conditions for their growth (e.g. Huntley and Webb, 1988, 1989; Webb and Bartlein, 1992). Those effects can for instance be seen along the two phytogeographical gradients across Europe. Ranges of variations in the specimens suggest a major influence of ecological constrains and geographical gradients at population level. Migrationism has often been assumed as framework for the explanation of a wide range of spatial patterns related to both biogeographical and cultural processes. Both Ellenberg's and Brown's approaches advocate in fact that the distribution and abundance under certain environmental conditions reflect at the same time: the response of a local population of a given species to local conditions and the extent to which local habitats meet the primary niche requirement of the given species. Both processes imply a certain degree of autocorrelation in space between environments in close proximity. This is not always true in fragmented landscapes. A monotonic inference on the climatic significance of vegetation communities can therefore become dubious, invalidating any realistic temporal reconstruction of local vegetation history. Thus, although functional relationships with biotic and abiotic environmental conditions should mark the local success of plants (Schulze and Chapin III, 1987; Keddy, 1990; Solbrig, 1993), any impact of human settlements in the past has to be taken into account carefully. Several efforts have arisen out of these observations and will be discussed briefly.

4.1. Photosynthesis and respiration as proxies for temperature, rainfall and CO_2

Long-term field experiments on the response of plant specimens, plant populations and plant communities are necessary for reliable environmental risk assessments and ecological modelling. Most prognostic models of the response of the vegetation to disturbance, pollution and global change are focused on the biophysical feedback of the landscape (i.e., the ecophysiological constraints defined by the character states of Table 1). They try to take into account a very large list of key environmental variables (UV-B, O_3, CO_2 elevation, SO_2, NO_2-deposition, temperature, and rainfall). These forcing factors have to be considered not only singly, but also – or even especially – in combination as extreme events. Aside from geological and technological CO_2 production, vegetation removes CO_2 from the atmosphere, stores it in organic forms and releases CO_2, methane and other trace gases back to the atmosphere (Shugart et al., 1992). Therefore, increasing CO_2-pollution can be regarded as one of the greatest sources of disturbance in changing atmosphere composition, due to its role in both the respiration and the photosynthesis. Most General Circulation Models forecast a global warming of several degrees Celsius as a result of increased concentrations of greenhouse gases (Prentice, 1990; Schneider, 1992). At the same time, the (estimated) flux of volatile organic compounds from natural vegetation could have damped or even amplified past climate changes during the last 20,000 years (Adams et al., 2001). During this time span, continental biomes like the Sahara-Gobi desert belt shifted from a sink for 218 ~ 283 Gt of atmospheric C to a source of carbon, showing the extent of a still unknown carbon reservoir dynamic (Lioubimtseva et al., 1998).

Therefore, a central problem in mapping and monitoring the vegetation all over the world concerns the stability of the vegetation structure within the general framework of global climate change and man disturbance. The increased atmospheric CO_2 concentration has almost immediate effects on two crucial ecophysiological processes of terrestrial plants, the photosynthesis rate and the rate at which plant leaves lose water. The attractivity of monitoring photosynthetic processes rests on their ability to elucidate and quantify the current climate. Plants can be divided roughly in three photosynthetic pathways according to their biochemistry (C_4, C_3 and Crassulacean Acid Metabolism (CAM)) and leaf anatomy (Kranz or non-Kranz). Although C_4 biota account for less than one fifth of the total global productivity (Ehleringer et al., 1997), two dominant C_4 photosynthesis subtypes coupled with the Kranz anatomy, namely NAD-*me* and NADP-*me* can be easily used to yield information on climate (e.g., Ehleringer and Vogel, 1993; Schulze et al., 1996). C_4 photosynthesis does not only confer benefit on plants growing under conditions with low CO_2 concentrations (Fig. 5), it also controls the loss of stomata water under warm and dry conditions. The above stated C_4 subtypes differ in terms of morphological and biochemical details. In particular, an evident, significant sensitivity for the minimal annual rainfall has been put into evidence in Namibian grasslands (Fig. 6) whereas the relative dominance of Kranz-grasses with either a NAD-*me* or a NADP-*me* metabolism growing along a continentality gradient across the Kalahari show remarkable opposite trends (Mulder and Ellis, 2000). Similar approaches can also rely entirely upon geochemical analyses. Huang et al. (1999) provide a valuable example of high-resolution time-variation of individual lignin phenols and leaf-wax *n*-alkanes. Beside the CO_2 starvation of the Last Glacial Maximum, their study confirms the important role of aridity in controlling the abundance of C_4 plants. The abundance records of certain C_4 bioindicators may assess quantitatively climate-dependent (or man-induced) dynamic changes in the plant's potential relative growth rate (*sensu* Grime, 1979).

Leaf anatomy differences associated with water-use efficiency strongly influence the final competitivity of single species and the actual resilience of the given vegetation they belong to (Pyankov et al., 1998; Mulder and Ellis, 2000). This drought habitat–response relation can contribute to the evaluation of the sustainability of land-use in developing countries and can be monitored easily in the field as well as in the laboratory. Yet, microscopic observations of plant leaves offer encouraging possibilities for monitoring not only water stress, but also air pollution. In fact the largest fraction of terrestrial evaporative water flux to the atmosphere passes through leaf stomata. Since the density and geometry of these pores determines the stomatal control of leaf conductance, this mechanism balances the conflicting priorities of C-gain for photosynthesis and water conservation in terrestrial plants (Mansfield, 1998). The stomata closure induced by SO_2 is of primary importance for ecological evaluation and risk assessments. Stomatal conductance can be inferred easily from needles and the so-obtained statistical results can be related to the air concentration of SO_2, the soil concentration of SO_4^{2-} and to the health of forest canopies. Slovik et al. (1996) demonstrated a statistically highly significant correlation between damaged spruce canopies and air SO_2 pollution. Adverse airborne-pollutant effects on the photosynthetic apparatus are mostly considered to result from the reduction in SO_2 uptake through reduction in stomatal conductance or even stomatal closure at high SO_2 concentrations. However, interactions between CO_2 and SO_2 have to be taken into account (Lee et al., 1997, Robinson et al., 1998).

The initial enthusiasm for environmental modelling of past concentrations of greenhouse gases, which relies entirely on computer-aided determination of stomatal parameters, was based on studies with serious methodological problems (Indermühle et al., 1999). Interaction and covariance between gases and leaf stomata further affect the possible use of stomatal conductance for prognostic purposes, since the buffer capacities of soils seem unpredictable. In fact, although the effects of the acid rain in Europe decreased during the last two decades and can be monitored easily by epiphytic lichens (see Wolterbeek et al., 2002, this volume), the present-day recovery of vascular plants shows still a consistent delay due to the high soil concentration of SO_4^{2-}. Therefore, the forest management is to some extent forced to counteract the nutrient depletion in soils with a frequent liming (Alewell et al., 2000). Yet, recent pilot results are encouraging. Although most extrapolations from short-term observation failed to predict

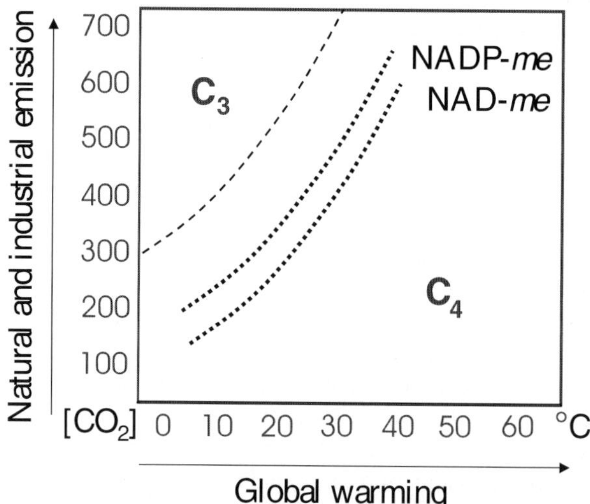

Figure 5. Crossover daytime average air-temperature during the growing season of monocots (cf. Fig. 6) of the correspondent CO_2-uptake quantum yield as a function of the CO_2 atmospheric concentration (ppmv). Changes in the CO_2-uptake quantum yield due to photorespiration, and to a lesser extent to pollution. C_4-types split in NAD-*me* (Nicotinamide-adenine dinucleotide-malic enzyme) and NADP-*me* (Nicotinamide-adenine dinucleotide phosphate-malic enzyme); PEP-*ck* curve (phosphoenolpyruvate carboxykinase) undrawn. Thresholds redrawn from Edwards and Walker (1983) and Ehleringer et al. (1997).

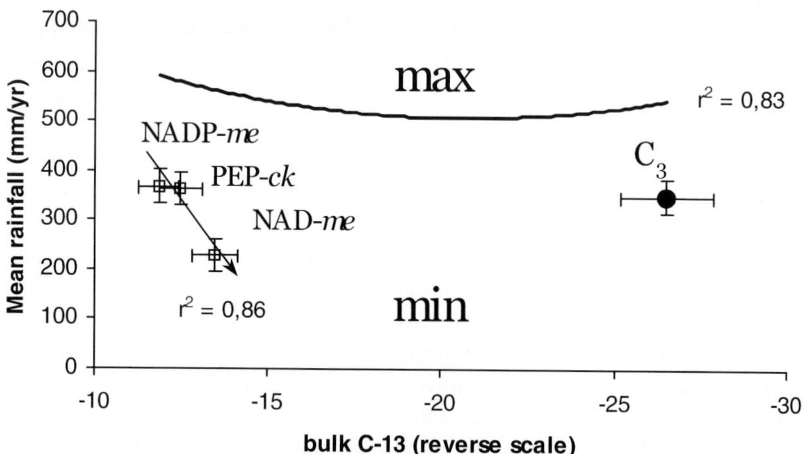

Figure 6. Crossover mean annual precipitation in arid landscapes (cf. Fig. 5) as a function of the CO_2-uptake quantum yield (here inferred from the bulk C-13). Abbreviations as in Fig. 5. Raw data from Mulder and Ellis (2000). It is evident that the investigated Namibian plants are more sensitive for the minimum precipitation values. These rainfall minima play a crucial role in the desertification process.

functional patterns as mere response to CO_2 stress, Anderson et al. (2001) have put into evidence that the stomatal conductance in perennial plants showed significant curvilinear decreases with the atmospheric CO_2 concentration, suggesting historical trends.

4.2. Plant remains as proxies for vegetation history and human influence

Plant macro-remains (especially tree rings and seeds) and micro-remains (dispersed pollen, spores, and biogenic silica particles that precipitate in plant tissues) provide unique tools for an environmental reconstruction of past landscapes all over the world. The study of plant remains occurs within four disciplines, namely dendrochronology (tree rings), archaeobotany (seeds in archaeological settings), palynology (pollen and spores in sediments and surface samples) and micro-palaeontology (plant opal and diatoms). Particularly the study of growth rings of trees (dendrochronology) helps to improve understanding of spatial and temporal variability of climatic, geomorphic, and ecological systems. The growth of tree rings is due to external conditions when they were formed. Therefore, all analysed tree rings can be dated to their formation period to improve understanding of past environmental conditions. Tree rings are not merely counted, but cross-dated by matching the ring-width across many samples to identify the correct year of formation of each ring. Dendrochronology comprises various sub-disciplines, from dendroclimatology (reconstruction of past temperature and rainfall), dendroecology (assessment of fire hazards, insect and fungal pests, forest dynamics), dendrohydrology (study of snowfall and melt timing, floods, desertification), dendro-chemistry (interpretation of chemical environmental changes through the analysis of inorganic elements, relationships between ring-width and nutrient availability through time) to archaeology (dating past human cultures).

Valuable monitoring of Mediterranean oaks continuously exposed to elevated CO_2 concentrations show that in an enriched atmosphere the growth response of trees varies over their entire lifetime (Hattenschwiler et al., 1997), and the most significant CO_2-effect occurs probably during the first year of growth (Telewski et al., 1999). Aside from the widespread archaeological applications, the tree rings prove to be very reliable bioindicators in the other subdisciplines, from dendroecology through dendro-hydrology up to dendrochemistry. For instance, a fine quantitative assessment of the global fire histories between 1650 and 1990 and their teleconnections with El Niño-Southern Oscillation and the switch with La Niña conditions has recently been provided by Kitzberger et al. (2001). According to these authors, the tree rings allowed to establish a long-term record of fire history in Arizona, New Mexico (USA) and Patagonia (Argentina) where El Niño events enhance the production of fine fuels, while La Niña events create the conditions for a further widespread wildfire. These results are encouraging for prognostic purposes, and tree rings offer the opportunity for a better land management. Another successful example is provided by Antonic et al. (2001), who applied dendrochronology to the modelling of acceptable groundwater regime for a floodplain forest characterised by a very sensitive pedunculate oak. The important role of oak rings as bioindicators for groundwater disturbance is confirmed by Edmands et al. (2001). After a preliminary study on the uptake and mobility of uranium in bark, sapwood and heartwood tree rings they suggested that the monitoring of U-concentrations in oak bark and sapwood is a rather inexpensive alternative to standard drilling wells. Furthermore, the $\delta^{13}C$ record of annual tree rings seems to be a powerful tool for the evaluation of the actual impact of air pollution on cedar forests (Sakata and Suzuki, 1998) and confirms that SO_2 is a major cause in the forest decline.

Palynology embraces the study of pollen grains and spores, studying the spore dispersal of respectively the male gametophyte of seed plants and the resting phase of cryptogams (algae, mosses and fungi included). The study of pollen grains has multi-disciplinary applications, ranging from melissopalynology (the pollen contained in honey points to the foraging source of the bees and consequently rates the product quality) through forensic palynology (bioindicators can help to resolve complex legal disputes, e.g. whether or not *Cannabis* samples were obtained from the same cultivation) up to aerobiology (the monitoring of airborne pollen and spore concentration in relation to air temperature and hay fever). The most widespread application is to trace the vegetation history in species and communities, assigning tentative dates to correlate deposits and assessing the global change (Moore et al., 1991). All distributional trends of pollen and spores in space (transects) and in time (cores) show either regional or local dispersal (Mulder and Janssen, 1999 and Odgaard, 1999, respectively). Therefore, pollen captured by surface samples or traps can provide quantitative values for a climatic assessment of the distinct local vegetation zones (e.g., Hicks, 1999), while pollen recovered offshore from marine sediments can help to reconstruct the regional vegetation history at biome level (e.g., Shi et al., 2001).

For each study area, the basic stages in tracing the human impact on the local vegetation with a palynological approach is the establishment of which pollen types are the most valuable bioindicators. Several weeds have recognisable pollen grains at species level and have proven to be sensitive bioindicators of human disturbance and land management (agriculture, clearances, etc.). *Plantago lanceolata* is typical for European wet meadows and pastures, and a continuous, low presence of this pollen type proves evidence for grazing (Behre, 1981) and increased use of artificial fertilisers (Joosten, 1985). In winter croplands *Centaurea cyanus*, *Fallopia convolvulus* and *Scleranthus annuus* are widespread indicators, while Polygonaceae (knotweed and dock) are more common in fallow land (cf. Moore et al., 1991 and Behre and Kučan, 1994). The first human attempts to consolidate sand dunes along the Northwest European coasts can be detected between 1681 (Holland) and 1690 (England) by recognising the introduced and successively naturalised mesem *Carpobrotus edulis* (Mulder, 2002). However, the resolution of palynological investigation can be often much lower (i.e., it is difficult or even impossible to identify the palynomorphs at species level). Still, the habitat-response relationship and the eventual niche specialisation of a given plant community can be inferred easily by the trends in the pollen assemblage along a soil profile. For instance, bioindicators for metal-polluted soils like *Noccanea caerulescens*, *Armeria maritima* ssp. *halleri* or *Viola calaminaria* cannot be recognised at species level by pollen analyses (the pollen grains of these taxa belong – together with a large number of other species – to the *Raphanus raphanistrum*-type, *Limonium binervosum*-type and the *Viola palustris*-type, respectively). Still, their combined occurrence allows for an ecotoxicological evaluation at larger scales of the past landscape, since these pollen grains may refer to the same coenon, namely *Violetalia calaminariae* (Br.-Bl. et Tüxen 43). The same arguments can be used for an assessment of the land-use or a refined reconstruction of the vegetation biomes.

Therefore multivariate analyses became so popular in palynology during the last decade (e.g., Gaillard et al., 1992; Birks, 1998; Brostrom et al., 1998; Hjelle, 1998, 1999; Mulder et al., 2002), although the importance of numerical techniques has been previously claimed by Birks and Gordon already in 1985.

Although the main goal of palynology is the reconstruction of the past landscape, archaeobotany focused on the economic and social behaviour of man (e.g., trade, diet, etc.). In Central Europe a high amount of archaeobotanical investigations provide a fine-scaled assessment of prehistoric and protohistoric land-use (e.g. Willerding, 1986). The human impact during the Neolithic was different according to the landscape. Most forest were not used intensively and permanent open land was rare (Rösch, 1996). In open areas, farming was performed by shifting cultivation with slash and burn, mobilising nutrients and hampering most weeds (cf. Huttunen, 1980), enabling a re-development of forest. Only much later the practice of coppicing or even forest-clearance changed the European landscape dramatically, facilitating the migration of many aggressive weeds like the chenopods. A multivariate analysis of weed occurring in 274 archaeological sites allows to state that the synanthropism of crucial taxa like chenopods (at least during the whole prehistory and protohistory north of the Alps) is rather weak (Mulder, 1999). Although some chenopods are evidently correlated with more-or-less man-induced desertification processes, Chenopodiaceae-rich communities bear evidence for their natural migration during the Holocene (cf. Pyankov et al., 2002). Thus, in palaeoethnobotany the overlapping between climate-induced migration and man-induced migration is subtle.

On the other hand, in the Southern Hemisphere the overlapping between natural and man-induced migration is much clearer. For instance, hunt coupled with slash and burn agriculture has also been recognised during the Iron Age of south-eastern savanna biomes, but only during the last millennium southern Africa was marked by dramatic agricultural development and significant changes in the vegetation (Hoffman, 1997). The increasing importance of cattle led to several ecological problems, among which a selective form of grazing with consequent patchiness of the vegetation structure. In fact, tannin-like substances are present in many grasses (Ellis, 1980), affecting their digestibility to stock, especially ruminants. Some of these grasses are so specialised to high nitrate and phosphate soils that they can still be used a bioindicators of past cattle holding. According to Denbow (1979), the blue buffalo grass *Cenchrus ciliaris* is consistently associated with vitrified dung deposits and his populations are easily discernible as "bald spots" on aerial photographs. Yet, although the man-induced changes on the landscape are strikingly evident even in Africa (Acocks, 1988), the environmental impact of man on and in the temperate biome during his recent history is highly debated (Vera, 2000).

References

Aarssen, L.W., 2001. On correlations and causations between productivity and species richness in vegetation: predictions from habitat attributes. Basic Appl. Ecol. 2, 105–114.

Acocks, J.P.H, 1988. Veld Types of South Africa, 3rd edn. Memoirs van die botaniese opname van Suid-Afrika 57, pp. 1–146.

Adams, J.H., Constable, J.V.H., Guenther, A.B., Zimmermann, P., 2001. An estimate of natural volatile organic compounds emissions from vegetation since the last glacial maximum. Chemosphere – Glob. Change Sci. 3, 73–91.

Alewell, C., Manderscheid, B., Gerstberger, P., Matzner, E., 2000. Effects of reduced atmospheric deposition on soil solution chemistry and elemental contents of spruce needles in NE-Bavaria, Germany. J. Plant Nutr. Soil Sci. 163, 509–516.

Alkemade, J.R.M., van Grinsven, J.J.M., Wiertz, J., Kros, J., 1998. Towards integrated modelling with particular reference to the environmental effects of nutrients. Environ. Poll. 102, 101–105.

Anderson, L.J., Maherali, H., Johnson, H.B., Polley, H.W., Jackson, R.B., 2001. Gas exchange and photosynthetic acclimation over subambient to elevated CO_2 in a C_3–C_4 grassland. Glob. Change Biol. 7, 693–707.

Antonic, O., Hatic, D., Krian, J., Bukovec, D., 2001. Modelling groundwater regime acceptable for the forest survival after the building of the hydro-electric power plant. Ecol. Model. 138, 277–288.

Bakkenes, M., de Zwart, D. Alkemade, J.R.M., 2002. MOVE: nationaal Model voor de Vegetatie versie 3.2. Achtergronden en analyse van modelvarianten. RIVM report nr. 408657 006, (2 vols), Bilthoven.

Bassler, G., Lichtenecker, A., Karrer, G., 2000. Gliederung der extensiven Grünlandtypen im Transekt von Oppenberg bis Taupliz. MAB-Forschungsbericht: "Landschaft und Landwirtschaft im Wandel", pp. 51–96.

Behre, K.-E., 1981. The interpretation of anthropogenic indicators in pollen diagrams. Pollen Spores 23, 225–245.

Behre, K.-E., Kučan, D., 1994. Die Geschichte der Kulturlandschaft und des Ackerbaus in der Siedlungskammer Flögeln, Niedersachsen, seit der Jungsteinzeit. Probleme der Küstenforschung im Südlichen Nordseegebiet 21.

Bijlma, R.J., Lambers, H., 2000. A dynamic whole-plant model of integrated metabolism of nitrogen and carbon. 2. Balanced growth driven by C fluxes and regulated by signals from C and N substrate. Plant Soil 220, 71–87.

Bio, A.M.F., 2000. Does vegetation suit our models? Data and model assumptions and the assessment of species distribution in space. Netherlands Geographical Studies 265.

Birks, H.J.B., 1998. Numerical tools in palaeolimnology – progress, potentialities and problems. J. Paleolimnol. 20, 307–332.

Birks, H.J.B., Gordon, A.D., 1985. Numerical Methods in Quaternary Pollen Analysis. Academic Press, London.

Borhidi, A., 1995. Social behaviour types, the naturalness and relative ecological indicator values of the higher plants in the Hungarian flora. Acta Bot. Hung. 39, 97–181.

Braun-Blanquet, J., 1951. Pflanzensoziologie. Grundzüge der Vegetationskunde (2nd rev. edn). Springer Verlag, Wien.

Braun-Blanquet, J., Jenny, H., 1926. Vegetationsentwicklung und Bodenbildung in der alpinen Stufe der Zentralalpen (Klimaxgebiet des *Caricion curvulae*). Neue Denkschrift der Schweizerischen Naturforschungs Gesellschaft 63, 183–349.

Brej, T., 1998. Heavy metal tolerance in *Agropyron repens* (L.) P. Bauv. Populations from the Legnica copper smelter area, Lower Silesia. Acta Soc. Bot. Pol. 67, 325–333.

Brostrom, A., Gaillard, M.J., Ihse, M., Odgaard, B., 1998. Pollen-landscape relationships in modern analogues of ancient cultural landscapes in southern Sweden – a first step towards quantification of vegetation openness in the past. Veg. Hist. Archaeobot. 7, 189–201.

Brown, J.H., 1995. Macroecology. University of Chicago Press, Chicago.

Brown, J.H., 1999. Macroecology: progress and prospect. Oikos 87, 3–14.

Carlson, R.E., Foley, T.A., 1991. The parameter R^2 in multiquadratic interpolation. Computer & Mathematics with Applications 21, 29–42.

Clay, K., 1990. Fungal endophytes of grasses. Ann. Rev. Ecol. Syst. 21, 275–297.

Collins, S.L., Glenn, S.M., Gibson, D.J., 1995. Experimental analysis of immediate disturbance and initial floristic composition: decoupling cause and effect. Ecology 76, 486–492.

Cottingham, K.L., Brown, B.L., Lennon, J.T., 2001. Biodiversity may regulate the temporal variability of ecological systems. Ecol. Lett. 4, 72–85.

Danin, A., Orshan, G., 1999. Vegetation of Israel. I. Desert and coastal vegetation. Backhuys, Leiden.

Delvaux, B., Kruyts, N., Cremers, A., 2000. Rhizospheric mobilization of radiocesium in soils. Environ. Sci. Technol. 34, 1489–1493.

Denbow, J.R., 1979. *Cenchrus ciliaris*: an ecological indicator of Iron Age middens using aerial photography in eastern Botswana. S. Afr. J. Sci. 75, 405–408.

De Ruiter, P.C., van Veen, J.A., Moore, J.C., Brussaard, L., Hunt, H.W., 1993. Calculation of nitrogen mineralization in soil food webs. Plant Soil 157, 263–273.

Du Rietz, G.E., 1931. Life forms of terrestrial flowering plants. Acta Phytogeogr. Suec. 3, 1–95.

ECNC (European Centre for Nature Conservation), 2001. Species richness in Europe. [online at: http://www.ecnc.nl/doc/data/species.html]

Edmands, J.D., Brabander, D.J., Coleman, D.S., 2001. Uptake and mobility of uranium in black oaks: implications for biomonitoring depleted uranium-contaminated groundwater. Chemosphere 44, 789–795.

Edwards, G., Walker, D.A., 1983. C_3, C_4: mechanisms, and cellular and environmental regulation, of photosynthesis. Blackwell, Oxford.

Ehleringer, J.R., Cerling, T.E., Helliker, B.R., 1997. C_4 photosynthesis, atmospheric CO_2, and climate. Oecologia 112, 285–299.

Ehleringer, J.R., Vogel, J.C., 1993. Historical aspects of stable isotopes in plant carbon and water relations. In: Ehleringer, J.R., Hall, A.E., Farquhar, G.D. (Eds), Stable Isotopes and Plant Carbon–Water Relations. Academic Press, San Diego, pp. 9–18.

Ellenberg, H., Weber, W., Düll, R., Wirth, V., Werner, W., Paulißen, D., 1992. Zeigerwerte von Pflanzen in Mitteleuropa, 2nd edn. Scripta Geobotanica 18. Goltze, Göttingen.

Ellis, R.P., 1980. Tannin-like substances in grass leaves. Memoirs van die botaniese opname von Suid-Afrika, 59, 1–80.

Entry, J.A., Vance, N.C., Hamilton, M.A., Zabowski, D., Watrud, L.S., Adriano, D.C., 1996. Phytoremediation of soil contaminated with low concentrations of radionuclides. Water, Air, Soil Poll. 88, 167–176.

Entry, J.A., Watrud, L.S., Reeves, M., 1999. Accumulation of ^{137}Cs and ^{90}Sr from contaminated soil by three grass species inoculated with mycorrhizal fungi. Environ. Poll. 104, 449–457.

Eriksson, O., 2000. Functional roles of remnant plant populations in communities and ecosystems. Global Ecol. and Biogeography 9, 443–449.

Ernst, W.H.O., 2002 (2003). The use of higher plants as bioindicators. In: Markert, B.A., Breure, A.M., Zechmeister, H.G. (Eds), Bioindicators and Biomonitors. Elsevier, Oxford, pp. 423–463.

Fægri, K., van der Pijl, L., 1979. The Principles of Pollination Ecology, 3rd rev. edn. Pergamon Press, Oxford.

Franzaring, J., van der Eerden, L.J.M., 2000. Accumulation of airborne persistent organic pollutants (POPs) in plants. Basic Appl. Ecol. 1, 25–30.

Füll, Ch., Jung, S., Schulte, Ch., 2000. Plant protection products: assessing the risk for terrestrial plants. Chemosphere 41, 625–629.

Gaillard, M.-J., Birks, H.J.B., Emanuelsson, U., Berglund, B.E., 1992. Modern pollen/land-use relationships as an aid in the reconstruction of past land-uses and cultural landscapes: an example from south Sweden. Veg. Hist. Archaeobot. 1, 3–17.

Galli, U., Schuepp, H., Brunold, C., 1994. Heavy metal binding by mycorrhizal fungi. Physiol. Plant. 92, 364–368.

Gibbs Russell, G.E., Watson, L., Koekemoer, M., Smook, L., Barker, N.P., Anderson, H.M., Dallwitz, M.J., 1991. Grasses of southern Africa. Memoirs van die botaniese opname van Suid-Afrika 58.

Gore, A.J.P., (Ed.) 1983. Mires: Swamp, Bog, Fen and Moor. Ecosystems of the Worlds 4A. Elsevier, Amsterdam, 440 pp.

Grime, J.P., 1973. Competitive exclusion in herbaceous vegetation. Nature 242, 344–347.

Grime, J.P., 1979. Plant Strategies and Vegetation Processes. Wiley, Chichester.

Grime, J.P., 1997. Biodiversity and ecosystem function: the debate deepens. Science 277, 1260–1261.

Hattenschwiler, S., Miglietta, F., Raschi, A., Korner, C., 1997. Thirty years of in situ tree growth under elevated CO_2: a model for future forest responses? Glob. Change Biol. 3, 463–471.

Hicks, S., 1999. The relationship between climate and annual pollen deposition at northern tree-lines. Chemosphere: Glob. Change Sci. 1, 403–416.

Hjelle, K.L., 1998. Herb pollen representation in surface moss samples from mown meadows and pastures in western Norway. Veg. Hist. Archaeobot. 7, 79–96.

Hjelle, K.L., 1999. Modern pollen assemblages from mown and grazed vegetation types in western Norway. Rev. Palaeobot. Palynol. 107, 55–81.

Hoffman, M.T., 1997. Human impact on vegetation. In: Cowling, R.M., Richardson, D.M., Pierce, S.M. (Eds), Vegetation of Southern Africa. Cambridge University Press, Cambridge, pp. 507–534.

Horváth, F., Rapcsák, T., Fölsz, F., Hoffer, J., Lõkös, L., Peregovits, L., Rajczy, M., Samu, F., Szép, T., Szilágyi, G., 1997. A Biodiverzitás-monitorozó Program metaadatbázisának terve (TERMET). In: Horváth, F., Rapcsák, T., Szilágyi, G. (Eds), Nemzeti Biodiverzitásmonitorozó Rendszer I. Informatikai alapozás. Budapest, Magyar Természettudományi Múzeum, pp. 88–105.

Huang, Y., Freeman, K.H., Eglinton, T.I., Street-Perrott, F.A., 1999. $\delta^{13}C$ analyses of individual lignin phenols in Quaternary lake sediments: a novel proxy for deciphering past terrestrial vegetation changes. Geology 27, 471–474.

Humphry, R.W., Mortimer, M., Marrs, R.B., 2001. The effect of plant density on the response of *Agrostema githago* to herbicide. J. Appl. Ecol. 38, 1290–1302.

Huntley, B., Webb, T. III, 1988. Vegetation History. Kluwer, Dordrecht.

Huntley, B., Webb, T. III, 1989. Migration: species' response to climatic variations caused by changes in the earth's orbit. J. Biogeogr. 16, 5–19.

Huttunen, P., 1980. Early land-use, especially the slash-and-burn cultivation in the comune of Lammi, southern Finland, interpreted mainly using pollen and charcoal analyses. Acta Bot. Fenn. 113, 1–45.

Indermühle, A., Stauffner, B., Stocker, T.F., Raynaud, D., Barnola, J.-M., Birks, H.H., Eide, W., Birks, H.J.B., Wagner, W.S., Kürschner, W.M., Visscher, H., Bohnke, S.J.P., Dilcher, D.L., van Geel, B., 1999. Early Holocene atmospheric CO_2 concentrations. Science 286, 1815a (Technical Comments).

Jones, C.G., Lawton, J.H., Shachak, M., 1994. Organisms as ecosystem engineer. Oikos 69, 373–386.

Joosten, J.H.J., 1985. A 130 year micro- and macrofossil record from regeneration peat in former peasant peat pits in the Peel, The Netherlands: a palaeoecological study with agricultural and climatological implications. Palaeogeogr. Palaeoclimatol. Palaeoecol. 49, 277–312.

Kahn, A.G., Kuek, C., Chaudhry, T.M., Khoo, C.S., Hayes, W.J., 2000. Role of plants, mycorrhizae and phytochelators in heavy metal contaminated land remediation. Chemosphere 41, 197–207.

Karrer, G., 1992. Österreichische Waldboden-Zustandsinventur, Teil VII: Vegetationsökologische Analysen. Mitteilungen der forstlichen Bundesversuchsanstalt Wien 168/II, pp. 193–226.

Karrer, G., Kilian, W., 1990. Standorte und Waldgesellschaften im Leithagebirge Revier Sommerein. Mitteilungen der forstlichen Bundesversuchsanstalt Wien 165.

Kastinger, Ch., Weber, A., 2001. Bee-flies (*Bombylius* spp., Bombyliidae, Diptera) and the pollination of flowers. Flora 196, 3–25.

Keddy, P.A., 1990. Competitive hierarchies and centrifugal organization in plant communities. In: Grace, J.B., Tilman, D. (Eds), Perspectives in Plant Competition. Academic Press, San Diego, pp. 256–290.

Kitzberger, T., Swetnam, T.W., Veblen, T.T., 2001. Inter-hemispheric synchrony of forest fires and the El Niño-Southern Oscillation. Glob. Ecol. Biogeogr. 10, 315–326.

Kluth, S., Kruess, A., Tscharntke, T., 2001. Interactions between the rust fungus *Puccinia punctiformis* and ectophagous and endophagous insects on creeping thistle. J. Appl. Ecol. 38, 548–556.

Landolt, E., 1977. Ökologische Zeigerwerte zur Schweizer Flora. Veröff. Geobot. Inst. ETH, Stiftung Rübel, Zürich, 64.

Latour, J.B., Reiling, R., 1993. A multiple stress model for vegetation ("MOVE"): a tool for scenario studies and standard-setting. Sci. Tot. Environ. (Suppl.) Part 2, 1513–1526.

Leake, J.R., Shaw, G., Read, D.J., 1989. The role of ericoid mycorrhizas in the ecology of ericaceous plants. Agric. Ecosyst. Environ. 29, 237–250.

Lee, E.H., Pausch, R.C., Rowland, R.A., Mulchi, C.L., Rudorff, B.F.T., 1997. Response of field-grown soybean (cv. Essex) to elevated SO_2 under two atmospheric CO_2 concentrations. Environ. Exp. Bot. 37, 85–93.

Lijzen, J.P.A., Mesman, M., Aldenberg, T., Mulder, Ch., Otte, P.F., Posthumus, R., Roex, E., Swartjes, F.A., Versluijs, C.W., van Vlaardingen, P., van Wezel, A.P., van Wijnen, H., 2002. Evaluatie onderbouwing Bodem Gebruiks Waarden. RIVM report no. 711701 029, Bilthoven.

Lioubimtseva, E., Simon, B., Faure, H., Faure-Denard, L., Adams, J.M., 1998. Impact of climatic change on carbon storage in the Sahara-Gobi desert belt since the Last Glacial maximum. Glob. Planet. Change 16–17, 95–105.

Mansfield, T.A., 1998. Stomata and plant water relations: does air pollution create problems? Environ. Poll. 101, 1–11.

Markert, B.A., Breure, A.M., Zechmeister, H.G. 2002 (2003). Definitions, strategies and principles for bioindication/biomonitoring of the environment. In: Markert, B.A., Breure, A.M., Zechmeister, H.G. (Eds), Bioindicators and Biomonitors. Elsevier, Oxford, pp. 3–39.

Marschner, P., Godbold, D.L., Jentschke, G., 1996. Dynamics of lead accumulation in mycorrhizal and non-mycorrhizal Norway spruce (*Picea abies* (L.) Karst). Plant Soil 178, 239–245.

Masters, G.J., McNeill, S., 1996. Evidence that plant varieties respond differently to NO_2 pollution as indicated by resistance to insect herbivores. Environ. Poll. 91, 351–354.

Mathijssen-Spiekman, E.A.M., Wolters-Balk, M.A.H., 2001. The integrated monitoring area Lhee-broekerzand – The Netherlands: data of 1997, 1998, 1999. RIVM report no. 607165 001, Bilthoven.

May, D.H., 1995. Tertiäre Vegetationsgeschichte Europas. Gustav Fischer, Jena.

McCune, B. 1997. Influence of noisy environmental data on canonical correspondence analysis. Ecology 78, 2617–2623.

Moore, P.B., Webb, P.A., Collinson, M.E., 1991. Pollen Analysis, 2nd edn. Blackwell Scientific, Oxford.

Mueller-Dombois, D., Ellenberg, H., 1974. Aims and methods of vegetation ecology. Wiley, New York.

Mulder, Ch., 1999. Biogeographic re-appraisal of the Chenopodiaceae of Mediterranean drylands: a quantitative outline of their general ecological significance in the Holocene. Palaeoecol. Africa 26, 161–188.

Mulder, Ch., 2002. Aizoaceae. In: Punt, W., Blackmore, S., Hoen, P.P., Stafford, P.J. (Eds), The Northwest European Pollen Flora 61/VIII. Elsevier, Amsterdam.

Mulder, Ch., Breure, A.M., Joosten, J.H.J., 2002. Fungal functional diversity inferred along Ellenberg's abiotic gradients: Palynological evidence from different soil microbiota. Grana 41 (in press).

Mulder, Ch., Ellis, R.P., 2000. Ecological significance of South-West African grass leaf phytoliths: a climatic response of vegetation biomes to modern aridification trends. In: Jacobs, S.W.L., Everett, J. (Eds), Grasses – Systematics and Evolution. Proceedings 2nd International Conference on the Comparative Biology of the Monocotyledons, Sydney 2, CSIRO, Melbourne, pp. 246–256.

Mulder, Ch., Janssen, C.R., 1998. Application of Chernobyl Caesium-137 fallout and naturally occurring Lead-210 for standardization of time in moss samples: recent pollen-flora relationships in the Allgäuer Alpen, Germany. Rev. Palaeobot. Palynol. 103, 23–40.

Mulder, Ch., Janssen, C.R,. 1999. Occurrence of pollen and spores in relation to present-day vegetation in a Dutch heathland area. J. Veg. Sci. 10, 87–100.

Odgaard, B.V., 1999. Fossil pollen as a record of past biodiversity. J. Biogeogr. 26, 7–17.

Posthuma, L., Suter, G.W. II, Traas, T.P. (Eds), 2002. Species sensitivity distributions in ecotoxicology. Lewis Publishers, Boca Raton.

Prentice, K.C., 1990. Bioclimatic distribution of vegetation for General Circulation Model studies. J. Geophys. Res. 95, 11,811–11,830.

Pyankov, V.I., Ivanova, L.A., Lambers, H., 1998. Quantitative anatomy of photosynthetic tissues of plant species of different functional types in a boreal vegetation. In: Lambers, H., Poorter, H., van Vuuren, M.M.I. (Eds), Inherent variation in plant growth. Physiological mechanisms and ecological consequences. Backhuys, Leiden, pp. 71–87.

Pyankov, V., Black. C., Stichler, W., Ziegler, H., 2002. Photosynthesis in *Salsola* species (Chenopodiaceae) from Southern Africa relative to their C_4 syndrome origin and their African–Asian arid zone migration pathways. Plant Biol. 4, 62–69.

Raunkiaer, C., 1910. Statistik der Lebensformen als Grundlage für die biologische Pflanzengeographie. Beih. Bot. Centbl. 27, 171–206.

Raunkiaer, C., 1934. The Life Forms of Plants and Statistical Plant Geography. Oxford University Press, Oxford.

Robinson, M.F., Heath, J., Mansfield, T.A., 1998. Disturbance in stomatal behaviour caused by air pollutants. J. Exp. Bot. 49 (Special Issue), 461–469.

Rösch, M., 1996. New approaches to prehistoric land-use reconstruction in south-western Germany. Veg. Hist. Archaeobot. 5, 65–79.

Salemaa, M., Vanha-Majamaa, I., Gardner, P.J., 1999. Compensatory growth of two clonal dwarf shrubs, *Arctostaphylos uva-ursi* and *Vaccinium uliginosum* in a heavy metal polluted environment. Plant Ecol. 141, 79–91.

Sanità di Toppi, L., Gabrielli, R., 1999. Response to cadmium in higher plants. Environ. Exper. Bot. 41, 105–130.

Sakata, M., Suzuki, K., 1998. Assessment method for environmental stresses in trees using delta C-13 records of annual growth rings. Geochem. J. 32, 331–338.

Schaminée, J.H.J., Stortelder, A.H.F., Westhoff, V., Barkman, J.J., Doing, H., van Duuren, L., 1995. Inleiding tot de plantensociologie. Grondslagen, methoden en toepassingen. Opulus Press, Uppsala.

Schneider, S. H., 1992. The climatic response to greenhouse gases. Adv. Ecol. Res. 22, 1–32.

Schulze, E.-D., 1982. Plant life forms and their carbon, water and nutrient relations. In: Lange, O.L., Nobel, P.S., Osmond, C.B., Ziegler, H. (Eds), Physiological Plant Ecology. II. Water Relations and Carbon Assimilation. Springer Verlag, Berlin, pp. 615–676.

Schulze, E.-D., Chapin, F.S. III, 1987. Plant specialisation to environments of different resource availability. In: Schulze, E.-D., Zwölfer, H. (Eds), Potentials and Limitations to Ecosystem Analysis. Ecol. Stud. 61, 120–148.

Schulze, E.-D., Ellis, R.P., Schulze, W., Trimborn, P., Ziegler, H., 1996. Diversity, metabolic types and $\delta^{13}C$ carbon isotope ratios in the grass flora of Namibia in relation to growth form, precipitation and habitat condition. Oecologia 106, 352–369.

Sharples, J.M., Meharg, A.A., Chambers, S.M., Cairney, J.W.G., 2001. Arsenate resistance in the ericoid mycorrhizal fungus *Hymenoscyphus ericae*. New Phytol. 151, 265–270.

Shi, N., Schneider, R., Beug, H.-J., Dupont, L.M., 2001. Southeast trade wind variations during the last 135 kyr: evidence from pollen spectra in eastern South Atlantic sediments. Earth Planet. Sci. Lett. 187, 311–321.

Shugart, H. H., Smith, T. M., Post W. M., 1992. The potential for application of individual-based simulation models for assessing the effects of global change. Ann. Rev. Ecol. Syst. 23, 15–38.

Slovik, S., Siegmund, A., Fuhrer, H.W., Heber, U., 1996. Stomatal uptake of SO_2, NO_x and O_3 by spruce crowns (*Picea abies*) and canopy damage in central Europe. New Phytol. 132, 661–676.

Solbrig, O. T., 1993. Plant traits and adaptive strategies: their role in ecosystem function. In: Schulze, E.-D., Mooney, H.A. (Eds), Biodiversity and Ecosystem Function. Ecol. Stud. 99, 97–116.

Soó, R., 1980. Synopsis systematico-geobotanica florae vegetationisque Hungariae. VI. Akadémia Kiadó, Budapest.

Steiner, M., Linkov, I., Yoshida, S., 2002. The role of fungi in the transfer and cycling of radionuclides in forest ecosystems. J. Environ. Radioact. 58, 217–241.

Stuessy, T.F., 1990. Plant taxonomy: the systematic evaluation of comparative data. Columbia University Press, New York.

Telewski, F.W., Swanson, R.T., Strain, R., Burns, J.M., 1999. Wood properties and ring width responses to long-term atmospheric CO_2 enrichment in field-grown loblolly pine (*Pinus taeda* L.). Plant, Cell Environ. 22, 213–219.

USDA–ARS (United States Department of Agriculture–Agricultural Research Service), 1991. Pollination Workshop Proceedings. Denver.

Van der Maarel, E. 1996. Vegetation dynamics and dynamic vegetation science. Acta Bot. Neerl. 45, 421–442.

Vera, F.W.M., 2000. Grazing Ecology and Forest History. CAB International, Wallingford.

Waide, R.B., Willig, M.R., Steiner, C.F., Mittelbach, G., Gough, L., Dodson, S.I., Juday, G.P., Parmenter, R., 1999. The relationship between productivity and species richness. Ann. Rev. Ecol. Syst. 30, 257–300.

Walter, H., 1951. Grundlagen der Pflanzenverbreitung. I. Standortslehre. Eugen Ulmer, Stuttgart.

Walter, H., 1964. Die Vegetation der Erde in öko-physiologischer Betrachtung, 2nd rev. edn, Vol. 1. Gustav Fischer Verlag, Jena.

Wamelink, G.W.W., van Dobben, H.F., van der Eerden, L.J.M., 1998. Experimental calibration of Ellenberg's indicator values for nitrogen. Environ. Poll. 102, 371–375.

Watson, L., Dallwitz, M.J., 1994. The Grass Genera of the World, 2nd edn. CAB International, Wallingford.

WCED (World Commission on Environment and Development), 1987. Our Common Future. Oxford University Press, London.

Webb, T. III, Bartlein, P.J., 1992. Global changes during the last 3 million years: climatic controls and biotic responses. Ann. Rev. Ecol. Syst. 23, 141–173.

Weiss, P., Offenthaler, I., Öhlinger, R., Wimmer, J., 2002 (2003). Higher plants as accumulative bioindicators. In: Markert, B.A., Breure, A.M., Zechmeister, H.G. (Eds), Bioindicators and Biomonitors. Elsevier, Oxford, pp. 465–500.

Wilkins, D.A., 1991. The influence of sheating (ecto-)mycorrhizas of trees on the uptake and toxicity of metals. Agri. Ecosyst. Environ. 35, 245–260.

Willerding, U., 1986. Zur Geschichte der Unkräuter Mitteleuropas. Karl Wachholtz Verlag, Neumünster.

Woodward, F.I., 1986. Climate and Plant Distribution. Cambridge University Press, Cambridge.

Wolterbeek, H.T., Garty, J., Reis, M.A., Freitas, M.C., 2002 (2003). Biomonitors in use: lichens and metal air pollution. In: Markert, B.A., Breure, A.M., Zechmeister, H.G. (Eds), Bioindicators and Biomonitors. Elsevier, Oxford, pp. 377–419.

Zobel, M., 1997. The relative role of species pools in determining plant species richness: an alternative explanation of species coexistence. Trends Ecol. Evol. 12, 266–269.

IIe

Invertebrates

Chapter 15

Nematodes

Sebastian Höss and Walter Traunspurger

Abstract

In most sediments and soils, nematodes are the dominant organism group among the meta-zoans and play an important role for the food web. In biomonitoring studies, these endobenthic organisms provide several advantages over most macrobenthic organisms, including their omnipresence, high abundance and diversity of species and trophic groups. Single species and community level bioassays, as well as assessments of in situ nematode communities proved to be suitable tools for evaluting the quality of sediments and soils.

Various nematode species were used as test organisms for single species bioassays, however mainly one species, Caenorhabditis elegans, is used for ecotoxicological assessments of sediments, soils, and waste water. C. elegans has shown to be a suitable test organisms for solid and liquid substrates, using various toxicity parameters, such as lethality, growth, reproduction, and behaviour.

Community assessments showed that nematode assemblages were changed by various types of pollution. Investigated disturbances include oil spills, sewage outfall, mechanical disturbances, and pollutants such as heavy metals, organo-metals, phenols, and hydrocarbons. Pollution-induced shifts in community structure have been detected at the level of species, genus, family and ratio of Secernentea/Adenophorea or, ataxonomically, by characterizing feeding types by their buccal cavities. Field studies, as well as experiments in model ecosystems showed that univariate and multivariate measures were able to detect pollution induced changes in nematode communities. Analyses of communities in terms of composition of species (multivariate analysis, k-dominance), trophic groups or life history strategists (Maturity Index) appeared to be most promising. However, it became obvious that, particularly for freshwater systems, there is need for further research on nematode communities in unpolluted and polluted ecosystems, to improve the significance of nematode community assessments in biomonitoring studies.

Keywords (alphabetical order): bioassay, *Caenorhabditis elegans*, cluster analysis, community structure, feeding types, heavy metals, *k*-dominance, maturity index, meiobenthos, model ecosystem, multidimensional scaling, nematoda, oil spill, pentachlorphenol, pollution, sediment, soil

1. Ecological relevance of meiofauna and nematodes for biomonitoring

In ecological studies meiobenthic communities (intermediate size class between micro- and macrobenthos) have been neglected for a long time. In 1984, Bruce Coull pointed out that the "meiofaunal ecologists have been slower to utilize the experimental field approach than their macrofaunal colleagues" (Coull and Palmer, 1984).

As a consequence, also in pollution assessments meiofauna ranked behind macrofauna, as ecological knowledge is required for assessing pollution induced changes of communities. The generally small body size of meiofauna (pass through a 500 μm sieve, are retained by a 42 μm sieve; Fenchel, 1978) and their uniformity in shape leads to taxonomic and experimental problems, explaining the delayed progress of meiofaunal research. However, activity in meiofaunal taxonomy and ecology increased considerably in the last three decades, increasing the importance of using meiofauna in pollution impact assessment (reviewed e.g. by Coull and Chandler, 1992; Coull and Palmer, 1984; Ferris and Ferris, 1979; Kennedy and Jacoby, 1999; Moore and Bett, 1989). The motivation for using meiobenthos in environmental studies is based on several advantages of meiofauna over macrofauna (Heip, 1980; Kennedy and Jacoby, 1999; Trett et al., 2000):

- *High abundance*: Meiofaunal organisms represent the most abundant metazoans in soils and sediments (e.g. Traunspurger, 1996a; Yeates, 1981). They can reach densities up to 40 million m^2 (Heip et al., 1985). Due to these high densities, statistically valid sampling can be achieved more easily than with macrofauna, even with small, easily processed samples.
- *High biodiversity*: Compared to macrofauna, the numbers of meiofaunal species belonging to a single phylum in a given habitat can be one order of magnitude higher. Because of this diversity, meiofauna covers a broad range of physiologies and feeding types, and thus provide a balanced assessment of effects of prevailing conditions on food webs and community processes.
- *Pervasiveness and tolerance*: Representatives of meiofauna are found in all environments examined so far. This includes such extremes as hot, volcanic springs, anoxic sediments, sea ice, and polluted sediments. This is possible as meiofauna include species that are tolerant to a range of different environmental stresses. However, the group also includes sensitive, stress-intolerant species. Thus, changes throughout a wide spectrum of stress conditions can be assessed for a large number of different ecosystems.
- *Low mobility*: Communities of meiofauna are continuously exposed to harmful materials that enter their environment, due to their limited ability to escape. Therefore, the community structure is more directly related to the physicochemistry of the habitat sampled than in the case of macrofaunal communities.
- *Generation time*: Effects of short-term as well as longer term influences become apparent in meiofaunal communities, as their life-cycles cover a broad range of times spreads (from as little as six days to over two years). Most species have a generation time of about 1–3 months (Traunspurger, 2002).

Within most meiofaunal communities nematodes are the dominant organism group. This was found for freshwaters (lentic habitats: e.g. Prejs, 1977; Traunspurger, 1996a; Traunspurger, 1996b; lotic habitats e.g. Anderson, 1992; Traunspurger, 2000), marine sediments (e.g. Heip et al., 1985) and soils (e.g. Yeates, 1981). The number of valid species of free-living nematodes described so far is about 11,000 (Andrassy, 1992), but there is general agreement, that most of the nematode species still remain to be described. Nematodes have both interstitial and burrowing representatives that occupy many different trophic levels: There are species feeding on detritus, bacteria, algae, fungi, and higher plants, as well as omnivorous and predatory species (Traunspurger, 1997; Yeates et al., 1993). Thereby, nematodes occupy important positions in benthic food webs, substantially influencing the material and energy flux in sediments. Bacterivorous nematodes, for example, are able to enhance abundance and biomass of benthic bacteria (Traunspurger et al., 1997), thus stimulating microbial metabolism in sediments.

The ecological importance of nematodes and the general advantages of meiofauna for pollution assessments as listed above, have led to an increasing attention of environmental scientists in this organism group. In this chapter, we will give an overview of the use of nematodes in pollution assessments, demonstrating the possibilities of using meiofauna studies in biomonitoring programmes. To do so, we structured the present studies by the type of investigation (field or laboratory approach), the type of the investigated habitat (aquatic, terrestrial), the type of evaluation (statistics, indices), and the type of contamination (e.g. organics, heavy metals, oil spill). This structure (see Table 1) should help the reader to find out where deficiencies lie in this area and which approaches seem promising and applicable for biomonitoring. On the basis of this information, we will give an outlook, how further research on and application of meiofauna could contribute to more reliable ecosystem risk assessments.

2. Single species bioassay

Laboratory single species tests are an important part of ecotoxicological assessments. Although already discussed elsewhere in this volume (Altenburger and Schmitt-Jansen, 2002; Ratte et al., 2002), we want to give a short overview of bioassays using nematodes as test organisms, to present a complete picture of the use of this organism group in ecotoxicology.

For two or three decades, nematodes have been used as test organisms for laboratory bioassays (e.g. Boroditsky and Samoiloff, 1973; Haight et al., 1982; Samoiloff et al., 1980). Various nematode species were used to assess the potential toxicity of various compounds in aqueous medium (e.g. Haight et al., 1982; Traunspurger ct al., 1997; Williams and Dusenbery, 1990b), on agar (Popham and Webster, 1979; Vranken et al., 1985), as well as in more complex matrices as sediments and soils (Donkin and Dusenbery, 1993; Höss et al., 1997). Mainly free living, bacterivorous nematodes were chosen as test organisms: Besides the marine *Monhystera disjuncta* (e.g. Vranken and Heip, 1986) and species of the genus *Panagrellus* (e.g. Haight et al., 1982; Samoiloff et al., 1980; Sherry et al., 1997), ecotoxicology with nematodes has focused on the intensively studied *Caenorhabditis elegans*. This soil dwelling species has proven to be an adequate test organism for various substrates, so that recently methods have been standardized for the assessment of waste water (Hitchcock et al., 1997), sediment (Höss et al., 1999; Traunspurger et al., 1997) and soil (Freeman et al., 2000; Peredney and Williams, 2000). For these purposes, a variety of toxicity parameters were studied for *C. elegans*: Lethality (Donkin and Dusenbery, 1993; Tatara et al., 1997; Williams and Dusenbery, 1990b), growth (Höss et al., 1997; Traunspurger et al., 1997), reproduction

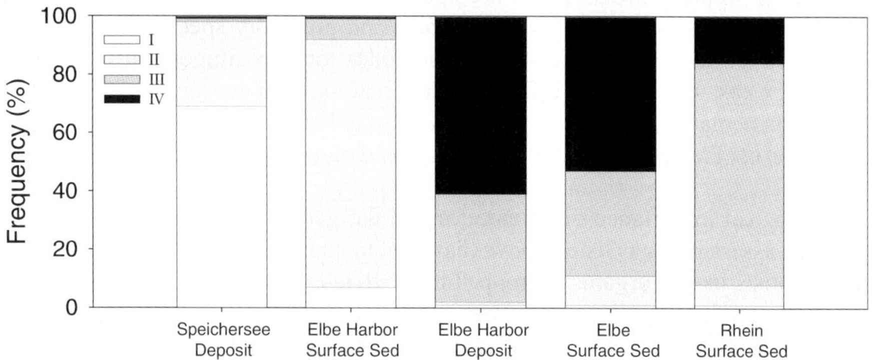

Figure 1. Frequency (%) of toxicity classes (I–IV) occurring in various deposited and surface sediments from a lake (Speichersee), a harbour of the river Elbe (Elbe Harbour), the river Elbe and the river Rhine; toxicity was assessed as growth inhibition of *Caenorhabditis elegans* (Nematoda); I = lowest toxicity, IV = highest toxicity; data from Henschel (2002).

(Dhawan et al., 1999; Traunspurger et al., 1997) and behaviour (Boyd et al., 2000; Williams and Dusenbery, 1990a). Also, *C. elegans* was used as a bioindicator utilizing the ability to express stress proteins in the presence of heavy metals (Cioci et al., 2000; Guven et al., 1994; Stringham and Candido, 1994), as test organism for bioconcentration studies (Haitzer et al., 1999a; Haitzer et al., 1999b) and for screening mutagen substances (Lew et al., 1983).

The need of infaunal organisms as test organisms for whole sediment or soil bioassays is beyond doubt (Burton, 1991), and various sediment and soil dwelling organisms have been used for assessments of solid phases (Ingersoll et al., 1995; Keddy et al., 1995; Traunspurger and Drews, 1996). However, although nematodes are one of the most important organisms in sediments and soils, the use of this organism group in ecotoxicology has always been by far underrepresented.

Henschel (2002) compared different studies that used a nematode sediment bioassay with *C. elegans* (Ahlf and Gratzer, 1999; Maaß et al., 1997). For the purpose of better comparability, toxicity data were graduated in four toxicity classes. Surface sediments from two rivers (Elbe, Rhein) and a harbour of the river Elbe (Hamburger Hafen), as well as sediment deposits from the Elbe harbour and from a contaminated lake (Ismaninger Speichersee) were compared in terms of frequency of toxicity classes. Figure 1 shows that the results of nematode bioassays (sublethal toxicity parameter: growth) differed between the systems investigated. Surface sediments from Elbe and Rhine, as well as the sediment deposits from Elbe harbour showed a more alarming toxicity pattern than the surface sediments from Elbe harbour and the lake sediment deposits. This is obvious from the higher proportion of sediments in toxicity classes 3 and 4. Moreover, the nematodes indicated a much higher toxicity in the sediment deposits of Elbe harbour, compared to the surface sediments. Ecotoxicological assessment of sediments or soils using nematodes as test organisms is still at its beginning. However, particularly the nematode *C. elegans* seems to be a promising test organism for sediment and soil risk assessments.

3. Community assessments

Assessments of *in situ* benthic communities or assemblages are an ecologically more relevant approach than single species bioassays, as they integrate interactions of all biotic and abiotic factors of the benthos, including pollution. But, this approach also contains more uncertainties. Information about the ecology of nematodes in unpolluted aquatic environments is absolutely required for estimating pollution-induced changes in community structure. In 1979, Ferris and Ferris reviewed knowledge about the ecology of nematodes in both "normal" and polluted habitats, emphasizing the importance of obtaining baseline data on natural variability within habitats. Various factors, regardless of sediment pollution, are able to influence nematode communities, so that it is nearly impossible to demonstrate unequivocal effect-cause relationships with single factors. Food availability, particle size, and salinity were found to have considerable influence on the structure of marine nematode communities (e.g. Jensen, 1984; Soetaert et al., 1995; Tietjen, 1980; Vanreusel, 1991; Yeates and Coleman, 1982) and, therefore, these factors are able to mask effects of pollutants. Moreover, the response of communities is dependent on the environmental conditions that the communities normally experience. It could be found that nematode assemblages were most affected by the kind of disturbance that they do not normally experience naturally (Schratzberger and Warwick, 1999).

Thus, it is not possible to define a universally valid baseline for nematode communities, which evokes one of the major problems in pollution assessments: the negative control. Relative changes of communities, as implied by the term itself, can only be measured by comparing one set of data to another. Usually, the community of a polluted site is compared to that of an unpolluted site, the control. An ideal control site comprises "natural" sediment or soil properties comparable to the polluted site except for contamination, allowing any changes in community structure to be traced back to the pollution. These ideal conditions are rarely found in the field. However, suitable statistical techniques, combined with knowledge gathered from laboratory studies (model ecosystems) can help accurately to interpret community data in biomonitoring studies.

In the following, we will give a short summary of available methods for evaluating differences in nematode communities, followed by examples dealing with marine, estuarine, freshwater and terrestrial nematode communities and various types of evaluation methods. These examples include laboratory experiments and field studies, Table 1 gives an overview of relevant studies, that deal with nematode communities in pollution ecology.

3.1. Evaluation methods

Changes in community structure can be evaluated using various features of nematode communities. Abundance or biomass, feeding or life history strategies, presence or absence of key species, biodiversity, and species composition are useful parameters for characterizing nematode communities. To visualize differences of nematode assemblages in terms of these parameters, univariate and multivariate techniques can be used.

Table 1. Nematode community assessments in experimental and field studies.

Reference	Type of habitat	Evaluation method		Disturbance
		Univariate	Multivariate	
Experimental approach				
Austen and McEvoy, 1997a	Aquatic, estuarine	X	X	Tributyltin
Austen and McEvoy, 1997b	Aquatic, estuarine	X	X	Heavy metals
Austen and Somerfield, 1997	Aquatic, estuarine	X	X	Heavy metals
Austen et al., 1994	Aquatic, estuarine	X	X	Heavy metals
Bongers et al., 2001	Aquatic, terrestrial	X		Cu
Cantelmo and Rao, 1978	Aquatic, marine	X		Pentachlorphenol
Carman and Todaro, 1996	Aquatic, marine	X		PAH
Carman et al., 1995	Aquatic, marine	X		PAH
Carman et al., 1997	Aquatic, marine	X		Diesel fuel
Delaune et al., 1984	Aquatic, marine	X		Crude oil
Fiscus and Neher, 2002	Terrestrial	X	X	Physical and chemical disturbance
Freckman and Ettema, 1993	Terrestrial	X	X	Different agricultural management practices
Fuller et al., 1997	Terrestrial	X	X	Trichlorethylene, toluene
Gee and Warwick, 1985	Aquatic, marine	X	X	Organic enrichment
Korthals et al., 1996	Terrestrial	X		Cu, pH
Korthals et al., 1996	Terrestrial	X		Heavy metals
Korthals et al., 2000	Terrestrial	X		Cu, Zn
Lenz and Eisenbeis, 2000	Terrestrial	X		Tillage
Millward and Grant, 1995	Aquatic, estuarine	X		Heavy metals
Parmelee et al., 1993	Terrestrial	X		Cu, trinitotoluol, p-nitrophenol
Ruess et al., 1993	Terrestrial	X		Acidification
Schratzberger and Warwick, 1999	Aquatic: estuarine	X	X	Biological, physical, organic enrichment
Schratzberger et al., 2000	Aquatic: estuarine	X	X	Heavy metals
Warwick et al., 1988	Aquatic, marine	X	X	Cu, hydrocarbons
Weiss and Larink, 1991	Terrestrial	X		Heavy metals, sewage sludge

Field approach

Reference	Habitat			Impact
Beier and Traunspurger, 2001	Aquatic: freshwater	X		Organic pollution
Bloemers et al., 1997	Terrestrial	X		Forest clearing
Bongers and van de Haar, 1990	Aquatic: estuarine	X	X	
Bongers et al., 1991	Aquatic: marine, estuarine	X		Organic pollution, oil spill, heavy metals
Boucher, 1985	Aquatic: marine, estuarine	X		Oil spill
Danovaro et al., 1995	Aquatic: marine	X	X	Oil spill
Essink and Romeyn, 1994	Aquatic: estuarine	X		Organic pollution
Fricke et al., 1981	Aquatic: marine	X		Oil spill
Gyedu-Ababio et al., 1999	Aquatic: marine	X	X	Heavy metals
Hennig et al., 1983	Aquatic: marine	X		Chemical effluent, mechanical perturbation, oil, organic enrichment
Hodda and Nicholas, 1986	Aquatic: estuarine	X	X	Heavy metals, ammonia, orthophosphate
Hodda et al., 1997	Terrestrial	X		Forest clearing, land-use
Lambshead et al., 1983	Aquatic: marine	X		
Lambshead, 1984	Aquatic: marine, estuarine	X		Organic contamination
Neilson et al., 1996	Aquatic: estuarine	X		Sewage outfall, heavy metals
Newell et al., 1990	Aquatic: marine, freshwater			Liquid waste
Newell et al., 1991	Aquatic: marine, freshwater	X	X	Metallic residues
Panesar et al., 2000	Terrestrial	X	X	Forest harvesting systems
Raffaelli, 1982	Aquatic: marine	X		Organic contamination
Sandullini and de Nicola-Guidici, 1990	Aquatic: marine		X	Heavy metals, organics
Somerfield et al., 1994a	Aquatic: estuarine	X	X	Heavy metals
Somerfield et al., 1994b	Aquatic: estuarine	X	X	Heavy metals
Somerfield et al., 1995	Aquatic: marine	X	X	Heavy metals
Trett et al., 2000	Terrestrial	X	X	Heavy metals
Urzelai et al., 2000	Terrestrial	X		Waste
Vidakovic, 1983	Aquatic: marine	X		Raw domestic sewage
Zullini and Peretti, 1986	Aquatic: freshwater	X		Lead
Zullini, 1976	Aquatic: freshwater	X		

Univariate methods generally reduce the information on communities to single values (indices; Table 2), so that differences can be analyzed using univariate statistical tests, such as one-way ANOVAs. Some of these indices are relatively easy to obtain, as no or only limited taxonomic skills are required. Moreover, these indices are generally easy to interpret, which is of advantage when e.g. administrative decisions by regulating authorities have to be taken. However, taxonomic or functional identities are not retained during univariate analysis. These methods might indicate the occurrence of similar community structures in places which have a totally different species composition. Table 2 summarizes commonly used univariate methods that are addressed in the examples of this chapter. Although k-dominance and feeding type distribution can not strictly be classified as univariate in the mathematical sense, both measures were included in the univariate group, because, like the indices, they are species-identity independent. For a more detailed discussion of biodiversity index see e.g. Magurran, (1988).

Table 2.   Univariate measures for analyzing nematode communities.

Measures	Information	References
Total abundance or biomass	Total number or biomass of indiviuals in a defined sampling unit	
Species Richness: S	Number of species in a defined sampling unit	Magurran, 1988
Shannon Index: H′	Distribution of species abundances: gives more weight to rare species; higher index indicates higher diversity	Pielou, 1977
Simpson's Diversity Index: D	Distribution of species abundances: gives more weight to common species; higher index indicates higher dominance	Pielou, 1977
Evenness: Pilou's J′	Distribution of species abundances: E = H′/logS; higher index (between 0 and 1) indicates higher diversity	Pielou, 1977
k-dominance curves	Distribution of species abundances: plotting percentage of cumulative abundance (k-dominance) against species rank (k) of two nematode assembages A and B, B is more diverse than A if the curve is everywhere below or touching that of A	Lambshead et al., 1983
Index of trophic diversity (ITD)	Distribution of feeding types	Heip et al., 1985
Maturity Index (MI)	Distribution of different life history strategies of nematodes: based on colonizers to persisters scale of 1 to 5; a low index indicates disturbance	Bongers, 1990

Multivariate methods consider the identity of each species or functional group combined with data on abundance or biomass to compare assemblages (for detailed information see Clarke, 1993). Similarities or dissimilarities of nematode species composition are used as a measure for differences in community structure. Non-metric multidimensional scaling ordination (MDS) or cluster analysis are employed to visualize similarities or dissimilarities between different assemblages. To test for differences in community structure multivariate statistics (MANOVA) have to be performed.

3.2. Experimental approach

In controlled experiments all factors but the one of interest can be held (more or less!) constant, allowing to set up a control, to which treatments can be compared. For this purpose, natural sediments or soils usually are transferred into experimental containers, to obtain small and well-defined representatives of ecosystems. These model ecosystems (micro- or mesocosms, depending on their size) are then manipulated experimentally for a certain period of time. These experiments can also be regarded as community level bioassays. Nematodes are suitable organisms for microcosm experiments, because of the relatively short generation time of most nematode species community changes can be measured over the time scale at which such experiments can be realistically maintained (several months). Also, there are no problems of recruitment within the microcosm systems since nematodes have direct benthic development (unlike macrobenthic groups where many species have planctonic larval recruitment).

In this section, we will give some examples of aquatic and terrestrial laboratory studies with nematodes, demonstrating the possibility of studying effects of single substances on nematode communities in controlled laboratory experiments using various types of pollutants. In addition, some evaluation methods for measuring differences in nematode assemblages are compared.

Cantelmo and Rao (1978) studied the effect of the biocide pentachlorphenol (PCP) on estuarine nematode communities in aquaria that contained artificial sediments, made up of clean (contaminant free) sand. The aquaria received PCP at three concentrations (7–622 µg/L) in a flow-through system over nine weeks of exposure. While diversity indices, such as the Shannon index and evenness were not influenced by any PCP concentration, the distribution of feeding types (classified according to Wieser, 1953), was distinctly affected by the PCP treatment. The authors found a shift from communities dominated by epistrate feeders, as present in the controls and aquaria exposed to low concentrations of PCP, to communities dominated by a detritus feeder (*Diplolaimella punicea*) at the highest concentration of PCP (Fig. 2). The decline in the abundance of epistrate feeders in the highly contaminated treatments could be explained by reduced food availability, as also algal growth was inhibited by PCP. Other studies have indicated that primary production influences densities of plant feeding nematodes (Ingham et al., 1985; Yeates, 1987). This simple experimental design allowed the observed direct or food chain effect to be attributed to the added substance, as in the artificial control sediment the presence of contaminants could be excluded. However, a control sediment consisting of clean sand does not reflect

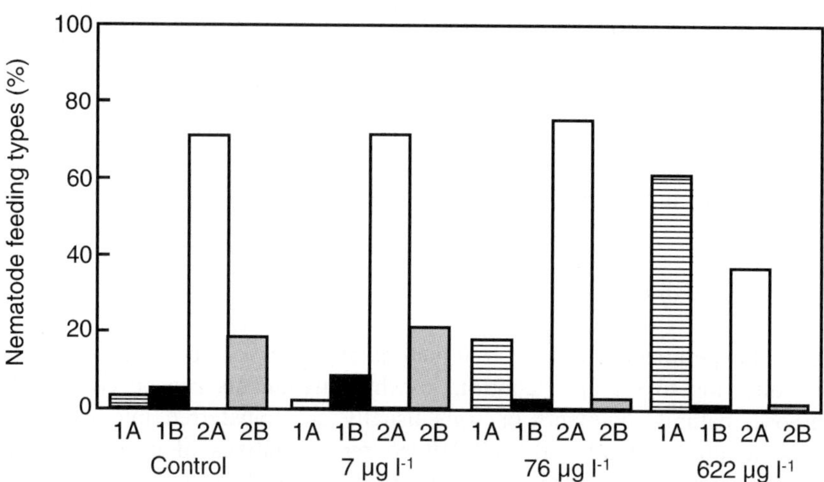

Figure 2. Distribution of nematodes, classified according to their feeding types, in control aquaria and experimental aquaria exposed to pentachlorphenol (PCP); 1A: Selective deposit feeder; 1B: non-selective deposit feeder; 2A: epistrate feeder; 2B: predator/omnivore; reproduced from Cantelmo and Rao (1978), with kind permission of Springer-Verlag, Heidelberg.

real conditions in natural sediments, as natural occurring binding sites such as organic matter, sulphides, or metal oxides that strongly influence the bioavailability of contaminants are absent.

Instead of an artificial sediment, Austen et al. (1994) used defaunated natural sediments for a microcosm study which is an ecologically more relevant approach. However, natural sediment also contains more confounding factors than a defined artificial sediment, complicating the assignment of any effect to a specific cause. In this study two different estuarine sediments (Exe: sandy sediment; Lynher: muddy sediment) were spiked with copper (Cu), zinc (Zn), and cadmium (Cd) and nematode assemblages were analyzed using univariate and multivariate techniques. While Cd did not alter the nematode community structures in any sediment, Cu and Zn distinctly influenced meiobenthic communities. All applied methods showed distinct changes in nematode community structures due to Zn and Cu contamination, with multidimensional scaling ordination (MDS) showing the clearest effects. Results of MDS (using the Bray-Curtis similarity measure with √√ transformed data; Fig. 3) combined with statistical analysis (ANOSIM) indicate that the meiobenthic communities in the copper and zinc treatments differed both from the control and from each other. In the Lynher mud sediment nematode assemblages, dissimilarity between control and contaminated treatments and between different metal treatments was greatest at the highest metal concentrations indicating some dose effect. For the Exe sand nematodes copper treatment communities showed greatest dissimilarities to those in the controls, but the difference between high doses and other dose levels was not as pronounced as for the muddy sediments. The stronger effects of Cu and Zn in the sandy than in the muddy sediments was explained with differences in bioavailability,

Lynher estuary: muddy sediment

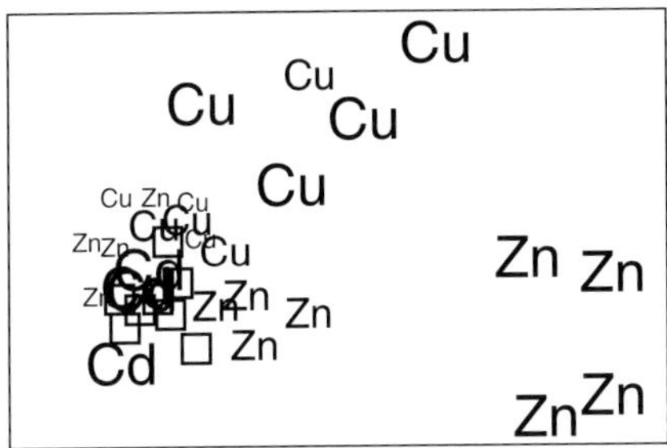

Exe estuary: sandy sediment

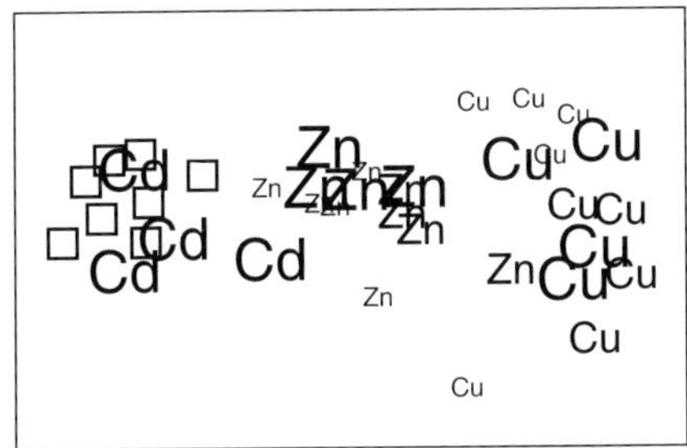

Figure 3. Multidimensional scaling ordination (Bray-Curtis similarity measure with $\sqrt{\sqrt{}}$ transformed data) of nematode assemblages in microcosms containing zinc (Zn), copper (Cu) and cadmium (Cd). Separation on axes indicates level of similarity between samples; Sediments derived from Exe and Lynher estuaries (south west England); squares = controls; size of symbols represent dosage level: low (small), medium, high (large); reproduced from Austen et al. (1994), with kind permission of Elsevier Science Ltd.

as the latter sediment contains more metal-binding components (organic matter, iron oxides), reducing concentrations of readily available metal ions in the porewater. In another study the authors found that the sediment composition strongly influenced the effect of the organo-metal tributyltin on nematode communities (Austen and McEvoy, 1997a).

A short-term laboratory experiment with natural soils also showed distinct changes in nematode community structure, when soils were spiked with heavy metals (Korthals et al., 1996). While Cd did not influence nematodes (note that concentrations were an order of magnitude lower than concentrations of the other metals), copper, nickel, and zinc affected the relative abundance of nematode life history groups, reflected by a dose-dependent decrease of the Maturity Index (Bongers, 1990; Bongers et al., 1991; Bongers and Bongers, 1998; Fig. 4).

This decrease was driven by relative changes in the colonizer–persister (c–p) group distribution among the non plant feeding nematodes. Mainly, some taxa with c–p values of 1 to 2, e.g. Rhabditidae (including dauer-larvae), *Aphelenchoides* and *Pseudhalenchus,* were relatively insensitive to higher Cu, Ni and Zn concentrations compared to most other taxa. Besides the Maturity Index, the distribution of feeding

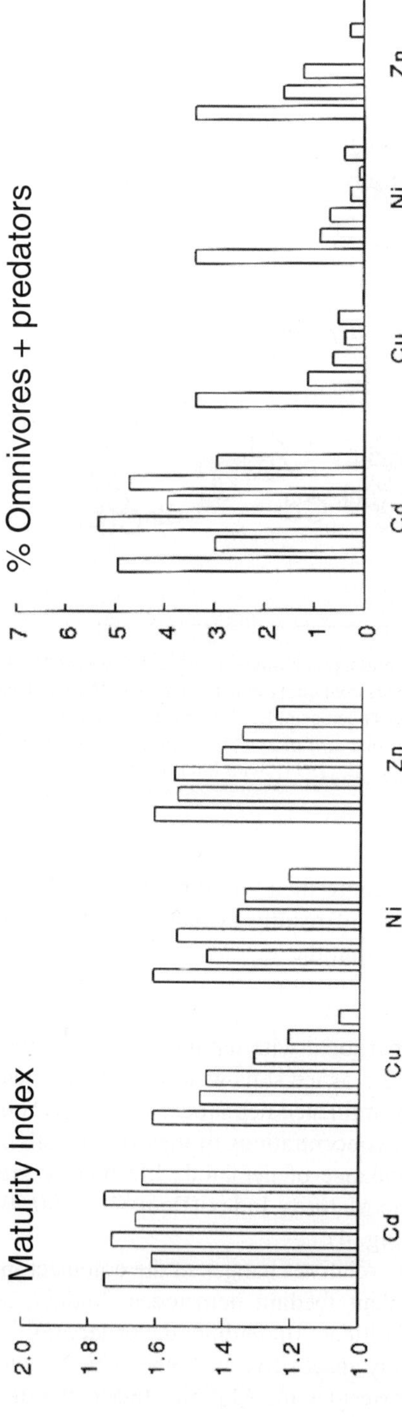

Figure 4. Effects of Cd (0, 10, 20, 40, 80, and 160 mg/kg), Cu, Ni and Zn (0, 100, 200, 400, 800 and 1600 mg/kg) on maturity index and feeding types of nematode assemblages after 1–2 weeks exposure; concentration increase from left to right; reproduced from Korthals et al. (1996), with kind permission of Elsevier Science Ltd.

types, classified according to Yeates et al. (1993), altered in the presence of metals. Omnivorous and predatory nematodes showed to be quite sensitive to Cu, Ni and Zn (first effects at 100 mg/kg, Fig. 4). The relative abundance of bacterial, plant and hyphal feeding nematodes were only affected at the highest Cu concentration (1600 mg/kg). These results are consistent with those of another study, where also omnivorous and predatory nematodes showed to be the most sensitive to Cu (Parmelee et al., 1993). Korthals et al. (1996) found that differences in sensitivity to metals for different taxa exposed to the same metal were much larger than between the sensitivities to Cu, Ni or Zn within the same taxon. These results point out the need of laboratory studies to investigate the characteristics that define the sensitivity of nematodes to heavy metals. Also, the authors stress the meaningfulness of a "natural soil method", as an ecologically realistic approach, because bioavailability is an important issue in risk assessment studies.

The conditions of controlled experiments allow interactions of single compounds with nematode communities to be investigated. The examples show that both, aquatic and terrestrial nematode communities, react on specific disturbances of single pollutants, in terms of composition of species, feeding types or life history groups. In controlled experiments it is possible to get information about the sensitivity of families or even species of nematodes to different pollutants and dose-response relationships of pollutants with nematode communities, as quality and quantity of the tested compound can be chosen arbitrarily. Moreover, information about the bioavailability of different pollutants can be gathered by manipulating sediment or soil properties of the investigated model ecosystems. These data are very important for the interpretation of changes in nematode communities and for the identification of causes for these changes in field studies.

3.3. Field approach

In field studies, the investigator can neither control sediment properties, nor quantity and quality of contaminants. Hence, it is much more difficult to recognize pollution induced changes of community structures. Investigations of communities in polluted sediments require control sites to which polluted sites can be compared. As mentioned above, the sediment or soil at the control site should have similar properties, but without contamination. Of course, these criteria can only be roughly achieved in field studies. In principle, there are two possibilities how control sites are separated from the polluted sites: temporally or spatially. The former case occurs, if a site where data on benthic communities is available is affected by an acute pollution event (e.g. oil spill), so that the community structures before and after the accident can be compared. In addition, the progression of recreation can be monitored over time. In the case of spatial separation, a site that is remote from the source of pollution (e.g. sewage effluent) is selected as control.

In the following section we want to give examples of field studies dealing with nematode communities in aquatic and terrestrial ecosystems. Studies of aquatic systems are clearly dominated by investigations of marine and estuarine nematode assemblages. Until now, only a few studies exist that deal with disturbances of freshwater nematode communities.

The first example demonstrates the use of nematodes for monitoring benthic disturbances caused by an acute pollution event, an oil spill (Danovaro et al., 1995). In a monitoring study the dynamics of the meiobenthic assemblages of the affected site (Golfo Marconi, Ligurian Sea) had been investigated in advance of the oil spill, so that a control was available. Three months after the beginning of the study about 30,000 t of crude oil were released in front of the Livorno harbour (Ligurian Sea) as a consequence of an oil tanker accident. Because of the general circulation of the Ligurian Sea the oil slicks drifted towards the study site. The meiofaunal assemblage was immediately investigated after the oil contamination and compared it to pre-pollution (control) and post-pollution conditions, using univariate and multivariate methods. Despite a clear decline in total density, neither meiofauna analysis at the level of major groups nor univariate indices were able to detect oil-induced disturbances. Conversely, the analysis of nematodes to genus level proved to be highly efficient. Response of the nematode community to oil pollution was found to be extremely rapid. Significant changes occurred immediately after the oil spill and recovered after only two weeks. Changes in community structure appeared particularly clear from k-dominance curves, but also were apparent when using cluster analysis and MDS ordination. The k-dominance curve of the sample with heavily contaminated oil pollution (Apr. II in Figure 5) was clearly above the control curve (before the oil spill: Jan. 91–Apr. in Figure 5),

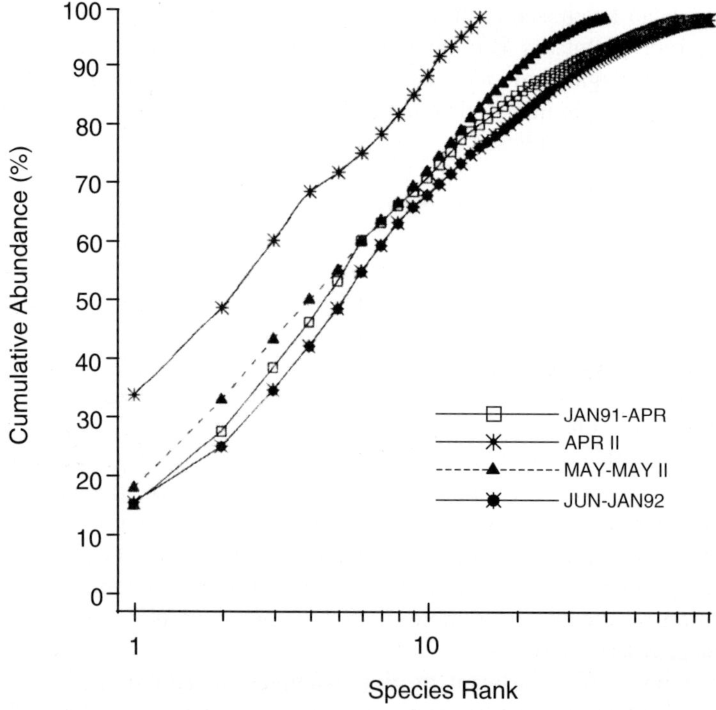

Figure 5. K-dominance curves from nematode communities (1) before the oil spill (Jan. 91–Apr.), (2) immediately after the oil spill (Apr II), (3) a month later (May–May II), and (4) 2 to 9 month after the oil spill (Jun.–Jan. 92); reproduced from Danovaro et al. (1995), with kind permission of Elsevier Science Ltd.

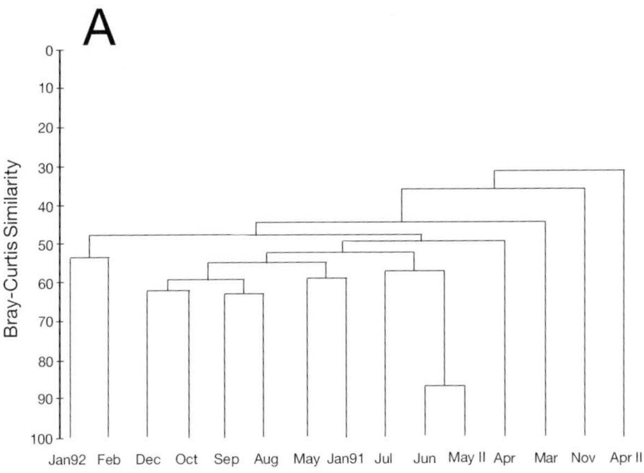

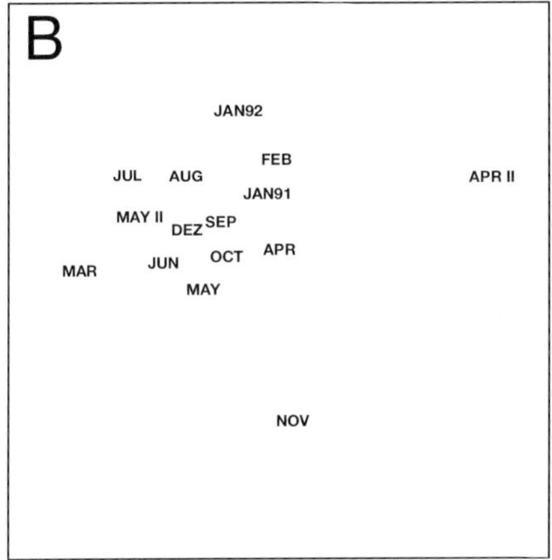

Figure 6. Cluster analysis (A) and MDS ordination (B), using Bray-Curtis similarity measure of $\sqrt{\sqrt{}}$ transformed data; stress for MDS = 0.07; reproduced from Danovaro et al. (1995), with kind permission of Elsevier Science Ltd.

whereas the curve of the samples one month after the oil spill, was at an intermediate position (May–May II in Figure 5). Finally, the *k*-dominance curve of the samples from two to nine month after the accident was very close to the control.

Similar results were found for nematode species composition. Also cluster analysis and MDS ordination at genus level separated the sample with the highest oil concentration (Apr II; Fig. 6) from the other samples. The difference between the disturbed and the pre- and post-oiled conditions was due to the dominance of four important genera: *Daptonema, Viscosia, Prochromadorella* and *Microlaimus*. Some of the genera present throughout the year (such as *Chromaspirina, Monoposthia, Paracanthonchus* and *Setosabatieria*) disappeared after the pollution event to recover one month later.

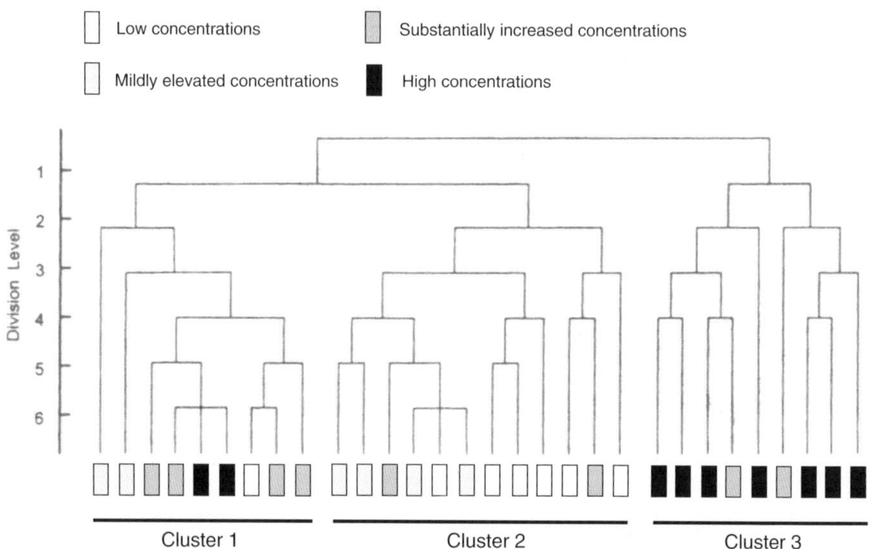

Figure 7. Dendrogram showing the three main clusters of nematode assemblages for sites of the explosive burning ground survey, combined with the classification of these sites in terms of heavy metal concentrations; squares represent the classes of metal contamination with shading increasing with metal concentrations; combined from Table 2 and Fig. 5 in Trett et al. (2000), permission to reprint is granted by the American Chemical Society.

The next example deals with inherited heavy metal pollution in soils from the former burning grounds at ICI Explosives (ICI Nobel Enterprises) in Scotland (Trett et al., 2000). In contrast to the first example, here nematode communities were exposed to chronic pollution, allowing the community structure to adapt to the respective contamination. Trett et al. (2000) compared data on nematode communities with those of heavy metal contamination, to explain differences in benthic community structures and found a reasonably good correlation of community diversity with heavy metal loadings. Cluster analysis based on nematode community data could clearly distinguish three clusters of nematode assemblages (Fig. 7).

The bacterial feeding species were important diagnostic features in the classification of the nematode communities. Species such as *Acrobeles ciliatus* and some members of the family Rhabditidae were absent from cluster III. In contrast, when present, the densities of two plant-parasitic species, *Rotylenchus buxophilus* and a Tylenchid species, were frequently high in cluster III. It is obvious from Figure 7 that in the most highly contaminated sites nematode assemblages from cluster III were present, whereas cluster II assemblages were mainly found in low or mildly contaminated sites. K-dominance curves of moderately contaminated sites could clearly be distinguished from those of heavily contaminated sites (Fig. 8), which also agreed with nematode species richness at these sites. The high dominance at the highly contaminated sites was attributed to more tolerant or resistant species, including non-specialist feeding types, such as the non-selective deposit feeders and detrivores. Differential

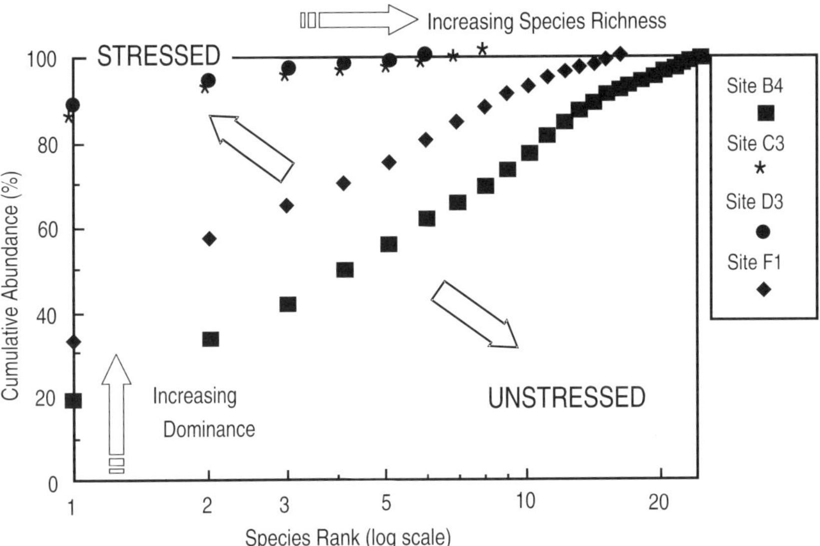

Figure 8. K-dominace curves for selected nematode assemblages present in the explosive burning ground survey. Examples shown include curves for sites with varrying heavy metal contamination (B4: low concentrations; F1: substantially increased concentrations; C3 and D3: high concentrations); B4 and D3 represent species rich (24 species) and species poor (6 species) communities, respectively; reproduced from Trett et al. (2000), permission to reprint is granted by the American Chemical Society.

effects on (selective) bacterial-feeding nematodes in the highly contaminated soils were explained by heavy metal induced alterations of the soil microflora. In this study, the maturity index failed to distinguish between the more heavily contaminated site communities and those from the least contaminated sites.

Beier and Traunspurger (2001) studied nematode communities of two small German streams, taking sediments from two "unpolluted" sites with relatively low anthropogenic impact, as well as from two organically polluted sites, highly influenced by sewage effluents. As in the latter example, benthic communities were exposed to chronic pollution, however, this study deals with organic pollution that continuously disturbed the "polluted" sites. The authors found no substantial difference in the distribution of feeding types (as classified by Traunspurger, 1997) of polluted and unpolluted sites. All communities were clearly dominated by deposit-feeders, relative abundances ranging from 70 to 75 %. In order to uncover changes within the deposit feeders, taxonomic identification to family level was required. At the polluted sites, the "niche" of the deposit-feeding Monhysteridae and Plectidae was taken over by the deposit feeding Diplogasteridae, Diploscapteridae and Rhabditidae. The change of the relative abundance within the dominant nematode families led to a distinct shift in the ratio of the two major nematode "classes", Secernentea and Adenophorea, which can be used as a pollution index (S/A ratio). Zullini (1976) showed that in polluted and disturbed habitats of river Seveso (Italy) the relative abundance of Secernentea is higher than in unpolluted habitats. Beier and Traunspurger (2001) found a distinctly

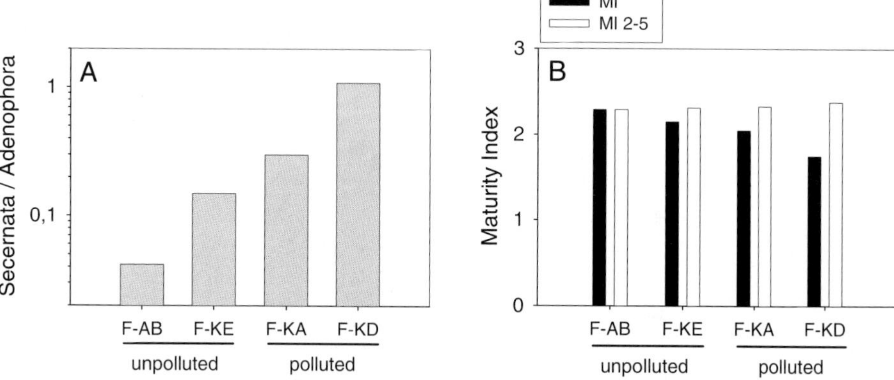

Figure 9. (A) Ratios of Secernata to Adenophora and (B) Maturity Indices (B; MI, MI2–5) for nematode communities from sediments of two unpolluted and two polluted sites in small streams; data from Beier and Traunspurger (2001).

higher S/C ratio at the polluted compared to the unpolluted sites (Fig. 9A). In addition, the Maturity Index (MI: including c–p 1 nematodes) showed a clear decrease towards the polluted sites, indicating a more disturbed nematode community at the polluted than at the unpolluted sites (Fig. 9B). This decrease is mainly driven by a higher relative abundance of nematodes belonging to the c–p 1 group ("colonizers"), rather than by a distinction of "persisters" (c–p 3–5). Excluding the c–p 1 group from calculations of the Maturity Index (MI2–5), the difference between the various sites disappeared (Fig. 9B). The additional use of the MI2–5 can provide supplementary information on the nature of the underlying changes in the habitat. In this case it indicated that the disturbance of the nematode community is related to eutrophication, rather than to pollution (De Goede et al., 1993).

Zullini and Peretti (1986) studied nematode communities in mosses, a special freshwater habitat that shelter several animal species in biocoenoses relatively isolated from surrounding habitats. The southern plain of the river Po (Italy), was considerably contaminated by lead (Pb), due to heavy industrial input of factories producing ceramic tiles. The authors found Pb concentrations to be correlated with the number of nematode species, the Shannon-Wiener diversity index (H') and nematode biomass (Fig. 10). Mosses inhabit a totally different nematode species composition than stream sediments (see latter example: Beier and Traunspurger, 2001), with epistrate feeders dominating the nematode community. The higher abundance of relatively sensitive species (members of Dorylaimidae or Monhysteridae) may provide a useful monitoring tool. The nematode suborder Dorylaimina was shown to be sensitive to pollution, as its percentage in the nematode community was negatively correlated with Pb concentrations (Fig. 10). The authors observed a cut-off level at 300 ppm Pb, separating biotopes with high from those with very low dorylaim percentage.

The examples demonstrate that the analysis of nematode communities can be a valuable tool for detecting pollution-induced disturbances of benthic habitats. Acute short-term, as well as chronic long-term pollution events were shown considerably to

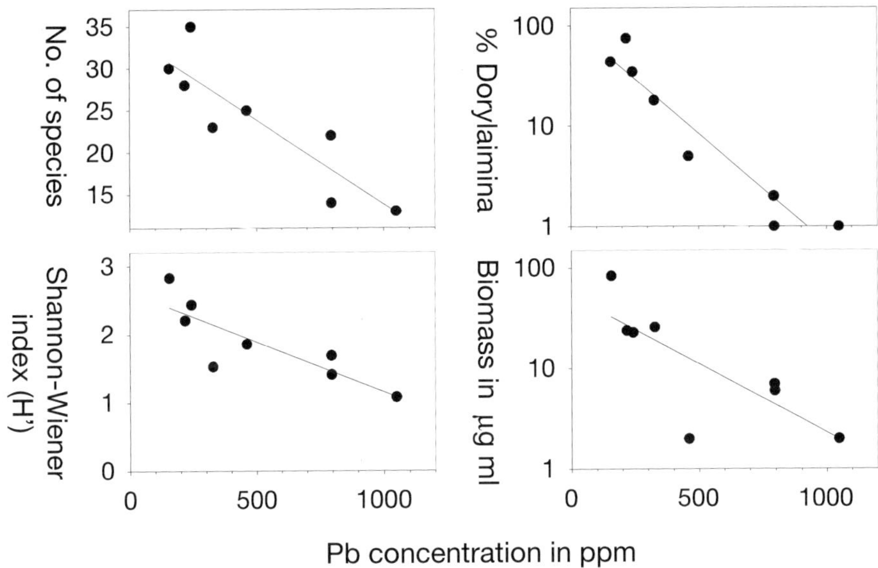

Figure 10. Number of species, percentage of Dorylaimina, Shanon-Wiener diversity index, and biomass of nematodes in stations with different Pb content; adapted from Zullini and Peretti (1986), with kind permission of Kluwer Academic Publishers.

influence nematode community structures of marine, estuarine, freshwater, and terrestrial ecosystems. Nematode communities have been affected by various types of pollution, including crude oil, heavy metals and organic compounds. Both, univariate and multivariate techniques were found to be able to detect contaminant-induced changes in nematode assemblages. However, analyses of communities in terms of composition of species (multivariate analysis, *k*-dominance), trophic groups or life history strategists (Maturity Index) appeared to be most promising. In some cases, simple diversity indices failed to detect a change in nematode communities, although a shift in species composition occurred. It has been shown previously that species-identity-independent measures of community structure are less sensitive than multivariate methods (Warwick and Clarke, 1991). Community analysis on the genus or species level can provide essential information for the understanding of interactions between pollutants and ecosystems.

4. Conclusions and perspectives

Nematodes are the most abundant and species richest organism group among benthic metazoans. They occupy important positions in benthic food webs, forming many different feeding types, including bacterivorous, detrivorous, fungivorous, omnivorous, and predatory nematodes. Also, nematodes evolved species with various life strategies, ranging from relatively tolerant species with short generation times that could easily adapt to new environmental conditions to more sensitive species with long generation times. All these features of nematodes were already used in biomonitoring studies and turned out to be suitable tools for measuring pollution induced disturbances of benthic ecosystems. For this purpose, nematodes were used in single-species bioassays, community level bioassays (micro- and mesocosms), and *in situ* community assessments. However, nematodes are still underrepresented in routine applications of biomonitoring programmes.

In single species bioassays mainly one species, *Caenorhabditis elegans*, is used for ecotoxicological assessments of sediments, soils, and waste water. *C. elegans* has shown to be a suitable test organisms for solid and liquid substrates, using various toxicity parameters, such as lethality, growth, reproduction, and behaviour. Readily standardized guidelines for bioassays with *C. elegans* are available, making this organism a promising candidate for entering the ecotoxicological routine application.

It could be shown that nematode communities react on various types of pollution with changes in their community structure. Investigated disturbances include oil spills, sewage outfall, mechanical disturbances, and pollutants such as heavy metals, organo-metals, phenols and hydrocarbons. Pollution-induced shifts in community structure have been detected at the level of species, genus, family and ratio of Secernentea/Adenophorea or, ataxonomically, by characterizing feeding types by their buccal cavities. It depends on many factors at which systematic or functional level a pollution effect can be detected. Quality and quantity of pollution may decide, how severely a community is disturbed. Specific effects of contaminants on food organisms, such as algae or bacteria, can alter the distribution of feeding types in a characteristic manner. The specific sensitivity of different species or higher taxonomic levels for various contaminants plays an important role for the community structure. These factors also decide, if changes in community structure can be detected by univariate measures, or if a multivariate analysis of species composition is required to visualize changes. Thus, it is not possible to give a universally valid recommendation, how to measure changes in nematode community structure. It has to be decided individually for each case, how much effort (costs, taxonomic skills, etc.) can be invested. To decide this, all available information about the type of the investigated habitat and pollution should be taken into account.

However, the examples shown in this chapter, demonstrate that essential information would be lost, if nematode community data at species or genus level were not considered. Multivariate methods, such as MDS ordination, and *k*-dominance curves seemed to be more sensible to nematode community changes than most univariate measures. Nevertheless, classifications of nematodes considering ecological characteristics, such as feeding types or life history strategies, can be considered as helpful tools for estimating disturbed communities. Measures like the Maturity Index, already include some interpretation of the community data, as the classification (c–p scaling) is made upon ecological information, including sensitivity to pollution. In the case of comparing communities on the basis of similarity matrices (e.g. MDS), data still need to be interpreted. For administrative decisions, simple indices may thus be preferable.

It is obvious that there is still a need for further research on the ecology of nematode communities in unpolluted and polluted habitats. This is particularly true for freshwater nematode communities. In contrast to marine and terrestrial systems, freshwater nematodes so far have not been taken into consideration for ecological (pollution) studies. It is necessary to put more effort in investigations of nematode communities in freshwater model ecosystems, to increase the knowledge about the interactions of freshwater nematode communities with pollutants in different types of sediments. If this basic knowledge can be improved (not only for freshwater), this ubiquitous organism group could considerably contribute to biomonitoring programmes in nearly all kinds of ecosystems.

Acknowledgements

We want to thank Matthias Höss for correcting the English language and for helpful comments on the manuscript.

References

Ahlf, W., Gratzer, H., 1999. Erarbeitung von Kriterien zur Ableitung von Qualitätszielen für Sedimente und Schwebstoffe – Entwicklung methodischer Ansätze, UBA-FB 98–119, Umweltbundesamt, Berlin.

Altenburger, R., Schmitt-Jansen, M., 2002 (2003). Preduicting toxic effects of contaminants in ecosystems using single species investigations. In: Markert, B.A., Breure, A.M., Zechmeister, H.G. (Eds), Bioindicators and Biomonitors. Elsevier, Oxford, pp. 153–198.

Anderson, R.V., 1992. Free-living nematode associations in Pool 19, Mississippi river. Journal of Freshwater Biology 7, 243–250.

Andrassy, I., 1992. A short census of free-living nematodes. Fundamental and Applied Nematology 15, 187–188.

Austen, M.C., McEvoy, A.J., 1997a. Experimental effects of tributyltin (TBT) contaminated sediment on a range of meiobenthic communities. Environmental Pollution 96, 435–444.

Austen, M.C., McEvoy, A.J., 1997b. The use of offshore meiobenthic communities in laboratory microcosm experiments: response to heavy metal contamination. Journal of Experimental Marine Biology and Ecology 211, 247–261.

Austen, M.C., McEvoy, A.J., Warwick, R.M., 1994. The specifity of meiobenthic community responses to different pollutants: results from microcosms experiments. Marine Pollution Bulletin 28, 557–563.

Austen, M.C., Somerfield, P.J., 1997. A community level sediment bioassay applied to an estuarine heavy metal radient. Marine Environmental Research 43, 315–328.

Beier, S., Traunspurger, W., 2001. The meiofauna community of two small German streams as indicators of pollution. Journal of Ecosystem Stress and Recovery 8, 387–405.

Bloemers, G.F., Hodda, M.E., Lambshead, P.J.D., Lawton, J.H., Wanless, F.R., 1997. The effects of forest disturbance on diversity of tropical sp nematodes. Oecologia 111, 575–582.

Bongers, T., 1990. The maturity index: an ecological measure of environmental disturbance based on nematode species composition. Oecologia 83, 14–19.

Bongers, T., Alkemade, R., Yeates, G.W., 1991. Interpretation of disturbance-induced maturity decrease in marine nematode assemblages by means of the Maturity Index. Marine Ecology Progress Series 76, 135–142.

Bongers, T., Bongers, M., 1998. Functional diversity of nematodes. Applied Soil Ecology 10, 239–251.

Bongers, T., Ilieva-Makulec, K., Ekschmitt, K., 2001. Acute sensitivity of nematode taxa to $CuSO_4$ and relationships with feeding-type and life-history classification. Environmental Toxicology and Chemistry 20, 1511–1516.

Bongers, T., van de Haar, J., 1990. On the potential of basing an ecological typology of aquatic sediments on the nematode fauna: an example from the River Rhine. Hydrobiological Bulletin 24, 37–45.

Boroditsky, J.M., Samoiloff, M.R., 1973. Effects of growth inhibitors on development of the reproductive system of the free-living nematode *Panagrellus redivivus* (Cephalobidae). Canadian Journal of Zoology 51, 483–492.

Boucher, G., 1985. Long term monitoring of meiofauna densities after the *Amoco Cadiz* oil spill. Marine Pollution Bulletin 16, 328–333.

Boyd, W.A., Anderson, G.L., Dusenbery, D.B., Williams, P.L., 2000. Computer tracking method for assessing behavioral changes in the nematode *Caenorhabditis elegans*. In: Price, F.T., Brix, K.V., Lane, N.K. (Eds), Environmental Toxicology and Risk Assessment: Recent Achivements in Environmental Fate and Transport, Ninth Volume, ASTM STP 1381. American Society for Testing and Materials, West Conshohocken, pp. 225–238.

Burton, G.A., 1991. Assessing the toxicity of freshwater sediments. Environmental Toxicology and Chemistry 10, 1585–1627.

Cantelmo, F.R., Rao, K.R., 1978. Effects of pentachlorphenol (PCP) on meiobenthic communities established in an experimental system. Marine Biology 46, 17–22.

Carman, K.R., Fleeger, J.W., Means, J.C., Pomarico, S.M., McMillin, D.J., 1995. Experimental investigation of the effects of polynuclear aromatic hydrocarbons on an estuarine sediment food web. Marine Environmental Research 40, 289–318.

Carman, K.R., Fleeger, J.W., Pomarico, S.M., 1997. Response of a benthic food web to hydrocarbon contamination. Limnology and Oceanography 42, 561–571.

Carman, K.R., Todaro, M.A., 1996. Influence of polycyclic aromatic hydrocarbons on the meiobenthic-copepod community of a Louisiana salt marsh. Journal of Experimental Marine Biology and Ecology 198, 37–54.

Cioci, L.K., Qiu, L., Freedman, J.H., 2000. Transgenic strains of the nematode *Caenorhabditis elegans* as biomonitors of metal contamination. Environmental Toxicology and Chemistry 19, 2122–2129.

Clarke, K.R., 1993. Non-parametric multivariate analyses of changes in community structure. Australian Journal of Ecology 18, 117–143.

Coull, B.C., Chandler, G.T., 1992. Pollution and meiofauna: field, laboratory and mesocosm studies. Oceanography and Marine Biology Annual Reviews 30, 191–271.

Coull, B.C., Palmer, M.A., 1984. Field experimentation in meiofaunal ecology. Hydrobiologia 118, 1–19.

Danovaro, R., Fabiano, M., Vincx, M., 1995. Meiofauna response to the *Agip Abruzzo* oil spill in subtidal sediments of the Ligurian Sea. Marine Pollution Bulletin 30, 133–145.

De Goede, R.G.M., Bongers, T., Ettema, C.H., 1993. Graphical presentation and interpretation of nematode community structure: c–p triangles. Mededelingen Faculteit Landbouwwetenschappen Universiteit Gent 58, 743–750.

Delaune, R.D., Smith, C.J., Patrick, W.H., Fleeger, J.W., Tolley, M.D., 1984. Effect of oil on salt marsh biota: methods for restoration. Environmental Pollution (Series A) 36, 207–227.

Dhawan, R., Dusenbery, D.B., Williams, P.L., 1999. Comparison of lethality, reproduction, and behavior as toxicological endpoints in the nematode *Caenorhabditis elegans*. Journal of Toxicology and Environmental Health 58, 451–462.

Donkin, S.G., Dusenbery, D.B., 1993. A soil toxicity test using the nematode *C.elegans* and an effective method of recovery. Archives of Environmental Contamination and Toxicology 25, 145–151.

Essink, K., Romeyn, K., 1994. Estuarine nematodes as indicators of organic pollution: an example from the Ems Estuary (The Netherlands). Netherlands Journal of Aquatic Ecology 28, 213–219.

Fenchel, T., 1978. The ecology of micro and meiobenthos. Annual Review of Ecology and Systematics 9, 99–121.

Ferris, V.R., Ferris, J.M., 1979. Thread worms (Nematoda). In: Pollution Ecology of Estuarine Invertebrates. Academic Press, London, pp. 1–33.

Fiscus, D.A., Neher, D.A., 2002. Distinguishing sensitivity of free-living soil nematode genera to physical and chemical disturbances. Ecological Applications 12, 565–575.

Freckman, D.W., Ettema, C.H., 1993. Assessing nematode communities in agroecosystems of varying human intervention. Agriculture, Ecosystems and Environment 45, 239–261.

Freeman, M.N., Peredney, C.L., Williams, P.L., 2000. A soil bioassay using the nematode *Caenorhabditis elegans*. In: Henshel, D.S., Black, M.C., Harrass, M.C. (Eds), Environmental Toxicology and Risk Assessment: Standardization of Biomarkers for Endocrine Disruption and Environmental Assessment, Eighth Volume, ASTM STP 1364. American Society for Testing and Materials, West Conshohocken, pp. 305–318.

Fricke, A.H., Hennig, H.F.-K.O., Orren, M.J., 1981. Relationship between oil pollution and psammolittoral meiofauna density of two South African beaches. Marine Environmental Research 5, 59–77.

Fuller, M.E., Scow, K.M., Lau, S.S., Ferris, H., 1997. Trichloroethylen (TCE) and toluene effects on the structure and function of the soil community. Soil Biology and Biochemistry 29, 75–89.

Gee, J.M., Warwick, R.M., 1985. Effects of organic enrichment on abundance and community structure of in sublittoral soft sediments. Journal of Experimental Marine Biology and Ecology 91, 247–262.

Guven, K., Duce, J.A., de Pomerai, D.I., 1994. Evaluation of a stress-inducible transgenic nematode strain for rapid aquatic toxicity testing. Aquatic Toxicology 29, 119–137.

Gyedu-Ababio, T.K., Furstenberg, J.P., Blaird, D., Vanreusel, A., 1999. Nematodes as indicators of pollution: a case study from the Swartkops River system, South Africa. Hydrobiologia 397, 155–169.

Haight, M., Mudry, T., Pasternak, J., 1982. Toxicity of seven heavy metals on *Panagrellus silusiae*: the efficacy of the free-living nematode as an in vivo toxicological bioassay. Nematologica 28, 1–11.

Haitzer, M., Burnison, B.K., Höss, S., Traunspurger, W., Steinberg, C.E.W., 1999a. Effects of quantity, quality, and contact time of dissolved organic matter on bioconcentration of Benzo[a]pyrene in the nematode *Caenorhabditis elegans*. Environmental Toxicology and Chemistry 18, 459–465.

Haitzer, M., Höss, S., Traunspurger, W., Steinberg, C., 1999b. Relationship between concentration of dissolved organic matter (DOM) and the effect of DOM on the bioconcentration of benzo[a]pyrene. Aquatic Toxicology 45, 147–158.

Heip, C., 1980. Meiobenthos as a tool in the assessment of marine environmental quality. Rapport et Proces Verbaux des Reunins Conseils Permanant International pour l'Exploration de la Mer 179, 182–187.

Heip, C., Vincx, M., Vranken, G., 1985. The ecology of marine nematodes. Oceanography and Marine Biology Annual Reviews 23, 399–489.

Hennig, H.F.-K.O., Eagle, G.A., Fielder, L., Fricke, A.H., Gledhill, W.J., Greenwood, P.J., Orren, M.J., 1983. Ratio and population density of psammolittoral meiofauna as a perturbation indicator of sandy beaches in South Africa. Environmental Monitoring and Assessment 3, 45–60.

Henschel, T., 2002. Ökotoxikologisches Effektmonitoring und integrierte Gefahrenabschätzung schadstoffbelasteter Gewässersedimente: Feldstudien von Systemen unterschiedlicher Kontaminationsmuster. Ph.D. Thesis, Humbold University, Berlin.

Hitchcock, D.R., Black, M.C., Williams, P.L., 1997. Investigations into using the nematode *Caenorhabditis elegans* for municipal and industrial wastewater toxicity testing. Archives of Environmental Contamination and Toxicology 33, 252–260.

Hodda, M.E., Bloemers, G.F., Lawton, J.H., Lambshead, P.J.D., 1997. The effect of clearing and subsequent land-use on abundance and biomass of soil nematodes in tropical forest. Pedobiologia 41, 279–294.

Hodda, M.E., Nicholas, W.L., 1986. Nematode diversity and industrial pollution in the Hunter River estuary, NSW, Australia. Marine Pollution Bulletin 17, 251–255.

Höss, S., Haitzer, M., Traunspurger, W., Gratzer, H., Ahlf, W., Steinberg, C., 1997. Influence of particle size distribution and content of organic matter on the toxicity of copper in bioassays using *Caenorhabditis elegans* (Nematoda). Water, Air and Soil Pollution 99, 689–695.

Höss, S., Haitzer, M., Traunspurger, W., Steinberg, C.E.W., 1999. Growth and fertility of *Caenorhabditis elegans* (Nematoda) in unpolluted freshwater sediments – response to particle size distribution and organic content. Environmental Toxicology and Chemistry 18, 2921–2925.

Ingersoll, C.G., Ankley, G.T., Benoit, D.A., Brunson, E.L., Burton, G.A., Dwyer, F.J., Hoke, R.A., Landrum, P.F., Norberg-King, T.J., Winger, P.V., 1995. Toxicity and bioaccumulation of sediment-associated contaminants using freshwater invertebrates: a review of methods and applications. Environmental Toxicology and Chemistry 14, 1885–1894.

Ingham, R.E., Trofymow, J.A., Ingham, E.R., Coleman, D.C., 1985. Interactions of bacteria, fungi, and their nematode grazers: effects on nutrient cycling and plant growth. Ecological Monographs 55, 119–140.

Jensen, P., 1984. Ecology of benthic and epiphytic nematodes in brackish water. Hydrobiologia 108, 201–217.

Keddy, C.J., Greene, J.C., Bonnell, M.A., 1995. Review of whole-organism bioassays: soil, freshwater sediment, and freshwater assessment in Canada. Ecotoxicology and Environmental Safety 30, 221–251.

Kennedy, A.D., Jacoby, C.A., 1999. Biological indicators of marine environmental health: meiofauna – a neglected benthic component? Environmental Monitoring and Assessment 54, 47–68.

Korthals, G.W., Alexiev, A.D., Lexmond, T.M., Kammenga, J.E., Bongers, T., 1996. Long-term effects of copper and pH on the nematode community in an agroecosystem. Environmental Toxicology and Chemistry 15, 979–985.

Korthals, G.W., Bongers, M., Fokkema, A., Dueck, T.A., Lexmond, T.M., 2000. Joint toxicity of copper and zinc to a terrestrial nematode community in an acid sandy soil. Ecotoxicology 9, 219–228.

Korthals, G.W., Van de Ende, A., Van Megen, H., Lexmond, T.M., Kammenga, J.E., Bongers, T., 1996. Short-term effects of cadmium, copper, nickel, and zinc on soil nematodes from different feeding and life-history strategy groups. Applied Soil Ecology 4, 107–117.

Lambshead, P.J.D., 1984. The nematode/copepod ratio. Some anomalous results from the Firth of Clyde. Marine Pollution Bulletin 15, 256–259.

Lambshead, P.J.D., Platt, H.M., Shaw, K.M., 1983. The detection of differences among assemblages of marine benthic species based on an assessmnet of dominance and diversity. Journal of Natural History 17, 859–874.

Lenz, R., Eisenbeis, G., 2000. Short-term effects of different tillage in a sustainable farming system on nematode community structure. Biology and Fertility of Soils 31, 237–244.

Lew, K.K., Nichols, D.G., Kolber, A.W., 1983. In vivo assay to screen for mutagens/carcinogens in the nematode *C.elegans*. In: Kolber, A.R., de Woskin, R.S., Hughes, T.J. (Eds), In vitro Toxicity Testing of Environmental Agents: A Survey of Test Systems. Plenum, New York, pp. 139–150.

Maaß, V., Schmidt, C., Lüschow, R., Leitz, Th., 1997. Sedimentuntersuchungen im Hamburger Hafen 1994/1995, Strom und Hafenbau Hamburg, Hamburg.

Magurran, A.E., 1988. Ecological Diversity and its Management. Croom Helm, London.

Millward, R., Grant, A., 1995. Assessing the impact of copper on nematode communities from a chronically metal-enriched estuary using pollution-induced community tolerance. Marine Pollution Bulletin 30, 701–706.

Moore, C.G., Bett, B.J., 1989. The use of meiofauna in marine pollution impact assessment. Zoological Journal of the Linnean Society 96, 263–280.

Neilson, R., Boag, B., Palmer, L.F., 1996. The effect of environment on marine nematode assemblages as indicated by the maturity index. Nematologica 42, 232–242.

Newell, R., Maughan, D., Trett, M.W., Newell, P., Seiderer, M., 1991. Modification of benthic community structure in response to acid-iron wastes discharge. Marine Pollution Bulletin 22, 112–118.

Newell, R., Newell, P., Trett, M.W., 1990. Assessment of the impact of liquid wastes on benthic invertebrate assemblages. The Science of the Total Environment 97/98, 855–867.

Panesar, T.S., Marshal, V.G., Barclay, H.J., 2000. The impact of clearcutting and partial harvesting systems on population dynamics of soil nematodes in coastal Douglas-fir forest. Pedobiologia 44, 641–665.

Parmelee, R.W., Wentsel, R.S., Phillips, C.T., Simini, M., Checkal, R.T., 1993. Soil microcosm for testing the effects of chemical pollutants on soil fauna communities and trophic structure. Environmental Toxicology and Chemistry 12, 1477–1486.

Peredney, C.L., Williams, P.L., 2000. Utility of *Caenorhabditis elegans* for assessing heavy metal contamination in artificial soil. Archives of Environmental Contamination and Toxicology 39, 113–118.

Pielou, P.C., 1977. Ecological diversity and its measurements. In: Pielou, P.C. (Ed.), Mathematical Ecology. Wiley, New York, pp. 291–311.

Popham, J.D., Webster, J.M., 1979. Cadmium toxicity in the freeliving nematode, *Caenorhabditis elegans*. Environmental Research 20, 183–191.

Prejs, K., 1977. The species diversity, numbers and biomass of benthic nematodes in central part of lakes with different trophy. Ecologia Polska 25, 31–44.

Raffaelli, D., 1982. An assessment of the potential of major meiofauna groups for monitoring organic pollution. Marine Environmental Research 7, 151–164.

Ratte, H.T., Hammers-Wire, M., Cleuvers, M., 2002 (2003). Ecotoxicity testing. In: Markert, B.A., Breure, A.M., Zechmeister, H.G. (Eds), Bioindicators and Biomonitors. Elsevier, Oxford, pp. 221–256.

Ruess, L., Funke, W., Breunig, A., 1993. Influence of experimental acidification on nematodes, bacteria and fungi: soil microcosms and field experiments. Zoologische Jahrbücher (Systematik) 120, 189–199.

Samoiloff, M.R., Schulz, S., Jordan, Y., Denich, K., Arnott, E., 1980. A rapid simple long-term toxicity assay for aquatic contaminants using the nematode *Panagrellus redivivus*. Canadian Journal of Fisheries and Aquatic Science 37, 1167–1174.

Sandullini, R., de Nicola-Guidici, M., 1990. Pollution effects on the structure of meiofaunal communities in the Bay of Naples. Marine Pollution Bulletin 21, 144–153.

Schratzberger, M., Rees, H.L., Boyd, S.E., 2000. Effects of simulated deposition of dredged material on structure of nematode assemblages – the role of contamination. Marine Biology 137, 613–622.

Schratzberger, M., Warwick, R.M., 1999. Differential effects of various types of disturbances on the structure of nematode assemblages: an experimental approach. Marine Ecology Progress Series 181, 227–236.

Sherry, J., Scott, B., Dutka, B., 1997. Use of various acute, sublethal and early life stage tests to evaluate the toxicity of refinery effluents. Environmental Toxicology and Chemistry 16, 2249–2257.

Soetaert, K., Vincx, M., Wittoek, J., Tulkens, M., 1995. Meiobenthic distribution and nematode community structure in five European estuaries. Hydrobiologia 311, 185–206.

Somerfield, P.J., Gee, J.M., Warwick, R.M., 1994a. Benthic community structure in relation to an instantaneous discharge of waste water from a tin mine. Marine Pollution Bulletin 28, 363–369.

Somerfield, P.J., Gee, J.M., Warwick, R.M., 1994b. Soft sediment meiofaunal community structure in relation to a long-term heavy metal gradient in the Fal Estuary system. Marine Ecology Progress Series 105, 79–88.

Somerfield, P.J., Rees, H.L., Warwick, R.M., 1995. Interrelationships in community structure between shallow-water marine meiofauna and macrofauna in relation to dredging disposal. Marine Ecology Progress Series 127, 103–112.

Stringham, E.G., Candido, E.P.M., 1994. Transgenic hsp 16-lacZ Strains of the soil nematode *Caenorhabditis elegans* as biological monitors of environmental stress. Environmental Toxicology and Chemistry 13, 1211–1220.

Tatara, C.P., Newman, M.C., McCloskey, J.T., Williams, P.L., 1997. Predicting relative metal toxicity with ion characteristics: *Caenorhabditis elegans* LC 50. Aquatic Toxicology 39, 279–290.

Tietjen, J.H., 1980. Population structure and species distribution of the free living nematodes inhabiting sands of the New York Bight apex. Estuarine and Coastal Marine Science 10, 61–73.

Traunspurger, W., 1996a. Distribution of benthic nematodes in the littoral of an oligotrophic lake (Königssee, Nationalpark Berchtesgaden, FRG). Archiv für Hydrobiologie 135, 393–412.

Traunspurger, W., 1996b. Distribution of benthic nematodes in the littoriprofundal of an oligotrophic lake (Königssee, Nationalpark Berchtesgaden, FRG). Archiv für Hydrobiologie 135, 557–575.

Traunspurger, W., 1997. Bathymetric, seasonal and vertical distribution of feeding types of nematodes in an oligotrophic lake. Vie Milieu 47, 1–7.

Traunspurger, W., 2000. The biology and ecology of lotic nematodes. Freshwater Biology 44, 29–45.

Traunspurger, W., 2002. Freshwater nematodes: biology and ecology. In: Rundle, S.D., Robertson, A., Schmid-Araya, J. (Eds), Freshwater Meiofauna: Biology and Ecology. Backhuys, London.

Traunspurger, W., Bergtold, M., Goedkoop, W., 1997. The effect of nematodes on bacterial activity and abundance in a freshwater sediment. Oecologia 112, 118–122.

Traunspurger, W., Drews, C., 1996. Toxicity analysis of freshwater and marine sediments with meio- and macrobenthic organisms: a review. Hydrobiologia 328, 215–261.

Traunspurger, W., Haitzer, M., Höss, S., Beier, S., Ahlf, W., Steinberg, C., 1997. Ecotoxicological assessment of aquatic sediments with *Caenorhabditis elegans* (nematoda) – a method for testing in liquid medium and whole sediment samples. Environmental Toxicology and Chemistry 16, 245–250.

Trett, M.W., Calvo-Urbano, B., Forster, S.J., Hutchinson, J.D., Feil, R.L., Trett, S.P., Best, J.G., 2000. Terrestrial meiofauna and contaminated land assessment. Environmental Science and Technology 34, 1594–1602.

Urzelai, A., Hernandez, A.J., Pastor, J., 2000. Biotic indices based on soil nematode communities for assessing soil quality in terrestrial ecosystems. The Science of the Total Environment 247, 253–261.

Vanreusel, A., 1991. Ecology of free-living marine nematodes in the Voordelta (southern bight of the north sea). 2. Habitat preferences of the domonant species. Nematologica 37, 343–359.

Vidakovic, J., 1983. The influence of raw domestic sewage on density and distribution of meiofauna. Marine Pollution Bulletin 14, 84–88.

Vranken, G., Heip, C., 1986. Toxicity of copper, mercury and lead to a marine nematode. Marine Pollution Bulletin 17, 453–457.

Vranken, G., Vanderhaeghen, R., Heip, C., 1985. Toxicity of cadmium to free-living marine and brackish water nematodes (*Monhystera microphtalma*, *Monhystera disjuncta*, *Pellioditis marina*). Diseases of Aquatic Organisms 1, 49–58.

Warwick, R.M., Carr, M.R., Clarke, K.R., Gee, J.M., Green, R.H., 1988. A mesocosm experiment on the effects of hydrocarbon and copper pollution on a sublittoral soft-sediment meiobenthic community. Marine Ecology Progress Series 46, 181–191.

Warwick, R.M., Clarke, K.R., 1991. A comparison of some methods for analysing changes in benthic community structure. Journal of the Marine Biology Association, 71, 225–244.

Weiss, B., Larink, O., 1991. Influence of sewage sludge and heavy metals on nematodes in an arable soil. Biology and Fertility of Soils 12, 5–9.

Wieser, W., 1953. Die Beziehung zwischen Mundhöhlengestalt, Ernährungsweise und Vorkommen bei freilebenden marinen Nematoden. Arkiv for Zoologi 2, 439–484.

Williams, P.L., Dusenbery, D.B., 1990a. A promising indicator of neurobehavioral toxicity using the nematode *Caenorhabditis elegans* and computer tracking. Toxicology and Industrial Health 6, 425–440.

Williams, P.L., Dusenbery, D.B., 1990b. Aquatic toxicity testing using the nematode *Caenorhabditis elegans*. Environmental Toxicology and Chemistry 9, 1285–1290.

Yeates, G.W., 1981. Nematode populations in relation to soil environmental factors: a review. Pedobiologia 22, 312–338.

Yeates, G.W., 1987. How plants affects nematodes. Advances in Ecological Research 17, 61–113.

Yeates, G.W., Bongers, T., De Goede, R.G.M., Freckman, D.W., Georgieva, S.S., 1993. Feeding habits in soil nematode families and genera – an outline for soil ecologists. Journal of Nematology 25, 315–331.

Yeates, G.W., Coleman, D.C., 1982. Role of nematodes in decomposition. In: Freckman, D.W. (Ed.), Nematodes in Soil Ecosystems. University of Texas Press, Austin, pp. 55–80.

Zullini, A., 1976. Nematodes as indicators of river pollution. Nematologia Mediterranea 4, 13–22.

Zullini, A., Peretti, E., 1986. Lead pollution and moss-inhabiting nematodes of an industrial area. Water, Air and Soil Pollution 27, 403–410.

Bioindicators and biomonitors
B.A. Markert, A.M. Breure, H.G. Zechmeister, editors
© 2003 Elsevier Science Ltd. All rights reserved.

Chapter 16

Oligochaeta

Wim Didden

Abstract

Among oligochaetes bioindicators and -monitors have been identified at all levels of biological organization, from the molecular to the ecosystem level. For these various levels an overview is given of current and developing biomonitors and -indicators. On the molecular and cellular level these include rapidly evolving techniques that make use of the sensitivity of genetic material to stress factors, induction of stress proteins, changes in functioning of the nervous system and the immune system, and in (ultra)structure of cells and organelles. On the individual and population level use is being made of stress-induced changes in abundance and biomass, of the various degrees to which accumulation and bioconcentration of contaminants occurs in oligochaete species, and of easily measurable changes that may be induced in reproduction, growth, behaviour and morphology. On the community and ecosystem level community composition may change as a result of stress, potentially leading to changes in ecosystem functioning. The various bioindicators and -monitors are outlined, and their applicability is discussed. It is concluded that oligochaetes have a high potential as bioindicators and -monitors, that may be even more realized if the link between low-level processes and ecosystem-level processes would be clarified further.

Keywords: Lumbricidae, Enchytraeidae, Tubificidae, ecotoxicology, stress, bioindicator, biomonior, biomarker, review

1. Introduction

Bioindicators and biomonitors (sensu Markert et al., 1999) are used as representatives of (part of) an ecosystem or one or more of its functions. Their use is restricted to the detection or (at its best) prediction of changes in composition or functioning of the system (Reinecke, 1998; Lancaster, 2000), but the basic consideration behind their use is that (living) organisms provide the best reflection of the actual suitability of their habitat and of changes therein. Although it is common practice to determine habitat factors (for instance by chemical analyses of soil or sediment), their relation with ecosystem functioning is often difficult to interpret without knowledge on the ways in which the various organisms come into contact and react with these factors. Moreover, the number of measurable parameters is practically infinite, making it impossible to measure them all at any location. Because organisms may be regarded as integrating and reflecting the current habitat characteristics, it is, in principle, possible to detect a change in habitat characteristics by looking at the organisms, without having to be able to identify the exact nature of the change. Moreover, using organisms it is possible to establish the presence of unknown or unexpected stressors that would otherwise be left unnoticed. This is why the use of biological indicators and monitors, both for assessment of toxicological risk and to monitor soil remediation processes, is becoming increasingly popular (cf. Ramade, 1987; Spurgeon et al., 1996; LABO, 1998; Kille et al., 1999; Sayles et al., 1999; Haimi, 2000; Prygiel et al., 2000; Sijm et al., 2000).

Indicators and monitors may be used to test the response to specific stresses, which is mainly performed under laboratory conditions. Alternatively, general information on the level of stress may be collected by assessing the state of the indicator or monitor in the field. There are, however, several factors that may complicate the interpretation of the results:

- The common problem of the translation of laboratory data to the field situation, mainly caused by the inherently larger complexity in the field (e.g. Bogomolov et al., 1996). A commonly used approach to this problem is the use of micro- and mesocosms as an intermediate between laboratory and field (e.g. Parmelee et al., 1993; Römbke et al., 1994, Marinussen, 1997, Martikainen et al., 1998; Moser et al., 1999; Hodge et al., 2000).
- The occurrence of adaptation to stress factors. Spurgeon et al. (1994), for instance, has shown that *Eisenia fetida* was able to survive in soil with concentrations of heavy metals that were higher than the LC_{50} and EC_{50} values determined for this species.
- The involvement of regulatory mechanisms, that may obscure the relation between stress level and response.
- The occurrence of avoidance behaviour, by which a biomonitor may avoid heterogeneously distributed stress factors.
- The effect of trade-off of defence mechanisms on, for instance, growth or reproductivity. In this way it is possible for an organism to survive, but the effect of the stressor may be expressed at the population level.

Oligochaeta are generally regarded as highly suitable bioindicators or biomonitors (e.g. Stenersen et al., 1992; Goven et al., 1994b; Abdul Rida and Bouché, 1995a; Bunn et al., 1996; Reinecke and Reinecke, 1998; Cortet et al., 1999; Paoletti, 1999; Prygiel et al., 2000; Didden and Römbke, 2001). Their main advantages are:

- Oligochaeta contain key species for ecosystem functioning, notably for decomposition and soil structure maintenance.
- Oligochaete worms are widespread and abundant, both in terrestrial and aquatic environments.
- Due to their behaviour and morphology they are in contact with both the aqueous and the solid phase of the substrate.
- Most species are relatively large and therefore easy to handle and to culture.
- Most species are not extremely sensitive to low levels of contamination.
- Their reactions towards stress are measurable and reproducible at various levels of organization.
- They can be used both in laboratory and field conditions.
- There is a vast and growing body of knowledge on their biology, ecology and ecotoxicology.
- Their internal organization is not highly complex, yet they possess strongly differentiated organs and tissues.

- The chemical composition of their body is fairly constant, providing a homogeneous reference for toxic substances.
- Oligochaetes are non-controversial as test animals.

Earthworms, in particular Lumbricidae, are, due to their size and relatively easy handling, by far the most intensively studied oligochaete group in the field of ecotoxicology and biomonitoring. Their occurrence, however, is largely limited to terrestrial habitats, which restricts their use. For studies involving sediments rather intensive use is being made of species from the Lumbriculidae and Tubificidae, while in recent years increasing attention is being devoted to the Enchytraeidae, a widespread group which occurs both in terrestrial and aquatic environments (Didden and Römbke, 2001).

It should be stressed here that the effect of any stressor on oligochaetes depends on its bioavailability. Bioavailability of xenobiotics to oligochaetes depends on a number of factors, of physical, chemical, biological, temporal and spatial nature, thus making it difficult – if not impossible – to estimate from the concentration in soil or sediment. Some examples may illustrate this point:

Many data support the importance of the concentration of a toxicant in the soil solution. Zweers (1996), for instance, found in a variety of field soils a good correlation of the Zn concentration in the pore-water with concentration in *Enchytraeus crypticus*, but no correlation with the total concentration in soil. This relation may not be true for all species and chemical substances, however, due to physiological and behavioural differences. Sjögren et al. (1995), for instance, found that *Cognettia sphagnetorum* accumulated Zn linearly proportional to the total concentration in the substrate while Cu was accumulated only to a certain level. This indicates that in this species an appreciable part of Zn and Cu intake takes place via food, and that it is able to regulate the accumulation of Cu to a certain extent. Belfroid et al. (1995) maintained, based on model calculations, that earthworms take up most organic hydrophobic chemicals from the interstitial water. Uptake via food might be significant only in the case of high organic matter contents ($\geq$ 20%) of the soil. However, this is not necessarily true in all cases. Callahan et al. (1994), studying the effects of a number of chemicals on four earthworm species, found that in some cases earthworm sensitivity was much higher in soil tests than in contact tests, suggesting that in these cases uptake of the chemicals was both via the cuticula and via food. Notably in the case of hydrophobic contaminants uptake via food may be an important uptake route as was shown with pyrene in the aquatic oligochaete *Lumbriculus variegatus* (Leppänen and Kukkonen, 1998; Conrad et al., 2000).

Effects of physico-chemical soil characteristics such as organic matter content, moisture and pH on bioavailability have repeatedly been demonstrated (e.g. Ma, 1982, 1984; Ma et al., 1983; Puurtinen and Martikainen, 1997). Such effects may also be related to the increased complexity of the test system, as Salminen and Haimi (1997) found an effect of PCP on the enchytraeid *C. sphagnetorum* in field lysimeters at a concentration that caused no effects in laboratory microcosms.

The bioavailability of a substance in soil may decrease with ageing (Gevao et al., 2000; Reid et al., 2000), decreasing the uptake by oligochaetes as was shown, for instance, by Kelsey et al. (1997) for phenanthrene. Organic xenobiotics may be detoxified by biotransformation via cytochrome P450, but as these enzymes are probably not inducible in oligochaetes (with a possible exception for PCB's), such substances will accumulate (Stenersen, 1992). The way of living of earthworm species also influences bioavailability, as, due to slow desorption and diffusion processes, bioavailability will be less for species that live in permanent burrows (Stenersen, 1992).

It may be concluded that bioavailability of a substance depends not only on characteristics of the substance itself and of the soil, but also those of the species under study. Although bioavailability is, therefore, only measurable using organisms, specific chemical extraction procedures may give an estimate for specific species (Conder and Lanno, 2000).

Bioindicators and -monitors may be identified at various levels of biological organization, ranging from the molecular level to the ecosystem level, and the type of information they confer differs accordingly. Figure 1 gives a schematic overview of the various levels, their general characteristics and the advantages and disadvantages associated.

In general, it may be stated that there will be an inverse relation between the level of organization at which an indicator is determined and its predictive potential (e.g. Booth et al., 2000). Also, at lower organizational levels (until the individual level) there may exist a direct relation between occurrence of stress and the reaction of the indicator, suggesting the use of such indicators in an early warning system (sentinel species). Yet, an effect occurring on this level does not necessarily indicate an effect on higher levels (Cortet et al., 1999), and may therefore not have ecological relevance. Higher-level indicators, on the other hand, may indicate the occurrence of environmental changes without giving information on their causes (Cortet et al., 1999). It should be understood, however, that indirect effects of stressors (such as on food availability, predation level or soil structure) may well be of more importance for ecosystem functioning than direct effects (Haimi, 2000).

With Oligochaeta, bioindicators and -monitors are in use on all organization levels mentioned above. In the following overview these will be treated according to their organizational level.

2. The cellular and subcellular level

2.1. Genetic methods

Because of its general relevance for risk assessment the detection of genotoxic substances may be regarded an important topic, and several tests using earthworms have been developed. Walsh et al. (1995, 1997), for instance, demonstrated increased presence of DNA-adducts in *Lumbricus terrestris* and *E. fetida* after 1 to 2 weeks' exposure to PAH contaminated soil. Applying a comet assay to *E. fetida* from metal polluted soil Salagovic et al. (1996) recorded significantly more single strand breaks in coelomocytes.

A response to stress conditions often consists in the induction or upregulations of specific enzymes and proteins. Willuhn et al. (1996b) found that exposure to Cd

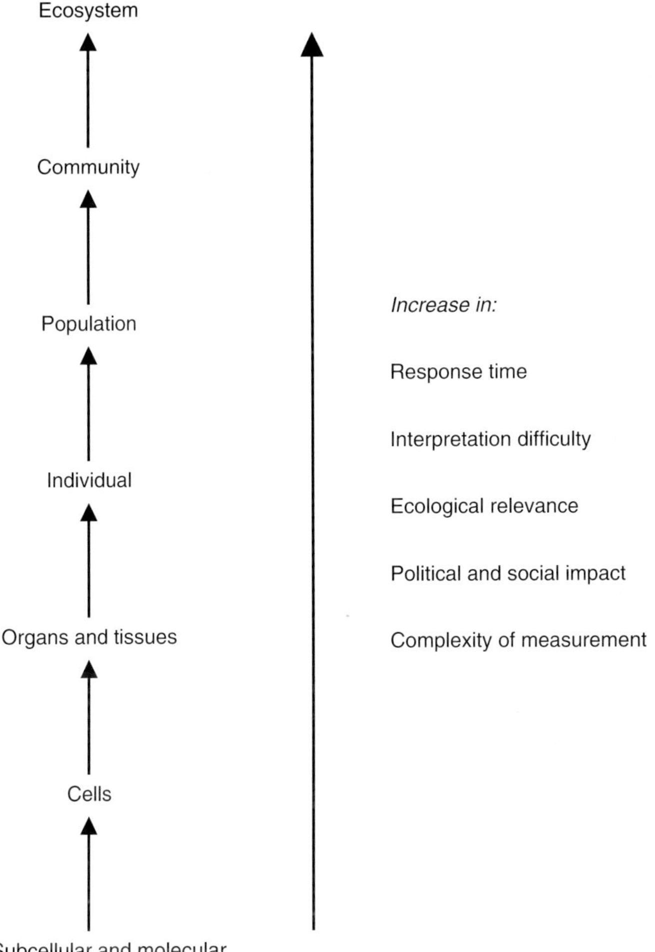

Figure 1. The organizational level at which bioindicators and -monitors may be identified and their general characteristics, advantages and disadvantages. Adapted from Morgan et al., 1999a.

induced a mRNA species in *Enchytraeus buchholzi* coding for a cystein-rich nonmetallothionein protein that probably was engaged in the binding of Cd. Stürzenbaum et al. (1999) established the occurrence of two isoforms of cyclophilin in the earthworm *Lumbricus rubellus*, one that was an invariable housekeeping gene, but the other was inducible by heavy metal (Zn, Pb, Cu and Cd, possibly especially Zn) stress. This cyclophilin A may be involved in the regulation of transcription processes and therefore in the induction of metallothioneins (see Section 3.2). Kille et al. (1999) used a number of recently developed and relatively fast and inexpensive techniques to determine the level of gene expression for stress related responses in metal (Pb, Cu, Zn, Cd) exposed earthworms. They found a clear upregulation for the mitochondrial function (reflecting increased energy demand as a general indication of stress), lysosomal glycoprotein (indicating loss of lysosomal integrity, see Section 2.4) and for metallothionein (specifically pointing to metal stress).

Given the rapid development in economically feasible genetic methods and the apparent sensitivity of oligochaete genomes to stress factors, such approaches may well lead to standard screening tools in toxicological risk evaluations.

2.2. Stress proteins

Induction of stress proteins that are engaged in biotransformation or detoxification may be regarded as a direct response to stress factors, and will be produced rather quickly. Mariño et al. (1999) applied an immunostaining technique to determine heat shock protein (hsp) levels in *L. rubellus* under metal (Pb and Zn) stress. These hsp's can be mainly found in the chloragocytes, where oligochaetes sequester a number of heavy metals (see Section 3.2). They reported a dose-related hsp expression, and as the response occurred within 3 weeks after transferring worms to metal contaminated soil they suggested this approach to be suitable for screening and monitoring purposes. As there are a number of different hsp's that as a group may react in a stress-specific way, this approach could also be used to identify the type of stress present. However, Mariño et al. (1999) also found indications for metal tolerance in worms from contaminated sites, restricting the possible use for monitoring.

2.3. Neurological indicators and monitors

The nerve system is essential for survival in any higher animal, and its impairment will have consequences on all levels of the animals' functioning. Therefore, elements of the nervous system may provide an early warning system for potentially hazardous stress factors. One approach in this field uses the activity of cholinesterases (ChE) that are involved in neurotransmission, in earthworms as a biomonitor. Stenersen et al. (1992) studied the effects of the ChE inhibiting pesticides carbaryl and paraoxon on ChE activity in *Eisenia andrei*, *Eisenia veneta* and *E. fetida* and found clear effects on the nervous system even at low doses. Probably due to the occurrence of various different ChE's there was an interspecific difference in lethal toxicity, with *E. veneta* being most sensitive. O'Halloran et al. (1999) and Booth et al. (2000), however, reporting a clear ChE inhibition in *Aporrectodea caliginosa* by sublethal application of organophosphorous and carbamate pesticides in laboratory experiments, found no response in mesocosm or field test. This may indicate decreased bioavailability under these conditions either by heterogeneous distribution of the pesticides or by a behavioral response of the worms (see Section 3.5).

Another approach was applied by Drewes (1997), who used the functioning of the escape reflex. This parameter has relevance because the escape reflex has a direct bearing on survival of an individual. Measurable effects on this mechanism may be found in the sensory functions involved, conduction of the stimulus and in the functioning of the motorneurons. In laboratory experiments it was shown that many xenobiotic substances produced reproducible changes in the functioning of the escape reflex in *L. terrestris*, *E. fetida* and *L. variegatus*, indicating that this approach is potentially useful as a predictive tool for effects of xenobiotics in the field.

2.4. *Immunoassays*

The immune system of oligochaetes, which is located in the coelomocytes and the coelomic fluid, is in many respects analogous to that found in other animals, including mammals (Goven et al., 1994b; Bunn et al., 1996). Thus, any stress factor that affects the functioning of their immune system may be regarded a potential hazard to higher animals also. Moreover, as the immune system constitutes a vital defence mechanism against infections, impairment will lead to increased sensitivity towards infections, which will also have consequences on the population and community level. Accordingly, the effects of environmental stressors on the earthworm immune system are receiving increasing attention as potential indicators of sublethal exposure and bioavailability of xenobiotics (e.g. Goven et al., 1994b; Bunn et al., 1996; Weeks and Svendsen, 1996; Svendsen and Weeks, 1997a,b; Giggleman et al., 1998).

A number of aspects of the earthworm immune system have been used as an endpoint, and several of these appear to be rather promising. Table 1 shows an overview of the various approaches.

From a comparison of 20 bioassays with Cd, phenol, PCP and trifluralin Bierkens et al. (1998) reported earthworm immunoassays to be among the most sensitive. For the neutral red retention assay, that has been applied both in laboratory and field conditions, Svendsen and Weeks (1997a,b) performed experiments to link the effects with effects on the individual and population level. They exposed *E. andrei* and *L. rubellus* to a series of Cu concentrations and found that effects on neutral red retention occurred in two distinct stages and only at the second stage effects on the individual and population level occurred. With *L. rubellus* the strength of the effects was possibly related to exposure time. They concluded that this assay has potential as an early warning system. It appears that neutral red retention time may also be more sensitive than coelomocyte activity, as Eason et al. (1999) found effects on neutral red retention time in the same experiments were no effects on coelomocyte activity were recorded. Moreover, Reinecke and Reinecke (1999) reported reduced neutral red retention in Cd-tolerant *E. fetida* living in Cd-contaminated soil, indicating that this assay may be less susceptible to adaptation of the worms.

2.5. *Chloragocytes*

Chloragocytes have important functions in oligochaetes because they function as storage organs for glycogen, lipids, phosphate and a number of xenobiotics such as polyphenols and flavines, and also play an important role in metabolic and respiratory processes (Prentø, 1994). Chloragosomes may act as ion-exchangers (both for cations and anions) and, because of the occurrence of lipids, apolar interactions may take place. Moreover, storage of substances in chloragocytes may last up to several months (Prentø, 1994). Fischer and Molnár (1992) contended that in *E. fetida* the chloragocytes showed clear responses to several different forms of environmental stress. Water stress, temperature stress, respiratory stress, salt stress and the pesticides paraquat and carbaryl all induced a significant enlargement of the chloragocyte nucleus. In the case of paraquat there also was at first an enlargement of the whole cell, after which the chloragocytes disappeared almost completely. In the case of temperature stress and

Table 1. Earthworm immunoassays used or proposed as bioindicator or biomonitor.

Endpoint	Stressor	Species	Reference
Coelomocyte activity	Various pesticides, PCB	*E. veneta*; *L. terrestris*	Goven et al., 1994b; Bunn et al., 1996
Fagocytosis	PCP	*L. terrestris*	Giggleman et al., 1998
	NaN$_3$	*L. terrestris*	Brousseau et al., 1997
	PCB	*E. fetida, Eisenia hortensis, L. terrestris*	Ville et al., 1995
Coelomocyte counts	PCB	*L. terrestris*	Goven et al., 1994b
Lysozyme activity	Cu	*L. terrestris*	Goven et al., 1994a
	PCB	*E. fetida, E. hortensis, L. terrestris*	Ville et al., 1995
NBT dye reduction	Fly ash	*L. terrestris*	Goven et al., 1994b
Coelomocyte cytotoxicity	PCB	*L. terrestris*	Suzuki et al., 1995
Neutral red retention	Effects of plastics fire	*Lumbricus castaneus*	Svendsen et al., 1996
	Cu	*E. andrei, L. rubellus*	Svendsen and Weeks, 1997a,b
	Chlorpyrifos	*E. andrei*	Eason et al., 1999
	BaP	*E. andrei*	Eason et al., 1999
	Soil from gas plant	*E. andrei*	Eason et al., 1999
	Pb, Zn and Cd	*E. fetida*	Reinecke and Reinecke, 1999
Wound healing	PCB	*E. fetida, E. hortensis, L. terrestris*	Ville et al., 1995
Graft rejection	PCB	*E. fetida, E. hortensis, L. terrestris*	Ville et al., 1995
Hemolysis	PCB	*E. fetida, E. hortensis, L. terrestris*	Ville et al., 1995
Proteolysis	PCB	*E. fetida, E. hortensis, L. terrestris*	Ville et al., 1995

salt stress there was an enhanced pigmentation in the chloragosomes, accompanied by a decrease of the concentrations of P, Ca, S, K and Cl in the chloragocytes. It was also established that Cr was stored in the chloragocytes. This could imply that chloragocytes in earthworms are potentially suitable for general stress indication, and possibly even could point to the type of stress involved.

2.6. Ultrastructure of cells and organelles

Hagens and Westheide (1987) found that sublethal exposure to the insecticide parathion caused ultrastructural changes in gut epithelial cells and chloragocytes of *Enchytraeus doerjesi*. Reinecke et al. (1995) reported that the pesticide dieldrin affected the sperm ultrastructure in *Eudrilus eugeniae*, and Reinecke and Reinecke (1997) found similar effects of Pb and Mn on *E. fetida*. Such effects may have ecological relevance, because even if cells or organs recover from such damage this may trade off to reproductive capacity and lifespan, as was demonstrated by Purschke et al. (1991). In the case of damaged sperm cells the effect on reproduction is even more direct and has a direct bearing to the population and ecosystem level.

3. The individual and population level

3.1. Abundance and biomass

Abundance and biomass of worm species or worm communities may be considered to reflect their response to habitat characteristics. For a large number of substances lethal doses for oligochaete species have been determined in single species laboratory tests (see e.g. Edwards and Bohlen, 1996 and Didden and Römbke, 2001 for overviews) that can be used for the interpretation of abundance data in relation to physico-chemical soil characteristics. Although several acute and chronic test systems are widely applied, there still is a need for further standardization of these tests to provide better comparability (Abdul Rida and Bouché, 1997).

From the work of Abdul Rida and Bouché (1994, 1995b) it is clear that there exist appreciable specific differences in earthworm sensitivity to heavy metals: they found that the genus *Scherotheca* was extremely sensitive, especially to Pb and Cu, and disappeared from contaminated sites. Similarly, Sturmbauer et al. (1999) reported clear intraspecific differences in sensitivity towards Cd in the aquatic oligochaete *Tubifex tubifex*. These differences were related to the occurrence of different mitochondrial lineages within this "species". A difficulty in the interpretation of abundance and biomass data from the field may be the development of resistance to a contaminant, as was shown by Langdon et al. (1999) for *L. rubellus* and arsenic and by Spurgeon and Hopkin (2000) for *E. fetida* and Zn.

In many cases there has been found a clear response of the abundance of oligochaete worms to toxicant stress. Salminen et al. (1996) and Salminen and Haimi (1996, 1997) have shown that the abundance of the enchytraeid *C. sphagnetorum* reacted negatively to terbuthylazine and PCP. Generally, it has been found that tests with mortality as endpoint are less sensitive than tests applying sublethal endpoints, but the advantage of mortality tests is that they are less time-consuming. Possibly, their sensitivity may be enhanced by including sublethal parameters such as the condition index (a semi-quantitative parameter based on muscle tone and response to stimulation) as was done by Langdon et al. (1999) with *L. rubellus*.

3.2. Accumulation and bioconcentration

The amount of a substance accumulated by an organism may be regarded as an indication of the bioavailability of this substance in the environment. The extent to which accumulation occurs is generally represented with the bioconcentration factor (BCF), which is the ratio of the concentration in the organism to the concentration in the substrate. It should be kept in mind, however, that the concentration in an organism is not necessarily related to toxic effects, as both the mechanisms involved in the uptake process and the action of a substance may vary widely and involve metabolic, excretion and detoxification pathways (Morgan and Morgan, 1993). BCF-values for risk evaluation or monitoring should therefore be interpreted against the background of knowledge on these mechanisms.

Because of the relative ease of bioaccumulation measurements these are frequently part of monitoring programmes, and often involve oligochaetes. To enable inference of the bioaccumulation of a toxicant from the concentration in the substrate, its potential bioconcentration should be determined. This, however, may cause problems, notably in laboratory experiments, as accumulation in the animal depends both on the rate at which a substance is taken up and on the rate at which depuration occurs. Sheppard et al. (1997) studied uptake and depuration kinetics of a number of elements in *L. terrestris* and found that these processes occurred rather slowly, notably for I and Cd. The consequence would be that the time needed to reach a steady state would exceed the experimental time that is normally applied in laboratories. Notably for Cd they expected accumulation to continue throughout the lifespan of *L. terrestris*. This may mean that accumulation in the field may easily be underestimated when laboratory determinations are applied. Bioaccumulation in oligochaetes is a function of the substance involved, soil characteristics, concentration, weather and climate, the species considered, age and condition of the specimens, uptake route for the substance etc. Determination of BCFs may only be useful in cases were no regulatory mechanism is involved (Cortet et al., 1999), but in the absence of such mechanisms accumulators may serve as sentinel species by exposing individuals in micro- or mesocosms in the field (e.g. Marinussen, 1997).

Bioaccumulation in oligochaetes has been studied for a fairly large number of substances. For PAHs very high BCFs (ranging from 452 to 2390) were reported in *L. variegatus* by Sheedy et al. (1998) and for the pesticide diazinon BCFs of 0.7 to 2.3 were found in *L. terrestris* in the field (Stephenson et al., 1997).

Bioaccumulation of heavy metals has been studied extensively, and by and large the available data show a clear relation between total soil concentration and concentration in the worm, although the form of this relation depends on metal and worm species. Figure 2 summarizes the results from a number of field studies on this subject. In general, BCFs tend to be highest at low soil levels, and to decline below unity at high levels. Cd was always found to be accumulated, although a BCF of 1 is

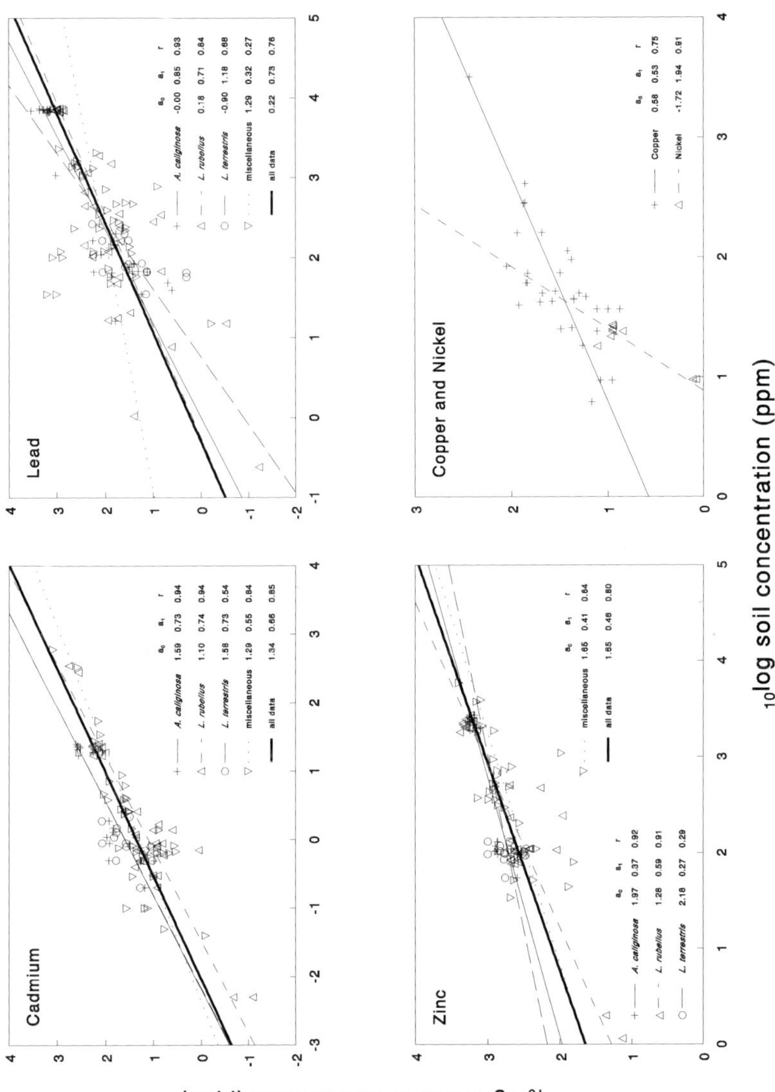

Figure 2. The relation between total concentration in the soil and concentration in worms for earthworms and enchytraeids, based on a (arbitrarily selected) number of field studies. Regression equations were calculated for various species and for the complete set of data, and include intercept (a_0), slope (a_1) and correlation coefficient (r). Data from Carter et al., 1980; Roth-Holzapfel and Funke, 1989; Morgan and Morgan, 1993; Roth, 1993; Abdul Rida and Bouché, 1994, 1995, 1996; Heck et al., 1995; Emmerling et al., 1996; Spurgeon and Hopkin, 1996; Mariño et al., 1998, 1999; Kille et al., 1999; Koeckritz et al., 1999; Reinecke and Reinecke, 1999; Cortet et al., 2000.

approached at very high soil levels. The figure suggests that Ni may be an exception in that this metal appears to be excluded at low soil levels (as was found by Roth-Holzapfel and Funke., 1989, Abdul Rida and Bouché, 1994 and Neuhauser et al., 1995), but might be concentrated at higher soil levels. Due to the lack of data at higher levels, however, such a conclusion is not warranted. For Cr, BCFs of up to 3 were recorded for enchytraeids and lumbricid earthworms (Roth-Holzapfel and Funke, 1989). From Figure 2 it appears that, although interspecific differences do occur, species of oligochaetes exhibit the same trend, suggesting that they may have a common strategy towards heavy metals.

From laboratory experiments comparable results were reported. For the tubificid *T. tubifex* Bouché et al. (2000) reported extremely high BCFs (up to 18430) for Cd. Bogomolov et al. (1996) found a good relationship between Cu total soil concentration and accumulation in *Aporrectodea tuberculata*. Reinecke and Reinecke (1997) reported exclusion of Pb and accumulation of Mn by *E. fetida*, and suggested the existence of different regulatory mechanisms for both metals. Veltz et al. (1996) concluded that *L. variegatus* would be a suitable monitor for Pt, because this species was rather tolerant and accumulated Pt in proportion to concentration, temperature and time. Physico-chemical characteristics of the substrate may certainly also play a role, as Pt-toxicity was clearly inversely related to $CaCO_3$-concentration (Veltz et al., 1996). For Pb there may also be a regulatory mechanism involving Ca, as Heck et al. (1995) found clearly lower BCFs for this metal in lumbricids and enchytraeids after liming.

The occurrence of adaptation mechanisms is apparent, as Neuhauser et al. (1995) found that worms from polluted areas eliminated Cd and Zn quicker than worms from an unpolluted area. This was not the case for other metals, however. Koeckritz et al. (1999) suggested a physiological regulation of Cu and Pb by *A. caliginosa*. Sjögren et al. (1995) proposed that the enchytraeid *C. sphagnetorum* possesses a regulatory mechanism for Cu. Resistance to Cd has been demonstrated in various oligochaete species (e.g. Willuhn et al., 1996c for *E. buchholzi*; Wallace et al., 1998 for *Limnodrilus hoffmeisteri*; Reinecke et al., 1999 for *E. fetida*), and this may potentially affect conclusions drawn from bioindicator tests with such species. Cd resistance may be associated with the induction of metallothioneines, as found in *L. rubellus* by Mariño et al. (1998), or other Cd-binding proteins as was found by Willuhn et al. (1996c) in *E. buchholzi*. Although in neither case Cu induced the proteins, it appeared that induction by Cd reduced copper toxicity (Willuhn et al., 1996a; Mariño et al., 1998). Tolerance for Zn was reported for the enchytraeid *Marionina clavata* from Zn-polluted soil (Notenboom et al., 1997) and for *E. fetida* in a laboratory experiment (Spurgeon and Hopkin, 2000). The origin for metal resistance or tolerance is not always clear. Aziz et al. (1999) reported tolerance for Pb and Zn in *L. rubellus* populations, but could find no clear indications for genetic adaptation to be involved and suggested the occurrence of differences in detoxification mechanisms. The results of Spurgeon and Hopkin (2000), on the other hand, suggest that in *E. fetida* genetically based resistance to Zn (and Cu) may develop in as few as two generations, probably involving a number of loci responsible for physiological responses in the worms.

Though a relative measure, bioaccumulation of a substance may be regarded to give an integrated picture of its bioavailability and, moreover, also is a prerequisite for biomagnification, in which a substance is taken up in the food-chain (cf. Spurgeon and Hopkin, 1996; Stephenson et al., 1997) and becomes available to higher trophic levels. This was clearly demonstrated for Cd accumulation by the aquatic oligochaete *L. hoffmeisteri* (Wallace et al., 1998) and for diazinon accumulation in *L. terrestris* (Stephenson et al., 1997).

3.3. Reproduction

Many stressors affecting sublethal parameters in worms will also affect the reproduction capability, either directly or as a trade-off of defence mechanisms (e.g. Morgan et al., 1999b). Reproduction is a rather easy and straightforward measure, and has generally been proven to be more sensitive to environmental stress than mortality (Spurgeon et al., 1994). Therefore, the reproductive potential of worms is often regarded as a good bioindicator both for risk assessment and monitoring of remediation measures. Moreover, there is much information available from standard laboratory toxicity tests with specific toxicants, allowing predictions to be made regarding their effects under field conditions. Achazi et al. (1997) found that in PAH contaminated soil effects on reproduction in enchytraeids occurred at clearly lower levels than effects on mortality. Reduced cocoon production in *E. fetida* when exposed to heavy metals such as Cu, Pb, Ni, Zn and especially Cd has repeatedly been reported (Neuhauser et al., 1984; Spurgeon et al., 1994). It also appeared that worms only partly recovered when transferred to control soil (Neuhauser et al., 1984). Reduced cocoon production and fertility was also found in Cd-resistant *E. fetida* (Reinecke et al., 1999). Fischer and Molnár (1997) conducted a series of experiments with *E. fetida*, in which the worms were exposed to sublethal concentrations of a range of metal chlorides. They found that in all cases reproduction (cocoon production) was retarded, whereas growth of the worms was only affected by Co, Cu, Ni, Li and Tl.

3.4. Growth

As reproduction, growth may be regarded a more sensitive indicator than abundance. Neuhauser et al. (1984) found clear growth reduction in *E. fetida* when exposed to Cd, Cu, Pb, Ni and Zn. After transfer to control soil the effect disappeared, suggesting this to be an adequate bioindicator for metal stress. Bogomolov et al. (1996) showed for *A. tuberculata* that growth was retarded at a Cu-concentration of 200 mg kg^{-1}, whereas acute toxicity occurred only at 800 mg kg^{-1}. There also may be strong species-specific differences, as Streit (1984) found a much higher sensitivity in *Octolasion cyaneum*. Eason et al. (1999) found reduced growth in *E. andrei* when exposed to chlorpyrifos and BaP at low doses, growth being a much more sensitive parameter than mortality. From a comparison of 20 bioassays with Cd, phenol, PCP and trifluralin Bierkens et al. (1998) reported earthworm weight loss (with *E. fetida*) to be among the most sensitive.

3.5. Behaviour

Changes in behaviour may be used to quantify effects of stress on individuals and populations. Avoidance behaviour of oligochaetes towards xenobiotics have repeatedly been reported (e.g. Achazi et al., 1999; Gunn and Sad, 1994; Salminen and

Sulkava, 1996; Yeardley et al., 1996) and avoidance may possibly be used as an indicator in the assessment of soil toxicity. A limitation would be that an effect might only be measured with irritating substances (Yeardley et al., 1996). In experiments by Slimak (1997) with 10 pesticides and *L. terrestris*, a consistent and dose-related response was observed, suggesting good applicability for this parameter. However, Hodge et al. (2000) found no avoidance response in *A. caliginosa* towards the organophosphate pesticides chlorpyrifos and diazinon, neither in laboratory, microcosm or field tests, indicating a limited potential of avoidance as a bioindicator. In a forest soil contaminated with Cu and Ni Salminen and Haimi (1999) found no clear relation between metal contamination and the abundance of *C. sphagnetorum*, but they suggested that this may have been caused by the worms avoiding contaminated patches. Eason et al. (1999) found changes in behaviour (hypersensitivity and coiling) in *E. andrei* when exposed to chlorpyrifos and BaP at low doses. Achazi et al. (1999) reported that tests with *E. crypticus* and PAHs were more sensitive with avoidance than with reproduction as an endpoint.

3.6. Morphology

Some oligochaete species, notably Tubificidae (Prygiel et al., 2000), possess a morphological response to environmental stress; the proportion of worms without hair setae being positively related to the effect of micro-pollution. This could become a simple and fast type of bioindicator for aquatic ecosystems, the more so because the determination of morphology may indicate the type of stress involved (Prygiel et al., 2000). Unfortunately, such indicators are not known in terrestrial worms.

Sjögren et al. (1995) suggested for the enchytraeid *C. sphagnetorum* that autotomy of posterior segments in which metals have accumulated, followed by regeneration, might act as an effective detoxification mechanism. Clear indications that this may be a more common strategy were presented by Nakamura and Shiraishi (1999) for nickel intoxication in *E. buchholzi*, by Lucan-Bouché et al. (1999) for Cu and Pb in *T. tubifex* and by Bouché et al. (2000) for Cd in the same species. Bouché et al. (2000) proposed the use of autotomy as a sublethal endpoint in *T. tubifex*, as they found that the response was dose-related. In view of the more common occurrence of this detoxification mechanism this may be worth exploring.

4. The community and ecosystem level

4.1. Community composition

Graefe (1994, 1997) defined biological soil quality as a condition of the soil that emanates from the interaction between all soil organisms, which is equivalent to the composition and activity of the soil community. As it is virtually impossible to describe this complex system completely, an indicator should be used consisting of a part of the soil community that exhibits sufficient differentiation. For this, he proposed to make use of the composition of annelid communities, as these comprise both earthworms as key species in ecosystem functioning and smaller annelids (in terrestrial ecosystems mainly enchytraeids) that by their higher abundance and diversity allow a more sensitive differentiation between situations. Applying an adaptation of the system of plant indicator values (Ellenberg, 1979) to annelids, both a qualitative and quantitative description of the annelid community as an indicator was possible (e.g. Graefe 1997, 1998, 1999, Graefe and Belotti, 1999, Graefe et al., 1998). Springett et al. (1996) found comparable responses in annelid communities in New Zealand soils. Likewise, Irmler (1999) was able to characterize earthworm communities from various habitats based on the abiotic soil parameters moisture, pH, calcium, carbon and C:N. Such data may be used for the development of an indicator system because they allow, at least in principle, a comparison of the potentially occurring community with the actual one (cf. Irmler, 2000).

Yet, because of the inherent complexity in field studies of oligochaete communities as regards the relations between community composition, soil characteristics and management practices it is often problematic, if not impossible, to ascribe changes recorded to any particular factor or factors (e.g. Tarrant et al., 1997).

4.2. Ecosystem functioning

From a societal point of view the most important parameter measured by any monitoring programme would be the sustainability of ecosystems. After all, the future of society depends on the functioning of the worlds' ecosystems, providing food, oxygen and drinking water. To meet this need two main approaches are currently being developed. The first of these makes use of micro- or mesocosms with multi-species communities, the second attempts to monitor the performance of ecosystems directly in the field. Both these approaches also employ oligochaetes, which is not surprising in view of the fact that they may be considered a key group in ecosystem functioning. Therefore, any effects on e.g. burrowing activity of earthworms may eventually have profound effects on the system as a whole. Apart from that, a stress factor negatively affecting oligochaete species may have an adverse effect on other organisms as well, resulting in changed ecosystem performance.

Parmelee et al. (1993) used microcosms with field communities of soil invertebrates to evaluate the effects of various chemical stressors on the soil food web structure. They found clear indications that direct effects on one trophic category (in this case predatory nematodes) produced indirect effects on another (herbivorous nematodes), indicating also effects on ecosystem functioning. Salminen et al. (1995, 1996) and Salminen and Haimi (1996, 1997, 1998) performed a series of experiments in which microcosms with field soil and more or less complex communities (including enchytraeid worms) were exposed to various types of chemical stress. Besides the effects on microbes and fauna they also measured effects on higher-level processes, using soil respiration, ATP-content, nutrient leaching and primary production as parameters. PCP, for instance, negatively affected both the abundance of *C. sphagnetorum* and primary production in a microcosm experiment (Salminen and Haimi, 1996). It was also found that effects on faunal elements and on soil processes did not necessarily coincide and could occur at different moments during the experiments, as illustrated in Figure 3. This could indicate that measurement of effects on the level of ecosystem processes may provide useful information that cannot be obtained through

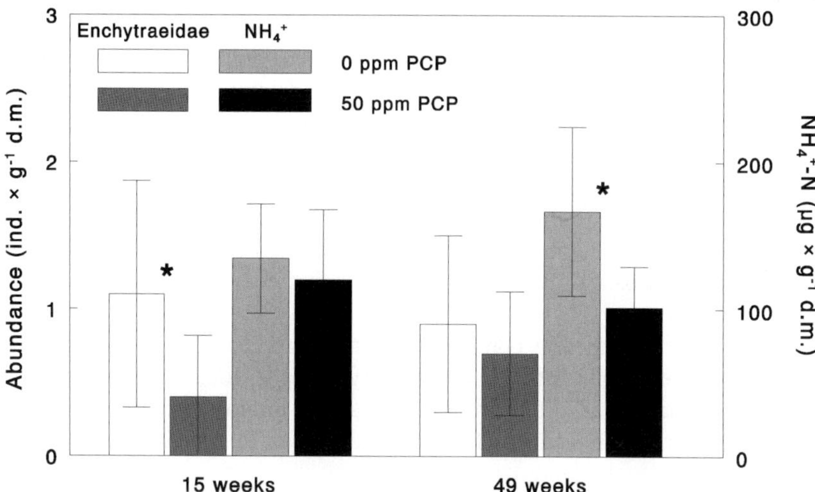

Figure 3. The effect of PCP on abundance of Enchytraeidae and nitrogen mineralization in a lysimeter experiment performed in the field. Significant differences between two treatments denoted by an asterisk above the bars. Data from Salminen and Haimi, 1997.

measurements on lower levels. Yet, as long as the link between lower and higher-level effects remains obscure, they are difficult to use for predictions.

For monitoring of the performance of ecosystems in the field it is important to identify key ecosystem processes, because the inherent complexity of any ecosystem makes it impossible to monitor all processes and organisms involved. Schouten et al. (2000) started the development of a biological indicator system of soil ecosystem processes that was intended to produce an integrated view of the ecological state of the soil relative to a desired situation. They termed such a key process "life support function" (LSF) and selected a number of these as a starting point for the development of the system. Oligochaetes (earthworms and enchytraeids) were considered to act as an important functional group in several of these LSFs, viz.:

- Fragmentation and degradation of organic material.
- Recycling of nutrients.
- Soil structure evolution (bioturbation and aggregate formation).

Among others, diversity and abundance of oligochaetes were used as indicative variables in the indicator system. A pilot project involving grasslands and horticultural farms (Schouten et al., 1999) demonstrated that these variables were clearly discriminative between soil types and land-use forms, and that it was possible to use the measured variables to create an "indicator ruler" that distinguished between land-use intensities.

Important issues in this approach are, of course, the exact nature of the link between the measured variables and the LSFs on the one hand, and the establishment of a desired situation that may serve as a reference.

5. Conclusions

It is evident that oligochaetes may be used profitably as indicator and monitoring organisms. On all organizational levels they may provide bioindicators and -monitors, a number of which have been proven to be suitable. There also are, of course, a number of practical and theoretical questions that have to be solved to make full use of their potential. The most important of these is the problem of the link between processes on the physiological and individual level, and the ecosystem level. Knowledge of the exact nature of this link may on the one hand open the possibility to apply fast, low level monitoring techniques with a high ecological relevance, and may on the other hand help in the interpretation of complex field surveys.

Because there will, however, never be a single ideal indicator organism or test procedure (Bierkens et al., 1998), it will remain inevitable that tiered test-systems should be used, preferably including various types of organisms on various levels of biological organization. Which type of test or monitoring programme should be used is, of course, dependent on the aim of the research in question: if the focus is on assessment of soil or sediment quality in large scale surveys use must be made of indicators on higher organizational levels, whereas questions on the suitability of a certain soil may best be answered applying lower level methods.

References

Abdul Rida, A.M.M., Bouché, M.B., 1994. A method to assess chemical biorisks in terrestrial ecosystems. In: Donker, M.H., Eijsackers, H., Heimbach, F. (Eds), Ecotoxicology of Soil Organisms. Lewis, Boca Raton, pp. 383–394.

Abdul Rida, A.M.M., Bouché, M.B., 1995a. Earthworm contribution to ecotoxicological assessments. Acta Zool. Fennica 196, 307–310.

Abdul Rida, A.M.M., Bouché, M.B., 1995b. The eradication of an earthworm genus by heavy metals in southern France. Appl. Soil Ecol. 2, 45–52.

Abdul Rida, A.M.M., Bouché, M.B., 1997. Earthworm toxicology: from acute to chronic tests. Soil Biol. Biochem. 29, 699–703.

Achazi, R.K., Chroszcz, G., Mierke, W., 1997. Standardization of test methods with terrestrial invertebrates for assessing remediation procedures for contaminated soils. ECO-Informa 12, 284–289.

Achazi, R.K., Fröhlich, E., Henneken, M., Pilz, C., 1999. The effect of soil from former irrigation fields and of sewage sludge on dispersal activity and colonizing success of the annelid *Enchytraeus crypticus* (Enchytraeidae, Oligochaeta). Newsletter on Enchytraeidae 6, 117–126.

Aziz, N.A., Morgan, A.J., Kille, P., 1999. Metal resistance in earthworms; genetic adaptation or physiological acclimation. Pedobiologia 43, 594–601.

Belfroid, A.C., Seinen, W., van Gestel, C.A.M., Hermens, J.L.M., van Leeuwen, C.J., 1995. Modelling the accumulation of hydrophobic organic chemicals in earthworms – application of the equilibrium partitioning theory. ESPR – Environ. Sci. and Pollut. Res. 2, 5–15.

Bierkens, J., Klein, G., Corbisier, P., van den Heuvel, R., Verschaeve, L., Weltens, R., Schoeters, G., 1998. Comparative sensitivity of 20 bioassays for soil quality. Chemosphere 37, 2935–2947.

Bogomolov, D.M., Chen, S.K., Parmelee, R.W., Subler, S., Edwards, C.A., 1996. An ecosystem approach to soil toxicity testing: a study of copper contamination in laboratory soil microcosms. Appl. Soil Ecol. 4, 95–105.

Booth, L.H., Heppelthwaite, V., McGlinchy, A., 2000. The effect of environmental parameters on growth, cholinesterase activity and glutathione S-transferase activity in the earthworm (*Aporrectodea caliginosa*). Biomarkers 5, 46–55.

Booth, L.H., Hodge, S., O'Halloran, K., 2000. Use of cholinesterase in *Aporrectodea caliginosa* (Oligochaeta; Lumbricidae) to detect organophosphate contamination: comparison of laboratory tests, mesocosms, and field studies. Environ. Toxicol. Chem. 19, 417–422.

Bouché, M.-L., Habets, F., Biagianti-Risbourg, S., Vernet, G., 2000. Toxic effects and bioaccumulation of cadmium in the aquatic oligochaete *Tubifex tubifex*. Ecotoxicol. Environ. Safety 46, 246–251.

Brousseau, P., Fugère, N., Bernier, J., Coderre, D., Nadeau, D., Poirier, G., Fournier, M., 1997. Evaluation of earthworm exposure to contaminated soil by cytometric assay of coelomocytes phagocytosis in *Lumbricus terrestris* (Oligochaeta). Soil Biol. Biochem. 29, 681–684.

Bunn, K.E., Thompson, H.M., Tarrant, K.A., 1996. Effects of agrochemicals on the immune systems of earthworms. Bull. Environ. Contam. Toxicol. 57, 632–639.

Callahan, C.A., Shirazi, M.A., Neuhauser, E.F., 1994. Comparative toxicity of chemicals to earthworms. Environ. Toxicol. Chem. 13, 291–298.

Carter, A., Hayes, E.A., Laukulich, L.M., 1980. Earthworms as biological monitors of changes in heavy metal levels in an agricultural soil in British Columbia. In: Dindal, D.L. (Ed.), Soil Biology as Related to Land Use Practices. EPA, Washington, pp. 344–357.

Conder, J.M., Lanno, R.P., 2000. Evaluation of surrogate measures of cadmium, lead, and zinc bioavailability to *Eisenia fetida*. Chemosphere 41, 1659–1668.

Conrad, A.U., Comber, S.D., Simkiss, K., 2000. New method for the assessment of contaminant uptake routes in the oligochaete *Lumbriculus variegatus*. Bull. Environ. Contam. Toxicol. 65, 16–21.

Cortet, J., Gomot-De Vauflery, A., Poinsot-Balaguer, N., Gomot, L., Texier, C., Cluzeau, D., 1999. The use of invertebrate soil fauna in monitoring pollutant effects. Eur. J. Soil Biol. 35, 115–134.

Didden, W., Römbke, J., 2001. Enchytraeids as indicator organisms for chemical stress in terrestrial ecosystems. Ecotoxicol. Environ. Safety 50, 25–43.

Drewes, C.D., 1997. Sublethal effects of environmental toxicants on oligochaete escape reflexes. Amer. Zool. 37, 346–353.

Eason, C.T., Svendsen, C., O' Halloran, K., Weeks, J.M., 1999. An assessment of the lysosomal neutral red retention test and immune function assay in earthworms (*Eisenia andrei*) following exposure to chlorpyrifos, benzo-a-pyrene (BaP), and contaminated soil. Pedobiologia 43, 641–645.

Edwards, C.A., Bohlen, P.J., 1996. Biology and ecology of earthworms. Chapman & Hall, London.

Ellenberg, H., 1979. Zeigerwerte der Gefäßpflanzen Mitteleuropas. 2. Aufl.. Scripta Geobotanica 9, 1–122.

Emmerling, C., Krause, K. and Schröder, D., 1996. Regenwürmer als Bioindikatoren für Schwermetallbelastungen von Böden unter Freilandbedingungen. Z. Pflanzenern. Bodenkd. 160, 33–39.

Fischer, E., Molnár, L., 1992. Environmental aspects of the chloragogenous tissue of earthworms. Soil Biol. Biochem. 24, 1723–1727.

Fischer, E., Molnár, L., 1997. Growth and reproduction of *Eisenia fetida* (Oligochaeta, Lumbricidae) in semi-natural soil containing various metal chlorides. Soil Biol. Biochem. 29, 667–670.

Gevao, B., Semple, K.T., Jones, K.C., 2000. Bound pesticide residues in soils: a review. Environ. Pollut. 108, 3–14.

Giggleman, M.A., Fitzpatrick, L.C., Goven, A.J., Venables, B.J., 1998. Effects of pentachlorophenol on survival of earthworms (*Lumbricus terrestris*) and phagocytosis by their immunoactive coelomocytes. Environ. Toxicol. Chem. 17, 2391–2394.

Goven, A.J., Chen, S.C., Fitzpatrick, L.C., Venables, B.J., 1994a. Lysozyme activity in earthworm (*Lumbricus terrestris*) coelomic fluid and coelomocytes – enzyme assay for immunotoxicity of xenobiotics. Environ. Toxicol. Chem. 13, 607–613.

Goven, A.J., Fitzpatrick, L.C., Venables, B.J., 1994b. Chemical toxicity and host defense in earthworms – an invertebrate model. Ann. N. Y. Acad. Sci. 712, 280–300.

Graefe, U., 1994. Gibt es bodentyp-spezifische Tiergesellschaften? Mitteilgn. Dtsch. Bodenkundl. Gesellsch. 75, 11–14.

Graefe, U., 1997. Bodenorganismen als Indikatoren des biologischen Bodenzustandes. Mitteilgn. Dtsch. Bodenkundl. Gesellsch. 85, 687–690.

Graefe, U., 1998. Annelidenzönosen nasser Böden und ihre Einordnung in Zersetzergesellschaften. Mitteilgn. Dtsch. Bodenkundl. Gesellsch. 88, 109–112.

Graefe, U., 1999. Die Empfindlichkeit von Bodenbiozönosen gegenüber Änderungen der Bodennutzung. Mitteilgn. Dtsch. Bodenkundl. Gesellsch. 91, 609–612.

Graefe, U., Belotti, E., 1999. Strukturmerkmale der Bodenbiozönose als Grundlage für ein natürliches System der Humusformen. Mitteilgn. Dtsch. Bodenkundl. Gesellsch. 89, 181–184.

Graefe, U., Elsner, D.-C., Necker, U., 1998. Monitoring auf Boden-Dauerbeobachtungsflächen: bodenzoologische Parameter zur Kennzeichnung des biologischen Bodenzustandes. Mitteilgn. Dtsch. Bodenkundl. Gesellsch. 87, 343–346.

Gunn, A., Sadd, J.W., 1994. The effect of ivermectin on the survival, behaviour and cocoon production of the earthworm *Eisenia fetida*. Pedobiologia 38, 327–333.

Hagens, M., Westheide, W., 1987. Subletale Schädigungen bei *Enchytraeus minutus* (Oligochaeta, Annelida) durch das Insektizid Parathion: Veränderungen in der Ultrastruktur von Chloragog- und Darmzellen in Abhängigkeit von der Belastungsdauer. Verh. Ges. Ökol. 16, 423–426.

Haimi, J., 2000. Decomposer animals and bioremediation of soils. Environ. Pollut. 107, 233–238.

Heck, M., Rink, U., Weigmann, G., 1995. Blei- und Cadmiumbelastung von Bodentieren in einem immissionsbeeinflußten Forst in der Nähe von Berlin. Z. Ökologie u. Naturschutz 4, 75–85.

Hodge, S., Webster, K.M., Booth, L., Hepplethwaite, V., O' Halloran, K., 2000. Non-avoidance of organophosphate insecticides by the earthworm *Aporrectodea caliginosa* (Lumbricidae). Soil Biol. Biochem. 32, 425–428.

Irmler. U., 1999. Die standörtlichen Bedingungen der Regenwürmer (Lumbricidae) in Schleswig-Holstein. Faun.-Ökol. Mitt. 7, 509–518.

Irmler. U., 2000. Environmental characteristics of ground beetle assemblages in northern German forests as basis for an expert system. Z. Ökologie u. Naturschutz 8, 227–237.

Kelsey, J.W., Kottler, B.D., Alexander, M., 1997. Selective chemical extractants to predict bioavailability of soil aged organic chemicals. Environ. Sc. Techn. 31, 214–217.

Kille, P., Stürzenbaum, S.R., Galay, M., Winters, C., Morgan, A.J., 1999. Molecular diagnosis of pollution impact in earthworms – towards integrated biomonitoring. Pedobiologia 43, 602–607.

Koeckritz, T., Irmler, U., Weppen, P., 1999. Schwermetallbelastung von *Aporrectodea caliginosa* (Oligochaeta, Lumbricidae) im Raum Kiel. J. Plant. Nutr. Soil. Sci. 162, 477–482.

LABO (Ad-hoc-Arbeitsgruppe Boden-Dauerbeobachtung der Bund/Länder-Arbeitsgemeinschaft Bodenschutz), 1998: Boden-Dauerbeobachtung – Einrichtung und Betrieb von Boden-Dauerbeobachtungsflächen.

Lancaster, J., 2000. The ridiculous notion of assessing ecological health and identifying the useful concepts underneath. Hum. Ecol. Risk Assess. 6, 213–222.

Langdon, C.J., Piearce, T.G., Black, S., Semple, K.T., 1999. Resistance to arsenic-toxicity in a population of the earthworm *Lumbricus rubellus*. Soil Biol. Biochem. 31, 1963–1967.

Leppänen, M.T., Kukkonen, J.V.K., 1998. Relative importance of ingested sediment and pore water as bioaccumulation routes for pyrene to oligochaete (*Lumbriculus variegatus*, Müller). Environ. Sci. Technol. 32, 1503–1508.

Lucan-Bouché, M.L., Biagianti-Risbourg, S., Arsac, F., Vernet, G., 1999. An original decontamination process developed by the aquatic oligochaete *Tubifex tubifex* exposed to copper and lead. Aquat. Toxicol. 45, 9–17.

Ma, W.C., 1982. The influence of soil properties and worm-related factors on the concentration of heavy metals in earthworms. Pedobiologia 24, 109–119.

Ma, W.C., 1984. Sublethal toxic effects of copper on growth, reproduction and litter breakdown activity in the earthworm *Lumbricus rubellus*, with observations of the influence of temperature and soil pH. Environ. Pollut. A 33, 207–219.

Ma, W.C., Edelmann, T., van Beersum, I., Jans, T., 1983. Uptake of cadmium, zinc, lead, and copper by earthworms near a zinc-smelting complex: influence of soil pH and organic matter. Bull. Environ. Contam. Toxicol. 30, 424–427.

Mariño, F., Stürzenbaum, S.R., Kille, P., Morgan, A.J., 1998. Cu-Cd interactions in earthworms maintained in laboratory microcosms: the examination of a putative copper paradox. Comp. Biochem. Physiol. Part C 120, 217–223.

Mariño, F., Winters, C., Morgan, A.J., 1999. Heat shock protein (hsp60, hsp70, hsp90) expression in earthworms exposed to metal stressors in the field and laboratory. Pedobiologia 43, 615–624.

Marinussen, M.P.J.C., 1997. Heavy metal accumulation in earthworms exposed to spatially variable soil contamination. Ph.D. Thesis, Agricultural University, Wageningen.

Martikainen, E., Haimi, J., Ahtiainen, J., 1998. Effects of dimethoate and benomyl on soil organisms and soil processes – a microcosm study. Appl. Soil Ecol. 9, 381–387.

Markert, B., Wappelhorst, O., Weckert, V., Herpin, U., Siewers, U., Friese, K., Breulmann, G., 1999. The use of bioindicators for monitoring the heavy-metal status of the environment. J. Radioanal. Nucl. Chem. 240, 425–429.

Morgan, J.E., Morgan, A.J., 1993. Seasonal changes in the tissue-metal (Cd, Zn and Pb) concentrations in 2 ecophysiologically dissimilar earthworm species: pollution-monitoring implications. Environmental Pollution 82, 1–7.

Morgan, A.J., Stürzenbaum, S.R., Kille, P., 1999a. A short overview of molecular biomarker strategies with particular regard to recent developments in earthworms. Pedobiologia 43, 574–584.

Morgan, A.J., Stürzenbaum, S.R., Winters, C., Kille, P., 1999b. Cellular and molecular aspects of metal sequestration and toxicity in earthworms. Invert. Reprod. Devel. 36, 17–24.

Moser, T., Förster, B., Römbke, J., 1999. Overview on the use of enchytraeidae in terrestrial model ecosystems (or "microcosm") investigations. Newsletter on Enchytraeidae 6, 111–116.

Nakamura, Y., Shiraishi, H., 1999. The nodule nickel in posterior segments of *Enchytraeus buchholzi* (Enchytraeidae: Oligochaeta). Edaphologia 62, 93–96.

Neuhauser, E.F., Malecki, M.R., Loehr, R.C., 1984. Growth and reproduction of the earthworm *Eisenia fetida* after exposure to sublethal concentrations of metals. Pedobiologia 27, 89–97.

Neuhauser, E.F., Cukic, Z.V., Malecki, M.R., Loehr, R.C., Durkin, P.R., 1995. Bioconcentration and biokinetics of heavy metals in the earthworm. Environ. Pollut. 89, 293–301.

Notenboom, J., Römbke, J., Folkerts, A-J., de Groot, A., 1997. Effects of heavy metals on enchytraeid worms: comparison between observations in the field and laboratory data. Poster, SETAC Amsterdam.

O'Halloran, K., Booth, L.H., Hodge, S., Thomsen, S., Wratten, S.D., 1999. Biomarker responses of the earthworm *Aporrectodea caliginosa* to organophosphates: hierarchical tests. Pedobiologia 43, 646–651.

Paoletti, M.G., 1999. The role of earthworms for assessment of sustainability and as bioindicators. Agr. Ecosyst. Environ. 74, 137–155.

Parmelee, R.W., Wentsel, R.S., Phillips, C.T., Simini, M., Checkai, R.T., 1993. Soil microcosm for testing the effects of chemical pollutants on soil fauna communities and trophic structure. Environ. Toxicol. Chem. 12, 1477–1486.

Prentø, P., 1994. Uptake and long-time storage of natural and synthetic dyes by earthworm chloragocytes. In vivo and in vitro investigations. Comp. Biochem. Physiol. 109A, 805–816.

Prygiel, J., Rosso-Darmet, A., Lafont, M., Lesniak, C., Durbec, A., Ouddane, B., 2000. Use of oligochaete communities for assessment of ecotoxicological risk in fine sediment of rivers and canals of the Artois-Picardie water basin (France). Hydrobiologia 410, 25–37.

Purschke, G., Hagens, M., Westheide, W., 1991. Ultrahistopathology of enchytraeid oligochaetes (Annelida) after exposure to pesticides – a means of identification of sublethal effects? Comp. Biochem. Phys. 100C, 119–122.

Puurtinen, H.M., Martikainen, E.A.T., 1997. Effect of soil moisture on pesticide toxicity to an enchytraeid worm, *Enchytraeus* sp. Arch. Env. Cont. Toxicol. 33, 34–41.

Ramade, F., 1987. Ecotoxicology. John Wiley, Chichester.

Reid, B.J., Jones, K.C., Semple, K.T., 2000. Bioavailability of persistent organic pollutants in soils and sediments – a perspective on mechanisms, consequences and assessment. Environ. Pollut. 108, 103–112.

Reinecke, A.J., 1998. Die ekologiese relevansie van chemiese stres op die "gesondheid" van die biologiese omgewing. S.A. Tydskr. Natuurwet. Tegnol. 17, 142–145.

Reinecke, A.J., Reinecke, S.A., 1998. The use of earthworms in ecotoxicological evaluation and risk assessment: new approaches. In: Edwards, C.A. (Ed.), Earthworm Ecology. St. Lucie, Boca Raton, pp. 273–293.

Reinecke, S.A., Prinsloo, M.W., Reinecke, A.J., 1999. Resistance of *Eisenia fetida* (Oligochaeta) to cadmium after long-term exposure. Ecotoxicol. Environ. Safety 42, 75–80.

Reinecke, S.A., Reinecke, A.J., 1997. The influence of lead and manganese on spermatozoa of *Eisenia fetida* (Oligochaeta). Soil Biol. Biochem. 29, 737–742.

Reinecke, S.A., Reinecke, A.J., 1999. Lysosomal response of earthworm coelomocytes induced by long-term experimental exposure to heavy metals. Pedobiologia 43, 585–593.

Reinecke, S.A., Reinecke, A.J., Froneman, M.L., 1995. The effects of dieldrin on the sperm ultrastructure of the earthworm *Eudrilus eugeniae* (Oligochaeta). Environ. Toxicol. Chem. 14, 961–965.

Römbke, J., Knacker, Th., Förster, B., Marcinkowski, A., 1994. Comparison of effects of two pesticides on soil organisms in laboratory tests, microcosms, and in the field. In: Donker, M.H., Eijsackers, H., Heimbach, F. (Eds), Ecotoxicology of Soil Organisms. Lewis, Boca Raton, pp. 229–240.

Roth, M., 1993. Investigations on lead in the soil invertebrates of a forest ecosystem. Pedobiologia 37, 270–279.

Roth-Holzapfel, M., Funke, W., 1989. Elementflüsse im Ökosystem "Fichtenforst" unter besonderer Berücksichtigung wirbelloser Tiere. KfK-PEF 50, 93–103.

Šalagovič, J., Gilles, J. Verschaeve, L., Kalina, I., 1996. The comet assay for the detection of genotoxic damage in the earthworms: a promising tool for asessing the biological hazards of polluted sites. Folia Biol. (Praha) 42, 17–21.

Salminen, J., Eriksson, I., Haimi, J., 1996. Effects of terbuthylazine on soil fauna and decomposition processes. Ecotoxicol. Environ. Safety 34, 184–189.

Salminen, J., Haimi, J., 1996. Effects of pentachlorophenol in forest soil: a microcosm experiment for testing ecosystem responses to anthropogenic stress. Biol. Fertil. Soils 23, 182–188.

Salminen, J., Haimi, J., 1997. Effects of pentachlorophenol on soil organisms and decomposition in forest soil. J. Appl. Ecol. 34, 101–110.

Salminen, J., Haimi, J., 1998. Responses of the soil decomposer community and decomposition processes to the combined stress of pentachlorophenol and acid precipitation. Appl. Soil Ecol. 9, 475–481.

Salminen, J., Haimi, J., 1999. Horizontal distribution of copper, nickel and enchytraeid worms in polluted soil. Environ. Pollut. 104, 351–358.

Salminen, J., Haimi, J., Sironen, A., Ahtiainen, J., 1995. Effects of pentachlorophenol and biotic interactions on soil fauna and decomposition in humus soil. Ecotoxicol. Environ. Safety 31, 250–257.

Salminen, J., Sulkava, P., 1996. Distribution of soil animals in patchily contaminated soil. Soil Biol. Biochem. 28, 1349–1355.

Sayles, G.D., Acheson, C.M., Kupferle, M.J., Shan, Y., Zhou, Q., Meier, J.R., Chang, L., Brenner, R.C., 1999. Land treatment of PAH contaminated soil: performance measured by chemical and toxicity assays. Environ. Sci. Technol. 33, 4310–4317.

Schouten, T., Bloem, J., Didden, W.A.M., Rutgers, M., Siepel, H., Posthuma, L., Breure, A.M., 2000. Development of a biological indicator for soil quality. SETAC Globe 1, 30–32.

Schouten, A.J., Breure, A.M., Bloem, J., Didden, W., de Ruiter, P.C., Siepel, H., 1999. Life Support Functions of the Soil: Operationalization for the Policy. RIVM Report 607601003 National Institute of Public Health and the Environment, Bilthoven (in Dutch with English summary).

Sheedy, B.R., Mattson, V.R., Cox, J.S., Kosian, P.A., Phipps, G.L., Ankley, G.T., 1998. Bioconcentration of polycyclic aromatic hydrocarbons by the freshwater oligochaete *Lumbriculus variegatus*. Chemosphere 36, 3061–3070.

Sheppard, S.C., Evenden, W.G., Cornwell, T.C., 1997. Depuration and uptake kinetics of I, Cs, Mn, Zn and Cd by the earthworm (*Lumbricus terrestris*) in radiotracer-spiked litter. Environ. Toxicol. Chem. 16, 2106–2112.

Sijm, D., Kraaij, R., Belfroid, A., 2000. Bioavailability in soil or sediment: exposure of different organisms and approaches to study it. Environ. Pollut. 108, 113–119.

Sjögren, M., Augustsson, A., Rundgren, S., 1995. Dispersal and fragmentation of the enchytraeid *Cognettia sphagnetorum* in metal polluted soil. Pedobiologia 39, 207–218.

Slimak, K.M., 1997. Avoidance response as a sublethal effect of pesticides on *Lumbricus terrestris* (Oligochaeta). Soil Biol. Biochem. 29, 713–715.

Springett, J.A., Gray, R.A.J., Bakker, L., 1996. Influence of agriculture on Enchytraeidae fauna of soils in the south-west of the North Island of New Zealand. Pedobiologia 40, 461–466.

Spurgeon, D.J., Hopkin, S.P., 1996. Risk assessment of the threat of secondary poisoning by metals to predators of earthworms in the vicinity of a primary smelting works. Sci. Total Envir. 187, 167–183.

Spurgeon, D.J., Hopkin, S.P., 2000. The development of genetically inherited resistance to zinc in laboratory-selected generations of the earthworm *Eisenia fetida*. Environ. Pollut. 109, 193–201.

Spurgeon, D.J., Hopkin, S.P., Jones, D.T., 1994. Effects of cadmium, copper, lead and zinc on growth, reproduction and survival of the earthworm *Eisenia fetida* (Savigny): assessing the environmental impact of point-source metal contamination in terrestrial ecosystems. Environ. Pollut. 84, 123–130.

Spurgeon, D.J., Sandifer R.D., Hopkin, S.P., 1996. The use of macroinvertebrates for population and community monitoring of metal contamination – indicator taxa, effect parameters and the need for a soil invertebrate prediction and classification scheme (SIVPACS). In: van Straalen, N.M., Krivolutsky, D.A. (Eds), Bioindicator Systems for Soil Pollution. Kluwer, Dordrecht, pp. 95–110.

Stenersen, J., 1992. Uptake and metabolism of xenobiotics by earthworms. In: Greig-Smith, P.W., Becker, H., Edwards, P.J., Heimbach, F. (Eds), Ecotoxicology of Earthworms. Intercept, Andover, pp. 129–138.

Stenersen, J., Brekke, E., Engelstad, F., 1992. Earthworms for toxicity testing: species differences in response towards cholinesterase inhibiting insecticides. Soil Biol. Biochem. 24, 1761–1764.

Stephenson, G.L., Wren, C.D., Middelraad, I.C.J., Warner, J.E., 1997. Exposure of the earthworm, *Lumbricus terrestris*, to diazinon, and the relative risk to passerine birds. Soil Biol. Biochem. 29, 717–720.

Streit, B., 1984. Effects of high copper concentrations on soil invertebrates (earthworms and oribatid mites). Oecologia 64, 381–388.

Sturmbauer, C., Opadiya, G.B., Niederstätter, H., Riedmann, A., Dallinger, R., 1999. Mitochondrial DNA reveals cryptic oligochaete species differing in cadmium resistance. Molec. Biol. Evol. 16, 967–974.

Stürzenbaum, S.R., Morgan, A.J., Kille, P., 1999. Characterisation and quantification of earthworm cyclophilins: identification of invariant and heavy metal responsive isoforms. BBA – Gene Struct. Express. 1489, 467–473.

Suzuki, M.M., Cooper, E.L., Eyambe, G.S., Goven, A.J., Fitzpatrick, L.C., Venables, B.J., 1995. Polychlorinated biphenyls (PCBs) depress allogeneic natural cytotoxicity by earthworm coelomocytes. Environ. Toxicol. Chem. 14, 1697–1700.

Svendsen, C., Meharg, A.A., Freestone, P., Weeks, J.M., 1996. Use of an earthworm lysosomal biomarker for the ecological assessment of pollution from an industrial plastics fire. Appl. Soil Ecol. 3, 99–107.

Svendsen, C., Weeks, J.M., 1997a. Relevance and applicability of a simple earthworm biomarker of copper exposure. I. Links to ecological effects in a laboratory study with *Eisenia andrei*. Ecotox. Environ. Saf. 36, 72–79.

Svendsen, C., Weeks, J.M., 1997b. Relevance and applicability of a simple earthworm biomarker of copper exposure. II. Validation and applicability under field conditions in a mesocosm experiment with *Lumbricus rubellus*. Ecotox. Environ. Saf. 36, 80–88.

Tarrant, K.A., Field, S.A., Langton, S.D., Hart, A.D.M., 1997. Effects on earthworm populations of reducing pesticide use in arable crop rotations. Soil Biol. Biochem. 29, 657–661.

Veltz, I., Arsac, F., Biagianti-Risbourg, S., Habets, F., Lechenault, H., Vernet, G., 1996. Effects of platinum (Pt^{4+}) on *Lumbriculus variegatus* Müller (Annelida, Oligochaetae): acute toxicity and bioaccumulation. Arch. Environ. Contam. Toxicol. 31, 63–67.

Ville, P., Roch, P., Cooper, E.L., Masson, P., Narbonne, J.F., 1995. PCBs increase molecular-related activities (lysozyme, antibacterial, hemolysis, proteases) but inhibit macrophage-related functions (phagocytosis, wound healing) in earthworms. J. Invertebr. Pathol. 65, 217–224.

Wallace, W.G., Lopez, G.R., Levinton, J.S., 1998. Cadmium resistance in an oligochaete and its effect on cadmium trophic transfer to an omnivorous shrimp. Mar. Ecol. Progr. Ser. 172, 225–237.

Walsh, P., El Adlouni, C., Mukhopadhyay, M.J., Viel, G., Nadeau, D., Poirier, G.G., 1995. ^{32}P-postlabeling determination of DNA adducts in the earthworm *Lumbricus terrestris* exposed to PAH-contaminated soils. Bull. Environ. Contam. Toxicol. 54, 654–661.

Walsh, P., El Adlouni, C., Nadeau, D., Fournier, M., Coderre, D., Poirier, G.G., 1997. DNA adducts in earthworms exposed to a contaminated soil. Soil Biol. Biochem. 29, 721–724.

Weeks, J.M., Svendsen, C., 1996. Neutral red retention by lysosomes from earthworm (*Lumbricus rubellus*) coelomocytes: a simple biomarker of exposure to soil copper. Environ. Toxicol. Chem. 15, 1801–1805.

Willuhn, J., Otto, A., Koewius, H., Wunderlich, F., 1996a. Subtoxic cadmium-concentrations reduce copper-toxicity in the earthworm *Enchytraeus buchholzi*. Chemosphere 32, 2205–2210.

Willuhn, J., Otto, A., Schmitt-Wrede, H.P., Wunderlich, F., 1996b. Earthworm gene as indicator of bio-efficacious cadmium. Biochem. Biophys. Res. Commun. 220, 581–585.

Willuhn, J., Schmitt-Wrede, H.-P., Otto, A., Wunderlich, F., 1996c. Cadmium-detoxification in the earthworm *Enchytraeus*: specific expression of a putative aldehyde dehydrogenase. Biochem. Biophys. Res. Commun. 226, 128–134.

Yeardley, R.B., Lazorchak, J.M., Gast, L.C., 1996. The potential of an earthworm avoidance test for evaluation of hazardous waste sites. Environ. Toxicol. Chem. 15, 1532–1537.

Zweers, P., 1996. Opname kinetiek van metalen in de potworm *Enchytraeus crypticus* bij blootstelling aan verschillende metaalverontreinigde grond. Studentenverslag Landbouwuniversiteit, vakgroep Terrestrische Oecologie en Natuurbeheer, Wageningen.

Bioindicators and biomonitors
B.A. Markert, A.M. Breure, H.G. Zechmeister, editors
© 2003 Elsevier Science Ltd. All rights reserved.

Chapter 17

Molluscs as bioindicators

Jörg Oehlmann and Ulrike Schulte-Oehlmann

Abstract

The ecological role and importance of molluscs as one of the most species-rich phyla of the animal kingdom are briefly summarised with special emphasis on certain characteristics, which make them especially suited for monitoring programmes in the field. The advantages, perspectives and limitations for the use of terrestrial and aquatic molluscs for the monitoring of chemical stressors in their specific environment are compared. Furthermore, examples of bioaccumulation and biological effect monitoring surveys are given with a differentiation of sub-organism, organism and community level effects for the latter. Finally, the possibilities for monitoring of tributyltin compounds in coastal and freshwater ecosystem are demonstrated as a case study.

Keywords: molluscs, accumulation, biological effects, biomarker, monitoring, indication, endocrine disrupters, tributyltin

1. Introduction: The "molluscan gap" – ecological relevance of molluscs and their role in ecotoxicology

The molluscs represent one of the most diverse and species-rich phyla of the animal kingdom. With more than 130,000 known recent species they are only second to the arthropods (Gruner, 1993; Gruner et al., 1993). Of the seven molluscan classes, gastropods make up more than 80% of the species with bivalves constituting the major part of the rest (15%). The other five classes, in decreasing species numbers, are the Polyplacophora, the highly evolved cephalopods and finally the scaphopods, Aplacophora and Monoplacophora (Table 1). Especially the cephalopods exhibit a comparative degree of physiological and neuronal complexity and organisation as vertebrates. Cephalopods from the genus *Architeuthis* represent with an overall length of more than 20 m the largest living invertebrates, but also the smallest metazoans can be found within the molluscs, some Aplacophora with a size of less than 1 mm.

Although molluscs are basically a marine group of animals, gastropods and bivalves have also expanded their distribution to various freshwater environments. Gastropods have additionally penetrated into a huge variety of terrestrial habitats so that molluscs can be found today from the abysses of the sea to mudflats, from lakes and rivers and their banks to forests, alpine mountains, but also in steppes and desserts; they occur on nearly all latitudes of the planet from polar to tropical temperatures (Hyman, 1967; Purchon, 1968).

Table 1. Overview of the different classes and subclasses within the phylum Mollusca indicating their species numbers and geographical distribution.

Class Subclass	Number of species	Distribution
Aplacophora	250	Exclusively marine benthic organisms; from low tide level to depths > 6000 m
Polyplacophora	1000	Exclusively marine benthic organisms; mainly in the eulittoral zone, but also in depths up to 4000 m
Monoplacophora	20	Exclusively marine benthic organisms in depths between 170 and 6500 m; reported from the Pacific, south Atlantic, the Antarctic region and the Indian Ocean (Gulf of Aden)

Gastropoda
  Prosobranchia
  Pulmonata
  Opisthobranchia
	110,000
	60,000
	44,000
	6000
		Cosmopolitan in all terrestrial, freshwater and marine environments, including steppes, deserts, alpine mountains, polar regions, the deep sea and the pelagic zone

Bivalvia
  Protobranchia
  Filibranchia
  Eulamellibranchia
  Septibranchia
	20,500
	550
	2200
	17,500
	250
		Cosmopolitan in all freshwater and marine environments from the eulittoral to the abyssal zone and from tropical to polar regions

Scaphopoda	350	Exclusively marine cosmopolitans; in sediments from the eulittoral zone to depths of 7000 m
Cephalopoda	760	Exclusively marine cosmopolitans, as benthic or pelagic organisms from surface waters to the deep sea

Due to their ubiquitous distribution and enormous species number, molluscs play important ecological roles in the different aquatic and terrestrial ecosystems of the world. They provide key species for ecosystem functioning, e.g. for litter decomposition but also because they contribute to huge amounts of the biomass on the different trophic levels in ecosystems (from primary consumers to top predators). Many other groups feed on molluscs, like echinoderms, fish, birds and mammals. Molluscs act as vectors for a number of human-relevant parasites and diseases, e.g. as intermediary hosts of trematodes, and a number of mollusc species live as endoparasites themselves (Purchon, 1968; Götting, 1996).

In contrast to their ecological importance, the contribution of molluscs to ecotoxicological research and routine measurements in the laboratory is much smaller. This is especially true for the standard testing of chemicals, where species from other invertebrate groups, like arthropods, mainly insects and crustaceans, but also nematodes and annelids are much more considered. This is particularly due to the fact that life cycles of molluscs are normally longer than of most other invertebrate test species and that the maintenance of healthy mollusc brood stocks requires more technical and thus financial efforts, namely for those aquatic species which have a planktonic larval phase. The exception from this general rule is the field of bioindication and biomonitoring where molluscs have been successfully used to obtain information on the quality of terrestrial, marine and freshwater ecosystems and to quantify the exposure to and effects of contaminants in their environment (Markert et al., 1999). This is particularly the case for the two most diverse classes of molluscs, gastropods and bivalves, while cephalopods play a more secondary role, and representatives from the remaining four classes have not been used as bioindicators according to the published literature.

As canaries already warned miners in the European coal mines of the 19th century of "bad air", methane occurrence and the danger of firedamp explosions, it has long been known that molluscs are indicators of poor water quality. More than 90 years ago, Ortmann (1909) described for streams in North America and Kolkwitz and Marsson (1909) for German surface waters that the lack or decline of freshwater mussel populations is an effect of pollutants. In 1976, the "Mussel Watch" was initiated in the United States of America as one of the first geographical large scale environmental surveillance programmes which made use of living organisms (Goldberg, 1975; Goldberg et al., 1978). The original scheme covered a coordinated and standardised sampling and measurement of pollutants in four bivalve species (two *Mytilus* species, *Crassostrea virginica* and *Ostrea equestris*) at more than 100 sample sites on the coast of North America. The mollusc tissues were analysed for heavy metals, radionuclides, halogenated hydrocarbons, and petroleum hydrocarbons and provided useful data on baseline levels of these substances. The mussel watch programme was adopted by further countries in the following years and was employed almost on a global scale so that information on coastal ecosystem pollution with certain substances is available today worldwide (cf. 3.1).

Molluscs are for a number of reasons well suited as bioindicators or biomonitors. Although these aspects are also shared with other systematic groups of the animal kingdom, it is the unique combination of these different features which characterises molluscs as ideal bioindicators. The most important characteristics are:

- Gastropods and bivalves are widespread and abundant in all marine and freshwater ecosystems worldwide. Additionally, gastropods can also be found in almost all terrestrial environments.
- Although some terrestrial gastropods are endemic with a rather limited distribution, most molluscs, especially those living in the aquatic environment, exhibit a broad distribution within and even between continents, facilitating their use in geographical large scale surveys. Furthermore, a number of species and genera are even cosmopolitans (e.g. mussels of the genus *Mytilus* with the two species *M. edulis* and *M. galloprovincialis* being the most widespread).
- Many molluscs are key species for the functioning of marine, freshwater and terrestrial ecosystems so that it is likely that a pollutant that affects such a mollusc population will also exhibit a negative impact for the entire ecosystem. Examples of pollution effects on the ecosystem level, caused by an interference with mollusc populations, will be provided in the Section 3.3.
- The majority of gastropod and bivalve species exhibit an extremely limited mobility or are completely sessile as adults. Therefore, these molluscs represent

the contamination of their habitat ideally. The only exceptions are pelagic snails from the prosobranch genus *Janthina*, the Heteropoda (*Atlanta, Carinaria, Pterotrachea*) and Pteropoda (*Hyalela, Creseis, Styliola*) (Fioroni, 1981). Most of the aquatic mollusc species, especially in the temperate, subtropical and tropical region, have a planktonic larval stage which guarantees a high dispersal potential and allows a recruitment of populations even in those habitats where sexually mature adults might have become extinct due to the high level of contamination.

- Molluscs represent a broad variety of reproductive modes, like simultaneous and consecutive hermaphroditism, gonochory and parthenogenesis, each of them combined with semelparity or iteroparity so that effects of contaminants affecting specifically these types of reproduction can be monitored. Furthermore, molluscs exhibit an extraordinary variation of life-cycle-strategies, especially with respect to their longevity. While the majority of the cephalopods, the marine opisthobranch snails and most of the freshwater and terrestrial gastropods are short living species with a maximum life span of one year, the marine prosobranch snails and many bivalves are long-living so that they can integrate contaminations of their environment over long time periods. Approximately 40% of the marine bivalve species and more than 20% of the marine prosobranches attain maximum ages of more than 14 years according to Heller (1990). For single species even longer life spans have been reported, like for example more than 50 years for the abalone *Haliotis cracherodii* by Powell and Cummins (1985) and more than 100 years for some marine (120 years for *Panope generosa*, 150 years for *Crenomytilus grayanus* and 220 years for *Arctica islandica* according to Jones, 1983) and freshwater bivalves (116 years for *Margaritifera margaritifera* according to Bauer, 1987).

- Most gastropod and bivalve species used for biomonitoring and bioindication purposes are relatively large and therefore easy to handle. Consequently, they can be used both under laboratory and field conditions, for active and passive biomonitoring.

- Due to the lack of an exoskeleton, as it is present in arthropods, molluscs are in direct contact with the ambient medium (water or soil). Therefore, chemicals can be taken up not only from the diet (via the gastro-intestinal tract) but also additionally from ambient water or soil via the integument, including the respiratory organs in aquatic species, resulting in a greater accumulation potency for contaminants.

- Compared with other invertebrate groups like arthropods and especially vertebrates, molluscs exhibit only a limited ability to excrete pollutants directly via their kidneys or other excretory organs and tissues, to metabolise organic chemicals, and physiologically to inactivate toxic heavy metals, e.g. by the formation of and binding to metallothioneins (Lee, 1985; Berger et al., 1995a; Legierse et al., 1998). As a consequence, molluscs attain higher bioaccumulation or bioconcentration factors for many toxicants than other systematic groups. Therefore, pollutants might exhibit negative impacts on molluscs at lower environmental concentrations than on other invertebrates or vertebrates, facilitating their use as a kind of ecological early warning system.

- The high sensitivity of molluscs to environmental chemicals is also represented by the fact that they contribute in a disproportionate extent to the "red lists" of endangered wildlife species worldwide. Recent data from Germany indicate that not less than 204 (= 61%) of the 333 occurring freshwater and terrestrial mollusc species are rated as threatened or already extinct (Jungbluth and von Knorre, 1995). Similar data are only available for Sweden and Madeira, where 25 of 133 and 72 of 190 terrestrial gastropods are endangered or already exterminated (Waldén, 1986). Unfortunately, for the marine and coastal environment no comparable surveys have been made in the past but there are numerous reports of population declines for a number of marine molluscan species which have attained a specific economic or scientific interest. For some of these reports it has been shown that environmental chemicals are the causative agent (cf. 3.4).

- The internal organisation, especially the normal morphological and histological structure of the different organs and tissues, and the physiology of the most abundant gastropod and bivalve species used for biomonitoring is characterised quite well. Our knowledge on the biology and ecology of these species has improved considerably in recent decades.
- Consequently, biological effects of environmental stress in general and of contaminant exposure in particular are measurable at various levels of biological organisation (from molecules to communities).
- Molluscs are non-controversial as organisms for ecotoxicological research, especially as test animals and for environmental monitoring.

2. Terrestrial bioindication and biomonitoring with molluscs

Gastropods represent the only molluscan class in terrestrial ecosystems and consequently, snails are the only molluscs which can be used for bioindication and biomonitoring purposes in these environments. Most terrestrial gastropods belong to the class pulmonates (order Stylommatophora), but also the members of the two prosobranch subtropical and tropical families Hydrocenidae and Helicinidae and the highly endangered European littorinid snail *Pomatias elegans* are living in terrestrial environments (Gruner, 1993).

For the time being, molluscs play clearly a secondary role in the surveillance of terrestrial ecosystems when compared with lichens, bryophytes, vascular plants and other invertebrate groups like nematodes, annelids and arthropods. Nevertheless, biomonitoring attempts with snails have found an increasing interest during the last decade and a number of promising projects have already been conducted. Most of them made use of the snails' bioaccumulation potential for metals and organic contaminants, but there are also examples for surveys which assessed the biological effects of soil contaminants on different levels of biological organisation.

2.1. Bioaccumulation

A number of invertebrate species are known to be efficient accumulators of trace elements (Dallinger, 1994). Generally, metal accumulation by such organisms is favoured by their limited ability to excrete these contaminants directly after their uptake and also by efficient physiological inactivation mechanisms, such as intracellular compartmentalisation, or metal inactivation by binding to metallothioneins (cf. 2.2). Such biological accumulators have often been used as accumulation indicators of environmental metal pollution. Ideally, metal concentrations in the animal's body reflect environmental pollution levels quantitatively. In reality, however, many factors like the nutritional, physiological and reproductive status, the sex and age of the animals influence such quantitative relationships. Therefore, these factors have to be considered carefully before invertebrates in general and particularly terrestrial snails can be utilised as accumulation indicator for metal pollution. Coughtrey and Martin (1977) compared the concentrations of Cd, Cu, Pb and Zn in the garden snail *Helix aspersa* collected from sites of varying degree of metal contamination. Already in this early study the authors found a positive linear relationship between metal uptake and body weight and concluded that molluscs of similar weight and/or size should be used for monitoring purposes. The different patterns of metal uptake by different organs were interpreted as a result of the organ-specific physiological activity. Even today with our more detailed knowledge on the physiology of pollutant accumulation and the role of detoxification enzymes and metallothioneins (cf. 2.2), the demand of Coughtrey and Martin (1977) for a comprehensive study of the physiological aspects of heavy metal uptake in molluscs is still valid.

There are numerous reports on the use of pulmonate snails as accumulation bioindicators so that only a limited number of examples can be presented here. A common characteristic of all these programmes is that their geographical focus is local or regional at its best, but that no attempts have been made so far to perform larger geographical surveys. One of the few exceptions is the use of slugs (*Arion rufus, A. ater, Limax cinereoniger, L. maximus*) within the "Ecological Effect Cadaster Baden-Württemberg" in Germany, which is designed as a long range monitoring programme and now running since several years (Spang, 1995). Adult snails of comparable size are sampled in autumn each year at an extended number of permanent surveillance sites all over the country with a sample size of 5 to 10 specimens. After a defecation period of three days in the laboratory the entire snails are frozen, the tissues homogenised, freeze-dried and analysed for a number of metals and organic pollutants.

Gomot de Vaufleury and Pihan (2000) used young garden snails (*Helix aspersa*) for an active biomonitoring of metals in France. About two months old snails from the laboratory were caged and exposed for four weeks on the soil in different areas, including a forest remote from human activities as a field control, urban and industrial sites, including waste disposal dumps. Additionally, laboratory controls were analysed in parallel. At the end of the experiment, mortality, growth (cf. 2.3) and accumulation of Cd, Cu, Pb and Zn in the foot and the visceral complex were measured. While the Cu concentrations where almost identical in the foot and visceral complex at all analysed sites indicating a strong homeostasis ability of the snails for this essential element, the tissue concentrations of Cd, Pb and Zn in the visceral complex where 2- to 50-fold higher compared to the foot in the different areas. Furthermore, the snail accumulated significantly higher amounts of Pb and Zn at the contaminated sites when compared with the laboratory and field controls (Fig. 1).

The main advantage of such an active biomonitoring approach is that it can be easily modified to consider also many other groups of contaminants, like pesticides, polycyclic aromatic hydrocarbons (PAH), etc. (e.g. Coeurdassier et al., 2001), although

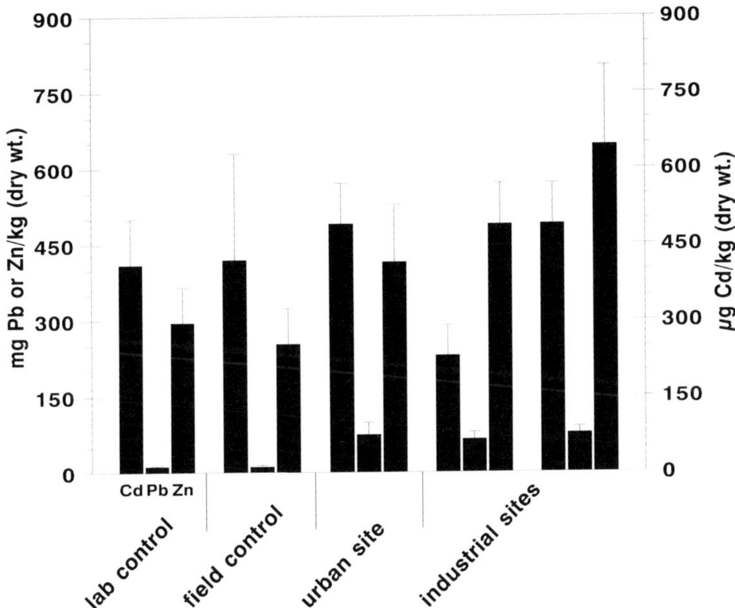

Figure 1. Concentrations of cadmium, lead and zinc in the visceral complex of young garden snails (*Helix aspersa*) transplanted from the laboratory to different exposure sites in France (values from Gomot de Vaufleury and Pihan, 2000).

heavy metals and other trace elements have gained much more interest than organic pollutants in the past. Bertani et al. (1994) used the land snail *Eobania vermiculata* to monitor the mercury exposure in Italy in an area affected by a chlor-alkali complex. Additionally, Hg residues were also assessed in plants, soils and air from the same region. The results indicate a low bioavailability of Hg for the snails as the mercury levels within the gut exceeded with values between 0.2 and 5.97 mg/kg (dry wt.) those in the soft tissues of the snails with 0.03 to 0.72 mg/kg (dry wt.). The latter concentrations were in the same range as measured in plants at the same sampling sites indicating a low bioaccumulation of Hg in this snail species. Gaso et al. (1995) conducted a study of radionuclide pollution with ^{226}Ra, ^{137}Cs and ^{40}K in the surroundings of a storage centre for radioactive wastes in Mexico. The parallel measurements of radionuclide residues in *Helix aspersa* and in soil samples from the same stations allowed a calculation of bioaccumulation factors, which were generally lower than 1 (5.5×10^{-2} and 2.0×10^{-3} for ^{226}Ra and ^{137}Cs in soft tissues, respectively). Nevertheless, the ^{226}Ra levels in these snails exceeded those measured at a reference site 100 km away by a factor of 9.

A further advantage of these types of active surveys is that they can also be used to evaluate the contamination of industrial waste dump sites and to monitor the efficiency of in situ-remediation techniques. An example is provided by Pihan and Gomot de Vaufleury (2000) who utilised two terrestrial snail species, *Helix pomatia* and *H. aspersa*, to assess the intensity of heavy metal pollution around the waste dump site before and after the remediation process. Active biomonitoring approaches with specimens taken from the laboratory avoid additionally the natural variability in autochthonous land snail populations. Due to the live-long exposure of wild snails to pollutants present in their natural habitat, they might have produced specific detoxification enzymes (e.g. MFO system) or metal-binding molecules (e.g. metallothioneins) resulting in a marked modification of pollutant accumulation. Consequently, the measured contaminant concentrations in the tissues of the indigenous bioindicators will not necessarily reflect the concentrations of organic compounds or trace elements in their environment. This might have contributed to the results reported by Richmond and Beeby (1992), who found lower lead body burdens in a *Helix aspersa* population from a heavily polluted area than in low contaminated reference regions of England.

One of the main problems with accumulation monitoring studies is the evaluation of the results and the development of assessment criteria. Berger and Dallinger (1993) have established a classification system for heavy metal contamination in the pulmonate snail *Arianta arbustorum*. They distinguish three levels of contamination, named as classes. Class 1 represents the background or control level, class 2 represents slightly contaminated areas with heavy metal body burdens in snails, which are typically found near major traffic routes and in cities, while the highest concentrations (class 3) are attained in the vicinity of smelters and mines.

Land snails have been successfully used as bioaccumulation indicators or monitors in the past and will also play a prominent role in this area of environmental surveillance in the future. Nevertheless, it has to be considered that such studies can only offer rather limited insights into the ecological and ecotoxicological relevance of the actual pollutant exposure in the environment. Although they provide information on the bioavailability of contaminants in ecosystems, it is hardly possible to derive any predictions of biological effects of these pollutants at the given level of exposure; such predictions are the main perspectives of biological effect monitoring.

2.2. Biological effects on the sub-organism level

Terrestrial snails offer the possibility to assess the effects of environmental pollutants on the sub-organism level using a wide range of structural and physiological endpoints, which are generally referred to as biomarkers or biological markers. In most studies a differentiation between biomarkers of exposure and biomarkers of effect is made (e.g. Handy and Depledge, 1999). The advantages and limitations of the application of biomarkers in terrestrial invertebrates with respect to the ecotoxicological risk assessment in soils and the monitoring of soil contamination were recently reviewed by Kammenga et al. (2000). Such studies raise in general a problem, which has been already addressed by Markert and Oehlmann (1998), the ecological relevance for higher biological integration levels (populations, ecosystems, etc.). Organisms, populations, biocenoses and ultimately entire ecosystems are naturally subject to a number of biotic and abiotic stress factors (e.g. climatic fluctuations, varying radiation, food availability, predation, parasites, diseases, competition), including pollutants. This stress situation is of existential importance at every level of biological organisation. The ability to react to stressors is therefore a vital attribute of all living systems. Conversely, no evolutive development of individual species and thus of the ecosystem as a whole is possible without natural stressors. Many of the biomarker studies raise the question, whether the investigated sub-organismic reactions of pollutants are really indicating an adverse biological effect in the sense that they are damaging individuals or even the population or if they are just indicating that the individual is coping with this stress. Workers in this field may have been too interested in expanding the merits of their various techniques without giving full regard to the question "What does it mean, how relevant and applicable is it?". Thus, the present situation is that a number of scientifically sound techniques are available which, whilst being very sensitive, yield little information on the impact of pollution on the health of the community. It has to be pointed out that the ultimate objective of ecotoxicology is not the protection of the individual (as it is for human toxicology), but of the population. From a biological point of view it can therefore be argued that unless an effect has consequences at the population level it is insignificant. An alternative view, which may be related to the precautionary principle, is that preventive action should be taken when effects on the biomarker level are detected in individual animals. But nevertheless, it is this relationship between observed changes on the sub-organismic level and the consequences for the population which has been and is still lacking in most studies, but is required for biological effects in indicator species to be of value protecting the health of the environment.

2.2.1. Ultrastructural alterations

The assessment of cytotoxic effects of environmental pollutants by ultrastructural analyses of target cells and tissues is extremely time consuming and expensive. Therefore, such studies play clearly a secondary role for biomonitoring purposes in general. This is also due to the fact that alternative approaches, like the measurement of membrane stability (e.g. neutral red retention time assay) are available today. Köhler and Triebskorn (1998) have investigated in a comparative study the effects of heavy metal exposure on a number of soil invertebrates, including the slug *Deroceras reticulatum*. Ultrastructural alterations in the cells of the midgut gland epithelium were recorded qualitatively and quantitatively by transmission electron microscopy. The authors propose an "impact index" which itself is easy to handle for the assessment of ultrastructural damage by heavy metals in the indicator species, but also in this study the problem is the data collection with enormous technical and time efforts.

2.2.2. Metallothioneins (MTs)

Metallothioneins (MTs) are low-molecular-weight, cytosolic proteins with a high cysteine content, showing a strong affinity toward certain essential and nonessential trace elements, such as Cd, Cu and Zn (Kägi and Schäffer, 1988). So far, MTs have been identified in a large number of tissues and species throughout the animal kingdom, including a number of terrestrial snails (Dallinger, 1994; Dallinger et al., 2000). Although a variety of biochemical data prove MTs to be structurally well defined and highly conserved proteins, their biological function is still under discussion (Cherian and Chan, 1993). A number of studies demonstrated that the synthesis of MTs can be induced by certain trace elements, but also by organic chemicals and other non-chemical stress factors, like infections, starvation and injuries (Cherian and Chan, 1993, Berger et al., 1995a). Nevertheless, it has been shown that Cd, Cu, Hg and other trace elements are the most potent inducers of MT synthesis and it has therefore been speculated that the detoxification of metals is the primary biological function of these proteins, although further physiological roles such as the involvement in the cellular homeostasis of essential trace elements, gene regulation and protection against oxidant stress are evident in invertebrates (Engel and Brouwer, 1987; Sato and Bremner, 1993).

The involvement of MTs in cadmium detoxification of terrestrial gastropods has been proven in detail for a number of species, e.g. the Roman snail *Helix pomatia* (Dallinger, 1993; Berger et al., 1995a; Dallinger et al., 2000), slugs from the genus *Arion* (Dallinger et al., 1989) and *Arianta arbustorum* (Berger et al., 1995b). Compared to other invertebrate taxa such as earthworms, land snails have the advantage that their MTs are by far less unstable, particularly under conventional conditions of preparation, facilitating their use in biomonitoring surveys (Dallinger et al., 2000).

A promising perspective is the finding that MTs in metal-loaded organisms can be present in different isoforms that are specifically synthesised in response to different metals. The latter fact may have implications for the use of specific MT isoforms as potential element-specific exposure biomarkers for metal stress in invertebrates as proposed by Dallinger (1994). One possible strategy to achieve this objective may be to assess parameters of MT synthesis at the molecular or biochemical level. In any case care must be taken to consider intrinsic physiological parameters, such as nutritional or developmental factors, which could also interfere with MT synthesis (Dallinger, 1994).

2.2.3. Heat shock proteins (HSPs)

Heat shock proteins (HSPs) represent a second class of cellular stress molecules next to the MTs described previously, which both offer some protection from cellular damage. The term "heat shock protein" was coined by Ritossa (1962), who first described these proteins in *Drosophila melanogaster* following an exposure to high temperatures. In the following years, a range of environmental stressors have been shown to induce HSPs (compare Table 2), including trace metals, organic chemicals, temperature variations, changes in osmolarity, oxygen deficiency and UV radiation (Sanders, 1990; Schlesinger, 1990; Bauman et al., 1993; Myrmel et al., 1994). A number of HSP families are distinguished, classified by their molecular weight: HSP90, HSP70, chaperonin (= HSP60) and the so-called LMWs (with low molecular weights). The primary biological function they are involved in is the ensuring of the correct spatial arrangement and folding of cellular proteins (Hartl, 1996). Although all HSPs seem to be highly conserved in evolutionary terms and allow therefore the investigation of stress effects in the broadest sense in a variety of species, independent of their systematic status, the HSP70 family plays the most prominent role in this respect. HSP70 has been identified in archaeobacteria, several species of algae, vascular plants, many terrestrial and aquatic invertebrate taxa, including molluscs, and all chordate classes (Lewis et al., 1999). This protein family exhibits the largest specific activity compared to other HSPs and is thus easier to detect. An increase in the total specific

Table 2. Selection of chemicals and non-chemical environmental stressors that induce heat shock proteins in various biological taxa (modified from Lewis et al., 1999).

Environmental stress	Taxa studied
Cadmium	Nematodes, molluscs, crustaceans, fish, mammals
Copper	Algae, nematodes, molluscs, crustaceans
Lead	Nematodes, molluscs, crustaceans, mammals
Mercury	Nematodes
Silver	Molluscs, crustaceans
Zinc	Nematodes, molluscs
Arsenite	Protozoa, nematodes, fish, amphibians, mammals
Mixtures of metals	Molluscs, crustaceans, fish
Lead	Nematodes, molluscs, crustaceans, mammals
Tributyltin (TBT)	Molluscs
Fluorothane	Molluscs, crustaceans, fish
Organophosphates and carbamates	Crustaceans, fish
Thermal stress	Nematodes, molluscs, crustaceans, fish, amphibians, reptiles, mammals
UV radiation	Bacteria

activity of HSP70 within an organism can be used as a non-specific indicator of stress. The various detection methods for HSPs with their advantages and limitations have recently been reviewed by Lewis et al. (1999).

The induction of HSP is generally slower, but persists longer, during exposure of organisms to chemicals compared to thermal stress, probably due to the uptake and clearance kinetics of the particular trace elements or compounds. A number of authors have found no relationship between tissue concentrations of contaminants and the HSP response, e.g. Lundebye et al. (1997). This can be at least partly explained by the fact that HSP induction may not always be involved in the cellular detoxification of pollutants: Paraquat is one of the few herbicides that induce HSPs in animals, while MTs seem to be more important and effective at low heavy metal exposure levels.

HSPs have been used as biomarkers also in terrestrial snails for biomonitoring purposes. Köhler et al. (1992) conducted a comparative evaluation of HSP70 induction in three diplopod, one isopod and two slug species (*Deroceras reticulatum* and *Arion ater*) both, in the field and in the laboratory. They tested the effects of elevated temperatures and the exposure to heavy metals and organic compounds (molluscicides) on the gastropods, but were unable to find any differences in the HSP70 levels in slugs exposed to the molluscicide cloethocarb and the control groups. Furthermore, Köhler et al. (1996) reported that *Deroceras reticulatum* exposed to sublethal concentrations of Cd, Pb or Zn for three weeks, showed an increased expression of HSP70.

With respect to HSPs, there are many contradictory studies available, as also pointed out by De Pomerai (1996) and Lewis et al. (1999) in their recent reviews. They identified a number of areas of so-far neglected research, like spatial and seasonal variability in baseline specific activities and inducibility of HSPs. Furthermore, it should be recognised that normal or background protein quality and content of the tissues may vary considerably between tissues, specimens and sampling sites and consequently, specific HSP activities in bioindicators should be interpreted with caution.

2.2.4. *Lysosomal stability and membrane integrity*

The attraction of using cytological techniques are that they allow pollution effects to be detected at the earliest stage where the functioning of the cell may be impaired. They have gained wide interest and acceptance in many invertebrate studies, although to a lesser extent for terrestrial gastropods. Especially, parameters such as lysosomal stability (or integrity) can distinguish between specimens from polluted and "clean" sites though not always reflecting a dose response relationship. The sensitivity of these parameters to general environmental stressors such as temperature and nutrition of the animals and the apparent over-riding effect of the reproductive status restricts the periods and conditions, under which they can be applied. The assessment of lysosomal stability as a general measure of membrane stability may require the transplantation of caged animals for monitoring purposes to control at least some of the intrinsic biological parameters, which affect this biomarker.

Lysosomes have the ability to concentrate a wide range of environmental pollutants, both organic and inorganic, resulting in an increase of membrane permeability and loss of acid hydrolases into the cytosol, eventually causing cellular damage (Moore, 1990). The assessment of lysosomal stability is therefore a parameter that correlates significantly with the environmental conditions. First experimental evidence for molluscs was provided by Bayne et al. (1979) using marine bivalves. On a routine basis, the damage to biological cell membranes with a resulting efflux of hydrolases from the lysosomes into the cytosol is measured by the neutral red retention (NRR) assay. The vital cell stain neutral red is readily adsorbed by cells and then actively accumulated by the lysosomes resulting in an intense red colouring of these organelles. Any damage to the membranes impedes the ability of the lysosomes to accumulate and/or retain the stain.

The NRR assay has been used extensively in a number of surveys with haemocytic and digestive gland cell lysosomes from a number of marine bivalves, including *Mytilus edulis*, *M. galloprovincialis*, *Ostrea edulis* and *Crassostrea virginica* (e.g. Lowe and Pipe, 1994). Like already stated for MTs and HSPs before, the NRR assay responses not only to pollutants but also to natural stressors, like temperature and osmolarity changes (Ringwood et al., 1998). So far, only few attempts have been made to apply this biomarker also for terrestrial biomonitoring with snails. Recently, Snyman et al. (2000) have utilised this biomarker with haemocytes in the garden snail *Helix aspersa* in South Africa to monitor the effects of the agrochemical copper oxychloride, which is commonly used as a broad-range fungicide on a variety of fruits and vegetables, especially in vineyards. They found a significant and concentration dependent decrease of the NRR time by up to 80% compared to the control.

2.2.5. Oxidative metabolism

The occurrence of oxidative enzymes is a general characteristic of all living systems. One of the most important enzyme systems is the cytochrome P-450-dependent monooxygenase system, which is also named the mixed- or multi-function oxidase (MFO) system. The ancestral gene for cytochrome P-450, the terminal component of the MFO system, is thought to have originated over 2 billion years ago, and subsequent divergent evolution has produced many different forms or isoenzymes (Nebert et al., 1989). In the animal kingdom, the MFO system has been detected in almost all systematic taxa, including the molluscs (Livingstone et al., 1990). It functions in the transformation of endogenous and exogenous compounds and serves as an important bio-catalyst in numerous and diverse biochemical pathways. In the animal kingdom, the roles played by cytochrome P-450 in endogenous pathways include the synthesis and degradation of steroids, prostaglandins, fatty acids and a broad spectrum of further biological molecules (Schenkman and Kupfer, 1982). In metabolism of xenobiotics, cytochrome P-450 plays key roles in the transformation of natural toxins and anthropogenic pollutants (e.g. aflatoxins, aldrin, biphenyls, nitroanisole, benzo[a]pyrene and other PAHs, hydrocarbons mixtures as diesel oil and crude oil), but also in the metabolic activation and inactivation of many chemical carcinogens. The nature and significance of these functions in an organism will depend largely on the complement of particular cytochrome P-450 proteins present, their catalytic function and regulation.

In molluscs, the MFO is localised mainly in the microsomes of the digestive gland, although it was also found in other tissues like gills and haemocytes. The levels of microsomal MFO components and activities are similar in different species and populations of the same species from around the world, but are generally up to two orders of magnitude lower than in most vertebrate species and a number of other invertebrate taxa such as arthropods (Lee, 1985; Livingstone et al., 1990; Dauberschmidt et al., 1997).

The activity of cytochrome P-450 and of the entire MFO system may be increased by exposure to a wide range of organic xenobiotics. This induction of the MFO system has been widely used as a biomarker of exposure in aquatic biomonitoring programmes with molluscs (cf. 3.2), but surprisingly, no attempts have been made to use this sensitive response to organic pollutant exposure also in terrestrial snails so far. Increases of the MFO activity by a factor of 6 to 10 in areas contaminated with PAHs and other hydrocarbons are reported in aquatic studies (e.g. Baumard et al., 1998), although the inducibility of the molluscan MFO system seems to be apparently more limited compared to vertebrates and insects (Livingstone et al., 1990).

A general problem limiting the applicability of this biomarker for monitoring studies is the marked seasonal variation of the MFO activities in most species analysed yet. In *Mytilus edulis* for example, the MFO activity and the cytochrome P-450 specific content of the digestive gland decline with the approach of spawning and the eventual release of gametes with a consequent increase during autumn when food reserves are built-up again (Livingstone, 1988; Fig. 2). For mussels and other bivalve species also sex-related differences were described with females showing generally higher activities than males (Livingstone et al., 1989).

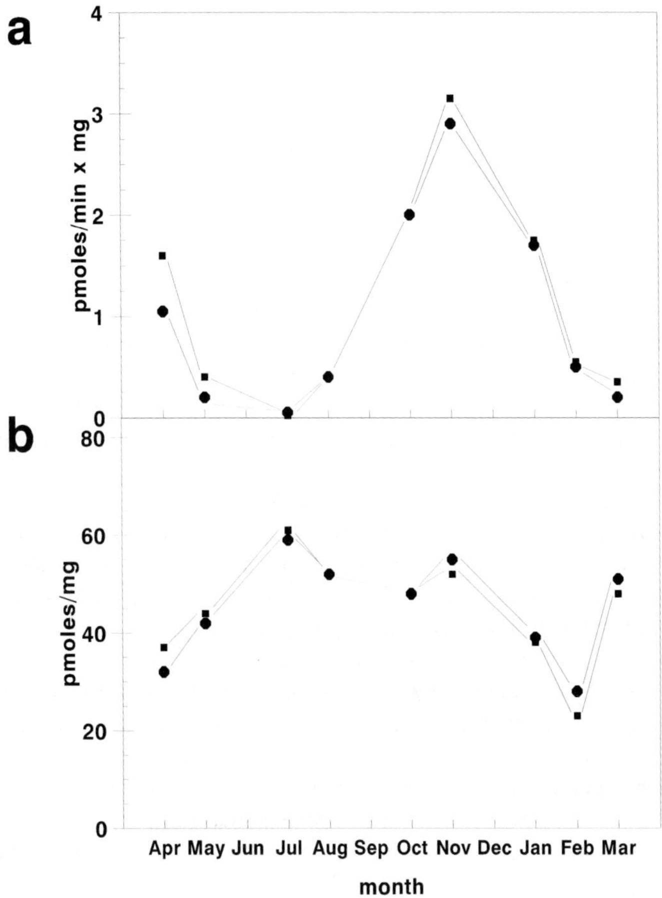

Figure 2. Seasonal variations in NADPH-independent 7-ethoxycoumarin O-deethylase (ECOD) specific activity as a measure of MFO activity (a) and cytochrome P-450 specific content in digestive gland microsomes of *Mytilus edulis* (b). Mean values of six samples, each consisting of the pooled tissue of six mussels, are presented: ■, males; ●, females (values from Livingstone et al., 1989).

2.2.6. *Acetylcholinesterase (AChE) activity*

A special group of enzymes, the esterases, have been widely utilised in the past two decades to assess the exposure of wildlife populations to two important classes of agricultural pesticides, the organophosphorus and carbamate insecticides. The class "B esterases" (according to Aldridge, 1953) represent serine hydrolases, which are inhibited by organophosphates such as paraoxon and include the acetylcholinesterase (AChE), the target molecule for the majority of these compounds. The natural substrate for AChE is the neurotransmitter acetylcholine (ACh), which is used by almost all taxa of the animal kingdom to transmit neuronal impulses across the synapse. An incoming (afferent) impulse results in an ACh release into the synaptic cleft, is followed by the diffusion of the transmitter through the cleft and a binding to postsynaptic ACh receptors, which finally leads to a depolarisation of the postsynaptic membrane causing a further conduction of the impulse. An inhibition of the AChE induces a permanent depolarisation of the postsynaptic membrane, resulting in a prolonged transmission of impulses with tetani and finally the death of the exposed specimen as ultimate effects (Oehlmann and Markert, 1997).

Many organophosphorus pesticides are applied in an inactive form and have to be activated by monooxygenases (e.g. the MFO system) in vivo to the AChE-inhibiting form. An example is the (more or less inactive) pesticide parathion, which is metabolised to the active agent paraoxon. The organophosphate inhibits the esterase through reaction with the serine at the active site of the enzyme at very low concentrations; thus the insecticide acts as a suicide substrate. Carbamates exhibit a comparable mode of action with a carbamylation of the serine moiety of the enzyme, but this interaction is less strong and carbamylated esterase can spontaneously reactivate while a reactivation of organophosphate-inhibited AChE is very slow (Thompson, 1999).

There are a number of factors apart from inhibition by organophosphates or carbamates, which affect the esterase activity of animal specimens and need to be taken into account when monitoring programmes are designed or their results are interpreted. A reduced AChE activity can be due to an exposure to other xenobiotics and even metals, it can be caused by starvation, parasite infections and other diseases. Additionally, there are also a number of natural sources of variation which can affect AChE activities (Rattner and Fairbrother, 1991). The most important are species-specific and even interindividual differences within a species, age related, diurnal and even seasonal changes.

In contrast to the broad use of AChE activity measurements as a biological marker of organophosphate and carbamate exposure in aquatic molluscs (cf. 3.2), only few attempts have been made with this assay for terrestrial monitoring. Two of the rare examples are the studies of Schuytema et al. (1994) and Coeurdassier et al. (2001), who investigated the effects of a dietary uptake of various organophosphorus insecticides and the carbamate carbaryl in the garden snail *Helix aspersa*. The inhibition of AChE activity was next to clinical effects the most sensitive parameter with a decrease of 80% in activity at an insecticide concentration of 250 µg/g in the food of the gastropods (Fig. 3a).

In land molluscs, Rorke and Gardner (1974) reported a strong inhibition of AChE activity in the haemolymph of *Helix aspersa*, caused by a carbamate or by the active oxon of fenitrothion. Young and Wilkins (1989) did not observe any inhibition of this activity in the slug *Deroceras reticulatum* when exposed to methiocarb. They explained this resistance with a difference in the sensitivity of the species' five AChE isoenzymes.

2.3. Biological effects on the organism and community level

Biological effects of pollutants or other environmental stressors on the organism or community level have been used only in a very limited number of monitoring studies with terrestrial snails. Honek (1993) investigated the value of shell banding in the

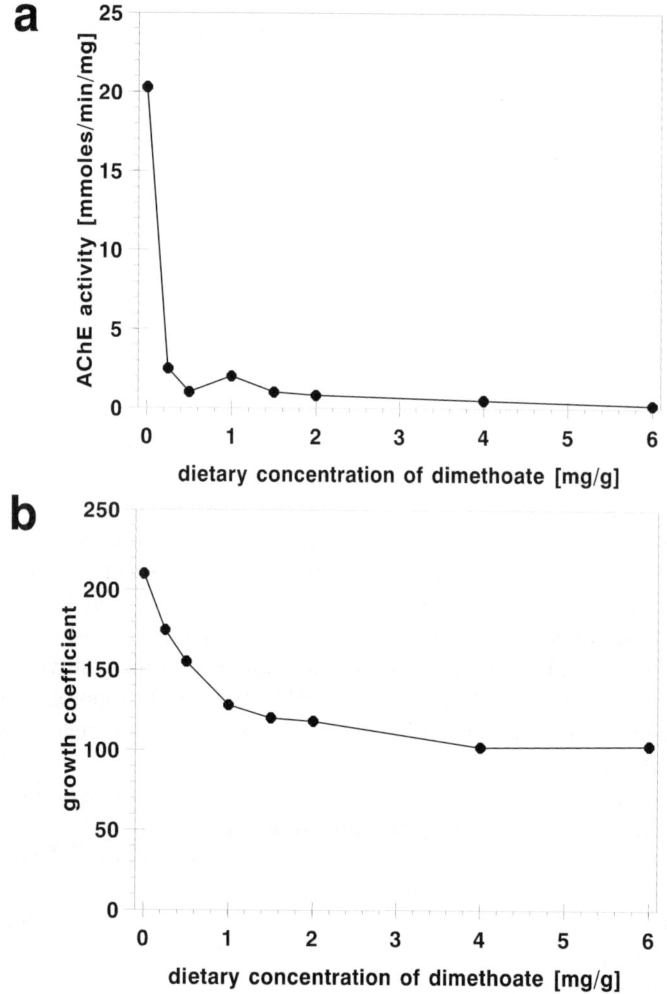

Figure 3.   Effects of the organophosphorus insecticide dimethoate on acetylcholinesterase (AChE) activity (a) and growth in the garden snail *Helix aspersa* (b) after four weeks of dietary exposure (values from Coeurdassier et al., 2001).

pulmonate *Helicella candicans* in the Czech Republic as an indicator of industrial air pollution. He found a significant positive correlation between the incidence of melanistic shell phenotypes in more than 180 analysed populations and a decrease of incident sun exposure. Although some of these areas with a high proportion of dark-shelled snails are characterised by a high degree of industrial air pollution, in other regions natural environmental conditions seemed to be the underlying cause for the observed melanism, like meteorological aspects (fog and clouds) or high and dense vegetation covers. The author concluded on the basis of his field investigations and parallel experiments in the laboratory that in areas of reduced sunshine duration and/or intensity, dark snail specimens may be at an advantage, especially during the breeding season in autumn. Because the above mentioned natural factors affect the incidence of melanistic forms in the populations, the value of this biological marker as an indicator of air pollution is rather limited.

The structure of terrestrial snail biocenoses is analysed as part of the "Ecological Effect Cadaster Baden-Württemberg" in Germany. The objective for the assessment of this synecological marker is to detect shifts in the species composition, which might be attributed to environmental pollution or non-chemical stressors (Spang, 1995). The effect cadaster is designed as a long range monitoring programme and includes an annual sampling of molluscs at approximately 200 permanent observation areas in forest, agricultural and urban ecosystems, mainly in autumn. The quantitative sampling of snail species with a body length of more than 5 mm is performed in an area of 3×2 m at these stations. Smaller species are sampled in two adjacent areas of 0.25 m^2 each from the vegetation cover, the litter layer and the upper 10 cm of the soil. The smaller gastropods are generally analysed in the laboratory under dissection microscopes because it is imperative for a characterisation of the recent biocenosis to distinguish living snails from dead shells.

Pollutant effects on the level of single organisms such as survival (or mortality), growth, weight gain, reproduction, morphological and histological alterations have been assessed in a number of terrestrial biomonitoring surveys with terrestrial gastropods. Gomot de Vaufleury and Pihan (2000) used the garden snail *Helix aspersa* as an active effect biomonitor to detect the biological impact of metals in soils on growth and mortality. During the four weeks of their study the mortality was 10.7% in the field control and 0% in the laboratory control, but up to 47.6% in those groups transplanted to industrial and urban sites. Snails transplanted from the laboratory to metal-polluted sites in the field exhibited statically significant lower shell and body weight at the end of the experiment. The latter effect, which was assessed separately for the foot and the visceral complex, was due to dry weight differences and not to the moisture differences of the tissues. A comparable approach has been made with the same bioindicator species by Coeurdassier et al. (2001), but with a focus on organophosphorus insecticides as a model class of organic contaminants. A statistically significant increase of the mortality was observed in snails exposed to dietary dimethoate concentrations of 250 μg/g after four weeks. Regardless of the criterion used to assess growth, like total fresh or dry weight gain, growth coefficients of soft body or shell diameter (Fig. 3b), a concentration dependent growth inhibition was always detectable at $\geq$250 μg dimethoate/g. Additionally, the authors found a number of typical morphological alterations in the snails following an exposure to the pesticide like a dropping of the eyestalks and a characteristic extension of the body with a swelling of the foot and the mantle edge forming a fold, which protrudes from the shell. Comparable results were also reported by Schuytema et al. (1994) for *Helix aspersa* in experiments with a number of further organophosphorus insecticides and the carbamate carbaryl (Table 3). These observations indicate that in situ encagement of land snails can be applied as a valuable active biomonitoring approach to assess the degree of soil pollution with metals and organic pollutants.

A very interesting example of the potential of terrestrial gastropods as biomonitoring tools to assess anthropogenic effects on ecosystems was published by Graveland

Table 3.   Comparison of lethal and sublethal effects of different organophosphorus insecticides and the carbamate carbaryl in the garden snail *Helix aspersa* following a dietary exposure (values from Schuytema et al., 1994; Coeurdassier et al., 2001).

Compound	Exposure time (d)	Effects	LC_{50} (μg/g)	NOEC (μg/g)	LOEC (μg/g)
Azinphosmethyl	10	Mor; Es, Sb	188	34	155
Aminocarb	10	Mor; Es, Sb	313	40	154
Trichlorfon	10	Mor; Es, Sb	664	127	501
Dimethoate	10	Mor; Es, Sb	>6,000	100	250
	28	Mor; Es, Sb	3,700	100	250
Fenitrothion	10	Mor; Es, Sb	7,060	138	797
Methylparathion	10	Mor; Es, Sb	>10,000	39	156
Carbaryl	10	Es, Sb	>10,000	626	2,500

Es = eyestalk dropped; LC_{50} = lethal concentration for 50% of tested specimens; LOEC = lowest observed effect concentration; Mor = mortality; NOEC = no observed effect concentration; Sb = swollen body.

et al. (1994) and Graveland and van der Wal (1996). They report that on poor soils in the Netherlands, an increasing number of great tits, *Parus major*, and other forest passerines produce eggs with defective shells, resulting in low reproductive success. This effect was attributed to calcium deficiency and has been observed in Germany and Sweden, too. Snail shells were the main calcium source for birds in forests where defective eggshells did not occur, but were very rare in areas where tits often had eggshell defects. The authors investigated whether a decrease in snail abundance on poor soils could be responsible for the decline in eggshell quality and the underlying reasons for this decline. There was a marked and highly significant negative correlation of the snail abundance and the incidence of eggshell defects in the birds. Furthermore, the gastropod density correlated with the calcium content and also with pH of the litter layer. The liming of a calcium-poor forest soil with a low snail abundance resulted in a marked recovery of gastropod densities and also in an improved reproductive success of *Parus* populations within the next four years. An extended survey indicated that snail densities had declined on calcium-poor soils over the last two decades, but not on calcium-rich soils. The authors concluded that acid deposition was responsible for the reduced calcium content on poor soils. This anthropogenic acidification caused a decline of terrestrial mollusc populations depriving the bird populations of their most important calcium source.

3. Aquatic bioindication and biomonitoring with molluscs

Although six of the seven classes within the molluscs comprise exclusively aquatic species and even the majority of species of the remaining class, the gastropods, are non-terrestrial, it is evident, that snails and bivalves are the only two molluscan groups which have been widely used for bioindication and biomonitoring purposes in the past. Representatives from the remaining classes found some interest for ecotoxicological testing of chemicals in the laboratory but no attempts have been made so far to consider them for field surveys.

It is virtually impossible to give a complete overview of the use of molluscs in aquatic bioindication and biomonitoring programmes within the scope of this book chapter. As an alternative approach examples will be presented showing that marine and freshwater gastropods and bivalves are widely applied as accumulation indicators and monitors of effects of contaminants on different levels of biological integration. If appropriate, we will also refer to review publications, which provide more detailed information in special areas of interest. The section will end with the group of tributyltin compounds as a special case study: Molluscs have not only played a prominent role in monitoring programmes for these organometallic pollutants in the past but were also affected by tributyltin from the lowest to the highest levels of biological integration. Consequently, the case of tributyltin effects in molluscs is one of the rare examples which show that contaminant responses can be assessed by a variety of techniques and that comparatively trivial biochemical changes induced by xenobiotics may have even consequences on the ecosystem level.

3.1. Bioaccumulation

The so-called mussel watch, initiated in the United States of America in 1976, was one of the first environmental surveillance programmes, which made use of living organisms in an extended geographical area (Goldberg, 1975; Goldberg et al., 1978). In the beginning, the mussel watch comprised a coordinated, standardised sampling and measurement of heavy metals, radionuclids, halogenated and petroleum hydrocarbons in four marine bivalve species (*Mytilus edulis, M. californicus, Crassostrea virginica* and *Ostrea equestris*) at more than 100 sample sites on the coast of North America.

In these early years of accumulation monitoring with bivalves, a number of regional surveys were conducted to investigate the dynamics of heavy metal uptake in the indicator species and to analyse seasonal influences of environmental factors. Frazier (1975, 1976) observed in his studies that Mn and Fe concentrations, but not levels of Zn and Cu in the soft tissues of the oyster *Crassostrea virginica* correlated with shell deposition rates. The concentrations of the two latter heavy metals increased gradually during spring and early summer in the tissues, followed by a rapid loss during late summer and autumn, if ambient concentrations in sea water were low. Under high environmental exposure conditions to heavy metals, the author found a rapid and concentrations dependent uptake of Fe, Zn and Cu in summer and autumn and a delayed accumulation in early spring. At the same time, the potential of freshwater bivalves was also analysed in first investigations as by Anderson (1977), who conducted a comparative evaluation of heavy metal accumulation in the shell and soft tissues of six clam species in the Fox River in Illinois and Wisconsin.

Since 1976, bivalves have been used to assess the levels of contamination in marine ecosystems, and certain systematic groups, notably mussels and oysters, have been extensively studied worldwide (for review Phillips, 1977; Rainbow and Phillips, 1993; Boening, 1999). Their main advantage is that contamination levels in these organisms provide a time-integrated measure of pollutant bioavailability, responding essentially to that fraction of the total environmental load, which is of direct ecotoxicological relevance. Bioavailability and thus uptake of metals and other pollutants are highly dependent on chemical and biological factors. Among biological factors, there are major differences in bioaccumulation between bivalve species (Boening, 1999). Within a single species, accumulation can be a function of age, size, sex, genotype, nutritional and reproductive status. Chemical factors that influence bioaccumulation are organic carbon, water hardness, temperature, pH, dissolved oxygen, sediment grain size and hydrologic features of the system (Elder and Collins, 1991). Tissue analysis is influenced by sediments and detritus in the digestive tract at the time of collection, so that a depuration time of 24–48 h in clean water is generally necessary. However, excessive depuration times may cause inaccurate results due to a partial elimination of contaminants from body tissues. Under most circumstances, depuration is initiated with a rapid initial clearance of the pollutant, which is followed by a greatly decelerated loss after reaching a certain level (McKinney and Rogers, 1992). Uptake and accumulation in deposit-feeders would be expected to correlate to contaminant concentrations in sediments, whereas accumulation in filter-feeders would most likely reflect ambient concentrations in water (Boening, 1999).

Mussels (e.g. *Mytilus edulis*, *M. galloprovincialis*, *M. californicus*), oysters (e.g. *Crassostrea virginica*, *Ostrea edulis*) and clams (e.g. *Mercenaria mercenaria*, *Venerupis* spec., *Macoma balthica*) are the most commonly used bivalve groups for accumulation monitoring studies, while other species than those mentioned were considered only occasionally if they offer specific advantages, such as high abundance, or if habitat conditions in the study area were unsuited for the established biomonitors. Especially in tropical and subtropical regions other bivalve species have been employed as indicators. Metal accumulation in the sediment dwelling mussel *Donax trunculus* was analysed by Fishelson et al. (1999) on the Mediterranean coast of Israel, where *D. trunculus* attains densities of up to 2000 specimens/m^2. The authors found a site and age specific accumulation of Cd, Pb, Cu and Hg, with the highest values for Hg in the vicinity of a PVC producing plant and elevated Cd concentrations at a station polluted with oil and waste from the petrochemical industry. The residuals of Cd, Cu and Hg were relatively high in young and noticeably low in medium sized mussel specimens.

Gregory et al. (1999) confirmed the efficient bioaccumulation of Hg in their study with the mussel *Perna perna*. The same species was also employed by Avelar et al. (2000) as an indicator of heavy metal pollution in Brazil. Although the main objective of this study was the analysis of seasonal variations in the concentrations of Cd, Cu, Cr, Pb and Zn in the bivalves, the authors detected higher values of Cd (up to 4.57 μg/g dry wt.), Cr (up to 48.5 μg/g dry wt.) and Pb (up to 60.0 μg/g dry wt.) compared to other biomonitoring studies in tropical areas. The results showed furthermore an elevated metal accumulation during July compared to spring and autumn. Sarkar et al. (1994) suggested the two bivalve species *Crassostrea cucullata* and *Anadara granosa* as suitable bioindicators for heavy metal pollution in India and other Asian countries and Szefer et al. (1998) proposed the mussel *Mytella strigata* for the same purpose in central and south America. Further potential accumulation indicators for tropical regions are presented in the literature review of Avelar et al. (2000).

Especially if new species are introduced as accumulation monitors, the question of species- and contaminant-specific time integration capacities is crucial for the interpretation of assessed concentrations in molluscs. This objective was already addressed for the well established bivalve monitors, for example by Boisson et al. (1998) and Regoli and Orlando (1999) for *Mytilus galloprovincialis*. Boisson et al. (1998) examined lead accumulation from Pb-labelled seawater. An equilibrium was reached after 21 days. The elimination of Pb after transfer into clean sea water was biphasic with a rapid loss in the initial phase (half life: 1.4 ± 0.3 days) and a very slow excretion in the second phase (half life: 2.5 ± 0.7 months). Regoli and Orlando (1999) transplanted *Mytilus galloprovincialis* from a clean to a heavy metal polluted site. The authors describe that metal concentrations reached a steady state after only two weeks, indicating that mussels can rapidly equilibrate with enhanced environmental levels of these pollutants. After a transfer to clean sea water in the laboratory, the authors found a contaminant specific excretion rate, resulting in extreme diverse biological half life times. Due to the relatively fast uptake and long depuration half life times of the various metals, the ability of mussels accurately to record short term variations in trace element concentrations in the surrounding waters is limited, a fact which should be taken into consideration in order to define the appropriate sampling frequency for mussels used in biomonitoring programmes. Far less is known with regard to the toxicokinetics of organic contaminants in these well established monitor species and for the various non-classical mussels, which are used worldwide as surrogates.

The original idea of the mussel watch programme, which primarily aimed on the site-integrated assessment of coastal marine pollution was further developed in the past. Special emphasis was laid on the analysis of temporal contamination trends, but also on the use of mussels as sentinels after the occurrence of accidents with a release of larger amounts of pollutants. Zatta et al. (1992) conducted a survey of heavy metal and As contamination with *Mytilus galloprovincialis* in the Venetian lagoon in 1988. A comparison with data from studies, which were carried out in the same area a decade earlier revealed that the degree of heavy metal pollution in the lagoon has improved. The investigations published in Hung et al. (2001) show that the copper contamination in Taiwanese oysters increased from the time period 1980–1985 (174 ± 71 µg/g dry wt., $n = 142$) to 1986–1993 (513 ± 369 µg/g dry wt., $n = 188$) with a consequent drop in the period 1994–1996 (214 ± 89 µg/g dry wt., $n = 207$) and 1997/1998 (185 ± 110 µg/g dry wt., $n = 58$). An example for the use of two bivalves, *Crassostrea angulata* and *Scrobicularia plana*, as sentinels to monitor the extent of a chemical spill was provided by Blasco et al. (1999) after the Aznalcollar accident in Spain. In April 1998, the holding pool of the Aznalcollar mine burst its banks and 5 million m² of sludge poured into the nearby River Guadiamar, a tributary of the River Guadalquivir. Within hours of the accident, approximately 2.5 Hm² acidic water with high concentrations of metals, especially Zn, had entered the river. Within a biomonitoring programme to evaluate the impact of this spill on the fauna in the two rivers and the Guadalquivir estuary, a number of aquatic species, including the two molluscs, were analysed for accumulated metals. In the oyster *Crassostrea angulata*, Cd and Cu concentrations were far above the human consumption limits for shellfish. An increase of the Zn levels was observed, as Zn concentrations in *C. angulata* were higher than those reported 30 years ago (Fig. 4).

Mussel watch surveys were also conducted with special emphasis on organic contaminants, mainly PCBs, DDT and petroleum hydrocarbons (e.g. Geyer et al., 1984;

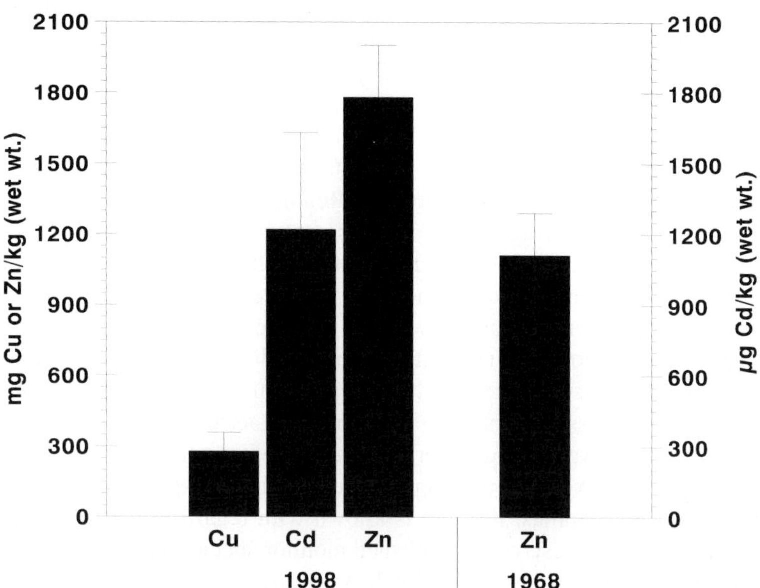

Figure 4.   Concentrations of Cu, Cd and Zn (means ± standard deviation) in the soft tissues of the oyster *Crassostrea angulata* from the estuary of the River Guadalquivir, Spain, from April to September 1998 after the Aznalcollar accident in 1998 compared to Zn residues measured in oysters from the same station in 1968 (values from Blasco et al., 1999).

Stephenson et al., 1995; Villeneuve et al., 1999). In a 15 year time span, from 1977–1992, mussels were collected at 378 stations in California. From these stations, 47 were chosen to conduct statistical analyses based on the criteria that they had been sampled at least six times for total DDT, total PCBs, and total chlordanes. Declines of total DDT and chlordanes were noted at approximately half of the stations (Stephenson et al., 1995). Villeneuve et al. (1999) reported for the Mediterranean mussel watch programme that the residues of DDT and PCBs decreased by more than 80% between 1973 and 1989 in the mussels. This decrease is in agreement with the ban on DDT implemented in 1975 in western Europe and gradual cessation of PCB production in the 1970s and 1980s. Nevertheless, the residues of these compounds measured in mussels confirm the well known long persistence of DDT and PCBs, which are still present in relatively high concentrations in some regions.

In many areas, marine gastropods have been considered as additional accumulation monitors, transforming the original mussel watch into a mollusc watch. Perhaps the most extensive comparative investigation was conducted by Hung et al. (2001), who analysed interspecific differences of trace metal accumulation in 30 mollusc species in coastal waters of Taiwan. The authors showed that the highest concentrations of Cu, Zn, Cd, Pb and Cr were measured in gastropod species, while bivalves were better accumulators of Ni, As and Sn. Szefer et al. (1999) compared the metal accumulation

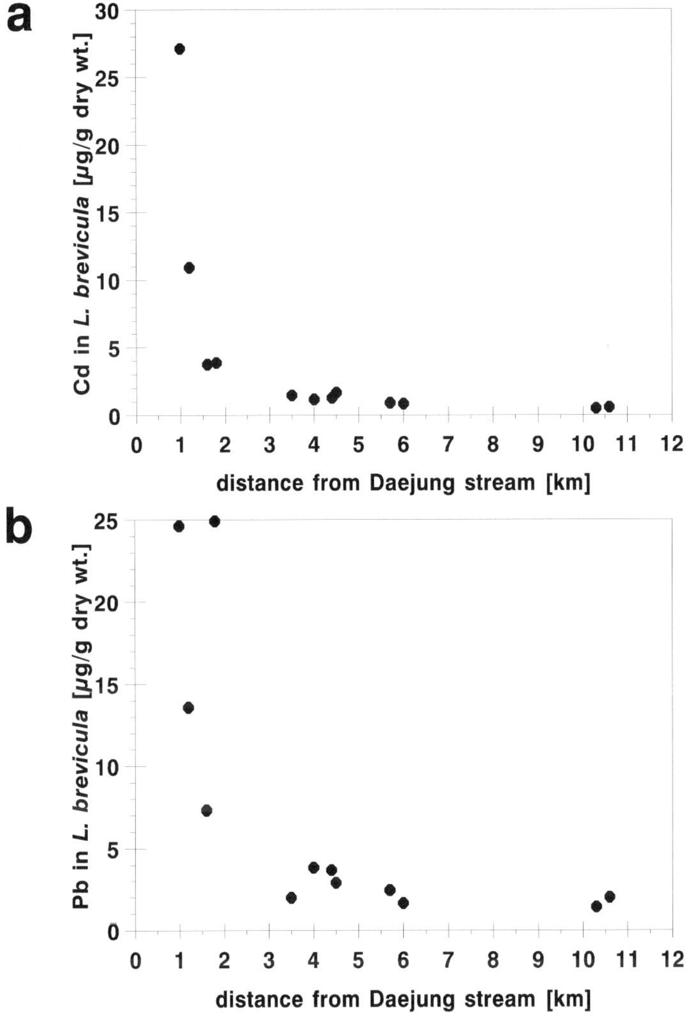

Figure 5.   Decrease of Cd (a) and Pb concentrations (b) in the tissues of the winkle *Littorina brevicula* with distance from mouth of the highly contaminated Daejung stream in the southeast of Korea (values from Kang et al., 1999).

in four mollusc species, the bivalves *Ostrea cucullata* and *Pitar* spec., the gastropod *Turbo coronatus* and the polyplacophore *Acanthopleura haddoni* in the Gulf of Aden and rated the two latter species as well suited biomonitors.

The use of marine gastropods as accumulation monitors for heavy metals was reviewed by Bryan et al. (1977, 1983, 1985). Especially the edible winkle, *Littorina littorea*, has been employed as an indictor of trace metal pollution on both sides of the Atlantic and further species from the same family in other areas of the world (e.g. *Littorina brevicula* by Kang et al., 1999, 2000; compare also Fig. 5). Campanella et al. (2001) proposed two other marine gastropods, *Monodonta turbinata* and *Patella caerulea*, as cosmopolitan biomonitors. Furthermore, marine snails were successfully used as bioaccumulators of organic contaminants in the past (e.g. Walsh et al., 1994, 1995).

Freshwater molluscs have also been frequently used as bioaccumulator organisms (compare review of Elder and Collins, 1991). Two important advantages of snails and bivalves over most other freshwater organisms for biomonitoring research are their large size and limited mobility. In addition, they are abundant in many types of freshwater environments and are relatively easy to collect and identify. At concentrations of metals and organic pollutants that are within ranges common to natural waters, they are generally effective bioaccumulators. Biomonitoring studies with freshwater molluscs have covered a wide diversity of species, contaminants, and environments. The principal generalisation that can be drawn from this research is that bioaccumulation and toxicity are extremely situation dependent; hence, it is difficult to extrapolate results from any particular study to other situations where the biological species or environmental conditions are different. Under some conditions, the bioconcentration factors can be in the range of 10^3 to 10^6, relative to water. Most studies that provide comparisons among taxonomic groups indicate that bioaccumulation in molluscs is greater than that in fish (Elder and Collins, 1991).

One of the most frequently employed freshwater bivalve accumulator species is the zebra mussel *Dreissena polymorpha*, which can easily be caged and transplanted to a number of sites within a river catchment, facilitating a spatial and temporal trend monitoring of pollutant exposure (e.g. Camusso et al., 1994). For the same purpose also a number of other freshwater bivalves (e.g. Renaud et al., 1995), prosobranch gastropods, like *Bithynia tentaculata*, and pulmonates such as *Physa gyrina* and *Biomphalaria glabrata* were used (Abd Allah et al., 1999; Flessas et al., 2000). In most of these studies heavy metals were in the centre of interest, but there are also examples of biomonitoring approaches for radionuclides, organometallic compounds like methyl mercury and organic pollutants (e.g. Hameed et al. 1996, Desy et al., 2000).

Bivalve molluscs also play an important role in the environmental specimen banking (ESB) programmes in a number of countries, e.g. the zebra mussel *Dreissena polymorpha* for freshwater and the blue mussel *Mytilus edulis* in marine ecosystems in the German ESB (Klein, 1999). The ESB allows not only a monitoring of current contaminant levels in various ecosystems but offers furthermore the possibility for a retrospective detection of pollutants in stored samples given the case that a certain compound will be identified as a threat for the environment in the future (Wise et al., 1993; Rossbach and Kniewald, 1997).

Although bioaccumulation monitoring with aquatic molluscs can provide some insights into the exposure to and bioavailability of compounds in marine and freshwater ecosystems, it does not allow predictions to be derived on the biological effects of these substances on individuals or populations. This issue has been identified as a major limitation in the national ESB and mussel watch programmes. Currently, there are a number of attempts on their way, to consider also measures of biological effects in these programmes, as proposed by Klein (1999) for the German ESB and Goldberg and Bertine (2000) for the mussel watch.

3.2. Biological effects on the sub-organism level

The effects of pollutants on aquatic molluscs can be assessed using a variety of structural, physiological and functional endpoints on the sub-organism level. The main advantage of such an approach with respect to biological effect monitoring programmes is the capability to consider large sample numbers at comparatively low costs and in a short period of time, facilitating geographical large scale surveys. But the general problem of these so-called biomarker studies, as already stated in Section 2.2 is the ecological relevance of the assessed effects for higher biological integration levels.

3.2.1. Ultrastructural alterations

The general applicability of ultrastructural alteration assessment for biomonitoring purposes has been characterised in Section 2.2.1. Recently, Orbea et al. (1999) proposed to use the peroxisomal structure together with activity measurements of the peroxisomal enzyme catalase as biomarkers of environmental organic pollution in mussels from estuarine ecosystems. They sampled mussels in monthly intervals for over one year in two estuaries in the Bay of Biscay with different degrees of pollution. Stereological procedures were applied to detect changes in peroxisome structure and microspectrophotometry was used to quantify changes in catalase activity. The animals from the two studied sampling sites were characterised by different polycyclic aromatic hydrocarbon (PAH) burdens with mussels from Plentzia generally showing lower total PAH contents than mussels from Galea. The peroxisome structure in animals from the two estuaries was characterised by site-specific seasonal variations: A strong peroxisome proliferatory response was found in mussels sampled in Plentzia during the summer months, while mussels from Galea showed few variations over the year. It appeared that mussels exposed chronically to PAHs and other pollutants, such as those from Galea, lost their ability to respond to this exposure in terms of peroxisome proliferation. In contrast, mussels collected in Plentzia effectively responded to an increased bioavailability of organic pollutants during the summer by increasing peroxisome volume and surface and numerical densities in digestive epithelial cells. However, these increases were transient because elevated PAH body burdens detected in mussels sampled in Plentzia in autumn were not accompanied by a peroxisome proliferatory response.

3.2.2. Metallothioneins (MTs)

Metallothioneins (MTs) and other selectively metal binding proteins have found a comparable attention and application for aquatic studies like in terrestrial surveys (cf. 2.2.2). Their use as a tool in biomonitoring programmes has recently been reviewed by Viarengo et al. (1999), considering also the latest knowledge on MT gene regulation and inducibility. It appears that in fish MTs should be considered as a kind of general stress protein, which is particularly responsive to heavy metals. In aquatic molluscs and especially snails and mussels, MTs seem to be more specifically involved in responses to heavy metals and can thus be considered as a biomarker of exposure to metal contamination.

Viarengo et al. (1997) have developed a comparably easy to handle spectrophotometric method to evaluate MT concentrations in tissues of marine organisms and have shown its applicability as a biomarker of Cd, Cu and Zn exposure in the digestive gland of Mediterranean and Antarctic mussels. The authors describe a three-fold induction in exposed mussels compared to uncontaminated controls. MT levels have also been used successfully as a biomarker of heavy metal exposure in marine snails, such as the netted whelk *Nassarius reticulatus* (Andersen et al., 1989), the periwinkle *Littorina littorea* (Bebianno and Langston, 1995) and recently also for freshwater mussels. Couillard et al. (1995) employed an active biomonitoring approach using transplants of the bivalve *Pyganodon grandis*, while High et al. (1997) utilised the zebra mussel *Dreissena polymorpha*.

3.2.3. Heat shock proteins (HSPs)

The potential and limitations of the use of heat shock proteins (HSPs) as a biological marker of pollutant exposure in molluscs were already addressed in Section 2.2.3. HSPs have proven useful as part of a suite of biochemical markers of xenobiotic exposure in aquatic molluscs, although it has to be considered that by themselves, HSP induction is a marker of multiple stress exposure (Snyder et al., 2001). Therefore, HSPs cannot indicate exposure to any specific stressor without direct control or knowledge of almost all environmental conditions. When combined with additional physiological observations, HSPs can, however, be indicative of the severity of the stress exposure. For example, Steinert and Pickwell (1993) demonstrated that *Mytilus edulis* gill tissue showed 12-fold induction of HSP70 following an exposure to tributyltin and this induction correlated directly with a reduction in mussel filtration rates. Clayton et al. (2000) found a more stressor- and HSP family-specific effect in an investigation with the freshwater mussel *Dreissena polymorpha*. The concentration response curve for HSP60 expression in Cu exposed mussels was biphasic, with a return to control or lower levels after a maximum expression of three times control levels. In contrast, HSP60 and HSP70 levels were elevated at all tributyltin concentrations, and HSP70 concentrations increased in Cu exposed mussels beyond the induction threshold.

3.2.4. Lysosomal stability and membrane integrity

The application of parameters like lysosomal stability, measured by the neutral red retention (NRR) time assay, and size have gained widest acceptance in studies with aquatic bivalves, particularly in the blue mussel *Mytilus edulis* (e.g. Moore, 1982). Major problems of these techniques, like the general lack of a concentration response relationship and the over-riding effect of other environmental stressors next to pollutants such as temperature, salinity and nutrition, which are likely to limit the use of lysosomal stability as a biological marker for monitoring purposes (Stickle et al., 1985), have already been addressed in Section 2.2.4.

Matozzo et al. (2001) studied the effects of heavy metal exposure in the clam *Tapes philippinarum* on NRR capacity and found that Cd and Cu has a marked effect on this measure of membrane integrity. Fishelson et al. (1999) applied in their study with the marine bivalve *Donax trunculus* in the eastern Mediterranean an alternative to the NRR assay, the acridine orange (AO) assay. Both vital stains are accumulated in viable lysosomes so that any disturbance of the accumulation can act as an indicator of disturbed lysosomal function, a decrease of membrane stability and therefore also as a marker for cell viability. The fluorescent cationic probe AO penetrates through the plasma membrane via diffusion and is actively accumulated by lysosomes of intact cells, where it acquires a red fluorescence (Bresler and Yanko, 1995). The data of the investigation of Fishelson et al. (1999) demonstrated a site-dependent trend of decrease in the capability of the lysosomes to accumulate AO with the most prominent effect at sites of highest pollutant exposure. The decrease in the lysosomal accumulation of AO in gills and mantle epithelia correlated with a higher frequency of cytopathological processes such as lysosomal enlargement and secondary lysosome formation.

3.2.5. Phase I and phase II enzymes of pollutant metabolism

Of the wide range of biochemical measurements, which have been proposed as indicators of stress, those which appear to offer most potential are based on proteins which can be induced by specific stimuli, most notably the enzymes of xenobiotic metabolising systems. The so-called phase I enzymes catalyse a functionalisation reaction with the introduction of a polar moiety such as a hydroxyl group into the organic, often lipophilic contaminant by oxidation, epoxidation, reduction or hydrolysis. The product is then characterised by a lower (metabolic deactivation) or occasionally also by an enhanced toxicity (metabolic activation). In the following phase II, the metabolites of phase I are conjugated to various endogenous substrates such as peptides (e.g. glutathione), carbohydrates (e.g. activated glucuronic acid), sulphate or amino acids to further increase the water solubility of the metabolites (Fig. 6).

The primary oxidative enzymes involved in phase I reactions belong to a number of different enzyme families, including the cytochrome P-450-dependent monooxygenase (MFO) and the flavine-dependent monooxygenase (FMO) systems, prostaglandin synthetase (PGS) and other peroxidases, monoamine oxidase (MAO), and a number of dehydrogenases. The MFO system has found the most interest for environmental monitoring but since this system does also metabolise endogenous substrates such as steroid hormones, prostaglandines and fatty acids next to xenobiotic compounds, the interpretation of MFO induction should be made with caution (for review Livingstone, 1988; Livingstone et al., 1989; see also Section 2.2.5.).

MFO induction as an exposure response for dioxins, PCBs, PAH and further groups of organic contaminants (Livingstone and Farrar, 1985), can be visualised by means of immunocytochemical techniques in native tissues or by electrophoretic separation of homogenates and microsomal preparations (Segner and Braunbeck, 1998). Alternatively, artificial substrates have been employed to assess MFO activities or of other enzymes of the phase I group such as 7-ethoxyresorufin-O-deethylase (EROD) or benzo(a)pyrene hydroxylase (B(a)PH). In general, it has been found that phase I enzyme levels correlate well with measured pollutant levels in the field for a number of aquatic vertebrates and invertebrates, including molluscs. EROD activities and cytochrome P-450 concentrations demonstrated the greatest sensitivity within the

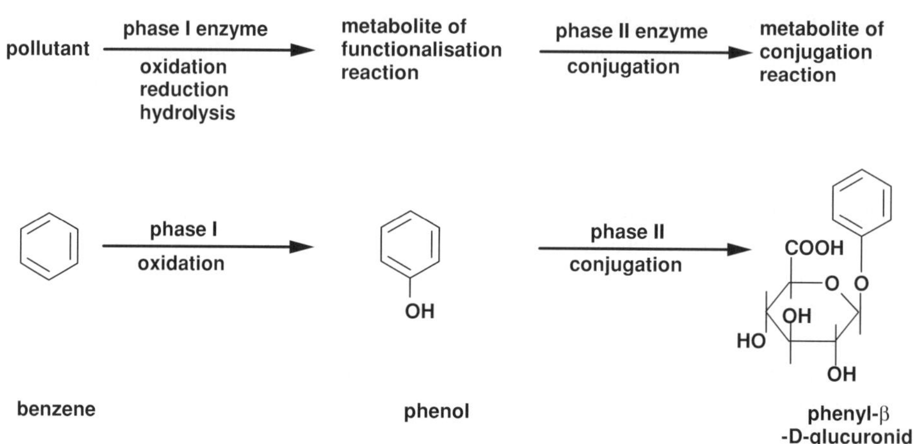

Figure 6. General metabolism scheme of xenobiotics (above) with the example of a specific metabolism of benzene (below) (modified from Oehlmann and Markert, 1997).

different parameters, which can be principally used. Further phase I enzymes like the epoxide hydrolase and phase II enzymes such as the glutathione-S-transferase have received by far less attention for biomonitoring purposes.

In contrast to the numerous studies on MFO activities and MFO induction in marine molluscs, particularly bivalves, the application of this biochemical measure of organic contaminant exposure has only occasionally been employed in freshwater environments. Wilbrink et al. (1991) provide one of the rare studies on MFO activities in a freshwater gastropod, the pond snail *Lymnaea stagnalis*, while Dauberschmidt et al. (1997) investigated the MFO system in the zebra mussel *Dreissena polymorpha*.

3.2.6. *Acetylcholinesterase (AChE) activity*

Although acetylcholinesterase (AChE) activities are considered in a number of biomonitoring programmes as a specific marker of organophosphorus or carbamate pesticide exposure, it has recently been shown by Labrot et al. (1996) that AChE activities can be modulated by metals in the freshwater bivalve *Corbicula fluminea* in vivo. Escartin and Porte (1997) investigated the use of cholinesterase and carboxylesterase activities in the mussel *Mytilus galloprovincialis* as a marker of pesticide exposure and effects in the Ebro Delta in Spain, an area of intense rice culture with a marked application of insecticides. They determined body burdens of organophosphates and seasonal variations of the activities of AChE and of two further esterases in the gills and digestive glands of mussels. The gill esterases were found to be more sensitive to in vitro inhibition than the same enzymes in the digestive glands. The carboxylesterase (CbE) was more sensitive than AChE and the authors concluded that CbE may play a protective role by removing a significant amount of the activated metabolite prior to reaching the target site AChE during an organophosphate intoxication. In mussels collected from the delta, gill AChE activity varied with season. The maximum activity was recorded in January and minimum values in April and May. The gill CbE exhibited its maximal activity from September to February and low activity from April to August. The low activity period coincided with the opening of paddy field irrigation channels with a consequent inflow of contaminated water into the delta. This minimum activity also coincided with maximum residues of fenitrothion and vamidothion in mussel tissues. The activity changes could be related to changes in water temperature, changes in salinity or normal seasonal variations. However, mussels collected from a less polluted reference area did not show such variations suggesting these are directly due to the effects of exposure to the pesticide in the run-off from the paddy fields.

Comparable attempts as described here for *Mytilus galloprovincialis* have been made also with freshwater mussels by Fleming et al. (1995) and Moulton et al. (1996).

3.2.7. Genetical markers

Pollutants can effectively influence the genetic stock in a population at two levels. They can select for certain characteristics in a population and thus alter the allele frequency as indicated in the study of Sultan et al. (2000) with the bivalve *Donax trunculus*. In addition, they can damage the DNA giving rise to gene mutations or chromosomal aberrations. A number of techniques exist for detecting genetic damages and these have been applied also to a number of mollusc species, primarily marine mussels. Chromosome mitotic abnormalities have been used to indicate differences in damage as a result of pollution exposure in mussel and oyster embryos and the sister chromatid exchange (SCE) system has been employed to demonstrate genetic damage in juvenile and adult *Mytilus edulis*, including early life stages. For these techniques molluscs have found wider application than for example fish, because the latter have larger numbers of small chromosomes, which give rise for a number of methodological problems. The comparatively fast rate of cell division in mussels, irrespective of their age, is of advantage compared to adult fish, as it allows shortened exposure periods to colchicine to accumulate dividing cells at metaphase and also to 5-bromodesoxyuridine (Brdu) as a marker, which itself can cause DNA damage. Recently, a simpler method, the micronucleus (MN) test, has been proposed for monitoring contamination exposure and effects in marine molluscs. This test, which seems to be predictive of the reproductive success, has the advantage of being karyotype independent and is thus less expensive to conduct on samples without extended pretreatments, although Burgeot et al. (1996) concluded on the basis of their comparative study that the MN test has to be improved before it can be applied for monitoring as a routine test. The incidence of micronuclei and other nuclear abnormalities in gill cells and haemocytes of Mediterranean mussels were investigated by Venier et al. (1997). The authors reported a concentration dependent increase for both markers with respect to benzo[a]pyrene exposure. Mersch and Beauvais (1997) transplanted caged zebra mussels, *Dreissena polymorpha*, to six monitoring sites in France, receiving industrial effluents suspected of containing genotoxic chemicals. After an exposure period of two months, the induction of MN in haemocytes was successfully determined as a criterion for genetic damage.

The SCE technique and its applicability to mussels such as *Mytilus galloprovincialis* is described in detail by Martinez-Expositio et al. (1994) and Pasantes et al. (1996).

The latter study described additionally seasonal variations and inter-population differences of SCE frequencies. Jha et al. (2000a) applied SCEs and further chromosomal aberrations as endpoints in a study of genotoxic effects of UK harbour sediments using embryos and larval stages of *Mytilus edulis*. The evaluation of the genotoxicity gave a positive response for all considered endpoints, emphasising the need for the assessment of short and long term genotoxic impacts of dredged disposal on marine biota.

One of the best studied markers of DNA damage in aquatic molluscs and particularly in marine mussels is the so-called comet assay, which is capable to determine DNA single strand breaks. Steinert et al. (1998) conducted a survey of DNA damage in San Diego Bay, which was determined in haemocytes collected from transplanted and autochthonous *Mytilus edulis* at stations in and around the Naval Station San Diego. Transplanted mussels were exposed at selected stations for approximately 30 days in plastic mesh bags. Those stations exhibiting the extremes of contaminant exposure, both highest and lowest concentrations, were easily identified by the comet assay results. The assay and in particular germ cell DNA damage determinations were found to respond rapidly to station contaminants. The robustness of the assay was investigated by Wilson et al. (1998) for *Mytilus edulis*. The authors studied baseline levels of single strand breaks in isolated gill cells and how they were affected by age or size of animals, time since collection and the applied feeding regime. Comet assay results in untreated controls were found to be highly variable over time. Fluctuations between low and very high DNA damage occurred over just 14 days post collection. No differences were observed between age or size and feeding regime of the mussels but a vitamin E supplementation in the diet of the organisms resulted in a marked reduction in the levels of DNA damage in the controls and in an increased sensitivity of the comet assay at the lower end of the concentration range.

3.2.8. Immunotoxicity

The use of immunosuppression or -stimulation as an indicator of pollution stress is complicated by the range of the different factors involved in the immune response of intact organisms. Whilst the immunological mechanisms of invertebrates are believed to be less complex than those of vertebrates, the various assays for measuring immune responses particularly in mussels have found wide application, especially the phagocytosis assay for *Mytilus edulis* (for review Pipe et al., 1995a). The immune system of molluscs depends largely on circulating haemocytes present in sinuses, which are able to migrate throughout the tissues to protect against potential pathogens and undertake immunosurveillance. These haemolymph cells can be divided into a number of subpopulations on the basis of functional and staining characteristics (Noel et al., 1994). The primary defence strategies involve phagocytosis, incorporating release of oxygen metabolites and degradative enzymes, and the secretion of agglutinating and cytotoxic compounds (Pipe et al., 1995a). The conventional methodology for assessing invertebrate phagocyte activation and phagocytosis has generally relied upon labour intensive and time consuming microscopic assessment or agarose plate assays. One of the major difficulties encountered when trying to assess overall immunocompetence is the interindividual variability, which arises due to the polygenic nature of natural mussel populations. The generation of statistically significant data requires large numbers of replicates. To overcome the logistical problems associated with performing numerous assays with sufficient replicates, a number of new techniques, based on optical density values obtained with a 96-well microtitre plate reader, have been developed and already applied for marine mussels such as *Mytilus edulis* and *Mercenaria mercenaria*, but also for freshwater clams like *Anodonta cygnaea* (Anderson and Mora, 1995).

Biomonitoring examples considering measures of immunomodulation are provided by Pipe et al. (1995b), Dyrynda et al. (1998) and more recently by Matozzo et al. (2001). Pipe et al. (1995b) sampled *Mytilus galloprovincialis* at three times of the year from various sites within the Venice Lagoon and a reference site in the north Adriatic. The immune response of the mussels was assessed using a range of assays, including total and differential cell counts, phagocytosis, degradative enzyme levels and release of reactive oxygen metabolites. Chlorinated hydrocarbons, including lindane, DDT and PCBs, together with trace metal levels were measured in digestive gland tissues from the mussels. The measurements of immune response and the contaminant levels showed seasonal fluctuations. However, the results demonstrated significant differences in a number of immunotoxicity assays, which showed some correlation with the levels of tissue-bound contaminants. Dyrynda et al. (1998) used a comparative approach with the blue mussel *Mytilus edulis* from six sites within the UK, comparing stations with histories of severe contamination problems with relatively uncontaminated reference sites. The results showed that significant differences in immunocompetence were evident between mussels from contaminated and reference sites. Haemocytes of mussels from contaminated sites showed enhanced superoxide production and reduced activity of degradative enzymes. Nevertheless, the results indicated that not all immune parameters are affected by contamination and that the type and extent of effects on immune defences vary with the nature or concentration of the contaminants.

Watermann et al. (1996) have introduced the assessment of inflammatory processes as an additional parameter of immunocompetence, measured as the incidence and/or intensities of haemocyte infiltrations and granulocytomas in the midgut gland, mucosa of the intestine and in storage tissues of the mussel *Mytilus edulis* and the winkle *Littorina littorea*. The authors found a significant positive correlation of these responses with the exposure of mollusc populations to xenobiotics in their environment.

3.3. Biological effects on the organism and community level

Due to the large number of relevant studies reported in the literature, only a limited example selection of monitoring programmes can be presented in this chapter, which employ biological effects in aquatic molluscs on the organism level, such as mortality or survival, growth, biological fitness (often termed as "condition"), structural alterations and interference with development.

3.3.1. Mortality or survival

In the 1960s and 1970s it was a standard procedure to expose caged mussels, oysters, clams or other bivalves and even snails for a certain period at different sites in coastal waters and to assess the percentage of dead specimens when the transplants were recollected at the end of the survey. These mortality data were often a kind of by-product of bioaccumulation studies. The same technique was later adopted for the monitoring of freshwater ecosystems and even for the surveillance of sewage outfalls or of other pollution point sources. The interpretation of such transplant results is not as easy as it looks at first glance because it remained under certain circumstances doubtful whether the exposed molluscs died due to the existing level of contamination or just because other environmental factors were unsuited (e.g. salinity, temperature, lack of food or hypoxia following the fouling of the exposure bags). Additionally, mortality is a rather rigid endpoint for the assessment of environmental quality in aquatic ecosystems and can therefore not compete with more sophisticated and modern biomarker techniques, speaking in terms of sensitivity. Nevertheless, even today mortality can be of some worth as an endpoint, especially in monitoring programmes with mussels, as demonstrated by Viarengo et al. (1995). The authors describe that a short exposure of *Mytilus edulis* to sublethal concentrations of pollutants such as Cu, Arochlor 1254 or other organic contaminants reduced the capacity of the animals to survive in air significantly. The effect was markedly concentration dependent and was strongly increased by pollutant mixtures. This parameter exhibits a sensitivity, which is in the same range as commonly used biomarkers, like the NRR time assay. Because this methodology is simple, inexpensive and does require only modest equipment, it can be integrated also in monitoring programmes for contaminated coastal areas.

3.3.2. Growth – scope for growth index

Chronic toxicities arising from exposure to sublethal concentrations of toxicants are often integrated and reflected in key life-history parameters (e.g., growth rates and reproductive efforts), or patterns of energy allocation of the organism. The rate of growth is one of the most sensitive measures of stress in an organism and responses also to chemical stressors in the environment. The quantification of growth is not as simple as it could be probably expected, especially in aquatic molluscs. Although molluscs offer the unique opportunity to use not only morphometric measures of their soft tissues, like weight or length, but additionally also shell parameters to assess growth, the problem remains, that many species exhibit an intermittent growth pattern. The shell is formed primarily at certain times of the year and body weight changes exhibit a marked seasonality due to the reproductive cycle with a considerable weight gain during the formation phase of the sexual products and a consequent loss of up to 80% of the individual biomass during spawning (Giese, 1959; Giese and Pearse, 1977). Part of the difficulties of quantifying and also interpreting growth measures can be overcome by the determination of the energy available for growth and reproduction in an organisms, also named scope for growth (SfG). SfG analyses have become particularly popular for monitoring studies with aquatic molluscs in the past, which might reflect the above mentioned problems of growth assessments in this special group of invertebrates. SfG is an integrative measure of the energy status of an organism at a particular time. Three critical components (food consumption, egestion and respiration) are required in order to calculate the overall energy value, represented by the index P according to Winberg (1960):

$$P = A - (R + U)$$

Where A is the energy absorbed from food, R is the equivalent energy used for respiration and U is the energy dispensed during egestion. Among those who favoured this concept for monitoring purposes with molluscs are Widdows et al. (1982) and Bayne et al. (1985). Bayne and Worral (1980) demonstrated a close relationship between this index and tissue growth in two populations of *Mytilus edulis*.

The sensitivity of SfG as a parameter for environmental monitoring, but also the limitations of such studies were recently demonstrated by Sobral and Widdows (1997). *Ruditapes decussatus* specimens from southern Portugal were exposed to a sublethal copper concentration of 10 µg/l for 20 days. The experiment showed two phases. Initially, Cu was rapidly accumulated, clearance rates declined markedly and respiration rates increased, resulting in a rapid SfG decline, which showed a negative value after five days. From day 9 on, the rate of Cu uptake declined and physiological responses were more stable with positive SfG values, but still significantly reduced when compared to the control (ca. 23% of the control values). This indicates that though animals partially recovered through detoxifying mechanisms, Cu caused sustained impairment of physiological functions. The experiment confirmed that though integrated SfG measurements are a sensitive methodology to detect deviations from normal performance and assess stress at environmental realistic pollutant concentrations, the result is also affected by the adaptation of animals to their specific environmental conditions. Cranford et al. (1999) exposed adult sea scallops, *Placopecten magellanicus*, in the laboratory under environmentally relevant conditions to different types and concentrations of drilling fluids. The authors observed a close relation between SfG values and actual growth measurements. Their results showed furthermore that chronic intermittent exposure of sea scallops to dilute concentrations of drilling wastes, which were characterised by acute lethal tests as practically nontoxic, can affect growth.

Wo et al. (1999) investigated in a comparative study the effects of sublethal Cd concentrations on different growth parameters of the intertidal marine gastropod *Nassarius festivus*. Their sensitivity based assessment suggests that SfG is the most sensitive growth biomarker, followed by the RNA/DNA ratio, and then the conventional growth measurement based on shell size and body weight. At all tested Cd concentrations, the authors determined negative SfG values (Fig. 7).

SfG measurements were performed by Widdows et al. (1995) for *Mytilus edulis* specimens, collected from 26 coastal sites from the Shetland Islands to the Thames estuary and eight offshore light vessels, to monitor changes in environmental quality along the British North Sea coastline. SfG values declined from north to south, reflecting both the major inflow of clean water from the North Atlantic via the north of Scotland, and the overall increase in environmental contamination with increasing urbanisation and industrialisation towards the south. There were coastal regions (e.g. Humber–Wash area and the Thames estuary) as well as specific sites, which showed markedly reduced SfG. The authors showed that at more than half of the sites the reduced SfG could be entirely explained by the recorded concentrations of contaminants in the tissues with polyaromatic hydrocarbons being one of the major responsible pollutant classes. At none of the stations were metals accumulated to concentrations

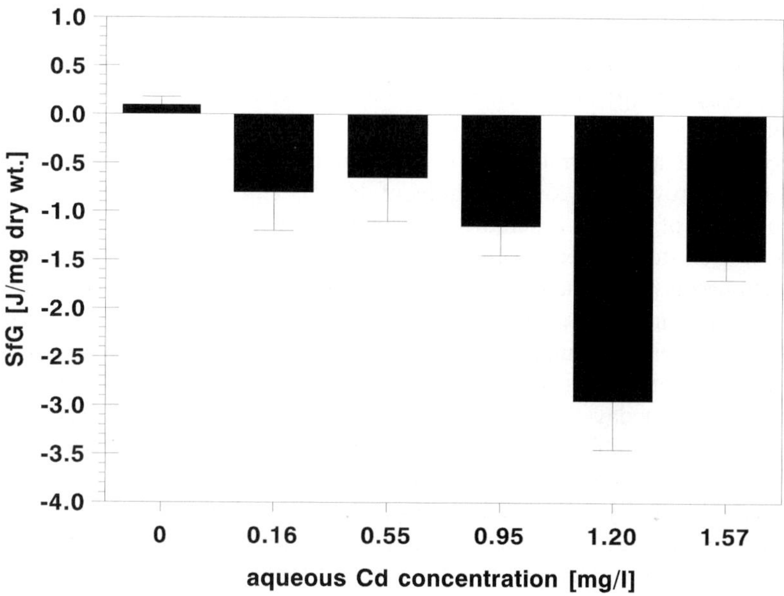

Figure 7. Effect of aqueous Cd exposure on the scope for growth (SfG) in *Nassarius festivus* (mean values ± standard deviation; values from Wo et al., 1999).

that could cause a significant reduction in SfG values. A comparative approach was performed by Widdows et al. (1997) with *Mytilus galloprovincialis* from the Venice lagoon. The authors assessed significant negative correlations between SfG values and tissue concentrations of petroleum hydrocarbons, PCBs, DDT and HCH, but found no significant correlations between SfG and metal tissue concentrations for Cd, Co, Cr, Cu, Fe, Hg, Mn, Ni, Pb and Zn.

Din and Ahamad (1995) utilised caged blood cockles, *Anadara granosa*, to monitor the effects of a highly polluted industrial discharge along a transect with eight stations on growth rates in the Juru area, Malaysia. In this region a marked decline in fisheries was reported since the early 1970s and the authors could demonstrated a massive reduction of SfG values with the highest effects at the stations in direct vicinity of the effluent. The applicability of the SfG concept for tropical freshwater snails was recently demonstrated by Lam (1996).

3.3.3. Fitness – condition index

Integrated physiological measurements of stress can generally be estimated by determinations of the body condition index, which can be calculated in different ways: (1) ratio of dry tissue to shell weight; (2) ratio of wet tissue to shell weight; (3) ratio of wet tissue volume to internal shell volume. These responses are indicative of multiple stressors on the animal, as toxicants deplete energy reserves which were potentially destined for growth and reproduction, thus reducing also the reproductive success and the overall fitness of the organism (Lucas and Beninger, 1985). Reserves are alternatively channelled into energy consuming detoxication processes and animals from contaminated waters have retarded growth and poor tissue condition. Condition indices from mussels are, therefore, potentially very sensitive to pollution. However, condition indices are affected by a number of additional environmental stressors next to pollutants, such as salinity, temperature, infestation with parasites and food availability.

Nicholson (1999) investigated in his comparative study in Hong Kong a range of cytological and physiological biomarker responses in caged green mussels, *Perna viridis*, for their suitability to indicate pollutant exposure, including condition indices. Specimens from the contaminated locations had significantly lower condition indices than reference site individuals. Condition indices integrate stress responses on somatic growth, but are not always indicative of stress as they can be affected by seasonal changes and associated nutritional and reproductive states at spatially different sites (Roesijadi et al., 1984; Leavitt et al., 1990). Gold-Bouchot et al. (1995) reported a significant negative correlation between condition indices in *Crassostrea virginica* populations in Mexico and ambient concentrations of Cd and Zn.

Juvenile *Mytilus edulis* specimens were transplanted by Grout and Levings (2001) along a pollution gradient of acid mine drainage from an abandoned Cu mine in Canada. Cages, each containing 75 mussels, were placed at a total of 15 stations and were exposed to concentrations of dissolved Cu in surface waters ranging from 5 to 1009 μg/l for a period of 41 days. Declines in survival and condition index occurred in mussels that bioaccumulated more than 40 μg Cu/g (dry wt.). The reduced survival of transplanted mussels was supported by an absence of natural mussels in the contaminated areas.

Veldhuizen-Tsoerkan et al. (1991) collected *Mytilus edulis* specimens from a relatively unpolluted area of the Eastern Scheldt in the Netherlands and transplanted them along contaminated sites of the Western Scheldt for 2.5 and 5 months. The authors determined established stress indices, including the condition index, and found a negative effect on condition index development at polluted sites in their study area.

3.3.4. Structural alterations – histopathology

The assessment of structural alterations on the cell and tissue level has a long and successful history in environmental monitoring programmes. They provide an integrative measure for the intactness of key organs in an organism and are a powerful tool to detect a broad range of different contaminant effects. The main disadvantage of using histopathological measures for such purposes is that they are time and cost intensive and require experienced, well-trained personnel to avoid misinterpretations of the findings.

Couch (1984) proposed the atrophy of the diverticular epithelium in *Crassostrea virginica* as a non-specific pollution index, based on a correlation of this effect with the contaminant exposure level at a number of analysed stations. Weis et al. (1993) reported the same histopathological alteration in oysters transplanted from a reference site to a station where they were exposed to the wood preservative chromated copper arsenate. Bowmer et al. (1994) analysed growth, reproductive performance, structural changes and heavy metal accumulation in *Cerastoderma edule* exposed to pulverised fuel ash. Although no apparent bioaccumulation was observed, the authors found histopathological and reproductive effects, which correlated well with mortality, implying that the occurrence of relatively slight effects, particularly in the digestive gland, could be an indicator of mortality. Berthou et al. (1987) conducted a study on the European flat oyster, *Ostrea edulis*, and the Pacific oyster, *Crassostrea gigas*, after the "Amoco Cadiz" oil spill in Brittany. The histopathological analyses revealed the highest incidence of lesions in the digestive tract, followed by the interstitial tissue and the gills. These effects were accompanied by an increased mortality in the populations during the first three months after the accident.

Gold-Bouchot et al. (1995) reported in *Crassostrea virginica* tissues from oyster populations in three coastal lagoons in Mexico, that more than half of all specimens showed histopathological lesions, which could be related to salinity but also to the concentrations of cadmium and hydrocarbons in their environment. Gregory et al. (1999) investigated the possibility that changes in the surface morphology of gill filaments of the mussel *Perna perna* may be used to indicate the relative toxicity of pollutants in the marine environment. In Hg exposed specimens, the authors described a gradual increase in a number of histopathological parameters, such as diameters of microvilli, depletion of abfrontal cilia, increase in necrotic cells and of cilia on the lateral gill surfaces. These results were recently confirmed by Bigas et al. (2001).

Vaschenko et al. (1997) analysed histological alterations in the gonads of the scallop *Mizuhopecten yessoensis* sampled from six stations in the Sea of Japan and demonstrated a retardation of gametogenesis, oocyte resorption, autolysis of spermatozoa and their phagocytosis in the gonads of scallops from polluted sites. Additionally, the percentage of gonadial hermaphrodites was about 6% against 0.3–0.4% in the scallop populations from clean areas.

3.3.5. Developmental effects

From an ecological perspective, one of the most important responses of organisms to pollution is the impairment of reproduction due to either direct effects on the formation of germ cells (as described above) or to interference with normal development processes in animal early life stages. Additionally, it is well documented that early life stages of many species are several orders of magnitude more sensitive to pollutants than the adults. Among the molluscan embryonic or larval biotests, the so-called oyster embryo bioassay (OEB) with early life stages of *Crassostrea gigas* is in widespread use (Beiras and His, 1994; His et al., 1997b), especially for the assessment of toxicity in coastal sediments (Chapman and Morgan, 1983). Even in international programmes this bioassay has been validated as a useful biological tool for marine environmental quality assessment (Williams et al., 1986; Butler et al., 1992; Chapman et al., 1992). One of the main problems of this test is the availability of more or less homogeneous biological material during the entire year. Because cryopreserved, cloned, or established laboratory strains of marine test species are currently not available, techniques for long term preservation of environmental samples are required. Beiras et al. (1998) investigated the effects of temperature and duration of storage on the toxicity of estuarine sediments using the OEB. Sediments ranging from unpolluted (controls) to extremely polluted with heavy metals (>100 mg/kg Hg, Cu, Zn, and Pb) and total hydrocarbons (>1000 mg/kg) were collected from sites in southwest France and northern Spain. Development of oyster embryos was significantly reduced in polluted sediments. A prolonged storage of fresh sediments at 4°C resulted in a loss of toxicity, which was more rapid in the less contaminated sediments. Deep frozen sediments (−196°C) were highly toxic regardless of origin and storage time. Because deep freezing caused spurious toxicity in the control samples, the authors recommended not to consider a freezing of sediments at such low temperatures for biomonitoring studies. Should the sediments require prolonged storage, freezing at −20°C appeared to be the best choice.

His et al. (1997a) utilised the OEB for the identification of polluted sediments in mud flats of Arcachon Bay in western France, where oyster farmers discarded old coal-tarred material with a resulting PAH contamination of up to 10.5 mg PAH/kg. Vaschenko et al. (1997) recorded marked effects in the development of offspring of the scallop *Mizuhopecten yessoensis* sampled in Peter the Great Bay, Sea of Japan. At polluted sites, a decrease in fertilisation success, diminution in percentage of normal trochophores, D-veligers, veligers, and a retardation of larval growth were recorded. Those scallop populations inhabiting the most polluted areas of Peter the Great Bay seemed to be incapable of normal reproduction. Dixon and Pollard (1985) assessed embryo abnormalities in the ovoviviparous snail *Littorina saxatilis* as an indictor of pollutant exposure in coastal waters and Koster and van den Biggelaar (1980) described impairment of development in the scaphopod genus *Dentalium* after the "Amoco Cadiz" oil spill in northern France.

Dregolskaya (1993) investigated the effects of environmental pollutant exposure on development of early life stages in two freshwater gastropod species, the pulmonate *Lymnaea stagnalis* and the prosobranch snail *Bithynia tentaculata*. In both molluscs, the most vulnerable development stage to toxicant exposure was the last stage before metamorphosis, when the reduction of the protonephridium occurred and a definitive kidney had not yet been built up. The author described furthermore a species-specific difference in the sensitivity to Cu in the snails, but because pond snail embryos were obtained from an unpolluted region, while *Bithynia* eggs came from a region receiving industrial water discharges, the observed differences might be due to specific adaptations in these populations.

3.3.6. Community effects

Analysis of biological communities with respect to environmental changes was primarily focussed on the effects of introduced exotic species in a certain habitat with the zebra mussel *Dreissena polymorpha* being in the centre of interest in the past (e.g. Dermott and Kerec, 1997; Karatayev et al., 1997; Strayer et al., 1998). Nevertheless, benthic community analyses have been shown to be a sensitive measure to reflect the impact of pollution on marine and freshwater life, changes in community structure being directly related to the ecological "health" of the environment and were thus recognised in the Joint Assessment and Monitoring Programme (JAMP) of the Oslo and Paris Commissions (OSPARCOM). The established role of benthic community analysis in monitoring pollution has already provided a number of species, particularly molluscs, which are recognised as stress indicators.

McCarthy et al. (1997) published a case study conducted at the Slave River in Canada between 1990 and 1994. The catchment of this river receives discharges from industrial and agricultural processes providing a certain background pollution level. The main objective of the five-year programme was to establish a baseline data set in future monitoring programmes. Special emphasis was laid on the assessment of benthic invertebrate populations, since it was recognised that such a survey could be important for biomonitoring purposes. Abundance of organisms, taxon diversity, and presence or absence of sentinel species were used to assess environmental contamination. The study concluded that the abundance of benthic invertebrates at the numerous sites examined in the Slave River was very low and organisms that had been used in other biomonitoring studies, especially bivalve molluscs or large oligochaetes were rare or totally absent. Over 90% of the invertebrates collected from the Slave River were chironomids or small oligochaetes and comparisons of benthic invertebrate communities in the Slave River Delta indicated that few changes in percentage composition or diversity had occurred over a 10-year period. The authors reported that molluscs were the rarest invertebrate group in the entire river system, indicating their dedicated susceptibility to pollutants in freshwater systems.

Statzner et al. (2001) investigated benthic invertebrate communities in European freshwater systems to illustrate how multiple biological traits could provide a measure for the large scale biomonitoring of the functional composition of communities. Their measure considered the relative abundance of 63 categories of 11 biological traits (such as size, reproductive and dispersal potential, food and feeding habits) that indicate various ecological functions. Comparing this measure for 10 French reference regions with 37 other most natural stream types scattered across Europe demonstrated an extremely high spatial and temporal stability of the functional composition of natural invertebrate communities at the European scale. The authors identified highly significant differences between natural reference and human-impacted communities, especially in regulated stream sites below dams and in rivers receiving sewage inputs.

Gardner (2000) investigated the bivalve community composition in the Cook Strait region, New Zealand and found evidence that environmental parameters, particularly the seston quality, were crucial for the distribution of the mussels *Aulacomya maoriana*, *Mytilus galloprovincialis* and *Perna canaliculus*.

3.4. Tributyltin as an integrated case study

Tributyltin (TBT) compounds are mainly used as biocides in antifouling paints, but also in various other formulations. They produce a variety of malformations in aquatic animals with molluscs as one of the most TBT-sensitive groups of invertebrates (for review Bryan and Gibbs, 1991; Fent, 1996). As the impact of TBT on nontarget organisms became apparent in the early 1980s, France was the first European country to draw up regulations to control TBT emission and banned the use of TBT antifoulings on small boats (length <25 m) in 1982. The French legislation was adopted by other countries since 1987 almost worldwide, but in the following years TBT pollution of coastal waters was found to have remained on a high level or even increased further in many regions (e.g. Oehlmann et al., 1993; Minchin et al., 1995, 1996, 1997; Huet et al., 1996). This was the main reason for the decision of the International Maritime Organization (IMO) in autumn 2001 to ban the application of TBT-based paints on all boats by January 2003 and the presence on ship hulls by January 2008.

The first adverse effects of TBT on wildlife molluscs were observed in the Bay of Arcachon on the west coast of France, one of the main centres of oyster aquaculture in Europe. Alzieu et al. (1980, 1986) described ball-shaped shell deformations in adult oysters, *Crassostrea gigas*, and found that the annual spatfall also declined dramatically. Both effects resulted in a break-down of oyster production in the bay with marked economic consequences. Later, comparable effects were also observed in southern England and Ireland. Detailed laboratory and field investigations revealed that TBT, leached from antifouling paints of yachts and other smaller vessels in marinas and moorings near the oyster fields, was the causative agent for the adverse responses in oysters with trace concentrations as low as 10 to 20 ng TBT/l in ambient water being already effective (Bryan and Gibbs, 1991). The induction of shell deformities in oysters, but also in further bivalve groups, has been successfully applied as a biological marker of TBT effects in the following years (e.g. Alzieu et al., 1986; Dyrynda, 1992; Page et al., 1989, 1996; Phelps and Page, 1997). Another effect of TBT in molluscs was first described in a number of regions worldwide in the early 1970s without identifying the organotin compound as the responsible cause at that time: A virilisation of female prosobranchs, which has been termed as imposex (Smith, 1971) or pseudohermaphroditism (Jenner, 1979). The imposex phenomenon of prosobranchs, i.e. the formation of a penis and/or vas deferens on females of these gonochoristic species is induced at lower concentrations than all other described TBT effects. Furthermore, it is a specific response of organotin compounds, so that the use of imposex offers the unique possibility for a highly sensitive biological effect monitoring for a special group of man-made chemicals in the environment.

Molluscs are effective bioaccumulators of organotin compounds, a favouring aspect for their extreme sensitivity compared to other systematic groups. They exceed the highest reported TBT bioaccumulation factors (BAFs) for algae (3.0×10^4), annelids (3.0×10^3), crustaceans (4.4×10^4) and fish (3.0×10^4) by at least one order of magnitude (for review Oehlmann, 1994). BAFs of 1.5×10^3 to 3.0×10^5 were measured in marine molluscs and values of 8.3×10^3 to 4.5×10^5 for freshwater bivalves and snails (Bryan and Gibbs, 1991; Schulte-Oehlmann et al., 1995; Oehlmann et al. 1996a, b, 1998b). Due to this fact, molluscs and especially freshwater and marine bivalves have widely been used as bioaccumulation indicators of TBT contamination in the past. Examples for such approaches are the investigations of Becker et al. (1992) and Becker-van Slooten and Tarradellas (1995) in Swiss lakes, using the zebra mussel *Dreissena polymorpha*, and the analyses of Short and Sharp (1989), Higashiyama et al. (1991), Page (1995), Phelps and Page (1997), Morcillo et al. (1999), Jacobsen and Asmund (2000) with different marine bivalves.

A variety of biological markers on the sub-organism level in molluscs were applied for the assessment of TBT contamination, especially in the coastal environment. It is virtually impossible, to give a representative overview of all the different techniques and endpoints, which have been used in the past for this purpose, but at least some examples will be provided additionally to those already covered by the reviews of Bryan and Gibbs (1991) and Fent (1996). Brick and Deutsch (1993) and Sundermann et al. (1998) used ultrastructural endpoints for marine prosobranch snails. HSPs and other stress protein levels were investigated by Lundebye et al. (1997), Clayton et al. (2000) and Smith et al. (2000). Jha et al. (2000b) applied genetic markers like the SCE assay in mussels and Cima et al. (1999) used markers of immunotoxicity in the clam *Tapes philippinarum.* The majority of TBT-related biochemical marker studies was dedicated to the assessment of effects on key enzymes of the phase I and phase II metabolism and the use of these endpoints in biomonitoring studies. Special emphasis was laid on the cytochrome P-450-dependent MFO system (e.g. Livingstone et al., 1989; Ronis and Mason, 1996; Morcillo et al., 1998a, b, 1999). In contrast to other organic contaminants, TBT does not induce the MFO system of molluscs, but causes a marked reduction of both MFO activities and cytochrome P-450 content. These findings are in line with the results of Spooner et al. (1991), Schulte-Oehlmann et al. (1995), Bettin et al. (1996), Oehlmann and Bettin (1996), Morcillo and Porte (1997) and Morcillo et al. (1998a, b, 1999), who described an increase of testosterone concentrations in TBT exposed molluscs, most probably due to an inhibition of the cytochrome P-450-dependent aromatase.

The imposex response is known today for more than 150 prosobranch species. The gradual virilisation of imposex affected females can be described by a development scheme with six stages, furthermore divided in up to three different types (a-c) (Gibbs et al., 1987, further developed by Fioroni et al., 1991), which has the advantage to be applicable for all affected species worldwide. Females are sterilised in the imposex stages 5 and 6 by one of the following mechanisms: (a) blockade of the pallial oviduct (Fig. 8), as in *Nucella lapillus* at ambient TBT concentrations above 2.0 ng as Sn/l or (b) by a split bursa copulatrix and capsule gland as in *Ocenebra erinacea* (threshold concentration 8 ng TBT as Sn/l). The first possibility prevents the deposition of egg capsules, resulting in an accumulation of abortive capsular material in the pallial oviduct (stage 6); the second mechanism prevents copulation and capsule formation. In young and sexual immature specimens of some muricid species a protogyne sex-change can be induced by TBT concentrations, e.g. above 10 ng as Sn/l in *N. lapillus* (Gibbs et al., 1988; Oehlmann et al., 1991) and above 2 ng as Sn/l in *Ocinebrina aciculata* (Oehlmann et al., 1996a)

The classification in six different stages is the basis of the VDS (vas deferens sequence) index, calculated as the mean imposex stage of a population. This parameter allows the assessment of imposex intensities in natural populations and laboratory groups. It has been shown that imposex intensities, measured as the VDS index in a range of affected prosobranch species, show a highly significant correlation with TBT concentrations in ambient sea water, as demonstrated for the dog whelk, *Nucella lapillus* in Figure 9a. Consequently, the degree of coastal TBT pollution can be assessed with high precision by a determination of imposex intensities in prosobranch populations. A further advantage of the VDS index as a measure of imposex is the possibility to perform interspecific comparisons of TBT sensitivities of different species and that the index is also a measure of the reproductive capability of a given population (for details cf. Oehlmann et al., 1996b). The imposex phenomenon has been successfully used in TBT biomonitoring studies in Scotland (e.g. Bailey and Davies, 1988a, b, 1989, 1991), England (e.g. Gibbs et al., 1990, 1991), Ireland (e.g. Minchin et al., 1995, 1996, 1997), France (Oehlmann et al., 1993; Huet et al., 1996), the

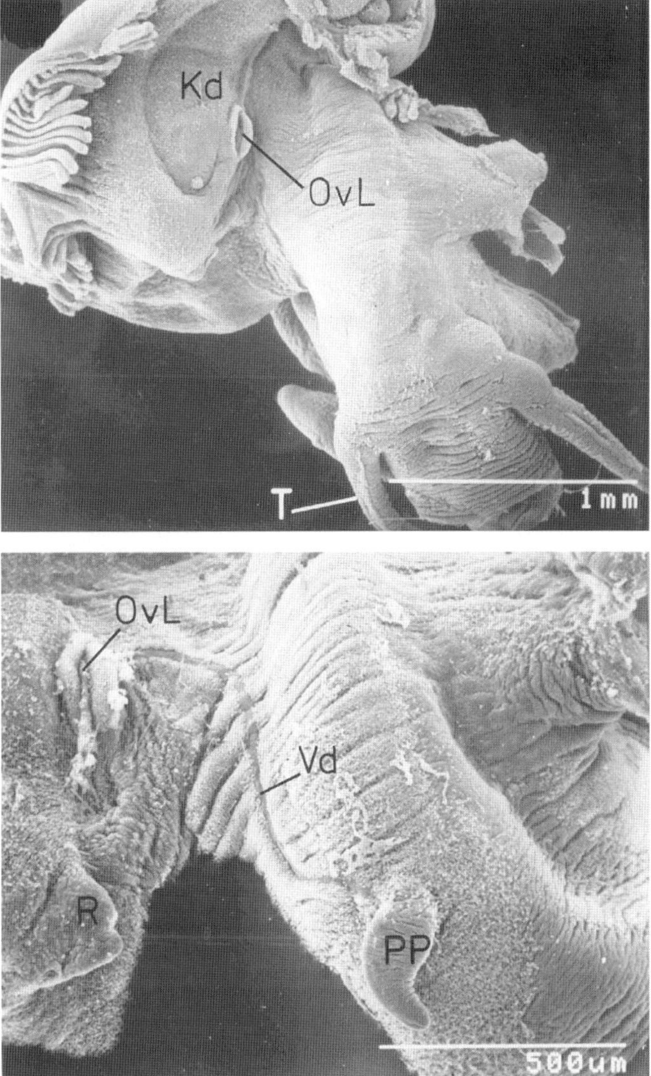

Figure 8. Hydrobia ulvae. Scanning electron micrographs of female specimens with their mantle cavity opened. Above: normal female without imposex; below: sterilised female in the final stage of imposex development with blocked oviduct. Abbreviations: Kd, capsule gland; OvL, Ooparous opening of oviduct (open above; blocked below); PP, Penis; R, rectum; T, tentacle; Vd vas deferens.

Mediterranean (e.g. Terlizzi et al., 1999) and outside Europe, e.g. in the United States (e.g. Short et al., 1989; Saavedra Alvarez and Ellis, 1990), Canada (e.g. Bright and Ellis, 1990), southeast Asia (e.g. Ellis and Pattisina, 1990; Horiguchi et al., 1997, 1998), New Zealand (e.g. Stewart et al., 1992) and Australia (e.g. Kohn and Almasi, 1993; Wilson et al., 1993).

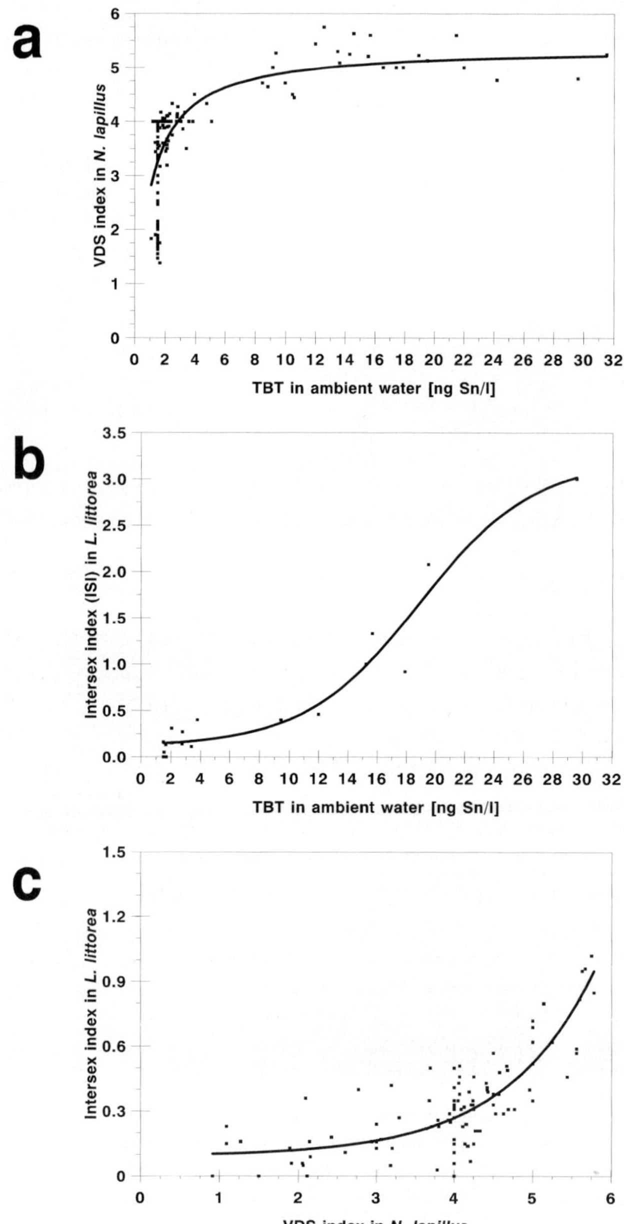

Figure 9.   Relationship between ambient TBT concentrations in water and imposex intensities in *Nucella lapillus* populations (a) or intersex intensities in *Littorina littorea* populations (b). In (c), the relationship between the intersex index in *L. littorea* populations and VDS indices in sympatrically living oopulations of *N. lapillus* is presented. (a) $y = 5.54x/(1.12 + x)$; $n = 151$ population samples from 81 stations; $r = 0.688$; $p < 0.0005$; (b) $y = 3.1/(1 + e^{-0.254(x-18.9)}) + 0.111$, $n = 18$ population samples from 11 stations, $r = 0.827$, $p < 0.0005$. (c) $y = 63.2/(1 + e^{-0.89(x-10.4)}) + 0.092$, $n = 103$ population samples from 12 French and 91 Irish stations.

In German coastal waters, but also in other European regions, the established TBT biomonitoring species are absent. The periwinkle *Littorina littorea* does not develop imposex but especially in direct proximity to harbours and marinas malformations of the female genital tract were found, which were termed as intersex (Bauer et al., 1995, 1997). The female specimens affected by intersex were either characterised by the development of male features on female pallial organs (inhibition of the ontogenetic closure of the pallial oviduct) or female sex organs were supplanted by the corresponding male formations. The intersex phenomenon of *L. littorea* is a gradual transformation of the female pallial tract, which can be described by an evolutive scheme with four stages (Bauer et al., 1995).

Intersex development causes restrictions of the reproductive capability of females. In stage 1, a loss of sperm during copulation is possible and consequently the reproductive success is reduced. Females in stages 2–4 are definitively sterile because the capsular material is spilled into the mantle cavity (stage 2) or the glands responsible for the formation of the egg capsule are missing (stages 3 and 4). Due to female sterility, populations of *Littorina littorea* can be in decline but are not likely to become extinct because of the planktonic veliger larvae of the species. Veligers produced by populations with lower intersex intensities can guarantee a minimum abundance of periwinkles even at sites suffering from high TBT contamination and reproductive failure.

The assessment of intersex intensities in periwinkle populations bases on the same principle as described for the VDS index. The intersex index (ISI) is the average intersex stage in a population. A value of 0.0 indicates that only normal females (stage 0) occur and no restrictions of the reproductive capability have to be expected. ISI values above 0 show that intersex affected females can be found and that reproductive success may be reduced. Intersex intensities are highly significantly correlated to ambient TBT concentrations (Fig. 9b) and can therefore be used together with or as an alternative to imposex assessments for the determination of the degree of coastal TBT pollution especially in regions with a relatively high level of contamination. In these areas periwinkles are very common and can be sampled in sufficient numbers because *L. littorea* (a) is tolerant of high TBT levels, (b) recruits from the plankton and (c) can occur in areas where dog whelks have expired.

Imposex in dog whelks and intersex in periwinkles have been used as combined biological markers for the convention-wide biological TBT effect monitoring of OSPARCOM (Oslo and Paris Commissions, 1996). For such large-scale surveys the interspecific comparison of intersex and imposex and the geographical uniformity of these two responses have to be clarified as prerequisites. Gibbs et al. (1991) demonstrated the geographical uniformity of the imposex response for *Nucella lapillus* and Oehlmann et al. (1998a) for the same species and additionally for intersex in *Littorina littorea*. The interspecific comparison of ISI values in periwinkle populations with VDS indices in sympatrically living dog whelk populations was performed by Oehlmann et al. (1998a) and is demonstrated in Figure 9c. The relationship between the ISI and VDS index showed that the use of intersex and imposex has specific advantages over different contamination ranges. Imposex in *N. lapillus* should be utilised for the assessment of lower TBT exposure levels in only slightly and moderately contaminated areas (ambient TBT concentrations <2.0 ng as Sn/l) whereas ISI values in

 J. Oehlmann, U. Schulte-Oehlmann

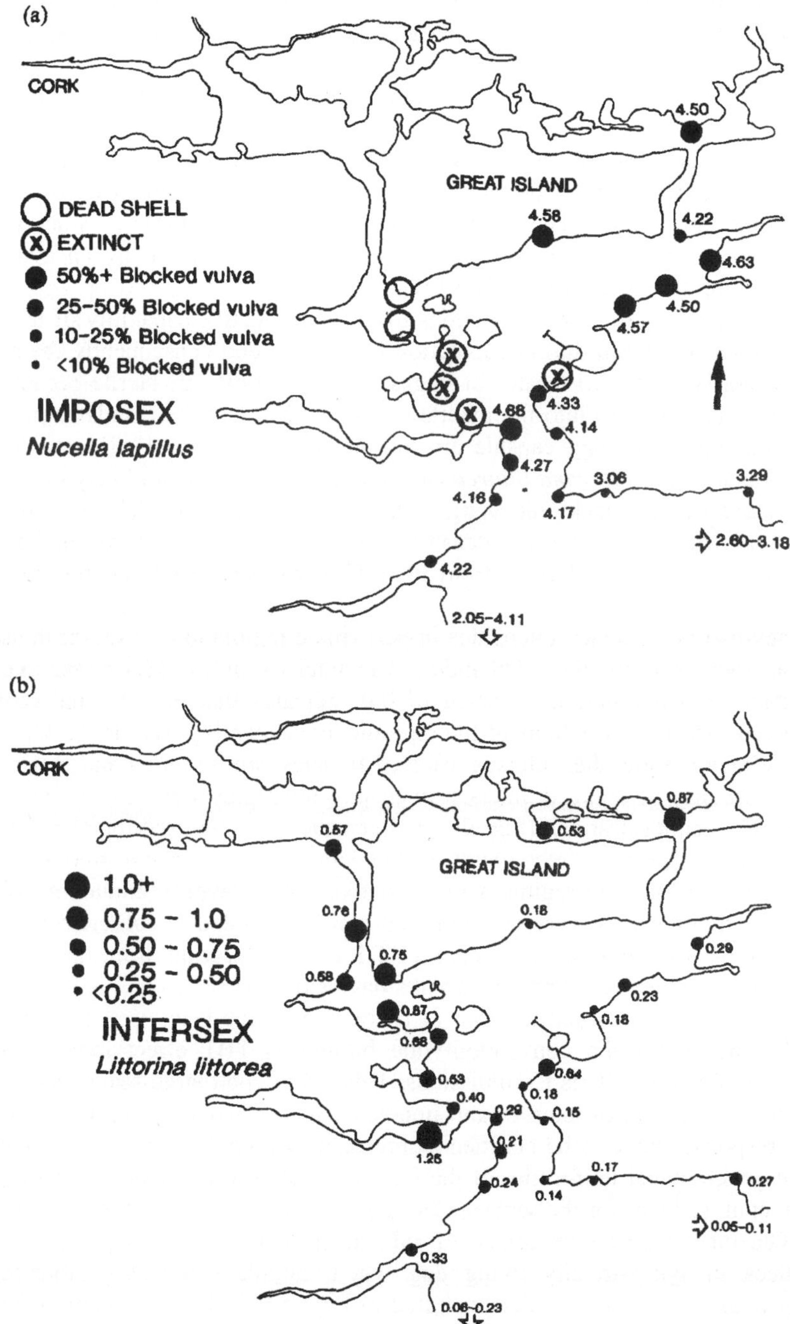

Figure 10. Biological parameters for (a) imposex intensities, measured as the VDS index and percentage of sterilised females in *Nucella lapillus* populations, and (b) intersex intensities in *Littorina littorea* populations, measured as the intersex index, in the Cork harbour area, Ireland (from Minchin et al., 1996).

L. littorea exhibit a kind of background noise with values between 0 and 0.4. At highly contaminated sites with ambient TBT concentrations above 2.0 ng TBT-Sn/l, VDS indices in dog whelks exceed 4.0 and the population becomes endangered due to progressive female sterilisation. It is in this contamination range that ISI values begin to increase. Should periwinkle populations exhibit an ISI of 0.5 or more, all females in a dog whelk population (which lives sympatrically) will have become sterile and the *Nucella* population will expire.

An example of the common use of both TBT bioindicators, *Nucella lapillus* and *Littorina littorea*, and their specific biological TBT responses is the analysis of Minchin et al. (1996) in the Cork harbour area in southern Ireland (Fig. 10). In the inner harbour, dog whelks had become extinct in the two decades before the analysis due to the existing level of contamination, but the second monitoring species, the periwinkle, was still available. Towards the open sea with decreasing TBT concentrations, where *Nucella* could survive, dog whelks were better suited as bioindicators and allowed a more appropriate differentiation of the contamination levels compared to periwinkles.

The biological TBT effect monitoring within the Joint Assessment and Monitoring Programme (JAMP) of OSPARCOM is one of the few examples of established quality assurance and quality control measures in biological environmental monitoring. The QUASIMEME (Quality Assurance Laboratory Performance Studies for Environmental Measurements in Marine Samples) office at the Marine Laboratory in Aberdeen organises regular training workshops and conducts exercises where identical subpopulations of dog whelks and periwinkles are analysed for imposex and intersex, respectively (Davies et al., 1999). Two reference laboratories, the marine laboratory in Aberdeen and our own laboratory in Frankfurt, are responsible for the training of participants and for the fixation of biological samples, which can be used as a standard reference material in all participating laboratories.

A monitoring of TBT effects on the community level and even beyond the scale of populations was performed by Rees et al. (1999) and Waldock et al. (1999). The authors conducted a number of subsequent surveys of the inter- and subtidal fauna of the River Crouch in southeast England between 1987 – the year of implementation of TBT restrictions in the UK – and 1992 and compared their results with older reports before the introduction of TBT-based antifouling paints. Overall, directional trends in community level attributed at a number of analysed stations suggested a moderate improvement in environmental conditions over the sampling period, which was coincident with a marked decline in TBT concentrations at the same stations. However, reference to historical data indicated that certain taxa that were previously frequent or common, especially the snails species *Doto coronata*, *Facelina* spec., *Archidoris pseudoargus*, *Lacuna crassior*, *Nassarius reticulatus* and *Nucella lapillus*, were only rarely recorded or still absent in the 1992 survey.

In contrast to these already institutionalised biological TBT effect monitoring programmes in coastal waters, by far less attention has been paid to comparable attempts in freshwater ecosystems, although even here an increasing pollution can be assessed due to the wide application of organotin compounds. Especially high TBT concentration in raw sewage, sewage sludge and even in effluents from sewage treatment plants are a cause for concern (e.g. Fent et al., 1991; Donard et al., 1993). The

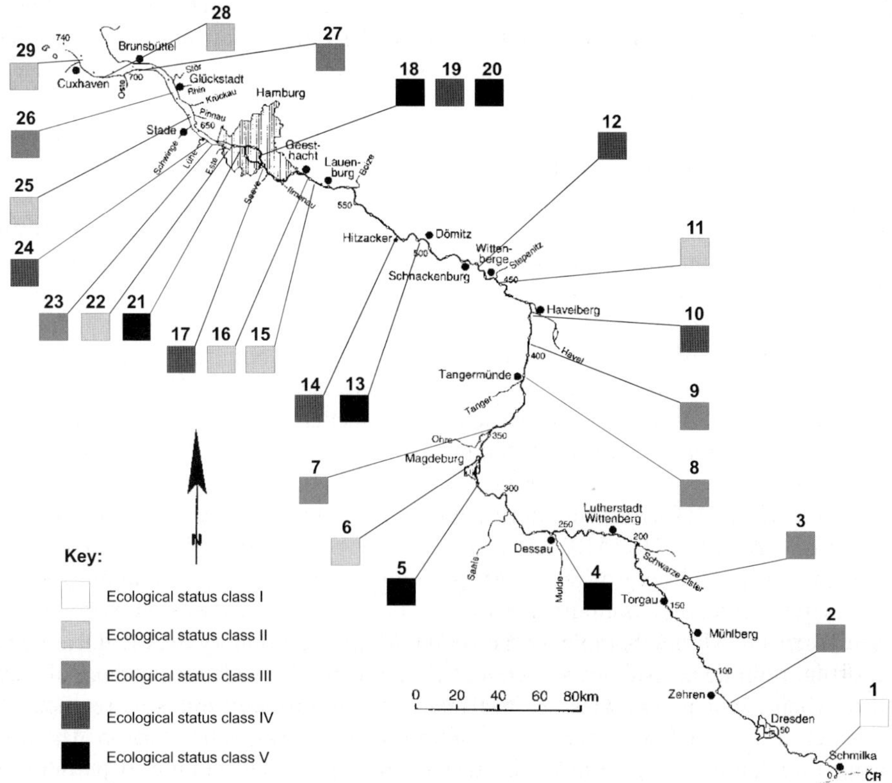

Figure 11.   Graphical representation of androgenic activities in sediments of the River Elbe in the year 2000, expressed as ecological status classes according to the European Water Framework Directive. The activities were assessed by using the imposex response in the netted whelk, *Nassarius reticulatus*, in the laboratory (Schulte-Oehlmann et al., 2001).

ramshorn snail *Marisa cornuarietis* was proposed by Schulte-Oehlmann et al. (1995) and the hydrobiid snail *Potamopyrgus antipodarum* by Schulte-Oehlmann (1997) as suited biological effect monitor species for freshwater environments. The latter and the netted whelk *Nassarius reticulatus* were employed by Schulte-Oehlmann et al. (2001) for a monitoring of androgenic activities in sediments of the River Elbe (Fig. 11). The majority of sediments exhibited marked androgenic activities and some of them, assigned to the ecological status classes IV and V according to the European Water Framework Directive, caused a maximum increase of imposex intensities in the netted whelk within four weeks. Although TBT was the main responsible pollutant for imposex development, an even stronger biological effect was observed as could be expected on the basis of analytically measured TBT concentrations for a number of sediments. This indicates that next to TBT further, not yet identified compounds may have contributed to the androgenic activities in these sediments.

Recently, it has been shown that marine and freshwater prosobranchs respond to an exposure of other endocrine active compounds such as oestrogens, further androgens and antiandrogens in the laboratory with a characteristic set of toxicological effects (Oehlmann et al., 2000; Schulte-Oehlmann et al., 2000; Tillmann et al., 2001) so that it can be assumed that in the near future these endpoints will also be used for biological effect monitoring in the field. Oestrogens caused primarily an induction of so-called "superfemales" resulting in an increased female mortality by the enhancement of spawning mass and egg production. The main effects of androgens were – as already demonstrated for tributyltin – a virilisation of females by imposex development and a marked decrease of the fecundity. Antiandrogens induced a growth reduction of the penis and of accessory male sex organs (e.g. penis sheath, prostate).

4. Conclusions

Molluscs have been successfully used as bioindicators in monitoring programmes in the past. Terrestrial ecosystems were much less considered than the aquatic environment but even for the latter it can be stated that bivalves, gastropods and especially the other molluscan classes have not yet received the attention they probably deserve, speaking in terms of their ecological importance. Major emphasis was laid in the past on the assessment of bioaccumulation in aquatic ecosystems, whereas the monitoring of biological effects in marine and freshwater environments has plaid a clearly secondary role. The example of the already institutionalised biological TBT effect monitoring programme within the JAMP of OSPARCOM with its extraordinary sensitivity for a special environmental pollutant and the relevance of effect measures for populations and higher levels of biological integration demonstrates the potential of such attempts with molluscs. It may be assumed that there is no reason to suppose that such far reaching effects of pollutants on mollusc populations are in any sense unique. The limited number of comparable successful monitoring programmes with molluscs is partially due to the fact that this special group of invertebrates was very much neglected in basic ecotoxicological research in the last decades and that therefore deleterious effects in the field, following an exposure to pollutants, may easily be missed or simply be unmeasurable at present.

References

Abd Allah, A.T., Thompson, S.N., Borchardt, D.B., Wanas, M.Q.A., 1999. *Biomphalaria glabrata*: A laboratory model illustrating the potential of pulmonate gastropods as freshwater biomonitors of heavy metal pollutants. Malocologia 41, 345–353.

Aldridge, W.N., 1953. Serum esterases. Biochemistry Journal 53, 110–117.

Alzieu, C., Thibaud, Y., Héral, M., Boutier, B., 1980. Évaluation des risques dus a l'emploi des peintures anti-salissures dans les zones conchylicoles. Revue des Travaux de 'Institut des Péches Maritimes 44, 306–348.

Alzieu, C., Sanjuan, J., Deltreil, J.-P., Borel, M., 1986. Tin contamination in Arcachon Bay: effects on oyster shell anomalies. Marine Pollution Bulletin 17, 494–498.

Andersen, R.A., Eriksen, K.D.H., Bakke, T., 1989. Evidence of presence of a low molecular weight, non metallothionein-like metal-binding protein in the marine gastropod *Nassarius reticulatus* L. Comparative Biochemistry and Physiology 94B, 285–291.

Anderson, R.S., Mora, L.M. 1995. Phagocytosis: A microtiter plate assay. In: Stolen, J.S., Fletcher, T.C., Smith, S.A., Zelikoff, J.T., Kaattari, S.L., Anderson, R.S., Söderhäll, K., Weeks-Perkins, B.A. (Eds), Techniques in Fish Immunology, Vol. IV. Immunology and Pathology of Aquatic Invertebrates. SOS Publications, Fair Haven, pp. 109–112.

Anderson, R.V., 1977. Contamination of cadmium, copper, lead, and zinc in six species of freshwater clams. Bulletin of Environmental Contamination and Toxicology 18, 492–496.

Avelar, W.E.P., Mantelatto, F.L.M., Tomazelli, A.C., Silva, D.M.L., Shuhama, T., Lopes, J.L.C., 2000. The marine mussel *Perna perna* (Mollusca, Bivalvia, Mytilidae) as an indicator of contamination by heavy metals in the Ubatuba Bay, Sao Paulo, Brazil. Water, Air and Soil Pollution 118, 65–72.

Bailey, S.K., Davies, I.M., 1988a. Tributyltin contamination in the Firth of Forth (1975–87). The Science of the Total Environment 76, 185–192.

Bailey, S.K., Davies, I.M., 1988b. Tributyltin contamination around an oil terminal in Sullom Voe (Shetland). Environmental Pollution 55, 161–172.

Bailey, S.K., Davies, I.M., 1989. The effects of tributyltin on dogwhelks (*Nucella lapillus*) from Scottish coastal waters. Journal of the Marine Biological Association of the UK 69, 335–354.

Bailey, S.K., Davies, I.M., 1991. Continuing impact of TBT, previously used in mariculture, on dogwhelk (*Nucella lapillus* L.) populations in a Scottish sea loch. Marine Environmental Research 32, 187–199.

Bauer, G., 1987. Reproductive strategy of the freshwater pearl mussel *Margaritifera margaritifera*. Journal of Animal Ecology 56, 691–701.

Bauer, B., Fioroni, P., Ide, I., Liebe, S., Oehlmann, J., Stroben, E., Watermann, B., 1995. TBT effects on the female genital system of *Littorina littorea*: a possible indicator of tributyltin pollution. Hydrobiologia 309, 15–27.

Bauer, B., Fioroni, P., Schulte-Oehlmann, U., Oehlmann, J., Kalbfus, W., 1997. The use of *Littorina littorea* for tributyltin (TBT) effect monitoring – results from the German TBT survey 1994/1995 and laboratory experiments. Environmental Pollution 96, 299–309.

Bauman, J.W., Liu, J., Klaassen, C.D., 1993. Production of metallothionein and heat-shock proteins in response to metals. Fundamental and Applied Toxicology 21, 15–22.

Baumard, P., Budzinski, H., Garrigues, P., 1998. PAHs in Arcachon Bay, France: origin and biomonitoring with caged organisms. Marine Pollution Bulletin 36, 577–586.

Bayne, B.L., Brown, D.A., Burns, K., Dixon, D.R., Ivanovici, A., Livingstone, D.R., Lowe, D.M., Moore, M.N., Stebbing, A.R.D., Widdows, J., 1985. The effects of stress and pollution on marine animals. Praeger Press, New York.

Bayne, B.L., Moore, M.N., Widdows, J., Livingstone, D.R., Salkeld, P., 1979. Measurement of the responses of individuals to environmental stress and pollution: Studies with bivalve molluscs. Philosophical Transactions of the Royal Society of London, Section B 286, 563–581.

Bayne, B.L., Worral, C.M., 1980. Growth and production of mussels (*Mytilus edulis*) from two populations. Marine Ecology Progress Series 3, 317–328.

Bebianno, M.J., Langston, W.J., 1995. Induction of metallothionein synthesis in the gill and kidney of *Littorina littorea* exposed to cadmium. Journal of the Marine Biological Association of the United Kingdom 75, 173–186.

Becker, K., Merlini, L., de Bertrand, N., de Alencastro, L.F., Tarradellas, J., 1992. Elevated levels of organotins in Lake Geneva: bivalves as sentinel organism. Bulletin of Environmental Contamination and Toxicology 48, 37–44.

Becker-van Slooten, K., Tarradellas, J., 1995. Organotins in Swiss lakes after their ban: assessment of water, sediment, and *Dreissena polymorpha* contamination over a four-year period. Archives of Environmental Contamination and Toxicology 29, 384–392.

Beiras, R., His, E., 1994. Effects of dissolved mercury on embryogenesis, survival, growth and metamorphosis of *Crassostrea gigas* oyster larvae. Marine Ecology Progress Series 113, 95–103.

Beiras, R., His, E., Seaman, M.N.L., 1998. Effects of storage temperature and duration on toxicity of sediments assessed by *Crassostrea gigas* oyster embryo assay. Environmental Toxicology and Chemistry 17, 2100–2105.

Berger, B., Dallinger, R., 1993. Terrestrial snails as quantitiative indicators of environmental metal pollution. Environmental Monitoring and Assessment 25, 65–84.

Berger, B., Dallinger, R., Thomaser, A., 1995a. Quantification of metallothionein as a biomarker for cadmium exposure in terrestrial gastropods. Environmental Toxicology and Chemistry 14, 781–791.

Berger, B., Hunziker, P.E., Hauer, C.R., Birchler, N., Dallinger, R., 1995b. Mass-spectrometry and amino-acid sequencing of 2 cadmium-binding metallothionein isoforms from the terrestrial gastropod *Arianta arbustorum*. Biochemical Journal 311, 951–957.

Bertani, R., Cosimi, C., Deliso, A., Ferrara, R., Maserti, B.E., Trifoglio, M., Zuccarelli, D., 1994. Mercury in a primary consumer (*Eobania vermiculata*) collected near a chlor-alkali complex. Environmental Technology 15, 1095–1100.

Berthou, F., Balouet, G., Bodennec, G., Marchand, M., 1987. The occurrence of hydrocarbons and histopathological abnormalities in oyster for seven years following the wreck of the *Amoco Cadiz* in Brittany (France). Marine Environmental Research 23, 103–133.

Bettin, C., Oehlmann, J., Stroben, E., 1996. TBT-induced imposex in marine neogastropods is mediated by an increasing androgen level. Helgoländer Meeresuntersuchungen 50, 299–317.

Bigas, M., Durfort, H., Poquet, M., 2001. Cytological effects of experimental exposure to Hg on the gill epithelium of the European flat oyster *Ostrea edulis*: Ultrastructural and quantitative changes related to bioaccumulation. Tissue and Cell 33, 178–188.

Blasco, J., Arias, A.M., Sáenz, V., 1999. Heavy metals in organisms of the River Guadalquivir estuary: possible incidence of the Aznalcollar disaster. The Science of the Total Environment 242, 249–259.

Boening, D.W., 1999. An evaluation of bivalves as biomonitors of heavy metals pollution in marine waters. Environmental Monitoring and Assessment 55, 459–470.

Boisson, F., Cotret, O., Fowler, S.W., 1998. Bioaccumulation and retention of lead in the mussel *Mytilus galloprovincialis* following uptake from seawater. The Science of the Total Environment 222, 55–61.

Bowmer, C.T., Jenner, H.A., Foekema, E., van der Meer, M., 1994. The detection of chronic biological effects in the marine intertidal bivalve *Cerastoderma edule* in model ecosystem studies with pulverised fuel ash: reproduction and histopathology. Environmental Pollution 85, 191–204.

Bresler, V., Yanko, V., 1995. Acute toxicity of heavy metals for benthic foraminifera *Pararotalia spinigera* (Le Calvez) and influence of seaweed-derived DOC. Environmental Toxicology and Chemistry 14, 1687–1695.

Brick, M., Deutsch, U., 1993. Ultrastructural investigations of the penis epithelia cells of three neo-gastropods, collected from TBT (tributyltin)-polluted areas. Aquatic Toxicology 27, 113–132.

Bright, D.A., Ellis, D.V., 1990. A comparative survey of imposex in northeast Pacific neogastropods (Prosobranchia) related to tributyltin contamination, and a choice of a suitable bioindicator. Canadian Journal of Zoology 68, 1915–1924.

Bryan, G.W., Gibbs, P.E., 1991. Impact of low concentrations of tributyltin (TBT) on marine organisms: a review. In: Newman, M.C., McIntosh, A.W. (Eds), Metal ecotoxicology: concepts and applications. Lewis, Ann Arbor, pp. 323–361.

Bryan, G.W., Langston, W.J., Hummerstone, L.G., Burt, G.R., Ho, Y.B., 1983. An assessment of the gastropod, *Littorina littorea*, as an indicator of heavy-metal contamination in United Kingdom estuaries. Journal of the Marine Biological Association of the United Kingdom 63, 327–345.

Bryan, G.W., Langston, W.J., Hummerstone, L.G., Burt, G.R., 1985. A guide to the assessment of heavy-metal contamination in estuaries using biological indicator. Occasional Publications of the Marine Biological Association of the United Kingdom 4, 1–92.

Bryan, G.W., Potts, G.W., Forster, G.R., 1977. Heavy metals in the gastropod mollusc *Haliotis tubercu-lata* (L.). Journal of the Marine Biological Association of the United Kingdom 57, 379–390.

Burgeot, T., Woll, S., Galgani, F., 1996. Evaluation of the micronucleus test on *Mytilus galloprovincialis* for monitoring applications along French coasts. Marine Pollution Bulletin 32, 39–46.

Butler, R., Chapman, P.M., van den Hurk, P., Roddie, B., Thain, J.E., 1992. A comparison of North American and West European oyster embryo-larval toxicity tests on North Sea sediments. Marine Ecology Progress Series 91, 245–251.

Cajaraville, M.P., Olabarrieta, I., Marigomez, I., 1996. In vitro activities of mussel hemocytes as biomarkers of environmental quality: a case study in the Albra Estuary (Biscay Bay). Ecotoxicology and Environmental Safety 35, 253–260.

Campanella, L., Conti, M.E., Cubadda, F., Sucapane, C., 2001. Trace metals in seagrass, algae and molluscs from an uncontaminated area in the Mediterranean. Environmental Pollution 111, 117–126.

Camusso, M., Balestrini, R., Muriano, F., Mariani, M., 1994. Use of freshwater mussel *Dreissena poly-morpha* to assess trace metal pollution in the lower river Po (Italy). Chemosphere 29, 729–745.

Chapman, P., Morgan, J.D., 1983. Sediment bioassays with oyster larvae. Bulletin of Environmental Contamination and Toxicology 31, 438–444.

Chapman, P., Swartz, R.C., Roddie, B., Phelps, H.L., van den Hurk, P., Butler, R., 1992. An international comparison of sediment toxicity tests in the North Sea. Marine Ecology Progress Series 91, 253–264.

Cherian, M.G., Chan, H.M., 1993. Biological functions of metallothionein – a review. In: Suzuki, K.T., Imura, N., Kimura, M. (Eds), Metallothionein III – biological roles and medical applications. Birkhäuser Verlag, Basel, pp. 87–109.

Cima, F., Marin, M.G., Matozzo, V., Da Ros, L., Ballarin, L., 1999. Biomarkers for TBT immunotoxicity studies on the cultivated clam *Tapes philippinarum* (Adams and Reeve, 1850). Marine Pollution Bulletin 39, 112–115.

Clayton, M.E., Steinmann, R., Fent, K., 2000. Different expression patterns of heat shock proteins hsp 60 and hsp 70 in zebra mussels (*Dreissena polymorpha*) exposed to copper and tributyltin. Aquatic Toxicology 47, 213–226.

Coeurdassier, M., Saint-Denis, M., Gomot-de Vaufleury, A., Ribera, D., Badot, P.-M., 2001. The garden snail (*Helix aspersa*) as a bioindicator of organophosphorus exposure: Effects of dimethoate on survival, growth, and acetylcholinesterase activity. Environmental Toxicology and Chemistry 20, 1951–1957.

Couch, J., 1984. Atrophy of diverticular epithelium as an indicator of environmental irritants in the oyster, *Crassostrea virginica*. Marine Environmental Research 14, 525–526.

Coughtrey, P.J., Martin, M.H., 1977. The uptake of lead, zinc, cadmium, and copper by the pulmonate mollusc, *Helix aspersa* Müller, and its relevance to the monitoring of heavy metal contamination of the environment. Oecologia 27, 65–74.

Couillard, Y., Campbell, P.G.C, Pellerin-Massicotte, J., Auclair, J.C., 1995. Field transplantation of a freshwater bivalve, *Pyganodon grandis*, across a metal contamination gradient. 2. Metallothionein response to Cd and Zn exposure, evidence for cytotoxicity, and links to effects at higher levels of biological organization. Canadian Journal of Fisheries and Aquatic Sciences 52, 703–715.

Cranford, P.J., Gordon, D.C., Lee, K., Armsworthy, S.L., Tremblay, G.H., 1999. Chronic toxicity and physiological disturbance effects of water- and oil-based drilling fluids and some major constituents on adult sea scallops (*Placopecten magellanicus*). Marine Environmental Research 48, 225–256.

Dallinger, R., 1993. Strategies of metal detoxification in terrestrial invertebrates. In: Dallinger, R., Rainbow, P.S. (Eds), Ecotoxicology of metals in invertebrates. Lewis, Boca Raton, pp. 245–289.

Dallinger, R., 1994. Invertebrate organisms as biological indicators of heavy-metal pollution. Applied Biochemistry and Biotechnology 48, 27–31.

Dallinger, R., Janssen, H.H., Bauer-Hilty, A., Berger, B., 1989. Characterization of an inducible cadmium-binding protein from hepatopancreas of metal-exposed slugs (Arionidae, Mollusca). Comparative Biochemistry and Physiology C92, 355–360.

Dallinger, R., Berger, B., Gruber, C., Hunziker, P., Sturzenbaum, S., 2000. Metallothioneins in terrestrial invertebrates: structural aspects, biological significance and implications for their use as biomarkers. Cellular and Molecular Biology 46, 331–346.

Dauberschmidt, C., Dietrich, D.R., Schlatter, C., 1997. Investigations on the biotransformation capacity of organophosphates in the mollusc *Dreissena polymorpha* P. Aquatic Toxicology 37, 283–294.

Davies, I.M., Minchin, A., Bauer, B., Harding, M.J.H., Wells, D.E., 1999. QUASIMEME laboratory performance study of the biological effects of tributyltin (imposex and intersex) on two marine gastropod molluscs. Journal of Environmental Monitoring 1, 233–238.

De Pomerai, D.I., 1996. Heat-shock proteins as biomarkers of pollution. Human and Experimental Toxicology 15, 279–285.

Dermott, R., Kerec, D., 1997. Changes to the deepwater benthos of eastern Lake Erie since the invasion of *Dreissena*: 1979–1993. Canadian Journal of Fisheries and Aquatic Sciences 54, 922–930.

Desy, J.C., Archambault, J.F., Pinel-Alloul, B., Hubert, J., Campbell, P.G.C., 2000. Relationships between total mercury in sediments and methyl mercury in the freshwater gastropod prosobranch *Bithynia tentaculata* in the St. Lawrence River, Quebec. Canadian Journal of Fisheries and Aquatic Sciences 57, 164–173.

Din, Z.B., Ahamad, A., 1995. Changes in the scope for growth of blood cockles (*Anadara granosa*) exposed to industrial discharge. Marine Pollution Bulletin 31, 406–410.

Dixon, D.R., Pollard, D., 1985. Embryo abnormalities in the periwinkle, *Littorina 'saxatilis'*, as indicator of stress in polluted marine environments. Marine Pollution Bulletin 16, 29–33.

Donard, O.F.X., Quevaullier, P., Bruchet, A., 1993. Tin and organotin speciation during wastewater and sludge treatment processes. Water Research 27, 1085–1089.

Dregolskaya, I.N., 1993. Sensitivity of embryos of freshwater molluscs from various habitats to a rise in concentration of copper ions in the environment. Russian Journal of Ecology 24, 139–143.

Dyrynda, E.A., 1992. Incidence of abnormal shell thickening in the Pacific oyster *Crassostrea gigas* in Poole Harbour (UK), subsequent to the 1987 TBT restrictions. Marine Pollution Bulletin 24, 156–163.

Dyrynda, E.A., Pipe, R.K., Burt, G.R., Ratcliffe, N.A., 1998. Modulations in the immune defences of mussels (*Mytilus edulis*) from contaminated sites in the UK. Aquatic Toxicology 42, 169–185.

Elder, J.F., Collins, J.J., 1991. Freshwater molluscs as indicators of bioavailability and toxicity of metals in surface-water systems. Reviews of Environmental Contamination and Toxicology 122, 36–79.

Ellis, D.V., Pattisina, L.A., 1990. Widespread neogastropod imposex: a biological indicator of global TBT contamination? Marine Pollution Bulletin 21, 248–253.

Engel, D.W., Brouwer, M., 1987. Metal regulation and molting in the blue crab, *Callinectes sapidus*: Metallothionein function in metal metabolism. Biological Bulletin 173, 237–249.

Escartin, E., Porte, C., 1997. The use of cholinesterase and carboxylesterase activities from *Mytilus galloprovincialis* in pollution monitoring. Environmental Toxicology and Chemistry 16, 2090–2095.

Fent, K., 1996. Ecotoxicology of organotin compounds. Critical Reviews in Toxicology 26, 1–117.

Fent, K., Hunn, J., Renggli, D., Siegrist, H., 1991. Fate of tributyltin in sewage sludge treatment. Marine Environmental Research 32, 223–231.

Fioroni, P., 1981. Einführung in die Meereszoologie. Wissenschaftliche Buchgesellschaft, Darmstadt.

Fioroni, P., Oehlmann, J., Stroben, E., 1991. The pseudohermaphroditism of prosobranchs; morphological aspects. Zoologischer Anzeiger 226, 1–26.

Fishelson, L., Bresler, V., Manelis, R., Zuk-Rimon, Z., Dotan, A., Hornung, H., Yawetz, A., 1999. Toxicological aspects associated with the ecology of *Donax trunculus* (Bivalvia, Mollusca) in a polluted environment. The Science of the Total Environment 226, 121–131.

Fleming, W.J., Augspurger, T.P., Alderman, J.A., 1995. Freshwater mussel die-off attributed to anticholinesterase poisoning. Environmental Toxicology and Chemistry 14, 877–879.

Flessas, C., Couillard, Y., Pinel-Alloul, B., St-Cyr, L., Campbell, P.G.C., 2000. Metal concentrations in two freshwater gastropods (Mollusca) in the St. Lawrence River and relationships with environmental contamination. Canadian Journal of Fisheries and Aquatic Sciences 57, 126–137.

Frazier, J.M., 1975. The dynamics of metals in the American oyster, *Crassostrea virginica*. I. Seasonal effects. Chesapeake Science 16, 162–171.

Frazier, J.M., 1976. The dynamics of metals in the American oyster, *Crassostrea virginica*. II. Environmental effects. Chesapeake Science 17, 188–197.

Gardner, J.P.A., 2000. Where are the mussels on Cook Strait (New Zealand) shores? Low seston quality as a possible factor limiting multi-species distributions. Marine Ecology Progress Series 194, 123–132.

Gaso, M.I., Cervantes, M.L., Segovia, N., Abascal, F., Salazar, S., Velazques, R., Mendoza, R., 1995. [137]Cs and [226]Ra determination in soil and land snails from a radioactive waste site. The Science of the Total Environment 173, 41–45.

Geyer, H., Freitag, D., Korte, F., 1984. Polychlorinated biphenyls (PCBs) in the marine environment, particularly in the Mediterranean. Ecotoxicology and Environmental Safety 8, 129–51.

Gibbs, P.E., Bryan, G.W., Pascoe, P.L., Burt, G.R., 1987. The use of the dog-whelk, *Nucella lapillus*, as an indicator of tributyltin (TBT) contamination. Journal of the Marine Biological Association of the U.K. 67, 507–523.

Gibbs, P.E., Bryan, G.W., Pascoe, P.L., Burt, G.R., 1990. Reproductive abnormalities in female *Ocenebra erinacea* (Gastropoda) resulting from tributyltin-induced imposex. Journal of the Marine Biological Association of the UK 70, 639–656.

Gibbs, P.E., Bryan, G.W., Pascoe, P.L., 1991. TBT-induced imposex in the dogwhelk, *Nucella lapillus*: Geographical uniformity of the response and effects. Marine Environmental Research 32, 79–87.

Gibbs, P.E., Pascoe, P.L., Burt, G.R., 1988. Sex change in the female dog-whelk, *Nucella lapillus*, induced by tributyltin from antifouling paints. Journal of the Marine Biological Association of the UK 68, 715–731.

Giese, A.C., 1959. Comparative physiology: annual reproductive cycles of marine invertebrates. Annual Reviews in Physiology 21, 547–576.

Giese, A.C., Pearse, J.S. (Eds), 1977. Reproduction of Marine Invertebrates. Vol. 4. Molluscs: Gastropods and Cephalopods. Academic Press, London.

Götting, K.-J., 1996. Mollusca, Weichtiere. In: Westheide, W., Rieger, R. (Eds), Spezielle Zoologie. 1. Teil: Einzeller und Wirbellose Tiere. Gustav Fischer Verlag, Stuttgart, pp. 276–330.

Goldberg, E.D., 1975. The mussel watch – a first step in global marine monitoring. Marine Pollution Bulletin 6, 111.

Goldberg, E.D., Bertine, K.K. 2000. Beyond the mussel watch – new directions for monitoring marine pollution. The Science of the Total Environment 247, 165–174.

Goldberg, E.D., Bowen, V.T., Farrington, J.W., Harvey, G., Martin, J.H., Parker, P.L., Riseborough, R.W., Robertson, W., Schneider, E., Gamble, E., 1978. The mussel watch. Environmental Conservation 5, 101–126.

Gold-Bouchot, G., Sima-Alvarez, R., Zapata-Perez, O., Güemez-Ricalde, J., 1995. Histopathological effects of petroleum hydrocarbons and heavy metals on the American oyster (*Crassostrea virginica*) from Tabasco, Mexico. Marine Pollution Bulletin 31, 439–445.

Gomot de Vaufleury, A., Pihan, F., 2000. Growing snails used as sentinels to evaluate terrestrial environment contamination by trace elements. Chemosphere 40, 275–284.

Graveland, J., van der Wal , R., 1996. Decline in snail abundance due to soil acidification causes eggshell defects in forest passerines. Oecologia 105, 351–360.

Graveland, J., van der Wal , R., van Balen, J.H., van Noordwijk, A.J., 1994. Poor reproduction in forest passerines from decline of snail abundance on acidified soils. Nature 368, 446–448.

Gregory, M.A., George, R.C., Marshall, D.J., Anandraj, A., McClug, T.P., 1999. The effects of mercury exposure on the surface morphology of gill filaments in *Perna perna* (Mollusca: Bivalvia). Marine Pollution Bulletin 39, 116–121.

Grout, J.A., Levings, C.D., 2001. Effects of acid mine drainage from an abandoned copper mine, Britannia Mines, Howe Sound, British Columbia, Canada, on transplanted blue mussels (*Mytilus edulis*). Marine Environmental Research 51, 265–288.

Gruner, H.-E. (Ed.), 1993. Lehrbuch der Speziellen Zoologie. Vol. 1: Wirbellose Tiere. Vol. 1, Part 3: Mollusca, Sipunculida, Echgiurida, Annelida, Onychophora, Tardigrada, Pentastomida, 5th edn. Gustav Fischer Verlag, Jena.

Gruner, H.-E., Hannemann, H.-J., Hartwich, G., Kilias, R., 1993. Urania-Tierreich. Wirbellose Tiere. Urania-Verlag, Leipzig.

Hameed, P.S., Shaheed, K., Iyengar, M.A.R., 1996. [228]Radium in the Kaveri river ecosystem. Current Science 70, 1076–1080.

Handy, R.D., Depledge, M.H., 1999. Physiological responses: their measurement and use as environmental biomarkers in ecotoxicology. Ecotoxicology 8, 329–349.

Hartl, F.U., 1996. Molecular chaperones in cellular protein folding. Nature 381, 571–580.

Heller, J., 1990. Longevity in molluscs. Malacologia 31, 259–295.

Higashiyama, T., Shiraishi, H., Otsuki, A., Hashimoto, S., 1991. Concentrations of organotin compounds in blue mussels from the wharves of Tokyo Bay. Marine Pollution Bulletin 22, 585–587.

High, K.A., Barthet, V.J., McLaren, J.W., Blais, J.S., 1997. Characterization of metallothionein-like proteins from zebra mussels (*Dreissena polymorpha*). Environmental Toxicology and Chemistry 16, 1111–1118.

His, E., Budzinski, H., Geffard, O., Beiras, R., 1997a. Effects of a hydrocarbon polluted sediment on *Crassostrea gigas* (Thunberg) metamorphosis. Comptes Rendus de l'Academie des Sciences. Serie III – Sciences de la Vie 320, 797–803.

His, E., Seaman, M.N.L., Beiras, R., 1997b. A simplification the bivalve embryogenesis and larval development bioassay method for water quality assessment. Water Research 31, 351–355.

Honek, A., 1993. Melanism in the land snail *Helicella candicans* (Gastropoda, Helicidae) and its possible adaptive significance. Malacologia 35, 79–87.

Horiguchi, T., Shiraishi, H., Shimizu, M., Morita, M., 1997. Imposex in sea snails, caused by organotin (tributyltin and triphenyltin) pollution in Japan: a survey. Applied Organometallic Chemistry 11, 451–455.

Horiguchi, T., Hyeon-Seo, C., Shiraishi, H., Shibata, Y., Soma, M., Morita, M., Shimizu, M., 1998. Field studies on imposex and organotin accumulation in the rock shell, *Thais clavigera*, from the Seto Inland Sea and the Sanriku region, Japan. The Science of the Total Environment 214, 65–70.

Huet, M., Paulet, Y.M., Glémarec, M., 1996. Tributyltin (TBT) pollution in the coastal waters of west Brittany as indicated by imposex in *Nucella lapillus*. Marine Environmental Research 41, 157–167.

Hung, T.C., Meng, P.J., Han, B.C., Chuang, A., Huang, C.C., 2001. Trace metals in different species of Mollusca, water and sediments from Taiwan coastal area. Chemosphere 44, 833–841.

Hyman, L.H., 1967. The Invertebrates, Vol. VI, Mollusca I. McGraw-Hill, New York.

Jacobsen, J.A., Asmund, G., 2000. TBT in marine sediments and blue mussels (*Mytilus edulis*) from central-west Greenland. The Science of the Total Environment 245, 131–136.

Jenner, M.G., 1979. Pseudohermaphroditism in *Ilyanassa obsoleta* (Mollusca: Neogastropoda). Science 205, 1407–1409.

Jha, A.N., Cheung, V.V., Foulkes, M.E., Hill, S.J., Depledge, M.H., 2000a. Detection of genotoxins in the marine environment: adoption and evaluation of an integrated approach using the embryo-larval stages of the marine mussel, *Mytilus edulis*. Mutation Research 464, 213–228.

Jha, A.N., Hagger, J.A., Hill, S.J., 2000b. Tributyltin induces cytogenetic damage in the early life stages of the marine mussel, *Mytilus edulis*. Environmental and Molecular Mutagenesis 35, 343–350.

Jones, D., 1983. Sclerochronology: reading the record of the molluscan shell. American Scientist 71, 384–391.

Jungbluth, J.H., von Knorre, D., 1995. Rote Liste der Binnenmollusken [Schnecken (Gastropoda) und Muscheln (Bivalvia)] in Deutschland. Mitteilungen der Deutschen Malakozoologischen Gesellschaft 56/57, 1–17.

Kang, S.G., Choi, M.S., Oh, I.S., Wright, D.A., Koh, C.H., 1999. Asessment of metal pollution in Onsan Bay, Korea using Asian periwinkle *Littorina brevicula* as a biomonitor. The Science of the Total Environment 234, 127–137.

Kang, S.G., Wright, D.A., Koh, C.H., 2000. Baseline metal concentration in the Asian periwinkle *Littorina brevicula* employed as a biomonitor to assess metal pollution in Korean coastal water. The Science of the Total Environment 263, 143–153.

Kägi, J.H.R., Schäffer, A., 1988. Biochemistry of metallothionein. Biochemistry 27, 8509–8515.

Kammenga, J.E., Dallinger, R., Donker, M.H., Köhler, H.R., Simonsen, V., Triebskorn, R., Weeks, J.M., 2000. Biomarkers in terrestrial invertebrates for ecotoxicological soil risk assessment. Reviews of Environmental Contamination and Toxicology 164, 93–147.

Karatayev, A.Y., Burlakova, L.E., Padilla, D.K. 1997. The effects of *Dreissena polymorpha* (Pallas) invasion on aquatic communities in eastern Europe. Journal of Shellfish Research 16, 187–203.

Klein, R., 1999. Retrospektive Wirkungsforschung mit lagerfähigen Umweltproben. In: Oehlmann, J., Markert, B. (Eds), Ökotoxikologie – ökosystemare Ansätze und Methoden. Ecomed, Landsberg, pp. 285–293.

Köhler, H.R., Rahman, B., Graff, S., Berkus, M., Triebskorn, R., 1996. Expression of the stress-70 protein family (HSP70) due to heavy metal contamination in the slug, *Deroceras reticulatum*: An approach to monitor sublethal stress conditions. Chemosphere 33, 1327–1340.

Köhler, H.R., Triebskorn, R., 1998. Assessment of the cytotoxic impact of heavy metals on soil invertebrates using a protocol integrating qualitative and quantitative components. Biomarkers 3, 109–127.

Köhler, H.R., Triebskorn, R., Stöcker, W., Kloetzel, P.M., Alberti, G., 1992. The 70 kD heat shock protein (hsp 70) in soil invertebrates: a possible tool for monitoring environmental toxicants. Archives of Environmental Contamination and Toxicology 22, 334–338.

Kohn, A.J., Almasi, K.N., 1993. Imposex in Australian *Conus*. Journal of the Marine Biological Association of the UK 73, 241–244.

Kolkwitz, R., Marsson, M., 1909. Ökologie der tierischen Saprobien. Internationale Revue der gesamten Hydrobiologie 2, 126–152.

Koster, A.S., van den Biggelaar, J.A.M., 1980. Abnormal development of *Dentalium* due to Amoco Cadiz oil spill. Marine Pollution Bulletin 11, 166–169.

Labrot, F., Ribera, D., Saint Denis, M., Narbonne, J.F., 1996. In vitro and in vivo studies of potential biomarkers of lead and uranium contamination: lipid peroxidation, acetylcholinesterase, catalase and glutathione peroxidase activities in three non-mammalian species. Biomarkers 1, 21–28.

Lam, P.K.S., 1996. Sublethal effects of cadmium on the energetics of a freshwater snail, *Brotia hainanensis* (Brot, 1872). Environmental Toxicology and Water Quality 11, 345–149.

Leavitt, D.F., Lancaster, B.A., Lancaster, A.S., Capuzzo, J.M., 1990. Changes in the biochemical composition of a subtropical bivalve, *Arca zebra*, in response to contaminant gradients in Bermuda. Journal of Experimental Marine Biology and Ecology 138, 85–98.

Lee, R.F., 1985. Metabolism of tributyltin oxide by crabs, oysters and fish. Marine Environmental Research 17, 145–148.

Legierse, K.C.H.M., Sijm, D.T.H.M, van Leeuwen, C.J., Seinen, W., Hermens, J.L.M., 1998. Bioconcentration kinetics of chlorobenzenes and the organophosphorus pesticide chlorthion in the pond snail *Lymnaea stagnalis* – a comparison with the guppy *Poecilia reticulata*. Aquatic Toxicology 41, 301–323.

Lewis, S., Handy, R.D., Cordi, B., Billinghurst, Z., Depledge, M.H., 1999. Stress proteins (HSPs): methods of detection and their use as an environmental biomarker. Ecotoxicology 8, 351–368.

Livingstone, D.R., 1988. Responses of the detoxication/toxication enzyme systems of molluscs to organic pollutants and xenobiotics. Marine Pollution Bulletin 16, 158–164.

Livingstone, D.R., Farrar, S.V., 1985. Responses of the mixed function oxidase system of some bivalve and gastropod molluscs to exposure to polynuclear aromatic and other hydrocarbons. Marine Environmental Research 17, 101–105.

Livingstone, D.R., Kirchin, M.A., Wiseman, A., 1989. Cytochrome P-450 and oxidative metabolism in molluscs. Xenobiotica 19, 1041–1062.

Livingstone, D.R., Arnold, R., Chipman, K., Kirchin, M.A., Marsh, J., 1990. The mixed-function oxygenase system in molluscs: metabolism, responses to xenobiotics and toxicity. Océanis 16, 331–347.

Lowe, D.M., Pipe, R.K., 1994. Contamination induced lyosomal membrane damage in marine mussel digestive cells: an in vitro study. Aquatic Toxicology 30, 357–365.

Lucas, A., Beninger, P.G., 1985. The use of physiological condition indices in marine bivalve aquaculture. Aquaculture 44, 187–200.

Lundebye, A.-K., Langston, W.J., Depledge, M.H., 1997. Stress proteins and condition index as biomarkers of tributyltin exposure and effect in mussels. Ecotoxicology 6, 127–136.

Markert, B., Oehlmann, J., 1998. Ecotoxicology. In: Ambasht, R.S. (Ed.), Modern trends in ecology and environment. Backhuys, Leiden, pp. 37–53.

Markert, B., Wappelhorst, O., Weckert, V., Herpin, U., Siewers, U., Friese, K., Breulmann, G., 1999. The use of bioindicators for monitoring the heavy-metal status of the environment. Journal of Radioanalytical and Nuclear Chemistry 240, 425–429.

Martinez-Expositio, M.J., Pasantes, J.J., Mendez, J., 1994. Proliferation kinetics of mussel (*Mytilus galloprovincialis*) gill cells. Marine Biology 120, 41–45.

Matozzo, V., Ballarin, L., Pampanin, D.M., Marin, M.G., 2001. Effects of copper and cadmium exposure on functional responses of hemocytes in the clam *Tapes philippinarum*. Archives of Environmental Contamination and Toxicology 41, 163–170.

McCarthy, L.H., Robertson, K., Hesslein, R.H., Williams, T.G., 1997. Baseline studies in the Slave River, NWT, 1990–1994. 4. Evaluation of benthic invertebrate populations and stable isotope analyses. The Science of the Total Environment 197, 111–125.

McKinney, J.D., Rogers, R., 1992. Metal bioavailability. Environmental Science and Technology 26, 1298–1299.

Mersch, J., Beauvais, M.N., 1997. The micronucleus assay in the zebra mussel, *Dreissena polymorpha*, to in situ monitor genotoxicity in freshwater environments. Mutation Research 393, 141–149.

Minchin, D., Oehlmann, J., Duggan, C.B., Stroben, E., Keatinge, M., 1995. Marine TBT antifouling contamination in Ireland, following legislation in 1987. Marine Pollution Bulletin 30, 633–639.

Minchin, D., Stroben, E., Oehlmann, J., Bauer, B., Duggan, C.B., Keatinge, M., 1996. Biological indicators used to map organotin contamination in Cork Harbour, Ireland. Marine Pollution Bulletin 32, 188–195.

Minchin, D., Bauer, B., Oehlmann, J., Schulte-Oehlmann, U., Duggan, C.B., 1997. Biological indicators used to map organotin contamination from a fishing port, Killybegs, Ireland. Marine Pollution Bulletin 34, 235–243.

Moore, M.N., 1982. Lysosomes and environmental stress. Marine Pollution Bulletin 13, 42–43.

Moore, M.N., 1990. Lysosomal cytochemistry in marine environmental monitoring. Histochemistry Journal 22, 189–191.

Morcillo, Y., Albalat, A., Porte, C., 1999. Mussels as sentinels of organotin pollution: Bioaccumulation and effects on P450-mediated aromatase activity. Environmental Toxicology and Chemistry 18, 1203–1208.

Morcillo, Y., Porte, C., 1997. Interaction of tributyl- and triphenyltin with the microsomal monooxygenase system of molluscs and fish from the Western Mediterranean. Aquatic Toxicology 38, 35–46.

Morcillo, Y., Ronis, M.J.J., Porte, C., 1998a. Effects of tributyltin on the Phase I testosterone metabolism and steroid titres of the clam *Ruditapes decussata*. Aquatic Toxicology 42, 1–13.

Morcillo, Y., Ronis, M.J.J., Solé, M., Porte, C., 1998b. Effects of tributyltin on the cytochrome P450 monooxygenase system and sex steroid metabolism in the clam *Ruditapes decussata*. Marine Environmental Research 46, 583–586.

Moulton, C.A., Fleming, W.J., Purnell, C.E., 1996. Effects of two cholinesterase-inhibiting pesticides on freshwater mussels. Environmental Toxicology and Chemistry 15, 131–137.

Myrmel, T., McCully, J.D., Malkin, L., Krukenkamp, I.B., Levitsky, S., 1994. Heat shock protein 70 mRNA is induced by anaerobic metabolism in rat hearts. Circulation 90, 299–305.

Nebert, D.W., Nelson, D.R., Adesnik, M., Coon, M.J., Eastbrook, R.W., Gonzales, F.J., Guengerich, F.P., Gunsalus, I.C., Johnson, E.F., Kemper, B., Levin, W., Phillips, I.R., Sato, R., Waterman, M.R., 1989. The P450 gene superfamily. Update on the naming of new genes and nomenclature of chromosomal loci. DNA 8, 1–13.

Nicholson, S., 1999. Cytological and physiological biomarker responses from green mussels, *Perna viridis* (L.) transplanted to contaminated sites in Hong Kong coastal waters. Marine Pollution Bulletin 39, 261–268.

Noel, D., Pipe, R.K., Elston, R., Bachère, E., Mailhe, E., 1994. Antigenic characterization of hemocyte subpopulations in the mussel *Mytilus edulis* by means of monoclonal antibodies. Marine Biology 119, 549–556.

Oehlmann, J., 1994. Imposex bei Muriciden (Gastropoda, Prosobranchia), eine ökotoxikologische Untersuchung zu TBT-Effekten. Cuvillier Verlag, Göttingen.

Oehlmann, J., Bauer, B., Minchin, D., Schulte-Oehlmann, U., Fioroni, P., Markert, B., 1998a. Imposex in *Nucella lapillus* and intersex in *Littorina littorea*: interspecific comparison of two TBT-induced effects and their geographical uniformity. Hydrobiologia 378, 199–213.

Oehlmann, J., Bettin, C., 1996. TBT-induced imposex and the role of steroids in marine snails. Malacological Review Supplement 6 (Molluscan Reproduction), 157–161.

Oehlmann, J., Fioroni, P., Stroben, E., Markert, B., 1996a. Tributyltin (TBT) effects on *Ocinebrina aciculata* (Gastropoda: Muricidae): imposex development, sterilization, sex change and population decline. The Science of the Total Environment 188, 205–223.

Oehlmann, J., Markert, B., 1997. Humantoxikologie. Eine Einführung für Apotheker, Ärzte, Natur- und Ingenieurwissenschaftler. Wissenschaftliche Verlagsgesellschaft, Stuttgart.

Oehlmann, J., Schulte-Oehlmann, U., Tillmann, M., Markert, B., 2000. Effects of endocrine disruptors on prosobranch snails (Mollusca: Gastropoda) in the laboratory. Part I: Bisphenol A and octylphenol as xeno-estrogens. Ecotoxicology 9, 383–397.

Oehlmann, J., Stroben, E., Fioroni, P., 1991. The morphological expression of imposex in *Nucella lapillus* (Linnaeus) (Gastropoda: Muricidae). Journal of Molluscan Studies 57, 375–390.

Oehlmann, J., Stroben, E., Fioroni, P., 1993. Fréquence et degré d'expression du pseudohermaphrodisme chez quelques Prosobranches Sténoglosses des côtes françaises (surtout de la baie de Morlaix et de la Manche). 2. Situation jusqu'au printemps de 1992. Cahiers Biologie Marine 34, 343–362.

Oehlmann, J., Stroben, E., Schulte-Oehlmann, U., Bauer, B., 1998b. Imposex development in response to TBT pollution in *Hinia incrassata* (Ström, 1768) (Prosobranchia, Stenoglossa). Aquatic Toxicology 43, 239–260.

Oehlmann, J., Stroben, E., Schulte-Oehlmann, U., Bauer, B., Fioroni, P., Markert, B., 1996b. Tributyltin biomonitoring using prosobranchs as sentinel organisms. Fresenius Journal of Analytical Chemistry 354, 540–545.

Orbea, A., Marigomez, I., Fernandez, C., Tarazona, J.V., Cancio, I., Cajaraville, M.P. 1999. Structure of peroxisomes and activity of the marker enzyme catalase in digestive epithelial cells in relation to PAH content of mussels from two Basque estuaries (Bay of Biscay): seasonal and site-specific variations. Archives of Environmental Contamination and Toxicology 36, 158–166.

Ortmann, A.E., 1909. The destruction of the fresh-water fauna in western Pennsylvania. Proceedings of the American Philosophical Society 48, 90–100.

Oslo and Paris Commissions (Eds), 1996. Agenda Item 9 of the Environmental Assessment and Monitoring Committee (ASMO). OSPARCOM, Vila Franca do Campo (= ASMO 96/9/8-E).

Page, D.S., 1995. A six-year monitoring study of tributyltin and dibutyltin in mussel tissues from the Lynher River, Tamar Estuary, UK. Marine Pollution Bulletin 30, 746–749.

Page, D.S., Gilfillan, E.S., Foster, J., Widdows, J., 1989. Tributyltin in *Mytilus edulis* from coastal locations in Devon and Cornwall (UK) and Maine (US) and its effect on shell morphology. Marine Environmental Research 28, 539–540.

Page, D.S., Dassanayake, T.M., Gilfillan, E.S., 1996. Relationship between tissue concentrations of tributyltin and shell morphology in field populations of *Mytilus edulis*. Bulletin of Environmental Contamination and Toxicology 56, 500–504.

Pasantes, J.J., Martinez-Expositio, M.J., Torreiro, A., Mendez, J., 1996. The sister chromatid exchange test as an indicator of marine pollution: some factors affecting SCE frequencies in *Mytilus galloprovincialis*. Marine Ecology Progress Series 143, 113–119.

Phelps, H.L., Page, D.S., 1997. Tributyltin biomonitoring in Portuguese estuaries with the Portuguese oyster (*Crassostrea angulata*). Environmental Technology 18, 1269–1276.

Pihan, F., Gomot de Vaufleury, A., 2000. The snail as a target organism for the evaluation of industrial waste dump contamination and the efficiency of its remediation. Ecotoxicology and Environmental Safety 46137–46147.

Pipe, R.K., Coles, J.A., Farley, S.R., 1995a. Assays for measuring immune response in the mussel *Mytilus edulis*. In: Stolen, J.S., Fletcher, T.C., Smith, S.A., Zelikoff, J.T., Kaattari, S.L., Anderson, R.S., Söderhäll, K., Weeks-Perkins, B.A. (Eds), Techniques in Fish Immunology, Vol. IV, Immunology and Pathology of Aquatic Invertebrates. SOS Publications, Fair Haven, pp. 93–100.

Pipe, R.K., Coles, J.A., Thomas, M.E., Fossato, V.U., Pulsford, A.L., 1995b. Evidence for environmentally derived immunomodulation in mussels from the Venice Lagoon. Aquatic Toxicology 32, 59–73.

Phillips, D.J.H., 1977. The use of biological indicator organisms to monitor trace metal pollution in marine and estuarine environments – a review. Environmental Pollution 13, 281–317.

Powell, E.N., Cummins, H., 1985. Are molluscan maximum life spans determined by long-term cycles in benthic communities? Oecologia 67, 177–182.

Purchon, R.D., 1968. The Biology of the Mollusca. Pergamon Press, Oxford.

Rainbow, P.S., Phillips, D.J.H., 1993. Cosmopolitan biomonitors of trace metals. Marine Pollution Bulletin 26, 593–601.

Rattner, B.A., and Fairbrother, A., 1991. Biological variability and the influence of stress on cholinesterase activity. In: Mineau, P. (Ed.), Cholinesterase-inhibiting Insecticides. Elsevier, Amsterdam, pp. 89–107.

Rees, H.L., Waldock, R., Matthiessen, P., Pendle, M.A., 1999. Surveys of the epibenthos of the Crouch Estuary (UK) in relation to TBT contamination. Journal of the Marine Biological Association of the UK 79, 209–223.

Regoli, F., Orlando, E., 1999. Accumulation and subcellular distribution of metals (Cu, Fe, Mn, Pb and Zn) in the Mediterranean mussel *Mytilus galloprovincialis* during a field transplant experiment. Marine Pollution Bulletin 28, 592–600.

Renaud, C.B., Kaiser, K.L.E., Comba, M.E., Metcalfe-Smith, J.L., 1995. Comparison between lamprey ammocoetes and bivalve mollusks as biomonitors of organochlorine contaminants. Canadian Journal of Fisheries and Aquatic Sciences 52, 376–282.

Richmond, L., Beeby, A., 1992. A comparative study of lead uptake by three populations of the snail *Helix aspersa* Müller. Polish Journal of Environmental Studies 1, 9–13.

Ringwood, A.H., Conners, D.E., Hoguet, J., 1998. Effects of natural and anthropogenic stressors on lysosomal destabilization in oysters *Crassostrea virginica*. Marine Ecology Progress Series 166, 163–171.

Ritossa, F., 1962. A new puffing pattern induced by temperature shock and DNP in *Drosophila*. Experientia 18, 571–573.

Roesijadi, G., Young, Y.S., Drum, A.S., Gurtisen, J.M., 1984. Behavior of trace metals in *Mytilus edulis* during a reciprocal transplant field experiment. Marine Ecology Progress Series 18, 155–170.

Ronis, M.J.J., Mason, A.Z., 1996. The metabolism of testosterone by the periwinkle (*Littorina littorea*) in vitro and in vivo: effects of tributyl tin. Marine Environmental Research 42, 161–166.

Rorke, M.A., Gardner, D.R., 1974. Lethality and behavioural symptoms produced by some organophosphorous compounds in the snail (*Helix aspersa*). Bulletin of Environmental Contamination and Toxicology 11, 417–424.

Rossbach, M., Kniewald, G., 1997. Concepts of marine specimen banking. Chemosphere 34, 1997–2010.

Saavedra Alvarez, M.M., Ellis, D.V., 1990. Widespread neogastropod imposex in the northeast Pacific: implications for TBT contamination surveys. Marine Pollution Bulletin 21, 244–247.

Sanders, B.M., 1990. Stress-proteins: potential as multitiered biomarkers. In: Shugart, L., McCarthy, J. (Eds), Environmental Biomarkers. Lewis, Boca Raton, pp. 165–191.

Sarkar, S.K., Bhattachaya, B., Debnath, S., 1994. The suitability of tropical marine bivalves as biomonitors of heavy metals in deltaic Sundarbans, northeast India. Chemosphere 29, 759–770.

Sato, M., Bremner, I., 1993. Oxygen free radicals and metallothionein. Free Radical Biology and Medicine 14, 325–337.

Schenkman, J.B., Kupfer, D. (Eds), 1982. Hepatic cyctochrome P-450 monooxygenase system. Pergamon Press, Oxford.

Schlesinger, M.J., 1990. Heat-shock proteins: a mini review. Journal of Biological Chemistry 265, 12111–12114.

Schulte-Oehlmann, U., 1997. Fortpflanzungsstörungen bei Süß- und Brackwasserschnecken – Einfluß der Umweltchemikalie Tributylzinn. Wissenschaft und Technik Verlag, Berlin.

Schulte-Oehlmann, U., Bettin, C., Fioroni, P., Oehlmann, J., Stroben, E., 1995. *Marisa cornuarietis* (Gastropoda, Prosobranchia): a potential TBT bioindicator for freshwater environments. Ecotoxicology 4, 372–384.

Schulte-Oehlmann, U., Duft, M., Tillmann, M., Markert, B., Stachel, B., Oehlmann, J., 2001. Biologisches Effektmonitoring an Sedimenten der Elbe mit *Potamopyrgus antipodarum* und *Hinia (Nassarius) reticulata* (Gastropoda: Prosobranchia). ARGE Elbe, Hamburg.

Schulte-Oehlmann, U., Watermann, B., Tillmann, M., Scherf, S., Markert, B., Oehlmann, J., 2000. Effects of endocrine disruptors on prosobranch snails (Mollusca: Gastropoda) in the laboratory. Part II: Triphenyltin as a xeno-androgen. Ecotoxicology 9, 399–412.

Schuytema, G.S., Nebeker, A.V., Griffis, W.L., 1994. Effects of dietary exposure to forest pesticides on the brown garden snail *Helix aspersa*. Archives of Environmental Contamination and Toxicology 26, 23–28.

Segner, H., Braunbeck, T., 1998. Cellular response profile to chemical stress. In: Schüürmann, G., Markert, B. (Eds), Ecotoxicology. Ecological Fundamentals, Chemical Exposure, and Biological Effects. Wiley, Chichester, and Spektrum Akademischer Verlag, New York, pp. 521–569.

Short, J.W., Sharp, J.L., 1989. Tributyltin in bay mussels (*Mytilus edulis*) of the Pacific coast of the United States. Environmental Science and Technology 23, 740–743.

Short, J.W., Rice, S.D., Brodersen, C.C., Stickle, W.B., 1989. Occurrence of tri-n-butyltin-caused imposex in the North Pacific marine snail *Nucella lima* in Auke Bay, Alaska. Marine Biology 102, 291–297.

Smith, B.S., 1971. Sexuality in the American mud snail, *Nassarius obsoletus* Say. Proceedings of the Malacological Society of London 39, 377.

Smith, K.L., Galloway, T.S., Depledge, M.H., 2000. Neuro-endocrine biomarkers of pollution-induced stress in marine invertebrates. The Science of the Total Environment 262, 185–190.

Smith, P.J., McVeagh, M., 1991. Widespread organotin pollution in New Zealand coastal waters as indicated by imposex in dogwhelks. Marine Pollution Bulletin 22, 409–413.

Snyder, M.J., Girvetz, E., Mulder, E.P., 2001. Induction of marine mollusc stress proteins by chemical or physical stress. Archives of Environmental Contamination and Toxicology 41, 22–29.

Snyman, R.G., Reinecke, S.A., Reinecke, A.J., 2000. Hemocyctic lysosome response in the snail *Helix aspersa* after exposure to the fungicide copper oxychloride. Archives of Environmental Contamination and Toxicology 39, 480–485.

Sobral, P., Widdows, J., 1997. Effects of copper exposure on the scope for growth of the clam *Ruditapes decussatus* from southern Portugal. Marine Pollution Bulletin 34, 992–1000.

Spang, W.D., 1995. Terrestrische Gastropoden (Landschnecken) als Reaktions- und Akkumulationsindikatoren. Umweltsystem-wissenschaften und Schadst-offorschung – Zeitschrift für Umweltchemie und Ökotoxikologie 7, 251–253.

Spooner, N., Gibbs, P.E., Bryan, G.W., Goad, L.J., 1991. The effect of tributyltin upon steroid titres in the female dogwhelk, *Nucella lapillus*, and the development of imposex. Marine Environmental Research 32, 37–49.

Statzner, B., Bis, B., Doledec, S., Usseglio-Polatera, P., 2001. Perspectives for biomonitoring at large spatial scales: a unified measure for the functional composition on invertebrate communities in European running waters. Basic and Applied Ecology 2, 73–85.

Steinert, S.A., Pickwell, G.V., 1993. Induction of HSP70 proteins in mussels by ingestion of tributyltin. Marine Environmental Research 35, 89–93.

Steinert, S.A., Steib-Montee, R., Leather, J.M., Chadwick, D.B., 1998. DNA damage in mussels at sites in San Diego Bay. Mutation Research 399, 65–85.

Stephenson, M.D., Martin, M., Tjeerdema, R.S., 1995. Long-term trends in DDT, polychlorinated biphenyls, and chlordane in California mussels. Archives of Environmental Contamination and Toxicology 28, 443–450.

Stewart, C., de Mora, S.J., Jones, M.R.L., Miller, M.C., 1992. Imposex in New Zealand neogastropods. Marine Pollution Bulletin 24, 204–209.

Stickle, W.B., Moore, M.N., Bayne, B.L., 1985. Effects of temperature, salinity and aerial exposure on predation and lysosomal stability on the dogwhelk *Thais* (*Nucella*) *lapillus* (L.). Journal of Experimental Marine Biology and Ecology 93, 235–258.

Strayer, D.L., Smith, L.C., Hunter, D.C., 1998. Effects of the zebra mussel (*Dreissena polymorpha*) invasion on the macrobenthos of the freshwater tidal Hudson River. Canadian Journal of Zoology 76, 419–425.

Sultan, A., Abelson, A., Bresler, V., Fishelson, L., Mokady, O., 2000. Biomonitoring marine environmental quality at the level of gene-expression – testing the feasibility of a new approach. Water Science and Technology 42, 269–274.

Sundermann, G., Bauer, B., Oehlmann, J., 1998. Ultrastructure of prostate gland tissue in males and females with intersex phenomena of *Littorina littorea* L. Hydrobiologia 378, 227–233.

Szefer, P., Geldon, J., Ali, A.A., Osuna, F.P., Ruiz-Fernandes, A.C., Galvan, S.R.G., 1998. Distribution and association of trace metals in soft tissue and byssus of *Mytella strigata* and other benthal organisms from Mazatlan harbour, Mangrove lagoon of the northwest coast of Mexico. Environment International 24, 359–374.

Szefer, P., Ali, A.A., Ba-Haroon, A.A., Rajeh, A.A., Geldon, J., Nabrzyski, M., 1999. Distribution and relationships of selected trace metals in molluscs and associated sediments from the Gulf of Aden, Yemen. Environmental Pollution 106, 299–314.

Terlizzi, A., Geraci, S., Gibbs, P.E., 1999. Tributyltin (TBT)-induced imposex in the neogastropod *Hexaplex trunculus* in Italian coastal waters: morphological aspects and ecological implications. Italian Journal of Zoology 66, 141–146.

Thompson, H.M., 1999. Esterases as markers of exposure to organophosphates and carbamates. Ecotoxicology 8, 369–384.

Tillmann, M., Schulte-Oehlmann, U., Duft, M., Markert, B., Oehlmann, J., 2001. Effects of endocrine disruptors on prosobranch snails (Mollusca: Gastropoda) in the laboratory. Part III. Cyproterone acetate and vinclozolin as antiandrogens. Ecotoxicology 10, 373–388.

Vaschenko, M.A., Syasina, I.G., Zhadan, P.M., Medvedeva, L.A., 1997. Reproductive function state of the scallop *Mizuhopecten yessoensis* Jay from polluted areas of Peter the Great Bay, Sea of Japan, Hydrobiologia 352, 231–240.

Veldhuizen-Tsoerkan, M.B., Holwerda, D.A., de Bont, A.M., Smaal, A.C., Zandee, D.I., 1991. A field study on stress indices in the sea mussel, *Mytilus edulis*: application of the "stress approach" in bio-monitoring. Archives of Environmental Contamination and Toxicology 21, 497–504.

Venier, P., Maron, S., Canova, S., 1997. Detection of micronuclei in gill cells and haemocytes of mussels exposed to benzo[a]pyrene. Mutation Research 390, 33–44.

Viarengo, A., Burlando, B., Dondero, F., Marro, A., Fabbri, R., 1999. Metallothionein as a tool in bio-monitoring programmes. Biomarkers 4, 455–466.

Viarengo, A., Canesi, L., Pertica, M., Mancinelli, G., Accomando, R., Smaal, A.C., Orunesu, M., 1995. Stress on stress response – a simple monitoring tool in the assessment of a general stress syndrome in mussels. Marine Environmental Research 39, 245–248.

Viarengo, A., Ponzano, E., Dondero, F., Fabbri, R., 1997. A simple spectrophotometric method for metallothionein evaluation in marine organisms: an application to Mediterranean and Antarctic molluscs. Marine Environmental Research 44, 69–84.

Villeneuve, J.P., Carvalho, F.P., Fowler, S.W., Cattini, C., 1999. Levels and trends of PCBs, chlorinated pesticides and petroleum hydrocarbons in mussels from the NW Mediterranean coast: comparison of concentrations in 1973/1974 and 1988/1989. The Science of the Total Environment 237–238, 57–65.

Waldén, H.W., 1986. Endangered species of land molluscs in Sweden and Madeira. In: Kay, E.A. (Ed.), The conservation biology of molluscs. Proceedings of a symposium held at the 9th International

Malacological Congress, Edinburgh, Scotland. International Union for Conservation of Nature and Natural Resources (IUCN) and Species Survival Commission (SSC), New York, pp. 19–24.

Waldock, R., Rees, H.L., Matthiessen, P., Pendle, M.A., 1999. Surveys of the benthic infauna of the Crouch Estuary (UK) in relation to TBT contamination. Journal of the Marine Biological Association of the UK 79, 225–232.

Walsh, K., Dunstan, R.H., Murdoch, R.N., Conroy, B.A., Roberts, T.K., Lake, P., 1994. Bioaccumulation of pollutants and changes in population parameters in the gastropod mollusk *Austrocochlea constricta*. Archives of Environmental Contamination and Toxicology 26, 367–373.

Walsh, K., Dunstan, R.H., Murdoch, R.N., 1995. Differential bioaccumulation of heavy metals and organopollutants in the soft tissue and shell of the marine gastropod, *Austrocochlea constricta*. Archives of Environmental Contamination and Toxicology 28, 35–39.

Watermann, B., Ide, I., Liebe, S., Witten, E.P., 1996. Hämocyteninfiltrationen und Granulocytome in Gastropoden und Bilvalviern, Indikatoren für Umweltbelastungen? In: Hoffmann, R., Bernoth, E.M. (Eds), Tagung der Fachgruppe "Fischkrankheiten" in Verbindung mit der EAFP/Deutsche Sektion. Deutsche Veterinärmedizinische Gesellschaft, Königswartha, pp. 218–229.

Weis, P., Weis, J., Couch, J., 1993. Histopathology and bioaccumulation in oysters *Crassostrea virginica* living on wood preserved with chromated copper arsenate. Diseases of Aquatic Organisms 17, 41.

Widdows, J., Bakke, T., Bayne, B.L., Donkin, P., Livingstone, D.R., Lowe, D.M., Moore, M.N., Evans, S.V., Moore, S.L., 1982. Responses of *Mytilus edulis* L. on exposure to the water accommodated fraction of North Sea oil. Marine Biology 67, 15–31.

Widdows, J., Donkin, P., Brinsley, M.D., Evans, S.V., Salkeld, P.N., Franklin, A., Law, R.J., Waldock, M.J., 1995. Scope for growth and contaminant levels in North Sea mussels *Mytilus edulis*. Marine Ecology Progress Series 127, 131–148.

Widdows, J., Basci, C., Fossato, V.U., 1997. Effects of pollution on the scope for growth of mussels (*Mytilus galloprovincialis*) from the Venice lagoon, Italy. Marine Environmental Research 43, 69–79.

Wilbrink, M., Groot, E.J., Jansen, R., de Vries, Y., Vermeulen, N.P.E., 1991. Occurrence of a cytochrome P-450-containing mixed-function oxidase system in the pond snail, *Lymnaea stagnalis*. Xenobiotica 21, 223–233.

Williams, L.G., Chapman, P.M., Ginn, T.C., 1986. A comparative evaluation of marine sediment toxicity using bacterial luminiscence, oyster embryo and amphipod sediment bioassay. Marine Environmental Research 19, 225–249.

Wilson, J.T., Pascoe, P.L., Parry, J.M., Dixon, D.R., 1998. Evaluation of the comet assay as a method for the detection of DNA damage in the cells of a marine invertebrate, *Mytilus edulis* L. (Mollusca: Pelecypoda). Mutation Research 399, 87–95.

Wilson, S.P., Ahsanullah, M., Thompson, G.B., 1993. Imposex in neogastropods: an indicator of tributyltin contamination in eastern Australia. Marine Pollution Bulletin 26, 44 48.

Winberg, G.G., 1960. Rate of metabolism and food requirements of fishes. Transaction Series of the Fisheries Research Board of Canada 194, 1–202.

Wise, S.A., Koster, B.J., Langland, J.K., Zeisler, R., 1993. Current activities within the National Biomonitoring Specimen Bank. The Science of the Total Environment 139–140, 1–12.

Wo, K.T., Lam, P.K.S., Wu, R.S.S., 1999. A comparison of growth biomarkers for assessing sublethal effects of cadmium on a marine gastropod, *Nassarius festivus*. Marine Pollution Bulletin 39, 165–173.

Young, A.G., Wilkins, R.M., 1989. The response of the invertebrate acetylcholinesterase to molluscicides. In: Henderson, I.F. (Ed.), Slugs and snails in world agriculture. London, British Crop Protection Council Monograph 41, pp. 121–128.

Zatta, P., Gobbo, S., Rocco, P., Perazzolo, M., Favarato, M., 1992. Evaluation of heavy metal pollution in the Venetian lagoon by using *Mytilus galloprovincialis* as biological indicator. The Science of the Total Environment 119, 29–41.

Vertebrates

Bioindicators and biomonitors
B.A. Markert, A.M. Breure, H.G. Zechmeister, editors

Chapter 18

Fish as bioindicators

Andreas Chovanec, Rudolf Hofer and Fritz Schiemer

Abstract

In this article, the role of fish as bioindicators is discussed. The comprehensive knowledge of taxonomy, habitat requirements, and physiology of fish is a key prerequisite of using fish as indicators. No other aquatic organism is suitable for the application of so many different methods which allow the evaluation of the severity of toxic impacts by determining the accumulation of toxicants in tissues, by using histological and haematological approaches or by detecting morphological anomalies. Due to its complex habitat requirements the fish fauna is a crucial indicator of the ecological integrity of aquatic systems at different scales, from microhabitat to catchment. The fitness of fish species both at the individual level (e.g. growth performance) and at population level (e.g. population structure) is determined by the connectivity of different habitat elements in a broad spatial-temporal context. Thus bioindication using fish represents a good monitoring tool especially with regard to both pollution aspects and to river engineering, e.g. river restoration and management.

Keywords: Toxicant, Pollutant, Accumulation, Ecological Integrity, Connectivity, Habitat Structures, Water Framework Directive

1. Introduction

Over the last 150 years, aquatic systems worldwide have been impacted by a wide array of anthropogenic factors (e.g. Falkenmark and Allard, 1991; Dynesius and Nilsson, 1994; Rahel, 2000). Human activities may alter the physical, chemical or biological processes associated with water resources and thus modify the resident biological community. Karr (1991) identified five primary classes of environmental factors, that, when affected by human activities, result in ecosystem degradation:

- food/energy source: e.g. type, amount, and particle size of organic material; seasonal pattern of available energy;
- water quality: e.g. dissolved oxygen, nutrients, toxic substances;
- habitat structure: e.g. substrate type, sinuosity, channel morphology, connectivity aspects;
- flow regime: e.g. temporal distribution of floods and low flows;
- biotic interactions: e.g. competition, parasitism.

In most cases biological communities are sound and precise indicators of the status of the aquatic system as they are subject to the full range of chemical and physical influences, additive and synergistic effects included. In this context fish play a crucial role as bioindicators in water resource management and applied limnological research:

fish serve as "ecological indicators", "keystones", "umbrellas", "flagships" and "vulnerables" (Noss, 1990).

According to Markert (1994) a bioindicator is an organism (or a part of an organism or a community of organisms) that contains information on the quality of the environment. Thus, the use of bioindicators should help to describe the natural environment, to detect and assess human impacts and to evaluate restoration or remediation measures; in all these cases fish are intensively used for indication purposes.

The spatial changes of fish communities along the course of river systems and the use of fish zonation patterns for river classification are examples of some of the most traditional bioindication approaches (Fritsch, 1872; Thienemann, 1912, 1925). Fish have also been traditionally used for classifying different types of standing waters. The nature of the fauna in stagnant water bodies reflects their morphometry, trophic status, thermal and oxygen stratification and the extent of littoral development.

There are several reasons why fish are widely used to describe natural characteristics of aquatic systems and to assess habitat alterations (Noss, 1990; Cairns et al., 1993; Chovanec and Spindler, 1997; Boon et al., 2000; Schiemer, 2000; Schmutz et al., 2000):

- A long tradition of ecological, physiological and ecotoxicological research on fish has led to an advanced knowledge of the ecological requirements of a large number of fish species. The effectiveness of bioindication approaches depends on the sound knowledge of the indicators' ecological demands and physiology (Schiemer et al., 2001).

- A large number of abiotic environmental variables at different spatio-temporal scales are linked to the complex habitat requirements of particular species and their ontogenetic stages. Due to the specific habitat requirements and habitat shifts during the larval and juvenile stages, 0+ fish for example are suitable indicators of the ecological status of river systems (Schiemer et al., 1991; Keckeis et al., 1996).

- As migratory organisms fish are suitable indicators of habitat connectivity or fragmentation (e.g. Jungwirth, 1998; Chovanec et al., 2002). Waidbacher and Haidvogl (1998) point out that members of the families Clupeidae (*Alosa pontica, Alosa caspia nordmanni*) and Acipenseridae (*Huso huso, Acipenser gueldenstaedti, Acipenser stellatus*) migrate distances > 300 km in the Danube River system. Species typical of the upper reaches of the Danube like the highly abundant barbel *Barbus barbus* and nase *Chondrostoma nasus* migrate distances of more than 50 km.

- Due to the size of fish (and their organs) a great variety of analytical procedures can be carried out. Pathological results concerning fish illustrate the effects of water pollution to the scientific community, water management and the public. Some methods, such as haematological and histo-pathological approaches, are taken from human medicine.

- Due to the longevity of fish certain indication effects, e.g. accumulation processes, are increased.

- As primary and secondary consumers at different levels fish reflect trophic conditions in aquatic systems.

- The reconstruction of pristine reference communities is possible due to the existence of historical information (Muhar et al., 2000).

- Fishery and sport fishing have a long history, in which fish play an important role as indicators of water quality; because of the use of fish by man particularly as food resource, the condition of fish communities is an important factor in water resource management.
- Depending on the problem and the indication approach selected, bioindication by using fish often meets the requirements of both top-down approaches (assessing changes in communities in the natural environment and testing for sources and causes of possible problems) and of bottom-up assessments (using laboratory data to model changes in the more complex natural ecosystems).
- The number of species is relatively small and species are already determinable in the field.

When using fish as bioindicators problems may arise:

- Fishery-caused alterations, such as species transfer, stocking, overfishing, make it more difficult to discuss other man-induced degradations of aquatic ecosystems.
- The mobility of many species makes it difficult to identify not only the exact source of pollution, but also the time and duration of exposure.

Bioindication based on the use of fish generally satisfies the criteria against which biological monitoring programmes should be judged (Halbwachs and Arndt, 1992; Cairns et al., 1993; Yoder and Rankin, 1995; Chovanec and Koller-Kreimel, 1999; Karr and Chu, 1999; Chovanec et al., 2000a):

- Sensitivity to stressors: The range of responses must be suitable for the intended application; factors of different strength should lead to reactions of different intensity (no all or none response, no extreme natural variability).
- The range of response has to be sensitive to the environmental factors and conditions being observed.
- Methods have to be broadly applied in a wide range of stressors and sites.
- The results obtained by bioindication programmes have to be representative of many parts of the aquatic communities.
- Information has to be provided fast enough to initiate effective management action before unacceptable damage has occurred.
- Standardised methods are necessary for obtaining comparable results.
- Bioindicators should be cost-effective to collect and identify.
- The application of bioindicators should be possible at a local scale as well as at a regional or landscape scale.
- Bioindicators should be at least one of the following three major types (Cairns et al., 1993):
  - compliance indicators, which are chosen to assess the attainment and maintenance of ecosystem objectives related to the restoration and maintenance of environmental quality;
  - diagnostic indicators, which provide insight into the cause of noncompliance;
  - early warning indicators, which allow for management actions to be implemented before conditions have deteriorated to the point where compliance indicators become relevant.
  In many bioindication programmes fish meet the requirements of all three types.

This chapter provides an overview of different assessment approaches using fish as bioindicators. Special attention is paid to the indication of different pollution aspects and to the assessment of running water habitat degradation and the loss of connectivity. The chapter describes indication approaches in the field, laboratory tests under controlled conditions are not discussed in detail. Recent developments concerning the role of fish as indicators in integrated assessment procedures are also considered, such as the requirements of the new EU Water Framework Directive.

2. Fish as indicators of environmental pollution

Despite rising efforts of many industrialised countries to reduce toxicants from industrial and motor vehicle exhausts and to purify industrial and communal waste waters, our ecosystems still contain harmful concentrations of an increasing number of chemicals. They accumulate in soils and sediments from which they can be remobilised after changing their physico-chemical condition, and many of these substances persist for decades (e.g. DDTs, PCBs). Concentrations of heavy metals in sediments may exceed those of the overlying water by a factor of one to ten thousand (Bryan and Langston, 1992). Even remote areas such as high mountains and arctic regions receive significant amounts of pollutants by atmospheric deposition after transport over long distances (Wania, 1999). Relatively small quantities of toxicants may threaten these highly vulnerable ecosystems (Skjelkvale and Wright, 1998).

The water quality of many rivers and lakes has improved significantly due to the increasing number of purification plants. However, the treatment of waste water reduces not only the concentration of toxic substances but also that of non-toxic organic compounds. This may lead to changes in the bioavailability of chemicals and their toxicity, in particular of those entering the water by run-off and atmospheric deposition. Suspended inorganic and organic particles have a large surface area and thus a high capacity for physically absorbing toxicants. Toxic chemicals have been shown to interact with dissolved or colloidal organic matter by various modes of binding and absorption (Spry and Wiener, 1991). Many of these complexes are too large or too polar to diffuse across the gill membrane (Haitzer et al., 1998). Some metal cations can form lipophilic complexes with specific organic compounds used in agriculture, forestry and industry (e.g. ditiocarbamates, diethyldithiophosphate) which easily pass the gill membrane. This leads to both higher levels of metal accumulation than expected from water concentrations and an altered distribution pattern, with the highest increase in the brain and eyes of fish (Tjälve and Gottofry, 1991). Uptake and toxicity of mercury strongly depends on methylation by bacterial activity (Boening, 2000). Due to its lipophilic character, methyl mercury is absorbed about ten times faster than the ionic form. On the other hand, several studies have shown that selenium may reduce mercury toxicity (Cuvin-Aralar and Furness, 1991).

Bioavailability and toxicity of metals are controlled not only by suspended particles and dissolved organic matter but also by water parameters such as hardness, alkalinity, pH, temperature, and oxygen concentration (Köck and Hofer, 1998). Some of these factors modulate the speciation of trace metals (Davies et al., 1976). Hydrate ions and hydroxo-complexes are the most bioavailable forms of metals absorbed by fish gills

(Erickson et al., 1994). However, metals behave differently in natural waters: The speciation of Pb, Cu, Hg, and Al is highly affected by pH, whereas that of Cd and Zn is only slightly sensitive to pH alterations (Campbell and Stokes, 1985). The calcium concentration of the water has a major impact on metal speciation and the permeability of gill membranes. Competition between divalent metal ions and calcium for binding sites on the gill surface and the passage through ion-sensitive channels reduce the uptake and toxicity of metals in hard water (Wicklund and Runn, 1988; Köck et al., 1995). Uptake of chemicals across gill membranes is also a function of water flow along the gills (McKim and Erickson, 1991). As a consequence, rising temperature, oxygen depletion and metabolic stimulation (e.g. during reproduction or stress) accelerate gill ventilation and thus the uptake of toxicants.

These examples demonstrate that simple approaches of chemical water analyses often fail to detect environmental changes that are harmful for aquatic organisms. The complex situation in natural waters, with their synergistic and antagonistic effects, makes it difficult to predict the impact of toxicants on the ecosystem. In many cases, the input of toxicants is not constant but intermittent and may remain undetected.

Bioindicators and, in particular, long-living organisms such as fish are sensitive to the impact of a complex mixture of chemicals on a specific aquatic ecosystem, integrating the environmental load over time and space. Pollutants usually cause a wide spectrum of effects and responses in organisms ranging from the cellular and biochemical level to the level of behaviour, growth and reproduction. During low and limited exposure to toxicants, fish respond at a sub-cellular level, but usually organisms can compensate for the toxic effect, and their health is not seriously affected. Prolonged and severe exposure, however, may induce a sequence of functional and structural changes which impair vital functions. Tissue concentrations of chemicals are excellent indicators of the environmental load of a specific toxicant but usually do not directly reflect the physiological and ecological consequences. Most of the biomonitoring techniques, however, focus on different kinds of stress responses which are often more or less general responses and cannot be attributed to specific toxicants. Permanent stress – even if it is moderate – interferes with hormonal and biochemical processes leading to increased metabolism, immunosuppression, disturbed osmoregulation, failure of reproduction or tissue damages. The low toxicant specificity of many stress responses is not just a disadvantage, it increases the value of bioindicators for monitoring the general environmental load in natural water bodies which may contain several out of hundreds of different harmful chemicals. For practical use in the field, biomonitoring methods based on fish should be insensitive to the stress of capture which may mask the effects of toxicants. The biological parameters analysed in the assay should be well understood and their modulation induced by endogenous and exogenous factors other than toxicants should be known. Data on commercially manipulated fish species should be handled with caution, and possible loads of geogenic origin (e.g. metals) have to be considered.

2.1. Toxicant accumulation in fish tissues

Tissue concentrations of chemicals are a function of uptake, storage, and excretion. In fish, two different routes of uptake are important, (1) directly from the water, in

Table 1.   Definition of terms used in ecotoxicology

Bioaccumulation (BA)	The accumulation of contaminants in organisms resulting from water or food uptake.
Bioconcentration (BC)	The accumulation of water-borne contaminants directly from the water by a non-dietary route.
Biomagnification	The accumulation of toxicants resulting from ingestion of contaminated diet.
Bioconcentration factor (BCF)	Quotient of the concentration of a chemical in an aquatic organism and in the water. The BCF can be predicted from the concentration of a lipophilic chemical in the water and its K_{ow}.
log K_{ow}	Octanol-water partition coefficient: In most cases, the BCF is proportional to the logarithm of K_{ow} (Banjeree and Baughman, 1991).

freshwater fish almost exclusively via the gills, in marine species at a low percentage also through the drinking of water, and (2) the oral uptake and assimilation of contaminated food. Hydrophilic molecules are unlikely to pass the gill membrane unless they are very small (diffusion along an osmotic gradient) or transported by ionic pumps or channels. Lipophilic compounds, however, are soluble in biological membranes and cross all barriers. The relatively low oxygen solubility in water requires an extremely large respiratory surface and a high pumping rate of water. Consequently, the direct uptake of water-borne toxicants (whose concentration is two orders of magnitude higher than in the air) is the main route in fish (bioconcentration; see Table 1).

Diet, as the most significant source of toxicants leading to biomagnification along the food chain, is usually restricted to lipophilic compounds which are almost insoluble in water (compounds with log K_{ow} values of 5–8) and slowly metabolised, e.g. halogenated contaminants and pesticides which resist biotransformation (Thomann, 1989; Suedel et al., 1994; Mackay and Fraser, 2000). Lipophilic contaminants are predominantly stored in lipids including biological membranes and muscles, thus being of major concern for human nutrition. On the other hand, lipids serve as a protective reservoir for lipophilic chemicals, and therefore their toxicity decreases with rising lipid content of fish (Geyer et al., 1994). Top predators (e.g. piscivorous fish) and species with high lipid contents have been shown to be the most sensitive indicators for environmental contamination with lipophilic compounds (Geyer et al., 1997).

Liver and kidney are the main sites of accumulation for most toxicants including metals. These organs are rich in metallothioneins with high affinities to Cd, Hg, Zn, and Cu. The liver is also involved in a variety of detoxification processes transforming harmful compounds into less toxic and water-soluble metabolites which are excreted into the bile. These metabolites are either eliminated with the faeces or reabsorbed from the gut and returned to the liver by enterohepatic circulation which may increase the half-life of toxicants in the fish. In the bile of trout exposed to several labelled organic substances Statham et al. (1976) found concentrations between 11 and 10,000 times higher than in the water. Even under field conditions it has been shown that bile analysis is a useful tool to evaluate the environmental load of xenobiotics (Pointet and Milliet, 2000).

The proportion of accumulated toxicants between different tissues of the fish largely depends on dynamic processes between uptake, storage, and elimination. After short-term exposure, gills or the digestive tract and the liver usually show a high load of toxicants, whereas concentrations in kidney, bones (Pb, Zn), and muscles (lipophilic substances) increase more slowly after a time-lag, but the accumulated chemicals are more persistent than in other organs (Olson et al. 1978; Köck et al., 1996). Due to active regulation tissue accumulation of essential metals (Cu, Zn) is saturated at low levels, and thus a relatively weak indicator of environmental contamination (McGeer et al., 2000). Acute intoxication stimulates mucus secretion which can act as a chelator (Shephard, 1994). This may explain at least some of the elevated metal concentrations observed in fish gills (Felts and Heath, 1984). Strongly varying proportions of in-organic and organic contaminants between tissue concentrations in wild captured fish presented in Kime (1998) are not only due to different environmental conditions and exposure times but also to species- or family-specific patterns. Salmonids, e.g., have higher copper concentrations in the liver than other families.

2.2. Detoxification and cell protection

2.2.1. Biotransformation

The liver of fish is the major site of the biotransformation of lipophilic chemicals into more hydrophilic compounds (phase I reaction), followed by their conjugation with endogenous substances such as glucuronic acid, sulfate, and glutathione (phase II; Goksoyr and Husoy, 1999). Phase I reaction includes oxidation catalysed by multiple forms of enzymes known under various names (mixed function oxidase, monoxi-genase, cytochrome P-450, CYP1A), located on the endoplasmatic reticulum of cells (microsomal fraction). In most cases, these metabolites are less toxic, but there are also examples where toxicity develops through this transformation (e.g. parathione). The conjugated products are water-soluble and easily excreted into bile, urine or through the gills.

Under normal conditions the mixed function oxygenases are involved in the meta-bolism of steroid hormones. Specific forms of these enzymes cover a broad substrate specificity including several kinds of lipophilic xenobiotics such as PAHs, PCBs, dioxins, and halogenated compounds which are able to induce the synthesis of these enzyme complexes. It has been shown that in fish from areas polluted with oil, indus-trial or domestic wastes, hepatic microsomal cytochrome P-450 activity is highly elevated and can be used as an early dose indicator (Payne et al.,1987; Bucheli and Fent, 1995; Goksoyr, 1995). Meanwhile, the cytochrome P-450 system in fish has been incorporated as a routine method in many monitoring programmes. However, the degree of induction varies with species, sex, hormonal status, season, and temperature (Kleinow et al., 1987). Some toxicants, such as organotin compounds, inhibit the activity of cytochrome P-450 and glutathione S-aryl-transferase, and thus impair metabolism and elimination of other toxicants (Fent, 1999). Distinct changes in the activity of cytochrome P-450 may also interfere with the steroid metabolism of fish and affect their hormonal balance.

2.2.2. Oxidative stress

The metabolism of xenobiotics frequently results in the formation of reactive oxygen species (radicals) which significantly contribute to their toxicity. Due to their high reactivity, radicals damage all kinds of biological structures including nucleic acids (mutation, carcinogenesis), proteins, and membranes (lipid peroxidation). Peroxidation of unsaturated fatty acids is the most common phenomenon induced by reactive oxygen species. Peroxidation of fatty acids is a chain reaction modifying many molecules until it is interrupted by either dimerisation of two radicals or by the reaction with "radical scavengers" such as tocopherol and β-carotene. In a cascade of reactions, low molecular weight antioxidants including ascorbic acid, uric acid and glutathione take up or release single electrons and form relatively stable radicals with low reactivity (Lackner, 1998). Key enzymes for the detoxification of reactive oxygen species are superoxide dismutase, glutathione peroxidase, peroxidase, and catalase, in conjunction with enzymes providing reducing equivalents (glucose-6-phosphate dehydrogenase, glutathione reductase). Glutathione (GSH), a tripeptide, plays a central role in detoxification processes: Not only does it act as an antioxidant but it is also the substrate of glutathione peroxidase and the co-factor of glutathione-S-transferase involved in the conjugation of xenobiotics, and because of its sulfhydryl group, GSH can also bind metals. When GSH scavenges a radical or is oxidised it undergoes a dimerisation to the oxidised glutathione disulfide (GSSG) or mixed disulphides with proteins. In particular, the GSH/GSSG ratio is a sensitive measure of oxidative stress (Lackner, 1998).

Most components of this complex detoxification system are induced by intracellular radicals, toxicants and stress, and can be used as an unspecific but powerful early warning system for environmental pollution (Filho, 1996; Lackner, 1998).

2.2.3. Metallothioneins (MTs)

MTs are low molecular weight proteins with many sulfhydryl groups binding a variety of metals. MTs are found in all tissues, particularly in the liver and kidney, and play an important role in the intracellular regulation of the essential metals Zn and Cu. However, the affinity of Cd and Hg to MTs is even higher so that they may displace essential metals. MT concentrations in tissues increase by exposure of fish to Cu, Zn, Cd, or Hg. The changed pattern of metals bound to the protein may serve as an indicator for specific metal pollution (Hamilton and Mehrle, 1986; Dallinger et al., 1997). However, tissue levels of MTs are also affected by other factors such as reproduction and different kinds of stress (Overnell et al., 1987; Hyllner et al., 1989). For routine work tissue concentrations of metals are easier to obtain and more informative than MT analyses.

2.2.4. Heat-shock proteins (HSPs)

Cells contain a family of various low molecular weight proteins (28–100 kDa) with a high degree of identity among different organisms. Heat-shock proteins (HSPs) guarantee that cellular protein integrity and function are maintained, in particular under stressful conditions. HSP 70 is the most commonly known representative. Several studies with cell lines and cell cultures from fish tissues have shown that stressors such as heat shock and various environmental contaminants induce the *de novo* synthesis of HSPs (Iwama et al., 1998, 1999). Fewer investigations have been performed with the whole fish and they mainly refer to effects of heat shock. It has been shown that metals, bleached kraft mill effluents but also bacterial and viral diseases increase the cellular level of HSPs in various tissues. Although HSPs studies in fish are still at an early stage they may serve as a sensitive but unspecific indicator of the stress status of fish. Handling stress, however, does not induce HSP70 expression (Vijayan et al., 1997), but there is only limited knowledge on the relation between stress hormone levels and the formation of HSPs.

2.3. Inhibition of enzyme activities

2.3.1. δ Aminolevulinic acid dehydratase (ALA-D):

ALA-D condenses two molecules of aminolevulinic acid to one molecule of porphobilinogen, a component of haemoglobin and haem-containing enzymes involved in detoxification (cytochromes, catalases, and peroxidases). ALA-D activity in erythrocytes is extremely sensitive and highly specific to inorganic lead intoxication, but the more toxic organic lead compounds have only little effect (Johansson-Sjöbeck and Larsson, 1979; Hodson et al., 1984). Therefore, simultaneous analyses of lead and ALA-D in the blood may give valuable information on the proportion of alkyl-lead in the environment (high blood levels of lead but little inhibition of ALA-D).

Although ALA-D is a sensitive and highly specific indicator for pollution with inorganic lead, even severe inhibition by more than 50% of its activity is not associated with any signs of anaemia, probably due to the large reserve capacity of this enzyme. "Normal" ALA-D activity varies considerably between species and size classes of fish, and is activated by zinc and tetrachloro-1,2-benzoquinone, a component of bleached kraft pulp mill effluents (Andersson et al., 1988).

2.3.2. Acetylcholinesterase (AchE)

AchE has a key function in the regulation of nerve impulse transmission by hydrolysing the accumulated acetylcholin in cholinergic synapses. Insecticides, in particular organophosphates and carbamate insecticides, but also toxins of blue green algae inhibit the activity of this enzyme, leading to behavioural dysfunctions (Brewer et al., 2001). Fish die when more than 70–80 % of the activity is lost (Dembele et al. 2000). Both the highest activity of AchE and the highest sensitivity to inhibition were found in the brain. However, the activity is not uniformly distributed within the brain, and the sensitivity to inhibition is species-specific (Perkins and Schlenk, 2000). Other tissues suitable for environmental monitoring are muscles and blood (cholinesterase).

2.4. Histology

Increased levels of stress proteins and specific detoxification systems are typical responses of organisms to toxicant exposure reflecting their compensatory potential. Alterations at molecular levels manifest themselves rapidly and do not necessarily imply impaired cellular functions. Changes at a structural level, however, reflect the gradual impact on the metabolism ranging from adaptive to degenerative responses with severe consequences for the survival of fish. Due to the limited set of structural responses of cells and organelles histopathological changes have been considered rather unspecific, but they often specify responses of specific physiological functions. It has been shown that at least at an ultrastructural level toxicants induce a highly specific pattern of cellular alterations (Braunbeck, 1998). However, the knowledge of these responses is restricted to a limited number of toxicants investigated. Confronted with a variety of different potential toxicants, often acting simultaneously, the situation in the field is much more complex. In contrast to analytical methods the majority of histopathological investigations is of a descriptive nature, and the number of semi-quantitative and quantitative studies is small (e.g. Rocha et al., 1995, Braunbeck, 1998, Bernet et al., 1999).

Gills, liver, and kidney are the organs mostly used for histopathological investigations. As these organs are also impaired during infections by bacteria, viruses, and parasites, a comprehensive assessment of the health status of fish is necessary to differentiate between toxicant-induced changes and diseases. As tissues start to degrade very quickly after the death of fish, particularly at high temperatures, only samples of freshly killed fish can be used for histological evaluations. In this respect, gills are among the most delicate organs.

As gills have a large and permeable surface directly exposed to the water, they are highly sensitive to toxicants and are the main site for their uptake. During acute exposure to high concentrations, gills are the first and often most severely damaged organs of the fish. Typical responses to acute and chronic exposure to toxicants are lifting and necrosis of the lamellar epithelia and fusion of gill lamellae (Mallatt, 1985). More chronic effects include hyperplasia (proliferation) of the filamental epithelia which gradually reduces the respiratory surface, and the rise of mucus cell numbers caused by physical and chemical irritants and pathogens. Among many other functions, the mucus plays an important role in trapping toxicants as it provides a renewable surface and anything that attaches to mucus is rapidly removed (Shephard, 1994). Hypertrophy of gill epithelia, in particular the activation and proliferation of chloride cells, indicates a disturbance of the ionic balance induced by toxicants or simply by all kinds of prolonged stress as a result of increased corticosteroid levels (Perry and Laurent, 1993).

The liver is of high diagnostic value as it is not only the key organ for the basic metabolism of fish but also the major site for biotransformation, accumulation, and excretion of toxicants. Pollutant-induced changes in the liver have been reviewed by Couch (1975), Hinton and Lauren (1990), and Braunbeck (1998). Severe pathological changes induced by both toxicants and diseases include focal inflammations (hepatitis, perivasculitis, pericholangitis), necroses, and fibroses. The liver of fish is the most sensitive target organ of carcinogenic toxicants which may induce tumors (hepatoma and cholangioma; Bailey et al., 1996). Cloudy swelling of hepatocytes followed by hydropic degeneration and vacuolisation are the result of a disturbed Na/K membrane pump. An increased number of pycnotic hepatocytes reflects a high cell turnover, which is not only observed in intoxicated fish but also in females during vitellogenesis. Glycogen depletion in the liver often indicates a high metabolic status as it is found in females during vitellogenesis as well as in fish involved in detoxification processes and under chronic stress, but it might also be the result of starvation. In species with glycogen as the main energy source of the liver (e.g. salmonids), various pesticides may cause a lipid vacuolisation of hepatocytes (lipid degeneration, steatosis), found also in cultured fish due to dietary imbalance or aflatoxin contaminated diet. The diagnostic value of the abundance of melanomacrophages in the liver is equivocal (Haaparata et al., 1996). Besides species and age dependence, melanomacrophages have also been correlated with detoxification processes (binding of metals and radicals, phagocytosis of cell material) during intoxication, diseases, and starvation (Agius, 1985, Meinelt et al., 1997).

Light microscopical alterations mainly describe degenerative processes, but ultrastructural changes also reflect fast-responding adaptive processes of high indicative value in which all cell organelles are involved (Braunbeck, 1998). The proliferation of the smooth endoplasmatic reticulum is associated with the induction of xenobiotic biotransformation processes. Differences in numbers and staining activity of peroxisomes indicate changes in oxidative stress responses, and the formation of autophagic vacuoles reflects the turnover of cellular components. Substantial alterations are also seen in the structures of the rough endoplasmatic reticulum and mitochondria.

The supplementation of routine histological investigations by the more timeconsuming ultrastructural analysis of the liver from a selected number of specimens is recommended.

2.5. Haematology

Any kind of acute stress leads to an immediate release of stress hormones (catecholamines followed by corticosteroids) which causes a general activation of the metabolism and, consequently, a better chance of survival in extreme situations (Pickering, 1981). Prolonged stress, however, leads to exhaustion, osmotic imbalance, and immunosuppression. The use of haematological stress responses as biomarker for acute or chronic intoxication is limited as they are difficult to separate from effects induced by handling stress during the capture of wild fish. Within only a few minutes of stress, blood concentrations of glucose and lactate, and the number and volume of red blood cells increase significantly (Soivio and Oikari, 1976; Nikinmaa, 1992). Slower responses include rising numbers of immature erythrocytes released from the erythropoietic tissues (indicating an increased turnover of red blood cells), and the decline of lymphocyte numbers (lymphopenia). The high abundance of neutrophilic granulocytes, however, is a sign of disease rather than intoxication. Suitable biomarkers are more independent of acute stress and include anaemia, micronuclei (small nuclear segments constricted from the erythrocyte nucleus), amitotic divisions and deformations of erythrocytes (Nikinmaa, 1992). The latter should be evaluated with caution as they might also be the result of artefacts, in particular when blood smears are prepared under unfavourable conditions in the field. The amitotic division of erythrocytes is probably the response to severe respiratory stress caused either by oxygen depletion or toxicants (Murad et al., 1993). Increased levels of tissue enzymes in the blood plasma such as GOT (glutamate oxalacetate transaminase) and GPT (glutamate-pyruvate-transaminase) indicate lesions in liver and kidney during acute intoxication or diseases rather than chronic exposure to toxicants (Bucher and Hofer, 1990). Reduced activities of ALA-D caused by lead (see above) and the oxidation of haemoglobin to methaemoglobin in fish exposed to oxidants such as nitrite or mono-chloramine have proved to be useful indicators of specific intoxication (Buckley, 1982; Jensen, 1990). In general, blood composition is modified by natural environmental parameters including temperature and season, and is often sex- and species-specific (Luscova, 1997).

2.6. Endocrine disrupture

Many natural and synthetic substances interfere with the hypothalamus-pituitary-gonadal axis leading to alterations in secondary sexual characteristics and in the development of gonads affecting sexual behaviour and fertility (Crews et al., 2000). These substances mimic or block the action of hormones or affect their synthesis, metabolism, and transport. Specific responses depend on the exposure during sensitive periods of development, in particular during embryonic stages. Effects on reproduction are thus often delayed and may not be observed for years. Domestic and industrial effluents may contain a variety of endocrine disruptors such as natural and synthetic oestrogens originating from anti-baby pills and farm animals, natural phytosterols (pulpmill effluents), alkylphenols (non-ionic surfactants, e.g. p-nonylphenol), and organochlorine compounds (e.g. PCBs; Matthiessen and Sumpter, 1998). Many of these substances are highly persistent with a tendency to accumulate in organisms. Waste waters usually contain a mixture of hormone-like substances leading to additive or synergistic effects on the reproduction of fish which are difficult to predict.

The vitellogenin induction in males, analysed in the blood plasma using antibodies, is the most sensitive indicator of the presence of many kinds of endocrine disrupting substances in the water (Purdom et al., 1994; Allen et al., 1999). Vitellogenin is the protein precursor of yolk synthesised in the liver under the control of oestrogen, a process which does not occur in normal males. On the other hand, vitellogenin synthesis *in vitro* has been found to be depressed by polycyclic and halogenated aromatic hydrocarbons (Anderson et al., 1996). Alternative biomarkers for endocrine disruptors in fish include changes in the secondary sexual characteristics or the occurrence of intersexes which are seen just after dissection or may require histological techniques to detect more subtle changes in the testes (Ankley et al., 1998). Gonado-somatic indices (reduced testicular growth) and liver-somatic indices (increased relative liver weight) as well as sex ratios are other sensitive indicators of the presence of endocrine disruptors, but they are less specific as they may also depend on other environmental factors. Although the feminisation of fish exposed to endocrine disrupting compounds is the most common effect, many substances also interact with androgen receptors (e.g. pp'DDE) and may lead to the masculinisation of females, e.g. in the vicinity of kraft mill effluent (Taylor and Harrison, 1999). The effect of toxicants on thyroid hormone receptors (e.g. certain hydroxylated PCBs) has not been thoroughly investigated (Tyler at al., 1998).

2.7. Morphological anomalies

The rising abundance of deformed fish larvae and fry may indicate the presence of teratogenic substances, often without any visible effects on adult fish. As a consequence, fish populations gradually decline without any apparent reason. Selenium, progressively accumulating in food chains due to human activities, is one of these substances leading to a massive reproduction failure and is predicted to be a major threat to fish populations in several regions (Lemly, 1999). Most morphological anomalies found in fish concern bone malformations including skeletal anomalies, deformations of opercula, gill arches and finrays, and scale disorientation (Bengtsson et al., 1985; Janssens de Bisthoven, 1999). Substances affecting neuromuscular functions (e.g. organochlorine pesticides, zinc) may also cause spinal deformations and skeletal fractures due to an elevated muscle tone, even in adult fish. Non toxicant-induced deformations are the result of mechanical damages or vitamin C deficiency. Morphological anomalies are easily detectable indicators of a disturbed development or genetic defects and are often only one of many symptoms including physiological handicaps. Thus, with rising severity of deformation, larvae and fry will be eliminated rapidly from the population due to natural mortality, predation, and competition which considerably mask contamination effects. Fluctuating asymmetry, the random deviation from perfect symmetry in an otherwise bilaterally symmetric trait, has proved to be an excellent indicator of developmental instability being associated with reduced fitness (Möller, 1997). Females of the brook stickleback *Culaea inconstans* with symmetric pectoral fin ray numbers have about 15% more eggs per clutch than females with asymmetric fin ray counts (Hechter et al., 2000). In fish, however, the few data available do not support the theory that fluctuating asymmetry is a useful tool to monitor environmental pollution (Parsons, 1990).

2.8. Active biomonitoring

Ecotoxicological laboratory experiments under controlled conditions yield valuable basic data on threshold concentrations, dose-specific responses ranging from the molecular to the behavioural level, detoxification processes, mechanisms of adaptation, and modulation of toxicity caused by environmental factors. Although the body of knowledge of sublethal effects of pollutants is huge and rapidly growing, the toxicological status under complex field conditions cannot be predicted adequately. In contrast, studies on natural fish populations reflect the real situation, but, due to the mobility of fish, their individual life history and, in particular, the duration of exposure are unknown. Active biomonitoring using fish cages exposed to both a control site and the contaminated area may combine the advantages of laboratory and field studies (Grizzle et al., 1988). However, the use of cages may cause other serious problems: Destruction of the experimental set-up by storms, floods, and vandalism make such experiments risky. In experiments with fish sensitive to social stress, e.g. salmonids, the resulting immunosuppression combined with skin lesions in fish pushing against the net promote fungal infections and fish mortality. As caged fish are usually fed with pellets, the dietary input of toxicants is excluded. As a consequence, data obtained by active monitoring in fish cages have to be evaluated very carefully and potential artefacts caused by social stress or injuries have to be excluded.

The use of fish as sensors has been recommended for actively monitoring the water quality of industrial effluents and drinking water systems in continuously running and automatically working early warning systems, with the focus on short-term changes of toxicity. Testing under controlled and exposed conditions consists of a flow-through system supplying experimental chambers and recording units measuring a specific physiological or behavioural activity of the fish. The biological responses are converted into electronic signals feeding a data analysis system which discriminates between "normal conditions" and abnormally high deviations caused by increasing levels of toxicants. An alarm system should indicate rising toxic conditions. Automatic registration methods for monitoring rheodactil or locomotory behaviour, the frequency of opercular ventilation, heart rates, coughing rates as a response to irritation of the gill surface, and electric organ discharges of a weakly electric fish have been developed or adapted (Cairns and Van der Schalie, 1980; Geller, 1984; Vogl et al., 1999). Although these methods are useful and sensitive for short-term experiments, they create serious problems when applied for routine monitoring in an automatic alarm system. First, it is difficult to discriminate between the wide range of natural fluctuations of fish behaviour or metabolism and the specific responses due to intoxication. Second, some methods, e.g. the non-invasive registration of gill ventilation and heart rate, do not give stable bioelectric signals over a long period of time.

2.9. *Fish macroparasites as pollution indicators*

Heavy metals play an important role as substances affecting aquatic organisms. Their impact, particularly on fish, is receiving considerable attention. Investigations on chronic exposure to sublethal concentrations of pollutants and their effects on the host-parasite interrelationship and the parasites in particular are often neglected (e.g. Overstreet, 1997). The number of quantitative studies of heavy metals in both fish and their parasites has increased recently (e. g. Bucher et al., 1992; MacKenzie et al., 1995; Sures, 1999; Sures et al., 1999). Most investigations focus on either comparative studies of parasite community changes subsequent to environmental stress (e.g. Marcogliese and Cone, 1997; Dušek et al., 1998), or on single investigations of heavy metal concentrations in fish and their parasites (Sures and Siddall, 1999; Sures et al., 1999). Chubb (1997) confirms the results of Sures and co-workers (summarised in Sures et al., 1997; Sures et al., 1999) where especially acanthocephalans accumulate high burdens of lead and cadmium and therefore may be used as sensitive indicators for monitoring heavy metal contaminations in aquatic ecosystems (see also Kuperman, 1995; Zimmermann et al., 1999).

If fish parasites are to be used as indicators of pollutants they must meet several requirements to be comparable with free-living organisms. Kennedy (1997) suggested that the following conditions are necessary if fish parasites are likely to be indicators for pollution: the fish host must be abundant and easily accessible; parasite species, despite their overdispersed distribution, must show a high prevalence and abundance in the fish host population; parasites should be easily identified and not laborious to remove and count; information on the ecology and biology of both fish host and parasite should be available.

3. Fish as indicators of the ecological integrity of running waters

3.1. Habitat structures and connectivity aspects

Fish are used not only for the indication of pollution aspects but also to evaluate compound and complex structural properties of the environment. It is well established that river engineering has led to drastic changes in the ecological conditions with regard to longitudinal continuity, lateral interactions with the bordering riparian zones and the structural subunits within the instream channel. River engineering has led to deficiencies in functional processes, in habitat diversity, and in the diversity of characteristic biota (Noss, 1990; Townsend, 1996; Ward, 1998; Ward and Stanford, 1995). Modern, ecologically orientated river engineering attempts to develop management and restoration schemes in order to maintain and restore the major properties and functional processes and the characteristic biodiversity. An appropriate indication and monitoring system has to be developed to achieve this goal. Finding relevant descriptors requires the merging of geomorphological, hydrological and biological parameters. The physical indication system has to include the range and hierarchy of spatial scales – from the scale of the catchment area to bed sediments, and to combine the spatial properties with the relevant biological requirements of individual species and the different stages during ontogeny.

Fish have proved to be of significance as bioindicators of the so-called ecological integrity (Karr, 1991; Schiemer, 2000) because during their life cycle, the various guilds integrate a wide range of riverine conditions including the properties of bed sediments relevant for egg development and the longitudinal integrity for spawning migrations (Copp, 1989; Gaudin, 2001; Persat et al., 1995; Schiemer et al., 1991; (Schiemer et al., 2001a).

3.1.1. Structural properties of rivers at different scales

Over the past 20 years several concepts for understanding functional processes and biodiversity patterns of rivers have been developed emphasizing their four-dimensional nature and hierarchical structure with regard to spatial and temporal properties (Frissell et al., 1986; Noss, 1990; Gregory et al., 1991; Ward and Stanford, 1995; Townsend, 1996; Ward, 1998; Ward et al., 1999).

It is well known that the key factors for understanding running water ecosystems are the hydrological connectivity between the river and its environment and – over a longer time span – geomorphic processes within the river corridor controlled by

hydrology. Fluvial "disturbances" create characteristic patch dynamics, spatial heterogeneity, and the mosaic structure of densely packed ecotones and an array of
successions over a range of different scales. They also provide the habitat diversity
and the specific habitat conditions for characteristic species and result in high levels
of local species richness (alpha-diversity), habitat diversity and – differences between
habitats (beta-diversity) and consequently, overall species richness (Naiman et al.,
1988; Junk et al., 1989; Naiman and Décamps, 1990; Schiemer and Zalewski, 1992 ;
Bayley, 1995; Ward, 1998).

The floodpulse concept (Junk et al., 1989) refers specifically to these lateral interactions and exchange processes between the river and the semi-terrestrial adjoining
area, which are inundated by regular or irregular floods. An "inshore retention concept"
recently formulated (Schiemer et al., 2001b) highlights the significance of hydraulic
storage as a function of channel morphology, heterogeneity and the sinuosity of the
littoral zone. There is supportive evidence that the retentivity of a river reach is significant for the local microclimate, production processes, and the ability of channels to
retain particulate organic matter. With regard to fish, the production of food and refuge
capacity to withstand wash-out effects and population losses at early stages of their
life history is particularly relevant (Hildrew, 1996). In terms of hydraulics, inshore
retention also varies according to spatial scales: small storage zones will enhance the
growth and production of riverine phyto- and zooplankton while littoral diversification along a river reach will be required as microhabitats and refugia for 0+ fish. The
model thus proposes that hydraulic retention can be related to biological functions and
forms a general framework for the understanding of the ecology of large rivers.

A fish habitat is defined by structural conditions at various orders of magnitude.
Aspects of the catchment-scale, i.e. longitudinal integrity of the river (the upstream–
downstream connectedness), are very relevant for fish with regard to population
genetics, but also with regard to discharge and transport processes of nutrients and
organic matter. Sediment transport of a river and siltation are expressions of catchment processes (but can also indicate the extent and quality of local buffer strips, see
below).

On the macroscale of the reach of a river, the overall form of the river, the riffle-
pool sequence, the relief of the floodplain, the connectedness between river and
sidearm habitats, longitudinal connectivity between spawning and nursery areas and
flood refugia are decisive. On the mesoscale, structural properties will define instream
habitat availability. The form and sinuosity of inshore structure (e.g. structure of gravelbanks) for example and the way it interacts with the hydrological dynamics of a river
determines microhabitat availability and the configuration and steepness of ecological
gradients.

At a microscale, the relief of gravelbanks for example defines the physical nature
of the quality of bed sediments as spawning substrate and the ontogenetically changing
microhabitats.

Because of the complex array of their habitat requirements, fish react sensitively to
the various aspects of ecological integrity, which refers to the functional entity of a
system with regard to its original state (Schiemer, 2000), which can be described by
the physical and abiotic conditions calibrated by the requirements of characteristic
species.

3.1.2. Fish assemblages and the identification of guilds

Fish can be classified according to habitat requirements, e.g. rheotopic or stagnotopic, or more specifically according to their requirements during specific life phases e.g. for spawning. Balon (1975, 1981) has developed a classification scheme of reproductive types based on spawning site selection, adaptive features of eggs and embryos and the degree of parental care. The main groups distinquished are (a) non-guarders, (b) guarders, and (c) bearers. Within the "non-guarders" different types are distinquished according to spawning substrate, e.g. benthic spawners on coarse bottom substrate (lithophilous), sand (psammophilous) or plants (phytophilous) and brood hiders, e.g. cave spawners or spawners on invertebrates (e.g. *Rhodeus sericeus amarus*).

The various reproductive styles are obviously associated with specific habitat requirements. The distinct sequence of fish assemblages occurring along the longitudinal course of a river system can be related to river gradient and physical habitat conditions: habitat structure, substrate types, current velocity temperature and oxygen regime (Huet, 1959). Fish species richness generally increases with stream order (Lotrich, 1973; Horwitz, 1978) along with increasing habitat diversity and complexity and an enhanced trophic and energetic basis for the fish populations. The sequence of assemblages along a river system can be characterised by the array of ontogenetic and seasonal requirements of the characteristic species (e.g. Jungwirth et al., 2000). They define the ecological integrity of aquatic systems with regard to the connectedness of supplementary and complementary habitats (see Schlosser, 1995) and source and sink areas of populations.

Large rivers usually contain different guilds of fish according to life cycle ranges. As far as spawning and nursery sites are concerned floodplain rivers provide a great diversity of habitats which allows the existence of a variety of ecological guilds. For European rivers we have distinguished five major guilds (Schiemer and Waidbacher, 1992) according to the preferred zones of occurrence of adults and the spawning and nursery grounds (Fig. 1).

1. Riverine species dependent on the connectivity of the river with its tributaries. This group requires rhithral conditions for spawning and during the early life history stages (e.g. *Hucho hucho*).
2. Riverine species with spawning grounds and nurseries in the inshore zone of the river itself (majority of species, e.g. *Barbus barbus* or *Chondrostoma nasus*) Group 1 and 2 are now frequently referred to as Rheophilic A.
3. Riverine species with a preference for low-flow conditions (e.g. connected backwaters) during certain periods in the adult stage (e.g. feeding grounds or winter refuge), but with spawning grounds and nurseries in the river (e.g. *Aspius aspius*, *Leuciscus idus*). Such species are referred to as Rheophilic B.
4. Eurytopic species (habitat generalists found both in rivers and various types of stagnant water bodies. Some of these species require flooded vegetation as spawning area, e. g. *Esox lucius*).
5. Limnophilic species confined to various microhabitats of the floodplain (e.g. disconnected former river branches) with vigorous development of submerged vegetation.

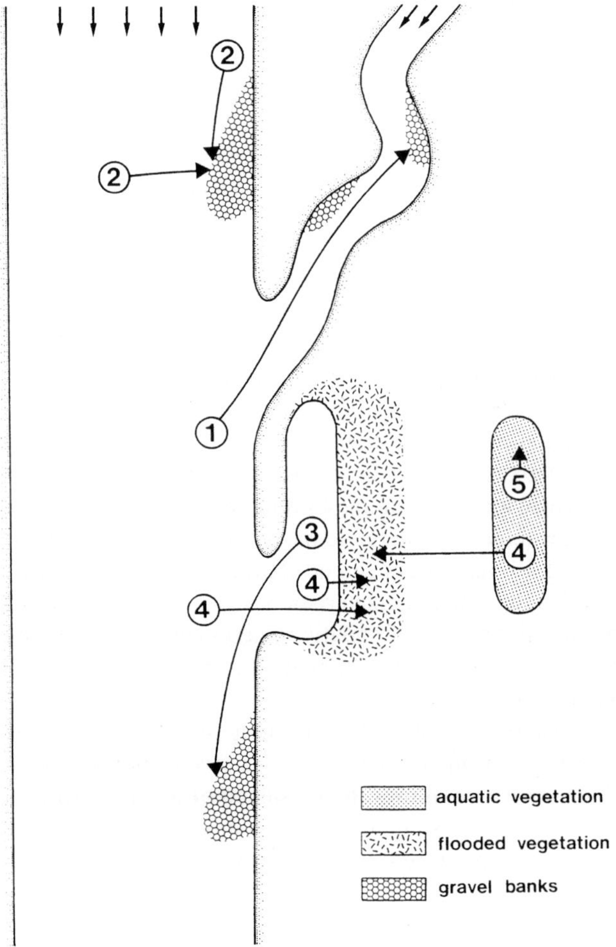

Figure 1. Schematic presentation of main habitat requirements of five guilds of fish. Circles: preferred habitats of adults; arrows: spawning and nursery sites (from Schiemer & Waidbacher, 1992).

Riverine species are either linked to the river only, or they spend phases in stagnant connected water bodies. Stagnotopic forms are found in highly vegetated disconnected water bodies of the floodplain. Stagnotopic species are characteristic of well disconnected, small water bodies and eurytopic forms are found in a wide range of river-floodplain habitats.

Various structural properties are essential for the various guilds. Some stagnotopic species such as *Misgurnus fossilis* are found exclusively in strongly fragmented and vegetated pools in the wetlands. Such species belong to a highly endangered guild which indicates that the wetland habitats of these species that border the floodplains have been reduced successively due to land reclamation and river restoration. For such disconnectance specialists, ecological integrity means the long-term sustained existence of the required habitat type as a result of floodplain development over the scale of a century or more.

In the large rivers in Europe and North America, the rheophilic guilds contain the highest number of species, but also the highest proportion of endangered taxa. This endangerment and the population decline are strong indications of habitat loss and the loss of the ecological integrity of aquatic systems. For large anadromous migrators such as sturgeon, the integrity of the catchment area is required. Some of the species require the connectance between the river and the floodplains, as complementary habitats, e.g. *Abramis ballerus* in the Danube system, which is a riverine spawner but enters connected backwaters for feeding and as a winter refuge. Such species are excellent indicators of lateral connectivity between lotic and lenitic habitat types. In smaller rivers, pools temporarily isolated along the shoreline have proved to be favourable for larval fish and may function as local hot spots (Ulmann and Peter, in manuscript).

3.1.3. Complex arrays of habitat requirements and critical phases

Fish vary dramatically in size and requirements during their ontogeny. Associated with this variation is a complex array of habitat-use patterns being mediated by migratory processes. Autecological requirements of the individual species and the match between these requirements and environmental conditions are decisive for the survival of a population. The habitat-template and niche concept provides the basis for this approach (e.g. Winemiller, 1992).

The following scheme (Fig. 2) – based on stream fish – specifies these ontogenetical and seasonal requirements for physical habitats. Interconnectivity of the various habitat patches is required for spawning and embryonic development as well as a mosaic of

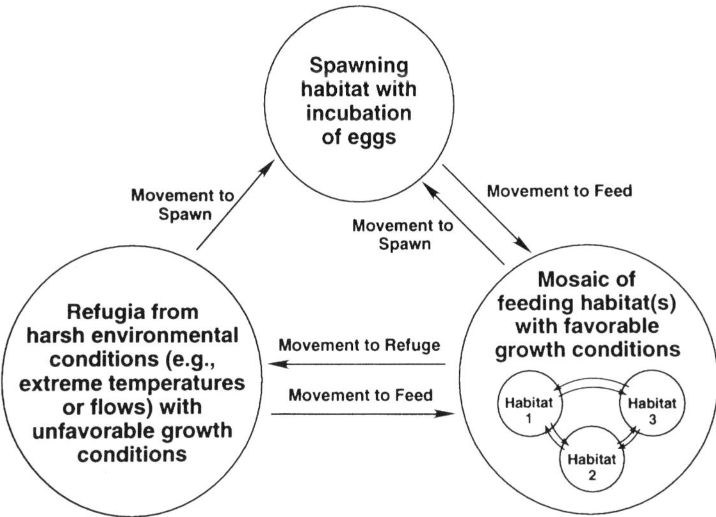

Figure 2. The basic life cycle of stream fish with emphasis on patterns of habitat use and movement (from Schlosser, 1995).

feeding habitats allowing for favourable growth conditions in the course of the life cycle and the availability of refugia under harsh environmental conditions, e.g. during the winter or in the case of environmental disturbances. Floods, for example, and successful migration between the various subhabitat types are decisive for the completion of the life cycle. Suboptimal conditions and the restricted availability of habitats or refugia will lead to reduced individual performances in growth or reproduction and population losses and ultimately to changes in the composition of the fish community.

The success or failure of a species is largely determined during a few critical stages of the life history. The value of fish as indicators is largely dependent on the match or mismatch between environmental conditions and requirements during these critical phases. At population level, the success of a species will be determined primarily by reproductive effort and mortality rates during the embryonic and early larval phases. The quality of the spawning substrates is critical especially in the lithophilous group. The spawning substrate quality is crucial for the reproductive success as are the water percolation in the bed sediments and the oxygen supply for the developing eggs. The high sensitivity during the embryonic stage towards oxygen conditions has been documented for many species (see Kamler, 1992; Keckeis et al., 1996). Low oxygen levels retard development and lead to a high egg mortality. Fish therefore act as sensitive indicators of the quality of spawning grounds with respect to the content of fine sediment and siltation. Thus, the recruitment of a fish population is a good indicator of conditions at the landscape level and the nature and quality of buffer strips (Berkman and Rabeni, 1987; Bisson et al., 1992; Rabeni and Smale, 1995).

An especially critical phase – (particularly relevant for recruitment and year-class strength) is the larval period. Mortality is high due to narrow niche dimensions. During the larval and early juvenile phases, drastic changes in microhabitat requirements of major environmental variables, food, water velocity, structural elements, etc. are observed (Schiemer and Spindler, 1989; Sempeski and Gaudin, 1995; see overview in Gaudin, 2001).

During the spawning and early life history period most riverine species are bound to the inshore zone of the river. Their occurrence or disappearance is indicative of the structural properties of the inshore zone. Inshore structure combines and integrates the following elements required for successful recruitment on the micro- and mesoscale:

1. Appropriate stream bed structures for spawning must be in close proximity and connectance to larval microhabitats. Emerging larvae drift passively to these nursery zones. Population losses are generally higher in channelised rivers with lower flow diversification.
2. Microhabitat gradients are required to cover ontogenetic niche shifts with regard to the velocity of the water current, substrate type and food. The nurseries of riverine species are almost exclusively found in richly structured littoral zones of the river. Artificial linear shorelines with steeper slopes are inappropriate microhabitats for larval and juvenile fish. The ontogenetic niche profiles are the result of changing microhabitat requirements. The larvae are bound to sheltered bays along the river shoreline, where velocities are low despite changing water levels. With increasing age and size, early juveniles shift into deeper water strata or to adjoining shallow gravel banks (Fig. 3, Fig. 4).

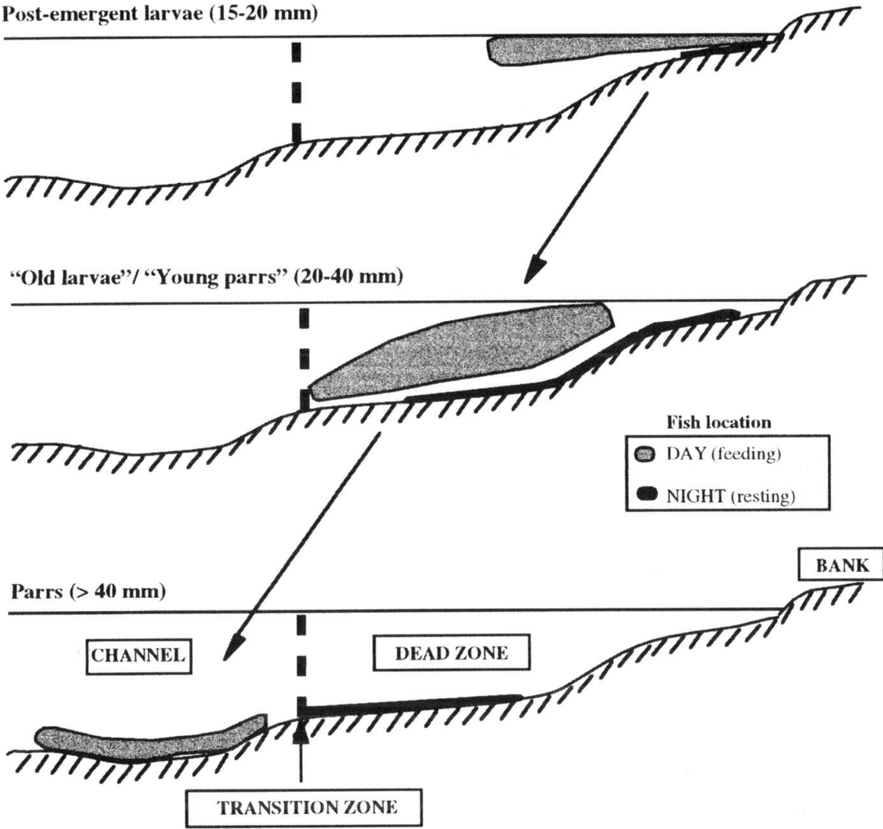

Figure 3. Dynamics of habitat use in young grayling. The different compartments are schematically represented by grey and black surfaces if the habitat is used by day or by night, respectively. Arrows show the tendency of grayling to shift to the river channel with increasing size (from Sempeski & Gaudin, 1995).

3. An important quality of the littoral zone is the proximity to backwaters openly connected with the river; such backwaters serve as a production zone of fine particulate organic food, e.g. zooplankton (in the sense of the "inshore retention concept" see above).
4. Refugia to reduce wash-out effects at sudden flood pulses. Shallow sloping embankments and littoral diversification can function as buffer zones for 0+ fish against wash-out effects in the event of floods.

Inshore structure is a significant requirement for the 0+ stages, as far as the extent of refugia under conditions of stochastic water level fluctuations and the connectivity of complementary microhabitats are concerned. These conditions are particularly relevant with regard to population losses especially in the early stages – when the ability of the fish to search for adequate conditions or to escape critical conditions is limited. The structural properties of the inshore constitute an important determinant for 0+

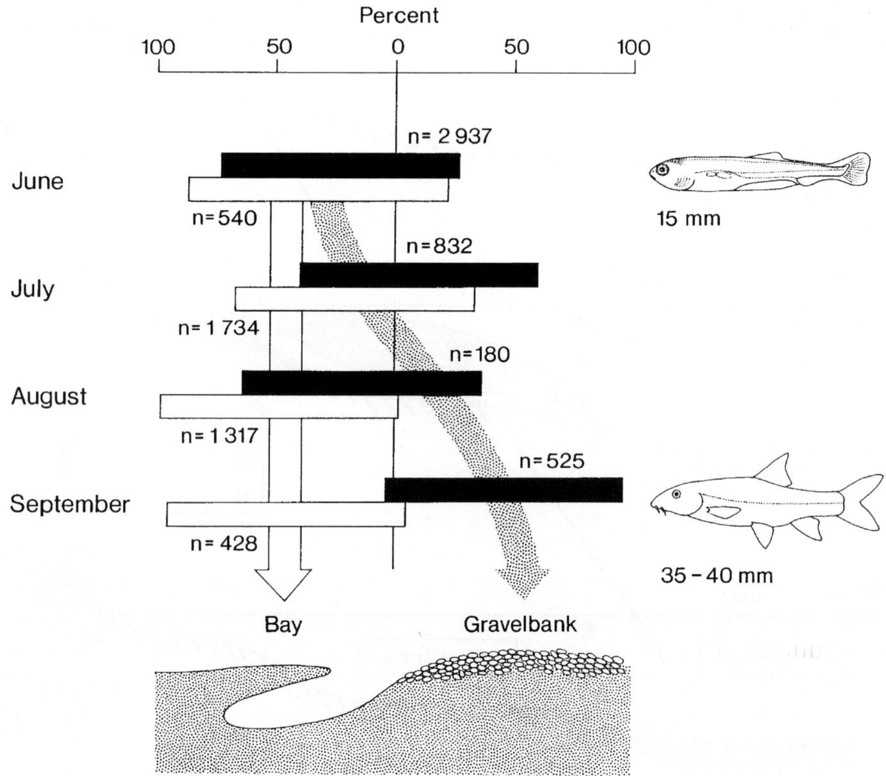

Figure 4. Habitat separation between eurytopic (white) and rheophilic (black) fish of the Danube River downstream of Vienna (Austria) during the first months of life. The columns indicate the portion of the total sample of fry found in adjacent bays and gravel banks. The average size of fish in June and September is indicated (from Schiemer & Spindler, 1989).

fish with regard to food, temperature and water velocity and will thus be decisive for individual growth performance during the critical stages.

Ecological quality is not merely dependent on the structural properties but also on the interaction between geomorphology and hydrology. Water level fluctuations lead to a continuous shift of microhabitat positions and determine the availability, connectance, and quality of microhabits and refugia. The quality of inshore zones depends on the degree to which two dynamic processes are matched: (a) the onto-genetic change in requirements and (b) the hydrological dynamics of the river which result in the continuous change of microhabitat locations and conditions. Considering the strong diurnal hydrological fluctuations occurring in large rivers, the inshore zones represent a highly stochastic environment for the early life history stages. For example in the Danube at Vienna the average daily amplitude lies between 20–40 cm. Structural heterogeneity of the shoreline is a buffer again to population losses (Schiemer et al., 2001a).

Thus, the interaction between structural properties and water level defines ecological qualities and the likeliness of disturbances at a microhabitat scale. Abrupt changes of conditions, as often encountered in artificial shoreline constructions represent disturbances for fish. In this context, disturbances are considered to be abrupt changes of habitat availability or of critical environmental conditions such as water velocity, dry falling or strongly rising temperatures. Such conditions lead to population loss if refugia are not available and appropriately connected.

Population loss is determined by a combination of 3 conditions:

(a) The hydrological extent and timing of water level fluctuations.
(b) Spatial heterogeneity determines the robustness against hydrological changes.
(c) Fish size, determining the escape range: small fish require higher spatial heterogeneity than larger ones.

For a detailed insight into bioindication and a causal understanding of environmental constraints in the early life history period, the functional response towards major environmental parameters has to be studied.

3.1.4. Individual fitness and habitat profitability in 0+ riverine fish

The interaction of the geomorphology and the hydrological dynamics of the river is not only significant for the population dynamics of riverine fish, but must be addressed when mentioning habitat quality, which determines individual growth. 0+ fish have to grow fast through a narrow corridor of environmental constraints and threats, and the match or mismatch between requirements and environmental conditions is particularly important (Wieser et al., 1988).

Suboptimal conditions are critical because they lead to prolonged duration of larval development, which makes it more likely that fish have to endure unfavourable or even lethal conditions. Fast growth through critical periods increases the probability of survival. A synoptic approach to defining the niche dimensions, environmental hardships and limitations especially during the critical larval phase, is to a large extent determined by the balance between energetic gains and costs. The knowledge of the functional response to major environmental conditions during the critical early life history phase – (temperature, current velocity and food availability) provides insight into the causes behind the endangerment of individual species and improves the indication value of fish. For *Chondrostoma nasus*, a characteristic species of European epipotamal rivers, such studies have shown the tight energetic balance espcially in the larval phase. In a continuously shifting, stochastic environment as is the inshore zone of large rivers, the costs of station maintenance are obviously high and such additional costs or lower consumption will constrain growth and extend the duration of the critical period. With increasing body size and more advanced developmental stages, the energetic scope widens and the escape range is enhanced.

3.1.5. The indication value of riverine fish

Fish are not only used as indicators of environmental conditions but individual species are targets for habitat quality assessments. Several models have been developed to

Table 2. The significance of various elements of "ecological integrity" of floodplain rivers for community structure, population dynamics, and performance in fish (from Schiemer, 2000).

Biotic indicators	Ecological parameters			
	Habitat diversity	Connectance of complementary habitat elements	Refugia	Physiographic and trophic habitat condition
Community structure				
Biodiversity	++	+		
Assemblage structure	+	++	++	
Endangeredness	+	++	++	
0+ assemblage		++	++	++
Population dynamics				
Year class strength		+	++	
Recruitment/mortality rates		+	++	+
Population exchange processes (e.g. migration, drift)		+		
Bioenergetics				
Growth				++
Condition				++
Egg size/egg quality				++

define river conditions and habitat quality for individual salmonid species with regard to fisheries and the conservation of species stocks (e.g. Wesche et al., 1987; Barnard and Wyatt, 1995; Bowlby and Roff, 1996).

The main value of fish, however, is their role within an overall environmental assessment. The development of management criteria and restoration strategies for river systems in order to improve ecological properties and functions requires appropriate biological indicators to analyse environmental deficiencies and to document changes and improvements. In this respect fish are the most significant single indicator group. Fish can be used within an integrative assessment of "biotic integrity" (Karr, 1991; Angermeier and Karr, 1994) and overall river corridor surveys. To use fish successfully as indicators, the monitoring aspects have to be clearly defined and the scope of indication has to be identified.

The matrix (Table 2) relates various elements of ecological conditions and integrity and their significance for community organisation, population dynamics, and bioenergetic performances of fish. It is quite obvious that the structure of the assemblage, the presence or absence of individual species of fish, and their state of endangeredness provide a broad overview and a first overall indication of the situation. A comparison with the original situation ("reference standard", "Leitbild") and characteristic fish assemblages then (which is relatively well assessed for many river types in Europe) reveals deficiencies.

For a more detailed evaluation the local distribution pattern of the fish fauna, its population structure, seasonality of occurrence, and growth performance must be assessed. Particularly the 0+ stages of fish are excellent indicators of the diversity and function of riparian ecotones and the structure of the shoreline. Simple indices of shoreline diversification, e.g. the littoral development per unit river length, highly correlate with species number, diversity and population density of 0+ fish. Such parameters are valuable criteria for an ecologically oriented river engineering at a local scale (Copp, 1989; Schiemer et al., 1991).

A detailed understanding of the requirements of a wide range of target species is necessary to further enhance the value of the fish fauna as indicators. Fitness has to be addressed at the level of both population dynamics and the ecophysiology of critical stages especially in the early life history period (Elliott, 1994; Keckeis and Schiemer, 2001). The subject of a recent workshop was the state of the art of using 0+ fish as indicators of the ecological integrity of large rivers (Schiemer and Keckeis, 2001).

3.2. The Index of Biotic Integrity

The Index of Biotic Integrity (IBI) was developed by Karr (1981) for use in non-salmonid, small warmwater streams in Illinois and Indiana. IBI incorporates many attributes of fish communities to evaluate biological conditions at sites influenced by different human activities (Karr and Chu, 1999). The original version of IBI included 12 metrics reflecting species richness and composition, trophic condition, and fish abundance and condition (Karr, 1981; Karr et al., 1986): number of fish species, number of darter species, number of sunfish species, number of sucker species, number of intolerant species, relative abundance of green sunfish, relative abundance of omnivores, relative abundance of insectivorous cyprinids, relative abundance of top carnivores, number of individuals (*), relative abundance of hybrids (*), relative abundance of diseased individuals (asterisks indicate metrics that have been considered not reliable).

Each metric is assigned a score based on expectations for that metric at minimally disturbed sites for that region and stream size. Metrics that approximate what experts would expect at reference sites are assigned a score of 5; those that deviate somewhat from such sites receive a score of 3; those that deviate strongly are scored 1. The final index is the sum of all the metrics' scores. The total IBI score determines the ranking of the site in one of the integrity classes (excellent, good, fair, poor, very poor).

As the IBI became more widely used different versions were developed for different regions and different types of ecosystems. Although these new versions had a multimetric structure, they differed from the original IBI in number, identity, and scoring of metrics (e.g. Oberdorff and Hughes, 1992; Simon and Lyons, 1995; Gammon and Simon, 2000).

3.3. The role of fish in the EU "Water Framework Directive"

3.3.1. General principles of the Water Framework Directive

The European water management has to be fundamentally reorganised on the basis of the new "Directive 2000/60/EC of the European Parliament and of the Council establishing a framework for Community action in the field of water policy" (Water Framework Directive – WFD; European Union, 2000). One major goal of the WFD is to achieve the "good ecological status" and "good chemical status" in all surface water bodies of the EU. The term ecological status corresponds to the general philosophy of an integrated approach for evaluating the ecological integrity of water bodies (Chovanec et al., 2000b).

The main principles for the assessment of the ecological status of surface waters and, thus, the basis of the application of fish-oriented methods are:

- Ecosystem approach: As opposed to other water-relevant EU Directives, the key issue of the WFD is not to guarantee the different aspects of water utilisation but to maintain or restore the health of aquatic ecosystems. In this context fish play an essential role as indicators within integrated assessment procedures.
- Type-specific approach: The assessment procedure is based on a comparison between a type-specific pristine or near-pristine reference condition and the present-day status of the water bodies (Muhar et al., 2000; Wimmer et al., 2000).
- Bioindication approach: According to the WFD, assessment has to be based on the investigation of the aquatic communities, algae (phytoplankton, phytobenthos), macrophytes, benthic macroinvertebrates, and fish.
- Classification: The degree of deviation has to be ranked in a five-class-system (high, good, moderate, poor and bad ecological status).
- Heavily modified and artificial water bodies: The WFD permits the member states to identify artificial and heavily modified water bodies, for which the objective is the achievement of a good ecological potential. For these types of water bodies not the type-specific natural condition but a so called "maximum ecological potential" has to be established as the reference condition. Heavily modified water bodies are bodies of surface water which are substantially changed in their character as a result of physical alterations by human activities. Migratory fish species indicating habitat fragmentation and landscape connectivity patterns will play an important role in the assessment of e.g. channelised and / or impounded rivers.
- Monitoring: A monitoring network has to be designed in each country which provides a comprehensive and representative overview of the ecological status within each river basin.

The WFD provides general definitions of three of the five classes of ecological status concerning biotic and abiotic parameters; definitions for the fish fauna are given in Table 3. Fish also play an important role in the setting of chemical quality standards. Where possible, both acute and chronic data as well as accumulation data shall be obtained for different taxa (e.g. fish) which are relevant for the water body type concerned as well as any other aquatic taxa for which data are available.

Table 3. Definitions for the high, good and moderate ecological status of river and lakes according to fish-based investigations (European Union, 2000).

High status	Good status	Moderate status
Species composition and abundance correspond totally or nearly totally to undisturbed conditions. All the type-specific disturbance-sensitive species are present.	There are slight changes in species composition and abundance from the type-specific communities attributable to anthropogenic impacts on physico-chemical and hydromorphological quality elements.	The composition and abundance of fish species differ moderately from the type-specific communities attributable to anthropogenic impacts on physico-chemical or hydromorphological quality elements.
The age structures of the fish communities show little sign of anthropogenic disturbance and are not indicative of a failure in the reproduction or development of any particular species.	The age structures of the fish communities show signs of disturbance attributable to anthropogenic impacts on physico-chemical or hydromorphological quality elements, and, in a few instances, are indicative of a failure in the reproduction or development of a particular species, to the extent that some age classes may be missing.	The age structure of the fish communities shows major signs of disturbance, attributable to anthropogenic impacts on physico-chemical or hydromorphological quality elements, to the extent that a moderate proportion of the type-specific species are absent or of very low abundance.

Table 4. Classification scheme for fish-based assessment of ecological integrity (Schmutz et al., 2000).

Criteria	Levels of ecological integrity				
	High	Good	Moderate	Poor	Bad
Type-specific species	None or nearly none missing	Some species missing	Several species missing	Many species missing	Most species missing
Self-sustaining species	None or some missing	Several species missing	Many species missing	Most species missing	Nearly all species missing
Fish region	No shift	No shift	Shift	Shift	Shift
Number of guilds	No guild missing	No guild missing	Single guilds missing	Many guilds missing	Most guilds missing
Guild composition	No alteration	Slight alteration	Substantial alteration	Complete alteration	Complete alteration
Biomass and density	No or nearly no changes	Slight changes	Substantial changes	Heavy changes	Extremely changed
Population age structure	No or nearly no changes	Slight changes	Substantial changes	Heavy changes	Extremely changed

3.3.2. Implementation of the Water Framework Directive: need for action

Karr (1981) pointed out that "a key problem in classification is defining the baseline", and the definition of reference conditions for river classification is no exception. The implementation of the WFD requires the elaboration of a comprehensive abiotic (e.g. Wimmer et al., 2000) and biotic typology of surface waters including the definition of type-specific reference communities for fish as well as a sound scheme for ranking the ecological status in a five-class system.

For lakes, a first approach of developing reference fish communities by analysing historical data from between 1500 to 1940 was published by Gassner and Wanzenböck (1999). The communities are described using factors such as species number and composition, ecological guilds and abundance.

A five-tiered classification scheme for a fish-based assessment of the ecological integrity of running waters was developed by Schmutz et al. (2000) and is close to the requirements of the WFD (Table 4). The weighting of the criteria in an epirhithralic stream will be different from that in a large potamalic river/floodplain-system. In the first case, fish species diversity is low even under natural conditions and both diversity and species composition are, thus, not sensitive criteria. On the other hand, quantitative information on the population size and age structure of each species is attainable at acceptable costs and without methodological sampling problems. In river/floodplain-systems, species diversity is usually high: the sensitivity of the diversity and species composition criteria is, therefore, much higher and quantitative information becomes less important (Schmutz et al., 2000). As pristine reference sites of rivers are generally rare in Europe and other parts of the world, the definition of river-type-specific fish communities has to be based also on other information, such as historical fish data, historical abiotic data (Muhar et al., 2000) and reference models (Milner et al., 1993; Barnard and Wyatt, 1995).

4. Concluding remarks

Fish are one of the most frequently used group of bioindicators in ecotoxicological field studies. The advantage of a comprehensive basic knowledge of toxicology, physiology, and histology exceeds the disadvantage of fish mobility. No other aquatic organism is suitable for the application of so many different methods which allow the evaluation of the severity of toxic impacts ranging from compensatory responses at a molecular and an ultrastructural level (serving as an early warning indicator) to sublethal and pathological changes as alarm signals for population declines and irreversible consequences for the whole ecosystem (Figure 5). Some biomarkers are indicators of unspecific stress, others respond to a group of toxicants with comparable attributes, and only a few biomarkers are highly substance-specific (Table 5). The bioindication of the occurrence of specific substances and their impact on specific biota and the ecosystem are the main focuses of ecotoxicological studies. Several methods of ecological and toxicological relevance with varying specificity have to be applied simultaneously to evaluate the ecotoxicological situation under the complex environmental conditions in the field.

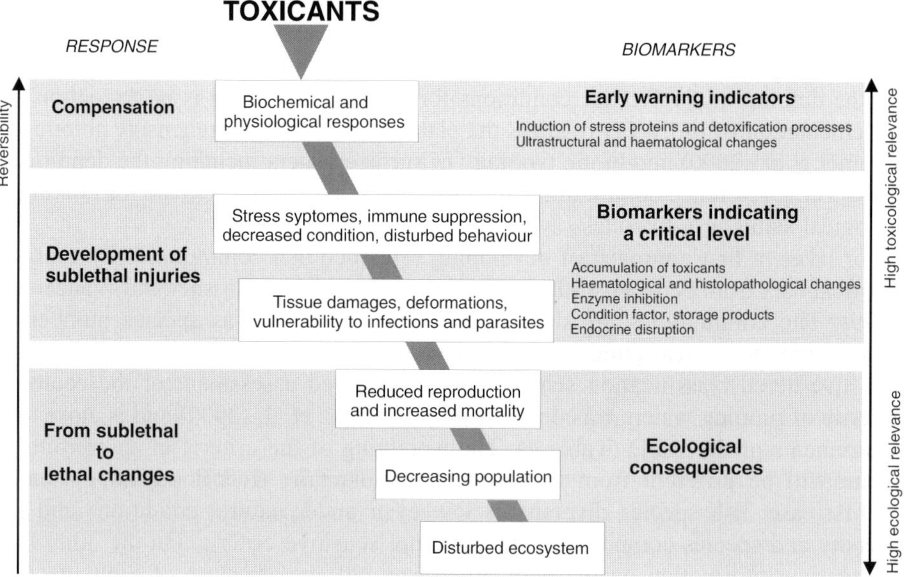

Figure 5. Different levels of responses of fish exposed to toxicants and the use of fish as a bioindicator of environmental pollution.

The use of macroparasites as indicators of heavy metal contamination is of increasing relevance in environmental control. There is a need to develop combined approaches including both parasite community aspects and accumulation aspects.

Due to its complex habitat requirements the fish fauna is a crucial indicator of the ecological integrity of aquatic systems at different scales, from microhabitat to catchment. The fitness of fish species both at the individual level (e.g. growth performance) and at population level (e.g. population structure) is determined by the connectivity of different habitat elements in a broad spatio-temporal context. Thus bioindication using fish represents a good monitoring tool especially with regard to river engineering, e.g. river restoration and management.

In order to further strengthen the role of fish as valuable indicators of the ecological integrity of aquatic systems, research is required ranging from the ecological demands of certain target species to ecosystem processes.

References

Agius, C., 1985. The melanomacrophage centres of fish: a review. In: Manning, M.J., Tatner, M.F. (Eds), Fish Immunology. Academic Press, London, pp. 85–105.

Allen, Y., Matthiessen, P., Scott, A.P., Haworth, S., Feist, S., Thain, J.E., 1999. The extent of oestrogenic contamination in the UK estuarine and marine environments – further surveys of flounder. Sci. Total Environ. 233, 5–20.

Anderson, M.J., Miller, M.R., Hintom, D.E., 1996. In vitro modulation of 17ß-estradiol- induced vitellogenin synthesis: effects of cytochrome P4501A1 inducing compounds on rainbow trout (*Oncorhynchus mykiss*) liver cells. Aquat. Toxicol. 34, 327–350.

Table 5. Biomarkers used in fish ecotoxicology.

Biomarker	Indicates pollution with specific chemicals	Specific of certain biological processes	Impact of non-toxic factors
Accumulation	Yes	Less specific	Environmental factors
Biotransformation	Several groups of lipophilic xenobiotics	Detoxification	Temperature, sex, species
Oxidative stress	No	Detoxification	None
Metallothioneins	Cd, Hg, Zn, Cu	Detoxification	Stress
Heat-shock proteins	No	Protein protection	Stress, temperature, diseases
ALA-D	Inorganic Pb	Decreased haemo-synthesis at very high Pb concentrations	Environment, species, fish size
Acetylcholin-esterase	Most insecticides	Nervous system	Species, environment
Histology	No	Specific tissue damages, compensation	Diseases, age
Haematology	Most parameters no (methaemoglobin for oxidants; ALA-D for Pb)	Stress response, anaemia, tissue damages	Stress, diseases
Endocrine disruption	Hormones and chemicals with hormone-like attributes	Reproduction	None
Skeletal anomalies	Teratogenic substances, neurotoxins	Decreased survival	Genetic impact, mechanical damages, Vit. C

Andersson, T., Förlin, L., Härdig, J., Larsson, A., 1988. Physiological disturbances in fish living in coastal water polluted bleached kraft pulp mill effluents. Can. J. Fish. Aquat. Sci. 45, 1525–1536.

Angermeier, P. L., Karr, J. R., 1994. Biological integrity versus biological diversity as policy directives: protecting biotic resources. Bioscience 44, 690–697.

Ankley, G., Mihaich, E., Stahl, R., Tillitt, D., Colborn, T., McMaster, S., Miller, R., Bantle, J., Campbell, P., Denslow, N., Dickerson, R., Folmar, L., Fry, M., Giesy, J., Earl Gray, L., Guiney, P., Hutchinson, T., Kennedy, S., Kramer, V., LeBlanc, G., Mayes, M., Nimrod, A., Patino, R., Peterson, R., Purdy, R., Ringer, R., Thomas, P., Touart, L., Van der Kraak, G., Zacharewski, T., 1998. Overview of a workshop on screening methods for detecting potential (anti-) estrogenic/androgenic chemicals in wildlife. Environ. Toxicol. Chem. 17, 68–87.

Bailey, G.S., Williams, D.E., Hendricks, J.D., 1996. Fish models for environmental carcinogenesis: the rainbow trout. Environ. Health Perspec. 104, 5–21.

Balon, E. K., 1975. Reproductive guilds of fishes: a proposal and definition. J. Fish. Res. Bd. Can. 32, 821–864.

Balon, E. K., 1981. Additions and amendments to the classification of reproductive styles in fishes. Env. Biol. Fish. 6, 377–389.

Banjeree, S., Baughman, G.L., 1991. Bioconcentration factors and lipid solubility. Environ. Sci. Technol. 25, 536–539.

Barnard, S., Wyatt, R. J., 1995. An analysis of predictive models for stream salmonid population. Bull. Fr. Peche Piscicult. 337/338/339, 365–373.

Bayley, P. B., 1995. Understanding large river-floodplain ecosystems. BioScience 45 (3), 153–158.

Bengtsson, B.E., Bengtsson, A., Himberg, M., 1985. Fish deformities and pollution in some Swedish waters. Ambio 14, 32–35.

Berkman, H.E. and Rabeni, C.F., 1987. Effect of siltation on stream fish communities. Envir. Biol. Fishes 18, 285–294.

Bernet, D., Schmidt, H., Meier, W., Burkhardt-Holm, P., Wahli, T., 1999. Histopathology in fish: proposal for the protocol to assess aquatic pollution. J. Fish Dis. 22, 25–34.

Bisson, P.A., Quinn, T.P., Reeves, G.H., Gregory, S.V., 1992. Best management practices, cumulative effects, and long-term trends in fish abundance in Pacific northwest river systems. In: Naiman, R.J. (Ed.), Watershed Management. Springer Verlag, New York, pp. 189–232.

Boening, D.W., 2000. Ecological effects, transport, and fate of mercury: a general review. Chemosphere 40, 1335–1351.

Boon, P.J., Davies, B.R., Petts, G.E., 2000. Global perspectives on river conservation. Science, policy and practice. Wiley, Chichester.

Bowlby, J.N., Roff, J.C., 1996. Trout biomass an habitat relationships in southern Ontario streams. Trans. Am. Fish. Soc. 115, 503–514.

Braunbeck, T., 1998. Cytological alterations in fish hepatocytes following in vivo and in vitro sublethal exposure to xenobiotics – structural biomarkers of environmental contamination. In: Braunbeck, T., Hinton, D.E., Streit, B. (Eds), Fish Ecotoxicology. Birkhäuser Verl., Basel, pp. 61–140.

Brewer, S.K., Little, E.E., DeLonay, A.J., Beauvais, S.L., Jones, S.B., Ellersieck, M.R., 2000. Behavioral dysfunctions correlate to altered physiology in rainbow trout (*Oncorhynchus mykiss*) exposed to cholinesterase-inhibiting chemicals. Arch. Environm. Contam. Toxicol. 40, 70–76.

Bryan, G.W., Langston, W.J., 1992. Bioavailability, accumulation and effects of heavy metals in sediments with special reference to United Kingdom estuaries: a review. Environ. Poll. 76, 89–131.

Bucheli, T.D., Fent, K., 1995. Induction of cytochrome P450 as a biomarker for environmental contamination in aquatic ecosystems. Crit. Rev. Environ. Sci. Techn. 25, 201–268.

Bucher, F., Hofer, R., 1990. Effects of domestic wastewater on serum enzyme activities of brown trout (*Salmo trutta*). Comp. Biochem. Physiol. 97C, 381–385.

Bucher, F., Hofer, R., El-Matbouli, M., 1992. The prevalence and pathology of Zschokkella nova (Klokacewa) (Myxosporea) in the liver of Cottus gobio (L.) from a polluted river. Dis. aquat. Org. 14, 137–143.

Buckley, J. A., 1982. Hemoglobin-glutathione relationships in trout erythrocytes treated with monochloramine. Bull. Environ. Contam. Toxicol. 29, 637–644.

Cairns, J., Van der Schalie, W.H., 1980. Biological monitoring. Part 1 – Early warning systems. Water Res. 14, 1179–1196.

Cairns, J., McCormick, P.V., Niederlehner, B.R., 1993. A proposed framework for developing indicators of ecosystem health. Hydrobiologia 263, 1–44.

Campbell, P.G.C., Stokes, P.M., 1985. Acidification and toxicity of metals of aquatic biota. Can. J. Fish. Aquat. Sci 42, 2034–2049.

Chovanec, A., Koller-Kreimel, V., 1999. Indikatoren einer nachhaltigen Nutzung von Oberflächengewässern. In: Federal Environment Agency Vienna Umweltbundesamt Wien (Ed.), Umweltindikatoren für Österreich – Regionale und nationale Maßzahlen zur Dokumentation der Umweltsituation auf dem Weg zu einer nachhaltigen Entwicklung. Conference Papers, Tagungsberichte Vol. 26, pp. 38–52.

Chovanec, A., Spindler, T., 1997. Zur Verwendung von Fischen als Bioindikatoren in Österreich. In: Spindler, T., Fischfauna in Österreich. Ökologie – Gefährdung – Bioindikation – Fischerei – Gesetzgebung. Monographien des Umweltbundesamtes, Band 87, Wien, pp. 76–86.

Chovanec, A., Schiemer, F., Cabela, A., Gressler, S., Grötzer, C., Pascher, K., Raab, R., Teufl, H., Wimmer, R., 2000a. Constructed inshore zones as river corridors through urban areas – the Danube in Vienna: preliminary results. Regul. Rivers: Res. Mgmt. 16, 175–187.

Chovanec, A., Jäger, P., Jungwirth, M., Koller-Kreimel, V., Moog, O., Muhar, S., Schmutz, S., 2000b. The Austrian way of assessing the ecological integrity of running waters – a contribution to the EU Water Framework Directive. Hydrobiologia 422/423, 445–452.

Chovanec, A., Schiemer, F., Waidbacher, H., Spolwind, R., 2002. Rehabilitation of a heavily modified river section of the Danube in Vienna (Austria): biological assessment of landscape linkages on different scales. Internat. Rev. Hydrobiol. 87 (2/3), 183–195.

Chubb, J.C., 1997. Fish parasites as indicators of environmental quality: a second perspective. Parassitologia 39 (3), 255.

Copp, G.H., 1989. The habitat diversity and fishing reproductive function of floodplain ecosystems. Environ. Biol. Fish. 26, 1–27.

Couch, J.A., 1975. Histopathological effects of pesticides and related chemicals on the liver of fishes. In: Ribelin, W.E., Migaki, G. (Eds), The Pathology of Fishes, University of Wisconsin Press, Madison, pp. 559–584.

Crews, D., Willingham, E., Skipper, J., 2000. Endocrine disrupters: present issues, future directions. The Quarterly Review of Biology 75/3, 243–260.

Cuvin-Aralar, M.L.A., Furness, R.W., 1991. Mercury and selenium interaction: a review. Ecotoxicol. and Environ. Safety 21, 348–364.

Dallinger, R., Egg, M., Köck, G., Hofer, R., 1997. The role of metallothionein in cadmium accumulation of Arctic char (*Salvelinus alpinus*) from high mountain lakes. Aquat. Toxicol. 38, 47–66.

Davies, P.H., Goettl, J.P., Sinley, J.R., Smith, N.F., 1976. Acute and chronic toxicity of lead to rainbow trout, *Salmo gairdenri*, in hard and soft water. Wat. Res. 10, 199–206.

Dembele, K., Haubruge, E., Gaspar, C., 2000. Concentration effects of selected insecticides on brain acetylcholinesterase in the common carp (Cyprinus carpio L.). Ecotoxicol. Environ. Saf. 45, 49–54.

Dušek, L., Gelnar, M., Šebelová, Š., 1998. Biodiversity of parasites in a freshwater environment with respect to pollution: metazoan parasites of chub (*Leuciscus cephalus* L.) as a model for statistical evaluation. Int. J. Parasitol. 28, 1555–1571.

Dynesius, M., Nilsson, C., 1994. Fragmentation and flow regulation of river systems in the northern third of the world. Science 266, 753–762.

Elliott, J. M., 1994. Quantitative Ecology and The Brown Trout. Oxford Series in Ecology and Evolution. Oxford University Press, New York.

Erickson, R.J., Bills, T.D., Clark, J.R., Hansen, D.J., Knezovich, J.P., Mayer, F.L., McElroy, A.E., 1994. Synopsis of discussion session on physico-chemical factors affecting toxicity. In: Hamelink, J.L., Landrum, P.F., Bergman, H.L., Benson, W.H. (Eds), Bioavailability: Physical, Chemical, and Biological Interactions. Lewis, Boca Raton, pp. 31–38.

European Union, 2000. Directive 2000/60/EC of the European Parliament and of the Council establishing a framework for Community action in the field of water policy.

Falkenmark, M., Allard, B., 1991. Water quality genesis and disturbances of natural freshwaters. In: Hutzinger, O. (Ed.), The Handbook of Environmental Chemistry, Vol. 5, Part A, Water Pollution. Springer Verlag, Berlin, pp. 45–78.

Felts, P.A., Heath, A.G., 1984. Interactions of temperature and sublethal environmental copper exposure on the energy metabolism of bluegill, *Lepomis macrochirus*. J. Fish Biol. 25, 445–453.

Fent, K., 1999. Effects of organotin compounds in fish: from the molecular to the population level. In: Braunbeck, T., Hinton, D.E., Streit, B. (Eds), Fish Ecotoxicology. Birkhäuser Verl., Basel, pp. 259–302.

Filho, D.W., 1996. Fish antioxidant defences – a comparative approach. Braz. J. Med. Biol. Res. 29, 1735–1742.

Frissell, C.A., Liss, W.J., Warren, C.E., Hurley, M.D., 1986. A hierarchical framework for stream habitat classification: viewing streams in a watershed context. Env. Mgmt. 10, 199–214.

Fritsch, A.J., 1872. Die Wirbeltiere Böhmens. Ein Verzeichnis aller bisher in Böhmen beobachteten Säugetiere, Vögel, Amphibien und Fische. Archiv für die naturwissenschaftliche Landesdurchforschung von Böhmen 2 (2), 1–152.

Gammon, J.R., Simon, T.P., 2000. Variation in a Great River Index of Biotic Integrity over a 20-year period. Hydrobiologia 422/423, 291–304.

Gassner, H., Wanzenböck, J., 1999. Fischökologische Leitbilder fünf ausgewählter Salzkammergutseen. Limnologica 29, 436–448.

Gaudin, P., 2001. Habitat shifts in juvenile riverine fishes. Archiv Hydrobiol. Suppl. 135/2, Large Rivers 12, 393–408.

Geller, W., 1984. A toxicity warning monitor using the weakly electric fish, *Gnathonemus petersi*. Water Res. 18, 1285–1290.

Geyer, H.J., Scheunert, I., Brüggemann, R., Matthies, M., Steinberg, C.E.W., Zitko, V., Kettrup, A., Garrison, W., 1994. The relevance of aquatic organisms' lipid content to the toxicity of lipophilic chemicals: toxicity of lindane to different fish species. Ecotoxical. Environ. Safety 28, 53–70.

Geyer, H.J., Scheunert, I., Brüggemann, R., Langer, D., R., Korte, F., Kettrup, A., Mansour, M., Steinberg, C.E.W., Nyholm, N., Muir, D.C.G., 1997. Half-lives and bioconcentration of lindane (δ-HCH) in different fish species and relationship with their lipid content. Chemosphere 35, 343–351.

Goksoyr, A., 1995. Use of cytochrome P450 1A (CYP1A) in fish as a biomarker of aquatic pollution. Arch. Toxicology Suppl. 17, 80–95.

Goksoyr, A., Husoy, A.M., 1999. Immunochemical approaches to studies of CYP1A localization and induction by xenobiotics in fish. In: Braunbeck, T., Hinton, D.E., Streit, B. (Eds), Fish Ecotoxicology. Birkhäuser Verl., Basel, pp. 165–202.

Gregory, S.V., Swanson, F.J., McKee, W.A., Cummins, K.W., 1991. An ecosystem perspective of riparian zones. BioSience 41, 540–551.

Grizzle, J.M., Horowitz, S.A., Strength, D.R., 1988. Caged fish as monitors of pollution: effects of chlorinated effluents from a wastewater treatment plant. Water Resour. Bull. 24, 951–959.

Haaparata, A., Valtonen, E.T., Hoffmann, R., Holmes, J., 1996. Do macrophage centres in freshwater fishes reflect the differences in water quality? Aquat. Toxicol. 34, 253–272.

Haitzer, M., Höss, S., Traunspurger, W., Steinberg, C. 1998. Effects of dissolved organic matter (DOM) on the bioconcentration of organic chemicals in aquatic organisms – a review. Chemosphere 37, 1335–1362.

Halbwachs, G., Arndt, U., 1992. Möglichkeiten und Grenzen der Bioindikation. VDI-Berichte 901, 7–15.

Hamilton, S.J., Mehrle, P.M., 1986. Metallothionein in fish: a review of its importance in assessing stress from metal contaminants. Trans. Am. Fish. Soc. 115, 596–609.

Hechter, R.P., Moodie, P.F., Moodie, G.E.E., 2000. Pectoral fin asymmetry, dimorphism and fecundity in the brook stickleback, *Culaea inconstans*. Behaviour 137, 999–1009.

Hildrew, A. G., 1996. Whole river ecology: spatial scale and heterogeneity in the ecology of running waters. Arch. Hydrobiol. Suppl. 113, Large Rivers 10, 25–43.

Hinton, D.E., Lauren, D.J., 1990. Liver structural alterations accompanying chronic toxicity: potential biomarkers of exposure. In: McCartney, J.F., Shugart, L.R. (Eds), Biomarkers of environmental contamination. Lewis, Boca Raton, pp. 15–57.

Hodson, P.V., Blunt, B.R., Whittle, D.M., 1984. Monitoring lead exposure on fish. In: Cairns, V.W., Hodson, P.V., Nriagu, J.O. (Eds), Contaminant Effects on Fisheries. John Wiley, New York, pp. 87–98.

Horwitz, R.J., 1978. Temporal variability patterns and the distributional patterns of stream fishes. Ecol. Monogr. 48, 307–321.

Huet, M., 1959. Profiles and biology of Western European streams as related to fish management. Trans. Am. Fish. Soc. 88, 153–163.

Hyllner, J.S., Haux, C., Andersson, T., Olsson, P.E., 1989. Cortisol induction of metallothionein in primary cultures of rainbow trout hepatocytes. J. Cell Physiol. 139, 24–28.

Iwama, G.K., Thomas, M.M., Vijayan, M.M., Forsyth, R.B., 1998. Stress protein expression in fish. Rev. Fish Biol. Fish. 8, 35–56.

Iwama, G.K., Vijayan, M.M., Forsyth, R.B., Ackerman, P.A., 1999. Heat shock proteins and physiological stress in fish. Am. Zoologist 39, 901–909.

Janssens de Bisthoven, L., 1999. Biomonitoring with morphological deformaties in aquatic organisms. In: Gerhardt, A. (Ed.), Biomonitoring of Polluted Water. Environmental Research Forum 9, Trans Tech, pp. 65–94.

Jensen, F.B., 1990. Nitrite and red blood cell function in carp: control factors for nitrite entry, membrane potassium ion permeation, oxygen affinity and methaemoglobin formation. J. Exp. Biol. 152, 149–166.

Johansson-Sjöbeck, M.L., Larsson, A., 1979. Effects of inorganic δ-amino levulinic acid dehydratase activity and haematological variables in rainbow trout (Salmo gairdnery). Arch. Environ. Contam. Toxicol 8, 419–431.

Jungwirth, M., 1998. River continuum and fish migration – going beyond the longitudinal river corridor in understanding ecological integrity. In: Jungwirth, M., Schmutz, S., Weiss, S. (Eds), Fish Migration and Fish Bypasses. Fishing News Books, Oxford, pp. 19–32.

Jungwirth, M., Muhar, S., Schmutz, S., 2000. Fundamentals of fish ecological integrity and their relation to the extended serial discontinuity concept. Hydrobiologia 422/423, 85–97.

Junk, W.J., Bayley, P.B., Sparks, R.E., 1989. The flood-pulse concept in river floodplain systems. In Dodge, D.P. (Ed.), Proceedings of the International Large River Symposium, Canadian Special Publication of Fisheries and Aquatic Sciences 106, 110–127.

Kamler, E., 1992. Early Life History of Fish. An Energetics Approach. Chapman & Hall, London.

Karr, J.R., 1981. Assessment of biotic integrity using fish communities. Fisheries 6 (6), 21–27.

Karr, J.R., 1991. Biological integrity: a long neglected aspect of water resource management. Ecol. Appl. 1, 66–84.

Karr, J.R., Chu, E.W., 1999. Restoring Life in Running Waters. Better Biological Monitoring. Island Press, Washington, DC.

Karr, J.R., Fausch, K.D., Angermeier, P.L., Yant, P.R., Schlosser, I.J., 1986. Assessment of biological integrity in running waters: a method and its rationale. Illinois Nat. Hist. Surv. Spec. Publ. 5.

Keckeis, H., Schiemer, F., 2001. The ecology of the early life history stages of riverine fish: new perspectives in conservation and river management. Archiv Hydrobiol. Suppl. 135/2, Large Rivers 12, 517–522.

Keckeis, H., Frankiewicz, P., Schiemer, F., 1996. The importance of inshore areas for spawning nase *Chondrostoma nasus* (Cyprinidae) in a free-flowing section of a large river (Danube, Austria). Arch. Hydrobiol. Suppl. 113, Large Rivers 10, 51–64.

Kennedy, C.R., 1997. Freshwater fish parasites and environmental quality: an overview and caution. Parassitologia 39 (3), 249–254.

Kime, D.E., 1998. Endocrine Disruption in Fish. Kluwer Academic, Dordrecht.

Kleinow, K.M., Melancon, M.J., Lech, J.J., 1987. Biotransformation and induction: implications for toxicity, bioaccumulation and monitoring of environmental xenobiotics in fish. Environ. Health Perspect. 71, 105–119.

Köck, G., Hofer, R., 1998. Origin of cadmium and lead in clear softwater lakes of high-altitude and high-latitude, and their bioavailability and toxicity to fish. In: Braunbeck, T., Hinton, D.E., Streit, B. (Eds), Fish Ecotoxicology. Birkhäuser Verl., Basel, pp. 225–257.

Köck, G., Hofer, R., Wögrat, S., 1995. Accumulation of trace metals (Pb, Cd, Cu, Zn) in Arctic char (Salvelinus alpinus) from oligotrophic Alpine lakes: relation to lake alkalinity. Can. J. Fish. Aquat. Sci. 52, 2367–2376.

Köck, G., Triendl, M., Hofer, R., 1996. Seasonal patterns of metal accumulation in Arctic char (Salvelinus alpinus) from oligotrophic Alpine lakes related to temperature. Can. J. Fish. Aquat. Sci. 53, 780–786.

Kuperman, B.I., 1995. Parasites as bioindicators for estimation of aquatic pollution. In: Thurston, R.V. (Ed.), Fish Physiology, Toxicology, and Water Quality. Proc. 4th Intern. Symp., Bozeman, Montana, USA, Ecosystems Res. Div., Athens, Georgia 30605, pp. 51–64.

Lackner, R., 1998. "Oxidative stress" in fish by environmental pollution. In: Braunbeck, T., Hinton, D.E., Streit, B. (Eds), Fish Ecotoxicology. Birkhäuser Verl., Basel, pp. 203–224.

Lemly, A.D., 1999. Selenium impacts on fish: an insidious time bomb. Human and Ecological Risk Assessment 5, 1139–1151.

Lotrich, V.A., 1973. Growth, production and community composition of fishes inhabiting a first, second and third order stream of eastern Kentucky. Ecol. Monogr. 43, 377–97.

Luscova, V., 1997. Annual cycles and normal values of hematological parameters in fishes. Acta Sc. Nat. Brno 31/5.

Mackay, D., Fraser, A., 2000. Bioaccumulation of persistent organic chemicals: mechanisms and models. Environ. Pollut. 110, 375–391.

MacKenzie, K., Williams, H.H., Williams, B., McVicar, A.M., Siddall R., 1995. Parasites as indicators of water quality and the potential use of helminth transmission in marine pollution studies. Advances in Parasitology 35, 86–144.

Mallatt, J., 1985. Fish gill structural changes induced by toxicants and other irritants: a statistical review. Can. J. Fish. Aquat. Sci. 42, 630–648.

Marcogliese, D.J., Cone, D.K., 1997. Parasite communities as indicators of ecosystem stress. Parassitologia 39 (3), 227–232.

Markert, B., 1994. Biomonitoring – Quo vadis. UWSF-Z. Umweltchem. Ökotox. 6 (3), 145–149.

Matthiessen, P., Sumpter, J.P., 1998. Effects of estrogenic substances in the aquatic environment. In: Braunbeck, T., Hinton, D.E., Streit, B. (Eds), Fish Ecotoxicology. Birkhäuser Verl., Basel, pp. 319–335.

McGeer, J.C., Szebedinszky, C., McDonald, D.G., Wood, C.M., 2000. Effects of chronic sublethal exposure to waterborne Cu, Cd or Zn in rainbow trout. 2. Tissue specific metal accumulation. Aquat. Toxicol. 50, 245–256.

McKim, J.M., Erickson, R.J., 1991. Environmental impact on the physiological mechanisms controlling xenobiotic transfer across fish gills. Physiol. Zool. 64, 39–67.

Meinelt, T., Krüger, R., Pietrock, M., Osten, R., Steinberg, C., 1997. Mercury pollution and macrophage centres in pike (*Esox lucius*) tissues. Environ. Sci. & Pollut. Res. 4, 32–36.

Milner, N., Wyatt, R.J., Scott, M., 1993. Variability in the distribution and abundance of stream salmonids, and the associated use of habitat models. J. Fish. Biol. 43, 103–119.

Möller, A.P., 1997. Developmental stability and fitness: a review. The American Naturalist 149, 916–932.

Muhar, S., Schwarz, S., Schmutz, S., Jungwirth, M., 2000. Identification of rivers with high and good habitat quality: methodological approach and applications in Austria. Hydrobiologia 422/423, 343–358.

Murad, A., Everill, S., Houston, A., 1993. Division of goldfish erythrocytes in circulation. Can. J. Zool. 71, 2190–2198.

Naiman, R.J., Décamps, H. (Eds), 1990. The Ecology and Management of Aquatic–Terrestrial Ecotones. UNESCO, Paris, and Parthenon, Carnforth.

Naiman, R.J., Décamps, H., Pastor, J., Johnson, C.A., 1988. The potential importance of boundaries to fluvial ecosystem. J. North Am. Benthol. Soc. 7, 289–306.

Nikinmaa, M., 1992. How does environmental pollution affect red cell function in fish? Aquatic Toxicol. 22, 227–238.

Noss, R.F., 1990. Indicators for monitoring biodiversity: a hierarchical approach. Conserv. Biol. 4, 355–364.

Oberdorff, T., Hughes, R.M., 1992. Modification on an index of biotic integrity based on fish assemblages to characterize rivers of the Seine-Normandie basin, France. Hydrobiologia 228, 117–130.

Olson, K.R., Squibb, K., Crusin, R.J., 1978. Tissue uptake, cellular distribution, and metabolism of $^{14}CH_3HgCl$ and $CH_3{}^{203}HgCl$ by rainbow trout, Salmo gairdneri. J. Fish. Res. Board Can. 35, 381–390.

Overnell, J., McIntosh, R., Fletcher, T.C., 1987. The levels of metallothionein and zinc in plaice, Pleuronecta platessa L., during the breeding season, and the effect of oestradiol injection. J. Fish Biol. 30, 539–546.

Overstreet R.M., 1997. Parasitological data as monitors of environmental health. Parassitologia 39 (3), 169–175.

Parsons, P.A., 1990. Fluctuation asymmetry: an epigenetic measure of stress. Biol. Rev. 65, 131–145.

Payne, J.F., Fancey, L.L., Rahimtula, A.D., Porter, E.L., 1987. Review and perspective on the use of mixed-function oxygenase enzymes in biological monitoring. Comp. Biochem. Physiol. 86C, 223–245.

Perkins, E.J. Jr., Schlenk, D., 2000. In vivo acetylcholinesterase inhibition, metabolism, and toxicokinetics of aldicarb in channel catfish: role of biotransformation in acute toxicity. Toxicol. Sci. 53, 308–315.

Perry, S.F., Laurent, P., 1993. Environmental effects on fish gill structure and function. In: Ranking, J.C., Jensen, F.B. (Eds), Fish Physiology, Fish and Fisheries Series 9, Chapman & Hall, London, pp. 231–264.

Persat, H., Olivier, J.-M., Bravard, J. P., 1995. Stream and riparian management of large braided Mid-European rivers, and consequences for fish. In Armantrout, N.B. (Ed.), Condition of the World's Aquatic Habitats. Proc. of the World Fisheries Congress, Theme 1, Oxford and IBH, New Delhi, pp. 139–169.

Pickering, A.D. (Ed.), 1981. Stress and Fish. Academic Press, London.

Pointet, K., Milliet, A., 2000. PAHs analysis of fish whole gall bladders and livers from the Natural Reserve of Camargue by GC/MS. Chemosphere 40, 293–299.

Purdom, C.E., Hardiman, P.A., Bye, V.J., Eno, N.C., Tyler, C.R., Sumpter, J.P., 1994. Estrogenic effects of effluents from sewage treatment works. Chem. Ecol. 8, 275–285.

Rabeni, C.F., Smale, M.A., 1995. Effects of siltation on stream fishes and the potential mitigating role of the buffering riparian zone. Hydrobiologia 303, 211–219.

Rahel, F.J., 2000. Homogenization of fish faunas across the United States. Science 288, 854–856.

Rocha, E., Monteiro, R.A.F., Rereira, C.A., 1995. Microanatomical organization of hepatic stroma of the brown trout, *Salmo trutta fario* (Teleostei, Salmonidae): a qualitative and quantitative approach. J. Morphol. 223, 1–11.

Schiemer, F., 2000. Fish as indicators for the assessment of the ecological integrity of large rivers. Hydrobiologia 422/423, 271–278.

Schiemer, F., Keckeis, H. (Eds), 2001. 0+ fish as indicators of the ecological status of large rivers. Riverine 0+ fish workshop. Arch. Hydrobiol. Suppl. 135/2–4, Large Rivers 12, 105–522.

Schiemer, F. Spindler, T., 1989. Endangered fish species of the Danube river in Austria. Regul. Rivers: Res. Mgmt., 4, 397–407.

Schiemer, F., Waidbacher, H. 1992. Strategies for conservation of the Danubian fish fauna. In Boon, P.J., Calow, P. and Petts, G.E. (Eds), River Conservation and Management. John Wiley, Chichester, pp. 364–384.

Schiemer, F., Zalewski, M., 1992. The importance of riparian ecotones for diversity and productivity of riverine fish communities. Neth. J. Zool. 42, 323–335.

Schiemer, F., Spindler, T., Wintersberger, H., Schneider, A., Chovanec, A., 1991. Fish fry associations: important indicators for the ecological status of large rivers. Verh. Internat. Verein. Limnol. 24, 2497–2500.

Schiemer, F., Keckeis, H., Winkler, G., Flore, L., 2001a. Large rivers: the relevance of ecotonal structure and hydrological properties for the fish fauna. Archiv Hydrobiol. Suppl. 135/2, Large Rivers 12, 487–508.

Schiemer, F., Keckeis, H., Reckendorfer, W., Winkler, G., 2001b. The "inshore retention concept" and its significance for large rivers. Archiv Hydrobiol. Suppl. 135/2, Large Rivers 12, 509–516.

Schlosser, I.J., 1995. Critical landscape attributes that influence fish population dynamics in headwater streams. Hydrobiologia 303, 71–81.

Schmutz, S., Kaufmann, M., Vogel, B., Jungwirth, M., Muhar, S., 2000. A multi-level concept for fish-based, river-type-specific assessment of ecological integrity. Hydrobiologia 422/423, 279–289.

Sempeski, P. and Gaudin, P., 1995. Size-related changes in dial distribution of young grayling (*Thymallus thymallus*). Canad. J. Fish. Aquat. Sci. 52, 1842–1848.

Shephard K.L., 1994. Functions for fish mucus. Rev. in Fish Biol. and Fisheries 4, 401–429.

Simon, T. P., Lyons, J., 1995. Application of the Index of Biotic Integrity to evaluate water resource integrity in freshwater ecosystems. In: Davis, W.S., Simon, T.P. (Eds), Biological Assessment and Criteria – Tools for Water Resource Planning and Decision Making. Lewis, Boca Raton, pp. 245–262.

Skielkvale, B.L., Wright, R.F., 1998. Mountain lakes, sensitivity to acid deposition and global climate change. Ambio 27, 280–286.

Soivio, A., Oikari, A., 1976. Haematological effects of stress on a teleost, *Esox lucius* L. J. Fish Biol. 8, 397–411.

Spry, D.J., Wiener, J.G., 1991. Metal bioavailability and toxicity to fish in low-alkalinity lakes: a critical review. Environ. Pollut. 71, 243–304.

Statham, C.N., Melancon, M.J., Lech, J.J., 1976. Bioconcentration of xenobiotics in trout bile: a proposed monitory aid for some waterborne chemicals. Science 193, 680–681.

Suedel, B.C., Boraczek, J.A., Peddicord, R.K., Clifford, P.A., Dilton, T.M., 1994. Trophic transfer and biomagnification potential of contaminants in aquatic ecosystems. Rev. Environ. Contam. Toxicol. 136, 21–89.

Sures, B., Siddall, R., 1999. *Pomphorhynchus laevis*: the intestinal acanthocephalan as a lead sink for its fish host, chub (*Leuciscus cephalus*). Exp. Parasitol. 93, 66–72.

Sures, B., Siddall, R., Taraschewski, H., 1999. Parasites as accumulation indicators of heavy metal pollution. Parasitology Today 15, 16–21.

Sures, B., Taraschewski, H., Rydlo, M., 1997. Intestinal fish parasites as heavy metal bioindicators: a comparison between *Acanthocephalus lucii* (Palaeacanthocephala) and the Zebra Mussel, *Dreissena polymorpha*. Bull. Environ. Contam. Toxicol. 59, 14–21.

Taylor, M.R., Harrison, P.T.C., 1999. Ecological effects of endocrine disruption: current evidence and research priorities. Chemosphere 39, 1237–1248.

Thienemann, A., 1912. Der Bergbach des Sauerlandes – Faunistische Untersuchungen. Internat. Revue ges. Hydrobiol. Suppl. 4, 1–125.

Thienemann, A., 1925. Die Binnengewässer Mitteleuropas. Die Binnengewässer, Band 1. Scheizerbart'sche Verlagsbuchhandlung, Stuttgart.

Thomann, R.V., 1989. Bioaccumulation model of organic chemical distribution in aquatic food chains. Environ. Sci. Technol. 23, 699–707.

Tjälve, H., Gottofry, J. 1991. Effects of lipophilic complex formation on the uptake and distribution of some metals in fish. Pharmacol. & Toxicol. 69, 430–439.

Townsend, C. R., 1996. Concepts in river ecology: pattern and process in the catchment hierarchy. Arch. Hydrobiol. Suppl. 113, 3–21.

Tyler, C.R., Jobling, S., Sumpter, J.P., 1998. Endocrine disruption in wildlife: a critical review of the evidence. Crit. Rev. Toxicol. 28, 319–361.

Ulmann P., Peter, A. (in manuscript). The importance of temporarily isolated pools in constrained river reaches as nurseries for 0+ rheophilic fishes.

Vijayan, M.M., Pereira, C., Forsyth, R.B., Kenedy, C.J., Iwama, G.K., 1997. Handling stress does not affect the expression of the hepatic heat shock protein 70 and conjugation enzymes in rainbow trout treated with ß-naphthoflavone. Life Sci. 61, 117–127.

Vogl, C., Grillitsch, B., Wytke, R., Soieser, O.H., Scholz, W., 1999. Qualification of spontaneous undirected locomotor behavior of fish for sublethal toxicity testing. Part 1 – Variability of measurement parameters under general test conditions. Environ. Toxicol. Chem. 18, 2736–2742.

Waidbacher, H.G., Haidvogl, G., 1998. Fish migration and fish passage facilities in the Danube: past and present. In: Jungwirth, M., Schmutz, S., Weiss, S. (Eds), Fish Migration and Fish Bypasses. Fishing News Books, Oxford, pp. 85–98.

Wania, F., 1999. On the origin of elevated levels of persistent chemicals in the environment. Environ. Sci. & Pollut. Res. 6, 11–19.

Ward, J.V., 1998. Riverine landscapes: biodiversity patterns, disturbance regimes, and aquatic conservation. Biological Conservation 83, 269–278.

Ward, J.V., Stanford, J.A., 1995. Ecological connectivity in alluvial river ecosystems and its disruption by flow regulation. Regul. Rivers: Res. Mgmt. 11, 105–119.

Ward, J.V., Tockner, K., Schiemer, F., 1999. Biodiversity of floodplain river ecosystems: ecotones and connectivity. Regul. Rivers: Res. Mgmt. 15, 125–139.

Wesche, T.A., Goertler, C.M., Hubert, W.A., 1987. Modified habitat suitability index model for brown trout in southeastern Wyoming. N. A. J. Fish. Manage. 7, 232–237.

Wicklund, A., Runn, P., 1988. Calcium effects on cadmium uptake, redistribution and elimination in minnows, *Phoxinus phoxinus*, acclimated to different calcium concentrations. Aquat. Toxicol. 13, 109–122.

Wieser, W., Forstner, H., Schiemer, F., Mark, W., 1988. Growth rates and growth efficiencies in larvae and juveniles of Rutilus rutilus and other cyprinid species: effects of temperature and food in the laboratory and in the field. Can. J. Fish. Aquat. Sci. 45, 943–950.

Wimmer, R., Chovanec, A., Moog, O., Fink, M.H., Gruber, D., 2000. Abiotic stream classification as a basis for a surveillance monitoring network in Austria in accordance with the EU Water Framework Directive. Acta hydrochim. hydrobiol. 28 (4), 177–184.

Winemiller, K.O., 1992. Life-history strategies and the effectiveness of sexual selection. Oikos 63, 318–327.

Yoder, C.O., Rankin, E.T., 1995. Biological criteria program development and implementation in Ohio. In: Davis, W.S., Simon, T.P. (Eds), Biological Assessment and Criteria – Tools for Water Resource Planning and Decision Making. Lewis, Boca Raton, pp. 109–144.

Zimmermann, S., Sures, B., Taraschewski, H., 1999. Schwermetallanreicherung bei parasitischen Würmern in Abhängigkeit von verschiedenen Umweltfaktoren – ein Beitrag zum Einsatz von Endoparasiten als Bioindikatoren für den aquatischen Lebensraum. In: Oehlmann, J., Markert, B. (Eds), Ökotoxikologie – ökosystemare Ansätze und Methoden. Ecomed Verlag, Jena, pp. 335–340.

Chapter 19

Biomonitoring with birds

Peter H. Becker

Abstract

Birds play an important role as bioindicators: birds are conspicuous, relatively easy to observe, one of the best studied groups of organisms, and in the focus of public interest and care. Top predators like raptors and seabirds accumulate toxic chemicals, which affect parameters like physiology, reproduction and even cause death, all of which lead to population declines and endangering, and which have often in the past been early warning of environmental change. Therefore birds are an attractive choice as biomonitors, indicating specific environmental change such as contamination with chemicals, marine pollution, fish stocks, and any other environmental change. The most striking examples of the value of birds as biomonitors originate from their use as qualitative and quantitative accumulative indicators of pesticides and heavy metals, based on logistically convenient and non-destructive avian matrices such as eggs, feathers or blood, and on high biomagnification rates in dose-dependent responses. Furthermore, large samples of avian museum material permit retrospective analyses, e.g. of heavy metal pollution history. In addition, birds are also of value as sensitive indicators owing to their responding to environmental change in terms of physiology, reproduction (e.g. egg-shell thickness, reproductive success) or demography. Avian biomonitors have been successfully introduced into current monitoring programmes with the following aims (Table 4): to indicate temporal and spatial trends in chemical pollution in terrestrial and aquatic ecosystems; to monitor marine oil pollution; to detect diverse environmental changes such as habitat alteration or fragmentation, and climate change by monitoring bird populations (abundance, distribution, demography). However, despite their undoubted advantages as biomonitors, birds are not being used as often or as effectively as they could be. Better use of ornithological knowledge, better co-operation between researchers doing biomonitoring is called for, and the use of birds as accumulative indicators should be combined with their use as sensitive indicators. Funding is also a problem despite the cost effectiveness of obtaining information via birds – not least because of the support of many bird monitoring volunteers. Another difficulty arises with the long-term perspectives in the monitoring programmes, which is necessary if adequate use is to be made of birds as an early warning. As birds are of interest to and popular with the general public, their choice as biomonitors also furnishes a good instrument for increasing the acceptance and understanding of environmental monitoring on the part of the public.

Keywords: Birds; reproductive success; population monitoring; environmental change; chemicals; oil pollution; top predators; raptors; seabirds; early warning; sensitive indicators; accumulative indicators; feathers; eggs; egg-shell thickness; contamination trend monitoring

1. Introduction: what's so special about birds as bioindicators?

Increasing human populations, greater demands on resources, advancing technological and industrial developments and the input of substantial quantities of various foreign substances into the environment both intentionally and unintentionally, result in serious impacts on the natural environment, with manifold negative consequences on the ecosystems, species and on mankind itself. Consequently, the need for environmental monitoring has never been greater. Furthermore, the avifauna has suffered greatly from the anthropogenic pressure on the natural environment, and many bird species or populations are endangered and can be found on the latest red lists (World Conservation Monitoring Centre, 1990; Nordheim et al., 1996; Köppel et al., 1998). In central Europe, the four main factor combinations contributing to bird population declines are: agriculture and fisheries, hunting and direct persecution, environmental chemicals, and climatic change (Bauer and Berthold, 1996).

Birds have often played an important role as indicators of environmental problems. To mention just two prominent examples: again and again the mass mortality of marine birds through oiling draws public attention to marine oil pollution, as recently the incidents involving the PALLAS (Germany, October 1998, Fleet et al., 1999), the ERIKA (Brittany, December, 1999), the TREASURE (South Africa, June 2000), and the JESSICA (Galapagos Islands, January 2001). High mortality in several species of raptor and seabird during the 1950/60ies in the industrialised countries, together with reproductive failures, running into population declines, led to public awareness of the contamination problems caused by pesticides (for details see 3.1.1).

Why do birds so often provide the first warning of environmental problems? Birds are conspicuous organisms and relatively easy to observe. They are the object of considerable public attention, and so discernible changes in their biology seldom go unremarked. Birds are one of the best-studied and best-known groups of organisms. Furthermore, especially those bird species being at a high position in the food chain are threatened by toxic chemicals, as persistent compounds accumulate to such high levels that they affect physiology, reproduction or, even, cause death, and lead to population declines.

For all of these reasons birds are an attractive choice as bioindicators. Wild birds have been used in biomonitoring, as passive biomonitors with the role of either accumulative and/or sensitive indicators. The most important bird groups in biomonitoring are species in the top positions of the food chains in terrestrial and aquatic ecosystems which accumulate environmental chemicals to high concentrations in their bodies: raptors and fish consuming birds, especially seabirds.

There exists a huge amount of literature on birds as bioindicators which is hard to keep track of and even harder to review. One important basis for this chapter was the comprehensive and meritorious review of birds as monitors of environmental change by Furness and Greenwood (1993), but it is outside the scope and intention of this chapter to summarise this book. Rather, I will especially address the following aspects:

- why birds are suitable and favoured bioindicators;
- the kinds of environmental changes which according to our present knowledge birds are able to indicate;
- current monitoring programmes using birds successfully;
- and perspectives of appropriate use of birds in future monitoring programmes.

This chapter is primarily intended for non-ornithologists. Seabirds tend to predominate in the examples presented for two reasons: first, the fact is that already many aspects of seabirds' lives have been evaluated with a view to indicating the state of the marine environment (e.g. Furness, 1987; Furness and Greenwood, 1993; Furness and Camphuysen, 1997; ICES, 1999). The second reason is that I personally am a seabird ecologist with specific experience and skills in planning and performing of biomonitoring projects involving marine birds.

2. Why are birds suitable as bioindicators?

Birds have often been chosen as bioindicators, but are they really suitable as indicators, and do they adequately meet the requirements of an bioindicator (see also Section 3)?

The following advantages advocate the use of birds in biomonitoring:

- Birds are easy to identify, and classification and systematics are well established. Consequently, there is little risk of monitoring being confounded by uncertainties regarding the identities of, or relationships between, the species studied.
- The general biology and ecology of birds are well known, which enhances their usefulness as biomonitors, by reducing the risk of misinterpretations. Predominately species whose biology is especially well known have been used for biomonitoring.
- Birds occupy various positions in the food chain, especially the higher trophic levels, indicating chemical contamination in various compartments of ecosystems by biomagnification of persistent chemicals.
- Sampling of tissues is relatively easy, and non-destructive techniques are available that do not harm the bird, which is necessary in protected or endangered species. Samples such as feathers or eggs are easy to collect which reduce the costs of sampling. Furthermore birds are superior to other animals with respect to the ease of gathering population data such as numbers or demographic parameters like reproductive success. At bird breeding colonies many samples can be taken or data recorded in a comparatively short time.
- In *k*-selected species like seabirds or raptors, adult survival is high and reproductive output low (e.g. Croxall and Rothery, 1991). Populations are relatively stable over time, and their changes occur delayed but indicate changes in the environment more distinctly than is the case in species with r-strategy.

Other advantages accrue with respect to scale:

- The long life-span of many bird species means that birds integrate the effects of environmental stresses over time.
- Many birds are widespread species allowing comparisons between different ecosystems, countries or even continents.
- The mobility of birds implies an integrative value in bioindication over broad spatial scales.
- Avian mobility and quick (re)settling capacity allow their use as first indicators of positive environmental change, for example through regeneration or institution of habitats.

There are also a number of pragmatic and public-relations reasons which plead for avian biomonitors:

- As birds are relatively easy to study, huge amounts of data have already been gathered, archived and published, and large samples have been collected and stored in museums and specimen banks. This offers a unique data, sample and knowledge base for current and retrospective biomonitoring.
- There is great interest in birds by well informed "amateurs": Many bird species specialists with great knowledge are among them, and without this vast army of volunteers the impressive collections of world wide data on bird populations would be inconceivable. Those amateur efforts can be directed into useful monitoring programmes.
- Birds are conspicuous: Striking biological events like mass mortality or wrecks do not go undetected.
- Public interest in birds is high, and people are emotionally involved with birds. With birds as sentinel organisms it is relatively easy to initiate political action for environmental protection.

However, along with these avian advantages come a number of drawbacks and limitations, depending on the monitoring aims:

- The longevity of birds makes it more difficult to establish any short-term perturbations.
- The mobility of birds can hinder their site-specific use as indicators. The occurrence of different populations during migration or staging at one site may obscure local environmental stress and reduce their value as indicators.
- Bird numbers are regulated by density-dependent processes, and so their population sizes may be somewhat buffered against the impacts of environmental changes. A multitude of factors affects demographic parameters and stages, so that the effects of specific factors are difficult to isolate.
- With wild birds or some species groups, laboratory bio-effect studies are often difficult or impossible.
- Most bird species are protected or even endangered, and licences are required for investigations or sampling.

Other, more specific advantages and limitations of birds or of the parameters of avian biology in biomonitoring are elucidated below, when specific factors are addressed which are indicated by birds. If a monitoring project involving birds is planned, the advantages and limitations of using birds should be carefully identified and assessed. The choice of a suitable sampling design can often eliminate or minimise possible disadvantages.

3. What environmental changes can birds indicate ?

3.1. Environmental change caused by specific factors

3.1.1. Environmental chemicals

The clearest examples of the value of birds as monitors of the environment originate from their use over the last four decades as qualitative and quantitative indicators of

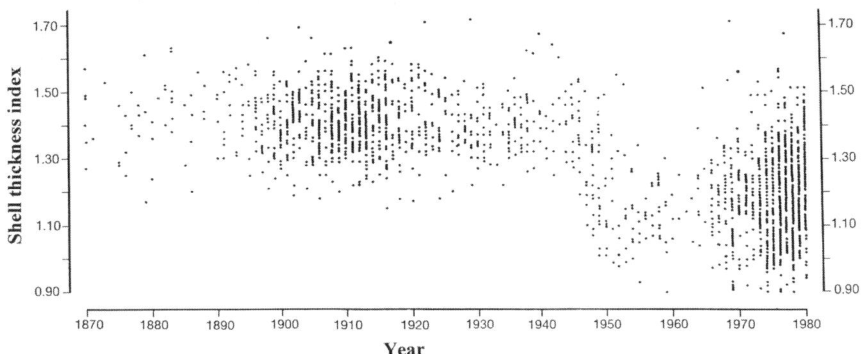

Figure 1. Egg-shell thickness index of British Sparrowhawks (1870–1980) dropped abruptly from 1947, coincident with the widespread introduction of DDT in agriculture. Shells available in museums and private collections have permitted comparisons of recent and historical samples. Each dot represents the mean shell index of a clutch; nearly 2000 clutches are included from all regions of Britain. Shell thickness index = Shell weight (mg)/length × breadth (mm). After Newton (1986).

pesticides in food webs (reviews e.g. Risebrough, 1986; Moriarty, 1990; Newton, 1979, 1986; Newton et al., 1993; Becker, 1991; Furness, 1993; Grasman et al., 1998). Lipophilic and highly persistent substances are accumulated in the food chains through bioconcentration and biomagnification. Top predators like raptors and seabirds have been found to accumulate the highest concentrations of those pollutants, and to respond sensitively to toxic chemicals. The conspicuous death of many birds of prey and seabirds during the 1960s in the industrialised world, eggshell thinning, reproductive failure and population crashes alarmed ornithologists and induced research into the possible causes. Thus the qualitative signs of change in bird populations were recognised as warnings and they were finally discovered to have been caused by pesticides. In other words, the birds acted as sentinel organisms sounding the alarm at signs of environmental damage through organochlorines, and permitting diagnostic investigation through a combination of copiously available background information, widespread interests in birds with the resource of turning to museum collections to investigate retrospectively long-term changes in eggshells (Fig. 1). Once their environmentally deleterious effects became known, the use of many of these compounds was restricted or forbidden by law, leading to a reduction of the chemicals in the environment and to a subsequent recovery of the bird populations.

Birds of prey: As detected through surveys by ornithologists in the early 1960s, formerly common and widespread raptors like the Sparrowhawk *Accipiter nisus* and the Peregrine *Falco peregrinus* – as well as rare species such as White-tailed Sea Eagles *Haliaetus albicilla* in Sweden – suddenly showed marked declines in breeding numbers or disappeared completely from some regions of the UK, of other European countries, and of North America. In Europe, their population crash followed the widespread introduction of new organochlorine pesticides, the cyclodiene compounds aldrin and dieldrin. Detailed analyses over several years revealed the cyclodienes to cause mortality. Eggshell thinning as a major factor yielding a significant reduction of the breeding success in raptors in Europe and North America (Peregrine, Bald Eagle *Haliaetus leucocephalus*, Grasman et al., 1998) was then traced back to DDE residues in the eggs, resulting from metabolic conversion of the bioaccumulated insecticide DDT in widespread use since the late 1940s (e.g. Ratcliffe, 1967; Hickey and Anderson, 1969; Peakall, 1970; Peakall et al., 1975; Enderson et al., 1982; Helander et al., 1982; Newton, 1986; Newton et al., 1989). Many detailed field and laboratory studies led to the elucidation of the raptor-pesticide-syndrome (Ratcliffe, 1970; Cooke, 1973).

In the light of strong, albeit circumstantial, evidence, the use of these organochlorine pesticides was progressively reduced and finally forbidden by law by the 1970s. The subsequent recovery of the affected raptor populations, the increase of eggshell thickness (Fig. 1) and reproductive outcome associated with the reduction in contamination of the environment is well documented for several species and countries (e.g. Sparrowhawk: Newton, 1986; Newton and Willie, 1992; Sibly et al., 2000; Denker et al., 2001; Peregrine: Ratcliffe, 1980; Newton et al., 1989; Schilling and König, 1980; Schilling and Rockenbauch, 1985; Wegner, 2000; White-tailed Sea Eagle at the Baltic Sea: Helander, 1982; further species of raptors: Ellenberg, 1981; Risebrough, 1986).

During the mid-1950s, declines in the Swedish avifauna, in particular in raptors like Goshawk *Accipiter gentilis,* and the conspicuous deaths of various bird species have pointed to environmental contamination with mercury. Multiple increases in mercury levels in predatory birds between the 1950s and 1960s were found (Berg et al., 1966). The problems were caused by alkyl mercury used as fungicides in seed protection, and in the pulp and paper industry. Large amounts of mercury had been released into the air, mainly from the chlorine-alkali industry (Johnels et al., 1979; Lindquist et al., 1991). After 1966 the use of alkyl mercury as seed disinfectant was banned, after 1990 no use of seed dressing containing mercury was allowed, and industry emissions were eliminated or reduced. These actions led to drastic reductions of the emissions and environmental contamination with mercury and thence to bird population recoveries (Helander, 1999).

Seabirds: During the mid-1960s, chicks, fledglings, and adults of Sandwich Terns (*Sterna sandvicensis*) belonging to a colony in the Dutch Wadden Sea were seen dying in tremors and convulsion. Also in other coastal bird species like terns, Herring Gulls *Larus argentatus* and Eiders *Somateria mollissima*, enormous mortality was observed from 1964–1968. Koeman et al. (1967) found considerable concentrations of endrin, dieldrin and telodrin – organochlorine pesticides of high toxicity in the tissues of birds and in their diets, and were able to show that the pesticides originated from effluents of a factory near the estuary of the river Rhine. That the levels were lethal was verified in laboratory experiments.

The consequence of this pollution was a steep decline in the Dutch seabird populations (Duincker and Koeman, 1978), especially in Sandwich Terns (Fig. 2) and Eiders. In Eiders, only females were affected as a consequence of marked weight loss during incubation. Mortality diminished with increasing distance of the breeding site from the pollution source (Swennen, 1972). Also in many species of coastal birds breeding in the German part of the Wadden Sea, striking decreases in breeding

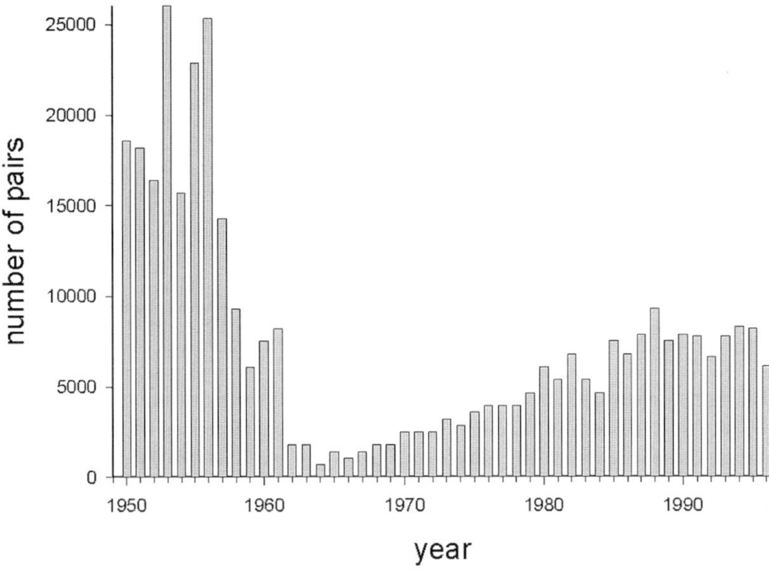

Figure 2. The crash of the Sandwich Tern population in the Dutch Wadden Sea during the 1960s was caused by heavy pollution of the North Sea with toxic cyclodiens. After legal prohibition, the tern population was able to recover but to date still has not reached its former size. Data from Brenninkmeijer and Stienen (1992) and Rasmussen et al. (2000).

populations were obvious during the mid-1960s as a consequence of this pollution event; hence, tern eggs from Germany contained concentrations of cyclodienes comparable to those in the Dutch seabirds' eggs (Becker, 1991). Measures were taken to stop the discharges, resulting in significant decreases in the amounts of telodrin in the North Sea coastal environments, and bird populations were able to recover on the Dutch and German North Sea coasts (Duincker and Koeman, 1978; Becker 1991; Stienen, 1998). From North America, there are other examples of seabirds signalling of environmental change through pollution (3.1.1.2).

The particular avian involvement in these environmental hazards occasioned many short-term investigations into the problems and elicited a number of bird-based contaminant monitoring programs, some of which are still active today (see 4). In these studies the value of birds as bioindicators of chemicals has repeatedly been proved and highlighted, and many environmental problems were first brought to light by the use of birds, both by their function as accumulative and as sensitive indicators of environmental chemicals.

3.1.1.1 Birds as accumulative indicators

High biomagnification rates, dose-dependent response: Birds are good indicators of those environmental chemicals which tend to biomagnify through food chains and which are accumulated in lipid-rich tissues, e.g. organochlorines and methylmercury.

The relevance of contaminants' levels in organisms in high trophic positions like many birds as warnings of human health risks may be greater than their concentrations in water, sediment or soil (Table 1). In particular, the ability to "integrate pollutant signals over time and space" by bioaccumulating pollutants in tissues means that to obtain a given level of accuracy measurements are required of a smaller number of animal samples than of physical samples (Furness et al., 1993) thus increasing the power of trend analyses. This is especially true of birds: together with the biomagnification effect (Tables 1 and 2, Figs 3 and 4), the lower within-sample variance especially in the egg as matrix (Gilbertson et al., 1987, further examples in ICES, 1999) means that spatial or temporal differences in contamination are indicated more conspicuously than in physical samples or in samples of the birds' prey (Tables 1 and 2).

Table 1. Industrial chemicals in abiotic and biotic compartments of the aquatic environment in 1996/1997 at three sites of the Wadden Sea (cf. Fig. 21) characterised by different degrees of contamination. These chemicals are discharged in high amounts by the river Elbe into the Wadden Sea. The resulting intersite differences are indicated most distinctly by the birds, especially by the fish eating Common Tern with the highest biomagnification factor (cf. Fig. 3).

		Location		
Contaminant	Matrix	Wadden Sea Niedersachsen	Elbe Estuary	Wadden Sea Schleswig-Holstein
Mercury	Water (l)	30–493	1100	3–77
	Sediment (dw)	0.2	0.3	0.2
	Mussel (dw)	0.1	0.2	0.1
	Flounder (fw)	74	78	65
	Ostercatcher egg (dw)	1040	908	1568
	Common Tern egg (dw)	1368	7304	1652
PCBs	Water (l)	1.5–9	12	< .05
(6 congeners)	Sediment (dw)	11	–	–
	Mussel (dw)	64	154	41
	Flounder (fat w)	10200	3000	5100
	Oystercatcher egg (dw)	1512	1604	952
	Common Tern egg (dw)	1352	4384	956
HCB	Water (l)	< 0.2	6	< 0.5
	Sediment (dw)	0.5	–	–
	Mussel (dw)	0.4	8.3	0.4
	Flounder (fat w)	1600	1900	< 80
	Oystercatcher egg (dw)	36	172	16
	Common Tern egg (dw)	84	1488	64

Species: Blue Mussel *Mytilus edulis*; Flounder *Platichthys flesus*; Oystercatcher *Haematopus ostralegus*; Common Tern *Sterna hirundo*. Concentrations: arithmetic means (water: ng l^{-1}; Flounder, mercury: ng g^{-1} fresh muscle (fw); PCBs, HCB: ng g^{-1} fat of liver (fat w); other matrices ng g^{-1} dry weight (dw), bird egg concentrations calculated from fresh weight values (75% water content); – : not determined. Sources: Bakker et al. (1999), Becker et al. (1998), Wassergütestelle Elbe, BLMP (2000), Broeg et al. (1999).

Table 2. Biomagnification factors between organochlorines in food and eggs of Oystercatchers (*Haematopus ostralegus*), Herring Gull (*Larus argentatus*) and Common Tern (*Sterna hirundo*) from Spiekeroog, German Wadden Sea, in 1993 (Mattig et al., 1996).

		Biomagnification factors		
Species	Food	PCBs	DDTs	HCB
Oystercatcher	Benthic animals[1]	4–14	3–23	12–21
Herring Gull	Benthic animals[1]	5–19	6–46	17–30
	Fish[2]	2–3	2	2–3
Common Tern	Fish[2]	3	3	3–5

1 *Cardium, Mytilus*; 2 Herring, Plaice
See Fig. for PCB-concentrations.

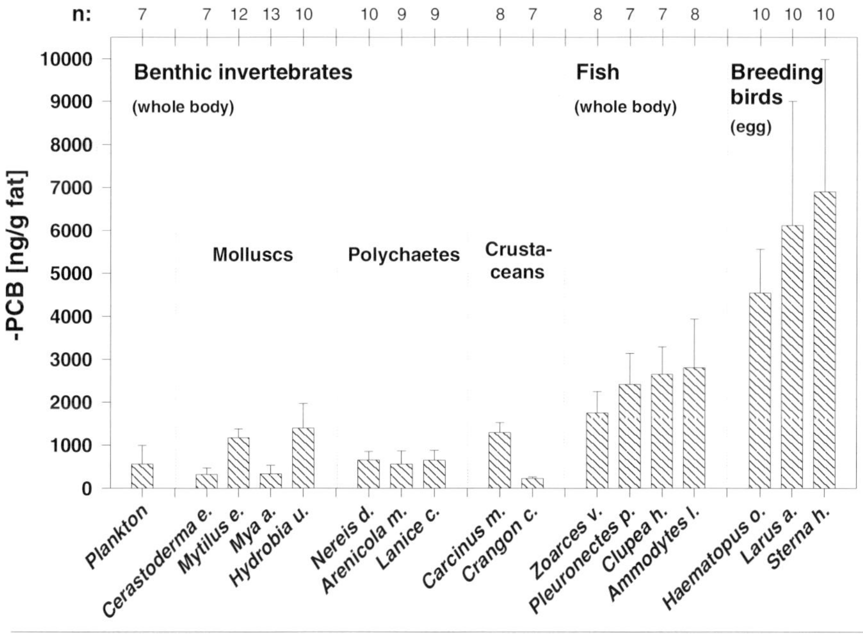

Figure 3. Bioaccumulation of PCBs in the food web of the Wadden Sea. Sum of concentrations of 8 PCB-congeners (means ± 1 standard deviation on fat weight basis) is presented for plankton, nine benthic invertebrates (*Cerastoderma edule, Mytilus edulis, Mya arenaria, Hydrobia ulvae, Nereis diversicolor, Arenicola marina, Lanice conchilega, Carcinus maenas, Crangon crangon*), 4 fish (juvenile stages: *Zoarces viviparus, Pleuronectes platessa, Clupea harengus, Ammodytes lancea*) and three coastal bird species (*Haematopus ostralegus, Larus argentatus, Sterna hirundo*). Mattig et al. (1996). See also Table 2 for biomagnification factors.

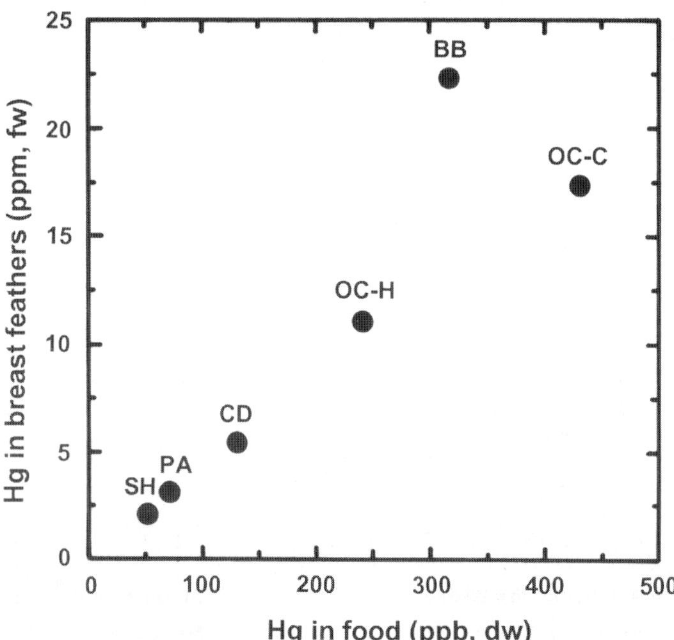

Figure 4.   Relationship between mean mercury concentrations in breast feathers of seabirds from the Azores and in the food of these birds sampled during the breeding season in the colony. Biomagnification factors were between 129–225. BB: *Bulweria bulweria* (Bulwer's Petrel); CD: *Calonectris diomedea* (Cory's Shearwater); OC: *Oceanodroma castro* (Madeiran Storm-Petrel, H = June breeders, C = November breeders); PA: *Puffinus assimilis* (Little Shearwater); SH: *Sterna hirundo* (Common Tern). After Monteiro et al. (1998).

Both laboratory experiments and oral dosing of birds in the field with mercury and other chemicals have shown that concentrations in tissues like liver, kidney, feathers and eggs are dose-dependent (e.g. mercury: Tejning, 1967; Lewis and Furness, 1991, Fig. 5). This being so, birds indicate the current environmental burden with a chemical, and also react relatively quickly to its change, as is clearly shown by the long-term data of Common Terns on the Elbe river (Fig. 6), of Guillemots *Uria aalge* in the Baltic (Fig. 7), as well as by small-scale spatial patterns in bird contamination (see below).

Birds' contamination reflects that of the food they eat (Figs 4, 5, 7; Nisbet and Reynolds, 1984; Dirksen et al. 1995). In coastal birds of the Wadden Sea, the fish-feeding species like terns and the Herring Gull were found to be more contaminated than benthic feeders (Fig. 3, Becker, 1991; Becker et al., 1991; Mattig et al., 2000). Among birds in terrestrial food chains, raptors have been found to show the highest levels of organochlorine pesticides (e.g. Conrad, 1977).

Choice of the matrix: The concentration of a pollutant in the birds' tissues is the result of dynamic and complex processes of uptake, elimination, metabolisation, body distribution and accumulation (e.g. Walker, 1994). Which tissues are suitable for use

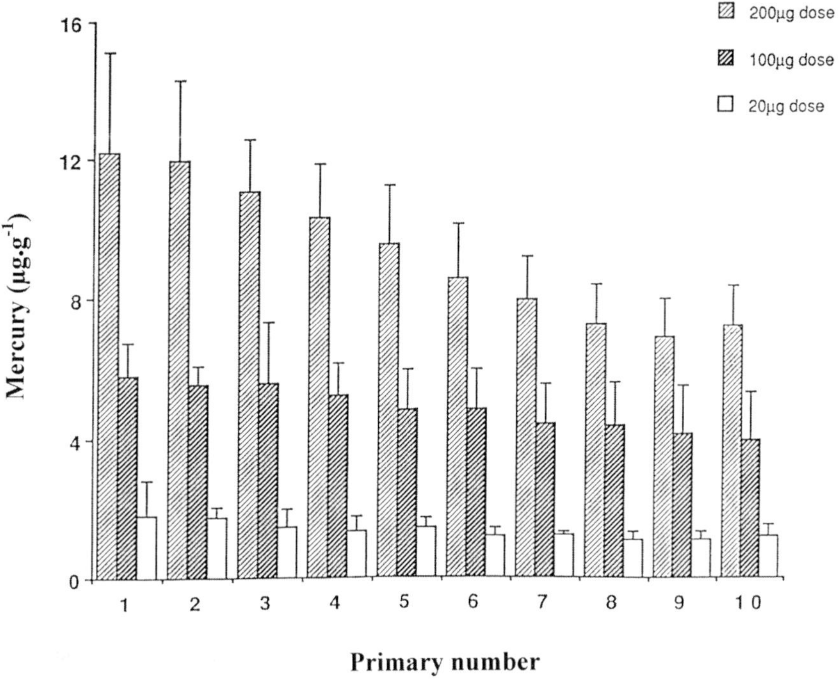

Figure 5. Mercury concentrations in individual primary feathers of Black-headed Gull chicks, fed with 20 µg, 100 µg and 200 µg of mercury, increase in a dose-dependent way. Additionally, a progressive and significant reduction in the mercury levels was evident as growth sequence progressed from the first to the tenth primary, owing to the depletion of mercury pools in the body. From Lewis and Furness (1991).

as samples in the analysis of the chemical load of a bird? Various matrices have been studied with respect to their indicative value, and each has advantages and disadvantages (Furness, 1993), depending for example on the aim of the study, the bird species in question or the chemicals to be analysed.

Residue levels are commonly positively correlated between tissue-types of an individual bird, and by means of known conversion factors the levels in other – unsampled – tissues can be estimated. The liver has been found to be of special indicative value, but to take internal tissues has the great disadvantage that the birds have to be killed. As this is not appropriate in scarce or protected species for both practical and ethical reasons, some studies instead used tissues from birds found dead. These included birds that had starved and mobilised fat reserves, thus adding the problem of an increase of fat soluble chemicals in blood and soft tissues, especially the liver, increasing the variance in the sample and obscuring trends. Nevertheless, important studies have been performed using internal tissues (e.g. Newton et al., 1993; Pain et al., 1995).

Advances in chemical analytical techniques over the last few years now make it possible to study pollutants in very small amounts of material, such as in small blood samples which can be obtained from caught individuals without harming them (e.g. Elliott and Shutt, 1993; Henriksen et al., 1998; Kahle and Becker, 1999), a method of

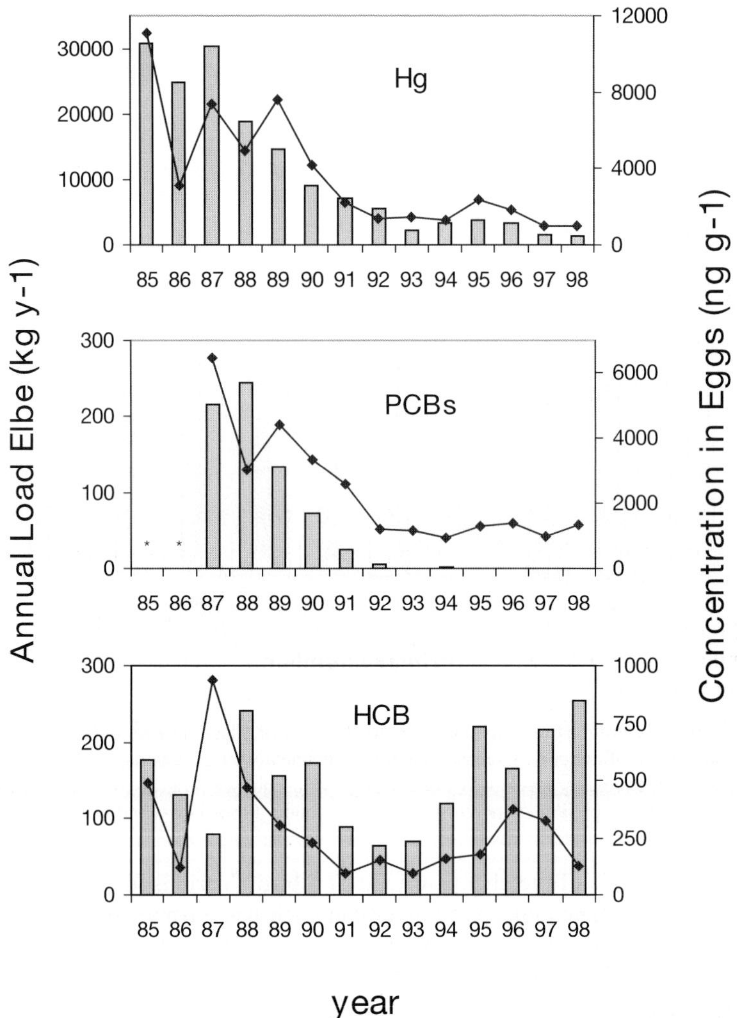

Figure 6. Concentrations of chemicals in Common Tern eggs (lines) reflect the annual loads of the river Elbe (columns). Mercury, PCBs and HCB are chemicals discharged in vast quantities by the Elbe (cf. Fig. 11, 22). Annual loads are derived from the sum of weekly discharges from June of the previous year before until May of the reference year when the eggs have been sampled. PCBs: sum of the congeners CB101, 138, 153 and 180. *: no data available; from 1993–1998, weekly water concentrations of PCBs were often below the detection limit; therefore, no loads have been calculated. Data from annual reports of the Wassergütestelle Elbe and Becker et al. (1998).

sampling that will gain in importance in years to come, as anlytical techniques are further refined. Other frequently applied methods involve sampling the birds "products" such as eggs or feathers.

The removal of eggs is less damaging than that of adults, and has been used as a method in numerous studies and monitoring projects (see 4). Intraclutch variation

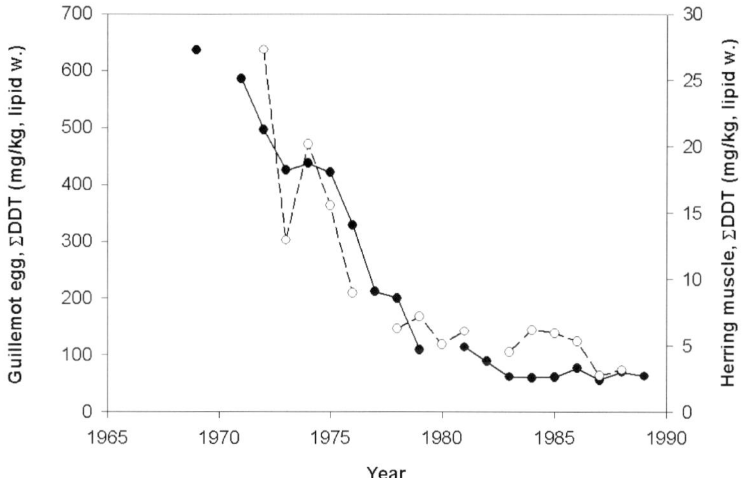

Figure 7. Baltic Guillemots' – egg levels of DDTs from 1969–1989 correspond to the respective levels in Herring, the main food of Guillemots. Biomagnification factor was about 20. After Bignert et al. (1995).

between egg chemical levels is low, so one egg taken from a clutch provides all the information needed. Several studies have shown seabird eggs to be good indicators of local pollutant contamination, even in migrating species like terns, since concentrations in eggs tend to reflect pollutant uptake by the female foraging close to the colony in the few days prior to egg laying. Advantages and disadvantages of eggs as a matrix are discussed e.g. by Gilbertson et al. (1987), Becker (1989), Furness (1993), Becker et al. (1998).

Advantages
- high lipid content and accumulation of lipophilic persistent compounds;
- consistent composition;
- originate in a defined area and year;
- reflection of the contamination of breeding females (healthy and reproductive part of the population);
- being restricted to the breeding season, reduced seasonal variability in chemicals' levels due to limitations of the breeding period;
- ease and speed of sampling;
- ease of handling and storing samples;
- sensitive reaction of birds during the egg and early chick stage to toxic chemicals, and relevance of egg residues to embryotoxic effects;
- feasibility of studies bearing on the relationships between contaminants, eggshell quality and hatching success.

Disadvantages
- relevance to only a part of the population (reproductive females) and of the year (breeding season);

- variation of pollutant levels with the laying sequence;
- failure of some heavy metals to accumulate in the egg (e.g. cadmium and lead).

In rare species, only addled eggs should be taken for reasons of conservation (e.g. Little Tern *Sterna albifrons*, Thyen et al., 2000a), with the problematical result that pollutant levels may not be representative for the whole female population (e.g. when produced by younger birds, or by more highly contaminated individuals), or when dehydrated or infected with bacteria.

Feathers are good indicators of metal pollution (Burger, 1993), especially for organic bound metals like mercury or tin, which accumulate in the plumage, and for which feathers are the main elimination route (Lewis and Furness, 1991; Lewis et al., 1993; Guruge et al., 1996). The quantities of a metal incorporated into the feather represent the body burden at the time of feather growth, during the development of the young or during moult (see Fig. 6). Some small body feathers are sufficient and can be removed from live birds with virtually no harm.

Indicating time trends in chemical pollution: Numerous cases of pollution – mainly in the industrial countries – in both terrestrial and aquatic ecosystems have come to light through being indicated by top-predatory birds accumulating environmental chemicals. The decline of pesticide contamination of raptors and their environment attendant on progressive reductions in the pesticides' use has been clearly demon-

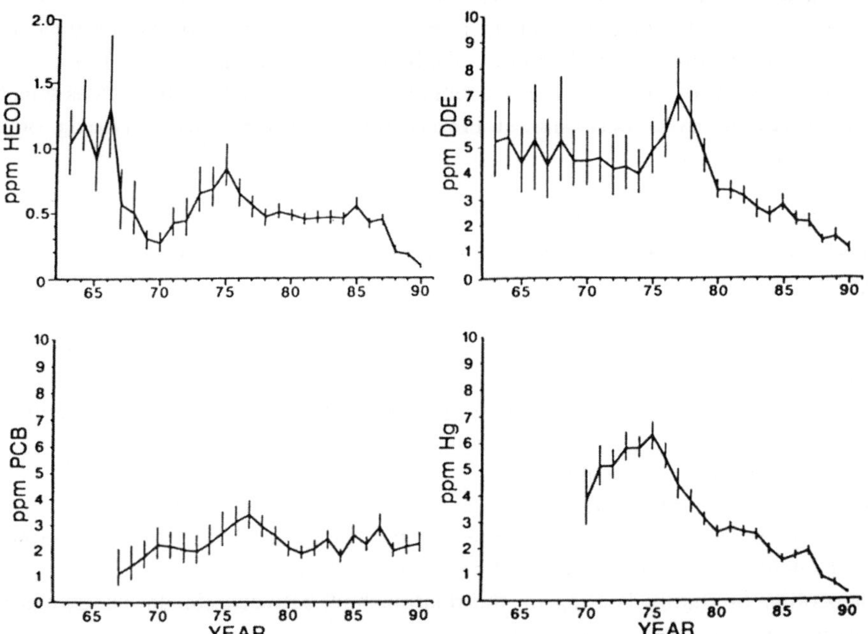

Figure 8.   Levels of chemical contaminants in the livers of British Sparrowhawks decreased significantly from 1963–1990 (except for PCBs, n.s.). Lines show three-year moving geometric means of residue levels, bars geometric standard errors. HEOD is the active ingredient in dieldrin and a metabolite of aldrin. From Newton et al. (1993).

strated by reference to their eggs or tissues taken from their carcasses (Fig. 8; e.g. Sparrowhawk: Newton, 1986; Newton and Wyllie, 1992; Newton et al., 1993; Denker et al., 2001; Peregrine: Schilling and Rockenbauch, 1985; Newton et al., 1989; Wegner, 2000; Goshawk: Baum and Conrad, 1978; White-tailed Sea Eagle: Helander et al., 1982; Kestrel *Falco tinnunculus*; Newton et al., 1993).

Birds in aquatic systems indicate the changes in pollutant loads in water and food chains: For running waters, Ormerod and Tyler (1993) show the value of Dipper *Cinclus cinclus* and Grey Wagtail *Motacilla alba* eggs as indicators of organochlorine pollution. The fish-eating Common Tern *Sterna hirundo* has proved to demonstrate pollutant loads in big rivers like the Elbe in Germany (Fig. 6): Contaminant levels were high and fluctuating during the 1980s, but decreased greatly during the 1990s as shown by water and bird analyses (Becker and Sommer, 1998; Becker et al., 1998). Especially high levels of egg contamination were found in years with high river spates, e.g. in 1987 (Fig. 6, Becker et al., 1991). In Britain, contamination levels in the livers of the fish-eaters Grey Heron *Ardea cinerea,* Kingfisher *Alcedo atthis* and Great-crested Grebe *Podiceps cristatus* confirmed the environmental declines in HEOD, DDE, PCBs and mercury from 1963–1990 (Newton et al., 1993).

Fish-consuming waterbird species of the Great Lakes (Canada) such as the Herring Gull, the Double-crested Cormorant *Phalacrocorax auritus* and the Common Tern have been studied since 1971 with respect to contaminant levels. The results show the declining chemicals' (mainly by DDE and PCBs) loads of these heavily polluted lakes (Fig. 14; e.g. Ryckman et al. 1997, 1998).

Seabirds have in the past often revealed temporal change in the chemical pollution of marine areas. In Britain Gannet (Newton et al., 1990) and Shag (*Phalacrocorax aristotelis*, Coulson et al., 1972) eggs, in Canada the eggs of several seabird species have revealed temporal trends (Chapdelaine et al., 1987; Elliott et al., 1994) just as Brown Pelican *Pelecanus occidentalis* eggs did in South Carolina (Blus, 1982). Guillemot (Fig. 7) and Little Tern eggs (Thyen et al., 2000a) have shown the declines of organochlorine pollution in the Baltic Sea. For the North American Atlantic coast, Nisbet and Reynolds (1984) documented decreases in contamination of Common Tern eggs during the 1970s. In Germany, long-term studies of the contamination of seabirds' eggs in the Wadden Sea revealed decreases in the levels of most organochlorines from 1981–1998, but predominantly during the 1990s (Fig. 22, Becker et al., 1992, 1998). In the early 1990s, Joiris et al. (1997) revealed increases in organochlorine and mercury levels in Guillemots wintering in the southern North Sea.

Birds as retrospective indicators of heavy metal pollution history: Avian museum material is available from as far back as the middle of the last century and has proved to be valuable as a matrix in the analysis of heavy metal levels, especially of mercury. Scandinavian studies were the first to show multiple increases in mercury levels in predatory birds between the 1950s and 1960s (Berg et al., 1966). A similar increase discovered in feathers from Guillemots and Black Guillemots *Uria lomvia* was attributed to the increased use of mercury seed dressing in Scandinavian countries in the 1960 (Applequist et al., 1985). These studies, however, measured total mercury levels in the feathers, and many museum samples are contaminated with inorganic mercury used in the preservation of skins. Accordingly, methods are to be preferred that

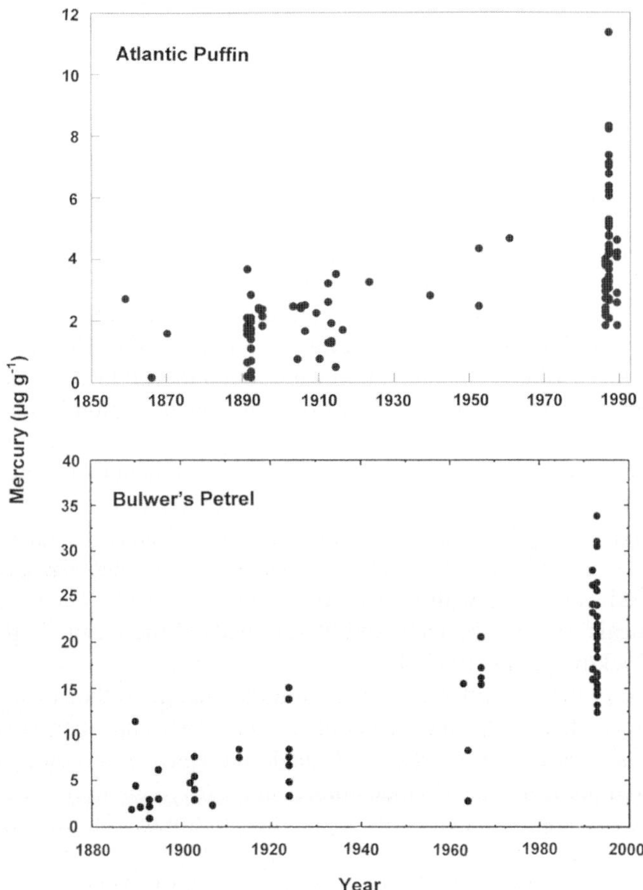

Figure 9.   Historical variation in body feather mercury concentrations of two seabirds, Atlantic Puffins from south-west Britain and Ireland (top), and Bulwer's Petrels from the Azores, feeding on mesopelagic fish (cf. Fig. 4). From Thompson et al. (1992) and Monteiro and Furness (1997).

measure only the metal amounts actually incorporated in the feather, in the case of mercury the organic methylmercury which is eliminated into the growing feather, and originating from the mercury uptake of the bird, and thus biologically relevant (Furness et al., 1986, 1995; Thompson and Furness, 1989).

As feathers reflect a bird's body burden during feather growth, body feathers from adult birds or chicks can be used to show long term and spatial trends in mercury contamination. Seabird feathers are the only matrix known to me permitting retrospective monitoring of mercury contamination in marine food webs, whereas in terrestrial systems it is possible to analyse mercury concentrations in ice cores, peat layers and sediment columns (Swain et al., 1992). Such seabird studies drawing on material collected in more than 100 years, have shown increases in mercury of 400% in seabirds from the UK coast and from the Azores region (Fig. 9), of 300% in seabirds from the

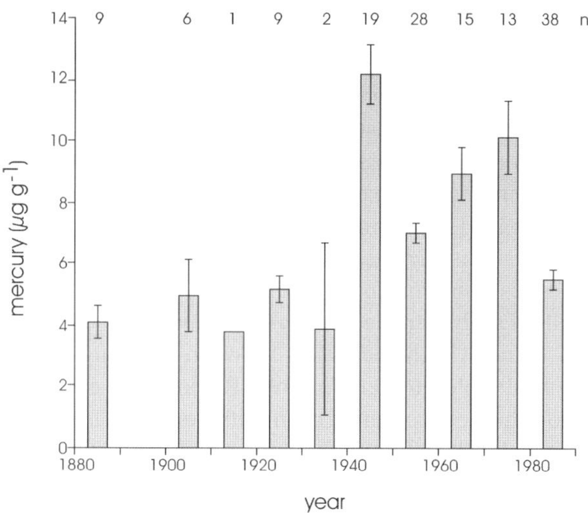

Figure 10.   Historical variation in body feather mercury concentrations of Herring Gulls from the German North Sea coast 1880–1990. The peaks are linked with World War II and the subsequent industrial development in central Europe. After Thompson et al. (1993b).

southern North Sea coast (Fig. 10), but no increases in seabirds from the southern hemisphere (Thompson et al., 1993a). These figures closely match with predictions from the modelling of the atmospheric transport of mercury from industrial sources (Mason et al., 1994; Fitzgerald, 1995). The temporal pattern of mercury in Herring Gull feathers from the German North Sea coast (Fig. 10) reflects mercury release occasioned by the Second World War and by the industrial activities of the 1960s and 1970s in central Europe (Thompson et al., 1993b). One hot spot for mercury contamination in North Sea seabirds is the German Bight as analyses of birds' feathers clearly indicated (Furness et al., 1995).

In the future, advances in chemical analytical techniques may provide new methods of performing retrospective analysis on other heavy metals and chemicals in museum skins. Since the 1980s, in some countries, contemporary bird samples have been stored frozen so as to be available for possible future retrospective analyses of environmental chemicals (see Kettrup, 2002).

Birds as indicators of spatial trends in chemical pollution: Spatial variation in contamination of raptors has often been the subject of study, and it has been possible to relate it to regional variation in the use or input of environmental chemicals into the environment. Conrad (1977) measured higher HCB-levels in the eggs of four raptor and owl species in agricultural areas of western Germany, whereas DDE was found in higher concentrations in regions where fruit growing predominated, and PCBs in industrial hot spots. In Sparrowhawks' eggs (Newton, 1986) and livers, and in the livers of Kestrels (Newton et al., 1993), higher levels of DDE and HEOD (from dieldrin) were detected in eastern, arable districts of Britain than elsewhere. In British Peregrine eggs, a declining gradient from south to north was found in the levels, fitting well with the amount of agricultural land (Newton et al., 1989). Mercury and PCB concentrations in eggs, however, were higher on coastal sites, presumably because coastal Peregrines eat more seabirds, which are heavily polluted by these chemicals. Similarly, White-tailed Sea Eagles were polluted by various chemicals much more heavily at sites along the Baltic coast than in Lapland (Helander et al., 1982). In Germany, Sparrowhawk eggs indicated different environmental pollution between eastern and western Germany with respect to industrial chemicals and pesticides during the 1990s (Denker et al., 2001; cf. Common Tern, see below).

Birds of aquatic ecosystems indicate distinctly the spatial trends in contamination on small or large scales, as numerous case studies demonstrate. In the Great Lakes for example, Herring Gulls, Double-crested cormorants and terns clearly indicate differences between Lakes – and within the Lakes between breeding-sites (Mineau et al., 1984; Weseloh et al., 1989; Weseloh et al., 1990; Ryckman et al., 1997, 1998). These fish-eating species show highest contamination with various chemicals at Lakes Erie, Ontario and Michigan, lowest at Lakes Superior and Huron characterised by lower chemical pollution. Custer et al. (1983) and Nisbet and Reynolds (1984) related differences in egg contamination between North American Atlantic coast colonies of Common Terns to local contamination from urban and agricultural sources. In several seabird species, mercury levels were high in north-west Iceland relative to levels in the same species from the British Isles or Norway, which was certainly not a result of local pollution but rather reflected patterns of atmospheric transport and deposition (Furness and Camphuysen, 1997). Renzoni et al. (1986) compared trace element and chlorinated hydrocarbons in the eggs and tissues of Cory's Shearwaters *Calonectris diomedea* between three Mediterranean and one Atlantic colony, the birds of which were found to be less polluted. Muñoz and Becker (1999) found low egg contamination of Kelp Gulls *Larus dominicanus* in Chile, displaying a geographical trend with declining levels from north to south, linked with decreasing human impacts.

In central Europe, Becker and Sommer (1998, Fig. 11) compared the chemical pollution of Common Terns breeding inland with that of those breeding on the coast by means of their egg contamination levels, indicating the contamination of freshwater and marine ecosystems over small to large scales. The distinct geographical patterns, differing between the chemicals, clearly revealed the areas of high and low pollution. In the 1990s, highest contamination with industrial chemicals was found on the rivers Rhine (HCB, PCBs), Elbe (mercury, HCB) and Weser (PCBs), whereas the highest pollution with pesticides was detected on the river Warta, Poland. A comparison of contaminants in Black-headed Gull *Larus ridibundus* eggs between those from the Baltic and those from the Wadden Sea (Kahle and Becker, 2000) along with recent studies in raptors (Denker, 2001, see above) have confirmed the higher pollution of eastern European birds and environment with DDT and metabolites, which used to be produced and employed in eastern Germany and Poland by the end of the 1980s (Heinisch et al., 1993). Studies not only of mercury egg levels (Fig. 11), but also those in the feathers of Herring Gull, Black-headed Gull, Common Gull *Larus canus* and Common Tern chicks clearly showed that seabirds breeding at the estuary of the river Elbe had much higher mercury levels than those from other areas on the German North Sea coast (Becker et al., 1993a; Kahle and Becker, 2000). In the gulls, this result was confirmed by a recent study using the blood of chicks (Kahle and Becker, 1999).

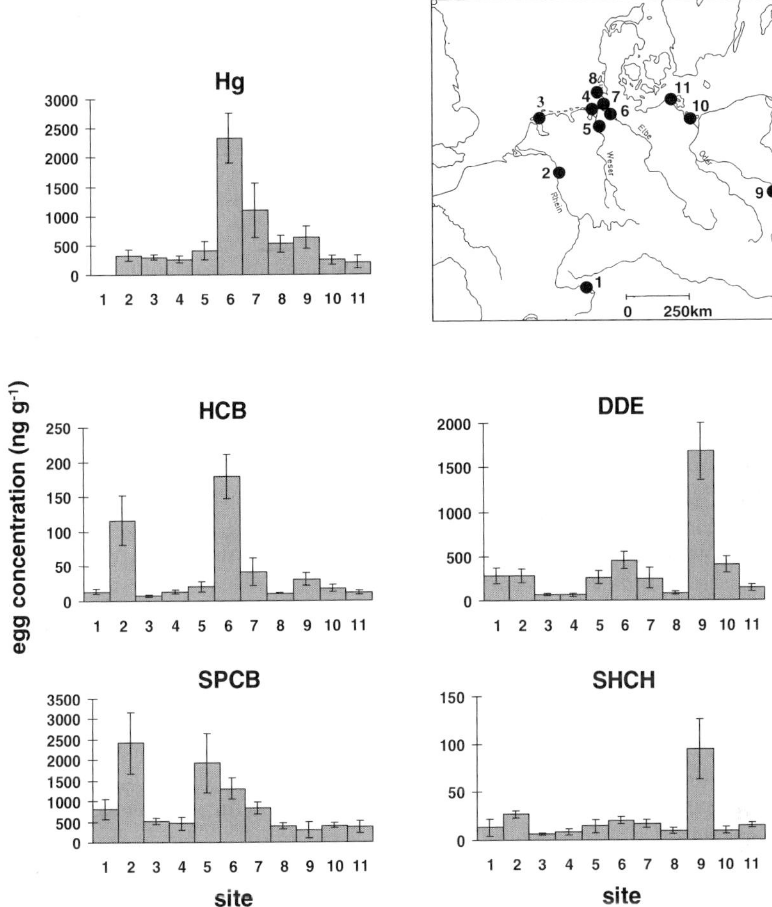

Figure 11. Spatial variation in mercury, PCBs (sum of six congeners), HCB, p,p′-DDE and HCH (SHCH, sum of α-, β- and γ-HCH) concentrations in Common Tern eggs sampled in Central Europe in 1995. Concentrations are given in ng g⁻¹ ± 95% confidence intervals (vertical lines). Eggs from area 9 (Warta, Poland) are from 1990. Sites: *Rhine*: (1) Konstanz, (2) Lower Rhine; *Wadden Sea*: (3) Griend, (4) Minsener Oldeoog, Jade, (7) Trischen, (8) Norderoog; *Weser*: (5) Bremen; *Elbe*: (6) Hullen; *Warta*: (9) Jeziorsko, Poland; *Baltic Sea*: (10) Böhmke and Werder, (11) Kirr. After Becker and Sommer (1998).

Bird studies can bring to light significant differences in egg contamination not only between regions but also on smaller scales. In the Wadden Sea, eggs from adjacent breeding sites in the inner German Bight (Fig. 11, Elbe estuary, Trischen island) were contaminated to much higher levels than those at western and northern breeding sites of the Wadden Sea, indicating the discharge of the industrial chemicals studied (especially mercury and HCB) and pesticides by the river Elbe into the North Sea. Considerable contamination by PCBs and lindane was also ascertained on Griend, Netherlands, in 1996 and 1997, which suggests that the Wadden Sea ecosystem is also influenced by pollutant loads from the river Rhine (Fig. 11, Becker and Sommer, 1998; Becker et al.,

1998; Oystercatcher: Exo et al., 1998). Thus, by the careful selection of sampling sites, the hot spots of pollution and the input sources of chemicals can be identified on the basis of the birds' samples. By reference to the degree of spatial differences, distinct local discharges of pollutants can be distinguished, as in the case for mercury or HCB transported by the Elbe into the North Sea (Fig. 11). In some chemicals detected in birds' eggs from the Wadden Sea (like the PCBs), similar and high concentrations without distinct spatial patterns indicate atmospheric deposition besides some riverine discharge (Fig. 11). The concentrations of HEOD (metabolite of dieldrin) in Dipper eggs from Welsh rivers were positively linked with sheep densities on the river catchments, suggesting that sheep dip, of which dieldrin used to be an active component, may have been responsible for at least some contamination (Ormerod and Tyler, 1993).

Further hints to pollution sources can be derived from the detailed analyses of the chemicals' mixtures in the birds. In the Great Lakes, discriminant analyses clearly separated Herring Gulls from various breeding sites by the patterns of chemicals' concentrations in eggs, with PCBs and HCB having the greatest power (Weseloh et al., 1990). Within-mixture patterns of the isomers of HCH, the metabolites of DDT or the congeners of PCBs are interesting in this respect. A high ratio of p,p'-DDT in the mixture may indicate recent application of this insecticide, and the composition of the HCH-mixtures can point to the use of lindane, or to loads by the technical mixture and β-HCH from industrial sources, as in the river Elbe (Becker et al., 1998). The composition of PCB-mixtures in birds depends on the degree of metabolism, which differs between species and sites (Beyerbach et al., 1993; Denker et al., 1994; Dietrich et al., 1997). In coastal birds of the Wadden Sea, the degree of metabolism and the percentage of highly chlorinated congeners within the mixture have been found to be higher in areas of stronger PCB-pollution, indicating an increased induction of metabolising enzymes (Beyerbach et al., 1993; Becker et al. 1998; see also 3.1.1.2). The metabolisation of PCB mixtures in combination with the varying PCB-levels in the environment and birds result in such discrete patterns in the eggs, clearly discriminating the breeding sites (Fig. 12).

The widespread distribution of some bird species allows comparisons of contamination on large scales, even between oceans or continents (e.g. Burger et al., 1992; Burger and Gochfeld, 1995), or between pollution at breeding and wintering grounds (Burger et al., 1992; Dietrich et al., 1997). By referring to systematically and ecologically similar species, such as the Herring Gull and the Kelp Gull, even differences in pollution of aquatic habitats between the northern and southern hemispheres can emerge (Table 3).

3.1.1.2 Birds as sensitive indicators of chemical pollution

Lack of knowledge and consideration of costs restrict the quantification of residue levels in birds to a few selected species and sites, and to a small number of chemicals of known toxicological relevance. However, thousands of anthropogenic chemicals are present in the environment, and each year new contaminants are introduced. Fox and Weseloh (1987) estimate that 30,000 different chemicals enter the Great Lakes, some of which are suspected of being toxic to birds. Another problem is that chemicals may interact in ways not foreseeable by reference to the measurable concentrations of individual substances. One approach to these problems is to combine regular

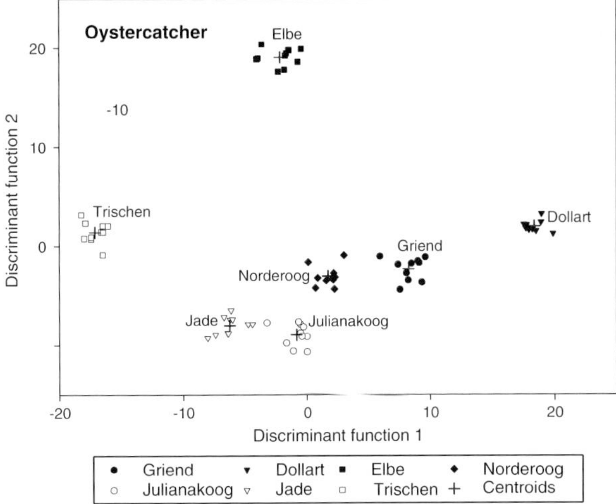

Figure 12. Diagram of the results of a discriminant analysis for separation of Oystercatchers' breeding sites on the Wadden Sea coast, based on concentrations of 62 PCB congeners in eggs collected in 1997. Each symbol represents one egg, + the group centroids. Major components of the discriminant function were the CBs 138, 153, 160, 180, 170 and 128, listed in order of decreasing importance. Griend, Julianapolder: Dutch Wadden Sea, the other sites: German Wadden Sea. From Becker et al. (1998).

Table 3. Intercontinental variation in the contamination of eggs (arithmetic means, ng g^{-1} egg fresh mass) of Herring Gulls (*Larus argentatus*) or Kelp Gulls (*L. dominicanus*) in 1995.

	Herring Gull		Kelp Gull
	Lake Ontario Canada	Wadden Sea Germany	Algarrobo Chile
pp′-DDE	2590	71	192
β-HCH	0	7	0
HCB	22	11	6
ΣPCB	14800	1462	245
Mercury	Not analysed	159	167
n	1 pooled sample	22	9

In gulls at the Great Lakes (data from Weseloh pers. comm.), concentrations of p,p′-DDE, HCB and ΣPCB had highest levels. In eggs from Germany (Kahle and Becker, 2000), β-HCH had highest, HCB and ΣPCB intermediate levels, and in Chile (Muñoz and Becker, 1999) p,p′-DDE values were intermediate.

monitoring of the contamination of birds with studies of possible effects, by making use of birds as sensitive indicators.

How chemicals can influence birds? Ecotoxicology is at the focus of many papers and reviews, but can addressed only very briefly here (see e.g. Fimreite, 1979, Ohlendorf et al., 1978; Scheuhammer, 1987; Peakall, 1973, 1986, 1992; Moriarty, 1990; Furness, 1993; Walker, 1994; Nisbet, 1994; Grasman et al., 1998). The following examples of well known and documented effects of environmental chemicals, which may be sublethal or lethal, have been ranked starting with primary effects on physiology, reproduction, individuals, and culminating with influences on bird populations and communities:

- *Enzymes and biochemical parameters* such as cytochrome p450 enzymes, porphyrins, vitamin A, thyroid or immune functions (e.g. Jefferies and Parslow, 1976; Peakall, 1986; Hoffman et al., 1987; Brouwer et al., 1990; Elliott et al., 1990; Walker, 1994; Murk et al., 1994; Bosveld et al., 1995; Henriksen et al., 1998; Grasman et al., 1998).
- *Hormones and feminisation* (pollutants acting as "endocrine disruptors", Peakall, 1986, 1992; Guilette, 1994; Fry, 1995; Umweltbundesamt, 1995).
- *Parasites* (intestinal nematodes: Sagerup et al., 2000).
- *Abnormalities, genotoxicity* (Ohlendorf et al., 1986; Ludwig et al., 1995; Grasman et al., 1998; Ryckman et al., 1998).
- *Behaviour of adults* (Ratcliffe, 1970; Cooke, 1973; Dobson, 1981).
- *Egg laying, clutch size* (Ratcliffe, 1970; Cooke, 1973).
- *Thickness and quality of the egg-shell.* Many studies in various bird species describe the thinning of egg-shells by DDE in a dose-dependent way, through inhibition of Ca^{2+} ATP-ase by p,p'-DDE (Fig. 13, Peakall, 1970; Cooke, 1973; Moriarty et al., 1986; Blus et al., 1997). A reduction of the shell quality results in egg breakage and reduces hatching success (Fig. 13). Birds of prey and some seabirds are particularly sensitive to this effect, e.g.Peregrine (Peakall et al., 1975; Newton et al., 1989); Sparrowhawk (Conrad, 1977; Newton, 1986; Newton and Willie, 1992, Figs 1, 13), Goshawk (Conrad, 1977), Barn Owl *Tyto alba* (Conrad, 1977); Guillemot (Bignert et al., 1995), Double-crested Cormorant: (Weseloh et al.,1995); Cormorant (Dirksen et al., 1995); Gannet (Parslow and Jefferies, 1977), Brown Pelican (Blus et al., 1997). Cormorants *Phalacrocorax carbo* and gannets are very sensitive to the effects of DDE because they incubate their eggs by wrapping the webs of their feet around them, and thinned shells have not the strength to withstand the weight of an incubating bird. Through comparison of recent egg shells with museum samples the change of shell-thickness with time and with the use of DDT could also be shown (e.g. in Peregrine and Sparrowhawk, Fig. 1, Newton, 1986; Newton et al., 1989; Brown Pelican, Blus, 1982). Shell-strength responds to chemicals more sensitively than does shell-thickness (e.g. Carlisle et al., 1986; Bennett et al., 1988).
- *Embryonic survival, hatchability.* Along with its dependence on eggshell quality, the hatching success of an egg depends on the survival of the embryo. The contaminant burden of chicks and sensitivity are highest during hatching before subsequent rapid growth dilutes the amount received in the egg. As many chemicals

have embryotoxic effects, a reduction of the hatching rate can indicate pollution problems (Hoffman et al., 1987; Grasman et al., 1998). Common Tern egg mortality was related to high PCB levels (Becker et al., 1993b), and high levels of PCBs in Forster's Tern *Sterna forsteri,* Caspian Tern *Sterna caspia*, Common Tern and Double-crested Cormorant and Cormorant eggs were associated with a steep reduction in hatchability (Kubiak et al., 1989; Tillitt et al., 1992; Becker et al., 1993b; Dirksen et al., 1995; Grasman et al., 1998).

- *Mortality and growth of young.* Chick growth (Burger and Gochfeld, 1988; Harris et al., 1993; Gochfeld and Burger, 1996) as well as chick survival (Koeman et al., 1967; Dirksen et al., 1995) may be impaired by contaminants.
- *Reproductive success* is the product of the success rates of the different reproductive stages, and used often to be linked with chemical pollution. Some examples of chemicals affecting the reproductive output are DDE in Brown Pelicans (Blus, 1982; Anderson and Gress, 1983), Gannets (Chapdelaine et al., 1987; Elliott et al., 1988), Cormorants (Dirksen et al., 1995), Double-crested Cormorants (Anderson and Gress, 1983; Weseloh et al., 1995, Fig. 14), Herring Gulls (Gilman et al., 1977; Mineau et al., 1984; Grasman et al., 1998), or White-tailed Sea Eagles (Helander et al., 1982). In Peregrines and Sparrowhawks, reproductive success was clearly negatively related to DDE residues, but not to levels of other chemicals (Fig. 13, Newton et al.,1989; Newton and Wyllie, 1992).

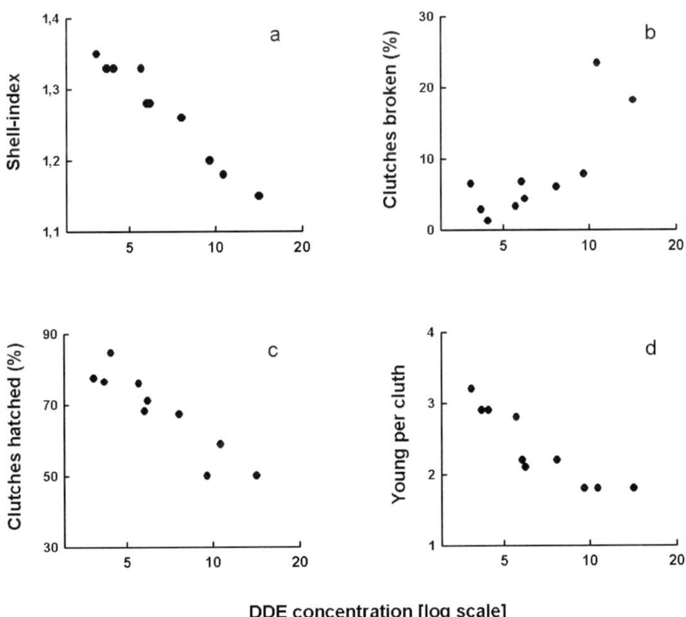

Figure 13. Relationships between DDE levels in addled eggs of Sparrowhawks from the Rockingham population (μg g^{-1} fresh weight) and (a) mean egg-shell thickness index, (b) percentage of clutches which failed through egg-breakage, (c) percentage of clutches which hatched, and (d) mean number of young per clutch. Each dot represents an annual value from 1980 to 1989, and all correlations are highly significant. From Newton and Wyllie (1992).

- *Survival of adults,* e.g. by reduction through direct poisoning (Koeman et al., 1967; Swennen, 1972; Conrad, 1979; Newton, 1986; Jenni-Eiermann,1996) and subadults (Sibly et al., 2000).
- *Population size.* Despite the indication of environmental damage through pesticides provided by the historical population crashes in raptors and seabirds have (Figs 2, 14), the information obtained by monitoring population size is less valuable as an early warning of the build-up of toxic pollutants than that obtained by monitoring breeding success (Furness and Monaghan, 1987; Becker, 1991; Thyen et al., 1998, 2000b). Hence, especially in long-lived "K-strategists" such as raptors or seabirds (see e.g. Croxall and Rothery, 1991), breeding population size itself gives a delayed response to changes in survival rates or reproductive output. However, population trends may be useful in detecting any changes in the quantity and quality of habitat (Morrison, 1986), and changes in numbers could trigger detailed research into the causes.

A conceptual framework showing how chemicals develop the disastrous chains of influences on bird life is presented e.g. by Hartner (1981) and Newton (1986). Effects and toxicity differ between chemicals, as for example in some well studied compounds: among the pesticides, *cyclodienes (Dieldrin, Aldrin, Telodrin)* have the highest toxicity for birds, causing direct poisoning and mortality. p,p′-*DDE (dichloro-diphenyl-trichlorethene)*, the main metabolite of the insecticide p,p′-DDT, disrupts hormones and has effects on egg production, eggshell-thinning and hatchability, leading to nest failure and declines in reproductive success. *PCBs (polychlorinated biphenyls)*: This industrial mixture of up to 209 congeners has a broad spectrum of negative influences on birds (e.g. Peakall, 1986), and can cause hormonal and enzymatic disturbance, organ damage, behavioural abnormalities, and is embryotoxic and genotoxic. Some congeners of very low concentrations are the most toxic ones, as they are related in structure to dioxin (TCDD-equivalents, e.g. Kubiak et al. (1989), Safe (1990), Grasman et al. 1998). *HCB (hexchlorobenzene)* used to be used as fungicide. Today HCB is an industrial chemical, highly toxic and carcinogenic. *Mercury:* Today mainly an industrial chemical, with high embryotoxicity and neurotoxicity.

The effects of chemicals on birds are not only known to be contaminant- but also species-specific: for example, DDE affects fish eating and raptorial birds much more than pigeons, ducks or chicken (Cooke, 1973), and critical levels differ widely between species (e.g. Ohlendorf et al., 1978; Pearce et al., 1979; Scheuhammer, 1987). In consequence, threshold levels derived for one bird species are not allowed to be transferable to other bird species without corroborative studies. This must be born in mind especially in transferring results from laboratory feeding trials, mostly performed in poultry, to wild birds of other orders.

Critical levels: Comparisons between sites or populations underlying various degrees of contamination may be helpful to detect threshold levels of a pollutant for the species in question. For example, Pearce et al. (1979) compared species-specific threshold levels of eggshell thinning by DDE for seabirds. Blus (1982) used the sample egg technique and identified 3 μg g^{-1} fresh egg mass for DDE as the value associated with substantially impaired reproductive success. Helander et al. (1982) estimated 500–600 μg g^{-1} of DDE and /or 800–900 μg g^{-1} of PCBs (lipid mass) as the critical

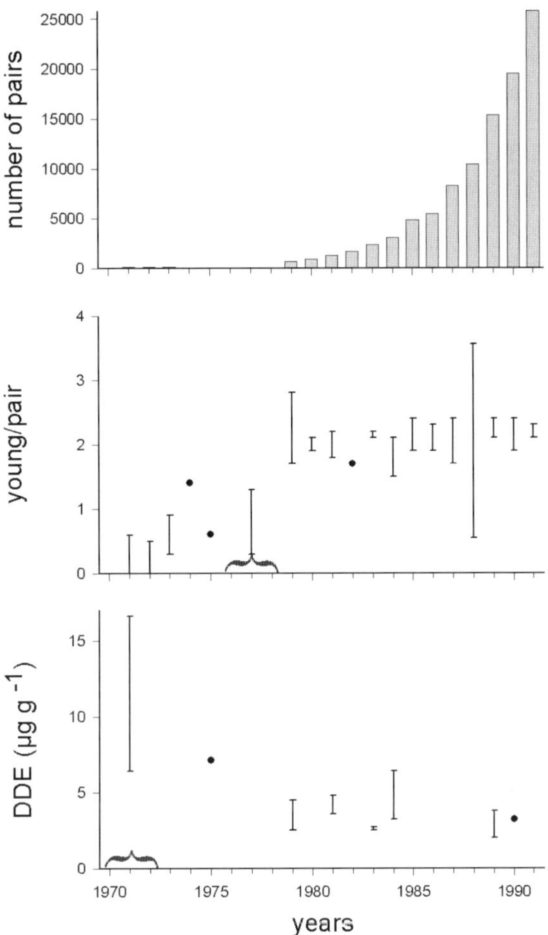

Figure 14. Increase of the Great-Lakes Double-crested Cormorants breeding population (totals) and of its reproductive success (young per pair) in relation to p,p′-DDE egg levels (fresh weight). DDE use was discontinued after 1972. Range of mean values of several study colonies is indicated by vertical lines, single values by dots. After data from Weseloh et al. (1995) and Ryckman et al. (1998).

range above which reproduction of White-tailed Sea Eagles may be assumed to be impaired. The range of PCBs levels critical for affecting the breeding success of birds was roughly estimated to be above 3–5 µg g^{-1} fresh egg mass (Lorenz and Neumeier, 1983; 7 µg g^{-1} in Forster's Terns, Harris et al., 1993, 5 µg g^{-1} in Common Terns, Becker et al., 1993b). Mercury levels as low as 0.5 µg g^{-1} can reduce hatchability (reviews Ohlendorf et al. 1978; Fimreite, 1979; Scheuhammer, 1987).

The historical pesticide disasters showed that bird populations are affected in various life stages and parameters by environmental chemicals (e.g. Sibly et al., 2000). Consequently, careful long-term investigations of bird populations are able to present information crucial to the discovery of environmental changes, not least those caused by chemicals, through monitoring parameters known to be affected by pollutants (see above). In this way birds can be sensitive indicators of pollution problems, and function as an early warning system as they did in the historical examples. Population size presents a delayed reaction to chemical effects, especially in long-lived species such as the top-predators, and the survival of adults and subadults are demographic parameters relatively difficult to measure. However, parameters related to reproduction respond immediately to chemicals' influences, as also to sublethal levels of pollution, and are especially suitable as "early warning" against detrimental chemical effects. Furthermore, reproductive parameters in birds are relatively easy to measure.

Through bird population studies (see 3.2.4, 4.4.4) possible environmental problems for the birds may be indicated by changes in sensitive demographic parameters. There are of course many other environmental factors besides contaminants affecting bird populations and possibly disguising the effects of chemicals. Furthermore, the effects of a single chemical may be obscured by positive intercorrelation between different pollutants, and synergistic or additive effects. But detailed careful studies of different life stages and population parameters may be able to isolate the crucial factors in any particular change, and provide warning of environmental hazards through chemicals as the historical examples have demonstrated. They are quite likely to reveal the toxic effects of pollutants in birds but may not be able to identify the causal agent because several chemicals may be involved.

3.1.2. Radionuclides

Brisbin (1993) reviewed the uptake of radionuclides by birds and discussed their usefulness as indicators of radionuclide contamination of the environment. The review showed that birds can provide valuable information on the extent over a wide range of spatial scales of radionuclide contamination, by radionuclides such as caesium which readily enter and move through natural food chains. This may be due to the high metabolic rates of birds relative to other similar sized biota. On small spatial scales, however, components of the physical environment, vegetation or less mobile fauna would be more appropriate indicators. There have been few studies that have attempted to survey the hazards to wildlife caused by radionuclides discharged into the environment.

3.1.3. Other foreign substances in the marine area

3.1.3.1. Oil pollution

Stranded (= beached) oiled seabirds have been used for nearly a century to demonstrate the effects of oil pollution on the marine environment (e.g. Bourne, 1976; Reineking and Vauk, 1982; Dunnet, 1982; Nisbet, 1994; Camphuysen, 1998; Fleet and Reineking, 2000). If coupled with chemical analysis, beached bird surveys can also be effective indicators of pollution of the seas by other lipophilic substances (Dahlmann et al., 1994; Camphuysen et al., 1999). Chemical fingerprinting the oil from carcasses permits identification of specific oil components and their distinct sources, and can even be used as evidence in prosecutions for the discharge of oil at sea (Dahlmann et al., 1994; Dahlmann and Sechehaye, 2000).

Total numbers of beached birds are subject to enormous fluctuations, being the result not only of changes in the amount of oil spilled in the marine environment, but also of natural factors like currents, the frequency of onshore winds and variation in the number of birds in a given area (e.g. Camphuysen, 1998). The fraction of oiled birds of the total stranded, the "oil rate", however, appears to be relatively constant and specific for a species and region, an adequate reflection of the risk of birds or corpses becoming oiled, an index of the pollution of the sea area with oil, and a useful tool for monitoring oil pollution. High oil rates are typical for mainly swimming, highly exposed seabirds in areas with frequent oil spills, low oil rates for mainly flying seabirds, in areas remote from the busiest shipping lanes or big harbours.

Beached bird surveys (BBS) provided clear evidence of long term trends in oiling rates of seabirds and of differences in oil impacts between regions (e.g. Camphuysen, 1998; Fleet and Reineking, 2000). For example, surveys in the Netherlands typically showed that 88% of stranded Guillemots were oiled, whereas in the Shetlands the oiling rate was only 18%. Gradual but significant decreases in oil rates were found over the last 15 years (Camphuysen, 1998). In the German Bight, however, over the last decade no clear temporal trend was evident (Bakker et al., 1999; Fleet and Reineking, 2000). Furthermore, BBS are a very dramatic way of demonstrating ongoing oil pollution offshore, which governments aim to mask, for example by preventing oil strandings.

3.1.3.2. Marine pollution with plastic particles

The use of seabirds as monitors of plastic pollution on the ocean surface (Prüter, 1987) has been discussed and advocated by a number of authors (e.g. Bourne, 1976; Furness, 1985, 1993; Nisbet,1994; Ryan, 1988; Spear et al., 1995; Blight and Burger, 1997). Petrels would probably make suitable indicators as they tend to accumulate plastic particles in their gizzards and their ability to process food may be impaired (Ryan, 1988). There is evidence that plastic particle burdens in pelagic seabirds are increasing, a warning signal that demands further attention. However, measuring quantities of ingested plastics requires sampling dead birds or killing birds, which latter is a deplorable method of monitoring. Taking beached seabird corpses (see above) involves the risk of biased samples. However, in order to study long term trends in plastic particle pollution it should be possible to obtain gizzard samples by sampling seabirds at colonies on a regular basis; birds may be killed by predators or in specific accidents, e.g. by entanglement in the vegetation or by attraction to lights (Furness, 1993; ICES, 1999).

3.1.4. Industrial impacts on ecosystems

Birds have been used to indicate the possible impacts of temporary industrial interference in ecosystems. For example, in 1994 a gas pipeline was built through the National Park "Niedersächsisches Wattenmeer" on the southern North Sea coast of Germany, between the islands of Baltrum and Langeoog. The programme to monitor the ecological effects involved using the Common Tern as indicator, a breeding seabird species integrating the food web around the site of action and apt to show fluctuations caused by the construction measures. The results from 1993–1995, however, with respect to diets, breeding success and chick growth provided no evidence of negative impacts (Frank, 1998).

3.1.5. Seabirds as indicators of marine prey stocks

In the last 10 years an increasing number of papers and reviews (Cairns, 1992; Montevecchi, 1993; Furness and Camphuysen, 1997; Greenstreet et al., 1999) have been addressing the usefulness of seabirds for monitoring prey stocks. The detailed knowledge available on seabird biology makes it possible to gain information about changes in prey stocks from the specific responses of seabird species if these are carefully studied and if other intrusive factors which may impede interpretation are taken into account.

The constellation of the fish community within foraging range and the availability of the prey species affect the diet composition of seabirds influencing seabirds in terms of average body mass, breeding output, growth and survival of young, all parameters which can be readily measured in seabirds. Their usefulness in monitoring fish stocks is reviewed in Montevecchi (1993). Foraging ranges vary among seabird species, but tend to be tens of kilometres at most during the breeding season. Thus, the prey stock related seabird parameters on the whole reflect the local fish communities and stocks. Possible ways to use seabirds to indicate fish stocks on a larger scale are presented by Greenstreet et al. (1999).

The diets of generalist fish feeders like for example Gannets (Fig. 15), and Cormorants or at specific sites also Common Terns (Becker et al., 1987) reflect the temporal variation in relative prey abundance at a given site. The prey taken by specialist feeders, e.g. the sandeel-dependent seabirds of Shetland (Monaghan et al., 1989) or the Herring *Clupea harengus* and Sprat (*Sprattus sprattus*) -dependent Puffins *Fratercula arctica* (Hislop and Harris, 1985; Anker-Nilssen, 1987) or terns in the Wadden Sea (Frank, 1992; Stienen et al., 2000) reflect the annual or seasonal fluctuations in stocks of particular species.

Diet composition can often be sampled fairly easily by means of analysing regurgitations, pellets or food fed to the chicks. Many studies have shown that diet is related to assessments made of fish communities or stock sizes, and is thus a valuable indicator of stock changes (e.g. Fig. 15, Montevecchi and Myers, 1992; Hislop and Harris, 1985; Becker et al., 1987; Greenstreet et al. 1999). Today the technique of stable isotope ratio determination (^{15}N/^{14}N and ^{13}C/^{12}C) in bird tissues is also applicable to investigations into long-term changes in seabirds' diet and trophic position (Thompson et al., 1995).

The reproductive period is the bottleneck in the biology of seabirds, and food availability is the most important environmental factor affecting the size of the breeding population, food provisioning and the growth of the young, and the number of fledged chicks (e.g. Furness and Monaghan, 1987; Greenstreet et al., 1999). Response to changes in food supply can be quite rapid and dramatic: For example, when sandeel availability decreased at the Shetlands in the mid 1980s (Bailey et al., 1991), Arctic Tern *Sterna paradisaea* breeding numbers and success fell dramatically; their breeding success correlated closely with estimates of sandeel stock sizes, as did that of other species as the Arctic Skua *Stercorarius parasiticus* (Philipps et al., 1996; Monaghan, 1992). When sandeel abundance recovered after 1991, Arctic Tern numbers increased again, in consequence of the recruitment of birds that had apparently refrained from breeding throughout about 7 "lean" years (Furness and Camphuysen, 1997). Furness

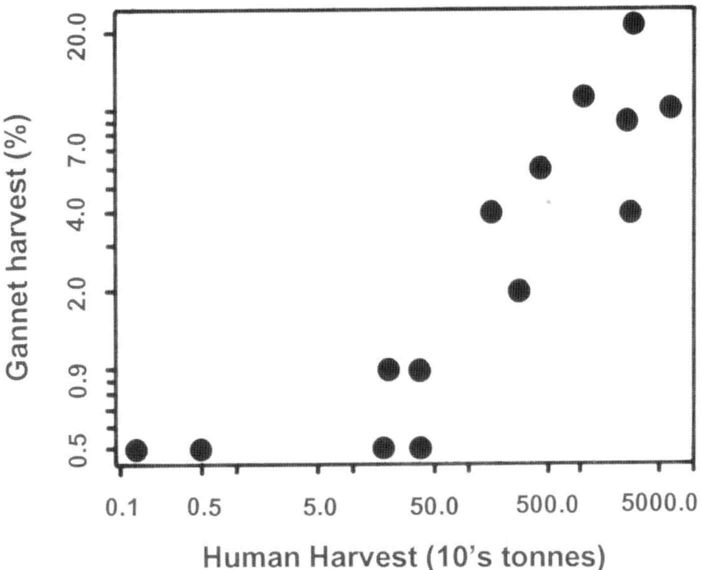

Figure 15.   Seabird diet as indicator of fish stocks: Relationship between Short-finned Squid in the prey harvests of Gannets (log percentage) and the commercial catch in Newfoundland (log), 1977–1991. After Montevecchi and Myers (1992).

and Ainley (1984) identified terns as particularly liable to act as sensitive indicators of food shortage, and this prediction has been supported by evidence from studies on the effects of sandeel shortage in Shetland (see above) as well as from several other investigations (e.g. Becker et al., 1987; Uttley, 1992; Becker, 1998; Greenstreet et al., 1999). Populations of seabirds have also been found to respond supplementary food supplies of anthropogenic origin e.g. from fishery (Fig. 16, Chapdelaine and Rail, 1997; Walter and Becker, 1997; Garthe et al., 1999; Oro and Ruiz 1997; Camphuysen and Garthe 2000).

As young fish make up most of the diet during the reproductive period, seabirds may constitute an additional means of sampling younger fish at a high temporal solution, and at a cost lower than that incurred in traditional fish monitoring methods. Using this additional information source provided by seabirds is unlikely to add to the fisheries' assessment costs. However, the integration of seabird research into interdisciplinary oceanographic programmes will aid in long-term programmes studying interactions within marine food webs and monitoring at least local fish stocks. This is likely to be important in view of the increasing pressures of the fisheries on marine fish resources expected over the coming decades.

3.1.6. Acidification and freshwater quality

The value of birds as indicators of freshwater pollution has already been mentioned (3.1.1.1). Increased acidification of the environment can affect birds in indirect ways, via other parts of the ecosystems (e.g. Zang, 1998). So plants, invertebrates or fish

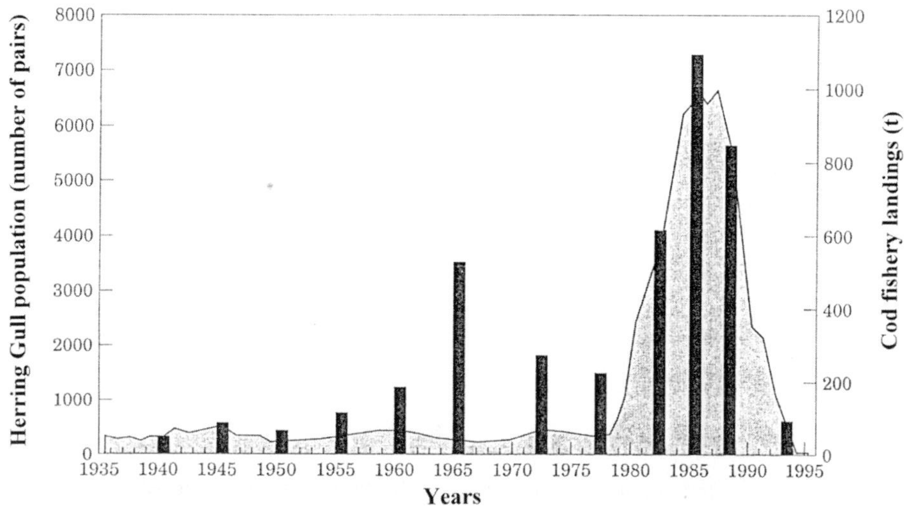

Figure 16. Relationship between the population of Herring Gulls (columns) and cod fishery landings in Quebec Canada (shaded area), 1935–1995. From Chapdelaine and Rail (1997).

may be more appropriate subjects for monitoring (Furness, 1993). In riparian birds such as Dippers the river pH affects breeding density and success, diet, egg mass and shell thickness (Ormerod and Tyler, 1993), which is an indication of the potential of freshwater bird species as a rudimentary early warning of changes in water acidification and in other parameters of water quality, e.g. nutrients, as reviewed by Ormerod and Tyler (1993). In recent years a number of papers have shown the deleterious effects on eggshell structure and quality of reduced calcium availability for egg shell production (Graveland et al., 1994; Beintema et al., 1997; Graveland and Drent, 1997; Weimer and Schmidt,1998).

3.2. Detecting effects of diverse environmental changes by monitoring bird populations

The preceding sections have argued that birds are useful as indicators of the state of the environment and as a warning of specific environmental change. But so many aspects of the environment are subject to change that we never will be able to monitor all of them specifically and in advance, through neither biological nor non-biological monitoring. Against the background of this dilemma, birds represent an alternative approach: The monitoring of the populations of a broad spectrum of bird species enables us to pick up the effects of environmental changes whose impact on ecosystems and wildlife would not otherwise be detected. Ornithology has often in the past provided the crucial hint of dangerous developments in the environment, like habitat alteration and fragmentation, introduced species as predators, or of general environmental quality. The likely causes of population change may be identified more easily in birds with their various advantages (see 2) than would be the case in less well studied organisms. The diversity of avian ecologies, habitats used and trophic relationships means that a broadly based population monitoring is likely to expose the effects of a wide range of environmental changes (Bohāc and Fuchs, 1991; Greenwood et al., 1993; Greenwood, 1999).

To return for a moment to the function of birds as indicators of general environmental change (Morrison, 1986), breeding bird surveys aimed at estimating abundances and population size are very common throughout the world, and as well as detecting temporal and spatial fluctuations in bird populations and communities provide valuable information on a great variety of environmental change. All this chapter can do is to present a small number of examples which may provide the interested reader with the initial access to the huge amount of relevant literature available.

3.2.1. Breeding bird census and atlas work

Most census work in the past was done on a national basis to collect data on the distribution and abundance of birds and used methods like grid mapping to cover the total areas of countries (e.g. SOVON, 1987; Rheinwald, 1993; Rheinwald et al., 1997), territory mapping on selected study plots and/or point counts (e.g. Flade and Schwarz, 1999). The extensive North American Breeding Bird Survey (BBS) was started as far back as 1966 (Sauer and Link, 1999; O' Connor, 1991). Järvinen and Väisänen's (1979) data from Finnish line transect counts over four decades reveal a decline of forest bird populations in consequence of a decrease of about 15% in the area of old forests in Finland. The British Trust for Ornithology (BTO) began the Common Bird Census in 1962. Annual changes in population size are calculated as percentages from summed territory counts for all plots that are covered in the same way in consecutive years. Today the BTO census in combination with other methods is moving towards an integrated population monitoring (see below). In Germany, in 1989 a national German Common Bird Census was started by the DDA using point stop counts and territory mapping, and now the first interesting results are beginning to emerge (Flade and Schwarz, 1999). For example, changes in European Community agricultural policies are clearly reflected in farmland birds: species living in meadows and pastures are decreasing in numbers, those related to large reserved areas show trends of recovery. All over the industrial countries such census work is indicating changing breeding bird communities and densities, and clearly pointing to the ongoing anthropogenic landscape deterioration (e.g. Marchant, 1999; Bauer and Berthold, 1996; Greenwood et al., 1993).

3.2.2. Breeding population size

Throughout the world, a great many studies on changing breeding population size in birds are conducted and present important information on population development (e.g. Figs 2, 13, 14, 17; Bauer and Berthold, 1996; Helbig and Flade, 1999). In the German and Dutch Wadden Sea, mainly since the Second World War, but for some important breeding sites even longer, data on the size of breeding populations have been collected and analysed, to show the impacts of changing breeding habitats on the Wadden Sea coast as well as those of environmental chemicals as important factors effecting population change (Fig. 2; Becker and Erdelen, 1987; Becker,1991; Südbeck, et al. 1998; Stienen and Brenninkmeijer, 1998; Hälterlein et al., 2000). Since 1990, the complete Wadden Sea coast has been included in the survey so that trend statements have been based on the total population (Melter et al., 1997; Rasmussen et al., 2000). Seabirds are monitored for example in Norway (Anker-Nilssen et al., 1996) and in Britain and Ireland, where a seabird colony register was established in 1984 by the UK Seabird Group and the Nature Conservancy Council which also including historical data and providing information about distribution and status, population trends and reasons for changes (Lloyd et al., 1991).

3.2.3. *Migrating and staging birds*

Through site-based counting or trapping birds during migration or staging in specific resting or wintering areas over years, data is accumulated on possible change in the populations which are staging at these sites or areas. At passerines trapping stations such as those at Ottenby (Sweden), Rybachi (Russia) or Heligoland (Germany) and many others, the birds caught are also ringed (3.2.4, 4.4.4). The annual catching figures at these stations have often been used to indicate population trends. However, interpretation is difficult if the origin of the populations is not clear: migrating and resting birds are often breeders in other, unknown parts of their annual ranges, and it may be here where the factors effecting change may be operating. Site-based ringing with standardised effort provides effective indices and surveillance of both numbers and survival, as also of productivity through catching the offspring in autumn (e.g. Greenwood, 1999).

Illustrative of the large number of studies in passerines are the studies in eastern Europe (Busse et al., 1995) and the Mettnau-Reit-Illmitz programme of the Vogelwarte Radolfzell in Germany (1974–2000): highly standardised techniques on three trapping sites produced findings clearly indicating wide fluctuations in the populations of the European avifauna (Berthold et al., 1999). The trapping totals yielded significant negative relationships between years and numbers for 20 of the 35 species. These results reflected important changes in the central European avifauna which were in general agreement with the current "Red Lists" (Bauer and Berthold, 1996). In Canada, too, population estimates derived from migration monitoring and from breeding population monitoring showed good correlations (Dunn and Hussell, 1995). In Finland population trends in wintering birds were studied from 1957–2000 (Väisänen and Solonen, 1997). Other projects estimating population size and trends in non-breeding birds, such as counting seabirds at sea in the Northeast Atlantic, and waterbird or wader populations are mentioned below (4.4.3).

3.2.4. *Integrated bird population monitoring*

Annual counts of birds tell us how numbers are changing, but other information and techniques are needed if we are to understand the mechanisms of the observed changes. The most important regulatory parameters of a bird population's size are the survival of adults, the survival of young, the age of first breeding, breeding success and the immigration- and emigration rates (Fig. 17). In consequence, once we understand the

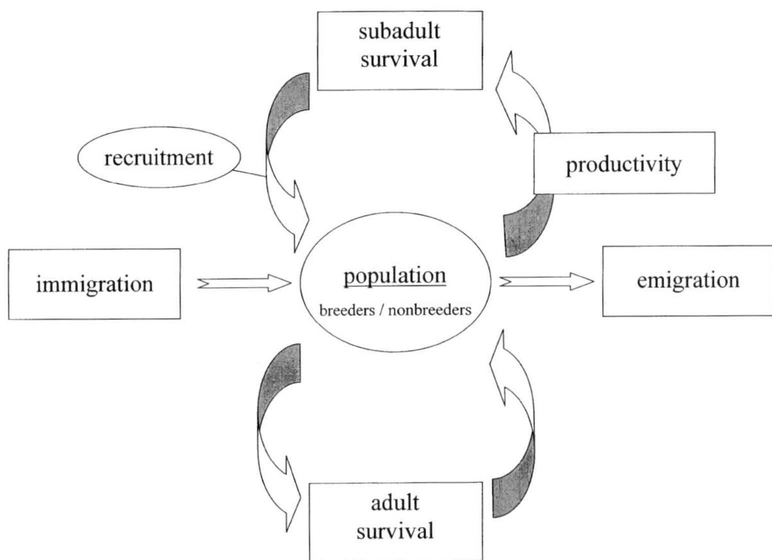

Figure 17. Scheme of important demographic parameters regulating bird populations.

life cycle stage at which a bird population is affected, it is possible to narrow down and identify factors which are causing the change.

The only way to measure survival and movements is to mark animals individually. In birds, marking individuals is a time-honoured and sophisticated technique, popularly known as "scientific bird ringing", which is used throughout the world in the investigation of many aspects of bird biology (Jenni et al., 1994; Baillie et al., 1999a,b; Jenni, 2001). In Europe, more than 10,000 – mostly amateur – ringers mark millions of birds each year, and thus form the basis for population studies. Scientific bird ringing is organised and directed by the ringing centres under scientific guidance, with the aim of studying not only demography but also migration and many other aspects of bird life. Today the classic metal bird ring is supplemented by new marking techniques which contribute to the success of population studies while obviating the necessity to recapture the birds, e.g. individual colour ringing, numbered rings readable at a distance as for instance used in the White Stork *Ciconia ciconia*, or in swans and geese, and transponders (microchips, Michard et al., 1994; Becker and Wendeln, 1997) which make it possible to record the individual birds automatically and to measure demographic parameters year after year (Common Tern, Becker et al., 2001b).

Intensive studies of bird populations have provided a detailed understanding of the mechanisms which cause changes in bird numbers (reviews in or by Perrins et al., 1991; Greenwood et al., 1993; Greenwood, 1999; Jenni, 2001; Helbig and Flade, 1999). Here only a few of the results can be highlighted.

Since 1988 in the UK, the British Trust for Ornithology (BTO) has been combining several national projects into an integrated monitoring system for common bird species

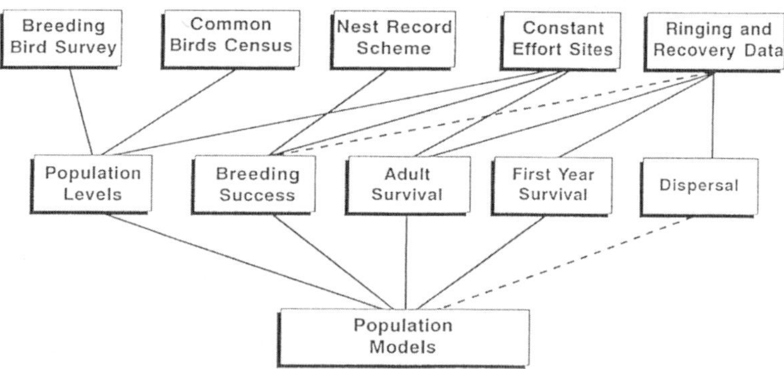

Figure 18. The BTO's Integrated Population Monitoring programme showing how demographic information is derived from various extensive surveys, is used to build population models. From Greenwood (1999).

(Fig. 18): The Common Bird Census, the nest record scheme to monitor annual changes in breeding performance, the "constant effort" sites scheme which uses summer mist netting of birds at woodland, scrub and wetland sites to provide indices of adult and juvenile abundance and survival; and the ringing scheme with sufficient data to make it possible to monitor the annual survival rates of at least 28 species (Baillie, 1990; Greenwood et al., 1993; Baillie, 2001). Such demographic information can be interpreted by modelling and used as indicator for environmental change (Greenwood et al., 1993).

Using this powerful tool, Baillie (1990) and Thomson et al. (1997), for example, were able to rule out changes in breeding output or survival of adult birds as the operating forces in the decline of British Song Thrushes *Turdus philomelos* and to identify insufficient survival of birds during their first year of life as the likely cause. Although weather certainly influences survival, it has not apparently been the cause of the long-term decline. For seed-eating birds declining in Great Britain's farmlands, it was suggested that the causes were changes in post-fledging survival and/or in the number of breeding attempts per year (Baillie et al., 1997; Siriwardena et al., 2000). Combined studies of reproductive output and other demographic parameters are likely to yield more robust data than those based solely on abundance (Paradis et al., 2000).

Besides trends in demographic parameters over time, key factor analyses are able to identify the influence of specific environmental factors (Greenwood et al., 1993). In some migrating species visiting the Sahel zone in Africa, it was found that the amount of rainfall was negatively correlated with overwinter survival in the White-throat *Sylvia communis* (Baillie and Peach, 1990; the Sedge Warbler *Acrocephalus schoenobaenus* (Fig. 19, Peach et al., 1991) and the White Stork (Kanyamibwa et al., 1993), and hence had an influence on the species' breeding population size in Europe.

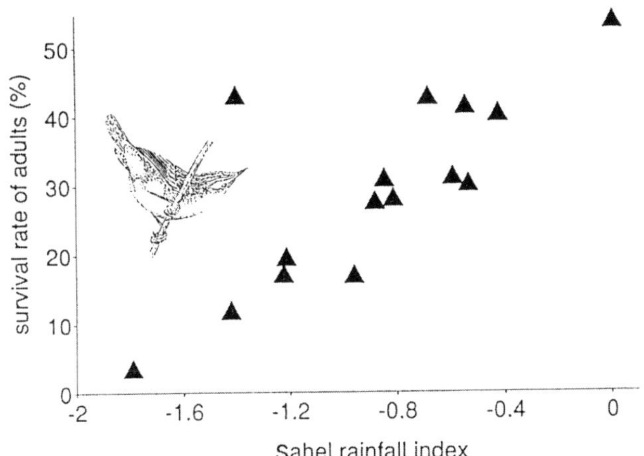

Figure 19. The survival rate of Sedge Warblers as measured by ringing decreases with increasing drought in the Sahel region. Falling survival rates have caused a long-term decline in the west European population which winters in this part of Africa. From Peach et al. (1991).

In consequence, these species indicate environmental factors such as the tropical weather conditions which obtained during their wintering period.

Reproductive performance like clutch and egg size, laying date, hatching and breeding success as well as chick growth are relatively easy to measure in birds. This can give important hints about the environmental conditions in the breeding area and may provide an early warning of an environmental problem like pollution, failing food supply or predation, e.g. through introduced predators. In consequence, the monitoring of reproduction is the first step towards an integrated population monitoring and is in operation e.g. in raptor monitoring (Mammen and Stubbe, 1999, see 4.4.4), seabirds monitoring in Great Britain (Walsh et al., 1991, 1995) and has been advocated and successfully tested in monitoring populations of coastal birds in the Wadden Sea (Becker, 1991; Exo et al., 1996; Bakker et al., 1998; Becker, 1998; Thyen et al., 1998, 2000b).

3.2.4.1. Long-term population studies

Many detailed long-term population studies of particular species have been undertaken in various parts of the world, uncovering vital evidence of environmental changes with major implications for conservation, and providing correlations that give strong indications of the causal mechanisms behind such changes. Some examples are the White Stork (Schulz, 1999; some populations studied for more than 70 years, Bairlein, 1991; Bairlein and Henneberg, 2000), seabirds (Wooller et al., 1992) like the Fulmar *Fulmarus glacialis* on the Orkneys (41 years, Ollason and Dunnet, 1988), Short-tailed Shearwater *Puffinus tenuirostris* (54 years, Bradley et al., 1991), the Snow Goose *Anser caerulescens* (Cooch and Cooke, 1991), the Mute Swan *Cygnus olor* (Coleman et al., 2001), the Sparrowhawk (37 years, Newton, 2001), the Kittiwake *Rissa tridactyla* in NE England (38 years, Coulson and Thomas, 1985, Aebischer et al., 1990), the Great Skua *Catharacta skua* in Shetland (nearly 30 years, Hamer

et al., 1991; Catry et al. 1999), the Puffin (30 years, Harris and Wanless, 1991) and
the Common Tern in the USA (30 years, Nisbet, 1978; Nisbet et al., 2002) and on the
Jade, German Wadden Sea (20 years, Becker, 1991, 1998), several long-term projects
on hole-breeding birds such as tits and Pied Flycatchers *Ficedula hypoleuca* (e.g.
Perrins, 1996; Perrins and McCleery, 2001; Dhondt, 2001; Winkel and Hudde, 1997).

3.2.4.2. Do birds indicate climate change?

Such long-term studies clearly reveal changes in bird biology during recent decades,
linked to global warming (Berthold, 1998; Bairlein and Winkel, 2001). Climate change
undoubtedly affects bird populations, though the effects may differ from species to
species and are difficult to predict. Climatic changes may affect the area of distribution, the reproductive biology, the migratory behaviour and mortality during winter.
In central Europe, some bird species breed earlier in spring, for example tits,
Nuthatches *Sitta europaea* (Fig. 20) and Pied Flycatchers (Crick and Sparks, 1999;
Bairlein and Winkel, 2001). The egg volume of this species is positively related to
temperature during the laying period, and has consequently increased during the recent
years (Järvinen, 1994). Several migrant species have been arriving at the breeding
grounds progressively earlier in spring and leaving progressively later in autumn.
Changes in oceanography possibly associated with global warming influence the distributions, movement patterns of the marine prey of seabirds, and thus affect foraging,
diet composition, breeding success and population development in a variety of species-
specific ways (Montevecchi and Myers, 1997; Kitaysky and Golubova, 2000).

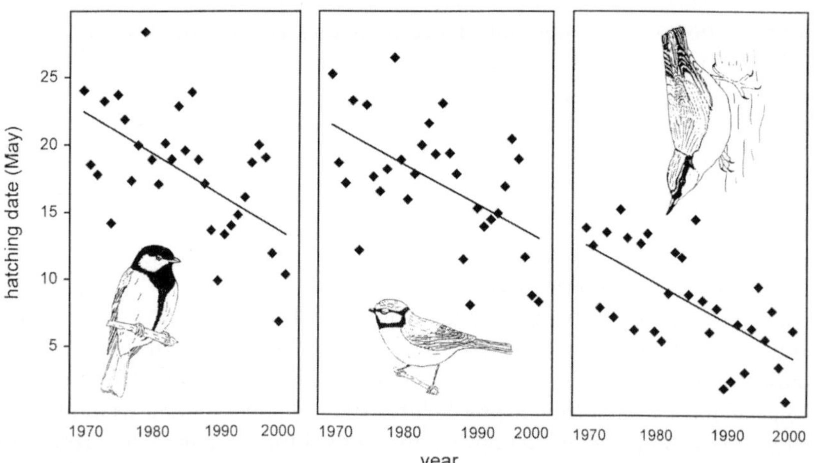

Figure 20.   Annual mean hatching dates advanced over the years in first broods of Great Tits *Parus major*,
Blue Tits *Parus caeruleus* and Nuthatches *Sitta europaea* near Brunswick, northern Germany, 1970–2000.
The earlier breeding corresponds with an increase in average air temperatures in March ($r = 0.23$, n.s.) and
April ($r = 0.68$, $p < 0.001$). From Bairlein and Winkel (2001).

4. Avian biomonitors in use

The value of birds for monitoring the environment having been recognised, over the past four decades birds have been introduced as components into a number of governmental run, established monitoring programmes on a national, and indeed international, scale (Table 4). Birds are in use as monitors to identify the trends and effects of specific factors like chemicals, or as sensitive indicators ("sentinel organisms") for the general state of the environment and its change. The projects listed highlight mainly international programmes, but the list is far from complete.

4.1. Birds as monitors of environmental chemicals

Because of the prominent role of birds as indicators of various aspects of environmental pollution with chemicals (3.1.1), biomonitoring of environmental chemicals with birds has been introduced in various countries since the 1960s. The main objectives of programmes utilising the avian value as accumulative indicators are as follows (e.g. Newton et al., 1993; Bignert et al., 1995, 1998; Bignert, 2001; TMAP, 1997; Becker et al., 2001):

- to provide long-term residue data on several representative species and sites;
- in time trend monitoring: to assess long-term temporal trends in chemical residues and to estimate the rate of changes;
- in spatial trend monitoring: to assess spatial variation in pollutants' levels, on small and large geographical scales;
- to explore site-specific and time-specific patterns in the composition of compound mixtures like PCBs, DDTs, HCHs, toxaphenes and organic tin compounds.

Such data can be used

- to detect incidents of local contamination discharges;
- to detect renewed application or discharge of banned contaminants;
- to assess the effectiveness of measures to reduce contamination, against the background of successive governmental restrictions on organochlorine and mercurial pesticide use;
- to assess the changing population status of affected species: the response of species to changes in pollution levels may affect the abundance or, physiology of species leading to structural changes in the ecosystem (TMAP, 1997). To achieve this aim, chemical monitoring has to be combined with bird population monitoring (4.4).

The most frequently used matrix is the bird egg (advantages see 3.1.1.1).

Since 1971, contamination of aquatic birds has been monitored by the Canadian Wildlife Service as an indicator of the pollution of the Great Lakes (Fig. 14; e.g. Mineau et al., 1984; Weseloh et al., 1988, 1990; Ryckman et al., 1997, 1998; CWS, 2001). Canadian seabirds are used as biomonitors of marine pollution with chemicals (Chapdelaine et al.,1987; Elliott et al., 1992). The Gannet and other seabirds are taken as biomonitors in Britain (Newton et al., 1990). The marine conditions of the Baltic Sea are monitored by means of various biota, including Guillemot eggs in Sweden

Table 4. Some examples of international or national Monitoring Schemes using birds.

Project Organization	Country Area	Species	Objectives	Samples	Reference (Examples)
• Census Atlas Work, EBCC, BBS[1]	Europe, USA	Landbirds	Distribution abundance	Territory, point-, transect-counts	Marchant (1999), Sauer and Link (1999)
• IWC Wetlands International	Worldwide	Waterbirds	Population size	Numbers	Rose (1994), Mooij (2000), Sudfeldt et al. (2000)
• CMMN	Canada, USA	Landbirds	Population size of migrants	Numbers	Dunn and Hussel (1995)
• Seabirds at Sea (JNCC)	North-East Atlantic	Seabirds	Abundance, distribution	Numbers	Tasker et al. (1994); Stone et al. (1995)
• BTO, German Ringing Centres	UK, Central Europe	Landbirds	Demography	Integrated population data	Baillie (1991) Greenwood et al. (1993)
• University of Halle	Central Europe	Birds of prey	Population	Numbers, reproductive success	Mammen and Stubbe (1999)
• TMAP	Wadden Sea: Germany, The Netherlands, Denmark	Coastal birds	• Population size of breeding birds	Numbers	• Rasmussen et al. (2000)
			• Population size of migratory birds	Numbers	• Poot et al. (1996), Günther and Rösner (2000)
			• Reproductive success[1]	Population data, chick growth	• Thyen et al. (1998, 2000b)
			• Contamination	Eggs	• Becker et al. (2001a)
			• Oil pollution	No. carcasses oiled	• Bakker et al. (1999), Fleet and Reineking (2000)

Table 4. (continued)

Project Organization	Country Area	Species	Objectives	Samples	Reference (Examples)
• BBS[2]	North-East Atlantic	Seabirds	Oiling rate	Numbers	Camphuysen (1998)
• CEH (ITE)	Britain	• Birds of prey • Seabirds	Contamination	• Eggs, tissues • Eggs	• Newton et al. (1993) • Newton et al. (1990)
• NSCMP	Sweden	Guillemot, White-tailed Eagle, Starling	Contamination, Population	Eggs, population data	Bignert et al. (1998), Helander (1999), Odsjö (2000)
• USFWS	USA	Gulls, Starlings, Ducks, eagles	Contamination	Eggs, tissues	Cain and Bunck (1983)
• CWS	Canada	Seabirds	Contamination	Eggs, tissues	Elliott et al. (1994)
• CWS	Canada, Great Lakes	Aquatic birds	Contamination, population	Eggs, population data	Grasman et al. (1998)

[1] not yet implemented.

Abbreviations: BBS[1] = Breeding Bird Survey; BBS[2] = Beached Bird Survey; CEH = Centre for Ecology and Hydrology, Monks Wood Huntingdon, UK (formerly ITE, Institute of Terrestrial Ecology); CMMN = Canadian Migration Monitoring Network; CWS = Canadian Wildlife Service; EBCC = European Bird Census Council; IWC = International Waterbird Census; NSCMP = National Swedish Contaminant Monitoring Programme; TMAP = Trilateral Monitoring and Assessment Programme; USFWS = US Fish and Wildlife Service; WI = Wetlands International since 1969 (Fig. 7, Bignert et al., 1995, 1998). Monitoring programmes for terrestrial biota include the Starling *Sturnus vulgaris* (Odsjö, 2000, Sweden; US National Biomonitoring Program).

In 1996/97, on the basis of projects carried out in the 1980s (Becker, 1989, 1991; Becker et al., 1991, 1992) and of a pilot study within the TMAP framework, monitoring chemicals in the Wadden Sea area using birds' eggs with generally agreed guidelines for monitoring design, sampling procedures and chemical analytical procedures was initiated (OSPAR, 1997; TMAP, 1997; Becker et al., 1998). Several sampling sites (11) along the Wadden Sea coast in the Netherlands, Germany (Niedersachsen, Schleswig- Holstein) and Denmark permit the evaluation of geographical trends and hot spots of contamination (Fig. 21). Ten eggs per species, site and year – one per nest, indicating the contamination of ten female birds – are sampled and allow statistical analyses of trends of the heavy metal mercury, pesticides and industrial chemicals including HCB and 62 PCB congeners (Sommer et al., 1997; OSPAR 1997; Becker et al., 2001a). By use of the data set available from the 1980s, contamination trends over a period of 20 years are now able to be analysed. Recent results indicate steep contamination decreases at the beginning of the 1990s and further reductions during the decade (Fig. 22), which is corroborated by the analyses of abiotic and other biotic samples in the Wadden Sea (Bakker et al., 1999). Yet despite the lowered inputs of chemicals into the North Sea by atmosphere and through rivers during this time, the contamination of the seabird eggs clearly indicates distinct geographical trends, with the hot spot at the Elbe estuary still persisting today (Fig. 21).

All these studies clearly showed that concentrations of most environmental chemicals have decreased continuously since the 1970s, in the Wadden Sea during the 1990s especially (Fig. 22). Obviously, many measures aimed at the protection of the environment have had measurable positive effects on the environment and on the birds as well. However, concentrations of some chemicals are still unacceptable high, especially the toxic PCBs (3.1.1.2). In Sweden, for example, the rate of decrease is slower in PCBs than in other compounds studied, indicating an ongoing pollution from sources like refuse tips (Bignert et al., 1998). In Germany, too, the bird data show further effluents of PCBs through the big rivers Rhine and Elbe (Figs 11, 22; Becker and Sommer, 1998; Becker et al., 2001a), even though production and use of PCBs have been forbidden since 1989.

Besides the use of birds in general as accumulative indicators of chemicals' levels, the capacity of avian top predators to act as sensitive indicators is utilised to monitor for possible biological effects of environmental chemicals. This approach (3.1.1.2) employs birds (i) as an early warning of new chemicals not covered by the regular monitoring programme (ii) to spot dangerous interactions of chemicals, (iii) to derive critical levels for toxic chemicals. In this way in Sweden, for example, the reproductive success of White-tailed Sea Eagles (Helander et al., 1998; Olsson et al., 1998; Helander, 1999) is a parameter of the national marine monitoring programme. In Canada, monitoring contamination of aquatic birds at the Great Lakes is supplemented by studies on reproductive performance, congenital anomalies, mutagenicity and on biochemical and physiological effects (Mineau et al., 1984; Ryckman et al., 1998; Grasman et al., 1998; CWS 2001), in order to receive an early warning of possible toxic effects of environmental chemicals. This approach also was recommended,

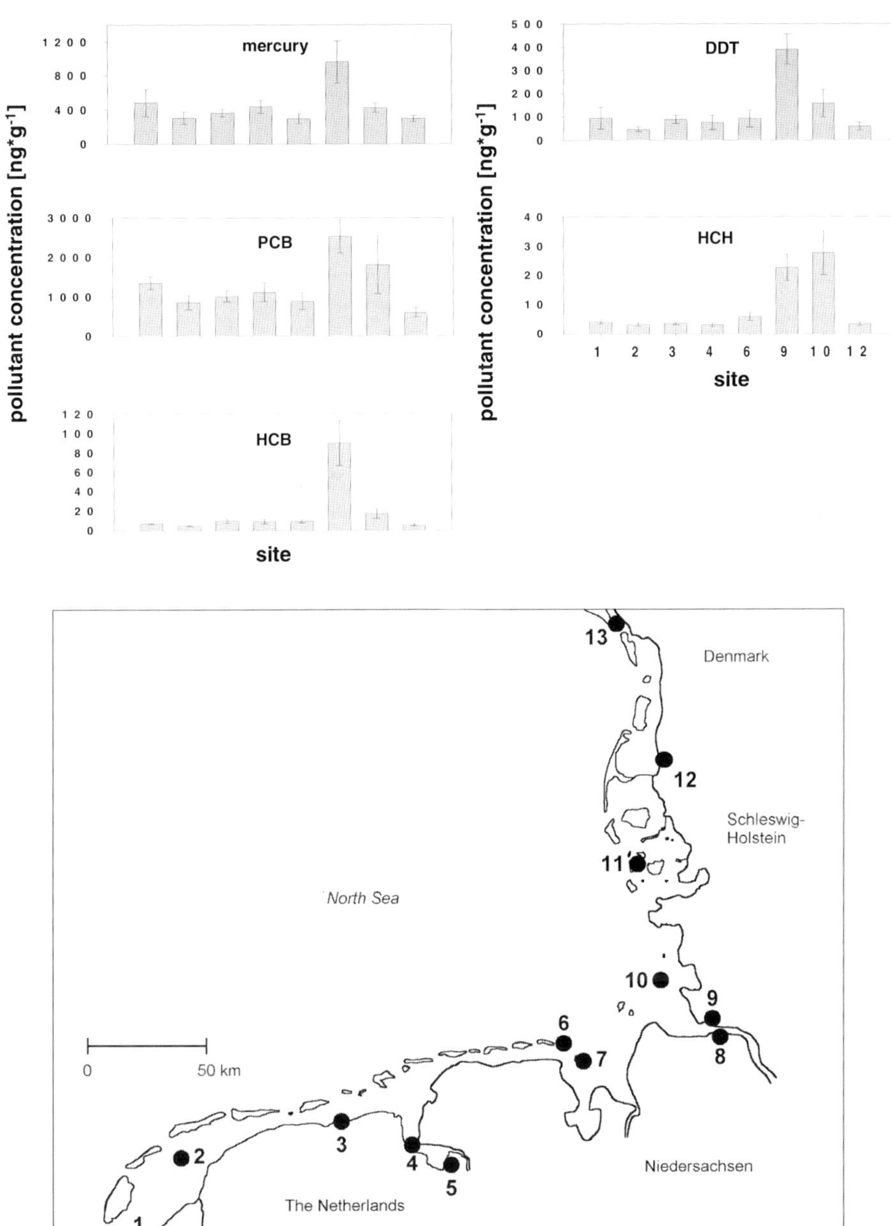

Figure 21. Spatial trends in concentrations of mercury, ΣPCB, HCB, ΣDDT and ΣHCH in eggs of Common Terns in the Wadden Sea in 2000 (means ± 95% confidence intervals). Recent data from the TMAP-project "Pollutants in bird eggs". Sites: 1 Balgzand, 2 Griend, 3 Julianakoog, 4 Delfzijl, 6 Jade, Minsener Oldeoog, 9 Elbe, Neufelderkoog, 10 Trischen, 12 Margrethekoog (Becker et al., 2001a).

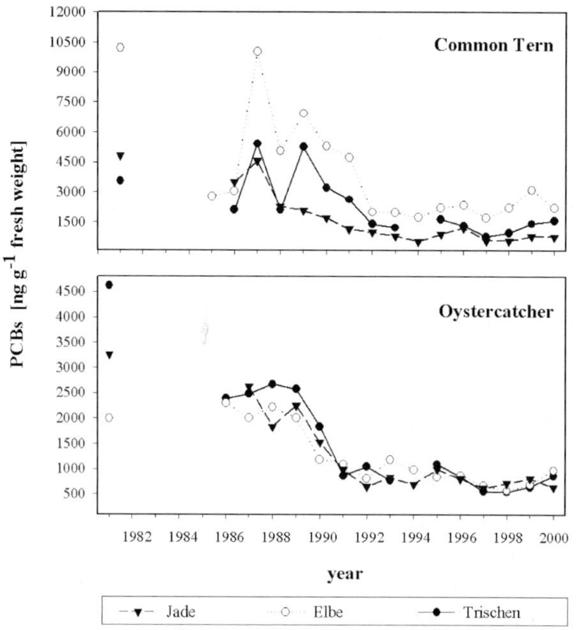

Figure 22. Declines of ΣPCB concentrations in eggs of Oystercatchers and Common Terns from three breeding sites in the German Wadden Sea, 1981–2000. See Fig. 21 for sites. From Becker (2001a).

successfully tested and adopted in the TMAP in the Wadden Sea (Thyen et al., 1998, 2000b) but has not yet been applied, because of a lack of funding.

4.2. Banking of avian samples

The storing of frozen environmental samples, including bird tissues, for future retrospective investigations is addressed in Kettrup (2002). Such invaluable material will make possible future retrospective studies of contaminants which defy analysis today as well as control analyses of suspected analytical errors or inter-calibration studies of new against old analytical methods and results. Such specimen banks for bird and other materials are run by Canada (CWS), Norway (NINA), and Sweden (NSCMP). In Germany, eggs of Herring Gulls from two breeding colonies in the Wadden Sea (since 1988) and one in the Baltic Sea have been put into storage (Umweltprobenbank, 1996; Kettrup and Marth, 1998), along with pigeons' eggs from inland sites.

4.3. Beached bird surveys to monitor marine oil pollution

The monitoring of beached seabirds (BBS) is carried out to assess (i) the existing level of chronic oil and chemical pollution in the marine areas studied, (ii) the effectiveness of measures taken to reduce chronic oil pollution, and (iii) the effect of oil incidents on coastal birds and seabirds. In Europe many countries on the North East Atlantic participate in these surveys (guidelines see OSPAR 1996). BBS are also part of the TMAP in the Wadden Sea (TMAP, 1997). The main European BBS have now been running for 25–35 years and are making a valuable contribution to the understanding of spatial and temporal trends in the pollution of European seas with oil (Camphuysen, 1998; Camphuysen and Heubeck, 2001). Such patterns of oiled marine birds reflect the risk for individual seabirds in different sea areas of encountering oil slicks and, hence, BBS are a useful tool for recording and measuring these differences and for studying the effectiveness of measures to reduce oil pollution at sea. BBSs are carried out predominately by amateurs guided and organised by professional staff who are also responsible for the data analysis. Compared to aerial surveillance for oil slicks at sea, BBSs are a very cost-effective method with a proven success rate and capable of application on a very wide range of scales.

4.4. Monitoring bird populations as indicators of environmental change

The results of bird population monitoring not only present important hints about environmental change, but also form the basis for the protection of bird populations and their environments, and to assess the success of measures of environmental protection and nature conservancy. That was at the back of every ornithologists' mind – amateur and professional alike – and reason enough for them all over the world to initiate bird population monitoring, to engage in field work, in collecting, analysing and interpreting the data. From this highly cost effective constellation has ensued the development of currently established biomonitoring programmes with birds.

4.4.1. Trends in breeding bird distribution, density, communities and population size: bird census and atlas work

Changes in numbers or distribution of species communities can act both as an alarm signal for problems in their habitats and as a pointer to the nature of those problems. That is the reason why bird atlas and census work is so commonly performed all over the world. In many countries it is a simple matter to organise extensive surveys by networks of observers, very often amateurs, to obtain data on distribution, numbers, reproductive success and survival in "common" forest, farmland, garden and urban birds. The advantage of such work lies in the comprehensive coverage, in the great quantity of data obtained permitting analyses of considerable detail, and in the large number of studied sites. Such surveys have mainly been carried out at local or national scales (Marchant et al., 1998; Flade and Schwarz, 1999; Sauer and Link, 1999), but extension to the European, international scale has both actual and potential benefits for international legislation and adds greater credibility to trend statements derived from national surveys. The EBCC (European Bird Census Council) has made considerable progress towards international bird monitoring in Europe (Hagemeijer and Blair, 1997; Greenwood, 1999; Marchant, 1999), but constant appeals for public funding need to be made on behalf of this important survey work.

Innumerable studies focus on the annual change of population size in birds world wide (3.2.2, 3.2.4). Within the framework of the TMAP for example, the monitoring of breeding birds in the Wadden Sea is carried out to assess the effects on occurrence, abundance, population size and community structure of environmental changes in, for instance, climate, pollution levels, fishery or human recreational activities and grazing on salt marshes. The counts are carried out following detailed guidelines (Hälterlein et al., 1995) and cover census areas annually, since 1990 with complete annual surveys of selected species, and at five-year intervals with counts of a larger range of species within the entire area in order to record the total population size (e.g. Hälterlein and Südbeck, 1996; Melter et al., 1997; Rasmussen et al., 2000; Hälterlein et al., 2000).

4.4.2. Migrating and wintering birds

The international Waterbird Census (IWC, started in 1964), under the auspices of "Wetlands International", organises world wide waterfowl monitoring, covering the most important wetlands and staging areas of waterbirds. Data are regularly analysed to provide information on population sizes, trends, seasonal distribution etc. of all waterfowl species. The data form an important basis for nature conservation and several international agreements and conventions (Rose and Scott, 1997; Mooij, 2000; Sudfeldt et al., 2000).

Seabirds at Sea: Since 1979, Ornithologists of all nations adjacent to the North Sea have been collecting data on seabird distribution at sea throughout the year using standardised methods (Tasker et al., 1984; Webb and Durinck, 1992; Garthe and Hüppop, 2000). Within year and inter sea-area fluctuations in numbers can be studied (Tasker et al., 1987; Camphuysen et al., 1993; Jensen et al., 1994; Garthe et al., 1995; Skov et al., 1995; Stone et al., 1995). The project, including a joint data base project, is leaded by The Seabirds at Sea Co-ordinating Group, Joint Nature Conservation Committee (JNCC), Aberdeen, UK.

Monitoring waterbirds staging in the Wadden Sea: The Wadden Sea is one of the world's most important wetlands for migratory waterbirds and has outstanding functions in the life cycle as staging, moulting and wintering area for at least 52 populations of 41 species using the East Atlantic Flyway (e.g. Exo, 1994; Meltofte et al., 1994). Since 1992, earlier counts of waterbirds in the Wadden Sea have been developed into a programme coordinated internationally within the framework of the TMAP. The objectives are (i) to detect changes in bird numbers and distribution (ii) to determine the total number of birds present in the Wadden Sea during the annual cycle (iii) to estimate total populations using the East Atlantic Flyway, and (iv) to collect data to explain the patterns and to indicate environmental changes. The project is organised according to methods described by Rösner et al. (1994) and Poot et al. (1996) and carries out midwinter counts, synchronous counts of goose species during their peak occurrence and spring-tide counts on selected sites (Poot et al., 1996; Rösner et al., 1999; Günther and Rösner, 2000).

4.4.3. Population development and demographic parameters

Integrated bird population monitoring, initiated by the BTO, UK (3.2.4), has since been adopted by several other European countries (e.g. the ringing centres Heligoland, Hiddensee and Radolfzell in Germany, Bairlein et al., 2000), to gather information on important demographic parameters besides bird numbers, such as productivity and survival rates in many common bird species. Such demographic information can be interpreted by modelling and used as an indicator for environmental change (Greenwood et al., 1993, 3.2.4).

Bird Ringing as a basic methodology for avian population studies is essential part of integrated bird monitoring programmes (3.2.4, Jenni et al., 1994; Baillie et al., 1999a,b; Jenni, 2001). Scientific bird ringing is organised by national ringing schemes involving hundreds of ringers. In Europe, all schemes are members of EURING (The European Union for Bird Ringing), which organises and standardises bird ringing in Europe, co-ordinating large scale research projects and promoting adequate methods of data analysis. Ringing data have been collected since the beginning of the 20th century, and hundreds of thousands of recovery records are stored in the EURING Data Bank, which is an important data source for diverse analyses of bird life, also with regard to the function of birds as biomonitors.

Monitoring reproductive success: In non-passerine bird species, a large-scale integrated population monitoring is more difficult to perform than it is in passerines and is thus restricted to specific studies (examples in 3.2.4). One important demographic parameter, however, the reproductive success, is relatively easy to measure in birds and offers important hints on the environmental situation in the breeding areas, using birds as sensitive indicators. In top predators, this monitoring ideally should be combined with their function as accumulative indicators and with the study of pollution with chemicals (4.1).

Monitoring reproductive success is part of some programmes monitoring the environment with bird species such as raptors in Europe (see below), aquatic birds in the Great Lakes (Mineau et al., 1984; Ryckman et al., 1997, 1998; Grasman et al., 1998), Swedish White-tailed Sea Eagles (4.1) and with six common coastal bird species in the Wadden Sea. Successfully tested in a two-year pilot study (Thyen et al., 1998, 2000b) and recommended to and adopted by the TMAP-Group (TMAP, 1997), the implementation is being held up by the lack of government funding. The main objectives are to assess (i) the response of bird species to changing pollutant levels which may affect abundance and physiology of species and leading to structural changes in the ecosystem Wadden Sea, and (ii) anthropogenic effects on the occurrence and populations like fisheries, agriculture, pollution, or recreational activities.

In the context of pollution problems, raptor and owl populations have been intensively studied in many countries since the 1950s. Besides the programmes already mentioned, a project was founded in 1988 to monitor raptors and owls in central Europe (Stubbe et al., 1996; Mammen and Stubbe, 1999), and by end of the 1990s it involved participants from 14 European countries. Data are collected of annual presence, density, reproduction and breeding success in many species of raptor and owl. The data base permits the detailed analyses of change in population size and reproduction of these top predators, which are important indicators of environmental change. This program is to be supplemented by regular analyses of environmental chemicals in raptor eggs in order to utilise the additional function of this group as accumulative indicator of terrestrial food webs.

5. Concluding remarks: obstacles, recommendations and perspectives

As argued in this chapter, birds are indeed useful indicators of the state of the environment in which they live, and have proved effective bioindicators as well as biomonitors of various kinds of environmental change. Consequently, avian bioindicators are already important components in current monitoring programmes in many countries and in a variety of ecosystems (Table 4).

However, the various advantages of birds in biomonitoring are not always obvious to everybody people dealing with environmental monitoring. Very often birds are simply forgotten – e.g. by marine ecologists unaware of the prominent and integrative data source presented by seabirds in monitoring the sea. One reason for this may be that marine birds – penguins aside – don't live "in" but "on" or "over" the sea. Nevertheless, they make use of the marine resources just like any other marine animals. For example, for the monitoring of populations of some economically important fish species as well as of marine pollution, birds often provide a cheaper alternative with at least the same indicative evidence as fish catches, for example, or water samples. These latter are much more costly to collect by ships or the fisheries, or to analyse in case of water samples dealing with low concentrations of chemicals, the techniques involved being very complex and expensive. Consequently, the amount of funding needed for biological indicators, especially birds, is small compared to that needed for water chemistry. However, over the last decade the International Council for the Exploration of the Sea (ICES) has established a working group on "Seabird Ecology", dealing with the question of the usefulness of seabirds as indicators of the state of marine environments (see e.g. Hunt and Furness, 1996; Reid, 1997; Furness and Tasker, 1999). Marine ornithology is an integral aspect of biological oceanography, and advances will be made if seabird research is integrated into marine research and monitoring programmes (Montevecchi, 1993).

The reservations of conservationists can also create problems, as bird species are often protected and/or live in protected natural areas such as national parks. However, the interference necessary when using birds as indicators is small compared with the great advantages they represent for monitoring a particular environment and as sentinel organisms, and therefore the risks are ethically justifiable. Usually only small samples of birds or their tissues are needed, and conservative methods have been developed whereby the birds are not harmed by the sampling procedures. Advanced methods of chemical analysis, of biomarker and bioassay techniques (Peakall, 1992) will be available and of great advantage in the future in the analysis of chemicals' levels and effects on physiology, methods involving merely the taking of a small sample of blood from a free-ranging bird without harming it. This will permit better assessment of the current contaminant loads or of hazardous pollution events, while evaluating their effects on physiology.

When monitoring is done with birds, the information available on their biology should be applied to reduce the risk of misinterpretation or poor performance of the monitoring programme (e.g. Brisbin, 1993). Thus, for studies of the contamination of birds with chemicals, not only the experience of analytical chemists is needed but also the input of experienced researchers into bird biology and ecology: Their knowledge of the advantages and limitations of birds as monitors is necessary both for planning the monitoring project and for scientific accompaniment and for interpretation of the results. The co-operation of researchers from various scientific fields should be encouraged in monitoring when a variety of expertise is necessary.

Along with the ornithological expertise comes the guarantee of the selection of an appropriate species, sample units and techniques with respect to the needs and aims of the specific monitoring program. Ornithology has in store reliable and proved methods for environmental monitoring with birds! Many researchers work hard to produce and test adequate methods of measuring environmental change with birds, but frequently the governments who initiate such investigations fail to provide the necessary support when the studies have to be translated into a monitoring programme, and to be supported by the public authorities.

Continuous, long-term, annual monitoring efforts are needed before the very useful instrument offered by biota samples to describe environmental processes can be effective (e.g. Bignert et al., 1998). Protracted series reveal the size of random between-year variation which is not part of a trend. This shows that it is risky to use small, scattered sets of data on occasionally collected samples for interpreting environmental issues. With respect to contamination, the need for more precise monitoring instruments will grow as the environmental concentrations of a pollutant decrease (Bignert et al., 1998), as has become clear in recent years with respect to pesticides.

Many of the above mentioned monitoring programmes using birds (4., Table 4) complain of poor or even non-existent public funding. Yet even if using birds as monitors costs much less than methods of environmental monitoring involving many other biota or techniques, adequate financial support is essential. Personnel costs are always heavy, even if many bird monitoring programmes are based largely on unpaid volunteers. In order to manage the projects, experienced researchers have to be employed on a long-term basis, for biomonitoring makes sense only if performed constantly over a long period. The efforts and costs of managing and evaluating the data gathered in increasing amounts year by year are often underestimated, and the failure to supply research groups or institutions with sufficient financial resources impedes the effective evaluation of the results, which is the ultimate objective of monitoring and absolutely essential if the findings of the monitoring are to be made available for public debate on the environment or for the pursuit of prospective aims, such as providing early warning of dangerous changes in the environment.

Programmes monitoring chemicals should be combined with studies of bird demography, in particular of possible effects on reproduction, in order to achieve the effect of birds acting as an early warning system (3.1.1.2, 3.2.4, 4.1, 4.4.4). Such studies should certainly include the hot spots of environmental pollution relevant also in the future: rivers, estuaries, industrial areas, where pollution sources and environmental chemicals become concentrated, and where the first signs of negative effects on nature are apt to become manifest. A focus of monitoring programmes on those sites may save costs, but for purposes of comparison, sites with basic pollution levels are also needed, in order to detect "source" (where natality is sufficient to maintain the bird population) and "sink" areas and to be able to derive critical levels and species' sensitivity towards chemicals.

This chapter was intended to develop an awareness of the relevance of birds and their ecology to biomonitoring the environment. As birds are a matter of public interest, their choice as biomonitors brings the added advantage of having an effective pedagogical instrument with which to increase public acceptance and understanding of environmental monitoring. People care about birds! Therefore with birds it is easier than with other animals to explain to the public the necessity of monitoring, the need for and adequate use of funding, to elucidate the biomonitoring findings and to win support against disastrous anthropogenic impacts on the environment.

Acknowledgements

My thanks go to Dipl.-Biol. Stefan Thyen to several colleagues for giving permission to reproduce figures, for comments on the draft of this chapter, and to Ken Wilson for checking the English. Tobias Dittmann, Anja Epding and Christiane Ketzenberg helped with the figures, Ingrid Campen, Martin Wagener and Elke Wiechmann with compiling the reference list. Dr Andreas Bignert presented recent information on the Swedish monitoring programme. RIKZ, Den Haag, Nationalparkverwaltung Niedersächsisches Wattenmeer, Wilhelmshaven, Landesamt für den Nationalpark Schleswig-Holsteinisches Wattenmeer, Tönning, and the Miliø- og Energiministeriet, Copenhagen, gave permit to present recent TMAP data.

References

Aebischer, N.J., Coulson, J.C., Colebrook, J.M., 1990. Parallel long-term trends across four marine trophic levels and weather. Nature 347, 753–755.

Anderson, D.W., Gress, F., 1983. Status of a northern population of California Brown Pelicans. Condor 85, 79–88.

Anker-Nilssen, T., 1987. The breeding performance of Puffins *Fratercula arctica* on Rost northern Norway in 1979–1985. Fauna norv. Ser. C, Cinclus 10, 21–38.

Anker-Nilssen, T., Erikstad, K.E., Lorentsen, S.H., 1996. Aims and effort in seabird monitoring: an assessment based on Norwegian data. Wildl. Biol. 2, 17–26.

Appelquist, H., Drabaek, I., Asbirk, S., 1985. Variation in mercury content of Guillemot feathers over 150 years. Mar. Pollut. Bull. 16, 244–248.

Bailey, R.S., Furness, R.W., Gauld, J.A., Kunzlik, P.A., 1991. Recent changes in the population of the sandeel (*Ammodytes marinus* Raitt) at Shetland in relation to estimates of seabird predation. ICES Mar. Sci. Symp. 193, 209–216.

Baillie, S.R., 1990. Integrated population monitoring of breeding birds in Britain and Ireland. Ibis 132, 151–166.

Baillie, S.R., 2001. The contribution of ringing to the conservation and management of bird populations: a review. In: Jenni, L. (Ed.), Proc. Conf. Bird Ringing 100 years, Helgoland 1999. Ardea Special issue.

Baillie, S.R., Gregory, R.D., Siriwardena, G.M., 1997. Farmland bird declines: patterns, processes and propects. In: Kirkwood, R.C. (Ed.), Biodiversity and Conservation in Agriculture. British Crop Protection Council Symposium Proceedings 69. British Crop Protection Council, Farnham, pp. 65–87.

Baillie, S.R., North, P.M., Gosler, A.G. (Eds), 1999a. Large-scale studies of marked birds. Proceedings of EURING 97 Conference. Bird Study (Supplement) 46, 1–308.

Baillie, S.R., Peach, W.J., 1990. Population limitation in Palaearctic-African migrant passerines. Ibis 134, 120–132.

Baillie, S.R., Wernham, C.V., Clark, J.A., 1999b. The conservation uses of ringing data. Proc. JNCC/BTO Workshop. Ringing and Migration (Suppl.) 19, 1–143.

Bairlein, F., 1991. Population studies of White Storks *(Ciconia ciconia)* in Europe. In: Perrins, C.M., Lebreton, J.D., Hirons, G.J.M. (Eds), Bird Population Studies: Relevance to Conservation and Management. Oxford University Press, New York, pp. 207–229.

Bairlein, F., Bauer, H.-G., Dorsch, H., 2000. Integrated monitoring of songbird populations. Vogelwelt 121, 217–220.

Bairlein, F., Henneberg, H.R., 2000. Der Weißstorch (*Ciconia ciconia*) im Oldenburger Land. Isensee (Oldenburger Forschungen; N. F., Bd. 12), Oldenburg.

Bairlein, F., Winkel, W., 2001. Birds and climate change. In: Lozan, J.L., Graßl, H., Hüpfer, P. (Eds), Climate of the 21st Century – Changes and Risks. Wissenschaftliche Auswertungen, Hamburg, pp. 278–282.

Bakker, J., Kellermann, A., Farke, H., Laursen, K., Knudsen, T., Marencic, H., Jong, F. de, Lüerßen, G., 1998. Implementation of the Trilateral Monitoring and Assessment Program (TMAP). Common Wadden Sea Secretariat, Wilhelmshaven.

Bakker, J. F., Bartelds, W., Becker, P. H., Bester, K., Dijkshuizen, D., Frederiks, B., Reineking, B., 1999. 4. Marine Chemistry. In: De Jong, F., Bakker, J.F., Berkel, C.J.M. van, Dankers, N.M.J.A., Dahl, K., Gätje, C., Marencic, H., Potel, P. (Eds), Wadden Sea Quality Status Report. Wadden Sea Ecosystem No. 9, Common Wadden Sea Secretariat, Trilateral Monitoring Group, Quality Status Group. Wilhelmshaven, pp. 85–117.

Bauer, H.G., Berthold, P., 1996. Die Brutvögel Mitteleuropas: Bestand und Gefährdung. Aula, Wiesbaden.

Baum, F., Conrad, B., 1978. Greifvögel als Indikatoren für Veränderungen der Umweltbelastung durch chlorierte Kohlenwasserstoffe. Tierärztliche Umschau 33, 661–679.

Becker, P.H., 1989. Seabirds as monitor organisms of contaminants along the German North Sea Coast. Helgoländer Meeresunters. 43, 395–403.

Becker, P.H., 1991. Population and contamination studies in coastal birds: the Common Tern *Sterna hirundo*. In: Perrins, C.M., Lebreton, J.D., Hirons, G.J.M. (Eds), Bird Population Studies: Relevance to Conservation and Management. Oxford University Press, Oxford, pp. 433–460.

Becker, P.H., 1998. Langzeittrends des Bruterfolgs der Flußseeschwalbe und seiner Einflußgrößen im Wattenmeer. Vogelwelt 119, 223–234.

Becker, P.H., Erdelen, M., 1987. Die Bestandsentwicklung von Brutvögeln der deutschen Nordseeküste 1950–1979. J. Ornithol. 128, 1–32.

Becker, P.H., Frank, D., Walter, U., 1987. Geographische und jährliche Variation der Ernährung der Flußseeschwalbe (*Sterna hirundo*) an der Nordseeküste. J. Ornithol. 128, 457–475.

Becker, P.H., Furness, R.W., Henning, D., 1993. The value of chick feathers to assess spatial and interspecific variation in the mercury contamination of seabirds. Environ. Monitoring Assessment 28, 255–262.

Becker, P.H., Heidmann, W.A., Büthe, A., Frank, D., Koepff, C., 1992. Umweltchemikalien in Eiern von Brutvögeln der deutschen Nordseeküste: Trends 1981–1990. J. Ornithol. 133, 109–124.

Becker, P.H., Koepff, C., Heidmann, W.A., Büthe, A., 1991. Schadstoffmonitoring mit Seevögeln. TEXTE 2/92, Umweltbundesamt, Berlin, 1–260.

Becker, P.H., Munoz Cifuentes, J., Behrends, R., Schmieder, K.R., 2001a. Contaminants in bird eggs in the Wadden Sea. Temporal and spatial trends 1991–2000. Wadden Sea Ecosystem No. 11. Common Wadden Sea Secretariat, Trilateral Monitoring and Assessment Group, Wilhemshaven, Germany, pp. 1–67.

Becker, P.H., Schuhmann, S., Koepff, C., 1993b. Hatching failure in Common Terns (*Sterna hirundo*) in relation to environmental chemicals. Environ. Pollut. 79, 207–213.

Becker, P.H., Sommer, U., 1998. Die derzeitige Belastung der Flußseeschwalbe *Sterna hirundo* mit Umweltchemikalien in Mitteleuropa. Vogelwelt 119, 243–249.

Becker, P.H., Thyen, S., Mickstein, S., Sommer, U., Schmieder, K.R., 1998. Monitoring pollutants in coastal bird eggs in the Wadden Sea. Final Report of the Pilot Study 1996–1997/Wadden Sea Ecosystem 8. Common Wadden Sea Secretariat Wilhelmshaven, 55–101.

Becker, P.H., Wendeln, H., 1997. A new application for transponders in population ecology of the Common Tern. Condor 99, 534–538.

Becker, P.H., Wendeln, H., González-Solís, J., 2001b. Population dynamics, recruitment, individual quality and reproductive strategies in Common Terns marked with transponders. In: Jenni, L. (Ed.), Proc. Conf. Bird Ringing 100 years, Heligoland 1999. Ardea 89 (Special issue), 241–252.

Beintema, A.J., Baarspul, T., Krijger, J.P. d., 1997. Calcium deficiency in Black Terns *Chlidonias niger* nesting on acid bogs. Ibis 139, 396–397.

Bennett, J.K., Ringer, R.K., Bennett, R.S., Williams, B.A., Humphrey, P.E., 1988. Comparison of breaking strength and shell thickness as evaluators of eggshell quality. Environ. Toxicol. Chem. 7, 351–357.

Berg, W., Johnels, A., Sjöstrand, B., Westermark, T., 1966. Mercury content in feathers of Swedish birds from the past 100 years. Oikos 17, 71–83.

Berthold, P., 1998. Vogelwelt und Klima: gegenwärtige Veränderungen. Naturw. Rdsch. 51, 337–346.

Berthold, P., Fiedler, W., Schlenker, R., Querner, U., 1999. Bestandsveränderungen mitteleuropäischer Kleinvögel: Abschlussbericht zum MRI-Programm. Vogelwarte 40, 1–51.

Beyerbach, M., Becker, P.H., Büthe, A., Denker, E., Heidmann, W.A., Yanes, G.S. d., 1993. Variation von PCB-Gemischen in Eiern von Vögeln des Wattenmeeres. J. Orn. 134, 325–334.

Bignert, A., 2001. Comments concerning the National Swedish Contaminant Monitoring Programme in Marine Biota. Report Contaminant Research Group Swedish Museum Natural History, 1–114.

Bignert, A., Litzen, K., Odsjö, T., Olsson, M., Persson, W., Reutergardh, L., 1995. Time-related factors influence the concentrations of sDDT, PCBs and shell parameters in eggs of Baltic Guillemot (*Uria aalge*), 1961–1989. Environ. Pollut. 89, 27–36.

Bignert, A., Olsson, M., Persson, W., Jensen, S., Zakrisson, S., Litzén, K., Eriksson, U., Häggberg, L., Alsberg, T., 1998. Temporal trends of organochlorines in Northern Europe, 1967–1995. Relation to global fractionation, leakage from sediments and international measures. Environm. Pollut. 99, 177–198.

Blight, L.K., Burger, A.E., 1997. Occurrence of plastic particles in seabirds from the eastern North Pacific. Mar. Pollut. Bull. 34, 323–325.

BLMP, 2000. Meeresumwelt 1994–1996. Bundesamt für Seeschiffahrt und Hydrographie, Hamburg und Rostock, pp. 1–151.

Blus, L.J., 1982. Further interpretation of the relation of organochlorine residues in Brown Pelican eggs to reproductive success. Environ. Pollut. Ser. A28, 15–33.

Blus, L.J., Wiemeyer, S.N., Bunck, C.M., 1997. Clarification of effects of DDE on shell thickness, size, mass, and shape of avian eggs. Environ. Pollut. 95, 67–74.

Bohac, J., Fuchs, R., 1991. The structure of animal communities as bioindicators of landscape deterioration. In: Jeffrey, D.W., Madden, B. (Eds), Bioindicators and Environmental Management. Academic Press, London.

Borg, K., Wanntorp, H., Erne, K., Hanko, E., 1969. Alkyl mercury poisoning in terrestrial Swedish wildlife. Viltrevy 6, 301–379.

Bosveld, A.T.C., Gradener, J., Murk, A.J., Brouwer, A., Kampen van, M., Evers, E.H.G., Berg van den, M., 1995. Effects of PCDDs, PCDFs and PCBs in Common Tern (*Sterna hirundo*) breeding in estuarine and coastal colonies in the Netherlands and Belgium. Environm. Toxicol. Chem. 14, 99–116.

Bourne, W.R.P., 1976. Seabirds and Pollution. In: Johnston, R. (Ed.), Marine Pollution. Bd. 6, Academic Press, London.

Bradley, J.S., Skira, I.J., Wooller, R.D., 1991. A long-term study of Short-tailed Shearwaters *Puffinus tenuirostris* on Fisher Island, Australia. Ibis 133, Suppl. I, 55–61.

Brenninkmeijer, A., Stienen, E.W.M., 1992. Ecologisch profiel van de grote stern (*Sterna sandvicensis*). RIN-rapport 92/17, 1–107.

Brisbin, L., 1993. Birds as monitors of radio-nuclide contamination. In: Furness, R.W., Greenwood, J.J.D. (Eds), Birds as Monitors of Environmental Change. Chapman and Hall, London.

Broeg, K., Zander, S., Diamant, A., Körting, W., Krüner, G., Paperna, I., Westernahgen v., H., 1999. The use of fish metabolic, pathological and parasitological indices in pollution monitoring. I. North Sea. Helgol. Mar. Res. 53, 171–194.

Brouwer, A., Murk, A.J., Koeman, J.H., 1990. Biochemical and physiological approaches in ecotoxicology. Funct. Ecol. 4, 275–281.

Burger, J., 1993. Metals in avian feathers: bioindicators of environmental pollution. In: Hodgson, E. (Ed.) Environ. Toxicol. Rev. 5. Toxicology Communications, Raleigh, NC, pp. 203–311.

Burger, J., Gochfeld, M., 1988. Effects of lead on growth in young Herring Gulls (*Larus argentatus*). J. Toxicol. Environ. Health 24, 227–236.

Burger, J., Gochfeld, M., 1995. Biomonitoring of heavy metals in the Pacific Basin using avian feathers. Environ. Toxicol. Chem. 14, 1233–1239.

Burger, J., Nisbet, I.C.T., Gochfeld, M., 1992. Metal levels in regrown feathers: assessment of contamination on the wintering and breeding grounds in the same individuals. J. Toxicol. Environ. Health 37, 363–374.

Busse, P., Baumanis, J., Leivits, A., Pakkala, H., Payevsky, V.A., Ojanen, M., 1995. Population number dynamics 1961–1990 of Sylvia species caught during autumn migration at some North and Central European bird stations. Ring 17, 11–30.

Cain, B.W., Bunck, C.M., 1983. Residues of organochlorine compounds in Starlings (*Sturnus vulgaris*), 1979. Environ. Monit. Assessm. 3, 161–172.

Cairns, D.K., 1992. Bridging the gap between ornithology and fisheries science: use of seabird data in stock assessment models. Condor 94, 811–824.

Camphuysen, C.J., Garthe, S., 2000. Seabirds and commercial fisheries: population trends of piscivorous seabirds explained? In: Kaiser, M.J., Groot, S.J. de (Eds), The Effects of Fishing on Non-Target Species and Habitats. Blackwell Science, Oxford, pp. 163–185.

Camphuysen, C.J., Heubeck, M., 2001. Marine oil pollution and beached bird survevys: the development of a sensitive monitoring instrument. Environ. Pollut. 112, 443–461.

Camphuysen, C.J.K., Ensor, K., Furness, R.W., Garthe, S., Hüppop, O., Leaper, G., Offringa, H., Tasker, M.L., 1993. Seabirds feeding on discards in winter in the North Sea. EC DG XIV research contract 92/3505, NIOZ Rapport 1993-8, Neth. Inst. for Sea Res., Texel, 140pp.

Camphuysen, K.C.J., 1998. Diurnal activity patterns and nocturnal group formation of wintering Common Murres in the central North Sea. Colonial Waterbirds 21, 406–413.

Camphuysen, K.C.J., Barreveld, H., Dahlmann, G., Franeker, J.A. van, 1999. Seabirds in the North Sea demobilized and killed by polyisobutylene (C4H8)n (PIB). Mar. Pollut. Bull. 38, 1171–1176.

Carlisle, J.C., Lamb, D.W., Toll, P.A., 1986. Breaking strength: an alternative indicator of toxic effects on avian eggshell quality. Environ. Toxicol. Chem. 5, 887–889.

Catry, P., Ruxton, G.D., Ratcliffe, N., Hamer, K.C., Furness, R.W., 1999. Short-lived repeatabilities in long-lived Great Skuas: implications for the study of individual quality. Oikos 84, 473–479.

Chapdelaine, G., Laporte, P., Nettleship, D.N., 1987. Population, productivity, and DDT contamination trends of Northern Gannets (*Sula bassanus*) at Bonaventure Island, Quebec, 1967–1984. Can. J. Zool. 65, 2992–2996.

Chapdelaine, G., Rail, J.F., 1997. Relationship between cod fishery activities and the population of Herring Gulls on the North Shore of the Gulf of St. Lawrence, Quebec, Canada. ICES J. Mar. Science 54, 708–713.

Coleman, A.E., Coleman, J.T., Coleman, P.A., Minton, C.D.T., 2001. A 39 year study of a Mute Swan *Cygnus olor* population in the English Midlands. In: Jenni, L. (Ed.), Proc. Conf. Bird Ringing 100 years, Helgoland 1999. Ardea 89 (Special issue), 123–133.

Conrad, B., 1977. Die Giftbelastung der Vogelwelt Deutschlands. Kilda-Verlag, Greven.

Conrad, B., 1978. Korrelation zwischen Embryonensterblichkeit und DDE-Kontamination beim Sperber (*Accipiter nisus*). J. Ornithol. 119, 109–111.

Conrad, B., 1979. Hohe Pestizidrückstände in tot aufgefundenen Sperbern (*Accipiter nisus*) als mögliche Todesursache. Vogelwarte 30, 21–28.

Cooch, E.G., Cooke, F., 1991. Demographic changes in a Snow Goose population: biological and management implications. In: Perrins, C.M., Lebreton, J.D., Hirons, G.J.M. (Eds), Bird Population Studies: Relevance to Conservation and Management. Oxford University Press, New York, pp. 168–189.

Cooke, A.S., 1973. Shell thinning in avian eggs by environmental pollutants. Environ. Pollut. 4, 85–150.

Cooke, A.S., 1979. Egg shell characteristics of gannets *Sula bassanus*,shags *Phalacrocorax aristotelis* and great black-backed gulls *Larus marinus* exposed to DDE and other environmental pollutants. Environ. Pollut. 19(1), 47–65.

Coulson, J.C., Thomas, C.S., 1985. Changes in the biology of the Kittiwake *Rissa tridactyla*: a 31-year study of a breeding colony. J. Anim. Ecol 54, 9–26.

Crick, H.Q.P., Sparks, T.H., 1999. Climate change related to egg-laying trends. Nature 399, 423–424.

Croxall, J.P., Rothery, P., 1991. Population regulation of seabirds: implication of their demography for conservation. In: Perrins, C.M., Lebreton, J.D., Hirons, G.J.M. (Eds), Bird Population Studies: Relevance to Conservation and Management. Oxford University Press, New York, pp. 272–296.

Custer, T.W., Erwin, R.M., Stafford, C., 1983. Organochlorine residues in Common Tern eggs from nine atlantic coast colonies,1980. Colonial Waterbirds 6, 197–204.

Dahlmann, G., Sechehaye, A., 2000. Verölte Seevögel an der deutschen Nordseeküste 1998/1999 – Ergebnisse der Ölanalysen. Seevögel 21, 11–12.

Dahlmann, G., Timm, D., Averbeck, C., Camphuysen, C.J., Skov, H., 1994. Oiled seabirds – comparative investigations on oiled seabirds and oiled beaches in the Netherlands, Denmark and Germany (1990–1993). Marine Pollution Bulletin 28, 305–310.

Denker, E., Becker, P.H., Beyerbach, M., Büthe, A., Heidmann, W.A., Staats de Yanes, G., 1994. Concentrations and metabolism of PCBs in eggs of waterbirds on the German North Sea Coast. Bull. Environ. Contam. Toxicol. 52, 220–225.

Denker, E., Büthe, A., Knüwer, H., Langgemach, T., Lepom, P., Rühling, I., 2001. Vergleich der Schadstoffbelastung in Eiern des Sperbers (*Accipiter nisus*) aus Brandenburg und Nordrhein-Westfalen, Deutschland. J. Ornithol. 142, 49–62.

Dhondt, A.A., 2001. Trade-offs between reproduction and survival in tits. In: Jenni, L. (Ed.), Proc. Conf. Bird Ringing 100 years, Helgoland 1999. Ardea 89 (Special issue), 155–166.

Dietrich, S., Büthe, A., Denker, E., Hötker, H., 1997. Organochlorines in eggs and food organisms of Avocets (*Recurvirostra avosetta*). Bull. Environ. Contam. Toxicol. 58, 219–226.

Dirksen, S., Boudewijn, T.J., Slager, L.K., Mes, R.G., Schaick, M.J.M., Voogt, P. de, 1995. Reduced breeding success of Cormorants (*Phalacrocorax carbo sinensis*) in relation to persistent organochlorine pollution of aquatic habitats in the Netherlands. Environ. Pollut. 88, 119–132.

Dobson, S., 1981. Auswirkungen chlorierter Kohlenwasserstoffe auf Physiologie und Verhalten bei Tauben im Experiment. Ökol. Vögel 3, 39–42.

Duinker, J.C., Koeman, J.H., 1978. Summary report on the distribution and effects of toxic pollutants (metals and chlorinated hydrocarbons) in the Wadden Sea. In: Essink, K., Wolff, W.J. (Eds), Pollution of the Wadden Sea area. Report 8. Wadden Sea Working Group, Leiden, pp. 45–54.

Dunnet, G.M., 1982. Oil pollution and seabird populations. Phil. Trans. R. Soc. Lond. B 316, 513–524.

Ellenberg, H. (Ed.), 1981. Greifvögel und Pestizide. Ökol. Vögel 3.

Elliott, J.E., Kenngy, S.W., Peakall, D.B., Won, H., 1990. Polychlorinated biphenyl (PCB) effects on hepatic mixed function oxidases and porphyria in birds. I. Japanese quail. Comp. Biochem. Physiol. 96, 205–210.

Elliott, J.E., Noble, D.G., Norstrom, R.J., Whitehead, P.E., Simon, M., Pearce, P.A., Peakall, D.B., 1992. Patterns and trends of organic contaminants in Canadian seabird eggs, 1968–90. In: Walker, C.H., Livingstone, D.R. (Eds), Persistent Pollutants in Marine Ecosystems. Pergamon Press, Oxford, pp. 181–194.

Elliott, J.E., Norstrom, R.J., Keith, J.A., 1988. Organochlorines and eggshell thinning in Northern Gannets (*Sula bassanus*) from eastern Canada, 1968–1984. Environ. Pollut. 52, 81–102.

Elliott, J.E., Shutt, L., 1993. Monitoring organochlorines in blood of Sharp-shinned Hawks (*Accipiter striatus*) migrating through the Great Lakes. Environ. Toxicol. Chem. 12, 241.

Endersen, J.H., Craig, G.R., Burnham, W.A., Berger, D.D., 1982. Eggshell thinning and organochlorine residues in Rocky Mountain Peregrines *Falco peregrinus*, and their prey. Can. Field. Nat. 96, 255–264.

Exo, K.M., 1994. Bedeutung des Wattenmeeres für Vögel. In: Lozan, J.L., Rachor, E., Reise, K., Westernhagen, H. v., Lenz, W. (Eds), Warnsignale aus dem Wattenmeer. Blackwell Wissenschafts-Verlag, Berlin, pp. 261–270.

Exo, K.M., Becker, P.H., Hälterlein, B., Scheufler, H., Hötker, H., Stiefel, A., Stock, M., Südbeck, P., Thorup, O., 1996. Bruterfolgsmonitoring bei Küstenvögeln. Vogelwelt 117, 287–293.

Exo, K.M., Becker, P.H., Sommer, U., 1998. Umweltchemikalien in Eiern von Binnenland- und Wattenmeerbrütern des Austernfischers (*Haematopus ostralegus*). J. Ornithol. 139, 401–405.

Fimreite, N., 1979. Accumulation and effects of mercury on birds. In: Niragu, J.O. (Ed.), The Biochemistry of Mercury in the Environment. Bd. 22. Amsterdam, Elsevier, pp. 601–627.

Fitzgerald, W.F., 1995. Is mercury increasing in the atmosphere? The need for an atmospheric mercury network. Water, Air and Soil Pollut. 80, 245–254.

Flade, M., Schwarz, J., 1999. Current status and new results from the German Common Birds Census (DDA Monitoring Program). Vogelwelt 120, Suppl., 47–51.

Fleet, D.M., Gaus, S., Hartwig, E., Potel, P., Reineking, B., Schulze-Dieckhoff, M., 1999b. PALLAS-Havarie und Seevogelsterben dominieren Spülsaumkontrollen im Winter 1998/99 – Ölopfer in der Deutschen Bucht im Zeitraum 01. Oktober 1998 bis 31. März 1999. Seevögel 20, 79–84.

Fleet, D.M., Reineking, B., 2000. Beached bird surveys at the German North Sea Coast. Wadden Sea Newsletter 2, 26–28.

Fox, G.A., Weseloh, D.V., 1987. Colonial waterbirds as bio-indicators of environmental contamination in the Great Lakes. In: Diamond, A.W., Filion, F.L. (Eds), The Value of Birds, Vol. 6. ICBP Technical Public, Cambridge, pp. 209–216.

Frank, D., 1992. The influence of feeding conditions on food provisioning of chicks in Common Terns *Sterna hirundo* nesting in the German Wadden Sea. Ardea 80, 45–55.

Frank, D., 1998. Bruterfolgsmonitoring an der Flußseeschwalbe *Sterna hirundo* als Instrument ökologischer Begleituntersuchungen zu einer Pipeline-Verlegung im Wattenmeer. Vogelwelt 119, 235–241.

Fry, D.M., 1995. Reproductive effects in birds exposed to pesticides and industrial chemicals. Environ. Health Persp. 103, 165–171.

Furness, R.W., 1985. Plastic particle pollution: accumulation by procellariiform seabirds at Scottish colonies. Mar. Pollut. Bull. 16, 103–106.

Furness, R.W., 1987. Seabirds as monitors of the marine environment. In: Diamond, A.W., Filion, F.L. (Eds), The Value of Birds. ICBP Technical Public. Vol. 6, 217–230.

Furness, R.W., 1993. Birds as monitors of pollutants. In: Furness, R.W., Greenwood, J.J.D. (Eds), Birds as Monitors of Environmental Change. Chapman and Hall, London, pp. 86–143.

Furness, R.W., Ainley, D.G., 1984. Threats to seabird populations presented by commercial fisheries. In: Croxall, J.P., Evans, P.G.H., Schreiber, R.W. (Eds), Status and Conservation of the World's Seabirds. ICBP Technical Publication No. 2, pp. 701–708.

Furness, R.W., Camphuysen, C.J., 1997. Seabirds as monitors of the marine environment. ICES J. Marine Science 54, 726–737.

Furness, R.W., Greenwood, J.J.D. (Eds), 1993. Birds as Monitors of Environmental Change. Chapman and Hall, London.

Furness, R.W., Greenwood, J.J.D., Jarvis, P.J., 1993. Can birds be used to monitor the environment? In: Furness, R.W., Greenwood, J.J.D. (Eds), Birds as Monitors of Environmental Change. Chapman and Hall, London, pp. 1–41.

Furness, R.W., Monaghan, P. (Eds), 1987. Seabird ecology. Blackie and Son, Glasgow and London.

Furness, R.W., Muirhead, S.J., Woodburn, M., 1986. Using bird feathers to measure mercury in the environment: relationships between mercury content and moult. Mar. Pollut. Bull. 17, 27–30.

Furness, R.W., Tasker, M.L. (Eds), 1999. Diets of seabirds and consequences of changes in food supply. ICES Coop. Research Report 232, 1–66.

Garthe, S., Alicki, K., Hüppop, O., Sprotte, B., 1995. Die Verbreitung und Häufigkeit ausgewählter See- und Küstenvogelarten während der Brutzeit in der südöstlichen Nordsee. J. Orn. 136, 253–266.

Garthe, S., Hüppop, O., 2000. Recent development of the seabirds at sea programme in Germany. Vogelwelt 121, 301–305.

Garthe, S., Walter, U., Tasker, M.L., Becker, P.H., Chapdelaine, G., Furness, R.W., 1999. Evaluation of the role of discards in supporting bird populations and their effects on the species composition of seabirds in the North Sea. In: Furness, R.W., Tasker, M.L. (Eds), Diets of Seabirds and Consequences of Changes in Food Supply. ICES Coop. Res. Rep. Bd. 232, pp. 29–41.

Gilbertson, M., Elliott, J.E., Peakall, D.B., 1987. Seabirds as indicators of marine pollution. In: Diamond, A.W., Filion, F.L. (Eds), The Value of Birds. ICBP Technical Public. Vol. 6, Cambridge, pp. 231–248.

Gilman, A.P., Fox, G.A., Peakall, D.B., Temple, S., Carroll, T.R., Haymes, G.T., 1977. Reproductive parameters and egg contaminant levels of Great Lakes Herring Gulls. J. Wildl. Mgmt. 41, 458–468.

Gochfeld, M., Burger, J., 1996. Effects of lead on growth and feeding behavior of young Common Terns (*Sterna hirundo*). Archives Environm. Contam. Toxic. 17, 513–517.

Grasman, K.A., Seanlon, P.F., Fox, G.A., 1998. Reproductive and physiological effects of environmental contaminants in fish-eating birds of the Great Lakes: a review of historical trends. Environ. Monit. Assessment 53, 117–145.

Graveland, J., Derwal, R. van, Balen, H. van, Nordwijk, A. van, 1994. Poor reproduction in forest passerines from decline in snail abundance. Nature 368, 446–448.

Graveland, J., Drent, R.H., 1997. Calcium availability limits breeding success of passerines on poor soils. J. Anim. Ecol. 66, 279–288.

Greenstreet, S.P.R., Becker, P.H., Barrett, R.T., Fossum, P., Leopold, M.F., 1999. Consumption of pre-recruit fish by seabirds and the possible use of this as an indicator of fish stock recruitment. In: Furness, R.W., Tasker, M.L. (Eds), Diets of Seabirds and Consequences of Changes in Food Supply. ICES Coop. Res. Rep., vol. 232, pp. 6–17.

Greenwood, J.J.D., 1999. Why conduct bird census and atlas work in Europe? Vogelwelt (Suppl.) 120, 11–23.

Greenwood, J.J.D., Baillie, S.R., Crick, H.P.Q., Marchant, J.H., Peach, W.J., 1993. Integrated population monitoring: detecting the effects of diverse changes. In: Furness, R.W., Greenwood, J.J.D. (Eds), Birds as Monitors of Environmental Change. Chapman and Hall, London, pp. 267–342.

Günther, K., Rösner, H.-U., 2000. Bestandsentwicklung der im schleswig-holsteinischen Wattenmeer rastenden Wat- und Wasservögel von 1988 bis 1999. Vogelwelt 121, 293–299.

Guillette, L.J., 1994. Endocrine-disrupting environmental contaminants and reproduction: lessons from the study of wildlife. In: Popkin, D.R., Peddle, L.J. (Eds), Women's Health Today: Perspectives on Current Research and Clinical Practice. Parthenon, New York, pp. 201–207.

Guruge, K.S., Tanabe, S., Iwata, H., Taksukawa, R., Yamagishi, S., 1996. Distribution, biomagnification, and elimination of butyltin compound residues in Common Cormorants (*Phalacrocorax carbo*) from Lake Biwa, Japan. Arch. Environ. Contam. Toxicol. 31, 210–217.

Hälterlein, B., Fleet, D.M., Henneberg, H.R., Mennebäck, T., Rasmussen, L.M., Südbeck, P., Thorup, O., Vogel, R., 1995. Anleitung zur Brutbestandserfassung von Küstenvögeln im Wattenmeerbereich. Seevögel 16, 3–24.

Hälterlein, B., Südbeck, P., Knief, W., Köppen, U., 2000. Population trends of coastal breeding birds of the German North and Baltic Sea coasts. Vogelwelt 121, 241–267

Hagemeijer, W.J.M., Blair, M.J., 1997. The EBCC atlas of European breeding birds: their distribution and abundance. Poyser, Academic Press, London.

Hamer, K.C., Furness, R.W., Caldow, R.W.G., 1991. The effects of changes in food availability on the breeding biology of great skuas *Catharacta skua* in Shetland. J. Zool., Lond. 223, 175–188.

Harris, H.J., Erdmann, T.C., Ankley, G.T., Lodge, K.B., 1993. Measures of reproductive success and polychlorinated biphenyl residues in eggs and chicks of Forster's Terns on Green Bay, Lake Michigan, Wisconsin – 1988. Arch. Environ. Contam. Toxicol. 25, 304–314.

Harris, M.P., Wanless, S., 1991. Population studies and conservation of Puffins *Fratercula arctica*. In: Perrins, C.M., Lebreton, J.D., Hirons, G.J.M. (Eds), Bird Population Studies: Relevance to Conservation and Management. Oxford University Press, New York, pp. 230–248.

Hartner, L., 1981. Wie schädigen die chlorierten Kohlenwasserstoffe die Vögel? Ökol. Vögel 3, 33–38.

Heinisch, E., Kettrup, A., Wenzel-Klein, S., (Eds), 1993. Schadstoffatlas Osteuropa: ökologisch-chemische und ökotoxikologische Fallstudien über organische Spurenstoffe in Ost-Mitteleuropa. Ecomed, Landsberg/Lech.

Helander, B., 1999. Havsörn och grasäl – fortsatt ökande bestand. Östersjä 99, 37–40.

Helander, B., Olsson, M., Reutergardh, L., 1982. Residue levels of organochlorine and mercury compounds in unhatched eggs and the relationships to breeding success in white-tailed sea eagles *Haliaeetus albicilla* in Sweden. Holarctic Ecol. 5, 349–366.

Helbig, A., Flade, M. (Eds), 1999. Bird Numbers 1999 – Where monitoring and ecological research meet. Vogelwelt (Suppl.) 120, 1–402.

Henriksen, E.O., Gabrielsen, G.W., Skaare, J.U., 1998. Validation of the use blood samples to assess tissue concentrations of organochlorines in Glaucous Gulls, *Larus hyperboreus*. Chemosphere 37, 2627–2643.

Henriksen, E.O., Gabrielsen, G.W., Skaare, J.U., Skjegstad, N., Jenssen, B.M., 1998. Relationships between PCB levels, hepatic EROD activity and plasma retinol in Glaucous Gulls, *Larus hyperboreus*. Mar. Environm. Res. 46, 45–49.

Hickey, J.J., Anderson, D.W., 1969. The Peregrine Falcon: life history and population. In: Hickey, J.J. (Ed.), Peregrine Falcon Populations: Their Biology and Decline. Univ. Wisconsin Press, Madison, pp. 3–42.

Hislop, J.R.G., Harris, M.P., 1985. Recent changes in the food of young Puffins *Fratercula arctica* on the Isle of May in relation to fish stocks. Ibis 127, 234–239.

Hoffman, D.J., Rattner, B.A., Sileo, L., Doucherty, D., Kubiak, T., 1987. Embryotoxicity, teratogenicity and arylhydro-carbon hydroxylase activity in Forster's Terns on Green Bay, Lake Michigan. Environ. Res. 42, 176–184.

Hunt, G.L., Furness, R.W. (Eds), 1996. Seabird/Fish Interactions, with particular reference to seabirds in the North Sea. ICES Coop. Research Report 216, 1–87.

ICES, 1999. Report of the Working Group on Seabird Ecology. ICES CM 1999/C:5, 1–58.

Järvinen, O., 1994. Global warming and egg size of birds. Ecography 17, 108–110.

Järvinen, O., Väisänen, R.A., 1979. Changes in bird populations as criteria of environmental changes. Holarctic Ecol. 2, 75–80.

Jefferies, D.J., Parslow, J.L.F., 1976. Thyroid changes in PCB -dosed Guillemots and their indication of one of the mechanisms of action of these materials. Environ. Pollut. 10, 293–311.

Jenni, L., Berthold, P., Peach, W., Spina, F. (Eds), 1994. Bird Ringing in Science and Environmental Management. EURING, Bologna.

Jenni-Eiermann, S., Bühler, U., Zbinden, N., 1996. Vergiftungen von Greifvögeln durch Carbofurananwendung im Ackerbau. Ornithol. Beob. 93, 69–77.

Jensen, H., Tasker, M.L., Coull, K., Emslie, D., 1994. A comparison of distribution of seabirds and prey fish stocks in the North Sea and adjacent areas. JNCC Report No. 207.

Johnels, A., Tyler, G., Westermark, T., 1979. A history of mercury levels in Swedish fauna. Ambio 8, 160–168.

Joiris, C.R., Tapia, G., Holsbeek, L., 1997. Increase of organochlorines and mercury levels in Common Guillemots *Uria aalge* during winter in the southern North Sea. Mar. Pollut. Bull. 34, 1049–1057.

Kahle, S., Becker, P.H., 1999. Bird blood as bioindicator for mercury in the environment. Chemosphere 39, 2451–2457.

Kahle, S., Becker, P.H., 2000. Die Belastung von Möwen mit Umweltchemikalien an der Deutschen Nord- und Ostseeküste in den Jahren 1995 und 1996. Seevögel 21, 47–53.

Kanyamibwa, S., Bairlein, F., Schierer, A., 1993. Comparison of survival rates between populations of the White Stork *Ciconia ciconia* in Central Europe. Ornis Scandinavica 24, 297–302.

Kettrup, A., Marth, P., 1998. Specimen banking as an environmental surveillance tool. In: Schüürmann, G., Markert, B. (Eds), Ecotoxicology: Ecological Fundamentals, Chemical Exposure, and Biological Effects. John Wiley, New York, pp. 413–450.

Kettrup, A.A.F., 2002 (2003). Environmental specimen banking. In: Markert, B.A., Breure, A.M., Zechmeister, H.G. (Eds), Bioindicators and Biomonitors. Elsevier, Oxford, pp. 775–796.

Kitaysky, A.S., Golubova, E.G., 2000. Climate change causes contrasting trends in reproductive performance of planktivorous and piscivorous alcids. J. Avian Ecol. 69, 248–262.

Koeman, J.H., Oskamp, A.A.G., Veen, J., Brouwer, E., Rooth, J., Zwart, P., 1967. Insecticides as a factor in the mortality of the Sandwich Tern (*Sterna sanvicensis*). Mededelingen Rijksfaculteit Landbourwetenschappen Gent 27, 841–854.

Köppel, C., Rennwald, E., Hirneisen, N., 1998. Rote Listen auf CD-Rom. Verlag für Interaktive Medien.

Kubiak, T.J., Harris, H.J., Smith, L.M., Schwartz, T.R., Stalling, D.L., Trick, J.A., Erdman, T.C., 1989. Microcontaminants and reproductive impairment of the Forster's Tern on Green Bay, Lake Michigan 1983. Arch. Environ. Contam. Toxicol. 18, 706–727.

Lewis, S.A., Becker, P.H., Furness, R.W., 1993. Mercury levels in eggs, internal tissues and feathers of Herring Gulls *Larus argentatus* from the German Wadden Sea. Environ. Pollut. 80, 293–299.

Lewis, S.A., Furness, R.W., 1991. Mercury accumulation and excretion in laboratory reared Black-headed Gull *Larus ridibundus* chicks. Arch. Environ. Contam. Toxicol. 21, 316–320.

Lindqvist, O. (Ed.), 1991. Mercury in the Swedish environment- recent research on causes, consequences and corrective methods. Water, Air, and Soil Pollution 55.

Lloyd, C., Tasker, M.L., Partridge, K., 1991. The status of seabirds in Britain and Ireland. Poyser, London.

Ludwig, J.P., Auman, H.J., Weseloh, D.V., Fox, G.A., Giesy, J.P., Ludwig, M.E., 1995. Evaluation of the effects of toxic chemicals in Great Lakes Cormorants: has causality been established? Colonial Waterbirds 18 (Special Public. I), 60–69.

Mammen, U., Stubbe, M., 1999. Monitoring of Raptors and Owls in Europe: a review of the first 10 years. Vogelwelt 120, Suppl., 75–78.

Marchant, J.H., 1999. Large-scale bird population monitoring in Europe: rationale, progress and prospects. Vogelwelt 120, Suppl., 25–29.

Marchant, J.H., Forrest, C., Greenwood, J.J.D., 1998. A review of large-scale generic population monitoring schemes in Europe. Bird Census News 10, 42–79.

Mason, R.P., Fitzgerald, W.F., Morel, F.F.M., 1994. The biochemical cycling of elemental mercury anthropogenic influences. Geochim. Cosmichim. Acta 58, 3191–3198.

Mattig, F.R., Bietz, H., Gießing, K., Becker, P.H., 1996. Schadstoffanreicherung im Nahrungsnetz des Wattenmeeres. UBA-Forschungsbericht 10802085/21, TPA 4.5, 1–390.

Mattig, F.R., Rösner, H.U., Gießing, K., Becker, P.H., 2000. Umweltchemikalien in Eiern des Alpenstrandläufers (*Calidris alpina*) aus Nordnorwegen im Vergleich zu Eiern von Brutvogelarten des Wattenmeeres. J. Ornithol. 141, 361–369.

Melter, J., Südbeck, P., Fleet, D.M., Rasmussen, L.M., Vogel, R.L., 1997. Changes in breeding bird numbers on census areas in the Wadden Sea 1990 until 1994. Wadden Sea Ecosystem 4, 7–93.

Meltofte, H., Blew, J., Frikke, J., Rösner, H.U., Smit, C.J., 1994. Numbers and distribution of waterbirds in the Wadden Sea. IWRB Publication 34, Wader Study Group Bull. 74, special issue, 1–192.

Michard, D., Ancel, A., Gendner, J.P., Lage, J., Le Maho, Y., Zorn, T., Gangloff, L., Schierrer, A., Struyf, K., Wey, G., 1994. Non-invasive bird tagging. Nature 376, 649–650.

Mineau, P., Fox, G.A., Norstrom, R.J., Weseloh, D.V., Hallen, D.J., Ellenton, J.A., 1984. Using the Herring Gull to monitor levels and effects of organochlorine contamination in the Canadian Great Lakes. In: Nriagu, J.O., Simmons, M.S. (Eds), Toxic Contaminants in the Great Lakes. John Wiley, New York, pp. 426–452.

Monaghan, P., 1992. Seabirds and sandeels: the conflict between exploitation and conservation in the northern North Sea. Biodiversity and Conservation 1, 98–111.

Monteiro, L.R., Furness, R.W., 1997. Accelerated increase in mercury contamination in North Atlantic mesopelagic food chains as indicated by time series of seabird feathers. Environ. Toxicol. Chem 16 2489–2493.

Monteiro, L.R., Granadeiro, J.P., Furness, R.W., 1998. Relationship between mercury levels and diet in Azores seabirds. Mar. Ecol. Prog. Ser. 166, 259–265.

Montevecchi, W.A., 1993. Birds as indicators of change in marine prey stocks. In: Furness, R.W., Greenwood, J.J.D. (Eds), Birds as Monitors of Environmental Change. Chapman and Hall, London, pp. 217–266.

Montevecchi, W.A., Myers, R.A., 1992. Monitoring fluctuations in pelagic fish availability with seabirds. CAFSAC Res. Doc. 92/94, 1–20.

Montevecchi, W.A., Myers, R.A., 1997. Centurial and decadal oceanographic influences on changes in Northern Gannet populations and diets in the north-west Atlantic: implications for climate change. ICES J. Mar. Science 54, 608–614.

Mooij, J.H., 2000. Ergebnisse des Gänsemonitorings in Deutschland und der westlichen Paläarktis von 1950 bis 1995. Vogelwelt 121, 319–330.

Moriarty, F., 1990. Ecotoxicology – The Study of Pollutants in Ecosystems. Academic Press, London.

Moriarty, F., Bell, A.A., Hanson, H., 1986. Does p,p'-DDT thin eggshells? Environ. Pollut. Ser. 40, 257–286.

Morrison, M.L., 1986. Bird populations as indicators of environmental change. In: Johnston, R.F. (Ed.), Current Ornithology. Plenum, New York, pp. 429–451.

Muñoz, J., Becker, P.H., 1999. The Kelp Gull as bioindicator of environmental chemicals in the Magellan region. A comparison with other coastal sites in Chile. Scienta Marina 63 (Suppl. 1), 495–502.

Murk, A.J., Bosveld, A.T.C., Berg, M. van den, Brouwer, A., 1994. Effects of polyhalogenated aromatic hydrocarbons (PHAHs) on biochemical parameters in chicks of the Common Tern. Aquatic Toxicol. 30, 91–115.

Newton, I. 1979. Population Ecology of Raptors. Poyser, Berkhamsted.

Newton, I. 1986. The Sparrowhawk. Poyser, Calton.

Newton, I., 2001. Causes and consequences of breeding dispersal in Sparrowhawk. In: Jenni, L. (Ed.), Proc. Conf. Bird Ringing 100 years, Helgoland 1999. Ardea 89 (Special issue), 143–154.

Newton, I., Bogan, J.A., Haas, M.B., 1989. Organochlorines and mercury in the eggs of British Peregrines *Falco peregrinus*. Ibis 131, 355–376.

Newton, I., Haas, M.B., Freestone, P., 1990. Trends in organochlorine and mercury levels in gannet eggs. Environ. Pollut. 63, 1–12.

Newton, I., Wyllie, I., 1992. Recovery of a Sparrowhawk population in relation to declining pesticide contamination. J. Appl. Ecol. 29, 476–484.

Newton, I., Wyllie, I., Asher, A., 1993. Long-term trends in organochlorine and mercury residues in some predatory birds in Britain. Environ. Pollut. 79, 143–151.

Nisbet, I.C.T., 1978. Population models for Common Terns in Massachusetts. Bird-Banding 49, 50–58.

Nisbet, I.C.T., 1994. Effects of pollution on marine birds. Bird Life Conserv. Series, 8–25.

Nisbet, I.C.T., Apanius, V., Friar, M.S., 2002. Breeding performance of very old Common Terns. J. Field Ornithol. 73, 117–240.

Nisbet, I.C.T., Reynolds, L.M., 1984. Organochlorine residues in Common Terns and associated estuarine organisms, Massachusetts, USA, 1971–81. Marine Environ. Research 11, 33–66.

O'Connor, R.J., 1991. Long-term bird population studies in the United States. Ibis 133, Suppl. 1, 36–48.

Ohlendorf, H.M., Hoffman, D.J., Saiki, M.K., Aldrich, T.W., 1986. Embryonic mortality and abnormalities of aquatic birds: apparent impacts of Selenium from irrigation drainwater. Sci. Total Environm. 52, 49–63.

Ohlendorf, H.M., Risebrough, R.W., Vermeer, K., 1978. Exposure of marine birds to environmental pollutants. Wildlife Research Report 9. Washington, DC.

Ollason, J.C., Dunnet, G.M., 1988. Variation in breeding success in fulmars. In: Clutton-Brock, T.H. (Ed.), Reproductive Success: Studies of Individual Variation in Contrasting Breeding Systems. The University of Chicago Press, Chicago, pp. 263–278.

Ormerod, S.J., Tyler, S.J., 1993. Birds as indicators of changes in water quality. In: Furness, R.W., Greenwood, J.J.D. (Eds), Birds as Monitors of Environmental Change. Chapman and Hall, London, pp. 179–216.

Oro, D., Ruiz, X., 1997. Exploitation of trawler discards by breeding seabirds in the north-western Mediterranean: differences between the Ebro Delta and the Balearic Islands areas. ICES J. Mar. Science 54, 695–707.

OSPAR, 1996. Working group on concentrations, trends and effects of substances in the marine environment (SIME): draft guidelines on standard methodology for the use of (oiled) beached birds as indicators of marine pollution. Document SIME 96/12/1, Oslo.

OSPAR, 1997. JAMP guidelines for monitoring contaminants in biota. 9/6/97, Oslo.

Pain, D.J., Sears, J., Newton, I., 1995. Lead concentrations in birds of prey in Britain. Environ. Pollut. 87, 173–180.

Paradis, E., Baillie, S.R., Sutherland, W.J., Dudley, C., Crick, H.Q.P., Gregory, R.D., 2000. Large-scale spatial variation in the breeding performance of Song Thrushes *Turdus philomelos* and Blackbirds *T. merula* in Britain. J. Appl. Ecol. 37, 73–87.

Parslow, J.L.F., Jefferies, D.J., 1977. Gannets and toxic chemicals. Brit. Birds 70, 366–372.

Peach, W.J., Baillie, S.R., Underhill, L.G., 1991. Survival of British Sedge Warblers *Acrocephalus schoenobaenus* in relation to west African rainfall. Ibis 133, 300–305.

Peakall, D., 1992. Animal Biomarkers as Pollution Indicators. Chapman and Hall, London.

Peakall, D.B., 1970. Pp′-DDT: effect on calcium metabolism and concentration of estradiol in the blood. Science 168, 592–594.

Peakall, D.B., 1973. Physiological effects of chlorinated hydrocarbons on avian species. In: Haque, R., Freed, V.H. (Eds), Environmental Dynamics of Pesticides, Plenum, New York, pp. 343–360.

Peakall, D.B., 1986. Accumulation and effects on birds. In: Waid, J.S. (Ed.), PCBs and the Environment, Vol. II. CRC Press, Boca Raton, FL.

Peakall, D.B., Boyd, H., 1987. Birds as bio-indicators of environmental conditions. In: Diamond, A.W., Filion, F.L. (Eds), The Value of Birds, Vol. 6. ICBP Technical Public. Cambridge, pp. 113–118.

Peakall, D.B., Cade, T.J., White, C.M., Haugh, J.R., 1975. Organochlorine residues in Alaskan Peregrines. Pestic. Monit. J. 8, 255–260.

Pearce, P.A., Peakall, D.B., Reynolds, L.M., 1979. Shell thinning and residues of organochlorines and mercury in seabird eggs, eastern Canada, 1970–1976. Pestic. Monitor. J. 13, 61–68.

Perrins, C.M., 1996. Eggs, egg formation and the timing of breeding. Ibis 138, 2–15.

Perrins, C.M., Lebreton, J.D., Hirons, J.M. (Eds), 1991. Bird population studies: relevance to conservation and management. Oxford University Press, New York.

Perrins, C.M., McCleery, R.H., 2001. The effect of fledging mass on the lives of Great Tits *Parus major*. In: Jenni, L. (Ed.), Proc. Conf. Bird Ringing 100 years, Helgoland 1999. Ardea Special issue.

Poot, M., Rasmussen, L.M., Roomen, M.V., Rösner, H.U., Südbeck, P., 1996. Migratory waterbirds in the Wadden Sea 1993/94. Wadden Sea Ecosystem No. 5, Common Wadden Sea Secretariat, Wilhelmshaven, pp.79.

Pruter, A.T., 1987. Sources, quantities and distribution of persistant plastics in the marine environment. Mar. Pollut. Bull. 18, 305–310.

Rasmussen, et al., 2000. Breeding birds in the Wadden Sea in 1996. Wadden Sea Ecosystem 10. Common Wadden Sea Secretariat, Wilhelmshaven, p. 122.

Ratcliffe, D.A., 1967. Decrease in eggshell weight in certain birds of prey. Nature 215, 208–210.

Ratcliffe, D.A., 1970. Changes attributable to pesticides in egg breakage frequency and eggshell thickness in some Britain birds. J. Appl. Ecol. 7, 67–115.

Ratcliffe, D.A., 1980. The peregrine falcon. Poyser, Calton.

Reid, J.B. (Ed.), 1997. Seabirds in the marine environment. ICES J. Marine Science 54, 503–739.

Reineking, B., Vauk, G., 1982. Seevögel- Opfer der Ölpest. Niederelbe, Otterndorf.

Renzoni, A., Focardi, S., Fossi, C., Leonzio, C., Mayol, J., 1986. Comparison between concentrations of mercury and other contaminants in eggs and tissues of Cory`s Shearwater (*Calonectris diomedea*) collected on atlantic and mediterranean islands. Environ. Pollut. 40, 17–35.

Rheinwald, G. 1993. Atlas der Verbreitung und Häufigkeit der Brutvögel Deutschlands: Kartierung um 1985. Schriftenreihe des Dachverband Deutscher Avifaunisten 12.

Rheinwald, G., Erhard, R., Wink, M., 1997. Untersuchungen zu Bestandsänderungen von Brutvögeln im Großraum Bonn durch Rasterkartierung und Punkt-Stopp-Erfassung. Charadrius 33, 179–195.

Risebrough, R.W., 1986. Pesticides and bird populations. In: Johnston, R.F. (Ed.), Current Ornithology. Plenum, New York, pp. 397–427.

Rösner, H.U., Fleet, D.M., Rasmussen, L.M., Roomen, M. van, Südbeck, P., Vogel, R., 1999. 5.1.1. Birds. In: De Jong, F., Bakker, J., Berkel, C.J.M. van, Dankers, N.M.J.A., Dahl, K., Gätje, C., Marencic, H., Potel, P. (Eds), 1999 Wadden Sea quality status report. Wadden Sea Ecosystem No. 9 Common Wadden Sea Secretariat, Trilateral Monitoring Group, Quality Status Group. Wilhelmshaven, Germany. pp. 85–117.

Rösner, H.-U., Günther, K., 1996. Monitoring von rastenden Wat- und Wasservögeln im Wattenmeer. Vogelwelt 117, 295–301.

Rösner, H.U., Roomen, M. van, Südbeck, P., Rasmussen, L.M., 1994. Migratory Waterbirds in the Wadden Sea 1992/93. Wadden Sea Ecosystem 2, pp. 1–72.

Rose, P.M., Scott, D.A., 1997. Waterfowl Population Estimates. 2nd edn. Wetlands International Publ. 44, Wageningen, Niederlande.

Ryan, P.G., 1988. Intraspecific variation in plastic ingestion by seabirds and the flux of plastic through seabird populations. Condor 90, 446–452.

Ryckman, D.P., Weseloh, D.V.C., Bishop, C.A., 1997. Contaminants in Herring Gull eggs from the Great Lakes: 25 years of monitoring levels and effects. In: Minister of Public Works and Goverment Services Canada Great Lakes. Fact Sheet, pp. 1–12.

Ryckman, D.P., Weseloh, D.V., Hamr, P., Fox, G.A., Collins, B., Ewins, P.J., Norstrom, R.J., 1998. Spatial and temporal trends in organochlorine contamination and bill deformities in Double-crested Cormorants (*Phalacrocorax auritus*) from the Canadian Great Lakes. Environ. Monit. Asses. 53, 169–195.

Safe, S., 1990. Polychlorinated biphenyls (PCBs), dibenzo-p-dioxins (PCDDs), dibenzofurans (PCDFs), and related compounds: environmental and mechanistic considerations which support the development of toxic equivalency factors (TEFs). CRC Crit. Reviews Toxicol. 21, 51–88.

Sagerup, K., Henriksen, E.O., Skorping, A., Skaare, J.U., Gabrielsen, G.W., 2000. Intensity of parasitic nematodes increases with organochlorine levels in the Glaucous Gull. J. Appl. Ecol. 37, 532–539.

Sauer, J.R., Link, W.A., 1999. Regional analysis of population trajectories from the North American Breeding Bird Survey. Vogelwelt (Suppl.) 120, 31–38.

Scheuhammer, A.M., 1987. The chronic toxicity of Aluminium, Cadmium, Mercury, and Lead in birds: a review. Environ. Pollut. 46, 263–295.

Schilling, F., König, C., 1980. Die Biozidbelastung des Wanderfalken (*Falco perigrinus*) in Baden-Württemberg und ihre Auswirkung auf die Populationsentwicklung. J. Ornithol. 121, 1–35.

Schilling, F., Rockenbauch, D., 1985. Der Wanderfalke in Baden-Württemberg gerettet. Beih. Veröff. Naturschutz Landschaftspflege Bad.-Württ. 46, 1–80.

Schulz, H. (Ed.), 1999. Weißstorch im Aufwind? – White Storks on the up? Proceedings Internat. Symp. on the White Stork, Hamburg 1996. NABU, Bonn.

Sibly, R.M., Newton, I., Walker, C.H., 2000. Effects of dieldrin on population growth rates of Sparrowhawks 1963–1986. J. Appl. Ecol. 37, 540–546.

Siriwardena, G.M., Baillie, S.R., Crick, H.Q.P., Wislon, J.D., 2000. The importance of variation in the breeding performance of seed-eating birds in determining their population trends on farmland. J. Appl. Ecol. 37, 128–148.

Skov, H., Durinck, J., Leopold, M.F., Tasker, M.L., 1995. Important birds areas for seabirds in the North Sea including the Channel and the Kattegat. BirdLife International, Cambridge.

SOVON, 1987. Atlas van de Nederlandse Vogels. SOVON, Arnhem.

Spear, L.B., Ainley, D.G., Ribic, C.A., 1995. Incidence of plastic in seabirds from the tropical Pacific, 1984–91: relation with distribution of species, sex, age, season, year and body weight. Mar. Environ. Res. 40, 123–146.

Stienen, E.W.M., Beers van, P.W.M., Brenninkmeijer, A., Habraken, J.M.P.M., Raaijmakers, M.H.J.E., Tienen van, P.G.M., 2000. Reflections of a specialist: patterns in food provisioning and foraging conditions in Sandwich Terns *Sterna sandvicensis*. Ardea 88, 33–49.

Stienen, E.W.M., Brenninkmeijer, A., 1998. Entwicklung des Brutbestandes der Flußseeschwalbe *Sterna hirundo* an der niederländischen Küste. Vogelwelt 119, 165–168.

Stone, C.J., Webb, A., Barton, C., Ratcliffe, N., Reed, T.C., Tasker, M.L., Camphuysen, C.J., Pienkowski, M.W., 1995. An atlas of seabird distribution in north-west European waters. JNCC, Peterborough.

Stubbe, M., Mammen, U., Gedeon, K., 1996. Das Monitoring-Programm Greifvögel und Eulen Europas. Vogelwelt 117, 261–267.

Südbeck, P., Hälterlein, B., Knief, W., Köppen, U., 1998. Bestandsentwicklung von Fluß- *Sterna hirundo* und Küstenseeschwalbe *S. paradisaea* an den Deutschen Küsten. Vogelwelt 119, 147–163.

Sudfeldt, C., Anthes, N., Wahl, J., 2000. Stand und Perspektiven des Wasservogelmonitorings in Deutschland. Vogelwelt 121, 307–317.

Swain, E.B., Engstrom, D.R., Brigham, M.E., Henning, T.A., Brezonik, P.L., 1992. Increasing rates of atmospheric mercury deposition in midcontinental North America. Science 257, 784–787.

Swennen, C., 1972. Chlorinated hydrocarbons attacked the Eider population in the Netherlands. TNO-nieuws 27, 556–560.

Tasker, M.L., Hope Jones, P., Dixon, T., Blake, B.F., 1984. Counting seabirds at sea from ships: a review of methods employed and a suggestion for a standardized approach. Auk 101, 567–577.

Tasker, M.L., Webb, A., Hall, A.J., Pienkowski, M.W., Langslow, D.R., 1987. Seabirds at the North Sea. Report Seabirds at Sea Project, Nature Conservancy Council.

Tejning, S., 1967. Biological effects of methyl mercury dicyandiamide-treated grain in the domestic fowl *Gallus gallus* L. Oikos Suppl. 8, 1–116.

Thompson, D.R., Becker, P.H., Furness, R.W., 1993b. Long-term changes in mercury concentrations in Herring Gulls *Larus argentatus* and Common Terns *Sterna hirundo* from the German North Sea coast. J. Appl. Ecol. 30, 316–320.

Thompson, D.R., Furness, R.W., 1989. Comparison of the levels of total and organic mercury in seabird feathers. Mar. Pollut. Bull. 20, 577–579.

Thompson, D.R., Furness, R.W., Lewis, S.A., 1993a. Temporal and spatial variation in mercury concentrations in some albatrosses and petrels from the sub-Antarctic. Polar. Biol. 13, 239–244.

Thompson, D.R., Furness, R.W., Lewis, S.A., 1995. Diets and long-term changes in $\delta15N$ and $\delta13C$ values in Northern Fulmars *Fulmarus glacialis* from two northeast Atlantic colonies. Mar. Ecol. Prog. Ser. 125, 3–11.

Thompson, D.R., Furness, R.W., Walsh, P.M., 1992. Historical changes in mercury concentrations in the marine ecosystem of the north and north-east Atlantic Ocean as indicated by seabird feathers. J. Appl. Ecol. 29, 79–84.

Thomson, D.L., Baillie, S.R., Peach, W.J., 1997. The demography and age-specific annual survival of song thrushes during periods of population stability and decline. J. Animal Ecol. 66, 414–424.

Thyen, S., Becker, P.H., 2000. Aktuelle Ergebnisse des Schadstoffmonitorings mit Küstenvögeln im Wattenmeer. Vogelwelt 121, 281–291.

Thyen, S., Becker, P.H., Behmann, H., 2000a. Organochlorine and mercury contamination of Little Terns (*Sterna albifrons*) breeding at the western Baltic Sea, 1978–1996. Environ. Pollut. 108, 225–238.

Thyen, S., Becker, P.H., Exo, K.-M., Hälterlein, B., Hötker, H., Südbeck, P., 1998. Monitoring breeding success of coastal birds. Final Report of the Pilot Study 1996–1997. Wadden Sea Ecosystem 8. Common Wadden Sea Secretariat, Wilhelmshaven, 7–55.

Thyen, S., Becker, P.H., Exo, K.-M., Hälterlein, B., Hötker, H., Südbeck, P., 2000b. Bruterfolgsmonitoring bei Küstenvögeln im Wattenmeer 1996 und 1997. Vogelwelt 121, 269–280.

Trilateral Monitoring and Assessment Group, 1997. TMAP Manual. The Trilateral Monitoring and Assessment Programme (TMAP). Common Wadden Sea Secretariat, Wilhelmshaven.

Umweltbundesamt, 1995. Umweltchemikalien mit endokriner Wirkung. UBA Texte 65/95.

Umweltprobenbank, 1996. Jahresbericht der Bank für Umweltproben 1995. Jülich.

Uttley, J.D., 1992. Food supply and allocation of parental effort in Arctic Terns *Sterna paradisaea*. Ardea 80, 83–91.

Väisänen, R.A., Solonen, T., 1997. Suomen talvilinnuston 40-vuotismuutokset (Population trends of 100 winter bird species in Finland in 1957–1996). Linnut-vuosikirja 1996, 70–97.

Walker, C.H., 1994. The ecotoxicology of persistent pollutants in marine fish-eating birds. In: Walker, C.H., Livingstone, D.R. (Eds), Persistent Pollutants in Marine Ecosystems. Pergamon Press, Oxford, pp. 211–232.

Walsh, P.M., Avery, M., Heubeck, M., 1991. Monitoring of seabirds numbers and breeding success. In: Stroud, D., Glue, D. (Eds), Britain's Birds in 1989–1990: The Conservation and Monitoring Review. British Trust for Ornithology, Norfolk, pp. 96–103.

Walsh, P.M., Halley, D.J., Harris, M.P., Nevo del A., Sim, I.M.W., Tasker, M.L., 1995. Seabird monitoring handbook for Britain and Ireland. Seabird monitoring handbook 1995.

Walter, U., Becker, P.H., 1997. Occurrence and consumption of seabirds scavenging on shrimp trawler discards in the Wadden Sea. ICES J. Marine Science 54, 684–694.

Webb, A., Durinck, J., 1992. Counting birds from ships. In: Komdeur, J., Bertelsen, J., Cracknell, G. (Eds), Manual for Aeroplane and Ship Surveys of Waterfowl and Seabirds. IWRB Spec. Publ. 19, 24–37.

Wegner, P., 2000. Die Biozidbelastung von Eiern des Wanderfalken (*Falco peregrinus*) aus Nordrhein-Westfalen und dem nördlichen Rheinland-Pfalz im Vergleich zu anderen Bundesländern. Charadrius 36, 113–125.

Weimer, V., Schmidt, H., 1998. Untersuchungen zur Eiqualität bei der Kohlmeise (*Marus major*) in Abhängigkeit von der Bodenbeschaffenheit. J. Ornithol. 139, 3–9.

Weseloh, D.V., Custer, T.W., Braune, B.M., 1989. Organochlorine contaminants in eggs of Common Terns from the Canadian Great Lakes, 1981. Environ. Pollut. 59, 141–160.

Weseloh, D.V., Elliott, J.E., Olive, J.H., 1988. Herring Gull surveillance program. Can. Wildl. Serv. D-2016 G.

Weseloh, D.V., Ewins, P.J., Struger, J., Mineau, P., Bishop, C.A., Postupalsky, S., Ludwig, J.P., 1995. Double-crested Cormorants of the Great Lakes: changes in population size, breeding distribution and reproductive output between 1913 and 1991. Colonial Waterbirds 18 (Special Publ. 1), 48–59.

Weseloh, D.V., Mineau, P., Struger, J., 1990. Geographical distribution of contaminants and productivity measures of Herring Gulls in the Great Lakes: Lake Erie and connecting channels. Sci. Total Environ. 91, 141–159.

Winkel, W., Hudde, H., 1997. Long-term trends in reproductive traits of Tits (*Parus major, P. caeruleus*) and Pied Flycatchers *Ficedula hypoleuca*. J. Avian Biology 28, 187–190.

World Conservation Monitoring Centre (Eds), 1990. 1990 IUCN red list of threatened animals. Gland, Cambridge.

Zang, H., 1998. Auswirkungen des "Sauren Regens" (Waldsterben) auf eine Kohlmeisen- (*Parus marjor*) Population in den Hochlagen des Harzes. J. Ornithol. 139, 263–268.

Chapter 20

Mammals as biomonitors

Frieda Tataruch and Horst Kierdorf

Abstract

Mammals represent useful organisms for biomonitoring purposes. Whereas plants as biomonitors are indicating the burden with environmental pollutants over a certain time span, animals can be used when both temporal and spatial information is required. By analysing the concentrations of accumulating contaminants in the body tissues of animals, information is obtained concerning the quality of the ecosystem of which the animal is an intrinsic part. Among the numerous members of the class of mammals, free ranging animals or "wildlife" fit best the requirements for a biomonitor. This is because they depend exclusively on the quality of food, water and air in their habitat (they consume flora or fauna that reflect local soil, water and air contamination). Contaminations will therefore influence the animal and potentially also its health.

A great number of environmental contaminants accumulate in animal organisms. Among these are the non-essential, potentially toxic trace elements such as lead, cadmium, mercury, arsenic and fluorine, radionuclides (like the caesium-isotopes 137 and 134, and strontium-90) as well as many organohalogen compounds such as pesticides and PCBs.

Most papers in the ecotoxicological literature deal with herbivorous species; the use of omnivorous and carnivorous species is limited by the fact that their diet composition is more variable and often influenced by food chain effects. Among the herbivores, species like roe deer, brown hare, moose, etc. fulfil the requirements for a good biomonitor and have the advantage that a sufficient amount of samples can be obtained by the regular hunting bags, making additional killing unnecessary.

An overview is given on the monitoring of lead, cadmium, mercury, fluorine and radiocaesium with different animal species and the temporal trends of environmental contamination are discussed. One section deals with the use of deer antlers (red deer and roe deer) for monitoring over very long periods.

Keywords: mammals, heavy metals, lead, cadmium, mercury, fluorine, radionuclides, radiocaesium, radiostrontium, organochlorines, PCBs, DDT, DDE, antler, deer, wild boar, brown hare, small mammals, dental fluorosis, osteofluorosis, historical monitoring

1. Introduction

Whereas plants, often used as biomonitors, are indicating the burden of environmental pollutants over a certain time span, animals can be used when both temporal and spatial information is required.

Some essential criteria are to be fulfilled by animal organisms selected for biomonitoring. These criteria have been previously discussed and formulated in detail by Ellenberg (1991):

1. Spatial representativeness
   - Distribution: The species should be wide spread, permitting large scale comparisons.
   - Abundance: Enough individuals must be available for field and laboratory studies.
   - Site fidelity: A prerequisite for making area- and site-related statements.
   - Euryoecia: If a species is only broadly limited by ecological conditions, this permits identification of site-specific characteristics and gradients.

2. Availability, practicability of specimen collection, and experimental suitability
   - Obtainable biomass: The size or number of individuals must be sufficiently large (may require statistical analysis) to permit taking of samples for laboratory analysis.
   - Experimental suitability: In order to corroborate findings made in the field and study special aspects (physiology, diet – related effects), it must be possible to manipulate the species for experiments in captivity under controlled conditions.
   - Longevity: Temporal changes can best be interpreted within the same individuals.

3. Ecological representativeness
   - Reference characteristics: The physiology and annual and daily rhythms of the species must be known, at least in principle, and the gender and age of individuals must be readily determinable.
   - Sensitivity: The population must be tolerant to the pollutants under study, yet at the same time react sensitively to them.
   - Similarity to humans: Given the role of bioindicators as sentinels for human health risks, a high degree of physiological similarity to human beings is necessary.

4. Trophic level and role of study animals within an ecosystem
   - Feeding habits: The trophic level occupied by the species in the ecosystem must be known. If the composition of the diet of individuals can be ascertained, then its specific role within the ecosystem can be identified with greater precision.

5. Reproducibility of specimen collection
   - Standardised specimen-collection procedures are a prerequisite if results from different areas are to be compared. They permit direct comparisons with studies that have used the same specimen structures at other locations.

Although these criteria were mainly formulated for birds as biomonitors they can also be used to mammals:

- Like all animals, mammals integrate their contaminant burden over time and area.
- The monitoring area depends on the habitat size of the species; by selection of different species, the monitoring area is variable.
- For many species, established physiological parameters are already available.
- The feeding preferences of most species are well known.
- Reliable methods for age determination exist.

- The body size of most species is big enough to provide adequate tissue samples for reliable chemical analyses.
- Many species are "harvested" regularly by hunting, so no additional killing is necessary.
- Animals accumulate trace metals, a parameter that can not always be measured using plants.
- The burden of an animal is influenced by all components of an ecosystem and therefore reflects the contamination of the total ecosystem.
- Humans are mammals too – results obtained on mammals therefore have a high degree of validity for humans.
- Due to their long life span, exposure effects over time can be demonstrated in many mammalian species – this again highlights the comparability of these species to the human situation.

Besides the above listed advantages of mammals there are also some disadvantages, which must be considered:

- The tissue level of pollutants depends on endogenic factors, like age, sex, and stress as well as on exogenic factors, e.g. season, habitat, etc. However, the influences of exogenic factors are one of the prerequisites for using a species as a biomonitor. A larger number of individuals is required to account for the influence of the diverse exogenic factors.
- High sample numbers increase the costs of monitoring.
- Sampling often requires the death of the animal, so only one sample (one set of samples) per animal can be obtained.
- Mammals are much slower to react on acidic pollutants than plants.

Within mammals, free ranging animals or "wildlife" best fit the requirements for a biomonitor. This is because they depend exclusively on the quality of food, water and air in their habitat (they consume flora or fauna that reflect the local contamination of soil, water and air). Any contamination present will therefore influence the animal and can have an effect on its health. Domestic animals, like cattle, pigs or sheep, can also be used for assessing environmental quality, providing they feed only on foodstuffs originating from their immediate environment. Although grazing sheep still fall in this category, due to modern agricultural practice pigs and most cattle no longer do. Their feed consists mainly of components which are produced not only outside of the original living area, but possibly even in other continents.

The same problem, which is even more pronounced in industrialised countries, exists for pets (e.g. cats and dogs). In spite of these disadvantages, some authors describe pets as a useful subgroup of biomonitors or "sentinels" (Buck, 1979; O'Brien et al., 1993).

By analysing the concentrations of accumulating contaminants in the body tissues of animals, information is obtained about the quality of the ecosystem of which the animal is an intrinsic part. As already mentioned, the degree of contamination of its habitat is the main exogenic factor influencing the contaminant burden of an animal. Among the contaminants that can be monitored using wildlife species as bioindicators are non-essential, potentially toxic trace elements such as lead, cadmium, mercury, arsenic and fluorine, radionuclides (like the caesium-isotopes 137 and 134, strontium-90) as well as many organohalogen compounds (e.g. pesticides and PCBs).

The monitoring of diverse ecosystems can be accomplished by utilising several different species of mammals depending on their habitat and size of home range. Although many wildlife species could in principle be used for biomonitoring, the available ecotoxicological literature concentrates on only a few species.

The suitability of a species for study is influenced by different factors: Herbivores accumulate metals, fluorine and radionuclides to a higher extent than carnivores, which in contrast are better monitors for lipophilic organohalogen compounds. Additionally, the size of the home range is important. Home range size should be relatively constant for all seasons and over the animal's life span, i.e. an individual that does not roam into distant habitats during mating should be chosen. This factor would therefore exclude wild boar (*Sus scrofa*) as a monitoring species since it shows a sex-dependent difference of home range size. Groups of female boar with their juveniles have a relatively constant home range of about 250 ha, whereas males live solitary with no stable territory and migrate over many kilometres.

The suitability of red deer (*Cervus elaphus*) is also limited by the fact that in undisturbed areas the winter ranges differ from the summer ranges as the animals move over many kilometres to areas with a better forage supply. Nowadays, especially in Central Europe, anthropogenic barriers like roads and human settlements inhibit this migration, so the animals have to stay in the same habitat during the entire year. This makes them in fact more useful for monitoring studies. However, if the food shortage in winter is compensated for by additional feeding provided by hunters this can alter the results, since the supplementary foodstuff is often imported from other areas.

Any population used for monitoring studies should not be under risk of extinction. The populations of the European brown hare (*Lepus europaeus*) have been declining for some decades and therefore the species is no longer hunted in many parts of its range. However, in Central Europe with fluctuating but more or less unthreatened populations, this species represents one of the best biomonitors for agriculturally used land.

A very good biomonitor is the roe deer (*Capreolus capreolus*), the species being abundant in nearly all parts of Europe in agricultural as well as in forest areas. The diversity of its habitat reaches from the plains to the timber line and even up to mountainous areas. Its home range normally has a size of less than 50 ha and is rather stable throughout lifetime. The physiology of this ruminant species has been under investigation for many years and is well known by now (Anderson et al., 1998). The roe deer's abundance has considerably increased during recent decades and hunting allows for the collection of enough tissue samples. The hunting period normally comprises 7 to 8 months, so that seasonal effects can also be investigated (Holm, 1993). The established methods for age determination are sufficiently reliable for studying age dependent influences.

Higher up in mountainous regions, the chamois (*Rupicapra rupicapra*), living in the higher forest zones and above the timber line, can also be used for biomonitoring. In boreal zones, like Scandinavia or North America, the moose (*Alces alces*) and the reindeer (*Rangifer tarandus*) have been used as biomonitors. In case of reindeer, however, the seasonal migration of many populations has to be taken into account.

Several taxa of small mammals, like shrews and rodents can also be used for biomonitoring of terrestrial ecosystems. Shrews (Soricidae), which would be very suitable biomonitors, are however threatened in many areas. Therefore they were replaced in monitoring studies by unthreatened species like voles (e.g. *Clethrionomys glareolus*) or white-footed and deer mice *(Peromyscus* spec.). The advantages of these species are that they are very abundant, do not migrate over long distances, have a widespread distribution and have generalised food habits. Their relatively short life span and high reproduction rate allow assessment of immediate effects, in that each new generation reflects the present type and amount of contaminant levels in the environment as well as an assessment of long term effects that may accrue over several generations of exposure (Beardsley et al., 1978).

The disadvantages of these animals are their small body size (so that very often the whole carcasses have to be analysed) as well as the fact that trapping techniques involve great expenses with relatively low success. Species like the insectivorous Soricidae are less abundant, threatened, and more difficult to trap than other species. Unfortunately, their low numbers make statistical analyses difficult, although they show higher accumulation rates of many contaminants.

2. Lead

Lead (Pb) is a naturally occurring heavy metal, which is abundant in the earth's crust. The main source of lead found in the environment is the combustion of leaded gasoline. This is because since the 1920s organolead compounds (tetramethyl and tetraethyl lead) have been added to gasoline as an anti-knocking agent. It has been estimated that emissions from internal combustion engines have been responsible for up to 90% of environmental lead pollution. Although the addition of lead compounds to gasoline is presently decreasing drastically in most parts of the Western world, leaded gasoline is still used in many other countries.

In addition to mining, smelting and refining of lead (including the manufacture of lead-containing products), the combustion of coal and oil also contributes to lead accumulation in, and pollution of the environment.

2.1. Distribution in plants

Atmospheric lead can be deposited onto plants. However, the translocation of this surface bound lead into the inner parts of the plant is negligible. Marked seasonal differences of lead pollution of plants have been demonstrated. In spring, plants have a relatively low lead contamination on the upper and lower surface areas of their leaves and shoots. During the growth period, more and more lead is deposited onto the plant surfaces, resulting in an increase of the total lead burden (Hecht et al., 1984). As a consequence, lead contents of feeding plants are normally higher in autumn than in spring.

Lead uptake from soils with natural concentrations (15–30 μg/g) is low and limited to the roots. This is due to the fact that lead is generally retained in the upper layers of soils, especially those with a high humus fraction and pH values above 5. Only

Table 1. Suitability of several mammalian species as biomonitors.

	Distribution	Habitat	Home range size	Feeding strategy	Availability of tissue samples
Roe deer (*Capreolus capreolus*)	Europe, Asia, low lands to high mountains	Woodland, fields	–50 ha, high fidelity	Herbivorous, concentrate selectors	Game species, stable to increasing populations
Red deer (*Cervus elaphus*)	Europe (exc. Finland)	Woodland	>5 km^2, very variable	Herbivorous, mixed feeder (grasses)	Game species
Brown hare (*Lepus europaeus*)	Europe, Asia, N. Africa	Fields, grassland, woodland	≈25 ha	Herbivorous	Game species, regionally declining populations
Chamois (*Rupicapra rupicapra*)	Mountains Europe, Asia	Sub-alpine, alpine, above timberline	high fidelity	Herbivorous, mixed feeder, partly concentrate selector	Game species, stable up to increasing populations
Moose (*Alces alces*)	N.E. Europe, N. America, N. Asia	Boreal forest	≈100 ha, (5–1000 ha), high fidelity	Herbivorous (trees, shrubs, heather, lichen), browser	Game species, stable to increasing populations
Wild boar (*Sus scrofa*)	Europe, Asia, N. Africa	Woodland, fields	≈250 ha, females high fidelity	Omnivorous	Game species, increasing populations
Shrews (*Soricidae*)	Europe, N. America	Woodland, grass-land, open areas	<3 ha	Insectivorous	Trapping (low success)
Voles (*Arvicolidae*)	America, Europe (exc. Ireland, S. Europe)	Woodland, grass-land, open areas	<3 ha	Mostly herbivorous, some species omnivorous	Trapping
Mice (*Muridae*)	Europe, America, Asia, Africa,	Woodland, grass-land, fields	<3 ha	Omnivorous	Trapping

plants growing on soils that contain very high amounts of lead accumulate larger amounts of Pb from the soil. The lead burden of a herbivore's forage is thus in general influenced more by air-borne than by soil-borne lead.

2.2. Absorption and distribution of lead within mammalian organs

In addition to uptake by feeding and drinking, animals can also absorb and accumulate lead by breathing. Absorption via the gastro-intestinal tract is relatively low (in the range of 5–10%) in comparison to absorption via the lungs which can reach up to 50%. However, in the latter case the amount of resorption depends on the size of the lead containing particles. Studies on domestic animals have shown how diet can influence lead absorption. Low calcium and vitamin D levels as well as an insufficient iron and protein supply increase the absorption rate of lead (Mahaffey, 1980).

The absorbed lead enters the bloodstream (up to 95% bound to erythrocytes) from where it is distributed to various organs and tissues. The target organs for lead are the skeleton, the liver and the kidneys, whereas muscle tissue has a remarkably lower affinity. For biomonitoring, usually the liver and kidneys are analysed as they are easily obtainable. Bones, which would in fact allow better analysis of the long term effects, are not often examined due to the more laborious sample collection and preparation techniques. Additionally, the half life of lead in different types of bone differs greatly. It varies from about 20 days in trabecular bone to 10 to 20 years in cortical bone (Ewers and Schlipköter, 1991). For small animals, whole-body samples were also analysed. The results of these analyses are difficult to interpret but are at least partly consistent with results from studies using liver and kidney samples. Blood lead values only give information on the current lead uptake and are not indicative of long term exposure. Some research groups also used hair samples for monitoring the lead burden of animals. The main disadvantages of hair analysis are the difficulties in removing external dust and dirt. There are no certified and standardised cleaning methods for hair samples available.

As previously mentioned, usually liver and kidney samples are analysed in order to monitor the lead exposure of an animal. The distribution pattern of lead within these organs is not as uniform as it is found with cadmium for example. This in turn also varies depending on the species. For ruminants like red and roe deer, many authors report variable ratios of lead between the liver and kidneys, whereas for brown hares, rabbits and wild boar higher levels are always found in the liver.

Any tissue sample obtained from animals which have been shot should be checked carefully for any damage by the projectile. Particles from the projectile, or abrasion from the lead-containing ammunition can cause a secondary lead contamination of the organ. This in turn may yield erroneous results (Moreth and Hecht, 1981; Hecht et al., 1984). Under such circumstances, samples suspected or known to be contaminated by projectile parts should be excluded from further analyses as even microscopically small particles can give rise to false and unrealistically high lead values. To avoid any such negative parameters, results should be based on at least two different and separately collected samples of the tissue in question. Several authors have recommended analysis of the kidneys. This is because the surrounding kidney fat protects them and therefore the risk of secondary contamination by the projectile is significantly lower than for the liver (Kleiminger and Holm, 1985; Tataruch, 1993b). The impact energy of lead shot that is used for hunting small game like brown hares and rabbits is remarkably lower than that of ball-cartridges. As a result, lead shot does not penetrate as deeply into the musculature and secondary contamination is normally limited to a small area of a few millimetres around the point of impact.

2.3. Influence of age and sex

Neither age nor sex have been proven to be significant factors for lead concentration in soft tissues but an age dependent accumulation is to be expected in the calcified tissues of the skeleton due to the long half life of lead in bone.

2.4. Influence of season

As already mentioned, the lead uptake of animals with their diet is higher in autumn than in spring. Corresponding seasonal variations in the lead content of organs have been found and have to be taken into consideration for monitoring programmes (Tataruch, 1984).

2.5. Influence of species and diet selection

One of the most suitable mammalian indicator species for lead is the brown hare, which accumulates lead very strongly. In comparison with other species living within the same habitat, the lead content of hares' organs is highest, followed by that of wild boars (Holm, 1984).

There are two probable explanations for this. First, that both species are monogastric with lower pH values in their stomach than e.g. in ruminants. This acidity results in a better resorption of minerals contained in their diet and therefore a higher accumulation of minerals. Second, they are in closer contact with the soil. The omnivorous wild boar feeds from the upper layers of soil. As a result, its diet contains considerable amounts of soil.

The lead contamination of the strictly herbivorous brown hare is not only influenced by the lead content of its feeding plants but also by the fact that this animal is much more exposed to soil particles and dust than many other herbivorous species. The reasons for this are as follows. It lives mainly on farmland, rests in hollows on bare soil and spends a large part of the day grooming its fur. Its feeding consists mainly of plants growing directly above ground, heavily contaminated by soil particles. This species-specific behaviour and feeding selection results in a higher uptake of soil, reflected by elevated concentrations of crude ash in the hares' stomach contents (Onderscheka et al., 1982). These same factors are also responsible for the higher lead content of the wild rabbit (Schinner, 1981; Hecht et al., 1984).

Ruminants generally have lower lead concentrations in the inner organs and the distribution pattern between the liver and kidneys varies. Several authors report higher lead levels for red deer than for roe deer from the same habitat (Table 3). A possible interpretation for this observation is that roe deer mainly feed on young shoots and buds which grow at a greater distance above the ground than the food items selected by red deer (Holm, 1984).

Table 2. Lead concentrations (mg/kg = ppm) in liver, kidneys and muscle of four species in the same biotope (Holm, 1984).

		n	med	x	$\pm s$	min	max
Roe deer	L	114	0.209	0.350	0.651	0.003	4.605
	K	113	0.294	0.454	0.599	0.053	4.450
	M	81	0.032	0.059	0.081	0.001	0.527
Red deer	L	187	0.689	1.016	1.053	0.018	7.241
	K	188	1.060	1.243	0.910	0.057	7.543
	M	134	0.041	0.061	0.086	0	0.729
Wild boar	L	26	0.910	1.337	1.167	0.356	4.949
	K	27	1.268	1.473	1.036	0.141	3.600
	M	16	0.040	0.060	0.044	0.020	0.185
Brown hare	L	17	5.824	6.038	3.185	1.031	12.698
	K	18	4.350	7.200	7.363	0.514	32.973
	M	16	0.142	0.210	0.195	0.040	0.795

med = median, x = mean, $\pm s$ = standard deviation, min = lowest value, max = highest value L = liver, K = kidney, M = muscle, 0: below detection limit

Table 3. Lead concentrations (mg/kg = ppm) in liver and kidneys of three ruminant species in an alpine habitat in Austria (Tataruch, unpublished).

		n	med	x	$\pm s$	min	max
Red deer	L	245	0.059	0.072	0.079	0.001	0.938
	K	223	0.057	0.070	0.053	0.002	0.355
Roe deer	L	165	0.040	0.048	0.041	0.004	0.362
	K	162	0.052	0.059	0.039	0.001	0.277
Chamois	L	278	0.042	0.052	0.048	0.001	0.311
	K	286	0.050	0.060	0.047	0.001	0.485

med = median, x = mean, $\pm s$ = standard deviation, min = lowest value, max = highest value L = liver, K = kidney

During the time when tetraethyl lead was a common gasoline additive, studies of environmental pollution by automobile exhausts were made using several species of small mammals. These studies were reviewed by Talmage and Walton in 1991. In general, the concentrations of lead in soils, vegetation, invertebrates and selected tissues of mammals correlated with traffic densities. There was also a gradient effect on tissue levels with distance from the road. In all species, specimens from the direct vicinity of heavily used highways exhibited the highest lead concentrations.

At mining sites and in smelter areas, members of different mammal species showed elevated tissue lead levels compared with specimens from control reference areas. In contaminated areas, insectivorous species, like *Sorex araneus*, had higher tissue levels than herbivorous species like wood mice (*Apodemus sylvaticus*) or field voles *(Microtus agrestis)*. The typically high lead concentrations found in members of the Soricidae are not only attributable to their different food selection but also to their high metabolic rate and comparatively high food intake, which equals their own body weight on a daily basis.

Ma (1989) investigated the transfer of lead from gunshot pellets to the food chain at a shooting range by analysing tissues of small mammals inhabiting the site. He found remarkably high levels of lead in specimens trapped on the shooting range compared to animals of the same species from control sites. Also in this study, lead concentrations were found to be higher in shrews than in bank voles (*Clethrionomys glareolus*) and wood mice (*Apodemus sylvaticus*).

Appleton et al. (2000b) demonstrated the usefulness of the teeth of bank voles (*Clethrionomys glareolus*) as sensitive and reliable indicators of environmental pollution by heavy metals. They found significantly different concentrations of heavy metals (including lead) in the teeth of animals from differently polluted sites.

2.6. Temporal trends

After the restrictions imposed on the addition of lead compounds to gasoline, a decrease in lead contamination of free living mammals could have been expected. So far only a few papers have been published dealing with lead levels in wildlife after the ban of lead addition. In a recent report by Hecht (2001), a survey on changes in lead contamination of roe deer in Germany during the period 1976 to 1997 and on the lead content of the respective summer feeding plants in the period 1976 to 1989 is given. The author reports a pronounced decline in the period up to 1989 attributed to the stepwise reduction of the amount of organolead additives to gasoline. For this period an ecological half life for lead of 8.2 to 8.5 years was calculated for contaminated inner organs of roe deer. Based on these calculations, in the late 1990s, i.e. after the addition of organolead additives had been completely banned for some time, the concentrations of lead found in animal tissues should have become very low (near to the detection limit of the analytical technique). However, this was not the case as easily detectable lead concentrations are presently still found in organ samples of roe deer. Based on these observations the half life of lead in the environment was estimated to lie between 14 and 15.5 years. Hecht (2001) attributes these findings to a long range atmospheric lead contamination, originating from neighbouring countries were leaded gasoline is still used. Similar results were reported by Tataruch (2000) for brown hares in Austria from 1976 to 1999. In the 1970s, high concentrations of lead were found in the livers of the animals. This was followed by a decline in the first half of the 1980s. In the late 1980s and consecutively in the 1990s, the values rose again due to the increase in traffic coming from those countries of Central and Eastern Europe where leaded gasoline is still in use.

In Norway the lead concentrations in moose livers declined probably due to a reduction of long-range atmospheric pollution (Frøslie et al., 2001). No comparable decline could be found for reindeer in south Norwegian mountains, which may be related with the slow turnover of lead in lichens, a principal food of this ruminant species.

3. Cadmium

Cadmium (Cd) is a widely distributed element in the earth's crust that is always found in combination with zinc. There has been a remarkable increase in Cd levels in ecosystems over the last 100 years (Korte, 1982). This has been due to its use in industrial processes such as plastic production, electroplating and the manufacture of alloys, batteries and fertilisers. In addition to this, Cd is present in the combustion products of fossil fuels, emissions from smelting and refining plants and is also deposited as a result of the wear of vehicle tyres.

3.1. Distribution in plants

Cadmium concentration in plants is mainly a result of uptake from soil. Atmospheric deposition on plant surfaces is of only minor importance, however Cd can penetrate into plants via the stomata. In general, the Cd concentration in a plant decreases from the roots to the shoots. However, the steepness of this gradient is strongly species dependent. Normally, fruits contain lower cadmium amounts than shoots, but in some plant species this barrier is not very effective, resulting in an enrichment of the metal in the fruits.

Cd uptake by plants not only depends on the total Cd content of the upper soil layers but also on several other soil characteristics: Cd binds strongly to clay minerals, humic acids and calcium. Therefore, in soils that contain high concentrations of these components the mobility of Cd is greatly reduced. pH values in the range of 7 lead to a diminished Cd release from chelates (humic acids) and clay minerals and in consequence to lower Cd levels in plants (Lorenz, 1979). In more acidic soils, the mobility of Cd is increased resulting in a higher transfer from soil into the plants.

3.2. Absorption and distribution among organs

Cd is a non-essential element for mammals. Under normal conditions, absorption is minimal at around 5–7% of the total oral uptake. When protein, calcium, zinc, iron, copper or vitamin D are low in the diet, the retention rate of Cd increases (Cooke and Johnson, 1996). In addition to dietary uptake, another exposure pathway for Cd is the inhalation of contaminated air. Because cadmium is better absorbed from inhaled air (10 to 40%, depending on particle size) than from ingested food, inhalation of cadmium containing particles is likely to be the most important exposure pathway in areas with elevated atmospheric Cd levels.

After absorption in the lung and gut, Cd is transported via the bloodstream to body stores, particularly the liver. In blood, more than 95 % of cadmium is bound to protein in the blood cells (Friberg et al., 1985). In the liver Cd is bound to metallothionein (Stoeppler, 1991) and the formed complex is transported to the kidneys, where it is almost completely absorbed at the proximal kidney tubules. Interestingly, approximately 50% of the total Cd burden of the body is found in the liver and kidneys although these organs account for only about 5% of the total body mass (Korte, 1987). Muscle tissue does not accumulate Cd, and even in the case of very high body burdens, Cd concentrations in muscles are low. As with lead, Cd concentrations in the blood are not useful as an indicator of the general body burden.

In the kidneys, Cd has a very long retention time. For cattle, Kreuzer et al. (1981) estimated a half life time of up to 12 years in the kidneys compared to only about 2 years in the liver. These long retention times in the kidneys result in a highly significant, positive correlation between the kidney concentrations of Cd and the age of the animal.

Since the kidneys are the main target organ for Cd, they are the best tissue samples for analysis of Cd burdens of mammals. There are remarkable differences in the concentrations of Cd between the cortex and the medulla and varying ratios of these parts in the analytical sample can give misleading results. It is therefore recommended that for the larger species, such as cervides, only the cortex of the kidney is used for analysis.

3.3. Influence of age and sex

Due to the extremely long half life of Cd in the kidneys, interpretation of analytical results have to take the age of the animal into consideration. Although age determination of wildlife is difficult and can yield imprecise results, only animals of the same age group should be compared to obtain a reliable comparison of Cd burdens. The influence of sex on Cd levels has been demonstrated for several species, in that females show higher concentrations than age matched males (Tataruch, 1984). This is most likely due to the fact that females consume more food per kilogramme body weight than males. Lactation is not an excretion pathway for Cd, as levels in milk are very low (Stoeppler, 1991).

3.4. Influence of season

Seasonal variations have a negligible effect on Cd distribution in animal tissue due to the long half-life of this element.

3.5. Influence of habitat

A significant influence on Cd burden of animals is exerted by the geological properties of the habitat, in particular the soil. Therefore, mammals living in habitats with low pH soils (e.g. primary rocks) show higher organ burdens compared with individuals of the same species living on limestone, as they feed on plants with elevated Cd concentrations.

3.6. Influence of species and diet selection

When comparing organ concentrations of cadmium between mammal species, the highest Cd levels are found in the omnivorous wild boar (Holm, 1984). This is attributed to the general rooting type of feeding habit typical for pigs, which leads to the uptake of a high portion (an average of 50–60%, sometimes up to 100%)

of subterranean components, like roots, bulbs, larvae and also a large amount of soil. This results in a high oral Cd uptake which in turn reflects the soil burden. Additionally, elevated Cd levels accumulate in animals like the wild boar in comparison to ruminants because in monogastric digestive systems, metals are absorbed to a higher extent compared to the ruminant situation (Kreuzer et al., 1978).

High Cd concentrations are also found in the herbivorous brown hare. This is most likely a consequence of its monogastric digestive system as well as its relatively high metabolic rate. Due to their comparatively small and stable habitat and the fact that they mainly live in agricultural ecosystems, brown hares are very suitable biomonitors in agricultural areas.

Among the ruminants, several authors found higher Cd concentrations in roe deer (Fig. 1) than in other species from the same habitat, e.g. red deer, chamois and moose (Holm, 1984; Frank, 1986; Tataruch, 1993a). Chemical analysis of the preferred feeding plants of these different ruminant species showed that this difference is due to the selective feeding habit of the roe deer (Anke et al., 1979; Tataruch, 1993a). Its diet consists of a high percentage of Cd accumulating plants like herbs, weeds, fungi and ferns (Seeger, 1978; Hatch et al., 1988). Together with its narrow home range, its feeding habits make the roe deer a very sensitive biomonitor, especially in forest areas. For boreal regions, Frank (1986) showed that the moose (*Alces alces*) is a very good indicator for the bioavailability of Cd, too. Both species show similar rates of accumulation.

For small mammals too, a positive correlation between the cadmium concentrations in soils of their habitat and the Cd concentrations in soft tissues was found (Talmage and Walton, 1991). As for other mammals, the highest values were found in the kidneys with average cadmium levels in this organ being tenfold higher than in the liver. Other tissues displayed much lower concentrations. Only in the case of *Sorex araneus* notable differences were reported. In animals from unpolluted areas, kidney levels were higher than liver levels. However, with high oral Cd exposure, kidney

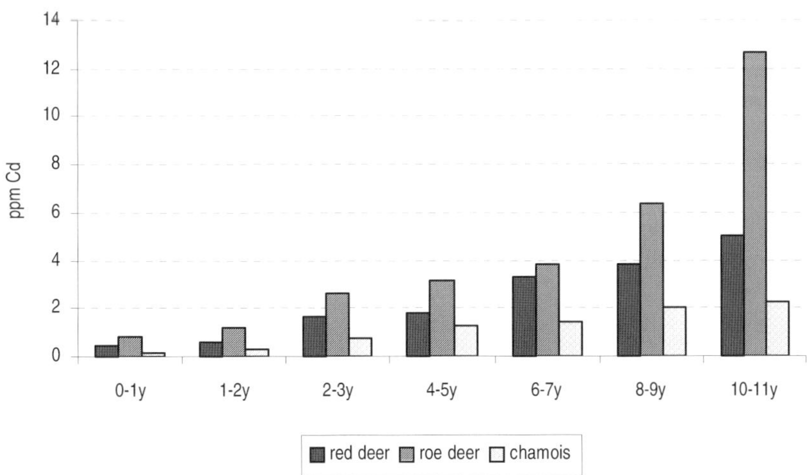

Figure 1. Cd-levels (median) in kidneys of three wild ruminant species in relation to age (in years).

Table 4. Cadmium concentrations (in mg/kg = ppm) in the kidneys (cortex) of roe deer.

	Age	*n*	med	min	max
Region A	−1 year	66	0.792	0.062	5.496
	1–2 years	62	1.184	0.281	3.594
	2–4 years	6	2.655	0.722	6.195
	4–6 years	17	3.115	0.145	31.400
	older	25	4.366	0.871	12.668
Region B	−1 year	3	0.359	0.140	1.808
	1–2 years	16	5.935	1.757	30.604
	2–4 years	13	8.917	1.684	36.288
	4–6 years	10	11.068	1.846	24.121
	older	2	19.185	10.040	28.331

Region A: no industrial depositions.
Region B: heavily polluted by depositions from metallurgical (heavy metals, mainly lead) industries.

concentrations quickly reached a maximum and even liver levels increased. In these animals the kidney/liver ratios, which are above 1 in practically all other mammal species and also in shrews from unpolluted sites, became lower than 1 (Hunter et al., 1987, 1989).

Talmage and Walton (1991) cite several publications reporting that shrews (Soricidae) show higher Cd levels in whole body samples than members of other families such as *Apodemus sylvaticus* or *Microtus agrestis*. These findings correlate very well with the dietary patterns of the species (Ma et al., 1991). First, shrews have a very high food intake, eating the equivalent of their own body weight each day. Second, *S. araneus* feeds primarily on invertebrates at the top of the invertebrate food web, whereas the other species are strictly herbivorous. Uptake of soil particles contaminating the surface of their prey or released from its gut content may also contribute to the dietary cadmium exposure of shrews. The higher Cd levels in *Sorex araneus* are, however, not only a result of higher dietary Cd exposure, but also reflect differences in the bioavailability of Cd from the diet. Additionally, basic physiological differences between species have also been shown to play a role in intestinal cadmium uptake.

Cd levels in the organs of larger carnivores are usually very low. This is explained by the very low accumulation level of Cd in muscle, which comprises the major part of a carnivorous meal.

4. Mercury

Mercury (Hg) is a ubiquitous, biologically non-essential metal. Human activities have resulted in the release of a wide variety of inorganic and organic forms of mercury. The main anthropogenic emission sources are electrical, chemical and paper industries as well as the combustion of fossil fuels. Until recently, the use of organomercury compounds for seed dressing was an important factor influencing wildlife on arable lands, but during the last two decades there has been an almost total world-wide ban on the use of these substances.

4.1. Distribution in plants

Hg present in soil has a very low availability to plants as the roots function as a barrier. Therefore, in general Hg concentrations in the environment are comparably low and are limited mostly to specific and/or localised aerogenous immissions. Significant accumulation of Hg has been observed in fungi, carrots and potatoes. When organomercurials were still in use as fungicides it was shown that in wheat, barley, oats and corn, Hg was transferred from the seed dressing into the new seedling (Von Burg and Greenwood, 1991). During growth, dilution of the mercury occurred and the concentration in the grain was reduced.

4.2. Absorption and distribution among mammalian organs

The level of gastrointestinal absorption of Hg in animals depends on its chemical form. Hg in its inorganic form is absorbed up to about 7% from food, which is similar to other heavy metals. On the other hand, the absorption of methyl mercury can be as high as 95%. Elemental Hg can not be absorbed orally, however, if its vapours are inhaled, around 80% will be absorbed.

The organ distribution of Hg in mammals follows the sequence of kidneys, liver, spleen, and brain, in descending order. This distribution pattern is also found after uptake of organomercury compounds, despite their highly lipophilic nature. Concentrations in blood and muscle are low.

4.3. Influence of species

Hg concentrations in the organs of terrestrial mammals have declined considerably in recent years. Only those species that feed on fungi, such as roe deer and wild boar, show higher levels in the kidney and liver. Fungi are known to accumulate metals, like mercury and cadmium. Roe deer feed intensively on growing fungi (Tataruch, 1993a). This results in a remarkable increase of mercury concentrations in the roe deer's inner organs (Fig. 2). Other species do not show such an extreme preference for fungi, resulting in a lower mercury burden.

The literature on mercury in small mammals indicates that concentration levels are once again influenced by feeding patterns. The seed feeding species *Apodemus sylvaticus* showed an increased mercury burden during the first weeks after fields were sown with seeds treated with an organomercury fungicide. After seeding, mercury levels declined with time. Other species, like *Clethrionomys glareolus* or *Microtus agrestis*, which feed on grasses, always showed comparably low mercury concentrations in their organs.

Species specific accumulation was found in the kidneys of small mammals trapped on a Hg contaminated floodplain. *Blarina brevicauda (soricidae),* whose food includes earthworms, insects, snails etc., showed a remarkably higher uptake compared to two herbivorous species (Talmage and Walton, 1991).

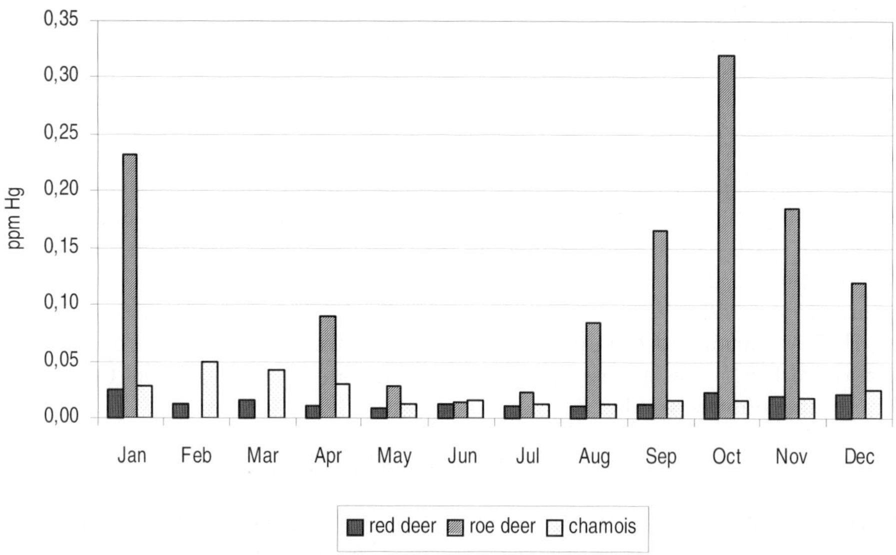

Figure 2.   Seasonal variations of Hg-levels (median) in kidneys of three wild ruminant species.

4.4. Temporal trends

Until the ban on mercury compounds for seed dressing was implemented, all species which fed on the easily accessible seedlings of winter crops, like roe deer or brown hare, were highly contaminated with Hg. This concerned brown hares in particular, for whom the seedlings were a very attractive food. It was shown by Tataruch and Onderscheka (1981) that due to the use of organomercury fungicides for seed dressing, Hg concentrations in the green leaves of cereal crops (on which the hares fed for a longer period in winter) were about 7 ppm dry matter. As concentrations in corn plants decreased as a result of dilution through growth and as the hares turned to other feeding plants, their Hg burden was reduced. After the legal banning of the use of Hg compounds in agriculture, the levels of Hg in the organs of hares became low during all seasons.

5. Fluorine

Fluorine is thirteenth among the elements in order of abundance in the earth's crust, and fluorides are apparently ubiquitous in the environment (WHO, 1984; Smith and Ekstrand, 1996). The halogen enters the biosphere by natural processes (e.g. volcanism, geothermal waters, rock weathering) as well as by numerous human activities such as the production of aluminium, steel, phosphate fertilisers, glass, cement, and brick products. Coal fired power plants and oil refineries are further sources of fluoride emissions. As already summarised by Roholm (1937), increased environmental levels of fluoride constitute a remarkable health risk to the affected organisms. Although single cases of acute toxicosis have been reported in wild mammals after ingestion of large doses of fluoride (Cooke et al., 1996), environmental fluoride is of particular concern because of its chronic toxicity to mammals. Due to the high affinity of the fluoride ion to calcium, chronic fluoride effects are mainly observed in the skeleton and the teeth of mammals. Fluoride selectively accumulates in bone and dental hard tissues, and higher plasma fluoride levels exert negative effects on their formation and mineralization (Shupe and Olson, 1983; Walton, 1988; Fejerskov et al., 1994).

5.1. Skeletal fluoride accumulation

The skeletal fluoride content of an individual provides a cumulative measure of net fluoride incorporation over lifetime (Robinson et al., 1996). Main factors determining the concentration of fluoride in bone are the magnitude and duration of exposure, the stage of skeletal development, the level of bone turnover and the degree of fluoride bioavailability (Turner et al., 1993; Whitford, 1996). Age related increase in bone fluoride content has been studied in different species of wild mammals, predominantly deer, collected from control ("unpolluted") habitats (Kay et al., 1976; Kierdorf et al., 1989; 1995; Machoy et al., 1995; Vikøren et al., 1996). Skeletal fluoride concentrations in mammals were also used to assess the degree of contamination in areas exposed to increased fluoride pollution from various sources (Karstad, 1967; Kay et al., 1975; Newman and Yu, 1976; Newman and Murphy, 1979; Kierdorf, 1988; Walton, 1988; Machoy et al., 1991; Boulton et al., 1994a; Cooke et al., 1996; Kierdorf et al., 1996 a, b; Vikøren et al., 1996; Schroder et al., 1999; Kierdorf and Kierdorf, 1999b; 2000c; Rafferty et al., 2000). From the above work it is evident that long-term monitoring of bone fluoride concentrations of wild mammals taken from the same habitat provides a means to reconstruct temporal changes in environmental levels of fluoride. Excluding work on antlers, that is discussed in a separate chapter, so far three studies used this approach to assess temporal trends of fluoride pollution. Suttie et al. (1987) compared mandibular bone fluoride concentrations in a local population of white-tailed deer (*Odocoileus virginianus*) before and after the onset of operation of an aluminium smelter in the USA. Kierdorf and Kierdorf (1999b) described the reduction of fluoride levels in skull bones of roe deer (*Capreolus capreolus*) from the vicinity of a brown coal-fired power plant in Germany over a 26 yr period. A third study (Kierdorf and Kierdorf, 2000c) presented data on temporal and geographical variation of mandibular fluoride content in roe deer from two industrialised areas of Germany. A pronounced decline in bone fluoride levels was recorded in both study areas over the sampling period (Fig. 3), indicating decreasing fluoride deposition into the wildlife habitats after the implementation of various emission control measures since the 1970s.

5.2. Dental fluorosis

The most obvious sign of elevated fluoride exposure in mammals is the occurrence of dental lesions, summarised as dental fluorosis, resulting from fluoride interference with

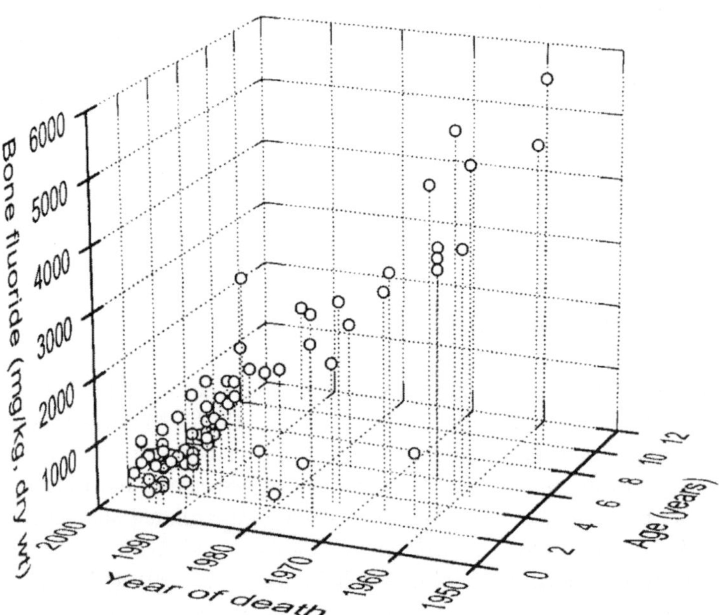

Figure 3. Relationship among the variables age, year of death, and bone fluoride content in roe deer mandibles (n = 76) collected between 1955 and 1998 in the southern Ruhr area (after Kierdorf and Kierdorf, 2000c).

the development of the dental hard tissues, especially dental enamel (Shupe and Olson, 1983; Fejerskov et al., 1994). On the basis of macroscopic and histopathological studies it was concluded that in wild mammals both the secretory and postsecretory stages of enamel formation can be affected by elevated plasma fluoride levels, leading to different pathological enamel changes (Shupe and Olson, 1983; Kierdorf and Kierdorf, 1989, 1997; Kierdorf et al., 1993, 1996a, 1997, 2000; Paranjpe et al., 1994; Vikøren and Stuve, 1996; Boulton et al., 1997). Macroscopically, fluorosed dental enamel is characterised by opacity, discoloration and the occurrence of enamel surface lesions (Shupe and Olson, 1983; Kierdorf, 1988; Kierdorf et al., 1993, 1996b, 2000a; Boulton et al., 1994a,b, 1997; Fejerskov et al., 1994; Paranjpe et al., 1994) (Figs 4a–d). The micromorphological alterations of fluorotic dental enamel were analysed in deer and wild boars (Kierdorf et al., 1993, 1996b, 1997, 2000; Kierdorf and Kierdorf, 1997). Changes attributed to a fluoride disturbance of the secretory stage of amelogenesis include an overall accentuation of the pattern of Retzius lines, the presence of enamel surface hypoplasias (Fig. 4d) and the occurrence of individual broad hypomineralized and structurally altered incremental bands deep to the hypoplastic lesions.

As a consequence of fluoride interference with the process of enamel maturation, fluorotic enamel of deer and wild boars exhibits subsurface hypomineralisation of different depth and extent (Kierdorf et al., 1993, 1996b, 1997, 2000). The pronounced hypomineralization of severely fluorotic enamel results in a markedly reduced mechanical stability of the tissue. As a consequence, increased dental attrition, posteruptive loss of larger areas of surface enamel, severe dental disfigurement and tooth fractures have been reported from fluorotic teeth of deer (Kierdorf et al., 1993; 1996 a; b) (Figs 4a–c). Pathological changes occurring in mandibular and alveolar bone of fluorotic red deer as a sequel of rapid tooth wear and tooth fractures (Fig. 4a) have been documented by Schultz et al. (1998). This study provided circumstantial evidence that the occurrence of severe periodontal disease and tooth loss can lead to a reduction in life expectancy of severely fluorotic deer.

So far only few studies have analysed the changes in dentinal structure present in fluorotic teeth of wild mammals (Kierdorf et al., 1993; Appleton et al., 2000a). According to these studies, fluorotic teeth of roe and red deer are characterised by the presence of larger areas of interglobular dentin and alternating bands of hypo- and hypermineralised dentin.

The spectrum of fluoride-induced dental changes and their distribution within the dentition are distinctive and to date no other agent or condition is known that causes identical symptoms in man or free-living mammals (Shupe et al., 1984; Fejerskov et al., 1988; Kierdorf, 1988). In human dental epidemiology, dental fluorosis has for decades been used as a highly sensitive indicator of increased fluoride exposure, and today several diagnostic scoring systems are available to assess the prevalence and severity of dental fluorosis in human populations (Fejerskov et al., 1988; 1994; Den Besten, 1994). More recently, the relationship between the degree of fluoride exposure, reflected by bone fluoride levels, and the occurrence of fluoride induced dental lesions has also been studied in wild mammals exposed to increased environmental levels of the halogen (Kierdorf, 1988; Boulton et al., 1994a; 1999; Kierdorf et al., 1996a; Kierdorf and Kierdorf, 1999c; Schroder et al., 1999; Rafferty et al., 2000). Scoring systems based on the frequency and severity of macroscopic signs of dental fluorosis were developed for deer and rodents (Kierdorf, 1988; Boulton et al., 1994a, 1999; Kierdorf et al., 1996a; Kierdorf and Kierdorf, 1999c) by modifying an established system used for assessing the degree of dental fluorosis in cattle (Shupe and Olson, 1983). Studies on free living deer (Kierdorf et al., 1996a, Kierdorf and Kierdorf, 1999c) exposed to regional, long-term fluoride pollution revealed a strong and highly significant positive relationship between skeletal fluoride levels and the severity of dental fluorosis (expressed as dental lesion index of fluorosis for the permanent cheek teeth) (Fig. 5), thereby proving that prevalence and severity of dental fluorosis can be used to assess environmental pollution by fluorides. A major advantage of this approach is that in many countries it can make use of an established and continuously operated system of specimen collection (mandibles of male deer) by hunters. The system, therefore, offers the opportunity for running large-scale monitoring programmes of fluoride pollution (involving high numbers of analysed individuals) at low cost. A first study using the biomarker dental fluorosis in roe deer to assess environmental pollution by fluorides on a regional scale was conducted by Kierdorf et al. (1999) in the Saxonian Ore mountains (Germany).

Since the "window of susceptibility" for fluoride action on deer teeth is the period of enamel formation, it must be stressed that dental fluorosis in deer is a biomarker of past fluoride exposure, reflecting elevated plasma fluoride levels during the period of tooth formation only (Kierdorf and Kierdorf, 1999c). The highly characteristic

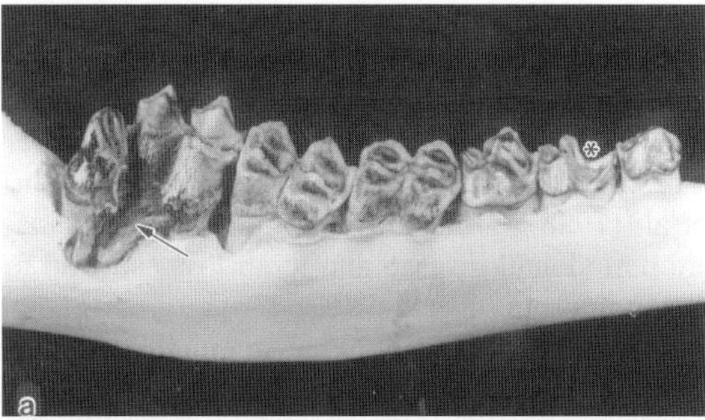
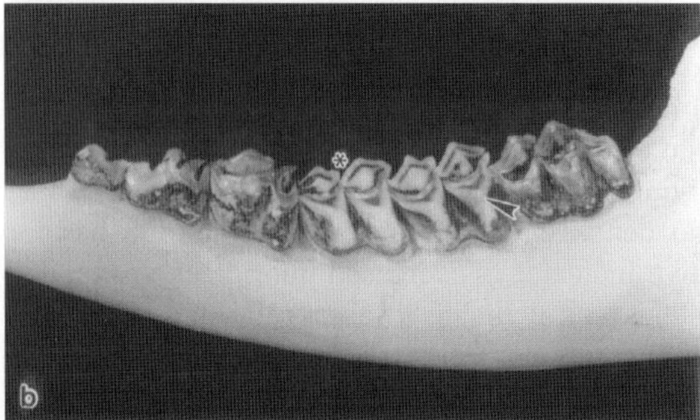
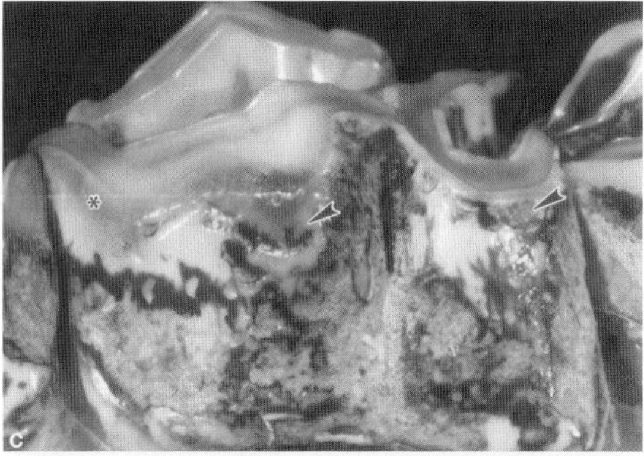
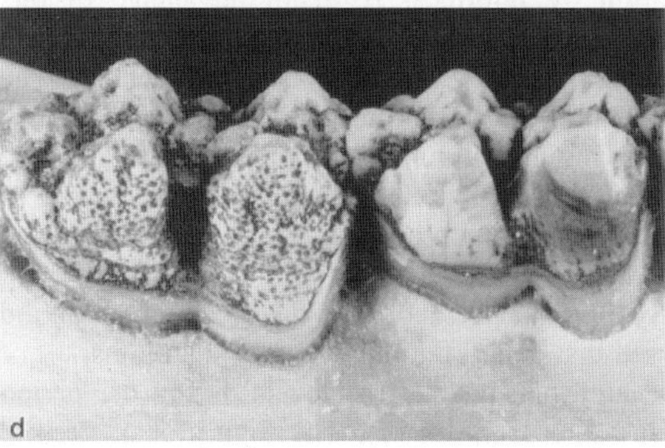

Figure 4.
Macroscopic symptoms of dental fluorosis: (a) Right hemimandible of a red deer stag showing severe dental fluorosis. Note fracturing of the M_3 (arrow) that has resulted in alveolar bone destruction. Abnormal wear of the permanent premolars (especially P_3, asterisk) and the M_2. Bucco-occlusal view. (b) Left hemimandible of a roe buck with severe dental fluorosis of the permanent premolars and the third molar. The M_1 (asterisk) shows no and the M_2 only slight (enamel discoloration, arrowhead) fluorotic alterations. Bucco-occlusal view. (c) P_4 of the fluorotic roe deer mandible depicted in "b", exhibiting typical signs of dental fluorosis, viz. pathologically increased attrition with disappearance of enamel ridges on the occlusal surface (asterisk), enamel discoloration and posteruptive loss of surface enamel (arrowheads). Bucco-occlusal view. (d) M_1 (right) and M_2 (left) of a fluorotic wild boar mandible. While the enamel of the first molar is characterized by opacity and discoloration only, the crown of the M_2 exhibits severe hypoplastic pitting. Linguo-occlusal view.

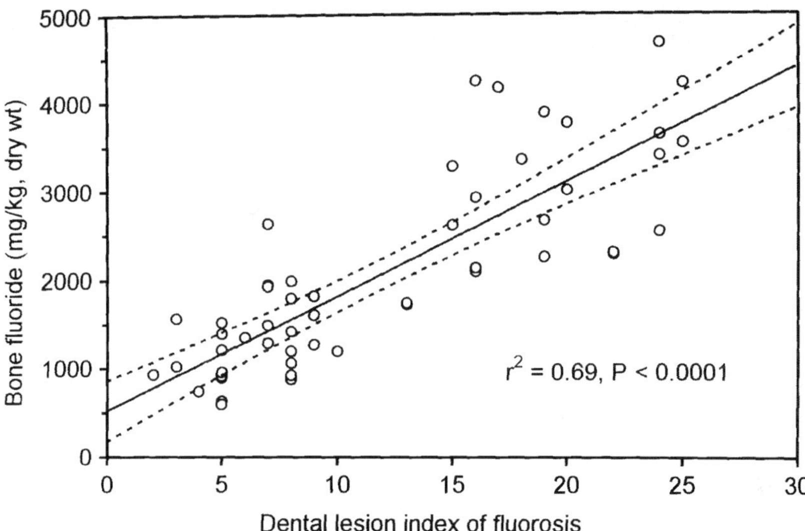

Figure 5. Relationship between mandibular bone fluoride content and degree of dental fluorosis in the permanent cheek teeth (P_2–M_3), expressed as dental lesion index of fluorosis, in a sample of 53 fluorotic red deer mandibles. Dashed lines indicate 95% confidence limits of the regression line (after Kierdorf and Kierdorf, 1999c).

variation in the prevalence and severity of fluorotic lesions within the row of permanent cheek teeth observed in roe and red deer (Figs 4a, b) can be related to the developmental sequence of the dentition, in particular the timing of crown formation of the different tooth types (Kierdorf and Kierdorf, 2000d). Furthermore, the action of several protective mechanisms during certain ontogenetic stages (foetal period and early postnatal period of milk feeding) has been postulated for these species (Kierdorf, 1988; Kierdorf et al., 1991, 1993, 1996a; Kierdorf and Kierdorf, 2000d). Differences in the timing of crown formation between homologous teeth of roe and red deer are thought to be responsible for the systematic variation in the prevalence and severity of fluorotic lesions between some homologous teeth of the two species (Kierdorf and Kierdorf, 2000d). A proper knowledge of the interdental and interspecific variation in fluorotic alterations of teeth is required when using different deer species as bioindicators of environmental pollution by fluorides.

A scoring scheme for the assessment of macroscopic signs of dental fluorosis in the teeth of short-tailed field voles (*Microtus agrestis*) was developed by Boulton et al. (1994a, b, 1999) for application in monitoring studies. Also in this species, the severity of the lesions in incisors and molars (expressed as lesion score) was positively correlated with skeletal and dental fluoride concentrations (Boulton et al., 1994 a,b, 1999). This relationship was not only observed in animals trapped at fluoride contaminated sites, but could also be established in controlled feeding experiments (Boulton et al., 1994b). A positive relationship between the degree of dental fluorosis and bone fluoride levels was also found in cotton rats (*Sigmodon hispidus*) from fluoride contaminated petrochemical waste sites (Schroder et al., 1999; Rafferty et al., 2000).

Contrary to deer teeth, incisors (and in some species also molars) of rodents grow throughout life. Therefore short-term (e.g. seasonal) changes in the degree of fluoride exposure during lifetime can be monitored by analysing variation in the prevalence and severity of fluorotic dental lesions in animals trapped at contaminated sites during different seasons (Kim et al., 2001).

5.3. Osteofluorosis

Osteofluorotic lesions have been reported from domestic animals, particularly cattle and wild ungulates after prolonged intake of fluoride (Shupe and Olson, 1983; Shupe et al., 1984). By contrast, no gross morphological changes have been recorded in the bones of various small mammal species originating from fluoride contaminated habitats (Walton, 1988; Cooke et al., 1996). Compared to the severe dental lesions developing as a consequence of elevated fluoride exposure, osteofluorotic lesions clearly do not play a major role in wild mammals and are of lesser importance with regard to biomonitoring studies.

6. Deer antlers as monitors of environmental pollution

Antlers are deciduous bony structures that develop on top of permanent frontal protuberances (the pedicles) of male deer (and females in *Rangifer*) and undergo periodic replacement (Goss, 1983). In deer from temperate and arctic regions, the timing of the antler cycle is strongly influenced by the photoperiod, and antler growth occurs during a seasonally fixed period of some months (Goss, 1983). After completion of growth and terminal intense mineralisation of antler bone, the velvet is shed from the antlers and the polished hard antlers are retained for some further month until they are cast and a new set of antlers is produced.

During their limited life span of some months, antlers accumulate potential environmental contaminants with a high affinity to mineralized tissues, such as fluoride (Karstad, 1967; Suttie et al., 1985; Walton and Ackroyd, 1988; Samujlo et al., 1994; Kierdorf et al., 1997b; Kierdorf and Kierdorf, 2000a,c, 2001), heavy metals, particularly lead (Sawicka-Kapusta, 1979; Kardell and Källman, 1986; Tataruch and Schönhofer, 1993; Medvedev, 1995; Tataruch, 1995; Kierdorf and Kierdorf, 1999; 2000 a,b), and strontium-90 (Schultz, 1964; Gelbke, 1972; Schönhofer et al. 1994; Strandberg and Strandgaard, 1995; Tiller and Poston, 2000). The concentrations of these contaminants in hard antlers constitute cumulative measures of exposure of the forming antler bone during a relatively well defined period. Antlers can thus be viewed as quite well "naturally standardized" environmental samples. Therefore, and because they are regularly collected and kept over long periods of time, antlers are well suited as monitoring units for studying temporal and/or geographical variation in environmental levels of "bone seeking" pollutants.

During antler growth, bone minerals are mobilized from the postcranial skeleton of deer by bone resorption, in order to meet the high mineral requirements of the forming antlers (Baxter et al., 1999). It may be hypothesized that along with these bone minerals also previously skeletally deposited contaminants are released, that can contribute to the exposure of forming antlers. This means that the concentrations of these substances in antler bone would not solely reflect uptake by the animal during the antler growth period, but would to some extent also be influenced by the previous (lifetime) exposure of an individual.

Larger series of antlers collected over longer periods in the same area can be used to monitor temporal trends of environmental pollution. Such studies have been conducted using antlers of roe deer (*Capreolus capreolus*) and red deer (*Cervus elaphus*) (Gelbke, 1972; Kardell and Källman, 1986; Tataruch and Schönhofer, 1993; Schönhofer et al., 1994; Tataruch, 1995; Kierdorf and Kierdorf 1999, 2000a,c, 2001). An example for this approach is shown in Figure 6 presenting data on the fluoride and

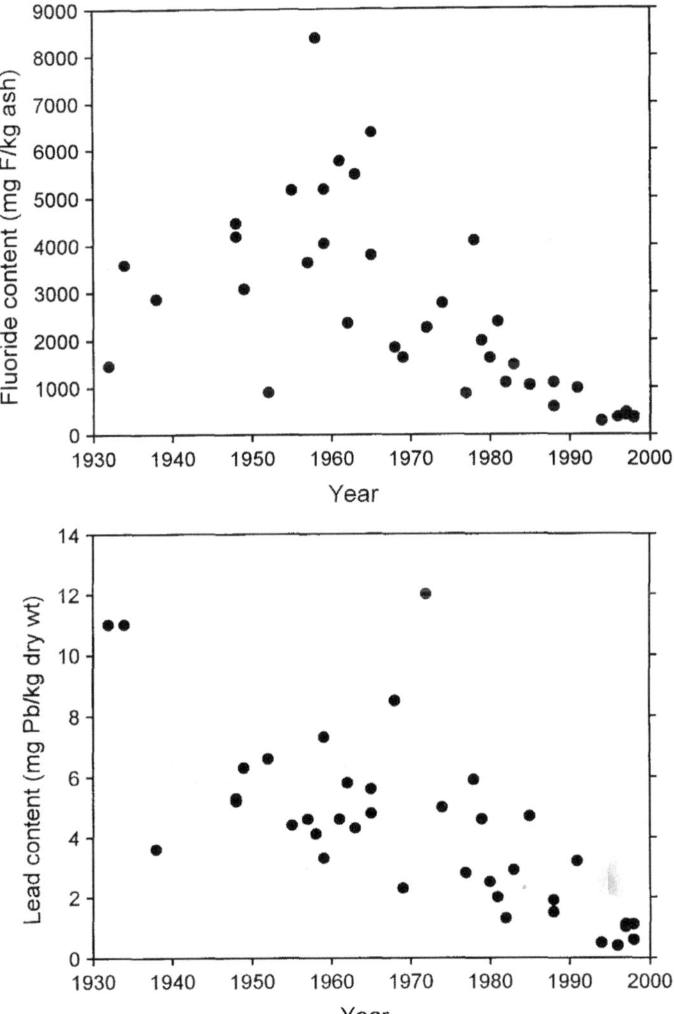

Figure 6. Fluoride and lead concentrations in antler bone of roe deer (*n* = 39) killed between 1932–1998 in a forest area of western Germany (after Kierdorf and Kierdorf, 2000a).

lead content of roe deer antlers obtained over a 67 year period in a small forest area of western Germany (Kierdorf and Kierdorf, 2000a). The decline in antler concentrations of both pollutants in the 1980s and 1990s indicates decreasing environmental contamination during this period.

Analysis of antlers grown prior to large scale industrialisation enables an assessment of environmental trace element concentrations that are only slightly affected by anthropogenic emissions. Thus, red deer antlers collected before 1860 in western Germany contained low fluoride concentrations ranging between 27.7 and 78.7 mg mg/kg ash (Kierdorf and Kierdorf, 2000e). These values were regarded as being quite close to natural background levels.

Antler samples obtained during the same period in different areas can be used to assess regional variation in environmental contamination (Walton and Ackroyd, 1988; Samujlo et al., 1994; Medvedev, 1995; Tataruch and Schönhofer, 1993; Tataruch, 1995; Kierdorf and Kierdorf, 1999a, 2000 b,e). In this way, significant regional differences of lead pollution were demonstrated among 13 study sites in western Germany using roe deer antlers collected in the 1990s (Fig. 7) (Kierdorf and Kierdorf, 2000b).

The effects of increased fluoride exposure on the formation and mineralisation of antler bone have been studied in red deer from fluoride polluted areas (Kierdorf et al., 2000b). Compared to controls, a reduced amount of mineralised bone and a lower mineralisation density was recorded in the antlers of the fluoride exposed red deer.

So far, only one experimental study on fluoride accumulation in antler bone has been conducted (Suttie et al., 1985). The data presented by these authors for white-tailed deer (*Odocoileus virginianus*) allow a rough estimation of the relationship between the fluoride content of food and antler fluoride concentrations also in other deer species.

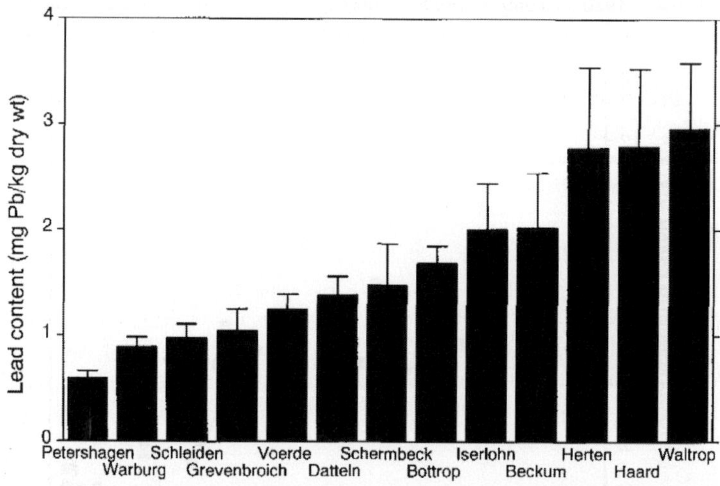

Figure 7. Lead concentrations (mean and SE) of roe deer antlers (*n* = 172) collected between 1990–1999 in 13 study sites of North Rhine-Westphalia, Germany (after Kierdorf and Kierdorf, 2000b).

7. Radionuclides

As a result of above ground testing of nuclear weapons there was a continuous atmospheric deposition of radionuclides onto many areas world-wide (mainly caesium-137 and strontium-90) in the years prior to 1986. In 1986, the Chernobyl accident resulted in a huge deposition of radionuclides (mainly caesium-137, caesium-134 and iodine-131) onto many parts of Europe. This incident in particular has lead to remarkably high radioactive contamination levels which exceed those during the weapons testing period by 50–100 times. The deposition in 1986 was largely governed by climatic conditions. Contamination was significantly higher in areas where rainfall was greatest at the time the Chernobyl plume passed over and also at higher altitudes. Combinations of altitude and high precipitation resulted in the greatest fallout.

The Chernobyl accident occurred at the end of April 1986, a few weeks before the start of the main hunting season of many game species. In early May, in consumer protection efforts, many European countries started to investigate the radioactive burden of game. As a result of these examinations, valuable information regarding the role of wildlife as biomonitors for radioactive contamination were obtained. Most of the research work done was undertaken in Sweden, Germany and Austria.

7.1. Distribution among different organs

In the mammalian organism caesium behaves like potassium, thus the highest concentrations of radiocaesium can be found in the muscles. The kidneys display variable concentrations depending on the relationship between excretion and absorption, in some cases showing higher activities of Cs-137 than the muscles. In the liver, the burden is significantly lower. Iodine-131 accumulates in the thyroid, but due to its short half-life of 8 days the activity declines rapidly. In 1986, iodine-131 was no longer detectable by end of June, even in areas highly polluted by the Chernobyl fallout. The radiocaesium components of the fallout are of more long term significance due to their long physical half-lives of 30.2 years for caesium-137 and 2.1 years for caesium-134.

7.2. Regional distribution

Until the end of June 1986, the radioactive contamination in wildlife, especially roe deer, showed a geographical distribution very similar to the precipitation pattern (Hecht, 1988; Tataruch et al., 1988). Areas with high precipitation levels during the passage of contaminated air masses showed severe contamination as a result of washout effects. Additionally, contamination of wildlife varied with the altitude of the habitat, with higher areas also exhibiting a higher contamination (Tataruch et al., 1988).

7.3. Influence of species and diet selection

Considerable differences have been found in radiocaesium activity concentrations of different animal species grazing in the same habitat. In the months following atmospheric fallout, higher contamination levels were detected in ruminants than in monogastric species such as wild boar, hares or game birds like grouse, ptarmigan, etc. (Howard et al., 1991). This is most likely caused by the larger absorption surface available in the rumen and the longer retention time of the diet within the intestine. Among ruminants, roe deer showed higher Cs-contamination than other species from the same area, e.g. red deer and chamois in Austria or moose in Sweden. These differences are due to several factors such as variations in food selection, metabolic rate of the animal and / or transfer coefficients. As already mentioned, roe deer feed mainly on herbs, buds, young shoots etc. Due to their greater surface area compared to grass, these parts are more heavily contaminated by direct fallout. This in turn results in a higher uptake of radiocaesium by roe deer. Transfer coefficients from feeding plants to the animal were higher in roe deer than in sheep or cattle (Howard et al., 1991).

In Scandinavia, very high radioactivity levels were measured in reindeer and this was shown to be the result of feeding on lichens (Eriksson, 1990). Due to their longevity and physiology, lichens accumulate high amounts of atmospheric pollutants.

7.4. Influence of age

Juveniles showed higher activity concentrations of radiocaesium in their muscles than adult individuals (Tataruch et al., 1988). This can be attributed to a relatively higher metabolic rate and a rapid growth in this age class.

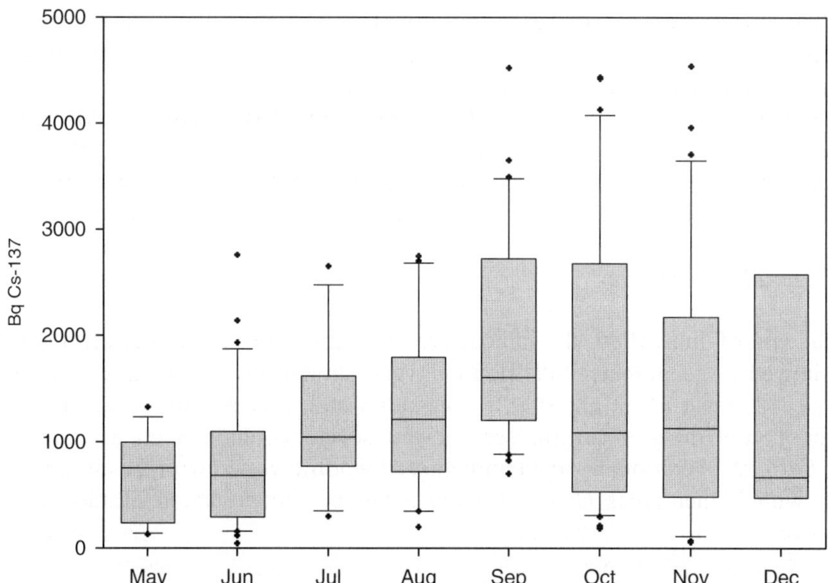

Figure 8.   Monthly changes of ^{137}Cs in muscles of roe deer.

The lower boundary of the box indicates the 25th percentile, the line in the box the median and the upper boundary of the box the 75th percentile. Whiskers above and below the box indicate the 90th and 10th percentiles. The dots indicate outlying values.

7.5. Influence of season

During the summer and autumn of 1986, the radioactivity levels in wildlife decreased. This was due to a reduction in contamination of feeding plants which was in turn a result of wash-off-effects and dilution of radionuclides by plant growth. However, in summer and autumn of 1987 a rise of the Cs-137 levels was observed especially in roe deer and moose (Tataruch et al., 1990; Howard et al., 1991; Karlen and Johanson, 1991). Thereafter, a seasonal increase starting around mid summer and lasting until October or November, was noticed every year in these species (Fig. 8). In reindeer a comparable peak load was observed in winter (Jones et al., 1989). The explanation for this seasonal variation was found in the species-specific feeding on specific forage plants which have a tendency to accumulate high concentrations of caesium. Many authors showed that for roe deer, a large uptake of fungi is the main reason for the increase in radiocaesium levels in summer (Hecht, 1988; Tataruch et al., 1990; Johanson et al., 1990; Avila et al., 1999). Kiefer et al. (1996) found a significant positive correlation between the precipitation in mid-summer, fungi growth, and caesium levels in roe deer. Additionally, other preferred feeding constituents, like *Vaccinum myrtillus* (except the berries) and ferns, also accumulate caesium. For moose, the uptake of dwarf-shrubs, in particular heather (*Calluna vulgaris*), accounted for the increased radiocaesium levels in animals shot in October (Bothmer et al., 1990). The peak load found in reindeer during winter was explained by an increasing amount of lichen uptake due to the lower availability of other more preferred food sources.

7.6. Long term changes in radiocaesium contamination of free living mammals

Levels of radiocaesium activity in wildlife living on agricultural land declined steadily throughout 1986. For roe deer, a biological half-life for Cs-137 was calculated in the range of 22–26 days in summer and about 40 days in winter (Hecht, 1988, Howard et al., 1991). In 1987, the activity level in the muscles of game animals was nearly as low as it had been before the Chernobyl accident. Due to the soil characteristics (high in potassium and organic matter, less acidic pH values, etc.) and agricultural practice in arable land (ploughing, harvesting, etc), the bioavailability of Cs became much lower in comparison to forest areas. However, despite the aforementioned decline it was also noted that the radiocaesium levels in roe deer in unbroken forest areas remained high (Fig. 9). The explanation for this observation is found in the increased bioavailability of Cs in soil types characteristic of forests. Such soils are often acidic, have high organic content, low potassium content and include clay minerals which bind caesium. Additionally, the penetration rate of radiocaesium into the soil was slower than expected and therefore it remained within the rooting zone. These factors resulted in longer than expected half-lives of radiocaesium in forest areas. Hecht (1992) calculated an ecological half-life of Cs for the roe deer-forest system of 3 years. Jones et al. (1989) predicted an effective half-life for the lichen-reindeer system of 10–20 years.

Contrary to the continuous reduction in contamination of roe deer during recent years, a remarkable elevation in the burden of wild boars was reported for several

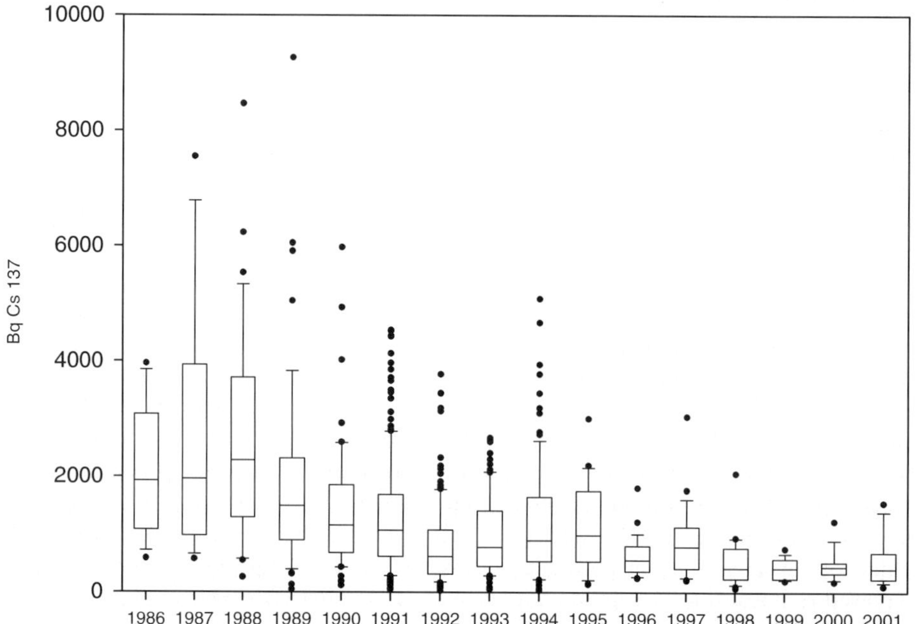

Figure 9. Cs-137 levels in roe deer (muscle) from an unbroken forest area in Austria in the period from 1986 to 2001.

The lower boundary of the box indicates the 25th percentile, the line in the box the median and the upper boundary of the box the 75th percentile. Whiskers above and below the box indicate the 90th and 10th percentiles. The dots indicate outlying values.

forest areas in Central Europe (Tataruch et al., 1996; Hecht, 1993). After an increase of contamination levels in wild boar in 1987 (at almost the same time as in the roe deer) the activity concentrations in the muscles of this omnivorous species remained high. In many areas that had been severely contaminated by direct fall out in 1986, like the Bayerischer Wald in Germany, the values are still increasing steadily. In some cases, peak levels which are many times higher than those recorded in 1986 are being reported. At present the most likely explanation for this observation is the uptake of fungi which accumulate radiocaesium to an extremely high extent.

8. Organohalogens

This group consists of a wide variety of chemicals, with the common property that they contain at least one halogen atom, mostly chlorine. They had (very few of them still have) manifold uses in industry and agriculture, mainly as pesticides. Some of them were distributed in the environment unintentionally as by-products or impurities of pesticides or other chemicals.

The common characteristics of organohalogens are their persistence and high lipophilia, responsible for their bioaccumulation and long-term effects in the environment. Because of their physicochemical properties (K_{OW}, lipophilia, vapour pressure) these compounds accumulate in the fat tissue of animals. In contrast to metals, the carry-over of parts of the body burden of female mammals to their offspring, by prenatal transplacental transfer and postnatal transfer via milk is significant. The long-term consequences of this behaviour are difficult to predict.

Residues of substances from two large groups of organohalogen compounds have become ubiquitous during the past decades, namely organochlorine pesticides and polychlorinated biphenyls. Over 30 organochlorine compounds have been used as pesticides, half of these widely. Some of them were stable enough to become distributed globally, mainly DDT, together with its isomers and metabolites (mainly DDE), dieldrin and HCB. Others, like aldrin, endrin or heptachlor, exerted acute deleterious effects on non-target animals, mainly birds, but were not accumulated in the environment due to their rapid break-down in biological systems and their short half-lives in animal tissue.

Polychlorinated biphenyls are a group of aromatic organic chemicals consisting of 209 congeners sharing a common basic two-ring structure. They differ only in the number and placement of chlorine atoms on these rings. Their manufacture began in the 1920s and continued until quite recently in some countries. They have been distributed widely around the globe and the concentrations of PCBs in living organisms have increased steadily over time, at least until the mid-1970s when PCBs were banned in some countries.

Animals may be exposed by ingestion of contaminated food or water, by inhalation or by absorption through skin and mucosa. Due to their lipophilia organohalogens are stored in the fatty tissue, liver and brain. For analyses of organohalogens mostly fat tissue or liver are used.

There are several problems for long time monitoring of organohalogen contamination of mammals. At the time, when the release of these substances was at its maximum, no appropriate analytical techniques were available to measure their impact on mammals respectively tissue concentrations of these substances. Later, with the improvement of analytical equipment, the global dispersal of these compounds could be demonstrated. At the beginning of the use of gas chromatography analytical standards of PCBs were not available. Therefore, most results until the 1980s are based on calibration with technical mixtures (e.g., Clophen®, Aroclor®). Later on, the development of capillary columns and the synthesis of single PCB congeners enabled congener specific determinations, although the total number of congeners analysed by the different research groups varies widely.

Today, many years after severe restrictions and prohibition of the application of organochlorine pesticides and the ban of PCBs' synthesis and use in open systems, the burden of terrestrial herbivorous mammals by these compounds is very low, if detectable at all, usually in the range of ppb. In omnivorous or carnivorous species, some persistent PCB congeners, like PCB-138, 153, 180 can be traced. In general, terrestrial mammals are no longer suitable biomonitors for organochlorines.

Marine mammals are important for monitoring long-term effects of organohalogen pollution of the marine environment world-wide. Due to their position at the end of the aquatic food chain and their rather long life-spans, they can be used as global pollution indicators. Their lipid reserves represent a large reservoir for accumulating lipophilic compounds, like the organochlorines. Mössner and Ballschmiter (1997) compared the concentrations of different organochlorines (seven indicator congeners of PCBs, three isomers of the hexachlorocyclohexanes and six components of the DDT-groups) in blubber samples of two seal species, three toothed whale species and one baleen whale species from different regions of the Northern hemisphere. The specimens from the western North Atlantic were significantly higher contaminated with organochlorines than the animals from the eastern North Pacific and the Bering Sea/ Arctic Ocean. The organochlorine burden, the 4,4′-DDE burden as well as the metabolic PCB patterns correlated with the different trophic levels of these specimens. Trophic levels seem to influence the organochlorine burden to a significant extent. Another problem for comparison of the results of different studies lies in the fact that the group of analysed substances varied remarkably between different studies. In members of six species of marine mammals from Alaska lower contamination levels were recorded than in specimens from the same species originating from areas around Greenland and Canada (Becker, 2000).

References

Andersen, R., Duncan, P, Linnell, J.D.C., 1998. The European roe deer. The biology of success. Scandinavian University Press, Oslo.

Anke, M., Grün, M., Briedermann, L., Missbach, K., Hennig, A., Kronemann, H., 1979. Die Mengen- und Spurenelementversorgung der Wildwiederkäuer. 1. Mitteilung: Der Kadmiumgehalt der Winteräsung und der Kadmiumstatus des Rot-, Dam-, Reh- und Muffelwildes. (The supply of wild ruminants with major and trace elements. 1. The cadmium content of winter grazing and the cadmium status of red deer, fallow deer, roes and mouflons). Archiv für Tierernährung 29, 829–844.

Appleton, J., Chesters, J., Kierdorf, U., Kierdorf, H., 2000a. Changes in the structure of dentine from cheek teeth of deer chronically exposed to high levels of environmental fluoride. Cells Tissues Organs 167, 266–272.

Appleton, J., Lee, K.M., Sawicka-Kapusta, K., Damek, M., Cooke, M., 2000b. The heavy metal content of the teeth of the bank vole (*Clethrionomys glareolus*) as an exposure marker of environmental pollution in Poland. Environmental Pollution 110, 441–449.

Avila, R., Johanson, K.J., Bergström, R., 1999. Model of the seasonal variations of fungi ingestion and ^{137}Cs activity concentrations in roe deer. Journal of Environmental Radioactivity 46, 99–112.

Baxter, B.J., Andrews, R.N., Barrel, G.K., 1999. Bone turnover associated with antler growth in red deer (*Cervus elaphus*). Anatomical Record 256, 14–19.

Beardsley, A., Vagg, M.J., Beckett, P.H.T., Sansom, B.F., 1978. Use of the field vole (*M. agrestis*) for monitoring potentially harmful elements in the environment. Environmental Pollution 16, 65–71.

Becker, P.R., 2000. Concentration of chlorinated hydrocarbons and heavy metals in Alaska arctic marine mammals. Marine Pollution Bulletin 40, 819–829.

Bothmer, S. von, Johanson, K.J., Bergström, R., 1990. Cesium-137 in moose diet; considerations on intake and accumulation. The Science of the Total Environment 91, 87–96.

Boulton, I.C., Cooke, J.A., Johnson, M.S., 1994a. Fluoride accumulation and toxicity in wild small mammals. Environmental Pollution 85, 161–167.

Boulton, I.C., Cooke, J.A., Johnson, M.S., 1994b. Experimental fluoride accumulation and toxicity in the short-tailed field vole (*Microtus agrestis*). Journal of Zoology 234, 409–421.

Boulton, I.C., Cooke, J.A., Johnson, M.S., 1997. Fluoride-induced lesions in the teeth of the short-tailed field vole (*Microtus agrestis*): a description of the dental pathology. Journal of Morphology 232, 155–167.

Boulton, I.C., Cooke, J.A., Johnson, M.S., 1999. Lesion scoring in field vole teeth: application to the biological monitoring of environmental fluoride contamination. Environmental Monitoring and Assessment 55, 409–422.

Buck, W.B., 1979. Animals as monitors of environmental quality. Veterinary and Human Toxicology 21, 277–284.

Cooke, J.A., Boulton, I.C., Johnson, M.S., 1996. Fluoride in small mammals. In: Beyer, W.N., Heinz, G.H., Redmon-Norwood, A.M. (Eds), Environmental Contaminants in Wildlife: Interpreting Tissue Concentrations. SETAC Special Publication Series, CRC Press, Boca Raton, pp. 473–482.

Cooke, J.A., Johnson, M.S., 1996. Cadmium in small mammals. In: Beyer, W.N., Heinz, G.H., Redmon-Norwood, A.W. (Eds), Environmental Contaminants in Wildlife: Interpreting Tissue Concentrations. SETAC Special Publication Series, CRC, Boca Raton, pp. 377–388.

Den Besten, P.K., 1994. Dental fluorosis: its use as a biomarker. Advances in Dental Research 8, 105–110.

Ellenberg, H., 1991. Bioindicators and Biological Monitoring. In: Ellenberg, H., Arndt, U., Bretthauer, R., Ruthsatz, B., Steubing, L. (Eds), Biological Monitoring: Signals from the Environment. Vieweg, Braunschweig, pp.13–74.

Eriksson, O., 1990. Cs-137 in forage plants vital to reindeer (*Rangifer tarandus tarandus* L.) in northern Sweden. In: Desmet, G., Nassimbeni, P., Belli, M. (Eds), Transfer of Radionuclides in Natural and Semi-Natural Environments. Elsevier Applied Science, London, pp. 194–201.

Ewers, U., Schlipköter, H.W., 1991. Lead. In: Merian, E. (Ed.), Metals and their Compounds in the environment. VCH, Weinheim, pp. 971–1014.

Fejerskov, O., Larsen, M.J., Richards, A., Baelum, V., 1994. Dental tissue effects of fluoride. Advances in Dental Research 8, 15–31.

Fejerskov, O., Manji, F., Baelum, V. Møller, I.J., 1988. Dental fluorosis – a handbook for health workers. Munksgaard, Copenhagen.

Frank, A., 1986. In search of biomonitors for cadmium: cadmium content of wild Swedish fauna during 1973–1976. The Science of the Total Environment 57, 57–65.

Friberg, L., Elinder, C.-G., Kjellstroem, T., Nordberg, G.F. (Eds), 1985. Cadmium and Health: A Toxicological and Epidemiological Appraisal, Vol. 1: Exposure, Dose and Metabolism. CRC Press, Boca Raton, FL.

Frøslie, A., Norheim, G., Rambæk, J.P., Steinnes, E., 1984. Levels of trace elements in liver from Norwegian moose, reindeer and red deer in relation to atmospheric deposition. Acta veterinaria Scandinavia 25, 333–345.

Frøslie, A., Sivertsen, T., Lochmiller, R., 2001. Perissodactyla and Artiodactyla. In: Shore, R.F., Rattner, B.A. (Eds), Ecotoxicology of wild mammals. John Wiley, Chichester, pp. 497–550.

Gelbke, W. 1972. Radioaktivität in Rotwild-Stangen und menschliche Strontium-90-Aufnahme. Zeitschrift für Tierphysiologie, Tierernährung und Futtermittelkunde 29, 178–195.

Goss, R.J., 1983. Deer antlers. Regeneration, Function, and Evolution. Academic Press, New York.

Hatch, D.J., Jones, L.H.P., Burau, R.G., 1988. The effect of pH on the uptake of cadmium by four plant species in flowing solution culture. Plant and Soil 105, 121–126.

Hecht, H., 1987. Unter welchen Bedingungen eignen sich freilebende jagdbare Tiere als Bioindikatoren? VDI Berichte 609, 101–122.

Hecht, H., 1988. Radioaktive Belastung von Wild- und Nutztieren nach dem Unfall von Tschernobyl (Radioactive contamination of game and farm animals after the Chernobyl disaster). Fleischwirtschaft 68, 508–513.

Hecht, H., 1992. Rehwild – ein Bioindikator zur Erfassung des Ausmaßes von Umweltbelastungen durch Störfälle in kerntechnischen Anlagen. VDI-Berichte 901: Bioindikation: ein wirksames Instrument der Umweltkontrolle. VDI-Verlag GmbH Düsseldorf, pp. 937–954.

Hecht, H., 1993. Die Radiocäsiumbelastung des Schwarzwildes nach dem Unfall von Tschernobyl. In: Honikel, K.O., Hecht, H. (Eds). Radiocäsium in Wald und Wild. Bundesanstalt für Fleischforschung, Kulmbach. pp. 90–104.

Hecht, H., 2001. Die Bleikontamination des Rehwildbrets in Deutschland ist weiter rückläufig (The decrease of the lead concentration of German roe deer continues). BAFF-Mitteilungsblatt 151, 1–6.

Hecht, H., Schinner, W., Kreuzer, W., 1984. Endogene und exogene Einflüsse auf die Gehalte an Blei und Cadmium in Muskel- und Organproben von Rehwild – 1. Mitteilung: Einfluß von Alter und Versuchsort (Endogenous and exogenous influences on contents of lead and cadmium in muscle and organ samples from roe deer – 1. The influence of age and experimental area). Fleischwirtschaft 64, 1–6.

Holm, J., 1984. Aufbau eines ursachenorientierten Monitoring-Systems für Schadstoffbelastungen beim Wild. 2. Belastungen von Wild mit Schwermetallen aus unterschiedlich strukturierten Herkunfts-regionen (Constructing a cause-oriented system for monitoring the contamination of game by harmful substances. 2. Contamination of game by heavy metals from differently structured regions of origin). Fleischwirtschaft 64, 613–619.

Holm, J., 1993. Investigation of roe deer – criteria for use as a bioindicator in specimen banking. The Science of the Total Environment 139/140, 237–249.

Howard, B.J., Beresford, N.A., Hove, K., 1991. Transfer of radiocaesium to ruminants in natural and semi-natural ecosystems and appropriate countermeasures. Health Physics 61, 715–725.

Hunter, B.A., Johnson, M.S., Thompson, D.J., 1987. Ecotoxicology of copper and cadmium in a contam-inated grassland ecosystem. III. Small mammals. Journal Applied Ecology 24, 601–614.

Hunter, B.A., Johnson, M.S., Thompson, D.J., 1989. Ecotoxicology of copper and cadmium in a contam-inated grassland ecosystem. IV. Tissue distribution and age accumulation in small mammals. Journal Applied Ecology 26, 89–99.

Johanson, K.J., Bergström, R., Bothmer, S., Karlen, G., 1990. Radiocaesium in wildlife of a forest in central Sweden. In: Desmet,G., Nassimbeni, P., Belli, M. (Eds), Transfer of Radionuclides in Natural and Semi-Natural Environments. Elsevier Applied Science, London, pp. 183–193.

Jones, B.E., Eriksson, O., Nordkvist, M., 1989. Radiocesium uptake in reindeer on natural pasture. The Science of the Total Environment 85, 207–212.

Kålås, J.A., Steinnes, E., Lierhagen, S., 2000. Lead exposure of small herbivorous vertebrates from atmos-pheric pollution. Environmental Pollution 107, 21–29.

Kardell, L., Källman, S., 1986. Heavy metals in antlers of roe deer from two Swedish forests, 1968–1983. Ambio 15, 232–235.

Karlen, G., Johanson, K.J., 1991. Seasonal variations in the activity concentration of ^{137}Cs in Swedish roe-deer and in their daily intake. Journal of Environmental Radioactivity 14, 91–103.

Karstad, L., 1967. Fluorosis in deer (*Odocoileus virginianus*). Bulletin of the Wildlife Disease Association 3, 42–46.

Kay, E., Tourangeau, P.C., Gordon, C.C., 1975. Industrial fluorosis in wild mule and white tail deer from western Montana. Fluoride 8, 192–191.

Kay, E., Tourangeau, P.C., Gordon, C.C., 1976. Populational variation of fluoride parameters in wild ungu-lates from the western United States. Fluoride 9, 73–90.

Kiefer, P., Pröhl, G., Müller, H., Lindner, G., Drissner, J., Zibold, G., 1996. Factors affecting the transfer of radiocaesium from soil to roe deer in forest ecosystems of southern Germany. The Science of the Total Environment 192, 49–61.

Kierdorf, H., Kierdorf, U., 1997. Disturbances of the secretory stage of amelogenesis in fluorosed deer teeth: a scanning electron microscopic study. Cell and Tissue Research 289, 125–135.

Kierdorf, H., Kierdorf, U., 1999a. Bleigehalte in Rothirschgeweihen aus Nordrhein-Westfalen: Ein Beitrag zum historischen Biomonitoring. (Lead content of red deer antlers from Northrhine-Westphalia: a contribution to historical biomonitoring). Zeitschrift für Jagdwissenschaft 45, 96–106.

Kierdorf, H., Kierdorf, U., 1999b. Reduction of fluoride deposition in the vicinity of a brown coal-fired power plant as indicated by bone fluoride concentration in roe deer (*Capreolus capreolus*). Bulletin of Environmental Contamination and Toxicology 63, 473–477.

Kierdorf, H., Kierdorf, U., 2000a. Roe deer antlers as monitoring units for assessing temporal changes in environmental pollution by fluoride and lead in a German forest area over a 67-year period. Archives of Environmental Contamination and Toxicology 39, 1–6.

Kierdorf, H., Kierdorf, U., 2000b. Vergleichende Untersuchungen zum Bleigehalt von Rehgeweihen aus verschiedenen Regionen Nordrhein-Westfalens (Deutschland) im Zeitraum 1990–1999. (Comparative analysis of the lead content of roe deer antlers from different regions in North Rhine-Westphalia (Germany) during the period 1990–1999). Zeitschrift für Jagdwissenschaft 46, 270–278.

Kierdorf, H., Kierdorf, U., Boyde, A., 1997a. A quantitative backscattered electron imaging study of hypomineralization and hypoplasia in fluorosed dental enamel of deer. Annals of Anatomy 179, 405–412.

Kierdorf, H., Kierdorf, U., Richards, A., Sedlacek, F., 2000a. Disturbed enamel formation in wild boars (*Sus scrofa* L.) from fluoride polluted areas in Central Europe. Anatomical Record 259, 12–24.

Kierdorf, H., Kierdorf, U., Sedlacek, F., 1999. Monitoring regional fluoride pollution in the Saxonian Ore mountains (Germany) using the biomarker dental fluorosis in roe deer (*Capreolus capreolus* L.). The Science of the Total Environment 232, 159–168.

Kierdorf, H., Kierdorf, U., Sedlacek, F., Erdelen, M., 1996a. Mandibular bone fluoride levels and occurrence of fluoride induced dental lesions in populations of wild red deer (*Cervus elaphus*) from Central Europe. Environmental Pollution 93, 75–81.

Kierdorf, U., 1988. Untersuchungen zum Nachweis immissionsbedingter chronischer Fluoridintoxikation beim Reh (*Capreolus capreolus*). (A study on chronic fluoride intoxication in roe deer (*Capreolus capreolus*) caused by immissions). Zeitschrift für Jagdwissenschaft 34, 192–204.

Kierdorf, U., Kierdorf, H., 1989. A scanning electron microscopic study on surface lesions in fluorosed enamel of roe deer (*Capreols capreolus*). Veterinary Pathology 26, 209–215.

Kierdorf, U., Kierdorf, H., 1999c. Dental fluorosis in wild deer: its use as a biomarker of increased fluoride exposure. Environmental Monitoring and Assessment 57, 265–275.

Kierdorf, U., Kierdorf, H., 2000c. Temporal and geographical variation in skeletal fluoride content of roe deer (*Capreolus capreolus*) from industrialized areas in Germany. Comparative Biochemistry and Physiology 126C, 61–68.

Kierdorf, U., Kierdorf, H., 2000d. Comparative analysis of dental fluorosis in roe deer (*Capreolus capreolus*) and red deer (*Cervus elaphus*): interdental variation and species differences. Journal of Zoology 250, 87–93.

Kierdorf, U., Kierdorf, H., 2000e. The fluoride content of antlers as an indicator of fluoride exposure in red deer (*Cervus elaphus*): a historical biomonitoring study. Archives of Environmental Contamination and Toxicology 38, 121–127.

Kierdorf, U., Kierdorf, H., 2001. Fluoride concentration in antler bone of roe deer (*Capreolus capreolus*) indicate decreasing fluoride pollution in an industrialized area of western Germany. Environmental Toxicology and Chemistry 20, 1507–1510.

Kierdorf, U., Kierdorf, H., Boyde, A., 2000b. Structure and mineralisation density of antler and pedicle bone in red deer (*Cervus elaphus* L.) exposed to different levels of environmental fluoride: a quantitative backscattered electron imaging study. Journal of Anatomy 196, 71–83.

Kierdorf, U., Kierdorf, H., Erdelen, M., Korsch, J.P., 1989. Mandibular fluoride concentration and its relation to age in roe deer (*Capreolus capreolus* L.). Comparative Biochemistry and Physiology 94A, 783–785.

Kierdorf, U., Kierdorf, H., Erdelen, M., Machoy, Z., 1995. Mandibular bone fluoride accumulation in wild red deer (*Cervus elaphus* L.) of known age. Comparative Biochemistry and Physiology 110A, 299–302.

Kierdorf, U., Kierdorf, H., Fejerskov, O., 1993. Fluoride-induced developmental changes in enamel and dentine of European roe deer (*Capreolus capreolus* L.) as a result of environmental pollution. Archives of Oral Biology 38, 1071–1081.

Kierdorf, U., Kierdorf, H., Krefting, E.R., 1991. Elektronenstrahl-Mikrosonden-Untersuchung des Fluorgehaltes und der Fluorverteilung in Schmelz und Dentin fluorotischer und nicht fluorotischer Backenzähne des Rehes (*Capreolus capreolus* L.). (Electron probe study on fluorine content and distribution in enamel and dentine of fluorosed and unfluorosed cheek teeth of roe deer (*Capreolus capreolus* L.)). Zeitschrift für Jagdwissenschaft 37, 232–239.

Kierdorf, U., Kierdorf, H., Sedlacek, F., Fejerskov, O., 1996b. Structural changes in fluorosed dental enamel of red deer (*Cervus elaphus* L.) from a region with severe environmental pollution by fluorides. Journal of Anatomy 188, 183–195.

Kierdorf, U., Richards, A., Sedlacek, F., Kierdorf, H. 1997b. Fluoride content and mineralization of red deer (*Cervus elaphus*) antlers and pedicles from fluoride polluted and uncontaminated regions. Archives of Environmental Contamination and Toxicology 32, 222–227.

Kim, S, Stair, E.L., Lochmiller, R.L., Rafferty, D.P., Schroder, J.L., Basta, N.T., Lish, J.W., Qualls, C.W., 2001. Widespread risks of dental fluorosis in cotton rats (*Sigmodon hispidus*) residing on petrochemical waste sites. Journal of Toxicology and Environmental Health A62, 107–125.

Kleiminger, J., Holm, J., 1985. Aufbau eines ursachenorientierten Monitoring-Systems für Schadstoffbelastungen beim Wild. 4. Wahl eines geeigneten Bioindikators für die Anzeige von Schadstoffen (Constructing a cause-oriented system for monitoring the contamination of game by harmful substances. 4. Choosing a suitable bioindicator of harmful substances). Fleischwirtschaft 65, 394–399.

Korte, F., 1982. Ecotoxicology of cadmium. Regulatory Toxicology and Pharmacology 2, 184–208.

Korte, F., 1987. Lehrbuch der Ökologischen Chemie. 2. Thieme, Stuttgart.

Kreuzer, W., Bunzl, K., Kracke, W., 1981. Zum Übergang von Cadmium aus dem Futter in Nieren, Lebern und Muskulatur von Schlachtrindern. (Transfer of cadmium from feed to the kidneys, livers and musculature of cattle). Fleischwirtschaft 61, 1886–1894.

Kreuzer, W., Kracke, W., Sansoni, B., Wißmath, P., 1978. Untersuchungen über den Blei- (Pb)- und Cadmium-(Cd)-Gehalt in Fleisch und Organen von Schlachtrindern. 1. Rinder aus einem wenig umweltbelasteten Gebiet. (Lead (Pb) and cadmium (Cd) contents in the meat and organs of slaughter cattle. 1. Cattle from an area with little environmental contamination. Fleischwirtschaft 58, 1022–1030.

Lorenz, H., 1979. Binding form of toxic heavy metals, mechanisms of entrance of heavy metals in the food chain and possible measures to reduce their level in foodstuff. Ecotoxicology Environmental Safety 3, 47–58.

Ma, W.-C., 1989. Effect of soil pollution with metallic lead pellets on lead bioaccumulation and organ/body weight alterations in small mammals. Archives of Environmental Contamination and Toxicology 18, 617–622.

Ma, W.-C., Denneman, W., Faber, J., 1991. Hazardous exposure of ground-living small mammals to cadmium and lead in contaminated terrestrial ecosystems. Archives of Environmental Contamination and Toxicology 20, 266–270.

Machoy, Z., Dabkowska, E., Nowicka, W., 1991. Increased fluoride content in mandibular bones of deer living in industrialized regions of Poland. Environmental Geochemistry and Health 13, 161–163.

Machoy, Z., Dabkowska, E., Samujlo, D., Ogonski, T., Raczynski, J., Gebczynska, Z., 1995. Relationship between fluoride content in bones and the age in European elk (*Alces alces* L.). Comparative Biochemistry and Physiology 111C, 117–120.

Mahaffey, K.R.,1980. Nutrient-lead interactions. In: Singhal, R.S., Thomas, J.A. (Eds), Lead Toxicity. Urban and Schwarzenberg, Baltimore, pp. 425–460.

Medvedev, N., 1995. Concentration of cadmium, lead and sulphur in tissues of wild, forest reindeer from north-west Russia. Environmental Pollution 90, 1–5.

Moreth, F., Hecht, H., 1981. Blei aus Geschossrückständen in Wildbret. Fleischwirtschaft 61, 1326.

Mössner, S., Ballschmiter, K.H., 1997. Marine mammals as global pollution indicators for organochlorines. Chemosphere 34, 1285–1296.

Newman, J.R., Murphy, J.J., 1979. Effects of industrial fluoride on black-tailed deer (preliminary report). Fluoride 12, 129–135.

Newman, J.R., Yu, M.H., 1976. Fluorosis in black-tailed deer. Journal of Wildlife Diseases 12, 39–41.

O'Brien, D.J., Kaneene, J.B., Poppenga, R.H., 1993. The use of mammals as sentinels for human exposure to toxic contaminants in the environment. Environmental Health Perspectives 99, 351–368.

Onderscheka, K. (Ed.), 1982. Normalwerte von Wildtieren. Teil 1: Feldhase (*Lepus europaeus P.*). (Reference values of wildlife. Part 1: European Brown Hare (*Lepus europaeus P.*)). Forschungsinstitut für Wildtierkunde, Wien.

Paranjpe, M.G., Chandra, A.M.S., Qualls, C.W., McMurry, S.T., Rohrer, M.D., Whaley, M.M., Lochmiller, R.L., McBee, K, 1994. Fluorosis in a wild cotton rat (*Sigmodon hispidus*) population inhabiting a petrochemical waste site. Toxicologic Pathology 22, 569–578.

Pirkle, J.L., Sampson, E.J., Needham, L.L., Patterson, D.G., Ashley, D.L., 1995. Using biological monitoring to assess human exposure to priority toxicants. Environmental Health Perspectives 103 (Suppl. 3), 45–48.

Pokorny, B., Ribaric-Lasnik, C., 2000. Lead, cadmium and zinc in tissues of roe deer (*Capreolus capreolus*) near the lead smelter in the Koroska region (northern Slovenia). Bulletin of Environmental Contamination and Toxicology 64, 20–26.

Rafferty, D.P., Lochmiller, R.L., Kim, S., Qualls, C.W., Schroder, J, Basta, N., McBee, K., 2000. Fluorosis risks to resident hispid cotton rats on land-treatment facilities for petrochemical wastes. Journal of Wildlife Diseases 36, 636–645.

Robinson, C., Kirkham, J., Weatherell J.A., 1996. Fluoride in teeth and bones. In: Fejerskov, O., Ekstrand, J., Burt, B.A. (Eds), Fluoride in Dentistry, 2nd edn. Munksgaard, Copenhagen, pp. 69–87.

Roholm, K., 1937. Fluorine intoxication, a clinical-hygienic study. H.K. Lewis, London.

Samujlo, D., Machoy-Mokrzynska, A., Dabkowska, E., Nowicka, W., Paterkowski, W., 1994. Fluoride accumulation in European deer antlers. Environmental Sciences 2, 189–194.

Sawicka-Kapusta, K., 1979. Roe deer antlers as bioindicators of environmental pollution in southern Poland. Environmental Pollution 19, 283–293.

Schinner, W., 1981. Untersuchungen über endogene und exogene Einflüsse auf den Blei (Pb) – und Cadmium (Cd) – Gehalt in Muskeln und Organen von Rehwild (*Capreolus capreolus L.*) und Wildkaninchen (*Lepus cuniculus L.*) (Endogeneous and exogenus influences on the lead and cadmium content in muscles and organs of roe deer and wild rabbits). Thesis Univ. Giessen.

Schönhofer, F., Tataruch, F., Friedrich, M., 1994. Strontium-90 in antlers of red deer: an indicator of environmental contamination by strontium-90. The Science of the Total Environment 157, 323–332.

Schroder, J.L., Basta, N.T., Rafferty, D.P., Lochmiller, R.L., Kim, S., Qualls, C.W., McBee, K., 1999. Soil and vegetation fluoride exposure pathways to cotton rats on a petrochemical-contaminated landfarm. Environmental Toxicology and Chemistry 18, 2028–2033.

Schultz, M., Kierdorf, U., Sedlacek, F., Kierdorf, H., 1998. Pathological bone changes in the mandibles of wild red deer (*Cervus elaphus*) exposed to high environmental levels of fluoride. Journal of Anatomy 193, 431–442.

Schultz, V., 1964. Sampling white-tailed deer antlers for strontium-90. Journal of Wildlife Management 28, 45–49.

Seeger, R., 1978. Cadmium in Pilzen. (Cadmium in mushrooms). Zeitschrift für Lebensmittel-Untersuchung und – Forschung 166, 23–34.

Shupe, J.L., Olson A.E., 1983. Clinical and pathological aspects of fluoride toxicosis in animals. In: Shupe, J.L., Peterson, H.B., Leone, N.C. (Eds), Fluorides – Effects on Vegetation, Animals and Humans. Paragon Press, Salt Lake City, pp. 319–338.

Shupe, J.L., Olson, A.E., Peterson, H.B., Low, J.B., 1984. Fluoride toxicosis in wild ungulates. Journal of the American Veterinary Medical Association 185, 1295–1300.

Smith, F.A., Ekstrand, J., 1996. The occurrence and the chemistry of fluoride. In: Fejerskov, O., Ekstrand, J, Burt, B.A. (Eds), Fluoride in Dentistry, 2nd edn. Munksgaard, Copenhagen, pp. 17–26.

Stoeppler, M., 1991. Cadmium. In: Merian, E. (Ed.), Metals and their Compounds in the Environment. Chemie, Weinheim, pp. 803–852.

Strandberg, M., Strandgaard, H., 1995. ^{90}Sr in antlers and bone of a Danish roe deer population. Journal of Environmental Radioactivity 27, 65–74.

Suttie, J.S., Dickie, R., Clay, A.B., Nielsen, P., Mahan, W.E., Baumann, D.P., Hamilton, R.J., 1987. Effects of fluoride emissions from a modern primary aluminum smelter on a local population of white-tailed deer (*Odocoileus virginianus*). Journal of Wildlife Diseases 23, 135–143.

Suttie, J.W., Hamilton, R.J., Clay, A.C., Tobin, M.L., Moore, W.G., 1985. Effects of fluoride on white-tailed deer (*Odocoileus virginianus*). Journal of Wildlife Diseases 21, 283–288.

Talmage, S.S., Walton, B.T., 1991. Small mammals as monitors of environmental contaminants. Reviews of Environmental Contamination and Toxicology 119, 47 145.

Tataruch, 1984. Untersuchungen zur Schwermetallbelastung der Feldhasen (*Lepus europaeus P.*) in Österreich. (Investigations on the heavy metal contamination of European brown hare in Austria). Habilitation thesis, University of Veterinary Medicine, Vienna.

Tataruch, F., 1991. Freilebende Wildtiere als Bioindikatoren der Schwermetallkontamination. (Free ranging wild animals as bioindicators of heavy metal contamination). VDI-Berichte 901, 925–936.

Tataruch, F., 1993a. Vergleichende Untersuchungen zur Schwermetallbelastung von Rot-, Reh- und Gamswild. (Comparable investigations on the heavy metal contamination of red deer, roe deer and chamois). Zeitschrift für Jagdwissenschaft 39, 190–200.

Tataruch, F. 1993b. Die Belastung freilebender Wildtiere mit Umweltschadstoffen. (Contamination of wildlife by environmental pollutants). Übersichten Tierernährung 21, 181–204.

Tataruch, F., 1995. Red deer antlers as biomonitors for lead contamination. Bulletin of Environmental Contamination and Toxicology 55, 332–337.

Tataruch, F., 2000. Monitoring of the heavy metal contamination of brown hares in Austria during the last decades. In: Fernandez de Luco, D. et al. (Eds) Proceedings of the Fourth Meeting of the European Wildlife Disease Association. SEDIFAS, Zaragoza, p. 24

Tataruch, F., Onderscheka, K., 1981. Belastung freilebender Tiere in Österreich mit Umweltschadstoffen (III): Gehalt an Quecksilber in Organen von Feldhasen. (Contamination of wild animals with environmental pollutants (III). Mercury content of organs of European brown hare). Zeitschrift für Jagdwissenschaft 27, 266–270.

Tataruch, F., Schönhofer, F., 1993. Reconstruction of environmental contamination of past decades by chemical analyses of red and roe deer antlers. In: Thompson, I.D. (Ed.), Proceedings of the XXI IUGB-Congress Vol. 2, Halifax, Canada, pp. 23–28.

Tataruch, F., Schönhofer, F., Klansek, E., 1990. Studies in levels of radioactivity in wildlife in Austria. In: Desmet, G., Nassimbeni, P., Belli, M. (Eds), The Transfer of Radionuclides in Natural and Seminatural Environments. Elsevier Appl. Science, London, pp. 210–216.

Tataruch, F., Schönhofer, F., Klansek, E., 1996. Radiocesium levels in roe deer and wild boar in two large forest areas in Austria. In: Gerzabek, M., Desmet, G., Howard,B.J., Heinrich, G., Schimmack, W., (Eds), Ten Years Terrestrial Radioecological Research Following the Chernobyl Accident. Proc. Int. Symp. Radioecology, pp. 285–293.

Tataruch, F., Schönhofer, F., Onderscheka, K., 1988. Untersuchungen zur radioaktiven Belastung der Wildtiere in Österreich. (Studies in levels of radioactivity in wild animals in Austria). Zeitschrift für Jagdwissenschaft 34, 22–35.

Tiller, B.L., Poston, T.M., 2000. Mule deer antlers as biomonitors of strontium-90 on the Hanford site. Journal of Environmental Radioactivity 47, 29–44.

Turner, C.H., Boivin, G., Meunier, P.J., 1993. A mathematical model for fluoride uptake by the skeleton. Calcified Tissue International 52, 130–138.

Vikøren, T., Stuve G., 1996. Fluoride exposure in cervids inhabiting areas adjacent to aluminum smelters in Norway. II. Fluorosis. Journal of Wildlife Diseases 32, 181–189.

Vikøren, T., Stuve, G., Frøslie, A., 1996. Fluoride exposure in cervids inhabiting areas adjacent to aluminum smelters in Norway. I. Residue levels. Journal of Wildlife Diseases 32, 169–180.

Von Burg, R., Greenwood, M.R., 1991. Mercury. In: Merian, E. (Ed.), Metals and their compounds in the environment. VCH, Weinheim, pp. 1045–1088.

Walton, K.C., 1988. Environmental fluoride and fluorosis in mammals. Mammal Review 18, 77–90.

Walton, K.C., Ackroyd, S., 1988. Fluoride in mandibles and antlers of roe and red deer from different areas of England and Scotland. Environmental Pollution 54, 17–27.

Whitford, G., 1996. The metabolism and toxicity of fluoride, 2nd ed. Monographs in Oral Science 16, Karger, Basel

WHO, 1984. Fluorine and fluorides. Environmental health criteria 36. World Health Organization, Geneva.

International programmes for biomonitoring purposes

Chapter 21

Environmental specimen banking

A.A.F. Kettrup

Abstract

Thousands of chemicals are traded on the market, but only in a few cases is full information available on their distribution and their effects on man and the environment. In addition to real time monitoring of chemicals, it is, thus, necessary to establish an environmental specimen bank (ESB) for the retrospective monitoring of chemicals in the future. The base of ESB are representative bioindicators of systematically collected biological and environmental samples. As a result of extremely low storage temperature $(T < -150°C)$ it is guaranteed that the samples are not subject to chemical changes during the long-term storage.

Keywords: environmental, specimen banking, bioindicator, bioavailability, environmental monitoring.

1. Introduction

According to the "European Inventory of Existing Commercial Substances" EINECS more than 100,000 different chemical substances are produced world-wide. One thousand to 2000 new chemicals are entering the market every year in addition to those already in circulation. For most of them we lack sufficient information about their effects on man, animals and plants and about their further reaction and fate in the environment; SRU, 1987). New technologies always produce unintended and unpredicted waste and impact. In most cases the introduction of chemicals into the environment represents an irreversible step. A considerable number of chemicals reaching the environment do not degrade at all or only very slowly. They accumulate in the environment and, having distributed, become ubiquitous of certain pollutants e.g. PCBs, chlorinated insecticides. Persistent biologically-active chemicals, sometimes at concentrations below our ability to analyse or detect, can pose serious pervasive and possible irreversible threats to human health and the integrity of the biosphere (Lewis, 1988).

Numerous industrial countries have passed laws in order to assess the hazards of chemicals for man and the environment. Investigations on forecasting the impact of chemicals are, however, afflicted by special costs and problems.

Problems are:

- around 10% of chemicals in the environment are actually monitored;
- there are knowledge deficiencies about toxicological properties, metabolism and behaviour of chemicals in the environment;
- there are deficiencies in analytical methods.

So, these problems are resulting in a lack of knowledge regarding damages in environment, human health risks, incorrect assessments and predictions. Persistent organic pollutants (POPs) are still under discussion, especially halogenated organic pollutants (Fiedler et al., 2000).

Effective damage protection and risk contaminant strategies can only be developed by policy makers if they have reliable basic information on chemical disposal.

The idea that organisms can provide an indication of the quality of their environment is widespread and at least as old as agriculture (Thalius, 1588). It is not possible to establish any clear definition for the term "bioindicator" considering the large number of literature published in the last decades. The following definitions are suggested in accordance with many European authors (Wittig, 1993):

- Bioindication is the use of an organism (a part of an organism or a society of organisms) to obtain information on the quality (of a part) of its environment. Organisms which are able to give information on the quality (of a part) of its environment are bioindicators.
- Biomonitoring is the continuous observation of an area with the help of bioindicators, which in the case may also be called biomonitors. With the aid of organisms trends in time and space concerning the distribution and ecological effects of environmental chemicals can be observed by a semi-quantitative evaluation of the results.

Biological samples from the environment are mainly used and analysed as representatives for larger entities or similar or related environmental compartments. This requires the selection of standardised (bio)indicatior systems, which react with known specifity and sensitivity to environmental chemicals and have the capability of spatial and/or temporal integration. Such indicator systems can be efficiently and reproducibly analysed and evaluated vicariously for the total entity of sensitive targets in the environment to be observed, which are often extremely variable with respect to the space, time and physiology. Bioindicator systems in such cases, where potential integral effects of complex or unknown immission types have to be detected and quantified. Such effects may occur on different levels from specific organs of single organism up to whole ecosystems. Bioindicators are also preferred in such cases, where they offer advantages due to their high sensitivity towards a broad spectrum of substances or because of their ability to accumulate a substance over an extended period of time or to integrate its influence in an area of known and relevant size. This is namely the case, if the sensitiveness of available analytical methods for dangerous substances is too low to find them in other environmental compartments like air, water and soils.

The most important property for using organisms as bioindicators is the bioavailibility of a chemical.

In the case of an earthworm (e.g. *Lumbricus terrestris*) the bioavailability is dependent on:

- properties of the chemical
  - Kow
  - dissociation constant
  - affinity to charged surfaces

- properties of the soil
  - organic matter content
  - pH
  - clay content

- properties of the organism
  - uptake kinetics
  - physiological condition

In the case of an earthworm it is relatively easy to determine the age and the physiological condition. Furthermore it is easy to sample an sufficient amount of the specimen. Earthworms like *Lumbricus terrestis* or *Aporrectodea longa* are good bioindicators for chlorinated hydrocarbons as can be shown in Figure 1.

Using earthworms as bioindicator it is possible to show that the soil contamination with chlorinated phenols is quite the same in the area of Leipzig (former DDR) and Saarbrücken near the French border.

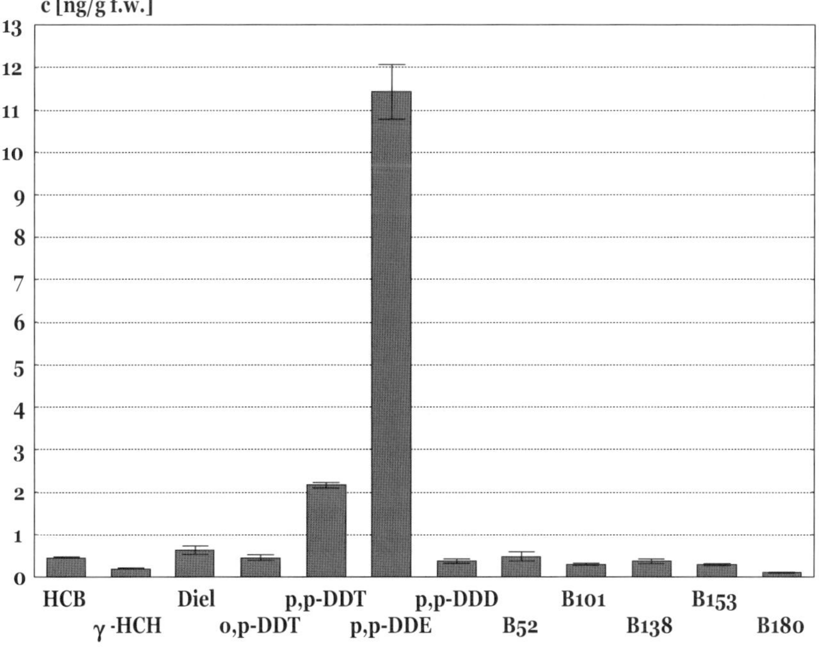

Figure 1. Sum of chlorinated hydrocarbons in *Lumbricus terrestris* sampled in the area of Leipzig.

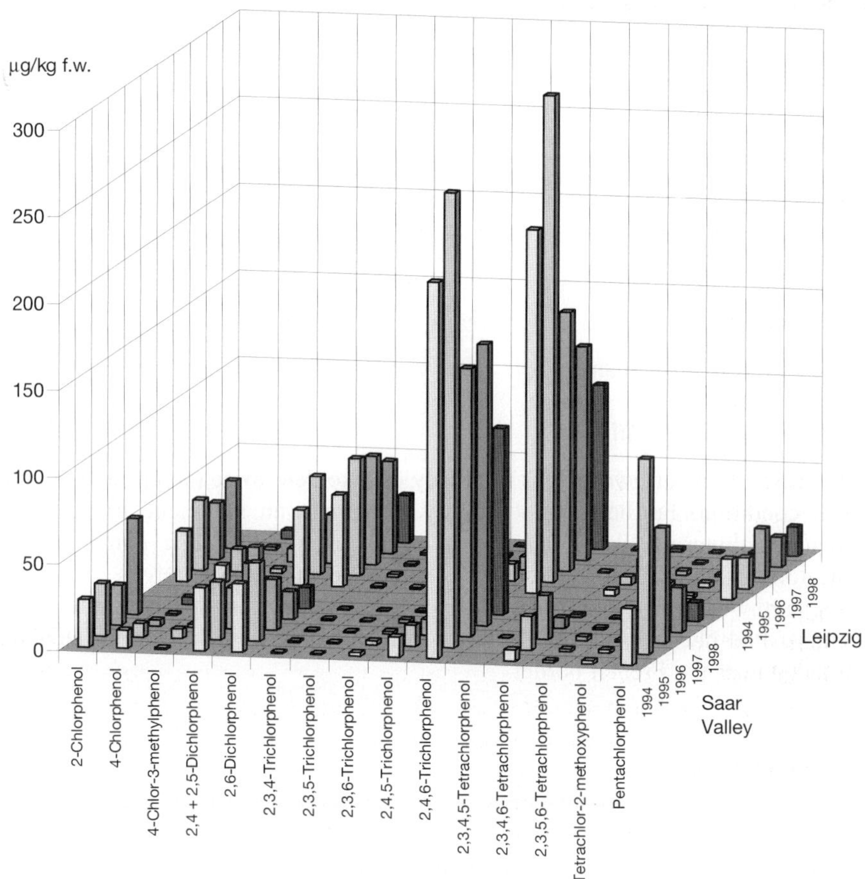

Figure 2. Amount of different chlorinated phenols in *Lumbricus terrestris.*

It is obvious that chlorophenols show the highest concentration in earthworms in both areas. Chlorphenols are used as an intermediate in the chemical industry but also as disinfectants, biocides, preservatives and pesticides.

In addition to the concentrations of toxic substances and their metabolites biological specimen can also be analysed for essential components and a broad spectrum of possible biochemical, physiological, morphological and/or genetical effects. Organisms and biological communities normally do not react to single components or substances in their environment. They show the effects of the totality of all the acting substances and environmental factors. Decisive for the use of biological specimen is their ecotoxicological relevance, that means the relevance or indicative function of the found effects for other living organisms and communities including men.

2. Idea of environmental specimen banking

With respect to effects of pollutants, their quantities and distribution under natural conditions the acquisition of reliable information requires a systematic programme of environmental monitoring in which concentrations of hazardous chemical substances are measured in suitable environmental specimen of various trophic stages and food chains. But actual monitoring of the environment can only be as good as our present knowledge and as analytical possibilities allow. From among the multitude of substances found in the environment only those can be monitored, which have already been recognised to be hazardous substances – and thus, the quality of regulatory decisions – suffers from the act that no results are available on pollutant burdens of former times or that the data which are available are ambiguous (Kayser et al., 1982).

Before this background at the beginning of the 1970s the idea of using biological samples as reference material to furnish proof of environmental pollution was put forward by Frederick Coulston of the Albany Medical College (Albany/NY), and Friedhelm Korte of the Institute of Ecological Chemistry, GSF-Forschungszentrum, Munich-Neuherberg.

In an environmental specimen bank (ESB) carefully selected, relevant environmental samples are stored systematically at temperatures below $-150°C$ immediately after collection. In this way no, or as small as possible, chemical changes occur over a long period of time. Baseline levels of contaminants in the environment can be established by taking samples at the present time for future demand in ecological-chemical research. Long-term storage of samples with indicator functions represents an necessary complement to the actual monitoring of the environment and a safety net in the assessment of chemical risk.

A systemically established archive of frequently collected representative environmental specimen samples fullfil the following important functions (Lewis, 1988; Kayser et al., 1982; Wise and Zeisler, 1984; Keune, 1993):

- They may be used for the determination of the environmental concentrations of those substances, which, at the time of storage, were not recognised to be hazardous or which at present cannot be analysed with adequate accuracy (retrospective monitoring)
- They may serve as reference samples for the documentation of the improvement of analytical efficiency and for the verification of previously obtained monitoring results.
- Early detection of environmental increases in hazardous chemicals thought to be under control is possible. Also, the effectiveness of restrictions, regulations, or management practices that have been applied to the community, the environmental or to the manufacture, distribution, disposal, or use of toxic chemicals can be assessed.
- Depending upon the analysis and evaluation of stored materials ESB can save considerable time and money when unexpected impacts are observed.
- Sources of chemical may be identified. Often, by the time a chemical is recognized as a health or environmental problem it is sufficiently wide-spread to defy identification of the principal sources or pathways.

- ESB can offset the lack of reliable data on pollutant burdens of earlier times because inconsistencies or ambiguities among available data usually limit assessments and regulatory decisions.

In Germany the Federal Minister for Research and Technology supported a comprehensive pre- and pilot phase of ESB between 1976 and 1984. During this period the technical feasibility regarding the sampling of different species, handling and shipping of samples, deep freezing, homogenisation, ultra trace analysis, packing materials, logistics, storage temperature and documentation was confirmed (Boehringer, 1988). The results were so encouraging that in 1985 the German government decided to set up a permanent ESB under the responsibility of the Federal Ministry for the Environment, Nature Conservation and Reactor Safety (BMU), coordinated by the Federal Environmental Agency (Umweltbundesamt). Two specimen banks are subsumed under the general heading of the German ESB:

- The Specimen Bank for Environmental Specimens at the Institute of Applied Physical Chemistry of the Research Centre Jülich (KFA) and since 2000 at the Fraunhofer Institute of Environmental Chemistry and Ecotoxicology/Schmallenberg.
- The Specimen Bank for Human Organ Specimens at the Institute of Pharmacology and Toxicology of the University of Münster.

The work was distributed among these and some other institutions depending on their special scientific capabilities:

- Institute of Biogeography/University of the Saarland.
- Institute of Ecological Chemistry of the Natl. Research Centre of Environment and Health/Munich-Neuherberg.
- Biochemical Institute for Environmental Carcinogens/Großhansdorf.
- ERGO/Hamburg.

In Germany environmental specimen banking has been successfully established as a permanent environmental surveillance tool.

The banking activities are focused on the preparation, characterisation and storage of representative samples from different ecosystems in Germany. Based on reliable and well-documented analytical procedures the obtained monitoring results offers the opportunity for long-term control of environmental pollution by spatial differences and time-dependent trends.

In the meantime an international cooperation of ESBs in the Federal Republic of Germany, USA, Canada, Japan, Finland, Sweden, Norway and Denmark has been established (Wise et al., 1988; Stoeppler and Zeisler, 1993).

3. Sampling areas and specimen types

Sampling areas have been chosen as to form a national network of Ecological Assessment Parks coordinating Environmental Specimen Banking with long-term ecological research and environmental monitoring (Klein et al., 1994; Lewis, 1987; Lewis et al., 1993). An overall concept has been developed by a committee of

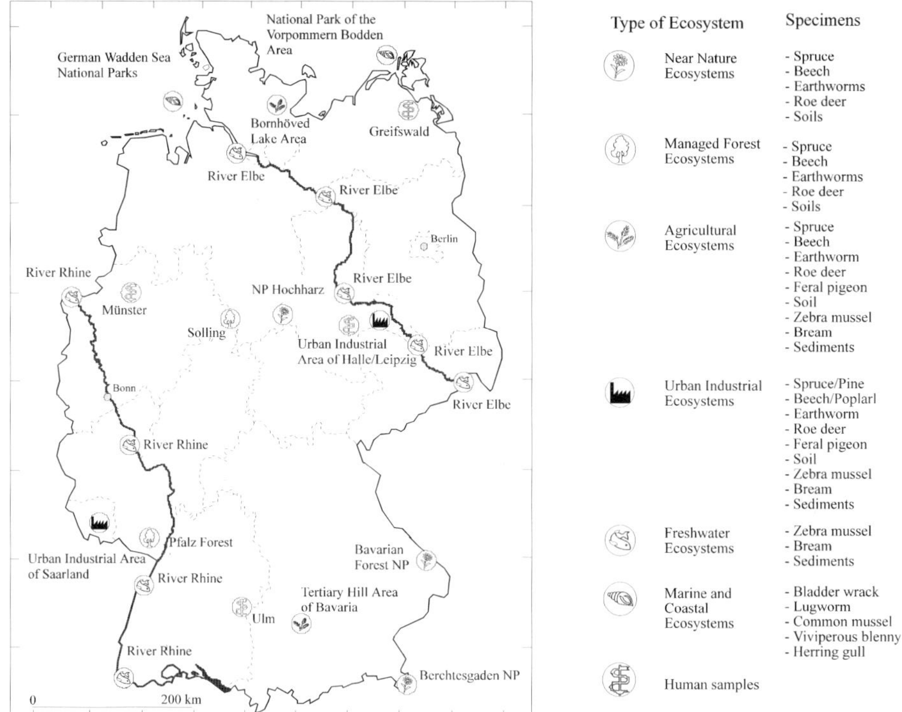

Figure 3. Sampling areas and specimens types (after Klein et al., 1994).

experts under the auspices of the BMU, taking into consideration different types of ecosystems with corresponding representative sampling areas according to the following criteria:

- stability of utilization;
- assured long-term-use;
- sufficient minimal size;
- availability of suitable samples;
- practicability e.g. accessibility, public ownership (National Park), no conflict with the protection of biotopes and species, high level of information, nearby suitable institutions for research.

The list of a present fourteen areas (see Fig. 3) comprises the major ecosystems and habitat types that occur within the Federal Republic of Germany including:

- limnic and marine ecosystems;
- urban industrial ecosystems;
- forest and agricultural ecosystems; and
- semi-natural ecosystems.

Continuous sampling is now performed in the following areas:

- the national parks of mud flats in Schleswig-Holstein and Lower Saxony (North Sea) and the Baltic Sea as marine ecosystems;
- River Elbe and River Rhine as limnic ecosystems;
- Saarland and the Halle/Leipzig/Bitterfeld area as urban-industrialized regions.

The selection and assignment of representative specimen of the terrestrial, limnic and marine ecosystems for the ESB was undertaken by a committee of experts in consideration of the above mentioned indicator functions so that a broad spectrum of different types of matrices (all trophic levels) and media (air, sediment, soil) with environmentally relevant concentrations of xenobiotics is available (see Fig. 3) (Klein et al., 1994; Lewis et al., 1993).

Furthermore, the following requirements must be fulfilled for using a matrix as bioindicator:

- The chemicals must be accumulated comparable to levels occurring in the environment.
- Contamination trends in the environment must correspond to those in the matrix.
- The matrix should have a widespread distribution and must be available in time and place to a sufficient extent.
- The organism should be sedentary and easy to identify.
- The species should accumulate the pollutants without being killed or rendered incapable of reproduction.

Standardized sampling guidelines in the sense of standard operating procedures (SOP) are the basis for the comparability, reliability, and repeatability of the banked samples.

They contain detailed instructions for the

- selection of sampling sites and specimens;
- sampling;
- providing cover for repeatability of sampling;
- area and sample characterisation;
- sample treatment and long-term storage;
- documentation of sampling and storage conditions;
- chemical analysis;
- data processing and evaluation; and
- quality assurance.

Nevertheless, sampling of biological and other environmental specimens is always influenced by factors which may modify the exposure as well as the accumulation behaviour of the specimen types in relation to xenobiotics, e.g. by climatic factors, weather conditions and changes in the population sampled or in the structure of the whole ecosystem (Wagner, 1995).

Ecological and biometrical sample characterisation provides basic information about changes in the quality of the sampled material and its comparability with previous and following samples from the same area or the same specimen type sampled in other areas. On the other hand biological sample characterisation can also give information about ecological and ecotoxicological effects to the population sampled.

In environmental specimen banking samples of different specimen types and ecosystems are frequently sampled, characterised, processed and stored with considerable effort in order to maintain the precautions necessary for deferred analysis on initially unknown substances or parameters. Quality assurance is therefore an absolute demand and an innovative challenge in ESB. Errors made during the sampling in the field, transportation and sample pre-treatment can seldom be recognised and never corrected afterwards during the following analytical measurements. Thus, the quality assurance system for ESB includes the whole process from planning, sampling, ecological and biometrical characterisation, packing, transportation, storage, homogenisation and sub-sampling up to the analytical procedures and the evaluation of the results (Emons, 1994; Grimmer et al., 1996; Jacob et al., 1996; Oxynos et al., 1992; Paulus et al., 1992, 1995; UPB, 1996).

An average 2.5 kg of material per specimen per sampling site was collected producing nearly 250 standardized sub-samples of approximately 10 g.

4. Analytical sample characterisation

The choice of pollutants or classes of pollutants for the analytical sample characterisation took place according to ecotoxicological importance as it was understood and decided two decades ago. Analytes chosen were:

- heavy metals;
- polycyclic aromatic hydrocarbons (PAHs);
- chlorinated hydrocarbons, especially POPs.

The analytical procedures for heavy metals in ESB samples were selected with emphasis on trace analysis capability and applicability for very complex biological matrices. During the pilot phase sample preparation was optimised in dependence on the matrix to be analysed and the detection technique to be used (Standard Operating Procedure of the German Environmental Specimen Bank). The following four groups of analytical methods are applied for trace analysis:

- Atomic Spectrometry (GF-AAS, CV-AAD, Hydride AAS, ICP-AES);
- Mass Spectrometry (IDMS, ICP-MS);
- Electrochemical methods (PSA, Voltammetry);
- Radiochemical methods gas-chromatography (INAA, PGCNAA).

An important aspect of the ESB consists in the long-term monitoring of heavy metals in biological samples between different regions in Germany. For example, a clear decline in mercury pollution in the estuary region of the Elbe in the past few years can be documented on the basis of samples of herring gull eggs (*Larus argentatus*) collected in the Trischen bird sanctuary (Grimmer, 1993). More than 90% of the mercury was present in the form of the highly toxic methyl mercury. Before the unification of the two German states (1988/89) the mercury concentrations in herring gull eggs from Trischen island was twice as high as for the subsequent years (1991–1995). The decreasing temporal trend is probably associated with the closure of industrial plants in the upper regions of the Elbe and its tributaries.

The determination of PAHs is of very high importance because they are considered to be the most relevant class of environmental carcinogens (Nesmerak, 1993). PAH immission sources results from incomplete fossile combustion, wood burning, waste gases of industrial and private combustion processes. The continuous measurement of PAH concentration in various selected environmental matrices representing the terrestrial, aquatic, and atmospheric ecosystem provides the opportunity to recognise trends in the environmental pollution.

For PAH determination sample homogenates are extracted with toluene or cyclohexane by Soxhlet. Interferences are removed by liquid/liquid distribution and chromatography on silica and Sephadex LH20. The detection of PAHs is carried out by gas chromatography equipped with FID or MS (SIM) detectors (Kettrup et al., 1999).

Spruce and pine sprouts are passive sampler reflecting the atmospheric pollutants by PAHs (Oxynos et al., 1993). For example, the Benzo(a)pyrene-concentration of spruce sprouts from the industrialised area of Saarland decreased by a factor of 3 within 10 years.

The same trend in pine sprouts has been observed for a sampling area in East Germany (Dübener Heide) during the period 1991–1995 with B(a)P declining from 3.5 µg/kg to 1.5 µg/kg. These findings show that the reduction of pollution by technical improvements such as modern vehicle conceptions as well as improved domestic and industrial devices in the past were successful.

Mussels *(Mytilus edulis)* as a bioindicator of the marine environment for the North Sea (Eckwarderhörne) exhibited likewise a decline of the PAH concentration although to a lesser extent (Oxynos et al., 1993). With the exception of 1988 when higher concentrations have been measured, the B(a)P decreased from 1.9 µg/kg in 1985 to 1.2 µg/kg in 1990 and remained practically constant. Long distance transfer by atmospheric pollutants from other countries and the dilution effect play an important role for the contamination of the North Sea.

For the determination of chlorinated hydrocarbons the samples are mixed with anhydrous sodium sulfate/seasand to form a free flowing product, which is extracted with n-hexane/acetone (2:1 v/v) in an extraction column. Clean up procedure is performed with gel-permeation and high performance liquid chromatography. Quantification is carried out by high resolution gas chromatography (HRGC) equipped with electron capture detection using two columns of different polarity. More details about the applied method are given in (Nesmerak, 1993; Oxynos et al., 1995; Schramm et al., 1996; Kettrup et al., 1999).

Due to the fact that the GSF-Institute of Ecological Chemistry was in charge for the analysis of chlorinated hydrocarbons some results are discussed in Section 5.

5. Chlorinated hydrocarbons as pollutants in ESB samples

Numerous chlorinated insecticides and industrial chlorinated hydrocarbons (e.g. polychlorinated biphenyls (PCBs) or polychlorinated dibenzo-p-dioxins and dibenzofurans (PCDD/F) are extremely resistant to degradation in the environment. Residues of these xenobiotics have been identified throughout the world although most of them have been banned since the 1970s. Because of their toxicological properties and accumulation effects, long-term studies on their residues levels are essential to understand the environmental contamination in the past and to predict future trends.

The former German Democratic Republic (GDR) was one of the major producers of elemental chlorine in East Europe (Heinisch, 1992). Due to economical reasons the production and application of low volatile CHC played a more important role in the chemical pest control in agriculture and forestry of East Germany than in the Western countries (Greenpeace, 1990). Under the pressure of worldwide application restrictions the production has ceased since the early 1970s (DDT) or late 1980s (lindane, toxaphene) respectively. Thus, the different economic and technical development of the two German states led to partly different patterns of environmental pollution. Reliable data of the pollution situation, especially of the industrialized areas, in the former GDR were not available on account of the strict political restriction regime. In order to detect the efficiency of legislative regulations for the environmental redevelopment, which were initiated directly after the reunification, and for the assessment of the environmental pollution a special sampling campaign was performed in representative areas of the former GDR in 1990/91 (Anon., 1994). These samples can be regarded as a basis for monitoring the development of spatial and temporal trends in Eastern Germany.

In the marine ecosystem herring gull eggs are a suitable bioindicator for chlorinated insectides, polychlorinated biphenyls (PCB) and polychlorinated dibenzo-p-dioxins and dibenzo-furans since they express the local and temporal environmental conditions (Oxynos et al., 1995).

The time series of herring gull eggs at the sampling locations on the islands of Mellum and Trischen are shown in Figure 4. The location influenced by the river Elbe (Trischen) exhibited higher concentrations than the location Mellum, which is influenced by fluxes of pollutants in the river Weser.

The significantly higher DDE concentrations in eggs from Trischen in 1989 can be explained by DDT applications in the former GDR in the mid 1980s.

One aim of the sampling campaign was the characterisation of the CHC burden along the German part of the River Elbe and to monitor the expected recovery of the river. For the past 20 years the River Elbe has been one of the most polluted rivers in Europe (Fraucke et al., 1994). Due to political changes in the Eastern part of Europe and especially in East Germany, various industries, who discharged their waste water into the river Elbe, were either closed or reduced their production activities. In order to characterise local variations of the CHC burden seven stations were selected with regard to various types and magnitudes of industrial pollution as shown in Figure 6 (Marth et al., 1999a).

In addition, bream samples were collected in nine lakes of Mecklenburg which were mainly influenced by agriculture, livestock fattening and forest industry. Figures 8–10 present a comparison of the main contaminants in bream muscle homogenates from the River Elbe and the lake area of Mecklenburg.

As shown in Figures 7–9 fish caught at the upper Elbe (Prossen and Löschwitz) exhibited the highest concentrations of DDE, OCS and PCB. This observation is probably a result of the considerable pollution of the River Elbe from industrialised areas (e.g. Pardubice, Neratovice, Usti) of the former CSFR (Marth et al., 1999b).

A significant decrease of these contaminants was found between stations Löschwitz and Vockerode. Between Vockerode and Heinrichsberg the levels remained nearly

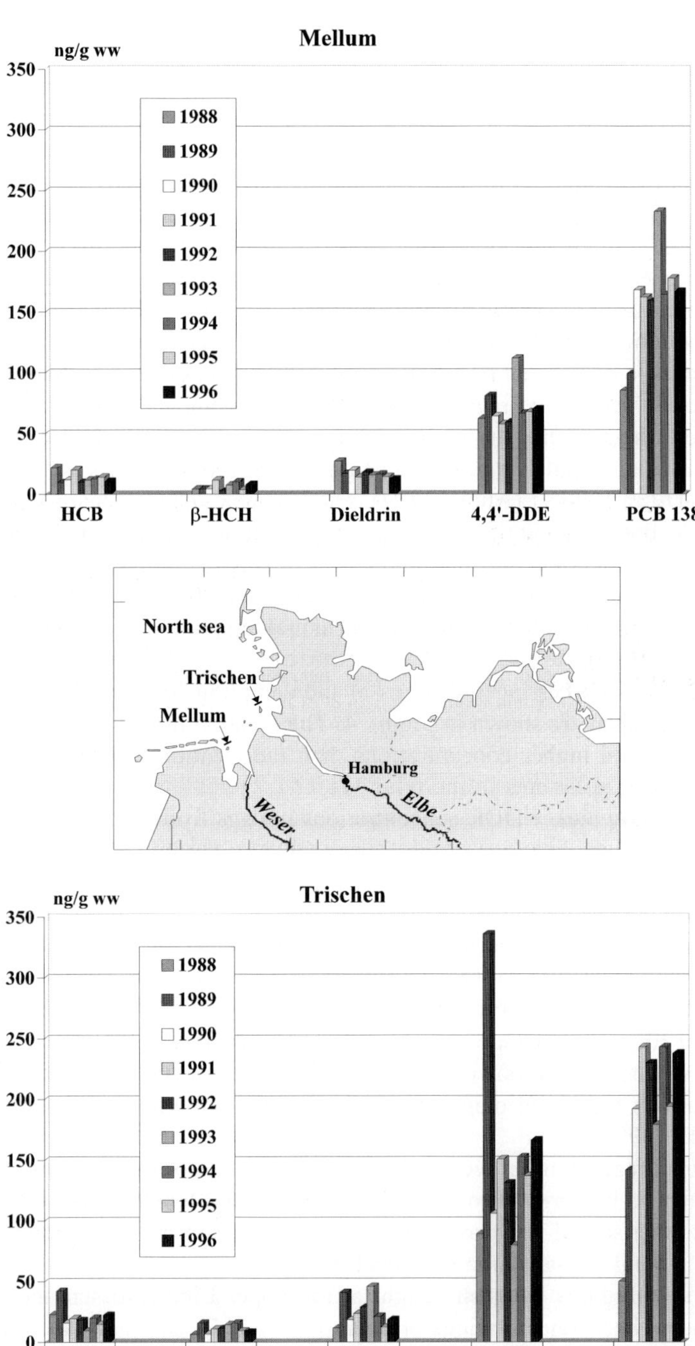

Figure 4. Time trend of selected chlorinated hydrocarbons in herring gull eggs from the islands Trischen and Mellum (ww: wet weight).

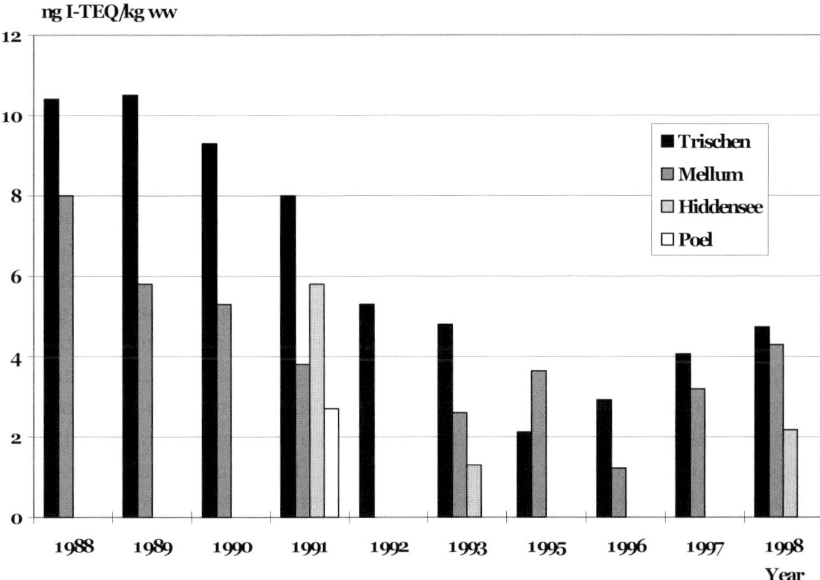

Figure 5. Time series of herring gull eggs from four sampling locations in the North and Baltic Sea (sub-sample of pooled sample, *n* = 1) (ng I-TEQ/kg ww : ng International Toxicity Equivalents/kg wet weight).

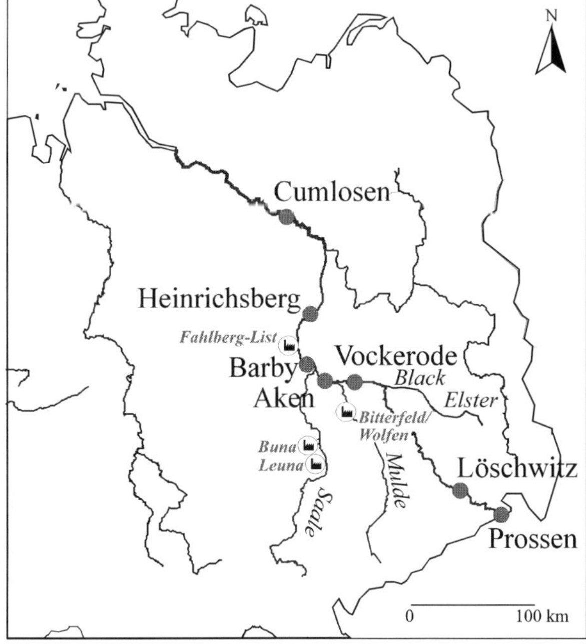

Figure 6. Location of the sampling stations of breams (*Abramis brama)* from the River Elbe and sites of chemical plants.

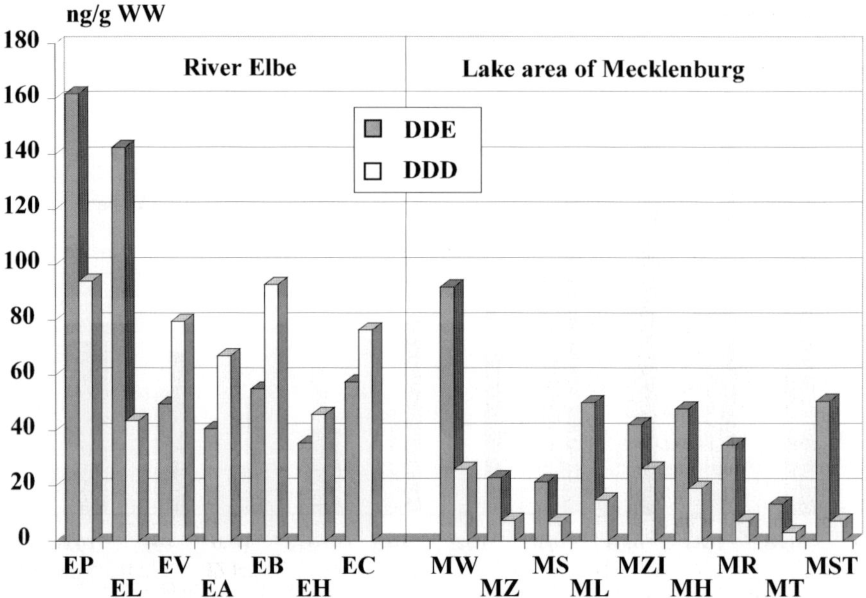

Figure 7.   Comparison of the DDE and DDD mean concentrations in bream muscle homogenates (*Abramis brama*) from the River Elbe and the lakes of Mecklenburg in 1990/91 (WW: wet weight).

Abbreviations of the sampling sites

River Elbe		*Lake area of Mecklenburg*	
EP	Prossen	MW	Wanzkaer See
EL	Löschwitz	MZ	Zwirn-See
EV	Vockerode	MS	Schweingarten-See
EA	Aken	ML	Lutowsee
EB	Barby	MZI	Zierker See
EH	Heinrichsberg	MH	Haussee
EC	Cumlosen	MR	Rödliner See
		MT	Großer See von Teterow
		MST	Stechlin-See

constant. The analytical data obtained for HCB concentrations in bream indicate a declining trend downstream of Barby. The high HCB contamination is considered to be an effect of the discharges from chemical plants in the former CSFR as well as of the influxes from the River Mulde (upstream from Aken) and River Saale (upstream from Barby). The increase of the HCH content in fish from these stations is also linked to the influence of both tributaries of the River Elbe (Fig. 9).

The major polluters of the River Mulde were the chemical plants in the Bitterfeld/ Wolfen area, although the production of chlorinated pesticides has stopped several years ago (Greenpeace, 1990). The CHC burden in the River Saale originated from Buna, Leuna and the pulp mill industry. Lindane and DDT containing wood preserving agents (e.g. Hylotox IP/Hylotox 59) were produced in Leuna. The high contamination levels of HCH in bream at the station Heinrichsberg might be explained by the

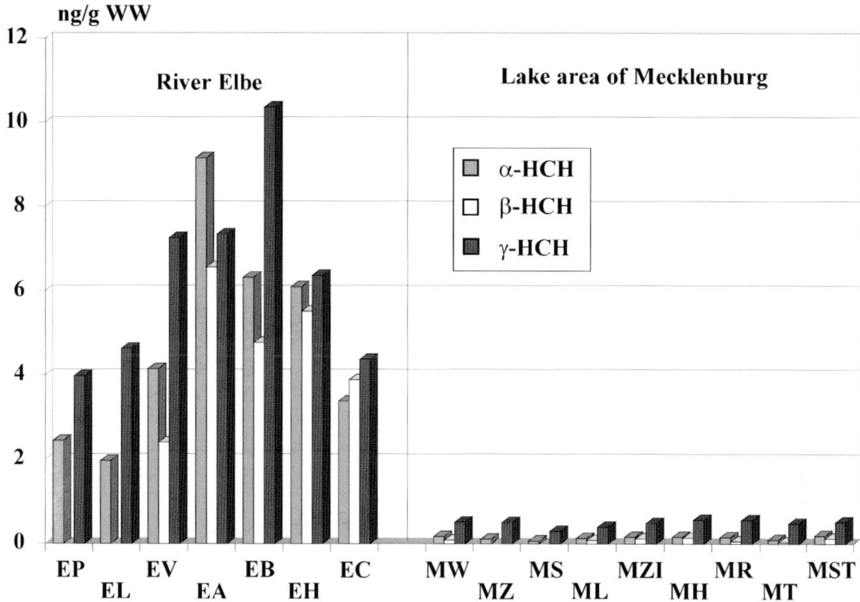

Figure 8.   Comparison of the HCH mean concentrations in bream muscle homogenates (*Abramis brama*) from the River Elbe and the lakes of Mecklenburg in 1990/91 (WW: wet weight) (Abbr.: cf. Fig. 7).

influence of discharges from the pesticides plant Fahlberg-List, located in Magdeburg. The results are consistent with waste water analysis of the plant and river sediment in 1990, which showed that both matrices were extremely contaminated by HCH (Fraucke et al., 1994). Although the production of lindane has ceased since 1981, the pesticides were highly accumulated in the river sediments.

Another interesting aspect can be seen from Figure 7. In the literature DDD is known as the dominant DDT metabolite in fish from the River Elbe (Anon., 1997). It is a degradation product of 4,4′-DDT by anaerobic conversion as well as a byproduct of the DDT synthesis. In sediment of the River Elbe and the River Mulde DDD levels were generally higher than those of DDE (Walkow et al., 1994). Interestingly, the recorded DDE values from the upstream stations (Prossen and Löschwitz) were twice as high as those of DDD. From Vockerode to Cumlosen DDD/DDE, ratios ranged between 1.3–1.7. This results seems to imply that the DDD burden of sediments is well reflected by bream.

With the exception of the DDT metabolites the CHC burden in bream from Mecklenburg was considerably lower in comparison with fish from the River Elbe as could be expected due to the absence of chemical industries. The DDE levels were comparable to those found in fish from the River Elbe on account of the intensive agriculture in this region. Slightly higher PCB concentrations were found in breams from the Lake "Stechlin-See" and the Lake "Zierker See". The Lake "Stechlin-See" is situated in a nature reserve which is influenced by the local recreation traffic from Berlin as well as by a former nuclear power station "Rheinsberg". The elevated PCB levels

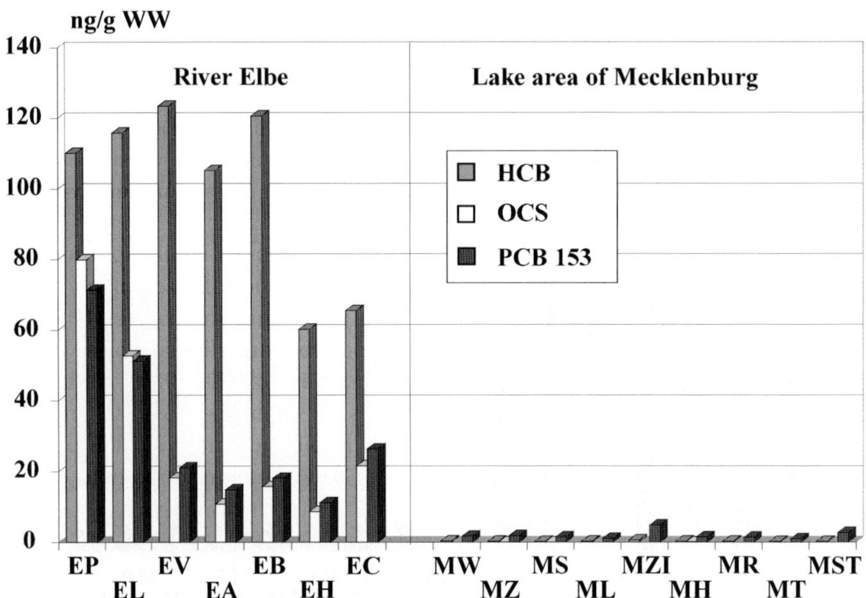

Figure 9. Comparison of the mean concentrations of selected CHC in bream muscle homogenates (*Abramis brama*) from the River Elbe and the lakes of Mecklenburg in 1990/91 (WW: wet weight) (Abbr.: cf. Fig. 7).

in the Lake "Zieker See" could be attributed to the long-term influx of untreated waste water from the city Neustrelitz.

Figures 10 and 11 illustrate results on the major CHC contaminates in bream livers from five different German rivers, obtained within the ESB routine programme in 1995 (Marth et al., 2000). Since 1994 bream samples have been periodically taken in the River Saar and since 1995 also in the River Mulde, River Saale and River Rhine (Fig. 11).

Samples were also obtained from stations along the River Elbe (Prossen, Zehren, Barby, Cumlosen, Blankenese), four station along the River Rhine (Weil am Rhein, Iffezheim, Koblenz and Bimmen), one station situated upstream of the River Saar (Güdingen) and Rehlingen located downstream under the industrialised area of Saarbrücken. The samples from the River Mulde were collected near Dessau and fish of the River Saale between Halle and Wettin.

It is apparent that the stations on the upper Elbe still exhibited the highest concentrations of HCB, DDE and PCB. High values for HCB and DDE were also found in samples from the River Mulde. This observation is mainly a result of the former industrial emissions of the chlorine industry and a leaching dumpsite at Bitterfeld. Wilken reported, that there had been further pesticide emitters in the area upstream of the River Mulde in Saxony, especially with respect to DDT. In West Germany the HCB and DDE concentrations were much lower in comparison with the levels in East Germany. Bream caught near the Rhine station Iffezheim contained HCB at higher levels because of a former chemical plant located upstream, which produced pentachlorophenol

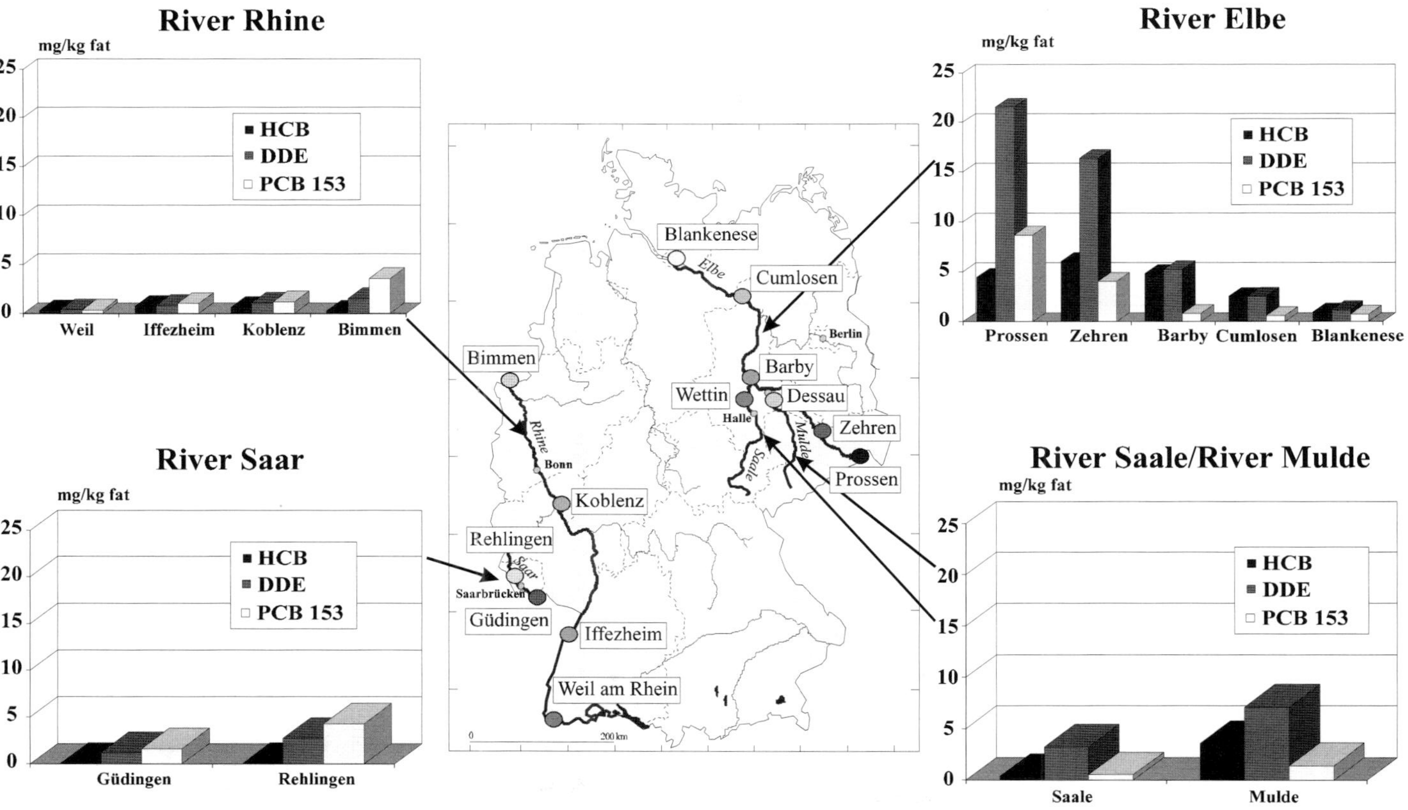

Figure 10. Comparison of the mean concentrations of selected CHC in bream liver homogenates (*Abramis brama*) from different German rivers in 1995.

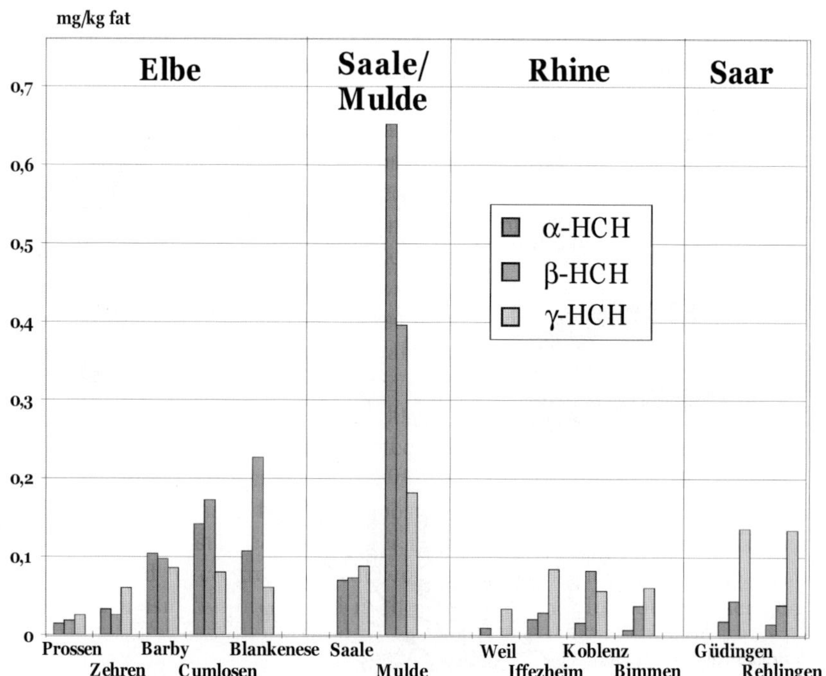

Figure 11.   Comparison of the mean HCH concentrations in bream liver homogenates (*Abramis brama*) from different German rivers in 1995.

(Marth et al., 2000). The highest DDT amount in fish from West Germany were found in the River Saar at Rehlingen, in contrast to its small HCB levels.

Elevated levels of PCB were detected in bream from the lower Rhine at Bimmen and from the lower Saar at Rehlingen. The increasing concentrations along the River Rhine are partly a result of the influx of the River Mosel (upstream of Koblenz), containing high amounts of PCB (Anon., 1995). The river Saar is strongly influenced by emissions from the mining industry situated in the Saar region.

The data shown in Figure 11 clearly indicate the HCH pollution of the River Mulde. Wilken et al. (Anon., 1994) reported that sediments and soils from River system Mulde are still highly contaminated by pesticides and PCDD/F although the industrial emission has stopped. These sediment-bound pollutants can be remobilised and may lead to a secondary contamination of the River Elbe by pollutant transport. Therefore, it remains to be seen how long the recovery of the River Elbe will take.

Generally HCH patterns of bream from East Germany differ from those found in fish from West-Germany. The γ-isomer predominates in the Western samples, whereas higher levels of the α and β-isomers are detected in the fish from the Eastern part of Germany.

The comparison of the CHC patterns in bream livers exhibits significant distinctions between the different ecosystems (Fig. 12). DDT metabolites and PCB contributed to one third each to the total CHC burden of bream livers from the limnic ecosystem of

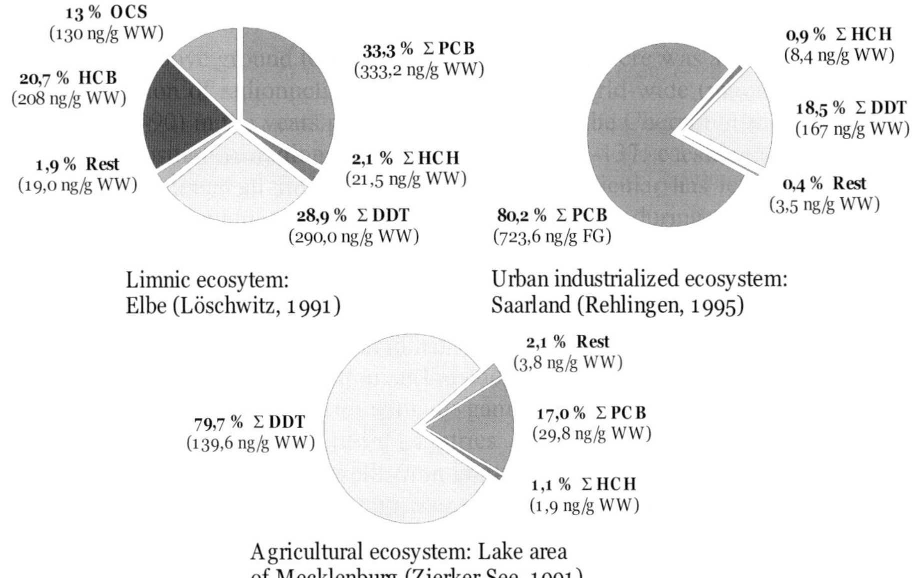

Limnic ecosytem:
Elbe (Löschwitz, 1991)

Urban industrialized ecosystem:
Saarland (Rehlingen, 1995)

Agricultural ecosystem: Lake area
of Mecklenburg (Zierker See, 1991)

Figure 12. Comparison of the proportion of different CHC pollutant classes in bream livers (*Abramis brama*) from East and West Germany.

the River Elbe. PCB were the major organochlorine contaminants (70%) in bream of a typical industrialized area (Saarland) in West Gemany in contrast to agricultural areas of East Germany, where DDT metabolites were the dominant pollutants (80%).

This surprising result that DDT metabolites show high amounts in environmental samples of Eastern Germany can also be demonstrated in pigeon eggs.

The collected specimens have been successfully utilized for the characterization of the environmental pollution by CHC in the former GDR.

The dominant position of the DDT group in nearly all Eastern samples clearly reflects the central role of DDT in the pest control of the former GDR. For this pollutants class significant spatial differences between East and West Germany were detected in terrestrial, atmospheric and limnic environments. The HCH concentrations in animal samples were considerably lower. With the expedition of local hot spots of pollution the lindane contamination of both German parts was in the same range, whereas the concentration of α- and β-HCH isomers, byproducts of the lindane production, appeared to be higher in samples from East Germany, especially in the vicinity of former pesticide plants. Relatively low PCB levels were observed in specimens from the East compared to West Germany due to low usage of PCB in the former GDR.

6. Conclusions

Twenty years of practical experience in environmental specimen banking have demonstrated that the concept of long-term storage of biological specimens for the retrospective analysis has contributed to traditional environmental pollution monitoring as an important complement. The ESB can serve as a valuable resource for the assessment of long-term trends of pollutants affecting human and environmental health, in particular for those pollutants that have been unnoticed.

The results fulfil the proposed goals of the ESB insofar as political decisions about the emission or other regulations on pollutants can be confirmed by their decrease in the environment (e.g. introduction of unleaded fuel, offshore incineration of hazardous waste). On the other hand illegal applications of chemical could have been identified e.g. DDT applications in the former GDR.

The generated Standard Operating Procedures for analytical and sampling procedure are well documented similar to the demands of Good Laboratory Practice or the European Standards for Analytical Work. Thus, in this field it became possible to identify and to avoid possible errors, in the future. Due to high quality standard of sampling and analysis the collection is of particular value for future analytical work on environmental contaminants of which we may know very little at present (UPB, 1996).

The increasing amount of generated data now allows the verification of many relationships discovered in the past in the environmental research with very high accuracy that are bioaccumulation, biomagnification, distribution, transport and at least degradation of chemicals. The history of the conceptional development of the ESB has resulted in a pool of knowledge which can be used for future decisions and recommendations for researchers and politicians.

As a new instrument for science, administration and management, ESB can support analytical and environmental research and monitoring generally in many ways and make it more effective and reliable, e.g. supply reference materials for environmental analysis, preservation of authentic records as an archive for long-term comparison of environmental change, securing and perpetuation of evidence in biology, medicine, forensic medicine, biotechnology, deposition or conversion of problematic wastes, environmental planning and risk assessment.

Acknowledgements

This work was supported by the Federal Ministry for the Environment, Nature Conservation and Nuclear Safety of Germany via the German Federal Environmental Agency. All co-workers of the Environmental Specimen Bank are gratefully acknowledged, especially D. Martens, P. Marth, V. Mayer, K. Oxynos, J.Schmitzer and K.W. Schramm.

References

Anon., 1994. Gewässergütebericht Elbe 1993 mit Zahlentafeln, Intern. Kommission zum Schutz der Elbe (IKSE), Hamburg.
Anon., 1995. Wasserbeschaffenheit von Mosel, Saar und Nebenflüssen im Jahre 1994, Intern. Kommission zum Schutz der Mosel und der Saar gegen Verunreinigungen (IKSMS), Trier.
Anon., 1997. Annual report of the Bank of environmental specimens, Umweltprobenbank, Jülich.
Boehringer, U.R.,1988. Umweltprobenbank: Bericht und Bewertung der Pilotphase. Bundesministerium für Forschung und Technologie, Umweltbundesamt Berlin, Springer-Verlag, Berlin.

Emons, H., 1994. Inorganic analysis within the German Environmental Specimen Bank. In: Aly, H.F., Nentwich, D. (Eds), German-Egyptian Seminar on Environmental Research Cairo, Aegypten, 21–23 March. Bilateral Seminars of the International Bureau, Forschungszentrum Jülich, 19, Umweltbundesamt, 1996, pp. 195–204.

Fiedler, H., Fürst, P., Malisch, R., Papke, O., Schrenk, D., 2000. State of the art: dioxins. Environ. Sci. Poll. Res. 7 (4), 239–242.

Fraucke, W., Franke, S., Hildebrandt, S., Link, M., Schwarzbauer, J., 1994. In: Die Belastung der Elbe-Nebenflüsse mit Schadstoffen, Forschungszentrum Karlsruhe, Karlsruhe, pp. 136–160.

Grimmer, G.,1993. Relevance of polycyclic hydrocarbons as environmental carcinogens. In Garrigues, P., Lamote, M. (Eds), Polycyclic Aromatic Compounds 13th Int. Symp. PAH 1–4 October 1991, Gordon and Breach, Bordeaux, pp. 31–41.

Grimmer, G., Hildebrandt, J., Jacob, J., Naujack, K.-W., 1996. Standard Operating Procedure for the Analysis of Polycyclic Aromatic Hydrocarbons (PAH) in Various Matrices, in press.

Greenpeace, 1990. The Elbe-results of the Measuring and Campaign Trip of the Beluga in the Spring of 1990. Ökopol-Institute of Ecology and Politics, Hamburg.

Heinisch, E., 1992. Umweltbelastung in Ostdeutschland – Fallbeispiele: Chlorierte Kohlenwasserstoffe Wiss. Buchgesellschaft, Darmstadt.

Jacob, J., Grimmer, G., Hildebrandt, J., 1996. Long-term decline of atmospheric and marine pollution by polycyclic aromatic hydrocarbons (PAH) in Germany. Submitted to Chemosphere.

Kayser, D., Boehringer, R. U., Schmidt-Bleek, F., 1982. The environmental specimen banking project of the Federal Republic of Germany (pilot-phase). Environmental Monitoring and Assessment 1, 241–255.

Keune, H., 1993. Environmental specimen banking (ESB): an essential part of integrated ecological monitoring on a global scale. The Sci. Tot. Environm. 139/140, 537–544.

Klein, R., Paulus, M., Wagner, G., Müller, P., 1994. Das ökologische Rahmenkonzept zur Qualitätssicherung in der Umweltprobenbank des Bundes. In: Paulus, M., Klein, R., Wagner, G., Müller, P. (Eds), Biomonitoring und Umweltprobenbank, Teil I. Beitragsserie in der UWSF-Z-Umweltchem. Ökotox. 6, pp. 223–221.

Kettrup A., Schramm, K.W., Marth, P., Oxynos, K., Schmitzer, J., 1999. Specimen banking as surveillance tool. Annali di Chmica 89, 489–498.

Lewis, R.A., 1987. Guidelines for Environmental Specimen Banking with Special Reference to the Republic of Germany: Ecological and Managerial Aspects. US Department of the Interior, National Park Service, US MAB Report, 12, 182.

Lewis, R.A., 1988. Remarks on the status of environmental specimen banking in relation to health and environment assessment. Proceedings of the 11th US–German Seminar of State and Planning on environmental specimen banking, Bayreuth, 1–3 May.

Lewis, R.A., Horras, C., Paulus, M., Klein, B., 1993. Auswahl ökologischer Umweltbeobachtungsgebiete in der Bundesrepublik Deutschland. In: Likens, G.E. (Ed.), An Ecosystem Approach to Aquatic Ecology, Springer-Verlag, New York.

Marth, P., Schramm, K.-W., Henkelmann, B., Wolf, A., Oxynos, K., Schmitzer, J., Kettrup, A., 1999a. Die Rolle der Umweltprobenbank in der Umweltüberwachung am Beispiel von chlorierten Kohlenwasserstoffen in ausgewählten Matrizes. UWSF-Z. Umweltchem. Ökotox., 11, 89–97.

Marth, P., Schramm, K.-W., Martens, D., Oxynos, K., Schmitzer, J., Kettrup, A., 1999b. Distribution of chlorinated hydrocarbons in different ecosystems in Germany. Intern. J. Environm. Anal. Chem. 75 (1–2), 229–249.

Marth, P., Martens, D., Schramm, K.-W., Schmitzer, J., Oxynos, K., Kettrup, A., 2000. Environmental specimen banking – Herring gull eggs and breams as bioindicators for monitoring long-term and special trends of chlorinated hydrocarbons. Pure Appl. Chem. 72, 1072–1034.

Nesmerak, I., 1993. Kontamination der Elbe aus dem Gebiet der Tschechischen Republik und der Moldau mit organischen Schadstoffen. In: Heinisch, E., Kettrup, A., Wenzel-Klein, S. (Eds), Schadstoffatlas Osteuropa, Ecomed-Verlag, Landsberg, pp. 167–170.

Oxynos K., Schmitzer J., Dürbeck H.W., Kettrup, A., 1992. Analysis of chlorinated hydrocarbons (CHC) in environmental samples. In: Rossbach, M., Schladot, J.D., Ostapczuk, P. (Eds), Specimen Banking, Springer-Verlag, Berlin, p. 127.

Oxynos K., Schmitzer J., Kettrup A., 1993. Herring gull eggs as bioindicator for chlorinated hydrocarbons. Sci. Tot. Environ. 139/140, 387–398.

Oxynos, K., Schramm, K.-W., Marth, P., Schmitzer, J., Kettrup, A., 1995. Chlorinated hydrocarbons-(CHC) and PCDD/F-levels in sediments and breams (*Abramis brama*) from the river Elbe (contribution to the German Environmental Specimen Bank). Fresenius J. Anal. Chem. 353, 98–100.

Paulus M., Altmeyer M., Klein, K, Klenke, R., Nentwich, K., Stegner, J., Wagner, G., Müller, P., 1992. Zustandsdokumentation der Belastungssituation auf dem Gebiet der ehemaligen DDR durch Umweltproben, UBA-Vorhaben Nr. 1080001.

Paulus, M., Klein, R., Zimmer, M., Jacob, J., Rossbach, M., 1995. Die Rolle der biometrischen Probencharakterisierung in der Umweltanalytik am Beispiel der Fichte (*Picea abies*). In: Paulus, M., Klein, R., Wagner, G., Müller, P. (Eds): Biomonitoring und Umweltprobenbank, Teil VI. Beitragsserie in der UWSF-Z-Umweltchem.Ökotox. 7, 236–244.

Schramm, K.-W., Kettrup, A., Schmitzer, J., Marth, P., Oxynos, K., 1996. Environmental Specimen Bank – a useful tool for prospective and retrospective environmental monitoring. TEN 3, 43–49.

SRU (Rat der Sachverständigen für Umweltfragen), 1987. Umweltgutachten 1987, Deutscher Bundestag, Drucksache 11/1569 und Verlag Kohlhammer, Stuttgart/Mainz.

Stoeppler, M., Zeisler, R. (Eds), 1993. Biological environmental specimen banking. The Sci. Tot. Environm., BESB special issue, 139/140.

Thalius, J., 1588. Sylvia hercynia, sive catalogus plantarum sponte nascentium in montibus. Frankfurt.

UPB, 1996. Jahresbericht der Umweltproben Bank Jülich von 1995 Forschungszentrum Jülich, Jülich, FRG.

Wagner, G., 1995. Basic approaches and methods for quality assurance and quality control in sample collection and storage for environmental monitoring. Sci. Tot. Environ. 176, 63–71.

Walkow, F., Wilken, M., Jager, E., Zeschmar-Lahl, B., 1994. Flooding area and sediment contamination of the River Mulde on Elbe-Influx, with PCDD/PCDF and other organic pollutants. Chemosphere 29, 2237–2252.

Wise, S.A., Zeisler, R., 1984. The pilot environmental specimen bank program. Environ. Sci. Technol. 18, 302A–307A.

Wise, S.A., Zeisler, R., Goldstein, G.M., (Eds) 1988. Progress in environmental specimen banking. NBS Special Publication 740, US Dept of Commerce, US Government Printing Office, Washington, DC.

Wittig, R., 1993. General aspects of biomonitoring heavy metals by plants. In: Markert, B. (Ed.), Plants as Biomonitors-Indicators for Heavy Metals in the Terrestrial Environment. VCH, Weinheim, pp. 3–27.

Chapter 22

Some concepts and future developments: developing agri-biodiversity indicators as a tool for policy makers

Kevin Parris

Abstract

Agriculture as the human activity occupying the largest share of the total land area for many countries, plays a key role with regard to biodiversity which is highly dependent on land-use. The expansion of farm production and intensification of input use are considered a major cause of the loss of biodiversity, while at the same time certain agro – ecosystems can serve to maintain biodiversity. Farming is also dependent on many biological services, such as the provision of genes to develop improved crop varieties and livestock breeds, crop pollination, and soil fertility provided by micro – organisms. In some cases non – native species cause damage to crops from alien pests and competition for livestock forage. The main focus of policy actions in the area of biodiversity has been to protect and conserve endangered species and habitats, but some countries have also begun to develop more holistic national biodiversity strategy plans. These plans usually incorporate the agricultural sector in biodiversity conservation. At the international level a range of agreements are also important in the context of agriculture and biodiversity, most notably, the Convention on Biological Diversity. This chapter examines recent efforts across OECD countries over the past 15 years to provide policy makers with indicators on the current state and changes in biodiversity associated with agricultural activities. Agri-environmental indicators are becoming an important tool for national policy makers to help them with policy monitoring, evaluation and forecasting. The chapter outlines in Section 1 the environmental context of the agri-biodiversity issue, followed in Section 2 by reviewing the national and international policy context. Section 3 defines agri-biodiversity indicators, covering genetic, wild species and ecosystem diversity. The chapter concludes in Section 4 by examining future challenges to further research in this area.

Keywords: biodiversity, agriculture, environment, indicators, policy, genetic diversity, wild species diversity, ecosystem diversity, OECD

1. Environmental context

The effects of agriculture on biodiversity are of considerable importance because farming is the human activity occupying the largest share of the total land area for many OECD countries. Even for countries where the share of agriculture in the total land area is smaller, agriculture can help by increasing the diversity of habitat types. The expansion of agricultural production and intensive use of inputs over recent decades in OECD countries is considered a major contributor to the loss of biodiversity. At the same time certain agricultural ecosystems can serve to maintain biodiversity, which may create conditions to favour species-rich communities, but that might be endangered by fallowing or changing to a different land-use, such as forestry (see Fjellstad and Dramstad, 1999; Hunziker, 1995; and Ihse, 1995). Agricultural food and fibre production is also dependent on many biological services. This can include, for example, the provision of genes for development of improved crop varieties and livestock breeds, crop pollination and soil fertility provided by micro-organisms. The importance of biodiversity for agriculture involves (Day, 1996; OECD, 1996, p. 20; and Pagiola and Kellenberg, 1997):

- facilitating the functioning of ecosystems, such as nutrient cycling, protection and enrichment of soils, pollination, regulation of temperature and local climates, and watershed filtration;
- providing the source of most of the world's food and fibre products, including the basis for crop and livestock genetic resources, their improvement, and the development of new resources; and,
- offering a range of scientific, health/medicinal, cultural, aesthetic, recreational and other intangible (and non-monetary values) and services from biodiversity richness and abundance.

Biodiversity, as it relates to agriculture, can be considered in terms of three levels, drawing on the Convention on Biological Diversity definition of biodiversity:

- *genetic diversity* ("within species"): the diversity of genes within domesticated plants and livestock species and wild relatives;
- *species diversity* ("between species"): the number and population of wild species (flora and fauna) affected by agriculture, including soil biota and the effects of non-native species on agriculture and biodiversity;
- *ecosystem diversity* ("of ecosystems"): the ecosystems formed by populations of species relevant to agriculture or species communities dependent on agricultural habitats.

The survival of these three levels of diversity is interdependent, as genetic diversity fosters the survival of species, enabling it to adapt to changing ecosystem conditions. A loss of species or the introduction of non-native species, can disturb the ecosystem and alter resilience to further changes (OECD, 1997).

Genetic diversity provides the means for agriculture to improve crop and livestock yields. Selective plant and animal breeding programmes in all most countries, drawing on a variety of genetic material, has helped to increase agricultural production with fewer inputs. In the *United States* it is estimated that over the past 60 years half of agriculture's productivity increases can be attributed to genetic improvements. Traditionally farmers have relied on 'landraces', that is, varieties of crops or livestock developed over many generations to raise yields. As these 'landraces' have been adapted for specific environmental conditions and farming systems, the genetic diversity is usually very high. With the advent of modern 'hybrid' breeding methods, which selects for specific desirable traits such as pest and disease resistance, maturation and stature, the yields of crops and livestock have been raised substantially. This process is likely to accelerate with developments in biotechnology, such as those involving genetic modification, cloning, and other such technologies. These 'new' technologies, however, have also raised concerns about their possible effects on human health, wild species, genetic erosion, the environment and development of genes resistant to pesticides (see Cunningham, 1999; Kate and Laird, 1999; and Spillane, 1999).

While estimates of the global total number of *wild species* vary greatly, it is clear that the number is very large, with estimates ranging from 5 to 100 million. Moreover, the richness of individual countries in biodiversity varies greatly according to the parameters chosen. In the context of agriculture, biodiversity 'richness' can differ according to specific climatic and agro-ecosystem conditions, and the type of farming management practices and systems adopted. Farming systems based on multiple crops and livestock with natural pasture areas are richer in biodiversity than monocultural farms. However, regardless of the type of farming system, agriculture by seeking to maximise the yield of a limited number of plant and animal species, inevitably weakens and reduces competition from other unwanted species (Debailleul, 1997). Species diversity and its relationship with agriculture is important in a number of different ways, including:

- *Species supporting agricultural production systems*, the so called "life-support-system", that is crypto-biota, including soil micro-organisms, worms, pest controlling species and pollinators.
- *Species related to agricultural activities*, covering (a) wild species using agricultural land as habitat ranging from marginal use to complete dependence on agro-ecosystems, and (b) wild species that use other habitats but are affected by farming activities, such as the impact of farm chemical run-off on marine life in coastal waters.
- *Non-native species* that can threaten agricultural production and agro-ecosystems, such as invasion of weeds and pests that are alien to indigenous biodiversity.

Ecosystem diversity and its relation to agriculture is manifest through:

- changes in farming practices and systems;
- changes in land-use between agricultural and other land-uses; and
- the interaction between agriculture and adjacent ecosystems.

In some cases agricultural land-use patterns and practices support the conservation and sustainable use of biodiversity, while in others they cause serious threats. In this context, agriculture generates both benefits and pressures on biodiversity, which vary across different regions and countries depending on local farming practices, biogeography, grazing periods, climate and other factors. Farming communities have an intrinsic interest in ensuring that land-use practices are sustainable and contribute to the conservation and sustainable use of biodiversity. Some semi-natural agricultural habitats can be preserved only if appropriate farming activities are continued. In many situations where agriculture production is a key element to sustain certain ecosystems, the change in land-use from agriculture to other uses can lead to the degradation of some ecosystems.

The negative impacts of agriculture on biodiversity must also be considered in terms of the benefits agriculture brings to society through providing food and fibre, employment and incomes. A better understanding of the processes and trade-offs involved between agricultural production, biodiversity loss and agriculture's role in some situations to maintain biodiversity are, nevertheless, critical to improving land-use decision making (Montgomery et al., 1999; and Wossink et al., 1999). The difficulty for scientists at present, is to quantify the critical thresholds of biodiversity resilience to stress and identify the measures and likely costs of restoring biodiversity stability. Equally the different forms in which agriculture impacts on biodiversity, while widely recognised, vary in their intensity and effects across countries. There are also other influences on biodiversity besides agricultural activity, such as from natural processes, for example, fires; non-indigenous species; other economic activities, for example, forestry and industry; and global climate change (Burns, 2000; European Commission, 1998; Mac et al., 1998). For policy makers to improve their responses in reducing biodiversity loss associated with agriculture, this will require a better understanding and measurement of the driving forces and state of biodiversity in agriculture.

2. Policy context

The preservation and enhancement of biodiversity poses a major challenge for agricultural policy decision makers, as world population and demand for food increase. It is estimated that, with current population trends, food production will have to increase by 24% by the year 2020 just to maintain the existing levels of food consumption and without any significant expansion of agricultural area. Policy makers will therefore need to find ways of minimising the conflicts between expanding production and biodiversity conservation, enhancing the many complementarities between agriculture and biodiversity, and finding ways to prevent the loss of biodiversity on agricultural land (Pagiola and Kellenberg, 1997).

Most agricultural policy affects, directly or indirectly, biodiversity. For a growing number of OECD countries, protecting and enhancing biodiversity is becoming an important part of their domestic and international agri-environmental policy objectives and actions. These policy actions are in response to a growing public concern over the increasing pressure and harmful impacts on natural and semi-natural ecosystems brought about through a variety of causes, including agricultural activity. There is also the perceived threat that damage to biodiversity could be highly detrimental to human welfare over the long term, although the consequences are complex and poorly understood (Smith, 1996).

In practice, implicitly or explicitly, government policy towards biodiversity involves balancing the trade-offs between socio-economic values and biodiversity conservation. Up to present the main focus of policy actions in the area of biodiversity has been to protect and conserve endangered species and habitats. Many OECD countries have introduced legislation for the protection of specific endangered species and habitats, and also designated certain areas as biosphere reserves, nature parks, and other protected sites. In moving towards a more holistic approach, some countries have begun to develop national biodiversity strategy plans, which usually incorporate the agricultural sector as a key player in biodiversity conservation. These strategy plans set out the relevant policy objectives and targets for managing and sustaining biodiversity. They also provide a starting point for establishing policy relevant biodiversity indicators to measure the performance of national policies and help monitor progress in fulfilling international obligations.

In most OECD countries a wide spectrum of organisations is also involved in the conservation of plant and animal genetic resources. However, the way these conservation efforts are organised varies across countries, ranging from involvement of governmental and non-governmental organisations, and from amateur collections to commercial companies. Some countries have national genebanks, others have several specialised agricultural research institutes responsible for the maintenance of agricultural genetic resources, while some countries work together in regional genebank networks.

At the international level a range of agreements and conventions are also important in the context of agriculture and biodiversity, most notably the International Convention on Biological Diversity (CBD) agreed at the UN Conference on Environment and Development at Rio in 1992. Recognition has been given by the CBD to the significance of biodiversity for agriculture. This has led the FAO to request member countries to negotiate, through the FAO inter-governmental Commission on Genetic Resources for Food and Agriculture (CGRFA), the revision of the international undertaking on plant genetic resources in agriculture in harmony with the CBD. In addition, in January 2000 within the overall context of the CBD, the Biosafety Protocol was agreed by 130 nations. This was the first major international agreement to control trade in genetically modified organisms (GMOs), covering food, animal feed and seeds.

Other related international conventions include, for example, the Convention on International Trade in Endangered Species of Wild Fauna and Flora (CITES, 1973), the Convention on Wetlands (Ramsar Convention, 1971), the Convention on Migratory Species of Wild Animals (Bonn, 1983), the North American Waterfowl Management Plan, and the Canada-United States Migratory Birds Convention (1995). The Commission for Environmental Cooperation, created by Canada, Mexico and the United States to examine the environmental provisions of the North American Free Trade Agreement, has begun to develop a strategy for improving biodiversity, including the role of agriculture (CEC, 2000).

3. Monitoring the impact of agriculture on biodiversity

The long-term challenge for agriculture is to produce food and industrial crops efficiently, profitably and safely, and to meet a growing world demand without degrading natural resources and the environment. To respond to this challenge and develop better policies, policy makers need indicators, which can help to monitor the environmental effects of agriculture and provide a tool for policy analysis. Agri-biodiversity indicators can provide a useful tool for the evaluation of domestic policies and international obligations related to biodiversity in agriculture. While the set of indicators to monitor biodiversity are potentially very large, a smaller and policy relevant set are being established by OECD, structured within the general framework of genetic, species, and ecosystem diversity (see Fig. 1). Together the indicators establish the initial steps in providing a coherent picture of biodiversity in relation to agriculture. It is the impact of agriculture on biodiversity which is the emphasis in this chapter, and not the effects on agriculture of biodiversity and related ecosystem services (OECD is also developing an indicator concerning the impacts of non-native species on agriculture and agro-ecosystems, not examined here, but see OECD, 2001).

Biodiversity level   **Indicators**

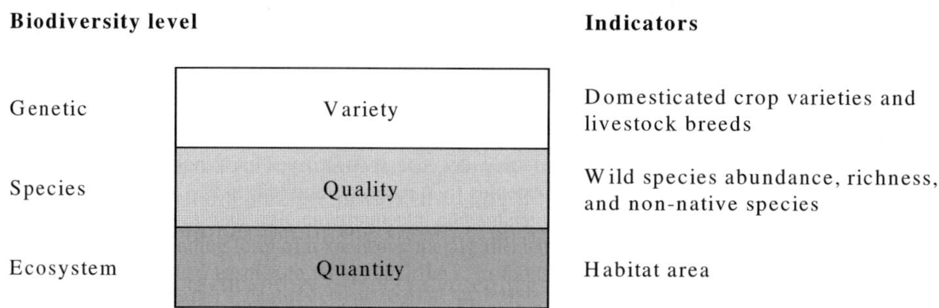

Figure 1. Coverage of biodiversity indicators in relation to agriculture.

Source: OECD (2001), Environmental Indicators for Agriculture, Volume 3 – Methods and Results, Publications Service, Paris, France.

3.1 Genetic diversity

3.1.1. Indicator definitions and methods of calculation

1. For the main crop/livestock categories (e.g. wheat, rice, cattle, pigs) the total number of crop varieties/livestock breeds that have been registered and certified for marketing.
2. The share of key crop varieties in total marketed production for individual crops (e.g. wheat).
3. The share of the key livestock breeds in respective categories of livestock numbers (e.g. the share of Friesian, Jersey, Charolais, etc., in total cattle numbers).
4. The number of national crop varieties/livestock breeds that are endangered.

The first three indicators track the extent of diversity in the range of crop varieties and livestock breeds used for agricultural production. These indicators require data covering the total registered or marketed number of crop varieties/livestock breeds, and total crop production/livestock numbers for the main categories of crops (e.g. wheat, etc.) and livestock (e.g. cattle, etc.). The fourth indicator, provides information on the extent of genetic erosion and loss of domesticated varieties/breeds from the much wider genetic pool than just those varieties/breeds marketed for production. Sources for species data include national genebanks and breeding organisations, although the FAO has begun to develop internationally co-ordinated databases for genetic resources in agriculture.

3.1.1. Recent trends

There seems broad consensus that global losses of genetic resources for food and agriculture have been substantial over the past 100 years. Even so, trends in the populations and numbers of 'wild' relatives of domesticated agricultural plants and livestock are poorly documented. To a larger extent national definitions, systems of classification and monitoring of the state and trends of genetic diversity in agriculture, are based on the approaches being developed through the Convention on Biological Diversity

(CBD), and the related work of the FAO inter-governmental Commission on Genetic Resources for Food and Agriculture. In addition, many OECD countries have major genebanks of crop and livestock genetic material (see FAO, 1996, 1998). For *European Union* countries EU Regulation No.1467/94 provides a programme for the conservation, characterisation, collection and utilisation of genetic resources in agriculture, while in principle conservation of agricultural genetic resources can be supported through EU Regulation No. 2078/92 (European Commission, 1998, pp. 48–50). The latter EU regulation is applied to promote conservation of threatened farm animal species through provision of support for farmers who undertake to rear local livestock breeds in danger of extinction and to cultivate crops threatened by genetic erosion (see ECNC, 2000; and the European Commission, 1999, p. 131).

Overall, there has been an *increasing number of crop varieties* registered for marketing and as a share in crop production over the past 13 years in OECD countries. This trend suggests that for many countries arable farming has improved its resilience to environmental change and risk through diversifying the number of crop varieties used in production. The trend in the share of the one to five dominant varieties in the total marketed production for specific crops has also declined in a large number of cases. The share of these dominant crop varieties, however, is still in excess of 70% for most crop categories, although for some countries the dominance of major varieties in crop production is lower (see Fig. 2). These trends are supported by research, that reveals, over a longer time period than shown in Figure 2, the share of the total area of wheat planted to the dominant cultivar has declined in France, Hungary, Italy, the Netherlands and the United Kingdom (Smale, 1997, p. 1261).

The most frequently cited cause of the *loss of genetic diversity* from country reports provided to the FAO (FAO, 1996, pp. 13–14), was the introduction of new varieties of crops leading to the replacement and loss of traditional, highly variable crop varieties. In Korea 74% of varieties of 14 crops grown on farms in 1985 had been replaced by 1993 (see Ahn et al., 1996). In the United States, a study drawing on information about varieties grown by US farmers in the 19th century revealed that most varieties can no longer be found either in commercial agriculture or any US genebank (FAO, 1996, p. 14).

The overall trend for livestock shows an *increasing number of breeds* registered for marketing and as a share of total livestock numbers. This indicates a growing diversity of the breeds used for livestock production for most categories of livestock and OECD countries. Changes from 1985 to 1998 in the number of livestock breeds, registered or certified for marketing, shows an increase for nearly all major livestock categories and for most countries, although data for poultry are extremely limited. These trends are also reflected in the reduction in the share of the three major livestock breeds in total livestock numbers for respective livestock categories (Fig. 3). Differences across OECD countries and between livestock categories exist, in particular, for sheep and cattle, while the dominance of a few breeds in total livestock numbers for respective categories is, in general, higher than for crops, in excess of 80% in most cases.

In the case of the *loss of livestock genetic diversity*, FAO estimates that globally for over 3800 breeds of cattle, goats, pigs, sheep, horses and donkeys that existed 100 years ago, 16% have become extinct and 15% are threatened. In cattle breeding, where the Holstein-Friesian breed has become the dominant breed for milk production world-

Change in the share of the one to five dominant varieties in total marketed production

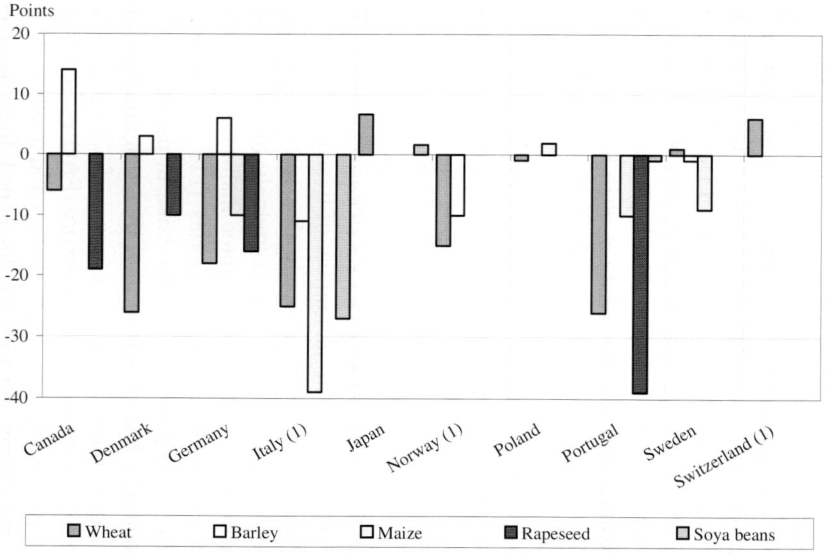

Percentage share of the one to five dominant varieties in total marketed production: 1998

	Wheat	Barley	Maize	Rapeseed	Soya beans
Canada	74	60	..	75	..
Denmark	70	71	..	82	
Germany	41	67	47	54	..
Italy	63	42	56	100	60
Japan	83	..	..	..	50
Norway	85	80	..	100	..
Poland	59	..	37	..	..
Portugal	32	..	20	26	4
Sweden	55	58	44	..	..
Switzerland	90	..	..	90	90

Notes: Data are not available for all crop categories and all countries.

1. Percentages are zero or close to zero per cent for Italy (rapeseed), Norway (rapeseed), Switzerland (rapeseed, soya beans).

Source: OECD (2001), Environmental Indicators for Agriculture, Volume 3 - Methods and Results, Publications Service, Paris, France.

Figure 2. Share of the one to five dominant varieties in total marketed crop production: 1985 to 1998.

Change in the share of the three major livestock breeds in total livestock numbers

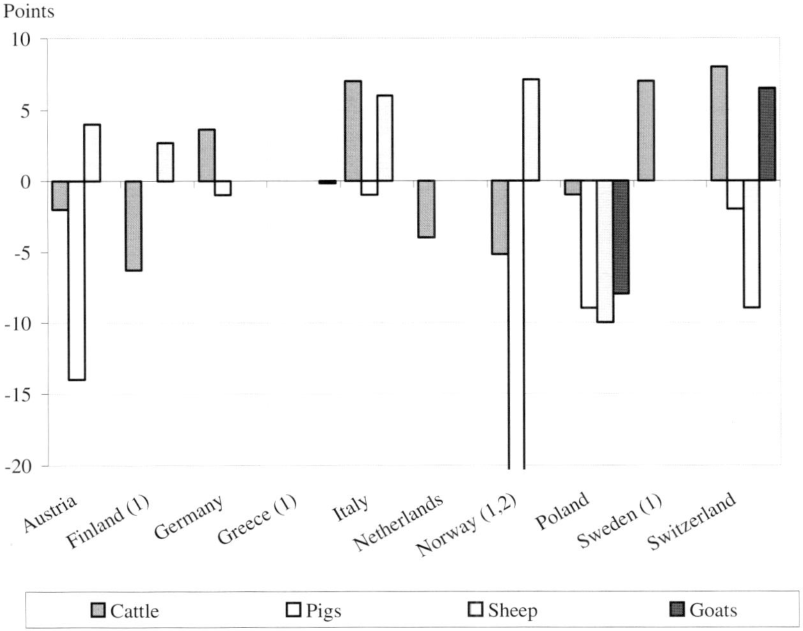

Percentage share of the three major livestock breeds in total livestock numbers: 1998

	Cattle	Pigs	Sheep	Goats
Austria	93	71	79	..
Finland	67	95	97	100
Germany	90	94	..	..
Greece	98	93	68	100
Italy	94	98	89	..
Netherlands	91	..	..	..
Norway	91	36	86	100
Poland	98	84	68	64
Sweden	92	95	..	95
Switzerland	98	98	82	74

Notes: Data are not available for all livestock categories and all countries.

1. Percentages are zero or close to zero per cent for Finland (pigs, goats), Greece (goats), Norway (goats), Sweden (goats).

2. Percentage is greater than -20% for Norway (pigs).

Source: OECD (2001), Environmental Indicators for Agriculture, Volume 3 - Methods and Results, Publications Service, Paris, France.

Figure 3. Share of the three major livestock breeds in total livestock numbers: 1985 to 1998.

wide, the number of sire-lines is decreasing and for the pig and poultry sectors only a small number of breeds dominate global production. Estimates for Germany show that in 1997 the number of endangered breeds was 12 out of a total of 77 for cattle, 14 out of 41 for sheep, 3 out of 16 for goats, 12 out of 103 for horses, but with no endangered breeds for pigs.

3.1.2. Indicator interpretation and limitations

Preventing the erosion of genetic diversity and dependence of agricultural production on a small number of varieties/breeds is important for agriculture. Genetic dependence on a small number of varieties/breeds can heighten the risks associated with changes in environmental conditions and susceptibility to pests and disease. Genetic erosion could impair the future potential to raise crop and livestock yields, as genetic material loss is generally irreversible. The baseline from which this loss should be measured is yet to be determined, although initially the early 1980s is being used as a baseline. Tracking in situ conservation of rare crop varieties/livestock breeds can be important for conservation of certain specific ecosystems.

In some cases the increase in particular national varieties/breeds, shown in Figures 2 and 3, is the consequence of the expanding international trade in varieties/breeds. The Hereford cattle breed, for example, while previously a dominant breed in the United Kingdom, is now becoming more common in Norway. Some caution is required, however, in using and interpreting indicators that measure genetic diversity by the trends in numbers of *crop varieties*. First, the genetic structure of the varieties in current use is likely to be similar, independent of the number of varieties grown. In other words, twenty main varieties grown in 1998, for example, may not have more genetic diversity than two main varieties grown in 1985. Second, varieties for certain crops are not registered in some OECD countries, in particular, this applies to fruit and vegetables and forage plants. Third, and perhaps most importantly, these indicators only account for what is grown or registered for marketing at any given time. The available gene pool is much wider.

For some countries the information on *livestock breeds* in Figure 3 may underestimate the "real" situation, as not all livestock are registered, and in some cases registered animals represent the elite breeding population and not "commercial" animals. These indicators could also be improved by providing a breakdown by sex, which registration statistics often neglect, and providing information on the number of livestock breeds considered threatened because of low population numbers. The FAO is now in the process of developing the international Domestic Animal Diversity Information System (DAD-IS) database to address the issue of the loss of animal genetic resources and their better use and development (FAO, 1998).

The genetic diversity between livestock breeds, as well as within breeds, is also important. So far exterior characteristics, e.g. coat colour, have been used to distinguish breeds. This is, however, a rather crude measure and does not sufficiently distinguish within breeds. Productivity levels may also be used as breed characteristics, but that would not represent genetic progress in production potential made, for example, in rare breeds. Moreover, productivity levels also fail to take account of other desirable traits in domestic animals such as hardiness to cold or drought, behavioural traits, meat quality (e.g. taste, etc.). Indicators of ecosystem diversity are also important in assessing genetic resources, because the plant varieties and livestock breeds have generally developed within specific agro-ecosystems. It is the adaptation of these breeds to these ecosystems that can make their conservation desirable.

3.2. Wild species diversity

3.2.1. Indicator definition and method of calculation

OECD countries have applied different approaches to describe and assess the state and trends in population distribution and numbers of wild species associated with agriculture. To a large extent this reflects differences in policy priorities, availability of data, and varying stages of scientific research on biodiversity issues. Thus, at this stage of the work it is not possible to develop a consistent method of calculation across OECD countries. Instead, it is only possible to develop an indicator that reveals the trends in population distributions and numbers of wild species related to agriculture, according to different country approaches, such as in terms of measuring species abundance, species richness, species distribution, key species, endangered species, or groups of species having similar functions (i.e. species guilds). These different approaches have varying advantages and disadvantages in terms of accuracy, sensitivity, feasibility and cost.

Most OECD countries do not have specific monitoring systems to track wild species populations and numbers on agricultural land. Background information is available, however, for a number of species and species groups related to agriculture, but usually this is not collected in a systematic manner. Many OECD countries, and some international organisations, however, report on a regular basis the total number of known and threatened species of mammals and birds, and to a lesser extent fish, reptiles, amphibians, invertebrates, vascular plants, mosses, lichens, fungi and algae, but none of this information relates specifically to agriculture (OECD, 1998a).

A few countries have begun to establish monitoring systems specifically to track wild species trends in agro-ecosystems. Canada, has started to examine the issue of monitoring wild species on agricultural land. Some countries also use hunting statistics as proxies for the likely impact of agriculture on wild species (e.g. Denmark, hares; Norway, roe deer, rooks, blackheaded gull and partridge, since the 1940s). Germany is developing a system that will involve monitoring the occurrence and frequency of 100 selected species through periodic sampling for defined ecological areas, including agro-ecosystems. In the *Netherlands*, for example, a monitoring programme exists, and is being further developed, covering plants, birds, butterflies, dragonflies, amphibians, mammals, fish, aquatic macro fauna and soil fauna (e.g. nematodes, worms, mites, fungi, etc.). Measurable species within these groups are selected, providing a representative cross-section of the agro-ecosystem, and consist of rare species as well as more common species. Species are measured under the Dutch system in various units, such as distribution, presence/absence, density, total numbers, breeding pairs, or area coverage, depending on what is feasible in sample areas or plots.

There remains an active discussion amongst biologists as to the merits of using the species abundance or species richness in monitoring biodiversity. *Species abundance* measures both the decline or increase of populations, which may result from human activities, such as agriculture. *Species richness* refers to the total number of a specific taxonomic group or functional groups associated with key ecosystems per site. Species richness measures presence/absence of species and is, therefore, a relatively insensitive variable compared to species abundance. The species richness of agricultural areas may rank in the order of tens of thousands of species, and it is not feasible to measure them all. Thus, choices will be necessary with regard to, specific taxonomic groups or functional groups, and spatial scales taking into account that species richness at the national level as an indicator has a different meaning than the average species richness per unit area of agricultural land. Moreover, data on the presence of species or taxonomic groups in the baseline state may be hard to find, especially on smaller spatial scales, while monitoring the current state on larger spatial scales is a costly activity and the sensitivity can be extremely low.

3.2.2. Recent trends

The information on the impact of agriculture on wild species, that either use agricultural land as habitat or use other habitats but are affected by farming activities, is limited for OECD countries, although two key points emerge from the data and research material that does exist. First, agricultural land provides an important habitat area for 'remaining' wild species (i.e. wild species that exist following the conversion of "natural" habitat to agricultural land-use), but especially birds and vascular plants (Fig. 4). Second, the population trends of wild species using agricultural land as habitat indicate in most cases a reduction over the period from 1985 to 1998, representing the continuation of a longer-term trend. There is some evidence, however, that the decline has slowed or even reversed over recent years in some countries, although from a low base, i.e. wild species population on agricultural land are increasing.

Concerning population trends for wild species using agricultural land as habitat, there is considerable work completed on the status of birds on agricultural land, especially in Europe (IUCN, 1999). A comparison of different habitat types (e.g. agriculture, forests, wetlands, etc.) reveals that across Europe agricultural habitats account for the highest proportion of birds with an unfavourable conservation status (Tucker and Heath, 1994; and Tucker and Evans, 1997). Much of the adverse impact of agriculture on bird populations has been attributed to pesticides (EEA, 1998, p. 166) and changing land-use patterns in agriculture, especially the loss of extensive grazing land (see OECD, 2001).

In Finland, over one-third of the country's vascular plants are found on pasture (MAF Finland, 1996, pp. 4–6). With the reduction in the pasture area in Finland over the past few decades and large-scale structural changes in the agricultural sector, it is estimated that these changes have threatened the disappearance of almost 290 species of flora and fauna, and, in addition, thousands of other species have declined (Table 1). A study has been undertaken in Germany, by the Federal Ministry for the Environment, to examine the various human and natural factors that have caused declines in plant species over the past 10 years, including the impact of agriculture. Intensified agricultural land-use, cessation of use, fallowing and natural succession, appears to have been the major cause of the decline in plant species, although the

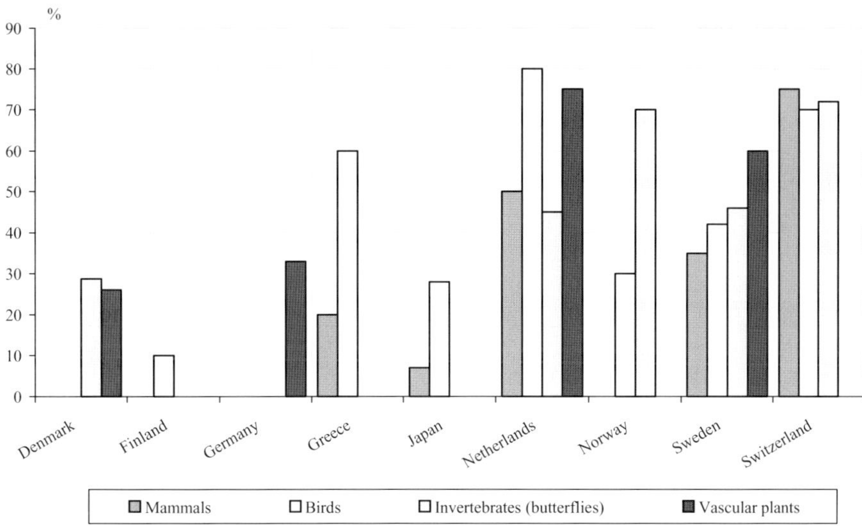

Notes: Data are not available for all categories of wild species for all countries.

1. This figure should be interpreted with care as definitions of the use of agricultural land as habitat by wild species can vary. Species can use agricultural land as "primary" habitat (strongly dependent on habitat) or "secondary" habitat (uses habitat but is not dependent on it).

Source: OECD (2001), Environmental Indicators for Agriculture, Volume 3 - Methods and Results, Publications Service, Paris, France.

Figure 4. Share of selected wild species categories that use agricultural land as habitat:[1] 1998

destruction of habitats and afforestation also has been important, although the causes of plant species decline in agriculture have had less effect in recent years. Monitoring of the Peregrine Falcon (*Falco peregrinus*) population in Ireland, showed a decline in numbers between the 1940s to the 1970s because of contamination by organochlorine pesticides. With restrictions on the use of these pesticide compounds there has been a recovery in the population from 225 breeding pairs in 1981 to 350 in 1991 (Environmental Protection Agency, 1999).

Almost a half of the bird species using farmland as habitat in the United Kingdom have declined in population size over the past 20 years (UK Department of the Environment, 1996, pp. 120–121). Within farmland habitats the decline in numbers of species was higher on cultivated arable land (about 60% of the bird species) than in grazing land (about 40%). While the decline in UK farmland birds is a cause of concern, at the same time, some bird species benefit from intensive farming and their populations are not decreasing, for example, the stock dove (*Columba oenas*) and jackdaw (*Coruus monedula*) (MAFF, 2000). However, some rare species, such as the corncrake (*Crex crex*), stone curlew (Burhinus oedicnemus) and cirl bunting (*Emberiza cirlus*), are responding well to conservation efforts (RSPB, 1999).

The marked reduction in UK bird populations can partly be linked to the declines in farmland habitat quality as a result of intensification of agricultural practices. Pasture (a good source of invertebrate food) has been lost from the arable areas in the east and

Table 1. Number of species threatened to disappear and dependent on agricultural habitats: Finland, early 1990s.

Agricultural habitat	Verte-brates[1]	Inverte-brates[2]	Vascular plants	Crypto-gams[3]	Total
Fresh meadows	1	18	8	6	33
Woodland pastures	0	34	13	22	69
Dry meadows	1	122	27	23	173
Fields	4	7	3	0	14
Total	6	181	51	51	289

Notes:
1. Mammals, birds, fish, etc.
2. Annelids, molluscs, butterflies, beetles, and other insects, arthropods, and invertebrates.
3. Non-flowering plants, such as algae, mosses, ferns, etc.

Source: MAF Finland (1996).

cereals from the pastoral areas in the west of the country. Most unimproved grassland has been lost since the 1930s, thus reducing the variety and numbers of birds, especially in the west. Most cereal crops are now planted in the autumn, not the spring. As a result there are substantially fewer stubble fields, which are a source of food for the bird population over the winter. Both hedgerow removal and the loss of other uncropped habitats have also reduced nesting and feeding opportunities for some species. Pesticide use is another factor implicated in the decline of farmland birds in the UK. Concern has focused on whether pesticides, by removing insect pests and weed species, may have an indirect effect on some bird populations by reducing food sources. The UK Ministry of Agriculture has recently commissioned a 5-year research project, involving collaboration with various conservation bodies, specifically to investigate the role of pesticides and other factors in the decline of farmland bird species.

Recent surveys by the United Kingdom Mammal Society, also reveal the reduction in mammal species that use agricultural land as habitat, such as voles (*Arvicola*), shrews (*Sorex*) and field mice (*Mus rylvaticus*). The main causes for these declines over the period since the 1970s have been attributed to the loss of rough grazing land and small habitat features on farmland such as ditches, hedges, etc.

Over the past 15 years in Canada, many farmers have begun to replace conventional tillage practices with conservation tillage, including no-till (see Neave et al., 2000). In Canada, studies have shown that wild species benefit from conservation tillage. Invertebrate numbers have been shown to rise as a result of the protection afforded by crop residue cover and reduced mortality caused by ploughing. Many species of birds become more common as their prey, invertebrates, grow in numbers. Deer mice (*Hesperomys*), too, may become more abundant, possibly because of higher survival rates or greater mobility of the population than is the case in conventionally tilled fields.

Agricultural activities in the United States are considered to affect 380 of the over 660 wild species listed as threatened or endangered in 1995 (USDA, 1997, pp. 17–18).

The main threats to wild species from agriculture in the US originate from converting land to cropland and grazing, with exposure to farm chemicals also important. The competition between agriculture and endangered species for land was heightened in the US with the introduction of the Endangered Species Act in 1973, which has the express objective of protecting ecosystems on which threatened and endangered species depend. Several agricultural programmes include measures that are designed to reduce the conflict between agriculture and biodiversity loss, including the Conservation Reserve Programme and the Wetlands Reserve Programme.

3.2.3. Indicator interpretation and limitations

Where agriculture is the dominant land-use activity, as is the case for many OECD countries, then it is to be expected that agriculture is likely to provide the major habitat area for wild species. In this context, Figure 4 needs to be interpreted with some care, as it is unclear if forests or other ecosystems were re-established on agricultural land what the relative share of wild species on different land-uses would be. The interpretation of wild species indicators is not straightforward, and care is required in relating species reductions or increases to agriculture, where other external factors, such as changes in the weather or populations of natural living organisms and predators, may have an important influence. It will also be necessary to take care in interpreting such indicators across countries, as the number of species will tend to be greater in large countries than for small countries, hence, the possibility of expressing the indicator according to a standard area unit could be considered.

Defining baselines is an important step in calibrating, comparing and interpreting indicators of biodiversity, but in practical terms baselines will usually be limited by available data. Baselines can be useful as objective measures of status at a given point in time against which changes in status can be compared. However, irreversible ecosystem and climate changes may prevent restoration of pre-existing species populations. In these cases progress towards agreed targets may be more useful for policy decision-making than measuring distance from baselines, especially when it is difficult to establish common baselines across OECD countries.

Setting baselines is a complex and often an arbitrary process, with many alternative baselines possible, and with each alternative generating different results and policy information. A number of baseline options with respect to wild species can be considered. These include first, setting the baseline at the time of the CBD's agreement in 1992; second, determining a baseline that represents the evolution of biodiversity in an ecosystem that has been unaffected by any significant human influence, i.e. the original 'natural' state; and third, establishing a baseline, in the case of agriculture, prior to the intensive use of inputs in agriculture, which for many OECD countries is around the early 1950s.

Measurement against the conditions at the time of CBD ratification is an alternative, but assessing biodiversity using 1992 as a baseline would be perceived as giving a biased result, because at that time OECD countries had already achieved a high level of socio-economic and agricultural development partly at the expense of biodiversity. Comparing an agricultural area with the original "natural" baseline, e.g. a forest or wetland, is of little value in that it will simply show that the majority of the original biodiversity has disappeared. However, the original natural state baseline is the relevant baseline in the case of clearing additional forests for agricultural use. It can also be of interest to potential resilience, if an area is no longer cultivated or if agriculture becomes less intensive. The so-called "climax" baseline, on the other hand, characterises the developing natural state after human activities on an area have ceased, and can be an important baseline in the case of a potential change in land-use from agriculture to another use.

Establishing a baseline for agriculture in terms of the period before the intensive use of inputs, also raises a number of questions. For example, how to define "intensification", at what point this is considered to begin (which can also vary for regions across a country), and what are the impacts of agriculture on biodiversity under different systems of input intensity (e.g. intensive use of machinery and chemicals, organic, and extensive "low" input farming systems)? For many countries, which have only just begun to establish wild species monitoring systems, the only practical baseline will be the first year of the monitoring programme.

3.3. Ecosystem diversity

Indicators of ecosystem diversity include the proportion of semi-natural and uncultivated natural habitats on agricultural land, and the extent of changes in agricultural land-use. Ecosystem diversity indicators represent the 'quantity' aspect of biodiversity shown at the base of the rectangle in Figure 1. There are four categories of indicators discussed here. The first three categories draw on the classification of agricultural land into intensively farmed, semi-natural, and uncultivated natural habitats. The fourth category involves developing an integrated approach by considering the entire agricultural landscape, including "man-made" features on agricultural land (e.g. hedges, farm buildings) which can provide wildlife habitat and combining this with information on biodiversity.

3.3.1. Intensively farmed agricultural habitats

3.3.1.1. Indicator definition and method of calculation
The indicator showing the *percentage share of the agricultural area covered by each crop type*, requires annual agricultural census data on the respective areas of individual crops and/or major cropping areas, i.e. arable crops (e.g. cereals), permanent crops (e.g. fruit trees), and pasture. To date no comprehensive cross country analysis of the environmental impact of changing crop compositions on intensively farmed agricultural habitat has been completed. While no generally agreed definition exists of the term 'intensively farmed agricultural habitats' it is widely interpreted as concerning agricultural areas which are used to produce arable crops and improved grassland for food, feed, and renewable raw materials (Tucker and Evans, 1997, pp. 267–325). These areas are commonly treated with fertilisers and pesticides and subject to farm management practices, such as ploughing, sowing, weeding, and harvesting.

Intensively farmed areas are artificial habitats subject to regular disturbance of the soil and dominated by annual/perennial crop species. The value of these areas as habitat is generally low, because of the paucity of non-crop vegetation combined with the use of pesticides, but they do provide habitat for some vascular plants, invertebrates, small mammals and birds. Often they are temporarily valuable habitats for migratory birds. The richness and abundance of wildlife on intensively farmed land will vary according to the:

- Type of crops cultivated: cereals, oilcrops, improved grassland, etc.
- Methods of production: farm management practices covering nutrients, soil, water, etc., and the farming system, such as "conventional", "integrated", "organic", etc.
- Spatial composition of the cultivated areas: field size, cropping patterns, etc.
- Proximity to other categories of habitats: semi-natural and uncultivated natural habitats.

In terms of the *type of crops* cultivated under an intensively farmed system, any type of crop can, to some extent, support wildlife, although the degree to which one crop is more supportive than another is unclear.

A Canadian study showed, however, that the diversity of vertebrates was higher in cereal crops relative to oilcrops (Fig. 5). The same study also showed that some crops may rank highly for a species in terms of a nesting site (for example fruit trees), but can be less important as a site for feeding. With regard to the *methods of production* used on intensively cultivated land, this is well documented as a critical factor affecting wildlife. Important in this context is the type of farming system used to produce crops ranging from "conventional" systems typified by widespread use of farm chemicals to "organic" systems where these inputs are not used (OECD is also developing an indicator share the share of organic farming in the total agricultural area, see OECD, 2001). There is also the extent to which farm management practices retain or remove non-farmed marginal features such as field margins, hedges, and ditches, which provide crucial habitat sites for wildlife.

The *spatial composition of cultivated areas*, concerns the effects on wildlife of the cropping patterns, ranging from monocultural systems to more diverse systems of cropping, and rotation patterns, with interspersed patches of non-crop vegetation. High levels of spatial crop heterogeneity, however, is not necessarily a beneficial indicator for all wildlife, for example, improved grassland habitats can provide a species rich and abundant habitat (Tucker and Evans, 1997). *Proximity to other categories of habitats* can be both beneficial and harmful depending on the proximity of intensive habitats to semi-natural and uncultivated habitats. Also important in this context are the types of farm management practices and systems under which each habitat type is maintained.

3.3.1.2. Recent trends

The overall decline in the total agricultural land area for most OECD countries since the early 1980s (and over a longer time scale for many countries) has been associated with the conversion of highly productive agricultural land usually to urban, industrial and road development, and a large share of marginal farming land converted to forest (Fig. 8). In many cases, such as in the European Union and the United States, the decrease in the area of intensively farmed land, i.e. for arable and permanent crops, has proceeded at a faster rate than for extensively farmed land, i.e. permanent pasture. At the same time agricultural production on the remaining intensively farmed agricultural land has increased through improving productivity by, for example, the greater use of farm chemicals and the removal of boundary features such as field border strips, to increase field size for larger farm machinery.

These developments in agricultural land-use over the past 20 years are largely recognised as having had a harmful impact on the environment in most OECD countries, both through the conversion of habitat features, such as shelterbelts, and the effects of diffuse pollution. This pattern of change, however, began to alter from a period around the late 1980s/early 1990s, with the introduction of agri-environmental and land diversion schemes in many countries. This started to improve farm management practices on intensively farmed land, such as the adoption of conservation tillage, and to lead to an increase in the area of land under more extensive farming practices, organic farming systems or land diverted to non-agricultural uses (e.g. long term fallow, forested). While evidence is still limited, these changes in farm management practices and the pattern of land-use, have led to the conservation and restoration of certain high nature value habitats on agricultural land. This has helped the recovery in some populations of wildlife species, and reduced diffuse pollution. Even so, it is still too early to be sure about the extent of these changes within or across OECD countries, or the permanence of the increase in some wildlife populations using agricultural land as habitat.

A study of the Northeast part of the United States, reflects the more general trends described above over the past 20 years (Mac et al., 1998, pp. 192–195). With the continuing expansion of urban communities, significant areas of prime agricultural land in the region were converted to housing, commercial development and roads. This led to the rise in land prices and encouraged more intensive farming on remaining land, including the removal of hedgerows, field-border strips, wetlands and woodland, and the greater use of farm chemicals. As a result of the removal of these habitats wildlife populations declined. An examination of the major sources of threats to endangered and threatened wild species in the United States found that agriculture was the major source of threat (USDA, 1997, pp. 17–18). Of the different sources of agricultural threats to endangered species, the conversion of land to agricultural production threatens the most species. Grazing and the use of pesticides are also important, but fertilisers less so. This pattern of different agricultural threats is broadly reflected for vertebrates, invertebrates and plants, although for invertebrates pesticide use is a key threat. The introduction of measures in the mid-1980s such as the US Conservation Reserve Programme and the encouragement of farm management practices beneficial to reducing soil erosion, has helped to maintain and restore certain habitats on farmland. Also the adoption of conservation tillage has increased the availability of crop residues for wildlife. The US study of the Northeast region has shown that numbers of wild turkeys and Canadian geese have increased because of the availability of crop residues in autumn and winter (Mac et al., 1998). Some mammals, such as possums, deer, racoons and skunks, have also taken advantage of residual maize and other crops.

In *Canada*, a recent study has examined the use of agricultural land by vertebrates (Neave and Neave, 1998). Using the Canadian prairies eco-zone as an example, which represents over 80% of the total national agricultural area, this revealed that all habitat types were used by some species (Fig. 5). However, while uncultivated habitats on agricultural land, especially wetlands, woodlands, and "natural" pasture, supported the greatest number of species for breeding, feeding, cover and wintering; cropland was

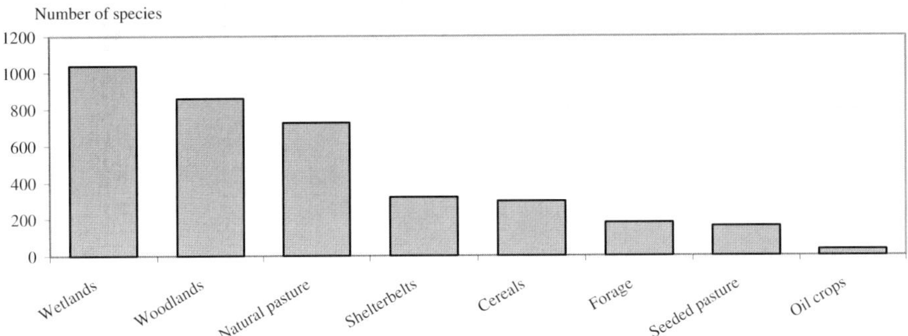

Figure 5.   Number of vertebrate[1] species using habitat[2] on agricultural land: Canadian Prairies, mid-1990s.

Notes:
1. Vertebrates, including birds, mammals, amphibians amd reptiles.
2. This includes the addition of species using prairies as a primary and secondary habitat for five activities, i.e. reproduction, feed, cover, wintering and staging (birds only).

Source: Adapted from Neave and Neave (1988).

primarily used for feeding purposes only. Moreover, within the cropland category, cereal crops supported larger numbers of species than oilcrops.

The changes outlined in the US study, have also been broadly reflected by similar developments in the European Union over the past two decades (European Commission, 1999). Crop production both increased and intensified, with an increased use of inputs and less diversified crop rotations, i.e. an increase in wheat, oilcrops and a reduction in secondary cereals, such as oats and rye. The area of permanent crops and pasture also declined, in some cases involving the ploughing up of meadows leading to the removal of habitat features such as hedges and other field boundaries. The overall consequence of these changes on the environment was an increase in diffuse pollution through the greater use of chemical inputs, and the removal of habitat, both to the detriment of wildlife. The reform of the EU's Common Agricultural Policy and introduction of agri-environmental measures in the early 1990s, has begun to encourage changes in farming practices, for example, the development of field margins on cropland and the maintenance of hedgerows. In addition, the policy of taking land out of production, 'set-aside', has resulted in an increase in fallow land from around 1 million hectares in the early 1980s up to over 4 million hectares by the mid-1990s (European Commission, 1999). While it is still too early to make any overall assessment of the impacts of these changes on the environment in the EU, evidence from some member states would suggest that some environmental improvements have been achieved, especially the restoration of habitats.

Field margins surrounding intensively cropped areas can provide and enhance wildlife habitats without altering cropping patterns or the intensity of output on the remaining cropped land. As part of the United Kingdom's Biodiversity Action Plan, farmers are being encouraged under a Habitat Action Plan, to maintain and restore grass margins, conservation headlands, and uncropped field margins (MAFF, 2000). Recent trends show that the estimated area under cereal field margin management, that contributes to the targets in the Habitat Action Plan, has increased substantially in recent years. A study of the corn bunting (Miliaria calandra) in Portugal concluded

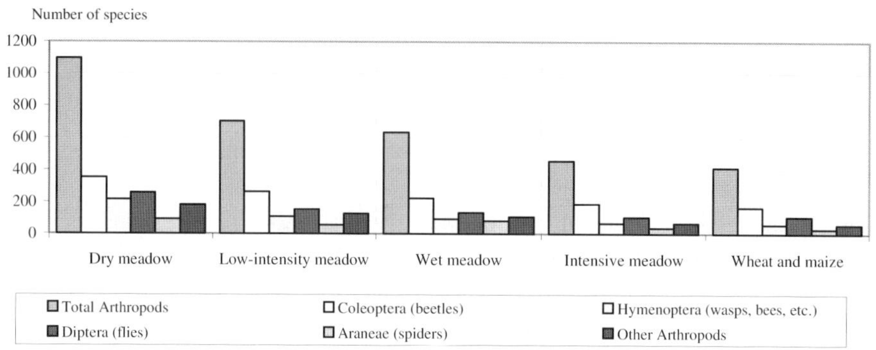

Figure 6.   Number of arthropod[1] species using intensive farmed land[2] compared to semi-natural habitats[3]: Switzerland, 1987.

Notes:
1. Anthropods recorded in this study were mainly spiders, beetles, flies, wasps and bees.
2. Intensively farmed land refers to wheat, maize and intensive meadow.
3. Semi-natural habitats refer to dry meadow, wet meadow and low-intensity meadow.
Source: Adapted from Duelli et al (1999).

that extensively managed farmland appears to offer a preferable habitat to this bird than intensively managed areas (Stoate et al., 2000). While the corn bunting is present on intensively managed areas, its population is in decline there. Research completed in Switzerland, drew similar conclusions to the Canadian study above. The Swiss study (Duelli et al., 1999), covering the use of agricultural land by arthropods (e.g. spiders, beetles, flies, bees), showed that the most beneficial crop for this species group was dry and low intensity meadows compared to wheat and maize crops (Fig. 6).

3.3.1.3. Indicator interpretation and limitations
Interpreting the effects on the environment of changes in cropping patterns on *intensively farmed land* requires some care, as this can vary considerably not only with the type of crop, but also with the management system used to produce the crop. For example, organic wheat production with field margins can be expected to have very different implications for the environment than wheat produced without field margins using a high intensity of chemical inputs. Moreover, wildlife on intensively farmed land will also be affected by the extent of the spatial distribution across the farmed landscape of semi-natural and uncultivated habitat, and the extent of field boundaries or openness of the landscape.

3.3.2. Semi-natural agricultural habitats

3.3.2.1. Indicator definition and method of calculation
The indicator of the share of the agricultural area covered by semi-natural agricultural habitats, requires data, from the annual agricultural census or national land inventory, covering the area of semi-natural agricultural habitats. A clear understanding and definition of the scope of semi-natural agricultural habitats is also necessary. Semi-natural agricultural habitats can be broadly defined as areas of land subject to "low intensity" farming practices, but this leaves open the difficulty of determining what is "low inten-

sity" farming. For practical purposes, semi-natural agricultural habitats have been broadly defined in this chapter as covering grazing marshes and water meadows, pastoral woodlands, alpine pastures, and dry meadows and pasture.

Semi-natural agricultural habitats can be characterised in terms of areas of farmland where the use of farm chemicals is either totally absent or they are applied at considerably lower rates per unit area than in more intensively cultivated areas. Also these habitats are relatively undisturbed by farming practices, such as from ploughing, mowing, and weeding. Typically, semi-natural habitats arise through interaction with other ecosystems, and can be broadly classified as follows (see Baldock et al., 1995; Baldock, 1999; Peco et al., 1999; and Tucker and Evans, 1997):

- Semi-natural habitats typical of agricultural ecosystems, such as extensive grassland and pasture; fallow land; extensive margins in cropped land; and "low intensity" permanent crop areas, including certain fruit orchards and olive groves.
- Semi-natural habitats arising from the interaction between agricultural and aquatic ecosystems, including some types of wetlands exploited for agricultural use, such as grazing in marshes and water meadows.
- Semi-natural habitats arising from the interaction between agricultural and forest ecosystems, including agro-forestry and pastoral woodland.
- Semi-natural habitats arising from the interaction between agricultural and mountain ecosystems, including alpine pastures and grass patches.
- Semi-natural habitats arising from the interaction between agricultural and steppe ecosystems, ranging from semi-arid to desert steppe, including dry meadows and pastureland.

The value of semi-natural agricultural habitats for wild flora and fauna varies according to the individual type of habitat. In general these habitats are considered to have systematically better conditions for wildlife than intensively farmed habitats. They also include some important sites for nature conservation, with frequently a high level of species richness of botanical and entomological value. Moreover, the interspersion of intensively farmed areas by semi-natural habitat can enhance the quality of the entire agricultural ecosystem, both from the viewpoint of biodiversity and a varied landscape.

3.3.2.2. Recent trends

Changes in the area of semi-natural habitats on agricultural land show considerable variation for the limited number of OECD countries for which data are available (Fig. 7). For some countries semi-natural habitats as a share of the total agricultural land area is in excess of 50%, but in Canada, Japan and Sweden the ratio is considerably lower than this. Also, for certain countries the area of semi-natural agricultural habitats has shown a modest increase since 1985, despite a reduction in the total agricultural land area (Fig. 7). In some cases this can be explained by land diversion schemes leading to a shift from arable land to fallow and pasture. Semi-natural agricultural habitats have been often converted to other land-uses because of their location, which is commonly in marginal farming areas, where both the physical terrain and climatic conditions result in low productivity and poor financial returns. This makes such areas especially unattractive for younger entrants into farming, particularly where higher income and less arduous employment opportunities exist.

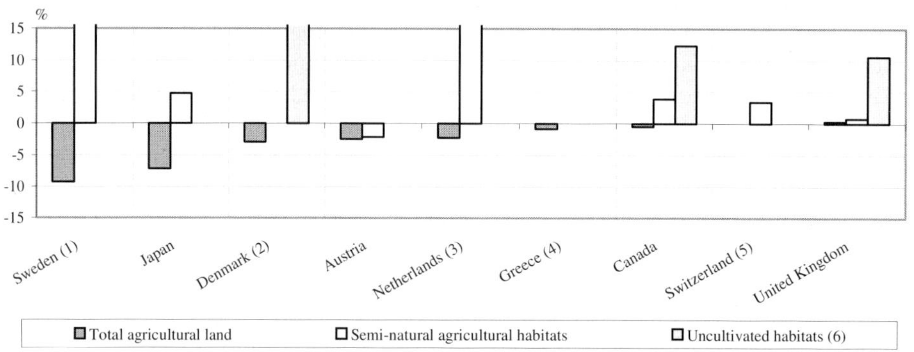

Figure 7. Area of total agricultural land, semi-natural agricultural habitats and uncultivated habitats: 1985 to 1998.

Notes:
1. Area of semi-natural habitat showed an increase of 33%.
2. Area of uncultivated land includes only woodland which showed an increase of 21%.
3. Area of semi-natural habitat showed an increase of 547%.
4. Negligible change in semi-natural habitat area.
5. No change in agricultural land area.
6. Uncultivated natural and man-made habitats on and/or bordering agricultural land, e.g. woodlands, small rivers, wetlands, farm buildings, etc.

Source: OECD (2001), Environmental Indicators for Agriculture, Volume 3 – Methods and Results, Publications Service, Paris, France.

A study of the Vejle county of *Denmark* (accounting for nearly two-thirds of the total national agricultural land area) from 1970–1995, revealed a number of important changes to semi-natural agricultural habitats and biodiversity in the area (IUCN, 1999, pp. 15–18). Over this period semi-natural grasslands decreased by over 40%, accompanied by a shift from low intensity pastoral farms to high intensity pig and cattle enterprises. Wild flora seed banks in arable fields declined by 60%, while there were significant reductions in areas of wet and dry heathland and peat bogs. The intensification of agriculture was recognised as a major influence on these changes, although measures introduced in the early 1990s, including agri-environmental management payments, are helping to maintain and restore semi-natural agricultural habitats. Under new legislation on environmental protection, the creation of ecological farmland areas was introduced in Poland in 1999 to help restore semi-natural agricultural habitats such as water meadows, and agro-forestry areas (FAO, 1999, pp. 204–205). Agriculture is also included under measures to provide for national parks and landscape reserves, accounting for over 40% of the total area of these parks/reserves in the late 1990s.

In the United Kingdom intensification of farm land-use was considered to be one of the main contributors to the reduction in the area of semi-natural agricultural habitats over the past 50 years. However, as a result of targeted agri-environment policies, reductions in price support, technological developments and consumer demand, trends towards extensification of agriculture may be emerging (Stott and Haines-Young, 1998). The UK has also set targets for the maintenance and restoration of priority semi-natural grassland habitats, under the UK Biodiversity Action Plan

(MAFF, 2000). The trend towards greater extensification in the UK is illustrated by the increase in area of unmanaged grassland/tall herb, reflecting the introduction of arable set-aside. Intensification was associated largely with the agricultural improvement of grasslands, but this was more than balanced, in terms of area, by the reversion of previously improved farm land to unmanaged grassland. There was also a minor net gain in area of wetlands, resulting from the conversion of intensive farm land to wetland, but small net losses in moorland and bogs. These results should be interpreted with caution, however, because of both sampling and possible observer error, and because it is unlikely that semi-natural habitats created by reversion or restoration compensate, in ecological terms, for the loss of semi-natural habitat of "high" nature conservation value.

3.3.2.3. Indicator interpretation and limitations

It is generally assumed that the larger the area covered by semi-natural agricultural habitats, the more beneficial are the effects on wildlife, as in general these habitats harbour a much greater variety and abundance of species than do intensively farmed agricultural areas. Semi-natural agriculture habitats are also characterised through their symbiotic relationship with surrounding habitats. However, the reversion of intensively farmed land to semi-natural habitat is unlikely, in ecological terms, to fully compensate for the loss of semi-natural habitat of "high" nature value. While research clearly points to the sharp decline in the extent of certain types of semi-natural agricultural habitats, such as traditional terraced olive groves, alpine meadows, heathland, and pastoral woodlands, it is extremely difficult to provide any systematic view due to the lack of consistent time series data (EEA, 1998, p. 164). Not only is data limited in terms of the extent of these habitats, but knowledge of their quality is also lacking. These problems are compounded by the absence of a clear definition of what constitutes a "traditional" semi-natural farmed habitat, although some attempts have been made to define low-intensity farming systems (Baldock et al., 1995).

3.3.3. Uncultivated natural habitats

3.3.3.1. Indicator definitions and method of calculation

No commonly accepted definition of "uncultivated natural habitats" exists, but it is generally considered to include those habitats that are on, crossing, or bordering agricultural areas. The main examples include: small ponds, lakes and rivers, unexploited wetlands, bogs and other aquatic habitats; natural woodlands and forests; and rocky outcrops. The indicators developed here focus on two of these habitat: aquatic ecosystems, in particular, wetlands; and "natural" forests, by tracking changes in the net area of these habitats converted to agricultural use.

The net area of *aquatic ecosystems*, such as unexploited wetlands, bogs, small ponds, lakes, and diverted rivers, converted to agricultural use gives an estimate of the loss of aquatic ecosystems through drainage or reclamation for farming offset by the restoration or reversion of these ecosystems from agricultural use. This approach is being used in the United States, for example, to help assess domestic wetland conservation policies (Heimlich et al., 1998). The conversion of agricultural land back into an aquatic ecosystem, may in some cases be part of efforts to help reduce flooding by the reclamation of farm land in order to increase the free flow of dammed rivers. In other instances this restoration has the objective of restoring the ecosystem as a valued aquatic environment.

The area of *natural forest* converted to agricultural use, encompasses both natural "primary" forest, such as areas of tropical rainforest in Australia and Mexico, and also "secondary" forests. Secondary forests are those forests which are, or have been commercially exploited, and in which the physical conditions and diversity closely resemble the natural state, having developed over a long time period. There is generally a contact zone with agriculture, where forests have been cleared for farming through cutting and burning. Farm land is also restored back to use as woodland or forest, and a "net change" approach might be appropriate in these cases. However, the main concern here is the destruction of 'natural' forest which even if restored could take hundreds of years, if not more, to return to its "original" state.

3.3.3.2. Recent trends

There is little comprehensive time series data on the net area of *aquatic ecosystems* converted to agricultural land across OECD countries, although more data on the total area converted exists (Fig. 8). For the limited of number countries where data is available, there has been, over the past ten years, a net conversion of agricultural land to aquatic ecosystems, i.e. more aquatic ecosystems are being restored that being converted to agricultural use. The two exceptions to this trend are in Japan and Korea, where the reverse is the case. However, proposals in Korea for large-scale reclamation of tidal flats for agriculture have been cut back drastically to safeguard estuarine habitats (OECD, 1998b).

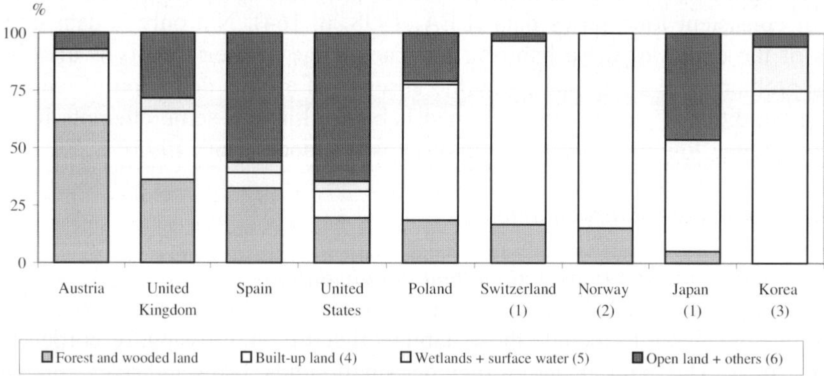

Figure 8. Share of different land use types in land converted from agriculture to other uses: mid-1980s to mid-1990s.

Notes:
~ for wetlands + surface water areas are not available.
~etlands + surface water areas and open land + others are not available.
~ded land not available.
~ mainly land used for urban or industrial development and transport infrastructure,
~ace water covers mainly small ponds, lakes and diverted rivers.
~sed for any of the above uses, such as barren land, exposed rocks and for
~n land but not forested.
~ronmental Indicators for Agriculture, Volume 3 – Methods and Results,
France.

It is evident that in some countries agricultural expansion has not been the major cause of the decrease in the area of aquatic ecosystems. In the United States, for example, between 1982 and 1992 agriculture accounted for only 20% of the total reduction in the area of wetlands, with conversion to urban development accounting for nearly 60% over the same period (Heimlich et al., 1998). Wetland conservation has been the focus of policy debate over recent years in the US. The share of wetlands converted to agricultural use dropped from more than 80 % in the period 1954 to 1974 to 20 % during 1982 to 1992 (Heimlich et al., 1998). The US appears to be reaching its goal, set through various wetland conservation measures, of "no net loss" of wetland area in the 1990s, that is conserving and restoring at least as much wetland as is lost. Heimlich et al. (1998) observe that while government policies are partly responsible for the decrease in wetland conversion, falling agricultural commodity prices also reduced the pressure on farmers to convert wetlands. Hence, it is difficult statistically to separate the policy and market factors responsible for decreased wetland conversion. It should be noted that the share of total US wetlands converted during the period 1954 to 1992 is relatively small, and there remains a considerable area of wetland that could be converted to agricultural production. In addition, the slower rate of wetland conversion over the 1984–1992 period may reflect that most of the wetlands suitable to conversion were already converted to farmland prior to 1982.

In the European Union the area of *wetlands* in coastal zones decreased by between less than 1 % in France and up to 16 % in Italy, during the period 1975 to 1990 (EUROSTAT, 1999, pp.50–51). Reduction in wetland areas are reported to be greater than this when comparing the period since the 1950s and 1960s, for example, a decrease of 60–65 % in Finland, 57 % in Germany (excluding marshes), and 55 % for the Netherlands. The conversion of agricultural land is identified as one of the major causes of wetland reduction in the EU, but other losses have occurred due to forestry, pollution and over-exploitation of aquifers.

For the few countries where data exist, the conversion of agricultural land to *woodland and forest* represents a significant share of agricultural land converted to other uses for a number of countries over the past decade (Fig. 8). Moreover, forest and woodland also represent the main land-use type of land converted to agricultural use. However, it is not clear whether these changes represent the conversion to 'natural' wooded areas or commercial forest.

3.3.3.3. Indicator interpretation and limitations

These indicators reveal the extent of loss or conservation of uncultivated natural habitats – aquatic ecosystems and "natural" forest – on and/or bordering agricultural land. Where such habitats are converted to agricultural use, this is usually associated with a high diminution of wildlife and amenity value, especially in cases where the habitats may have taken several thousand years to evolve. In any assessment of uncultivated habitat that has been restored from agricultural use, it is necessary to evaluate the conditions and type of restoration that has occurred. For example, whether agricultural land has been restored to a commercial or to a "natural" forest, and the period over which the change has occurred. Interpretation of these indicators may also require a regional and spatially disaggregated perspective. In some regions, an increase in the conversion of forest to agricultural land may involve the establishment of low intensity farming systems, which might have a positive impact on biodiversity depending on the type of forest converted, i.e. a "natural" or commercial area of forest.

3.4. Habitat matrix

3.4.1. Indicator definition and method of calculation

The habitat matrix identifies the ways in which various wild species use agricultural habitat types, ranging from cropped land to uncultivated habitat on agricultural land, and then relates this use to changes in the areas of these habitats. The indicator is then used to identify which habitat types on agricultural land support the most wildlife use and whether these habitat types are increasing, decreasing or remaining constant over time (this approach draws on Neave et al., 2000; and Neave and Neave, 1998). The methodology recognises that all farm land has some value as habitat. The matrix explicitly incorporates information on how various species use farmland to meet their habitat needs. It is also restricted to habitat change occurring within the agricultural land base only and not that due to other land-uses. The agricultural land base is defined to include areas of uncultivated natural habitat (e.g. marshes) and man-made features on agricultural land (e.g. farm buildings).

To construct the matrix it is necessary to identify how different species use various agricultural habitats. To accomplish this, habitat suitability matrices are developed individually for the main agricultural ecozones across a country (or bio-regions). These matrices incorporate information on all flora and fauna, or more partial information where detail for all taxonomic groups does not exist. The particular use each species makes of agricultural land habitats in each ecozone (see below) is then identified. Each "habitat use" is ranked according to how dependent a species is on a certain habitat for this use, including:

- Primary use, meaning that a species is dependent on, or strongly prefers, a certain type of habitat (also called critical habitat).
- Secondary use, meaning that a species uses a certain habitat (e.g. to obtain food in the case of fauna) but is not dependent on it.

Matrices for each specified ecozone are then collected. This information might be assembled from a range of sources, depending on the quality and quantity of data available in any given country, including written sources, expert judgements by wildlife and agricultural specialists and, ideally, actual field survey data of wildlife species. Once the matrices are completed, primary and secondary habitat use entries are summed separately into five main categories (this applies to fauna only), including first, breeding, nesting, and reproduction; second, feeding and foraging; third, cover, resting, roosting, basking, and loafing; fourth, wintering; and fifth, staging (for birds only). Each separate use of a habitat type by a species is recorded as a habitat use unit, that is not the number of species using the habitat, but the number of individual ways in which the habitat is used, such as, for feeding and nesting, these habitat use units are then summed by habitat type for each ecozone. The habitat types can correspond to any classification system for which data is available, but for many countries, at present, this will generally correspond to the main land-use categories defined in the annual Census of Agriculture. In general, the habitat matrix can provide an alternative or surrogate for the wild species diversity indicator. Where species population data is not available the matrix approach can provide an indirect measure for species diversity.

3.4.2. Recent trends

There are few attempts to provide an integrated holistic view of the impact on wild species of changes in the pattern of agriculture land-use. A number of countries are now beginning to examine the possibility of establishing a habitat matrix to examine the impact of all agricultural land-use changes (or the total land area including agriculture) on wildlife, for example, Mexico, Switzerland, and the United Kingdom. Possibly the most advanced work of this type has been developed in Canada, while some work has also been undertaken in Finland.

The results of the Canadian study concludes that all agricultural land offers a variety of habitats for wildlife, but some types are superior to others (Fig. 5 and Neave et al., 2000). The Canadian study suggests that changes in agricultural land-use from less intensive to more intensive practices, such as bringing marginal land into crop production, create pressures on wildlife by making one or more of the resources they depend on more scarce or otherwise unavailable. On the other hand, the study indicates that reductions in summer fallow, and conversion of marginal cropland to other uses such as Tame or Seeded Pasture, will benefit wildlife, although these findings may not be valid for other countries. In general, from 1981 to 1996 agricultural habitat for wildlife in Canada shows positive or neutral trends for wild species in all ecozones except, the Pacific Maritime and Mixedwood Plains (Fig. 9). These two regions are noted for the intensity of their agriculture, although they account for less than 6% of the total agricultural land area. The

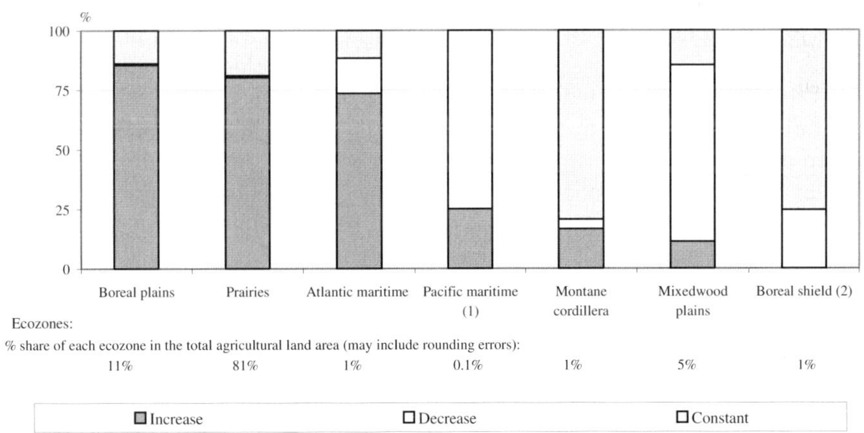

Figure 9. Share of habitat use units for which habitat area increased, decreased and remained constant: Canada, 1981 to 1996.

Notes:
1. The share for which habitat area remained constant equals 0%.
2. The share for which habitat areas increased equals 0%,

Source: Neave et al. (2000).

recent trend to reduce summer fallow in Canada and to convert cropland to permanent cover is a positive trend for wildlife, but one currently driven more by market forces, especially relatively low commodity prices, than by an apparent interest in wildlife.

Finland has begun to monitor the number of threatened species, flora and fauna, across different types of agricultural habitats (see Table 1). Although this research has not yet provided any time series results, it does show that dry meadows are the agricultural habitat where the most threatened species can be found. However, this could reflect that in general more species use dry meadows as habitat relative to crop land. A habitat matrix approach is also being used in Korea to examine the effects of different agricultural land-use patterns and farm management practices on wild species. Preliminary and unpublished research by the Korean National Institute of Environmental Research has revealed that species diversity of birds was higher on farmland than on other habitat types studied (e.g. forest edges, mountain areas, forests). However, for certain species (e.g. small mammals), cropland might act as a barrier to species dispersal, limiting them to forest edges and other habitats.

3.4.3. Indicator interpretation and limitations

The habitat matrix approach allows changes in area of habitat to be monitored and mapped, and identifies which species are most likely to benefit from, or be adversely affected by, the changes observed. The indicator is readily developed from standard agricultural land-use data that is available in most countries. The matrix is able to track the trends in habitat area over time, identify areas where critical habitats are threatened, and provide a link to the species making use of different agricultural habitats. Trends over time are calculated using land-use data obtained through national Agricultural Censuses. In most countries this data covers all farms, is spatially detailed, undergoes extensive testing with respondents, and is usually validated prior to publication. The matrices can be based on the biological and ecological literature and on interviews and consultations with field biologists, or, where data exists, from field survey information. Areas of different habitat types, and changes in those areas, can be mapped allowing policy efforts to target both valuable and/or vulnerable areas. Policy relevance is further enhanced because changes in habitat over time can be directly linked to the species making use of these habitats. In this way, species that may be affected by changes in land-use and habitat can be identified, including species at risk. Moreover, because of the link to land-use, the matrix can readily be linked to models which forecast agricultural land-use trends (Mac et al., 1998, pp. 55–57).

Notwithstanding the above points, there are several limitations to this approach, in particular, that the matrix records only information about the absence or presence of certain habitat uses, it does not reveal much about habitat and species quality. However, the matrix has the flexibility to use finer categories of habitats than provided from Census data where this data exists or where it is felt to be of value to collect such data. Related to this, the matrix does not always consider how successful is a particular habitat use, i.e. for reproduction or feeding. The indicator does not examine the effects of various land management practices that can differ significantly for the same habitat unit. Using broad land-use categories also does not account for biological factors that may limit a species' use of a particular habitat type. For example, a species may not be able to use a habitat because one need may be met (e.g. food) while others are not (e.g. water, cover), the habitat may be fragmented, there may be behavioural barriers to use, or the species may be too widely dispersed.

A further limitation to the habitat matrix is that it does not distinguish between the conservation value of different species. That is to say, the indicator might not reflect the replacement of rare, protected or endangered species. Moreover, the local extinction of a rare species, whose ecological niche is then occupied by other opportunistic or non-native species, would be reflected as a positive development in the matrix. To help better incorporate these aspects of species quality into the matrix, different classes of species could be weighted by using, for example, national information on threatened species. Hence the respective parts of agricultural habitats used by such species could be singled out to avoid biases in the calculation of the matrix.

4. Future challenges

Understanding the relationship between agriculture and biodiversity is still in an early phase of development and requires further research of the basic conceptual issues concerning the complex and multidimensional nature of biodiversity. This work will also benefit in the future from further co-operation internationally with efforts concerning biodiversity and agriculture underway in FAO, and more broadly through the Secretariat to the Convention on Biological Diversity. However, considerable research has been undertaken on the effects of agriculture on biodiversity, while there are now a range of databases established or being developed that are of relevance to the area.

While the emphasis of indicators of *genetic diversity* in agriculture, has been on the *in situ* diversity of domesticated crops and livestock, further work could examine *in situ* indicators of wild relatives for genetic improvement, especially for cultivated crops. As *in situ* indicators measure only a very small proportion of existing and available genetic diversity and can severely underestimate real available genetic diversity, *ex situ* indicators might be further developed (see FAO 1996; and 1998). Also, in the future, using molecular "fingerprint" genetic marker data to measure genetic diversity could allow more precise assessment of genetic diversity of domesticated species. For example, using different named varieties of maize could be misleading, as they may have very similar germplasm.

To improve monitoring of the state and trends in *wild species diversity* in agriculture across OECD countries, may require developing a standardised methodology for indicators of wild species on agricultural land. One possibility, being explored by some countries, is to develop species diversity indicators for agriculture through a Natural Capital Index (NCI) framework. The NCI is calculated as the product of the quantity of the ecosystem (e.g. agro-ecosystems) multiplied by the quality of the ecosystems (i.e. the average of changes in wild species numbers from a baseline period, see RIVM, 1998). This approach has similarities with the habitat matrix indicator. Comprehensive data on species distribution and population numbers are unavailable for most countries, although certain indicative wild species (e.g. birds) could serve as a useful proxy of biodiversity quality in agriculture. A pragmatic approach will be needed to choose indicative (endemic) species, or groups of species, that are important to the functioning of particular agricultural ecosystems. In this context, it will also be necessary to distinguish between indicators of wild species that help support agricultural production, such as pollinators and pest controlling organisms and predators, and wild species that use agricultural habitat or are affected by farming but use other habitats (especially for those farming systems that have been established over long periods of time).

A key element in further developing *ecosystem diversity* indicators may be the establishment across OECD countries of common *definitions* of the major types of habitat identified, namely: intensive, semi-natural and uncultivated habitats on agricultural land. This task might be helped by drawing on classification systems of different agricultural land-use types already developed for most countries in annual agricultural censuses. To help improve the analytical soundness and measurability of wildlife habitat indicators, further research could also be developed to examine the relationship between agricultural activity and habitats, covering *habitat fragmentation* (i.e. the degree to which a given habitat type is divided into separate patches), *heterogeneity* (i.e. average size of and variability of habitat types per monitoring area), and *vertical vegetation structure* (i.e. habitat strata, such as bushes and trees).

Baselines from which to interpret changes in biodiversity, can be important for valuing the state and trends in biodiversity. A number of baseline options can be considered for biodiversity, and setting such a baseline is a complex and often a relatively arbitrary process. Many countries are in the process of developing criteria and thresholds to interpret biodiversity indicators, and in many cases the only practical baseline will be the first year from the beginning of when programmes are monitored. However, given the difficulties in determining suitable baselines across OECD countries, it may be more useful for policy makers to measure progress towards agreed targets.

As targets and baselines are established it will also be useful for policy makers to improve understanding of the *spatial distribution of biodiversity in agriculture*. This may also require better understanding of the significance of particular species distribution patterns and how to interpret changes in these distributions over time. Knowledge is also poor of species numbers and distribution patterns in relation to different agricultural land-use types and farm management practices and systems. A feasible approach to this is to link biodiversity and agro-ecosystems into a matrix.

Biodiversity has an economic value to society operating at many different levels, but mainly in terms of biodiversity's use value, such as providing a life supporting system to agricultural production; and non-use values, for example, the knowledge of the continued existence of a particular species which others might enjoy or benefit (OECD, 1999). Placing a monetary value on biodiversity is especially difficult as in many instances no markets exist for biodiversity, and also market prices fail properly to reflect the many non-market benefits of biodiversity (Cooper, 1999; Smale, 1998; and Zohrabian and Traxier, 1999).

Research on the economic value of biodiversity is of considerable importance to policy makers and society in assessing the costs and benefits of biodiversity conservation, and in helping determine which policies might best achieve biodiversity goals in agriculture, as recognised in the CBD. While there is work underway in this area, further studies are required to estimate the economic benefits of biodiversity, and the costs and benefits of the trade-offs between increased agricultural production and biodiversity loss (OECD, 1997, pp. 42–45; Steffens and Hoehn, 1997).

Acknowledgements

The author wishes to thank Wilfrid Legg, Laetitia Reille, Theresa Poincet and other OECD colleagues for comments and assistance in preparing this paper. Any remaining errors in the paper are the responsibility of the author, and the views expressed do not necessarily reflect those of the OECD or its Member countries. This chapter draws extensively from the OECD report (2001), *Environmental Indicators for Agriculture Volume 3: Methods and Results*, Publications Service, Paris, France. For further information regarding the OECD work on agri-environmental indicators contact Kevin Parris at the address above or visit the OECD website at: http://www.oecd.org/agr/biodiversity/index.htm.

References

Ahn, W.S., Kang, J.H., Yoon, M.S., 1996. Genetic erosion of crop plants in Korea. In: Park Y.G., Sakamoto, S. (Eds), Biodiversity and Conservation of Plant Genetic Resources in Asia. Japan Scientific Societies Press, Tokyo, pp. 41–55.

Baldock, D, 1999. Indicators for high nature value farming systems in Europe. In: Brouwer F. and Crabtree B. (Eds), Environmental Indicators and Agricultural Policy. CAB International, Wallingford, pp. 121–135.

Baldock, D., Beaufoy, G., Clark, J., 1995. The Nature of Farming, Low Intensity Farming Systems in Nine European Countries. Institute for European Environmental Policy, London.

Burns, W., 2000. Bibliography: Impacts of Climate Change on Flora and Fauna Species and Associated Ecosystems. Pacific Institute for Studies in Development, Environmental Security, Oakland, CA. Available at: http://www.pacinst.org/ccbio.pdf.

CEC, 2000. Securing the Continent's Biological Wealth: Towards Effective Biodiversity Conservation in North America, Commission for Environmental Cooperation, Montreal. Available at: http://www.cec.org/programs_projects/conserv_biodiv/baseline.cfm?varlan=english.

Cooper, J.C. 1999. The Sharing of Benefits Derived from the Utilisation of Plant Genetic Resources for Food and Agriculture, Internal Memorandum, US Department of Agriculture, Washington DC.

Cunningham, E.P., 1999. Recent developments in biotechnology as they relate to animal genetic resources for food and agriculture, Background Study Paper No. 10, FAO Commission on Genetic Resources for Food and Agriculture, Eighth Session, Rome. Available at: http://www.fao.org/ag/cgrfa/docs8.htm.

Day, K., 1996. Agriculture's Links to Biodiversity, Agricultural Outlook, December, 1996. Economic Research Service, US Department of Agriculture, Washington, DC.

Debailleul, G., 1997. Economic incentives for biodiversity conservation in the agricultural sector. In OECD, Investing in Biological Diversity. The Cairns Conference, Paris, pp. 235–52.

Duelli, P., Obrist, M.K., Schmatz, D.R., 1999. Biodiversity evaluation in agricultural landscapes: above-ground insects. Agriculture, Ecosystems and Environment 74, 33–64.

ECNC [European Centre for Nature Conservation], 2000. Stimulating Positive Linkages between Agriculture and Biodiversity: Recommendations for the EC Agricultural Action Plan on Biodiversity, Tilburg, The Netherlands. Available at: http://www.ecnc.nl/.

EEA [European Environment Agency], 1998. Europe's Environment: The Second Assessment, Office for Official Publications of the European Communities, Luxembourg. Available at: http://themes.eea.eu.int/.

Environmental Protection Agency, 1999. Environment in Focus – A Discussion Document on Key National Environmental Indicators, Wexford, Ireland. Available at: http://www.epa.ie/pubs/default.htm.

European Commission, 1998. First Report on the implementation of the Convention on Biological Diversity by the European Community, Brussels.

European Commission, 1999. Agriculture, Environment, Rural Development: Facts and Figures – A Challenge for Agriculture. Office for Official Publications of the European Communities, Luxembourg. Available at: http://europa.eu.int/comm/dg06/envir/report/en/index.htm.

EUROSTAT [Statistical Office of the European Communities], 1999. Towards environmental pressure indicators for the EU, Environment and Energy Paper Theme 8, Luxembourg. Background documentation available at: http://e-m-a-i-l.nu/tepi/ and http://esl.jrc.it/envind/.

FAO [United Nations Food and Agriculture Organisation], 1996. Report on the State of the World's Plant Genetic Resources for Food and Agriculture, prepared for the International Technical Conference on Plant Genetic Resources, Leipzig. Available at: http://193.43.36.6/wrlmap_e.htm.

FAO, 1998. The State of the World's Animal Genetic Resources for Food and Agriculture. First session of the International Technical Working Group on Animal Genetic Resources for Food and Agriculture, Rome. Available at: http://www.fao.org/ag/cgrfa/docs8.htm.

FAO, 1999. Central and Eastern European Sustainable Agriculture Network, First Workshop Proceedings, REU Technical Series 61, FAO Subregional Office for Central and Eastern Europe, Rome. Available at: http://www.fao.org/regional/europe/public-e.htm.

Fjellstad, W.J., Dramstad, W.E., 1999. Patterns of change in two contrasting Norwegian agricultural landscapes. Landscape and Urban Planning 45 (4), 177–191.

Heimlich, R.E., Wiebe, K.D., Claassen, R., Gadsby, D., House, R.M., 1998. Wetlands and Agriculture: Private Interests and Public Benefits. Agricultural Economic Report No. 765, Economic Research Service, US Department of Agriculture, Washington, DC. Available at: http://www.ers.usda.gov/epubs/pdf/aer765/.

Hunziker, M., 1995. The spontaneous reafforestation in abandoned agricultural lands: perception and aesthetic assessment by locals and tourists. Landscape and Urban Planning 31 (3), 399–410.

Ihse, M., 1995. Swedish agricultural landscapes – patterns and changes during the last 50 years, studies by aerial photos. Landscape and Urban Planning 31 (1), 21–37.

IUCN [World Conservation Union], 1999. Background Study for the Development of an IUCN Policy on Agriculture and Biodiversity, report prepared by Nowicki P., Potter C. and Reed T., University of London,. Available at: http://www.iucn.org/places/europe/eu/docs/Agriculture_Biodiversity.pdf.

Kate, K. ten, Laird, S.A., 1999. The Commercial Use of Biodiversity – Access to Genetic Resources and Benefit Sharing, Earthscan Publications Ltd., London.

Mac, M.J., Opler, P.A., Haecker, C.E.P., Doran, P.D., 1998. Status and Trends of the Nation's Biological Resources, 2 Vols, United States Department of the Interior, United States Geological Survey, Reston, Virginia. Available at: http://biology.usgs.gov/pubs/execsumm/page2.htm.

MAF Finland, 1996. Renewable Natural Resources and Biological Diversity. Ministry of Agriculture and Forestry (MAF), Helsinki.

MAFF [Ministry of Agriculture, Fisheries and Food], 2000. Towards Sustainable Agriculture – A Pilot Set of Indicators. London. Available at: http://www.maff.gov.uk/ [Farming > Sustainable Agriculture].

Montgomery, C.A., Pollak, R.A., Freemark, K., White, D., 1999. Pricing biodiversity. Journal of Environmental Economics and Management, 38 (1), 1–19.

Neave, P., Neave, E., 1998. Agroecosystem Biodiversity Indicator – Habitat Component: Review and Assessment of Concepts and Indicators of Wildlife and Habitat Availability in the Agricultural Landscape – Concept Paper. Report No. 26, Agri-Environmental Indicator Project, Agriculture and Agri-Food Canada, Regina, Canada.

Neave, P., Neave, E., Weins, T., Riche, T., 2000. Availability of wildlife habitat on farmland. In McRae T., Smith C.A.S., Gregorich L.J. (Eds), Environmental Sustainability of Canadian Agriculture: Report of the Agri-Environmental Indicator Project. Agriculture and Agri-Food Canada (AAFC), Ottawa. Available at: http://www.agr.ca/policy/environment/publications/list.html.

OECD, 1996. Saving Biological Diversity. Paris.

OECD, 1997. Investing in Biological Diversity: The Cairns Conference. Paris.

OECD, 1998a. Towards Sustainable Development: Environmental Indicators. Paris.

OECD, 1998b. Environmental Performance Reviews: Korea. Paris.

OECD, 1999. Handbook of Incentive Measures for Biodiversity – Design and Implementation, Paris.

OECD, 2001. Environmental Indicators for Agriculture Volume 3: Methods and Results, Publications Service, Paris. Executive summary of this report available from the OECD agri-environmental indicators web site at: http://www.oecd.org/agr/env/indicators.htm (see "Publications").

Pagiola, S., Kellenberg, J., 1997. Mainstreaming biodiversity in agricultural development – toward good practice, World Bank Environment Paper No. 15, World Bank, Washington DC.

Peco, B., Malo, J.E., Onate, J.J., Suarez, F., Sumpsi, J., 1999. Agri-environmental indicators for extensive land-use systems in the Iberian Peninsula. In: Brouwer, F., Crabtree, B. (Eds), Environmental Indicators and Agricultural Policy. CAB International, Wallingford, pp. 137–156

RIVM [State Institute of Public Health and the Environment], 1998. Leefomgevingsbalans, voorzet voor vorm en inhoud (only in Dutch "Balance for the Natural Environment"). Bilthoven.

RSPB [Royal Society for the Protection of Birds], 1999. The State of the UK's Birds 1999, RSPB Annual Report, Sandy, Bedfordshire. Available at: http://www.rspb.org.uk/ [> Conservation Issues]

Smale, M., 1997. The Green Revolution and Wheat Genetic Diversity: Some Unfounded Assumptions, World Development 25 (8), 1257–1269.

Smale, M. (Ed.), 1998. Farmers, Gene Banks and Crop Breeding: Economic Analysis of Diversity in Wheat, Maize, and Rice. International Maize and Wheat Improvement Centre, Mexico, Kluwer Academic Publishers, Boston, MA.

Smith, F., 1996. Biological diversity, ecosystem stability and economic development. Ecological Economics 16 (3), 191–203.

Spillane, C., 1999. Recent developments in biotechnology as they relate to plant genetic resources for food and agriculture, Background Study Paper No. 9, FAO Commission on Genetic Resources for Food and Agriculture, Eighth Session, Rome. Available at: http://www.fao.org/ag/cgrfa/docs8.htm.

Steffens, K., Hoehn, J.P., 1997. Valuing Biodiversity: Issues and Illustrative Example. Staff Paper (97–7), February, Department of Agricultural Economics, Michigan State University, East Lansing, MI.

Stoate, C., Borralho, R., Araujo, M., 2000. Factors affecting corn bunting *Miliaria calandra* abundance in a Portuguese agricultural landscape. Agriculture, Ecosystems and Environment 74, 33–64.

Stott, A.P., Haines-Young, R., 1998. Linking land cover, intensity of use and botanical diversity in an accounting framework in the United Kingdom. In: Uno, K., Bartelmus, P., (Eds), Environmental Accounting in Theory and Practice. Kluwer Academic Publications, The Netherlands, pp. 245–262.

Tucker, G.M., Heath, M.F., 1994. Birds in Europe: Their Conservation Status. BirdLife Conservation Series No. 3, BirdLife International, Cambridge.

Tucker, G.M., Evans, M.I., 1997. Habitats for Birds in Europe: A Conservation Strategy for the Wider Environment. BirdLife Conservation Series No. 6 Birdlife International, Cambridge.

UK Department of the Environment, 1996. Indicators of Sustainable Development for the United Kingdom, London,. Available at: http://www.environment.detr.gov.uk/ [> Indicators of Sustainable Development for the UK].

USDA [United States Department of Agriculture], 1997. Agricultural Resources and Environmental Indicators, 1996–97. In: Agricultural Handbook No. 712. Natural Resources and Environment Division, Economic Research Service, Washington, DC. Available at: http://www.ers.usda.gov/ [Briefing Rooms > Agricultural Resources and Environmental Indicators].

Wossink, A., Wenum, J. van, Jurgens, C., Snoo, G. de, 1999. Co-ordinating economic, behavioural and spatial aspects of wildlife preservation in agriculture. European Review of Agricultural Economics 26 (4), 443–460.

Zohrabian, A., Traxier, G., 1999. Valuing Plant Genetic Resources: An Economic Model of Utilisation of the US National Crop Germplasm Collection. Paper presented to the Annual Meeting of the American Association of Agricultural Economics, 8–11 August, Nashville, TN.

Chapter 23

USEPA biomonitoring and bioindicator concepts needed to evaluate the biological integrity of aquatic systems

James M. Lazorchak, Brian H. Hill, Barbara S. Brown,
Frank H. McCormick, Virginia Engle, David J. Lattier,
Mark J. Bagley, Michael B. Griffith,
Anthony F. Maciorowski and Greg P. Toth

Abstract

This chapter presents the current uses, concepts and anticipated future directions of bio-monitoring and bioindicators in the regulatory and research programs of the United States Environmental Protection Agency (USEPA). The chapter provides a historical look on how biomonitoring and bioindicators evolved in the USEPA or its predecessor agencies from the 1960s – 1980s, then describes two current key biomonitoring and bioindicator programmes, the USEPA Office of Research and Development's Environmental Monitoring and Assessment Programme (EMAP) and USEPA's Office of Water's Biocriteria Programme. The remainder of the chapter is organized hierarchically beginning with concepts and moni-toring approaches using fish, macroinvertebrates, and periphyton assemblages, and functional ecosystem measures. The assemblage approaches are followed by current research and regu-latory use of whole organism toxicity testing assessments for measuring contamination in aquatic environments and remediation assessment. The chapter includes existing and proposed activities in the use of real-time biomonitoring to assess biological exposures to contaminants and other environmental changes. A new approach that uses small and large adult whole fish tissue as a bioindicator for assessing potential contaminant exposures to wildlife is presented, followed by a description of new research in molecular approaches to biomonitoring and bioindicators through measures of gene expression, use of microarrays and measures of genetic diversity.

Keywords: USEPA, Biomonitoring, Bioindicators, Marine, Freshwater, Fish, Macroinverte-brates, Algae, Molecular, Real-time

1. Overview of USEPA's current use of biomonitoring in regulatory and research programmes

This chapter presents the United States Environmental Protection Agency's (USEPA) current uses, concepts and anticipated future directions of biomonitoring in regulatory and research programmes. The terms biomonitoring and bioindicator used in this chapter will generally follow Markert et al. (1999): "Biomonitoring is a method of

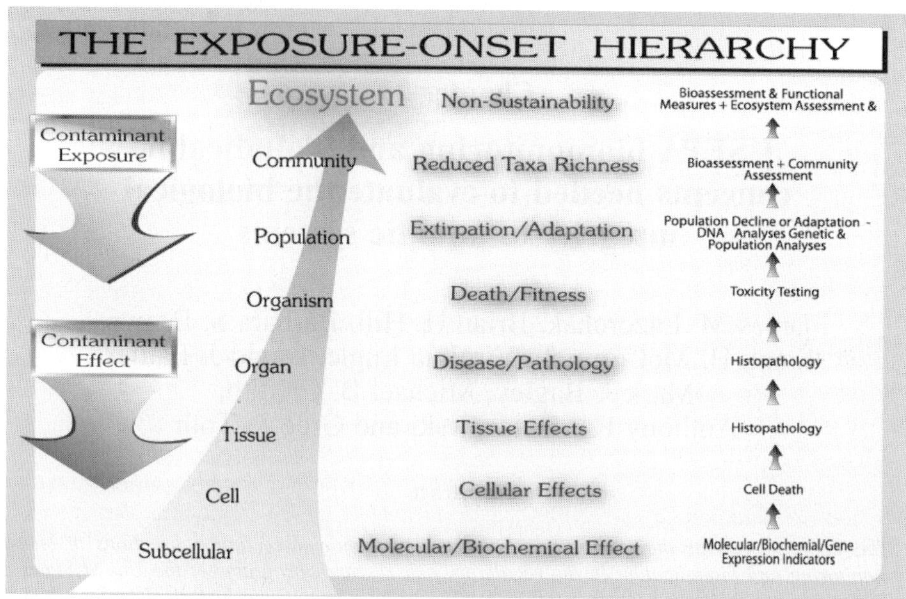

Figure 1. Exposure Onset Hierarchy. Modified from presentation made at the 20th Annual Meeting of the Society of Environmental Toxicology by Mary Haasch.

observing the impact of external factors on ecosystems and their development over a period, or of ascertaining differences between one location and another. A biomonitor or bioindicator is an organism (or a part of an organism or a community of organisms) that contains information on the *quality* of the environment (or part of the environment). A biomonitor, on the other hand, is an organism (or a part of an organism or a community of organisms) that contains information on the *quantitative* aspects of the quality of the environment." This chapter covers the USEPA use of (1) fish, macroinvertebrates, and periphyton surveys and assessments; (2) toxicity testing in laboratories and in the field; (3) fish tissue assessment; (4) microbial measures as functional ecosystem measures; (5) molecular measures; and (6) real-time biological monitoring as quantitative or qualitative measures of ecosystem health or ecosystem integrity. Figure 1 depicts an exposure paradigm that we will generally follow. The diagram starts on the left with how exposure moves to effects, starting at the molecular level and moving up to the ecosystem level. On the far right of the diagram are the different biological measures that are currently used to measure the effect at the corresponding hierarchal level.

The USEPA or its predecessor agencies began in the 1960s using fish and macroinvertebrate surveys to gather basic information on the existing fauna and to observe changes from year to year on large interstate navigable waters (Mason, et al., 1971). In these early years, information on water quality impacts in large rivers led to surveys in major waterways across the United States to document and quantify these impacts. Collection locations were usually selected upstream of major cities so that their faunal characteristics would reflect environmental conditions rather than specific sources of municipal or industrial pollution. In the late 1960s and early 1970s, the USEPA began examining interstate large rivers and smaller streams and testing effluents for toxicity. These studies were employed to (1) measure the toxicity of specific pollutants or effluents to individual species or communities of aquatic organisms under natural conditions; (2) detect violations of water quality standards; (3) evaluate the trophic status of waters; and (4) determine long term trends in water quality (Weber, 1973). USEPA field studies were mostly conducted by national enforcement programs and laboratory studies were conducted principally by national research laboratories (Weber, 1973). USEPA laboratory studies were undertaken to measure the effects of known or potentially deleterious substances on aquatic organisms, to estimate "safe" concentrations, and to determine some basic environmental requirements, such as temperature, pH, and dissolved oxygen, using important and sensitive species of aquatic organisms.

After the passage of the Federal Water Pollution Control Act Amendments of 1972 [referred to hereafter as the CWA (Clean Water Act)], the focus of biomonitoring was to collect information to assess the goal of restoring and maintaining the chemical, physical and biological integrity of the Nation's waters. In the years that followed the passage of these amendments, there were a number of deferring opinions on what was meant by integrity. Some felt that integrity of water meant, " the capability of supporting and maintaining a balanced, integrated, adaptive community of organisms having a composition and diversity comparable to that of the natural habitats of the region"(USEPA, 1977). Others defined integrity of water as " the maintenance of the community structure and function characteristic of a particular locale or deemed satisfactory to society" and " Integrity as used is intended to convey a concept that refers to a condition in which the natural structure and function of ecosystems is maintained" (USEPA, 1977). All these interpretations of integrity require some assessment of the biological health of an aquatic system. There was a basic change in the USEPA's emphasis of achieving better water quality from one solely based on numerical water quality standards for chemicals or physical conditions, to one that utilized a combination of numerical standards and a technology-based approach (permit-driven wastewater treatment technology) to achieving the integrity goal (USEPA, 1977). The CWA amendments of 1977 also directed the USEPA and States to collect biomonitoring information for a number of purposes: (1) basic water monitoring for assessing the status of water quality conditions meeting chemical and biological criteria and trend monitoring; (2) development of National Water Quality Criteria, which were to be used to set both chemical as well as biological water quality standards, and (3) compliance monitoring of permit conditions for effluents or non-point sources.

In the late 1980s, the USEPA began to restructure monitoring programs from one that emphasized compliance monitoring of permit conditions, to one that emphasized environmental results. The USEPA published qualitative and semi-quantitative bioassessment protocols designed to provide basic aquatic life data for planning and management purposes, such as screening, site ranking, and trend monitoring (USEPA, 1989). These protocols were fundamental assessment techniques to generate basic information on ambient physical, chemical and biological conditions.

2. Biological monitoring and the use of bioindicators in USEPA's biocriteria programme

In the 1990s, the trend toward measuring environmental results was enhanced with passage of the Governmental Performance and Results Act (GPRA) of 1993. Under GPRA, the USEPA established environmental performance objectives based on ecological outcomes, such as "By 2005, conserve and enhance the ecological health of the nation's waters and aquatic resources so that 75% of waters will support healthy aquatic communities." Such objectives required the development of broader indicators focused on assemblages of organisms and their supporting habitats, as well as "health" of individual organisms, i.e., shifting from organismal lethality to more subtle impacts. Therefore, bioindicators for monitoring and assessing environmental condition would be evaluated against "expected" (or reference) sites in the natural environment, as opposed to control (non-dosed) samples in the laboratory. During this same time frame the USEPA encouraged States first to adopt narrative biological standards into State water quality standards (Gibson, 1991). Biological criteria can be used by States to confirm impairment from known and unknown sources of impact, determine support of designated aquatic life use classifications and provide a tool to expand monitoring and assessment programmes expansion from source control to overall resource management (Yoder and Rankin, 1995). Some states use biological criteria better to delineate and protect aquatic life use classifications and in the enforcement of water quality standards (Yoder and Rankin, 1995). Further refinement of narrative criteria into numerical criteria or expectations of community structure and function in a least disturbed condition (or reference condition) were considered a next logical progression (Gibson, 1991). Patterns in community response to stress are then used to determine biological integrity and ecological function (Karr and Dudley, 1981). Biological criteria, therefore, supplement, rather than replace chemical and toxicological endpoints. They are based on the premise that the structure and function of an aquatic community within a specific habitat provide important information about the quality of surface waters (USEPA, 1989). The USEPA provided additional guidelines and standardized procedures for using benthic macroinvertebrates and fish in developing biocriteria (USEPA, 1990, 1993a). These procedures were more quantitative techniques for collecting, processing, and identifying specimens and also included taxonomic references.

3. Use of biomonitoring in risk assessments in the pesticide regulatory programmes

Ecological risk assessment methods and procedures Federal Insecticide, Fungicide, and Rodenticide Act (FIFRA) are detailed elsewhere (40 Code of Federal Regulations (CFR) 158.130; 40 CFR 158.145; Urban and Cook, 1986; SETAC, 1994; Touart and Maciorowski, 1997), and are briefly described in this section. Existing methods pre-date EPA's ecological risk assessment framework (USEPA, 1992) and guidelines (USEPA, 1996). However, two pesticide case studies (carbofuran, synthetic pyrethroids) were used in the Agency's state-of-the-practice for ecological risk assessment

Table 1. Sample assessment questions for the assessment of stream condition pertaining to stream fish assemblages.

What % of stream miles (and spatial distribution) have fish assemblages that differ from "reference" condition as measured by:

- Species richness?
- Number of species sensitive to human disturbance?
- Percentage of individuals tolerant of chemical or habitat disturbance?
- Percentage of non-indigenous individuals?
- Cumulative index of biotic integrity based on fish assemblage?

What % of stream miles support coldwater vs. warmwater fisheries as determined by the fish species?

(Modified from US EPA (1998a) and McCormick and Peck (2000).

prepared during the guidelines development process (USEPA, 1993b). Generally, ecological risk assessments for pesticide registration are prospective estimates based on single active ingredients and use-sites (e.g., corn, wheat, ornamental plants, etc.). The scope and complexity of pesticide risk assessments will vary with the specific chemical and its use, but a tiered, iterative approach is generally used. The tiers progress through simple risk quotients derived from laboratory fate, transport, and toxicity data in lower tiers, to a weight-of-the-evidence approach in higher tiers (Tables 2 and 3).

Exposure analysis consists of a preliminary or comprehensive fate and transport assessment (Table 1) based on registrant submitted data. The exposure analysis provides exposure profiles and estimated environmental concentrations (EEC) for the pesticide use (e.g. corn, cotton, wheat, etc.). Note that EEC's may be derived from four estimation procedures ranging from simple to complex . The ecological effects analysis (Table 2) is also tiered. Tier I provides an acute toxicity profile for birds, fish, mammals, and invertebrates. Tier II provides a sub-chronic and chronic toxicity (No-Observed-Effect Concentration or NOEC) profile and bioaccumulation potential for the same test species. Depending upon the hazard and exposure characteristics of a particular pesticide and use pattern, Tier II analyses may be conducted for all representative taxa, or may focus on either aquatic or terrestrial species. When warranted, Tier III effects analysis is used to refine NOEC and bioaccumulation estimates.

Following exposure and effects analysis, ecological risk is estimated as a function of ecotoxicological effects and environmental exposure using the quotient method (Table 2). A number of risk quotients are calculated (e.g., acute avian, acute fish, acute invertebrate, chronic avian, chronic fish, chronic invertebrate, etc.) and compared to regulatory risk criteria (e.g., presumption of acceptable risk, presumption of unacceptable risk, etc.). Traditionally, if regulatory criteria were exceeded, a high risk potential was assumed to exist for the pesticide-use combination. If registrants wished to refute a presumption of risk finding, Tier IV effects analysis consisting of field studies, simulated field studies, or other special studies could be conducted (Touart, 1988; Fite et al., 1988). The types of ecological studies that have been required include

Table 2. Generalized exposure analysis and assessment methods and procedures used in prospective ecological risk screens of pesticides.[a]

Preliminary Exposure Analysis includes simple laboratory tests and models to provide an initial fate profile for pesticide (Hydrolysis and photolysis in soil and water, aerobic and anaerobic soil metabolism, and mobility).

Fate and Transport Assessment provides a comprehensive profile of the chemical (persistence, mobility, leachability, binding capacity, degradates) and may include field dissipation studies, published literature, other field monitoring data; groundwater studies; and modelled surface water estimates.

Estimated Environmental Concentrations (EEC) are derived during the exposure analysis or comprehensive fate and transport assessment. There are four EEC estimation procedures.

Level 1: A direct application, high exposure model designed to estimate direct exposure to a non-flowing, shallow water (<15 cm) system.

Level 2: Adds simple drift or runoff exposure variables such as drainage basin size, surface area of receiving water, average depth, pesticide solubility, surface runoff, or spray drift loss which attenuate the Level 1 direct application model estimate.

Level 3: Computer runoff and aquatic exposure simulation models. A loading model (SWRBB-WQ[b], PRZM[c],etc.) is used to estimate field losses of pesticide associated with surface runoff and erosion, which then serves as a input to a partitioning model (EXAMS II[d]) to estimate sorbed and dissolved residue concentrations. Simulations are based on either reference environment scenarios or environmental scenarios derived from typical pesticide use circumstances.

Level 4: Stochastic modelling where EECS are expressed as exceedance probabilities for the environment, field, and cropping conditions.

---

[a] For additional details regarding environmental fate data requirements see 40 CFR §158.130, SETAC (1994); Touart (1995).
[b] Simulator for Water Resources in Rural Basins–Water Quality.
[c] Pesticide Root Zone Model.
[d] Exposure Analysis Modelling System.

qualitative avian mortality screening studies, pond studies, mesocosm studies, and microcosm studies.

Although field studies and simulated field studies represent one type of bio-monitoring, they are presently rarely used due to regulatory policy changes in 1992 (SETAC, 1994). Such studies provided useful information for evaluating short-term impacts of pesticides, but provided little information regarding longer-term impacts, latent effects, recovery of individuals and populations, or alterations of higher dimensional ecological phenomena. Further, many of the acute and chronic effects observed in field studies could be predicted from the acute and chronic laboratory data sed in risk assessment.

Table 3. Generalized ecological effects analysis and risk quotient methods and procedures used in prospective risk screens of pesticides.

Tier I Effects Analysis provides acute toxicity values and dose-response information (mammalian and avian acute oral LD_{50}; avian dietary LC_{50}; seedling emergence and vegetative vigor EC_{25}; honey bee acute contact LD_{50}; and additional wild mammal, estuarine, and plant tests depending on pesticide use category.

Tier II Effects Analysis provides sub-chronic and chronic toxicity values (NOEC) including avian reproduction studies; special avian or mammal studies; fish early life stage studies; invertebrate life cycle studies; and a fish bioaccumulation factor.

Tier III Effects Analysis provides refined NOEC estimates for chronic toxicity that may include a fish full life cycle test, aquatic organism accumulation, or food chain transfer tests

The Quotient Method is used to provide a set of acute and chronic risk quotients (RQ) for fish, birds, invertebrates, plants and endangered species. The RQs are calculated by dividing exposure (EEC) by hazard (LD_{50} or LC_{50} or NOEC). Risk quotients are then compared to regulatory risk criteria as follows.

Presumption of acceptable risk	Presumption of risk that may be mitigated by restricted use	Presumption of unacceptable risk	
		Nonendangered	Endangered species
Acute toxicity			
$EEC < 0.1\ LC_{50}$	$0.1\ LC_{50} \leqslant EEC \leqslant 0.5\ LC_{50}$	$EEC > 0.50\ LC_{50}$	$EEC > 0.05\ LC_{50}$ or $EC > 0.10\ LC_{10}$
Chronic toxicity			
$EEC < Chronic\ NOEC$	N/A	$EEC > NOEC$	$EEC > NOEC$

Tier IV Effects Analysis allows registrants to rebut a presumption of risk derived from laboratory studies by performing field or simulated field studies, including qualitative terrestrial field studies, farm pond studies, mesocosm studies, or other special studies.

[a] For additional details regarding ecological effects data requirements see 40 CFR §158.145 Subdivision E; Urban and Cook (1986); SETAC (1994); Touart (1995).

4. Biological monitoring and use of bioindicators in the USEPA Environmental Monitoring and Assessment Programme (EMAP) surface waters and estuarine programme

In 1989, the USEPA initiated the Environmental Monitoring and Assessment Programme (EMAP), an integrated multi-resource programme designed to develop methods to estimate the condition of the Nation's ecological resources at various geographic scales over long periods of time. (Messer et al., 1991). Within EMAP, an indicator was defined as any environmental measurement that can be used to quantitatively estimate the condition of ecological resources, the magnitude of stress, the exposure of biological components to stress, or the amount of change in condition. "Condition indicators" provided quantitative information on the state of ecological resources of interest, a subset of which were biotic indicators which estimated the condition of a biological component of a resource (Barber, 1994). Such indicators were specifically associated with previously identified environmental values of interest and the assessment endpoints that represented those environmental values.

For freshwater systems, EMAP initiated its first effort in the Mid-Atlantic Highlands ecoregion, (Whittier and Paulsen, 1992; USEPA, 1997). The study was identified as the Mid-Atlantic Highlands Assessment (MAHA) and was conducted to develop and demonstrate EMAP approaches such as probability-based survey designs and appropriate indicators of ecological condition as applied to address specific regional assessment questions of interest to the USEPA. The monitoring framework for MAHA used a regional-scale probability-based survey design to select sampling sites. This design permits unbiased inferences the subset of sites where samples and data are collected with known certainty to explicitly defined populations of ecological resource units (Larsen, 1995, 1997; Diaz-Ramos et al., 1996). For MAHA, populations were defined based on the total length of streams. For example, the design allows one to estimate the total length of streams in the target population (e.g., all permanent streams appearing on a particular scale of map) which meet some criteria (e.g., all first-order target streams, all target streams within a specific ecoregion, etc.). The distribution of indicator scores can then be examined for these defined populations to determine the estimated length of stream characterized by a particular set of indicator values, with associated uncertainty in these estimates represented by confidence bounds.

A similar study was undertaken in 1997 and 1998 in the Mid-Atlantic Estuaries, spanning the Delaware and Chesapeake Estuaries and the Atlantic coastal bays. The purpose of the study was to fill information gaps identified during preparation of the *Condition of the Mid Atlantic Estuaries* report (USEPA, 1998a) and to demonstrate how to integrate different institutional monitoring programmes. The study integrated monitoring programs from the Chesapeake Bay Programme, the Delaware River Basin Commission, the National Oceanic and Atmospheric Administration, and EMAP into a compatible design using a stratified probability-based design and a core suite of indicators which were measured by all programmes (USEPA, 1998a). The core suite of indicators included measures to characterize habitat (e.g., salinity, temperature, grain size), stressors (e.g., toxics, nutrients, dissolved oxygen), and biological response (e.g., benthic community, fish community, chlorophyll *a*). Consistent methods and sampling design over the programmes allowed the results across programs to be aggregated to gain a better understanding of the regional scale condition. For the estuarine component of the programme, the biotic indicators selected for use or development included benthic species composition and biomass, fish community composition, contaminants in fish flesh and shellfish, the gross pathology and histopathology of fish, and (for linkage between biology and stressor) sediment chemistry and toxicity (Holland, 1990). EMAP built on the pioneering work of Karr (1981) and others to develop multimetric indices of biotic integrity. Such indices would represent the response of biological communities to environmental stressors by not only quantifying the current condition of the ecosystem but also by integrating the effects of multiple anthropogenic and natural stressors over time.

5. Community and ecosystem measures

5.1. Fish monitoring and bioindicators (freshwater)

In support of the biocriteria programme, USEPA has conducted research on the development fish biotic indices and evaluated various components of such indices. Fish species exhibit diverse evolutionary, morphological, ecological, and behavioural adaptations to their natural habitat and thus are particularly effective indicators of the condition of aquatic systems (Karr et al., 1986; Fausch et al., 1990; Simon and Lyons, 1995). The biological characteristics of stream fish assemblages, including the capability to integrate the effects of a variety of stressors across different time scales and levels of ecological organization, and the importance and familiarity of fishes to the general public, make them conducive to the development of an indicator of ecological condition (Karr et al., 1986; Simon, 1991; Simon and Lyons, 1995, and USEPA, 1999).

5.1.1. Multimetric approaches

Multimetric indicators such as the Index of Biotic Integrity (IBI) represent a means to integrate various structural and functional attributes of an ecosystem and provide an overall assessment of ecosystem condition (Fausch et al., 1990; Karr, 1991; Karr and Chu, 1997). Structural and functional attributes of the fish assemblage (derived from species presence/absence and relative abundance data) are aggregated into metric categories (taxonomic composition, abundance and individual condition, trophic, and reproductive function) that are hypothesized to respond predictably to increasing intensities of human disturbance (Karr et al., 1986; Karr, 1991, Barbour et al., 1995; Hughes and Oberdorff, 1999). Candidate metrics are tested for responsiveness to biotic or abiotic conditions resulting from increasing human disturbance, and their biological importance (Hughes et al., 1998; McCormick and Peck, 2000; McCormick et al., 2001). The IBI was originally developed with 12 metrics (Karr, 1981), but IBI's have subsequently been developed with fewer and more metrics (Hughes et al., 1998; Halliwell et al., 1999; Moyle and Marchetti, 1999; McCormick et al., 2001). Response values for each metric selected are transformed to a metric score based on the degree of deviation of the response value from that expected at a similar site under conditions of minimal human disturbance. The individual metric scores are then aggregated to produce a multimetric index score in which a higher score indicates better ecological condition (i.e., closer to the expected condition when human disturbance is minimal). Possible causes of poor ecological condition may be identified (although specific cause-effect relationships cannot always be ascertained) by examining correlations between the index or its component metrics and various measures of ecosystem stress. More detailed descriptions of the general approach used to develop multimetric indices can be found in Hughes et al. (1998), US EPA (1999), McCormick and Peck (2000), and McCormick et al. (2001).

5.1.2. Indicator development

Several factors should be considered in the indicator development for use in biomonitoring and biocriteria programmes (Yoder and Rankin, 1995; McCormick and Peck, 2000).

5.1.2.1. Conceptual relevance of the indicator

The design of monitoring studies should be driven in part by a series of specific assessment questions related to the condition of stream resources. The indicator should be linked to identified assessment questions, should contribute information to address multiple assessment questions, and should complement other potential indicators (Table 1). The nature of the question suggests that an appropriate indicator would focus at the assemblage level and consist of multiple components to address the various aspects of the questions. The indicator is also useful in that the basic fish species and abundance data used to develop it can also be used with little or no additional effort to address other assessment questions of interest. These subsidiary questions are relevant to a separate societal value of interest to the MAHA study, fishery health. McCormick and Peck (2000) graphically represented conceptual relationships between major structural components and processes to illustrate possible routes of exposure from anthropogenic stressors (Fig. 2).

5.1.2.2. Feasibility of implementation

Collection of field data at an individual sampling site is based on standard approaches for stream fish assemblages (Ohio EPA, 1987; Lyons, 1992; McCormick, 1993). Fish assemblage sampling is conducted using a combination of gear types (electrofishing and seining), standardized sampling times and distances (Ohio EPA, 1987; McCormick and Hughes, 1998; US EPA, 1987). McCormick and Peck (2000) presented the results of a pilot study on wadeable streams in the Interior Highlands and Central Lowlands that showed that 90% of the species in a reach were sampled by single-pass, backpack electrofishing over a distance equal to 40 times the mean channel width (Fig. 4). EMAP documented protocols for sampling wadeable and non-wadeable streams that were developed in the Mid-Atlantic region of the United States but have been used, with some modifications, in the Southern Rocky Mountains, California's Central Valley, Interior Highlands and the Great Plains (Lazorchak et al., 1998). An appropriate quality assurance programme can be developed and implemented for the indicator and monitoring framework using available resources and techniques (e.g., Chaloud and Peck, 1994).

5.1.2.3. Sources of error

McCormick and Peck (2000) addressed the different types of errors that can affect either the measurement data or the development of indicator values from measurement data. Measurement-related errors of field collection data, in terms of number of species collected, species composition, and number of individuals, cannot be estimated directly for the indicator by collecting replicate samples during a single visit to a site

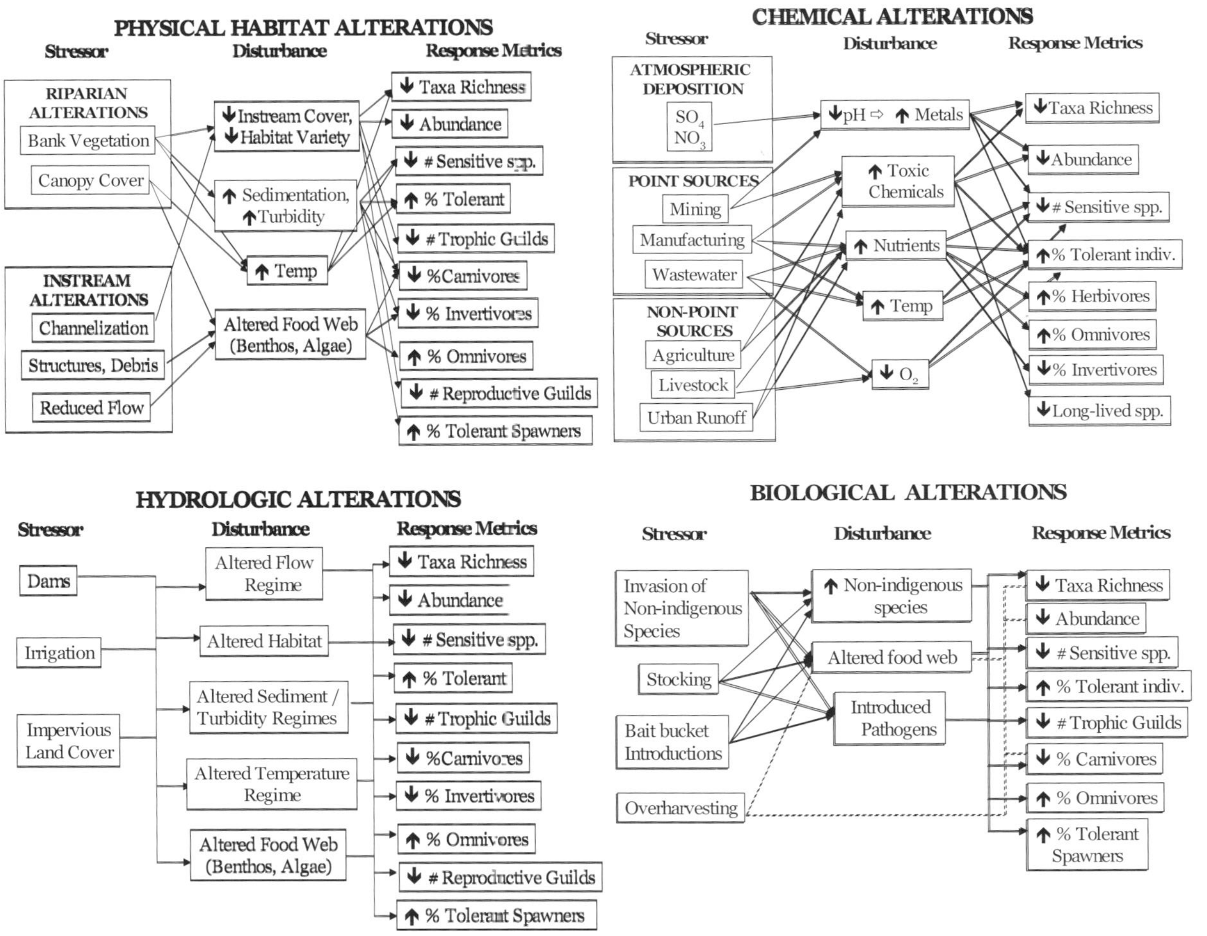

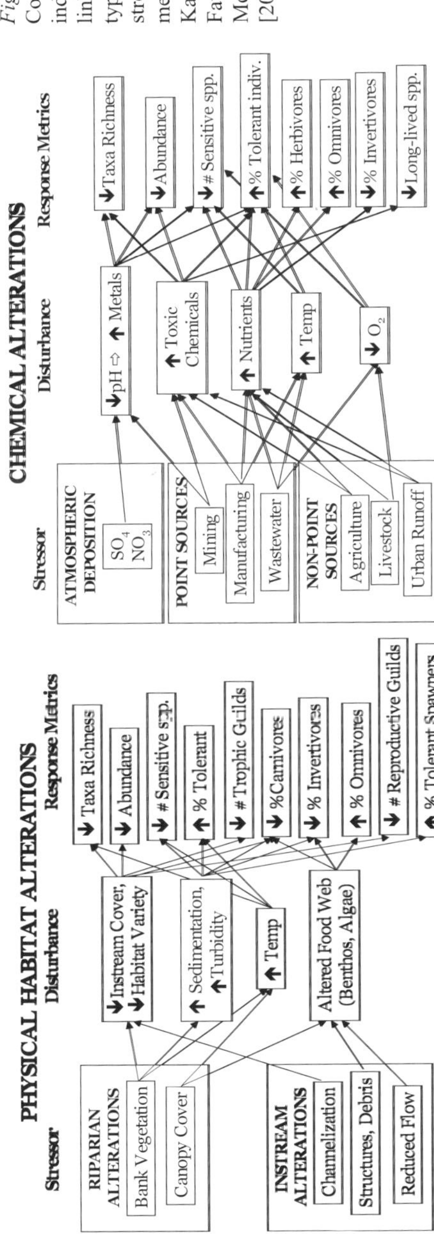

Figure 2. Conceptual model of indicator, showing linkages between various types and classes of stressors and component metrics (Derived from Karr et al., [1986], Fausch et al., [1990], and McCormick and Peck, [2000].

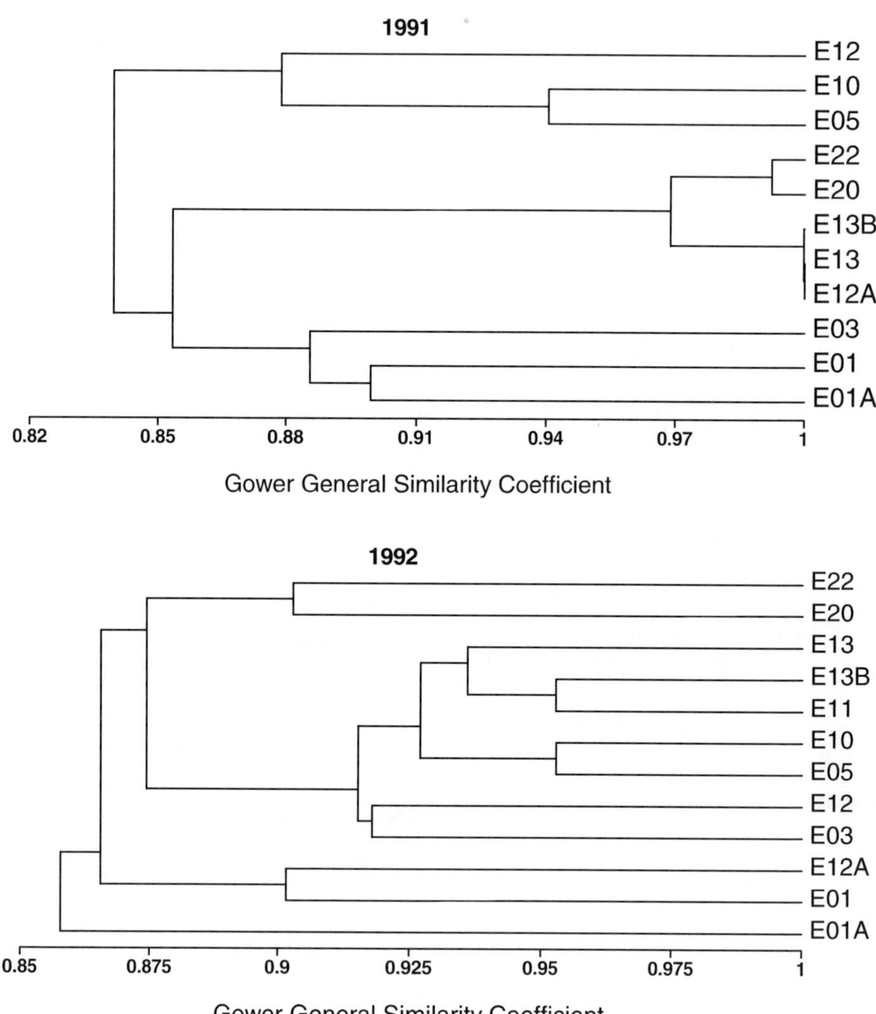

Figure 3. Community similarity dendrograms (based on for Eagle River periphyton assemblages in 1991 and 1992.

(Fore et al., 1996) but must be addressed by professional ichthyologists. The other critical source of error in measurement data is incorrect identifications of fish species. Various means of controlling this source of error include the collection and confirmation of voucher specimens, using personnel experienced in fish identification and additional training in field identification of regional fishes. McCormick and Peck (2000) presented a more quantitative evaluation of five types of errors related to field identification of fish species. Transcription errors occur when the wrong species (or species code) is recorded on the field data form. The remaining four relate to actual errors in species identification and include a cumulative estimate of errors for all

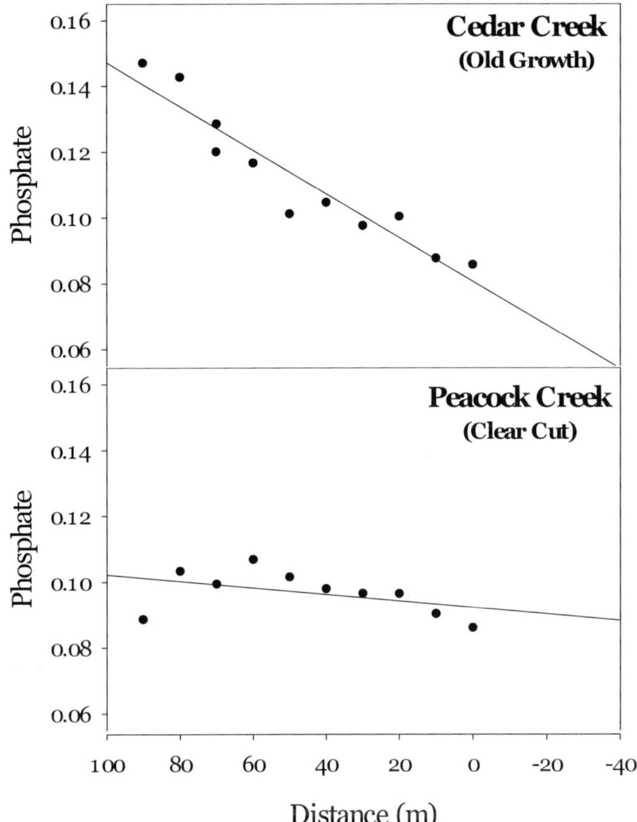

Figure 4. Regression of PO_4^{2-} concentrations against downstream distance for old growth (top) and harvested (bottom) watersheds in northwestern California.

species, errors specific to groups of fishes that are difficult to identify to the species level in the field (e.g., sculpins, genus *Cottus*, and a cyprinid genus *Nocomis*), and errors at the genus level.

5.1.2.4. Sources of variance

It is important to identify the components and magnitude of variance that affect the ability of the indicator to detect differences in condition among sites. Among-site variance is variation due to differences in the indicator value among a sample of stream sites. This component represents the environmental "signal" to be detected and interpreted with respect to an ecological condition. Extraneous variance consists of the remaining temporal and measurement-related variation. Collectively, these components represent "noise" that inhibit the ability to detect and interpret the environmental "signal" and include the extent to which regional-scale effects (e.g., climate, hydrology) and temporal variance affect the ability to detect differences among sites.

5.1.2.5. Geographic factors

Environmental assessment is potentially affected by selection of appropriate geographic and temporal scales (McCormick et al., 2000, 2001). Multi-metric indicators developed for a particular geographic area and scale of monitoring effort should not be applied to other scales of monitoring or other geographic areas without evaluation and modification. Within a biogeographic province, spatial and temporal variability is relatively low. However, some seasonal variability associated with spawning activity may confound assessments if no consistent index period (a specific time frame for sampling, i.e April 15–May 30 for spring low flow or July 1–September 1 for summer low flow) is selected for sampling. Numbers of species vary with ecoregion, drainage basin, and watershed size. Understanding the patterns of geographic variation in the structure of fish assemblages is crucial to developing a comprehensive assessment of stream conditions. Understanding the influence of geographic factors in structuring fish assemblages is crucial to developing a comprehensive assessment of stream conditions. The variability in responses at different spatial and temporal scales may affect the interpretation of bioassessment endpoints and has important implications for large-scale monitoring programmes. Establishing reference conditions for MAHA streams required identifying the local factors (e.g., stream size, gradient, temperature, substrate composition, and habitat complexity) that control fish assemblage structure in minimally disturbed streams. McCormick et al. (2000; 2001) found no substantial differences in the range or general distribution of fish assemblage response values across ecoregions.

5.2. Macroinvertebrate monitoring and bioindicators (freshwater)

Although Karr's (1981) IBI was originally developed for fish assemblages, the utility of macroinvertebrate assemblage structure for describing the integrity of aquatic ecosystems has been widely recognized (Kerans and Karr, 1994; DeShon, 1995; Barbour et al., 1996; Fore et al., 1996). Their role in aquatic food webs as primary consumers of producers (i.e., periphyton) and decomposers (i.e., heterotrophic bacteria and fungi) and as prey for secondary and tertiary consumers (i.e., fish) make macroinvertebrates important to the community's total integrity. As a result, measurements of macroinvertebrate assemblages have been an integral part of monitoring biological conditions of streams and lakes both in the Environmental Monitoring and Assessment Programme (EMAP) Mid-Atlantic pilot study, in regional EMAP (R-EMAP) studies, and in state monitoring programs.

In the multimetric approach, assemblage structure is summarized with simple numerical measures of an assemblage's attributes called metrics. To create an index, selected metrics are calculated and scored using a standardized scale (i.e., a continuous scale from 1–10 or a discrete scale of 1, 3, or 5). Then, the scores are summed. Many metrics have been proposed for use in macroinvertebrate IBIs that measure different categories of assemblage attributes (USEPA, 1999). Richness measures include the total number of species or genera or the number of species or genera in selected taxonomic groups, such as Ephemeroptera, Plecoptera, and Trichoptera (EPT) or Chironomidae (i.e., midges). Evenness measures assess the other component of diversity, the relative dominance of the most abundant taxa in the assemblage. Composition measures assess the abundance of a taxa group, such as EPT or Chironomidae, relative to total macroinvertebrate abundance or the abundance of a tribe, subfamily, or family relative to its larger taxonomic grouping (ex., % of tanytarisinid midges to Chironomidae or % of Hydropsychidae to Trichoptera). Tolerance measures assess taxa richness or relative abundance of macroinvertebrates considered tolerant or intolerant of different environmental stressors or the abundance weighted average tolerance of the macroinvertebrate assemblage to a stressor gradient (ex., Hilsenhoff's Biotic Index for organic pollution). Feeding, habitat, and life cycle measures assess taxa richness or relative abundances of macroinvertebrates with specific functional feeding (i.e., scrapers, shredders, predators), habitat use (i.e., clingers, burrowers), or life cycle (i.e., univoltine, semivoltine) adaptations to the habitat template of the ecosystem (Southwood, 1977; Townsend and Hildrew, 1994). Similarity measures assess compositional resemblance of the macroinvertebrate assemblage to that expected under reference conditions for a region. However, similarity measures have been underutilized because they require identification of an expected assemblage by comparison with data from a reference site or predicted by modeling.

One major objective of a bioassessment is diagnosis of the anthropogenic stressors at a site. Selection of metrics for incorporation into macroinvertebrate indices of biotic integrity has been largely based on general observations on the response of assemblages to increasing perturbation (Kerans and Karr, 1994; Fore et al., 1996; USEPA, 1999). However, few metrics have been tested for their relation to specific stressor gradients (Wallace et al., 1996; Carlisle and Clements, 1999). Testing is needed to apply the multimetric approach to diagnosis of the causes of decreased biotic integrity at individual sites (Griffith et al., 2001).

The development of field sampling designs that employ multiple reference and polluted sites has been proposed as an alternative to the traditional upstream versus downstream approach used in most biomonitoring studies (Clements et al., 2000). Spatially-extensive monitoring programmes can characterize ecological conditions within an ecoregion and provide the necessary background information to evaluate future changes in water quality. Clements, et al. (2000) used physicochemical characteristics, heavy metal concentrations, and benthic macroinvertebrate community structure data from 95 sites in the Southern Rocky Mountain ecoregion in Colorado collected in 1995–1996 as part of the USEPA R-EMAP programme. Most sites (82%) were selected using a systematic, randomized sampling design. Each site was placed into one of four metal categories (background, low, medium, and high metals), based on the cumulative criterion unit (CCU), which was defined as the ratio of the instream metal concentration to the USEPA criterion concentration, summed for all metals measured. A CCU of 1.0 represents a conservative estimate of the total metal concentration that, when exceeded, is likely to cause harm to aquatic organisms. Although the CCU was less than 2.0 at most (66.3%) sites, values exceeded 10.0 at 13 highly polluted stations. Differences among metal categories were highly significant for most measures of macroinvertebrate abundance and all measures of species richness. Clements et al. (2000) observed the greatest effects on several species of heptageniid mayflies (Ephemeroptera: Heptageniidae), which were highly sensitive to heavy metals and were reduced by >75% at moderately polluted stations. The influence of taxonomic aggregation on responses to metals was also greatest for mayflies. In general, total abundance of mayflies and abundance of heptageniids were better indicators of metal pollution than abundance of dominant mayfly taxa. Heavy metal concentration was the most important predictor of benthic community structure at these sites. Because of the ubiquitous distribution of heavy metal pollution in the Southern Rocky Mountain ecoregion, we concluded that potential effects of heavy metals should be considered when investigating large scale spatial patterns of benthic macroinvertebrate communities in Colorado's mountain streams.

Griffith et al. (2001) conducted multivariate analyses of R-EMAP data from the same mineralized belt of the Southern Rockies Ecoregion in Colorado and the Central Valley Ecoregion in California have suggested that various metrics respond differently to environmental stressor gradients. Richness and evenness measures were correlated with dissolved and sediment metal concentrations in Rocky Mountain streams variously affected by metal mining. Richness, evenness, composition, and feeding measures were generally uncorrelated with alterations in dominant taxa related to increased dissolved cation and anion concentrations associated with irrigation runoff in Central Valley streams (Griffith et al., 2002), but specifically designed tolerance or similarity measures are likely to be more sensitive to this chemical gradient. Richness and composition measures, particularly for EPT taxa, were correlated with alterations in substrates, in-stream habitats, and riparian structure and shading associated with agriculture (i.e., livestock grazing in Colorado or irrigated row crops in California) in both ecoregions (Griffith et al., 2001, 2002. These differences illustrate the potential to create indices of biotic integrity composed of diagnostic metrics for specific stressor gradients.

5.3. Macroinvertebrate monitoring and bioindicators (marine)

Benthic indices of environmental condition were developed, tested, and validated for each of the biogeographic regions that were defined for the EMAP estuarine monitoring programme (Table 4). The EMAP approach to development of multimetric

Table 4.   References for benthic indices of estuarine condition that were used by EMAP.

Biogeographic province	Geographic range	References
Virginian Province	Cape Cod, MA to Chesapeake Bay, VA	Weisberg et al. (1993) Schimmel et al. (1994) Paul et al. (1999)
Carolinian Province	Cape Henry, VA to St. Lucie Inlet, FL	Hyland et al. (1996) Hyland et al. (1998)
West Indian Province	Indian River, FL to Tampa Bay, FL	Macauley et al. (2002)
Louisianian Province	Anclote Key, FL to Rio Grande, TX	Summers et al. (1993) Engle et al. (1994) Engle and Summers (1999) Macauley et al. (1999)

benthic indices in estuaries was loosely adapted from Karr's (1981) IBI approach. A set of sites were identified as having known condition with regard to stressors that would potentially elicit a response in the benthic macroinvertebrate community. Although the critical levels of the stressors varied among provinces, typically, dissolved oxygen, sediment toxicity, and sediment contaminants were used to identify reference and degraded sites. In addition, sites were chosen to represent the range of natural conditions (e.g., salinity, sediment type) found within a province. A suite of benthic community components were then chosen to represent the range of potential responses to stressors. The list typically included measures of abundance, biomass, and diversity as well as proportional abundances of taxonomic, trophic, or functional groups. Various multivariate methods such as discriminant and multiple regression analysis were used to identify a subset of components from this list that best discriminated between the reference and degraded sites. An index was then created as a linear combination of the subset of components weighted by their proportional contribution to the multivariate model. The benthic index was used to classify sites of unknown condition. Using the EMAP design and analysis procedures, the proportion of estuarine area with reference or degraded benthic condition was calculated for each biogeographical province.

The benthic index for the Virginian Province comprised Gleason's D based upon infauna and epifauna (normalized for salinity), abundance of tubificid oligochaetes (normalized for salinity), and abundance of spionid polychaetes. The overall efficiency for correct classification using this index was 86% for both reference and degraded sites. A four-year assessment of benthic condition using this index indicated that $25 \pm 3\%$ of the estuarine area in the Virginian Province was impacted (Paul et al., 1999). A benthic IBI for the Carolinian Province included the following metrics: (1) mean abundance, (2) mean number of taxa, (3) 100 – % abundance of the top two numerical dominants, and (4) % abundance of pollution-sensitive taxa. This index correctly classified 93% of the development sites and 75% of the validation sites. In 1995, 21% of the estuarine area of the Carolinian Province was classified as degraded using this index (Hyland et al., 1998). The West Indian Province was only sampled in 1995; therefore, the benthic index that was developed is preliminary and no validation has occurred. The benthic index was composed of the abundance of gastropods and all molluscs, total abundance of all organisms, and the proportion of polychaetes that were spionids. Using this index, $33\pm11\%$ of the estuarine area was classified as degraded (Macauley et al., 2002). In the Louisianian Province the benthic index represented a linear combination of five metrics: proportion of expected diversity (Shannon-Wiener H′ normalized for salinity), mean abundance of tubificid oligochaetes, and the proportional abundances of capitellid polychaetes, bivalves, and amphipods. The average classification efficiency for this index was 74% for degraded sites and 77% for reference sites. Degraded benthic condition occurred in $23 \pm 7\%$ of the estuarine resource in the Louisianian Province using this index (Macauley et al., 1999).

The need is increasing for biological indicators that are diagnostic for multiple, combined, and often unidentified stressors. EMAP has evolved into Coastal 2000, a programme designed to transfer technology to the States to assist with monitoring design and indicator development. This technology transfer will enable the States to improve their reporting capability for 305(b) and 303(d) Clean Water Act requirements. In addition, EPA, through the STAR programme, is funding research on the development of tools to evaluate the health of estuaries and the Great Lakes with a particular emphasis on diagnostic indicators that are applicable over large geographic scales.

5.4. *Anticipated marine biomonitoring research activities*

One innovative measure to assess the health and integrity of coastal ecosystems is to identify and count the bottom-dwelling organisms living in the sediment. These animals, which are a major food source for many fish, create intricate tubes and tunnels in the sediment to depths as much as three feet. A healthy sediment is characterized by a high degree of tube and tunnel formation and, by contrast, an impacted sediment has fewer large, deep burrowing animal and their associated tubes and tunnels. Traditionally, sediment health is determined by collecting, identifying and counting these organisms, but this procedure requires specialized training and is labor-intensive and time-consuming. Computer Axial Tomography (CAT) imaging offers a rapid cost-effective alternative to this traditional method by quantifying the burrows and tunnels in sediment cores. Scientists first collect intact mud cores from an estuary, using cylindrical plastic tubes pushed into the sediment. The cores are tightly sealed at the top and bottom and transported to a hospital for CAT imaging. The resulting image data are stored on magnetic tape and may be analyzed on a personal computer back at the laboratory. A three-dimensional image of tubes and tunnels within the core can be quantified, and these measures can be used to identify, monitor and assess the effects of human activities on sediment habitats. Because medical CAT imaging scanners are located throughout the world, this technique could be widely available for environmental managers to evaluate the health of sediments.

5.5. *Measures of periphyton assemblage structure and ecosystem function*

Efforts to use periphyton assemblage structure and ecosystems functions for the biological monitoring of aquatic ecosystems fall into two broad categories: measures of assemblage structure (taxa richness and diversity, assemblage similarity, the relative abundances of indicator taxa, chlorophyll and biomass) and measures of community function, which can be further divided into organismal-level measures (cellular integrity, growth, photosynthesis, cellular respiration, enzyme activity) and community-level measures (primary productivity, community respiration, nutrient uptake). Researchers from the USEPA have used measures of periphyton assemblage structure and ecosystem function to monitor biological condition of streams and lakes under two programmes: Superfund and the Environmental Monitoring and Assessment Programme (EMAP). The USEPA researchers measured structural and functional responses of stream communities to elevated heavy metals related to mining activities in our Superfund assessments of the Eagle River, Colorado. USEPA also measured periphyton assemblage structure and ecosystem function in EMAP and regional EMAP (R-EMAP) studies of the ecological conditions of streams in the Appalachians, and used diatom assemblage structure in the assessment of ecological conditions in New Jersey lakes.

5.6. *Structural measures of periphyton assemblages*

Measurement and analysis of assemblage structure is the mainstay of biological monitoring programs. Assemblage structure may be measured as lists of the total number of species present within an assemblage, abundance of indicator species, or as an aggregate index derived from other attributes of the assemblage structure.

5.6.1. *Species richness and diversity and assemblage similarity*

Species diversity has two components: species richness and evenness. Several studies have used species richness to monitor stream assemblage responses to disturbance. It is generally assumed that richness is inversely related to environmental stressors, and several researchers have documented decreases in diatom species richness as a result of stream contamination by organic enrichment, metals, and pesticides (Lange-Bertalot, 1979; Crossey and La Point, 1988; Whitton et al., 1991). USEPA's work in support of Superfund and EMAP found poor correlations of species richness with human disturbance gradients (Hill et al., 2000a, b; 2001). Species better adapted to the prevailing environmental conditions will have an advantage resulting in an uneven distribution of individuals among taxa. Evenness is often reported as % dominance of the assemblage by single species, and results from studies employing diatom assemblages have indicated that dominance increases with nutrient enrichment (Stevenson and Pan, 1999) and metal contamination (Crossey and LaPoint,1988). Superfund and EMAP studies have found similar nutrient and metal relationships (Hill et al., 2000a, 2001), as well as correlations with watershed land-uses (Hill et al., 2000b; Hill and Kurtenbach, 2001). Assemblage similarity, the degree of compositional agreement among the species in two or more assemblages along an environmental gradient, are particularly suited for identifying changes in assemblage structure relative to the distance from the source of perturbation, and may be more sensitive to low level stress than are diversity indices (Hellawell, 1977). Medley and Clements (1998) reported that assemblage similarity was better related to metal concentrations in the stream than were diversity indices. Hill et al. (2000b) found similar results in Superfund studies on the Eagle River, Colorado.

5.6.2. *Relative abundance of indicator species*

Indicator species are used to assess current or recent environmental conditions. The relative abundances of indicator taxa provides a measure of not only how well those taxa tolerate existing environmental conditions, but also provides an indirect measure of those environmental conditions. Shifts in the relative abundances of diatom taxa have been used for monitoring aquatic ecosystem contamination by heavy metals (Medley and Clements, 1998) and watershed land-use changes (Kutka and Richards, 1996). USEPA used diatoms as indicators of these stressors in Superfund assessments of the Eagle River, Colorado (Hill et al., 2000b), and in regional assessments of stream and lake water quality (Pan et al., 1996; Hill et al., 2000a, 2001; Hill and Kurtenbach, 2001). Figure 3 is an example on how community similarity dendrograms were used to analyze Eagle River periphyton assemblages in 1991 and 1992.

5.6.3. Chlorophyll

Chlorophyll *a* concentration has been widely used to assess nutrient enrichment of streams, even in regional-scale studies (Leland, 1995; Pan et al., 1999, 2000). USEPA did not, however, find a significant relationship between chlorophyll a content of periphyton and stream chemistry, habitat, or watershed land-use in our EMAP studies of Appalachian streams (Hill et al., 2000a, 2001).

5.6.4. Biomass

One of the simplest measures of aquatic plant assemblage structure is standing crop of biomass. The relationship between standing crop and water quality, however, is not easily interpreted. Clark et al. (1979) compared methods of estimating periphyton biomass in response to chemical perturbations in stream mesocosms, and found that no one method was consistently better than any of the others in detecting the impact of copper, chromium, and chloride contaminations, but found that biomass colonizing clean substrates was depressed in response to these contaminants. Hill et al. (2000b) reported no significant effects of heavy metals on periphyton biomass in their Superfund assessment of the Eagle River, Colorado. They found only weak correlations between periphyton biomass and stream chemistry, channel substrate size, and watershed land-use in our EMAP studies of Appalachian streams (Hill et al., 2000a, 2001).

5.6.5. Cellular integrity

Measures of cellular integrity fall into two broad categories, those related to morphological changes in cell structure and those related to changes in cell membrane permeability. Few researchers have used changes in cellular structure to monitor physiological condition of the cells or to predict water quality. Analysis of five species of fossil diatoms collected from Mono Lake, California, revealed a large percentage of deformed individuals, possibly related to the transition of this lake from freshwater to alkaline, brackish waters (Solladay, 1994). USEPA reported significantly increased numbers of deformed *Fragilaria* frustules with increasing dissolved metal concentrations in our Superfund assessment of the Eagle River, Colorado (McFarland et al., 1997).

5.7. Functional measures of plant assemblages

Recent research has been critical of the reliance on structural measures of biotic conditions to assess aquatic ecosystem integrity. From an ecosystem management perspective structural measures are proving to be less reliable than previously thought. It has been argued that the high spatial and temporal variability exhibited by biotic assemblages preclude the use of population data alone as indicators of anthropogenic disturbances, and resource managers are urged to exercise caution in the use these data. Ecosystem research over the past several years has increasingly focused on functional parameters rather than the more traditional structural metrics. Emphasis on system-level functional roles may not answer population-level questions, but it permits clustering of genetically and taxonomically diverse groups into functional guilds. Functional indicators are less likely to be constrained by regionally restricted biota.

Thus, functional approaches lead to a more global view of stream ecosystems, a view that is much less variable than one based only on taxa inhabiting stream communities. Hunsaker et al. (1990) argued that for regional ecological risk assessments to be effective, the system must be functionally defined, with the spatio-temporal boundaries of the system set by functional attributes of the communities inhabiting the system. Assessments that are functionally based are likely to have greater applicability across regions (Hunsaker et al., 1990).

5.7.1. Photosynthesis

One of the most direct measures of plant physiology is photosynthesis. This non-taxonomic integrator of physiological condition is responsive to changes in environmental condition, and can be accurately measured either by carbon incorporation or oxygen evolution. The actual mechanism of photosynthetic inhibition varies by chemical, but most inhibitors fall into three categories, those that interrupt electron transport activity, those that alter the structure of chloroplasts, and those that reduce chlorophyll concentrations within the chloroplast. Most herbicides and organochlorine and organophosphate pesticides inhibit photosynthesis by blocking electron transport. Adjusting photosynthesis for chlorophyll *a* per unit mass allows for comparisons of communities with differing levels of biomass, resulting in lower variance components to these measures. Hill et al. (1997), found depressed periphyton photosynthesis in the metal-impacted Eagle River, Colorado, during an investigation for the USEPA Superfund programme. In their Superfund assessment of a metal-impacted river, Hill et al. (1997) also found significantly reduced quantum yield (photosynthesis adjusted for solar radiation) and assimilation ratio (chlorophyll-adjusted photosynthesis) in streams with elevated dissolved metal concentrations.

5.7.2. Respiration

Respiration in aquatic plants is often overlooked in the assessment of physiological condition and responses to perturbations. Respiration integrates most cellular functions and indirectly measures impacts to all cell systems. The mechanisms of respiratory inhibition or stimulation by chemical substances are poorly understood, but electron transfer within glycolysis and the Krebs cycle seems likely points of action. In regional-scale EMAP surveys of Appalachian, Rocky Mountain, and California Central Valley streams (Hill et al.,1998, 2000c), we found similar rates of respiration among these diverse regions, and reported significant correlations between respiration and stream chemistry and habitat variables.

5.7.3. Microbial enzyme activity

The use of microbial enzyme activity to assess the integrity of aquatic ecosystems a relatively new idea. The lack of a substantial microbial history in ecosystem assessments stems largely from the lack of understanding of the microbial assemblage within the ecosystem. Through its role in detritus processing, the microbial assemblage integrates carbon and nutrient cycling within the process of energy flow through ecosystems. Because of its role in ecosystem function, the microbial assemblage may be the best indicator of overall ecosystem process integrity and any change in microbial metabolic rates may be construed as an impact. Since microbial metabolic pathways are dependent on respiration, respiration should be a sensitive indicator of the condition of the stream microbial assemblage. Dehydrogenase activity has been used to measure the effects of metabolic activity of stream microbial communities and their responses to physical and chemical disturbances (Burton and Lanza, 1987; Burton et al., 1987; Blenkinsopp and Lock, 1990, 1992). Hill et al. (2002) compared the O_2 depletion method with DHA in EMAP studies of Appalachian streams. Hill et al. (2000a) reported that APA was positively correlated with riparian zone agriculture, and negatively correlated with indices of human disturbances in the riparian zones.

5.7.4. Community metabolism

Community metabolism (primary productivity and respiration) is a commonly measured functional attribute of stream ecosystems. That metabolism is not used more often in monitoring may be linked to the perception that its response to environmental conditions is too variable and thus is of limited use for assessing a streams response to environmental conditions. Hill et al. (1997) found significant differences in community metabolism between metal-impacted and reference sites in their Superfund research on the Eagle River, Colorado.

5.7.5. Nutrient uptake and spiraling

Nutrient spiralling, defined as spatially-dependent nutrient cycling in stream ecosystems (Elwood et al., 1983), links the concept of nutrient cycling with unidirectional flow. Nutrient cycles of ecosystems are viewed as either closed (i.e., an atom of nutrient is continuously recycled within the ecosystem) or open (i.e., an atom of a nutrient is cycled within the system, but is eventually exported from the system). Most ecosystems are considered to be open, though the degree of openness may depend on the relative time scale used to analyze nutrient cycles. Because of the unidirectional flow, nutrient cycling was never considered as an attribute of streams. A basic difference between spiraling and cycling in an open system is that spiraling moves the nutrient downstream within the same system rather than losing it from the system. That is, transport occurs as a part of the nutrient cycle rather than as an alternative to it. Newbold et al. (1981) developed an index of nutrient spiraling known as spiral length, defined as the average downstream distance associated with one complete cycle of a nutrient atom. Under steady-state conditions spiral length is expressed as the ratio of total downstream transport of a nutrient to nutrient utilization. It appears that uptake length accounts for as much as 98% of spiral length (Newbold et al., 1983; Mulholland et al., 1990), and that uptake length may be measures downstream depletion of pulse additions of non-radioactive nutrients (Stream Solute Workshop, 1990; Webster et al., 1991). Webster et al. (1991) reported decreased PO_4^{-3} retention in logged streams, resulting in longer uptake lengths, and attributed this to biotic and abiotic changes in the stream. As part of the Agency's stream monitoring methods development research. We found differences in NH_4^{+1} and PO_4^{-3} uptake in streams draining harvested and old growth watershed, and attributed these results to differences in biotic activity and transient channel storage. Figure 4 shows the regression of PO_4^{2-} concentrations against downstream distance for old growth and harvested watersheds in our northwestern California sites.

6. Toxicity assessments

6.1. Point source toxicity assessment

Whole effluent toxicity testing (WET) is defined as "the aggregate toxic effect of an effluent measured directly by an aquatic toxicity test" [USEPA Regulations, 54 FR 23868 at 23895; June 2, 1989]. Aquatic toxicity test methods designed specifically for measuring WET and receiving water toxicity have been codified in USEPA regulations (40 CFR part 136 [60 FR 53529; October 16, 1995]). These WET test methods employ a suite of standardized freshwater, marine, and estuarine tests using plants, invertebrates, and vertebrates to estimate acute and short-term chronic toxicity of effluents and receiving waters. Specific test procedures for conducting WET and receiving water tests are included in USEPA, 1993c and USEPA, 1994a. These three method manuals (WET method manuals) were incorporated by reference into USEPA 40 CFR part 136 in 1995. As regulations, use of these methods and adherence to the specific test procedures outlined in the WET method manuals is required when monitoring WET under the National Pollutant Discharge Elimination System (NPDES).

6.2. Receiving water toxicity assessment

The USEPA conducted a field toxicity study in order to determine if laboratory estimates of safe concentrations of pollutants were valid for protection of real streams (Geckler et al., 1976). The study was conducted on Shayler Run, in Clermont County, Ohio and examined at the effects of copper on stream biota. Copper was added to the stream for 33 months to maintain a concentration of 120 µg/L, a concentration that was expected to adversely affect some fish species but not others. The stream also received sewage effluent containing a variety of compounds known to affect acute copper bioavailability. All but one abundant species of fish and four of the five most abundant macroinvertebrate species were adversely affected by exposure to copper at this concentration. Direct effects on fish were death, avoidance, and restricted spawning. Acute and chronic tests with copper were also conducted in standard laboratory conditions and streamside with fathead minnows. This study concluded that laboratory derived data could be used to predict toxic effects in a natural stream situation. In general, the toxicity of copper was underestimated by the laboratory data because of avoidance of fish to copper was not measured by laboratory exposures (Geckler et al., 1976). Indirect effects on fish, as a result of the effects of copper on the aquatic food chain, could not be demonstrated. More recently the USEPA has used methods similar to the WET methods to assess toxicity in receiving waters. In the fall of 1995 and spring of 1997 the USEPA Region VIII collected physicochemical and toxicity information from the Clear Creek watershed in central Colorado. The purpose of this investigation was to evaluate the relative advantages of an ecotoxicological approach for identifying residual contaminant sources and evaluating established clean up goals within a watershed. *Ceriodaphnia* and fathead minnow 48-hr acute toxicity tests and metal analyses were performed on 32 stream samples collected in 1995 and 37 stream samples collected in 1997 from the Clear Creek watershed. Stream water was shipped overnight to the USEPA Aquatic Research Facility in Cincinnati for

Ceriodaphnia and fathead minnow toxicity testing and to the USEPA Region VIII Laboratory in Denver, Colorado for metal analyses. Both profile tests (100% stream water) and definitive toxicity tests (stream samples serially diluted with moderately hard reconstituted water) were performed. *Ceriodaphnia* toxicity results (LC50s and 100% stream water) in 1995 and 1997 showed a similar trend throughout the watershed; the upper 20 miles of the mainstem of Clear Creek had LC50s >20% stream water, while the lower 20 miles had LC50s <10% stream water. Converting the toxicity results to a No Observed Acute Effect Level (NOAEL), in metal concentrations, was demonstrated presented as an alternative approach for evaluating clean up goals within an entire watershed (Fig. 5). Figure 5 shows the results of zinc metal analyses, calculated USEPA National Water Quality Criteria and NOAELs calculated from *Ceriodaphnia* and fathead minnow tests. Although the NOAELs are higher than national criteria they can be used for interim clean-up goals and can be used to tract progress in eventually meeting the Criteria. As part of a Superfund Innovative Technology Evaluation (SITE) Programme, the USEPA evaluated a remediation technology that was put in place at the Summitville Mine Superfund Site in southern Colorado. The technology evaluated was a successive alkalinity producing system (SAPS) for removing high concentrations of metals (aluminum, copper, iron, manganese, and zinc). Two treated and one untreated water sample were evaluated using a series of acute aquatic toxicity tests with *Pimephales promelas*, the fathead minnow and *Ceriodaphnia dubia*, and chronic aquatic toxicity tests with *Oncorhynchus mykiss*, rainbow trout. All tests used moderately hard reconstituted water as the control and dilution water. The *P. promelas* used in this study were three days old, the *C. dubia* were <24 H old, and the rainbow trout *O. mykiss* used were 18 days old, 5 days post swimup, provided. The trout tests were conducted at 15 C, the two other species were tested at 20 C. *Ceriodaphnia* were more sensitive than rainbow trout, than the fathead minnow. Both treated samples reduced toxicity by 7–8-fold for *Ceriodaphnia*, 10-fold for rainbow trout, and about 5-fold for the fathead minnow. However, a substantial amount of toxicity remained. A 100-fold more reduction in the concentration of metals would need to be achieved to remove acute toxicity to rainbow trout, a 1000-fold reduction in metals in both treatments would be needed to remove acute toxicity to *Ceriodaphnia* and a 50-fold reduction for no acute effects to fathead minnows.

6.3. Sediment toxicity assessment

As part of the EMAP Surface Water Programme, sediment samples were collected to assess toxicity on a Regional scale in streams and rivers of the Mid-Atlantic U.S. in 1994, 1997 and 1998 and in the Southern Mineralized Zone Ecoregion in the Colorado Rocky Mountains in 1994 and 1995. Sample sites were selected randomly using a probability design so that the results could be inferred for the entire region. Sediments were collected from each site by scooping 11 small samples of surficial sediments within a 150–800 m long sample reach (reach length was proportional to stream width) and composited in a bucket. Samples were then placed in a plastic bag and shipped on ice back to the lab and stored at 4–6°C. A 7-day *Hyalella azteca*, amphipod, lethality and growth method was used to assess the toxicity of all sediment samples. During

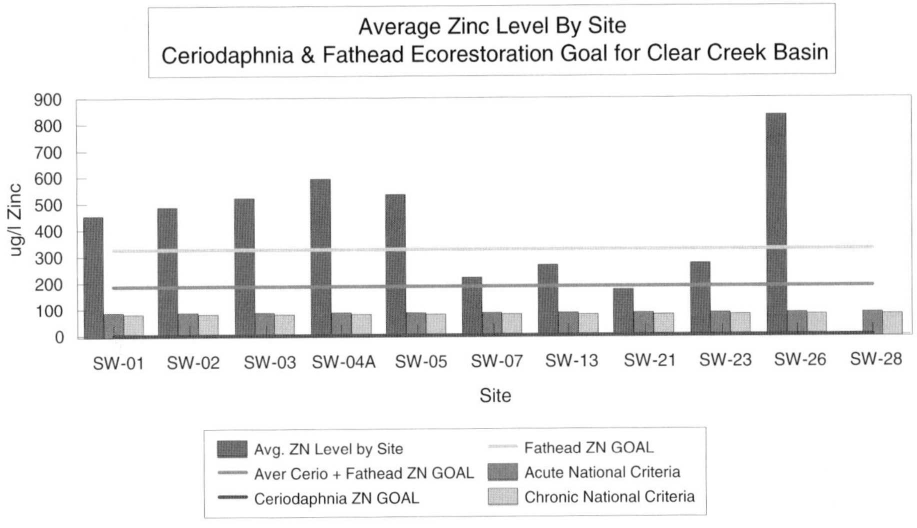

Figure 5. Figure demonstrating an approach to setting ecorestoration goals as interim goals for achieving site specific water quality standards.

the spring of 1994 in Mid-Atlantic streams, an estimated 3890 km of stream length (2.1% of the 188,700 km in the target population) were found to have toxic sediment (survival or growth significantly less than ($p = 0.05$)) the control). During the summers of 1997 and 1998 in Mid-Atlantic streams and river sediments, an estimated 5610 km (2.2%) of the 250,500 km of target length were found to be toxic. In the 1994/1995 Southern Mineralized Zone Ecoregion of the Colorado Rockies, an estimated 10.2% or 673 km of streams had toxic sediment.

Sediment toxicity assessments have also been used to evaluate the success of remediation of contaminated sediments (Tabak et al., 2000). Freshwater and marine sediment toxicity tests were used to measure baseline toxicity of sediment samples collected from New Jersey/New York Harbour (NJ/NY) (minimally contaminated) and East River (PAH-contaminated) sediment (ERC). Four freshwater toxicity tests were used: (1) amphipod *(Hyalella azteca)* mortality and growth tests (a standard 10-day USEPA method and two 7-day exposure methods (one using the standard amount of sediment, 100 ml; one using a reduced sediment volume, 17 ml) – the reduced volume freshwater amphipod test was developed and used in this study since existing volume requirements of the USEPA standard method exceeded the amounts available from enhanced or natural attenuation treatment); (2) a 7-day aquatic worm *(Lumbriculus variegatus)* mortality and budding test; (3) a 7/8-day fathead minnow (*Pimephales promelas*) embryo-larval survival and teratogenic test (FHM-EL); and (4) a 4-day vascular aquatic plant (*Lemna minor)* frond number/growth/chlorophyll *a* test (Duckweed). Two marine tests were also used: (1) an amphipod (*Ampelisca abdita*)10-day mortality test and a sheepshead minnow (*Cyprinodon variegatus*) embryo-larval sediment test (SHM-EL). ERC sediments were found to be highly toxic to all freshwater and marine organisms tested whereas the NJ/NY, sample showed no significant toxicity to the marine amphipod, but was slightly toxic to the freshwater worm and to freshwater and marine fish. For all tests with freshwater organisms and the one marine amphipod no survival was found in any of the tests except for one of the freshwater amphipod tests (55%). The ERC sediment significantly reduced frond production (-58.3%) and chlorophyll *a* levels (-35.4%) in the freshwater duckweed test. To determine the cause of toxicity in the sediments, five sediment manipulations were performed: (1) a sediment purge procedure, where 2 to 4 volumes of lab water were replaced over the sediment in a 24-h period; (2) a sediment dilution procedure where, grade 40 silica sand was mixed with PAH-contaminated sediments on a weight:weight basis; (3) a sediment aeration procedure, where sediment samples were aerated by adding 80 ml of sediment (140 g) to a 250 ml glass graduated cylinder and 120 ml of overlying water followed by aeration for 24–48 h; 4) an Ambersorb Treatment Procedure, where PAH-contaminated sediment samples were treated with two types of organic removal resins – Ambersorb 563 (AS 563) and Ambersorb 572 (AS 572); and (5) an Amberlite Treatment Procedure where IRC-718, an inorganic removal resin, was mixed with PAH-contaminated sediments. Results showed that freshwater amphipod survival was improved with the sediment aeration procedure and with 8% AS 563 and AS 572 treatments. Toxicity can also be reduced with the sediment dilution technique (100-fold). These manipulations revealed that hydrogen sulphide, organic compounds and inorganic compounds (metals) were factors in ERC sediment toxicity.

For the estuarine component of the EMAP programme, sediment toxicity tests included the 10-day acute test method on *Ampelisca abdita*, as well as *Ampelisca verilli,* and Microtox. For the four-year period from 1990–93 sediment toxicity (survival < 80%) using *Ampelisca abdita* was observed in 9±2% of the bottom sediments in estuarine area in the Virginian Province (Paul et al. 1999). In 1994, 3.2±6.2% of the estuarine area of the Carolinian Province showed sediment toxicity; in 1995, 0% was observed (Hyland et al., 1998). Sediment toxicity was also observed in the estuarine resource in the Louisianian Province (Macauley et al. 1999).

6.4. Tissue contaminants

The USEPA conducted a national screening-level investigation in 1987 to determine the prevalence of selected bioaccumulative pollutants in fish, and to correlate elevated fish tissue contaminant levels with pollutant sources. Game fish and bottom-dwelling fish were collected from 314 locations thought to be influenced by various point and nonpoint sources, and the fish tissue samples were analyzed to determine levels of selected contaminants. A list of 60 target analytes was developed for the study, including dioxins and furans, PCBs, pesticides and herbicides, mercury, and several other organic compounds. Results of the 1987 study indicated that target analytes were present in fish tissue at many of the sampling sites, and some of the contaminants occurred at levels posing potential human health risks.

In 1992–1994 the EMAP-Surface Waters programme conducted a survey of 167 lakes in the Northeastern United States and analyzed whole fish composite samples for contaminants, including Al, As, Cd, Cr, Cu, Fe, Hg, Ni, Pb, Se, and Zn. Using values for fish tissue contaminant levels that pose consumption risk derived from the literature and hazards assessment models, methylmercury (MeHg) was determined to be the metal contaminant of regional concern to fish consumers: 26% of lakes contained fish with MeHg exceeding a critical value of 0.2 µg/g, which implies risk to human consumers. Compared to USEPA wildlife values, 54% and 98% of lakes contain fish with MeHg exceeding critical values (0.1 µg/g; 0.02 µg/g) which implies risk to piscivorous mammals and birds, respectively. The other metals analyzed appeared to be at safe levels on a regional scale, and of only localized concern with regard to human health (Yeardley et al., 1998). In the EMAP 1993/1994 Mid-Atlantic Region assessment, fish assemblages from first- through third-order streams were dominated by small, short-lived fish (minnows (Cyprinidae), darters (Percidae), and sculpins (Cottidae)) that were more widely distributed and abundant than the large species typically chosen for tissue contaminant studies (suckers (Catostomidae), trout (Salmonidae), bass and sunfish (Centrarchidae), and carp (Cyprinidae)). Whole fish homogenate concentrations from small fish species exceeded the detection limits for contaminants such as mercury, DDT and PCBs in 50 to 65% of the stream length assessed and yielded greater regional estimates of wildlife exposure to mercury, DDT, dieldrin and chlordane than large fish species. Whole fish homogenate residues in large species exceeded the detection limits in only 32 to 38% of the stream length but regional estimates of wildlife exposure to PCBs were higher in larger fish. To make regional estimates of wildlife exposure to whole fish contaminants, USEPA developed wildlife values to reflect a potential threshold for toxic effects from chlordane, DDT and metabolites, dieldrin, endrin, mercury and PCBs exposure, using the approach described in USEPA's Great Lakes Water Quality Initiative. Whole fish homogenate concentrations of mercury, PCBs, DDT and metabolites, chlordane and dieldrin exceeded one or more of the wildlife values in both small and large fish species. This study concluded that, their greater distribution and higher estimates of contamination indicate that small, short-lived fish may be an excellent choice as target species for conducting regional fish contaminant studies (Lazorchak et al., 2002). USEPA also used the same EMAP data but looked at all 56 analytes and used toxicological bench-mark values for the belted Kingfisher (USEPA, 2001). Twenty two of the 56 analytes had median values that were above detection limits for either small or large fish. All sites from which samples were taken showed exposure to at least one contaminant. Seventy sites (100%) exceeded at least one of the 16 benchmark values and 22 sites (31.4%) exceeded four or more benchmark values (USEPA, 2001).

In 2000, the USEPA initiated work on three year national study of chemical residues in fish tissue, designed to expand the scope of the 1987 study. The new study is statistically designed and provided screening level data on fish tissue contaminants from a greater number of water bodies than were sampled in 1987. The 2000 study expanded the scope of the 1987 study which focused on chemical residues in fish tissue near point source discharges. The year 2000–2002 study intended: (1) to provide information on the national distribution of selected Persistent Bioaccumulative Toxics (PBT) residues in game fish and bottom-dwelling fish in lakes and reservoirs of the continental United States (excluding the Great Lakes); (2) to include lakes and reservoirs selected according to a probability design; (3) to involve the collection of fish from those randomly-selected lakes and reservoirs over a three year survey period; (4) will not be used to set fish consumption advisories; and (5) to include the analysis of fish tissue for PBT chemicals selected from the USEPA's multi-media candidate PBT list of 451 chemicals and a list of 130 chemicals from several contemporary fish and bioaccumulation studies. Lakes and reservoirs were chosen as the target population because they are accumulative environments where contamination is detectable, provide important sport fisheries nationwide, offer other recreational (non-fishing) access and opportunities, and occur in agricultural, urban, and less-developed areas, so that associations with each primary use may be determined. Lakes and reservoirs were the focus rather than other water body types due to the fact that fish consumption advisories represent 16.5% of the Nation's total lake acres (plus 100% of the Great Lakes), compared to 8.2% of the Nation's total river miles.

7. Real-time biological monitoring approaches

Assessing the environmental exposures of nonpoint sources, municipal and industrial point sources and storm water emanating from large metropolitan areas on large aquatic ecosystems presents many new and complex challenges. Unlike traditional point source discharges, storm water discharges and nonpoint sources of major metropolitan areas are sporadic and vary in intensity creating temporally and spatially variable shock loadings to receiving waters. Consequently, traditional assessment techniques which rely solely on sampling and characterization of the water column are ineffective in determining exposures to organisms in ecosystems that are temporally and spatially variable. In addition, the traditional acute and chronic water column and sediment toxicity tests (USEPA, 1993c; USEPA, 1994, a, b) used in toxicity testing are not without weaknesses when applied to variable exposures. Foremost among the weaknesses are concerns of how well, if at all, these methods mimic exposure of aquatic organisms during episodic events. Understanding interactions among toxicants (additivity, synergism, antagonism, etc.) is simply inadequate to predict biological responses (Waller et al., 1996).

Aquatic organisms have been shown to generate bioelectric signals which propagate into surrounding water. These signals can be recorded as rhythmic analog signals representative of specific movement activities (*e.g.* gill beats, heart rates, etc.). In addition, gape measurements (the degree to which a bivalve is open or closed) have been used with clams and mussels (bivalves) as a means to determine the status of these organisms. Utilizing appropriate statistical techniques and accompanying electronics, changes in bioelectric action potential (BAP) responses of fish and gape in bivalves can be detected, processed, and continuously recorded. They have also been used in detecting water quality induced stress in aquatic organisms (Waller et al., 1996). In response to toxic stress or unfavorable environmental conditions, bivalves (clams and mussels) may temporarily avoid exposure by closing their valves. Indeed this behaviour is the basis of current biomonitoring approaches where value-movement or "gape" is used as a measure of environmental stress or early warning to developing toxicity in aquatic systems (Waller et al., 1996 and Allen et al., 1996). Gape activity in live bivalves is the result of integrated responses of muscle/nerve functions, specifically adductor muscle contraction. Monitoring gape position as open or closed ("catch" state) provides an all-or-none quantal response that in the individual bivalve is a

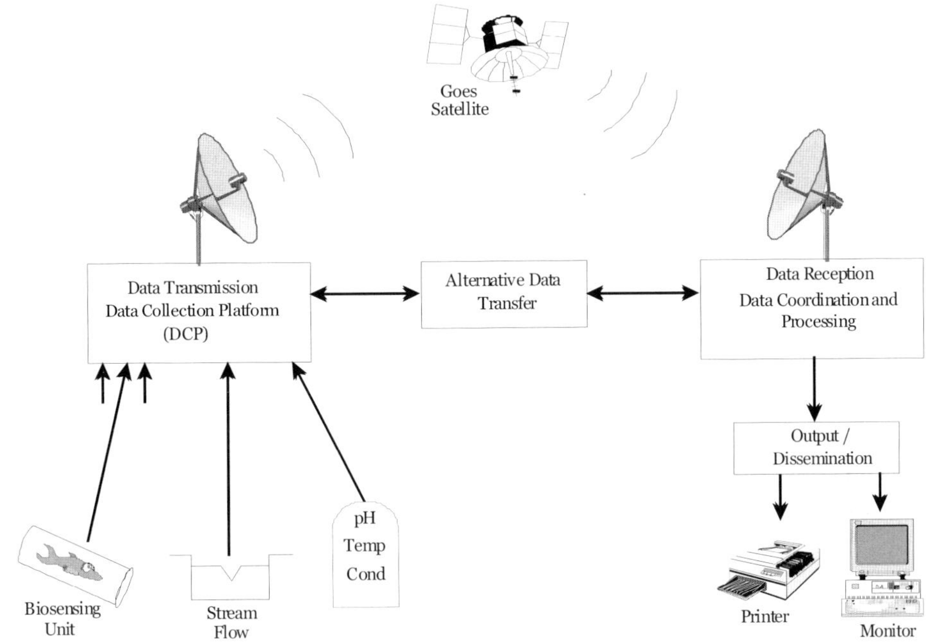

Figure 6. Schematic diagram of how real-time biological and chemical data are collected, transmitted and received.

qualitative phenomenon. Graded responses, such as a measured contraction of adductor, cardiac or foot muscles, are readily quantifiable, and provide higher levels of sensitivity to environmental stress analysis. Graded muscle responses are readily detected as bioelectric action potentials using electrocardiographic (ECG) techniques, but have limited utility in practice because they require that electrodes (probes) be inserted and are thus invasive and destructive to the bivalve.

A fish gill ventilatory activity monitor developed for remote field operations has been in use since the late 1970s and early 1980s (Morgan et al., 1978; Morgan et al., 1979a and b). Until the late 1980s, researchers were primarily interested in monitoring bioelectric events from aquatic animals in terms of the rate in which they occurred, i.e. the number of events per unit time. Although rate monitoring provided useful information on the state of the organism, considerable losses in sensitivity and information were experienced when looking only at heart and breathing rates. To account for these losses the system has been redesigned to digitally record the bioelectric signals generated by muscle and nerve responses and to process the data by digital processing (DSP) utilities.

The USEPA in cooperation with the University of North Texas, Denton, Texas and Tennessee Technical University, Cookeville, Tennessee installed a biological monitoring station on the Little Miami River at Miamiville, OH. The river is designated as a National and State Scenic River. The purpose of this station was to monitor the water quality of the Little Miami River near Cincinnati using physical/chemical and biological (fish and clams) sensors. The University of North Texas has placed clams in the river and has monitored their behaviour while Tennessee Technological University monitors caged fish "breathing" characteristics. The clam biomonitoring system was made up of three major components; Data Collection, Processing and Communication, and Power (Fig. 6). The data collection component consisted of both a biological and physical/chemical aspect. The biological aspect consisted of 15 clams. Gape was measured once per minute using proximity sensors and a stainless steel proxy attached to a shell of a clam. Physical/chemical data, temperature, pH, conductivity and dissolved oxygen, were also collected in real time using a multiprobe. Plots of these data over 24 hour periods were posted on a website (www.ias.unt.edu/~jallen/littlemiami/Clam_Page.html). Data were collected and processed using a microcomputer. The computer also telemetered the data to the University of North Texas Aquatic Toxicology Laboratory. A digital cellular modem was used to connect to the internet and pass the data for further analysis, display, and archiving. The field installation was powered using four, 90 amphr batteries. A solar array maintains the charge on the battery during the daylight hours. One finding of the study was that deployment in flowing systems with quickly fluctuating depths presented problems in operation and maintenance. In addition, uploading data from the two biosensors required more energy and cellular time than anticipated. However, there was success for up to 60 continuous days of operation and some baseline levels were established as well as an understanding of clam behavior under these conditions.

A USEPA Environmental Monitoring for Public Access and Community Tracking (EMAPCT) funded real-time biological monitoring project has been in operation in 2000 and 2001 on the Chicamacomico River, Maryland. The U.S. Army Centre for Environmental Health Research (USACEHR) developed an automated fish monitoring system, known as the Real Time Environmental Protection System (REPS). REPS was designed to detect harmful water quality conditions in the Chesapeake Bay and other waterways. In cooperation with the Maryland Department of Natural Resources, a portable REPS facility has been monitoring the water at a potential site of toxic Pfiesteria activity on the Chicamacomico River. REPS complements other on-going monitoring efforts to give early warning of potential risks to human and ecological health http://www.aquaticpath.umd.edu/empact/index.html . The REPS has been in continuous operation on the Chicamacomico River at Drawbridge, MD from July through November 2000, generating data on water quality and fish behaviour. The REPS did not detect severe stress resulting in fish death at any time during the monitoring period. Short-term fish stress events occurred occasionally, generally during storm events, but fish recovered quickly and no *Pfiesteria* were detected in water samples automatically collected during these events.

8. Molecular approaches

8.1. *Measures of gene expression*

Expression of certain genes in aquatic organisms is the cellular *reaction* to environmental *action*. Molecular biology offers sensitive and expedient tools for the detection of exposure to environmental stressors. Molecular approaches provide the means for detection of the "first cellular event(s)" in response to environmental changes – specifically, immediate changes in gene expression. Environmental exposure monitoring using gene activity as an indicator is based on the hypothesis that sub-cellular events resulting from an organism's contact with chemical milieus are manifested far in advance of those effects observed at higher levels of biological organization. Specifically, this approach involves detection of changes in gene transcription and relative levels of tissue-specific messenger RNA (mRNA), which occur as a result of direct contact with xenobiotic chemicals present in the environment. Protein products that are synthesized in response to environmental change represent the terminal aspect in a multi-step biochemical pathway that is replete with diverse cellular control mechanisms. Most of these studies in teleosts have centered on the single chemical/single gene response (i.e., β-naphthaflavone/P450IA1 and estradiol/vitellogenin). Recent USEPA studies describe the quantifiable induction of vitellogenin gene transcription in common carp (*Cyprinus carpio*) (Lattier, et al., 2001), and fathead minnows (*Pimephales promelas*) (Lattier et al., 2002) as indicators of environmental oestrogens. This scheme largely ignores the reality of environmental complexity, such as chemical fate and transport, synergism of chemical mixtures, multiple genes competing for limited intracellular pools of transcription co-factors, and gene induction profiles resulting from chronically exposed organisms. Emerging technologies, such as *differential display* and microarray DNA chips, provide a means to detect differences in inestimable gene products induced by the above scenarios. When applied judiciously, and in concert with higher order ecological analyses, molecular biology can provide the important link between immediate environmental exposure and long term biological, community and population effects.

Using the commonly found USEPA toxicological standard model, fathead minnow (*Pimephales promelas*), and the technique to detect changes in total cellular RNAs (*differential display*), we will establish the temporal global gene expression profile changes in embryos, resulting from exposure to several individual compounds. These will include a potent inducer of P450IA1, an identified metal, an oestrogenic compound, and a target pesticide. At the dose and time wherein each inducer causes the greatest alteration in the mRNA expression profile, a subtractive cDNA libraries will be constructed. Subtractive cloning is a powerful technique for isolating genes expressed or present in one cell population but not in another. Constitutively transcribed mRNAs, or those that are common between two cell or tissue types, are selectively removed. The remaining mRNA arises from genes that are uniquely expressed under specific environmental, developmental, or disease conditions. For the purpose of our indicator initiative, individual, inducer-specific subtractive libraries (ostensibly representing *only* those up- or down-regulated gene products resulting from a specific exposure) will then be combined, and arrayed in total on DNA chip(s), thus providing probes for detecting exposure induced changes in gene expression.

The platform for DNA arrayed on chips is accomplished through nanofabrication. Probe DNA is typically attached to standard $1'' \times 3''$ microscope slides by photolithography or other covalent chemistry, and spotted by "pins" at a density of 75–125 _m on-center with 0.1–2.0 nanograms of probe DNA per spot. Current technology offers the ability to place between 15,000 and 45,000 probe sequences on a single array.

Following fabrication of the chip, mRNA (cDNA) of exposed organisms and that of the control animals is isolated, and differentially labelled with the fluorescent dyes *Cy*3 [green] and *Cy*5 [red] in enzymatic reactions. Equal amounts of the fluorescent tagged nucleic acid, from both control and exposed animals, will be combined and simultaneously *hybridized* to complementary probes covalently bound to the chip(s). Following this competitive hybridization, the fluorescent dyes present on the arrayed chip are then induced to an excitation state by two independent lasers. The chip is then 'read' with a dual wavelength scanner at 632 and 532 nanometers for the red and green labels, respectively. In the resulting pseudo colour image, the green *Cy3* and red *Cy5* signals are overlaid. The range of spectral colour intensities, detected on the DNA spots, indicates whether a gene in question was expressed exclusively in one cell type (absolute red), or the other (absolute green). A yellow spot indicates equal intensity for the dyes, and suggests that a gene was expressed equally in both cell types. Comparison of patterns of gene expression between unexposed organisms and those exposed to environmental mixtures will permit investigators to determine genes that are activated, inhibited, or uncharacteristically modulated by multiple chemical stressors present in binary, ternary, and more complex mixtures.

This approach will facilitate understanding of the synergistic nature of expressed genes, and multi-gene pathways, when challenged with chemical mixtures, establish patterns of expressed genes that correlate with single or complex toxicant exposures, and provide information about the relative *bioavailable* concentrations of environmental stressors. This approach will also provide the ability to differentiate gene expression patterns influenced by chronic chemical exposure, when compared to genes that are activated by acute onset exposure. Embryo-specific expression patterns using individual compounds and environmental mixtures will then be indexed, providing 'expressed sequence tags', gene expression patterns, and a means to structurally characterized differentially expressed cDNAs. Early developmental stages of *Pimephales promelas* were chosen for analysis based on the following rationale: (i) This species is ubiquitous, and represents the USEPA toxicological standard model. (ii) The number of genes expressed during plastic early development far exceeds expression profiles in later life stages. (iii) Disruption of genes critical in early development could influence reproductive outcomes, and pose deleterious consequences to communities and populations. Gene expression analyses of early developmental stages provide unparalleled means by which to link patterns of gene expression with whole animal and population effects. (iv) Use of this developmental stage may comply with future Agency guidelines restricting vivisection and/or whole animal studies.

The fathead minnow is an excellent freshwater fish to be used as a model for testing gene expression for two reasons, (1) its wide distribution in North America, and (2) its long history of use in acute and chronic testing of contaminants, effluents and receiving waters by North American and European ecotoxicologists. The fathead minnow is a popular bait fish, and the ease with which it is propagated has led to its widespread introduction both within and outside the native range of the species. It has been so widely distributed in the eastern and southwestern United States by bait transportation that it is difficult to determine its original range. The presumed native distribution (Lee et al., 1980) extended from the Great Slave Lake in the northwest to New Brunswick, in eastern Canada, southward throughout the Mississippi valley in the United States, to southern Chihuahua in Mexico. Distribution records for this species also now include Oregon, and the Central Valley and other locations in California (Mount, 1968), but there are no records for British Columbia. This species is found in a wide range of habitats. It is most abundant in muddy brooks, streams, creeks, ponds, and small lakes, is uncommon or absent in streams of moderate and high gradients and in most of the larger and deeper impoundments, and is tolerant of high temperature and turbidity, and low oxygen concentrations.

The fathead minnow is unrivalled in toxicological research concerning the effects of pollution on freshwater resources. Tolerance to adverse conditions and ease of spawning makes the fathead minnow ideal for laboratory culture. Brood stock can be maintained in spawning condition year-round, ensuring a constant supply of larval fish for toxicity testing purposes. The fathead has been used since the 1960s (Mount, 1968; Norberg and Mount, 1985; Pickering and Lazorchak, 1995; Pickering et al., 1996) to generate acute life cycle, short-term chronic and embryo-larval toxicity information on various chemicals, municipal and industrial discharges and receiving waters. Since the 1970s more standardized tests have been developed and used for the fathead minnow throughout North America and Europe than any other freshwater fish. The USEPA has standardized acute, short-term chronic and embryo-larval survival and teratogenicity toxicity methods (USEPA, 1993c, 1994 a). The American Society of Testing and Materials has published a guide for acute and chronic fathead short-term chronic fathead minnow testing (ASTM, 1998). Standard methods have also been published describing acute and chronic methods for fathead minnows (APHA, 1998). Environment Canada has published a method for testing the larval growth and survival of fathead minnows (Environment Canada, 1992) and in Europe the Organization for Economic Cooperation and Development (OECD) has adopted acute and short-term chronic methods using fathead minnows (OECD, 1984, 1992).

8.2. Measures of genetic diversity

Molecular analysis of intraspecific genetic diversity provides a logical extension of previously described approaches to measure biological integrity with fish assemblage data. Although studies of molecular genetic diversity at the USEPA are in their infancy, there are a number of compelling reasons to believe that molecular genetic measures will ultimately provide highly useful bioindicators.

8.2.1. Rationale for development of a genetic diversity indicator

Biodiversity is usually defined in terms of three hierarchically related components: genetic diversity, species diversity, and ecosystem diversity. Genetic diversity is basal in this hierarchy, so erosion of biodiversity at this level will eventually impact diversity at species and ecosystem levels. For this reason, it has been argued that a fully developed indicator of genetic diversity may provide a sensitive and efficient measure of ecosystem health, especially when ecosystems are exposed to long-term, low-level chronic stressors (Bickham et al., 2000). Genetic diversity is fundamentally a trait of biological populations, and significant changes in genetic diversity reflect important population-level changes. Genetic erosion may involve loss of diversity within individual populations, often as a consequence of small (effective) population size, or it may involve loss of diversity among populations, which may reflect removal of previous barriers to effective migration. Both types of genetic erosion can increase extinction risk. Loss of diversity among populations is of particular concern where locally adapted populations are introgressed with foreign, nonadaptive genes, lowering population fitness (Waser, 1993). Loss of diversity within small populations may put them at immediate risk from inbreeding depression, but they are also at risk over the long term because of loss of adaptive potential (Lande, 1999). Current rates of anthropogenic change are unprecedented and populations with little genetic diversity may not be able to adapt in response to these changes quickly enough to avoid extinction (Lande and Shannon, 1996; Lynch, 1996; Lande, 1999).

In addition to its utility as a direct measure of ecosystem condition, a genetic diversity indicator will provide fundamental data that enhances the value and interpretation of other ecological assessment data, such as those obtained from landscape analyses and species assemblage data. An under-appreciated problem for ecological assessments is how to define the appropriate ecological unit of analysis. For any one species, the biological population is the most logical assessment unit for questions of biological integrity. Characterization of population genetic structure is one of the simplest applications of genetic diversity analyses, and delineation of distinct populations and their geographic boundaries is a tractable problem using genetic methods. For assessments of biological communities, a multi-species analysis of genetic diversity for different guilds and life-history types will provide the most meaningful information for defining assessment units.

The notion that genetic diversity should be responsive to changes in environmental condition is not new, and attempts to relate the variability of molecular genetic markers to specific aquatic stressors date back more than 30 years. These studies include both field surveys and controlled laboratory experiments of fish populations, and have evaluated the effects of metals, acidity, pesticides, radionuclides, and complex effluents (reviewed by Gillespie and Guttman, 1999). Taken as a whole, this body of studies presents overwhelming evidence that molecular genetic diversity can be substantially altered by deteriorated environmental conditions. Attempts to evaluate genetic diversity in relation to environmental quality at large, regional scales are noticeably absent from the literature. In fact, most field surveys have compared a relatively small number of populations in contaminant-exposed and "reference" sites. However, since reference sites are often chosen based on ecological, rather than genetic criteria, it is often unclear whether genetic differences between reference and exposed sites reflect recent changes or long-standing genetic differences.

8.2.2. USEPA genetic diversity research

Current EPA research is designed to assess the utility of incorporating a genetic diversity indicator into large-scale assessment and monitoring efforts. To avoid inherent problems in defining reference sites, a "landscape genetics" approach is used to make appropriate comparisons. Under this strategy, regional patterns of genetic similarities among populations are compared with current and historical geographic information to infer patterns of dispersal and evolutionary divergence among populations. Levels of genetic diversity within populations are compared to patterns of evolutionary divergence among populations to differentiate recent from historical genetic changes. Ecological data for each site is compared to inferred patterns of recent genetic change in order to evaluate possible environmental or anthropogenic factors that initiated genetic changes. An ongoing pilot analysis focuses on genetics of cyprinid stream fish, particularly the central stoneroller (*Campostoma anomalum*), creek chub (*Semotilus atromaculatus*), and blacknose dace (*Rhynichtys atratulus*) in the eastern half of the USA. Two different types of genetic assays are being evaluated. The first utilizes DNA fingerprinting technologies such as randomly amplified polymorphic DNA (RAPD; Welsh and McClelland, 1990; Williams et al., 1990) and amplified fragment length polymorphisms (AFLP; Vos et al., 1995) to identify differences between individuals at large numbers of anonymous DNA loci. The advantage of these techniques is that genetic profiles can be generated relatively quickly, with little developmental effort and modest molecular biological expertise. The second approach under evaluation entails sequence analysis of mitochondrial DNA combined with length-polymorphism analysis of nuclear DNA microsatellites (also called simple-sequence repeats). Microsatellite markers are extremely polymorphic, often segregating for dozens of alleles within populations, and therefore are extremely sensitive indicators of changes in genetic diversity within populations (Luikart et al., 1998). Mitochondrial DNA provides a relatively straightforward interpretation of evolutionary relationships among populations (Avise, 1994), so it should be highly complementary to microsatellite markers using the landscape genetics approach.

Both of these strategies target genetic loci that are expected to be selectively neutral, on average. A valid concern with both approaches is whether the patterns observed for these genetic markers are comparable to patterns at more ecologically and evolutionarily relevant (fitness-influencing) loci (Lynch, 1996; Rodriguez-Clark, 1999). Future research plans call for targeted analysis of large numbers of protein-coding and regulatory loci, building on an expanding database of knowledge of gene function resulting from the various fish genome projects (e.g., Detrich et al., 1999), and increased analytical capabilities resulting from USEPA gene expression research.

Acknowledgments

We would like to thank Teresa Ruby of CSC Graphics Support, c/o USEPA, Cincinnati, Ohio, for her preparation of Figure 1 and her patience in the many revisions the authors made to it.

References

Allen, H.J., Waller, W.T., Acevedo, M.F., Morgan, E.L., Dickson, K.L., Kennedy, J.H., 1996. Minimally invasive techniques to monitor valve-movement behavior in bivalves. Environ. Tech. 17, 501–507.

APHA, 1998. Standard Methods for the Examination of Water and Wastewater, 20th edn. American Public Health Association, Washington, DC.

ASTM, 1998. Biological effects and environmental fate; Section 11: water and environmental technology. In American Society for Testing and Materials Annual Book of ASTM Standards, 11.05, American Society for Testing and Materials, West Conshohocken, PA.

Avise, J.C., 1994. Molecular Markers, Natural History and Evolution. Chapman and Hall, New York.

Barber, M.C., 1994. Indicator Development Strategy. EPA/620/R-94/022. US Environmental Protection Agency, Office of Research and Development, North Carolina.

Barbour, M.T., Stribling, S.B., Karr, J.R., 1995. Multimetric approach for establishing biocriteria and measuring biological condition. In: Davis, W.S., Simon, T.P. (Eds), Biological Assessment and Criteria: Tools for Water Resource Planning and Decision Making. Lewis Publishers, Boca Raton, pp. 63–77.

Barbour, M.T., Gerritsen, J., Griffith, G.E., Frydenborg, R., Mccarron, E., White, J.S., Bastian, M.L., 1996. A framework for biological criteria for Florida streams using benthic macroinvertebrates. Journal of the North American Benthological Society 15, 185–211.

Bickham, J.W., Sandhu, S., Hebert, P.D.N., Chikhi, L., Athwal, R., 2000. Effects of chemical contaminants on genetic diversity in natural populations: implications for biomonitoring and ecotoxicology. Mutation Research 463, 33–51.

Blenkinsopp, S.A., Lock, M.A., 1990. The measurement of electron transport system activity in river biofilms. Water Research 24, 441–445.

Blenkinsopp, S.A., Lock, M.A., 1992. Impact of storm-flow on electron transport system activity in river biofilms. Freshwater Biology 27, 397–404.

Boston, H.L., Hill, W.R., Stewart, A.J., 1991. Evaluating direct toxicity and food-chain effects in aquatic systems using natural periphyton communities. In: Gorsuch, J.W., Lower, W.R., Wang, W., Lewis, M.A. (Eds), Plants for Toxicity Assessments, Vol. 2. ASTM STP 1115, American Society for Testing and Materials, Philadelphia, pp. 126–145.

Broberg, A., 1985. A modified method for studies of electron transport system activity in freshwater sediments. Hydrobiologia 120, 181–187.

Burton, G.A., Lanza, G.R., 1987. Aquatic microbial activity and macrofaunal profile of an Oklahoma stream. Water Research 21, 1173–1182.

Burton, G.A., Drotar, A., Lazorchak, J.M., Bahls, L.L., 1987. Relationship of microbial activity and *Ceriodaphnia* responses to mining impacts on the Clark Fork River, Montana. Archives for Environmental Contamination and Toxicology 16, 523–530.

Cairns, J., Niederlehner, B.R., Smith, E.R., 1992. The emergence of functional attributes as endpoints in ecotoxicology. In: Burton, G.A. (Ed.), Sediment Toxicity Assessment. Lewis Publishers, Chelsea, MI, pp. 111–128.

Carlisle, D.M., Clements, W.H., 1999. Sensitivity and variability of metics used in biological assessments of running waters. Environmental Toxicology and Chemistry 18, 285–291.

Carter, J., 1990. A new *Eunotia* and its great morphological variations under stress caused by a habitat loaded with copper salts. Ouvrage dedie a H. Germain, Koletz 1990, 13–17.

Chaloud, D.J., Peck, D.V. (Eds), 1994. Environmental Monitoring and Assessment Program: Integrated Quality Assurance Project Plan for the Surface Waters Resource Group. EPA/600/X-91/080, Revision 2.00. US Environmental Protection Agency, Las Vegas.

Chapman, D. (Ed.), 1992. Water Quality Assessment. Chapman and Hall, London.

Charles, D.F., 1985. Relationships between surface sediment diatoms assemblages and lakewater characteristics in Adirondack lakes. Ecology 66, 994–1011.

Clark, J.R., Dickson, K.L., Cairns, J., 1979. Estimating aufwuchs biomass. In: Weitzel, R.L. (Ed.), Methods and Measurement of Periphyton Communities: A Review. Special Technical Publication 690, American Society for Testing and Materials, Philadelphia, pp. 116–141.

Clements, W.H., Carlisle, D.M., Lazorchak, J.M., Johnson, P.C., 2000. The role of heavy metals in structuring benthic macroinvertebrate communities in Colorado's mountain streams. Eco. Appl. 10, 2, 626–638.

Crossey, M.J., LaPoint, T.W., 1988. A comparison of periphyton assemblage structural and functional responses to heavy metals. Hydrobiologia 162, 109–121.

Descy, J.P., 1979. A new approach to water quality estimation using diatoms. Nova Hedwigia 64, 305–323.

DeShon, J.E., 1995. Development and application of the invertebrate community index (ICI). In: Davis, W.S., Simon, T.P. (Eds), Biological Assessment and Criteria: Tools for Water Resource Planning and Decision Making. Lewis Publishers, Boca Raton, pp. 217–243.

Detrich, H.W., Westerfield, M., Zon, L.I., 1999. The Zebrafish: Genetics and Genomics. Methods in Cell Biology, 60. Academic Press, San Diego.

Diaz-Ramos, S., Stevens Jr., D.L., Olsen, A.R., 1996. EMAP Statistical Methods Manual. EPA 620/R-96/002. US Environmental Protection Agency, Office of Research and Development, Washington, DC.

Dixit, S.S., Smol, J.P., Kingston, J.C., Charles, D.F., 1992. Diatoms: powerful indicators of environmental change. Environmental Science and Technology 26, 22–33.

Elwood, J.W., Newbold, J.D., O'Neill, R.V., Van Winkle, W., 1983. Resource spiraling: an operational paradigm for analyzing lotic ecosystems. In: Fontaine III, T.D., Bartell, S.M. (Eds), Dynamics of Lotic Ecosystems. Ann Arbor Science Publishers, Ann Arbor, pp. 3–27

Engle, V.D., Summers, J.K., Gaston, G.R., 1994. A benthic index of environmental condition of Gulf of Mexico estuaries. *Estuaries* 17, 372–384.

Engle, V.D., Summers, J.K., 1999. Refinement, validation, and application of a benthic condition index for Gulf of Mexico estuaries. *Estuaries* 22, 624–635.

Environment Canada, 1992. Biological Test Method: Tests of Larval Growth and Survival Using Fathead Minnows. EPS 1/RM/22. Environmental Protection Series, Ottawa, Ontario.

Fausch, K.D., Lyons, J., Karr, J.R., Angermeier, P.L., 1990. Fish communities as indicators of environmental degradation. In: Adams, S.M. (Ed.), Biological Indicators of Stress in Fish. American Fisheries Society Symposium 8. Bethesda, pp. 123–144.

Fite, E.C., Turner, L.W., Cook, N.J., Stunkard, C., 1988. Guidance document for conducting terrestrial field studies. Hazard Evaluation Division Technical Guidance Document. Office of Pesticide Programs, US EPA, Washington, DC. EPA 540/09–88–109.

Fore, L.S., Karr, J.R., Wisseman, R.W., 1996. Assessing invertebrate responses to human activities: evaluating alternative approaches. Journal of the North American Benthological Society 15, 212–231.

Geckler, J.R., Horning, W.B., Neiheisel, T.M., Pickering, Q.H., Robinson, E.L., Stephan, C.E., 1976. Validity of Laboratory Tests for Predicting Copper Toxicity in Streams. US Environmental Protection Agency, Dulth, MN. EPA-600/3–76–116.

Geller, W., 1984. A toxicity monitor using the weakly electric fish *Gnathonemus petersi*. Wat. Res. 18, 1285–1290.

Gibson Jr, G.R., 1991 Biological criteria: research and regulation. Proceedings of a Symposium. US Environmental Protection Agency, Office of Water, Washington, DC EPA-440/5–91–005.

Gillespie, R.B., Guttman, S.I., 1999. Chemical-induced changes in the genetic structure of populations: effects on allozymes. In: Forbes, V.E. (Ed.), Genetics and Ecotoxicology. Taylor & Francis, Philadelphia, pp. 55–77.

Griffith, M.B., Kaufmann, P.R., Herlihy, A.T., Hill, B.H., 2001. Analysis of macroinvertebrate assemblages in relation to environmental gradients in Rocky Mountain streams. Ecological Applications 11, 489–505.

Griffith, M.B., Husby, P., Hall, R.K., Kaufmann, P.R., Hill, B.H., 2002. Analysis of macroinvertebrate assemblages in lotic habitats of California's Central Valley. Freshwater Biology, (in press).

Hall, R.I., Smol, J.P., 1992. A weighted-averaging regression and calibration model for inferrring total phosphorus concentration from diatoms in British Columbia (Canada) lakes. Freshwater Biology 27, 417–434.

Halliwell, D.B., Langdon, R.W., Daniels, R.A., Kurtenbach, J.P., Jacobsen, R.A., 1999. Classification of freshwater fish species of the Northeastern United States for use in the development of indices of biotic integrity, with regional applications. In: Simon, T.P. (Ed.), Assessing the Sustainability and Biological Integrity of Water Resources Using Fish Communities. CRC Press, Boca Raton, pp. 301–338.

Ham, K.D., Peterson, M.J., 1994. Effect of fluctuating low-level chlorine concentrations on valve-movement behavior of the Asiatic clam (*Corbicula fluminea*). Env. Tox. Chem. 13 (3), 493–498.

Healey, F.P., Henzel, L.L., 1979. Fluorometric measurements of alkaline phosphatase activity in algae. Freshwater Biology 9, 429–439.

Hellawell, J.M., 1977. Change in natural and managed ecosystems: detection, measurement and assessment. Proceedings of the Royal Academy of London, B. Biological Sciences 197, 31–56.

Hill, B.H., Hall, R.K., Husby, P., Herlihy, A.T., Dunne, M., 2000c. Interregional comparisons of sediment microbial respiration in streams. Freshwater Biology 44, 213–222.

Hill, B.H., Herlihy, A.T., Kaufmann, P.R., 2002. Benthic microbial respiration in Appalachian Mountain, Piedmont and coastal plains streams of the eastern U.S.A. Freshwater Biology 47, 185–194.

Hill, B.H., Herlihy, A.T., Kaufmann, P.R., Sinsabaugh, R.L., 1998. Sediment microbial respiration in a synoptic survey of mid-Atlantic region streams. Freshwater Biology 39, 493–501.

Hill, B.H., Herlihy, A.T., Kaufmann, P.R., Stevenson, R.J., McCormick, F.H., Burch Johnson, C., 2000a. Use of periphyton assemblage data as an index of biotic integrity. Journal of the North American Benthological Society 19, 50–67.

Hill, B. H., Kurtenbach, J.P., 2001. Correlations of sedimentary diatoms with watershed land-use and limnological conditions in New Jersey lakes. Lake and Reservoir Management, in review

Hill, B.H., Lazorchak, J.M., McCormick, F.H., Willingham, W.T., 1997. The effects of elevated metals on benthic community metabolism in a Rocky Mountain stream. Environmental Pollution 95, 183–190.

Hill, B.H., Stevenson, R.J., Pan, Y., Herlihy, A.T., Kaufmann, P.R., Burch Johnson, C., 2001. Comparison of correlations between environmental characteristics and stream diatom assemblages characterized at genus and species levels. Journal of the North American Benthological Society 20, 299–310.

Hill, B.H., Willingham, W.T., Parrish, L.P., McFarland, B.H., 2000b. Periphyton assemblage responses to elevated metal concentrations in a Rocky Mountain stream. Hydrobiologia 428, 161–169.

Holland, A.F. (Ed.), 1990. Near Coastal Program Plan for 1990: Estuaries. EPA/600/4–90/033. US Environmental Protection Agency, Environmental Research Laboratory, Office of Research and Development, Narragansett, Rhode Island.

Hughes, R.M., 1995. Defining acceptable biological status by comparing with reference conditions. In: Davis, W.S., Simon, T.P. (Eds), Biological Assessment and Criteria: Tools for Water Resource Planning and Decision Making. Lewis Publishers, Boca Raton, pp. 31–48.

Hughes, R.M., Kaufmann, P.R., Herlihy, A.T., Kincaid, T.M., Reynolds, L., Larsen, D.P., 1998. A process for developing and evaluating indices of fish assemblage integrity. Canadian Journal of Fisheries and Aquatic Sciences 55, 1618–1631.

Hughes, R.M., Oberdorff, T., 1999. Applications of IBI concepts and metrics towaters outside the United States and Canada. In: Simon, T.P. (Ed.), Assessing the Sustainability and Biological Integrity of Water Resources Using Fish Communities. CRC Press, Boca Raton, pp. 79–93.

Hunsaker, C.T., Graham, R.L., Suter, G.W., O'Neill, R.V., Barnthouse, L.W., Gardner, R.H., 1990. Assessing ecological risk on a regional scale. Environmental Management 14, 325–332.

Hyland, J.L., Herrlinger, T.J., Snoots, T.R., Ringwood, A.H., Van Dolah, R.F., Hackney, C.T., Nelson, G.A., Rosen, J.S., Kokkinakis, S.A., 1996. Environmental quality of estuaries of the Carolinian Province, 1994. Annual statistical summary for the 1994 EMAP-Estuaries Demonstration Project in the Carolinian Province. NOAA Technical Memorandum NOS ORCA 97. NOAA/NOS, Office of Ocean Resources Conservation and Assessment, Silver Spring, MD.

Hyland, J.L., Balthis, L., Hackney, C.T., McRae, G., Ringwood, A.H., Snoots, T.R., Van Dolah, R.F., Wade, T.L., 1998. Environmental quality of estuaries of the Carolinian Province, 1995. Annual statistical summary for the 1995 EMAP-Estuaries Demonstration Project in the Carolinian Province. NOAA Technical Memorandum NOS ORCA 123. NOAA/NOS, Office of Ocean Resources Conservation and Assessment, Silver Spring, MD.

Karr, J.R., 1981. Assessment of biotic integrity using fish communities. Fisheries 66, 21–27.

Karr, J.R., Fausch, K.D., Angermeier, P.L., Yant, P.R., Schlosser, I.J., 1986. Assessing biological integrity in running waters: a method and its rationale. Illinois Natural History Survey Special Publication 5, Champaign, IL.

Karr, J.R., 1991. Biological integrity, a long neglected aspect of water resource management. Ecological Applications 1, 66–84.

Karr, J.R., Dudley, D.R., 1981. Ecological perspective on water quality goals. Environmental Management 5, 55–68.

Karr, J.R., Heidinger, R.C., Helmer, E.H., 1985. Sensitivity of the index of biotic integrity to changes in chlorine and ammonia levels from wastewater treatment facilities. Journal of the Water Pollution Control Federation 57, 912–915.

Karr, J.R., Chu, E.W., 1997. Biological Monitoring and Assessment: Using Multimetric Indexes Effectively. EPA 235/R97/001. University of Washington, Seattle.

Kerans, B.L., Karr, J.R., 1994. A benthic index of biotic integrity (B-IBI) for rivers of the Tennessee Valley. Ecological Applications 4, 768–785.

Kutka, F.J., Richards., C., 1996. Relating diatom assemblage structure to stream habitat quality. Journal of the North American Benthological Society 15, 469–480.

Lande, R., 1999. Extinction risks from anthropogenic, ecological, and genetic factors. In: Landweber, L.F., Dobson, A.P. (Eds), Genetics and the Extinction of Species. Princeton University Press. Princeton, NJ, pp. 1–22.

Lande, R., Shannon, S., 1996. The role of genetic variability in adaptation and population persistence in a changing environment. Evolution 50, 434–437.

Lange-Bertalot, H., 1979. Pollution tolerance of diatoms as a criterion for water quality estimation. Nova Hedwigia 64, 285–304.

Lanza G.R., Cairns J., 1972. Physio-morphological effects of abrupt thermal stress on diatoms. Transactions of the American Microscopical Society 91, 276–298.

Larsen, D.P., 1995. The role of ecological sample surveys in the implementation of biocriteria. In: W.S. Davis and T.P. Simon (Eds), Biological Assessment and Criteria: Tools for Water Resource Planning and Decision Making. Lewis Publishers, Boca Raton, pp. 287–300.

Larsen, D.P., 1997. Sample survey design issues for bioassessment of inland aquatic ecosystems. Human and Ecological Risk Assessment 3 (6), 979–991.

Lattier, D.L., Gordon, D.A., Burks, D.J., Toth, G.P., 2001. Vitellogenin gene transcription: a relative quantitative exposure indicator of environmental estrogens. Environmental Toxicology and Chemistry 20, 1979–1985.

Lattier, D.L, Reddy, T.V., Gordon, D.A., Lazorchak, J.M, Smith, M.E., Williams, D.E., Wiechman, B., Flick, R.W., Miracle, A.L., Toth, G.P., 2002. 17α-ethynylestradiol induced vitellogenin gene transcription quantified in fathead minnow (*Pimephales promelas*) adult male livers, embryo larvae and gills. J. Environ. Toxicol. Chem. (in press).

Lazorchak, J.M., Klemm, D.J., Peck, D.V. (Eds), 1998. Environmental Monitoring and Assessment Program-Surface Waters: Field Operations and Methods for Measuring the Ecological Condition of Wadeable Streams. US Environmental Protection Agency, Cincinnati.

Lazorchak, J.M., McCormick, F.H., Herlihy, A.T., Henry, T.R., 2002. Fish tissue contamination in streams of the mid-atlantic region: an approach to regional indicator selection and assessment. Submitted J. Environ. Toxicol. Chem.

Lee, D.S., Shute, J.R., 1980. *Pimephales promelas*. In: Lee, D.S., Gilbert, C.R., Hocutt, C.H., Jenkins, R.E., McAllister, D.E., Stauffer Jr, R. (Eds), Atlas of North American Freshwater Fishes, Publication Number 1980–12, North Carolina State Museum of Natural History, Raleigh, North Carolina.

Leland, H.V., 1995. Distribution of phytobenthos in the Yakima River basin, Washington, in relation to geology, land-use, and other environmental factors. Canadian Journal of Fisheries and Aquatic Sciences 52, 1108–1129.

Leopold, A., 1941. Lakes in relation to terrestrial life patterns. In: Needham, J.G., Sears, P.B., Leopold, A. (Eds), A Symposium on Hydrobiology. University of Wisconsin Press, Madison, pp. 17–22.

Lowe, R.L., 1974. Environmental requirements and pollution tolerance of freshwater diatoms. EPA-670/4–74–005, United States Environmental Protection Agency, Cincinnati.

Luikart, G., Sherwin, W.B., Steele, B.M., Allendorf, F.W., 1998. Usefulness of molecular markers for detecting population bottlenecks via monitoring genetic change. Molecular Ecology, 7, 963–974.

Lynch, M., 1996. A quantitative-genetic perspective on conservation issues. In: Avise, J.C. (Ed.), Conservation Genetics: Case Histories from Nature. Chapman and Hall, New York, pp. 471–501.

Lyons, J.S., 1992. The length of stream to sample with a towed electrofishing unit when fish species richness is estimated. North American Journal of Fisheries Management 12, 198–203.

Macauley, J.M., Summers, J.K., Engle, V.D., 1999. Estimating the ecological condition of the estuaries of the Gulf of Mexico. Environ. Monit. Assess. 57, 59–83.

Macauley, J.M., Summers, J.K., Engle, V.D., Harwell, L.C., 2002. Ecological Condition of South Florida Estuaries. Environ. Monit. Assess. 75 (3), 253–269..

Markert, B., Wappelhorst, O., Weckert, V., Herpin, U., Siewers, U., Friese, K., 1999. The use of bio-indicators for monitoring the heavy-metal status of the environment. Journal of Radioanalytical Nuclear Chemistry 240 (2), 425–429.

Marshall, J.S., Mellinger, D.L., 1980. Dynamics of cadmium-stressed plankton communities. Canadian Journal of Fisheries and Aquatic Sciences 37, 403–414.

Mason Jr, W.T., Lewis, P.A., Anderson, J.B., 1971. Macroinvertebrate collections and water quality monitoring in the Ohio River Basin 1963–1967. US Environmental Protection Agency, Office of Technical Programs, Ohio Basin Region and Analytical Quality Control Laboratory, Water Quality Office. Cincinnati.

Matthews, R.A., Buikema Jr, A.L., Cairns Jr, J., Rodgers Jr, J.H., 1982. Biological monitoring: IIA. Receiving system functional methods, relationships, and indices. Wat. Res. 16, 129–139.

McCormick, F.H., Fish Communities as indicators of stream condition, 1993. In: Hughes, R.M. (Ed.), EMAP Streams Bioassessment Workshop. Report of the Proceedings. EPA/600/R-93/138. United States Environmental Protection Agency, Corvallis.

McCormick, F.H., Hughes, R.M., 1998. Aquatic vertebrate indicator. In: Klemm, D.J., Lazorchak, J.M., Peck, D.V. (Eds), Environmental Monitoring and Assessment Program-Surface Waters: Field Operations and Methods for Measuring the Ecological Condition of Wadeable Streams. US Environmental Protection Agency, Cincinnati.

McCormick, F.H., Peck, D.V., 2000. Fish indicator development: metrics and indices of biotic integrity. In: Jackson, L., Fischer, W., Kurtz, J. (Eds), EPA /620/R-99/005.

McCormick, F.H., Peck, D.V. Larsen, D.P., 2000. A comparison of ecological classification hierarchies for mid-Atlantic stream fish assemblages. Journal of the North American Benthological Society 19, 385–404.

McCormick, F.H., Hughes, R.M., Kaufmann, P.R., Herlihy, A.T., Peck, D.V., Stoddard, J.L., 2001. Development of an index of biotic integrity for the mid-Atlantic Highlands Assessment Project. Transactions of the American Fisheries Society 130, 857–877.

McFarland, B.H, Hill, B.H., Willingham, W.T., 1997. Abnormal *Fragilaria* spp. found in streams impacted by mine drainage. Journal of Freshwater Ecology 12, 141–149.

Medley, C.N., Clements, W.H., 1998. Responses of diatom communities to heavy metals in streams: the influence of longitudinal variation. Ecological Applications 8, 633–644.

Messer, J.J., Linthurst, R.A., Overton, W.S., 1991. An EPA program for monitoring ecological status and trends. Environmental Monitoring and Assessment 17, 67–78.

Morgan, E.L., et al., 1978. Biological water quality monitoring from remote stations and NOAA GOES Satellite. In: Proc. 4th Joint Conf. on Sensing Environmental Pollutants. American Chemical Society, New Orleans, November 1977, pp. 885–887.

Morgan, E.L., et al. 1979a. New developments in automated biosensing from remote water quality stations and satellite data retrieval for recourse management. In: Proc. 3rd World Congress on Water Resources, International Water Resources Association, Mexico City, 23–27 April, pp. 3534–3542.

Morgan, E.L. et al., 1979b. Automated biomonitoring applications to remote water quality stations and satellite data retrieval: new developments in achieving real-time biosensing for watershed management. Proceedings 9th Intersociety Conf. on Environ. Systems. Aerospace Div. Amer. Soc. Mech. Engineers, San Francisco, 16–19 July.

Morgan, E.L., Young, R.C., Wright Jr, J.R., 1988a. Developing portable computer-automated biomonitoring for a regional water quality surveillance network. In: Gruber D., Diamond, J. (Eds), Automated Biomonitoring. Ellis Horwood, Chichester.

Morgan, E.L., Wright Jr, J.R., Young, R.C., 1988b. Developing portable automated biomonitoring system for aquatic hazard evaluation. Special Technical Publication 988, American Society for Testing and Materials. Philadelphia.

Morgan, E.L., Yokley Jr, P., Rausina, G., Wright Jr, J.R., McFadden, J.F., Red, J.T., 1989. A toxicity test protocol for mature bivalve mussels using automated biological monitoring. In: Pesticides in Terrestrial and Aquatic Environments, Proceedings of a National Research Conference, Virginia Water Resources Research Center, Virginia Polytechnic Institute and State University, Blacksburg.

Morgan, E.L., et al., 1991. Remote biosensing applications in environmental assessment. Environmental Auditor 2 (4), 213–227.

Morgan, E.L., Waller, W.T., 1993. Detecting aquatic ecosystem change on regional scales: a proposed integrated remote-automated biosensing, GIS based network. Bull. Ecol. Soc. Am. 73 (4) (abstract).

Morgan, E.L. et al., 1995. Stress detection in bivalve mollusc using non-invasive bioelectric monitoring of myoneural behavior. In: 2nd World Congress of Soc. Environ. Toxicol. and Chem., Global Environmental Protection: Science, Politics, and Common Sense. 5–9 November 1995, Vancouver (abstract).

Mount, D.I., 1968. Chronic toxicity of copper to fathead minnows (*Pimephales promelas* Rafinesque). Water Research 2, 214–223.

Moyle, P.B., Marchetti, M.P., 1999. Applications of indices of biotic integrity to California streams and watersheds. In: Simon, T.P. (Ed.), Assessing the Sustainability and Biological Integrity of Water Resources Using Fish Communities. CRC Press, Boca Raton, pp. 367–380.

Mulholland, P.J., Rosemond, A.D., 1992. Periphyton response to longitudinal nutrient depletion in a woodland stream: evidence of upstream-downstream linkage. Journal of the North American Benthological Society 11, 405–419.

Mulholland, P.J., Steinman, A.D., Elwood, J.W., 1990. Measurement of phosphorus uptake length in streams: comparison of radiotracer and stable PO_4 releases. Canadian Journal of Fisheries and Aquatic Sciences 47, 2351–2357.

Newbold, J.D., Elwood, J.W., O'Neill, R.V., Van Winkle, W., 1981. Measuring nutrient spiraling in streams. Canadian Journal of Fisheries and Aquatic Sciences 38, 860–863.

Newbold, J.D., Elwood, J.W., O'Neill, R.V., Sheldon, A.L., 1983. Phosphorus dynamics in a woodland stream ecosystem: a study of nutrient spiraling. Ecology 64, 1249–1265.

Newman, R.M., Perry, J.A., Tam, E., Crawford, R.L., 1987. Effects of chronic chlorine exposure on litter processing in outdoor experimental streams. Freshwater Biology 18, 415–428.

Niemi, G.J., Detenbeck, N.E., Perry, J.A., 1993. Comparative analysis of variables to measure recovery rates in streams. Environmental Toxicology and Chemistry 12, 1541–1547.

Norberg, T.J., Mount, D.I., 1985. A new fathead minnow (*Pimephales promelas*) subchronic toxicity test. Environmental Toxicology and Chemistry 4, 711–718.

OECD, 1984. Guidelines for testing of chemicals. Section 2, Effects on Biotic Systems. Technical Guide No. 204, Fish, Prolonged Toxicity Test: 14-Day Study. Organization for Economic Cooperation and Development, Paris.

OECD, 1992. Guidelines for testing of chemicals. Section 2, Effects on Biotic Systems. Technical Guide No. 210, Fish, Early-Life Stage Toxicity Test. Organization for Economic Cooperation and Development, Paris.

Ohio EPA (Environmental Protection Agency), 1987. Biological Criteria for the Protection of Aquatic Life, Vol. 2. Users manual for biological field assessment of Ohio surface waters. Division of Water Quality Monitoring and Assessment, Columbus.

Omernik, J.M., 1987. Ecoregions of the conterminous United States. Annals of the Association of American Geographers 77, 118–125.

Omernik, J.M., Griffith, G.E., 1991. Ecological regions versus hydrologic units: frameworks for managing water quality. Journal of Soil and Water Conservation 46, 334–340.

Omernik, J.M., 1995. Ecoregions, a spatial framework for environmental management. In: Davis, W.S., Simon, T.P. (Eds), Biological Assessment and Criteria: Tools for Water Resource Planning and Decision Making. Lewis Publishers, Boca Raton, pp. 49–62.

Pan, Y., Stevenson, R.J., Hill, B.H., Herlihy, A.T., Collins, G.B., 1996. Using diatoms as indicators of ecological conditions in lotic systems: a regional assessment. Journal of the North American Benthological Society 15, 481–495.

Pan, Y., Stevenson, R.J., Hill, B.H., Kaufmann, P.R., Herlihy, A.T., 1999. Spatial patterns and ecological determinants of benthic algal assemblages in mid-Atlantic streams, USA. Journal of Phycology 35, 460–468.

Pan, Y., Stevenson, R.J., Hill, B.H., Herlihy, A.T., 2000. Ecoregions and benthic diatom assemblages in mid-Atlantic Highland streams, USA. Journal of the North American Benthological Society 19, 518–540.

Paul, J.F., Gentile, J.H., Scott, K.J., Schimmel, S.C., Campbell, D.E., Latimer, R.W., 1999. EMAP-Virginian Province Four-Year Assessment Report (1990–93). EPA/600/R-99/004. US Environmental Protection Agency, Atlantic Ecology Division, Narragansett, Rhode Island.

Pickering, Q.H., Lazorchak, J.M., 1995. Evaluation of the robustness of the fathead minnow, *Pimephales promelas*, larval survival and growth test, USEPA method 1000.0. Environmental Toxicology and Chemistry 14, 653–659.

Pickering, Q.H., Lazorchak, J.M., Winks, K.L., 1996. Subchronic sensitivity of one-, four-, and seven-day-old fathead minnow (*Pimephales promelas*) larvae to five toxicants. Environmental Toxicology and Chemistry 15, 353–359.

Rodriguez-Clark, K.M., 1999. Genetic theory and evidence supporting current practices in captive breeding for conservation. In: Landweber, L.F., Dobson, A.P. (Eds), Genetics and the Extinction of Species. Princeton University Press. Princeton, pp. 47–74.

Sayler, G.S., Sherrill, T.W., Perkins, R.E., Mallory, L.M., Shiaris, M.P., Pedersen, D., 1982. Impact of coal-coking effluent on sediment microbial communities: a multivariate approach. Applied and Environmental Microbiology 44, 1118–1129.

Schimmel, S.C., Melzian, B.D., Campbell, D.E., Benji, S.J., Rosen, J.S., Buffum, H.W., 1994. Statistical summary: EMAP – Estuaries Virginian Province – 1991. EPA/620/R–94/005. USEPA, Office of Research and Development, Rhode Island.

SETAC, 1994 Final Report: Aquatic Risk Assessment and Mitigation Dialogue Group. Pensacola, FL. Society of Environmental Toxicology and Chemistry, SETAC Foundation for Environmental Education.

Simon, T.P., 1991. Development of ecoregion expectations for the index of biotic integrity. I. Central Corn Belt Plain. EPA 905/9–90–005. US Environmental Protections Agency, Region V Environmental Sciences Division, Chicago, Illinois.

Simon, T.P., Lyons, J., 1995. Application of the index of biotic integrity to evaluate water resources integrity in freshwater ecosystems. In: Davis, W.S., Simon, T.P. (Eds), Biological Assessment and Criteria: Tools for Water Resource Planning and Decision Making. Lewis Publishers, Boca Raton, pp. 245–262.

Sloof, W., 1979. Detection limits of a biological monitoring system based on fish respiration. Bull. Environ. Contam. Toxicol. 23, 517–523.

Slooff, W., De Zwart, D., Marquenie, J.M., 1983. Detection limits of a biological monitoring system for chemical water pollution based on mussel activity. Bull. Environ. Contam. Toxicol. 30, 400–405.

Solladay, J.D., 1994. Morphological variation in fossil diatoms from Mono Lake. In: Kociolek, J.P. (Ed.), Proceedings of the 11th International Diatom Symposium, Memoirs of the California Academy of Sciences 17, pp. 337–348.

Southwood, T.R.E., 1977. Habitat, the templet for ecological strategies. Journal of Animal Ecology 46, 337–365.

Stevenson, R.J., Pan, Y., 1999. Assessing ecological conditions in rivers and streams with diatoms. In: Stoermer, E.F., Smol, J.P. (Eds), The Diatoms: Applications to the Environmental and Earth Sciences. Cambridge University Press, Cambridge, pp. 11–40.

Stream Solute Workshop, 1990. Concepts and methods for assessing solute dynamics in stream ecosystems. Journal of the North American Benthological Society 9, 95–119.

Summers, J.K, Macauley, J.M., Heitmuller, P.T., Engle, V.D., Adams, A.M., Brooks, G.T., 1993. Statistical summary: EMAP – Estuaries Louisianian Province – 1991. EPA/620/R–93/007. USEPA, Office of Research and Development, Washington, DC.

Tabak, H.H., Lazorchak, J.M., Khodadoust, A.P., Anita, J.E., Suidan, M.T., 2000. Studies on in-situ bio-remediation of PAH contaminated sediments: bioavailability, biodegradation, and toxicity issues. Proceedings of the Second Battelle International Conference on Remediation of Chlorinated and Recalcitrant Compounds, 22–25 May, Monterey, California, (in press).

Touart, L.W., 1988. Aquatic mesocosm tests to support pesticide registration. Hazard Evaluation Division Technical Guidance Document. Office of Pesticide Programs, US EPA, Washington, DC. EPA 540/09–88–035.

Touart, L.W., 1995. The Federal Insecticide, Fungicide and Rodenticide Act. In: Rand, G.M. (Ed.), Fundamentals of Aqatic Toxicology, Taylor & Francis, New York, pp. 657–668.

Touart, L.W., Maciorowski, A.F., 1997. Information needs for pesticide registration in the United States. Ecol. Appl. 74, 1086–1093.

Townsend, C.R., Hildrew, A.G., 1994. Species traits in relation to a habitat templet for river systems. Freshwater Biology 31, 265–275.

Tu, C.M., 1981. Effects of pesticides on activities of enzymes and microorganisms in clay soil. Journal of Environmental Science and Health B 16, 179–191.

Tyler, G., 1976. Heavy metal pollution, phosphatase activity, and mineralization of organic phosphorus in forest soils. Soil Biology and Biochemistry 8, 327–332.

Urban, D.J., Cook, J.N., 1986. Hazard Evaluation Division standard evaluation procedure. Office of Pesticide Programs, US Environmental Protection Agency, Washington, DC. EPA 540/19–83–001.

USEPA, 1977. The Integrity of Water. Proceeedings of a Symposium. US Environmental Protection Agency. Office of Water and Hazardous Materials. Washington, DC Stock Number 055–001–01068–1, pp 1–230.

USEPA, 1978. The *Selenastrum capricornutum* Printz Algal Assay Bottle Test. US Environmental Protection Agency, Corvallis, Oregon. EPA-600/9–78–018,

USEPA, 1989. Rapid Bioassessment Protocols for Use in Streams and Rivers. Benthic Macroinvertebrates and Fish. US Environmental Protection Agency Office of Water. Washington, DC. EPA 440/4–89/001.

USEPA, 1990. Macroinvertebrate Field and Laboratory Methods for Evaluating the Biological Integrity of Surface Waters. US Environmental Protection Agency. Washington DC. EPA/600/4–90/030.

USEPA, 1991. Technical Support Document for Water Quality-based Toxics Control, 2nd edn. US Environmental Protection Agency. Office of Water, Washington, DC. Report 505/2–90/001.

USEPA, 1992. Framework for Ecological Risk Assessment. Risk Assessment Forum, Office of Research and Development, Washington, DC. EPA/600/R-92/001.

USEPA, 1993a. Fish Field and Laboratory Methods for Evaluating the Biological Integrity of Surface Waters. US Environmental Protection Agency. Washington DC. EPA/600/R-92/111.

USEPA, 1993b. A Review of Ecological Assessment Case Studies from a Risk Assessment Perspective. Risk Assessment Forum, Office of Research and Development, US Environmental Protection Agency, Washington, DC. EPA/630/R-92/005.

USEPA, 1993c. Methods for Measuring the Acute Toxicity of Effluents and Receiving Waters to Freshwater and Marine and Marine Organisms. US Environmental Protection Agency Report. Office of Research and Development, Washington, DC. EPA/600/4–90/027F.

USEPA, 1994a. Short-term Methods for Estimating the Chronic Toxicity of Effluents and Receiving Waters to Freshwater Organisms (Lewis, P.A., Klemm, D.J., Lazorchak, J.M., Eds). US Environmental Protection Agency Environmental Monitoring Systems Laboratory, Cincinnati. Report EPA/600/4–91/002

USEPA, 1994b. Methods for Measuring the Toxicity and Bioaccumulation of Sediment-associated Contaminants with Freshwater Invertebrate. US Environmental Protection Agency. Office of Research and Development, Washington DC. Report 600/R-94–024.

USEPA, 1996. Proposed guidelines for ecological risk assessment. Federal Register 61(175), 47552–47631.

USEPA, 1997. Surface waters: Field operations and methods for measuring the ecological condition of wadeable streams. EPA/620/R–94/004F. Office of Research and Development, Washington, DC.

USEPA, 1998a. Condition of the Mid Atlantic Estuaries. Washington, DC. EPA/600/R-98/147.

USEPA, 1998b. National Water Quality Inventory: 1996 Report to Congress. Washington, DC. EPA/841/R-97/008.

USEPA, 1999. Rapid Bioassessment Protocols for Use in Streams and Wadeable Rivers: Periphyton, Benthic Macroinvertebrates and Fish, 2nd edn. Office of Water, US Environmental Protection Agency, Washington, DC. EPA 841-B-99–002.

USEPA, 2001. A Survey of Fish Contamination in Small Wadeable Streams in the Mid-Atlantic Region. Office of Research and Development, Washington DC. February. EPA/600/R-00/107.

Vos, P., Hogers, R., Bleeker, M.R., van de Lee, T., Hornes, M., Fritjers, A., Pot, J., Peleman, J., Kuiper, M., Zabeau, M., 1995. AFLP: a new technique for DNA fingerprinting. Nucleic Acids Research 23, 4407–4414.

Wallace, J.B., Grubaugh, J.W., Whiles, M.R., 1996. Biotic indices and stream ecosystem processes: results from an experimental study. Ecological Applications 6, 140–151.

Waller, W.T., Acevedo, M.F., Allen, H.J., Schwalm, F.U., 1996. University of North Texas, Institute of Applied Sciences, Denton. Technical Report No. 172. Texas Water Resources Institute, The Texas A&M University System College Station.

Waser, N.M., 1993. Population structure, optimal outcrossing, and assortative mating in angiosperms. In: Thornhill, N.W. (Ed.), The Natural History of Inbreeding and Outbreeding: Theoretical and Empirical Perspectives. University of Chicago Press, Chicago, pp. 1–13.

Weber, C.I. (Eds), 1973. Biological Field and Laboratory Methods for Measuring the Quality of Surface Waters and Effluents. US Environmental Protection Agency, National Enforcement Research Center, Cincinnati. EPA-670/4-73-001.

Webster, J.R., 1975. Analysis of potassium and calcium dynamics in ecosystems on three southern Appalachian watersheds of contrasting vegetation. Ph.D. Dissertation, University of Georgia, Athens.

Webster, J.R., D'Angelo, D.J., Peters, G.T., 1991. Nitrate and phosphate uptake in streams at Coweeta Hydrologic Laboratory. Verhandlungen der Internationalen Vereinengen Limnologie 24, 1681–1686.

Weisberg, S.B., Frithsen, J.B., Holland, A.F., Paul, J.F., Scott, K.J., Summers, J.K., Wilson, H.T., Heimbuck, D.G., Gerritsen, J., Schimmel, S.C., Latimer, R.W., 1993. Vrginian Province Demonstration Project Report, EMAP – Estuaries, 1990. EPA/620/R–93/006. USEPA, Office of Research and Development, Washington, DC.

Welsh, J., McClelland, M., 1990. Fingerprinting genomes using PCR with arbitrary primers. Nucleic Acids Research 18, 7213–7218.

Whittier, T.R., Paulsen, S.G., 1992. The surface waters component of the Environmental Monitoring and Assessment Program (EMAP): an overview. Journal of Aquatic Ecosystem Health 1, 119–126.

Whitton, B.A., Rott, E., Friedrich, G. (Eds), 1991. Use of Algae for Monitoring Rivers. Institut für Botanik, Universität Innsbruck, Innsbruck.

Williams, J.G., Kubelik, A.R., Livak, K.J., Rafalski, J.A., Tingey, S.V., 1990. DNA polymorphisms amplified by arbitrary primers are useful as genetic markers. Nucleic Acids Research 18, 6531–6535.

Yoder, C.O., Rankin, E.T., 1995. Biological criteria program development and implementation in Ohio. In: Davis, W.S., Simon, T.P. (Eds), Biological Assessment and Criteria: Tools for Water Resource Planning and Decision Making. Lewis Publishers, Boca Raton, pp. 109–144.

Yeardley Jr, Roger B. Lazorchak, J.M., Paulsen, S., 1998. Elemental fish tissue contamination in northeastern US lakes: evaluation of an approach to regional assessment. J. Envir. Tox. Chem. 17 (9), 1875–1884.

Bioindicators and biomonitors
B.A. Markert, A.M. Breure, H.G. Zechmeister, editors
© 2003 Elsevier Science Ltd. All rights reserved.   875

Chapter 24

IAEA approaches to assessment of chemical elements in atmosphere

Borut Smodiš

Abstract

The International Atomic Energy Agency (IAEA) was formed in 1957 as a specialised agency within the United Nations system for scientific and technical co-operation in the nuclear field. Its Statute authorises it, among others, "... to accelerate and enlarge the contribution of atomic energy to peace, health and prosperity throughout the world". Since its foundation, therefore, the IAEA has been supporting the use of nuclear analytical techniques such as neutron activation analysis, particle induced X-ray emission and X-ray fluorescence, for undertaking environmental monitoring and assessment activities ever since founded. Research work on trace element air pollution has been systematically supported since 1992. The activities supported include mainly (1) collection and analysis of coarse (2.5–10 μm diameter) and fine (<2.5 μm diameter) airborne particulate matter using standardized low-volume stacked filter air samplers and (2) trace element atmospheric pollution studies using plants as biomonitors. The IAEA has two main support tools for its Member States: the Research Contract Programme for encouraging research of interest to the IAEA's programme of work based on its mandate, and the Technical Co-operation Programme focused on solving technical problems by providing the necessary know-how upon countries' requests. Basic mechanisms and functions of both tools are presented. In supporting assessment of chemical elements in atmosphere the IAEA is implementing environmental metrology in practice: First, available environmental monitoring and assessment tools are adopted, upgraded and tested within co-ordinated research projects. Once tested, these tools are transferred to developing countries through technical co-operation projects. Projects on studying element atmospheric pollution, emphasising the one on "Validation and application of plants as biomonitors of trace element atmospheric pollution, analysed by nuclear and related techniques", are presented. In the framework of this project, the participants are harmonizing approaches for using plants for the assessment of atmospheric element deposition. Production and characterization of reference materials, and organization of quality control exercises form an important part of IAEA continuous support to its Member States in achieving high quality measurement results. Recent activities pertaining to the assessment of chemical elements in the atmosphere are presented. These activities include production of reference lichen material IAEA-336, and organisation of five proficiency testing and analytical quality control exercises involving air filters and plant materials, coded NAT–3 to NAT–7.

Keywords: airborne particulate matter, atmospheric pollution, bioindicator, biomonitoring, heavy metals, International Atomic Energy Agency, lichen, moss, nuclear analytical techniques, trace elements

1. Introduction

Air pollution has become a matter of global concern, particularly in developing countries, mainly due to various anthropogenic activities. The significant contribution of air pollution to the diminished health status of the exposed human populations, forest decline, loss of agricultural productivity, etc., has been a cause of increasing public concern on this issue throughout the world. Therefore, it has become a topic of intense scientific, governmental, and also industrial interest.

A wide array of air pollutants including particulates, liquids, and gases is being emitted both from natural and anthropogenic sources. The pollutants may include any natural or artificial composition of matter capable of being airborne. They may occur as solid particles, liquid droplets, gases, or in various admixtures of these forms. Natural sources include volcanic emissions, accidental fires in forests and on prairies, dust storms, soil particles, salt particles emitted from the oceans, and various products given off by plants. Anthropogenic sources of air pollution comprise emissions due to industrial activities (e.g. manufacturing products from raw materials, industries that convert products to other products, etc.), power generation (e.g. fossil fuel combustion), traffic, agriculture, waste incineration, residential heating, and many others. The main contaminants include sulphur dioxide, airborne particulate matter (APM), carbon monoxide, reactive hydrocarbon compounds, nitrogen oxides, and ozone. Among them, heavy metals and other toxic elements, mostly associated with APM, represent an important group to be considered. After emission, the pollutants are subjected to physical, chemical, and photochemical transformations, which ultimately decide their fate and atmospheric concentrations.

Increased urbanization and industrialization, and the rapid growth of transportation, in connection with population growth, have altered the nature of air quality problems and the focus of air pollution research during recent years. In addition to air pollution found around large, uncontrolled point sources, problems associated with regional-scale elevated levels of pollutants have been identified. Air pollutants do not remain confined near the source of emission, but spread over large distances, depending upon topography and meteorological conditions, especially wind direction and speed, and vertical and horizontal thermal gradients. Hence, air pollution issues are raising a high public interest also in remote areas.

There are two conceptual approaches to collecting samples relevant for air and atmospheric deposition related pollution studies: (1) the direct collection of APM, precipitation, and total deposit, and (2) the use of suitable air pollution biomonitors. The first approach is aimed at quantitative surveys of local, short-range, medium-range or global transport of pollutants, including human health-related studies when collecting size fractionated APM. It requires continuous sampling on a long-term basis at a large number of sites, in order to ensure the temporal and spatial representativeness of measurements. The application of such direct measurements on a large scale is extremely costly and person-power intensive. Furthermore, it is often not possible, due to logistic problems, to install instrumental equipment at all needed locations. Therefore, the use of plants or other biological organisms, either already present at the investigation points or to be brought into the system, may be more practical. This approach is considered as a non-expensive but yet reliable means of air quality status assessment in a country or a region. Certain types of biological organisms integrate pollution over time, reducing the need for continuous chemical monitoring, thus avoiding the difficulty of interpreting "snapshot" measures, and offering the potential of a retrospective monitoring. Many of them enrich the substance to be determined so that the analytical accessibility is improved and the measurement uncertainty reduced. By observing and measuring the changes in an appropriately selected organism a conclusion as to the kind of pollution (e.g. heavy metal), its source (e.g. from a smelter) and possibly its intensity (given as the relationship between degree of pollution and observable or measurable changes in the organism) can be drawn. In this context, biomonitoring may be defined as a continuous observation of a geographical area with the help of suitable organisms that reflect changes over space and time (e.g. by their elemental content). Numerous searches for appropriate species of detecting pollution have taken place, resulting in the use of properties of an organism or its part to obtain information on pollution levels in the air. The pioneer work conducted by Rühling and Tyler in the late 1960s (Rühling and Tyler, 1968; Tyler, 1970) offered the potential to trace the release and to track the movement of individual heavy metals over large areas. Using this method, point-source emitters could be detected and long-distance transport of individual metals proven. This procedure appeared to be cost-effective because it did not require installation of a sampler or measuring instrument at each measuring point and, consequently, no maintenance was required. Nowadays, the use of organisms, mainly plants, for assessing the levels of airborne pollution is widely accepted.

However, to be acceptable as a biomonitor of atmospheric pollution, an organism has to meet certain general and specific requirements (Sloof, 1993). It is also important that it should not take up appreciable amounts of elements from sources other than atmospheric and should average element content over a suitable time period as a result of integrated time exposure. The plant species chosen most frequently as receptors of atmospheric trace element pollution are (Pakeman et al., 1998; Markert, 1993; Sloof, 1993; Steinnes, 1989; Jervis and Qureshi, 1994; Lippo et al., 1995; Loppi et al., 1997; Wyttenbach et al., 1997): mosses, lichens, coniferous trees (bark, needles, shoots) and deciduous trees (bark, leaves). Mosses and lichens are considered as the most appropriate ones, because both (1) lack roots and therefore obtain most of their nutrient supply directly from precipitation and dry deposition of airborne particles with mostly insignificant risk of taking up metals from the substrate, (2) have no, or only a very reduced cuticle, so that ions retained on their surface have direct access to exchange sites on the cell walls.

1.1. Terminology used

Biological monitoring can be broadly defined as the measurement of the response of living organisms (often called bioindicators or biomarkers) to changes in their environment (Pakeman et al., 1998). The term "change" may, however, include changes in the chemical composition of biota, such as the accumulation of pollutants. A *bioindicator* is an organism that reveals the presence of a substance in its surroundings with observable and measurable changes, which can be distinguished from the effects of other natural or anthropogenic stresses. The term "bioindicator" is therefore usually reserved for an organism that can be used to obtain only *qualitative* information of its immediate environment, while the term *"biomonitor"* refers to an organism that can be used to obtain *quantitative* information (Markert, 1993). In both bioindication and biomonitoring, an organism responds to variations in the environmental conditions caused (e.g. by pollutant(s)) by changing its mode of life in the morphological and/or metabolic-physiological way. Such changes may be monitored by: (i) observation or (ii) measurement (e.g. by chemical determination of trace elements). *Biomonitoring* is the continuous observation of an area with the help of bioindicators (e.g. by repeated measurement of their responses in a manner which reveals changes over space and time), which are in this case called biomonitors. *Passive* bioindicators are organisms that already exist in the area under investigation, and *active* bioindicators are organisms that are brought into the area under investigation and exposed for a defined period of time (Pakeman et al., 1998; Markert, 1993; Markert, 1996).

1.2. Multielement analytical techniques

For the assessment of element pollution levels and identification of their sources, which are a prerequisite for studying effects of contaminants on the environment and human health, a multivariate data base containing as many pollutant elements should be generated. Therefore, multielement methods are usually used for such studies. The analysis of samples for their element content is governed by the sample type, the elements of interest, the sensitivity, precision and accuracy needed and the availability of (or access to) the technique. The choice of multielement methods available includes inductively coupled plasma atomic emission spectrometry (ICPAES), inductively coupled plasma mass spectrometry (ICPMS), X-ray fluorescence spectrometry (XRF), ion beam analysis (IBA) [i.e. particle induced X-ray emission (PIXE) and proton induced gamma-ray emission (PIGE)], nuclear activation analysis [neutron activation analysis (NAA), prompt gamma neutron activation analysis (PGNAA), charged particle activation analysis (CPAA)], and several other methods, which are seldom used on a routine basis. Some of these methods can be complemented by the use of single-element techniques such as anodic stripping voltammetry (ASV) or atomic absorption spectrometry (AAS).

The ability of nuclear and related analytical techniques (NAA, IBA, and XRF) to analyze solid phase samples for many elements without the need for sample dissolution or digestion with the high degree of sensitivity and selectivity, makes them particularly suitable for the elemental analysis of biomonitor samples. This fact brings the issue of biomonitoring trace element atmospheric deposition within the scope of programmes of the International Atomic Energy Agency (IAEA).

1.3. The role of nuclear and nuclear related analytical techniques

For handling solid samples, nuclear activation analysis, XRF and PIXE are the most appropriate as they are nondestructive methods, so the probability of contaminating a sample with the element to be determined is negligible, and eventual losses during the sample destruction and preparation are excluded. Among the three methods, NAA in its instrumental (INAA) or radiochemical mode is often used as one of the analytical methods in the process of certifying new reference materials. This is due to its good accuracy, selectivity and sensitivity for many elements.

Environmental samples such as soils, airborne particulate matter (APM) or plant materials, often contain an insoluble component, e.g. a mineral fraction, which could be only partially soluble in acids routinely used for sample destruction applied in many laboratories. Therefore, the accuracy and comparability of measurements obtained using different procedures and/or methods for determination of total elemental content should be validated and taken into account. It has been shown, for instance, that several elements in National Institute of Standards and Technology (NIST) leaf standard reference materials (SRMs) are only partially soluble in $HNO_3/HClO_4$ (Ihnat, 1982), that about 20% of the uranium in NIST SRMs 1571, 1575, and 1572 (Orchard Leaves, Pine Needles, and Citrus Leaves, respectively) resides in a phase that is not made soluble by conventional wet-ashing procedures (Byrne and Benedik, 1988), and that even in NIST SRMs of the new generation such as 1515 Apple Leaves and 1547 Peach Leaves, which were prepared using dedicated equipment and modern technology, several per cent of the total element content is contained in the mineral grit phase (Lindstrom et al., 1990). It was also shown (Greenberg et al., 1990) that incomplete solubilization can occur even when well-accepted, vigorous regimens are used. So, the analysis of any botanical sample (this is also valid for soil and APM material) by solution-based analytical techniques for the total element content must ensure that the mineral as well as the bulk material are dissolved.

Methods intercomparisons using different multielement analytical techniques (Quevauviller et al., 1993; Steinnes et al., 1993; Frontasyeva et al., 1995; Quevauviller et al., 1996; Steinnes et al., 1997) confirm the conclusions drawn above. The quality of destructive analysis of biomonitors is influenced by specific matrix effects that may not be matched by the matrix of other plant reference materials. The level of the mineral fraction contained in a biomonitor may depend on the level of pollution, so the recovery of an element during the dissolution may not be the same for the same type of material analysed. One should also be careful when using reference materials or samples, since in some cases the values are reported on an acid-soluble basis, so misleading conclusions may be drawn by their analysis (Steinnes et al., 1997).

These facts and problems encountered by using destructive methods make nuclear and nuclear-related analytical techniques (NATs) indispensable in laboratory intercomparisons to improve the quality of trace element determinations in environmental materials. NATs also proved to be a highly effective tool for the analysis of various materials used in studying trace element air pollution, including plants and other organisms, by providing multi-element measurement results. NAA has been extensively used for such studies and numerous references and review papers can be found in the literature (e.g. Steinnes, 1980; Sloof et al., 1993; Sloof and Wolterbeek, 1993; Svetina Gros et al., 1993; Jeran et al., 1993; Steinnes et al., 1994; Kuik and Wolterbeek, 1995; Freitas, 1995; Ribeiro Guevara et al., Jeran et al., 1996; Svetina Gros et al., 1996; Jeran et al., 1996a). The k_0-based INAA (Smodiš, 1992) has been found to be a particularly useful tool, offering the possibility of determining all the radionuclides appearing in the accumulated gamma spectra of the neutron-irradiated sample. It has also been shown (e.g. Stropnik et al., 1993; Takala et al., 1994) that relatively simple

Energy Dispersive X-ray Fluorescence Spectrometry (EDXRF) using radionuclide excitation sources can be efficiently used in air pollution research. In addition, the samples analysed by this technique can subsequently be submitted to analysis by other techniques, which is not always the case when using other methods.

2. International Atomic Energy Agency

The International Atomic Energy Agency (IAEA) was formed in 1957 as a specialized agency within the United Nations system. It serves as the world's central inter-governmental forum for scientific and technical cooperation in the nuclear field. Organizationally, it comprises six departments: Department of Nuclear Sciences and Applications, Department of Safeguards, Department of Technical Cooperation, Department of Nuclear Energy, Department of Nuclear Safety, and Department of Management. More than 2200 staff members in the Secretariat coming from many of its 130 Member States (as of the end of 1999) supervise projects within six major programmes: 1. Nuclear Power and Fuel Cycle, 2. Nuclear Applications, 3. Nuclear, Radiation and Waste Safety, 4. Nuclear Verification and Security of Material, 5. Management of Technical Cooperation for Development, and 6. Policy-making, Coordination and Support.

The opening articles of the Statute define the IAEA objectives, among others:

"The Agency shall seek to accelerate and enlarge the contribution of atomic energy to peace, health and prosperity throughout the world."

and its functions, among others:

"The Agency is authorized:

To encourage and assist research on, and development and practical application of, atomic energy for peaceful uses throughout the world; and, if requested to do so, to act as an intermediary for the purposes of securing the performance of services or the supplying of materials, equipment, or facilities by one member of the Agency for another; and to perform any operation or service useful in research on, or development or practical application of, atomic energy for peaceful purposes;

To foster the exchange of scientific and technical information on peaceful uses of atomic energy;

To encourage the exchange of training of scientists and experts in the field of peaceful uses of atomic energy;

Allocate its resources in such a manner as to secure efficient utilization and the greatest possible general benefit in all areas of the world, bearing in mind the special needs of the under- developed areas of the world.

The IAEA has been undertaking environmental monitoring and protection activities ever since founded in 1957. In accordance with its above statutory mandates, the Major Programmes for Nuclear Applications and Management of Technical

Cooperation for Development aim to further enhance the contribution of nuclear applications for peaceful purposes in order to contribute to sustainable development through, inter alia, the protection of marine and terrestrial environments.

2.1. IAEA's support mechanisms

2.1.1. The research contract programme

The Research Contract Programme (RCP) has been designed to fulfil Article III of the IAEA Statute which authorizes the Agency to encourage and assist research on, and development and practical application of, atomic energy for peaceful purposes throughout the world and to foster the exchange of scientific and technical information, as well as the exchange of scientists in the field of peaceful uses of atomic energy. Therefore, the RCP is one of the scientific arms of the IAEA. Its primary function is to support the research of interest to the IAEA's programme of work and its principle aim is to obtain scientific results in an effective manner. However, the Agency does not support purely basic research (which aims to improve knowledge *per se* rather than generate results or technologies which are likely to benefit society in the foreseeable future) unless this can provide a stimulus or support for more "downstream" or end-beneficiary oriented research. Research Contract activities provide opportunities for scientists and institutions in Member States to conduct the more "upstream" types of research in which strategic and applied issues with their subsequent opportunities for spillovers drive the research agenda. Technical Co-operation, on the other hand, focuses on the transfer of well-tried and tested technologies which nevertheless often require significant adaptive research in the process of adoption by end-users or end-beneficiaries, as well as the involvement of an often complex array of stakeholders.

Research efforts supported by the Agency are normally carried out within the framework of Co-ordinated Research Projects (CRPs). These CRPs are developed in relation to well defined research topics on which an appropriate number of institutes are invited to collaborate and represent an effective means of bringing together researchers in both developing and industrialized countries to solve a problem of common interest. The research supported by the Agency relates to the tasks included in the approved Programme, with the Agency supporting and co-ordinating the research through research contracts and cost-free research agreements and Research Co-ordination Meetings (RCMs). The funding provided by the IAEA usually represents "seed money" which often invites additional funding for participating institutes in Member States. Thus, a relatively small IAEA contribution can attract much larger funding for a project than the initial amount provided by the IAEA.

Once a CRP is formed, research teams from an average of ten to twelve institutes are normally selected for participation in the programme, which lasts three to five years. Regardless of the field of research, each CRP involves certain common aspects. Institutes in developed countries work in close co-ordination with those in developing countries and all participants are encouraged to conduct work which will produce new research results and to apply these results to needs in their countries. Co-operation between institutes is strongly encouraged and supported by RCMs that are financed by the IAEA for contractors and agreement holders and held at appropriate intervals, usually at intervals of around 18 months, for each CRP. At these meetings, the progress of the CRP is reviewed in detail and the future direction of work is established. At the final RCM, results are reviewed and evaluated by all participants and, where appropriate, recommendations are made for future work in the field.

The results of research supported under the RCP are shared by all Member States, and are disseminated through national, international and Agency scientific and technical publications and other communications media. Where feasible, the knowledge gained is also used to enhance the quality of projects delivered to Member States through the Agency's Technical Co-operation Programme.

The present programme (1999 statistics) provides funding of about US $7.5 million, which is used to support approximately 160 CRPs in many different subject areas, comprising about 2000 individual research contracts and research agreements. Research on trace element atmospheric pollution form only a small section of the IAEA's total research programme, but nevertheless encompass four recent and ongoing CRPs and annual funding of approximately $200,000.

2.1.2. The technical co-operation programme

The purpose of the Technical Co-operation Programme is to accelerate and enlarge the contribution of atomic energy to peace, health and prosperity throughout the IAEA Member States. For achieving this goal, the Programme is based on assessment of the development priorities and conditions in each specific country or region, and the project requests received from Member States. Within the framework of its Technical Co-operation Programme (TCP), the IAEA assists developing Member States in solving technical rather than scientific problems, at the same time providing the necessary know-how. Technical Co-operation (TC) projects are therefore carried out with a view to increasing the recipient institution's self-reliance. Projects may comprise one or more of the following components: experts, provision of equipment and materials, fellowships and scientific visits, training courses, meetings, and workshops. The projects provide expertise through scientists who are already well versed in particular techniques (experts) and who visit institutions that are carrying out such projects in order to train, advise or otherwise assist local scientists in conducting the work plan of the project.

By these means, TC projects aim at building up the research capability of institutions in developing countries to the point where scientists are able to conduct such work more-or-less independently and without further substantial outside funding when the project terminates.

The TCP comprises national projects, regional and interregional projects. A national project derives from a formal request for co-operation by a Member State. For administrative purposes, Member States engaged in technical cooperation with the IAEA are grouped in geographical areas or regions. Regional projects are (1) those proposed by the IAEA or a group of Member States in response to the expressed needs of several Member States of a region; and (2) those proposed by Member States of a region collaborating within a Regional Co-operative Agreement. These projects seek to achieve greater implementation efficiency or to improve effectiveness by consolidating common objectives and modalities between multiple Member States. Such projects employ greater use of group activities, distance learning materials, technical networking, integrated expert missions, technical co-operation between developing countries and in-kind contributions from host facilities. Interregional projects are established by the IAEA to serve the common needs of several Member States in different geographical regions. They use similar mechanisms as those used for regional projects.

Over the past four decades, the IAEA has planned and delivered projects worth more than US $800 million in many fields, including energy, safety, agriculture, industry, medicine, water and environmental studies. In its early stage, the TCP was mostly aimed at **building up scientific and technical capacity** and the supporting infrastructure. In that period, many countries were keen on establishing the foundation for using the new technology in a variety of areas, both power and non-power. At the same time, most countries faced a major constraint: the lack of infrastructure, particularly the lack of human resources trained in the field of peaceful uses of nuclear energy. Therefore, a key during this first phase was the emphasis on human resources development.

In the early 1990s, after decades of capacity building, the TCP became less supply-oriented and more demand-driven, focused on using built capacity in meeting needs for sustainable development. The core concept of this approach was embodied in the motto *"Partners in Development"*. Because the new strategy was meant to guide the transition from a programme of technology-driven assistance to one aimed at problem solving, new partnerships had to be forged with actual "problem-holders". These are the end-users of nuclear technologies, the main link in the chain connecting national counterparts with the ultimate beneficiaries, the public at large. In the second half of the 1990s a great deal of effort went into helping the nuclear research establishment reach out to mainstream development ministries such as health, agriculture, natural resources and environment. The focus on problem solving led logically to the need to collaborate with other parties, such as development financing and development co-operation organizations, working on the same problems from their own perspective. A third type of partnership is between the more advanced developing countries having know-how and establishments that equal those of developed ones, and the least developed countries.

On the turn of the centuries, the TCP is oriented towards **targeted programming**, where projects are results-based and focused on specific problems. Due to funding constraints, the emphasis is on establishing priorities. TC projects are being linked to existing programmes of national investments in sustainable development. Evidence-based priority setting is becoming a key means of targeting programming by finding good opportunities and avoiding dead-ends, providing that TC projects deliver real impact. Priority setting is widely based on forty years of experience; where this experience is not available, the priority-setting process is based on careful feasibility and cost-effectiveness analyses under local conditions. The TCP is now on the way of playing more of a facilitating and monitoring role, as well as strengthening the management of nuclear technology for development.

3. IAEA support to applied research on air pollution

3.1. Current and recent IAEA projects on air pollution

The IAEA has been systematically supporting work on air pollution since 1992 through various co-ordinated research and technical co-operation projects (Table 1). Since the most serious and frequently occurring air pollution problems in many of the world's largest cities are thought to be associated with airborne particulate matter (APM), the support has been focused on APM. Scientific interest in possible health effects has, until recently, mainly centred on particles in the size range below 10 _m (so-called PM_{10}) which, because of their small size, are able to penetrate deep into the lungs of people exposed to them. Studies designed to disaggregate the effects of different particle sizes show that the strongest relationships are with the "fine" particles smaller than 2.5 _m diameter (so-called $PM_{2.5}$). This is hardly surprising since, not only do *coarse particles* (2.5–10 μm effective aerodynamic diameter) and *fine particles* (<2.5 μm) differ in their deposition characteristics in the lungs (coarse particles generally cannot penetrate further than the trachea and primary bronchi whereas fine particles can penetrate all the way to the alveoli), but they also generally come from different sources. Coarse particles mainly comprise windblown dust from agricultural fields, unpaved roads and, in some regions, from nearby deserts, volcanoes and oceans. Fine particles, on the other hand, mainly come from anthropogenic activities, i.e. from industrial and residential combustion, from vehicle exhausts, and from chemical transformations acting on sulphur dioxide, nitrogen oxides and volatile organic compounds emitted from combustion activities.

When these IAEA projects were started in 1992, the importance of being able to distinguish between coarse and fine particles was not fully realized. Therefore, the main objectives of this work have been to obtain comparative data on coarse and fine particle levels in the air in areas of high pollution (e.g. a city center or a populated area downwind of a large pollution source) and low pollution (e.g. a rural area), and to identify major sources of air pollution affecting each of the participating countries.

There were two CRPs on applied research on air pollution using NATs: a global one with 19 participants (1992–1997), and a regional one for the participants from East Asia and Pacific region (RCA) with 13 participants (1995–1998). The latter project was supported by the United Nations Development Programme (UNDP). There were two important common themes to this work: (1) a common design of air sampler, and (2) the common use of nuclear analytical techniques.

An effort was made in setting-up an appropriate *environmental metrology* tools by selecting the standard air sampler which characteristics would make it well suited for the purpose. A so-called Ghent stacked filter unit, using the principle of sequential filtration has been used for collecting APM (Maenhaut et al., 1994). A schematic diagram of the sampler as typically deployed in the field including the pump, flow control system and rain shield, is shown in Fig. 1. This sampler, designed at the University of Ghent, Belgium, is now being used by all participants in the IAEA's CRPs on air pollution research and related projects. The sampler uses an "open face" type stacked filter unit, in which two polycarbonate filters (one filter of 8 μm pore size and the other of 0.4 μm pore size) are employed for the collection of APM. The filter

Table 1. Recent and current major IAEA projects on studying element atmospheric pollution.

Dates	Title	Participating countries
1992–1997	Applied Research on Air Pollution Using Nuclear-Related Analytical Techniques (CRP)	Argentina, Australia, Bangladesh, Belgium, Brazil, Chile, China, Czech Republic, Hungary, India, Iran, Jamaica, Kenya, Paraguay, Portugal, Slovenia, Turkey, Thailand, USA
1995–1998	Applied Research on Air Pollution Using Nuclear Related Analytical Techniques in the Asia and Pacific Region (CRP)	Bangladesh, China, Indonesia, Korea, Malaysia, Mongolia, Myanmar, Pakistan, Philippines, Singapore, Sri Lanka, Thailand and Vietnam
1996–2000	Assessment of Levels and Health-Effects of Airborne Particulate Matter in Mining, Metal Refining and Metal Working Industries Using Nuclear and Related Analytical Techniques (CRP)	Brazil, China, Czech Republic, Denmark, India, Indonesia, Kenya, Portugal, Russia, South Africa, Slovenia
1997–2002	Validation and application of plants as biomonitors of trace element atmospheric pollution, analysed by nuclear and related techniques (CRP)	Argentina, Brazil, Chile, China, Germany, Ghana, India, Israel, Jamaica, The Netherlands, Norway, Portugal, Romania, Russia
1998–2002	Air pollution and its trends (regional East Asia and the Pacific TC project)	Australia, Bangladesh, China, India, Indonesia, Korea, Malaysia, Myanmar, New Zealand, Pakistan, Philippines, Singapore, Sri Lanka, Thailand, Vietnam
1999–2001	Determining the content of atmospheric contamination (regional Latin America TC project)	Argentina, Brazil, Chile, Mexico

unit is inserted in a cylindrical container, which is provided with a pre-impaction plate for the collection of particles larger than 10 μm. The sampler is designed to operate at a flow rate of 16 L/min and collects the coarse and the fine particles. Comparative tests of the Ghent sampler against other commonly used commercial samplers show good agreement for PM_{10} and a good correlation, but slightly lower collection efficiency, for $PM_{2.5}$ (Hopke et al., 1998).

A variety of analytical techniques are being used for the analysis of the air filters. However, the main emphasis in the IAEA's programmes is on nuclear and related techniques. These techniques have characteristics that make them highly suitable for conducting non-destructive multi-element analyses of airborne particulate matter on filters. The usefulness of these techniques lies not only in their ability to determine individual elements but also to identify specific *sources* of pollution. This may be done by using groups of elements as "fingerprints" for different pollution sources (motor vehicles, industry, soil, etc.) and then by employing chemometric data evaluation methods (factor analysis, chemical mass balance, etc.) to quantify each specific fingerprint.

A relatively new feature of the IAEA's projects on APM pollution studies is to have central data evaluation whereby one expert is collecting data from all of the IAEA's projects participants, and evaluating them in a way that permits a single consistent style of presentation.

Metrological approach and procedures developed within the framework of the CRPs shown in Table 1 have been successfully applied in several national and regional

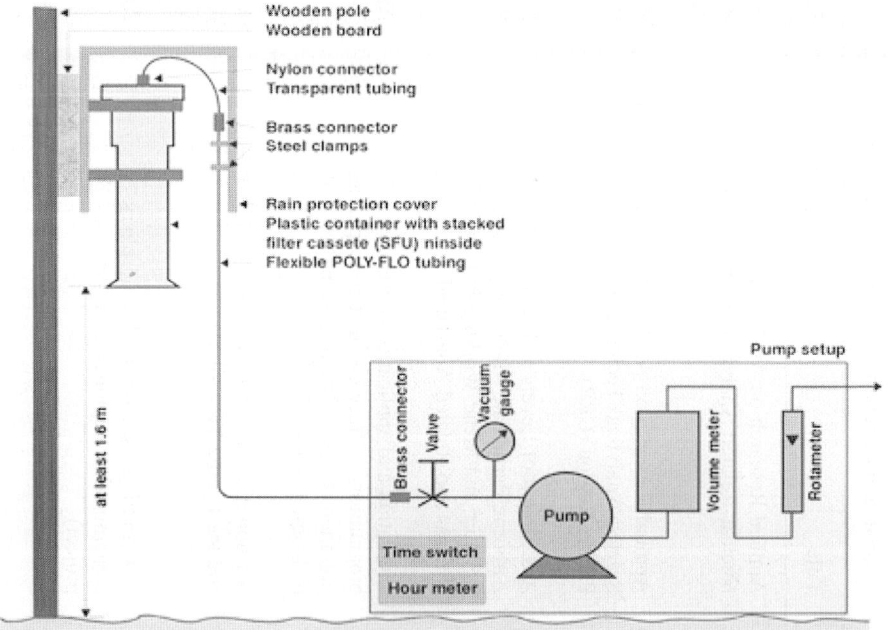

Figure 1.　　Schematic diagram of the Ghent sampler.

Table 2. National TC projects on trace element atmospheric pollution during the period 1997–2000.

Title	Participating country
Monitoring of trace element air pollution	Portugal
Diagnosis and monitoring of environmental pollution	Ecuador
Nuclear techniques in monitoring industrial pollution	Sri Lanka
Air pollutant characterization and source identification	Philippines
Trace element methods for workplace monitoring	Nigeria
Evaluation of air pollution and its effects on health	Uruguay
Nuclear analytical measurements of size fractionated APM	China

Technical Co-operation Projects. Two regional TC projects are shown in Table 1 and national TC projects are shown in Table 2.

The project on "Air pollution and its trends", carried out within the Regional Co-operative Agreement (RCA), is supported by the United Nations Development Programme, UNDP. Participants are assessing and comparing air pollution levels in strategically chosen areas within East Asia and the Pacific region, identifying and quantifying the critical air pollution sources, and collecting data for future work on transboundary movement of APM. One of important outputs of this project is preparedness for identifying regional air pollution events such as haze due to uncontrolled biomass burning. All the participants are using the same type of Ghent sampler, analytical results are centrally evaluated, processed and modeled. The use of non-destructive multielement nuclear analytical techniques by all the participants enables fingerprinting of pollution sources, thus giving possibility for their identification and apportionment.

The project on "Determining the content of atmospheric contamination", carried out within the Regional Co-operative Arrangements for the Promotion of Nuclear Science and Technology in Latin America (ARCAL), is focused on "mega dirty mega cities" in Latin America. Participants are identifying the toxic heavy metals levels in the air

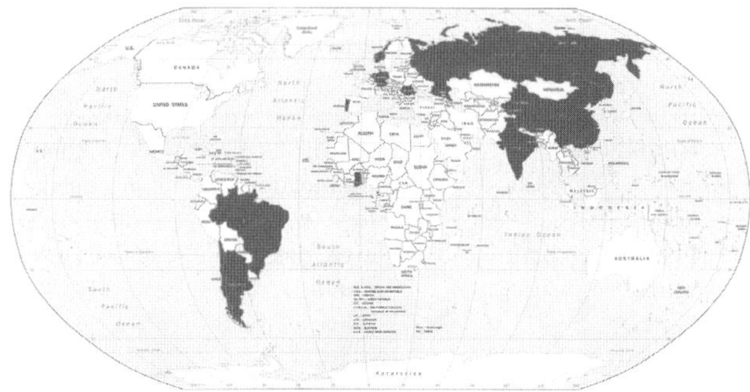

Figure 2. Countries operating the Ghent sampler with IAEA's assistance.

and characterising APM collected in the four countries capitals during critical (cold, dry, hot, humid) periods, using the same harmonized methodology as in the above-mentioned project.

Since 1992, more than 80 Ghent samplers have been delivered to more than 40 countries, mostly with the IAEA support, thus forming a comprehensive network for collecting APM data (Fig. 2).

3.2. The harmonized approach to using plants as biomonitors of trace element atmospheric pollution

In the two IAEA CRPs on applied research on air pollution using nuclear-related analytical techniques (1992–1998), participants from 29 countries had focused mainly on harmonized collection on airborne particulate matter. However, participants from six countries had included the collection and analysis of various bioindicators as a supplementary part of their studies, using both passive and active monitoring around point sources as well as larger areas (Table 3). As evident from Table 3, NAA was the main analytical tool used.

During a TC project on monitoring of trace element air pollution carried out in Portugal, the idea arose of organizing a workshop on biomonitoring atmospheric pollution. Indeed in September 1997, the International Workshop on Biomonitoring of Atmospheric Pollution (with emphasis on trace elements), BioMAP, was organized in Lisbon, Portugal, by Instituto Tecnológico e Nuclear (ITN), Sacavém, Portugal, in co-operation with the IAEA. Altogether 54 participants presented 35 contributions with results from 16 different countries (Table 4). The workshop re-emphasized the importance of using biomonitors for estimating atmospheric trace element pollution. Although there are still many problems to be studied and solved, it is hardly possible by any other approach to obtain a detailed picture of variations in time and space within reasonable limits of costs. The problems addressed most frequently included sample preparation, interspecies calibration, and the establishment of the relationship between atmospheric deposition of a pollutant and its corresponding content in the biomonitor. From the papers presented and the discussions held, it was evident that the most appropriate organisms for biomonitoring atmospheric pollution are lichens and mosses (IAEA, 2000). It was also clear that nuclear and related analytical techniques are

Table 3. IAEA member states that have participated in relevant CRPs during the period 1992–1998, using bioindication/biomonitoring.

Country	Bioindicator	Type of bioindication	Analytical method used
Argentina	lichen, tree bark	active, passive	NAA
Bangladesh	moss	passive	NAA, PIXE
Jamaica	lichen	passive	NAA, XRF
Portugal	lichen	passive	NAA, PIXE
Slovenia	lichen	passive	NAA
Vietnam	lichen, moss	passive	NAA

Table 4. Countries for which bioindication/biomonitoring results were reported at the BioMAP workshop.

Country	Bioindicator	Type of bioindication	Analytical method used
Argentina	lichen, tree bark	active, passive	NAA
Brazil	lichen, wood	passive	NAA, XRF
Canada	lichen	passive	ICPAES, ICPMS
Czech Rep.	deciduous tree bark	passive	NAA, XRF
France	moss	passive	ICPMS, NAA
Germany	coniferous tree, bark, lichen, moss, standard grass culture	active, passive	ICPAES, ICPMS
Italy	moss	passive	gamma-spectrometry
Israel	lichen	active, passive	ICPMS
Norway	coniferous tree bark	passive	ICPAES, ICPMS
Poland	coniferous tree bark	passive	ICPAES, ICPMS
Portugal	lichen, moss tree rings	active, passive	AAS, NAA, PIXE, *SXRF
Russian Fed.	coniferous tree bark, lichen	passive	ICPAES, ICPMS
Slovenia	lichen	active	NAA
Sri Lanka	lichen	passive	ICPAES, ICPMS
Yugoslavia	lichen	passive	NAA

*SXRF: synchrotron radiation X-ray fluorescence spectrometry

playing an important role in these studies. In more than half of presentations, NATs were used for obtaining analytical results.

Encouraged by the large number of participants and increased interest for biomonitoring in its Member States, the IAEA again co-operated in organizing the 2nd International Workshop on Biomonitoring of Atmospheric Pollution (BioMAP), held in Praia da Vitória, Azores islands, Portugal, 2000. A number of high quality contributions were presented and the main conclusions of the workshop was that in future years more emphasis and research efforts should be focused to linking atmospheric deposition data with the health-related information available. Many good quality data have been collected in air pollution biomonitoring surveys, which now should be compared with epidemiological surveys in order better to understand sources and causes of pollution-related diseases.

Stemming from the positive experience from the previous CRPs on air pollution and the results presented at the 1st International Workshop, the IAEA decided in 1998 to start a new CRP on "Validation and application of plants as biomonitors of trace element atmospheric pollution analysed by nuclear and related techniques" (Table 5). Additional evidence to support research work on such metrology practice was (1) the positive experience in using biomonitors, especially lower plants such as mosses and lichens in several developed countries (Markert, 1993, Rühling and Steinnes, 1998),

Table 5.   Countries participating in the CRP on Validation and application of plants as bio-monitors of trace element atmospheric pollution, analyzed by nuclear and related techniques.

Country	Bioindicator used	Type of bioindication	Analytical method used
Argentina	lichens	passive, active	NAA
Brazil	lichens	passive	NAA
Chile	lichens	passive	NAA
China	lichens, leaves	passive	NAA
Germany	mosses	passive, active	ICPAES, ICPMS
Ghana	lichens, leaves	passive	NAA, XRF
India	mosses, shrubs	passive	ICPMS, NAA, PIXE
Israel	lichens	active	ICPMS
Jamaica	lichens, mosses, tillandsia	passive	NAA, XRF
The Netherlands	lichens, mosses, bark	passive	NAA
Norway	mosses, pits	passive	ICPMS
Portugal	lichens, bark	active	NAA, PIXE
Romania	mosses	passive	AAS, NAA
Russian Fed.	mosses, pits	passive	AAS, NAA

(2) the continuous use of biomonitoring in several Member States (Freitas et al., 1997; Wolterbeek and Bode, 1995; Jeran et al., 1996), and (3) the fact that nuclear and related analytical techniques have been shown to be particularly appropriate for the analysis of air pollution biomonitors. The countries, participating in this CRP, are listed in Table 5.

The specific objective of the CRP is twofold: (1) to identify suitable biomonitors of atmospheric pollution for local and/or regional application (e.g. moss and/or lichen), and (2) to validate, whenever possible, these bioindicators for general environmental monitoring. In order to achieve this objective, the participants are:

- developing sampling design and guidelines for sample collection, sample preparation and analysis, including quality control;
- collecting, preparing and analysing samples in accordance with the guidelines;
- carrying out statistical analysis and creating graphical plots showing the geographical distributions of the elements, the levels of selected environmental pollutants, as well as identifying pollution sources and collecting the data on time trends;
- studying calibration of elemental content in biomonitoring species against absolute data for bulk deposition or air concentrations, whichever approach, or against another biomonitoring species already validated.

Stemming from the evidenced use of biomonitors (Markert, 1993, Rühling et al., 1996), the participants from 14 countries are concerned with searching for appropriate organisms in geographically and climatically very diverse parts of the world. Based on expertise already available and general requirements needed to be met by a bio-indicator, as already presented in the introductory part of this paper, mosses and lichens are the most frequently used species. After carrying out national or regional biomonitoring surveys, the participants are creating graphical plots reflecting the geographical distributions of the deposited toxic and other elements within the investigated areas, establishing pollution time trends, and applying statistical tools for identifying pollution sources. The emphasis is on the development and adoption of appropriate ***environmental metrology in practice***. The participants are therefore harmonizing the procedures used for surveys and quantification of the biomonitors applied, i.e. calibrating elementa content in the chosen biomonitoring species against the data for bulk deposition, air concentrations, or another biomonitoring species already validated.

The main expected outputs of the CRP are:

- established procedures for conducting environmental monitoring campaigns pertained to air pollution using biomonitors, thereby providing national governments with some of the capabilities they need in order to implement national and regional programmes of sustainable development;
- improved ability to conduct environmental measurements in conformity with modern quality management concepts;
- reliable information obtained on potential pollutants and their sources;
- appropriately presented data, over space and time, on trace elements of environmental and health concern such as As, Cu, Ni, Sb, Se, V, Cd, Cr, Hg, Pb.

There have been two RCMs held for the project: In September 1998 and in March 2000, both in Vienna. During these meetings, progress was reviewed and plans for future investigations were presented. Particular emphasis was on harmonizing research approaches and biomonitoring techniques. The report of the first meeting is available from the IAEA (IAEA, 1999).

During the first RCM the participants focused their attention on general strategies in carrying out biomonitoring surveys and appropriate procedures for sample collection and preparation. It was agreed that there should be a means to assess the quality of a survey. This may be measured by analysis of the signal-to-noise ratio, which can be determined by the ratio between the survey and local variance (Wolterbeek et al., 1996). This approach implies that the quality can hardly be determined before the start-up of the survey. However, knowledge of the local variance may lead to an increased survey signal-to-noise ratio. Therefore, attention should not only be focused on analytical variances, but especially the total local variances, including biological ones. It was presented that a high bioaccumulation factor, which may be translated into a better analytical precision, has only limited value as a criterion for the selection of the biomonitor species, and consequently, attention should be focused on other than analytical uncertainties. Sample preparation was raised as a very important issue.

Namely the proportion of airborne particulate matter (APM) in the total atmospheric deposit varies substantially with the geographic distribution of sampling sites, and dust should be removed from the biological material prior to analysis, raising the probability of unwanted leaching of certain elements. Therefore, it was recommended that short rinsing (for several seconds) with distilled water, rather than washing, should be applied, if necessary. In the case of visible contamination with APM (e.g. in deserts or in places with very low precipitation rate), washing by dipping the material into the distilled water consecutively until it remains clear was suggested. For collecting biomonitors, procedures similar as applied by Nordic environmental specimen banking were recommended (Giege et al., 1995).

The participants agreed to focus their research work on the following scientific aspects:

Quantification: Assessment of a quantitative relationship between the elemental content of the biomonitor species and the (wet or bulk) deposition or the atmospheric concentrations (quantification aspects).

Time resolution: Assessment of the element accumulation rate in the biomonitor to permit estimates of the time needed for the monitor to reflect new elementa atmospheric / deposition conditions (time resolution aspects).

Geographical resolution: Resolution-strength of the biomonitor in a spatial sense involving local variability in biomonitor responses, survey design (grid density) and spatial variability in deposition and atmospheric concentrations that should be considered simultaneously (geographical resolution aspects).

Survey: Assessment of geographical differences and/or time trends in deposition and/or atmospheric concentrations by the determination of the elemental content of the sampled biomonitor species (biomonitoring survey aspects).

Mapping: Graphical presentation of the biomonitoring results, when local and/or regional surveys are carried out (mapping).

Impact: Assessment of the changes in biomonitor parameters as a result of ambient and/or internal conditions. This means that selected physiological/biochemical parameters are to be quantified in relationship with varying extent of deposition and/or atmospheric concentrations. Furthermore, changes in selected parameter values are studied in order to get insight in the consequences: changed values may lead to changed relationships between monitor and deposition and/or atmospheric concentrations (impact aspects).

Within the lifetime of CRP the participants had prepared: (1) written protocols on sample collection procedures, (2) written protocols on sample preparation procedures, and (3) written protocols on sample analysis. All the protocols were prepared on the basis of guidelines agreed upon and adopted during the first RCM (IAEA, 1999). The following sample preparation procedure for lichens and mosses, for instance, was proposed: cleaning, separation of samples from substrate (lichens), rinsing in distilled water for 5–10 seconds, drying at 40°C for 48 hours or freeze-drying, grinding using liquid nitrogen if available. The participants had also agreed upon harmonised guidelines for the selection of sampling sites and types of samples to be collected, sampling techniques and equipment to be used, data processing and interpretation, as well as quality assurance and quality control procedures in all phases of the studies concerned.

All the participants have tested at least one biomonitor species with respect to sample collection, preparation and analysis. Many participants have already carried out monitoring surveys and presented maps showing geographical distribution of selected toxic elements.

One of the conclusions to-date is an experience that due to diverse meteorological and other environmental conditions in different parts of the world (e.g. temperature, amount of dust in air, amount of precipitation, etc.), the possibility of harmonizing operating procedures is limited. If the full potential of biomonitoring approach is to

Figure 3.   Tillandsia recurvata: composite ball aged 1–3 years.

be utilized, the procedures optimised to the local or regional conditions are to be applied.

Another important finding is that several epiphytic species from the genus *Tillandsia* (family Bromeliaceae) have a potential of becoming a regional biomonitor for tropical and subtropical parts of Americas (Fig. 3). Studies of its applicability are presently carried out in several Latin American and Caribbean countries.

Validation of the selected biomonitors against bulk deposition or air concentration of the elements studied still remains to be done in most of the participating countries. At this stage of the project implementation, it is not yet clear to what extent such validation is feasible for all experimental situations encountered in the programme.

3.3. Analytical quality control and proficiency testing

For most of the past 40 years the IAEA has been supporting programs aimed at improving analytical quality assurance (QA) and quality control (QC) in its Member States. Nuclear analytical techniques are well established as important tools in a wide variety of different kinds of environmental studies. One of their particular strengths is in analytical quality assurance, including the validation of other analytical methods and the development of new analytical reference materials. These methodologies are therefore helping to harmonize the analytical data produced. This topic was discussed in an IAEA symposium on "Harmonisation of health related environmental measurements using nuclear and isotopic techniques" (IAEA, 1997). It was concluded that the methods considered are playing an important role in the application of newly emerging quality management and quality assurance standards (e.g. ISO-25 and ISO-9000) in environmental analytical laboratories. At the same time they are helping to meet some of the goals of Agenda 21, of the United Nations Conference on Environment and Development, relating to the monitoring and control of environmental pollutants.

In all its projects the IAEA is giving special attention and high priority to QA/QC. The main activities include the organization of intercomparison exercises and proficiency tests, the production and distribution of reference materials, and the provision of training. Much of this work has been concerned with analyses of biological and environmental samples for radionuclides, trace elements, organometallic compounds, stable isotopes, and organic contaminants. The primary "end users" are participants from Member States that have access to (1) techniques for measuring radioactivity and/or (2) nuclear and related analytical techniques. Arising out of these activities, the IAEA has, over the years, produced a wide variety of reference materials, intercomparison samples, and proficiency testing materials.

A lichen reference material IAEA-336 has been issued through co-operation of the IAEA and the Instituto Tecnológico e Nuclear, Portugal. Lichen collected from areas presumed to be of low contamination by trace elements was used in preparing the material, which was submitted to an international intercomparison exercise, involving 42 laboratories from 26 countries (Heller-Zeisler et al., 1999; Freitas et al., 1993; Stone et al., 1995). Evaluation of the analytical data resulted in recommended values for 19 elements and information values for 14 elements. A second lichen material, coded IAEA-338 is now being prepared and evaluated, having more elevated levels for trace elements.

Despite the several decades of analytical research in APM characterisation, the determination of chemical composition of APM is still not a trivial task; chemical measurement results are often not of acceptable quality due to the complex nature of the determinations. Therefore, attention has recently focused on the use of reference air filter samples, prepared in the IAEA's laboratory, made from two different dusts. About 250 sets of reference air filters were prepared, consisting of two polycarbonate filters loaded with two urban particulate matter collected in Vienna and Prague and one blank. These materials have been thoroughly characterized for proficiency testing purposes and may later be used for a worldwide intercomparison exercise (Kučera et al., 2000).

Quality assurance activities of one kind or another are included in both Research Contract and Technical Co-operation projects on atmospheric pollution. To this end, a variety of analytical quality control materials (Table 6) has been used in world-wide intercomparison exercises, some of them supplied as bulk material, and some as particulate matter deposited on filters. For all intercomparison materials mentioned in Table 6, the degree of inhomogeneity, and for many also the element mass concentration target values were determined before sending them out (Stone et al., 1995; Heller-Zeisler et al., 1998; Zeisler et al., 1998, Kučera et al., 2000).

Besides preparing written analytical protocols, participants in all CRPs and many TC projects on atmospheric pollution are requested to participate in analytical quality control exercises organized by the IAEA.

The NAT-3 interlaboratory comparison (IC) was organized with a simulated filter-based APM test material. The material had been previously tested for the degree of inhomogeneity (which had been found to be in the 5% range, in terms of relative

Table 6.   Materials used for recent proficiency testing and analytical quality control exercises.

Test code	Material used	Date
NAT–3	Vienna dust loaded on filter	1998
NAT–4	Welding dust loaded on filter	1998
NAT–5	Lichen (2 samples)	1998
NAT–6	Moss (2 samples)	1999
NAT–7	Prague and Vienna dust loaded on filter	1999

standard deviation), and characterized for 14 elements using INAA (Heller-Zeisler et al., 1998). The filters, along with a portion of bulk material used for loading them, were distributed among 37 laboratories, and 34 actually reported results. The laboratories submitted 41 sets of results for 53 elements, and the results were evaluated with an AQCS-PC programme, used by the IAEA for such intercomparisons (Szopa et al., 1996). It was found out that most laboratories performed satisfactorily, and no statistically significant differences among the analytical methods used were found (Bleise and Smodiš, 1999). The whole sets of results from only two laboratories were statistically rejected.

The NAT–4 IC was focused on the determination of four elements in welding fume dust loaded filters. The exercise was limited to participants concerned with workplace environmental monitoring, so no further details are given in this paper.

Within the scope of NAT–5 IC two lichen materials: *Evernia prunastri (L) Ach.* species from unpolluted area in Portugal, and *Pseudevernia furfuracea* species from a mining area in Austria were distributed among 16 participating laboratories. The one from Portugal was actually the IAEA reference material IAEA-336 Lichen, under different code name, so that the participants could not reveal its origin. Results for both materials were reported by 15 participating laboratories. They submitted 17 sets of results for 47 elements. The results for the Austrian lichen sample were evaluated in the standard way for IAEA intercomparisons, whereas the results for IAEA-336 were evaluated as a proficiency test according to draft ISO 13528 (ISO, 1998). For both materials, only about 5% of the results were detected as outliers by statistical data evaluation. No laboratory having all results out of the acceptable uncertainty range was found (Bleise and Smodiš, 1999a). The relative deviation of the NAT-5 mean values from the target ones were lower than 10 % for all but two analytes, Hg and Sb (Fig. 4). With the exception of Hg and Sb, all IC mean values were within the assigned confidence intervals.

The NAT–6 IC comprised two moss samples, *Pleurozium schreberi* species, supplied by the Finnish Forest Research Institute. Analogically to the NAT–5 exercise, one moss sample, M2, was from a contaminated site and the other, M3, from an unpolluted area. Both test materials were distributed under different names, so that the participants could not reveal their original code names. In total, 18 institutions were invited to participate in this exercise and 16 participants reported results. Analytical techniques applied comprised AAS, ICPAES, ICPMS, INAA, isotope dilution time-of-flight mass spectrometry and voltammetry. A maximum number of 21 results could

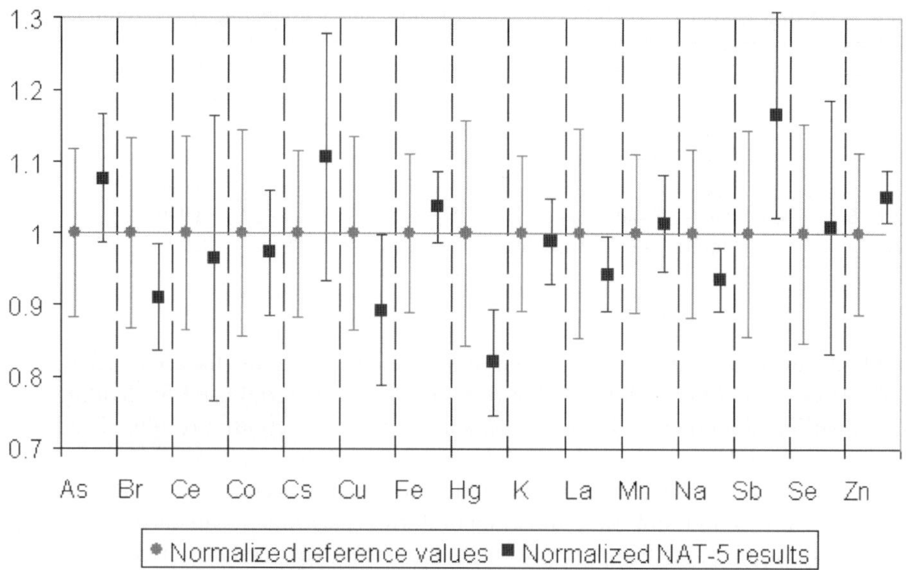

Figure 4.   Comparison of 15 element reference values with NAT-5 results for IAEA-336. Both reference (adopted from Heller-Zeisler et al. (1999) and IC values are normalised to unity.

consequently be received for an element. Results for the critical ten elements, recommended for analysis in monitoring surveys, are summarised in Tables 7 and 8. When comparing the IC results with the literature values, it should be emphasized that the intercomparison participants were determining total elemental content in the test samples whilst the literature values are referred to as "nitric-acid-soluble concentrations of the elements studied" (Steinnes et al., 1997). Also the approach for deriving the values was different. The IAEA used its AQCS-PC programme that applies four different outlier tests: Grubb's, Dixon's, coefficient of skewness and coefficient of kurtosis. Any laboratory mean thereby identified as an outlier by one of these tests is subsequently excluded from further consideration. The remaining accepted laboratory means are then combined in the usual statistical way, without weighting. As an output, values for the overall mean and its associated standard deviation, standard error and 95% confidence interval are provided. Following this procedure, only about 6% of laboratory means were statistically rejected. Differently, Steinnes and co-workers based rejection of outliers on a stepwise comparison of single values with the mean value for the same element, where the rejection criteria was two standard deviations from the mean value. Uncertainties associated with the recommended values are expressed in terms of standard deviations. An additional acceptance criterion was that at least two analytical techniques based on different physical principles should be applied for each element. Such procedure resulted in rejection of many submitted results and consequently relatively small uncertainties for some elements (Tables 7 and 8). Agreement of the IC results with the recommended values is good, providing that the first values reflect the total elemental content, and the latter values reflect the

Table 7. Summary results for the moss M2. The uncertainty quoted in the "IC value" column presents 95% confidence limits.

Element	No. of accepted laboratory means	IC value [mg/kg]	Recommended value [mg/kg] (Steinnes et al., 1997)
As	11	0.93 ± 0.14	0.98 ± 0.07
Cd	15	0.466 ± 0.035	0.454 ± 0.019
Cr	16	1.40 ± 0.20	0.97 ± 0.17
Cu	20	69.6 ± 2.9	68.7 ± 2.5
Fe	17	310 ± 17	262 ± 35
Hg	6	0.058 ± 0.017	0.058 ± 0.005
Ni	13	17.7 ± 1.6	16.3 ± 0.9
Pb	10	7.69 ± 0.46	6.37 ± 0.43
V	14	1.52 ± 0.08	1.43 ± 0.17
Zn	19	38.3 ± 1.6	36.1 ± 1.2

nitric acid soluble content. Nevertheless, the IC mean values tends to be slightly higher for most of elements, except for As, though still within the uncertainty limits. Systematically higher values are only for Cr (44% higher mean value), Fe (18%), Pb (21%) in M2, and for Fe (26%) in M3.

In the NAT-7 proficiency test, consisting of two filters loaded with Vienna and Prague dust (Kučera et al., 2000), 47 institutions took part. Since the evaluation has not been completed yet, the results will be presented elsewhere.

As a result of all these quality control exercises the performance of the laboratories participating in the IAEA projects on air pollution studies improved significantly over these years and most participants are now able obtaining reliable analytical results (Smodiš and Bleise, 2000). Analytical measurements of plant materials, in particular, showed to be of very high quality.

Table 8. Summary results for the moss M3. The uncertainty quoted in the "IC value" column presents 95% confidence limits.

Element	No. of accepted laboratory means	IC value [mg/kg]	Recommended value [mg/kg] (Steinnes et al., 1997)
As	10	0.111 ± 0.012	0.105 ± 0.007
Cd	13	0.103 ± 0.013	0.106 ± 0.005
Cr	15	0.69 ± 0.13	0.67 ± 0.19
Cu	12	3.87 ± 0.27	3.76 ± 0.23
Fe	16	174 ± 9	138 ± 12
Hg	5	0.043 ± 0.007	0.035 ± 0.004
Ni	12	1.23 ± 0.23	0.95 ± 0.08
Pb	10	3.66 ± 0.20	3.33 ± 0.25
V	14	1.28 ± 0.07	1.19 ± 0.15
Zn	20	24.8 ± 1.3	25.4 ± 1.1

3.4. Planned future activities

- The CRP on "Validation and application of plants as biomonitors of trace element atmospheric pollution analysed by nuclear and related techniques" is expected to be continued until 2002.
- A new CRP on correlating toxic element atmospheric deposition data with diseases is expected to be initiated in 2003. Namely, many areas in the IAEA Member States affected by airborne emissions of toxic elements have recently been identified and characterized within the recent and on-going projects. It is expected that by the year 2003 sufficient data will be collected to enable starting research on its correlation with the available epidemiological data on pulmonary and other element-related diseases. This would be a definitive step ahead in practical exploitation of the accumulating experimental evidence on air pollution, obtained either by analyzing APM or using biomonitors. It is expected that the toxic element atmospheric data will be linked with health-related information such as mortality, pulmonary diseases, cancer incidence, etc.
- The regional TC project for East Asia and the Pacific region on "Air pollution and its trends" will continue until 2002. In the future, more emphasis will be put into receptor modelling of large-scale data, examining time trends and transboundary source/receptor relationships. Atmospheric chemistry, the nature of source emissions, and the physical processes that give rise to regional and global scale atmospheric transport will be studied. Subject to approval, the project might be continued also during the 2003/2004 cycle.
- The Latin America regional TC project on "Application of biomonitors, nuclear and related techniques in atmospheric pollution studies" will be initiated in 2002, expected to be concluded in 2004. The participating countries Argentina, Bolivia, Brazil, Chile, Cuba, Ecuador, El Salvador, Guatemala, Mexico, Paraguay, Peru, Uruguay, and Venezuela will directly benefit from the outcomes of the CRP on "Validation and application of plants as biomonitors of trace element atmospheric pollution analysed by nuclear and related techniques". Biomonitoring metrology tools developed and validated within the CRP will be put into practice for assessing the degree of atmospheric pollution within the region.
- One moss and two spruce shoots reference materials are planned to be prepared and characterised for analytical quality control purposes.

4. Conclusions

The International Atomic Energy Agency is following, in its approaches to assessment of chemical elements in atmosphere, a concept of implementing environmental metrology in practice. This means that proven environmental monitoring and assessment tools are adopted and applied in its developing Member States. Methodological and technological gaps due to requirements needed to be met under specific environmental and/or climatic conditions are filled in with applied research work within co-ordinated research projects. Once these metrological tolls are tested, they are transferred to developing countries by the mechanism of technical co-operation projects. Such an approach has been successfully implemented for measuring elemental composition of both $PM_{2.5}$ and PM_{10} components of airborne particulate matter. Reliable air samplers that have been extensively tested have been delivered in more than 40 countries worldwide, where they may provide information on air pollution, its trends and transport. A similar approach is now being implemented for biomonitoring trace element atmospheric pollution. Appropriate plants that would fit for purpose are searched for and validated worldwide within a CRP. The selected species, when applied for regular biomonitoring surveys implemented through TC projects, will provide comparative and compatible data on regional scales. An important part of implementing this environmental metrology is a continuous effort for improving analytical measurements by organizing quality control exercises.

Altogether 75 counterparts from 40 IAEA Member States have participated within the six major projects on studying element atmospheric pollution during the period 1992–2000 (Table 9). Biomonitoring surveys, using either active or passive bioindicators, are already carried out on a regular basis in 18 countries. The number of these countries is expected to increase substantially in the years to come. This metrology will be further harmonized, thus providing comparable data on a regional scale.

In all of these activities, proficiency testing and analytical quality control are important issue that merits special attention. Since 1998, 145 participants from 60 countries have taken part in quality control exercises concerned with air quality measurements, organised by the IAEA. The participants have submitted 156 sets of results for up to 57 elements in one sample. Due to these exercises, the awareness of good laboratory practice has been built in many developing countries and the quality of environmental measurements in the participating laboratories is continuously improving. Regular quality control exercises proved to be important tools in obtaining high quality measurement results for biomonitors and other environmental materials. Therefore, the IAEA plans to continue organising interlaboratory comparison studies and proficiency testing for samples relevant to atmospheric pollution studies.

Table 9. Countries that have participated in major projects on studying element atmospheric pollution during the period 1992–2000. The numbers indicate in how many projects each country has participated.

Argentina	3	Ghana	1	Mexico	1	Romania	1
Australia	2	Hungary	1	Mongolia	1	Russian Fed.	2
Bangladesh	3	India	4	Myanmar	2	Singapore	2
Belgium	1	Indonesia	3	The Netherlands	1	Slovenia	2
Brazil	4	Iran	1	New Zealand	1	South Africa	1
Chile	3	Israel	1	Norway	1	Sri Lanka	2
China	5	Jamaica	2	Pakistan	2	Thailand	2
Czech Rep.	2	Kenya	2	Paraguay	1	Turkey	1
Denmark	1	Korea	2	Philippines	2	USA	1
Germany	1	Malaysia	2	Portugal	3	Vietnam	2

Acknowledgements

The author wish to acknowledge dedicated scientific efforts of all the numerous participants in the concerned projects and quality control exercises; without them this contribution would not be possible. Thanks are also due to colleagues in the Nutritional and Health-Related Studies Section of the IAEA and the Chemistry Unit of the Agency's Laboratories Seibersdorf for their continuous help and support.

References

Bleise, A., Smodiš, B., 1999. Report on the Intercomparison Run NAT-3 For the Determination of Trace and Minor Elements in Urban Dust Artificially Loaded on Air Filters. NAHRES-43, International Atomic Energy Agency, Vienna, Austria.

Bleise, A., Smodiš, B., 1999a. Report on the Intercomparison Run NAT-5 For the Determination of Trace and Minor Elements in Two Lichen Samples. NAHRES-46, International Atomic Energy Agency, Vienna, Austria.

Byrne, A.R., Benedik, L., 1988. Determination of uranium at trace levels by radiochemical neutron-activation analysis employing radioisotopic yield determination. Talanta 35, 161–166.

Englund, H.M., Berry, W.T. (Eds), Proceedings of the Second International Clean Air Congress. Academic Press, New York.

Freitas, M.C., Catarino, F.M., Branquinho, C., Maguas, C., 1993. Preparation of a lichen reference material. Journal of Radioanalytical and Nuclear Chemistry 169, 47–55.

Freitas, M.C., 1995. Elemental bioaccumulators in air pollution studies. Journal of Radioanalytical and Nuclear Chemistry 192, 171–181.

Freitas, M.C., Reis, M.A., Alves, L.C., Wolterbeek, H.Th., Verburg, T., Gouveia, M.A., 1997. Bio-monitoring of trace-element air pollution on Portugal: qualitative survey. Journal of Radioanalytical and Nuclear Chemistry 217, 21–30.

Frontasyeva, M.V., Grass, F., Nazarov, V.M., Steinnes, E., 1995. Intercomparison of moss reference material by different multi-element techniques. Journal of Radioanalytical and Nuclear Chemistry 192, 371–379.

Giege, B., Barikmo, J., Juha-Pekka, H., Odsjö, T., Petersen, H., Petersen, Æ., Wallentinus, H.-G., 1995. Nordic Environmental Specimen Banking – Methods in Use in ESB. TemaNord, 543, Nordic Council of Ministers, Copenhagen.

Greenberg, R.R., Kingston, H.M., Watters, R.L., Pratt, K.W, 1990. Dissolution problems with botanical reference materials. Fresenius Journal of Analytical Chemistry 338, 394–398.

Heller-Zeisler, S.F., Fajgelj, A., Bernasconi, G., Tajani, A., Zeisler, R., 1998. Examination of the procedure for the production of a simulated filter-based air particulate matter reference material. Fresenius Journal of Analytical Chemistry 360, 435–438.

Heller-Zeisler, S.F., Zeisler, R., Zeiller, L., Parr, R.M., Radecki, Z., Burns, K.I., De Regge, P., 1999. Report on the Intercomparison Run for the Determination of Trace and Minor Elements in Lichen Material IAEA-336. NAHRES-33 (IAEA/AL/79), International Atomic Energy Agency, Vienna, Austria.

Hopke, P.K., Xie, Y., Raunemaa, T., Biegalski, S., Landsberger, S., Maenhaut, W., Artaxo, P., Cohen, D., 1998. Characterization of the Gent stacked filter unit PM_{10} sampler. Aerosol Science and Technology 27, 726–735.

IAEA, 1997. Proceedings of the Symposium on Harmonisation of Health Related Environmental Measurements Using Nuclear and Isotopic Techniques. IAEA-STI/PUB/1006, International Atomic Energy Agency, Vienna, Austria.

IAEA, 1999. Report on the First Research Co-ordination Meeting for the Co-ordinated Research on Validation and Application of Plants as Biomonitors of Trace Element Atmospheric Pollution, Analysed by Nuclear and Related Techniques. NAHRES-45, International Atomic Energy Agency, Vienna, Austria.

IAEA, 2000. Proceedings of the International Workshop on Biomonitoring of Atmospheric Pollution – BioMAP. IAEA-TECDOC-1152, International Atomic Energy Agency, Vienna, Austria.

Ihnat, M., 1982. Importance of acid-soluble residue in plant analysis for total macro and micro elements. Communications in Soil Science and Plant Analysis 13, 969–979.

Jeran, Z., Smodiš, B., Jaćimović, R., 1993. Multielemental analysis of transplanted lichens *(Hypogymnia Physodes, L. Nyl.)* by instrumental neutron activation analysis. Acta Chimica Slovenica 40, 289–299.

ISO, 1998. Draft International Standard ISO 13528. Statistical Methods for Use in Proficiency Testing by Interlaboratory Comparison. International Standardization Organization, Geneva.

Jeran, Z., Jaćimović, R., Batič, F., Smodiš, B., Wolterbeek, H.Th., 1996. Atmospheric heavy metal pollution in Slovenia derived from results for epiphytic lichens. Fresenius Journal of Analytical Chemistry 354, 681–687.

Jeran, Z., Jaćimović, R., Smodiš, B., Batič, F., 1996a. The use of lichens in atmospheric trace element deposition studies in Slovenia. Phyton 36, 91–94.

Jervis, R.E., Qureshi, R.Y, 1994. Effectiveness of oak leaves as bioindicators of environmental pollution. Journal of Radioanalytical and Nuclear Chemistry 188, 149–155.

Kučera, J., Parr, R.M., Smodiš, B., Fajgelj, A., Mattiuzzi, M., Havránek, V., 2000. Use of INAA, PIXE and XRF in homogeneity testing of new IAEA reference air filters. Journal of Radioanalytical and Nuclear Chemistry 244, 121–126.

Kuik, P., Wolterbeek, H.Th., 1995. Factor analysis of atmospheric trace-element deposition data in the Netherlands obtained by moss monitoring. Water, Air, and Soil Pollution 84, 323–346.

Lindstrom, R.M., Byrne, A.R., Becker, D.A., Smodiš, B., Garrity, K.M., 1990. Characterization of the mineral fraction in botanical reference materials and its influence on homogeneity and analytical results. Fresenius Journal of Analytical Chemistry 338, 569–571.

Lippo, H., Poikolainen, J., Kubin, E., 1995. The use of moss, lichen and pine bark in the nationwide monitoring of atmospheric heavy metal deposition in Finland. Water, Air, and Soil Pollution 85, 2241–2246.

Loppi, S., Nelli, L., Ancora, S., Bargagli, R., 1997. Passive monitoring of trace elements by means of tree leaves, epiphytic lichens and bark substrate. Environmental Monitoring and Assessment 45, 81–88.

Maenhaut, W., François, F., Cafmeyer, J., 1994. The "Gent" Stacked Filter Unit (SFU) sampler for the collection of atmospheric aerosols in two size fractions: description and instructions for installation and use. In Report on the First Research Co-ordination Meeting for the Co-ordinated Research Project on Applied Research on Air Pollution Using Nuclear-related Analytical Techniques, NAHRES-19, International Atomic Energy Agency, Vienna, Austria, pp. 249–263.

Markert, B. (Ed.), 1993. Plants as biomonitors – Indicators for heavy metals in the terrestrial environment. VCH, Weinheim.

Markert, B., 1996. Instrumental element and multi-element analysis of plant samples, Methods and applications. John Wiley, New York (first published 1993).

Pakeman, R.J., Hankard, P.K., Osborn, D., 1998. Plants as biomonitors of atmospheric pollution: their potential for use in pollution regulation. Reviews of Environmental Contamination and Toxicology 157, 1–23.

Quevauviller, Ph., Van Renterghem, D., Muntau, H., Griepnik, B., 1993. Intercomparison to improve the quality of trace element determination in lichen. International Journal of Environmental Analytical Chemistry 53, 233–242.

Quevauviller, Ph., Herzig, R., Muntau, H., 1996. Certified reference material of lichen (CRM 482) for the quality control of trace element biomonitoring. Science of the Total Environment 187, 143–152.

Ribeiro Guevara, S., Arribére, M.A., Calvelo, S., Román Ross, G., 1995. Elemental composition of lichens at Nahuel Huapi national park, Patagonia, Argentina. Journal of Radioanalytical and Nuclear Chemistry 198, 437–448.

Rühling, Å., Tyler, G., 1968. An ecological approach to the lead problem. Botaniska Notiser 122, 248–342.

Rühling, Å, Steinnes, E., Berg, T. (Eds), 1996. Atmospheric Heavy Metal Deposition in Northern Europe 1995. Nord 1996, 37. Nordic Council of Ministers, Copenhagen, Denmark.

Rühling, Å., Steinnes, E. (Eds), 1998. Atmospheric Heavy Metal Deposition in Europe. Nord 1998, 15. Nordic Council of Ministers, Copenhagen, Denmark.

Sloof, J.E., 1993. Environmental lichenology: biomonitoring trace-element air pollution. Ph.D. Thesis, Delft University of Technology.

Sloof, J., Wolterbeek, B.Th., 1993. Interspecies comparison of lichens as biomonitors of trace-element air pollution. Environmental Monitoring and Assessment 25, 149–157.

Smodiš, B., 1992. Reactor neutron activation analysis using the k_0-standardization method. Part I: Theory and experimental considerations. Vestnik Slovenskega Kemijskega Društva 39, 503–519.

Smodiš, B., Bleise, A., 2000. Neutron activation analysis in the IAEA projects on air pollution. Journal of Radioanalytical and Nuclear Chemistry 244, 97–102.

Steinnes, E., 1980. Atmospheric deposition of heavy metals in Norway studied by analysis of moss samples using neutron activation analysis and atomic absorption spectrometry. Journal of Radioanalytical Chemistry 58, 387–391.

Steinnes, E., 1989. Biomonitors of air pollution by heavy metals. In: Pacyna, J.M., Ottar, B. (Eds), Control and Fate of Atmospheric Trace Metals. Kluwer Academic Publishers, pp. 321–338.

Steinnes, E., Hanssen, J.E., Rambæk, J.P., Vogt, N.B., 1994. Atmospheric deposition of trace elements in Norway: Temporal and spatial trends studied by moss analysis. Water, Air, and Soil Pollution 74, 121–140.

Steinnes, E., Johansen, O., Røyset, O., Ødegård, M., 1993. Comparison of different multielement techniques for analysis of mosses used as biomonitors. Environmental Monitoring and Assessment 25, 87–97.

Steinnes, E., Rühling, Å., Lippo, H., Mäkinen, A., 1997. Reference materials for large-scale metal deposition surveys. Accreditation and Quality Assurance 2, 243–249.

Stone, S.F., Freitas, M.C., Parr, R.M., Zeisler, R., 1995. Elemental characterization of a candidate lichen research material – IAEA-336. Fresenius Journal of Analytical Chemistry 352, 227–231.

Stropnik, B., Byrne, A.R., Smodiš, B., 1993. Air pollution monitoring by ED XRF in the Šalek valley: Part 1. Acta Chimica Slovenica 40, 301–330.

Svetina Gros, M., Smodiš, B., Jeran, Z., Jaćimović, R., 1996. Trace elements in the water cycle of the Šalek valley, Slovenia, using INAA. Journal of Radioanalytical and Nuclear Chemistry 204, 45–55.

Svetina Gros, M., Smodiš, B., Pirc, S., 1933. Following the water cycle in the Šalek valley by the INAA method. Acta Chimica Slovenica 40, 243–253.

Szopa, Z., Jaszczuk, J., Dybczynski, R., 1996. A new multifunctional PC program for evaluation of interlaboratory comparison results. Nukleonika 41, 117–127.

Takala, K., Olkkonen, H., Salminen, R., 1994. Iron content and its relations to the sulphur and titanium contents of epiphytic and terricolous lichens and pine bark in Finland. Environmental Pollution 84, 131–138.

Tyler, G., 1970. Moss analysis – a method for surveying heavy metal deposition. In Englund, H.M., Berry, W.T. (Eds), Proceedings of the Second International Clean Air Congress, Academic Press, New York.

Wolterbeek, H.Th., Bode, P., 1995. Strategies in sampling and sample handling in the context of large-scale plant biomonitoring surveys of trace element air pollution. Science of the Total Environment 176, 33–40.

Wolterbeek, H.Th., Bode, P., Verburg, T.G., 1996. Assessing the quality of biomonitoring via signal-to-noise ration analysis. Science of the Total Environment 180, 107–116.

Wyttenbach, A., Bajo, S., Furrer, V., Langenauer, M., Tobler, L., 1997. The accumulation of As, Br and I in needles of Norway spruce (Picea abies [L.] karst.) at sites with low pollution. Water, Air, and Soil Pollution 94, 417–430.

Zeisler, R., Heller-Zeisler, S.F., Fajgelj, A., 1998. On achieving measurement quality in air pollution studies employing nuclear techniques. Journal of Radioanalytical and Nuclear Chemistry 233, 15–19.

Bioindicators and biomonitors
B.A. Markert, A.M. Breure, H.G. Zechmeister, editors

Chapter 25

Bio-indicators and the indicator approach of the European Environment Agency

Peter Bosch and Ulla Pinborg

Abstract

The European Environment Agency produces assessments based on biodiversity indicators in a broad sense. These include indicators not only focused on species or species groups, but also indicators on changes in the distribution and function of habitats and indicators on protected areas. Bio-indicators as based on species thus fall within the overall work on biodiversity indicators. The use of bio-indicators by the European Environment Agency is characterised firstly by the nature of these indicators. In countries with advanced environmental policies like in west and central Europe, indicators on pressures and driving forces play the most important role in showing progress for the majority of environmental issues, and bio-indicators have a limited role in actual policy making. The exception is, of course, in nature and biodiversity management, and in situations where organisms and their functions are a part of the solution to environmental problems. For these policies state of the environment indicators and bio-indicators specifically, come in the forefront of attention. For the majority of the environmental problems, the main role of bio-indicators is to monitor if current environmental policies have a beneficial effect.

Secondly the EEA is depending on the monitoring carried out by institutions in its member countries, which limits the range of species, or functionalities of species to be included in its indicators.

Keywords: state indicators, pressure indicators, environmental reporting

1. The European Environment Agency

The European Environment Agency (EEA) is a European Community body with the aim of serving the Community and the Member States with information to support policy making for environmental protection in the perspective of sustainable development. The main activities of the EEA are compiling and assessing data and information on the current and foreseeable state of the environment.

The EEA was founded by a regulation of the European Community (Council regulation EEC/12120/90), and it started its activities in 1993. The EEA has a membership of the fifteen European Union countries, plus Norway, Iceland, Liechtenstein. The Accession Countries to the EU are expected to become member of the EEA soon. Although the EEA office in Copenhagen houses only 70 people, many more experts in many institutions in all European countries, are involved in its work. Simultaneous with the foundation of the EEA, also the EIONET (the European Environmental

Information and Observation Network) was created, a huge network of 500 institutions in Europe to help the EEA with bringing together, harmonising, and reporting on environmental data and information in Europe. EEA reports and databases are increasingly made available to everyone in Europe via the Internet (http://www.eea.eu.int).

With the development in attention going from "environment", to "the environmental pillar in sustainable development" to "environmentally sustainable development" the EEA is in a constant process of creating clear lines for our clients focusing on the essentials in the overload of environmental and sustainability information.

The primary clients are policy-making agents and politicians at EU level in the European Commission, in the European Parliament, in the Council and in the Member states. These clients are increasingly aware of the use of and usefulness of indicators in their processes. Indicators can play an important role within the policy preparation and the evaluation stages of the policy cycle (Fig. 1). It is also at these points that the two major reports that the EEA produces regularly aim to support policy: a State of the Environment/Environmental Outlook report of which the last edition (EEA, 1999) served to prepare the 6th Environmental Action Programme of the European Commission, and the regular Environmental signals indicators report series (EEA, 2000, 2001), which aims to support the regular evaluation of the policies set out in the Action programme.

However, our primary clients are not the only actors driving policies and able to bring along changes. Informed citizens, NGOs, companies, lower levels of governments are EEA's secondary target groups.

But, regardless of the users of the indicators, a few principal characteristics seem needed to make them effective communication tools:

1. Indicators should report progress over time and must go with an assessment of the reasons for their development.
2. They should be few in number, and users should get used to their presentation.
3. They become more powerful when linked with formal targets or informal or indicative (sustainable) reference values. Linked with targets, indicators become tools for management and help to make policy makers accountable;

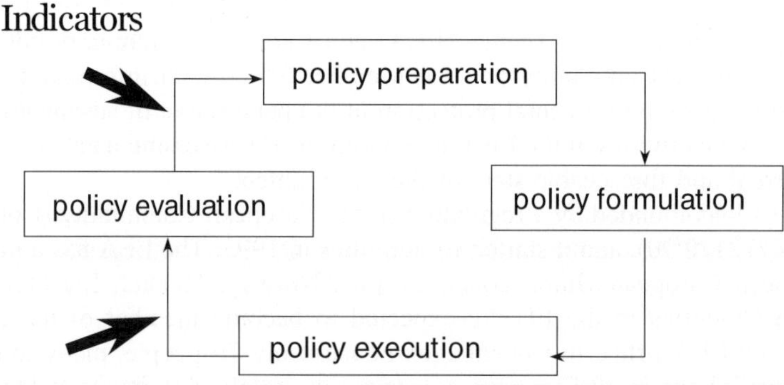

Figure 1.   Indicators and the policy cycle.

4. With or without targets, using indicators to compare or benchmark individual sectors or countries or companies with each other is another way to make decision makers accountable and to foster progress as both failure and success stories become evident. The question why one sector/country/company is doing better than another is a good entrance to explore still unknown opportunities to do better. At the same time exposing this kind of information to the outside world could lead to 'peer pressure' to do better.

2. Indicators and support to the policy process

As an institution founded to support policy, the EEA pays a lot of attention to the proper link of indicators and the policy processes they serve. For its assessments the EEA uses the concept of biodiversity indicators, which also includes indicators not specifically focused on species or species groups, such as indicators on changes in the distribution of habitats and indicators on protected areas. Bio-indicators as based on species fall within the overall work on biodiversity indicators. Results of bio-monitoring are used in assessments of various environmental issues. For both categories a clear link to policy making will always be strived for.

Since the European Council in Gothenburg in June 2001, the EU sustainable development strategy provides an umbrella for a number of interlinked and mutually supporting policies on the EU level. Regarding biodiversity the Commission's Communication on the Sustainable Development Strategy (CEC, 2001) mentions explicitly the objective to protect and restore habitats and natural ecosystem and to halt the loss of biodiversity by 2010. As one of the implementation measures the Commission will establish a system of biodiversity indicators by 2003.

The most important EU policy lines elaborating this objective are the 6th Environmental Action Programme and the so-called Cardiff process for the integration of environment in sector and other policies.

The proposal for the decision of the European Parliament and the Council on the 6th Environmental Action Programme (EU Council, 2001) reiterates and details the above-mentioned 2010 objective. As implementation measures it lists the implementation and monitoring of the Community's biodiversity strategy (1998) with its accompanying Biodiversity Action Plans for environment, agriculture, fisheries and development (all adopted in 2001); the continued implementation of the Natura 2000 network; additional measures to protect species covered by the Habitats and Birds directives outside Natura 2000 areas; and the development of measures aimed at prevention of invasion of alien species. In all these plans development of indicators is receiving more and more attention.

The Cardiff process for the integration of environment and other policies started as a European Council initiative in 1998. Councils with a sector focus, such as the Energy Council and the transport Council were asked to prepare strategic documents on how they would integrate environment in their policy making. As most of the threats to nature and biodiversity are coming from the development of transport infrastructure, agriculture, fisheries, tourism and other sector developments, the integration of environment in sector and other policies is of utmost importance for biodiversity

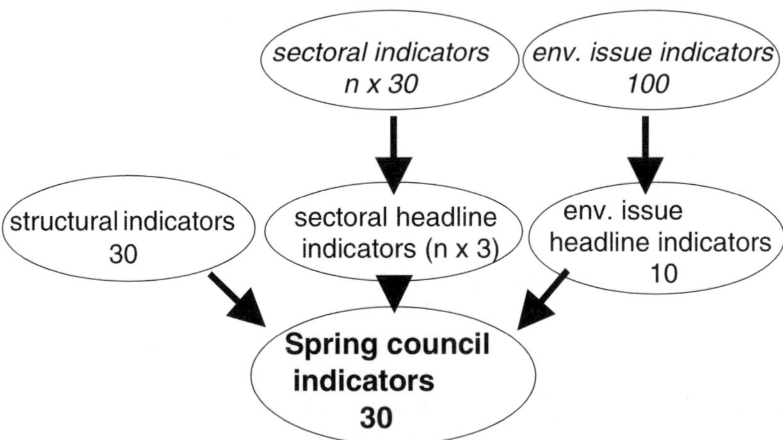

Figure 2. Indicator architecture with tentative number of indicators per group.

protection and nature management. The threats arise from land-use and landscape fragmentation, climate change, air and water pollution, leading to eutrophication, acidification, use and introduction of species. Apart from EU agriculture and fisheries policies, however, the EU sector strategies developed under the Cardiff initiative are still paying little attention to biodiversity issues (Kraemer, 2001).

To report on progress and assess the effects and effectiveness of these policies an accompanying set of interlinked indicators is needed. Furthermore, there should be a clear structure to communicate to policy makers how each part of the information is related to the various policy processes. To achieve this clarity clusters of environmental indicators and assessment and reporting mechanisms are being developed (Fig. 2).

Behind the 6th Environmental Action Programme, and the envisaged more detailed action programmes for environmental issues, indicator-based reporting is necessary on the state of the environment for the various issues and the driving forces and responses influencing it. The EEA is developing for the issue Nature and Biodiversity a set of indicators, which will be regularly used in all EEA assessments. These indicators concern species, habitats and their functions, as well as societal responses to biodiversity degradation.

For use in reports to high-level policy makers, a selection has been made of the hundreds of indicators for environmental issues, resulting in eleven EU Environmental Headline Indicators (Table 1) (EEA, 2001). For nature and biodiversity the current 'stand-in' indicator is foreseen to be substituted by a species and habitat related indicator in the future.

Behind the sector integration strategies, there should be regular reporting mechanisms based on indicators, as requested by the various Councils since Cardiff.

Following on the example of the successful Transport and Environment Reporting Mechanism (TERM, see Box 1), the EEA is developing with its partners similar indicator based reporting on Environment and Energy, Environment and Agriculture and, if resources are made available, will develop it for tourism and fisheries. Similar to

Table 1.　Environmental headline indicators for the EU (status January 2001).

Issue	Current indicators	Proposals for ideal indicators
6th Environmental Action Programme theme: Climate change		
1.　Climate change	Aggregated emissions of three main greenhouse gases	Aggregated emissions of six greenhouse gases of the Kyoto Protocol
6th Environmental Action Programme theme: Nature and biodiversity		
2.　Nature and biodiversity	Designated "Special Protection Areas" (Birds Directive)	Biodiversity index, or conservation status of key species and habitats
3.　Air quality: acidification	Aggregated emissions of acidifying substances	Same
6th Environmental Action Programme theme: Environment and human health		
4.　Air Quality: summer smog	Aggregated emissions of ozone precursor substances	Same, and: number of days of pollution exceeding standards
5.　Urban air quality	Number of days of exceedance (several pollutants)	Urban air quality indicators or index; urban transport indicators
6.　Water quality	Phosphate and nitrate concentration in large rivers	European index for the status of water bodies
7.　Chemicals	Production of hazardous chemicals	Production of hazardous chemicals, weighted
6th Environmental Action Programme theme: Waste and resources		
8.　Waste	Municipal and hazardous waste generated and landfilled	Resource use in line with the waste strategy
9.　Resource use	Gross inland energy consumption	Material balance indicator
10.　Water quant.	Total fresh water abstraction	Intensity of water use
11.　Land-use	Land-use by selected categories	Land-use change matrix

the 'environmental headline indicators' (see below) a limited set of main indicators can be selected from the currently around 30 available integration indicators per sector

Finally, the Gothenburg summit requested the European Commission to report each year at the Spring European Council on progress under the sustainable development strategy, based on indicators. For the EEA the process of regular reporting to spring European Councils is a unique opportunity to deliver key indicators and assessments on environmental aspects of sustainable development, including progress

---

Box 1. The TERM process and concept: a model to be followed by other sectors?

The European Council, at its Summit in Cardiff in 1998, requested the Commission and the transport ministers to focus their efforts on developing integrated transport and environment strategies. At the same time, and following initial work by the European Environment Agency on transport and environment indicators, the joint Transport and Environment Council invited the Commission and the EEA to set up a Transport and Environment Reporting Mechanism (TERM), which should enable policy-makers to gauge the progress of their integration policies.

The main output of TERM is a regular indicator-based report through which the effectiveness of transport and environment integration strategies can be monitored. The first indicator report was published in 2000. TERM-2001 is currently under preparation (publication expected in September 2001).

The 30 TERM indicators were selected and grouped to address the seven key questions:

1. Is the environmental performance of the transport sector improving?
2. Are we getting better at managing transport demand and at improving the modal split?
3. Are spatial and transport planning becoming better coordinated so as to match transport demand to the needs of access?
4. Are we optimising the use of existing transport infrastructure capacity and moving towards a better-balanced intermodal transport system?
5. Are we moving towards a fairer and more efficient pricing system, which ensures that external costs are internalised?
6. How rapidly are improved technologies being implemented and how efficiently are vehicles being used?
7. How effectively are environmental management and monitoring tools being used to support policy and decision-making?

The indicators cover all the most important aspects of the transport and environment system (Driving forces, Pressures, State of the environment, Impacts, and societal Responses – the so-called DPSIR framework) and include eco-efficiency indicators.

---

towards integration of environment in sectoral activities. The EEA will re-orient its Environmental signals indicator report to support this reporting process, while retaining the reports own structure and logic to serve multiple clients.

3. Frameworks and typologies for indicators

With sustainable development now a main line in EU policy making, the EEA sees environmental indicators placed within a system aiming to describe developments

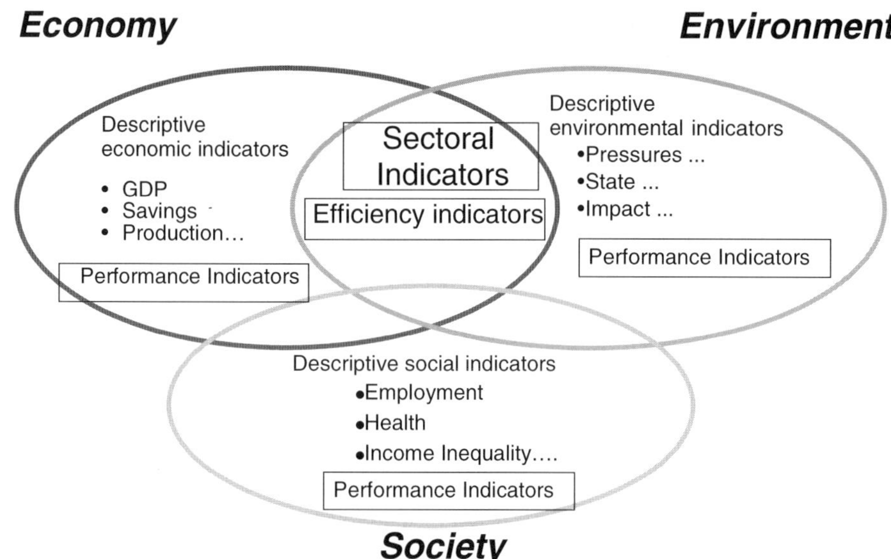

Figure 3. General model sustainability indicators.

towards sustainable development, which makes clear where trade-offs are to be made and where win-win situations may occur (Fig. 3).

The descriptions in the boxes in Figure 4 refer to a typology of indicators. A first category of indicators answers the question: "how are pressures on the environment and how is the quality of the environment developing". These are called *Descriptive indicators*, and are usually presented as a line diagram showing the development of a variable over time, for example "cadmium contents in blue mussels", or "number of indigenous species in biogeographical regions".

A second category of indicators answers the question "and is that relevant?" *Performance indicators* use generally the same variables as descriptive indicators but are connected with target values. "The number of days in which ozone levels are exceeding WHO standards" is clearly an example of a performance indicator. The "designation of Natura 2000 sites compared with an estimate of important natural areas or an area target per country" is also a performance indicator.

The third category, which can be found in between the environment and the economy circle are *Efficiency indicators*. These answer the question "have we become more efficient in our economic processes?". Eco-efficiency indicators have proven to be useful communication tools in benchmarking exercises: a "2% eco-efficiency improvement in a given year" is a common language, whatever the economic structure of a country or whatever business sector considered.

Classifying the types of indicators helps in identifying the questions on which policy makers need an answer, and brings a balance in indicator sets.

Another important tool to structure a collection of indicators and to communicate their application is the analytical framework for the assessment. The EEA uses a

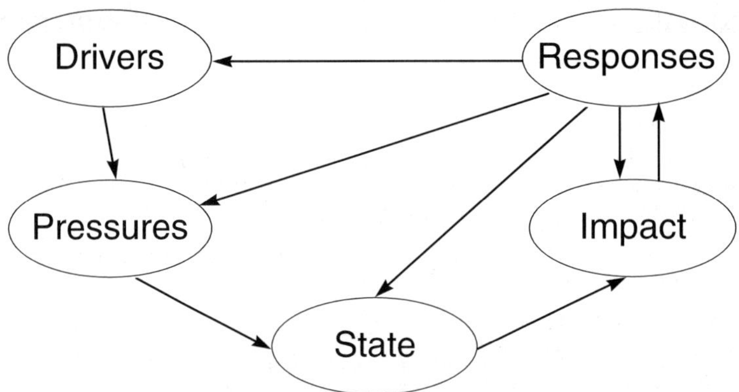

Figure 4.   The DPSIR framework for reporting on environmental issues.

slightly extended version of the well known OECD-model, which is called the **D**riving forces – **P**ressures – **S**tate – **I**mpact – **R**esponses (DPSIR) framework (Fig. 4).

This extended framework is used because it makes clear what the difference is between the *driving force*, like development of industry or number of vehicle kilometres of passenger cars, and the *pressures*, for example the emission of carbon dioxide by passenger cars. From the viewpoint of an ecosystem the last are often called "stresses", which focuses more precisely than the concept of pressures on the disturbances of a particular system. The *state* of the environment is expressed in quality parameters for air, water and soil. *Impact* is a more difficult concept: it stands for the effects of a changed environment on the health of human beings and other organisms and on the effects on nature and biodiversity. All those impacts give rise to *responses* of society.

Sometimes indicators can be placed in between the DPSIR elements or consist of combinations of the boxes. Eco-efficiency indicators like "emission coefficients" and "energy productivity" (or its inverse "energy intensity") show what happens in between the driving forces and pressures. The combination in one diagram of the pressure: "release of nutrients from agriculture" and the state: "development of nitrate concentration in surface waters" tells a story of time delay in natural processes and the "time bombs" created in the environment. Together with indicators on the impacts of eutrophication on nature, as appears from a change in species, this small set of indicators provides a concise picture of a problem, it hints to possible solutions, but also shows some of the complexities of the system.

According to the apparent role in the policy process environmental indicators can be classified in four classes:

1. Those indicators that are fully integrated in the policy process in the sense that a change in the indicator leads to a reaction from the policy side. The emission of greenhouse gases and the distance to the targets of the Kyoto protocol has in many countries led to a redirection of current policies, a reformulation of national targets and it is regularly published in the newspapers. There are a number of similar

examples on the European and national scale, such as the recycling of packaging waste, the amount of animal manure in areas with severe eutrophication problems, and of course fish stocks. More recently the designation percentage of the national territory for Natura 2000 has received this status.

2. Those indicators that are clearly relevant to a policy process, but lack the precise link to policies to lead to an immediate reaction from the side of policy makers. Concentrations of nitrogen and phosphorus in surface waters have a role to maintain the attention for the eutrophication problem, but as the parameters are not immediately linked with certain policy actions, the publication of the indicator will not immediately give rise to a lot of policy attention.

3. Those indicators that have only a limited relevancy to policy makers due to their definition in units not relevant for the policy makers' actions, or because the indicators describe processes that cannot or can only to a very limited extend be influenced by policy measures. For instance, the percentage of the population linked to waste water treatment facilities does not appeal to policy makers when they are working in a political process that does not aim at 100% connection, but at the provision of waste water treatment to agglomerations with more than 2000 inhabitants. The EEA tends to maintain a number of indicators from this category, with the motivation that they are the best available proxy indicators. In some cases that maybe true, but in other cases the use of these indicators distracts us from giving better and more relevant signals on progress.

4. And finally, there is a group of indicators important for showing how the environmental system works, but that are not designed to be particularly policy relevant. For instance, as long as we are not able to entangle the complex of factors including weather conditions that influence the occurrence of wetland birds on national or even EU level (thus not for a specific area) the resulting indicator is mainly important as a background to other indicators that indeed have an action perspective. With increased knowledge or a redirection of environmental policies, some of the background indicators may turn into a policy relevant indicator in the future.

4. Requirements from bio-indicators for supporting policy making

The DPSIR scheme provides an explanation of the overall role of bio-indicators in policy making. For problems that are in the beginning of their policy life cycle (Winsemius, 1986), that is, in the stage of problem identification, quality indicators and especially bio-indicators and bio-monitors identify developments in the state of the environment that may be alarming. The most well known cases of "state"-indicators that gave rise to policy reactions are those on the sudden decline of selected species (fish in acidified Scandinavian lakes, seals in the Dutch Wadden Sea, for instance), surface water quality (salt in the river Rhine which was used for irrigation in horticulture, for example) and on air quality in cities (summer smog in Paris, Athens). This function of "state"-indicators is thus limited in time: as soon as a problem is politically accepted and measures are being designed, the attention shifts to pressure indicators.

There is, however, a long period in which state indicators support the process of getting political acceptance of policy responses. Greenhouse gas policies provide clear examples where indicators on climate change and its impacts in terms of average temperatures, movement of the tree line, or species distribution are being used to gather political support for signing the Kyoto protocol.

In the next and longer stages of the policy cycle (formulation of policy responses, implementation of measures and control) bio-indicators and bio-monitors have another role, less in the forefront of policy making. Policymakers focus in these phases on what they can influence, the driving forces through volume measures, the pressures with technical measures and educational projects. The state of the environment, which is communicated through bio-indicators, is only a derived result of activities in society and policy reactions. The exception is, of course, management of biodiversity as such or when organisms play a role in the solution of environmental problems. In these situations indicators such as biomass production, forest-types as carbon dioxide sinks and forest composition are important measures of progress.

In the control phase of the policy cycle, bio-indicators become important again to watch the recovery of the environment and a limited number of indicators will be used to continuously monitor the state of the environment. A current example of this are the indicators on forest health related to main tree groups, with regard to the acidification problem, which were included in every national indicator report in the 1980s (in the problem identification phase) and nowadays (in the phase of the implementation of measures) can be found in technical reports only.

Taking into account the exceptions mentioned above, one might conclude that the more a policy topic is developed the lesser the role for State of the environment information and bio-indicators and the larger the role for driving force/pressure indicators.

Still indicators on the state of the environment in general, bio-diversity and bio-indicators in particular can potentially play an important role in the policy making process. But therefore they need to comply with a number of requirements:

1. *They must refer to policy processes*
   But what are relevant ones? Following from the observations on the policy life cycle of environmental problems, it will be clear that bio-indicators have a limited role in the phase of the implementation of measures or the formulation of laws or regulations. More often bio-indicators will play a role in the preparation and evaluation of plans and strategies. As almost all environmental problems or activities of societal sectors in the end impact on the natural environment, in principle bio-indicators could be relevant to almost all types of planning. However some of the strategies aim more explicitly at a good quality of the environment, such as water quality and quantity management, chemicals policy, spatial planning, and of course nature and biodiversity policies.

   Bio-indicators can also play a role in the development of integrated sectoral plans (transport, agriculture, fisheries) and ultimately in strategies for sustainable development. The more precise the targets of these plans are being or have been formulated, the better a bio-indicator can be linked to the process. Although it will as a rule seldom occur that a target is so explicitly formulated as "before the year 2000 the salmon must be back in the river Rhine", as once happened in the

Netherlands, in many other cases there are opportunities to promote a bio-indicator as an appealing spearhead for policies. The present ambitious EU target stating that by 2010 no more species shall be lost, calls for a proper bio-indicator.

2. *They must be responsive to changes*
Being responsive to changes in the outside world is a requirement for any indicator, but for bio-indicators to play a role in policy making, this requires extra attention. The bio-indicators need to function in a dynamic environment together with emission indicators which are yearly updated and react immediately on changes in consumption patterns. This requirement comes down to making a good selection of species or characteristics of species or by combining information on occurrence and abundance of species with information on the habitat area or another variable that shows a fast response to changes in pressures.

3. *They must be easily understandable in their relation with driving forces or pressures*
Or, in other words: the indicators must fit into a story. Before any selection of, or even discussion on indicators can take place, the context of their use should be clear. A graph on "the spreading of imposex in dogwhelks", for instance, does not easily raise interest unless the relation with the use of a certain endocrine disrupting chemical is explained. When combined with general information on the use of tributyl tin in vessels, the indicator indeed gets an action perspective. The selection of species, species groups, habitats and their functionalities clearly needs to take into account their significance with respect to driving forces. One should be aware that "the story" reflects messages and therewith values that are communicated with the indicators.

5. The use of bio-indicators in EEA-assessments

The European Environment Agency, is, as an international institution, limited by the data collection by its Member countries. Essentially two type of indicators are being used: indicators based on biomonitoring, and bio-indicators based on the observation of changes in species population distribution.

For bio-monitors data limitations mean an almost exclusive reliance on the few species-based monitoring exercises set up by other national and international organisations, such as the secretariats of the marine conventions (e.g. Ospar/Mon, 1998) and NGOs, or incidental international studies such as the heavy metals in moss study (Rühling, 1994).

An example of an indicator based on biomonitoring is given in Figure 5. (taken from EEA, 2001). The development of hazardous substance concentrations is presented as a performance indicator related to the Ecological Reference Index (ERI) for the blue mussel. The ERI is the ratio between the mean concentration of a pollutant in an organism and the OSPAR Background/Reference Concentration (BRC) or the upper value of the range of OSPAR Ecotoxicological Assessment Criteria (EAC) for that pollutant and that organism. BRC-values have been set for heavy metals in

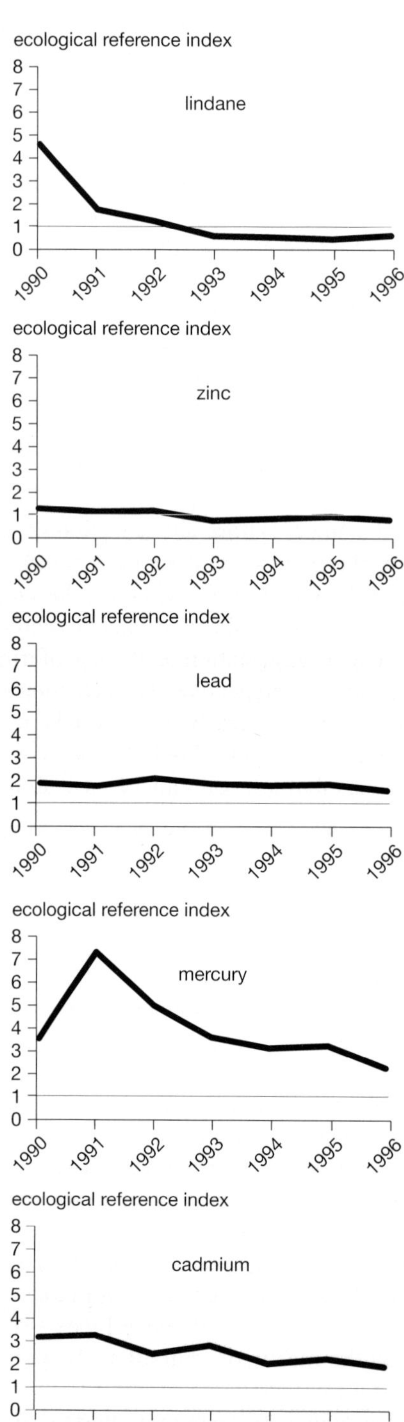

Figure 5.

Hazardous substances in blue mussels in the North East Atlantic

Note: The ratio between mean concentrations of hazardous substances in the blue mussel and values for Background/Reference concentrations (BRC-values) or Ecotoxicological Assessment Criteria (EAC-values) for these substances has been used to present the development on a comparable scale. The BRC/EAC-values have no legal significance and are only related to risks for marine ecosystems and not to human health risks (from consumption of mussels). The ratio is called Ecological Reference Index (ERI). BRC/EAC-values have an upper and lower limit; the upper value has been used to calculate the ERI. ERI = 1 for the upper limits of the BRC/EAC-range.

Data source: OSPAR/MON 1998

mussels and an EAC-value has been set for lindane on the basis of toxicity tests with marine and freshwater organisms and includes a safety factor. An ERI value above '1' may indicate an ecological risk for the species concerned.

Mean concentrations of lindane, cadmium and zinc in blue mussels in the northeast Atlantic fell in the period 1990 to 1996, which is consistent with falling inputs into the area. Inputs of mercury and lead also fell during this period, but no clear trends are apparent for their mean concentrations in the blue mussel. A high natural (seasonal) variability of concentrations in mussels, variability in age and sex of collected mussels, the relatively small number of time series for blue mussels and differences in the bio-availability of hazardous substances, as well as uncertainties on the representativeness of the samples may influence the results. Statistically significant trends are found mainly at locations in estuaries and fjords, which are closer to the sources of pollution.

The bio-indicators based on population distribution that the EEA can use refer so far to well-monitored species/species groups: birds, higher plants, and large mammals. Sources have been, for example, Wetlands International, Birdlife International and the forest health monitoring programme.

Most of the bio-indicators developed are based on indexes for the occurrence of a group of species. Figure 6 taken from (EEA, 2000) uses the data from the International Waterbird Census project, covering 23 open water species. The increase shown by the index is probably as much the result of the recovery after the cold winters of 1982, 1985, and 1987, as of the increased wetland conservation activities in the period.

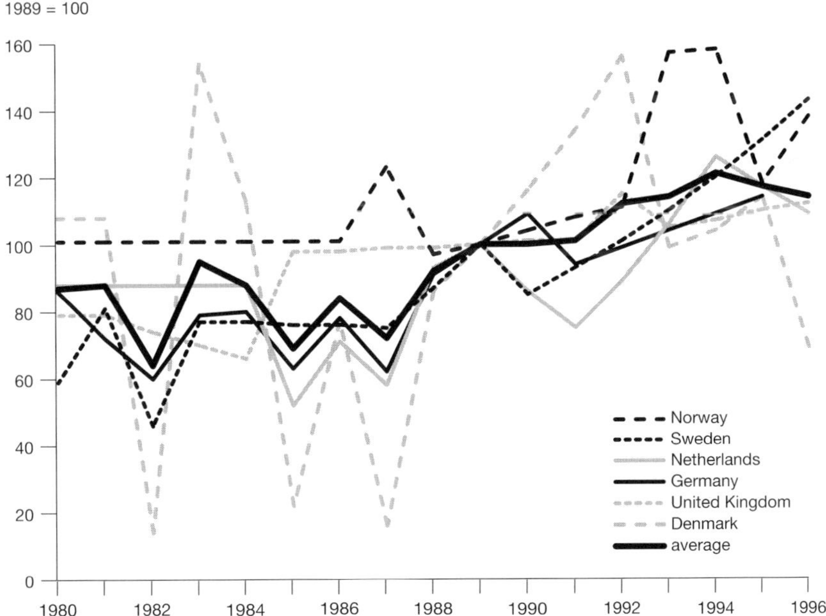

Figure 6. Index of wintering waterbirds in six north-western European countries.

Source: Ramsar Bureau, Wetlands International, EEA.

Clearly this indicator does not comply with the responsiveness requirement given above and no policy use of the indicator has been seen yet.

The use of species groups that can be linked to categories of human activities or habitat management, however, is a promising development.

The possibilities for a European indicator that links species with ecosystems such as agricultural land or forest, after the example of the UK indicator for the population of farmland and woodland birds (H13 in DETR,1999), is being explored. Based on UK experiences, the potential for the use of this indicator are quite high at the national level. At European level it may be useful as a headline indicator (see Table 1), if it can be related to main ecosystems and to groups of species clearly associated with changes in these ecosystems.

References

CEC, 2001. A Sustainable Europe for a Better World: A European Union Strategy for Sustainable Development. Commission's proposal to the Gothenburg European Council. COM(2001)264 Final.

DETR, 1999. Quality of Life Counts. Indicators for a Strategy on Sustainable Development for the United Kingdom: A Baseline Assessment. Department of the Environment, Transport and the Regions. London.

EEA, 1999. Environment in the European Union at the turn of the century. Environmental Assessment Report No. 2. European Environment Agency, Copenhagen.

EEA, 2000. Environmental signals 2000. Environmental Assessment Report No. 6. European Environment Agency, Copenhagen.

EEA, 2001. Environmental signals 2001. Environmental Assessment Report No. 8. European Environment Agency, Copenhagen.

EU Council, 2001. Proposal for a Decision of the European Parliament and of the Council laying down the Community Environmental Action Programme 2001–2010 (6th EAP). 9775/01, Brussels.

Kraemer, R.A., 2001. Results of the "Cardiff-processes" – assessing the state of development and charting the way ahead. Report to the German Feral Environment Agency and the German Federal Ministry for the Environment, Nature Conservation and Nuclear Safety. UFOPLAN Report No. 299-19-120. Ecologic, Berlin.

Ospar/Mon, 1998. Summary Report of the Ad Hoc Working Group on Monitoring (MON). Copenhagen, 23–27 February.

Rühling, Å. (Ed.), 1994. Atmospheric heavy metal deposition in Europe – estimations based on moss analysis. Nordic Council of Ministers. Nord 1994, 9.

Winsemius, P., 1986. Gast in eigen huis, beschouwingen over milieumanagement. [Guest at home, reflections on environmental management]. Samson H.D. Tjeenk Willink, Alphen aan den Rijn.

Chapter 26

Critical assessment of international marine monitoring programmes for biological effects of contaminants in the North-East Atlantic area

Peter Matthiessen

Abstract

This chapter describes the use of techniques for monitoring the biological effects of contaminants by means of international environmental monitoring programmes in the north-east Atlantic, with particular focus on the work of the Helsinki Commission (HELCOM) and the Oslo and Paris Commission (OSPAR). Some biological monitoring has been included in such programmes since they began in the early 1980s, but the development and introduction of these techniques has lagged behind the use of chemical analytical methods. Even today, chemical and biological monitoring have not been fully integrated in the programmes of most north-east Atlantic coastal states, despite encouragement from international organisations such as OSPAR and the International Council for the Exploration of the Sea (ICES). This has led to a number of significant shortcomings in our understanding of marine environmental quality: (1) Much data obtained on the presence of contaminants in north-east Atlantic marine ecosystems has been impossible to interpret reliably in terms of likely biological impacts, either on individuals or on populations; (2) Coverage of the area by biological monitoring has been very patchy, with the best data being available from the North Sea; (3) We have, as yet, little information on whether reductions in the concentrations of many marine contaminants are being reflected in similar reductions in their effects on wildlife; (4) With some exceptions, we have rarely obtained data which diagnose the chemical causes of observed biological effects; (5) In broad terms, it has not yet been possible to use data on the effects of contaminants on individual organisms to make confident predictions about possible ecosystem-level impacts. Having said all that, there is no doubt that both organic and inorganic contaminants are having undesirable effects on marine organisms in many areas. Such effects range from mortality to subtle impacts on reproduction and normal endocrine function, and are particularly marked in many urbanised estuaries and other semi-enclosed marine waters. Effects of persistent substances have also been observed further afield in coastal and offshore organisms, particularly in those associated with sediments or higher trophic levels where such contaminants tend to accumulate. Some of these effects are known to have caused population declines or impoverished communities in some instances, but it is currently not possible to state with confidence that present contaminant inputs to the North-east Atlantic are a threat to the marine ecosystem as a whole. The chapter concludes with a number of recommendations which address the problems identified in marine biological effects monitoring programmes.

Keywords: Marine monitoring; biological effects; North-east Atlantic; OSPAR; HELCOM

1. Introduction

Until the mid-twentieth century, the world's oceans and coastal seas were regarded as effectively limitless resources which could be exploited with little fear that their capacity to supply food and materials, or to absorb wastes, would be exceeded. In particular, the idea that wastes discharged to the seas could be adversely affecting the marine ecosystem would have seemed just as fanciful as the notion that human activities could alter the atmosphere and change climate. This is not to say that some apparently localised contaminant-related effects in the marine environment had not been noticed earlier, merely that widespread systemic impacts were considered unlikely due to the vastness of the oceans in comparison with the seemingly small scale of polluting inputs. Thus, for example, the biological effects of polluting discharges to the Thames estuary in the United Kingdom were specifically addressed by Albert Günther of the British Museum (Natural History) in London in a report to the Metropolitan Sewage Discharge Commission in 1883 (Günther, 1883, cited in Wheeler, 1979). The essentially local nature of marine pollution at that time was due to the relatively small scale of discharges combined with the absence of significant numbers of persistent synthetic chemicals which only came on the scene in later years.

Human perceptions of the global environmental impacts associated with pollution only started to crystallise in the early 1960s as a result of, for example, Rachel Carson's seminal book *Silent Spring* (Carson, 1962), leading up to the United Nations Conference on the Human Environment in 1972 which precipitated the United Nations Environment Programme (UNEP). This resulted, *inter alia*, in the establishment of the Global Investigation of Pollution in the Marine Environment (GIPME), the Global Environment Monitoring System (GEMS), and the production of holistic reports on various environmental compartments (e.g. Meybeck *et al.*, 1989, dealing with freshwater systems; Goldberg, 1976, on marine waters). The IMO/ FAO/ UNESCO/ WMO/ IAEA/ UN/ UNEP Joint Group of Experts on the Scientific Aspects of Marine Pollution (GESAMP) has been particularly active in this area, producing a first review of the health of the oceans in 1982 (GESAMP, 1982), and reports on the state of the marine environment in 1990 and 1997 (GESAMP, 1990, 1997). These reviews were short on hard data and fairly general, because co-ordinated marine monitoring programmes targeted on contaminants and their biological effects had only been started in the 1980s, although some radiological monitoring had got underway in the 1950s. For example, GESAMP (1990) only briefly mentioned that there was an association between organochlorine residues and declining predatory seabird populations, between petroleum hydrocarbons and adverse effects in molluscs, sea grasses and crustaceans, between polychlorinated biphenyls (PCBs) and sterility in Baltic and Wadden Sea seals, and between various contaminants and diseases in fish and marine mammals. Apart from these examples of impacts seen in the field, the approach taken was to look at concentrations of chemicals which had been reported to be present in the marine environment, and to compare these with concentrations known to be toxic to various marine organisms, largely on the basis of acute, lethal toxicity tests. This was the best that could be done at the time, but there was not much detail, and little attempt to focus attention on particular areas of concern. Only subsequently did GESAMP publish guidelines for marine environmental assessments (GESAMP, 1994), and for the use of biological indicators in marine monitoring programmes (GESAMP, 1995), publications which were influential in the development of some of the international programmes described below.

2. Early chemistry-based marine monitoring programmes

The oceans are a truly international resource, and marine pollutants and polluters do not respect international boundaries. This perception is especially acute in the countries bordering the North, Baltic and Mediterranean Seas, and is probably a major reason why international efforts to assess the quality of the North Sea were some of the first such activities to get under way. This process has been conducted primarily at the instigation of a series of International Conferences on the Protection of the North Sea convened by the coastal states and reporting direct to ministers responsible for the environment. Each of the conferences has considered reports on North Sea quality, the first being submitted to the conference held in Bremen in 1984, and an updated version of it to the London conference in 1987 (NSC, 1987). However, even before the North Sea Conferences, the International Council for the Exploration of the Sea (ICES) had set up its Working Group on the Pollution of the North Sea which reported in considerable detail on the presence of chemical contaminants and on early attempts to measure the toxicity of some of these substances to marine biota (e.g. ICES, 1974), and the ICES Working Group on Marine Pollution Baseline and Monitoring Studies in the North Atlantic was also active (e.g. Pearce *et al.*, 1986).

The early North Sea Quality Status Reports (QSRs) were more detailed than those produced by GESAMP, but although they contained some biological data, its collection was not closely coordinated and the biological information did not derive from an international monitoring programme. However, much of the chemical monitoring data available at this time was produced by the Oslo and Paris Commission's Joint Monitoring Group (JMG) which started to develop its Joint Monitoring Programme (JMP) in 1978. The purpose of the JMP, working entirely through the governments of coastal states, was to assess:

- possible hazards of marine pollution to human health;
- harm to living resources and marine life;
- existing levels and spatial extent of marine pollution;
- the effectiveness of measures being taken to reduce pollution from land-based sources i.e. temporal trends.

In theory, the JMP covered the entire OSPAR Convention area (i.e. a huge segment of ocean stretching from the North Pole south to 36°N, bounded in the west by longitude 42°W and in the east by 50°E, excluding the Baltic and Mediterranean), but sampling coverage was very uneven and focused mainly on coastal areas, with a few open sea sites in the Irish Sea, North Sea and Skagerrak. The substances to be monitored were as follows:

- in seawater: Cd, Cu, Hg and Zn;
- in organisms: Cd, Cu, Pb, Hg, Zn and PCBs;
- in sediments: Cd, Cu, Pb, Hg and Zn.

There was also a request for gamma-hexachlorocyclohexane (γ-HCH) to be measured voluntarily in organisms and sediments, and PCBs to be measured in sediments. Yet other substances were recorded on an ad-hoc basis, and provisions were put in place for analytical quality assurance of chemical determinands.

As with the GESAMP reports (GESAMP, 1982, 1990), the main emphasis in the early QSRs lay in the reporting of contaminant concentrations in waters and their interpretation in the light of data from supposed clean areas, laboratory-based toxicity data, and various water quality standards derived from such data. It was recognised that this was not ideal because it took no account of real environmental mixtures which might exert toxicity not shown by individual components. It also failed to account for the reduced bioavailability experienced by contaminants in the presence of various mitigating factors such as suspended particulates and organic carbon, and had difficulty in accounting for possible long-term effects on the basis of short-term acute toxicity data. Nevertheless, it was concluded that metal levels in seawater from the North Sea were almost all below the various quality standards available at the time, although it was pointed out that most such standards had been developed as guides to maximum permissible levels below regulated discharges, not as quality targets for the open sea.

Information on residues in sediments could not even be compared with toxicity data as there were few such data available. Furthermore, the data on residues in biota were simply compared with JMG 'upper', 'medium' and 'lower' categories which bore no relation to levels of environmental or public health significance, having been derived purely for statistical convenience. The 1987 North Sea QSR, for example, highlighted the fact that some fish and mussels from the Oslo Fjord, Firth of Forth, English northeast coast, Humber, central and southern North Sea, German Bight, and Belgian and Dutch coasts contained elevated concentrations of various metals, but it was unable to come to a conclusion about whether these were a problem.

The 1987 North Sea QSR did, however, make some remarks on the probable effects or otherwise of contaminants on wildlife. For example:

- The report indicated that the diversity of benthic invertebrate communities had been adversely affected near various point sources such as offshore oil and gas installations in the central northern North Sea and in areas of limited water exchange such as some Norwegian fjords. It made the point, however, that more subtle contaminant influences on far-field communities were difficult to distinguish from physical habitat disturbance and natural changes.
- Changes in plants (both attached algae and higher plants) and increases in algal blooms in some areas were tentatively ascribed to eutrophication.
- Declines in certain fisheries were firmly ascribed to fishing intensity, although it was noted that some fisheries had been damaged because contamination had led to fishery closures in order to protect human health.
- The circumstantial evidence for a link between contaminants and certain fish diseases such as lymphocystis in dab was considered to be fairly strong, but far from proven.
- In the light of decreasing organochlorine (OC) contamination, the greatest contaminant-related threat to North Sea seabirds was held to be oil spills.

- In marine mammals, a firm link was made between elevated levels of OCs (especially PCBs) and population declines in Wadden Sea and Baltic seals, due to interference with normal reproduction. However, the possible effects of contaminants in other marine mammals were unknown.
- Finally, it was indicated that contaminants can cause "stress" at concentrations below those that are acutely toxic. Although few field data were available at the time, it was recognised that bioassays[1] and what are now called biomarkers[2] could measure various aspects of stress, and might be able to provide early warnings of more serious biological impacts. There was, however, an understandable desire to avoid the automatic equation of 'effect' with 'harm'. The report nevertheless called for the development of more sensitive measures of 'stress', and this clearly prefigured developments reflected in subsequent QSRs.

In summary, the North Sea QSR in 1987 was able to draw few clear conclusions concerning links between the presence of contaminants in marine environmental compartments and their possible biological effects, and hence, it was difficult to make decisions about appropriate pollution prevention measures. This was partly due to the paucity of suitable biological monitoring techniques at that time; nevertheless, it was a significant shortcoming because the mere presence of contaminants is not in itself a cause for concern – contamination does not equal pollution. Marine chemists already had a wide range of exquisitely sensitive analytical methods available to them, so it is understandable that marine monitoring should have been essentially chemistry-driven. Another factor which limited the scope for interpreting marine pollution in biological terms was that, for institutional and organisational reasons, chemical and biological marine monitoring were generally conducted as entirely separate activities, often on different research cruises. The attribution of cause-effect relationships was therefore almost impossible. Finally, the 1987 QSR was not able to draw on a body of biological effects data derived from an international monitoring programme, and so was inevitably based on unco-ordinated information culled from published data and national submissions. This reduced the scope for reaching conclusions on a truly regional basis. The 1987 North Sea QSR was nevertheless a significant milestone on the road to the development of a modern marine monitoring programme in the northeast Atlantic area. It led directly to the establishment of the North Sea Task Force, the first international programme to include serious consideration of the biological effects of contaminants in the marine environment.

Biological effects studies were also not included in early international monitoring of the Baltic Sea. Assessments published to date by the Helsinki Commission (HELCOM – the body responsible for controlling polluting inputs to the Baltic) of the state of the Baltic marine environment (HELCOM 1986, 1990, 1996) do not report biological effects measurements as such, but they make it clear that there have been severe anthropogenic changes due to polluting discharges into this almost landlocked sea. Some of the major issues covered by the HELCOM monitoring programme have been as follows:

- Increasing levels of nutrients in many areas (winter surface concentrations of up to 12 μmol/l of nitrate+nitrite in the Belt Sea, and up to 2 μmol/l of phosphate in the Gulf of Finland, in 1989–1993) have led to more frequent algal blooms (mean spring chlorophyll-*a* concentrations of up to 7–18 mg/m³ in surface waters of the Gulfs of Riga and Finland in 1979–1993, although 1–6 mg/m³ more generally) which on decay have produced regular or continuous oxygen deficits and elevated hydrogen sulphide concentrations (>150 μmol/l) in deep waters of the central and western Baltic. This has caused mass die-offs of entire benthic communities over very wide areas (up to 70,000 km²), and has also affected the food resources and spawning areas of some fish and shellfish species (e.g. cod, crab and lobster). A seawater inflow (310 km³) from the North Sea in 1993 caused temporarily improved conditions, but both main Baltic basins have largely returned to anoxia.

- Eutrophication has probably also increased the incidence of harmful algal blooms (especially some cyanobacteria and dinoflagellates) which can cover up to 75,000 km² in the summer and whose toxins kill marine organisms and make seafood unfit for human consumption.

- Among the metals, cadmium, copper and nickel in seawater have been present at levels 2–3 times greater than in the North Sea and 5–10 times greater than in the North Atlantic, but other heavy metals are not seriously elevated, and all appeared to be static or on the decrease in 1980–1993.

- However, suspended particulates have contained 10->200 times the metal concentrations found in the North Atlantic, and this has been reflected in deposited sediments in many areas.

- Metal levels in fish and bivalves have generally not been very elevated, although there have been a number of localised exceptions, with moderately high levels of zinc, cadmium, mercury and lead in some bivalves, and elevated levels of cadmium in fish from the northern Baltic (up to 3 μg Cd/g). Lead levels in fish have been on the decline, and mercury levels have shown no consistent trend, but cadmium in herring liver from several areas was increasing at the rate of 5–8% per year during 1981–1994.

- In some areas, organotin concentrations are causing adverse effects in molluscs.

- There have been high but generally decreasing levels of PCBs and other organochlorines (e.g. DDT and HCH) in birds, fish and seals. For example, by 1995, total PCB levels in herring muscle had dropped to 0.5–1.5 ng/g (from levels up to 3.5 ng/g in 1978), with parallel decreases in total DDT (down to 0.1–0.7 ng/g). Concentrations of lindane residues in fish and mussels are also decreasing. Rates of decrease in most organochlorine residues in various biota during 1972–1995 ranged from 4 to 13% per year. However, some organohalogens such as brominated flame retardants are known to be increasing.

- Total hydrocarbons in water have not been especially high (up to 2 μg/l), although birds and coastal benthic communities have been damaged by chronic oil pollution (inputs of 20,000–70,000 t/year). For example, in the winter of 1994–1995, more than 25,000 dead oiled ducks were found along the coasts of Sweden, Poland, Lithuania and Latvia. There are about three big oil spills per year in the Baltic, each averaging about 225 tonnes, but 600–700 smaller spills are also recorded.

- Microbial populations appear to have adapted to some extent to high contaminant levels and have acquired the ability to degrade toxicants such as PCBs and PAHs.

Unlike the early North Sea QSRs, the first three Baltic Periodic Assessments made little attempt to interpret contaminant concentrations in terms of potential biological impact. One of the few major biological effects measurements concerned benthic invertebrate communities which had clearly been devastated over a huge area by depleted oxygen concentrations in the deeper waters. The distribution and species diversity of macrophytobenthos also appears to have been affected by eutrophication, with decreases in eelgrass and bladder wrack beds (*Zostera marina* and *Fucus vesiculosus*) and increases in fast-growing algae (e.g. *Cladophora* spp.).

PCBs have been identified as the main cause of population declines in Baltic seals and mink, and DDE has been responsible for eggshell thinning in guillemots and white-tailed sea eagles, but many of these populations are now beginning to recover. For example, nesting success in sea eagles dropped from 72% before 1950 to 25% in the 1960s and 1970s, but then recovered to 68% in 1994. Grey seals from the northern Baltic have increased in numbers from 400 in 1982 to about 1400–1700 in 1990–1994, although some still have pathological deformities related to PCBs, and mortality in the young is >50% in the southern Baltic. Finally, the fish disease lymphocystis (an indicator of stress, including that caused by pollution) increased in flounder during 1986–1993, and the prevalence of flounder liver nodules (pre-neoplastic change that is probably related to contaminants such as PAHs) was as high as 10% in older animals.

In summary, data from the Baltic for the period to 1993 suggest that many contaminants (apart from nutrients) were on the decrease, as were some of their effects, but detailed biological effects monitoring was not generally being conducted at that time.

3. International marine biological effects monitoring programmes in the North-East Atlantic

The rest of this chapter encompasses international marine biological effects monitoring in the whole OSPAR area i.e. OSPAR Region I (Arctic), Region II (Greater North Sea), Region III (Celtic Seas), Region IV (Bay of Biscay and Iberian Coast), and Region V (Wider Atlantic), as well as in the area of the Helsinki Commission (HELCOM) (Baltic Sea). However, much of the focus will inevitably be on the North and Irish Seas where the majority of biological effects monitoring has been conducted to date. The paper will include consideration of the effects of contaminants, but generally excludes eutrophication and the effects of radioactivity.

3.1. North Sea Task Force and the 1993 North Sea Quality Status Report

The Ministerial Declaration which emanated from the Second International Conference on the Protection of the North Sea (London, 24–25 November 1987) requested (paragraph 52) "the International Council for the Exploration of the Sea (ICES) and the Oslo and Paris Commissions (OSPAR) to consider together the optimal means to achieve [harmonised methods for monitoring] including the possible benefits of a joint Working Group (or Task Force) . . . and to organise a harmonized programme of studies . . .". Annex G of the Declaration stated that "A co-ordinated scientific programme needs to be developed in the North Sea, to provide more consistent and dependable data and to permit links between inputs, concentrations and effects to be established with greater confidence". This request led to the creation of the North Sea Task Force (NSTF) which began work in 1988, and which ultimately produced the 1993 North Sea Quality Status Report (OSPAR, 1993).

In 1990, the Third International Conference on the Protection of the North Sea held in the Netherlands invited the NSTF to address a number of specific topics, including the impact of fishing activities on the ecosystem, surveillance of chemicals not covered in routine monitoring programmes, the environmental impact of persistent chemicals, the role of atmospheric inputs as a source of contaminants, and an assessment of existing damage. The practical monitoring work required to address these issues was still conducted by the individual North Sea coastal states, but henceforth was to be part of a Monitoring Master Plan which specified the data to be collected, the sampling stations and, increasingly, data quality. Furthermore, the North Sea was divided up into 10 sub-regions, in each of which one or two countries were to take the lead in producing a separate sub-regional report. As with any international collaborative programme, the amount of effort devoted by coastal states to the NSTF varied considerably according to available funding and expertise.

Although a much more slick and professional production than the report to the Second North Sea Conference, the 1993 QSR broadly followed a similar format, with chapters covering a general description of the North Sea, physical oceanography, marine chemistry, marine biology, and man's impact on ecosystems. The report ended with an overall scientific assessment and some conclusions. However, only six pages were devoted to the effects of contaminants, and despite the earlier requirement to provide information on cause-effect relationships, relatively little progress had been made in this direction. The following chemical stressors were discussed.

3.1.1. Metals

The report concluded that the greatest metal-related risks to the North Sea ecosystem (mainly in estuaries and coastal areas) stemmed from copper which was thought to be affecting algae, cadmium and mercury threatening some top predators such as seals, porpoises and seabirds, and lead perhaps posing a risk to some consumers of shellfish including man. However, the available information was based solely on measured concentrations in water and tissues, from which potential impacts were inferred. Cause-effect links were only made through correlation and were rather tenuous in some cases due to the fact that a proportion of metal in seawater is adsorbed to dissolved organic matter and is therefore less bioavailable.

3.1.2. Tributyltin

In this case, it was possible to attribute widespread damage to bivalve and gastropod mollusc populations to the presence of tributyltin (TBT) derived from antifouling paints. In some coastal and estuarine areas, TBT had completed eliminated the dogwhelk *Nucella lapillus* through the induction of masculine characteristics in females (imposex), and seriously interfered with normal breeding and shell growth in oysters. Furthermore, edible whelks *Buccinum undatum* caught near offshore shipping lanes were also affected. This was the first well-documented example of a particular synthetic chemical causing widespread ecological damage in the marine environment.

3.1.3. Oils and Polycyclic Aromatic Hydrocarbons (PAH)

Hydrocarbons derived from offshore drilling activities were clearly identified as having caused localised damage to benthic invertebrate abundance and diversity, over a total estimated area of 1900 to 8000 km^2. There was also some information that oil in water reduced scope-for-growth in caged bivalves and had caused localised tainting of fish in some areas. Oil slicks were of course associated directly with fouling of birds and some marine mammals. An association was also made between PAHs in sediments, elevated cytochrome P4501A activity and pre-neoplastic liver lesions in dab in the German Bight, although the causative links were not firmly established.

3.1.4. Synthetic organic chemicals

Firm links were made between polychlorinated biphenyls (PCBs), reproductive effects and declining populations in Dutch seals, and organochlorines in general were experimentally associated with reduced hatching success in whiting. Other than these few examples, there was little information presented on the effects of synthetic organics.

3.1.5. Combined effects of chemicals

The NSTF reported on results obtained with more or less non-specific bioassays and biomarkers, and represented one of the first concerted attempts to use such techniques in an international monitoring programme. Results from the use of three main techniques were discussed: incidence of fish disease, induction of cytochrome P4501A in dab, and oyster embryo bioassays of seawater. Attention was drawn to the high prevalence of epidermal hyperplasia/papilloma found in dab caught in certain polluted areas, although at least a proportion of these cases are triggered by natural changes. More significantly, elevated levels of liver tumours in flatfish were associated with PAH contamination in a number of coastal and offshore areas, although once again, cause-effect relationships were not firmly established. An early response to PAH and PCB exposure in fish is the induction of the cytochrome P4501A enzyme system, measured as increased activity of ethoxyresorufin-o-deethylase (EROD). Elevated EROD activity in dab was reported from several locations including the central/northern North Sea oil exploration area, the Dogger Bank, and a number of estuaries. Unfortunately, there were problems in comparing EROD data from different laboratories, indicating the need for much stricter quality control in the future. Finally, oyster embryo bioassay data showed that seawater and sediment elutriate quality was good in most coastal and offshore areas, but were lethally and sub-lethally toxic at some stations in a number of industrialised estuaries. A number of other observations made with non-NSTF methods were also reported (e.g. widespread abnormalities in pelagic fish embryos in North Sea coastal areas), but none were firmly linked with contaminants.

It will be apparent from the foregoing that some progress was beginning to be made in the deployment of methods for studying the biological effects of specific chemicals or groups of chemicals. This particularly applied to the surveys of imposex and shell-thickening in molluscs which were organotin-specific, and the measurements of EROD induction in fish which identified the presence of planar organic molecules such as PAHs, PCBs and polychlorinated dibenzofurans (PCDFs). A number of other observations of biological effects (e.g. effects of oils on benthic communities) were linked fairly closely to causative substances, but it was clear that the causality issue was still only being addressed in a patchy and incomplete manner. Also, a big issue which began to arise was how to interpret the ecological significance of effects observed with biomarkers and bioassays – it was clearly not enough to say that an organism had been exposed to contaminants, it was also desirable to know whether it had been damaged (i.e. polluted), and whether this damage had knock-on effects for populations and communities. The evidence available at the time suggested that species diversity in benthic invertebrate communities, for example, was only seriously reduced near certain point sources (e.g. offshore oil installations) and in some urbanised estuaries where organic enrichment was suspected of being the main culprit. Finally, it became apparent as a result of the NSTF that future international biological effects monitoring programmes would need to pay much more attention to quality assurance and control if data from different countries were to be sufficiently comparable to identify those areas most and least affected by pollution.

Nevertheless, on the basis of the NSTF observations, a number of recommendations were made, including:

- Existing goals for reduction of metal inputs to the North Sea should continue to be implemented, and cadmium emissions more strictly controlled.
- Further reductions should be made in the use of TBT.
- Reduction policies for PAHs should be considered, and the potential risks posed by other organic chemicals assessed.
- Strategies should be developed to prevent the entry of PCBs and other organochlorines into the sea.
- Reduction of oil in discharged cuttings and produced water should be continued, and additional measures considered to reduce accidental oil pollution by shipping.

From the point of view of the present paper, perhaps the main conclusion of the NSTF (articulated by the NSTF Vice-Chairman Dr John Portmann at the Scientific Symposium on the North Sea Quality Status Report – Andersen *et al.*, 1996) was that, (a) the only way to decide if environmental media are harmful is to undertake biological effects studies, and (b) there was a clear need for the development of better biological effects methods. The 1993 North Sea QSR was nevertheless the first such assessment to start applying these techniques in a co-ordinated way in the north-east Atlantic.

3.2. The HELCOM monitoring programme (COMBINE)

In the 1990s, the HELCOM Monitoring and Assessment Group developed a Manual for Marine Monitoring (available on the HELCOM website – www.helcom.fi) to support its new COMBINE monitoring programme. The programme covers all the Baltic coastal states, although their contributions vary even more widely than in the OSPAR area. This is mainly due to the political history of the region, and the great disparity in resources and expertise available to the various HELCOM members. The objectives of COMBINE are:

- to compare contaminant levels in selected biota in order to detect spatial trends and "hotspots";
- to measure temporal trends in contaminants in selected biota;
- to measure contaminants in biota in order to assess whether they pose a threat to these species or to higher trophic levels.

It will be noticed that although this programme is focused on biota, it is still very much 'chemistry-led', although in the detail there is limited provision for 'supporting' biological effects studies and for residue measurements in waters and sediments. COMBINE specifies the locations where monitoring should occur, and the species which should be studied.

In the open sea, Baltic monitoring species include pelagic organisms:

- Herring
- Cod
- Guillemot.

In the coastal zone, they include:

- Macroalgae – bladder wrack
- Invertebrates – mussels, clams and isopods
- Fish – flounder, perch, viviparous blenny
- Birds – common tern, white-tailed sea eagle
- Seals – grey, ringed and common.

There is also a detailed specification of the contaminants to be measured in various tissues and media; these include heavy metals, DDTs, PCBs, hexachlorobenzene (HCB), α- and γ-HCH, PAHs, total hydrocarbons, brominated flame retardants, and tributyltin (TBT). Of course, not all contaminants are measured in each matrix, and not all HELCOM member states have undertaken to make all the recommended measurements in their areas. However, the emphasis is on the monitoring of substances which are likely to be persistent, toxic and bioaccumulative.

The supporting programme of biological measurements is in theory fairly well integrated with the collection of chemical data, but as with the chemical determinands, there is great disparity between the member states (Table 1). Only two measures are used which could be considered as diagnostic of the effects of chemical contaminants – EROD induction in perch and blenny (diagnostic of exposure to planar PCBs, PAHs and PCDFs), and imposex in whelk (diagnostic of effects of triorganotins), and neither is applied in more than one or two countries. There is also some encouragement to measure depression of acetylcholinesterase activity as an indicator of exposure to organophosphates and carbamate pesticides, but this is not part of the core programme. All the other biological variables are indicative of general health, but do not respond solely to contaminants.

Table 1. Biological effects monitoring conducted in various countries as part of the HELCOM COMBINE monitoring programme in the Baltic Sea.

Species	Variables	Denmark	Estonia	Finland	Germany	Latvia	Lithuania	Poland	Russia	Sweden
Whelk *Buccinum*	Imposex (caused by organotins)	+								
Perch*	Physiology, population parameters, reproduction, biomarkers			+	+			(+)		+
Viviparous blenny*	Physiology, population parameters, reproduction, biomarkers							+		+
Fish community**	Population parameters			+				+		+
Seals***	Population dynamics, reproduction, biomarkers			+				+		+
White-tailed eagle	Population dynamics, reproduction			+						+

Notes:

* Variables may include growth, gonad weight, fecundity, condition factor, external disease, gonadosomatic index, hepatosomatic index, haematocrit, leucocrit, plasma ions, EROD activity, blood lactate and tissue glycogen
** Variables may include species composition, catch per unit effort and age composition
*** The biomarkers are unspecified in the COMBINE manual

Unfortunately, little further comment can be made here as the first results of the COMBINE programme covering 1994–1998 were not published until late 2001 (as part of the 4th HELCOM Periodic Assessment). It is clear that HELCOM has made a tacit commitment to integrated monitoring that includes both chemical and biological determinands, but the national take-up of biological effects monitoring techniques is still very poor. Indeed, the only effects technique which has yet been formally adopted by HELCOM as part of its core programme of studies in the coastal zone is reproductive success and population stability in the viviparous blenny (*Zoarces viviparus*) (K. Grip, HELCOM, pers. comm. 2001).

3.3. The OSPAR Joint Assessment and Monitoring Programme (JAMP) and the OSPAR Quality Status Report 2000

Subsequent to the establishment of the revised OSPAR Convention in 1992, the Oslo and Paris Commission decided in 1994 to develop a new joint monitoring programme for the entire convention area. It was the intention to include, as far as possible, both chemical and biological monitoring techniques in an integrated fashion. The OSPAR member states also agreed to work towards the production of a new QSR by 2000, which would be the first to encompass all five OSPAR regions, each of which would also report separately. The new programme (the Joint Assessment and Monitoring Programme – JAMP) was adopted in 1995, and within it there is now a Coordinated Environmental Monitoring Programme (CEMP) which is that part of the JAMP where the national contributions overlap and are co-ordinated (SIME, 2000; Stagg, 1998). The main monitoring focus has moved away from seeking spatial variations in contaminants and their effects (although this continues to some extent), towards identifying temporal trends in order to assess, for example, whether various regulatory actions are being successful.

It is worth pointing out that OSPAR does not contribute financially to the implementation of the JAMP, but is reliant on the inputs of national monitoring programmes. As such, OSPAR cannot insist that monitoring should be conducted in a particular way, and each country's efforts are dictated by its own budgets and priorities. This leads to considerable geographic unevenness in the JAMP as a whole.

The general strategy of the JAMP is to address a series of issues which are perceived as being of concern. These include both chemical and biological questions, and those of direct relevance to biological effects measurement are listed in Table 2. They are clearly very ambitious objectives, and they have not yet all been achieved, but in order to begin the monitoring process, OSPAR and ICES convened a joint workshop in 1995 to agree on the required biological monitoring tools (OSPAR/ICES, 1995). It was agreed that two parallel (and partly overlapping) biological monitoring programmes were needed, one to monitor effects that are not necessarily chemical-specific, and a second to address the effects of particular chemicals. Within the non-specific or general biological effects programme, there was a need both for techniques to monitor the quality status of areas where effects had not been previously suspected, and for tools to monitor areas of known impact. On the other hand, the contaminant-specific effects programme required interlocking suites of techniques which, taken together, would be diagnostic of particular causes.

Table 2.   Issues involving the biological effects of contaminants to be addressed by the OSPAR Joint Assessment and Monitoring Programme (JAMP).

Issue of concern	Biological questions to be answered
Tributyltin (TBT)	• To what extent do biological effects occur in the vicinity of major shipping routes, offshore installations, marinas and shipyards?
Polychlorinated biphenyls (PCBs)	• Do high concentrations in marine mammals disturb enzyme systems? • Do high concentrations pose a risk to the marine ecosystem? • Do high concentrations of non-ortho and mono-ortho chlorobiphenyls in seafood pose a risk to human health?
Polycyclic aromatic hydrocarbons (PAHs)	• Do PAHs affect fish and shellfish?
Biological effects of pollutants	• Where do pollutants cause deleterious biological effects?
Oil	What are the effects of oil on benthic communities? • What are the effects of aromatics discharged with production water?
Chemicals used in mariculture	• In which areas do pesticides and antibiotics affect marine biota?
Ecosystem health	• How can ecosystem health be assessed in order to determine the extent of human impact?

The OSPAR/ICES workshop went on to elaborate a series of biological effects monitoring techniques which would be suitable for the purposes described above. Those finally approved by OSPAR in 1997 are listed in Table 3, together with appropriate chemical methods, showing the programmes within which they were proposed for use. It will be apparent that the methods range from early-warning indicators of damage, through indicators of longer-term change, to ultimate population and community responses. Furthermore, some are chemical-specific while others are simply indicators of general stress which can respond both to chemicals and to natural stressors.

It is vital to recognise that these techniques (including the recommended chemical analyses) were intended to be applied to a given objective as integrated suites, and it was generally not considered sensible for users to 'cherry-pick' a subset of the methods. It was, however, recognised that a few methods might not be applicable in all areas (e.g. reproduction in blenny). The concept of integrated monitoring had been originated by ICES whose Working Group on the Biological Effects of Contaminants had been instrumental in developing, and pointing out the benefits of, biological effects monitoring methods deployed in a programme which was fully co-ordinated with chemical

Table 3. Biological effects and chemical monitoring techniques approved in 1997 by the OSPAR Working Group on Assessment and Monitoring (ASMO) for integrated use in the JAMP, showing the purposes for which they are required.

Method	General biological effects monitoring		Contaminant-specific effects monitoring		
	General quality status	Areas of known or suspected impact	PAHs	Mercury, cadmium and lead	TBT
EROD induction in fish* liver	+	+	+		
Lysosomal stability in fish* liver	+	+			
Fish* liver pathology	+	+	+	+	
Reproductive success in viviparous blenny	+	+			
External diseases in fish*	+	+			
Benthic invertebrate community structure	+	+			
Sediment and water column bioassays		+			
PAHs in sediment			+		
PAH metabolites in fish* bile			+		
DNA adduct induction in fish* liver			+		
Metals in sediment and fish* liver				+	
Metallothionein induction in fish* liver				+	
ALA-D** inhibition in fish* blood				+	
Antioxidant defences in fish* liver				+	
TBT in molluscs***					+
Imposex/intersex in gastropods or shell thickening in oyster***					+

Notes:
* Species not specified, but dab *Limanda limanda* most often used in northern waters.
** δ-amino levulinic acid dehydratase.
*** Gastropods *Nucella lapillus* or *Littorina littorea*; oyster *Crassostrea gigas*.

measurements. In brief, these benefits include the use of organisms as sentinels which integrate effects of all the contaminants present in environmental media, the use of organisms to discriminate between bioavailable and non-bioavailable fractions of contaminants, and the use of specific biomarkers and combined biological and chemical approaches which have the potential to diagnose the causes of particular effects. A fuller discussion of these issues can be found in Addison (2002).

Subsequently, in 1998, OSPAR issued detailed guidelines for general and contaminant-specific biological effects monitoring in the JAMP (OSPAR, 1998 a and b). Earlier experience in the NSTF and elsewhere had highlighted the problems that arose if biological effects methods used in an international programme were not quality assured. OSPAR therefore decided that all the recommended methods must be subject to quality assurance (QA) before they were given full approval. They were consequently given Category II status, implying that they could be used immediately, but that caution should be exercised when comparing data from different OSPAR member states. As a result, a European Union-funded international research project entitled Biological Effects Quality Assurance in Monitoring Programmes (BEQUALM) was initiated at the end of 1998 (Matthiessen, 1999; www.cefas.co.uk/bequalm). The purpose of this was to develop QA procedures for as many of the OSPAR-recommended biological monitoring techniques as possible. The BEQUALM programme was completed at the end of 2001, whereupon a self-funding infrastructure for regular intercomparisons etc. was begun to be put in place.

By 1998, production of the QSR 2000 was already in progress, so there was little opportunity for the fruits of the new JAMP biological effects monitoring methods to be incorporated in the report. Indeed, it is immediately apparent on reading the QSR 2000 that it has not resulted from an integrated programme. Nevertheless, there has been some application of biological techniques in OSPAR member states since the 1993 QSR, and this has been reflected in the QSR 2000 (OSPAR, 2000).

The QSR 2000 was organised in a fairly similar way to earlier reports, with sections on Geography/Hydrography/Climate, Human Activities, Chemistry, and Biology, and a final chapter giving an overall assessment. The chemistry data, having largely been collected in a separate exercise from biological measurements, could only be assessed in the traditional way i.e. by comparison with predicted 'safe' concentrations based mainly on laboratory toxicity data but with the application of large safety factors. OSPAR had been through a protracted process of developing so-called ecotoxicological assessment criteria (EAC) which consist of an upper and lower value for each contaminant and matrix, and had also derived background/reference concentrations (BRC) for contaminants that are partly natural in origin (metals and PAHs). It is likely that the EACs are over-precautionary in at least some cases, so OSPAR recommends that exceedances of the upper values should act as triggers for further investigation, not necessarily for regulatory action. OSPAR (2000) lists the EACs and BRCs, and it should be noted that many of them are provisional due to inadequate toxicity datasets. Table 4 details those matrices and regions where EACs or BRCs were reported to have been exceeded in some places.

For the reasons outlined above, it is difficult to evaluate these data in any meaningful way, although the QSR did indicate that concentrations of several metals and organics were generally on the decline. In at least one case (copper in seawater), the

Table 4. OSPAR regions* in which some samples exceed upper ecotoxicological assessment criteria (EAC) or background/reference concentrations (BRC), as reported in the QSR 2000.

	Water	Sediment	Biota
Cadmium	Region II estuaries exceed EAC	Regions II, III and IV estuaries, and Region II coast/open sea exceed EAC	Levels in mussels from Region II fjords are much higher than BRCs
Mercury	Region IV nearshore exceeds EAC	Region II and III estuaries and Region IV coast and estuaries exceed EAC	Levels in mussels from Region IV coasts are much higher than BRCs. Elevated levels in Region III coastal fish and seals
Copper	Almost everywhere that was monitored exceeds EAC	Regions I, III and IV, coasts and estuaries exceed EAC	None
Lead	None	All regions except Region I, estuaries, coast and open sea exceed EAC	Levels in mussels from Region II fjords are much higher than BRCs. Elevated levels in Region III seals
Tributyltin	EAC exceeded in many estuaries	All regions monitored exceeded EAC	EAC in mussels exceeded in many places
Polychlorinated biphenyls (total ICES 7 congeners)	None	Regions I, II, III and IV estuaries, and Regions II and IV coastal areas exceed EAC	All except Region V mussels exceed EAC. Fish from all regions exceed EAC, with even higher levels in marine mammals
Dioxins and furans	–	Dioxins much higher than background in Regions I and II	–
Hexachloro-benzene	None	Elevated levels in Region II estuaries	–
Lindane	Region II open sea – some exceedances	–	–
DDE	–	–	Region III mussels and fish – some EAC exceedances
Total PAH	None	Region II estuaries exceed EAC	–

Notes:
* Region I – Arctic waters (Greenland, Barents, Norwegian and Iceland Seas).
   Region II – Greater North Sea (including English Channel).
   Region III – Celtic Seas (Irish and Celtic Seas and Atlantic margin).
   Region IV – Bay of Biscay and Iberian coast.
   Region V – Wider Atlantic.

upper EAC value (0.05 μg/l) was exceeded almost everywhere, but as this is below the natural background (0.05–0.36 μg/l), one is not inspired to draw apocalyptic conclusions. On the other hand, the widespread high concentrations of PCBs in many biota probably *are* a cause for concern given their properties as endocrine disrupters, as indeed are the levels of TBT in the sediments and biota of many estuaries. The elevated levels of PAHs in many estuarine sediments are also of potential concern, although their bioavailability was not measured.

In addition to the substances mentioned above, the QSR 2000 refers to a number of other anthropogenic materials which are present and may be of concern in the marine environment although their effects are largely unknown: brominated flame retardants, chlorinated paraffins, synthetic musks and alkylphenols. Indeed, OSPAR now has an agreed strategy for reducing so-called 'hazardous' substances in present discharges to near-background levels in the marine environment by 2020. It has elaborated a list of such chemicals for priority action and a further list of candidate substances for consideration (OSPAR, 1998c), but there is little knowledge at present about whether the majority of these substances pose a real threat to marine ecosystems.

Given the huge cost of collecting the chemical data reported in the QSR 2000, it is encouraging to note that OSPAR is now more committed to integrated chemical/biological monitoring which is likely to produce results that are more cost-effective and easier to interpret. Because of this, it is surprising that the QSR 2000 only devoted about two pages to the biological effects of contaminants. The report does, however, list a number of effects which can be attributed to the action of contaminants, although not all derive from observations made as part of the core JAMP programme:

- Effects of TBT (imposex) on gastropod molluscs are still widespread and evident at all but the most remote coastal sites. Such effects are also still present offshore in the vicinity of shipping lanes, and are attributed to the continuing use of TBT as an antifoulant on vessels with a waterline length in excess of 25 m.
- PCBs are thought to have severely damaged otter populations along the Skagerrak coast, and seals in the Wadden Sea. Although there is no absolute proof of this, there are good supporting experimental data obtained with seals.
- Elevated levels of EROD activity have been observed in a number of flatfish in UK marine waters, especially near some sewage sludge disposal grounds (now no longer used) and in some urbanised estuaries. The precise causes and wider biological significance of these observations are unclear, although PAHs and PCBs are strong EROD inducers and it is known that EROD induction can be an early warning of neoplastic liver damage.
- Oestrogenic endocrine disruption is reported to be occurring in flounder caught in many industrialised UK estuaries, with induction of yolk protein (vitellogenin) and ovotestis (egg cells in the testis) in male fish. It is not yet known if these effects pose a threat to fish populations.
- Although metal levels are elevated in some places, there are no recent reports of resulting biological effects.
- Oyster embryo bioassays of waters and bioassays of sediments with amphipod crustaceans and polychaete worms have shown that there are acutely or sublethally toxic concentrations of contaminants in some urbanised UK estuaries. The benthic

invertebrate communities in some of these locations are known to be impoverished, although these ecosystem-level impacts may be largely due to organic enrichment rather than poisoning as such.

- Measurements in coastal mussels around the UK have shown that scope-for-growth, a measure of generalised pollution stress, is depressed in the more urbanised regions (southern North Sea, central Irish Sea). The majority of the effect can be attributed to aromatic hydrocarbons.
- Toxicity Identification and Evaluation (TIE) techniques using a harpacticoid copepod bioassay have identified some of the causes of the toxicity noted in some urbanised UK estuaries. These include alkylphenols, alkyl-substituted naphthalenes and fluorenes, and dimethyl benzoquinone.

It appears from this list that most biological effects of contaminants are occurring in UK waters alone, but this is an artefact caused by the very active biological monitoring programme in that country (the National Marine Monitoring Programme – NMMP). Further details of biological monitoring results obtained in the UK are described by Matthiessen and Law (2002). Serious marine biological effects monitoring is only now getting started in most other OSPAR member states, so it will be a while before a holistic picture is available for the entire OSPAR area. As stated above, it is essential that chemical and biological techniques are fully integrated in the future.

In summary, the use of biological effects monitoring in the OSPAR area, while further advanced than in HELCOM, is still very patchy. More importantly, few of the techniques used to date are truly diagnostic of causes or can be interpreted in ecological terms. Nevertheless, despite the fact that concentrations of many contaminants are on the decline as a result of increasing controls on discharges, there is no doubt that contaminants (usually acting as complex mixtures rather than singly) are causing adverse effects in many estuaries and some coastal waters. The real challenge for the future will be to understand the extent to which such changes pose a threat to the long-term sustainability of marine ecosystems.

4. Conclusions and recommendations

It is worth considering briefly why marine pollution biomonitoring is undertaken. In essence, the objective is to provide a sort of safety net which will detect the effects of problem contaminants that have slipped through the environmental risk assessment of chemicals which takes place before manufacture and use, or that have not been properly controlled by the regulation of industrial and domestic discharges. Of course, it is desirable that such a safety net is sensitive to early warning signs of serious polluting impacts, and is not merely able to tell us that significant ecological damage has already taken place. Until recently, marine monitoring programmes in the north-east Atlantic area were only able to perform this early-warning function very imperfectly, being largely focused on short-term acute changes with poor long-term predictivity, or on ecosystem-level impacts which merely show that serious damage has been done. Fortunately, this state of affairs is now improving.

It is fair to say that, although some biological effects of contaminants in the northeast Atlantic area have been reported since the earliest international marine monitoring programmes began in the 1980s, the systematic use of biological responses as holistic measures of the effects of environmental contamination is a much more recent phenomenon which even now is a long way from being fully implemented. Most observers are agreed that chemical or biological observations made in isolation from each other will be unlikely to establish cause-effect relationships and will therefore not provide environmental managers with the information they need to make cost-effective improvements to marine environmental quality. However, the implementation of truly integrated monitoring is still in its infancy.

This is partly due to the fact that the development of validated and sensitive biological effects monitoring techniques has lagged well behind equivalent developments in analytical chemistry. Indeed, even now, there is a lack of sensitive and diagnostic biological techniques which can reliably give an early warning of adverse effects at the population or community levels, and which can therefore produce results that are readily understood by ecologists and administrators. This is partly due to our still rather poor understanding of the way in which marine ecosystems work, and to the fact that there are many contingencies which intervene between the responses of an individual and possible consequent impacts on the ecosystem. In other words, there is no simple proportionality between responses at different levels of biological organisation, and this hampers our ability to interpret individual-level biological effects monitoring data in ecological terms. On the other hand, surveys of community-level change will only provide information on ecological damage after it has occurred, and they are unlikely to provide a diagnosis of the causes of such change. Monitoring must therefore take a middle road by using integrated suites of techniques with some predictive and diagnostic capability, and the OSPAR JAMP has made a good start, although only a few countries have achieved even partial implementation.

These and other problems have been usefully discussed by the ICES Advisory Committee on the Marine Environment (ACME) (Annex 9 of ICES, 2001), which evaluated the structure, process and limitations of marine environmental assessments and quality status reports. The ACME discussion highlights the enormous complexity of managing activities such as the JAMP and production of the QSR 2000, and indicates that part of the difficulty in implementing truly integrated monitoring on an international scale is caused by logistic problems. For example, the expertise and resources currently available for implementing the JAMP vary widely across OSPAR member states, many national sampling programmes still do not integrate the collection of biological and chemical data into a single activity, and even such apparently simple matters as common data reporting and processing formats have not been properly agreed, let alone adopted. OSPAR is moving towards better co-ordination of monitoring activities, but if this fails to materialise, a case could be made for more centralised organisation and funding, perhaps at the European Union level. Finally, the need for acceptable quality assurance of biological data has delayed the full application of the JAMP until 2002 at the earliest.

In the face of these difficulties, it is perhaps not surprising that the conclusions which can be drawn about the impacts of chemicals on the environment of the north-east Atlantic are very limited and tentative. Unfortunately, this lack of precise knowledge combined with the influence of environmental pressure groups has led to the OSPAR policy of reducing inputs to the sea of all 'hazardous' substances to negligible levels by 2020. It is questionable whether this is even achievable, but it is a good example of taking a large hammer to crack a possibly very small nut. Biological and chemical data from an integrated monitoring programme would provide much more precise and selective information to underpin the control of environmental chemicals.

It is worth considering a number of recommendations as a result of this review:

1. It is unwise to conduct marine environmental monitoring programmes without the use of sensitive biological tools which can respond to the mixture of contaminants that are present and bioavailable.

2. Using chemical monitoring data alone, it is generally difficult or impossible to interpret the likely biological implications of the presence of individual anthropogenic substances in the marine environment, due to the confounding effects of variable bioavailability and complex environmental mixtures. For that reason, the use of chemical-based environmental quality standards in isolation from direct measures of biological effect should be discouraged.

3. Marine monitoring programmes should be designed in such a way that the requisite chemical and biological data are gathered in a fully co-ordinated fashion, allowing proper interpretation of likely causes of the observed effects. There are signs that the existing international marine monitoring bodies (OSPAR and HELCOM) may lack sufficient power or influence to achieve this without financial and organisational changes.

4. Both chemical and biological effects monitoring data should be collected in internationally standardised reporting and processing formats which are interchangeable between laboratories across the entire area of interest.

5. If more than one laboratory is involved, monitoring data must all be subject to agreed quality assurance procedures.

6. If significant biological effects are observed, a promising approach to the interpretation of their causes is to use Toxicity Identification and Evaluation (TIE) techniques which employ bioassays to pinpoint progressively simpler sample fractions that contain the active substance(s).

7. There is a continuing need for the development of sensitive biological effects monitoring techniques which are diagnostic of causes and/or predictive of effects at higher levels of biological organisation. It is likely that genomic techniques such as micro-arrays, which are currently being developed, will be able to provide highly specific and diagnostic monitoring tools.

8. Improved understanding of marine ecosystem structure and function is required to allow better interpretation of changes observed at the level of the population, individual and cell. This is essential for holistic understanding of the importance of pollutant impacts.

9. There is a need for continued funding to support the generation of long-term chemical and biological datasets which can be used to predict the effects of contaminants and other stressors.

Acknowledgements

This paper would not have been possible without discussions held over many years with a multitude of colleagues in CEFAS, ICES and OSPAR. I would particularly like to thank the members of the ICES Working Group on the Biological Effects of Contaminants for their stimulating ideas. I am also indebted to John Portmann and Paul Leonard who provided helpful and constructive comments on an early draft. The UK work referred to herein was largely supported by the UK Department of Environment, Food and Rural Affairs (DEFRA – www.defra.gov.uk).

Notes

1 Bioassays may be defined as toxicity tests of an environmental medium such as water or sediment using the holistic responses of experimental animals (e.g. mortality, growth) as the endpoint.
2 Biomarkers are biological responses to contaminants of wild or caged organisms at the sub-organism level (e.g. biochemical or cellular changes).

References

Addison, R.F., 2002. Whither "biological effects monitoring"? In: Anderson, E.D. (Ed.), Proceedings of the ICES History Symposium, Helsinki, Finland, 1–4 August 2000, ICES Marine Science Symposia 215, 184–194.

Andersen, J., Karup, H., Nielsen, U.B. (Eds), 1996. Scientific Symposium on the North Sea Quality Status Report 1993, 18–21 April 1994, Ebeltoft, Denmark. Proceedings. Danish Environmental Protection Agency, Copenhagen.

Carson, R., 1962. Silent Spring. Houghton Mifflin, Boston, MA.

GESAMP, 1982. Review of the health of the oceans. GESAMP Reports and Studies (15). Published also as UNEP Regional Seas Reports and Studies (16).

GESAMP, 1990. The state of the marine environment. GESAMP Reports and Studies (39). Published also as UNEP Regional Seas Reports and Studies (115).

GESAMP, 1994. Guidelines for marine environmental assessments. GESAMP Reports and Studies (54), International Maritime Organisation, London.

GESAMP, 1995. Biological indicators and their use in the measurement of the condition of the marine environment. GESAMP Reports and Studies (55), United Nations Environment Programme, Nairobi.

GESAMP, 1997. The State of the Marine Environment. Blackwell Scientific, Oxford.

Goldberg, E.D., 1976. The Health of the Oceans. UNESCO, Paris.

HELCOM, 1986. First periodic assessment of the state of the marine environment of the Baltic Sea area, 1980–1985. Baltic Sea Environment Proceedings (17A and B), Helsinki Commission, Helsinki.

HELCOM, 1990. Second periodic assessment of the state of the marine environment of the Baltic Sea area, 1984–1988. Baltic Sea Environment Proceedings (35A and B), Helsinki Commission, Helsinki.

HELCOM, 1996. Third periodic assessment of the state of the marine environment of the Baltic Sea area, 1989–1993. Baltic Sea Environment Proceedings (64A and B), Helsinki Commission, Helsinki.

ICES, 1974. Report of the working group for the international study of the pollution of the North Sea and its effects on living resources and their exploitation. International Council for the Exploration of the Sea, Cooperative Research Report (39).

ICES, 2001. Report of the ICES Advisory Committee on the Marine Environment, 2000. ICES Cooperative Research Report (in press), International Council for the Exploration of the Sea, Copenhagen.

Matthiessen, P., 1999. Biological effects quality assurance in monitoring programmes (BEQUALM). QUASIMEME Bulletin (6), 6–7.

Matthiessen, P., Law, R.J., 2002. Contaminants and Their Effects on Estuarine and Coastal Organisms in the United Kingdom in the late Twentieth Century. Environmental Pollution 120, 739–757.

Meybeck, M., Chapman, D., Helmer, R. (Eds), 1989. Global Freshwater Quality. A First Assessment. World Health Organization and United Nations Environment Programme, Blackwell, Oxford.

NSC, 1987. Quality status of the North Sea. Report to the Second International Conference on the Protection of the North Sea by its Scientific and Technical Working Group.

OSPAR, 1993. North Sea quality status report 1993. A report of the North Sea Task Force, Oslo and Paris Commissions, London.

OSPAR, 1998a. JAMP guidelines for general biological effects monitoring. Technical Annexes 1–3, Oslo and Paris Commission, London.

OSPAR, 1998b. JAMP guidelines for contaminant-specific biological effects monitoring. Technical Annexes 1–10. Oslo and Paris Commission, London.

OSPAR, 1998c. OSPAR strategy with regard to hazardous substances. OSPAR ref. 1998–16, Oslo and Paris Commissions, London.

OSPAR, 2000. Quality Status Report 2000. Oslo and Paris Commission, London.

OSPAR/ICES, 1995. Report of the Workshop on Biological Effects Monitoring Techniques, Aberdeen, 2–6 October 1995, Oslo and Paris Commission, London.

Pearce, J.B., Dybern, B.I., Portmann, J., Uthe, J., 1986. A review of the North Atlantic Working Group and its Continuation of Activities. International Council for the Exploration of the Sea, Copenhagen, ICES C.M. 1986/E:48.

SIME, 2000. OSPAR Coordinated Environmental Monitoring Programme (CEMP). Summary Record of the OSPAR Working Group on Concentrations, Trends and Effects of Substances in the Marine Environment (SIME), Annex 5, OSPAR Commission, London.

Stagg, R.M., 1998. The development of an international programme for monitoring the biological effects of contaminants in the OSPAR convention area. Marine Environmental Research 46 (1–5), 307–313.

Wheeler, A., 1979. The Tidal Thames. The History of a River and its Fishes. Routledge and Kegan Paul, London.

Subject index